THE MAMMALS
OF
NORTH AMERICA

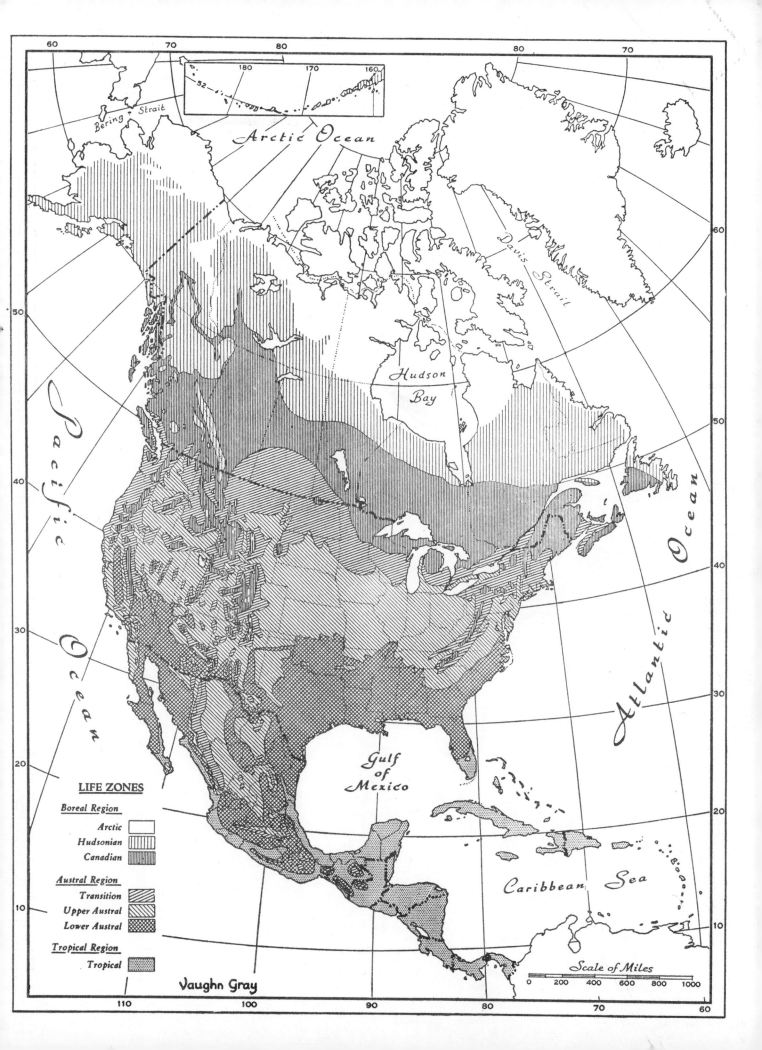

LIFE ZONES

Boreal Region

Arctic

Hudsonian

Canadian

Austral Region

Transition

Upper Austral

Lower Austral

Tropical Region

Tropical

Vaughn Gray

Scale of Miles

0   200   400   600   800   1000

THE MAMMALS

OF

NORTH AMERICA

E. RAYMOND HALL, Ph.D.

RESEARCH ASSOCIATE, MUSEUM OF NATURAL HISTORY,
AND PROFESSOR OF ZOOLOGY (EMERITUS)
THE UNIVERSITY OF KANSAS

Volume II
SECOND EDITION

A WILEY-INTERSCIENCE PUBLICATION

JOHN WILEY & SONS, New York · Chichester · Brisbane · Toronto

Published April 3, 1981

Library of Congress Cataloging in Publication Data

Hall, Eugene Raymond.
   The mammals of North America.

   "A Wiley-Interscience publication."
   Bibliography: p.
   Includes indexes.
   1. Mammals—North America. 2. Mammals—West
Indies. I. Title.
QL715.H15   1979   599′.097   79-4109
   ISBN 0-471-05443-7 (v. 1)
   ISBN 0-471-05444-5 (v. 2)

Printed in the United States of America

10 9 8 7 6 5 4 3 2 1

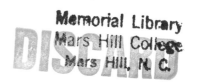
CONTENTS

Volume II

THE MAMMALS
OF
NORTH AMERICA

Family CASTORIDAE—Beavers

Rostrum broad and deep; braincase narrow; basioccipital region with conspicuous pitlike depression; cheek-teeth not ever-growing but so excessively hypsodont that the slightly reduced heptamerous pattern changes little with age and rarely if ever wears out; external form highly modified for aquatic life; tail paddlelike and caudal vertebrae flattened (after Miller and Gidley, 1918:435). Insicors strongly developed; baculum present. Dentition, i. $\frac{1}{1}$, c. $\frac{0}{0}$, p. $\frac{1}{1}$, m. $\frac{3}{3}$.

Genus **Castor** Linnaeus—Beavers

1758. *Castor* Linnaeus, Syst. Nat., ed. 10, 1:58. Type, *Castor fiber* Linnaeus, 1758.

Largest rodent north of Panamá, many adults weighing more than 60 pounds; body thickset and compact; legs short; ears small; hind feet large, pentadactyl, unguiculate, toes webbed; tail broad, flat, nearly hairless and covered with large scales. Pelage consisting of unusually dense, fine underfur but overlaid with many coarse guard-hairs.

Castor canadensis
Beaver

× $\frac{1}{8}$

Upper parts a rich glossy brown, the shade varying seasonally and geographically; underparts paler brown to tawny; tail and feet black.

According to Lavrov and Orlov (1973:742), "The karyotype of C. fiber is more archaic and that of C. canadensis [is] derived from the former. Craniological differences between these species are distinct. Asiatic populations of beavers are not intermediate between European and American beavers and belong to the species C. fiber."

Beavers were extirpated by unrestricted trapping for fur in many areas of their natural range. After 1900 some of these areas were stocked with live animals, not always of the subspecies that had been extirpated. Some streams far outside the natural range of the species were stocked; a stream in the San Bernardino Mountains of southern California is an example (Booth, 1968:46).

Castor canadensis acadicus V. Bailey and Doutt

1942. *Castor canadensis acadicus* V. Bailey and Doutt, Jour. Mamm., 23:87, February 14, type from Nepisiquit River, New Brunswick.

MARGINAL RECORDS.—Quebec (Harper, 1961:50, 51): Richmond Gulf; Clearwater Lake; Matamek River, thence down Atlantic seaboard (*Ingonish, Cape Breton Id., Nova Scotia*, introduced) to Maine: Mount Desert Island (Manville, 1960:416, as *C. canadensis* only), thence probably to eastern Connecticut. New Hampshire: Charleston [? = Charlestown]. New York: Raquette Lake. Quebec: East Main River (Harper, 1961:50).

Castor canadensis baileyi Nelson

1927. *Castor canadensis baileyi* Nelson, Proc. Biol. Soc. Washington, 40:125, September 26, type from Humboldt River, 4 mi. above Winnemucca, Humboldt Co., Nevada.

MARGINAL RECORDS.—Oregon: Silvies River; Kiger Creek. Nevada: Marys River, 25 mi. N Deeth, 5800 ft.; *Deeth;* Toyn Creek, 7000 ft., near summit Harrison Pass; type locality. Oregon: Big Fish Creek.

Castor canadensis belugae Taylor

1916. *Castor canadensis belugae* Taylor, Univ. California Publ. Zool., 12:429, March 20, type from Beluga River, Cook Inlet region, Alaska.

MARGINAL RECORDS.—Yukon: *Porcupine River drainage;* mouth Waters River, $\frac{1}{2}$ mi. WSW Lapierre House (Youngman, 1975:78, as *C. c. canadensis*); Mile 286, Macmillan Pass. British Columbia: Lower Stikine River; Terrace; Wistaria; Anahim Lake; W branch Homathko River; thence northward along Pacific Coast, including several coastal islands, to Alaska: Snug Harbor; between lower end Savonoski River and Grosvenor Lake, Katmai National Monument

(Cahalane, 1959:205); *between base Mt. Katolinat and lower Ukak River, Katmai National Monument* (*ibid.*); Mt. Peulik (Osgood, 1904:33), thence northward including all of Alaska northward to *Brooks Range* except west coastal plains; east of Ladd Air Force Base.

Castor canadensis caecator Bangs

1913. *Castor caecator* Bangs, Bull. Mus. Comp. Zool., 54:513, July, type from near Bay St. George, Newfoundland.
1942. *Castor canadensis caecator*, G. M. Allen, Extinct and vanishing mammals of the Western Hemisphere, p. 62, December 11.

MARGINAL RECORDS.—Newfoundland: Port Saunders; Gander River; *Red Indian Lake;* type locality. Not found, Bay d'Espoir.

Castor canadensis canadensis Kuhl

1820. *Castor canadensis* Kuhl, Beiträge zur Zoologie und vergleichenden Anatomie, Abth. 1, p. 64, type locality, Hudson Bay.

MARGINAL RECORDS.—Yukon Territory: *northern Yukon.* Mackenzie: Mackenzie Delta. Keewatin: about 6 mi. above mouth Little River (Harper, 1956:24). Manitoba: Seal River. Ontario: Lake Abitibi region; Algonquin Park; thence southeastward to Atlantic Coast possibly in *western Connecticut* and down coast to Virginia: Brunswick County; *Nottoway County;* Nelson County; Wythe County. West Virginia: upper end Long Shoal, a few miles below Kanawha Falls. Pennsylvania: Greene County; Erie County. Ontario: Bothwell; Nepigon. Minnesota: Grand Portage Indian Reservation (Timm, 1975:20). Iowa: *possibly*

extreme northeastern Iowa. North Dakota: Mouse River Valley. Saskatchewan: Long Creek. Montana: Glacier National Park. British Columbia: *East Pine* (Cowan and Guiguet, 1965:174); Little Prairie (*ibid.*); vic. Laurier Pass. Yukon: 138 mi. N Watson Lake and 5 mi. E Little Hyland River, 6000 ft. (Youngman, 1968:75).

Castor canadensis carolinensis Rhoads

1898. *Castor canadensis carolinensis* Rhoads, Trans. Amer. Philos. Soc., n.s., 19:420, September, type from Dan River, near Danbury, Stokes Co., North Carolina.

MARGINAL RECORDS.—Indiana: Laporte County; Steuben County; thence eastward *presumably throughout Ohio except extreme northeastern part.* Tennessee: Holston River between Kingsport and Three Springs Ford. Virginia: Franklin County; *Bedford County;* Campbell County; *Charlotte County,* thence down coast to Georgia: 1 mi. E Lloyd's Island, Okefinokee Swamp. Louisiana (Lowery, 1974:220, 224): *Honey Island swamp, NE Pearl River;* Gonzales; *near Baton Rouge on Airline Hwy.;* Solitude; Beauregard Parish; Bienville Parish, thence northward through Mississippi Valley probably into *Illinois.*

Castor canadensis concisor Warren and Hall

1939. *Castor canadensis concisor* Warren and Hall, Jour. Mamm., 20:358, August 14, type from Monument Creek, SW of Monument, El Paso Co., Colorado.

MARGINAL RECORDS.—Wyoming: 8 mi. N, 16 mi. E Encampment (Long, 1965a:623). Colorado: Morgan County; Arapahoe County; Lake Moraine, 10,250

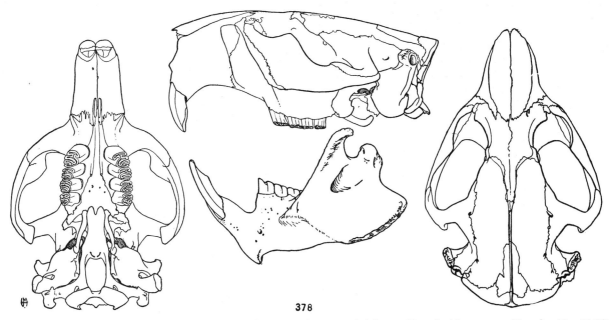

378

Fig. 378. *Castor canadensis repentinus*, Colorado River, ½ mi. N California–Nevada Monument, Nevada, No. 61670 M.V.Z., ♀, X ½.

ft.; 4 mi. W La Veta (Armstrong, 1972:185). New Mexico: *Cimarroncito Creek;* Rayado Canyon. Colorado: Costilla River (Armstrong, 1972:185); Navajo River (*ibid.*); Mancos River (Anderson, 1961:48); Gunnison River, 4750 ft., *ca.* 29 mi. NW Delta; Trappers Lake. Wyoming: 7½ mi. N, 18½ mi. E Savery.

Castor canadensis duchesnei Durrant and Crane

1948. *Castor canadensis duchesnei* Durrant and Crane, Univ. Kansas Publ., Mus. Nat. Hist., 1(20):413, December 24, type from Duchesne River, 5600 ft., 10 mi. NW Duchesne, Duchesne Co., Utah.

MARGINAL RECORDS.—Utah: Currant Creek, 6000 ft., Strawberry Valley; type locality. Colorado: 9½ mi. SW Pagoda Peak.

Castor canadensis frondator Mearns

1897. *Castor canadensis frondator* Mearns, Preliminary diagnoses of new mammals of the genera *Sciurus, Castor, Neotoma,* and *Sigmodon,* from the Mexican border of the United States, p. 2, March 5 (preprint of Proc. U.S. Nat. Mus., 20:502, January 19, 1898), type from Río San Pedro, Sonora, near Monument 98 of Mexican boundary line.

MARGINAL RECORDS.—New Mexico: Liberty; *Blanco;* headwaters Zuñi River; headwaters Gila River; Gila River near Bedrock. Sonora: Cañon de Guadalupe; type locality. Arizona: Fort Verde.

Castor canadensis idoneus Jewett and Hall

1940. *Castor canadensis idoneus* Jewett and Hall, Jour. Mamm., 23:87, February 15, type from Foley Creek, tributary to Nehalem River, Tillamook Co., Oregon.

MARGINAL RECORDS.—Washington: Puget Island. Oregon: type locality; Blaine.

Castor canadensis labradorensis V. Bailey and Doutt

1942. *Castor canadensis labradorensis* V. Bailey and Doutt, Jour. Mamm., 23:86, February 14, type from 5 mi. above Grand Falls, Hamilton River, Labrador.

MARGINAL RECORDS (Harper, 1961:49, 50).—Labrador: Hopedale, thence down coast to Charles River, W and N to Quebec: George River. Not found: Labrador: near Namaycush Lake; Winokapau Lake.

Castor canadensis leucodontus Gray

1869. *Castor canadensis leucodonta* Gray, Ann. Mag. Nat. Hist., ser. 4, 4:293, October, type from Vancouver Island, British Columbia.
1898. *Castor canadensis pacificus* Rhoads, Trans. Amer. Philos. Soc., n.s., 19:422, September, type from Lake Keechelus, about 3000 ft., Cascade Mts., Kittitas Co., Washington. Regarded as inseparable from *leucodontus* by Dalquest, Univ. Kansas Publ., Mus. Nat. Hist., 2:322, April 9, 1948.

MARGINAL RECORDS.—British Columbia: Tlell, Queen Charlotte Islands (Cowan and Guiguet, 1965:174, introduced in 1950); Quatsino, Vancouver Island; Alta Lake; Okanagan Landing; Bush River; Golden; Fort Steele; Newgate. Idaho: Coolin; Bonner County; Shoshone County. Oregon: Pine Creek near Pine; Jordan Creek near Rome; *headwaters Deschutes River;* Willamette Valley. Washington: Bear Prairie, 3 mi. S Longmire, Mount Rainier National Park, thence along coast to British Columbia: Beaver Creek, near Alberni, Vancouver Island (Cowan and Guiguet, 1965:174).

Castor canadensis mexicanus V. Bailey

1913. *Castor canadensis mexicanus* V. Bailey, Proc. Biol. Soc. Washington, 26:191, October 23, type from Ruidoso Creek, 6 mi. below Ruidoso, Lincoln Co., New Mexico.

MARGINAL RECORDS.—New Mexico: Culebra Mts., Costilla River, 9400 ft.; headwaters Pecos River; 6 mi. below Ruidoso on Ruidoso Creek (Findley, *et al.*, 1975:188). Texas: Pecos River Canyon; Starr County (Davis, 1966:174, as *C. canadensis* only); Brownsville. Tamaulipas: *Rio Grande, 12 mi. below Matamoros.* Nuevo León: China. Coahuila: Sabinas. Texas: Johnson Ranch, Big Bend National Park, 2060 ft. Chihuahua: Camargo (Anderson, 1972:321). Texas: Rio Grande, El Paso Co. New Mexico: Garfield; near Albuquerque.

Castor canadensis michiganensis V. Bailey

1913. *Castor canadensis michiganensis* V. Bailey, Proc. Biol. Soc. Washington, 26:192, October 23, type from Tahquamenan River, 5 mi. above falls, Luce Co., Michigan.

MARGINAL RECORDS.—Michigan: Isle Royale. Ontario: Pancake Bay. Michigan: Monroe County; Kalamazoo County. Wisconsin (Jackson, 1961:203): Babcock; Basswood Lake, 10 mi. SE Iron River.

Castor canadensis missouriensis V. Bailey

1919. *Castor canadensis missouriensis* V. Bailey, Jour. Mamm., 1:32, November 28, type from Apple Creek, 7 mi. E Bismarck, Burleigh Co., North Dakota.

MARGINAL RECORDS.—Alberta: Battle Creek. Saskatchewan: Rocky Creek, S of Wood Mtn. North Dakota: Antelope Creek near Goodall; near Sawyer; type locality. Iowa: Sac County. Missouri: *northwestern part.* Oklahoma (Glass, 1960:24, as *C. canadensis* only): near McAlester; El Reno; Ellis County; Beaver County; Texas County. New Mexico: 34⅔ mi. N, 2½ mi. E Clayton (Best, 1971:210, as *C. canadensis* only). Kansas: St. Francis. Wyoming (Long, 1965a:623): Little Laramie River Valley; 7 mi. S South Pass City; Fort Bridger; Teton Basin, N Fork Teton River; *Teton Canyon;* Swan Lake Flat. Alberta: Milk River, N of Sweet Grass Hills.

Castor canadensis pallidus Durrant and Crane

1948. *Castor canadensis pallidus* Durrant and Crane, Univ. Kansas Publ., Mus. Nat. Hist., 1:409, December 24, type from Lynn Canyon, 7500 ft., Boxelder Co., Utah.

MARGINAL RECORDS.—Utah: type locality; *Raft River, 5 mi. S Yost, Raft River Mts., 6000 ft.*

Castor canadensis phaeus Heller

1909. *Castor canadensis phaeus* Heller, Univ. California Publ. Zool., 5:250, February 18, type from Pleasant Bay, Admiralty Island, Alaska.

MARGINAL RECORDS.—Alaska: Chichagof Island; Mole Harbor, Admiralty Island.

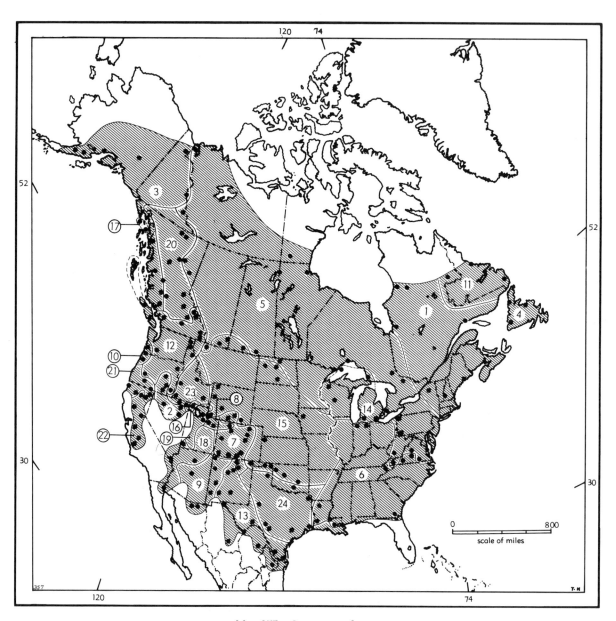

Map 357. *Castor canadensis.*

1. *C. c. acadicus*	7. *C. c. concisor*	13. *C. c. mexicanus*	19. *C. c. rostralis*
2. *C. c. baileyi*	8. *C. c. duchesnei*	14. *C. c. michiganensis*	20. *C. c. sagittatus*
3. *C. c. belugae*	9. *C. c. frondator*	15. *C. c. missouriensis*	21. *C. c. shastensis*
4. *C. c. caecator*	10. *C. c. idoneus*	16. *C. c. pallidus*	22. *C. c. subauratus*
5. *C. c. canadensis*	11. *C. c. labradorensis*	17. *C. c. phaeus*	23. *C. c. taylori*
6. *C. c. carolinensis*	12. *C. c. leucodontus*	18. *C. c. repentinus*	24. *C. c. texensis*

Castor canadensis repentinus Goldman

1932. *Castor canadensis repentinus* Goldman, Jour. Mamm., 13:266, August 9, type from Bright Angel Creek, 4000 ft., Grand Canyon, Arizona.

MARGINAL RECORDS.—Utah: *presumably Colorado River drainage S of Tavaputs Plateau.* Arizona: Cameron, Little Colorado River (Hoffmeister, 1971:124). Sonora: Colorado River, 30 mi. N of mouth. Nevada: Searchlight Ferry. Utah: Magotsu Creek, *ca.* 30 mi. NW St. George (Stock, 1970:431).

Castor canadensis rostralis Durrant and Crane

1948. *Castor canadensis rostralis* Durrant and Crane, Univ. Kansas Publ., Mus. Nat. Hist., 1:411, December 24, type from Red Butte Canyon, 5000 ft., Fort Douglas, Salt Lake Co., Utah.

MARGINAL RECORDS.—Utah: type locality; Kansas, 5500 ft.; *Charleston, Heber Valley, 5500 ft.; Millcreek Canyon, 7000 ft., 6 mi. above mouth.*

Castor canadensis sagittatus Benson

1933. *Castor canadensis sagittatus* Benson, Jour. Mamm., 14:320, November 13, type from Indianpoint Creek, 3200 ft., 16 mi. NE Barkerville, British Columbia.

MARGINAL RECORDS.—British Columbia: Lower Liard Crossing; Fort Grahame; *Indianpoint Lake and vic.* (Cowan and Guiguet, 1965:175); Isaacs Lake; Murtle River (Cowan and Guiguet, 1965:175); North Fork of Eagle River (*ibid.*); Chezacut; Wistaria; Tetana Lake (Cowan and Guiguet, 1965:174); Fort Halkett.

Castor canadensis shastensis Taylor

1916. *Castor subauratus shastensis* Taylor, Univ. California Publ. Zool., 12:433, March 20, type from Cassel, Hot Creek, near Pit River, Shasta Co., California.
1933. *Castor canadensis shastensis*, Grinnell, Univ. California Publ. Zool., 40:166, September 26.

MARGINAL RECORDS.—Oregon: Yamsey Mts.; Goose Lake. California: *head Lassen Creek;* N fork Pit River, above Alturas; type locality; Scott River; Klamath River above Requa.

Castor canadensis subauratus Taylor

1912. *Castor subauratus* Taylor, Univ. California Publ. Zool., 10:167, May 21, type from Grayson, San Joaquin River, Stanislaus Co., California.
1933. *Castor canadensis subauratus*, Grinnell, Univ. California Publ. Zool., 40:166, September 26.

MARGINAL RECORDS.—California: McCloud; American River near Fair Oaks; *Merced River near Snelling;* Kings River, Fresno Co.; Tulare Lake; *San Joaquin River near Mendota;* Westley, near Grayson; *Cache Slough, 10 mi. N Rio Vista;* N of Marysville Buttes.

Castor canadensis taylori Davis

1939. *Cator canadensis taylori* Davis, The Recent mammals of Idaho, Caxton Printers, Caldwell, Idaho, p. 273, April 5, type from Big Wood River, near Bellevue, Blaine Co., Idaho.

MARGINAL RECORDS.—Idaho: Salmon (City?); Portneuf River, 10 mi. NW Pocatello. Nevada: Goose Creek, 10½ mi. W Utah boundary, 5150 ft.; Jarbidge River. Idaho: S fork Owyhee River, 12 mi. N Nevada line; Boise River, 5 mi. W Boise.

Castor canadensis texensis V. Bailey

1905. *Castor canadensis texensis* V. Bailey, N. Amer. Fauna, 25:122, October 24, type from Cummings Creek, Colorado Co., Texas.

MARGINAL RECORDS.—Texas: Canadian River not far from Canadian. Oklahoma: near Tishomingo (Glass, 1960:24). Texas: Texarkana; Neches River near Beaumont; Cummings Creek; 40 or 45 mi. NW Uvalde; Edwards County (Davis, 1966:174, as *C. canadensis* only); Colorado River 10 mi. below Colorado City.

SUBORDER MYOMORPHA

Medial masseter muscle traverses infraorbital canal and exits on side of rostrum; infraorbital canal much enlarged; zygomatic plate (horizontal, narrow and beneath infraorbital canal) as in Dipodoidea, or (relatively broad and inclined) as in Muroidea but not so broad and inclined as in advanced sciuromorphs; postorbital processes never present. Dentition, i. $\frac{1}{1}$, c. $\frac{0}{0}$, p. $\frac{1-0}{1-0}$, m. $\frac{3-2}{3-2}$ (p. $\frac{0}{0}$, m. $\frac{3}{3}$ in most species, after Wilson, 1949:120).

KEY TO NORTH AMERICAN FAMILIES OF MYOMORPHA

1. Tail at least 1½ times length of head and body, naked and scaly; hind legs much elongated; infraorbital foramen much enlarged. . . . . Zapodidae, p. 840
1'. Tail less than 1½ times length of head and body, either haired or naked and scaly; hind legs not greatly elongated; infraorbital foramen moderately large. . Muridae, p. 605

FAMILY MURIDAE—Murids

Infraorbital foramen not conspicuously wider below than above (in all post-Oligocene genera); zygomatic plate broadened and tilted upward; masseter lateralis extends its line of attachment onto zygomatic plate; head of masseter lateralis superficialis distinct from zygoma; infraorbital foramen transmitting muscle but never extremely enlarged (usually less so than in Dipodoidae);

fibula reduced and fused high on leg with tibia; cheek-teeth cuspidate, prismatic, or laminate. Dentition, i. $\frac{1}{1}$, c. $\frac{0}{0}$, p. $\frac{0}{0}$, m. $\frac{3}{3}$ (modified from Ellerman, *et al.*, 1941:1, 2). Lineage can be traced back to early Oligocene.

KEY TO NORTH AMERICAN SUBFAMILIES OF MURIDAE

1. Molar teeth with tubercles arranged in 3 longitudinal series. . . . . . .Murinae,   p. 838
1′. Molar teeth flat-crowned or with tubercles arranged in 2 longitudinal series.
  2. Skull comparatively smooth and unspecialized; molars having 2 longitudinal rows of tubercles or, if not tuberculate, prisms not arranged as alternating triangles and rostrum tapering anteriorly (see Fig. 401).   Cricetinae,   p. 606
  2′. Skull comparatively angular and sculptured; molars with enamel pattern composed of alternating triangles; rostrum short (see Figs. 464–487).
                                      Microtinae,   p. 777

SUBFAMILY CRICETINAE

Slender to thickset rodents with pointed muzzles; limbs not hidden to any marked extent by integument; pentadactyl; tail usually relatively long; palatine bones not especially thickened; molars brachydont or hypsodont and, when hypsodont, not possessing alternating triangular prisms; molars with discrete roots; skull not becoming much modified by ridges for jaw-muscle arrangement, not developing squamosal crest nor median interorbital crest (modified from Ellerman, *et al.*, 1941:327). Hershkovitz (1966:81–149) reviews the phallic characters of the New World Cricetinae.

KEY TO NORTH AMERICAN GENERA OF CRICETINAE

(Modified from Ellerman, 1941:332–340)

1. External form much modified for aquatic life (hind toes fringed with stiff hairs; underfur dense; webbing present between some toes); braincase much flattened. . . . .*Rheomys,*   p. 774
1′. External form not modified for aquatic life (hind toes not fringed with stiff hairs; underfur not especially dense; no webbing between toes); braincase not markedly flattened.
  2. Upper cheek-teeth specialized, their normal cuspidate pattern not apparent at any time (young *Neotomodon* a possible exception); molars prismatic and flat-crowned.
    3. Third upper molar simplified, without inner fold. . . . .*Nelsonia,*   p. 773
    3′. Third upper molar with inner fold.
      4. Folds of upper molars in subadults widely open (dental pattern suggestive of that of Microtinae).
        5. Palate extending posteriorly at least to end of tooth-rows. . . . . . .*Neotomodon,*   p. 745
        5′. Palate extending no farther posteriorly than level of anterior part of last molar.
          6. Supraorbital ridge much elevated and strongly beaded; tympanic bullae much enlarged; their long axes parallel. . . . .*Xenomys,*   p. 773
          6′. Supraorbital ridge relatively little elevated and not beaded; tympanic bullae little enlarged, their long axes converging anteriorly. . . . .*Neotoma,*   p. 746
      4′. Folds of upper molars in subadults not widely open and areas of dentine less sharply projecting, making molars appear compressed. . . . .*Sigmodon,*   p. 735
  2′. Upper cheek-teeth not markedly specialized, the cuspidate pattern usually apparent; molars usually not flat-crowned.
    7. Upper incisors conspicuously grooved. . . . .*Reithrodontomys,*   p. 632
    7′. Upper incisors not conspicuously grooved.
      8. Cheek-teeth complex with subsidiary ridges normally present in main outer folds of M1 and M2; teeth cuspidate (secondary ridges tend to disappear in some species of *Peromyscus*).
        9. Cusps of upper molars alternate; anterointernal cusp reduced to absent; M3 notably reduced.
          10. Ears dusky or dusky edged with whitish, in slight contrast to color of head and body; posterior palatine foramina approx. midway between anterior palatine foramina and posterior edge of palate; dentine spaces of molars mostly confluent. . . . .*Peromyscus,*   p. 655
          10′. Ears bright ochraceous, same color as body; posterior palatine foramina nearer to posterior edge of palate than to anterior palatine foramina; dentine spaces of molars mostly closed. . . . .*Ochrotomys,*   p. 721
      9′. Cusps of upper molars opposite or nearly so; anterointernal cusp not obliterated; M3 normally not reduced.

Genus **Oryzomys** Baird—Rice Rats

Revised by Goldman, N. Amer. Fauna, 43:1–100, 6 pls., 11 figs., September 23, 1918.

1858. *Oryzomys* Baird, Mammals, *in* Repts. Expl. Surv. . . . , 8(1):458, July 14. Type, *Mus palustris* Harlan.

External measurements: 216–387; 101–235; 23–40. Form mouselike; pelage coarse but not bristly or spiny; tail usually long, annulations showing through sparse hairs. Skull lightly constructed and not markedly sculptured; zygomatic arches depressed to near level of tooth-row; palate extending posteriorly beyond tooth-rows; palatal pits present; tympanic bullae moderately inflated. Crowns of molars with well-developed cusps, styles, or tubercles, and main cusps opposite in two longitudinal series.

KEY TO NORTH AMERICAN SUBGENERA OF ORYZOMYS

(Adapted from Goldman, 1918:17)

Subgenus **Oryzomys** Baird—Rice Rats

1858. *Oryzomys* Baird, Mammals, *in* Repts. Expl. Surv. . . . , 8(1):458, July 14. Type, *Mus palustris* Harlan.

1948. *Micronectomys* Hershkovitz, Proc. U.S. Nat. Mus., 98:55, June 30. Type, *Nectomys dimidiatus* Thomas 1905. *Micronectomys* was a *nomen nudum* according to Hershkovitz, Jour. Mamm., 51:789–794, November 30, 1970, who there also (Jour. Mamm.) possibly left the name standing as a *nomen nudum*.

1948. *Macruroryzomys* Hershkovitz, Proc. U.S. Nat. Mus., 98:56, June 30. Type *Nectomys hammondi* Thomas 1913. *Macruroryzomys* was a *nomen nudum* according to Hershkovitz, Jour. Mamm., 51:789–794, November 30, 1970, who there also (Jour. Mamm.) possibly left the name standing as a *nomen nudum*.

Upper parts usually contrasting strongly with underparts; tail about equal to, or longer than, head and body; anterior border of lachrymal articulating about equally with maxilla and frontal; supraorbital and temporal ridges usually prominent; secondary parastyle (of Goldman, 1918: Fig. 1) well developed; slightly worn crown of 2nd upper molar with central enamel island elongated or absent; upper incisors decidely curved backward near points (after Goldman, 1918:17).

KEY TO SPECIES OF SUBGENUS ORYZOMYS

1. Hind foot without prominent tufts of digital bristles projecting beyond ends of 3 median claws.
    2. Occurring on María Madre Island. . . . .*O. nelsoni*, p. 612
    2′. Not occurring on María Madre Island.
        3. Known only from eastern Nicaragua; there differs from *O. palustris* in grayish as opposed to buffy venter and tail always shorter than length of head and body instead of approx. equal length. . . . .*O. dimidiatus*, p. 612
        3′. Known from New Jersey southward into Panamá (possibly including *O. fulgens* from Valley of Mexico). . . . .*O. palustris*, p. 608
1′. Hind foot with prominent tufts of digital bristles projecting beyond ends of three median claws.
    4. Hind foot less than 33.
        5. Supraorbital vibrissae less than 40 mm.
            6. Ears clothed internally with fine blackish hairs.
                7. Second upper molar with central enamel island present; 3rd lower molar with outer re-entrant angle extending approx. halfway across crown.
                    8. Upper parts dark ochraceous buff, ochraceous tawny, or even brownish, usually with blackish on head and mid-dorsally; tail bicolored, but dusky above and below distally; braincase vaulted. . . . .*O. alfaroi*, p. 615
                    8′. Upper parts buffy brown mixed with black; tail unicolored; braincase flattened. *O. caudatus*, p. 615
                7′. Second upper molar with central enamel island absent; 3rd lower molar with outer re-entrant angle extending more than halfway across crown. . . . .*O. capito*, p. 618
            6′. Ears clothed internally with buffy or rusty reddish hairs. . . . .*O. melanotis*, p. 613
        5′. Supraorbital vibrissae more than 50 mm. . . . .*O. bombycinus*, p. 616
    4′. Hind foot 33 or more.
        9. Tail longer than 215; hind foot longer than 39; supraorbital ridges prominent.*O. aphrastus*, p. 618
        9′. Tail shorter than 215; hind foot shorter than 39; supraorbital ridges absent. *O. albigularis*, p. 617

Oryzomys palustris
Marsh Rice Rat

x ¼

External measurements: 226–332; 108–182; 28–40. Upper parts grizzled grayish brown to ochraceous-tawny, mixed with blackish; sides paler, with less black; underparts white to ochraceous buff; tail varying from brownish above and whitish below to evenly dusky. Skull large;

braincase high; rostrum short; frontal region broad, with trenchant lateral margins that are slightly upturned and project as supraorbital ridges; temporal ridges usually well developed; interparietal small, subtriangular; elongated palatal slits approx. as long as palatal bridge; palatal pits prominent; sphenopalatine vacuities large to absent; inner re-entrant angles in upper molars and outer re-entrant angles in lower molars reaching less than halfway across moderately worn teeth.

Oryzomys palustris albiventer Merriam

1901. *Oryzomys albiventer* Merriam, Proc. Washington Acad. Sci., 3:279, July 26, type from Ameca, 4000 ft., Jalisco.
1960. *Oryzomys palustris albiventer*, Hall, Southwestern Nat., 5:173, November 1.

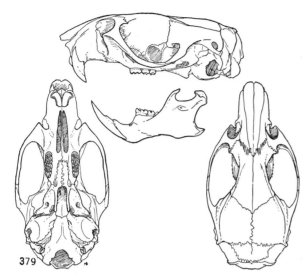

Fig. 379. *Oryzomys palustris palustris*, Goose Creek, Charleston Co., South Carolina, No. 97182 M.V.Z., ♂, X 1½.

1903. *Oryzomys molestus* Elliot, Field Columb. Mus., Publ. 71, Zool. Ser., 3(8):145, March 20, type from Ocotlán, 5000 ft., Jalisco.

MARGINAL RECORDS.—Jalisco: 2 mi. N, ½ mi. W Guadalajara, 5200 ft.; La Barca; type locality.

Oryzomys palustris antillarum Thomas

1898. *Oryzomys antillarum* Thomas, Ann. Mag. Nat. Hist., ser. 7, 1:177, February, type from Jamaica. Known only from Jamaica.
1966. *Oryzomys palustris antillarum*, Hershkovitz, Ectoparasites of Panama, Field Mus. Nat. Hist., p. 736, November 22.

Oryzomys palustris aquaticus J. A. Allen

1891. *Oryzomys aquaticus* J. A. Allen, Bull. Amer. Mus. Nat. Hist., 3:289, June 30, type from Brownsville, Cameron Co., Texas.
1960. *Oryzomys palustris aquaticus*, Hall, Southwestern Nat., 5:173, November 1. See addenda.

MARGINAL RECORDS.—Tamaulipas: Camargo. Texas: Lomita Ranch; *type locality;* 14$\frac{7}{10}$ mi. E Brownsville (Hall, 1960:172), thence southward to Tamaulipas (Alvarez, 1963:435): near Ciudad Tampico; *5 mi. N, 5 mi. W Altamira; 6 mi. N, 6 mi. W Altamira;* Sierra de Tamaulipas, 2 mi. S, 10 mi. W Piedra. Nuevo León: Río Ramos, 20 km. NW Montemorelos, 1000 ft.

Oryzomys palustris aztecus Merriam

1901. *Oryzomys crinitus aztecus* Merriam, Proc. Washington Acad. Sci., 3:282, July 26, type from Yautepec, Morelos.
1960. *Oryzomys palustris aztecus*, Hall, Southwestern Nat., 5:173, November 1.

MARGINAL RECORDS.—Morelos: Cuernavaca. Puebla: Tepanco, 1800 m.; Tehuacán, 1700 m. Oaxaca: Teotitlán, 920–950 m.; 3 km. NNE Cuicatlán, 570 m.; *1 km. S Cuicatlán, 590 m.; 2 km. SSW Cuicatlán, 570 m.* Guerrero: Tlalixtaquilla; Balsas. Morelos: Puente de Ixtla.

Oryzomys palustris azuerensis Bole

1937. *Oryzomys azuerensis* Bole, Sci. Publs., Cleveland Mus. Nat. Hist., 7:165, August 31, type from Paracoté, 1½ mi. S mouth Río Angulo, Mariato–Suay lands, Veraguas, Panamá.
1966. *O[ryzomys]. p[alustris]. azuerensis*, Handley, Ectoparasites of Panama, Field Mus. Nat. Hist., p. 781, November 22.

MARGINAL RECORDS (Handley, 1966:781).— Panamá: type locality; Guánico.

Oryzomys palustris coloratus Bangs

1898. *Oryzomys palustris coloratus* Bangs, Proc. Boston Soc. Nat. Hist., 28:189, March, type from Cape Sable, Monroe Co., Florida.
1901. *Oryzomys natator floridanus* Merriam, Proc. Washington Acad. Sci., 3:277, July 26, type from Everglade, Collier Co., Florida.

MARGINAL RECORDS.—Florida: Eden; thence southward around tip of peninsula to Everglade; Lake Okeechobee [= Ritta].—Layne (1974:392) records two specimens as only *Oryzomys* from *Cudjoe Key, approx. 23 mi. ENE of Key West.*—See addenda.

Oryzomys palustris couesi (Alston)

1877. *Hesperomys couesi* Alston, Proc. Zool. Soc. London, for 1876, p. 756, April, type from Cobán, Guatemala.
1960. *Oryzomys palustris couesi*, Hall, Southwestern Nat., 5:173, November 1. See addenda.
1897. *Oryzomys jalapae* J. A. Allen and Chapman, Bull. Amer. Mus. Nat. Hist., 9:206, June 16, type from Jalapa, 4400 ft., Veracruz.
1901. *Oryzomys jalapae rufinus* Merriam, Proc. Washington Acad. Sci., 3:285, July 26, type from Catemaco, 1000 ft., Veracruz.
1901. *Oryzomys teapensis* Merriam, Proc. Washington Acad. Sci., 3:286, July 26, type from Teapa, Tabasco.
1901. *Oryzomys goldmani* Merriam, Proc. Washington Acad. Sci., 3:288, July 26, type from Coatzacoalcos, Veracruz.
1904. *Oryzomys jalapae apatelius* Elliot, Field Columb. Mus., Publ. 90, Zool. Ser., 3:266, March 7, type from San Carlos, Veracruz.
1910. *Oryzomys richardsoni* J. A. Allen, Bull. Amer. Mus. Nat. Hist., 28:99, April 30, type from Peña Blanca, Nicaragua.

MARGINAL RECORDS.—Yucatán: 13 km. WSW Sisal (92231 KU). Quintana Roo: 4 mi. NNE Felipe Carrillo Puerto (92233 KU). Honduras: Yaruca; Catacamas. Nicaragua: Río Tuma; Chontales (Goldman, 1918:31). Costa Rica: San Juanillo. Nicaragua: 3 mi. NNW Diriamba (Jones and Genoways, 1970:14, as *O. palustris* only); San Antonio

(Jones and Genoways, 1970:7, as *O. palustris* only). Honduras: La Ciénaga; Las Flores. Guatemala: Finca Ciprés; Hda. California. Chiapas: Finca Esperanza, 250 m. Guatemala: Jacaltenango. Chiapas: Tumbalá; 10 km. S Solusuchiapa, 395 m. (Baker, *et al.*, 1973:78, 85). Oaxaca: mtn. near [= 15 mi. NE] Santo Domingo; Ixcuintepec (Goodwin, 1969:151); Comaltepec (*ibid.*). Veracruz: Orizaba. Puebla: Huauchinango; Metlaltoyuca. Veracruz: 5 km. N Jalapa, 5000 ft.; San Carlos. Not found: Nicaragua: Río San Juan del Norte (Goldman, 1918:31).

Oryzomys palustris cozumelae Merriam

1901. *Oryzomys cozumelae* Merriam, Proc. Biol. Soc. Washington, 14:103, July 19, type from Cozumel Island, Quintana Roo. Known only from Cozumel Island.
1965. *Oryzomys palustris cozumelae*, Jones and Lawlor, Univ. Kansas Publ., Mus. Nat. Hist., 16:413, April 13.

Oryzomys palustris crinitus Merriam

1901. *Oryzomys crinitus* Merriam, Proc. Washington Acad. Sci., 3:281, July 26, type from Tlalpan, Distrito Federal, México.
1960. *Oryzomys palustris crinitus*, Hall, Southwestern Nat., 5:173, November 1.

MARGINAL RECORDS.—Distrito Federal: type locality. Tlaxcala: 8 km. SW Tlaxcala, 7500 ft.

Oryzomys palustris gatunensis Goldman

1912. *Oryzomys gatunensis* Goldman, Smiths. Miscl. Coll., 56(36):7, February 19, type from Gatún, Canal Zone, Panamá.
1966. *O[ryzomys]. p[alustris]. gatunensis*, Handley, Ectoparasites of Panama, Field Mus. Nat. Hist., p. 781, November 22.

MARGINAL RECORDS (Handley, 1966:781).— Panamá: type locality; Pacora.

Oryzomys palustris lambi Burt

1934. *Oryzomys couesi lambi* Burt, Proc. Biol. Soc. Washington, 47:107, June 13, type from San José de Guaymas, Sonora. Known only from type locality.
1960. *Oryzomys palustris lambi*, Hall, Southwestern Nat., 5:173, November 1.

Oryzomys palustris mexicanus J. A. Allen

1897. *Oryzomys mexicanus* J. A. Allen, Bull. Amer. Mus. Nat. Hist., 9:52, March 15, type from Hda. San Marcos, 3500 ft., Tonila, Jalisco.
1960. *Oryzomys palustris mexicanus*, Hall, Southwestern Nat., 5:173, November 1.
1897. *Oryzomys bulleri* J. A. Allen, Bull. Amer. Mus. Nat. Hist., 9:53, March 15, type from Valle de Banderas, Nayarit.
1901. *Oryzomys rufus* Merriam, Proc. Washington Acad. Sci., 3:287, July 26, type from Santiago, 200 ft., Nayarit.

MARGINAL RECORDS.—Sinaloa: Mazatlán; *Chele;* Rosario. Nayarit: Santiago. Colima: Hda. Magdalena. Michoacán: 10 km. W Apatzingán, 1040 ft.; La Huacana. Guerrero: Río Aquacatillo, 30 km. N Acapulco, 1000 ft.; Ometepec. Oaxaca (Goodwin, 1969:151, unless otherwise noted): 6 km. S Putla, 840 m. (Webb and Baker, 1971:144); Juquila; Jalapa; Reforma (Hall and Kelson, 1959:558); Zanatepec; Tapanatepec, thence northward along coast to point of beginning.

Oryzomys palustris natator Chapman

1893. *Oryzomys palustris natator* Chapman, Bull. Amer. Mus. Nat. Hist., 5:44, March 17, type from Gainesville, Alachua Co., Florida.

MARGINAL RECORDS.—Florida: Anastasia Island; Micco; Lake Kissimmee; Tarpon Key (Bush Key); Crystal River, Citrus Co.; type locality.

Oryzomys palustris palustris (Harlan)

1837. *Mus palustris* Harlan, Amer. Jour. Sci., 31:385. Type locality, "Fast Land," near Salem, Salem Co., New Jersey.
1858. *Oryzomys palustris* Baird, Mammals, in Repts. Expl. Surv.... , 8(1):459, July 14.
?1854. *Arvicola oryzivora* Bachman, *in* Audubon and Bachman, The viviparous quadrupeds of North America, 3:214, type from St. Johns Parish, South Carolina.

MARGINAL RECORDS.—Pennsylvania: Tinicum. New Jersey: Cape May, thence down coast to Florida: Burnside Beach; New Berlin; St. Marks. Mississippi: Biloxi, thence along E bank Mississippi River through Tennessee to Missouri: Marble Hill. Illinois: western Franklin County (Klimstra and Scott, 1956: Fig. 1). Kentucky: Barbourville. Tennessee: Highcliff. South Carolina: Easley. North Carolina: Raleigh. Virginia: Brunswick County. Maryland: Nanjemoy Creek, 30 mi. SW Oxon Hill; 2 mi. NW Oxon Hill.

Guilday and Mayer-Oakes (1952:253, 254) record bones of *Oryzomys* from Indian archaeological sites from several places including the following. Ohio: near Cincinnati; near Lebanon; near Chillicothe; near Portsmouth. West Virginia: 2 mi. NNE Oglebay Park. Pennsylvania: Fayette County; Somerset County. These occurrences are not shown on Map 358.

Oryzomys palustris peninsulae Thomas

1897. *Oryzomys peninsulae* Thomas, Ann. Mag. Nat. Hist., ser. 6, 20:548, December, type from Santa Anita, Baja California.

MARGINAL RECORDS.—Baja California: type locality; *San José del Cabo*. On October 10, 1969, E. R. Hall examined the specimens used by Goldman (1918:45, 46), found them to agree with Goldman's description, but found the degree of difference between *O. peninsulae* and *O. palustris mexicanus* of the eastern coast of the Gulf of California to be less than between some pairs of intergrading subspecies of *O.*

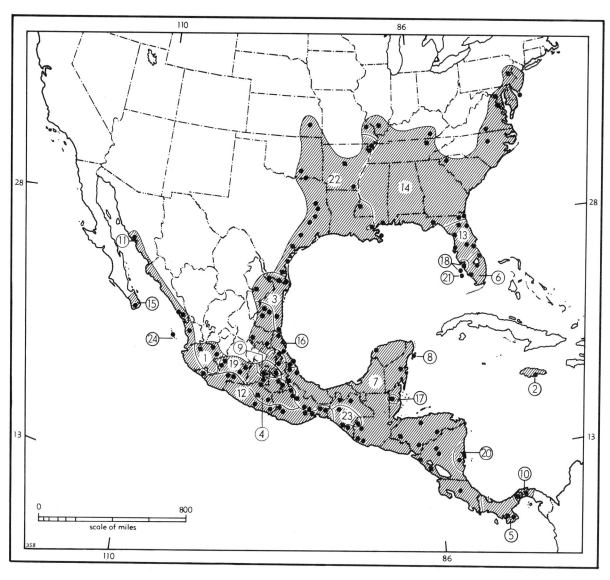

Map 358. *Oryzomys palustris* group.

1. *O. p. albiventer*
2. *O. p. antillarum*
3. *O. p. aquaticus*
4. *O. p. aztecus*
5. *O. p. azuerensis*
6. *O. p. coloratus*

7. *O. p. couesi*
8. *O. p. cozumelae*
9. *O. p. crinitus*
10. *O. p. gatunensis*
11. *O. p. lambi*
12. *O. p. mexicanus*

13. *O. p. natator*
14. *O. p. palustris*
15. *O. p. peninsulae*
16. *O. p. peragrus*
17. *O. p. pinicola*
18. *O. p. planirostris*

19. *O. p. regillus*
20. *O. p. richmondi*
21. *O. p. sanibeli*
22. *O. p. texensis*
23. *O. p. zygomaticus*
24. *O. nelsoni*

palustris (e.g., *O. p. texensis* and *O. p. aquaticus*). Therefore, *O. peninsulae* is here arranged as a subspecies of *O. palustris*.

Oryzomys palustris peragrus Merriam

1901. *Oryzomys mexicanus peragrus* Merriam, Proc. Washington Acad. Sci., 3:283, July 26, type from Río Verde, San Luis Potosí.

1960. *Oryzomys palustris peragrus*, Hall, Southwestern Nat., 5:173, November 1.

MARGINAL RECORDS.—Tamaulipas (Alvarez, 1963:436): 36 km. N, 10 km. W Ciudad Victoria; Río Corana. Veracruz: 5 mi. S Tampico (Hall and Dalquest, 1963:288); 3 km. SW San Marcos, 200 ft.; 4 km. W Tlapacoyan, 1700 ft.; *Miauapa* (Hall and Dalquest, 1963:288); 5 km. S Tihuatlán, 700 ft. San Luis Potosí: 3

km. N Tamazunchale; type locality. Tamaulipas: Juamave, 2400 ft. (Alvarez, 1963:436).

Oryzomys palustris pinicola A. Murie

1932. *Oryzomys couesi pinicola* A. Murie, Occas. Pap. Mus. Zool., Univ. Michigan, 245:1, June 9, type from a pine ridge, 12 mi. S El Cayo, Belize. Known only from type locality.
1960. *Oryzomys palustris pinicola*, Hall, Southwestern Nat., 5:173, November 1.

Oryzomys palustris planirostris Hamilton

1955. *Oryzomys palustris planirostris* Hamilton, Proc. Biol. Soc. Washington, 68:83, August 3, type from 1 mi. W 3rd bridge that spans Matachla Pass, Pine Island, Lee Co., Florida.

MARGINAL RECORDS.—Florida: 2 mi. N Fort Myers; type locality.

Oryzomys palustris regillus Goldman

1915. *Oryzomys couesi regillus* Goldman, Proc. Biol. Soc. Washington, 28:129, June 29, type from Los Reyes, Michoacán.
1960. *Oryzomys palustris regillus*, Hall, Southwestern Nat., 5:173, November 1.

MARGINAL RECORDS.—Michoacán: 1 mi. N Zamora, 5450 ft.; Queréndaro; 1 mi. E, 6 mi. S Tacámbaro, 4000 ft.; type locality; *4 mi. W Zamora, 5450 ft.*

Oryzomys palustris richmondi Merriam

1901. *Oryzomys richmondi* Merriam, Proc. Washington Acad. Sci., 3:284, July 26, type from Escondido River, 50 mi. above Bluefields, Nicaragua.
1960. *Oryzomys palustris richmondi*, Hall, Southwestern Nat., 5:173, November 1.

MARGINAL RECORDS.—Nicaragua: type locality; *Escondido River, 16 mi. above Bluefields.*

Oryzomys palustris sanibeli Hamilton

1955. *Oryzomys palustris sanibeli* Hamilton, Proc. Biol. Soc. Washington, 68:85, August 3, type from 4 mi W of lighthouse on Sanibel Island, Lee Co., Florida. Known only from type locality.

Oryzomys palustris texensis J. A. Allen

1894. *Oryzomys palustris texensis* J. A. Allen, Bull. Amer. Mus. Nat. Hist., 6:177, May 31, type from Rockport, Aransas Co., Texas.

MARGINAL RECORDS.—?Kansas: Neosho Falls. Arkansas: Fourche Bayou, near Little Rock. Missouri: Kennett; Portageville. Arkansas: Wilmot. Mississippi: Fayette. Louisiana: Lake Catherine, thence westward along coast, including *Brush Island* (Lowery, 1974:230) and Breton Island (*ibid.*), to Texas: *Padre Island*; $4\frac{1}{10}$

mi. W Port Mansfield (Hall, 1960:172); Corpus Christi; Victoria; Colorado County (Davis, 1966:195, as *O. palustris* only); Walker County (*ibid.*, as *O. palustris* only); Trinity County (*ibid.*, as *O. palustris* only); Nacogdoches County (*ibid.*, as *O. palustris* only); Rusk County (*ibid.*, as *O. palustris* only). Oklahoma: 5 mi. N Colbert; $3\frac{3}{8}$ mi. N Madill (McCarley, 1960:131).

Oryzomys palustris zygomaticus Merriam

1901. *Oryzomys zygomaticus* Merriam, Proc. Washington Acad. Sci., 3:285, July 26, type from Nentón, Guatemala.
1960. *Oryzomys palustris zygomaticus*, Hall, Southwestern Nat., 5:173, November 1.

MARGINAL RECORDS.—Chiapas: Ocuilapa. Guatemala: type locality. Chiapas: 1 km. S Mapastepec, 46 m.

Oryzomys nelsoni Merriam
Nelson's Rice Rat

1898. *Oryzomys nelsoni* Merriam, Proc. Biol. Soc. Washington, 12:15, January 27, type from María Madre Island, Tres Marías Islands, Nayarit. Known only from type locality.

External measurements of two adults, female and male, are: 320, 344; 185, 191; 37, 39. Upper parts rich ochraceous buff, richest on rump, darkened dorsally by black hairs; underparts white; tail dark brown, except on proximal third to half of underside, which is pale yellowish. Skull massive, long, narrow; rostrum massive, especially anteriorly, and strongly decurved; zygomata heavy. See Map 358.

Oryzomys fulgens Thomas
Thomas' Rice Rat

1893. *Oryzomys fulgens* Thomas, Ann. Mag. Nat. Hist., ser. 6, 11:403, May, type from "México." Probably in or near the Valley of Mexico. No other referred specimens.

Measurements of dry skin of type: head and body, 160; tail, 151; hind foot, 37.5. This species has not been critically compared with other species of *Oryzomys* by recent workers, and its status is doubtful. To judge from the original characterization, salient characteristics are: bright fulvous upper parts; anterior half of upper parts paler and duller than posterior half [molt?]; broad skull; inner wall of orbit forms one even curve; premaxillary processes barely reaching posterior level of nasals. Not shown on a distribution map.

Oryzomys dimidiatus (Thomas)
Nicaraguan Rice Rat

1905. *Nectomys dimidiatus* Thomas, Ann. Mag. Nat. Hist., ser. 7, 15:586, June, type from Río Escondido, 7 mi. below Rama, Nicaragua.

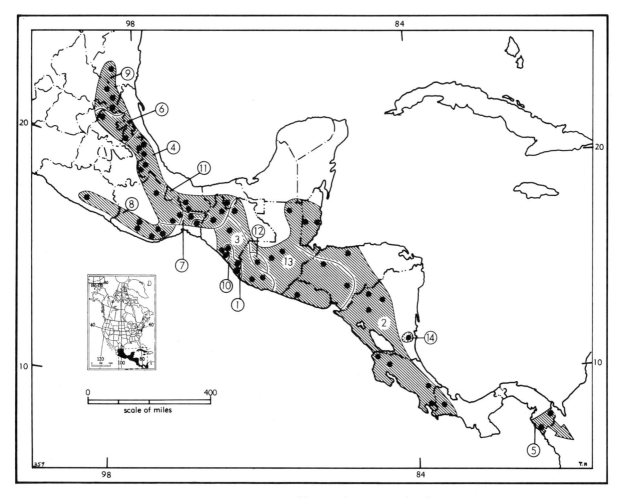

Map 359. *Oryzomys alfaroi* and *Oryzomys dimidiatus*.

Guide to kinds	5. *O. a. dariensis*	10. *O. a. hylocetes*
1. *O. alfaroi agrestis*	6. *O. a. dilutior*	11. *O. a. palatinus*
2. *O. a. alfaroi*	7. *O. a. gloriaensis*	12. *O. a. rhabdops*
3. *O. a. angusticeps*	8. *O. a. guerrerensis*	13. *O. a. saturatior*
4. *O. a. chapmani*	9. *O. a. huastecae*	14. *O. dimidiatus*

1970. *Oryzomys dimidiatus*, Hershkovitz, Jour. Mamm., 51:789, November 30. See also, Hershkovitz, Proc. U.S. Nat. Mus., 98:49, June 30, 1948, for placement of *dimidiatus* in *Oryzomys* without employing the name combination.

External measurements (holotype and a "young adult" male): 240, 228; 115, 110; 27 (dry), 28. Dorsum brownish to blackish; venter clay color to grayish; tail slaty gray above, whitish below. See Hershkovitz (1948:55; 1970:789–794) for detailed description, and Genoways and Jones (1971:833, 834) for additional details including comparison with the sympatric *O. palustris*.

MARGINAL RECORDS.—Nicaragua: type locality; *El Recreo*, 25 *m.*, *Zelaya* (Genoways and Jones, 1971:833).

Oryzomys melanotis
Black-eared Rice Rat

Hooper's (Occas. Pap. Mus. Zool., Univ. Michigan, 544:8, March 25, 1953) view that *O. rostratus* and *O. melanotis* are subspecies of one species accounts for four of the seven name combinations that follow.

External measurements: 216–277; 116–145; 26–33. Upper parts varying from ochraceous-buff to slightly paler, in worn adults sometimes becoming rusty red; underparts white with buffy suffusion, the basal gray usually showing through; pale-colored subauricular spot present; tail brownish above, paler below except at tip. Skull differs from that of *O. palustris* as follows: rostrum longer, anterior palatine foramina shorter in relation to palatal bridge, auditory bullae smaller, and m3 more deeply incised by outer re-entrant angle.

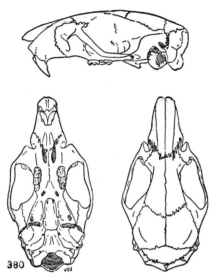

Fig. 380. *Oryzomys melanotis rostratus*, Potrero Llano, 350 ft., Veracruz, No. 30540 K.U., ♂, X 1½.

Oryzomys melanotis carrorum Lawrence

1947. *Oryzomys rostratus carrorum* Lawrence, Proc. New England Zool. Club, 24:101, May 29, type from Rancho Santa Ana, about 8 mi. SW Padilla, Río Soto la Marina, Tamaulipas.

MARGINAL RECORDS.—Tamaulipas: type locality; Sierra de Tamaulipas, 2 mi. S, 10 mi. W Piedra (Alvarez, 1963:437).

Oryzomys melanotis colimensis Goldman

1918. *Oryzomys melanotis colimensis* Goldman, N. Amer. Fauna, 43:51, September 23, type from Armería, Colima.

MARGINAL RECORDS.—Jalisco: 5 mi. NE Barra de Navidad, 200 ft.; *20 mi. SSE Autlán, 6500 ft.* Colima: Hda. Magdalena. Michoacán: 16 mi. S Arteaga, 800 ft., thence northward along coast to point of beginning.

Oryzomys melanotis megadon Merriam

1901. *Oryzomys rostratus megadon* Merriam, Proc. Washington Acad. Sci., 3:294, July 26, type from Teapa, Tabasco.

MARGINAL RECORDS.—Quintana Roo: 4 km. NNE Felipe Carrillo Puerto (92211 KU); Esmeralda

(Hall and Kelson, 1959:561, as *O. m. yucatanensis*). Guatemala: 11 km. NE Flores (Ryan, 1960:13, as *O. rostratus yucatanensis*). Tabasco: type locality. Campeche: Champotón; Apazote.

Oryzomys melanotis melanotis Thomas

1893. *Oryzomys melanotis* Thomas, Ann. Mag. Nat. Hist., ser. 6, 11:404, May, type from Mineral San Sebastián, Jalisco.

MARGINAL RECORDS.—Sinaloa: Los Limones. Nayarit: Santiago. Jalisco: type locality; Ixtapa.

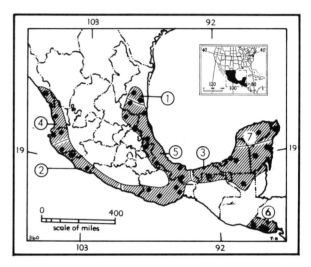

Map 360. *Oryzomys melanotis.*

Guide to subspecies	
1. *O. m. carrorum*	4. *O. m. melanotis*
2. *O. m. colimensis*	5. *O. m. rostratus*
3. *O. m. megadon*	6. *O. m. salvadorensis*
	7. *O. m. yucatanensis*

Oryzomys melanotis rostratus Merriam

1901. *Oryzomys rostratus* Merriam, Proc. Washington Acad. Sci., 3:293, July 26, type from Metlaltoyuca, Puebla.
1953. *Oryzomys melanotis rostratus*, Hooper, Occas. Pap. Mus. Zool., Univ. Michigan, 544:8, March 25.

MARGINAL RECORDS.—Tamaulipas: 70 km. by highway S Ciudad Victoria, 2 km. W El Carrizo; Altamira. Veracruz: 3 km. E San Andrés Tuxtla, 1000 ft.; 35 km. SE Jesús Carranza (Hall and Dalquest, 1963:289). Oaxaca (Goodwin, 1969:156, unless otherwise noted): mts. near [12 mi. NE] Santo Domingo; Kilometer 183, 36½ km. N San Gabriel Mixtepec (Schaldach, 1966:293); San Vicente; Jaltepec. Veracruz: Cautlapan, 4000 ft. (Hall and Dalquest, 1963:289). Puebla: Rancho El Ajenjibre (Warner and Beer, 1957:19); type locality. San Luis Potosí: 3 km. N Tamazunchale; El Salto.

Oryzomys melanotis salvadorensis Felten

1958. *Oryzomys rostratus salvadorensis* Felten, Senckenbergiana Biol., 39:3, March 31, type from Hda. San Antonio, El Salvador.

MARGINAL RECORDS (Burt and Stirton, 1961:61).—El Salvador: type locality; Kilometer 80 between San Salvador and San Miguel; Puerto del Triunfo.

Oryzomys melanotis yucatanensis Merriam

1901. *Oryzomys yucatanensis* Merriam, Proc. Washington Acad. Sci., 3:294, July 26, type from Chichén-Itzá, Yucatán.
1918. *Oryzomys rostratus yucatanensis*, Goldman, N. Amer. Fauna, 43:55, September 23.

MARGINAL RECORDS.—Quintana Roo: Puerto Morelos. Yucatán: type locality.

Oryzomys caudatus Merriam
Ixtlán Rice Rat

1901. *Oryzomys chapmani caudatus* Merriam, Proc. Washington Acad. Sci., 3:289, July 26, type from Comaltepec, 3500 ft., Oaxaca.
1969. *Oryzomys caudatus*, Goodwin, Bull. Amer. Mus. Nat. Hist., 141:157, April 30.

External measurements of six adults, including type, from Oaxaca are: 218–257; 97–116; 25–30. Pelage buffy brown mixed with black; tail unicolored. Skull large and heavy; braincase flattened.

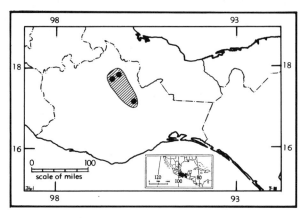

Map 361. *Oryzomys caudatus.*

MARGINAL RECORDS (Goodwin, 1969:157, 158).—Oaxaca: 13 mi. NE Llano de las Flores; Totontepec; type locality; *4 mi. SW Llano de las Flores; N of Llano de las Flores.*

Oryzomys alfaroi
Alfaro's Rice Rat

External measurements: 174–265; 94–114; 23–29.5. Upper parts varying from dark ochraceous buff to ochraceous tawny, or even brownish, usually with a mixture of black, especially on head and mid-dorsally; underparts white, usually with buffy suffusion, plumbeous basal color usu-ally showing through; tail sparsely haired, brownish above, paler below, becoming dusky above and below distally. Skull relatively elongated; braincase vaulted. (See Map 359.)

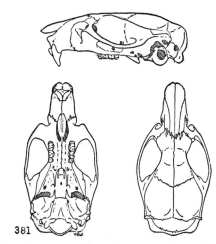

Fig. 381. *Oryzomys alfaroi chapmani*, 4 km. W Tlapacoyan, 1700 ft., Veracruz, No. 24107 K.U., ♀, X 1½.

Oryzomys alfaroi agrestis Goodwin

1959. *Oryzomys alfaroi agrestis* Goodwin, Amer. Mus. Novit., 1967:7, October 29, type from 12 mi. SE Tapachula, Chiapas, México, 1000 ft. [if this locality is correct, actual elevation would be less than 100 m.]. Known only from type locality.

Oryzomys alfaroi alfaroi (J. A. Allen)

1891. *Hesperomys (Oryzomys) alfaroi* J. A. Allen, Bull. Amer. Mus. Nat. Hist., 3:214, April 17, type from San Carlos, Costa Rica.
1894. *Oryzomys alfaroi* J. A. Allen, Abstr. Proc. Linnaean Soc. New York, 1893–1894, p. 36, July 20.
1908. *Oryzomys alfaroi incertus* J. A. Allen, Bull. Amer. Mus. Nat. Hist., 24:655, October 13, type from Río Grande, Zelaya, Nicaragua.

MARGINAL RECORDS.—Guatemala: Uaxactún. Belize: near Mountain Pine Ridge, 12 mi. S El Cayo; Bokowina. Honduras: Yaruca. Nicaragua: Tuma. Costa Rica: Santa Teresa Peralta. Panamá: Boquete. Costa Rica: Agua Buena, Puntarenas; 3½ mi. E Tilarán (No. 221 of David G. Huckaby in Louisiana State Univ. Mus. Zool.); Hda. Santa María. Nicaragua: Jinotega; Jícaro.

Oryzomys alfaroi angusticeps Merriam

1901. *Oryzomys angusticeps* Merriam, Proc. Washington Acad. Sci., 3:292, July 26, type from Volcán Santa María, 9000 ft., Guatemala.
1918. *Oryzomys alfaroi angusticeps*, Goldman, N. Amer. Fauna, 43:62, September 23.

MARGINAL RECORDS.—Chiapas: San Cristóbal. Guatemala: San Lucas; *Volcán San Lucas;* Finca Ciprés. Chiapas: *Pinabete;* Triunfo, 1950 m.; Catarina, 1300 m.

Oryzomys alfaroi chapmani Thomas

1898. *Oryzomys chapmani* Thomas, Ann. Mag. Nat. Hist., ser. 7, 1:179, February, type from Jalapa, 4400 ft., Veracruz.
1918. *Oryzomys alfaroi chapmani,* Goldman, N. Amer. Fauna, 43:67, September 23.

MARGINAL RECORDS.—Veracruz: 4 km. W Tlapacoyan, 1700 ft.; *5 km. N Jalapa, 4500 ft.* (Hall and Dalquest, 1963:290); type locality; Mirador, 3500 ft. Oaxaca (Goodwin, 1969:158): Campamento Vista Hermosa; *Macuiltianguis;* Petepa [= Petapa]; Santiago Lachiguirí; *Tarabundí.* Veracruz: *Huatusco, 5000 ft.* (Hall and Dalquest, 1963:290); Coscomatepec, 5000 ft.; 1 km. E Jalacingo, 6500 ft.

Oryzomys alfaroi dariensis Goldman

1915. *Oryzomys alfaroi dariensis* Goldman, Proc. Biol. Soc. Washington, 28:128, June 29, type from Cana, 2000 ft., eastern Panamá.

MARGINAL RECORDS.—Panamá: Mt. Tacarcuna, thence into South America and back to Panamá: type locality.

Oryzomys alfaroi dilutior Merriam

1901. *Oryzomys chapmani dilutior* Merriam, Proc. Washington Acad. Sci., 3:290, July 26, type from Huauchinango, 5000 ft., Puebla.
1918. *Oryzomys alfaroi dilutior,* Goldman, N. Amer. Fauna, 43:68, September 23.

MARGINAL RECORDS.—Hidalgo: 16 km. NW Jacala, 1550 m. Puebla: type locality.

Oryzomys alfaroi gloriaensis Goodwin

1956. *Oryzomys alfaroi gloriaensis* Goodwin, Amer. Mus. Novit., 1757:8, March 8, type from La Gloria, rain forest, about 2500 ft., 10 km. S Santa María Chimalapa, Juchitán Dist., Oaxaca.

MARGINAL RECORDS.—Oaxaca: type locality. *O. a. gloriaensis* has been recorded (Baker, *et al.,* 1973:78, 85) from *8 km. N Berriozábal, 1065 m., Chiapas.*

Oryzomys alfaroi guerrerensis Goldman

1915. *Oryzomys guerrerensis* Goldman, Proc. Biol. Soc. Washington, 28:127, June 29, type from Omilteme, 8000 ft., Guerrero.
1959. *Oryzomys alfaroi guerrerensis,* Hall and Kelson, Mammals of North America, Ronald Press, p. 562, March 31.

MARGINAL RECORDS.—Guerrero: type locality. Oaxaca (Goodwin, 1969:158, unless otherwise noted):

Kilometer 183, 36½ km. N San Gabriel Mixtepec (Schaldach, 1966:294); Río Grande; Zapotitlán; Pluma; Lacháo.

Oryzomys alfaroi huastecae Dalquest

1951. *Oryzomys alfaroi huastecae* Dalquest, Jour. Washington Acad. Sci., 41:363, November 14, type from 10 km. E Platanito, San Luis Potosí.

MARGINAL RECORDS.—Tamaulipas: Rancho del Cielo. San Luis Potosí: type locality; 3 km. N Valles; Xilitla.

Oryzomys alfaroi hylocetes Merriam

1901. *Oryzomys hylocetes* Merriam, Proc. Washington Acad. Sci., 3:291, July 26, type from Chicharras, 3500 ft., Chiapas.
1949. *Oryzomys alfaroi hylocetes,* Villa, Anal. Inst. Biol., Univ. Nac. Autó. México, 19(2):518, June 30.

MARGINAL RECORDS.—Chiapas: Finca Liquidámbar, 1210 m.; type locality; *Finca Prusia, 1200 m.*

Oryzomys alfaroi palatinus Merriam

1901. *Oryzomys palatinus* Merriam, Proc. Washington Acad. Sci., 3:290, July 26, type from Teapa, 3000 ft., Tabasco.
1918. *Oryzomys alfaroi palatinus,* Goldman, N. Amer. Fauna, 43:65, September 23.

MARGINAL RECORDS.—Veracruz: 20 km. ENE Jesús Carranza, 200 ft. Tabasco: type locality. Chiapas: 6½ km. SE Rayón, 1675 m. (Baker, *et al.,* 1973:78, 85); Ocuilapa. Oaxaca (Goodwin, 1969:158): Sierra Madre, N of Zanatepec; *Río Blanco.* Veracruz: 35 km. SE Jesús Carranza (Hall and Dalquest, 1963:291).

Oryzomys alfaroi rhabdops Merriam

1901. *Oryzomys rhabdops* Merriam, Proc. Washington Acad. Sci., 3:291, July 26, type from Calel, 10,000 ft., Guatemala. Known only from type locality.
1918. *Oryzomys alfaroi rhabdops,* Goldman, N. Amer. Fauna, 43:63, September 23.

Oryzomys alfaroi saturatior Merriam

1901. *Oryzomys chapmani saturatior* Merriam, Proc. Washington Acad. Sci., 3:290, July 26, type from Tumbalá, 5000 ft., Chiapas.
1918. *Oryzomys alfaroi saturatior,* Goldman, N. Amer. Fauna, 43:66, September 23.

MARGINAL RECORDS.—Chiapas: type locality; *Laguna Ocotal.* Guatemala: Chimoxan. Honduras: La Mica; Cantoral. El Salvador: Chilata (Burt and Stirton, 1961:61). Guatemala: El Soche.

Oryzomys bombycinus
Long-whiskered Rice Rat

Revised by Pine, Jour. Mamm., 52:590–596, August 26, 1971.

External measurements, recorded by Pine (1971:593), of 40-some specimens of unspecified age: 198–264; 96–130; 27–33. Appendages long, slender, and nearly naked. Pelage of upper parts long (*ca.* 12 mm. on back); underfur not wooly (as in, *e.g.*, *O. palustris*); upper parts approx. cinnamon brown or russet; underparts washed with white; rostrum long, narrow; nasals extending posteriorly to tips of premaxillary tongues; braincase broad; zygomata slender; frontal region broad, flattened, somewhat depressed anteromedially; distinct but narrow supraorbital ridges; palatal pits small; anterior palatine slits short; sphenopalatine vacuities small to absent; tympanic bullae small; crowns of molars deeply incised by re-entrant angles; mandibular tooth-row conspicuously narrowed posteriorly. Pine (1971: 590) notes that this species differs from *O. capito* in longer vibrissae and in that the parietotemporal suture corresponds in position with the temporal ridge instead of having the suture positioned below the ridge posteriorly.

Gardner and Patton (1976:40, 41) note that Pine (1971) did not mention *Oryzomys rivularis* J. A. Allen, 1901, from northwestern Ecuador, in his review of the taxonomy of *O. bombycinus*, and suggest that *O. rivularis*, an earlier name, may be conspecific with *O. bombycinus*.

Oryzomys bombycinus alleni Goldman

1915. *Oryzomys nitidus alleni* Goldman, Proc. Biol. Soc. Washington, 28:128, June 29, type from Tuís, about 20 mi. E Cartago, Cartago, Costa Rica.
1918. *Oryzomys bombycinus alleni* Goldman, N. Amer. Fauna, 43:78, September 23.

MARGINAL RECORDS (Pine, 1971:592).—Nicaragua: Río Kurinwas, 12° 51′ 30″ N, 84° 05′ W, *ca.* 10 m. Costa Rica: Cariari, *ca.* 100 m. Panamá: Río Changena, 9° 06′ N, 82° 34′ W. Costa Rica: type locality; Guápiles.

Oryzomys bombycinus bombycinus Goldman

1912. *Oryzomys bombycinus* Goldman, Smiths. Miscl. Coll., 56(36):6, February 19, type from Cerro Azul, 2500 ft., near headwaters Chagres River, Panamá.

MARGINAL RECORDS (Pine, 1971:590, 591).—Panamá: Cerro Bruja, 1000 ft.; type locality; *6 mi. E El Valle*; El Valle.

Oryzomys bombycinus orinus Pearson

1939. *Oryzomys bombycinus orinus* Pearson, Not. Naturae, Acad. Nat. Sci. Philadelphia, 6:2, June 8, type from Mt. Pirri, near Río Limón, Darién, Panamá.

MARGINAL RECORDS (Pine, 1971:595).—Panamá: Cerro Tacarcuna, 4000 ft.; type locality; *Loma Caña, 4900 ft.*, thence into South America.

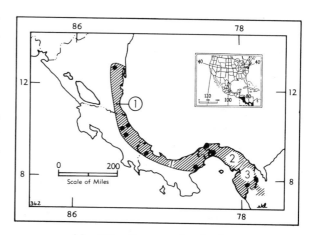

Map 362.   *Oryzomys bombycinus*.

1. *O. b. alleni*   2. *O. b. bombycinus*
             3. *O. b. orinus*

Oryzomys albigularis
Tomes' Rice Rat

External measurements: 309–360; 159–195; 33–38. Upper parts dark tawny, heavily mixed with black; throat whitish; remaining underparts whitish to ochraceous buffy; tail (skin) dark brown above, paler below. Skull large; rostrum long, heavy; nasals extending slightly posterior to premaxillaries; zygomata massive; frontals narrow; supraorbital ridges weak or absent; interparietal large; anterior palatine pits especially short; palatal pits large; sphenopalatine vacuities absent; crowns of molars deeply incised by re-entrant angles; mandibular tooth-row conspicuously narrow posteriorly.

Oryzomys albigularis devius Bangs

1902. *Oryzomys devius* Bangs, Bull. Mus. Comp. Zool., 39:34, April, type from Boquete, 5000 ft., Volcán de Chiriquí, Chiriquí, Panamá.
1966. O[*ryzomys*]. *a*[*lbigularis*]. *devius*, Handley, Ectoparasites of Panama, Field Mus. Nat. Hist., p. 779, November 22.

MARGINAL RECORDS.—Costa Rica: Lajas Villa Quesada; Estrella. Panamá: type locality. Costa Rica: El Muñeco.

Oryzomys albigularis pirrensis Goldman

1913. *Oryzomys pirrensis* Goldman, Smiths. Miscl. Coll. 60(22):5, February 28, type from near head of Río Limón, Mt. Pirri, 4500 ft., Panamá.
1961. *Oryzomys albigularis pirrensis*, Cabrera, Rev. Mus. Argentino de Cienc. Nat., 4:383, Agosta 25.

MARGINAL RECORDS.—Panamá: Mt. Tacarcuna, thence into South America and back to Panamá: type locality.

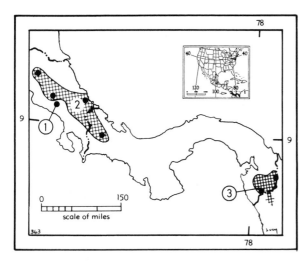

Map 363. *Oryzomys aphrastus* and *Oryzomys albigularis.*

1. *O. aphrastus*   2. *O. albigularis devius*
3. *O. albigularis pirrensis*

Oryzomys aphrastus Harris
Harris' Rice Rat

1932. *Oryzomys aphrastus* Harris, Occas. Pap. Mus. Zool., Univ. Michigan, 248:5, August 4, type from San Joaquín de Dota, about 4000 ft., San José, Costa Rica. Known only from type locality.

Measurements and description of the type (Harris, 1932:5, 6) are: total length, 387; tail, 235; hind foot, 40. Upper parts "very dark grizzled yellowish brown. Sides and flanks paler and more Ochraceous-Buff. . . . Tail blackish both above and below with tuft of black hairs at the tip. Feet dark brownish black. . . . Hairs of underparts lead gray at base tipped with Ochraceous-Buff." Skull long, narrow; supraorbital and temporal ridges prominent; auditory bullae small; nasals extend posterior to premaxillary tongues.

This "species" is provisionally placed between *O. albigularis* and *O. capito.*

Oryzomys capito
Large-headed Rice Rat

External measurements: 233–265; 115–133; 29–30.5. Color essentially as in *O. alfaroi.* Skull large; braincase somewhat flattened; maxillary arm of zygoma robust; nasals and premaxillary tongues approx. coterminous posteriorly; frontals broad, trenchant laterally, and depressed anteromedially; interparietal large; palatal pits small; anterior palatal pits much shorter than palatal bridge; tympanic bullae small; in molars, inner and outer re-entrant angles above and below deep. Also occurs in South America.

Oryzomys capito carrikeri J. A. Allen

1901. *Oryzomys carrikeri* J. A. Allen, Bull. Amer. Mus. Nat. Hist., 24:656, October 13, type from bank of Río Sicosla [= Sixaola], between Cuabre and mouth of Río Sixaola, Limón, Costa Rica.
1966. *O*[*ryzomys*]. *c*[*apito*]. *carrikeri,* Handley, Ectoparasites of Panama, Field Mus. Nat. Hist., p. 780, November 22.

MARGINAL RECORDS.—Costa Rica: type locality. Panamá (Handley, 1966:780): *Boca del Drago;* Almirante.

Oryzomys capito talamancae J. A. Allen

1891. *Oryzomys talamancae* J. A. Allen, Proc. U.S. Nat. Mus., 14:193, July 24, type from Talamanca, Limón, Costa Rica.
1966. *O*[*ryzomys*]. *c*[*apito*]. *talamancae,* Handley, Ectoparasites of Panama, Field Mus. Nat. Hist., p. 780, November 22.
1901. *Oryzomys panamensis* Thomas, Ann. Mag. Nat. Hist., ser. 7, 8:252, September, type from Panamá, Panamá.

MARGINAL RECORDS (Handley, 1966:780).—Panamá: Cerro Bruja; Mandinga; Armila; Tacarcuna Village; Cana; Jaqué; Cerro Campana; Cerro Hoya. Costa Rica: type locality. Panamá: Fort Sherman.

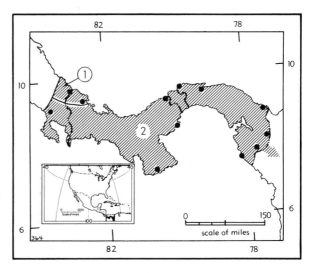

Map 364. *Oryzomys capito.*

1. *O. c. carrikeri*   2. *O. c. talamancae*

Subgenus Oecomys Thomas

Revised by Hershkovitz, Proc. U.S. Nat. Mus., 110:513–568, 6 figs., 12 pls., February 4, 1960.

1906. *Oecomys* Thomas, Ann. Mag. Nat. Hist., ser. 7, 18:444, December. Type, *Rhipidomys benevolens* Thomas [= *Rhipodomys phaeotis* Thomas = *Oryzomys bicolor phaeotis,* Hershkovitz 1960].

Tail long (averaging more than half of total length), and slightly penciled; hind foot short, and broad relative to its length; claws thick and recurved; anterior palatine foramina wide; sphenopalatine foramina normally absent (small in some specimens); anterior margin of maxillary plate viewed dorsally not projecting half so far anterior to remainder of maxilla as in subgenus *Oryzomys* (after Hershkovitz, 1960:515–532). Gardner and Patton (1976:12, 20) consider *Oecomys* to be a genus on basis of karyotypic features. Semiarboreal rice rats.

KEY TO NORTH AMERICAN SPECIES OF SUBGENUS OECOMYS

1. Hind foot less than 25.8. . . . *O. bicolor*, p. 619
1'. Hind foot more than 25.8. *O. concolor*, p. 619

Oryzomys bicolor
Bicolored Rice Rat

External measurements of holotype of *O. endersi* are 238, 124, 25. Upper parts approx. cinnamon brown heavily infused with black to bright ochraceous rufous, darkened, especially mid-dorsally, with black; underparts varying from white to ochraceous; tail brownish, unicolored or paler below but not sharply bicolored.

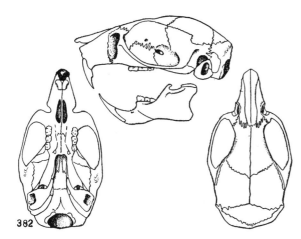

Fig. 382. *Oryzomys bicolor trabeatus*, Río Jesucito, Panamá, No. 19837 M.C.Z., holotype, ♂, X 1½.

Oryzomys bicolor trabeatus (G. M. Allen and Barbour)

1923. *Oecomys trabeatus* G. M. Allen and Barbour, Bull. Mus. Comp. Zool., 65:262, February, type from Río Jesucito, Panamá.
1960. *Oryzomys (Oecomys) bicolor trabeatus*, Hershkovitz, Proc. U.S. Nat. Mus., 110:533, February 4.
1933. *Oecomys endersi* Goldman, Jour. Washington Acad. Sci., 23:525, November 15, type from Barro Colorado Island, Canal Zone, Panamá.

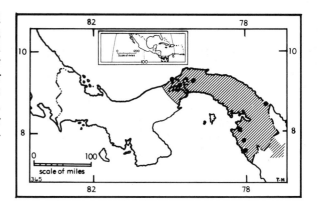

Map 365. *Oryzomys bicolor trabeatus*.

MARGINAL RECORDS (Handley, 1966:779).— Panamá: Fort Sherman; Cerro Azul; *Armila;* Puerto Obaldía; into and from South America into Panamá: Guayabo; Pelisa; type locality; Barro Colorado Island.

Oryzomys concolor
Concolored Rice Rat`

External measurements of specimens from Central America are 243–309, 123–162, 26–30. Coloration essentially as in *O. bicolor.* Also occurs in South America.

Oryzomys concolor tectus Thomas

1901. *Oryzomys tectus* Thomas, Ann. Mag. Nat. Hist., ser. 7,8:251, September, type from Bogava [= Bugaba], 800 ft., Chiriquí, Panamá.
1966. *O[ryzomys]. c[oncolor]. tectus*, Handley, Ectoparasites of Panama, Field Mus. Nat. Hist., p. 781, November 22.
1912. *Oryzomys frontalis* Goldman, Smiths. Miscl. Coll., 56(36):6, February 19, type from Corozal, Canal Zone, Panamá. Regarded as inseparable from *O. c. tectus* by Handley (1966:781).

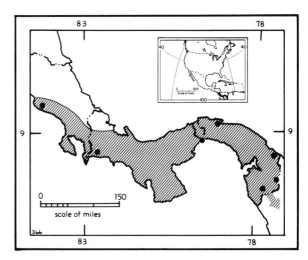

Map 366. *Oryzomys concolor tectus*.

MARGINAL RECORDS.—Costa Rica: San Gerónimo de Pirrís. Panamá (Handley, 1966:781): Mandinga; Armila; Tacarcuna Village, 1950 ft.; Cana; Corozal; type locality.

Subgenus **Oligoryzomys** Bangs

1900. *Oligoryzomys* Bangs, Proc. New England Zool. Club, 1:94, February 23. Type, *Oryzomys navus* Bangs.

Size small; hind foot usually less than 25; form slender and *Reithrodontomys*-like; tail markedly longer than head and body; 4 longer toes of hind feet bearing tufts of silvery bristles projecting beyond ends of claws. Skull delicate, smoothly rounded; supraorbital and temporal ridges absent; least interorbital constriction about equal to width of rostrum between antorbital foramina; molars with re-entrant angles usually broader than in subgenus *Oryzomys*, the salient angles formed by worn crowns of tubercles less evenly rounded; upper molars early exhibiting small, circular central enamel islands that persist until obliterated by wear in extreme old age (after Goldman, 1918:87, 88). Hershkovitz (1944:13) has noted that "The long-tailed *Microryzomys* [of South America] and *Oligoryzomys* represent hardly more than size gradations leading to the larger *Oryzomys*."

KEY TO SPECIES OF SUBGENUS OLIGORYZOMYS

1. Occurring on St. Vincent Island, Lesser Antilles. . . . .*O. victus,* p. 620
1'. Occurring in México and Central America. *O. fulvescens,* p. 620

Oryzomys victus Thomas
St. Vincent Rice Rat

1898. *Oryzomys victus* Thomas, Ann. Mag. Nat. Hist., ser. 7, 1:178, February, type from St. Vincent, Lesser Antilles. Known only from St. Vincent, Lesser Antilles.
(Not examined and group association not determined by Goldman, N. Amer. Fauna, 43:16, September 23, 1918.)

External measurements of type are: length of head and body, 96; tail, 121; hind foot without claws, 25. The characters of this species are not completely known and its systematic position is consequently problematical. If *Oligoryzomys* is accepted as a useful subgeneric grouping, and if *Oryzomys longicaudatus* (Bennett) is an *Oligoryzomys,* then it seems reasonable to assign *O. victus* to that subgenus because Thomas, in naming *O. victus,* compared it with *O. longicaudatus.*

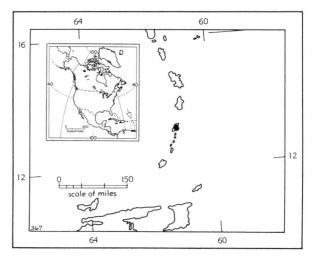

Map 367. *Oryzomys victus.*

Oryzomys fulvescens
Pygmy Rice Rat

External measurements: 168–235; 96–130; 20–25. Upper parts varying from ochraceous buff to tawny, most intense on rump, lightly mixed dorsally with blackish hairs; underparts white, sometimes strongly washed with buff; tail dark brown above, paler below, dusky above and below at tip.

Oryzomys fulvescens costaricensis J. A. Allen

1893. *Oryzomys costaricensis* J. A. Allen, Bull. Amer. Mus. Nat. Hist., 5:239, September 22, type from El General, 2150 ft., Puntarenas, Costa Rica.
1918. *Oryzomys fulvescens costaricensis,* Goldman, N. Amer. Fauna, 43:92, September 23.

MARGINAL RECORDS.—Costa Rica (Harris, 1943:12, unless otherwise noted): Hda. Santa Maria; San Gerónimo Pirrís (Goodwin, 1946:396); *Cartago;* El Copey de Dota; type locality; Buenos Aires (Goodwin, 1946:396). Panamá (Handley, 1966:781): *Cerro Punta, 4800–5600 ft.,* eastward to Barro Colorado Is-

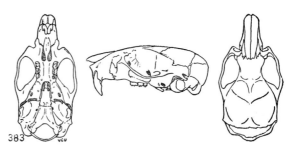

Fig. 383. *Oryzomys fulvescens fulvescens,* Cautlapan (= Ixtaczoquitlán), 4000 ft., Veracruz, No. 30575 K.U., ♂, X 1½.

land; Cerro Azul; Pacora, thence westward along Pacific Coast to point of beginning.

Oryzomys fulvescens creper Goodwin

1945. *Oryzomys fulvescens creper* Goodwin, Amer. Mus. Novit., 1293:2, July 20, type from Volcán Irazú, 9400 ft., Cartago, Costa Rica.

MARGINAL RECORDS.—Costa Rica: Cataratos San Carlos; Cervantes; *Buena Vista;* Escazú.

Oryzomys fulvescens engraciae Osgood

1945. *Oryzomys fulvescens engraciae* Osgood, Jour. Mamm., 26:300, November 14, type from Hda. Santa Engracia, NW of Ciudad Victoria, Tamaulipas.

MARGINAL RECORDS.—Nuevo León: 20 km. NW General Terán, 900 ft. Tamaulipas: type locality; *Altamira* (Alvarez, 1963:438); 7 km. N Tampico (*ibid.*); *Rancho Pano Ayuctle, 25 mi. N, 3 km. W El Mante* (*ibid.*); 10 km. N, 8 km. W El Encino.

Oryzomys fulvescens fulvescens (Saussure)

1860. *H[esperomys]. fulvescens* Saussure, Revue et Mag. Zool., Paris, ser. 2, 12:102, March, type from Veracruz; fixed by Merriam (Proc. Washington Acad. Sci., 3:295, July 26, 1901) at Orizaba.
1897. *Oryzomys fulvescens,* J. A. Allen and Chapman, Bull. Amer. Mus. Nat. Hist., 9:204, June 16.

MARGINAL RECORDS.—Tamaulipas: Rancho del Cielo. Veracruz: 5 km. S Tehuatlán, 700 ft.; 4 km. W Tlapacoyan, 1700 ft.; 3 km. E San Andrés Tuxtla, 100 ft.

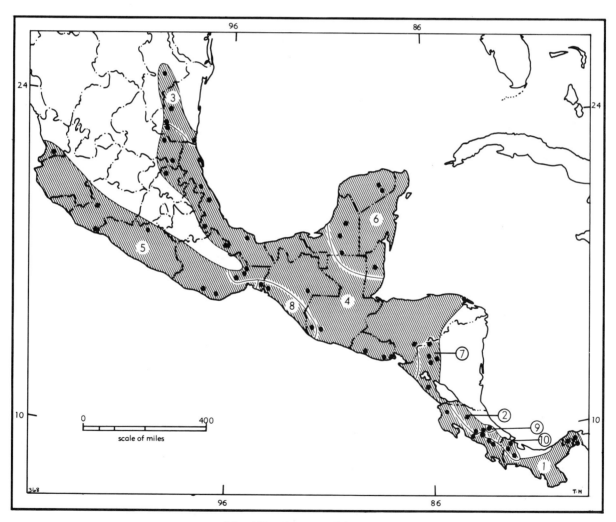

Map 368. *Oryzomys fulvescens.*

Guide	2. *O. f. creper*	5. *O. f. lenis*	8. *O. f. pacificus*
to subspecies	3. *O. f. engraciae*	6. *O. f. mayensis*	9. *O. f. reventazoni*
1. *O. f. costaricensis*	4. *O. f. fulvescens*	7. *O. f. nicaraguae*	10. *O. f. vegetus*

Honduras: Patuca; La Piedra de Jesús, Sabana Grande. El Salvador (Burt and Stirton, 1961:62): Lake Olomega; Puerto del Triunfo; Hda. Chilata. Guatemala: Finca Ciprés. Chiapas: Las Margaritas, 15 mi. E Comitán. Oaxaca (Goodwin, 1969:159): Santiago Lachiguirí; Matías Romero; Ubero; Tuxtepec. Veracruz: type locality. Hidalgo: 16 km. NW Jacala, 1550 m. San Luis Potosí: Xilitla; El Salto.

Oryzomys fulvescens lenis Goldman

1915. *Oryzomys fulvescens lenis* Goldman, Proc. Biol. Soc. Washington, 28:130, June 29, type from Los Reyes, Michoacán.

MARGINAL RECORDS.—Nayarit: vic. Santa Isabel. Michoacán: type locality. Guerrero: Chapa, 1470 m. Oaxaca (Goodwin, 1969:160): Pluma Hidalgo; San Gabriel Mixtepec. Michoacán: 22 km. S Arteaga (Alvarez, 1968:32), thence northward along coast to point of beginning.

Oryzomys fulvescens mayensis Goldman

1918. *Oryzomys fulvescens mayensis* Goldman, N. Amer. Fauna, 43:92, September 23, type from Apazote, near Yohaltún, Campeche.

MARGINAL RECORDS.—Yucatán: Tunkas; Chichén–Itzá. Belize: Glenwood Farm, less than 30 m. elev. (Kirkpatrick, *et al.*, 1975:331). Campeche: 103 km. SE Escárcega (93650 KU); 7 km. N, 51 km. E Escárcega; type locality.

Oryzomys fulvescens nicaraguae J. A. Allen

1910. *Oryzomys (Oligoryzomys) nicaraguae* J. A. Allen, Bull. Amer. Mus. Nat. Hist., 28:100, April 30, type from Vijagua [= Bijagua], Boaco, Nicaragua.
1946. *Oryzomys fulvescens nicaraguae*, Goodwin, Bull. Amer. Mus. Nat. Hist., 87:397, December 31.

MARGINAL RECORDS.—Nicaragua: 3½ km. S, 2 km. W Jalapa (Jones and Genoways, 1970:8, as *O. fulvescens* only); type locality; 3 mi. NNW Diriamba (Jones and Genoways, 1970:14, as *O. fulvescens* only); Santa María de Ostuma (Jones and Genoways, 1970:2, as *O. fulvescens* only); San Rafael del Norte.

Oryzomys fulvescens pacificus Hooper

1952. *Oryzomys fulvescens pacificus* Hooper, Proc. Biol. Soc. Washington, 65:23, January 29, type from Mapastepec, Chiapas.

MARGINAL RECORDS.—Oaxaca: Tapanatepec (Goodwin, 1969:160). Chiapas: Arriaga, 300 ft. Guatemala: Hda. California.

Oryzomys fulvescens reventazoni Goodwin

1945. *Oryzomys fulvescens reventazoni* Goodwin, Amer. Mus. Novit., 1293:3, July 20, type from Santa Teresa Peralta, Cartago, Costa Rica. Known only from type locality.

Oryzomys fulvescens vegetus Bangs

1902. *Oryzomys (Oligoryzomys) vegetus* Bangs, Bull. Mus. Comp. Zool., 39:35, April, type from Boquete, 4000 ft., Volcán de Chiriquí, Chiriquí, Panamá.
1918. *Oryzomys fulvescens vegetus*, Goldman, N. Amer. Fauna, 43:93, September 23.

MARGINAL RECORDS (Handley, 1966:781).— Panamá: upper Río Changena, 4800 ft.; Cerro Punta; type locality.

Subgenus Melanomys Thomas

1902. *Melanomys* Thomas, Ann. Mag. Nat. Hist., ser. 7, 10:248, September. Type, *Oryzomys phaeopus* Thomas.

Color dark; upper parts and underparts not strongly contrasted; form robust; tail approx. three-quarters length of head and body, black all around; hind feet broad, stout, digital bristles not projecting beyond ends of claws. Skull rotund; rostrum short, nearly straight; braincase large and inflated; frontals broad, their lateral margins projecting as supraorbital shelves; zygomata slender, but maxillary root decidedly expanded above along frontal and premaxillary sutures (after Goldman, 1918:94, 95).

Oryzomys caliginosus
Dusky Rice Rat

External measurements: 196–240; 85–105; 25–27.5. Upper parts varying from russet to tawny, finely but heavily mixed with black; underparts paler in most individuals, but neither sharply nor conspicuously so. Also occurs in South America.

Oryzomys caliginosus chrysomelas J. A. Allen

1897. *Oryzomys chrysomelas* J. A. Allen, Bull. Amer. Mus. Nat. Hist., 9:37, March 11, type from Suerre, near Jiménez, Limón, Costa Rica.
1918. *Oryzomys caliginosus chrysomelas*, Goldman, N. Amer. Fauna, 43:97, September 23.

MARGINAL RECORDS.—Honduras: Río Coco, within 1 mi. Waspam [= Huaspan] (Pine, 1969:643). Costa Rica: Port Limón; Talamanca, probably near Sipurio. Panamá (Handley, 1966:780): Boca del Drago; Isla Bastimentos; Bugaba. Costa Rica: Coto; San Gerónimo Pirrís; Cataratos San Carlos. Nicaragua: Peña Blanca; Tuma; Río Coco; *Río Grande*.

Oryzomys caliginosus idoneus Goldman

1912. *Oryzomys idoneus* Goldman, Smiths. Miscl. Coll., 56(36):5, February 19, type from Cerro Azul, 2500 ft., near headwaters of Chagres River, Panamá.

1'. Occurring north of Costa Rica.
    4. Tympanic bullae large and globose anteriorly, lacking marked anterior projection. . . . .*T. bullaris*,   p. 627
    4'. Tympanic bullae not markedly inflated, having conspicuous anterior projection.
        5. Maxillary tooth-row more than 9.1. . . . .*T. tumbalensis*,   p. 628
        5'. Maxillary tooth-row less than 9.1.   *T. nudicaudus*,   p. 627

Tylomys bullaris Merriam
Chiapan Climbing Rat

1901. *Tylomys bullaris* Merriam, Proc. Washington Acad. Sci., 3:561, November 29, type from Tuxtla, Chiapas. Known only from type locality.

External measurements of the immature type, the only known specimen, are: 324; 158; 37.5. Upper parts pale grayish plumbeous; upper lip and patch on side of nose white; underparts white; feet brown to toes which are white. Interparietal broad; zygomata widely spreading; auditory bullae markedly inflated, globose rather than pointed anteriorly.

Tylomys fulviventer Anthony
Fulvous-bellied Climbing Rat

1916. *Tylomys fulviventer* Anthony, Bull. Amer. Mus. Nat. Hist., 35:366, June 9, type from Tacarcuna, 4200 ft., Darién, Panamá.

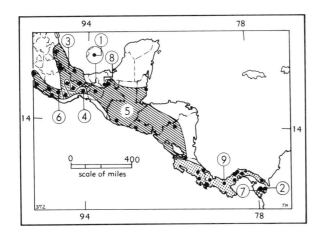

Map 372. *Tylomys.*

Guide to kinds	
1. *T. bullaris*	5. *T. nudicaudus nudicaudus*
2. *T. fulviventer*	6. *T. nudicaudus villai*
3. *T. nudicaudus gymnurus*	7. *T. panamensis*
4. *T. nudicaudus microdon*	8. *T. tumbalensis*
	9. *T. watsoni*

External measurements of the type are: 451; 233; 38. Upper parts dark brown, darkest medially; sides paler; underparts with russet median line extending from pectoral region to base of tail, ochraceous buff on either side of median stripe; tail black above basally, below gray, distally flesh colored; toes dusky. Anterior palatine foramina long, posteriorly attaining level of 1st molars, nasals exceeding premaxillae posteriorly.

MARGINAL RECORDS.—Panamá: type locality; *Tacarcuna Laguna* (Handley, 1966:782).

Tylomys nudicaudus
Peters' Climbing Rat

External measurements are: 400–500; 200–400; 38.6–45. Upper parts reddish brown fading to paler (clearer) on sides; underparts white or pale buffy fulvous; inguinal and axillary areas marked with white; legs and feet brown to russet (toes white in some specimens).

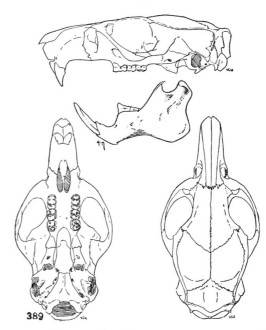

Fig. 389. *Tylomys nudicaudus gymnurus*, Río Atoyac, 8 km. NW Potrero, Veracruz, No. 18167 K.U., ♀, X 1.

Tylomys nudicaudus gymnurus Villa

1941. *Tylomis* [sic] *gymnurus* Villa, Anal. Inst. Biol. Univ. Nac. Autó. México, 12:763, November 18, type from Presidio, Veracruz.
1969. *Tylomys nudicaudus gymnurus*, Goodwin, Bull. Amer. Mus. Nat. Hist., 141:160, April 30.

MARGINAL RECORDS.—Puebla: 10 km. W Hueytamalco (Ramírez-P. and Sánchez-H., 1971:481). Veracruz: 8 km. NW Potrero; 3 km. E San Andrés Tuxtla; 25 km. SE Jesús Carranza. Oaxaca: Ixcuintepec; Lachivia (Goodwin, 1969:160). Veracruz: 3 km. N Presidio.

Tylomys nudicaudus microdon Goodwin

1955. *Tylomys nudicaudus microdon* Goodwin, Amer. Mus. Novit., 1738:1, June 10, type from La Gloria, rain forest at approx. 2500 ft., 10 km. SE Santa María Chimalapa, Isthmus of Tehuantepec, Oaxaca.

MARGINAL RECORDS.—Oaxaca: 4 mi. S Valle Nacional (Baker and Peterson, 1965:694); type locality.

Tylomys nudicaudus nudicaudus (Peters)

1866. *Hesperomys (Tylomys) nudicaudus* Peters, Monatsb. preuss. Akad. Wiss., Berlin, p. 404, pl. 1, figs. 1–4, type from Guatemala; La Primavera, 3200 ft., about 10 mi. SW Cobán, Alta Verapaz, regarded as probable type locality by Goodwin, Bull. Amer. Mus. Nat. Hist., 68:48, December 12, 1934.
1897. [*Tylomys*] *nudicaudus*, Trouessart, Catalogus mammalium . . . , fasc. 3, p. 520.

MARGINAL RECORDS.—Chiapas: Palenque. Belize: Silkgrass. Nicaragua: Río Coco. El Salvador: Chilata (Burt and Stirton, 1961:59). Guatemala: La Primavera. Chiapas: *Tuxtla Gutiérrez*; 8 km. N Berriozábal, 1065 m. (Baker, *et al.*, 1973:78, 85).

Tylomys nudicaudus villai Schaldach

1966. *Tylomys nudicaudus villai* Schaldach, Säugetierk. Mitteil., 14:294, October, type from Kilometer 183, 1600 m., 36½ km. N San Gabriel Mixtepec, Oaxaca.

MARGINAL RECORDS.—Guerrero: Cueva del Cañón del Zopilote, 13 km. S puente de Mexcala, 720 m. (Ramírez-P. and Sánchez-H., 1974:107, 108). Oaxaca: 11 km. S Chicahuaxtla, 1435 m. (Webb and Baker, 1971:144); type locality; 8 km. E Río Grande, 30 m. (Webb and Baker, 1971:144). Guerrero: Pedrera de Cajeles, 30 km. S Chilpancingo (Ramírez-P. and Sánchez-H., 1971:481).

Tylomys panamensis (Gray)
Panamá Climbing Rat

1873. *Neomys panamensis* Gray, Ann. Mag. Nat. Hist., ser. 4, 12:417, November, type from Panamá.
1897. [*Tylomys*] *panamensis*, Trouessart, Catalogus mammalium . . . , fasc. 3, p. 520.

"Mouse-coloured; back blackish-washed, with longer hairs; sides of head and body rather paler; throat, chest, underside of body, and inside of legs white; feet brownish with white hairs over the claws; tail slender, naked, glossy black, white at the end. Size of a common rat (*Mus decumanus* [= *Rattus norvegicus*])." (Gray, 1873:417.)

MARGINAL RECORDS.—Panamá: *Boca de Río Paya* (Handley, 1966:782); Cana.

Tylomys tumbalensis Merriam
Tumbalá Climbing Rat

1901. *Tylomys tumbalensis* Merriam, Proc. Washington Acad. Sci., 3:560, November 29, type from Tumbalá, Chiapas. Known only from type locality.

External measurements of the type, a young adult, are: 448; 234; 46. Upper parts dark gray to markedly blackish medially; sides brownish; chin, chest, part of inguinal region white; venter plumbeous washed with buff; toes dark brown; proximal half of tail blackish, distal part yellow. Skull large, long, flattened. According to Merriam (1901:560, 561), in comparison with the skull of *T. nudicaudus*, that of *T. tumbalensis* is: "less massive and more slender; rostrum and nasals decidedly more slender . . . ; bullae slightly larger; upper incisors weaker and more slender; molar series very large and heavy . . . (measuring 9.5 mm.)."

Tylomys watsoni Thomas
Watson's Climbing Rat

1899. *Tylomys watsoni* Thomas, Ann. Mag. Nat. Hist., ser. 7, 4:278, October, type from Bogava [= Bugaba], 800 ft., Volcán de Chiriquí, Chiriquí, Panamá.

External measurements of an adult male and the type are, respectively: 475, 493; 250, 243; 35, 38. Upper parts cinnamon buff with pronounced admixture of brown; underparts whitish; feet brown with white toes; tail mummy brown for basal two-thirds, whitish distally. Skull elongated; braincase flattened, depressed in interorbital region; premaxillae extending posteriorly slightly beyond nasals.

MARGINAL RECORDS.—Panamá (Handley, 1966:782): Cerro Brujo, 1000 ft.; Armila; Cerro Azul; 6 mi. E El Valle; type locality; *Boquerón*. Costa Rica: Palmar; *San Joaquín de Dota*; La Carpintera; Río Sixaola. Panamá (Handley, 1966:782): Boquete; Santa Fé; Salud.

Genus Ototylomys Merriam
Big-eared Climbing Rats

1901. *Ototylomys* Merriam, Proc. Washington Acad. Sci., 3:561, November 29. Type, *Ototylomys phyllotis* Merriam.

Closely resembles *Tylomys* in majority of features, but smaller; skull flattened, interparietal large; supraorbital ridge well developed and forming shelf over orbit; tympanic bullae much enlarged and inflated; coronoid process much reduced; pattern of molars diverging from basic oryzomyine pattern; cusps more nearly symmetrical than usual, folds in crown opposite one another, deep, and nearly meeting in middle of tooth; subsidiary ridges in main folds of upper molars much reduced.

Ototylomys phyllotis
Big-eared Climbing Rat

External measurements: 242–370; 119–191; 21–33. Dorsally dark grayish-brown tinged with cinnamon in highland areas and Tabasco to dusky brown in Yucatán, grayish-brown in Costa Rica, and blackish-brown in the wet lowlands of Nicaragua; ventrally, from creamy or dull white to slate (Tabasco, and highlands of Chiapas and Guatemala). Skull: greatest length, 39.0–41.8; zygomatic breadth, 18.4–20.4; length of incisive foramina, 6.9–8.1; palatal length, 5.4–6.1; length of maxillary tooth-row, 6.4–7.3.

Ramírez-P. and Sánchez-H. (1974:110, 111) record parts of two skulls recovered from owl pellets in a cave 13 km. south of the bridge at Mexcala, 720 m. in Guerrero, and understandably do not identify this material to subspecies.

Ototylomys phyllotis australis Osgood

1931. *Ototylomys phyllotis australis* Osgood, Field Mus. Nat. Hist., Publ. 295, Zool. Ser., 18:145, August 3, type from San Gerónimo, near Pozo Azul de Pirrís, Costa Rica.

MARGINAL RECORDS.—Costa Rica (Lawlor, 1969:40): ½ mi. E Finca Jiminez (*sic*); Monte Rey, 22 km. S San José, 1100 m.; *San Ignacio de Acosta*; type locality; Cerros de San Juan, 8 mi. S Santa Cruz.

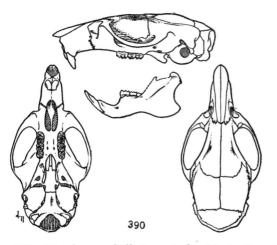

Fig. 390. *Ototylomys phyllotis australis*, Monte Rey, 22 km. S San José, Prov. San José, Costa Rica, No. 39255 K.U., ♀, X 1.

Ototylomys phyllotis connectens Sanborn

1935. *Ototylomys phyllotis connectens* Sanborn, Field Mus. Nat. Hist., Zool. Ser., 20(11):82, May 15, type from Cobán, Alta Verapaz, Guatemala.

MARGINAL RECORDS (Lawlor, 1969:38–40, unless otherwise noted).—Tabasco: 5 mi. SE Macuspana. Chiapas: ruins *ca.* 6 mi. SE Palenque. Guatemala: 20 km. NW Chinaja; Chimoxan; Senahu; Finca Concepción. Chiapas: Sabana de San Quintín, 215 m.; 8 km. N Berriozábal, 1065 m. (Baker, *et al.*, 1973:78, 85). Guerrero: Cueva del Cañón del Zopilote, 13 km. S puente de Mezcala [= Mexcala], 720 m. (Ramírez-P. and Sánchez-H., 1974:107, 110, 111). Tabasco: 5 mi. SW Teapa.

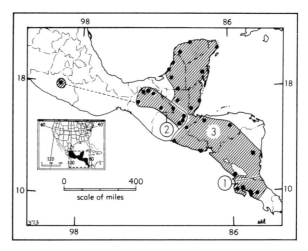

Map 373. *Ototylomys phyllotis*.

1. *O. p. australis*		2. *O. p. connectens*
			3. *O. p. phyllotis*

Ototylomys phyllotis phyllotis Merriam

1901. *Ototylomys phyllotis* Merriam, Proc. Washington Acad. Sci., 3:562, November 29, type from Tunkás, Yucatán.
1901. *Ototylomys phyllotis phaeus* Merriam, Proc. Washington Acad. Sci., 3:563, November 29, type from Apazote, near Yohaltún, Campeche (regarded as inseparable from *O. p. phyllotis* by Lawlor, Jour. Mamm., 50:36, February 26, 1969).
1908. *Ototylomys fumeus* J. A. Allen, Bull. Amer. Mus. Nat. Hist., 24:658, October 13, type from Matagalpa, Matagalpa, Nicaragua (regarded as inseparable from *O. p. phyllotis* by Lawlor, Jour. Mamm., 50:36, February 26, 1969).
1909. *Ototylomys guatemalae* Thomas, Abstr. Proc. Zool. Soc. London, p. 32, June 22, type from Tucuru, Río Polochic, about 50 mi. E Cobán, Guatemala (regarded as inseparable from *O. p. phyllotis* by Lawlor, Jour. Mamm., 50:36, February 26, 1969).
1953. *Ototylomys brevirostris* Laurie, Ann. Mag. Nat. Hist., ser. 12, 6:389, May, type from Kates Lagoon, 17° 57' N, 88° 30' W, Belize (regarded as inseparable from *O. p. phyllotis* by Lawlor, Jour. Mamm., 50:36, February 26, 1969).

1953. *Ototylomys brevirostris affinis* Laurie, Ann. Mag. Nat. Hist., ser. 12, 6:390, May, type from Chichén-Itzá, Yucatán (regarded as inseparable from *O. p. phyllotis* by Lawlor, Jour. Mamm., 50:36, February 26, 1969).

MARGINAL RECORDS (Lawlor, 1969:38–40).— Yucatán: 66 km. NE Mérida. Quintana Roo: La Vega; 60 km. N, 16 km. E Chetumal. Belize: 1 mi. S Pomona; 2 mi. W San Pedro Colombia. Honduras: Ilama; El Boquerón, Catamacas. Nicaragua: Río Curinguas, 12° 52′ N, 84° 03′ W; Río Javillo, 3 km. N, 4 km. W Sapoa, 40 m. El Salvador: Triumfo (*sic*). Guatemala: Astillero, 25 ft.; 3 mi. E Jocotán, 1400 ft.; Pacomón. Campeche: 103 km. SE Escárcega; 7½ km. W Escárcega; 5 km. S Champotón; 2 mi. S Campeche. Yucatán: Calcehtok.

Genus **Nyctomys** Saussure
Sumichrast's Vesper Rat

1860. *Nyctomys* Saussure, Revue et Mag. Zool., Paris, ser. 2, 12:106, March. Type, *Hesperomys sumichrasti* Saussure.

Braincase and frontals broad; supraorbital ridges well developed and extending across parietals to occiput; interparietal broad and large, completely separating parietals from supraoccipital; rostrum short; zygomatic plate narrow, straight anteriorly; infraorbital foramen prominent; tympanic bullae small; palate broad, ending in front of posterior part of tooth-row, and without lateral pits; cheek-teeth resembling those of *Oryzomys* but extremely complex; hind foot considerably modified for arboreal life; tail well haired and tufted terminally; hallux clawed (after Ellerman, *et al.*, 1941:375). The genus is monotypic.

The vesper rat is brightly colored, arboreal, builds outside nests of twigs and fibers much like those of the red squirrel, and only occasionally descends to the ground.

Nyctomys sumichrasti
Sumichrast's Vesper Rat

External measurements: 208–286; 85–156; 23–26. Upper parts buffy, cinnamon, or tawny with thin admixture of dark hairs, especially mid-dorsally; sides clearer, usually paler; underparts white or almost so; feet usually white or almost so, but in some subspecies hind feet dusky to brownish; tail almost or quite unicolored, usually brown. Additional characters as for the genus.

Nyctomys sumichrasti colimensis Laurie

1953. *Nyctomys sumichrasti colimensis* Laurie, Ann. Mag. Nat. Hist., ser. 12, 6:390, May, type from Juárez, Colima.

MARGINAL RECORDS.—Jalisco: vic. Chamela Bay (López-F., *et al.*, 1973:103). Colima: type locality.

Nyctomys sumichrasti costaricensis Goldman

1937. *Nyctomys sumichrasti costaricensis* Goldman, Jour. Washington Acad. Sci., 27:422, October 15, type from San Gerónimo de Pirrís, 100 ft., Puntarenas, Costa Rica.

MARGINAL RECORDS.—Costa Rica: type locality; Palmar.

Nyctomys sumichrasti decolorus (True)

1894. *Sitomys (Rhipidomys) decolorus* True, Proc. U.S. Nat. Mus., 16:689, February 5, type from Río de las Piedras, Honduras.
1916. *N[yctomys]. s[umichrasti]. decolorus*, Goldman, Proc. Biol. Soc. Washington, 29:155, September 6.

MARGINAL RECORDS.—Honduras: Catacombas; Yaruca, 1000 ft.; Catacamas; Cantoral; Las Flores.

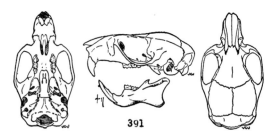

Fig. 391. *Nyctomys sumichrasti sumichrasti*, 20 km. E Jesús Carranza, 300 ft., Veracruz, No. 24137 K.U., ♀, X 1.

Nyctomys sumichrasti florencei Goldman

1937. *Nyctomys sumichrasti florencei* Goldman, Jour. Washington Acad. Sci., 27:421, October 15, type from Barra de Santiago, Ahuachapán, El Salvador.

MARGINAL RECORDS.—El Salvador: El Tablón (Burt and Stirton, 1961:60). Nicaragua: Ocotal (Goldman, 1937:422; on geographic grounds; referred by J. A. Allen, 1910:100, to *Rhipidomys salvini* (Tomes)); Santa María de Ostuma (Jones and Genoways, 1970:2, as *N. sumichrasti* only); 3 mi. NNW Diriamba (Jones and Genoways, 1970:14, as *N. sumichrasti* only); San Antonio (Jones and Genoways, 1970:7, as *N. sumichrasti* only). El Salvador (Burt and Stirton, 1961:60): Puerto El Triunfo; type locality.

Nyctomys sumichrasti nitellinus Bangs

1902. *Nyctomys nitellinus* Bangs, Bull. Mus. Comp. Zool., 39:30, April, type from Boquete, 4000 ft., Chiriquí, Panamá.
1916. *N[yctomys]. s[umichrasti]. nitellinus*, Goldman, Proc. Biol. Soc. Washington, 29:155, September 6.

MARGINAL RECORDS.—Costa Rica: La Carpintera. Panamá (Handley, 1966:782): Cayo Agua; Camp Chagres; Cerro Azul; type locality; Río Changena. Costa Rica: Cauhita.

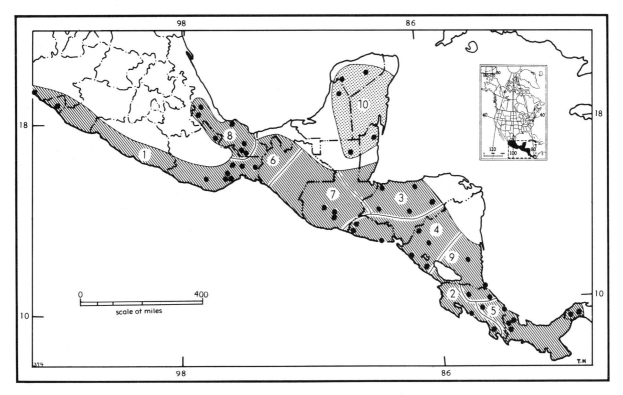

Map 374. *Nyctomys* and *Otonyctomys*.

1. *N. sumichrasti colimensis*
2. *N. sumichrasti costaricensis*
3. *N. sumichrasti decolorus*
4. *N. sumichrasti florencei*
5. *N. sumichrasti nitellinus*
6. *N. sumichrasti pallidulus*
7. *N. sumichrasti salvini*
8. *N. sumichrasti sumichrasti*
9. *N. sumichrasti venustulus*
10. *Otonyctomys hatti*

Nyctomys sumichrasti pallidulus Goldman

1937. *Nyctomys sumichrasti pallidulus* Goldman, Jour. Washington Acad. Sci., 27:420, October 15, type from Santo Domingo, 900 ft., 8 mi. W Lagunas, on Mexican National Railroad, Isthmus of Tehuantepec, Oaxaca.

MARGINAL RECORDS (Goodwin, 1969:163, unless otherwise noted).—Oaxaca: type locality; Las Conchas; Río Negro; San Antonio; Tenango; Río Jalatengo, 8 km. S Jalatengo, 1280 m. (124414 KU); Jalapa.

Nyctomys sumichrasti salvini (Tomes)

1862. *Hesperomys (Myoxomys) salvini* Tomes, Proc. Zool. Soc. London, p. 284, for 1861, April, type from Dueñas, Sacatepequez, Guatemala.
1916. *N[yctomys]. s[umichrasti]. salvini*, Goldman, Proc. Biol. Soc. Washington, 29:155, September 6.

MARGINAL RECORDS.—Guatemala: Panjachel [*sic*]; San Lucas; type locality.

Nyctomys sumichrasti sumichrasti (Saussure)

1860. *H[esperomys]. sumichrasti* Saussure, Revue et Mag. Zool., Paris, ser. 2, 12:107, March, type from eastern slope

mts. in Veracruz (fixed as Uvero, 20 kilometros al noroeste de Santiago Tuxtla, Veracruz, by Alvarez, Jour. Mamm., 44:583, 1963).
1902. *N[yctomys]. sumichrasti*, Bangs, Bull. Mus. Comp. Zool., 39(2):30, April.

MARGINAL RECORDS.—Veracruz: Jalapa; type locality; 20 km. E Jesús Carranza; 38 km. SE Jesús Carranza (Hall and Dalquest, 1963:295); 25 km. SE Jesús Carranza. Oaxaca: Chiltepec (Goodwin, 1969: 165). Veracruz: Coscomatepec, 5000 ft.

Nyctomys sumichrasti venustulus Goldman

1916. *Nyctomys sumichrasti venustulus* Goldman, Proc. Biol. Soc. Washington, 29:155, September 6, type from Greytown [= San Juan del Norte], Nicaragua.

MARGINAL RECORDS.—Nicaragua: Escondido River, 45 mi. from Bluefields; type locality. Costa Rica: Pacuare; Cataratos San Carlos.

Genus Otonyctomys Anthony
Yucatán Vesper Rat

1932. *Otonyctomys* Anthony, Amer. Mus. Novit., 586:1, November 16. Type, *Otonyctomys hatti* Anthony.

Resembling, and closely related to, *Nyctomys* but auditory bullae much larger, cheek-teeth smaller, and tarsus narrower. Anterior margin of zygomatic plate approx. perpendicular to palatal plane. Two pairs of inguinal mammae (after Anthony, 1932:1).

Otonyctomys hatti Anthony
Yucatán Vesper Rat

1932. *Otonyctomys hatti* Anthony, Amer. Mus. Novit., 586:1, November 16, type from Chichén-Itzá, Yucatán.

External measurements of five specimens are: 196–231; 97–127; 21–23; 14–15. Cranial measurements of four specimens are: condylobasal length, 24.4–26.9; zygomatic breadth, 14.6–15.7; length of upper tooth-row, 3.8–4.2; length of auditory bulla, 8.2–8.9. Upper parts nearly uniform russet to hazel, darkest on back; sides tawny to ochraceous-tawny; upper sides of feet whitish with wash of buff or tawny tones; underparts white with creamy wash; tail bone-brown above and below. See characters for the genus. See Map 374.

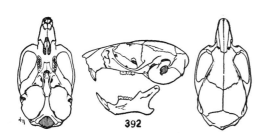

Fig. 392. *Otonyctomys hatti*, Chichén-Itzá, Yucatán, No. 91190 A.M.N.H., ♂, X 1.

MARGINAL RECORDS.—Yucatán: type locality. Belize: Rockstone Pond (Peterson, 1966:281). Guatemala: Tikal (Rick, 1965:335). Campeche: Dzibalchen (93870 KU). Yucatán: Actun Spukil (Peterson, 1966:281).

Genus **Reithrodontomys** Giglioli—Harvest Mice

Revised by A. H. Howell, N. Amer. Fauna, 36:1–97, 7 pls., June 5, 1914. Latin American species revised by Hooper, Miscl. Publ. Mus. Zool., Univ. Michigan, 77:1–255, 9 pls., January 16, 1952.

1874. *Reithrodontomys* Giglioli, Boll. Soc. Geogr. Ital., Roma, 11:326, May–July. Type by subsequent selection (A. H. Howell, N. Amer. Fauna, 36:13, June 5, 1914), *Reithrodon megalotis* Baird.
1874. *Ochetodon* Coues, Proc. Acad. Nat. Sci. Philadelphia, p. 184, December 15. No type selected; included species of both subgenera.

External measurements: 114–238; 48–145; 15–25. Tail slender, scaly, scantily haired; ears conspicuous, sometimes large; soles of hind feet with 6 tubercles; mammae, pec. 1, ing. 2 = 6. Skull comparatively smooth, little sculptured; zygomata slender; outer wall of infraorbital foramen a broad, thin lamina; anterior palatine foramina long and separated by a thin septum, terminating approx. at level of anterior end of tooth-row; posterior border of palate usually truncate, but often with small, median spine; tympanic bullae moderately inflated, arranged obliquely with relation to longitudinal axis of skull; upper incisors with deep median groove; molars tuberculate, rooted, tubercles arranged in 2 longitudinal series.

According to Hooper (1952a:13), mice of this genus molt at least twice before growing the adult pelage. There seems to be at least one annual molt thereafter. The juvenal pelage is dullest and more or less plumbeous. The adult pelage is brightest.

KEY TO NORTH AMERICAN SPECIES OF REITHRODONTOMYS

(Adapted from Hooper, 1952a:24–31)

1. Occurring in the United States.
    2. First primary fold of M3 at least as long as 2nd primary fold, each usually extending more than halfway across crown; worn occlusal surface of left m3 S-shaped. . . . . *R. fulvescens*, p. 644
    2'. First primary fold of M3 distinctly shorter than 2nd primary fold, extending less than halfway across crown; worn occlusal surface of left m3 C-shaped.
        3. Distinct labial ridge, often with cusplets, on m1–2; major fold and 2nd primary fold in M1–2 tend to coalesce, isolating anterior cusps from posterior cusps; occurring in southeastern United States. . . . . *R. humulis*, p. 637
        3'. No distinct labial ridge on m1–2; major fold and 2nd primary fold in M1–2 meet but do not coalesce; and thus isolate anterior cusps from posterior cusps; not occurring in southeastern United States.

4. Breadth of braincase not exceeding 9.6; rostrum short and broad; tail shorter than head and body; occurring in Great Plains and southwestern United States. . . .*R. montanus*,   p. 635

4′. Breadth of braincase more than 9.5; rostrum longer and narrower; tail approx. equal to, or longer than, head and body.

    5. Fur of upper parts long, dense, heavily pigmented; ears blackish; occurring in salt marshes, vic. San Francisco Bay, California. . . . .*R. raviventris*,   p. 641

    5′. Fur of upper parts shorter and less dense, varying from pale buff to reddish brown; ears buffy or fuscous. . . . .*R. megalotis*,   p. 637

1′. Occurring south of United States.

  6. Occurring in México.

    7. Second primary fold in m3 well developed; m3 resembling but smaller than m2; zygomatic plate little if any broader than mesopterygoid fossa (except in some *gracilis*); mesopterygoid fossa approx. as wide as either pterygoid fossa (except in some *gracilis*); occurring in eastern and southern México.

      8. Zygomata weak, especially anteriorly; braincase deep and barely narrower than zygomatic breadth; breadth of zygomatic plate less than 1.5; rostrum long, narrow; ear (from notch) barely shorter than hind foot; occurring usually above 8500 ft. in central and southern México. . . . .*R. microdon*,   p. 653

      8′. Zygomata strong, usually .5–1.0 broader than braincase; breadth of zygomatic plate more than 1.4; rostrum relatively short, broad; ear (from notch) usually 3–4 less than hind foot; occurring usually below 8500 ft.

        9. Occurring on Cozumel Island; total length more than 204. . . . . .*R. spectabilis*,   p. 652

        9′. Not occurring on Cozumel Island; total length less than 204.

          10. Length of skull usually less than 22; if more, then length of incisive foramina less than 4; frontals broad and flattened interorbitally; hind feet whitish or dusky above; occurring in tropical lowlands of southern México. . . . .*R. gracilis*,   p. 648

          10′. Length of skull usually more than 22; if less, then length of incisive foramina more than 4.0; frontals strongly constricted and not markedly flattened (but broad interorbitally); hind feet dusky above; occurring above 1000 ft.

                                                                  R. mexicanus,   p. 650

7′. Second primary fold in m3 faint or absent; m3 not resembling m2 in form; zygomatic plate broader than mesopterygoid fossa; mesopterygoid fossa not more than ¾ as wide as either pterygoid fossa; not restricted to eastern or southern México.

    11. First primary fold on M3 at least as long as 2nd primary fold, each extending more than halfway across crown; major fold well developed, sometimes continuous with 1st primary fold; worn occlusal surface of left M3 E-shaped.

      12. Zygomatic breadth more than 11.9; interorbital breadth more than 3.4; supraorbital shelf trenchant or beaded; tail unicolored or barely paler below; duskiness of tarsi extending to dorsum of hind feet. . . . .*R. hirsutus*,   p. 648

      12′. Zygomatic breadth less than 11.9; interorbital breadth less than 3.4; supraorbital shelf neither tranchant nor beaded, although sometimes slightly elevated; tail paler below than above, usually sharply bicolored; hind feet whitish or buffy above, not dusky. . . . .*R. fulvescens*,   p. 644

11′. First primary fold of M3 shorter than 2nd primary fold; major fold indistinct, not more than a shallow indentation on lingual face of tooth; worn occlusal surface of M3 C-shaped.

    13. Breadth of braincase more than 10.7; tail more than 90.

      14. No buffy hairs on inner surfaces of ears; ears 17–19 from notch to tip; rostrum long, narrow; interorbital region strongly constricted, hourglass-shaped; occurring above 9000 ft. in highlands of central México. . . . .*R. chrysopsis*,   p. 642

      14′. Some buffy hairs on inner surfaces of ears (magnification sometimes required to see same); ears less than 18; rostrum and interorbital region broader; occurring below 9000 ft. in central and southern México. . . . .*R. sumichrasti*,   p. 642

13′. Breadth of braincase less than 10.7; tail usually less than 85 (never more than 100).

    15. Tail shorter than head and body.

      16. Total length less than 140; tail less than 95% of length of head and body; breadth of braincase usually less than 9.8.

        17. Faint labial ridge on M1–2; tarsi whitish or whitish with thin longitudinal dusky line; pre- and postauricular areas bright buffy; occurring in coastal region of Sonora. . . . .*R. burti*,   p. 636

        17′. No labial ridge on M1–2; tarsi extensively dusky; head, pre- and postauricular areas grayish buff; not occurring in coastal region of Sonora. . . . .*R. montanus*,   p. 635

16'. Total length more than 140; tail more than 90% of length of head and
    body; breadth of braincase more than 9.6. . . . .*R. megalotis,* p. 637
15'. Tail longer than head and body.
    18. Breadth of mesopterygoid fossa more than 1.0, and approx. equal to dis-
        tance between posterior palatine foramina; tail more than 75, usually
        more than 85. . . . .*R. sumichrasti,* p. 642
    18'. Breadth of mestopterygoid fossa less than 1.3, and less than distance
        between posterior palatine foramina; tail usually less than 85.
                                                    R. megalotis, p. 637
6'. Occurring south of México.
  19. Second primary fold indistinct or absent in m3; m3 unlike m2 in form; zygomatic plate
     broader than mesopterygoid fossa; mesopterygoid fossa narrower than either pterygoid
     fossa.
     20. First primary fold in M3 at least as long as 2nd primary fold, each extending more than
        halfway across crown; major fold well developed, sometimes confluent with 1st pri-
        mary fold; in m3 worn occlusal surface of left tooth S-shaped. . . . .*R. fulvescens,* p. 644
     20'. First primary fold in M3 shorter than 2nd primary fold; major fold indistinct, not more
        than a shallow indentation on lingual surface of tooth; worn occlusal surface of M3
        C-shaped. . . . .*R. sumichrasti,* p. 642
  19'. Second primary fold well developed in m3, usually appearing as posterior 1 of 2 internal
     folds; m3 similar to but smaller than m2; zygomatic plate little if any broader than meso-
     pterygoid fossa; mesopterygoid fossa approx. as broad as either pterygoid fossa.
    21. Hind foot 22–26; length of molar tooth-row 3.9–4.5.
     22. Interorbital breadth more than 4.0; length of rostrum (measured from the shallow
        notch lying on superior orbital border of zygomatic arch [lateral to lachrymal
        bone] to tip of nasal on same side) more than 9.0; occurring in mts. of Costa Rica
        and Panamá above 7000 ft. . . . .*R. creper,* p. 655
     22'. Interorbital breadth less than 4.1; length of rostrum less than 9.2; occurring in mts.
        of Guatemala above 8000 ft. . . . .*R. tenuirostris,* p. 654
    21'. Hind foot less than 22; length of molar tooth-row less than 3.9.
     23. Greatest length of skull more than 24.
      24. Breadth of zygomatic plate less than 1.5; zygomatic breadth less than 12,
        approx. equal to breadth of braincase; depth of braincase more than 9.2.
                                           R. rodriguezi, p. 654
      24'. Breadth of zygomatic plate 1.5 or more; zygomatic breadth more than 12, and
        approx. .5–1.0 wider than braincase; depth of braincase less than 9.3.
                                           R. mexicanus, p. 650
     23'. Greatest length of skull less than 24.
      25. Braincase highly inflated; zygomatic breadth barely more than breadth of
        braincase; rostrum long, narrow; occurring in mts. of Guatemala above 8500
        ft. . . . .*R. microdon,* p. 653
      25'. Braincase moderately inflated; zygomatic breadth approx. .5–1.0 wider than
        braincase; rostrum broader; not restricted to mts. of Guatemala.
       26. Length of rostrum less than 7.
        27. Hind feet dusky above; occurring in lowlands of eastern Panamá.
                                      R. darienensis, p. 649
       27'. Hind feet whitish or dusky above; occurring north of Panamá.
                                      R. gracilis, p. 648
      26'. Length of rostrum more than 7.
       28. Length of molar tooth-row less than 3.2.
        29. Depth of braincase less than 8.6; dorsal surface of skull com-
          paratively flat.
         30. Hind feet dusky above; fur of upper parts long and
           dusky; zygomatic plate less than 1.5 wide; occurring in
           humid highlands of Costa Rica and uplands of
           Nicaragua.
          31. Upper parts near (*h*) Ochraceous-Tawny suffused
            with black; well-developed ectolophid on m1 and
            m2; incisive foramina terminating posterior to an-
            terior margins of first upper molars, and bony palate
            thus shorter (3.1–3.8). . . . . . .*R. brevirostris,* p. 652
         31'. Upper parts near (*j*) Buffy Brown lightly suffused

with ochraceous; ectolophid lacking on m1 and m2; incisive foramina terminating
    well anterior to 1st upper molars, and bony palate thus longer (3.9). . *R. paradoxus,*   p. 653
  30′. Hind feet whitish or lightly dusky above; fur of upper parts bright reddish buff and
    moderately short; zygomatic plate more than 1.5 wide; occurring in tropical lowlands.
                                                               R. gracilis,   p. 648
    29′. Depth of braincase more than 8.4; dorsal surface of skull convex. . . . . . . *R. mexicanus,*   p. 650
28′. Length of molar tooth-row more than 3.2.
    32. Length of skull less than 22.5; depth of skull less than 8.5; tail less than 105; occurring in
        tropical lowlands. . . . . *R. gracilis,*   p. 648
    32′. Length of skull more than 22; depth of braincase more than 8.3; tail rarely less than 100;
        occurring above 3000 ft. . . . .*R. mexicanus,*   p. 650

Subgenus **Reithrodontomys** Giglioli

1853. *Reithrodon* Le Conte, Proc. Acad. Nat. Sci. Philadel-
    phia, p. 413. Type, *Mus Le Contii* Audubon and Bachman
    [= *Mus humulis* Audubon and Bachman]. Not *Reithrodon*
    Waterhouse, 1837.
1874. *Reithrodontomys* Giglioli, Boll. Soc. Geogr. Ital., Roma,
    11:326, May–July. Type, *Reithrodon megalotis* Baird.
1903. *Rhithrodontomys* Elliot, Field Columb. Mus., Publ. 74,
    Zool. Ser., 3:164, May 7, an emendation. *Rhithrodontomys*
    peninsulae is the combination used.

Subgenus *Reithrodontomys* differs from subgenus *Aporodon* in: that part of skull anterior to
least interorbital constriction approx. equal in
length to the part posterior to interorbital constriction; braincase moderately inflated, extending laterally slightly beyond anterolateral limits
of zygomatic arches; zygomatic plate much
broader than mesopterygoid fossa (except in *R.
hirsutus*); pterygoid hamulae only slightly inflated, and barely, if at all, reflexed laterad.

Reithrodontomys montanus
Plains Harvest Mouse

External measurements: 107–143; 48–63;
14–17; 12–16. The color of this species varies
geographically in such a way that where it occurs
with another species the two may be almost indistinguishable on the basis of color. This is the case
in the Great Plains, where *R. montanus* and *R.
megalotis* occur together. Close inspection re-

veals, nevertheless, that *R. montanus* has a mid-
dorsal stripe (lacking in *R. megalotis* of the Great
Plains), narrower blackish dorsal stripe on tail,
and white instead of gray venter. In general the
upper parts are pale gray to buffy gray washed
with fulvous and sprinkled with black hairs, especially mid-dorsally; underparts white. Also, *R.
montanus* is smaller.

R. montanus is principally an upland species
and occurs in stands of short, sparse grass. After a
gestation period of 21 days, 2–5 (av. 2.9) young
are born.

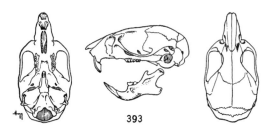

Fig. 393. *Reithrodontomys montanus albescens,* 2 mi. NE
Runnymede, Kansas, No. 12685 K.U., ♂, X 1½.

Reithrodontomys montanus albescens Cary

1903. *Reithrodontomys albescens* Cary, Proc. Biol. Soc. Wash-
    ington, 16:53, May 6, type from 18 mi. NW Kennedy,
    Cherry Co., Nebraska.
1911. *Reithrodontomys montanus albescens* Cary, N. Amer.
    Fauna, 33:110, August 17.

MARGINAL RECORDS.—South Dakota: 2 mi. N, 5
mi. W Ludlow (Andersen and Jones, 1971:378); ½ mi. W
Reva (*ibid.*); 15 mi. from mouth Belle Fourche River.
Nebraska (Jones, 1964c:193): *4 mi. N Bristow;* Nio-
brara; 1 mi. SW Neligh. Kansas: ½ mi. W Downs; 2 mi.
NE Runnymede; 17 mi. SW Meade. Colorado
(Armstrong, 1972:189, 190): Three Corners; 3 mi. W
Branson; Denver; 6 mi. S, 7 mi. W Fort Collins, 5600 ft.
Wyoming (Long, 1965a:626): 1 mi. S Pine Bluffs; be-
tween 5 mi. E and 18 mi. NW LaGrange (Maxwell and

Brown, 1968:144, as *R. montanus* only); 2 mi. S, ½ mi. E Lusk; *Rockypoint;* ⅝ mi. N, ³⁄₁₀ mi. E Rockypoint.

Reithrodontomys montanus griseus V. Bailey

1905. *Reithrodontomys griseus* V. Bailey, N. Amer. Fauna, 25:106, October 24, type from San Antonio, Bexar Co., Texas.
1935. *Reithrodontomys montanus griseus,* Benson, Jour. Mamm., 16:141, May 15.

MARGINAL RECORDS.—Nebraska (Jones, 1964c:194): 9 mi. NW Lincoln; London. Kansas: Lawrence. Missouri: *ca.* 7 mi. NW Jane (Long, 1961:417). Oklahoma: 3 mi. E Wainwright; 5 mi. N Colbert. Texas: 2 mi. N Sivells Bend; Bosque County (Davis, 1966:180, as *R. montanus* only); Robertson County (*ibid.,* as *R. montanus* only); 2 mi. SE College Station; type locality; Bear Creek, 5 mi. W Hunt; San Angelo; Martin County (Davis, 1966:180, as *R. montanus* only). New Mexico: 44 mi. NW Roswell; Santa Rosa. Oklahoma: 8 mi. E Keyes; Alva. Kansas: 3 mi. SE Arkansas City; 8 mi. NE Hutchinson; Concordia.

Reithrodontomys montanus montanus (Baird)

1855. *Reithrodon montanus* Baird, Proc. Acad. Nat. Sci. Philadelphia, 7:335, April, type from Rocky Mts., lat. 39°, probably near upper end San Luis Valley, Saguache Co., Colorado, according to J. A. Allen (Bull. Amer. Mus. Nat. Hist., 7:124, May 21, 1895), or very probably on Medano Creek, Alamosa Co., Colorado, according to Warren (Mammals of Colorado, p. 196, 1942).
1893. *Reithrodontomys montanus,* J. A. Allen, Bull. Amer. Mus. Nat. Hist., 5:80, April 28.

MARGINAL RECORDS.—Colorado: type locality. New Mexico: 3 mi. S Raton. Texas: Jeff Davis County (Davis, 1966:180, as *R. montanus* only); W of Glass Mts., Brewster Co. Chihuahua (Anderson, 1972:331): 8 mi. N, 11 mi. E Charco de Peña, 4700 ft.; 5 mi. W Jiménez, 4550 ft. Durango: 3 mi. SW Canutillo, 6300 ft. Chihuahua (Anderson, 1972:331): 25 mi. NNW Camargo; 35 mi. NW Dublán, 5300 ft. Sonora: 14 mi. S Nogales, 3500 ft. Arizona: 1½ mi. ENE Greaterville, Pima Co. (Hoffmeister, 1959:17). New Mexico (Findley, *et al.,* 1975:196): 6 mi. by road N Cloverdale; 3 mi. N Socorro; 1 mi. E Pena Blanca.

Reithrodontomys burti Benson
Sonoran Harvest Mouse

1939. *Reithrodontomys burti* Benson, Proc. Biol. Soc. Washington, 52:147, October 11, type from Rancho de Costa Rica, Río Sonora, Sonora.

External measurements: 116–136; 52–66; 16–17; ear, 14–17. Upper parts ochraceous-buff, hairs with black tips, most intense on sides and rump; underparts whitish with plumbeous color of base of hairs conspicuous; ankles white or with fine dusky line; tail-stripe narrow, indistinct. Tail

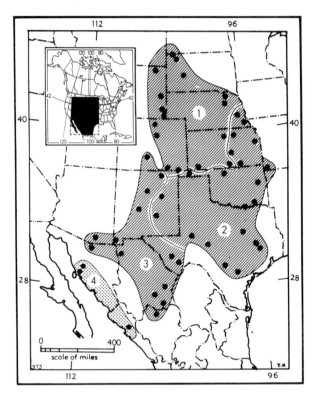

Map 375. *Reithrodontomys montanus* and *Reithrodontomys burti.*

1. *R. montanus albescens*   3. *R. montanus montanus*
2. *R. montanus griseus*   4. *R. burti*

relatively long (70–97% of length of head and body). Skull resembling that of *R. montanus,* but differing in wider interorbital area, broader zygomatic process of maxilla, larger infraorbital foramina and relatively longer nasals.

Hooper (1952:41) considers *R. burti* to be a relict species related to *R. montanus, R. megalotis,* and *R. humulis,* probably closest to *montanus* but in some characters related to *humulis.*

MARGINAL RECORDS.—Sonora: type locality. Sinaloa: 1 mi. S Pericos. Sonora: 11 mi. W Hermosilla, 600 ft.

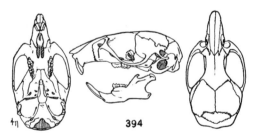

Fig. 394. *Reithrodontomys burti,* Rancho de Costa Rica, Río Sonora, Sonora, No. 82987 M.V.Z., ♀, X 1½.

Reithrodontomys humulis
Eastern Harvest Mouse

External measurements: 107–128; 45–60; 15–17; 8–9. Upper parts rich brown, sometimes faintly washed with grayish, dark mid-dorsal stripe usually present; paler on sides, with clear lateral line usually present; underparts ash colored, often with cinnamon or pinkish suffusion; tail bicolored, fuscous to dark brown above, grayish white below. Braincase narrow, highly arched; nasals and rostrum short, broad; zygomata parallel or slightly narrower anteriorly.

Reithrodontomys humulis humulis (Audubon and Bachman)

1841. *Mus humulis* Audubon and Bachman, Proc. Acad. Nat. Sci. Philadelphia, 1:97, type from Charleston, Charleston Co., South Carolina.
1907. *Reithrodontomys humulis*, Osgood, Proc. Biol. Soc. Washington, 20:49, April 18.
1842. *Mus carolinensis* Audubon and Bachman, Jour. Acad. Nat. Sci. Philadelphia, 8:306. Type locality, South Carolina (may refer to this species).
1842. *Mus Le Contii* Audubon and Bachman, Jour. Acad. Nat. Sci. Philadelphia, 8:307, type from Georgia.
1895. *Reithrodontomys humulis dickinsoni* Rhoads, Amer. Nat., 29:590, June, type from Willow Oak, Pasco Co., Florida.
1898. *Reithrodontomys lecontii impiger* Bangs, Proc. Biol. Soc. Washington, 12:167, August 10, type from White Sulphur Springs, 2000 ft., Greenbrier Co., West Virginia.

MARGINAL RECORDS.—Ohio: Sec. 23, Hocking Twp., Fairfield County. Virginia: Frederick County; Campbell County; Norfolk County. North Carolina: Currituck, thence down coast to Florida: Ritta; Tarpon Springs, thence westward along Gulf Coast to Louisiana: 1 mi. N Slidell (Lowery, 1974:233); University, East Baton Rouge Parish. Arkansas: ⁴/₅ *km. E Big Lake* (Sealander, 1977:149, as *R. humulis* only). Tennessee: 2 mi. E Open Lake. Ohio: Hamilton County; *Warren County* (Gottschang, 1965:48).

Reithrodontomys humulis merriami J. A. Allen

1895. *Reithrodontomys merriami* J. A. Allen, Bull. Amer. Mus. Nat. Hist., 7:119, May 21, type from Austin Bayou, near Alvin, Brazoria Co., Texas.

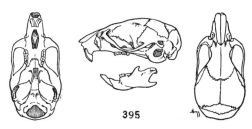

Fig. 395. *Reithrodontomys humulis humulis*, 2 mi. E Silver Springs, Marion Co., Florida, No. 27241 K.U., ♂, X 1½.

1914. *Reithrodontomys humulis merriami*, A. H. Howell, N. Amer. Fauna, 36:21, June 5.

MARGINAL RECORDS.—Arkansas: Fort Smith (6189 Louisiana State Univ. Mus. Zool.). Louisiana: Lafayette; Lake Arthur (Lowery, 1974:233). Texas: Labelle; type locality; Richmond; 1 mi. S New Boston (Packard, 1961:194). Oklahoma: 5 mi. N Colbert (Smith, 1964:204); 5 mi. N Wilburton (Jones and Anderson, 1959:153).

Reithrodontomys humulis virginianus A. H. Howell

1940. *Reithrodontomys humulis virginianus* A. H. Howell, Jour. Mamm., 21:346, August 13, type from Amelia, Amelia Co., Virginia.

MARGINAL RECORDS (Hooper, 1943:19, unless otherwise noted).—Maryland: Takoma Park. Virginia: Alexandria; vic. Triplett; *type locality;* Truxillo; 5½ mi. NW Chantilly (Peacock and Peacock, 1961:544).

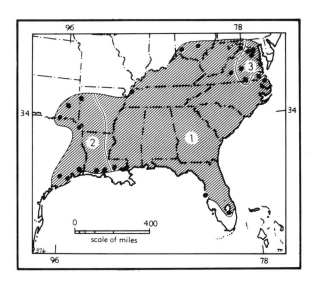

Map 376. *Reithrodontomys humulis*.

1. *R. h. humulis*   2. *R. h. merriami*
3. *R. h. virginianus*

megalotis-group
Reithrodontomys megalotis
Western Harvest Mouse

External measurements: 118–170; 55–96; 14–20; 10–16. Upper parts of various shades of buffy mixed with dark brown or blackish that sometimes predominates mid-dorsally; buffy tones purest on cheeks, shoulders, and flanks; underparts varying from dark buff to white; tail bicolored, darker above than below. Skull with broad zygomatic plate and broad pterygoid fossae.

There are no absolute characters by which this species can be distinguished from certain of its congeners; usually close comparison must be made and identification based on the summation of characters.

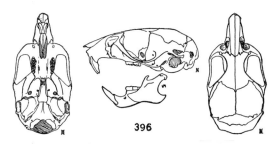

Fig. 396. *Reithrodontomys megalotis megalotis*, Crystal Spring, Lincoln Co., Nevada, No. 52998 M.V.Z., ♂, X 1½.

Reithrodontomys megalotis alticolus Merriam

1901. *Reithrodontomys saturatus alticolus* Merriam, Proc. Washington Acad. Sci., 3:556, November 29, type from Cerro San Felipe, 10,000 ft., Oaxaca.
1914. *Reithrodontomys megalotis alticolus*, A. H. Howell, N. Amer. Fauna, 36:37, June 5.

MARGINAL RECORDS.—Oaxaca: Tamazulapan, somewhere between 6500 and 7500 ft.; 3 mi. N Ozoletepec, 10,000 ft.; San Pedro Mixtepec (Goodwin, 1969:165); Río Molino (*ibid.*); San Andrés Chicahuaxtla (*ibid.*); Tlapacingo, 5200 ft. (?).

Reithrodontomys megalotis amoles A. H. Howell

1914. *Reithrodontomys amoles* A. H. Howell, N. Amer. Fauna, 36:40, June 5, type from Pinal de Amoles, approx. 7500 ft., Querétaro. Known only from type locality.
1952. *Reithrodontomys megalotis amoles*, Hooper, Miscl. Publ. Mus. Zool., Univ. Michigan, 77:64, January 16.

Reithrodontomys megalotis arizonensis J. A. Allen

1895. *Reithrodontomys arizonensis* J. A. Allen, Bull. Amer. Mus. Nat. Hist., 7:134, May 21, type from Rock Creek, 8000 ft., Chiricahua Mts., Cochise Co., Arizona. Known only from type locality.
1914. *Reithrodontomys megalotis arizonensis*, A. H. Howell, N. Amer. Fauna, 36:38, June 5.

Reithrodontomys megalotis aztecus J. A. Allen

1893. *Reithrodontomys aztecus* J. A. Allen, Bull. Amer. Mus. Nat. Hist., 5:79, April 28, type from La Plata, San Juan Co.,

New Mexico. (See J. A. Allen, Bull. Amer. Mus. Nat. Hist., 7:125, May 21, 1895.)
1914. *Reithrodontomys megalotis aztecus*, A. H. Howell, N. Amer. Fauna, 36:30, June 5.
1935. *Reithrodontomys megalotis caryi* A. H. Howell, Jour. Mamm., 16:143, May 15, type from Medano Ranch, 15 mi. NE Mosca, Alamosa Co., Colorado.

MARGINAL RECORDS.—Colorado (Armstrong, 1972:192, unless otherwise noted): Rifle (Hall and Kelson, 1959:585); Montrose; Coventry; 1 mi. W Mancos; 1 mi. S, 1½ mi. E Chimney Rock; Del Norte; 10 mi. S Saguache; Cañon City; Lamar. Kansas (Jones and Mursaloğlu, 1961:21): ½ mi. NW Bellefont; Meade County State Park, 14 mi. SW Meade. Oklahoma: 7 mi. S Turpin (*ibid.*). Texas: 9 mi. E Stinnett (*ibid.*). New Mexico (*ibid.*): 4 mi. SW Santa Rosa; 2 mi. S San Antonio; Apache Creek. Arizona: Canyon de Chelly. Utah: ½ mi. NW Bluff, 4500 ft.; Monticello. Colorado: Grand Junction.

Reithrodontomys megalotis catalinae (Elliot)

1904. *Rhithrodontomys catalinae* Elliot, Field Columb. Mus., Publ. 87, Zool. Ser., 3:246, January 7, type from Santa Catalina Island [near Avalon], Santa Barbara Islands, California. Known only from Santa Catalina Island.
1952. *Reithrodontomys megalotis catalinae*, Hooper, Miscl. Publ. Mus. Zool., Univ. Michigan, 77:23, January 16.

Reithrodontomys megalotis distichlis von Bloeker

1937. *Reithrodontomys megalotis distichlis* von Bloeker, Proc. Biol. Soc. Washington, 50:155, September 10, type from salt marsh at mouth Salinas River, Monterey Co., California.

MARGINAL RECORDS.—California: *Moss Landing, Elkhorn Slough;* type locality; *Seaside Lagoon.* See Blanks and Shellhammer (1968:730) for additional localities clustered around type locality.

Reithrodontomys megalotis dychei J. A. Allen

1895. *Reithrodontomys dychei* J. A. Allen, Bull. Amer. Mus. Nat. Hist., 7:120, May 21, type from Lawrence, Douglas Co., Kansas.
1914. *Reithrodontomys megalotis dychei*, A. H. Howell, N. Amer. Fauna, 36:30, June 5.
1895. *Reithrodontomys dychei nebrascensis* J. A. Allen, Bull. Amer. Mus. Nat. Hist., 7:122, May 21, type from Kennedy, Cherry Co., Nebraska.

MARGINAL RECORDS.—Alberta: Medicine Hat. Montana: 1 mi. N, 1 mi. W Malta, 2248 ft. North Dakota: Fort Clark; Fargo. Minnesota: Rosemount. Illinois: Mt. Carroll; 2½ mi. N Dekalb (Stupka, *et al.*, 1973:112, owl pellets); Barrington (*ibid.*). Indiana: 5 mi. N Morocco (Whitaker and Sly, 1970:381). Illinois: Illinois State Univ. Campus [in Normal] (Birkenholz, 1967:51); 7 mi. S Virginia (Stains and Turner, 1963:274, as *R. megalotis* only). Missouri: St. Louis. Arkansas: Leachville. Missouri: Thayer. Kansas: Neosho Falls; *16 mi. N, 4 mi. E Stafford* (Jones and Mursaloğlu,

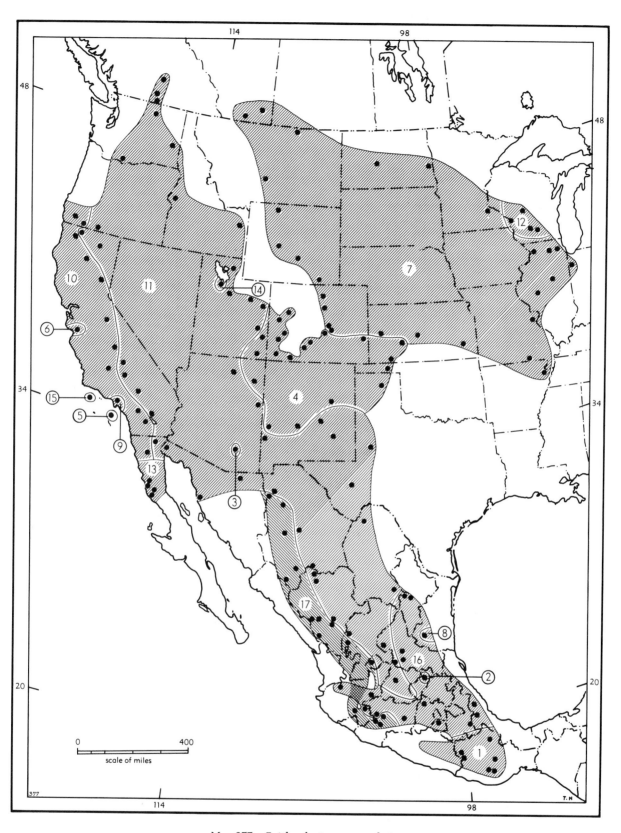

Map 377. *Reithrodontomys megalotis.*

Guide to subspecies
1. *R. m. alticolus*
2. *R. m. amoles*
3. *R. m. arizonensis*
4. *R. m. aztecus*

5. *R. m. catalinae*
6. *R. m. distichlis*
7. *R. m. dychei*
8. *R. m. hooperi*

9. *R. m. limicola*
10. *R. m. longicaudus*
11. *R. m. megalotis*
12. *R. m. pectoralis*

13. *R. m. peninsulae*
14. *R. m. ravus*
15. *R. m. santacruzae*
16. *R. m. saturatus*
17. *R. m. zacatecae*

1961:24); 3 mi. N, 2 mi. W Hoisington; 15 mi. W Scott City (Jones and Mursaloğlu, 1961:24). Colorado (Armstrong, 1972:193): 4 mi. S main gate, Camp Carson; Manitou; Golden; 16 mi. W Loveland, 6840 ft. Wyoming (Long, 1965a:626, 627): 1 mi. S, 3 mi. W Laramie; *Sun*; Splitrock; $\frac{3}{10}$ mi. NW Milford; 13 mi. N, 1 mi. E Cody. Montana: 11 mi. ENE Melville (Hoffmann, *et al.*, 1969:590, as *R. megalotis* only). Alberta: Mills River.

Reithrodontomys megalotis hooperi Goodwin

1954. *Reithrodontomys megalotis hooperi* Goodwin, Amer. Mus. Novit., 1660:1, May 25, type from Rancho del Cielo, 5 mi. NW Gómez Farías, 3500 ft., Tamaulipas. Known only from type locality.

Reithrodontomys megalotis limicola von Bloeker

1932. *Reithrodontomys megalotis limicola* von Bloeker, Proc. Biol. Soc. Washington, 45:133, September 9, type from Playa del Rey, Los Angeles Co., California.

MARGINAL RECORDS.—California: *Point Mugu;* type locality; *Anaheim Bay.*

Reithrodontomys megalotis longicaudus (Baird)

1858. *Reithrodon longicauda* Baird, Mammals, *in* Repts. Expl. Surv. . . . , 8(1):451, July 14, type from Petaluma, Sonoma Co., California.
1913. *Reithrodontomys megalotis longicauda*, Grinnell, Proc. California Acad. Sci., ser. 4, 3:303, August 28.
1893. *Reithrodontomys pallidus* Rhoads, Amer. Nat., 27:835, September, type from Santa Ysabel, San Jacinto Mts., San Diego Co., California.
1899. *Reithrodontomys klamathensis* Merriam, N. Amer. Fauna, 16:93, October 28, type from Big Spring (Mayten), Shasta Valley, Siskiyou Co., California.

MARGINAL RECORDS.—Oregon: Grants Pass; Ashland. California: Scott River Valley; Battle Creek Meadows; Portola; El Portal; Dunlap; Bakersfield; Hemet Lake, San Jacinto Mts.; Vallecito. Baja California: Rancho Viejo, 15 mi. E Alamos, thence northward along coast, except for ranges of *limicola* and *distichlis,* to point of beginning. Introduced on *San Clemente Island, California* (von Bloeker, 1967:255).

Reithrodontomys megalotis megalotis (Baird)

1858. *Reithrodon megalotis* Baird, Mammals, *in* Repts. Expl. Surv. . . . , 8(1):451, July 14, type from between Janos, Chihuahua, and San Luis Springs, Grant Co., New Mexico.
1893. *Reithrodontomys megalotis*, J. A. Allen, Bull. Amer. Mus. Nat. Hist., 5:79, April 28.
1895. *Reithrodontomys megalotis deserti* J. A. Allen, Bull. Amer. Mus. Nat. Hist., 7:127, May 21, type from Oasis Valley, Nye Co., Nevada.
1903. *Reithrodontomys megalotis sestinensis* J. A. Allen, Bull. Amer. Mus. Nat. Hist., 19:602, November 12, type from Río Sestín, 7500 ft., northwestern Durango.

1914. *Reithrodontomys megalotis nigrescens* A. H. Howell, N. Amer. Fauna, 36:32, June 5, type from Payette, Payette Co., Idaho.

MARGINAL RECORDS.—British Columbia: 2 mi. N Okanagan Landing (Munro, 1958:146, as *R. megalotis* only); NE end Skaha (Dog) Lake, just S of Penticton; Osoyoos Lake (Cowan and Guiguet, 1965:176). Washington: Timentwa; Colfax. Idaho: Crane Creek, 15 mi. E Midvale; Menan. Utah: Logan; Corner Canyon, near Draper Tunnel, 5000 ft.; 2 mi. E Duchesne; Willow Creek, 25 mi. S Ouray, 5250 ft.; Highway 160, 14 mi. N Moab, 4500 ft. Arizona: vic. Tuba; Zuni River. New Mexico: San Francisco River Valley, near Alma, Gallo Canyon, 35 mi. E Corona; Roswell. Texas: *ca.* 10 mi. E Lamesa (Blair, 1954:249, as *R. m. dychei.* If his specimen, No. 3999 Texas Nat. Hist. Coll., is correctly identified to species, it is subsp. *R. m. megalotis*, but in August 1973, when examined by E. R. Hall, the skull could not be found, and without examining it he could not be sure that No. 3999 is not *R. fulvescens canus*); 25 mi. W Fort Stockton. Coahuila: 35 mi. S, 14 mi. E Boquillas, 2350 ft.; Saltillo, 4800 ft. San Luis Potosí: 2 km. E Illescas; 2 km. NE Arriaga. Guanajuato: Santa Rosa, 8200 ft.?. Zacatecas: 1 mi. N, 8 mi. W Sombrerete, 7800 ft. Durango (Baker and Greer, 1962:106): 6 mi. S Morcillo, 6450 ft.; Río Sestín; Rosario. Chihuahua: 2 mi. W Parral, 6200 ft.; 8 mi. NE Laguna, *ca.* 3 mi. from Laguna de Bustillos (Anderson and Nelson, 1960:100, as *R. megalotis* only, in owl pellet); Casas Grandes, 4800 ft. Sonora: 5 mi. N Cananea, 4750 ft.; Puerto de Lobos, near sea level; Cienega Well, 50 ft.?. Baja California: Gardners Lagoon. California: Mecca; Victorville; Tehachapi; Bodfish; Bieber; Montague. Oregon: Klamath Lake Basin. Washington: Maryhill.

Reithrodontomys megalotis pectoralis Hanson

1944. *Reithrodontomys megalotis pectoralis* Hanson, Field Mus. Nat. Hist., Publ. 564, Zool. Ser., 29:205, October 26, type from Westpoint, Columbia Co., Wisconsin (Jackson, Mammals of Wisconsin, Univ. Wisconsin Press, 1961, p. 207, in recognizing *R. m. pectoralis*, made no attempt to identify as to subspecies specimens from the bordering states of Minnesota, Iowa, and Illinois).

MARGINAL RECORDS (Jackson, 1961:208).— Wisconsin: Sprague; *type locality;* Madison; *4 mi. W Middleton;* Mazomanie; La Crosse.

Reithrodontomys megalotis peninsulae (Elliot)

1903. *Rhithrodontomys peninsulae* Elliot, Field Columb. Mus., Publ. 74, Zool. Ser., 3(10):164, May 7, type from San Quintín, Baja California.
1914. *Reithrodontomys megalotis peninsulae*, A. H. Howell, N. Amer. Fauna, 36:35, June 5.

MARGINAL RECORDS.—Baja California: Socorro; Pozo Luciano, NW slope Sierra San Pedro Mártir Mts.; Rosario; type locality; *San Telmo.*

Reithrodontomys megalotis ravus Goldman

1939. *Reithrodontomys megalotis ravus* Goldman, Jour. Mamm., 20:355, August 14, type from N end Stansbury Island, 4250 ft., Great Salt Lake, Tooele Co., Utah.

MARGINAL RECORDS.—Utah: type locality; *Grantsville.*

Reithrodontomys megalotis santacruzae Pearson

1951. *Reithrodontomys megalotis santacruzae* Pearson, Jour. Mamm., 32:366, August 23, type from Prisoners Harbor, Santa Cruz Island, Santa Barbara Co., California. Known only from Santa Cruz Island.

Reithrodontomys megalotis saturatus J. A. Allen and Chapman

1897. *Reithrodontomys saturatus* J. A. Allen and Chapman, Bull. Amer. Mus. Nat. Hist., 9:201, June 16, type from Las Vigas, 8000 ft., Veracruz.
1914. *Reithrodontomys megalotis saturatus*, A. H. Howell, N. Amer. Fauna, 36:36, June 5.
1901. *Reithrodontomys saturatus cinereus* Merriam, Proc. Washington Acad. Sci., 3:556, November 29, type from near Chalchicomula (San Andrés), Puebla.

MARGINAL RECORDS.—Coahuila: *Diamante, 7500 ft.*; 12 mi. E San Antonio de las Alazanas, 9000 ft. Nuevo León: 12 mi. N Galeana, 7000 ft. Veracruz: 3 km. E Las Vigas, 8000 ft.; Volcán de Orizaba, timberline; Xuchil (Hall and Dalquest, 1963:296). Morelos: C. Cruz del Morelos, 2440 m. Michoacán: 10 mi. SE Pátzcuaro, 9200 ft.; 2 mi. NNW San Juan, 7700 ft. Jalisco: ½ mi. NW Mazamitla; 1 mi. WSW Ameca, 4000 ft.; vic. Ocotlán, 5000 ft. México: Atlacomulco, 8200 ft. San Luis Potosí: 3 km. SW San Isidro; Leoncito.

Reithrodontomys megalotis zacatecae Merriam

1901. *Reithrodontomys megalotis zacatecae* Merriam, Proc. Washington Acad. Sci., 3:557, November 29, type from Sierra de Valparaíso, Zacatecas.
1901. *Reithrodontomys megalotis obscurus* Merriam, Proc. Washington Acad. Sci., 3:558, November 29, type from Sierra Madre, near Guadalupe y Calvo, Chihuahua.

MARGINAL RECORDS.—Chihuahua (Anderson, 1972:331): 3 mi. SW Pacheco; 9 mi. WSW San Buenaventura; 2 mi. W Miñaca, 6900 ft.; Rancheria, 20 km. E Guachochic, 6250 ft. Durango: Laguna del Progreso (Baker and Greer, 1962:106); 1 mi. SW La Ciudad, 8300 ft. (*ibid.*); 28 mi. S, 17 mi. W Vicente Guerrero, *8350 ft. (ibid.).* Zacatecas: type locality. Aguascalientes: Sierra Fría. Jalisco: N slope El Nevado de Colima, 7300 ft. Michoacán: Nahuatzen, 8500 ft.; 9 mi. SE Pátzcuaro, 8000 ft.; Cerro Tancítaro, 1 mi. N Apo, 7000 ft. Durango: 9 mi. SW El Salto (Baker and Greer, 1962:106); 1½ mi. W San Luis (*ibid.*). Chihuahua: Sierra Madre near Guadalupe y Calvo, between 7000 and 9000 ft.

Reithrodontomys raviventris
Salt-marsh Harvest Mouse

Revised by Fisler, Univ. California Publ. Zool., 77:1–108, 23 figs., numerous tables, November 3, 1965.

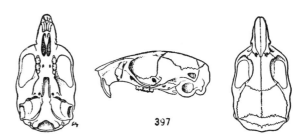

Fig. 397. *Reithrodontomys raviventris raviventris*, Palo Alto, Santa Clara Co., California, No. 3527 M.V.Z., ♀, X 1½.

External measurements: 135–162; 56–95; 15–21. Upper parts pinkish cinnamon interspersed with black hairs, especially mid-dorsally; underparts pinkish cinnamon or whitish; hind feet dark brown or whitish; tail dark brown or blackish above, paler, sometimes almost white, below. From *R. megalotis longicaudus*, differs as follows: pelage darker, thicker, longer; tail thicker (20 mm. from base, diameter 2.1–2.2 vs. 1.8–1.9 mm.); braincase longer.

This dark-colored harvest mouse is limited to the vicinity of San Francisco Bay, California. It

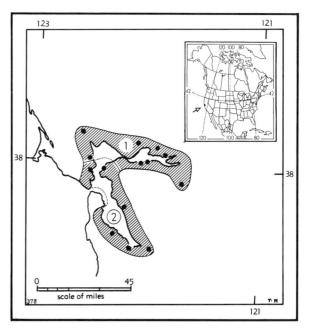

Map 378. *Reithrodontomys raviventris.*

1. *R. r. halicoetes*   2. *R. r. raviventris*

lives in the salt marshes where the ground is damp or wet and the marsh plants form a thick mat over a network of interstices.

Reithrodontomys raviventris halicoetes Dixon

1909. *Reithrodontomys halicoetes* Dixon, Univ. California Publ. Zool., 5:271, August 14, type from 3 mi. S Petaluma, Sonoma Co., California.
1914. *Reithrodontomys raviventris halicoetes*, A. H. Howell, N. Amer. Fauna, 36:42, June 5.

MARGINAL RECORDS (Fisler, 1965:100).— California: Petaluma and in other salt marshes on northern and western shores of San Pablo Bay and around Suisun Bay, as at Cordelia; Grizzly Island; Collinsville; Brentwood (*op. cit.*:7); ½ mi. W Avon; 1 mi. N, 2 mi. E Martinez; ¼ mi. NE Forbes.

Reithrodontomys raviventris raviventris Dixon

1908. *Reithrodontomys raviventris* Dixon, Proc. Biol. Soc. Washington, 21:197, October 20, type from salt marsh near Redwood City, San Mateo Co., California.

MARGINAL RECORDS (Fisler, 1965:99).— California: Salt marshes around San Francisco Bay, as at 2 mi. NE Larkspur; 1 mi. N mouth San Pablo Creek; Bay Farm Island; Alviso; Palo Alto; Belmont.

Reithrodontomys chrysopsis
Volcano Harvest Mouse

External measurements: 170–192; 89–108; 19–21; 17–19. Upper parts bright orange-buff mixed with black, black predominant on back and muzzle, buff tones purest on sides; underparts pinkish cinnamon; ocular ring black; ears black; tail sharply bicolored, blackish above, either slightly paler or silvery below. Skull with large, oval braincase (larger than in *tenuirostris*, *mexicanus*, or *hirsutus*); frontals small, strongly constricted interorbitally; rostrum long, broader than interorbital constriction; both anterior and posterior palatine foramina long, slender; sphenopalatine vacuities large, crescent-shaped; pterygoid fossa broader than mesopterygoid (interpterygoid) fossa; molars high-crowned.

Reithrodontomys chrysopsis chrysopsis Merriam

1900. *Reithrodontomys chrysopsis* Merriam, Proc. Biol. Soc. Washington, 13:152, June 13, type from Volcán Popocatépetl, 11,500 ft., México.
1901. *Reithrodontomys chrysopsis tolucae* Merriam, Proc. Washington Acad. Sci., 3:549, November 29, type from north slope Volcán de Toluca, 11,500 ft., México.
1901. *Reithrodontomys colimae* Merriam, Proc. Washington Acad. Sci., 3:551, November 29, type from near timberline, El Nevado de Colima, 12,000 ft., Jalisco.

MARGINAL RECORDS.—Michoacán: Cerro Patambán, 11,000 ft. Distrito Federal: La Venta, 10,200

ft. México: type locality. Morelos: 2 mi. W Huitzilac; *Cerro Zempoala, 3000 m.* (Ramírez-P., 1971:273). México: Salazar, 9000 ft. Michoacán: Cerro Tancítaro, 10,000–12,000 ft. Jalisco: NW slopes El Nevado de Colima, 9500–12,000 ft.

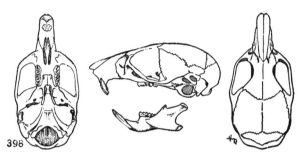

Fig. 398. *Reithrodontomys chrysopsis chrysopsis*, Monte Río Frío, 45 km. ESE Mexico City, México, No. 17979 K.U., ♂, X 1½.

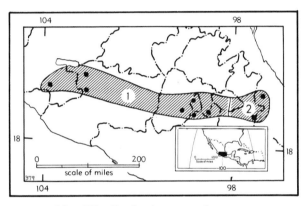

Map 379. *Reithrodontomys chrysopsis.*

1. *R. c. chrysopsis*   2. *R. c. perotensis*

Reithrodontomys chrysopsis perotensis Merriam

1901. *Reithrodontomys perotensis* Merriam, Proc. Washington Acad. Sci., 3:550, November 29, type from Cofre de Perote, 9500 ft., Veracruz.
1952. *Reithrodontomys chrysopsis perotensis*, Hooper, Miscl. Publ. Mus. Zool., Univ. Michigan, 77:89, January 16.
1901. *Reithrodontomys orizabae* Merriam, Proc. Washington Acad. Sci., 3:550, November 29, type from Volcán de Orizaba, 9500 ft., Puebla.

MARGINAL RECORDS.—Veracruz: N slope Cofre de Perote, Los Conejos, 10,600 ft. (Hall and Dalquest, 1963:297). Puebla: El Volcán de Orizaba, 9500 ft.

Reithrodontomys sumichrasti
Sumichrast's Harvest Mouse

External measurements: 143–206; 75–123; 17–22; 12–18. "In tone of coloration the races of *sumichrasti* may be arranged in 2 groups. Within each group the color deepens geographically from north to south. The races *sumichrasti, dorsalis, australis,* and *vulcanius* are dark-colored. . . . In each there is an abundance of black dorsally which largely obscures the cinnamon ground color. Thus, the upper parts are blackish and dull. The underparts are similarly somber. . . . The races *nerterus, luteolus,* and *modestus* comprise a group which is brighter in color. In them there is less black dorsally and the cinnamon-tinted bands of the hairs of the underparts obscure the blackish basal bands. The result is a distinctly brighter and on the average more buffy coat in those races." (Hooper, 1952a:68.) Skull with mesopterygoid fossa broad; tympanic bullae small; paracone of M2 not evenly rounded posteriorly but distinctly keeled on posterolabial margin, the keel projecting into 2nd primary fold.

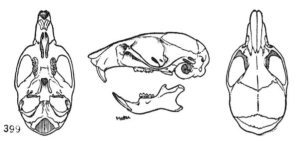

Fig. 399. *Reithrodontomys sumichrasti australis,* SW slope Volcán Irazú, 8500 ft., Prov. Cartago, Costa Rica, No. 26967 K.U., ♀, X 1½.

Reithrodontomys sumichrasti australis J. A. Allen

1895. *Reithrodontomys australis* J. A. Allen, Bull. Amer. Mus. Nat. Hist., 7:328, November 8, type from Volcán de Irazú, Costa Rica.
1952. *Reithrodontomys sumichrasti australis,* Hooper, Miscl. Publ. Mus. Zool., Univ. Michigan, 77:82, January 16.

MARGINAL RECORDS.—Costa Rica: type locality; *Altos Escazú, 4000 ft.*

Reithrodontomys sumichrasti dorsalis Merriam

1901. *Reithrodontomys dorsalis* Merriam, Proc. Washington Acad. Sci., 3:557, November 29, type from Calel, Guatemala.
1952. *Reithrodontomys sumichrasti dorsalis,* Hooper, Miscl. Publ. Mus. Zool., Univ. Michigan, 77:78, January 16.

MARGINAL RECORDS.—Chiapas: Tumbalá, 5000–5500 ft.; 28 mi. ESE Comitán, San José, 4900 ft. Guatemala: Cobán, 4500 ft.; Volcán Acatenango, Finca Montserrat, 5700 ft.; 2 mi. S, 7 mi. E La Unión (Anderson and Jones, 1960:522); Volcán Tajumulco, 10,400–13,200 ft. Chiapas (Baker, *et al.*, 1973:78, 85): *10 km. NNW Jitotol, 1645 m.; 6½ km. SE Rayón, 1675 m.*

Reithrodontomys sumichrasti luteolus A. H. Howell

1914. *Reithrodontomys rufescens luteolus* A. H. Howell, N. Amer. Fauna, 36:57, June 5, type from Juquila, 5000 ft., Oaxaca.
1952. *Reithrodontomys sumichrasti luteolus,* Hooper, Miscl. Publ. Mus. Zool., Univ. Michigan, 77:77, January 16.
1914. *Reithrodontomys alleni* A. H. Howell, N. Amer. Fauna, 36:59, June 5, type from mts. near Ozolotepec, 10,000 ft., Oaxaca.

MARGINAL RECORDS.—Guerrero: 3 mi. W Omilteme, 8200 ft. Oaxaca (Goodwin, 1969:165–166, unless otherwise noted): 16 km. SW Cuquila, 2380 m. (Webb and Baker, 1971:144); Santiago Lachiguirí; San Pedro Jilotepec; 3 mi. N Ozolotepec, La Cieneguía, 10,000 ft. (Hooper, 1952a:78); Río Molino; La Soldedad [= La Soledad]; Jamiltepec.

Reithrodontomys sumichrasti modestus Thomas

1907. *Reithrodontomys modestus* Thomas, Ann. Mag. Nat. Hist., ser. 7, 20:163, August, type from Jinotega, 4650 ft., Nicaragua.
1952. *Reithrodontomys sumichrasti modestus,* Hooper, Miscl. Publ. Mus. Zool., Univ. Michigan, 77:80, January 16.
1937. *Reithrodontomys dorsalis underwoodi* Goodwin, Amer. Mus. Novit., 921:2, May 3, type from Monte Verde, 4500 ft., 30 mi. NW Ocotepeque, Ocotepeque, Honduras.

MARGINAL RECORDS.—Honduras: Cementerio, 5300 ft.; Cerro Uyuca, 6100 ft. Nicaragua (Jones and Genoways, 1970:4): Savala; Santa María de Ostuma. El Salvador: Los Esesmiles, 6200–8000 ft. Honduras: Monte Verde, 4500 ft.

Reithrodontomys sumichrasti nerterus Merriam

1901. *Reithrodontomys colimae nerterus* Merriam, Proc. Washington Acad. Sci., 3:551, November 29, type from foothills of El Nevado de Colima, Jalisco.
1952. *Reithrodontomys sumichrasti nerterus,* Hooper, Miscl. Publ. Mus. Zool., Univ. Michigan, 77:74, January 16.
1901. *Reithrodontomys levipes otus* Merriam, Proc. Washington Acad. Sci., 3:555, November 29, type from foothills of El Nevado de Colima, Jalisco.
1949. *Reithrodontomys chrysopsis seclusus* Hall and Villa, Proc. Biol. Soc. Washington, 62:163, August 23, type from Mt. Tancítaro, 7800 ft., Michoacán.

MARGINAL RECORDS.—Jalisco: ½ mi. NW Mazamitla. Michoacán: 10 mi. NW Ciudad Hidalgo, Cerro San Andrés, 9400 ft.; 10 mi. ESE Zitácuaro, Macho de Agua, 8000 ft.; 9 mi. SE Pátzcuaro; *Dos Aguas, 7000 ft.* (Hooper, 1961:121); *Rancho Reparto, 6000 ft.* (ibid.); 20 min. SW Rancho Barolosa, 7600 ft. (ibid.). Jalisco: SE slope El Nevado de Colima, 9100 ft. (Baker and Phillips, 1965:692); 20 mi. SSE Autlán, 6500 ft.; *Sierra de Autlán.*

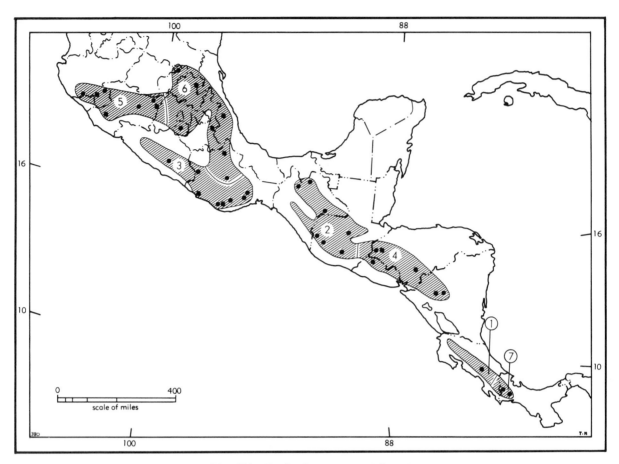

Map 380. *Reithrodontomys sumichrasti.*

1. *R. s. australis*	3. *R. s. luteolus*	6. *R. s. sumichrasti*
2. *R. s. dorsalis*	4. *R. s. modestus*	7. *R. s. vulcanius*
	5. *R. s. nerterus*	

Reithrodontomys sumichrasti sumichrasti (Saussure)

1861. *Reithrodon sumichrasti* Saussure, Revue et Mag. Zool., Paris, ser. 2, 13:3, type from México (restricted to Mirador, Veracruz, by Hooper (*infra*, p. 72).
1952. *Reithrodontomys sumichrasti sumichrasti*, Hooper, Miscl. Publ. Mus. Zool., Univ. Michigan, 77:71, January 16.
1897. *Reithrodontomys rufescens* J. A. Allen and Chapman, Bull. Amer. Mus. Nat. Hist., 9:199, June 16, type from Jalapa, 4400 ft., Veracruz.

MARGINAL RECORDS.—Querétaro: Amoles, somewhere between 7500 and 9500 ft. Hidalgo: Molango, 5400 ft. Veracruz: 5 km. N Jalapa, 4500 ft. (Hall and Dalquest, 1963:297). Oaxaca (Goodwin, 1969:165): Huehuetlán; San Felipe del Agua; *15 mi. SW Oaxaca de Juárez.* Puebla: Río Atlati, 8700 ft. Morelos: 1½ mi. SE Huitzilac; *Atzompa, 2825 m.* (Ramírez-P., 1971:272).

Reithrodontomys sumichrasti vulcanius Bangs

1902. *Reithrodontomys australis vulcanius* Bangs, Bull. Mus. Comp. Zool., 39:38, April, type from Volcán de Chiriquí, 10,300 ft., Chiriquí, Panamá.

1952. *Reithrodontomys sumichrasti vulcanius,* Hooper, Miscl. Publ. Mus. Zool., Univ. Michigan, 77:83, January 16.

MARGINAL RECORDS.—Panamá (Handley, 1966:783): Cerro Punta, 5000–7800 ft.; *type locality;* Boquete, 4000 ft.; *El Volcán.*

fulvescens-group
Reithrodontomys fulvescens
Fulvous Harvest Mouse

External measurements: 134–200; 72–116; 16–22; 11–17. Upper parts finely grizzled or "streaked" in a salt-and-pepper effect, resulting from a mixture of reddish brown and black; underparts varying geographically from pale buff to whitish; tail long (10–50% longer than head and body); blackish eye ring absent; lateral line present or absent. Skull with comparatively robust rostrum; premaxillary tongues broad; braincase elongate; frontals inflated in region of base of rostrum; zygomatic plate broader than mesopterygoid fossa; sphenopalatine vacuities oblong; length of incisive foramina only slightly more than breadth of rostrum.

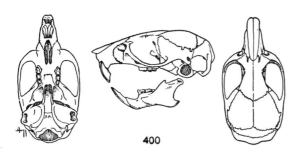

Fig. 400. *Reithrodontomys fulvescens intermedius*, Brownsville, Cameron Co., Texas, No. 1920 K.U., ♂, X 1½.

Reithrodontomys fulvescens amoenus (Elliot)

1905. *Rhithrodontomys amoenus* Elliot, Proc. Biol. Soc. Washington, 18:234, December 9, type from Reforma, about 500 ft., Oaxaca. Known only from type locality.
1952. *Reithrodontomys fulvescens amoenus*, Hooper, Miscl. Publ. Mus. Zool., Univ. Michigan, 77:120, January 16.

Reithrodontomys fulvescens aurantius J. A. Allen

1895. *Reithrodontomys mexicanus aurantius* J. A. Allen, Bull. Amer. Mus. Nat. Hist., 7:137, May 21, type from Lafayette, Lafayette Parish, Louisiana.
1914. *Reithrodontomys fulvescens aurantius*, A. H. Howell, N. Amer. Fauna, 36:48, June 5.
1899. *Reithrodontomys chrysotis* Elliot, Field Columb. Mus., Publ. 37, Zool. Ser., 1:281, May 15, type from Dougherty, Washita River, Murray Co., Oklahoma.

MARGINAL RECORDS.—Missouri: 8 mi. N, 3 mi. W Lamar (Long, 1965b:506); 5 mi. N Gainesville. Arkansas: Beebe. Mississippi (Kennedy, *et al.*, 1974:15): Bolivar County; Holmes County; Neshoba County; Forrest County. Texas (Davis, 1966:178, as *R. fulvescens* only, unless otherwise noted): Jefferson County; Brazoria County; Matagorda County; Eagle Lake (Hall and Kelson, 1959:592); 4 mi. NE San Marcos, 600 ft. (*ibid.*); Travis County; Dallas County. Oklahoma: Dougherty, 750 ft.; Noble, 1500 ft. Kansas: 1½ mi. SW Cedarvale.

Reithrodontomys fulvescens canus Benson

1939. *Reithrodontomys fulvescens canus* Benson, Proc. Biol. Soc. Washington, 52:149, October 11, type from 5 mi. SE Chihuahua, Chihuahua.

MARGINAL RECORDS.—New Mexico: 22 mi. S, 2 mi. E Rodeo (Findley and Pullen, 1958:306). Chihuahua (Anderson, 1972:329): 1½ mi. N San Francisco, 5100 ft.; Casas Grandes, 4800 ft.; Cañón del Potrero, 7 mi. W El Sauz, 5750 ft.; 8 mi. N, 11 mi. E Charco de Peña, 4700 ft. Texas: Limpia Canyon, 4300 ft., Davis Mts., 15 mi. N Fort Davis; Brewster County (Davis, 1966:178, as *R. fulvescens* only). Coahuila: Sierra del Carmen; 17 mi. N, 8 mi. W Saltillo, 5200 ft. Durango: San Juan, 3800 ft., 12 mi. W Gómez Palacio; Indé, 6100 ft. Chihuahua (Anderson, 1972:329): near Parral, *ca.* 10 mi. SE El Torreón, 5500 ft.; 2 mi. W Miñaca, 6900 ft.; Cherry Ranch, 11 mi. WNW Cocomorachic; Chuichupa.

Reithrodontomys fulvescens chiapensis A. H. Howell

1914. *Reithrodontomys fulvescens chiapensis* A. H. Howell, N. Amer. Fauna, 36:53, June 5, type from Canjob, about 5000 ft., Chiapas.

MARGINAL RECORDS.—Chiapas: near Bochil, 4300 ft. Guatemala: 6 mi. NNE Salamá, 5500 ft. Honduras: La Flor Archaga, 4500–5000 ft. El Salvador: Volcán San Miguel, 4300 ft. Guatemala: *2 mi. N, 2 mi. W Cuilapa* [= *Cuajiniquilapa*] (Anderson and Jones, 1960:522); 5 mi. S Guatemala City (*ibid.*); 4½ mi. W Sacapulas, 4300 ft. Chiapas: Jaltenango, 2300 ft.; Cintalapa, 1800 ft.

Reithrodontomys fulvescens difficilis Merriam

1901. *Reithrodontomys difficilis* Merriam, Proc. Washington Acad. Sci., 3:566, November 29, type from Orizaba, about 4500 ft., Veracruz.
1914. *Reithrodontomys fulvescens difficilis*, A. H. Howell, N. Amer. Fauna, 36:50, June 5.

MARGINAL RECORDS.—Puebla: Rancho El Ajengibre, Kilometer 264, México–Tuxpan road (3630 U. Minn.). Veracruz: *1½ mi. N Jalapa, 4500 ft.* (Hall and Dalquest, 1963:298); 4 mi. SE Jalapa, 4800 ft.; 9 mi. ENE Tlacotepec, 1500 ft.; Potrero Viejo, 1700 ft.; Río Blanco, 4200 ft. Hidaldo: San Agustín, 3500 ft.

Reithrodontomys fulvescens fulvescens J. A. Allen

1894. *Reithrodontomys mexicanus fulvescens* J. A. Allen, Bull. Amer. Mus. Nat. Hist., 6:319, November 7, type from Oposura, 2000 ft., Sonora.
1895. *Reithrodontomys fulvescens* J. A. Allen, Bull. Amer. Mus. Nat. Hist., 7:138, May 21.

MARGINAL RECORDS.—Arizona: Phoenix; mouth Miller Canyon, 5300 ft., Huachuca Mts. Sonora: type locality; La Estancia, 2150 ft., 6 mi. N Nacori; La Misión, 2900 ft., 2 mi. SW Magdalena. Arizona: 8 mi. SW Tucson, 2400 ft.

Reithrodontomys fulvescens griseoflavus Merriam

1901. *Reithrodontomys griseoflavus* Merriam, Proc. Washington Acad. Sci., 3:553, November 29, type from Ameca, 4000 ft., Jalisco.
1952. *Reithrodontomys fulvescens griseoflavus*, Hooper, Miscl. Publ. Mus. Zool., Univ. Michigan, 77:98, January 16.

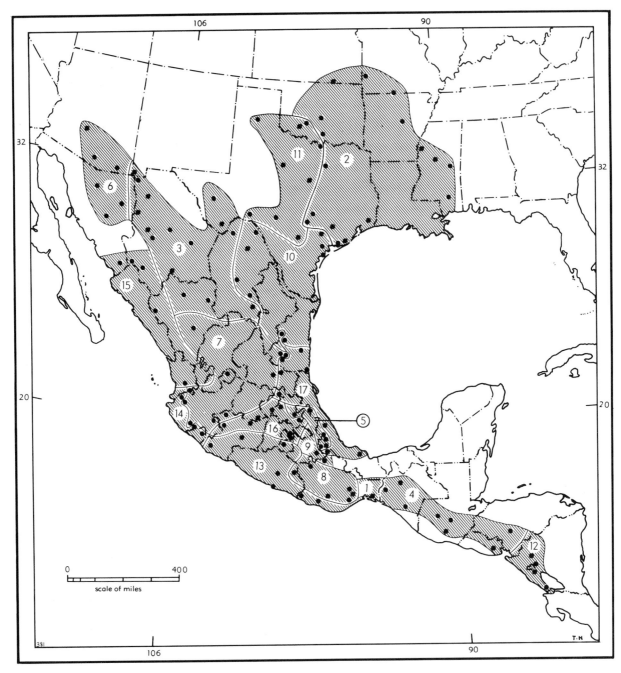

Map 381. *Reithrodontomys fulvescens.*

Guide to subspecies
1. *R. f. amoenus*
2. *R. f. aurantius*
3. *R. f. canus*
4. *R. f. chiapensis*
5. *R. f. difficilis*
6. *R. f. fulvescens*
7. *R. f. griseoflavus*
8. *R. f. helvolus*
9. *R. f. infernatis*
10. *R. f. intermedius*
11. *R. f. laceyi*
12. *R. f. meridionalis*
13. *R. f. mustelinus*
14. *R. f. nelsoni*
15. *R. f. tenuis*
16. *R. f. toltecus*
17. *R. f. tropicalis*

MARGINAL RECORDS.—Durango: 1 mi. N Chorro, 6400 ft. (Baker and Greer, 1962:105). Tamaulipas: Jaumave, 2700–3300 ft. San Luis Potosí: Río Verde, 3200 ft. Michoacán: vic. Zamora, 5500–6000 ft. Jalisco: ½ mi. NW Mazamitla; 6 mi. SSW Autlán, 4500 ft.; Talpa, 4000 ft. Nayarit: 2–4 mi. N Santa Isabel, 3800 ft. Aguascalientes: ½ mi. W Rincón de Romos. Durango: *Durango* (Baker and Greer, 1962:105); *5 mi. N Durango, 6400 ft. (ibid.).*

Reithrodontomys fulvescens helvolus Merriam

1901. *Reithrodontomys griseoflavus helvolus* Merriam, Proc. Washington Acad. Sci., 3:554, November 29, type from Oaxaca City, about 5000 ft., Oaxaca.
1914. *Reithrodontomys fulvescens helvolus*, A. H. Howell, N. Amer. Fauna, 36:52, June 5.

MARGINAL RECORDS.—Oaxaca: Huajuapan, 5000 ft.; San Pedro Jilotepec (Goodwin, 1969:166); La Concepción (*ibid.*); Arroyo Palmar (*ibid.*); Sola de Vega, 5000 ft. Guerrero: Tlapa, 3000 ft.

Reithrodontomys fulvescens infernatis Hooper

1950. *Reithrodontomys fulvescens infernatis* Hooper, Proc. Biol. Soc. Washington, 63:167, December 29, type from Teotitlán, 3100 ft., Oaxaca.

MARGINAL RECORDS.—Puebla: Tepanco, 6000 ft. Oaxaca: type locality.

Reithrodontomys fulvescens intermedius J. A. Allen

1895. *Reithrodontomys mexicanus intermedius* J. A. Allen, Bull. Amer. Mus. Nat. Hist., 7:136, May 21, type from Brownsville, Cameron Co., Texas.
1914. *Reithrodontomys fulvescens intermedius*, A. H. Howell, N. Amer. Fauna, 36:47, June 5.

MARGINAL RECORDS.—Texas: 14 mi. WSW Hallettsville; Victoria County (Davis, 1966:178, as *R. fulvescens* only); 6 mi. NE Rockport, thence southward along coast and coastal islands to Tamaulipas: Hda. Santa Engracia, 800 ft.; 29 mi. S Ciudad Victoria, 800 ft. (Alvarez, 1963:439). Coahuila: 2 mi. N San Lázaro, 64 mi. N, 22 mi. W Saltillo; 3 mi. NNW Cuatro Ciénegas; 6 mi. SW San Gerónimo; 1 mi. S, 9 mi. W Villa Acuña (Baker, 1956:252).

Reithrodontomys fulvescens laceyi J. A. Allen

1896. *Reithrodontomys laceyi* J. A. Allen, Bull. Amer. Mus. Nat. Hist., 8:235, November 21, type from Watsons Ranch, 15 mi. S San Antonio, Bexar Co., Texas.
1953. *Reithrodontomys fulvescens laceyi*, Russell, Texas Jour. Sci., 5:457, December.

MARGINAL RECORDS.—Oklahoma: Mt. Scott. Texas: 7 mi. N Gainesville; 6 mi. NW Walnut Springs; type locality; Val Verde County (Davis, 1966:178, as *R. fulvescens* only); 20 mi. W Mountain Home; 13½ mi. W Albany (Packard and Judd, 1968:537); Armstrong County (Davis, 1966:178, as *R. fulvescens* only). Oklahoma: 8 mi. NW Indiahoma.

Reithrodontomys fulvescens meridionalis Anderson and Jones

1960. *Reithrodontomys fulvescens meridionalis* Anderson and Jones, Univ. Kansas Publ., Mus. Nat. Hist., 9:522, January 14, type from 9 mi. NNW Estelí, Estelí, Nicaragua.

MARGINAL RECORDS.—Nicaragua: type locality; 11 mi. SE Darío (Jones and Genoways, 1970:5); Finca Amayo, 13 km. S, 14 km. E Rivas (*ibid.*); *11 km. S, 3 km. E Rivas* (*ibid.*); Kilometer 38 N Pan-American Hwy., 13½ km. N, 4½ km. E Tipitapa (*ibid.*); *8 mi. NNW Estelí* (*ibid.*).

Reithrodontomys fulvescens mustelinus A. H. Howell

1914. *Reithrodontomys fulvescens mustelinus* A. H. Howell, N. Amer. Fauna, 36:54, June 5, type from Llano Grande, 300 ft., Oaxaca.

MARGINAL RECORDS.—Michoacán: 1½–6 mi. S Tacámbaro, 4000–5700 ft. Morelos: 1 mi. W Tepoztlán, 6000 ft. Guerrero: vic. Chilpancingo, 4300–4500 ft. Oaxaca: type locality; Teotepec (Goodwin, 1969:167). Guerrero: Acapulco, near sea level. Michoacán: Coalcomán.

Reithrodontomys fulvescens nelsoni A. H. Howell

1914. *Reithrodontomys fulvescens nelsoni* A. H. Howell, N. Amer. Fauna, 36:53, June 5, type from Colima, Colima.

MARGINAL RECORDS.—Jalisco: San Sebastián; 2 mi. N La Resolana, 1200 ft. Colima: type locality, thence northward along coast to point of beginning.

Reithrodontomys fulvescens tenuis J. A. Allen

1899. *Reithrodontomys tenuis* J. A. Allen, Bull. Amer. Mus. Nat. Hist., 12:15, March 4, type from Rosario, Sinaloa.
1914. *Reithrodontomys fulvescens tenuis*, A. H. Howell, N. Amer. Fauna, 36:45, June 5.

MARGINAL RECORDS.—Sonora: vic. Alamos, 1500–4500 ft. Chihuahua: 1½ mi. SW Tocuina, 1500 ft. (Anderson, 1972:329); Las Guásimas, near Batopilas, 2800 ft. Durango: Chacala, 1800 ft. Nayarit: Tepic, 3000 ft., thence northward along coast to point of beginning.

Reithrodontomys fulvescens toltecus Merriam

1901. *Reithrodontomys levipes toltecus* Merriam, Proc. Washington Acad. Sci., 3:555, November 29, type from Tlalpan, Distrito Federal, México.
1914. *Reithrodontomys fulvescens toltecus*, A. H. Howell, N. Amer. Fauna, 36:51, June 5.
1903. *Reithrodontomys inexspectatus* Elliot, Field Columb. Mus., Publ. 71, Zool. Ser., 3(8):145, March 20, type from Pátzcuaro, Michoacán.

MARGINAL RECORDS.—Hidalgo: Zimapán, 6000–6400 ft.; 1 mi. W Tulancingo (6825 U. Minn.). México: Hda. Córdoba, 8300 ft.; 4 km. ENE Tlalmanalco, 2290 m. *Distrito Federal: 5 km. S Mexico City.* Michoacán: 4 mi. S Cuitzeo, 5900–6000 ft.; Los Reyes, 5000 ft. Guanajuato: Acámbaro, 6100 ft. Querétaro: Tequisquiapam, 6200 ft.

Reithrodontomys fulvescens tropicalis Davis

1944. *Reithrodontomys fulvescens tropicalis* Davis, Jour. Mamm., 25:393, December 12, type from Boca del Río, 8 km. S city of Veracruz, Veracruz.

MARGINAL RECORDS.—Tamaulipas (Alvarez, 1963:440): Hidalgo; *Rancho Santa Rosa, 25 km. N, 13 km. W Ciudad Victoria;* Sierra de Tamaulipas, 2 mi. S, 10 mi. W Piedra; *16 km. N Tampico;* 7 km. N Tampico, thence southward along coast to Veracruz: Catemaco, 1100 ft.; Presidio; 4 mi. NNW Cerro Gordo, 1500 ft. Querétaro: Jalpán, 2500 ft. San Luis Potosí; 2 mi. N, 10 mi. E Ciudad del Maíz. Tamaulipas (Alvarez, 1963:440): Rancho Pano Ayuctle; *2 km. W El Carrizo.*

Reithrodontomys hirsutus Merriam
Hairy Harvest Mouse

1901. *Reithrodontomys hirsutus* Merriam, Proc. Washington Acad. Sci., 3:553, November 29, type from Ameca, 4000 ft., Jalisco.
1901. *Reithrodontomys levipes* Merriam, Proc. Washington Acad. Sci., 3:554, November 29, type from San Sebastián, 3000 ft., Jalisco.

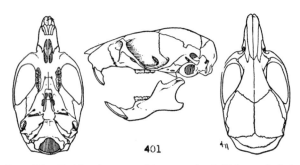

Fig. 401. *Reithrodontomys hirsutus,* 4 mi. N Santa Isabel, Nayarit, No. 94314 U.M.M.Z., ♂, X 1½.

External measurements: 175–202; 100–115; 20–22; 16–17. In coloration [*hirsutus*] resembles pale races of *R. fulvescens* and *R. gracilis.* The peculiar "streaked" appearance of the upper parts that is so characteristic of those species is accented even more in *hirustus.* The monocolored or faintly bicolored tail and slightly dusky hind feet are essentially matched in species of *Aporodon* and are seen in some specimens of *sumichrasti* and *chrysopsis.* . . . It is distinguishable from all species of the genus except *fulvescens* in characters of M3. In that tooth there are 2 long enamel folds, the major and the first primary. They are about equal in length (after Hooper, 1952a:123).

MARGINAL RECORDS.—Nayarit: 2–4 mi. N Santa Isabel, 3800 ft.; 1 mi. E Ixtlán del Río, 4000 ft. Jalisco: type locality; San Sebastián, 4000 ft. Nayarit: 1 mi. SW San José del Conde, 3000 ft.

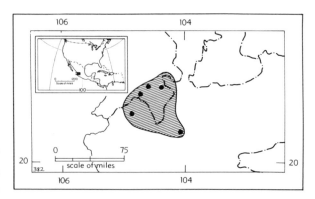

Map 382. *Reithrodontomys hirsutus.*

Subgenus **Aporodon** A. H. Howell[1]

1914. *Aporodon* A. H. Howell, N. Amer. Fauna, 36:63, June 5. Type, *Reithrodontomys tenuirostris* Merriam.

Certain differences from subgenus *Reithrodontomys* are as follows: that part of skull posterior to least interorbital constriction longer than anterior part; braincase elongate, large, greatly inflated, extending laterally beyond anterolateral limits of zygomatic arches; zygomatic plate narrower than mesopterygoid fossa; pterygoid hamulae well inflated, reflexed laterad, club-shaped from ventral aspect.

mexicanus-group
Reithrodontomys gracilis
Slender Harvest Mouse

External measurements: 152–191; 75–113; 15–20; approx. 14. Upper parts pale and with same "streaked" or "peppered" effect found in *fulvescens* and *hirsutus;* underparts varying from white to bright orange-cinnamon; lateral line sometimes present. Skull comparatively flat; zygomatic arches widespread and parallel, or almost so; frontals exceptionally broad and flat interorbitally; braincase broad, shallow, long; palate long; rostrum and incisive foramina short; molars with complete mesolophs.

Reithrodontomys gracilis anthonyi Goodwin

1932. *Reithrodontomys gracilis anthonyi* Goodwin, Amer. Mus. Novit., 560:3, September 16, type from Sacapulas, 4500 ft., El Quiché, Guatemala.

MARGINAL RECORDS.—Guatemala: type locality. El Salvador (Burt and Stirton, 1961:56, unless otherwise noted): 2 mi. SE San Cristóbal (Anderson and Jones, 1960:524); Monte Cristo Mine, 700 ft.; Amate de Campo. Guatemala: *2¼ mi. N 2½ mi. W San Cristóbal* (Anderson and Jones, 1960:524); Lago Atescatempo.

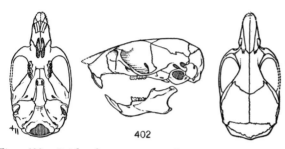

Fig. 402. *Reithrodontomys gracilis gracilis*, Mt. Pine Ridge, 12 mi. S Cayo, Belize, No. 63072 U.M.M.Z., ♂, X 1½.

Reithrodontomys gracilis gracilis J. A. Allen and Chapman

1897. *Reithrodontomys mexicanus gracilis* J. A. Allen and Chapman, Bull. Amer. Mus. Nat. Hist., 9:9, February 23, type from Chichén-Itzá, Yucatán.
1914. *Reithrodontomys gracilis*, A. H. Howell, N. Amer. Fauna, 36:76, June 5.

MARGINAL RECORDS.—Yucatán: 66 km. NE Mérida (93702 KU). Quintana Roo: Pueblo Nuevo X-Can (Jones, 1964b:124); 4 km. NNE Felipe Carrillo Puerto (*ibid.*); 27 km. NW Chetumal (93703 KU). Belize: Mountain Pine Ridge, 1000 ft., 12 mi. S Cayo. Guatemala: La Libertad. Campeche: 7½ km. W Escárcega (Jones, 1964b:124); San Juan.

Reithrodontomys gracilis harrisi Goodwin

1945. *Reithrodontomys harrisi* Goodwin, Amer. Mus. Novit., 1293:2, July 20, type from Hda. Santa María, 3200 ft., 18 mi. NE Liberia, Guanacaste, Costa Rica.
1952. *Reithrodontomys gracilis harrisi*, Hooper, Miscl. Publ. Mus. Zool., Univ. Michigan, 77:136, January 16.

MARGINAL RECORDS.—Nicaragua: 9 mi. NNW Estelí (Jones and Genoways, 1970:7). Costa Rica: type locality. Nicaragua (Jones and Genoways, 1970:7): Río Javillo, 3 km. N, 4 km. W Sapoá; *3 mi. SW Managua;* 4 mi. W Managua; San Antonio.

Reithrodontomys gracilis insularis Jones

1964. *Reithrodontomys gracilis insularis* Jones, Proc. Biol. Soc. Washington, 77:123, June 26, type from 8 mi. ENE Ciudad del Carmen, Isla del Carmen, Campeche.

MARGINAL RECORDS.—Campeche (Jones, 1964b: 124): *3 mi. E Ciudad del Carmen;* type locality; *1 km. SW Puerto Real.*

Reithrodontomys gracilis pacificus Goodwin

1932. *Reithrodontomys pacificus* Goodwin, Amer. Mus. Novit., 560:2, September 16, type from Hda. California, 6 mi. from Ocós, Guatemala.
1952. *Reithrodontomys gracilis pacificus*, Hooper, Miscl. Publ. Mus. Zool., Univ. Michigan, 77:135, January 16.

MARGINAL RECORDS.—Chiapas: Pijijiapan, 50 ft. Guatemala: Finca El Ciprés, 2000 ft.; San José, 15 ft. El Salvador: 1 mi. NW San Salvador (Anderson and Jones, 1960:525).

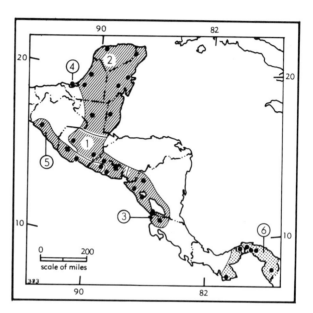

Map 383. *Reithrodontomys gracilis* and *Reithrodontomys darienensis*.

1. *R. g. anthonyi*   4. *R. g. insularis*
2. *R. g. gracilis*   5. *R. g. pacificus*
3. *R. g. harrisi*   6. *R. darienensis*

Reithrodontomys darienensis Pearson
Darién Harvest Mouse

1939. *Reithrodontomys darienensis* Pearson, Not. Naturae, Acad. Nat. Sci. Philadelphia, 6:1, June 8, type from [Santa Cruz de] Cana, 2000 ft., upper Río Tuyra, Darién, Panamá.

External measurements: 162–177; 100–112; 17–19; 14–15. Color as in *R. gracilis*. Selected features in which *R. darienensis* differs from *R. gracilis* are: tail actually and relatively longer (more than 148% of head–body length, vs. 131 in *R. gracilis*); braincase relatively shorter, and oval rather than obovoid; rostrum relatively shorter (82% of depth of rostrum, vs. 90%); zygomatic plate narrower (16% of cranial depth, vs. 21%); auditory bullae relatively smaller (*fide* Hooper, 1952a:137).

MARGINAL RECORDS.—Panamá: Gatún; Cerro Azul (Handley, 1966:782); 4 mi. W Chepo (*ibid.*); type locality; Pacora (*ibid.*); Cerro Hoya, 3000 ft. (*ibid.*).

Reithrodontomys mexicanus
Mexican Harvest Mouse

External measurements: 160–203; 92–126; 17–21; 14.5–18. Upper parts approx. tawny or orange-cinnamon, but varying in precise tone according to subspecies; underparts varying from whitish to pale cinnamon or orange-cinnamon; tail unicolored blackish or dark brown, or rarely paler below or white-tipped. Skull with broad braincase; zygomatic arch weak anteriorly; rostrum broad; palate usually approx. as broad as molar tooth-row, shorter than incisive foramina; molars comparatively large.

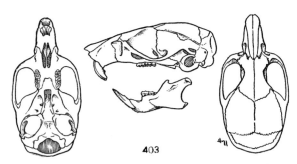

Fig. 403. *Reithrodontomys mexicanus cherrii*, 3 km. S Cartago, Prov. Cartago, Costa Rica, No. 26973 K.U., ♀, X 1½.

Reithrodontomys mexicanus cherrii (J. A. Allen)

1891. *Hesperomys (Vesperimus) cherrii* J. A. Allen, Bull. Amer. Mus. Nat. Hist., 3:211, April 17, type from San José, Costa Rica.
1914. *Reithrodontomys mexicanus cherrii*, A. H. Howell, N. Amer. Fauna, 36:73, June 5.
1895. *Reithrodontomys costaricensis* J. A. Allen, Bull. Amer. Mus. Nat. Hist., 7:139, May 21, type from La Carpintera, 5000 ft., Costa Rica.

MARGINAL RECORDS.—Costa Rica: Tapesco, 5000 ft.; slopes of Volcán de Irazú; El Copey de Dota, 6000–6500 ft.; vic. Alajuela, 3200–3500 ft.

Reithrodontomys mexicanus garichensis Enders and Pearson

1940. *Reithrodontomys mexicanus garichensis* Enders and Pearson, Not. Naturae, Acad. Nat. Sci. Philadelphia, 60:1, October 8, type from Río Gariché, 5 mi. SW El Volcán Post Office, 3200 ft., Chiriquí, Panamá.

MARGINAL RECORDS.—Panamá: Cerro Pando; Boquete, 3500–4000 ft.; near Boquerón.

Reithrodontomys mexicanus howelli Goodwin

1932. *Reithrodontomys mexicanus howelli* Goodwin, Amer. Mus. Novit., 560:1, September 16, type from Chichicastenango, 6500 ft. (Santo Tomás), El Quiché, Guatemala.

MARGINAL RECORDS.—Chiapas: Tumbalá, 4000 ft. Guatemala: Santa Clara, Sierra de las Miñas, 5500 ft.; 5 mi. N, 1 mi. W [Santa Cruz] El Chol (Anderson and Jones, 1960:527); 1 mi. WSW El Molino (*ibid.*); Jacaltenango, 4600 ft. Chiapas: Prusia, 3500–4500 ft.; Bochil, 4300 ft.

Reithrodontomys mexicanus lucifrons A. H. Howell

1932. *Reithrodontomys mexicanus lucifrons* A. H. Howell, Proc. Biol. Soc. Washington, 45:125, July 30, type from Cerro Cantoral (about 15 mi. by airline NNW Tegucigalpa), 6000 ft., Honduras.
1932. *Reithrodontomys mexicanus minusculus* A. H. Howell, Proc. Biol. Soc. Washington, 45:125, July 30, type from Comayaguela, 3000 ft., Honduras.

MARGINAL RECORDS.—Honduras: Hatillo, 4800 ft. Nicaragua (Jones and Genoways, 1970:10): 3½ km. S, 2 km. W Jalapa; Santa María de Ostuma; 5 mi. S, 2 mi. E Jinotega; 1 mi. NW Jinotega; Venecia, 7 km. N, 16 km. E Condega. Honduras: Sabana Grande, 3500 ft.; Humuya (= Muya, 4000 ft. ?).

Reithrodontomys mexicanus mexicanus (Saussure)

1860. R[eithrodon]. *mexicanus* Saussure, Revue et Mag. Zool., Paris, ser. 2, 12:109, type from mts. of Veracruz; restricted to Mirador, Veracruz, by Hooper, Miscl. Publ. Mus. Zool., Univ. Michigan, 77:140, January 16.
1914. *Reithrodontomys mexicanus mexicanus*, A. H. Howell, N. Amer. Fauna, 36:70, June 5. Not *Reithrodontomys mexicanus* (Saussure), being instead of J. A. Allen, 1895:135, which in part equaled *Reithrodontomys fulvescens difficilis*.
1901. *Reithrodontomys costaricensis jalapae* Merriam, Proc. Washington Acad. Sci., 3:552, November 29, type from Jalapa, 4000 ft., Veracruz.
1901. *Reithrodontomys goldmani* Merriam, Proc. Washington Acad. Sci., 3:552, November 29, type from Metlaltoyuca, 800 ft., Puebla.

MARGINAL RECORDS.—Tamaulipas: Rancho Pano Ayuctle, 6 mi. N Gómez Farías, 300 ft. (Jones and Anderson, 1958:447); *Rancho del Cielo, 3500 ft.* (Alvarez, 1963:440). Veracruz: 35 km. NW Tuxpan, 1000 ft. (Hall and Dalquest, 1963:299); Jalapa (*ibid.*). Oaxaca (Goodwin, 1969:171, unless otherwise noted): Tarabundí; Totontepec; Agua Zarca; *Cerro Pelón*; 2 mi. N Pluma Hidalgo, 4400 ft. (Jones and Anderson, 1958:447). Veracruz: Orizaba (Hall and Dalquest, 1963:299). Puebla: Huauchinango, 4900 ft. Hidalgo: Molango, 5200 ft. San Luis Potosí: above Xilitla at Miramar Grande, 5000 ft.; 10 km. E Platanito.

Reithrodontomys mexicanus ocotepequensis Goodwin

1937. *Reithrodontomys mexicanus ocotepequensis* Goodwin, Amer. Mus. Novit., 921:1, May 3, type from Monte Verde, 4500 ft., 30 mi. NE Ocotepeque, Ocotepeque, Honduras.

MARGINAL RECORDS.—Honduras: Monte Linderos, 5700 ft.; type locality. El Salvador: Los Esesmiles, 7500–8000 ft.

Reithrodontomys mexicanus orinus Hooper

1949. *Reithrodontomys mexicanus orinus* Hooper, Proc. Biol. Soc. Washington, 62:169, November 16, type from Hda. Chilata, 2000 ft., about 12 mi. SE Sonsonate, Sonsonate, El Salvador.

· MARGINAL RECORDS.—Guatemala: Finca San Rafael, 7000 ft. (Hooper, 1952a:149). El Salvador (Burt and Stirton, 1961:55): Hda. Montecristo; Los Esesmiles; Cerro Cacaguatique; Hda. Chilata. Guatemala: 7 mi. S, 6 mi. E Guatemala City (Anderson and Jones, 1960:528).

Reithrodontomys mexicanus potrerograndei Goodwin

1945. *Reithrodontomys mexicanus potrerograndei* Goodwin, Amer. Mus. Novit., 1293:1, July 20, type from Agua Buena (locally known as Cañas Gordas, but approx. 30 air line mi. N Cañas Gordas near Panamanian boundary), 3500 ft., Sabana de Potrero Grande, Puntarenas, Costa Rica.

MARGINAL RECORDS.—Costa Rica: type locality. Panamá: Río Chiriquí Viejo, Wald, 3800 ft.

Reithrodontomys mexicanus riparius Hooper

1955. *Reithrodontomys mexicanus riparius* Hooper, Occas. Pap. Mus. Zool., Univ. Michigan, 565:12, March 1, type from 2½ mi. SW Coalcomán, 3600 ft., Michoacán.

MARGINAL RECORDS.—Michoacán: Uruapan (Hooper, 1957:521, as *R. mexicanus* only); type locality.

Reithrodontomys mexicanus scansor Hooper

1950. *Reithrodontomys mexicanus scansor* Hooper, Jour. Washington Acad. Sci., 40:418, December 15, type from Villa Flores, 2000 ft., Chiapas.

MARGINAL RECORDS.—Oaxaca: mts. N of Zanatepec (Goodwin, 1969:172). Chiapas: Cintalapa, 1700 ft.; type locality.

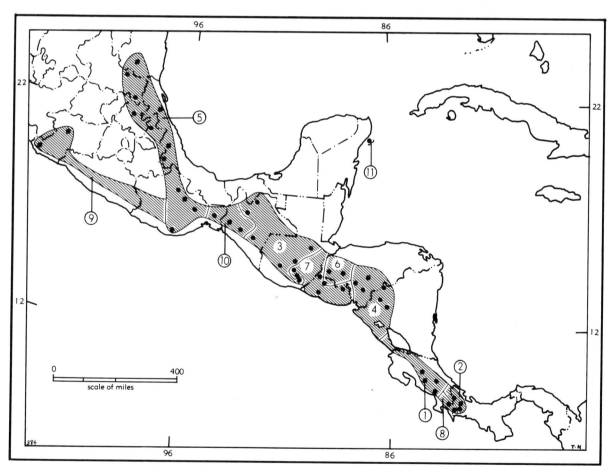

Map 384. *Reithrodontomys mexicanus* and *Reithrodontomys spectabilis*.

Guide to kinds
1. *R. mexicanus cherrii*
2. *R. mexicanus garichensis*
3. *R. mexicanus howelli*
4. *R. mexicanus lucifrons*
5. *R. mexicanus mexicanus*
6. *R. mexicanus ocotepequensis*
7. *R. mexicanus orinus*
8. *R. mexicanus potrerograndei*
9. *R. mexicanus riparius*
10. *R. mexicanus scansor*
11. *R. spectabilis*

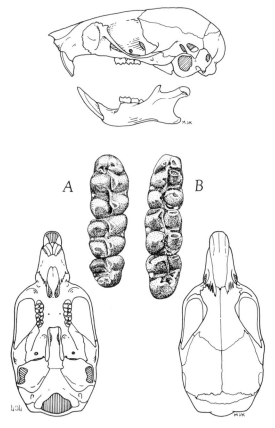

Fig. 404. *Reithrodontomys spectabilis*, 3½ km. N San Miguel, Isla Cozumel, Quintana Roo, No. 92293 K.U., ♀, X 2. Right upper (A) and left lower (B) molars X 10, after Jones and Lawlor (1965:414).

Reithrodontomys spectabilis Jones and Lawlor
Cozumel Island Harvest Mouse

1965. *Reithrodontomys spectabilis* Jones and Lawlor, Univ. Kansas Publ., Mus. Nat. Hist., 16:413, April 13, type from 2½ km. N San Miguel, Isla Cozumel, Quintana Roo.

Size large both externally and cranially; pelage short and relatively sparse; upper parts brownish ochraceous overall, brighter ochraceous on sides; underparts grayish white. Braincase relatively flattened and uninflated; zygomatic arches broad and strong; rostrum relatively short and broad; incisive foramina rarely reaching level of M1; teeth large.

External measurements: 205–221; 121–132; 20–22. Greatest length of skull, 24.6–26.2; zygomatic breadth, 11.8–12.7; interorbital breadth, 3.5–3.9; alveolar length of maxillary tooth-row, 3.7–3.9 (after Jones and Lawlor, 1965:413, 414).

MARGINAL RECORDS (Jones and Lawlor, 1965:415).—Quintana Roo: 3½ km. N San Miguel, Cozumel Island; *type locality*.

Reithrodontomys brevirostris
Short-nosed Harvest Mouse

External measurements: 159–203; 97–114; 16–20; 14–16. Upper parts bright ochraceous-tawny much darkened, especially mid-dorsally, by suffusion of black; underparts white, sharply delimited from upper parts; feet dull white with brown stripe extending to base of toes; tail unicolored, fuscous. Skull closely resembling that of *R. mexicanus cherrii* but smaller, braincase shallow and flattened; rostrum short; anterior part of frontals depressed and forming shallow sulcus at posterior border of nasals; zygomatic arches robust, parallel-sided; palate narrow; molar teeth small.

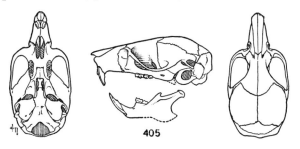

Fig. 405. *Reithrodontomys brevirostris brevirostris*, Laja Villa Quesada, 8 mi. off main road, San Carlos, Costa Rica, No. 139730 A.M.N.H., ♀, X 1½.

Reithrodontomys brevirostris brevirostris Goodwin

1943. *Reithrodontomys brevirostris* Goodwin, Amer. Mus. Novit., 1231:1, June 2, type from canyons above Villa Quesada, 5000 ft., Alajuela, Costa Rica.

MARGINAL RECORDS.—Costa Rica: type locality; Estrella de Cartago.

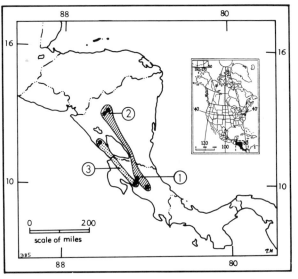

Map 385. *Reithrodontomys brevirostris* and *Reithrodontomys paradoxus*.

1. *R. b. brevirostris*   2. *R. b. nicaraguae*
3. *R. paradoxus*

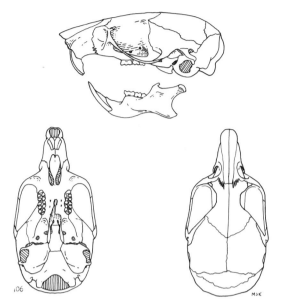

Fig. 406. *Reithrodontomys paradoxus*, 3 mi. NNW Diriamba, Carazo, Nicaragua, No. 71391 K.U., ♂, X 2.

Reithrodontomys brevirostris nicaraguae Jones and Genoways

1970. *Reithrodontomys brevirostris nicaraguae* Jones and Genoways, Occas. Pap., Western Found. of Vert. Zool., 2:10, July 20, type from Santa María de Ostuma, 1250 m., Matagalpa, Nicaragua.

MARGINAL RECORDS.—Nicaragua: Hda. La Trampa, 5½ km. N, 16 km. E Jinotega (Jones and Genoways, 1970:12); type locality.

Reithrodontomys paradoxus Jones and Genoways
Nicaraguan Harvest Mouse

1970. *Reithrodontomys paradoxus* Jones and Genoways, Occas. Pap., Western Found. of Vert. Zool., 2:12, July 20, type from 3 mi. NNW Diriamba, Carazo, *ca.* 660 m., Nicaragua.

External measurements: 167–179; 96–107; 18–19; 13–15. Dorsum near Buffy Brown lightly suffused with ochraceous, grading to ochraceous-buff on sides; underparts white; tarsus dusky, wedge-shaped dusky stripe extending to base of toes. Skull resembling that of *R. brevirostris* but braincase longer, less inflated, and dorsally more nearly flat; interorbital region lacking marked dorsal depression; incisive foramina terminating well anterior to 1st molars; bony palate longer; nasolachrymal capsules larger; lower molars lacking ectolophid (after Jones and Genoways, 1970).

MARGINAL RECORDS.—Nicaragua: type locality. Costa Rica: 5 mi. SW San Ramón (Jones and Genoways, 1970:14).

tenuirostris-group
Reithrodontomys microdon
Small-toothed Harvest Mouse

External measurements: 169–187; 101–117; 19–21; 16–17. Upper parts dark reddish brown; underparts vary from white to bright orange-cinnamon; eye ring blackish; hind feet blackish, sometimes rimmed with white; tail unicolored, blackish. Skull long, deep, broad; zygomata comparatively weak, markedly convergent anteriorly; incisors comparatively erect; incisive foramina long with reference to palatal length; auditory bullae moderately inflated.

The *tenuirostris*-group is the most specialized of the genus and the few specimens are all from cool, moist tropical forests. These mice probably are among the most arboreal of the harvest mice, which may account in some degree for their scarcity in collections.

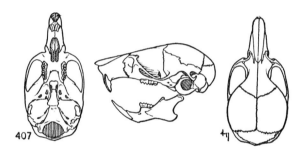

Fig. 407. *Reithrodontomys microdon microdon*, Todos Santos, Huehuetenango, Guatemala, No. 76922 U.S.N.M., ♂, X 1½.

Reithrodontomys microdon albilabris Merriam

1901. *Reithrodontomys microdon albilabris* Merriam, Proc. Washington Acad. Sci., 3:549, November 29, type from Cerro San Felipe, 10,000 ft., Oaxaca.

MARGINAL RECORDS.—Oaxaca: 6½ mi. SSW Vista Hermosa (Jones and Genoways, 1967:321); type locality; *Llano de las Flores* (Goodwin, 1969:172); *Cerro Pelón, 9200 ft.* (Musser, 1964:9).

Reithrodontomys microdon microdon Merriam

1901. *Reithrodontomys microdon* Merriam, Proc. Washington Acad. Sci., 3:548, November 29, type from Todos Santos, 10,000 ft., Guatemala.

MARGINAL RECORDS.—Chiapas: 4 mi. W Ciudad las Casas (= San Cristóbal) (Jones and Anderson, 1958:447); *6 mi. SE San Cristóbal, 7300 ft.* Guatemala: 2 mi. S San Juan Ixcoy (Anderson and Jones, 1960:526); Chichavac, 8800 ft., near Tecpán; Volcán Santa María; 3¼ mi. N, ¾ mi. E San Marcos (Anderson and Jones, 1960:526).

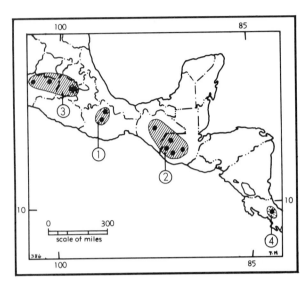

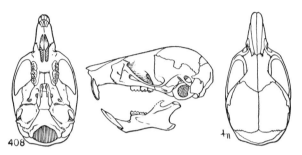

Fig. 408. *Reithrodontomys tenuirostris*, Todos Santos, Huehuetenango, Guatemala, No. 76996 U.S.N.M., ♀, X 1½.

Map 386. *Reithrodontomys microdon* and *Reithrodontomys rodriguezi*.

1. *R. microdon albilabris*
2. *R. microdon microdon*
3. *R. microdon wagneri*
4. *R. rodriguezi*

Reithrodontomys microdon wagneri Hooper

1950. *Reithrodontomys microdon wagneri* Hooper, Proc. Biol. Soc. Washington, 63:169, December 29, type from western flanks of Cerro San Andrés, 9400 ft., about 10 mi. NW of Ciudad Hidalgo, Michoacán.

MARGINAL RECORDS.—Michoacán: SE of Sirio, N slope Sierra Patamba, 9100 ft. (Jones and Anderson, 1958:447); type locality. Distrito Federal: Cañón Contreras, 15 mi. SW Mexico City, 9300–10,000 ft.

Reithrodontomys tenuirostris Merriam
Narrow-nosed Harvest Mouse

1901. *Reithrodontomys tenuirostris* Merriam, Proc. Washington Acad. Sci., 3:547, November 29, type from Todos Santos, 10,000 ft., Guatemala.
1901. *Reithrodontomys tenuirostris aureus* Merriam, Proc. Washington Acad. Sci., 3:548, November 29, type from Calel, SW of Momostenango, 10,200 ft., Quezaltenango, Guatemala.

External measurements of three specimens are: 200–231; 120–129; 23; 15–17. Upper parts tawny or cinnamon-orange heavily interspersed with long, black hairs; underparts pinkish-cinnamon, not sharply demarked from color of upper parts; eye ring and ears black; tail unicolored blackish brown, well haired. Braincase broad, deep; rostrum and incisive foramina long. Pelage long (approx. 10 mm. mid-dorsally) and lax.

MARGINAL RECORDS.—Guatemala: type locality; Cerro near Mataquescuintla, 8400 ft.; Volcán de Tajumulco, 8000–10,400 ft.

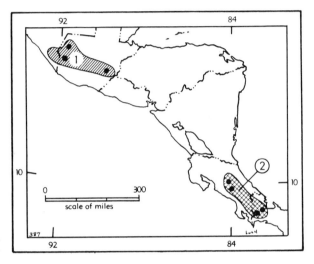

Map 387. *Reithrodontomys tenuirostris* (1) and *Reithrodontomys creper* (2).

Reithrodontomys rodriguezi Goodwin
Rodriguez's Harvest Mouse

1943. *Reithrodontomys rodriguezi* Goodwin, Amer. Mus. Novit., 1231:1, June 2, type from Volcán de Irazú, 9400 ft., Cartago, Costa Rica. Known only from type locality. See Map 386.

This species is known by only two specimens, both of which are subadults. Therefore adequate description of the species is not yet possible. The known specimens can be characterized as follows: external measurements: 193, 204; 116, 121; 20, 21; 14, 15. Pelage long, dense, soft. Upper parts cinnamon-buff darkened with long black hairs; underparts white; eye ring dusky; tail unicolored, dark brown except for white tip; feet white with dusky line extending to base of toes. Skull as in other species of *tenuirostris*-group; braincase large; rostrum long (length approx. 93% of cranial depth).

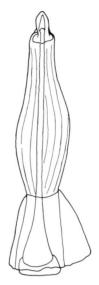

P. (Haplomylomys) californicus,
U.M.M.Z. 103726, × approx. 4.

P. (Peromyscus) maniculatus,
U.M.M.Z. 103718, × approx. 6.

P. (Megadontomys) thomasi,
M.V.Z. 113564, × approx. 4.

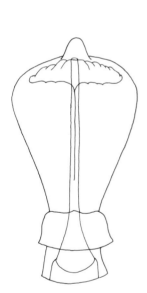

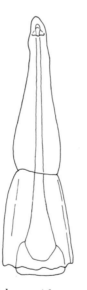

P. (Habromys) lepturus,
U.S.N.M. 68642, × approx. 9.

P. (Podomys) floridanus,
U.M.M.Z. 103733, × approx. 7.

P. (Isthmomys) pirrensis,
U.M.M.Z. P-3897, × approx. 3.

P. (Osgoodomys) banderanus,
A.M.N.H. 172068, × approx. 7.

Fig. 409c.   Ventral view of penis in seven subgenera of *Peromyscus* (after Hooper, 1958).

Peromyscus eremicus papagensis Goldman

1917. *Peromyscus eremicus papagensis* Goldman, Proc. Biol. Soc. Washington, 30:110, May 23, type from Sierra Pinacate, Sonora.

MARGINAL RECORDS.—Arizona: Pinacate Lava, Tule Desert 8 mi. NW U.S.–Mexican Monument 179 (Cockrum, 1961:173). Sonora: type locality.

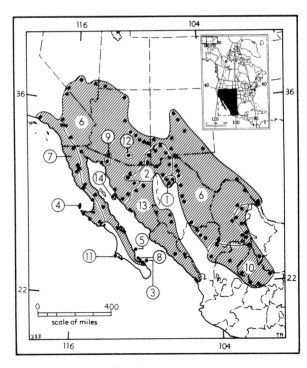

Map 388.   *Peromyscus eremicus.*

1. *P. e. alcorni*	8. *P. e. insulicola*
2. *P. e. anthonyi*	9. *P. e. papagensis*
3. *P. e. avius*	10. *P. e. phaeurus*
4. *P. e. cedrosensis*	11. *P. e. polypolius*
5. *P. e. cinereus*	12. *P. e. pullus*
6. *P. e. eremicus*	13. *P. e. sinaloensis*
7. *P. e. fraterculus*	14. *P. e. tiburonensis*

Peromyscus eremicus phaeurus Osgood

1904. *Peromyscus eremicus phaeurus* Osgood, Proc. Biol. Soc. Washington, 17:75, March 21, type from Hda. la Parada, San Luis Potosí.

MARGINAL RECORDS.—Coahuila: Sabinas. Nuevo León: Doctor Arroyo. San Luis Potosí: 10 mi. NW Ciudad del Maíz; San José de Albuquerque; 1 mi. W Bledos; Cerro Peñón Blanco. Zacatecas: Cañitas. Durango (Baker and Greer, 1962:114): 26 mi. SW Yerbanís, 6725 ft.; 2 mi. E Pedriceña, 4300 ft.; *2 mi. NW Chocolate, 4500 ft.* Coahuila (Baker, 1956:260): Sierra Encarnación; 10 mi. W Saltillo; 2 mi. S San Lázaro; 60 mi. N, 20 mi. W Saltillo; 4 mi. S, 9 mi. W San Buenaventura.

Peromyscus eremicus polypolius Osgood

1909. *Peromyscus eremicus polypolius* Osgood, N. Amer. Fauna, 28:248, April 17, type from Margarita Island, off West Coast southern Baja California. Known only from type locality.

Peromyscus eremicus pullus Blossom

1933. *Peromyscus eremicus pullus* Blossom, Occas. Pap. Mus. Zool., Univ. Michigan, 265:3, June 21, type from Black

Mtn., 10 mi. S Tucson, Pima Co., Arizona. Known only from type locality.

Peromyscus eremicus sinaloensis Anderson

1972. *Peromyscus eremicus sinaloensis* Anderson, Bull. Amer. Mus. Nat. Hist., 148:342, September 8, type from 26 mi. NE Choix, Sinaloa, 1300 ft.

MARGINAL RECORDS.—Sonora: Oposura (= Moctezuma). Chihuahua (Anderson, 1972:342): Carimechi; Barranca del Cobre, 23 mi. S, 1½ mi. E Creel; La Bufa. Durango: Santa Ana (Jones, 1964a:753, as *P. e. anthonyi*). Sinaloa: Culiacán. Sonora: Bahía San Carlos (Cockrum and Bradshaw, 1963:8, as *P. e. anthonyi*); *Ortiz*; Ures. See Anderson, 1972:342, who refers all occurrences from central Sonora south to southern Sinaloa to *P. e. sinaloensis*.

Peromyscus eremicus tiburonensis Mearns

1897. *Peromyscus tiburonensis* Mearns, Proc. U.S. Nat. Mus., 19:720, July 30, type from Tiburón Island, Gulf of California, Sonora. Known only from Tiburón Island.
1909. *Peromyscus eremicus tiburonensis*, Osgood, N. Amer. Fauna, 28:250, April 17.

Peromyscus eva
Eva's Desert Mouse

Revised by Lawlor, Occas. Pap. Mus. Zool., Univ. Michigan, 661:1–22, March 1, 1971.

External measurements: 185–218; 100–128; 20–21; 15.6–17.2. Cranial measurements: greatest length of skull, 24.1–26.5; zygomatic breadth, 12.0–13.3. Resembles *P. eremicus*, but tail relatively longer, length of skull and zygomatic breadth greater, nasals and rostrum longer, maxillary tooth-row longer and wider, number of tail vertebrae greater, phallus and baculum smaller and differing in shape. Dorsum varies from bright sandy-rufous in south to ochraceous-brown in north.

Peromyscus eva carmeni Townsend

1912. *Peromyscus eremicus carmeni* Townsend, Bull. Amer. Mus. Nat. Hist., 31:126, June 14, type from Carmen Island, Gulf of California, Baja California. Known only from type locality.
1971. *P[eromyscus]. eva carmeni*, Lawlor, Occas. Pap. Mus. Zool., Univ. Michigan, 661:17, March 1.

Peromyscus eva eva Thomas

1898. *Peromyscus eva* Thomas, Ann. Mag. Nat. Hist., ser. 7, 1:44, January, type from San José del Cabo, Baja California.

MARGINAL RECORDS.—Baja California (Lawlor, 1971:20, 21): Calmallí; San Ignacio; El Potrero; Puerto Escondida; Las Cruces; Bahía de los Muertos; type locality; Cabo San Lucas; Todos Santos; Matancita; San Jorge; El Patrocinio; 20 km. W San Ignacio; Aguaje de Santana.

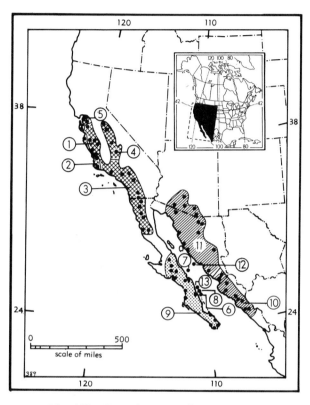

Map 389. Several species of *Peromyscus*.

1. *P. californicus benitoensis*
2. *P. californicus californicus*
3. *P. californicus insignis*
4. *P. californicus mariposae*
5. *P. californicus parasiticus*

6. *P. caniceps*
7. *P. dickeyi*
8. *P. eva carmeni*
9. *P. eva eva*
10. *P. merriami goldmani*
11. *P. merriami merriami*
12. *P. pembertoni*
13. *P. pseudocrinitus*

Peromyscus merriami
Merriam's Mouse

Revised by Hoffmeister and Diersing, Southwestern Nat., 18:354–357, October 5, 1973.

External measurements:183–223; 94–126; 20–24; 17–23. So closely resembles *P. eremicus* that the two are separated only with difficulty except when the differently shaped bacula are compared. Top of braincase of *P. merriami* more nearly flat, and the mice average larger than those of *P. eremicus* where the two species occur together.

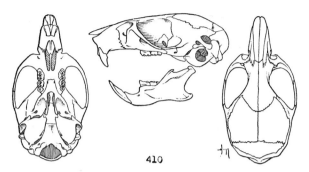

Fig. 410. *Peromyscus merriami merriami*, 8½ mi. N, 2½ mi. W Nogales, 3400 ft., Santa Cruz Co., Arizona, No. 22840 K.U., ♀, X 1½ [incorrectly identified as *Peromyscus eremicus anthonyi* in Hall and Kelson, 1959: Fig. 344 on p. 606].

Peromyscus merriami goldmani Osgood

1904. *Peromyscus goldmani* Osgood, Proc. Biol. Soc. Washington, 17:75, March 21, type from Alamos, Sonora.
1952. *Peromyscus merriami goldmani*, Hall and Kelson, Univ. Kansas Publ., Mus. Nat. Hist., 5:368, December 15.

MARGINAL RECORDS (Hoffmeister and Diersing, 1973:356, 357).—Sonora: 1 mi. W Alamos; *type locality; Vado Cuchijaqui, 9 mi. ESE Alamos.* Sinaloa: 2½ mi. N El Fuerte; 12 mi. N Culiacán, 400 ft.; 32 mi. SSE Culiacán; 6 mi. N, 1½ mi. E El Dorado; 1 mi. S Pericos; 4 mi. N Terrero; 13 mi. NNE Los Mochis. Sonora: 33 mi. SSE Navajoa.

Peromyscus merriami merriami Mearns

1896. *Peromyscus merriami* Mearns, Preliminary diagnoses of new mammals from the Mexican border of the United States, p. 2, May 25 (preprint of Proc. U.S. Nat. Mus., 19:138, December 21, 1896), type from Sonoyta, Sonora.

MARGINAL RECORDS.—Arizona (Hoffmeister and Lee, 1963:209): 1 mi. N, 10 mi. W Casa Grande; 4 mi. S, 1 mi. E Picacho; Wilmot Station, 13 mi. SE Tucson; 8½ mi. N, 1 mi. W Nogales. Sonora (Hoffmeister and Diersing, 1973:357, unless otherwise noted): 21 mi. SSE Nogales; *23 mi. S, 5 mi. E Nogales;* 9 mi. NNE Imuris; 10 mi. S Casa Blanca; Matape (Lawlor, 1971:20); 6 mi. NNE Ciudad Obregón; vic. Guaymas (Lawlor, 1971:20); Desemboque (Cockrum and Bradshaw, 1963:8); ½ mi. N Puerto Libertad; type locality. Arizona: Quitobaquito (Cockrum, 1961:174).

Peromyscus guardia
Angel Island Mouse

External measurements: total length, 189–223; tail, 93–123; length of skull, 25.4–27.2; width of interparietal, 7.0–8.8; length of palate, 4.1–4.6; length of nasals, 9.0–10.8. Color much as in *P. eremicus.* Skull markedly arched in dorsal profile; rostrum long; anterior palatine foramina not reaching posteriorly to plane of molars; interpterygoid fossae wide; tympanic bullae large. (Measurements after Banks, 1967a:215.)

Peromyscus guardia guardia Townsend

1912. *Peromyscus guardia* Townsend, Bull. Amer. Mus. Nat. Hist., 31:126, June 14, type from Angel de la Guarda Island, Gulf of California, Baja California. Known only from type locality.

Peromyscus guardia harbisoni Banks

1967. *Peromyscus guardia harbisoni* Banks, Jour. Mamm., 48:215, May 20, type from Isla Granite, 29° 33′ N, 113° 34′ W, Gulf of California, Baja California. Known only from type locality.

Peromyscus guardia mejiae Burt

1932. *Peromyscus guardia mejiae* Burt, Trans. San Diego Soc. Nat. Hist., 7:174, October 31, type from Mejía Island, lat. 29° 33′ N, long. 113° 35′ W, Gulf of California, Baja California. Known only from type locality.

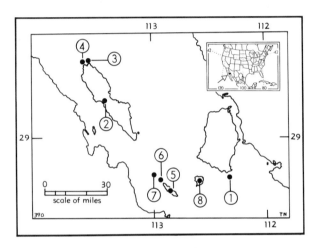

Map 390.   Four species of *Peromyscus*.

1. *P. collatus*
2. *P. guardia guardia*
3. *P. guardia harbisoni*
4. *P. guardia mejiae*
5. *P. interparietalis interparietalis*
6. *P. interparietalis lorenzi*
7. *P. interparietalis ryckmani*
8. *P. stephani*

Peromyscus interparietalis
San Lorenzo Deer Mouse

External measurements: total length, 182–215; tail, 90–117; length of skull, 24.4–27.2; width of interparietal, 8.1–8.9; length of nasals, 8.8–10.3 (after Banks, 1967a:215). Resembles *P. guardia* in size and color, but interpterygoid fossae narrower, bony palate shorter, incisive foramina reaching plane of 1st molars, and interparietal extending to suture between parietal and squamosal (after Burt, 1932:175). Banks (1967a:217) suggests that *P. guardia* and *P. inter-*

parietalis are two separate evolutionary lines from the original *P. eremicus* stock. See Map 390.

Peromyscus interparietalis interparietalis Burt

1932. *Peromyscus guardia interparietalis* Burt, Trans. San Diego Soc. Nat. Hist., 7:175, October 31, type from Isla San Lorenzo Sur, 28° 36′ N, 112° 51′ W, Gulf of California, Baja California. Known only from type locality.
1967. *Peromyscus interparietalis interparietalis*, Banks, Jour. Mamm., 48:216, May 20.

Peromyscus interparietalis lorenzi Banks

1967. *Peromyscus interparietalis lorenzi* Banks, Jour. Mamm., 48:216, May 20, type from Isla San Lorenzo Norte, 28° 42′ N, 112° 57′ W, Gulf of California, Baja California. Known only from type locality.

Peromyscus interparietalis ryckmani Banks

1967. *Peromyscus interparietalis ryckmani* Banks, Jour. Mamm., 48:216, May 20, type from Isla Salsipuedes, 28° 45′ N, 112° 59′ W, Gulf of California, ·Baja California. Known only from type locality.

Peromyscus collatus Burt
Turner Island Canyon Mouse

1932. *Peromyscus collatus* Burt, Trans. San Diego Soc. Nat. Hist., 7:172, October 31, type from Turners Island, lat. 28° 43′ N, long. 112° 19′ W, Gulf of California, Sonora. Known only from type locality.

Average external measurements of 10 specimens are: 174; 94; 20; 15. Externally closely resembling *Peromyscus eremicus tiburonensis* Mearns, especially in coloration and pelage; color darker than in *Peromyscus crinitus stephensi* Mearns; size about as in *tiburonensis* and *stephensi*. Skull with zygomata compressed anteriorly; rostrum depressed, short, and broad; 1st and 2nd upper molars with rudimentary accessory tubercles between outer primary tubercles as in *stephensi*, but differing from *stephensi* in broader, shorter rostrum with correlated broad incisors, and in having somewhat less inflated auditory bullae (after Burt, 1932:172, 173). Lawlor (1971:121) thought the correct name should be "*P. e. collatus*." See Map 390.

Peromyscus dickeyi Burt
Dickey's Deer Mouse

1932. *Peromyscus dickeyi* Burt, Trans. San Diego Soc. Nat. Hist., 7:176, October 31, type from Tortuga Island, lat. 27°21′N, long. 111°54′W, Gulf of California, Baja California. Known only from type locality.

External measurements of the type, an adult male, and average of 17 topotypes are, respectively: 185, 191; 82, 91; 23, 22; 18, 20. Upper parts dusky washed with pinkish cinnamon; ears dusky; lateral line present; underparts white, sometimes with buffy pectoral spot. "Skull broad and squarely built; zygomatic arches nearly parallel-sided; rostrum heavy; nasals wide anteriorly; premaxillae project well beyond posterior limits of nasals; incisors heavy; audital bullae small; lower jaw heavy, nearly as large as in *Peromyscus californicus insignis* Rhoads; ridges for muscle attachments prominent." (Burt, 1932:176.) See Map 389.

Peromyscus pembertoni Burt
Pemberton's Deer Mouse

1932. *Peromyscus pembertoni* Burt, Trans. San Diego Soc. Nat. Hist., 7:176, October 31, type from San Pedro Nolasco Island, lat. 27° 58′ N, long. 111° 24′ W, Gulf of California, Sonora. Known only from type locality.

External measurements of the type, an adult male, and average of 10 topotypes are, respectively: 210, 208; 100, 103; 24, 24; 18, 18. Color of upper parts "light vinaceous-cinnamon" lightly mixed with fine dusky lines on back, noticeably paler than in *dickeyi*; head somewhat paler than back; tail bicolored brownish above, white below; underparts white. Skull resembling that of *dickeyi*, but larger and heavier; zygomatic arches tapering anteriorly; sutures between frontals and parietals forming acute angle in median line; interparietal narrow anteroposteriorly; rostrum heavy as in *dickeyi*; premaxillae extending posteriorly beyond nasals. *Peromyscus pembertoni* differs from all species of subgenus *Haplomylomys* except *dickeyi* in relatively shorter tail (after Burt, 1932:177). See Map 389.

Peromyscus californicus
California Mouse

External measurements: 220–266; 117–148; 25–29; 20–25. Upper parts approx. russet mixed with dark brown, color varying with subspecies; underparts pale (approaching white), often having buffy pectoral spot; feet white or almost so; orbital ring dusky; tail bicolored, but usually not sharply so. Skull large; braincase well inflated; tympanic bullae large, well inflated; molars robust; two simple involutions apparent on labial sides of partly worn M1 and M2. Pelage long, lax; tail well haired but scaly annulations visible. See Map 389.

Peromyscus californicus benitoensis Grinnell and Orr

1934. *Peromyscus californicus benitoensis* Grinnell and Orr, Jour. Mamm., 15:216, August 10, type from near Cook P. O., 1300 ft., Bear Valley, San Benito Co., California.

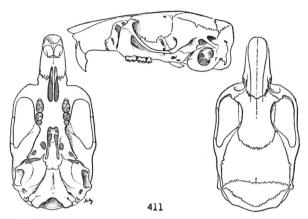

Fig. 411. *Peromyscus californicus parasiticus*, ½ mi. E Stadium, Berkeley, Alameda Co., California, No. 53372 M.V.Z., ♀, X 1½.

MARGINAL RECORDS.—California: Calaveras Valley, near Mt. Hamilton; vic. San Benito; vic. Hernandez; 2 mi. S San Miguel; Jolon; type locality.

Peromyscus californicus californicus (Gambel)

1848. *Mus californicus* Gambel, Proc. Acad. Nat. Sci. Philadelphia, 4:78, August, type from Monterey, Monterey Co., California.
1895. *Peromyscus californicus*, Rhoads, Proc. Acad. Nat. Sci. Philadelphia, 47:34, February 21.

MARGINAL RECORDS.—California: Seaside; Santa Margarita; Gaviota Pass, thence northward along coast to point of beginning.

Peromyscus californicus insignis Rhoads

1895. *Peromyscus insignis* Rhoads, Proc. Acad. Nat. Sci. Philadelphia, 47:33, February 21, type from Dulzura, San Diego Co., California.
1907. *Peromyscus californicus insignis*, Mearns, Bull. U.S. Nat. Mus., 56:429, April 13.

MARGINAL RECORDS.—California: Matilija; San Gabriel Mts.; 2 mi. E Strawberry Peak; San Gorgonio Pass; San Felipe Valley; Mountain Spring. Baja California: Los Pozos, 4200 ft., Sierra Juárez; El Rayo; El Valle de la Trinidad, 2500 ft.; Concepción, 6000 ft.; Rosarito; San Quintín, thence northward along coast to point of beginning.

Peromyscus californicus mariposae Grinnell and Orr

1934. *Peromyscus californicus mariposae* Grinnell and Orr, Jour. Mamm., 15:217, August 10, type from El Portal, 2500 ft., Mariposa Co., California.

MARGINAL RECORDS.—California: Pleasant Valley, 600 ft.; type locality; near Miramonte, 3200–3500 ft.; 12 mi. below Bodfish, 2000 ft.

Peromyscus californicus parasiticus (Baird)

1858. [*Hesperomys*] *parasiticus* Baird, Mammals, Repts. Expl. Surv. . . . , 8(1):479, July 14, type from Santa Clara Valley, Santa Clara Co., California.
1934. *Peromyscus californicus parasiticus,* Grinnell and Orr, Jour. Mamm., 15:213, August 10.

MARGINAL RECORDS.—California: vic. Berkeley; W side Mt. Diablo; Valencia Park Ranch.

Subgenus **Peromyscus** Gloger

Revised by Osgood, N. Amer. Fauna, 28:33–218, April 17, 1909.

1841. *Peromyscus* Gloger, Gemeinnütziges Hand- und Hilfsbuch der Naturgeschichte, 1:95. Type, *Peromyscus arboreus* Gloger [= *Mus sylvaticus noveboracensis* Fischer].
1894. *Trinodontomys* Rhoads, Proc. Acad. Nat. Sci. Philadelphia, 46:257, October. Type, *Sitomys insolatus* Rhoads [= *Hesperomys sonoriensis* Le Conte].

Pelage pattern usually bicolored; mammae, 3–3 (i. 2–2, a. 0–0, p. 1–1) or 2–2 (i. 2–2) in *crinitus, pseudocrinitus, caniceps, stephani,* and *megalops;* coronoid process of mandible usually small and only slightly elevated; accessory tubercles present in salient angles of 1st and 2nd upper molars but absent or rudimentary in the first four species named above as having only 2–2 mammae; outer accessory tubercles of m1 and m2 only slightly developed. Baculum one- to two-thirds as long as hind foot; glans penis rod-shaped in most species to vase-shaped in *P. hylocetes.* For variations in the phallus according to groups of species and even according to species see Hooper (1958) and Hooper and Musser (1964). Glands accessory to male reproductive tract essentially as in subgenus *Haplomylomys* except for rudimentary nature or absence of preputial glands (see Linzey and Layne, 1969).

GUIDE TO SPECIES OF SUBGENUS PEROMYSCUS

A. Occurring North of México

1. Two principal outer angles of M1 and M2 simple, without (or at most with rudimentary) accessory cusps or enamel loops; no pectoral mammae. . . . .*P. crinitus,* p. 690
1'. Two principal outer angles of M1 and M2 with more or less well-developed accessory tubercles or enamel loops; pectoral mammae, 1–1.
  2. Occurring on islands of coastal British Columbia and southern Alaska.
    3. Hind foot 25 or more. . . . .*P. sitkensis,* p. 683
    3'. Hind foot 25 or less. . . . .*P. maniculatus,* p. 670
  2'. Not occurring on islands of coastal British Columbia and southern Alaska.
    4. Ear (dry) 75% or more as long as hind foot.
      5. Ear (dry) usually more than 22; tail length equal to or more than length of head and body.
        P. truei, p. 701
      5'. Ear (dry) usually less than 22; tail length less than length of head and body in Colorado but south thereof longer than head and body. . . . .*P. difficilis,* p. 704
    4'. Ear (dry) less than 75% as long as hind foot.
      6. Tail usually less than 91.
        7. Total length less than 154.
          8. Occurring in southeastern United States south of northern Georgia.*P. polionotus,* p. 667
          8'. Not occurring in southeastern United States south of northern Georgia.
            P. maniculatus, p. 670
        7'. Total length more than 154.
          9. Greatest length of skull more than 28 or, if less, occurring in Florida. *P. gossypinus,* p. 689
          9'. Greatest length of skull less than 28.
            10. Black, narrow, sharply defined line on dorsum of tail; ventral surface of tail white: size averaging less than in *P. leucopus;* sides of body russet or warm brown; in some but not all areas where both *leucopus* and *maniculatus* occur, skulls differ. (See also *P. melanotis* of S. Ariz.) . . . *P. maniculatus.* p. 670
            10'. Less dark, wider, less sharply defined line on dorsum of tail; ventral surface of tail whitish or having faint grayish line; size averaging more than in *P. maniculatus;* sides of body or whole of upper parts distinctly cinnamon rufous; in some but not in all areas where both *leucopus* and *maniculatus* occur, cranial differences exist. . . . .*P. leucopus,* p. 685
      6'. Tail usually more than 91.
        11. Tail usually shorter than head and body or, if longer, occurring in the Pacific Northwest; tail not penicillate.
          12. Occurring on either Marthas Vineyard Island, Massachusetts, or in southwestern United States. . . . .*P. leucopus,* p. 685

12′. Not occurring on Marthas Vineyard Island, Massachusetts, nor in southwestern United
   States. . . . . *P. maniculatus*,   p. 670
11′. Tail usually longer than head and body, penicillate.
   13. Molar tooth-row less than 4.0; distally slender baculum ⅔ length of hind foot and supporting
       long cartilaginous spine on tip; ankles white. . . . . *P. pectoralis*,   p. 693
   13′. Molar tooth-row more than 4.0; baculum ½ length of hind foot and has a small knob-shaped tip
       capped by a minute cone of cartilage; ankles dusky.
       14. Hind foot ordinarily less than 23; mesolophid absent on m2 in 417 specimens and in 3
           possessing same much shorter than in *attwateri*; 2 large biarmed chromosomes.
                                                                       P. boylii,   p. 694
       14′. Hind foot ordinarily more than 23; mesolophid present on m2 (see Fig. 420); 6 large
           biarmed chromosomes. . . . . *P. attwateri*,   p. 697

B. Occurring in México

Owing to subspecific variation some species key out at more than one place in this Guide.

1. Occurring on islands in Gulf of California.
   a. Santa Catalina Island. . . . . *P. slevini*,   p. 683
   b. Santa Cruz and San Diego islands. . . . . *P. sejugis*,   p. 683
   c. San Pedro Nolasco Island. . . . . *P. boylii glasselli*,   p. 695
   d. San Esteban Island. . . . . *P. stephani*,   p. 698
   e. Coronados Island. . . . . *P. pseudocrinitus*,   p. 693
   f. Monserrate Island. . . . . *P. caniceps*,   p. 692
   g. Unnamed island in Gonzaga Bay. . . *P. maniculatus hueyi*, p. 677 and *P. crinitus pallidissimus*,   p. 692
1′. Not occurring on islands in Gulf of California (except *P. maniculatus hueyi* and *P. crinitis pallidissimus* both on same island; for distinguishing the two taxa see account of *P. c. pallidissimus*
   2. Occurring on Peninsula of Yucatán.
       3. Total length more than 206. . . . . *P. yucatanicus*,   p. 709
       3′. Total length less than 206. . . . . *P. leucopus*,   p. 685
   2′. Not occurring on Peninsula of Yucatán.
       4. Length of ear (dry) more than 75% of length of hind foot.
           5. Occurring in mountains of east-central Veracruz.
               6. Hind foot less than 24. *P. bullatus*,   p. 706
               6′. Hind foot 24 or more. . . . . *P. difficilis*,   p. 704
           5′. Not occurring in mountains of east-central Veracruz.
               7. Two principal outer angles of M1 and M2 simple, without (or at most with rudimen-
                   tary) accessory cusps or enamel loops; no pectoral mammae. . . . . . . *P. crinitus*,   p. 690
               7′. Two principal outer angles of M1 and M2 with more or less well-developed accessory
                   tubercles or enamel loops; pectoral mammae, 1–1.
                   8. Total length more than 212 and not in Baja California. *P. evides*, p. 699 and
                                                                       P. difficilis,   p. 704
                   8′. Less than 212 or, if more, in Baja California. . . . . *P. evides*, p. 699 or *P. truei*   p. 701
       4′. Length of ear (dry) less than 75% of length of hind foot.
           9. Total length more than 300. *P. zarhynchus*,   p. 716
           9′. Total length less than 300.
               10. Hind foot 30 or more.
                   11. Occurring north of 19° N lat.; nasals greatly expanded distally.   *P. furvus*,   p. 713
                   11′. Occurring south of 19° N lat.; nasals not greatly expanded distally.
                       12. Frontals distinctly beaded supraorbitally.
                           13. Upper parts blackish brown; feet dusky brown to base of toes; known
                               only from Oaxaca. . . . . *P. melanocarpus*,   p. 715
                           13′. Upper parts dark to bright tawny; feet not dusky brown to base of
                               toes; known only from Michoacán. . . . . *P. winkelmanni*,   p. 700
                       12′. Frontals only faintly beaded or not at all; feet not dusky brown to base of
                           toes.
                           14. Dusky wash on upper parts that are approx. tawny; tail coarsely
                               haired. . . . . *P. guatemalensis*,   p. 714
                           14′. Blackish-brown and tawny; tail scantily haired. . . . . *P. megalops*,   p. 715
               10′. Hind foot less than 30.
                   15. Hind foot less than 25.
                       16. Tail more than 150. *P. mekisturus*,   p. 709
                       16′. Tail less than 150.
                           17. Tail shorter than head and body.
                               18. Total length less than 210.

19. Tail usually sharply bicolored and penicillate; palatine slits usually long and nearly parallel-sided.
    20. Rostrum long (nasals about 11). . . . . *P. melanotis,*   p. 684
    20'. Rostrum shorter (nasals usually less than 11). . . . *P. maniculatus,*   p. 670
19'. Tail not sharply bicolored, slightly, if at all, penicillate; palatine slits variable. . . . . *P. leucopus,*   p. 685
18'. Total length more than 210; hind foot 25 or more. . . . .*P. hylocetes,*   p. 700
17'. Tail equal to or longer than head and body.
  21. Tail black or brown, unicolored, or slightly mottled on underside.*P. gymnotis,*   p. 712
  21'. Tail more or less bicolored.
    22. Supraorbital margin beaded or at least elevated. . . . . .*P. mexicanus,*   p. 710
    22'. Supraorbital border neither beaded nor elevated (sometimes trenchant).
      23. Ears relatively large (approx. 70% of length of hind foot). *P. polius,*   p. 698
      23'. Ears relatively small (less than 70% of length of hind foot).
        24. Hind foot 23 or more.
          25. Tarsal joint white. . . . . *P. polius,*   p. 698
          25'. Dusky coloration extending distally to and usually over tarsal joint.
            26. Tail usually less than 90. . . . .*P. leucopus,*   p. 685
            26'. Tail more than 90.
              27. Underparts brownish or ochraceous; tail scantily haired. . . . . *P. ochraventer,*   p. 709
              27'. Underparts not conspicuously brownish or ochraceous; tail more fully haired.
                28. Hind foot usually less than 25. . .*P. boylii,*   p. 694
                28'. Hind foot more than 25. . . . .*P. hylocetes,*   p. 700
        24'. Hind foot not more than 23.
          29. Tail equal to or barely longer than head and body, usually less than 90.
            30. Tail sharply bicolored, slightly penicillate.
                  P. maniculatus,   p. 670
            30'. Tail not sharply bicolored or barely penicillate.
                  P. leucopus,   p. 685
          29'. Tail always longer than head and body, usually more than 90.
            31. Dusky of hind leg extending distally to and usually over tarsal joint.
              32. Baculum ending in knob; molar tooth-row usually 4 mm. or more. . . . .*P. boylii,*   p. 694
              32'. Bacumlum slender (see Fig. 419); molar tooth-row usually less than 4 mm. . . .*P. pectoralis,*   p. 693
            31'. Tarsal joint white like upper side of foot.
              33. Total length 210–234. . . . .*P. polius,*   p. 698
              33'. Total length 180–210 (excluding *P. pectoralis collinus* which may reach 216—but it is southwest of geographic range of *P. polius*). . .*P. pectoralis,*   p. 693
15'. Hind foot 25 or more.
  34. Tarsal joint white like upper side of foot. *P. polius,*   p. 698
  34'. Dusky of hind leg extending at least to tarsal joint.
    35. Ear (dry) not more than 19.
      36. Dusky of hind leg extending over foot at least halfway to base of toes.
        37. Frontals with slight supraorbital bead.
          38. Molar tooth-row less than 5. . . . .*P. mexicanus,*   p. 710
          38'. Molar tooth-row 5 or more. . . . .*P. megalops,*   p. 715
        37'. Frontals lacking supraorbital bead.
          39. Color chiefly dusky; tail blackish all around or at least above and basally below. . . . . *P. furvus,*   p. 713
          39'. Color chiefly tawny or ochraceous; tail sharply and evenly bicolored.
                  P. boylii,   p. 694
      36'. Dusky of hind leg not extending so far over foot as halfway to base of toes.
        40. Supraorbital border slightly to strongly beaded or, if not, then tail not evenly bicolored.

C. Occurring in Central America

maniculatus-group
Peromyscus polionotus
Oldfield Mouse

External measurements: 122–153; 40–60; 15–19; ear (dry), 11.6–16.5. Upper parts varying from almost white to pale cinnamon or buffy; underparts white; tail bicolored. Skull much smaller than in neighboring subspecies of *P.* *maniculatus* or *P. gossypinus*, but resembling those of the smaller subspecies of *P. maniculatus*; palatine slits relatively shorter; auditory bullae relatively slightly larger.

Peromyscus polionotus albifrons Osgood

1909. *Peromyscus polionotus albifrons* Osgood, N. Amer. Fauna, 28:108, April 17, type from Whitfield, Walton Co., Florida.

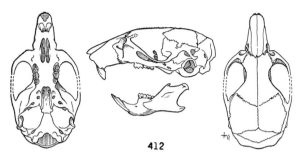

Fig. 412. *Peromyscus polionotus phasma*, Point Roma, Anastasia Island, St. Johns Co., Florida, No. 27076 K.U., ♂, X 1½.

MARGINAL RECORDS (Bowen, 1968:88).— Florida: Crestview; De Funiak Springs; type locality.

Peromyscus polionotus allophrys Bowen

1968. *Peromyscus polionotus allophrys* Bowen, Bull. Florida State Mus. (Biol. Sci.), 12(1):18, April 19, type from coastal dunes near Morrison Lake (about 10 mi. E Destin), Walton Co., Florida.

MARGINAL RECORDS (Bowen, 1968:88).— Florida: type locality; Eastern Lake; St. Andrews State Park.

Peromyscus polionotus ammobates Bowen

1968. *Peromyscus polionotus ammobates* Bowen, Bull. Florida State Mus. (Biol. Sci.), 12(1):16, April 19, type from sand bar W of Perdido Inlet (Alabama Point), Baldwin Co., Alabama.

MARGINAL RECORDS (Bowen, 1968:88).— Alabama: Fort Morgan; type locality.

Peromyscus polionotus colemani Schwartz

1954. *Peromyscus polionotus colemani* Schwartz, Jour. Mamm., 35:566, November 29, type from ¾ mi. E of Campton, Spartanburg Co., South Carolina (Campton is about 2½ mi. SE Inman, South Carolina).

MARGINAL RECORDS.—South Carolina: 2 mi. N Cleveland; 1½ mi. N Inman; 3 mi. N McCormick. Georgia: Whitehall. Alabama: Autaugaville; 4 mi. N Gainesville (Wolfe and Rogers, 1969:610, as *P. polionotus* only). Mississippi (Kennedy, *et al.*, 1974:16, as *P. polionotus* only): Lowndes County; Tishomingo County. Alabama: 13 mi. NE Centerville (Wolfe and Rogers, 1969:610, as *P. polionotus* only): extreme NE corner Jackson County; Sand Mountain, near Carpenter. Georgia: Tallulah Falls.

Peromyscus polionotus decoloratus A. H. Howell

1939. *Peromyscus polionotus decoloratus* A. H. Howell, Jour. Mamm., 20:363, August 14, type from Ponce Park, near Mosquito Inlet, Volusia Co., Florida.

MARGINAL RECORDS.—Florida: Bulow; type locality.

Peromyscus polionotus griseobracatus Bowen

1968. *Peromyscus polionotus griseobracatus* Bowen, Bull. Florida State Mus. (Biol. Sci.), 12(1): 16, April 19, type from about 5 mi. W Navarre, Santa Rosa Co., Florida.

MARGINAL RECORDS (Bowen, 1968:88).— Florida: Shalimar; type locality; 5 mi. E Pensacola.

Peromyscus polionotus leucocephalus A. H. Howell

1920. *Peromyscus leucocephalus* A. H. Howell, Jour. Mamm., 1:239, December 4, type from Santa Rosa Island, opposite Camp Walton, Okaloosa Co., Florida. Known only from Santa Rosa Island.
1926. *P[eromyscus]. p[olionotus]. leucocephalus*, Sumner, Jour. Mamm., 7:155, August 9.

Peromyscus polionotus lucubrans Schwartz

1954. *Peromyscus polionotus lucubrans* Schwartz, Jour. Mamm., 35:564, November 29, type from 4½ mi. E Savannah River on U.S. Highway 301, Allendale Co., South Carolina.

MARGINAL RECORDS.—South Carolina: "S. Bethune" (Biggers and Dawson, 1971:377, 380); Wisacky Field (*ibid.*); St. Matthews; 5 mi. W Orangeburg on State Highway 400; 7 mi. SE Ridgeland on State Highway 128; type locality; Ellenton; 4 mi. N Aiken, Shaw's Creek, on U.S. 1.

Peromyscus polionotus niveiventris (Chapman)

1889. *Hesperomys niveiventris* Chapman, Bull. Amer. Mus. Nat. Hist., 2:117, June 7, type from the east peninsula, opposite Micco, Brevard Co., Florida.
1909. *Peromyscus polionotus niveiventris*, Osgood, N. Amer. Fauna, 28:105, April 17.

MARGINAL RECORDS.—Florida: Canaveral; Hillsboro Inlet; *Hollywood Beach* (Layne, 1974:392).

Peromyscus polionotus peninsularis A. H. Howell

1939. *Peromyscus polionotus peninsularis* A. H. Howell, Jour. Mamm., 20:364, August 14, type from Saint Andrews Point Peninsula, Bay Co., Florida.

MARGINAL RECORDS (Bowen, 1968:88).— Florida: type locality; Port St. Joe; Money Bayou; Cape San Blas; St. Joseph's Point.

Peromyscus polionotus phasma Bangs

1898. *Peromyscus phasma* Bangs, Proc. Boston Soc. Nat. Hist., 28:199, March, type from Point Romo, Anastasia Island, St. Johns Co., Florida.
1909. *Peromyscus polionotus phasma*, Osgood, N. Amer. Fauna, 28:107, April 17.

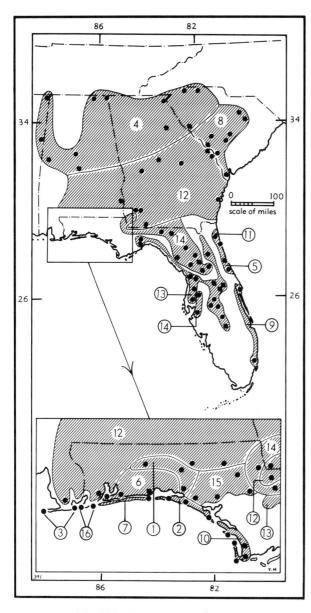

Map 391. *Peromyscus polionotus.*

1. *P. p. albifrons*
2. *P. p. allophrys*
3. *P. p. ammobates*
4. *P. p. colemani*
5. *P. p. decoloratus*
6. *P. p. griseobracatus*
7. *P. p. leucocephalus*
8. *P. p. lucubrans*
9. *P. p. niveiventris*
10. *P. p. peninsularis*
11. *P. p. phasma*
12. *P. p. polionotus*
13. *P. p. rhoadsi*
14. *P. p. subgriseus*
15. *P. p. sumneri*
16. *P. p. trissyllepsis*

MARGINAL RECORDS.—Florida: border of St. Johns and Duval counties; type locality.

Peromyscus polionotus polionotus (Wagner)

1843. *Mus polionotus* Wagner, Wiegmann's Arch. für Naturgesch., Jahrg. 9, 2:52, type from Georgia.

1907. *Peromyscus polionotus*, Osgood, Proc. Biol. Soc. Washington, 20:49, April 18.
1898. *Peromyscus subgriseus arenarius* Bangs, Proc. Boston Soc. Nat. Hist., 28:202, March, type from Hursman Lake, Scriven Co., Georgia. Not *P. eremicus arenarius* Mearns, 1896, type from near El Paso, Texas.
1898. *Peromyscus subgriseus baliolus* Bangs, Science, n. s., 8:215, August 19, a renaming of *arenarius* Bangs.

MARGINAL RECORDS.—Georgia: 16 mi. SE Augusta; 1 mi. W Savannah River, Hwy. 73; 6 mi. SW Townsend; Spring Hill Plantation, about 10 mi. SSW Thomasville. Florida: Rock Bluff (Bowen, 1968:87). Georgia: Sealy Camp (*ibid.*); W bank Ichwaynochaway Creek, ½ mi. below U.S. 91; 7½ mi. E Colquitt. Alabama: Abbeville (Bowen, 1968:87). Florida: 3 mi. S Chipley; Ponce de Leon (Bowen, 1968:87); Pensacola (*ibid.*). Alabama: 5 mi. N Gulf Shores (*ibid.*). Georgia: Butler; 7 mi. N Macon, Hwy. 129; Donovan.

Peromyscus polionotus rhoadsi Bangs

1898. *Peromyscus subgriseus rhoadsi* Bangs, Proc. Boston Soc. Nat. Hist., 28:201, March, type from head Anclote River, Hillsborough Co., Florida. According to Bowen (Bull. Florida State Mus. (Biol. Sci.), 12(1):88, April 19, 1968) the type locality possibly is near Lutz, Hillsborough Co., Florida.
1909. *Peromyscus polionotus rhoadsi*, Osgood, N. Amer. Fauna, 28:107, April 17.

MARGINAL RECORDS (Bowen, 1968:88).—Florida: 5 mi. S Tallahassee; Interlachen; Oviedo; Orlando; Loughman; Lake Pierce; Sebring; Hicoria; Auburndale; Zellwood; Inverness; Brooksville; type locality; Citronelle; Spring Creek.

Peromyscus polionotus subgriseus (Chapman)

1893. *Sitomys niveiventris subgriseus* Chapman, Bull. Amer. Mus. Nat. Hist., 5:341, December 22, type from Gainesville, Alachua Co., Florida.
1954. *P[eromyscus]. p[olionotus]. subgriseus*, Schwartz, Jour. Mamm., 35:562, November 29.

MARGINAL RECORDS (Bowen, 1968:87, unless otherwise noted).—Florida: Sneads; Greensboro; Greenville; Lee; Fort White; type locality; McIntosh; Ocala National Forest; Ocala; 4⅘ km. W Williston (M. H. Smith, 1968:455, as *P. polionotus* only); Eugene. Southern section: Dade City; Dug Creek.

Peromyscus polionotus sumneri Bowen

1968. *Peromyscus polionotus sumneri* Bowen, Bulls. Florida State Mus. (Biol. Sci.), 12(1): 20, April 19, type from Merial Lake, Bay Co., Florida.

MARGINAL RECORDS (Bowen, 1968:88).—Florida: Round Lake; Clarksville; type locality; Seminole Hills.

Peromyscus polionotus trissyllepsis Bowen

1968. *Peromyscus polionotus trissyllepsis* Bowen, Bull. Florida State Mus. (Biol. Sci.), 12(1):17, April 19, type from sand bar E of Perdido Inlet (Florida Point), Baldwin Co., Alabama.

MARGINAL RECORDS (Bowen, 1968:88).— Florida: Foster's Island (S side Big Lagoon). Alabama: type locality.

Peromyscus maniculatus
Deer Mouse

X²/₃

External measurements: 121–222; 46–123; 17–25; 12–20. Upper parts varying with subspecies but, in general, usually varying from pale grayish buff to deep reddish brown overlaid in varying degree by dusky; underparts white; tail clothed with short hairs, slightly penicillate, sharply bicolored, dark above, light below. Skull smooth and delicately built; braincase somewhat arched and well inflated; tympanic bullae moderate to small; rostrum slender, short, and tapered.

This species consists of a long series of intergrading populations. In a few instances, subspecies with adjoining geographic ranges do not intergrade directly but circuitously through other subspecies. Some subspecies (*e.g.*, *P. m. bairdii* and *P. m. gracilis*, in Michigan) occur in the same area, but are ecologically separated, and interbreeding is minimal. For these reasons it is not possible to characterize the species with any great degree of precision.

The species occupies a remarkable variety of habitats. Some subspecies are comparatively versatile in this respect and others more narrowly restricted in habitat.

Food is stored in important quantities. Both plant and animal materials are eaten. In most parts of its geographic range *P. maniculatus* is abundant and has great importance as a buffer species.

There are at least three pelages: juvenal, post-juvenal or subadult, and adult. The 1st pelage is distinctly gray and the color becomes progressively richer, that is, buffier, with succeeding molts. After adult pelage is attained, there is nor-

mally one annual molt. In temperate regions this molt usually takes place in late summer or early autumn, but varies in time of inception. Extreme wear in spring-taken specimens and alkaline bleaching of specimens from areas of strongly saline soils are sometimes mistaken for an additional molt.

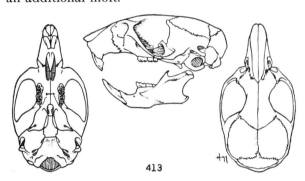

Fig. 413. *Peromyscus maniculatus bairdii*, 8.1 mi. E Arkansas City, Cowley Co., Kansas, No. 39268 K.U., ♀, X 1½.

Peromyscus maniculatus abietorum Bangs

1896. *Peromyscus canadensis abietorum* Bangs, Proc. Biol. Soc. Washington, 10:49, March 9, type from James River, Nova Scotia.
1909. *Peromyscus maniculatus abietorum*, Osgood, N. Amer. Fauna, 28:45, April 17.

MARGINAL RECORDS.—Quebec: Rivière du Loup, Gaspé Peninsula, thence along Atlantic Coast including Prince Edward Island and Nova Scotia to Maine: Big Deer Isle; Greenville.

Peromyscus maniculatus algidus Osgood

1909. *Peromyscus maniculatus algidus* Osgood, N. Amer. Fauna, 28:56, April 17, type from head of Lake Bennett (site of old Bennett City), British Columbia.

MARGINAL RECORDS.—Yukon (Youngman, 1975:82): Lake Laberge; Lake Marsh. Alaska: 1 mi. W Haines; *E side Chilkat River, 100 ft., 4 mi. N, 9 mi. W Haines.* British Columbia: Stonehouse Creek, 5½ mi. W jct. Kelsall River (Cowan and Guiguet, 1965:179). Yukon (Youngman, 1975:82): 1½ mi. S and 3 mi. E Dalton Post; Alaska Highway, Mile 1035.

Peromyscus maniculatus alpinus Cowan

1937. *Peromyscus maniculatus alpinus* Cowan, Proc. Biol. Soc. Washington, 50:215, December 28, type from Mt. Revelstoke, 6000 ft., 19 mi. NE Revelstoke, British Columbia.

MARGINAL RECORDS.—British Columbia: Kinbasket Lake; Glacier; type locality.

Peromyscus maniculatus anacapae von Bloeker

1942. *Peromyscus maniculatus anacapae* von Bloeker, Bull. Southern California Acad. Sci., 40(3):161, January 31, type from Fish Camp, West Anacapa Island, Ventura Co., California. Known from *West*, Middle, and *East* Anacapa islands, California (von Bloeker, 1941:161).

Peromyscus maniculatus angustus Hall

1932. *Peromyscus maniculatus angustus* Hall, Univ. California Publ. Zool., 38:423, November 8, type from Beaver Creek, 15 mi. NW Alberni, Vancouver Island, British Columbia.

MARGINAL RECORDS.—British Columbia: Campbell River (Cowan and Guiguet, 1965:180); Comox (*ibid.*); Hornby Island (*ibid.*); Parksville (*ibid.*); Newcastle Island; Saltspring Island; Sidney Island; Victoria (Cowan and Guiguet, 1965:180); Ucluelet (*ibid.*); Bare Island (*ibid.*); type locality.

Peromyscus maniculatus anticostiensis Moulthrop

1937. *Peromyscus maniculatus anticostiensis* Moulthrop, Sci. Publs., Cleveland Mus. Nat. Hist., 5(3):11, December 4, type from Fox Bay, E end Anticosti Island, Quebec. Known only from type locality.

Peromyscus maniculatus argentatus Copeland and Church

1906. *Peromyscus canadensis argentatus* Copeland and Church, Proc. Biol. Soc. Washington, 19:122, September 6, type from Grand Harbor, Grand Manan Island, New Brunswick. Known only from Grand Manan Island.
1909. *Peromyscus maniculatus argentatus*, Osgood, N. Amer. Fauna, 29:46, April 17.

Peromyscus maniculatus artemisiae (Rhoads)

1894. *Sitomys americanus artemisiae* Rhoads, Proc. Acad. Nat. Sci. Philadelphia, 46:260, October, type from Ashcroft, British Columbia.
1909. *Peromyscus maniculatus artemisiae*, Osgood, N. Amer. Fauna, 28:58, April 17.
1899. *Peromyscus texanus subarcticus* J. A. Allen, Bull. Amer. Mus. Nat. Hist., 12:15, March 4, type from Deerlodge Co., Montana.

MARGINAL RECORDS.—British Columbia: Indianpoint Lake; Field (Osgood, 1909:60); *Kootenay Park* (Cowan and Guiguet, 1965:180); Akamina Pass (*ibid.*). Montana: St. Mary Lake; Helena; Beartooth Mts. Wyoming (Long, 1965a:629): *28–31½ mi. N, 30–36 mi. W Cody*; SW slope Whirlwind Peak; *6 mi. S, 3 mi. W Whirlwind Peak*; Togwotee Pass; 10 mi. SE Afton; LaBarge Creek; *Border.* Idaho: S side Snake River, 3 mi. W Swan Valley; Crane Creek, 15 mi. E Midvale; 1 mi. N Heath. Oregon: *Blue Mountain section;* Sheep Creek, Wenaha Forest. Washington: College Place. Idaho: Cedar Mtn. Washington: Spokane Bridge; near Columbia River in northern Grant Co. (Sheppe,

1961:422); Duly Lake; Conconully. British Columbia: within a few miles of [E of] Allison Pass (Sheppe, 1961:432); Wright's Ranch, 3 mi. SW Princeton (*op. cit.*: 436); Lytton; Lillooet; Stuie (Sheppe, 1961:433); Anahim Lake; *Puntchezacut Lake* (Cowan and Guiguet, 1965:180); Bouchie Lake (*ibid.*).

Peromyscus maniculatus assimilis Nelson and Goldman

1931. *Peromyscus maniculatus assimilis* Nelson and Goldman, Jour. Mamm., 12:305, August 24, type from Coronadas Island, Baja California. Known only from type locality.

Peromyscus maniculatus austerus (Baird)

1855. *Hesperomys austerus* Baird, Proc. Acad. Nat. Sci. Philadelphia, 7:336, April, type from Old Fort Steilacoom, Pierce Co., Washington.
1909. *Peromyscus maniculatus austerus*, Osgood, N. Amer. Fauna, 28:63, April 17.
1899. *Peromyscus akeleyi* Elliot, Field Columb. Mus., Publ. 30, Zool. Ser., 1:226, February 2, type from Johnsons Ranch, Elwah River, Olympic Mts., Clallam Co., Washington.

MARGINAL RECORDS.—British Columbia: Loughborough Inlet; Bute Inlet (Cowan and Guiguet, 1965:181); Gambier Island (*ibid.*); Indian River (*ibid.*); Agassiz; within a few miles of [W of] Allison Pass (Sheppe, 1961:432). Washington: *Arlington;* Fall City; Bingen (Sheppe, 1961:431); Vancouver; mouth Kalama River; Cathlamet (Sheppe, 1961:430); Ilwaco (*ibid.*); Aberdeen (*ibid.*); 4 mi. N Shelton; Lake Cushman (Sheppe, 1961:429); Port Angeles, thence along coast to point of beginning.

Peromyscus maniculatus bairdii (Hoy and Kennicott)

1857. *Mus bairdii* Hoy and Kennicott, *in* Kennicott, Agricultural Report, U.S. Commissioner of Patents, 1856, p. 92, type from Bloomington, McLean Co., Illinois.
1909. *Peromyscus maniculatus bairdi*, Osgood, N. Amer. Fauna, 28:79, April 17.

MARGINAL RECORDS.—Manitoba: S end Lake Manitoba; Stony Mtn.; *Red River Valley at Winnipeg;* Rennie (Singh and McMillan, 1966:262, described these specimens as having certain "intergrading characters of *P. m. maniculatus*"). North Dakota: Pembina. Minnesota: Otter Tail County. Wisconsin (Jackson, 1961:215): *Wheeler;* Thorp Twp., Clark Co. Michigan: Menominee County; Emmet County; Huron County. Ontario: London. New York: Elba (Hamilton, 1950:100); North Pond (Whitaker and Goodwin, 1960:518); near Clayton (Lackey, 1977:194); *10 mi. NE Ithaca* (Brown and Welser, 1968:425); 8 mi. S Ithaca (Hamilton, 1950:100). Pennsylvania: vic. Wyalusing; Schuylkill County; York County. Maryland: 1⁹⁄₁₆ mi. N Bowie. Virginia (Peacock, 1967:243): 2¾ mi. NNW Chantilly; 6 mi. N Warrenton; Cedarville. West Virginia: Avalon. Tennessee: La Follette; 1 mi.

SE Morris Chapel (Beasley and Severinghaus, 1973:108); *6 mi. NW Selmer (ibid.)*; Memphis (*ibid.*). Arkansas: Forrest City; Pine Bluff. Missouri: Carthage. Kansas: Arkansas City; 8 mi. N, 1 mi. E Haven; 2½ mi. S Wilson; ½ mi. S, 3½ mi. W Beloit. Nebraska (Jones, 1964c:202): 2 mi. WSW Superior; *10 mi. S Hastings;* 10 mi. S Ord; *2 mi. S, 4 mi. E Ord;* Ewing; *6 mi. N Midway;* 5 mi. S, 2 mi. E Spencer; *5 mi. WNW Spencer.* North Dakota: Linton; Kenmare. Manitoba: Oak Lake. Note: In the northern part of the Lower Peninsula of Michigan the geographic ranges of *P. m. bairdii* and *P. m. gracilis* overlap, but the subspecies are ecologically separated; in Pennsylvania and New York the geographic ranges of *P. m. bairdii* and *P. m. nubiterrae* overlap, but the subspecies are ecologically separated.

Peromyscus maniculatus balaclavae McCabe and Cowan

1945. *Peromyscus maniculatus balaclavae*, McCabe and Cowan, Trans. Royal Canadian Inst., 25:197, February, type from Balaclava Island, British Columbia. Known from Hope and Balaclava islands, British Columbia.

Peromyscus maniculatus beresfordi Guiguet

1955. *Peromyscus maniculatus beresfordi* Guiguet, Rept. Provincial Mus. Nat. Hist. and Anthro., British Columbia, for 1954, p. B71, August 1, type from Beresford Island, British Columbia. Known only from Beresford Island.

Peromyscus maniculatus blandus Osgood

1904. *Peromyscus sonoriensis blandus* Osgood, Proc. Biol. Soc. Washington, 17:56, March 21, type from Escalón, Chihuahua.
1909. *Peromyscus maniculatus blandus* Osgood, N. Amer. Fauna, 28:84, April 17.

MARGINAL RECORDS.—New Mexico (Findley, *et al.*, 1975:206–210): 8½ mi. S San Jon; 20 mi. N Clovis; 4½ mi. N Clovis; 2 mi. S, 1½ mi. E Portales; 15 mi. W Hobbs. Texas: Winkler County Airport (Judd, 1970:278–279); Hermleigh (Packard and Garner, 1964:388, as *P. maniculatus* only); Tom Green County (Davis, 1966:186, as *P. maniculatus* only); Maxon Spring. Coahuila: Campo Centrale, 5300 ft., Sierra del Carmen, 30 mi. S Boquillas; 3 mi. S San Lázaro (Baker, 1956:264). Nuevo León: Horsetail Falls, near Monterrey. Tamaulipas: Miquihuana. San Luis Potosí: Santa María del Río. Jalisco: Lagos. Zacatecas: Plateado; Berriozábal; Cañitas. Durango: San Gabriel; 2 mi. S El Palmito, 4850 ft. (Baker and Greer, 1962:116); Río Sestín (*ibid.*); Rosario. Chihuahua (Anderson, 1972:346, 347, unless otherwise noted): 2 mi. W Parral, 6200 ft.; *Lago de Bustillos, Anahuac;* NE side Laguna de Bustillos, 6750 ft.; 20 km. N Galeana, 7000 ft. (Hall and Kelson, 1959:619); Casas Grandes Viejo, 4850 ft.; *1½ mi. W Casas Grandes Viejo, 5000 ft.; 4³/₁₀ mi. W Casas Grandes Viejo, 5500 ft.*; Llano de los Carretas, 27 mi. W Cuervo, 4300 ft.; 5½ mi. N, 2 mi. W San Francisco, 5100 ft. Arizona: *San Bernardino Ranch* (Cockrum, 1961:179); E of Portal. New Mexico (Findley, *et al.*, 1975:206–210): 10 mi. NW Lordsburg; 7 mi. SW Knutt; 1 mi. E U.S. Highway 84, Monticello Canyon; 17 mi. NW Carrizozo, Cerro Prieto; Carrizozo; Tularosa; Jarilla; Mesa; 13 mi. N, 3½ mi. W Fort Sumner.

Peromyscus maniculatus borealis Mearns

1890. *Hesperomys leucopus arcticus* Mearns, Bull. Amer. Mus. Nat. Hist., 2:285, February 21, type from Fort Simpson, Mackenzie, Northwest Territories. Not *Hesperomys arcticus* Coues, 1877 [= *Hesperomys maniculatus* Wagner], type locality, Labrador.
1911. *Peromyscus maniculatus borealis* Mearns, Proc. Biol. Soc. Washington, 24:102, May 15, a renaming of *arcticus* Mearns.

MARGINAL RECORDS.—Mackenzie: Fort Good Hope; Fort Rae; Fort Reliance. Alberta: Chipewyan. Manitoba: vic. Cormorant Lake, Mile 42 from The Pas, Churchill Railroad. Saskatchewan: Indian Head. Alberta: Crows Nest Pass; Mt. Assiniboine (Banfield, 1958:20). British Columbia: Thompson Pass (Cowan and Guiguet, 1965:183); Nukko Lake; Sinkut Mtn.; *Nulki Lake* (Cowan and Guiguet, 1965:183); Hazelton; Glenora; Sheslay River. Yukon (Youngman, 1975:82, 83): 12 mi. E Tagish; Lapie River, Canol Road, Mile 132; Nordenskiold River, 1 mi. NW Carmacks; Cultus Bay, Kluane Lake; Donjek River, Kluane Park; Dawson; 4½ mi. N Mayo.

Peromyscus maniculatus cancrivorus McCabe and Cowan

1945. *Peromyscus maniculatus cancrivorus* McCabe and Cowan, Trans. Royal Canadian Inst., 25:195, February, type from Table Island, Queen Charlotte Sound, British Columbia. Known only from type locality.

Peromyscus maniculatus carli Guiguet

1955. *Peromyscus maniculatus carli* Guiguet, Rept. Provincial Mus. Nat. Hist. and Anthro., British Columbia, for 1954, p. B72, August 1, type from Cox Island, British Columbia. In 1973, Thomas, Cytologia, 38:493, on basis of karyological data, placed *carli* in a "cohort" with *Peromyscus sitkensis*.

MARGINAL RECORDS.—British Columbia: Lanz Island; type locality.

Peromyscus maniculatus catalinae Elliot

1903. *Peromyscus catalinae* Elliot, Field Columb. Mus., Publ. 74, Zool. Ser., 3(10): 160, May 7, type from near Avalon, Santa Catalina Island, Santa Barbara Islands, Los Angeles Co., California. Known only from Santa Catalina Island.
1909. *Peromyscus maniculatus catalinae*, Osgood, N. Amer. Fauna, 28:97, April 17.

Peromyscus maniculatus cineritius J. A. Allen

1898. *Peromyscus cineritius* J. A. Allen, Bull. Amer. Mus. Nat. Hist., 10:155, April 12, type from San Roque Island, Baja California. Known only from type locality.
1909. *Peromyscus maniculatus cineritius*, Osgood, N. Amer. Fauna, 28:100, April 17.

Peromyscus maniculatus clementis Mearns

1896. *Peromyscus texanus clementis* Mearns, Preliminary diagnoses of new mammals from the Mexican border of the United States, p. 4, March 25 (preprint of Proc. U.S. Nat. Mus., 18:446, May 23, 1896), type from Pyramid Cove, SE end San Clemente Island, Santa Barbara Islands, California. Known only from San Clemente Island.

1909. *Peromyscus maniculatus clementis,* Osgood, N. Amer. Fauna, 28:96, April 17.

Peromyscus maniculatus coolidgei Thomas

1898. *Peromyscus leucopus coolidgei* Thomas, Ann. Mag. Nat. Hist., ser. 7, 1:45, January, type from Santa Anita, cape region of Baja California.

1909. *Peromyscus maniculatus coolidgei,* Osgood, N. Amer. Fauna, 28:94, April 17.

MARGINAL RECORDS.—Baja California: Rosario; San Francisquito, thence southward to tip of the peninsula, excepting the area occupied by *P. m. magdalenae.* Also Isla Smith (Lawlor, 1971:20, as *P. maniculatus* only).

Peromyscus maniculatus dorsalis Nelson and Goldman

1931. *Peromyscus maniculatus dorsalis* Nelson and Goldman, Jour. Washington Acad. Sci., 21:535, December 19, type from Natividad Island, Baja California. Known only from type locality.

Peromyscus maniculatus doylei McCabe and Cowan

1945. *Peromyscus maniculatus doylei* McCabe and Cowan, Trans. Royal Canadian Inst., 25:196, February, type from Doyle Island, Gordon Group, British Columbia. Known only from type locality. In 1973, Thomas, Cytologia, 38:493, on bais of karyological data, placed *doylei* in a "cohort" with *Peromyscus sitkensis.*

Peromyscus maniculatus dubius J. A. Allen

1898. *Peromyscus dubius* J. A. Allen, Bull. Amer. Mus. Nat. Hist., 10:157, April 12, type from Todos Santos Island, Baja California. Known only from type locality.

1909. *Peromyscus maniculatus dubius,* Osgood, N. Amer. Fauna, 28:98, April 17.

Peromyscus maniculatus elusus Nelson and Goldman

1931. *Peromyscus maniculatus elusus* Nelson and Goldman, Jour. Washington Acad. Sci., 21:533, December 19, type from Santa Barbara Island, Ventura Co., California. Known only from type locality and *Sutil Island* (von Bloeker, 1967:256).

Peromyscus maniculatus eremus Osgood

1909. *Peromyscus maniculatus eremus* Osgood, N. Amer. Fauna, 28:47, April 17, type from Pleasant Bay, Grindstone Island, Magdalen Islands, Quebec. Known only from Grindstone Island.

Peromyscus maniculatus exiguus J. A. Allen

1898. *Peromyscus exiguus* J. A. Allen, Bull. Amer. Mus. Nat. Hist., 10:157, April 12, type from San Martín Island, Baja California. Known only from San Martín Island.

1955. *Peromyscus maniculatus exiguus,* Miller and Kellogg, Bull. U.S. Nat. Mus., 205:485, March 3.

1931. *Peromyscus maniculatus martinensis* Nelson and Goldman, Jour. Washington Acad. Sci., 21:534, December 19, type from San Martín Island, Baja California.

Peromyscus maniculatus exterus Nelson and Goldman

1931. *Peromyscus maniculatus exterus* Nelson and Goldman, Jour. Washington Acad. Sci., 21:532, December 19, type from San Nicolas Island, Santa Barbara Co., California. Known only from type locality.

Peromyscus maniculatus fulvus Osgood

1904. *Peromyscus sonoriensis fulvus* Osgood, Proc. Biol. Soc. Washington, 17:57, March 21, type from city of Oaxaca, Oaxaca.

1909. *Peromyscus maniculatus fulvus* Osgood, N. Amer. Fauna, 28:86, April 17.

MARGINAL RECORDS.—Veracruz: Xuchil; Las Vigas. Oaxaca (Goodwin, 1969:172): 1 mi. E Tlacolula; Ejutla; Huajuapam [= Huajuapan]. México: Zempoala, 3200 m. Distrito Federal: Rancho La Noria, 1 mi. W Xochimilco, 2270 m.; *1 km. S San Mateo Xalpa, 2700 m.* Morelos: 5 mi. W Tepozlan [= Tepoztlán]. México: 23 km. E Mexico City. Hidalgo: Pachuca.

Peromyscus maniculatus gambelii (Baird)

1858. *Hesperomys gambelii* Baird, Mammals, *in* Repts. Expl. Surv. . . . , 8(1):464, July 14, type from Monterey, Monterey Co., California.

1909. *Peromyscus maniculatus gambeli,* Osgood, N. Amer. Fauna, 28:67, April 17.

1893. *Sitomys americanus thurberi* J. A. Allen, Bull. Amer. Mus. Nat. Hist., 5:185, August 18, type from Sierra San Pedro Mártir, Baja California.

1896. *Peromyscus texanus medius* Mearns, Preliminary diagnoses of new mammals from the Mexican border of the United States, p. 4, March 25 (preprint of Proc. U.S. Nat. Mus., 18:446, May 23, 1896), type from Nachogüero Valley, Baja California.

1925. *Peromyscus imperfectus* Dice, Carnegie Inst. Washington, Publ. 349:123, August, type from Rancho La Brea deposit no. 2051, Pleistocene age, Los Angeles Co., California. Considered to be a synonym of *P. maniculatus* [*gambelii?*], by Miller, W. E. (Nat. Hist. Mus. Los Angeles Co. Sci. Bull., 10:15, February 17, 1971).

MARGINAL RECORDS.—Washington: Chelan; 5 mi. N Coulee; Sprague; Steptoe Butte. Oregon: Pendleton; Burns; Lake Alvord. Nevada: 5 mi. N Summit Lake, 5900 ft.; 6 mi. N, 10½ mi. W Sulphur, 4000 ft.; *Fox Canyon, 6 mi. S Pahrum Peak, 4800 ft.; Sutcliffe, Pyramid Lake;* 8 mi. NE Reno; 3 mi. E Reno. California: Donner; *Emerald Bay, Lake Tahoe; Sonora Pass;*

Map 392.   Some subspecies of *Peromyscus maniculatus*. (See facing page for guide.)

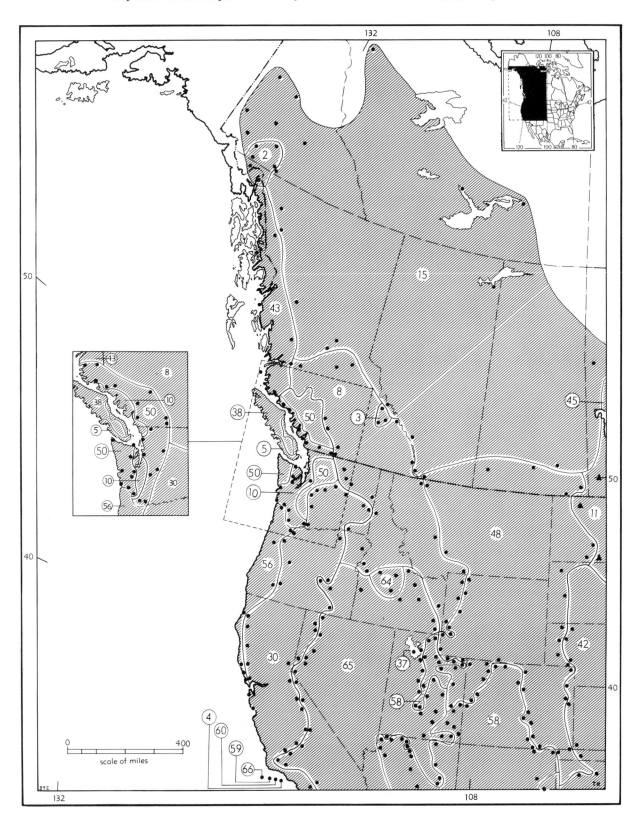

Map 393.   Some subspecies of *Peromyscus maniculatus*.

Guide to Maps 392 & 393
1. *P. m. abietorum*
2. *P. m. algidus*
3. *P. m. alpinus*
4. *P. m. anacapae*
5. *P. m. angustus*
6. *P. m. anticostiensis*
7. *P. m. argentatus*

8. *P. m. artemisiae*
10. *P. m. austerus*
11. *P. m. bairdii*
15. *P. m. borealis*
26. *P. m. eremus*
30. *P. m. gambelii*
33. *P. m. gracilus*
37. *P. m. inclarus*

38. *P. m. interdictus*
42. *P. m. luteus*
43. *P. m. macrorhinus*
45. *P. m. maniculatus*
48. *P. m. nebrascensis*
49. *P. m. nubiterrae*
50. *P. m. oreas*
51. *P. m. ozarkiarum*

53. *P. m. plumbeus*
56. *P. m. rubidus*
58. *P. m. rufinus*
59. *P. m. sanctaerosae*
60. *P. m. santacruzae*
64. *P. m. serratus*
65. *P. m. sonoriensis*
66. *P. m. streatori*

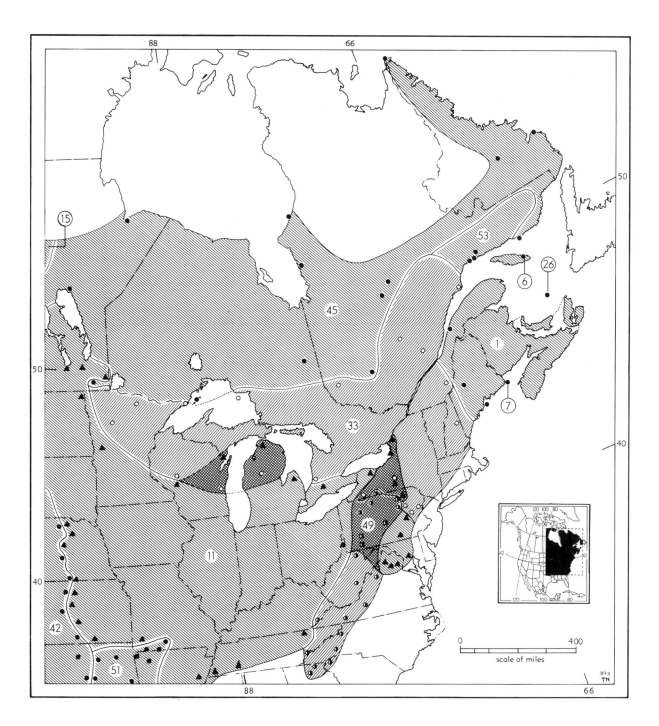

scale of miles

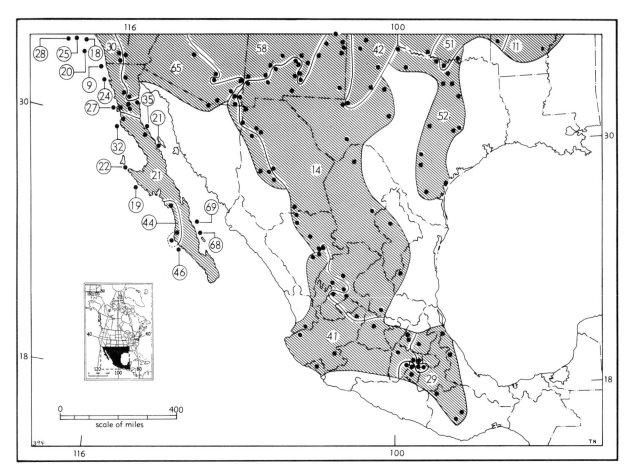

Map 394.   *Peromyscus maniculatus* (part), *Peromyscus sejugis*, and *Peromyscus slevini*.

9. *P. m. assimilis*	21. *P. m. coolidgei*	30. *P. m. gambelii*	51. *P. m. ozarkiarum*
11. *P. m. bairdii*	22. *P. m. dorsalis*	32. *P. m. geronimensis*	52. *P. m. pallescens*
14. *P. m. blandus*	24. *P. m. dubius*	35. *P. m. hueyi*	58. *P. m. rufinus*
18. *P. m. catalinae*	25. *P. m. elusus*	41. *P. m. labecula*	65. *P. m. sonoriensis*
19. *P. m. cineritius*	27. *P. m. exiguus*	42. *P. m. luteus*	68. *P. m. sejugis*
20. *P. m. clementis*	28. *P. m. exterus*	44. *P. m. magdalenae*	69. *P. m. slevini*
	29. *P. m. fulvus*	46. *P. m. margaritae*	

Mt. Dana; head San Joaquin River; Giant Forest, Sequoia National Forest; Orris; Buttonwillow; Riverside; Laguna Mts., San Diego Co.; Jacumba. Baja California: Piñon; Santa Rosa; San Felipe; Santa Eulalia; San Quentín [*sic*], thence up coast to California: Santa Cruz; *San Antonio;* Petaluma; Mt. St. Helena; Mt. Sanhedrin; Mad River, Trinity Co.; Hoopa Valley; *Horse Creek.* Oregon: Diamond Lake; Bend; Detroit; Mt. Hood. Washington: Bingen (Sheppe, 1961:431); Bethel Ridge, 30 mi. ESE Mt. Rainier, 3 mi. N Tieton Reservoir (Broadbooks, 1965:301, 302); Lodgepole Camp (Sheppe, 1961:435); near Easton (*ibid.*); Leavenworth.

Peromyscus maniculatus georgiensis Hall

1938. *Peromyscus maniculatus georgiensis* Hall, Amer. Nat., 72:455, September 10, type from Vananda, Texada Island, Georgia Strait, British Columbia.

MARGINAL RECORDS.—British Columbia: Savary Island; type locality; Thormanby Island; Bowen Island; Lasqueti Island.

Peromyscus maniculatus geronimensis J. A. Allen

1898. *Peromyscus geronimensis* J. A. Allen, Bull. Amer. Mus. Nat. Hist., 10:156, April 12, type from San Gerónimo Island, Baja California. Known only from type locality.
1909. *Peromyscus maniculatus geronimensis,* Osgood, N. Amer. Fauna, 28:99, April 17.

Peromyscus maniculatus gracilis (Le Conte)

1855. *H*[*esperomys*]. *gracilis* Le Conte, Proc. Acad. Nat. Sci. Philadelphia, 7:442, type from Michigan.
1909. *Peromyscus maniculatus gracilis,* Osgood, N. Amer. Fauna, 28:42, April 17.
1893. *Sitomys americanus canadensis* Miller, Proc. Biol. Soc.

Washington, 8:55, June 20, type from Peterboro, Madison Co., New York.

MARGINAL RECORDS.—Quebec: Godbout; Camp de la Roche (Wrigley, 1969:207, as *P. maniculatus* only); St. Sebastien (*ibid.*, as *P. maniculatus* only). New Hampshire: 4 mi. S Center Ossipee, thence southward to *highlands of western Massachusetts and northwestern Connecticut* (Waters, 1962:102, as *P. maniculatus* only). Pennsylvania: Monroe County; Carbon County; Sullivan County. New York: 5–10 mi. E Ithaca (Brown and Welser, 1968:425); Allegany State Park. Ontario: Port Franks. Michigan: Missaukee County. Wisconsin: Sheboygan Swamp, near Elkhart Lake (Jackson, 1961:212); Thorp Twp., Clark Co. Minnesota: Lake Itasca Forestry and Biological Station; Tower; 2 mi. N, 2½ mi. E Grand Portage (Timm, 1975:20). Ontario: Michipicoten Island, Lake Superior. Quebec: upper part valley of Gatineau; Lake St. John (Harper, 1961:53).

Peromyscus maniculatus hollisteri Osgood

1909. *Peromyscus maniculatus hollisteri* Osgood, N. Amer. Fauna, 28:62, April 17, type from Friday Harbor, San Juan Island, San Juan Co., Washington.

MARGINAL RECORDS.—Washington: Cypress Island; Blakeley Island; type locality.

Peromyscus maniculatus hueyi Nelson and Goldman

1932. *Peromyscus maniculatus hueyi* Nelson and Goldman, Trans. San Diego Soc. Nat. Hist., 7:51, April 15, type from small island in Gonzaga Bay, lat. about 29° 50′, Baja California. Known only from type locality.

Peromyscus maniculatus hylaeus Osgood

1908. *Peromyscus hylaeus* Osgood, Proc. Biol. Soc. Washington, 21:141, June 9, type from Hollis, Kasaan Bay, Prince of Wales Island, Alaska.
1909. *Peromyscus maniculatus hylaeus* Osgood, N. Amer. Fauna, 28:53, April 17.

MARGINAL RECORDS.—Alaska: Glacier Bay; Juneau; Taku Harbor; Petersburg, Mitkof Island; type locality; Lindenburg Peninsula, Kupreanof Island; near Killisnoo, Admiralty Island.

Peromyscus maniculatus inclarus Goldman

1939. *Peromyscus maniculatus inclarus* Goldman, Jour. Mamm., 20:355, August 14, type from Fremont Island, 4250 ft., Great Salt Lake, Weber Co., Utah. Known only from Fremont Island.

Peromyscus maniculatus interdictus Anderson

1932. *Peromyscus maniculatus interdictus* Anderson, Bull. Nat. Mus. Canada, 70:110, November 24, type from Forbidden Plateau, near E edge Strathcona Park, N of Mt. Albert Edward, about 17 mi. W Comox, 4200 ft., Vancouver Island, British Columbia.

MARGINAL RECORDS.—British Columbia (Cowan and Guiguet, 1965:185): Shushartie; Port Hardy; Sayward; type locality; Mt. Arrowsmith; Old Wolf Lake; Nootka; Cape Scott.

Peromyscus maniculatus isolatus Cowan

1935. *Peromyscus sitkensis isolatus* Cowan, Univ. California Publ. Zool., 40:434, November 14, type from Pine Island, Queen Charlotte Sound, N end Vancouver Island, British Columbia.
1945. *Peromyscus maniculatus isolatus*, McCabe and Cowan, Trans. Royal Canadian Inst., 25:194, February. In 1973, Thomas, Cytologia, 38:493, on basis of karyological data, placed specimens from Nigei Island in a "cohort" with *Peromyscus sitkensis*.

MARGINAL RECORDS.—British Columbia: type locality; Nigei Island.

Peromyscus maniculatus keeni (Rhoads)

1894. *Sitomys keeni* Rhoads, Proc. Acad. Nat. Sci. Philadelphia, 46:258, October, type from Massett, Graham Island, Queen Charlotte Islands, British Columbia.
1909. *Peromyscus maniculatus keeni*, Osgood, N. Amer. Fauna, 28:55, April 17.

MARGINAL RECORDS.—British Columbia: Rose Spit (Cowan and Guiguet, 1965:186); near Rose Harbor, Moresby Island; Cape Knox (Cowan and Guiguet, 1965:186).

Peromyscus maniculatus labecula Elliot

1903. *Peromyscus labecula* Elliot, Field Columb. Mus., Publ. 71, Zool. Ser., 3(8):143, March 20, type from Ocotlán, Jalisco.
1909. *Peromyscus maniculatus labecula*, Osgood, N. Amer. Fauna, 28:87, April 17.

MARGINAL RECORDS.—Durango: 26 mi. SW Yerbanís, 6725 ft. (Baker and Greer, 1962:116). Zacatecas: Sierra de Valparaíso; Zacatecas. Guanajuato: Santa Rosa. Hidalgo: Zimapán; Ixmiquilpan. Distrito Federal: Cerro de Santa Isabel, 8 km. N México, 2340 m. México: 4 km. ESE San Rafael, 2460 m. Puebla: Río Otlati, 15 km. NW San Martín, 8700 ft. Morelos: *1 ¹/₂ mi. SE Huitzilac; 2 mi. W Huitzilac; 5 km. N Tres Cumbres.* Distrito Federal: *Rancho La Noria, 1 mi. W Xochimilco, 2270 m.; ³/₄ mi. SSE San Andrés, 2620 m.; La Venta, 2780 m.* México: 6 mi. NNW Acambay. Jalisco: 1½ mi. N, ½ mi. NW Mazamitla. Colima: Hda. Magdalena. Jalisco: Mascota. Nayarit: Ojo de Aguas, near Amatlán. Durango: 8 mi. NW Durango, 6200 ft. (Baker and Greer, 1962:116); 6 mi. NW La Pila, 6150 ft. (*ibid.*).

Peromyscus maniculatus luteus Osgood

1905. *Peromyscus luteus* Osgood, Proc. Biol. Soc. Washington, 18:77, February 21, type from Kennedy, Cherry Co., Nebraska.

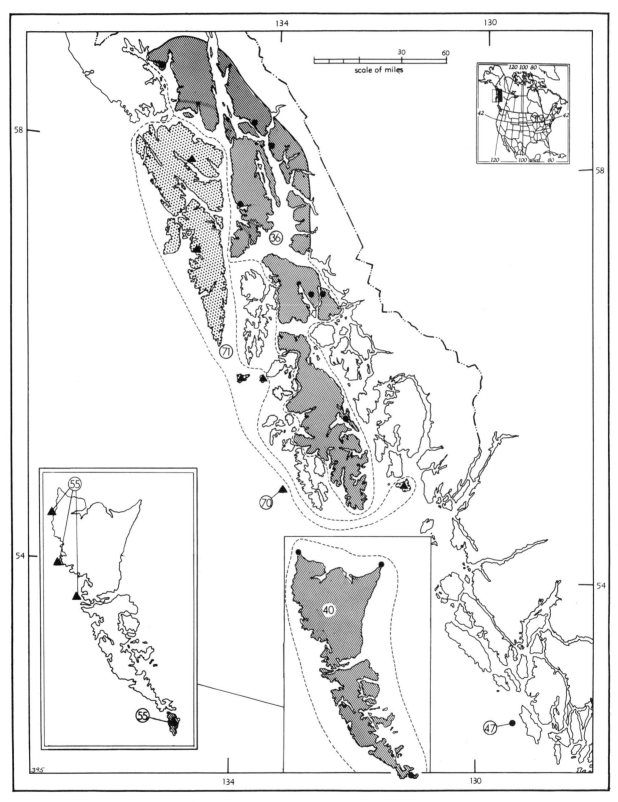

Map 395. *Peromyscus maniculatus* (part) and *Peromyscus sitkensis.*

36. *P. m. hylaeus*	47. *P. m. maritimus*	70. *P. s. oceanicus*
40. *P. m. keeni*	55. *P. m. prevostensis*	71. *P. s. sitkensis*

678

1909. *Peromyscus maniculatus luteus*, Osgood, N. Amer. Fauna, 28:77, April 17.

1896. *Peromyscus texanus nebrascensis*, J. A. Allen, Bull. Amer. Mus. Nat. Hist., 8:251 (part).

1911. *Peromyscus maniculatus nebrascensis*, Mearns, Proc. Biol. Soc. Washington, 24:102, May 15. From 1911 until 1958 authors applied the name *P. m. nebrascensis* to the subspecies *P. m. luteus* of the present work for reasons explained by Jones, Proc. Biol. Soc. Washington, 71:107–111, July 16, 1958.

MARGINAL RECORDS.—Nebraska (Jones, 1964c: 205): 10 mi. NE Stuart; *1 mi. S Atkinson; 10 mi. N Burwell; Comstock;* 1 mi. E Ravenna; *1³/₄ to 3³/₄ mi. S Kearney; Bladen;* Red Cloud. Kansas: ½ mi. W Downs; 3 mi. N, 2 mi. W Hoisington; 1 mi. N Harper. Oklahoma: Enid; Norman; Chattanooga. Texas (Judd, 1970:279): 7½ mi. E Crosbyton; Tom Good Ranch, Borden Co.; *4 mi. N Notrees;* 19 mi. E Kermit. New Mexico (*ibid.*): 17 mi. W Hobbs; 1 mi. S, 3 mi. W Caprock; Portales. Texas (*ibid.*): 4 mi. NW Hart; 27 mi. S Amarillo; 6 mi. W Stinnett; Canadian. Oklahoma: Tesquite Canyon. Kansas: Cimarron River, NW Rolla. Colorado (Armstrong, 1972:197): Burlington; 1½ mi. N, 3½ mi. W Wray. Nebraska (Jones, 1964c:205): *Windlass Hill, 2 mi. SE Lewellen;* ½ mi. S Oshkosh; 16 mi. NE Alliance; *Mirage Township;* Gordon.

Peromyscus maniculatus macrorhinus (Rhoads)

1894. *Sitomys macrorhinus* Rhoads, Proc. Acad. Nat. Sci. Philadelphia, 46:259, October, type from mouth Skeena River, British Columbia.

1909. *Peromyscus maniculatus macrorhinus*, Osgood, N. Amer. Fauna, 28:57, April 17.

MARGINAL RECORDS.—British Columbia: Docdaon Creek; W end Eutsuk Lake; Stuie; Kingcome Inlet; also the following islands: Calvert, *King, Swindle, Aristazabal, Princess Royal, North Estevan, Banks, Pitt, McCauley,* and *Porcher;* Metlakatla (Cowan and Guiguet, 1965:187). Alaska: Woronkofski Island.

Preomyscus maniculatus magdalenae Osgood

1909. *Peromyscus maniculatus magdalenae* Osgood, N. Amer. Fauna, 28:101, April 17, type from Magdalena Island, Baja California.

MARGINAL RECORDS.—Baja California: San Juanico Bay; Matancita; type locality.

Peromyscus maniculatus maniculatus (Wagner)

1845. *Hesperomys maniculatus* Wagner, Wiegmann's Arch. für Naturgesch., Jahrg. 11, 1:148. Type locality, the Moravian settlements in Labrador.

1898. *Peromyscus maniculatus*, Bangs, Amer. Nat., 32:496, July.

1877. [*Hesperomys*] *articus* Coues, *in* Coues and Allen, Monog. N. Amer. Rodentia, pp. 61 and 67, a *nomen nudum.*

1877. [*Hesperomys*] *bairdii* Coues, *in* Coues and Allen, Monog. N. Amer. Rodentia, pp. 61 and 67. Type locality, Labrador. Not *Mus bairdii* Hoy and Kennicott, 1857 [=

Peromyscus maniculatus bairdii], type from Bloomington, McLean Co., Illinois.

1897. *Peromyscus canadensis umbrinus* Miller, Proc. Boston Soc. Nat. Hist., 28:23, April 30, type from Peninsula Harbor, N shore Lake Superior, Ontario.

MARGINAL RECORDS.—Quebec: Port Burwell. Labrador: Gready (Harper, 1961:52). Quebec: Natashquan; Lake Albanel (Harper, 1961:52); Lake Mistassini (*ibid.*); 47°50'N, 75°30'W, S of Clova (MacLeod and Cameron, 1961:282). Ontario: Lowbush, Lake Abitibi. Michigan: Isle Royale. Manitoba: Marchand; Norway House; York Factory. Quebec: E James Bay Coast (Harper, 1961:52); Merry Island (Edwards, 1963:10, as *P. maniculatus* only). Labrador: Northwest River (Harper, 1961:52).

Peromyscus maniculatus margaritae Osgood

1909. *Peromyscus maniculatus margaritae* Osgood, N. Amer. Fauna, 28:95, April 17, type from Margarita Island, Baja California. Known only from type locality.

Peromyscus maniculatus maritimus McCabe and Cowan

1945. *Peromyscus maniculatus maritimus* McCabe and Cowan, Trans. Royal Canadian Inst., 25:199, February, type from largest of Moore Islands, British Columbia. Known only from type locality.

Peromyscus maniculatus nebrascensis (Coues)

1877. *Hesperomys sonoriensis* var. *nebrascensis* Coues, *in* Coues and Allen, Monog. N. Amer. Rodentia, U.S. Geol. Surv. Territories, 11:79, August, syntypes from Deer Creek, approx. 5 mi. from its mouth, Converse Co., Wyoming.

1909. *Peromyscus maniculatus nebrascensis*, Osgood, N. Amer. Fauna, 28:75, April 17.

1890. *Hesperomys leucopus nebrascensis*, Mearns, Bull. Amer. Mus. Nat. Hist., 2:285, February (a redescription of *nebrascensis* on p. 287, based on a specimen from Calf Creek, Custer Co., Montana).

1896. *Peromyscus texanus nebrascensis*, J. A. Allen, Bull. Amer. Mus. Nat. Hist., 8:251, November (part).

1911. *Peromyscus maniculatus osgoodi* Mearns, Proc. Biol. Soc. Washington, 24:102, May 15, type from Calf Creek, Custer Co., Montana. From 1911 until 1958 authors applied the name *P. m. osgoodi* to the subspecies *P. m. nebrascensis* of the present work for reasons explained by Jones, Proc. Biol. Soc. Washington, 71:107–111, July 16, 1958.

MARGINAL RECORDS.—Saskatchewan: Moosejaw; Glen Ewen; Estevan. North Dakota: Bismarck; Sioux County. South Dakota: ½ mi. W Reva (Andersen and Jones, 1971:379); Smithville. Nebraska (Jones, 1964c:207): 3 mi. E Chadron; *10 mi. S Chadron;* near Hemingford; Courthouse Rock; 1 mi. N, 2 mi. W Chappell. Colorado (Armstrong, 1972:200, 201): 5 mi. SW Julesburg; Tuttle; Craugh Ranch, Cimarron River. Oklahoma: Kenton. Colorado (Armstrong, 1972:199–201): Trinchera; 20 mi. E Walsenburg; 5 mi. W Pueblo; Cañon City; Minnehaha; Golden; *5 mi. W Boulder, 5600 ft.;* 3 mi. N, ¼ mi. E Livermore. Wyoming

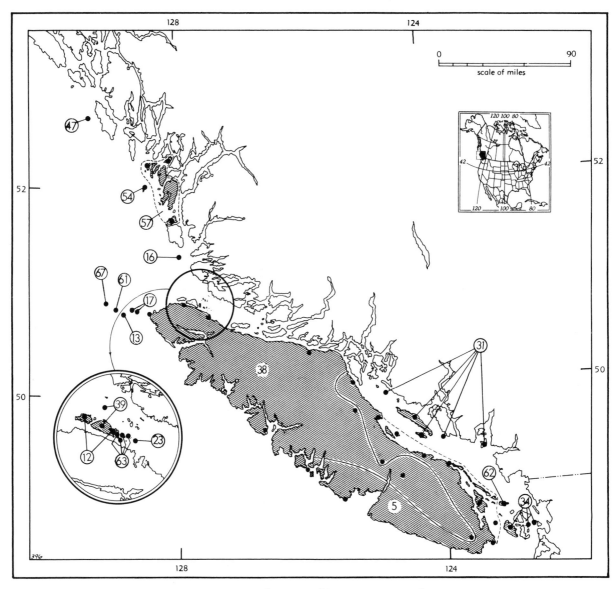

Map 396.   Some subspecies of *Peromyscus maniculatus*.

Guide to subspecies	16. *P. m. cancrivorus*	38. *P. m. interdictus*	61. *P. m. sartinensis*
5. *P. m. angustus*	17. *P. m. carli*	39. *P. m. isolatus*	62. *P. m. saturatus*
12. *P. m. balaclavae*	23. *P. m. doylei*	47. *P. m. maritimus*	63. *P. m. saxamans*
13. *P. m. beresfordi*	31. *P. m. georgiensis*	54. *P. m. pluvialis*	67. *P. m. triangularis*
	34. *P. m. hollisteri*	57. *P. m. rubriventer*	

(Long, 1965a:632): 6 mi. N, 15 mi. E Savery; *4 mi. N, 10 mi. E Savery*. Colorado (Armstrong, 1972:199, 200): Four Mile Creek, 6 mi. SW Baggs, Wyoming; Craig, 6100 ft.; Meeker, 6200 ft.; Grand Junction. Utah: 1 mi. N Moab Bridge; Willow Creek, 29 mi. S Ouray; Red Creek, 2 mi. N Fruitland; Ashley Creek, 10 (air line) mi. NW Vernal. Colorado: Ladore. Utah: Hideout Canyon, Ashley National Forest; *Sheep Creek, 7 mi. S Manila*; Henrys Park, Henrys Fork. Wyoming (Long, 1965a:632, 633): Evanston; *Bear River, 14 mi. N Evanston*; Cokeville; *Fontanelle*; 2 mi. S, 19 mi. W Big

Piney; 31 mi. N Pinedale; *Jackeys Creek, 3 mi. S Dubois; Needle Mtn.*; Valley; Clarks Fork. Montana: Jefferson River; St. Mary Lake. Alberta: McDonald Lake. Saskatchewan: Crane Lake.

Peromyscus maniculatus nubiterrae Rhoads

1896. *Peromyscus leucopus nubiterrae* Rhoads, Proc. Acad. Nat. Sci. Philadelphia, 48:187, April, type from summit Roan Mtn., 6370 ft., Mitchell Co., North Carolina.
1909. *Peromyscus maniculatus nubiterrae*, Osgood, N. Amer. Fauna, 28:47, April 17.

MARGINAL RECORDS.—New York: Allegany County. Pennsylvania: Bradford County; Sullivan County; Mifflin County. Virginia: Skyland; Bedford County; Mt. Rogers. North Carolina (Kirkland, 1975:226): Grandfather Mtn.; Cove Creek, Rutherford Co. South Carolina: near Caesars Head. Georgia: Neel['s] Gap (Golley, 1962:118); Blood Mtn. (ibid.); Brasstown Bald. Tennessee: Russell Field (Komarek and Komarek, 1938:154). Kentucky: Rosspoint, 1250 ft. West Virginia: Cold Knob Mts.; Morgantown. Pennsylvania: Fayette County; Westmoreland County; 10½ mi. NW Marienville (Bradshaw and George, 1969:823); Port Allegany. Note: In central Pennsylvania, P. m. nubiterrae and P. m. bairdii both occur. They are ecologically separated, P. m. nubiterrae occurring at higher elevations and P. m. bairdii generally in the valleys.

Peromyscus maniculatus oreas Bangs

1898. Peromyscus oreas Bangs, Proc. Biol. Soc. Washington, 12:84, March 24, type from Mount Baker Range, 6500 ft., British Columbia, near boundary of Whatcom County, Washington.
1909. Peromyscus maniculatus oreas, Osgood, N. Amer. Fauna, 28:51, April 17.—Sheppe's (Proc. Amer. Philos. Soc., 105:421–446, August, 1961) findings concerning the relation of P. oreas Bangs to four geographically allopatric, and in many places sympatric, subspecies of P. maniculatus (Wagner) are important for understanding evolution in these mice in the Pacific Northwest. He emphasized (p. 445) that "oreas is largely reproductively isolated from P. maniculatus," and noted that hybridization "may occur and there may be intergrade populations in some places, but in many places both species are found together in pure form." Consequently he regarded P. oreas as a species instead of a subspecies of P. maniculatus. His map (p. 421) shows the sympatry of oreas and adjoining subspecies of maniculatus in a useful fashion.

MARGINAL RECORDS.—Main (eastern) segment: British Columbia: Owikeno Lake; Lillooet; Wright's Ranch, 3 mi. SW Princeton (Sheppe, 1961:436); Copper Creek (ibid.). Washington: Grouse Creek; American Forks (Sheppe, 1961:435); Clear Lake (ibid.); Windriver C. C. C. Camp; Yacolt; La Wis Wis Camp (Sheppe, 1961:439); Bothell (op. cit.:430); Mt. Vernon (ibid.); Tomyhoi Lake. British Columbia; Lihumitson Park; Haney (Sheppe, 1961:433); Cypress Lake; Mt. Seymour; Alta Lake; head Bute Inlet; head Loughborough Inlet; Kingcome Inlet (Sheppe, 1961:431); Rivers Inlet (op. cit.:430). Western segment: Washington (Sheppe, 1961:429, 430, unless otherwise noted): Dungeness Hatchery; Lake Cushman; Tenino; near Chehalis; Kelso (Hall and Kelson, 1959:622); Cathlamet; Ilwaco; Aberdeen; Neah Bay (Sheppe, 1961:424).

Peromyscus maniculatus ozarkiarum Black

1935. Peromyscus maniculatus ozarkiarum Black, Jour. Mamm., 16:144, May 15, type from 3 mi. S Winslow, Washington Co., Arkansas.

MARGINAL RECORDS.—Missouri: Caney Fire-protection Tower, SE Ava (Brown, 1963:425). Arkansas: Huntsville; Mena. Oklahoma: 2 mi. NW Stapp, 1 mi. NE Zoe. Texas: Gainesville; Willow Creek, 6 mi. SW Gainesville; 2 mi. S Marysville. Oklahoma: 10 mi. SW Norman; Stillwater; Garnett. Arkansas: Gravette. Missouri (Brown, 1963:425): Roaring River State Park; 7 mi. SE Hollister.

Peromyscus maniculatus pallescens J. A. Allen

1896. Peromyscus michiganensis pallescens J. A. Allen, Bull. Amer. Mus. Nat. Hist., 8:238, November 21, type from San Antonio, Bexar Co., Texas.
1909. Peromyscus maniculatus pallescens, Osgood, N. Amer. Fauna, 28:83, April 17.

MARGINAL RECORDS.—Texas: Hardeman County (Dalquest, 1968:17, as P. m. cf. pallescens); Montague County (ibid., as P. m. cf. pallescens); Denton County. Oklahoma: 6 mi. NW Colbert; 5 mi. N Colbert. Texas (Davis, 1966:186, as P. maniculatus only, unless otherwise noted): Dallas County; Navarro County; Brazos County; Burleson County; about 4 mi. NE Rockport (Hall and Kelson, 1959:623); 12 mi. SW Alice (ibid.); Atascosa County; San Antonio (Hall and Kelson, 1959:623); 3 mi. NE Bertram (ibid.); Tarrant County; Baylor County (Baccus, 1971:182).

Peromyscus maniculatus plumbeus C. F. Jackson

1939. Peromyscus maniculatus plumbeus C. F. Jackson, Proc. Biol. Soc. Washington, 52:101, June 5, type from Pigou River, N shore Gulf of St. Lawrence, Saguenay Co., Quebec.

MARGINAL RECORDS.—Quebec: Matamek River (Harper, 1961:53); type locality; Bay of Seven Islands.

Peromyscus maniculatus pluvialis McCabe and Cowan

1945. Peromyscus maniculatus pluvialis McCabe and Cowan, Trans. Royal Canadian Inst., 25:199, February, type from northern island, Goose Island Group, lat. 52° N, long. 128° 31′ W, British Columbia. Known only from type locality.

Peromyscus maniculatus prevostensis Osgood

1901. Peromyscus prevostensis Osgood, N. Amer. Fauna, 21:29, September 26, type from Prevost [= Kunghit] Island, Queen Charlotte Group, British Columbia.
1965. Peromyscus maniculatus prevostensis, Cowan and Guiguet, British Columbia Provincial Mus., Handbook 11:187, October.

MARGINAL RECORDS (Cowan and Guiguet, 1965:188).—British Columbia: Frederick Island; Hippa Island; Marble Island; type locality.

Peromyscus maniculatus rubidus Osgood

1901. Peromyscus oreas rubidus Osgood, Proc. Biol. Soc. Washington, 14:193, December 12, type from Mendocino City, Mendocino Co., California.

1909. *Peromyscus maniculatus rubidus* Osgood, N. Amer. Fauna, 28:65, April 17.
1903. *Peromyscus perimekurus* Elliot, Field Columb. Mus., Publ. 74, Zool. Ser., 3(10):156, May 7, type from Goldbeach, Curry Co., Oregon.

MARGINAL RECORDS.—Washington: Puget Island. Oregon: W slope Mt. Hood; Wells; Prospect. California: Hoopa Valley; *Canyon Creek, Trinity Co.; Carlotta;* near Calpella; Camp Meeker; *Olema;* La Honda.

Peromyscus maniculatus rubriventer McCabe and Cowan

1945. *Peromyscus maniculatus rubriventer* McCabe and Cowan, Trans. Royal Canadian Inst., 25:196, February, type from Ruth Island, Hunter Islands, British Columbia.

MARGINAL RECORDS.—British Columbia: Chatfield Island; Hecate Island; Reginald Island (Cowan and Guiguet, 1965:188); *Townsend Island (ibid.); Smythe Island.*

Peromyscus maniculatus rufinus (Merriam)

1890. *Hesperomys leucopus rufinus* Merriam, N. Amer. Fauna, 3:65, September 11, type from San Francisco Mtn., 9000 ft., Coconino Co., Arizona.
1909. *Peromyscus maniculatus rufinus,* Osgood, N. Amer. Fauna, 28:72, April 17.

MARGINAL RECORDS.—Northwestern segment: Utah: 5 mi. SW Laketown, 6500 ft.; *Monte Cristo, 18 mi. W Woodruff, 8000 ft.;* Chalk Creek, 19 mi. NE Coalville; Hoop Lake, Ashley National Forest, 8000 ft.; jct. Deep and Carter creeks, 7900 ft.; *Beaver Creek, 19 mi. S Manila; Green Lake, 60 mi. N Vernal;* jct. Trout and Ashley creeks, 9700 ft.; Petty Mtn., 15 mi. N Mountain Home, 9500 ft.; Stockmore; *4 mi. E Mt. Alice, between Emery and Loa, 7450 ft.;* Elkhorn Guard Station, 14 mi. N Torrey, Fishlake Plateau, 9400 ft.; Fishlake, 8730 ft.; Maple Canyon; 1 mi. N Salt Lake City, 4500 ft.; Anderson's Ranch, 5700 ft., Blacksmiths Fork. Southeastern segment: Colorado (Armstrong, 1972:202–205): Pearl, 9000 ft.; *Owl Canyon;* Gold Hill; *Bear Creek, 8300 ft.; 5 mi. S Victor; 12 mi. W Pueblo;* East Spanish Peak, 10, 500 ft.; 1 mi. S, 7 mi. E Trinidad. New Mexico (Findley, et al., 1975:205 [Map 79]–210): Bear Canyon, Raton Range; 3 mi. W Kenton [Oklahoma]; 10 mi. N Cabra Springs; Capitan Mts.; 20 mi. S Cloudcroft; 5 mi. E Tularosa; Gallo Canyon, 35 mi. SE Corona; 4 mi E San Antonio; Cuchillo; Iron Canyon, 7 mi. N, 16 mi. E Santa Rita; 1 mi. W Redrock. Arizona: San Bernardino Ranch. Chihuahua (Anderson, 1972:347): 3 mi. S, 10 mi. E Pacheco; 7 mi. WSW Cuauhtémoc, 4 mi. NW San Francisco de Borja, 5700 ft.; 2 mi. W Miñaca; *Rancho San Ignacio, 4 mi. S, 1 mi. W Santo Tomás.* Arizona: Chiricahua Mts.; *NW slope Carr Peak, 8400 ft.;* N slope Mt. Lemmon, 8000–8600 ft. (Lange, 1960:448); Green Spring (Hoffmeister and Durham, 1971:38); *3 mi. N, 2 mi. W Mt. Dellenbaugh (ibid.);* Grand Gulch Mine (*ibid.*); 10 mi. W Wolf Hole

(*ibid.*); 8 mi. S St. George (*ibid.*); 25 mi. S Hurricane (*ibid.*); 1 mi. S, 2½ mi. W Fredonia (*ibid.*); Jacobs Lake (Cockrum, 1961:175); Nankoweap Campground, Mile 53 (Hoffmeister, 1971:172); *Kwagunt delta on Colorado River (ibid.); Palisades Creek, Mile 66 (op. cit.:173);* E side Cedar Mtn. (*ibid.*); type locality; Baker Butte; Keam Canyon; right fork Segie-ot-Sosie Canyon, ca. 11 mi. NW Kayenta (Cockrum, 1961:176). Utah: *Noland Ranch;* Riverview; Hatch Trading Post, 25 mi. SE Blanding, Montezuma Creek, 4500 ft.; Duck Lake, 1 mi. S Gooseberry Ranger Station, Elk Ridge, 8400 ft.; Moab; Castle Valley, 6000 ft., 18 mi. NE Moab. Colorado (Armstrong, 1972:202, 203): 7 mi. S Glade Park; De Beque; *Grand Hogback, 5 mi. S Meeker; 8 mi. NE Craig;* Elkhead Mountains, 20 mi. SE Slater.

Peromyscus maniculatus sanctaerosae von Bloeker

1940. *Peromyscus maniculatus sanctaerosae* von Bloeker, Bull. California Acad. Sci., 39(pt. 2):173, December 15, type from Elderberry Canyon, Santa Rosa Island, Santa Barbara Co., California. Known only from Santa Rosa Island.

Peromyscus maniculatus santacruzae Nelson and Goldman

1931. *Peromyscus maniculatus santacruzae* Nelson and Goldman, Jour. Washington Acad. Sci., 21:532, December 19, type from Santa Cruz Island, Santa Barbara Co., California. Known only from type locality.

Peromyscus maniculatus sartinensis Guiguet

1955. *Peromyscus maniculatus sartinensis* Guiguet, Rept. Provincial Mus. Nat. Hist. and Anthro., British Columbia, for 1954, p. B69, August 1, type from Sartine Island, British Columbia. Known only from Sartine Island.

Peromyscus maniculatus saturatus Bangs

1897. *Peromyscus texanus saturatus* Bangs, Amer. Nat., 31:75, January, type from Saturna Island, Gulf of Georgia, between Victoria and Vancouver City, British Columbia. Known only from Saturna Island.
1909. *Peromyscus maniculatus saturatus,* Osgood, N. Amer. Fauna, 28:61, April 17.

Peromyscus maniculatus saxamans McCabe and Cowan

1945. *Peromyscus maniculatus saxamans* McCabe and Cowan, Trans. Royal Canadian Inst., 25:198, February, type from Duncan Island, British Columbia.

MARGINAL RECORDS.—British Columbia: Hurst Island; Bell Island; Heard Island; type locality.

Peromyscus maniculatus serratus Davis

1939. *Peromyscus maniculatus serratus* Davis, The Recent mammals of Idaho, Caxton Printers, Caldwell, Idaho, p. 290, April 5, type from Mill Creek, 14 mi. W Challis, Custer Co., Idaho.

MARGINAL RECORDS.—Idaho: 5 mi. E Warm Lake; Alturas Lake; type locality.

Peromyscus maniculatus sonoriensis (Le Conte)

1853. *Hesp[eromys]. sonoriensis* Le Conte, Proc. Acad. Nat. Sci. Philadelphia, 6:413, type from Santa Cruz, Sonora.

1909. *Peromyscus maniculatus sonoriensis*, Osgood, N. Amer. Fauna, 28:89, April 17.

1890. *Hesperomys leucopus deserticolus* Mearns, Bull. Amer. Mus. Nat. Hist., 2:285 (described on p. 287), February 21, type from Mohave Desert, San Bernardino Co., California.

1894. *Sitomys insolatus* Rhoads, Proc. Acad. Nat. Sci. Philadelphia, 46:256, October, type from Oro Grande, Mohave Desert, San Bernardino Co., California.

1903. *Peromyscus oresterus* Elliot, Field Columb. Mus., Publ. 74, Zool. Ser., 3(10):159, May 7, type from Vallecitos, Sierra San Pedro Mártir, Baja California.

1937. *Peromyscus maniculatus gunnisoni* Goldman, Proc. Biol. Soc. Washington, 50:224, December 28, type from Gunnison Island, about 4300 ft., Great Salt Lake, Utah. Regarded as identical with *P. m. sonoriensis* by Durrant, Univ. Kansas Publ., Mus. Nat. Hist., 6:307, 310, August 10, 1952.

MARGINAL RECORDS.—Idaho: Lemhi; Birch Creek, Clark Co.; lava beds, 17 mi. W Idaho Falls; Crow Creek; Montpelier Creek. Utah: Ogden, 4400 ft.; Salt Lake City; Provo; Nephi; Manti; *Glenwood;* Loa; 5 mi. S Castle Dale, 5600 ft.; Christensen Ranch, Nine-mile Canyon, 10 mi. E summit, 6300 ft.; 7 mi. N Greenriver, 4100 ft.; San Rafael River, 15 mi. SW Greenriver; Mt. Ellen, Henry Mts.; Soldier Spring, Navajo Mtn.; Johns Canyon, 5150 ft., San Juan River; ½ mi. NW Bluff. Arizona: 7½ mi. N Adamana, 5337 ft. (Cockrum, 1961:178); Holbrook; *Painted Desert;* Kaibab Plateau; 6 mi. N Wolf Hole (Cockrum, 1961:178); Phoenix; Oracle; *10 mi. SE Oracle, 4600 ft.* (Lange, 1960:448); Ash Creek Ranch (Cockrum, 1961:178); Dos Cabezos; Huachuca Mts. Sonora: Ciénega Well; *opposite mouth Hardy River.* Baja California: La Grulla; Hanson Lagoon. California: Vallecito; Lythe Creek; Mt. Piños; Painted Rock, SE Simmler; Bakersfield; Kernville; Mt. Whitney; Mammoth; *Independence Creek; Leavett Meadows;* Markleeville. Nevada: *Zephyr Cove;* 3 mi. S Mt. Rose; 34 mi. E Reno; ½ mi. S Pyramid Lake; 30 mi. W, 4 mi. N Lovelock, 4300 ft.; 1¼ mi. N Sulphur, 4050 ft.; 1 mi. S Denio, Oregon, 4200 ft. Oregon: 2 mi. E Riley (Brown and Welser, 1968:425); S shore Malheur Lake. Idaho: Nampa; Craters of the Moon National Monument.

Peromyscus maniculatus streatori Nelson and Goldman

1931. *Peromyscus maniculatus streatori* Nelson and Goldman, Jour. Washington Acad. Sci., 21:531, December 19, type from San Miguel Island, Santa Barbara Co., California.

MARGINAL RECORDS.—California Channel Islands (von Bloeker, 1967:256): *Prince;* San Miguel.

Peromyscus maniculatus triangularis Guiguet

1955. *Peromyscus maniculatus triangularis* Guiguet, Rept. Provincial Mus. Nat. Hist. and Anthro., British Columbia, for 1954, p. B69, August 1, type from Triangle Island, British Columbia. Known only from Triangle Island. In 1973, Thomas, Cytologia, 38:493, on basis of karyological data, placed *triangularis* in a "cohort" with *Peromyscus sitkensis.*

Peromyscus sejugis Burt
Santa Cruz Island Mouse

1932. *Peromyscus sejugis* Burt, Trans. San Diego Soc. Nat. Hist., 7:171, October 31, type from Santa Cruz Island, lat. 25° 17′ N, long. 110° 43′ W, Gulf of California, Baja California. Confined, so far as known, to Santa Cruz Island and San Diego Island (25° 12′ N, 110° 42′ W), Gulf of California.

Color of upper parts grayish washed with avellaneous, giving general dull color to upper parts; underparts white; tail distinctly bicolored with narrow dorsal stripe (about 1.5 mm. wide in dry skin); ears distinctly dusky. Skull arched anteroposteriorly; rostrum heavy; nasals broad, tapering posteriorly and terminating beyond premaxillae; auditory bullae relatively small; length of shelf of bony palate greater than length of maxillary tooth-row (after Burt, 1932:171). Average measurements of 24 topotypes are: 173; 85; 22; 16. See Map 394.

Peromyscus slevini Mailliard
Slevin's Mouse

1924. *Peromyscus slevini* Mailliard, Proc. California Acad. Sci., ser. 4, 12:1221, July 22, type from Santa Catalina Island, 17 mi. NE Punta San Marcial, Gulf of California, Baja California. Known only from Santa Catalina Island, Baja California.

External measurements of the type, an adult male, are: 225; 120; 27; ear (dry), 16.5. Upper parts (worn pelage) pale cinnamon with admixture of dusky hairs dorsally; underparts white with pale wash of cinnamon pectorally; feet creamy white; forelegs pale cinnamon; tail bicolored, above darker than dorsum of body, below almost white. Skull with interparietal rhomboidal; nasals extending markedly posterior to premaxillae. See Map 394.

Peromyscus sitkensis
Sitka Mouse

External measurements: 205–230; 97–116; 25–27; ear (dry), 14.4–17.8. Upper parts rich russet to mars brown, shading to darker dorsally; underparts white or almost so; dusky tones well

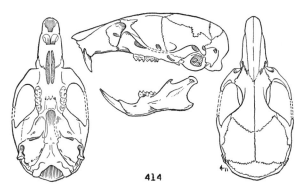

Fig. 414. *Peromyscus sitkensis sitkensis*, Sitka, Alaska, No. 433 K.U., ♂, X 1½.

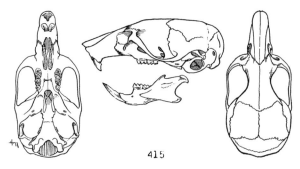

Fig. 415. *Peromyscus melanotis*, 6 km. SSE Altatongo, 9000 ft., Veracruz, No. 19479 K.U., ♀, X 1½.

developed on face, forelegs, and ankles; tail bicolored. Skull much as in *P. maniculatus*, but larger, more robust; nasals and rostrum relatively longer; auditory bullae relatively smaller.

Peromyscus sitkensis has a peculiarly disrupted distribution. Not only does it occur only on islands, but in some cases on islands some distance from each other with intervening islands occupied by *P. maniculatus*. The two species, *P. maniculatus* and *P. sitkensis*, are closely related and are not known to occur on the same island. In spite of its larger size, *P. sitkensis* is distributionally the weaker species. See Map 395.

Peromyscus sitkensis oceanicus Cowan

1935. *Peromyscus sitkensis oceanicus* Cowan, Univ. California Publ. Zool., 40:432, November 14, type from Forrester Island, Alaska. Known only from type locality.

Peromyscus sitkensis sitkensis Merriam

1897. *Peromyscus sitkensis* Merriam, Proc. Biol. Soc. Washington, 11:223, July 15, type from Sitka, Alaska.

MARGINAL RECORDS.—Alaska: Chichagof Island; Warren Island; Duke Island; Coronation Island; Baranof Island.

Peromyscus melanotis J. A. Allen and Chapman Black-eared Mouse

1897. *Peromyscus melanotis* J. A. Allen and Chapman, Bull. Amer. Mus. Nat. Hist., 9:203, June 16, type from Las Vigas, Veracruz.
1903. *Peromyscus cecilii* Thomas, Ann. Mag. Nat. Hist., ser. 7, 11:486, May, type from S slope Volcán de Orizaba, Puebla.
1904. *Peromyscus melanotis zamelas* Osgood, Proc. Biol. Soc. Washington, 17:59, March 21, type from Colonia García, Chihuahua.

External measurements: 132–175; 58–81; 17–22. Upper parts (winter-taken specimens) ochraceous tawny intermixed with dusky hairs, the

dusky hairs more numerous mid-dorsally; lateral line of ochraceous tawny fairly distinct; ears dusky brown with white edging; underparts and feet white; tail sharply bicolored, sooty brown above, white below. Summer pelage having dusky tones more pronounced. Skull resembling that of *P. maniculatus*, but with longer and slenderer rostrum and nasals; nasals more compressed posteriorly; braincase more rounded and inflated; auditory bullae relatively smaller.

Bowers (1974:720) identified as *P. melanotis* certain populations from southern Arizona previously identified as the species *P. maniculatus* by other systematists.

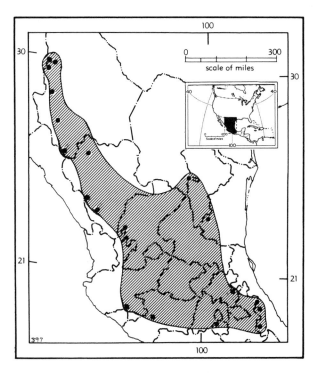

Map 397. *Peromyscus melanotis*.

MARGINAL RECORDS.—Chihuahua (Anderson, 1972:347, 348): Río Gavilán, 9 mi. SW Pacheco; *Water Canyon, 3 mi. S Colonia García, 7200 ft.; 9 mi. SE Colonia García, 8200 ft.; 10 mi. SE Colonia García;* 15 mi. SE Colonia García, 7500 ft. Durango: near Guanaceví. Coahuila: 12 mi. E San Antonio de las Alazanas, 9000 ft. Tamaulipas: Miquihana. Veracruz: 10 km. SW Jacales, 6500 ft. (Hall and Dalquest, 1963:302); 6 km. SSE Altotongo, 9000 ft.; 4 km. SE Las Vigas, 9500 ft. (Hall and Dalquest, 1963:302); Santa Bárbara Camp. Morelos: Huitzilac. Michoacán: Mt. Tancítaro, 12,000 ft. Jalisco: El Nevado del Colima, 9100 ft. (Baker and Phillips, 1965:692); *Sierra Nevada de Colima, 12,000 ft.* Zacatecas: Valparaíso Mts. Durango (Baker and Greer, 1962:117): 10 mi. SW El Salto; 1½ mi. W San Luis, 7750 ft. Chihuahua (Anderson, 1972:348): 10 mi. SW Guadalupe y Calvo; 2 mi. W Samachique, 7000 ft.; Yaguirachic, 130 mi. W Chihuahua; Chuhuichupa.

leucopus-group
Peromyscus leucopus
White-footed Mouse

External measurements: 156–205; 63–97; 19–24; 13–16. Upper parts variable, according to subspecies, from pale to rich reddish brown; feet and underparts white; tail bicolored but usually not sharply so, brownish above, whitish below; ears usually dusky, narrowly edged with whitish. Skull resembling that of *P. maniculatus.*

In areas where *P. maniculatus* and *P. leucopus* both occur, individual specimens are sometimes referred to the correct species only with considerable difficulty and identification is often a matter for the expert. Generally in such areas *P. leucopus* differs from *P. maniculatus* by larger size; longer, more sparsely haired, and less sharply bicolored tail; and larger hind feet. The habitat is chiefly wooded areas.

Peromyscus leucopus affinis (J. A. Allen)

1891. *Hesperomys (Vesperimus) affinis* J. A. Allen, Proc. U.S. Nat. Mus., 14:195, July 24, type from Barrio, Oaxaca. (See J. A. Allen and Chapman, Bull. Amer. Mus. Nat. Hist., 9:7, February 23, 1897.)
1909. *Peromyscus leucopus affinis,* Osgood, N. Amer. Fauna, 28:133, April 17.
1898. *Peromyscus musculoides* Merriam, Proc. Biol. Soc. Washington, 12:124, April 30, type from Cuicatlán, Oaxaca.

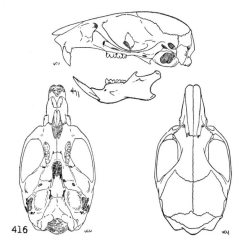

Fig. 416. *Peromyscus leucopus noveboracensis,* 4 mi. S Columbus, Cherokee Co., Kansas, No. 12360 K.U., ♂, X 1½.

MARGINAL RECORDS.—Veracruz: Pasa Nueva. Oaxaca: Santa Efigenia; Huilotepec; Cerro Marzorca (Goodwin, 1969:173); Cuicatlán; type locality.

Peromyscus leucopus ammodytes Bangs

1905. *Peromyscus leucopus ammodytes* Bangs, Proc. New England Zool. Club, 4:14, February 28, type from Monomoy Island, Barnstable Co., Massachusetts. Known only from Monomoy Island.

Peromyscus leucopus aridulus Osgood

1909. *Peromyscus leucopus aridulus* Osgood, N. Amer. Fauna, 28:122, April 17, type from Fort Custer, Big Horn Co., Montana.

MARGINAL RECORDS.—Saskatchewan: Mortlach (Beck, 1958:35). Minnesota: Browns Valley. Nebraska (Jones, 1964c:197, 198): 6 mi. N Midway; 1 mi. E Ravenna; *Logan Township;* 3 mi. S Red Cloud. Kansas (Fleharty and Stadel, 1968:232, 233): sec. 24, T. 15 S, R. 16 W; *3 mi. E Rush Center; 5½ mi. S, 4 mi. W Ness City;* 3 mi. N Alamota; *15½ mi. S, 1½ mi. W Gove;* Wallace County [1 mi. S, 2 mi. W Wallace, Fleharty, *in Litt.,* September 21, 1971]; 15 mi. N, 2 mi. W Bird City. Nebraska (Jones, 1964c:197, 198, unless otherwise noted): McCook; 1½ mi. S Brady; 4 mi. W Halsey; Hackberry Lake, Valentine National Wildlife Refuge; 10 mi. S Chadron; *Glen;* head Warbonnet Creek (Osgood, 1909:124). South Dakota: Custer. Wyoming (Long, 1965a:634): 2 mi. S Colony; *2 mi. N, 13 mi. W Hulett; 3 mi. S, 5 mi. E Rockypoint;* 3 mi. N, 3 mi. W Rockypoint. Montana (Hoffmann, *et al.,* 1969:591, as *P. leucopus* only, unless otherwise noted): *Lodge Grass;* Crow Agency (Hall and Kelson, 1959:629); 6 mi. NW Forsyth; 7 mi. NE Glendive. Alberta: Milk River; Eagle Butte.

Peromyscus leucopus arizonae (J. A. Allen)

1894. *Sitomys americanus arizonae* J. A. Allen, Bull. Amer. Mus. Nat. Hist., 6:321, November 7, type from Fairbank, Cochise Co., Arizona.
1909. *Peromyscus leucopus arizonae*, Osgood, N. Amer. Fauna, 28:126, April 17.

MARGINAL RECORDS.—New Mexico: Glenwood; Deming. Chihuahua (Anderson, 1972:344): Vado de Fusiles, 4000 ft.; 5 mi. N El Carmen; Cañón del Potrero, 7 mi. W El Sauz, 5750 ft.; 5 mi. N Chihuahua, 4700 ft.; *5 mi. SE Chihuahua, 5250 ft.*; jct. Río San Pedro and Río Conchos, 5 mi. N, 5 mi. E Meoqui, 3550 ft. Coahuila: 1 mi. SW San Pedro de las Colonias, 3700 ft. Durango: San Gabriel; Río Sestín (Baker and Greer, 1962:115, as *P. l. tornillo*); Rosario. Chihuahua (Anderson, 1972:344, 345): 2 mi. W Parral, 6200 ft.; 4 mi. NW San Francisco de Borja, 5700 ft.; 11 mi. NNW San Buenaventura; Río Piedras Verdes, 5 mi. WNW Colonia Juárez; 35 mi. NW Dublán, 5300 ft.; 5½ mi. N, 2 mi. W San Francisco. Sonora: San Bernardino Ranch; Santa Cruz; Sáric. Arizona: *8½ mi. N, 2½ mi. W Nogales, 3400 ft.*; Tumacacori Mission, near Tubac; Tucson; Calva, 2600 ft. (Cockrum, 1961:179); *York (ibid.)*. See Anderson, 1972:344, who refers all occurrences in Durango and one specimen from Coahuila to *P. l. arizonae*.

Peromyscus leucopus castaneus Osgood

1904. *Peromyscus texanus castaneus* Osgood, Proc. Biol. Soc. Washington, 17:58, March 21, type from Yohaltún, Campeche.
1909. *Peromyscus leucopus castaneus* Osgood, N. Amer. Fauna, 28:133, April 17.

MARGINAL RECORDS.—Yucatán: 2½ km. NW Dzitya (Birney, *et al.*, 1974:15); *4 km. E Dzitya (ibid.)*; Chichén-Itzá (Osgood, 1909:134, as *P. l. affinis* "approaching *castaneus*"). Quintana Roo: Santa Rosa. Campeche: type locality. Yucatán: 14 km. SW Muna (Birney, *et al.*, 1974:15).—Specimens that Osgood listed as *affinis* from Chichén-Itzá are here tentatively referred, perhaps incorrectly, to *castaneus* because Birney, *et al.* later referred all their specimens from the state of Yucatán to *P. l. castaneus*.

Peromyscus leucopus caudatus R. W. Smith

1939. *Peromyscus leucopus caudatus* R. W. Smith, Proc. Biol. Soc. Washington, 52:157, October 11, type from Wolfville, Kings Co., Nova Scotia.

MARGINAL RECORDS.—Nova Scotia: type locality; Newport; Halifax; South Milford; Digby.

Peromyscus leucopus cozumelae Merriam

1901. *Peromyscus cozumelae* Merriam, Proc. Biol. Soc. Washington, 14:103, July 19, type from Cozumel Island, Yucatán. Known only from type locality.
1909. *Peromyscus leucopus cozumelae*, Osgood, N. Amer. Fauna, 28:135, April 17.

Peromyscus leucopus easti Paradiso

1960. *Peromyscus leucopus easti* Paradiso, Proc. Biol. Soc. Washington, 73:21, August 10, type from 6⅜ mi. SE Pungo, Princess Anne Co., Virginia.

MARGINAL RECORDS (Paradiso, 1960:23).—Virginia: Virginia Beach; type locality.

Peromyscus leucopus fusus Bangs

1905. *Peromyscus leucopus fusus* Bangs, Proc. New England Zool. Club, 4:13, February 28, type from West Tisbury, Marthas Vineyard, Dukes Co., Massachusetts.

MARGINAL RECORDS (Waters, 1969:131).—Massachusetts: type locality; Nantucket Island.

Peromyscus leucopus incensus Goldman

1942. *Peromyscus leucopus incensus* Goldman, Proc. Biol. Soc. Washington, 55:157, October 17, type from Metlatoyuca, 800 ft., Puebla.

MARGINAL RECORDS.—Veracruz: Platón Sánchez, 800 ft. (Hall and Dalquest, 1963:304); San Andrés Tuxtla. Oaxaca (Goodwin, 1969:174): Nueva Raza; *San Juan Guichicovi*. Veracruz: Otatitlán. Puebla: type locality. Veracruz: *Chijal* (Hall and Dalquest, 1963:304).

Peromyscus leucopus lachiguiriensis Goodwin

1956. *Peromyscus leucopus lachiguiriensis* Goodwin, Amer. Mus. Novit., 1791:5, September 28, type from San José Lachiguirí, about 4000 ft., Oaxaca.

MARGINAL RECORDS.—Oaxaca: type locality; *San Juan Guivini* (Goodwin, 1969:174).

Peromyscus leucopus leucopus (Rafinesque)

1818. *Musculus leucopus* Rafinesque, Amer. Month. Mag., 3:446, October. Type locality, pine barrens of Kentucky.
1895. *Peromyscus leucopus*, Thomas, Ann. Mag. Nat. Hist., ser. 6, 15:192, February.
1939. *Peromyscus leucopus brevicaudus* Davis, Occas. Pap. Mus. Zool., Louisiana State Univ. Mus., 2:1, February 1, type from Huntsville, Walker Co., Texas. Regarded as inseparable from *P. l. leucopus* by McCarley, Texas Jour. Sci., 11:408, December 1959.

MARGINAL RECORDS.—Illinois: Flat Rock. Kentucky: vic. of Canmer; Mammoth. North Carolina: Highlands. Virginia: Dinwiddie County; Accomac County. North Carolina (Paul and Cordes, 1969:373): Currituck; Bertie County; Greenville; Atkinson. South Carolina: Calhoun County. Georgia: Richmond County (Golley, 1962:124). Alabama: Barachias; Greensboro. Mississippi (Kennedy, *et al.*, 1974:17): Wayne County; Harrison County. Louisiana (Lowery, 1974:241): 6 mi. SW Callender Naval Air Station; 1 mi. NE Montegut; Avery Island; 6 mi. SE Fields. Texas (Davis, 1966:188, as *P. leucopus* only, unless otherwise

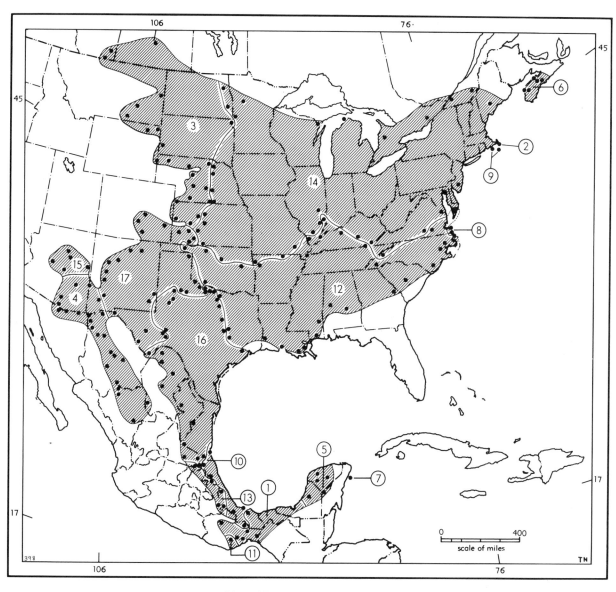

Map 398. *Peromyscus leucopus.*

Guide to subspecies	4. P. l. arizonae	9. P. l. fusus	14. P. l. noveboracensis
1. P. l. affinis	5. P. l. castaneus	10. P. l. incensus	15. P. l. ochraceus
2. P. l. ammodytes	6. P. l. caudatus	11. P. l. lachiguiriensis	16. P. l. texanus
3. P. l. aridulus	7. P. l. cozumelae	12. P. l. leucopus	17. P. l. tornillo
	8. P. l. easti	13. P. l. mesomelas	

noted): 6 mi. SE College Station; Robertson County; 3 mi. NE Leo (Hall and Kelson, 1959:630). Oklahoma: Chattanooga; 1½ mi. N Beaver; Alva; Fort Gibson. Arkansas: Huntsville. Tennessee: N end Reelfoot Lake. Illinois; Alto Pass; Shawneetown.

Peromyscus leucopus mesomelas Osgood

1904. *Peromyscus texanus mesomelas* Osgood, Proc. Biol. Soc. Washington, 17:57, March 21, type from Orizaba, Veracruz.

1909. *Peromyscus leucopus mesomelas* Osgood, N. Amer. Fauna, 28:132, April 17.

MARGINAL RECORDS.—San Luis Potosí: Huichihuayán; *3 km. W Axtla; 3 km. N Tamazunchale.* Veracruz: 7 km. W El Brinco, 800 ft. (Hall and Dalquest, 1963:304); 5 km. N Jalapa, 4500 ft.; 3 km. N Presidio, 1500 ft.; Río Blanco (Hall and Dalquest, 1963:304). San Luis Potosí: *Xilitla.*

Peromyscus leucopus noveboracensis (Fischer)

1829. [*Mus. sylvaticus*] δ *noveboracensis* Fischer, Synopsis mammalium, p. 318. Type locality, New York.
1897. *Peromyscus leucopus noveboracensis*, Miller, Proc. Boston Soc. Nat. Hist., 28:22, April 30.
1830. *Cricetus myoides* Gapper, Zool. Jour., 5:204, pl. 10. Type locality, between York and Lake Simcoe, Ontario.
1840. *Arvicola emmonsi* DeKay, *in* Emmons, Report on the quadrupeds of Massachusetts, p. 61. Based on animals from Massachusetts.
1841. *Peromyscus arboreus* Gloger, Gemeinnütziges Hand- und Hilfsbuch der Naturgeschichte, 1:95.
1842. *Mus michiganensis* Audubon and Bachman, Jour. Acad. Nat. Sci. Philadelphia, p. 304. Type locality, Erie Co., Michigan (= Ohio).
1853. *Hesperomys campestris* Le Conte, Proc. Acad. Nat. Sci. Philadelphia, 6:413. Type locality, New Jersey.
1901. *Peromyscus leucopus minnesotae* Mearns, Proc. Biol. Soc. Washington, 14:154, August 9, type from Fort Snelling, Hennepin Co., Minnesota.

MARGINAL RECORDS.—North Dakota: Manvel. Minnesota: vic. Lake Itasca. Michigan: Menominee County; Emmet County. Ontario: Mount Forest; Queen's University Biological Station, Chaffee's Locks (Sealander, 1961:57). Quebec: *along Ottawa River;* Mont St. Hilaire (Wrigley, 1969:208, as *P. leucopus* only); St. Sebastien (*ibid.*, as *P. leucopus* only). Maine: Oakland. Rhode Island: *Block Island* (Choate, 1972:216, as *P. leucopus* only). New Jersey: Wading River, thence southward along coast to Virginia: Richmond County; Luxemburg County. North Carolina: Weaverville. Kentucky: Black Mtn.; Eubank. Illinois: Charleston; Parkersburg; 6 mi. S Marion. Missouri: Williamsville. Arkansas: Hardy; 5 mi. SE Fayetteville. Oklahoma: Orlando. Kansas: 4 mi. SW Aetna; 3 mi. S, 1 mi. W Kingsley; Arkansas River, ½ mi. S, 1½ mi. W Ellinwood; *sec. 36, T. 13 S, R. 12 W* (Fleharty and Stadel, 1968:233); sec. 29, T. 9 S, R. 12 W (*ibid.*). Nebraska (Jones, 1964c:199): 2 mi. WSW Superior; 6 mi. S Grand Island; 1 mi. SW Neligh; 8 mi. SW Niobrara. Minnesota: S end Ten Mile Lake. North Dakota: Fargo.

Peromyscus leucopus ochraceus Osgood

1909. *Peromyscus leucopus ochraceus* Osgood, N. Amer. Fauna, 28:124, April 17, type from Winslow, Navajo Co., Arizona.

MARGINAL RECORDS.—Arizona (Cockrum, 1961:179): Turkey Tanks; type locality; St. Johns; *Springerville; N. Fork White River, 8200 ft.;* 10 mi. S Payson; Fort Whipple.

Peromyscus leucopus texanus (Woodhouse)

1853. *Hesperomys texana* Woodhouse, Proc. Acad. Nat. Sci. Philadelphia, 6:242, type probably from vic. Mason, Mason Co., Texas.
1909. *Peromyscus leucopus texanus*, Osgood, N. Amer. Fauna, 28:127, April 17.
1891. *Vesperimus mearnsii* J. A. Allen, Bull. Amer. Mus. Nat.

Hist., 3:300, June 30, type from Brownsville, Cameron Co., Texas.
1896. *Peromyscus canus* Mearns, Preliminary diagnoses of new mammals from the Mexican border of the United States, p. 3, March 25 (preprint of Proc. U.S. Nat. Mus., 18:445, May 23, 1896), type from Fort Clark, Kinney Co., Texas.

MARGINAL RECORDS.—Texas: Wichita County (Dalquest, 1968:18); Henrietta; Montague County (Dalquest, 1968:18); Decatur; Tarrant County (Davis, 1966:188, as *P. leucopus* only); 5 mi. N Waco; Milam County (Davis, 1966:188, as *P. leucopus* only); Dickinson Bayou, thence S along coast to Tamaulipas (Alvarez, 1963:443): Tampico; *2 mi. W Tampico.* San Luis Potosí: Ebano; *5 km. S Ebano;* Pujal; *vic. El Naranjo* (Brown and Welser, 1968:425, as *P. leucopus* only); 8 mi. W El Naranjo (Baker and Phillips, 1965:338); 1½ mi. E Río Verde; Presa de Guadalupe. Tamaulipas: Villa Mainero, 1700 ft. (Alvarez, 1963:442). Nuevo León: Santa Catarina. Coahuila (Baker, 1956:262): 6 mi. E Cuatro Ciénegas, 2200 ft.; Sabinas; 6 mi. N, 2 mi. E La Babia; 16 mi. N, 21 mi. E Piedra Blanca, 3200 ft. Texas: E base Burro Mesa, 3500 ft.; Fort Lancaster (Osgood, 1909:131); Dawson County (Davis, 1966:188, as *P. leucopus* only); 5 mi. N, 5 mi. E Draw (Garner, 1967:286, as *P. leucopus* only); Crosby County (Davis, 1966:188, as *P. leucopus* only).

Peromyscus leucopus tornillo Mearns

1896. *Peromyscus tornillo* Mearns, Preliminary diagnoses of new mammals from the Mexican border of the United States, p. 3, March 25 (preprint of Proc. U.S. Nat. Mus., 18:445, May 23, 1896), type from Rio Grande, about 6 mi. above El Paso, El Paso Co., Texas.
1909. *Peromyscus leucopus tornillo*, Osgood, N. Amer. Fauna, 28:125, April 17.
1903. *Peromyscus texanus flaccidus* J. A. Allen, Bull. Amer. Mus. Nat. Hist., 19:599, November 14, type from Río Sestín, Durango.

MARGINAL RECORDS.—Colorado: Broadmoor Golf Course (Armstrong, 1972:206); Lamar. Kansas (Fleharty and Stadel, 1968:232, 233, unless otherwise noted): *3 mi. N, 3 mi. W Syracuse;* 3 mi. N, 2 mi. E Lakin; *6 mi. N Kalvesta;* 10 mi. N, 8 mi. W Jetmore; 7 mi. S Kingsdown (Hall and Kelson, 1959:631); 11½ mi. E Meade (*ibid.*). Texas: Lipscomb. Oklahoma: 14 mi. S Olustee. Texas (Dalquest, 1968:18, unless otherwise noted): Wilbarger County; Archer County; King County; Dickens County. New Mexico (Findley, *et al.,* 1975:212–214, as *P. leucopus* only): 5 mi. N Tatum; 24 mi. E Loco Hills. Texas: lower Limpia Canyon, 15–16 mi. NE Fort Davis; Pecos County (Davis, 1966:188, as *P. leucopus* only); northeastern Terrell County; Presidio County (Davis, 1966:188, as *P. leucopus* only). Chihuahua (Anderson, 1972:345): Ciudad Juárez; *near Monument No. 1, Río Grande.* New Mexico (Findley, *et al.,* 1975:212–214, as *P. leucopus* only): Lake Valley; Datil; 16¼ mi. S, 17¼ mi. W Mush Mtn.; 2½ mi. S, 15 mi. W San Ysidro; Laguna; 4 mi. N El Rito; 5 mi. SW

Cimarron, Philmont Scout Ranch. Colorado: 1 mi. S, 2 mi. W Walsenburg (Armstrong, 1972:207); Cañon City.

Peromyscus gossypinus
Cotton Mouse

External measurements: 160–205; 68–97; 20–26; ear (dry), 15–16.5. Upper parts dark brown, often with strong fulvous or cinnamon wash; usually dusky admixture mid-dorsally; underparts white or creamy white; tail bicolored, brown above, whitish below, or uniformly brown; ears dusky brown, in some individuals faintly edged with white; feet white. Skull essentially similar to that of *P. leucopus* and *P. maniculatus*, but is usually larger than in sympatric or nearby subspecies of *P. leucopus* or *P. maniculatus;* molar teeth average larger than in *P. leucopus*. Useful criteria for distinguishing *P. gossypinus* from *P. leucopus* in southern Illinois have been listed by Hoffmeister (1977:222–224), and in Alabama by Linzey, *et al.* (1976:109–113).

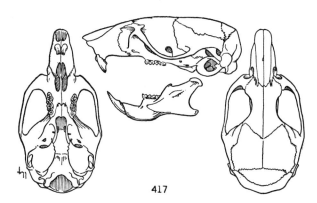

417

Fig. 417. *Peromyscus gossypinus gossypinus*, 2 mi. W Camp Cornelia (Okefinokee Ref. Headquarters), Ware Co., Georgia, No. 23360 K.U., ♀, X 1½.

Peromyscus gossypinus allapaticola Schwartz

1952. *Peromyscus gossypinus allapaticola* Schwartz, Jour. Mamm., 33:383, August 19, type from 12 mi. NE Rock Harbor, Key Largo, Monroe Co., Florida. Known only from Key Largo, until introduced on *Lignum Vitae Key* (Layne, 1974:393).

Peromyscus gossypinus anastasae Bangs

1898. *Peromyscus anastasae* Bangs, Proc. Boston Soc. Nat. Hist., 28:195, March, type from Point Romo, Anastasia Island, St. Johns Co., Florida.
1909. *Peromyscus gossypinus anastasae*, Osgood, N. Amer. Fauna, 28:141, April 17.
1898. *Peromyscus insulanus* Bangs, Proc. Boston Soc. Nat. Hist., 28:196, March, type from Cumberland Island, Camden Co., Georgia.

MARGINAL RECORDS.—Georgia: Cumberland Island. Florida: type locality.

Peromyscus gossypinus gossypinus (Le Conte)

1853. *Hesp[eromys]. gossypinus* Le Conte, Proc. Acad. Nat. Sci. Philadelphia, 6:411. Type locality, Georgia; probably the Le Conte Plantation, near Riceboro, Liberty Co. (See Bangs, Proc. Biol. Soc. Washington, 10:123, November 5, 1896.)
1896. *Peromyscus gossypinus*, Rhoads, Proc. Acad. Nat. Sci. Philadelphia, 58:189, April 21.
1831. *Hyp[udaeus]. gossipinus* Le Conte, *in* McMurtrie, The animal kingdom . . . by the Baron Cuvier . . . , 1 (App.):434, a *nomen nudum*.
1855. *Hesperomys cognatus* Le Conte, Proc. Acad. Nat. Sci. Philadelphia, p. 442. Type locality, Georgia.
1896. *Peromyscus gossypinus nigriculus* Bangs, Proc. Biol. Soc. Washington, 10:124, November 5, type from Burbridge, Plaquemines Parish, Louisiana.

MARGINAL RECORDS.—Virginia: just off State Route 10, N side Appomattox River, near Hopewell (Ulmer, 1963:273); Suffolk. North Carolina: Bertie. Florida: Summer Haven; Welaka; Gainesville; Whitfield. Mississippi: Bay St. Louis (Kennedy, *et al.,* 1974:17). Louisiana: Lake Charles; Le Compte; Tallulah. Alabama: Jackson. Georgia: Chattahoochie County (Golley, 1962:124); Butler. South Carolina (Golley, 1966:102): McCormick County; Lancaster County.—Lowery (1974:241–245) may be correct in referring specimens from northwestern Louisiana to *P. g. gossypinus* instead of to *P. g. megacephalus* as Osgood (1909:138, 139) did, but Lowery did not suggest a western boundary for *P. g. gossypinus*, and so left an unrealistic subspecific arrangement for more western animals (those from Texas). St. Romain (1976:79–88) has indicated a pattern of distribution for these two subspecies in Louisiana that is probably more nearly correct than is shown on Map 399, but he did not give a list of localities. Consequently, for the time being, Osgood's (*op. cit.*) classification is retained here.

Peromyscus gossypinus megacephalus (Rhoads)

1894. *Sitomys megacephalus* Rhoads, Proc. Acad. Nat. Sci. Philadelphia, 46:254, October, type from Woodville, Jackson Co., Alabama.
1909. *Peromyscus gossypinus megacephalus*, Osgood, N. Amer. Fauna, 28:138, April 17.
1896. *Peromyscus gossypinus mississippiensis* Rhoads, Proc. Acad. Nat. Sci. Philadelphia, 48:189, April 21, type from Samburg, Reelfoot Lake, Obion Co., Tennessee.

MARGINAL RECORDS.—Illinois: Ozark. Tennessee: Clarksville; Highcliff. Georgia (Golley, 1962:124): Dade County; Gordon County; Rabun County; Habersham County; Polk County. Alabama: Erin; Montgomery; Autaugaville. Louisiana: Waverly; Fishville; Leesville. Texas: Sour Lake; Grimes County (Davis, 1966:189, as *P. gossipinus* only); Brazos County (*ibid.,* as *P. gossypinus* only); Leon County (*ibid.,* as *P. gossypinus* only); Long Lake; northwestern Henderson

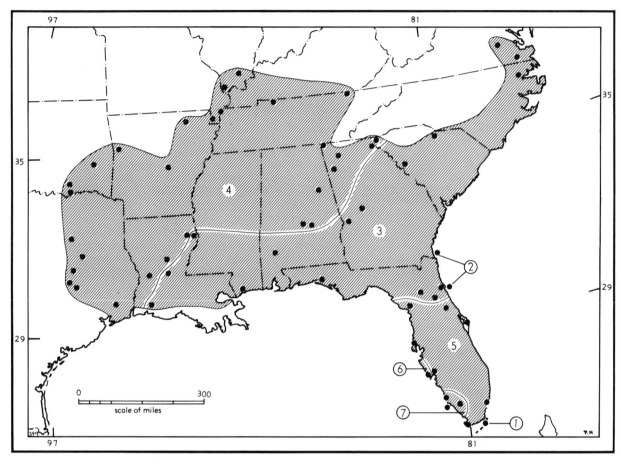

Map 399.   *Peromyscus gossypinus.*

1. *P. g. allapaticola*	3. *P. g. gossypinus*	6. *P. g. restrictus*
2. *P. g. anastasae*	4. *P. g. megacephalus*	7. *P. g. telmaphilus*
	5. *P. g. palmarius*	

County. Oklahoma: sec. 25, T. 7 S, R. 13 E (Taylor, 1965:640); sec. 2, T. 4 S, R. 12 E (Taylor, 1965:641); Red Oak. Arkansas: Barling; Fourche Bayou, near Little Rock; E side Black River, 3 mi. N Portia (Verts, 1965:100). Missouri: St. Francis River, W of Senath; Portageville. Illinois: Olive Branch (Hoffmeister, 1977:224). Not found: Missouri: Cushion Lake.

Peromyscus gossypinus palmarius Bangs

1896. *Peromyscus gossypinus palmarius* Bangs, Proc. Biol. Soc. Washington, 10:124, November 5, type from Oak Lodge, East Peninsula, opposite Micco, Brevard Co., Florida.

MARGINAL RECORDS.—Florida: Gulf Hammock; Glenwood; Cape Canaveral; Miami; *Goulds* (Layne, 1974:393); *Royal Palm Hammock* (*ibid.*); Flamingo (*ibid.*); *Chevalier Bay* (*ibid.*); Charlotte Harbor; Tarpon Springs.

Peromyscus gossypinus restrictus A. H. Howell

1939. *Peromyscus gossypinus restrictus* A. H. Howell, Jour. Mamm., 20:364, August 14, type from Chadwick Beach, near Englewood, Sarasota Co., Florida. Known only from type locality.

Peromyscus gossypinus telmaphilus Schwartz

1952. *Peromyscus gossypinus telmaphilus* Schwartz, Jour. Mamm., 33:384, August 19, type from Royal Palm Hammock, Collier Co., Florida.

MARGINAL RECORDS.—Florida: Naples; 8⅗ mi. E Monroe Station; Marco Island.

crinitus-group
Peromyscus crinitus
Canyon Mouse

Revised by Hall and Hoffmeister, Jour. Mamm., 23:51–56, February 14, 1942.

External measurements: 161–192; 82–118; 17.5–23; 15.3–21.5. Upper parts mixed ochraceous and brown or black, hairs plumbeous basally; underparts usually much paler, sometimes white. Tail longer than head and body (except in occasional individuals), bicolored (except in *delgadilli*); ear approx. as long as hind foot; mammae, i. ⅔. Skull with short tooth-row; width across anterior part of zygomatic arches less than greatest width of braincase; premaxillae not extending markedly posterior to nasals. M1 and M2 lack accessory cusps or folds except in occasional individuals.

Canyon mice are highly discontinuous in distribution because they are confined to rocky habitats. The nests are in crevices and clefts in rocks or, more commonly, in burrows under rocks. Altitudinally the species ranges from hot deserts below sea level up to cool heights of more than 10,000 ft.

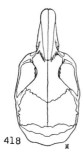

Fig. 418. *Peromyscus crinitus pergracilis*, 2 mi. W Smith Creek Cave, Nevada, No. 79179 M.V.Z., ♂, X 1½.

Peromyscus crinitus auripectus (J. A. Allen)

1893. *Sitomys auripectus* J. A. Allen, Bull. Amer. Mus. Nat. Hist., 5:75, April 28, type from Bluff City, San Juan Co., Utah.
1909. *Peromyscus crinitus auripectus*, Osgood, N. Amer. Fauna, 28:231, April 17.
1937. *Peromyscus crinitus peridoneus* Goldman, Jour. Mamm., 18:92, February 11, type from Bright Angel Trail, S side Grand Canyon, 4800 ft., Coconino Co., Arizona.

MARGINAL RECORDS.—Colorado (Armstrong, 1972:194): above Castle Park, Dinosaur National Monument; Lily; 8 mi. S, 4 mi. W Craig, 6400 ft.; 12 mi. above Glenwood Springs; Plateau Creek, 5 mi. E Tunnel; South Rim Headquarters, Black Canyon of the Gunnison National Monument, *ca.* 8000 ft.; Coventry, 6800 ft.; Balcony House, Mesa Verde National Park. New Mexico (Findley, *et al.*, 1975:200, as *P. crinitus* only): Los Pinos River, sec. 7, T. 30 N, R. 7 W; Chaco Canyon National Monument; 11 mi. N, ½ mi. E Prewitt. Arizona: Holbrook; 12 mi. N Deadman Flat, NE San Francisco Mtn.; *Supai Canyon*; Mile 156, N side Colorado River (Hoffmeister, 1971:173); *mouth Kanab Creek, Kaibab National Forst (ibid.)*; River Mile 21,

Glen Canyon, E bank (Durrant and Dean, 1959:90). Utah: Rainbow Bridge, 4000 ft.; White Canyon, River Mile 162, Glen Canyon, E bank (Durrant and Dean, 1959:90); 1 mi. E Hwy. 160, 6 mi. S Valley City, 4500 ft.; mouth Florence Canyon, 35 mi. N Greenriver, 4306 ft.

Peromyscus crinitus crinitus (Merriam)

1891. *Hesperomys crinitus* Merriam, N. Amer. Fauna, 5:53, July 30, type from Shoshone Falls, N side Snake River, Jerome Co., Idaho.
1899. *Peromyscus crinitus*, Bangs, *infra*.
1899. *Peromyscus crinitus scitulus* Bangs, Proc. New England Zool. Club, 1:67, July 31, type from Gardnerville, Douglas Co., Nevada.

MARGINAL RECORDS.—Oregon: Maupin; 3 mi. NE Huntington, 2100 ft. Idaho: 3 mi. NE Hammett; type locality; Salmon Creek, 8 mi. W Rogerson. Nevada: 18 mi. NE Iron Point, 4600 ft.; 4 mi. N Hot Creek, Hot Creek Range; *7 mi. W Tybo, Hot Creek Range, 6700–7200 ft.*; East Walker River, 2 mi. NW Morgans Ranch, 5050–5100 ft. California: 2 mi. NE Woodfords, 5600–5700 ft.; 1 mi. N Wendel; 5 mi. N Fredonyer Peak, 5700 ft.; Indian Well (Cave and Blue Grotto), Lava Beds National Monument.

Peromyscus crinitus delgadilli Benson

1940. *Peromyscus crinitus delgadilli* Benson, Proc. Biol. Soc. Washington, 53:1, February 16, type from 2 mi. S Crater Elegante, 34 mi. W Sonoita, Sierra del Pinacate, Sonora.

MARGINAL RECORDS.—Sonora: type locality; Río Sonoita, 30 mi. WSW Sonoita.

Peromyscus crinitus disparilis Goldman

1932. *Peromyscus crinitus disparilis* Goldman, Proc. Biol. Soc. Washington, 45:90, June 21, type from Tinajas Altas, 2000 ft., Gila Mts., Yuma Co., Arizona.
1940. *Peromyscus crinitus rupicolus* Benson, Proc. Biol. Soc. Washington, 53:2, February 16, type from Paso MacDougal, E end Sierra Hornaday, Sonora.
1940. *Peromyscus crinitus scopulorum* Benson, Proc. Biol. Soc. Washington, 53:2, February 16, type from Cerro La Cholla, 6 mi. WNW Punta Peñasca, Sonora.

MARGINAL RECORDS.—Arizona: 10 mi. W Wellton; 5 mi. S Wellton; Tule Tank. Sonora: Cerro La Cholla, 6 mi. WNW Punta Peñasca, 50 ft. Arizona: 20 mi. S Wellton.

Peromyscus crinitus doutti Goin

1944. *Peromyscus crinitus doutti* Goin, Jour. Mamm., 25:189, May 26, type from Antelope Canyon, 7200 ft., 20 mi. SE Duchesne, Duchesne Co., Utah.

MARGINAL RECORDS.—Wyoming: 4 mi. NE Linwood, Utah (Long, 1965a:627). Utah: Dinosaur Quarry, 6 mi. N Jensen, 5500 ft.; along Green River, 15

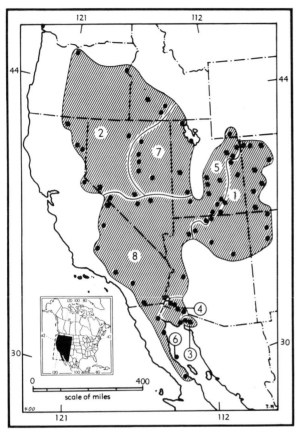

Map 400. *Peromyscus crinitus.*

1. *P. c. auripectus*	5. *P. c. doutti*
2. *P. c. crinitus*	6. *P. c. pallidissimus*
3. *P. c. delgadilli*	7. *P. c. pergracilis*
4. *P. c. disparilis*	8. *P. c. stephensi*

mi. (by airline) SW Ouray, 4500 ft.; 21 mi. out of San Rafael; Hite, Glen Canyon, W bank (Durrant and Dean, 1959:90); River Mile 137, Glen Canyon, W bank (*ibid.*); 8 mi. N Escalante; Sun View Forest Camp, Red Canyon, 3 mi. E Bicknell, 6900 ft.; Summit, E Sevier Co.; type locality. Wyoming: *1 mi. N Linwood* (Long, 1965a:627).

Peromyscus crinitus pallidissimus Huey

1931. *Peromyscus crinitus pallidissimus* Huey, Trans. San Diego Soc. Nat. Hist., 6:389, August 28, type from small island in Gonzaga Bay, 29° 50′ N, 114° 20′ W, Baja California. Known only from type locality. Differs as follows from *P. maniculatus hueyi*, which occurs on the same island: tail more instead of less than 90 per cent of length of head and body, upper parts paler, 2 instead of 3 pairs of mammae, and maxillary breadth inferred to be less than 11.

Peromyscus crinitus pergracilis Goldman

1939. *Peromyscus crinitus pergracilis* Goldman, Jour. Mamm., 20:356, August 14, type from S end Stansbury Island, 4250 ft., Great Salt Lake, Tooele Co., Utah.

MARGINAL RECORDS.—Utah: Kelton, 4225 ft.; type locality; White Valley; Beaver River, near Fort Cameron. Nevada: Water Canyon, 8 mi. N Lund; 8 mi. W Eureka; 4 mi. S Romano; Union; Elko.

Peromyscus crinitus stephensi Mearns

1897. *Peromyscus stephensi* Mearns, Proc. U.S. Nat. Mus., 19:721, July 30, type from 3 mi. E Mountain Spring, Imperial Co., California.
1909. *Peromyscus crinitus stephensi*, Osgood, N. Amer. Fauna, 28:232, April 17.
1904. *Peromyscus petraius* Elliot, Field Columb. Mus., Publ. 87, Zool. Ser., 3(14):244, January 7, type from Lone Pine, Inyo Co., California.

MARGINAL RECORDS.—Nevada: 14–15 mi. SW Sunnyside, White River Valley, 5800 ft. Utah: Parowan; *River Mile 113, Glen Canyon, W bank* (Durrant and Dean, 1959:90); River Mile 56, Glen Canyon, W bank (*ibid.*). Arizona: Lees Ferry, Glen Canyon, W bank (*ibid.*); lower end Toroweap Valley, 4200 ft.; Harquahala Mts., 3400 ft. California: Potholes, Colorado River. Baja California: El Mayor, Hardy River, 30 ft.; San Felipe; Bahía de los Angeles (Lawlor, 1971:20, as *P. crinitus* only). California: 1 mi. E Mountain Spring, 2000 ft.; Palm Springs, 452 ft.; Lovejoy Spring; mouth Kern River Canyon, 15 mi. NE Bakersfield; Bullfrog Lake, 10,600 ft.; White Mtn. (Dunmire, 1961:489, as *P. crinitus* only). Nevada: *Huntoon Valley, 5700 ft.*; Marietta, 4900 ft.; Old Mill, N end Reveille Valley, 6200 ft.

Peromyscus caniceps Burt
Burt's Deer Mouse

1932. *Peromyscus caniceps* Burt, Trans. San Diego Soc. Nat. Hist., 7:174, October 31, type from Monserrate Island, lat. 25° 38′ N, long. 111° 02′ W, Gulf of California, Baja California. Known only from type locality.

External measurements of type, an adult male, and the average of 16 topotypes are, respectively: 199, 202; 111, 112; 21, 22; 15, 15.5. Back and sides heavily washed with ochraceous-buff; head grayish, contrasting with back and sides; lateral line indistinct; underparts white washed with buff (in some specimens, "ochraceous-buff" of sides continuing ventrally, with no break in region of lateral line, and covering entire ventral surface except throat and chin); tail indistinctly bicolored, dusky above, whitish below; ears dusky. Skull more angular than in *pseudocrinitus;* zygomatic process of squamosal projects laterally noticeably beyond border of braincase, arches tapering anteriorly; 1st and 2nd upper molars with accessory tubercles in 13 out of 20 specimens; shelf of bony palate longer than maxillary tooth-row; auditory bullae small; nasals tapering slightly posteriorly (after Burt, 1932:174). See Map 389.

Peromyscus pseudocrinitus Burt
False Canyon Mouse

1932. *Peromyscus pseudocrinitus* Burt, Trans. San Diego Soc. Nat. Hist., 7:173, October 31, type from Coronados Island, lat. 26° 06′ N, long. 111° 18′ W, Gulf of California, Baja California. Known only from type locality.

Average external measurements of six specimens are: 194; 110; 21; 16. Darkest of the kinds of *Peromyscus* known from the islands in the Gulf of California; tail long, scantily haired, and proximal two-thirds distinctly bicolored. Upper parts plumbeous-black washed with cinnamon; underparts white. General outline of skull similar to that of *crinitus*, but skull larger and with relatively less inflated auditory bullae; nasals broad, parallel-sided, and bluntly rounded at posterior termination; premaxillae extend slightly behind nasals; accessory tubercles between outer primary tubercles in 1st and 2nd upper molars more prominent than in *crinitus*; shelf of bony palate shorter than maxillary tooth-row; interparietal divided in four of seven specimens (after Burt, 1932:173). See Map 389.

boylii-group
Peromyscus pectoralis
White-ankled Mouse

Revised by Schmidly, Southwestern Nat., 17:113–138, September 15, 1972.

External measurements: 185–219; 92–117; 19–23; 14.3–17.2; length of tail equals and usually exceeds length of head and body; ear shorter than hind foot; upper parts in north pale ochraceous buff or ochraceous buff overlaid with dusky, grading to much darker in south; underparts creamy white, some specimens having ochraceous buff pectoral suffusion; feet and ankles ordinarily white in northwest but ankles varying geographically to all dusky in southeast; soles of hind feet naked or with slight hairiness on heel; tail brown above, white to mottled with dusky below. Skull resembling that of *P. boylii* but differs (from *P. b. rowleyi* and especially *P. b. levipes*) as follows: smaller, braincase less vaulted, and upper tooth-row shorter.

In Texas differs from *P. attwateri* as follows: hind foot shorter than 23.5; upper parts paler; ankles white (not dusky); tail more sharply bicolored; nasals posteriorly truncate (not convex); upper tooth-rows and incisive foramina shorter.

Both Hooper (1952) and Schmidly (1972) concluded that *P. pectoralis* and *P. boylii* are two closely related species but could not discover any one cranial or external feature everywhere distinguishing the two. Hooper (1958:12, 13) noted that in *P. pectoralis* the distally slender baculum is two-thirds as long as the hind foot and that the tip ends in a long cartilaginous spine, whereas in *P. boylii* the baculum is half as long as the hind foot and has a small knob-shaped tip capped by a minute cone of cartilage. In a drawing, Clark (1953:190, Fig. 1) clearly showed the difference in shape of the terminal cartilage in *P. pectoralis* [= subspecies *laceianus*] and *P. boylii* [then the subspecies *attwateri*—now treated as a species closely allied to *P. boylii*].

In México south of the 25th parallel each of the three subspecies of *P. pectoralis* in several areas has dusky instead of white ankles, whereas farther north in México and in all of Texas the ankles of almost all specimens are white (Schmidly, 1972:124, Fig. 4).

Kilpatrick and Zimmerman (1975) suggested from their study of chromosomes and proteins "that the affinities of *P. pectoralis* may lie with other [than the *P. boylii* species group] forms of *Peromyscus*."

Peromyscus pectoralis collinus Hooper

1952. *Peromyscus pectoralis collinus* Hooper, Jour. Mamm., 33:372, August 19, type from San José, 2000 ft., Sierra San Carlos, 12 mi. NW San Carlos, Tamaulipas.

MARGINAL RECORDS (Schmidly, 1972:132).— Tamaulipas: Tamaulipeca; Marmolejo; Soto la Marina; 8 mi. S, 11 mi. W Piedras; *Rancho Acuña, 2650 ft.*; 8 km. NE Antiguo Morelos, 500 ft. San Luis Potosí: 2 mi. W Tamazunchale, 600 ft.; Hda. Capulín; Labor del Río. Guanajuato: 8 mi. S Ibarra, 8500 ft. Jalisco: 10 mi. NW Matanzas, 8000 ft. San Luis Potosí: Jesús María; 20 mi. W Antiguo Morelos, Tamaulipas. Tamaulipas: 14 mi. N, 6 mi. W Palmillas, 5500 ft.

Peromyscus pectoralis laceianus V. Bailey

1906. *Peromyscus pectoralis laceianus* V. Bailey, Proc. Biol. Soc. Washington, 19:57, May 1, type from Lacey Ranch, near Kerrville, Kerr Co., Texas.

MARGINAL RECORDS (Schmidly, 1972:134–137, unless otherwise noted).—Oklahoma: 4 mi. W Marietta (Kilpatrick and Caire, 1973:351). Texas: 10 mi. NNW Gainesville (Kilpatrick and Caire, 1973:351); 19 mi. NW Jacksboro; 9 mi. S Ranger; 3 mi. NW McNeil; Austin; Blanco River, 4 mi. W Kyle; San Antonio (Hall and Kelson, 1959:638). Coahuila: 6 mi. SW San Gerónimo; 9 mi. E Hermanas. Nuevo León: Ojo de Agua, 2½ mi. SW Sabinas Hidalgo; Río Ramos, 20 km. NE Montemorelos. Tamaulipas: Villagran; Ciudad Victoria; *20⁴/₅ mi. NE Juamave*; 35 km. SW Ciudad Victoria, Joya Verde. Nuevo León: 13 mi. N Matehuala, 1800 m. [Matehuala and 13 mi. due N thereof are in San Luis Potosí]. Zacatecas: Conception (sic) del Oro, 7680 ft. Coahuila: Parras; 22 mi. S, 5 mi. W Ocampo, 6000 ft.

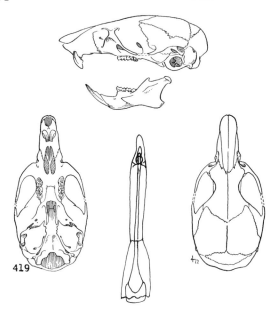

Fig. 419. *Peromyscus pectoralis laceianus*, 30 mi. SW Carlsbad, New Mexico, No. 7401 K.U., ♂, X 1½. [And baculum X approx. 3, ventral view, No. 101684 U.M.M.Z., as *P. pectoralis* only (after Hooper, 1958:Pl. 2).]

Chihuahua: 14 mi. SW Ciudad Camargo; *Chihuahua City*; 15 mi. NNW Chihuahua; 40 mi. E Gallego, 5000 ft. (Anderson, 1972:348). Texas: 11 mi. W Valentine; 7 mi. N Pine Springs, 5300 ft.; *1½ mi. N Nickel Creek*. New Mexico: 1½ mi. up Rattlesnake Canyon, 30 mi. SW Carlsbad; Carlsbad Caverns, 4 mi. W Whites City, 3700 ft. Texas: 11 mi. S, 2 mi. W Balmorhea; 7 mi. E Bakersfield; 15 mi. E Sheffield; Tennyson; 22 mi. SW Abilene; 5⅝ mi. N, 13½ mi. W Albany, 1500 ft.; 20 mi. SW Throckmorton (Packard and Judd, 1968:537).

Peromyscus pectoralis pectoralis Osgood

1904. *Peromyscus attwateri pectoralis* Osgood, Proc. Biol. Soc. Washington, 17:59, March 21, type from Jalpan, Querétaro.
1906. *Peromyscus pectoralis*, V. Bailey, Proc. Biol. Soc. Washington, 19:57, May 1.
1904. *Peromyscus attwateri eremicoides* Osgood, Proc. Biol. Soc. Washington, 17:60, March 21, type from Mapimí, Durango.

MARGINAL RECORDS (Schmidly, 1972:130).— Durango: 3 mi. E Las Nieves, 5400 ft.; 7 mi. N Campana, 3750 ft. Coahuila: 3 mi. SE Torreón; *Jimulco*. Tamaulipas: Miquihuana, 6200 ft.; Tula; *9 mi. SW Tula, 3900 ft.; 12 mi. SW Tula*. Aguascalientes: 1 km. S La Labor, 9 mi. N Calvillo. Jalisco: 9 mi. N, 2½ mi. E Encarnación de Díaz, 6650 ft. Querétaro: 3⁷⁄₁₀ mi. NW Jalpan, 2500 ft. Hidalgo: 3 mi. NE Jacala; San Agustín, 1100 m. Querétaro: Toliman, 1700 m. Jalisco: 3 mi. W Guadalajara. Nayarit: 1 mi. SW San José del Conde, 3000 ft. Durango: 15 mi. N Mezquital, 6700 ft.; 2 mi. S El Palmito, 4850 ft.; Indé.

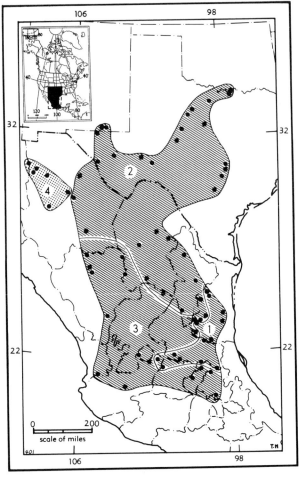

Map 401. *Peromyscus pectoralis* and *Peromyscus polius*.

1. *P. pectoralis collinus*
2. *P. pectoralis laceianus*
3. *P. pectoralis pectoralis*
4. *P. polius*

Peromyscus boylii
Brush Mouse

External measurements: 176–248; 90–132; 17–26; 14.7–19. Upper parts varying from dark, rich tawny or brownish to cinnamon, purest on sides; underparts white or creamy, pectoral spot buffy to ochraceous; tail long, well haired, somewhat penicillate, brownish above, white below; feet white. Size of skull highly variable among subspecies, rostrum depressed; zygomata narrowing anteriorly; auditory bullae less inflated than in *P. truei*. Plantar surface of hind foot haired.

See account of *P. pectoralis* on its resemblance to *P. boylii* and probable relationship of the two. Viewed dorsally postorbital constriction in skull of *P. boylii*, especially *P. b. levipes*, hourglass-shaped but in *P. aztecus* angular.

Peromyscus boylii ambiguus Alvarez

1961. *Peromyscus boylii ambiguus* Alvarez, Univ. Kansas Publ., Mus. Nat. Hist., 14:118, December 29, type from Monterrey, Nuevo León.

MARGINAL RECORDS (Alvarez, 1961:120).— Nuevo León: type locality; *Cerro de la Silla.* Tamaulipas: La Vegonia, Sierra San Carlos; 5 mi. S, 3 mi. W Victoria (Alvarez, 1963:443, as *P. b. levipes*); 2 mi. S, 10 mi. W Piedra (*ibid.*, as *P. b. levipes*); Acuña, Sierra de Tamaulipas (Schmidly, 1973:130); 5 mi. NW Gómez Farías, Rancho del Cielo, 3300 ft. (*ibid.*). San Luis Potosí: Villar (*ibid.*). Nuevo León: 17 mi. W, 9 mi. S Linares (*ibid.*). Coahuila: 12 km. E San Antonio de las Alazanas.

Peromyscus boylii beatae Thomas

1903. *Peromyscus beatae* Thomas, Ann. Mag. Nat. Hist., ser. 7, 11:485, May, type from Xometla Camp, Mt. Orizaba, Veracruz.
1961. *Peromyscus boylii beatae*, Alvarez, Univ. Kansas Publ., Mus. Nat. Hist., 14:116, December 29.

MARGINAL RECORDS (Alvarez, 1961:117, unless otherwise noted).—Veracruz: *1 km. E Jalancingo;* 5 km. E Las Vigas; *5 km. N Jalapa;* Orizaba (Hall and Dalquest, 1963:305); *Maltrata (ibid.); Xuchil; type locality; 10 km. SE Perote; 2 km. S Jalancingo.*

Peromyscus boylii boylii (Baird)

1855. *Hesperomys boylii* Baird, Proc. Acad. Nat. Sci. Philadelphia, 7:335, April, type from Middle Fork American River, near present town of Auburn, Eldorado Co., California.
1896. *Peromyscus boylii,* Mearns, Preliminary diagnoses of new mammals from the Mexican border of the United States, p. 3, May 25 (preprint of Proc. U.S. Nat. Mus., 19:139, December 21, 1896).
1893. *Sitomys robustus* J. A. Allen, Bull. Amer. Mus. Nat. Hist., 5:335, December 16, type from Lakeport, Lake Co., California.

MARGINAL RECORDS.—California: Beswick; Upper Lake; Lower Lake; Milford. Nevada: 3 mi. E (SE) Incline, 6250 ft., E shore Lake Tahoe. California:

Coarsegold; Dunlap; Scott Valley; Encinosa Creek, 3 mi. W Vacaville; Lakeport; Helena; Etna.

Peromyscus boylii cordillerae Dickey

1928. *Peromyscus boylii cordillerae* Dickey, Proc. Biol. Soc. Washington, 41:2, January 25, type from Mt. Cacaguatique, 3500 ft., Dept. San Miguel, El Salvador.

MARGINAL RECORDS (Burt and Stirton, 1961:56).—El Salvador: type locality; El Carmen. **See** addenda.

Peromyscus boylii glasselli Burt

1932. *Peromyscus boylii glasselli* Burt, Trans. San Diego Soc. Nat. Hist., 7:171, October 31, type from San Pedro Nolasco Island, lat. 27° 58′ N, long. 111° 24′ W, Gulf of California, Sonora. Known only from type locality.

Peromyscus boylii levipes Merriam

1898. *Peromyscus levipes* Merriam, Proc. Biol. Soc. Washington, 12:123, April 30, type from Mt. Malinche, 8400 ft., Tlaxcala.
1909. *Peromyscus boylei levipes,* Osgood, N. Amer. Fauna, 28:153, April 17.
1903. *Peromyscus sagax* Elliot, Field Columb. Mus., Publ. 71, Zool. Ser., 3(8):142, March 20, type from La Palma (not Pátzcuaro), Michoacán. (Regarded as inseparable from *levipes* by Hoffmeister, Jour. Mamm., 27:278, August 14, 1946).

MARGINAL RECORDS.—San Luis Potosí (Alvarez, 1961:116): 10 km. E Platanito; Xilitla. Veracruz: 3 km. N Zacualpan (*ibid.*); 6 km. SSE Altotonga (Hall and Dalquest, 1963:305). Tlaxcala: type locality. Oaxaca: Llano de los Flores (Goodwin, 1969:175); mts. near Ozolotepec. Chiapas: 10 km. NNW Jitotol, 1645 m. (Baker, *et al.*, 1973:78, 85); San Cristóbal; *6 mi. SE San Cristóbal* (5031 U. Minn.); *mts. near Comitán.* Guatemala: Hda. Chancol; Sacapulas; San Lucas; La Primavera; Zunil. Chiapas: Volcán de Tacaná. Oaxaca (Goodwin, 1969:175, unless otherwise noted): Río Mono Blanco; San Miguel [= San Miguel Caja de Agua]; *Lovene;* Río Molino (Musser, 1964:10); San Andrés Chicahuaxtla. Guerrero: mts. near Chilpancingo; *Cueva del Cañón del Zopilote, 13 km. S puente de Mexcala, 720 m.* (Ramírez-P. and Sánchez-H., 1974:107, 108). Michoacán: Pátzcuaro; Dos Aguas, 7000 ft. (Hooper, 1961:121). Jalisco: *Zapotlán;* El Nevado de Colima, 9100 ft. (Baker and Phillips, 1965:692); Ocotlán; 20 mi. SE Autlán, 8200 ft.; *6 mi. S Autlán* (31777 KU); 2 mi. W San Andrés. Guanajuato: vic. Ibarra, near León (Brown and Welser, 1968:425, as *P. boylii* only). San Luis Potosí: Jesús María (Schmidly, 1973:129). See Carleton (1977:40) for additional localities.

Peromyscus boylii madrensis Merriam

1898. *Peromyscus madrensis* Merriam, Proc. Biol. Soc. Washington, 12:16, January 27, type from María Madre Island, Tres Marías Islands, Nayarit.

1909. *Peromyscus boylei madrensis*, Osgood, N. Amer. Fauna, 28:152, April 17. According to Carleton's (Occas. Pap. Mus. Zool., Univ. Michigan, 675:37, 41, March 1, 1977) equivocal report on an unfinished study, *P. b. madrensis* may be a species instead of a subspecies.

MARGINAL RECORDS.—Nayarit: *Isla Juanito* (Carleton, 1977: 41); type locality; *María Magadalena Island; María Cleofa Island.*

Peromyscus boylii rowleyi (J. A. Allen)

1893. *Sitomys rowleyi* J. A. Allen, Bull. Amer. Mus. Nat. Hist., 5:76, April 28, type from Noland Ranch, N side San Juan River, San Juan Co., about 1½ mi. above Four Corners, Utah.
1896. *P[eromyscus]. b[oylii]. rowleyi*, Means, Preliminary diagnoses of new mammals from the Mexican border of the United States, p. 3, May 25 (preprint of Proc. U.S. Nat. Mus., 19:139, December 21, 1896).
1893. *Sitomys major* Rhoads, Amer. Nat., 27:831, September, type from Squirrel Inn, San Bernardino Co., California.
1893. *Sitomys rowleyi pinalis* Miller, Bull. Amer. Mus. Nat. Hist., 5:331, December 16, type from Granite Gap, Grant Co., New Mexico.
1903. *Peromyscus gaurus* Elliot, Field Columb. Mus., Publ. 74, Zool. Ser., 3(10):157, May 7, type from San Antonio, Sierra San Pedro Mártir, Baja California.
1904. *Peromyscus parasiticus* Elliot, Field Columb. Mus., Publ. 87, Zool. Ser., 3(14):244, January 7, type from Lone Pine, Inyo Co., California.
1904. *Peromyscus metallicola* Elliot, Field Columb. Mus., Publ. 87, Zool. Ser., 3(14):245, January 7, type from Providencia Mines, Sonora.

MARGINAL RECORDS.—Colorado (Armstrong, 1972:209): 1 mi. E Somerset, 6100 ft.; 3 mi. NE Cimarron, 7100 ft.; 2 mi. N Ridgeway, 7200 ft.; 1 mi. NW Dolores; 1 mi. N, 2 mi. W Juanita; Trinchera; 1 mi. S, 2 mi. W Walsenburg; *7 mi. SW Salida, 8300 ft.;* Salida; 10 mi. S Colorado Springs; Irwin's Ranch, T. 29 S, R. 52 W, 5200 ft.; Regnier. Texas: 20³⁄₁₀ mi. SW Canadian; 13 mi. E Canyon (Schmidly, 1973:128); 1 mi. S Post (*op. cit.:*124); The Bowl, 8000 ft.; 10 mi. S Toyahvale (Schmidly, 1973:128); 12 mi. E Alpine (*ibid.*). Coahuila: Tinaja de Chávez, 6400 ft., El Jardín Ranch, Sierra del Carmen. Chihuahua: 2 mi. SE Parral, 6300 ft. (Anderson, 1972:338). Durango: La Boquilla; San Gabriel. Aguascalientes: ½ mi. W Rincón de Romos. Durango: 28 mi. S, 17 mi. W Vicente Guerrero, 8350 ft. (Baker and Greer, 1962:112); 2 mi. N Pueblo Nuevo, 6000 ft. (*ibid.*); 18 mi. SW El Salto (*ibid.*); 1½ mi. W San Luis (*ibid.*); Arroyo de Bucy. Chihuahua (Anderson, 1972:338): *ca.* 10 mi. W Balleza; 4 mi. NW San Francisco be Borja, 5700 ft.; 12 mi. S Miñaca, 6900 ft.; La Polvosa, 6400 ft. Sonora: Providencia Mines; Sáric. Arizona (Cockrum, 1961:182–184): Weaver Camp, 4600 ft., 20 mi. NW Sasabe; Cuyama Valley, 9 mi. W Spanish Ranch [Chrystova], 1300 ft.; Harquahala Mts., 5000 ft.; Kingman. California: Providence Mts.; Mohave; San Gabriel Mts.; San Bernardino Mts.; San Jacinto Mts.; Mountain Spring. Baja California: Hanson Lagoon; San Matías Pass; Sierra San Pedro Mártir;

Rancho San Antonio. California: Dulzura; Santa Ana Mts.; Pozo; Seaside; Camp Badger; northern portion Panamint Mts. Nevada: Cold Creek, 6000 ft.; Ash Spring, Pahranagat Valley, 3800 ft. Utah: ¼ mi. S Enterprise Reservoir (Stock, 1970:431); Zion National Park (*ibid.*); River Mile 28, Glen Canyon, E bank (Durrant and Dean, 1959:92); River Mile 127, Glen Canyon, E bank (*ibid.*); mouth Nigger Bill Canyon, E side Colorado River, 4 mi. above Moab Bridge, 3995 ft. See Carleton, 1977:40, for additional localities.

Peromyscus boylii sacarensis Dickey

1928. *Peromyscus boylii sacarensis* Dickey, Proc. Biol. Soc. Washington, 41:3, January 25, type from San José del Sacare [= San José del Sacario of maps], 3600 ft., Chalatenango, El Salvador.

MARGINAL RECORDS.—Honduras: La Flor Archaga. El Salvador: type locality.

Peromyscus boylii simulus Osgood

1904. *Peromyscus spicilegus simulus* Osgood, Proc. Biol. Soc. Washington, 17:64, March 21, type from San Blas, Nayarit.
1909. *Peromyscus boylei simulus* Osgood, N. Amer. Fauna, 28:151, April 17.

MARGINAL RECORDS.—Sinaloa: 5 mi. S Copala, 750 ft. (Baker and Greer, 1962:113); *Chele;* Escuinapa. Nayarit: Navarrete; type locality. Sinaloa: Mazatlán. See Carleton, 1977:40, 41, for additional localities. According to Carleton's (1977:5, 41) equivocal report on an unfinished study, *Peromyscus boylii simulus* may be a species instead of a subspecies.

Peromyscus boylii spicilegus J. A. Allen

1897. *Peromyscus spicilegus* J. A. Allen, Bull. Amer. Mus. Nat. Hist., 9:50, March 15, type from Mineral San Sebastián, Mascota, Jalisco.
1909. *Peromyscus boylei spicilegus*, Osgood, N. Amer. Fauna, 28:149, April 17.

MARGINAL RECORDS.—Chihuahua (Anderson, 1972:338): Mojarachic, 6900 ft.; *N rim Barranca del Cobre, 23 mi. S, 1½ mi. E Creel, 7200 ft.;* bottom of Barranca del Cobre, 23 mi. S, 1½ mi. E Creel; Sierra Madre, 65 mi. E Batopilas. Zacatecas: Sierra Madre. Jalisco: La Laja. Zacatecas: Plateado. Jalisco: Etzatlán (Baker and Greer, 1962:113); 20 mi. SSE Autlán, 5500 and 6500 ft.; Estancia Jalisco. Colima: Hda. San Antonio. Jalisco: Talpa. Nayarit: Jalisco; Pedro Pablo. Sinaloa (Baker and Greer, 1962:113): Plomosas; 2 mi. SW Santa Lucía, 3750 ft. Durango (Baker and Greer, 1962:112): *Pueblo Nuevo, 5000 ft.;* Chacala. Chihuahua: *ca.* 10 mi. SW Guadalupe y Calvo (Anderson, 1972:338). Sinaloa: Sierra de Choix, 50 mi. NE Choix [probably in Sinaloa] (Baker and Greer, 1962:113, as *P. b. rowleyi*, but see Anderson, 1972:337, Fig. 326). Sonora: mts. near Alamos; Camoa. According to Carleton's (1977:5, 36, 41, 42) equivocal report on an unfinished study, *Peromyscus boylii spicilegus* may be

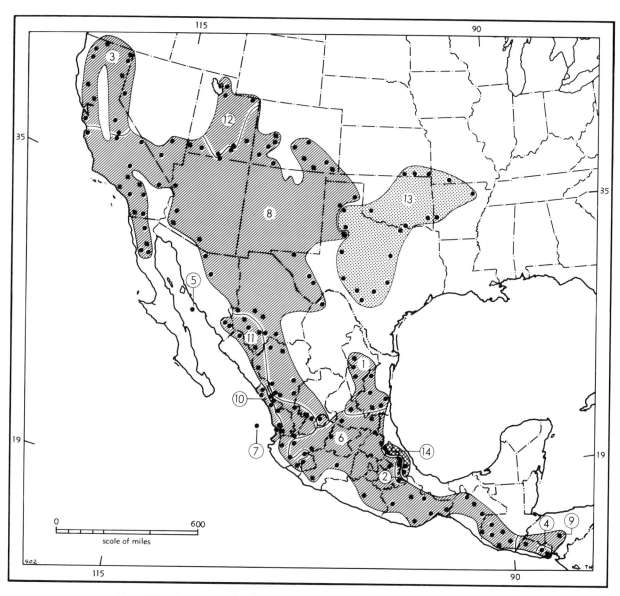

Map 402. *Peromyscus boylii*, *Peromyscus attwateri*, and *Peromyscus aztecus*.

Guide to kinds
1. *P. b. ambiguus*
2. *P. b. beatae*
3. *P. b. boylii*
4. *P. b. cordillerae*
5. *P. b. glasselli*
6. *P. b. levipes*
7. *P. b. madrensis*
8. *P. b. rowleyi*
9. *P. b. sacarensis*
10. *P. b. simulus*
11. *P. b. spicilegus*
12. *P. b. utahensis*
13. *P. attwateri*
14. *P. aztecus*

a species instead of a subspecies, may embrace an unnamed subspecies, and occurs in Michoacán. **See** addenda.

Peromyscus boylii utahensis Durrant

1946. *Peromyscus boylii utahensis* Durrant, Proc. Biol. Soc. Washington, 59:167, December 23, type from 5 mi. above lower power station, 5800 ft., Millcreek Canyon, Salt Lake Co., Utah.

MARGINAL RECORDS.—Utah: Ogden; Hideout [Canyon]; River Mile 148, Glen Canyon, W bank (Durrant and Dean, 1959:91); River Mile 34, Glen Canyon, W bank (*ibid.*); type locality.

Peromyscus attwateri J. A. Allen
Attwater's Mouse

Reviewed by Schmidly, Jour. Mamm., 54:111–130, April 26, 1973.

1895. *Peromyscus attwateri* J. A. Allen, Bull. Amer. Mus. Nat. Hist., 7:330, November 8, type from Turtle Creek, Kerr Co., Texas.

1896. *Peromyscus bellus* Bangs, Proc. Biol. Soc. Washington, 10:137, December 28, type from Stilwell, Adair Co., Oklahoma.

1905. *Peromyscus boylei laceyi* V. Bailey, N. Amer. Fauna, 25:99, October 24, type from Turtle Creek, Kerr Co., Texas.

1961. *Peromyscus boylii cansensis* Long, Univ. Kansas Publ., Mus. Nat. Hist., 14:101, December 29, type from 4 mi. E Sedan, Chautauqua Co., Kansas. (Inseparable from *P. b. attwateri* according to Choate, Phillips, and Genoways, Trans. Kansas Acad. Sci., 69:312, April 25, 1967.)

External measurements: 187–218; 96–112; 23–27; 15.4–17.3. Upper parts near Sayal Brown, darker and mixed with blackish along median dorsal area; sides Pinkish Cinnamon; fur on underparts pure white distally, but plumbeous basally. Skull resembling that of *P. boylii*, but averaging larger, anterior tip of nasals narrow instead of bulbous, infraorbital canal wider. Upper and lower molars alike in that M1 and M2 each has a mesoloph and m1 and m2 each has a metalophid. Hind foot longer than in *P. boylii*. Baculum closely resembling that of *boylii*. See Map 402.

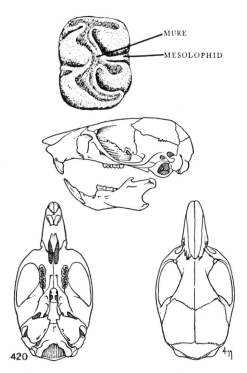

Fig. 420. *Peromyscus attwateri*, Ozark, Franklin Co., Arkansas, No. 10114 K.U., ♂, X 1½. And m2 X approx. 17 (after Schmidly, 1973:122, not No. 10114).

MARGINAL RECORDS (Schmidly, 1973:126, 127, unless otherwise noted).—Kansas: 4 mi. E Sedan; Schimmerhorn Park, Shoal Creek, 1 mi. S Galena. Missouri: *White and Elk river drainages* (Brown, 1964:196); 4 mi. N Branson. Arkansas: Batesville; 6 mi. N, 3 mi. E Mena. Oklahoma: 2 mi. W Smithville (Hall and Kelson, 1959:633); sec. 8, T. 8 S, R. 5 E (Taylor, 1965:641). Texas: 11 mi. NE Gainesville (Hall and Kelson, 1959:633); 7 mi. N Iredell; 4 mi. SW Austin; Boerne; Montell; 7 mi. W Rock Springs; Ozona; Big Spring; 10 mi. S Crosbyton; *3¹/₂ mi. W Post;* 20 mi. S, 5 mi. W Claude. Oklahoma: Quartz Mountain Park, Greer Co. Kansas: 5 mi. S, 2.5 mi. E Dexter.—On the basis of information provided by Diersing and Hoffmeister (1974:213), Schmidly (verbal information, February 11, 1975) regarded the specimen recorded by him (1973:127) from *Monahans Sand Hills, 4 mi. NE Monahans, Ward Co., Texas,* as probably incorrectly labeled as to locality. It likely was captured in the *Wichita Mts., Oklahoma.*

Peromyscus polius Osgood
Chihuahuan Mouse

1904. *Peromyscus polius* Osgood, Proc. Biol. Soc. Washington, 17:61, March 21, type from Colonia García, Chihuahua.

External measurements: 210–234; 111–120; 25–26; ear (dry), 17.2–18.5. Upper parts pinkish buff intermixed with dusky producing a dark brown overall tone; lateral line narrow, pinkish buff; ears grayish dusky, narrowly edged with grayish white; underparts white; feet, wrists, and ankles white; tail bicolored, brown above, whitish below. Skull "similar in general form to that of [*P. boylii*] *rowleyi*, but decidedly larger; molar teeth decidedly larger; palatine slits longer; audital bullae actually about same size, relatively smaller" (Osgood, 1909:178).

Osgood (*op. cit.*) placed *P. polius* in the *truei*-group. Hoffmeister (1951:22) tentatively placed *polius* in the *boylii*-group. Anderson (1972:349) noted that "detailed ecological study in the field and more detailed comparative studies of morphology are needed to clarify the relationships of *P. polius*." Kilpatrick and Zimmerman (1975: 143–162), from a study of chromosomes and proteins, favored placing *P. polius* in the *boylii*-group. See Map 401.

MARGINAL RECORDS (Anderson, 1972:349).— Chihuahua: 8 mi. W Altamirano; 10 mi. NE Colonia García, 6400 ft.; crest of Sierra del Arco, 10 mi. WSW San Buenaventura, 8250 ft.; Cañón del Potrero, 7 mi. W El Sauz, 5750 ft.; 2 mi. W Miñaca, 6900 ft.; Water Canyon, 3 mi. S Colonia García, 7200 ft.; *type locality.*

Peromyscus stephani Townsend
San Esteban Island Mouse

1912. *Peromyscus stephani* Townsend, Bull. Amer. Mus. Nat. Hist., 31:126, June 14, type from San Esteban Island, Gulf of California, Baja California. Known only from type locality.

Average external measurements of four specimens are: 195; 97; 22. In color close to typical *eremicus;* tail averaging shorter and hind foot larger. Skull decidedly shorter than in *eremicus*, dentition about the same; nasals more pointed posteriorly and reaching beyond premaxillae; frontals meet posteriorly at an angle on median line, instead of forming a curve as in *eremicus* (after Townsend, 1912:126).

Hooper and Musser (1964:12) tentatively placed this species in the *boylii*-group of the subgenus *Peromyscus* instead of in the subgenus *Haplomylomys* of which *eremicus* is the type species, and with which Townsend (1912:126) compared his material of *P. stephani*. See Map 390.

Peromyscus evides Osgood
Osgood's Deer Mouse

1904. *Peromyscus spicilegus evides* Osgood, Proc. Biol. Soc. Washington, 17:64, March 21, type from Juquila, Oaxaca.
1964. *Peromyscus evides*, Musser, Occas. Pap. Mus. Zool., Univ. Michigan, 636:9, June 17.

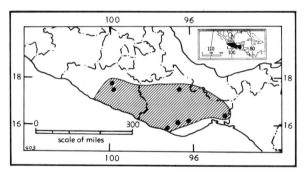

Map 403. *Peromyscus evides.*

1955. *Peromyscus hylocetes yautepecus* Goodwin, Amer. Mus. Novit., 1732:4, June 10, type from Santo Tomás Teipan, 7000 ft., 12 km. S San Bartolo Yautepec, Oaxaca.

External measurements: 171–230; 90–122; 20–26; 17–20. Upper parts rich tawny, dusky-tipped hairs forming mid-dorsal stripe; lateral line faint; black or nearly black orbital ring; soft, black hairs at anterior bases of ears; underparts creamy white, usually with tawny pectoral spot; tail blackish above, white below; forearm sooty to wrist, forefoot white; tarsal joint and proximal half of hind foot usually dusky except on sides.

Skull and teeth resembling those parts of *P. boylii spicilegus* but larger; supraorbital border shelf-like. For identifying specimens of *P. evides*, compare the account immediately above with those of *P. truei* and *P. difficilis*. See addenda.

MARGINAL RECORDS.—Guerrero: 12 mi. WSW Xochipala, 8200 ft. (Carleton, 1977:40). Oaxaca (Goodwin, 1969:175, unless otherwise noted): San Isidro Comaltepec; *Sierra Madre N of Zanatepec;* Cerro Baúl; Santo Tomás Teipan; Ríc Guajolote (Schaldach, 1966:296, as *P. hylocaetes* [sic] ssp., but see Goodwin, 1969:175, for assignment to *P. evides*); 10 mi. N Puerto Escondido (Carleton, 1977:40). Guerrero: 1 mi. SW Omilteme, 7260 ft. (*ibid.*).

Peromyscus aztecus (Saussure)
Aztec Mouse

1860. *H[esperomys]. aztecus* Saussure, Revue et Mag. Zool., Paris, ser. 2, 12:105, type probably from vic. Mirador, Veracruz, according to Osgood (N. Amer. Fauna, 28:156, 157, April 17, 1909).
1961. *Peromyscus aztecus*, Alvarez, Univ. Kansas Publ., Mus. Nat. Hist., 14:113, December 29.

External measurements: 215–238; 107–121; 24–26. Dorsal coloration near Sayal Brown; sides reddish; underparts Light Buff. Tail bicolored and about as long as head and body. Supraorbital border of skull angular, and bullae pointed anteriorly; anterior half of braincase nearly straight as viewed from above, thus making postorbital constriction angular instead of hourglass-shaped as in *P. boylii*. See Map 402. See addenda.

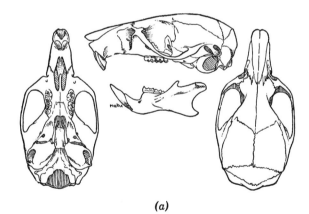

(a)

Fig. 421a. *Peromyscus aztecus*, Teocelo, 4500 ft., Veracruz, No. 30481 K.U., ♂, X 1½ [incorrectly identified as *Peromyscus simulatus* in Hall and Kelson, 1959:Fig. 359 on p. 647].

MARGINAL RECORDS.—Hidalgo: El Potrero, 13 mi. NE Metepec, 6600 ft. (Musser, 1964:2, 9). Puebla: Huachinango (Alvarez, 1961:115). Veracruz: *Jalapa, 5000 ft.* (Carleton, 1977:40); Teocelo (Musser, 1969:21); 2⅔ mi. SW Huatusco (Carleton, 1977:40).

Peromyscus oaxacensis Merriam
Oaxacan Deer Mouse

1898. *Peromyscus oaxacensis* Merriam, Proc. Biol. Soc. Washington, 12:122, April 30, type from Cerro San Felipe, 10,000 ft., Oaxaca.
1941. *Peromyscus hondurensis* Goodwin, Amer. Mus. Novit., 1121:1, June 9, type from Muya, a hill about 5 mi. N Chinacla, between 3000–4000 ft., La Paz, Honduras.

External measurements: 213–260; 102–135; 24–29; ear (dry), 15.8–17.5. Upper parts rich tawny to cinnamon brown, darkest along mid-dorsal line with admixture of black hairs; underparts white to creamy white; orbital ring dusky to black; tail bicolored, dusky to blackish above, white to whitish below. Skull resembling those of neighboring subspecies of *P. boylii*, but averaging larger and more robust; cheek-teeth averaging larger; relatively, auditory bullae same size as in *P. boylii;* palatine foramina averaging larger. **See** addenda.

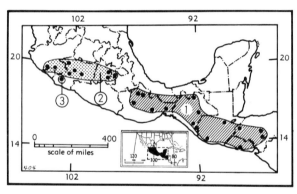

Map 404.　Three species of *Peromyscus*.

Guide to species
1. *P. oaxacensis*
2. *P. hylocetes*
3. *P. winkelmanni*

MARGINAL RECORDS.—Oaxaca: [Pápalo Santos] Reyes; San Isidro (Carleton, 1977:41). Chiapas: Triunfo, 1950 m.; Comitán Valley. Guatemala: Finca Concepción (Musser, 1969:7). Honduras (Musser, 1969:8): Cerro Pucca; Cantoral; Sabana Grande. El Salvador: Volcán de San Vicente (Burt and Stirton, 1961:57). Guatemala: La Primavera (Musser, 1968:7). Chiapas: Pinabete; Prusia, 1160 m. Oaxaca: Cerro Atravesado, 4000 ft.; San Pedro Jilotepec, 3000 ft. (Carleton, 1977:41); 15 mi. W Oaxaca.

Peromyscus winkelmanni Carleton
Forest Mouse

1977. *Peromyscus winkelmanni* Carleton, Occas. Pap. Mus. Zool., Univ. Michigan, 675:2, March 1, type from 6³⁄₁₀ mi. (by road) WSW Dos Aguas, 8000 ft., Michoacán.

External measurements of 12 specimens are: 235–265; 120–140; 27–29. Length of skull, 31.2–

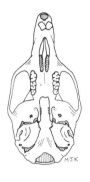 

(b)

Fig. 421b. *Peromyscus winkelmanni*, 6³⁄₁₀ mi. (by road) WSW Dos Aguas, 8000 ft., Michoacán, No. 110585 U.M.M.Z., holotype, ♂, X 1½ (after Carleton, 1977:3).

33.9; zygomatic breadth, 15.4–17.1; length of maxillary tooth-row, 5.1–5.6. Dorsum of a wet-season specimen tawny mixed with black; black hairs predominate on mid-dorsum; cheeks, sides, and flanks bright tawny, almost cinnamon, and transition from color of dorsum to venter abrupt; hair of underparts dark gray tipped with white; on hind foot dusky hair covering about half of metatarsal region and remainder white; tail thinly haired, slightly darker above than below, not sharply bicolored. Skull resembles those of *P. hylocetes* and *P. oaxacensis* but larger and with a distinct interorbital bead instead of a sharply angled or a shelflike interorbital region; bead weaker than in *P. banderanus* and *P. megalops;* accessory lophs and styles in upper and lower molars of most specimens. Glans penis broad with fluted surface and lacking dorsal and ventral lappets; baculum a simple rod capped with cone of cartilage usually less than 0.3 mm. long. Three pairs of mammae: 1 pair axillary and 2 inguinal (after Carleton, 1977:2, 3). See Map 404.

MARGINAL RECORDS (Carleton, 1977:2).—Michoacán: 2¹⁄₂ mi. SE *Dos Aguas;* type locality; 8²⁄₅ mi. *(by road) WSW Dos Aguas.*

Peromyscus hylocetes Merriam
Southern Wood Mouse

1898. *Peromyscus hylocetes* Merriam, Proc. Biol. Soc. Washington, 12:124, April 30, type from Pátzcuaro, 8000 ft., Michoacán.

External measurements: 220–238; 106–117; 25–27; ear (dry), 17.5–18.5. Upper parts pale ochraceous buff heavily mixed with dusky mid-dorsally, becoming tawny on sides; lateral line narrow, tawny; orbital ring and spot at base of whiskers sharply defined, black; underparts creamy with slaty undercoloring showing through; feet chiefly white, dusky extending to

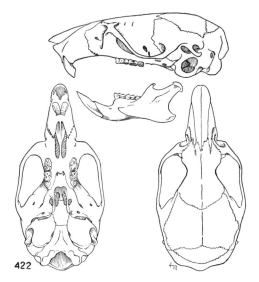

Fig. 422. *Peromyscus hylocetes,* Cerro Curitzaran, 3½ km. NNW San Juan, 2220 m., Michoacán, No. 28248 K.U., ♀, X 1½.

carpal joint and over tarsal joint; tail sharply bicolored, blackish above, white below. Skull essentially as in *P. boylii,* but with supraorbital border sharply angled.

This species is distinguished from its nearest relatives by the combination of blackish dorsum and sharply angled supraorbital border. **See addenda.**

MARGINAL RECORDS.—Michoacán: Patambán; *Cerro San Andreas, 10 mi. NW Hidalgo, 9400 ft.* (Carleton, 1977:41). Distrito Federal: 2 km. NW México, 3080 m. México: Amecameca. Morelos: Huitzilac. Michoacán: 9 mi. SE Pátzcuaro, 8000 ft.; Cerro Curitzaran, 3½ km. NNW San Juan, 2220 m.; Tancítaro, 10,000 ft. Jalisco: El Nevado de Colima, 9100 ft. (Baker and Phillips, 1965:692); 20 mi. SSE Autlán, 6500 ft.; *20 mi. SE Autlán, 8200 ft.; Sierra de Autlán, 7600 and 9000 ft.* (Carleton, 1977:41).

truei-group
Peromyscus truei
Piñon Mouse

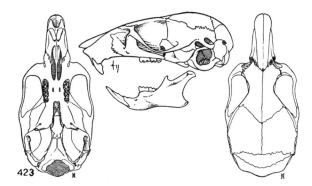

Fig. 423. *Peromyscus truei nevadensis,* ½ mi. W Debbs Creek, Nevada, No. 68476 M.V.Z., ♂, X 1½.

Subspecies reviewed by Hoffmeister, Illinois Biol. Monog., 21(4):ix + 104 pp., 5 pls., 24 figs. in text, November 12, 1951.

External measurements: 171–231; 76–123; 20–27; ear (dry), 18–25.9. Upper parts approx. grayish brown, but varying much geographically; underparts white or whitish; lateral line usually distinct; feet white; tail approx. as long as head and body, distinctly bicolored, brownish, or dusky above, whitish below. Ears large (usually longer than hind foot); fur long, silky. Skull medium in size for the genus; tooth-row usually 4.2–4.5; tympanic bullae large and well inflated; braincase vaulted; zygomata slightly convergent anteriorly. See accounts of *P. evides* and *P. difficilis* as an aid in identifying *P. truei.*

Peromyscus truei is found most often in rocky situations among pygmy conifers, especially in arid or semiarid regions. Occasionally it occurs at higher elevations.

Peromyscus truei chlorus Hoffmeister

1941. *Peromyscus truei chlorus* Hoffmeister, Proc. Biol. Soc. Washington, 54:131, September 30, type from Lost Horse Mine, S end Little San Bernardino Mts., 69 mi. E Riverside, Riverside Co., California.

MARGINAL RECORDS.—California: Hesperia, 2000 ft.; Barton Flats, 6400 ft.; Joshua Tree National Monument, 1½ mi. N Lost Horse Well, 4000 ft.; type locality; Santa Rosa Peak, 7500 ft.; Kenworthy, 4500 ft.; *Strawberry Valley, San Jacinto Mts., 6000 ft.;* Santa Ana River, 5500 ft.

Peromyscus truei comanche Blair

1943. *Peromyscus comanche* Blair, Contrib. Lab. Vert. Zool., Univ. Michigan, 24:7, July, type from Tule Canyon, Briscoe Co., Texas.
1973. *Peromyscus truei comanche,* Schmidly, Southwestern Nat., 18:276, October 5.

MARGINAL RECORDS (Packard and Judd, 1968:537, unless otherwise noted).—Texas: 17 mi. SE

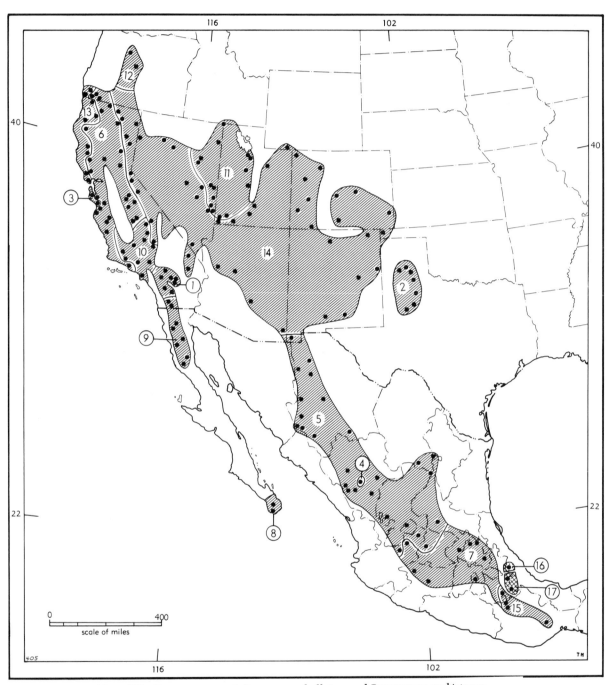

Map. 405.   *Peromyscus truei, Peromyscus bullatus,* and *Peromyscus mekisturus.*

Guide to kinds
1. *P. truei chlorus*
2. *P. truei comanche*
3. *P. truei dyselius*
4. *P. truei erasmus*
5. *P. truei gentilis*
6. *P. truei gilberti*
7. *P. truei gratus*
8. *P. truei lagunae*
9. *P. truei martirensis*
10. *P. truei montipinoris*
11. *P. truei nevadensis*
12. *P. truei preblei*
13. *P. truei sequoiensis*
14. *P. truei truei*
15. *P. truei zapotecae*
16. *P. bullatus*
17. *P. mekisturus*

Washburn; type locality; Quitaque (Schmidly, 1973:277); 1½ mi. E Dickens; 6 mi. E Justiceburg; 5 mi. N, 5 mi. E Draw (Garner, 1967:286, to species only); 21 mi. S Amarillo (Schmidly, 1973:277); *Palo Duro Canyon.*—In here placing *P. comanche* as a subspecies of *P. truei,* account has been taken of the evidence assembled by Johnson and Packard (1974:14), which led them to give *P. comanche* specific rank.

Peromyscus truei dyselius Elliot

1898. *Peromyscus dyselius* Elliot, Field Columb. Mus., Publ. 27, Zool. Ser., 1(10):207, April 16, type from Portola, San Mateo Co., California.
1951. *Peromyscus truei dyselius,* Hoffmeister, Illinois Biol. Monog., 21(4):63, November 12.

MARGINAL RECORDS.—California: Palo Alto: Berglund Ranch, 1450 ft., 5 mi. N Corralitos.

Peromyscus truei erasmus Finley

1952. *Peromyscus truei erasmus* Finley, Univ. Kansas Publ., Mus. Nat. Hist., 5:265, May 23, type from 8 mi. NE Durango, 6200 ft., Durango. Known only from type locality. Baker and Greer (1962:119) doubt that *P. t. erasmus* should be separated from *P. t. gentilis.*

Peromyscus truei gentilis Osgood

1904. *Peromyscus gratus gentilis* Osgood, Proc. Biol. Soc. Washington, 17:61, March 21, type from Lagos, Jalisco.
1909. *Peromyscus truei gentilis* Osgood, N. Amer. Fauna, 28:175, April 17.

MARGINAL RECORDS.—New Mexico: 1 mi. SW Aspen Springs, Animas Mts. (Findley, *et al.,* 1975:221, as *P. truei* only). Chihuahua: Casas Grandes; 9 mi. WSW San Buenaventura, 7450 ft. (Anderson, 1972:350); 8 mi. NE Laguna (*ibid.*); Parral. Durango: 26 mi. SW Yerbanís, 6725 ft. (Baker and Greer, 1962:119). Coahuila: Sierra Guadalupe (Baker, 1956:273); 12 mi. S, 2 mi. E Arteaga, 7500 ft.; Sierra Encarnación. San Luis Potosí: 1 km. N Arenal. Guanajuato: Silao. Jalisco: type locality; 7 mi. NNW Tepatitlán. Aguascalientes: ½ mi. W Rincón de Romos. Zacatecas: Valparaíso. Durango (Baker and Greer, 1962:119): 28 mi. S, 17 mi. W Vicente Guerrero, 8350 ft.; 30 mi. E El Salto; 3 mi. E Las Adjuntas; 1½ mi. W San Luis; 10 mi. SE Santiago Papasquiaro, 7100 ft. Chihuahua (Anderson, 1972:350): 103 mi. (by road) W Parral; 7 mi. SE Cerocahui; *Cerocahui;* 3 mi. NE Temoris, 5600 ft.; Mojárachic; Yaguirachic, 130 mi. W Chihuahua, 8500 ft.; 9 mi. SW Pacheco; *3 mi. SW Pacheco.*

Peromyscus truei gilberti (J. A. Allen)

1893. *Sitomys gilberti* J. A. Allen, Bull. Amer. Mus. Nat. Hist., 5:188, August 18, type from Bear Valley, San Benito Co., California.
1909. *Peromyscus truei gilberti,* Osgood, N. Amer. Fauna, 28:169, April 17.

MARGINAL RECORDS.—Oregon: Sams Valley; Ashland. California: East Ridge, Beswick; Montgomery Creek; Quincy; Middle Fork, American River, 1000 ft.; Dudley, 3000 ft.; Bass Lake; ½ mi. E Miramonte, 3500 ft.; Canyon Creek, 7 mi. E Orosi; Raymond, 940 ft.; 1 mi. W Coulterville, 1600 ft.; Marysville Buttes, 3–4 mi. NW Sutter, 300–500 ft.; 10 mi. W Gustine; 2 mi. NNE New Idria, 1900 ft.; *1 mi. S New Idria, 3700 ft.;* Waltham Creek, 4½ mi. SE Priest Valley, 1850 ft.; 2 mi. S San Miguel, 620 ft.; Matilija, thence up coast to 2

mi. S mouth Salinas River; near Gilroy; Alum Rock Park; Santa Rosa; Blue Lakes; South Fork Mtn.; Scott Valley, 1400 ft., 4 mi. S Fort Jones. Oregon: Grants Pass.

Peromyscus truei gratus Merriam

1898. *Peromyscus gratus* Merriam, Proc. Biol. Soc. Washington, 12:123, April 30, type from Tlalpan, Distrito Federal, México.
1909. *Peromyscus truei gratus,* Osgood, N. Amer. Fauna, 28:173, April 17.
1903. *Peromyscus sagax* Elliot, Field Columb. Mus., Publ. 71, Zool. Ser., 3(8):142, March 20, type from La Palma, Michoacán.
1903. *Peromyscus pavidus* Elliot, Field Columb. Mus., Publ. 71, Zool. Ser., 3(8):142, March 20, type from Pátzcuaro, Michoacán.
1904. *Peromyscus zelotes* Osgood, Proc. Biol. Soc. Washington, 17:67, March 21, type from Queréndaro, Michoacán.

MARGINAL RECORDS.—Jalisco: El Roble, 8 mi. NE Tepatitlán de Morales. Querétaro: Tequisquiapan. Hidalgo: Zimapán; Río Tasquillo, 26 km. E Zimapán, 5000 ft.; Pachuca. Distrito Federal: 200 m. E San Mateo Xalpa, 2390 m. Michoacán: 2 mi. W Pátzcuaro, 6700 ft.; 11 mi. W Zamora, 5750 ft.

Peromyscus truei lagunae Osgood

1909. *Peromyscus truei lagunae* Osgood, N. Amer. Fauna, 28:172, April 17, type from La Laguna, Sierra Laguna, Baja California.

MARGINAL RECORDS.—Baja California: La Chuparosa; Mt. Miraflores.

Peromyscus truei martirensis (J. A. Allen)

1893. *Sitomys martirensis* J. A. Allen, Bull. Amer. Mus. Nat. Hist., 5:187, August 18, type from Sierra San Pedro Mártir, 7000 ft., Baja California.
1909. *Peromyscus truei martirensis,* Osgood, N. Amer. Fauna, 28:171, April 17.
1903. *Peromyscus hemionotis* Elliot, Field Columb. Mus., Publ. 74, Zool. Ser., 3(10):157, May 7, type from Rosarito Divide, Sierra San Pedro Mártir, Baja California.

MARGINAL RECORDS.—California: Laguna Mts., 5500 ft.; ¾ mi. S Mountain Springs, 3300 ft. Baja California: Laguna Hanson, 5200 ft., Sierra Juárez; San Matías Spring; Santa Rosa; Rosarito Divide; La Grulla, Sierra San Pedro Mártir, 7200–7500 ft.; El Rayo.

Peromyscus truei montipinoris Elliot

1904. *Peromyscus montipinoris* Elliot, Field Columb. Mus., Publ. 90, Zool. Ser., 3(15):264, March 7, type from Lockwood Valley, near Mt. Piños, Ventura Co., California.
1951. *Peromyscus truei montipinoris,* Hoffmeister, Illinois Biol. Monog., 21(4):66, November 12.

MARGINAL RECORDS.—California: Whitney Creek, 10,650 ft.; *Freemans Canyon, E slope Walker Pass, 4900 ft.;* Mojave; Mint Canyon, 2100–2400 ft.;

Calabasas, 1200, 1300 ft.; Mt. Piños, 5500–8500 ft.; Cuyama Plain; Santiago Springs, Carrizo Plains; *divide W of McKittrick, 3000 ft.;* Kern River 12 mi. below Bodfish, 2000 ft.; Cannell Meadow, 7500 ft.; Jordan Hot Springs.

Peromyscus truei nevadensis Hall and Hoffmeister

1940. *Peromyscus truei nevadensis* Hall and Hoffmeister, Univ. California Publ. Zool., 42:401, April 30, type from ½ mi. W Debbs Creek, 6000 ft., Pilot Peak, Elko Co., Nevada.

MARGINAL RECORDS.—Utah: Raft River Mts.; Draper, 5000 ft.; Rock Canyon, 2 mi. N, 6 mi. E Provo; Nephi, 5095 ft.; near Salina; Cedar City; Pine Valley, 6400 ft. Nevada: 11 mi. E Panaca, 6500–6600 ft.; 2 mi. S Pioche, 6000 ft.; lat. 38° 17′ N, ¼ mi. W Utah–Nevada boundary, 7300 ft.; ½ mi. W Lehman Cave, 7500 ft.; Cleve Creek, 6900 ft., Shell Creek Range; Overland Pass, E slope Ruby Mts., 8 mi. S Elko Co. line; W side Ruby Lake, 3 mi. N White Pine Co. line, 6700 ft.; type locality.

Peromyscus truei preblei V. Bailey

1936. *Peromyscus truei preblei* V. Bailey, N. Amer. Fauna, 55:188, August 29, type from Crooked River, 20 mi. SE [= 12 mi. S, 6 mi. E] Prineville, Oregon.

MARGINAL RECORDS.—Oregon: Warmsprings; type locality; *Crooked River, 4 mi. W mouth Bear Creek.*

Peromyscus truei sequoiensis Hoffmeister

1941. *Peromyscus truei sequoiensis* Hoffmeister, Proc. Biol. Soc. Washington, 54:129, September 30, type from 1 mi. W Guerneville, Sonoma Co., California.

MARGINAL RECORDS.—Oregon: Galice, 500 ft. California: near Happy Camp; Taylor Creek, 5500 ft., Salmon Mts.; 1½ mi. S, ½ mi. E Willow Creek, 600 ft.; 3 mi. S Covelo; 3 mi. W summit Mt. Sanhedrin; Freestone; Ross, thence up coast to Oregon: Briggs Creek, 13 mi. SW Galice, 3000 ft.

Peromyscus truei truei (Shufeldt)

1885. *Hesperomys truei* Shufeldt, Proc. U.S. Nat. Mus., 8:407, September 14, type from Fort Wingate, McKinley Co., New Mexico.
1894. *P[eromyscus]. Truei,* Thomas, Ann. Mag. Nat. Hist., ser. 6, 14:365, November.
1890. *Hesperomys megalotis* Merriam, N. Amer. Fauna, 3:63, September 11, type from Black Tank, Little Colorado Desert, Coconino Co., Arizona.
1904. *Peromyscus lasius* Elliot, Field Columb. Mus., Publ. 90, Zool. Ser., 3(15):265, March 7, type from Hannopee Canyon, 7500 ft., Panamint Mts., Inyo Co., California.

MARGINAL RECORDS.—California: Crescent Butte, Lava Beds National Monument; near Lower Lake. Nevada: 17 mi. W Deep Hole, 4800 ft.; Eldorado Canyon, 6000 ft., Humbolt Range; S slope Granite Peak, East Range; Greenmonster Canyon, 7500 ft.,

Monitor Range; Water Canyon, 8 mi. N Lund; White River Valley, 14–15 mi. SW Sunnyside, 5500 ft.; Panaca. Utah: St. George; Zion National Park, 6500 ft.; Swamp Canyon, Bryce National Park; 5 mi. W Escalante, 6000 ft.; 13 mi. SE Price. Wyoming: Green River, 4 mi. NE Linwood, 5800 ft. Colorado: Escalante Hills, 20 mi. SE Ladore; McCoy; 12 mi. SE Rifle (Armstrong, 1972:210); North Rim, Black Canyon of Gunnison National Monument (*ibid.*); near Coventry, 6800 ft.; Piedra River, NW ¼ sec. 4, T. 32 N, R. 5 W (*ibid.*). New Mexico (Findley, *et al.,* 1975:221, 222, as *P. truei* only): 4½ mi. N El Rito; base Sierra Grande. Colorado: Fort Garland; Salida; 20 mi. S Colorado Springs; Rinehart's State Station, 20 mi. S Lamar. Oklahoma: Tesquite Canyon. New Mexico (Findley, *et al.,* 1975:221, 222, as *P. truei* only): 10 mi. S San Jon; 9 mi. N, 12 mi. W Fort Sumner; 15 mi. S Weed; Rope Springs, San Andreas Refuge. Arizona: Rock Creek Canyon, Chiricahua Mts., 7800 ft.; Walnut; Camp Verde, 3150 ft.; Pine Flat, Juniper Mts., 20 mi. NW Simmons. California: pass between Granite and Providence mts., 4100 ft.; Clark Mtn., 5000–7000 ft. Nevada: Clark Canyon (Deacon, *et al.,* 1964:404). California: Little Lake, 3100 ft.; Grays Meadow, Kearsarge Pass, Sierra Nevada, 6000 ft.; Williams Butte, 7000 ft.; Leavitts Meadow; ¼ mi. W Woodfords, 5700 ft.; Millford; Susanville; 1 mi. NNW Old Fort Crook.

Peromyscus truei zapotecae Hooper

1957. *Peromyscus truei zapotecae* Hooper, Occas. Pap. Mus. Zool., Univ. Michigan, 586:6, April 30, type from 1 mi. E Tlacolula, 5700 ft., Oaxaca.

MARGINAL RECORDS.—Puebla: 3 mi. N Tehuacán, 5500 ft. Oaxaca: type locality; Tlaxiaco (Goodwin, 1969:184); vic. Huajuapan, 4800 ft.

Peromyscus difficilis
Rock Mouse

External measurements: 180–260; 91–145; 22–28; 17.5–28. Upper parts brownish to black-

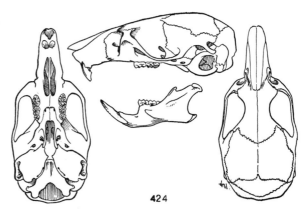

Fig. 424. *Peromyscus difficilis amplus,* Lago Salido, kilometer [post]253 E México; 8000 ft., Puebla, No. 14712 K.U., sex?, X 1½.

ish; underparts white to blackish suffused with silver; ears large, but when dry, shorter than hind foot; tail evenly bicolored, and ordinarily longer than head and body except in Colorado; rostrum long (nasals about 11); auditory bullae large; supraorbital border not beaded or sharply angled (after Hoffmeister and de la Torre, 1961:3–6). Compare the account immediately above with that of *P. evides* as an aid in distinguishing one species from the other. See also Diersing (1976:451–466) for differences between *Peromyscus difficilis* and some other species, and for comparisons of subspecies of *P. difficilis* one with another.

Peromyscus difficilis amplus Osgood

1904. *Peromyscus amplus* Osgood, Proc. Biol. Soc. Washington, 17:62, March 21, type from Coixtlahuaca, Oaxaca.
1909. *Peromyscus difficilis amplus* Osgood, N. Amer. Fauna, 28:181, April 17.

MARGINAL RECORDS.—Hidalgo: Tula; Marqués; Tulancingo. Veracruz: Perote; *1¹/₂ mi. S Perote, 8500 ft.* (Hall and Dalquest, 1963:307); *Maltrata (ibid.)*; 3 km. W Acultzingo, 7000 ft.; *4 km. W Acultzingo, 7500 ft.* (Hall and Dalquest, 1963:307). Oaxaca: type locality; Vanhuitlán (Goodwin, 1969:184); Tamazulapam. Tlaxcala: 8 km. SW Tlaxcala, 7500 ft. (Hoffmeister and de la Torre, 1961:12). Puebla: Río Otlati, 15 km. NW San Martín [Texmelucán], 8700 ft. México: Cerro La Caldera, 11 mi. SE México, 2350 m. (Hoffmeister and de la Torre, 1961:12); Atlacomulco, 2600 m. *(ibid.)*; *6 mi. NNW Acambay* (Hooper, 1955:18, as *P. d. felipensis*).

Peromyscus difficilis difficilis (J. A. Allen)

1891. *Vesperimus difficilis* J. A. Allen, Bull. Amer. Mus. Nat. Hist., 3:298, June 30, type from Sierra de Valparaíso, Zacatecas.
1897. [*Peromyscus*] *difficilis*, Trouessart, Catalogus mammalium . . . , fasc. 3, 518.

MARGINAL RECORDS.—Chihuahua (Anderson, 1972:338): Yaguirachic, 130 mi. W Chihuahua, 8500 ft.; 65 mi. E Batopilas. San Luis Potosí: Charcos; 3 km. SW San Isidro. Guanajuato: Santa Rosa. Zacatecas: Plateado; type locality. Durango: El Salto; 1½ mi. W San Luis; 11½ mi. E La Ciudad (Hoffmeister and de la Torre, 1961:7). Chihuahua (Anderson, 1972:338, 339): 10 mi. SW Guadalupe y Calvo; Churo, 7200 ft.

Peromyscus difficilis felipensis Merriam

1898. *Peromyscus felipensis* Merriam, Proc. Biol. Soc. Washington, 12:122, April 30, type from Cerro San Felipe, 10,200 ft., Oaxaca.
1909. *Peromyscus difficilis felipensis*, Osgood, N. Amer. Fauna, 28:182, April 17.

MARGINAL RECORDS.—Northern segment: México: Río Frío, 3000 m. Distrito Federal: Contreras, 2800 m. (Hoffmeister and de la Torre, 1961:13); *Ajusco (ibid.)*. México: Amecameca. Morelos: 2 mi. W Huit-

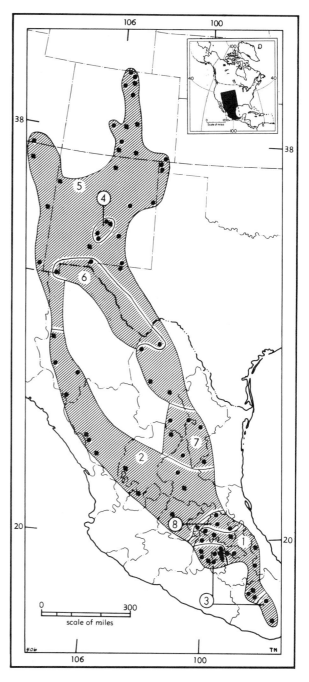

Map 406. *Peromyscus difficilis.*

1. *P. d. amplus*
2. *P. d. difficilis*
3. *P. d. felipensis*
4. *P. d. griseus*
5. *P. d. nasutus*
6. *P. d. penicillatus*
7. *P. d. petricola*
8. *P. d. saxicola*

zilac (Ramírez-P., 1971:276). México: Lago Zempoala, 45 km. SW Mexico City, 4500 ft.; Toluca Valley; Salazar. Southern segment: Oaxaca: type locality; *Mixtepec, 6000 ft.* (Goodwin, 1955a:2; see also Hoffmeister and de la Torre, 1961:12); Arroyo Agua Fría (Goodwin, 1969:185).

Peromyscus difficilis griseus Benson

1932. *Peromyscus nasutus griseus* Benson, Univ. California Publ. Zool., 38:338, April 14, type from the malpais, 3½ mi. W Carrizozo, 5150 ft., Lincoln Co., New Mexico.
1961. *Peromyscus difficilis griseus*, Hoffmeister and de la Torre, Jour. Mamm., 42:8, February 20.

MARGINAL RECORDS.—New Mexico: North Cerro Prieto, 6200 ft., 17 mi. NW Carrizozo; 4 mi. (by road) NW Carrizozo; *type locality;* 14 mi. W Three Rivers (Hoffmeister and de la Torre, 1961:9); Malpais Spring.

Peromyscus difficilis nasutus (J. A. Allen)

1891. *Vesperimus nasutus* J. A. Allen, Bull. Amer. Mus. Nat. Hist., 3:299, June 30, type from Estes Park, Larimer Co., Colorado.
1961. *Peromyscus difficilis nasutus*, Hoffmeister and de la Torre, Jour. Mamm., 42:7, February 20.

MARGINAL RECORDS.—Colorado (Armstrong, 1972:211, 212): Dale Creek, 1 mi. S Colorado–Wyoming boundary; 18 mi. N Fort Collins; 5 mi. S Fort Collins; Boulder; 14 mi. E Fountain; Trinidad; Regnier, 4500 ft. Oklahoma: 3 mi. N Kenton; Taft Canyon, Black Mesa (Hoffmeister and de la Torre, 1961:8). New Mexico: 8½ mi. S San Jon (Tamsitt, 1959:612); Santa Rosa; Hondo Canyon; 10 mi. S, 26 mi. W Carlsbad (Diersing and Hoffmeister, 1974:213); *4¾ mi. S, 11 mi. W White City (ibid.).* Coahuila (Hoffmeister and de la Torre, 1961:8): Carmen Mts., 6000 ft.; 2 mi. N, 18 mi. W Santa Teresa; *26 mi. W Santa Teresa, 7050 ft.;* 22 mi. S, 5 mi. W Ocampo; Sierra del Pino, 5 mi. N, 9 mi. W Acebuches. Texas: 2 mi. E Pine Springs (Diersing and Hoffmeister, 1974:213). New Mexico: 35 mi. NE Las Cruces (Diersing, 1976:464). Arizona: 1¼ mi. E Fly's Peak (*ibid.*); Springerville; Tuba City. Utah: Rainbow Bridge, 4000 ft. New Mexico (Findley, *et al.*, 1975:225, as *P. difficilis* only): Gallup; Arroyo Seco, 2 mi. N mouth Arroyo Hondo. Colorado (Armstrong, 1972:211, 212): 3 mi. W San Acacio; below Mosca Pass, ¾ mi. W Headquarters, Great Sand Dunes National Monument; 10 mi. SW Salida, 7100 ft.; 10 mi. N Cañon City; type locality; *Trail Creek Ranch, 15½ mi. NW Livermore.*

Peromyscus difficilis penicillatus Mearns

1896. *Peromyscus boylii penicillatus* Mearns, Preliminary diagnoses of new mammals from the Mexican border of the United States, p. 2, May 25 (preprint of Proc. U.S. Nat. Mus., 19:139, December 21, 1896), type from Franklin Mts., near El Paso, El Paso Co., Texas.
1976. *Peromyscus difficilis penicillatus*, Diersing, Proc. Biol. Soc. Washington, 89:462, October 12.

MARGINAL RECORDS (Diersing, 1976:463).—New Mexico: 16 mi. N El Paso. Coahuila: Sierra del Carmen (*15 mi. S, 25 mi. E Boquillas, 7300 ft.; Campo Madera, 8000 ft.; Oso Cañón; Botellas Cañón*). New Mexico: Dog Springs.

Peromyscus difficilis petricola Hoffmeister and de la Torre

1959. *Peromyscus difficilis petricola* Hoffmeister and de la Torre, Proc. Biol. Soc. Washington, 72:67, November 4, type from 12 mi. E San Antonio de las Alazanas, 9000 ft., Coahuila.

MARGINAL RECORDS.—Coahuila: 8 mi. S Bella Unión (Diersing, 1976:464); *type locality.* Nuevo León: 20 km. N Galeana, 7000 ft. (Hooper, 1947:51, as *P. d. difficilis*). Tamaulipas: Miquihuana (Hoffmeister and de la Torre, 1961:10); *20 mi. N Tula* (Alvarez, 1963:446). San Luis Potosí: Santa Ana, 2 mi. E Catorce (*ibid.*). Zacatecas: 4 mi. W Concepción del Oro (Diersing, 1976:464). Coahuila: Sierra Guadalupe (*ibid.*).

Peromyscus difficilis saxicola Hoffmeister and de la Torre

1959. *Peromyscus difficilis saxicola* Hoffmeister and de la Torre, Proc. Biol. Soc. Washington, 72:168, November 4, type from Cadereyta, 2100 m., Querétaro.

MARGINAL RECORDS (Hoffmeister and de la Torre, 1961:11, unless otherwise noted).—Hidalgo: vic. Jacala (Brown and Welser, 1968:425, as *P. difficilus* only); *Encarnación; San Agustín [Metzquititlán].* Veracruz (Hall and Dalquest, 1963:308): *6 km. WSW Zacualpilla, 6500 ft.;* 10 km. SW Jacales. Hidalgo: Ixmiquilpan. Querétaro: type locality.

Peromyscus bullatus Osgood
Perote Mouse

1904. *Peromyscus bullatus* Osgood, Proc. Biol. Soc. Washington, 17:63, March 21, type from Perote, Veracruz.

External measurements of the type, an adult female, are: 200; 93+; 23; 25 (dry). The tail of the type is broken. Upper parts rich ochraceous tawny heavily washed with dusky, especially mid-dorsally; underparts creamy white; feet white; tail brownish above, white below; "audi-

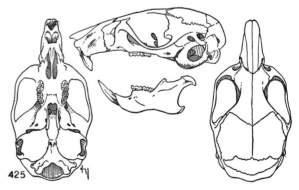

Fig. 425. *Peromyscus bullatus*, 2 km. W Limón, 7500 ft., Veracruz, No. 30478 K.U., ♀, X 1½.

tory bullae greatly enlarged, and nearly 20 per cent larger than in *P. truei*" (Hoffmeister, 1951:25). See Map 405.

MARGINAL RECORDS.—Veracruz: type locality; *2 km. W Limón* (Hall and Dalquest, 1963:306); *3 km. W Limón.*

melanophrys-group
Peromyscus perfulvus
Marsh Mouse

External measurements: 208–254; 110–138; 23–26. Upper parts bright cinnamon rufous with thinly scattered dusky hairs, which are more numerous on back than sides; underparts creamy; forefeet white and lacking dusky on wrists; hind feet white with brownish metatarsal area of variable extent; tail long, well haired, penicillate, unicolored sepia; soles of hind feet hairy on proximal quarter. Skull resembles that of *P. boylii;* braincase relatively long (in proportion to rostrum), little inflated; supraorbital border sharply angled but not beaded; interparietal large; tympanic bullae small.

Peromyscus perfulvus chrysopus Hooper

1955. *Peromyscus perfulvus chrysopus* Hooper, Occas. Pap. Mus. Zool., Univ. Michigan, 565:18, March 31, 1955, type from ½ mi. N Barro de Navidad, 50 ft., Jalisco.

MARGINAL RECORDS.—Jalisco: 4 mi. NNE Puerto Vallarta, 50 ft. (Genoways and Jones, 1973:14); shores of Bahía Tenacatita; *type locality; 3 mi. E Navidad.*

Peromyscus perfulvus perfulvus Osgood

1945. *Peromyscus perfulvus* Osgood, Jour. Mamm., 26:299, November 14, type from 10 km. W Apatzingán, 1040 ft., Michoacán.

MARGINAL RECORDS.—Michoacán: 12 mi. by road S Tzitzio, 3500 ft. Guerrero: Apaxtla, 4500 ft. Michoacán: type locality; *Apatzingán, 1500 ft.;* 6 mi. S, 1 mi. E Tacámbaro, 4000 ft.

Peromyscus melanophrys
Plateau Mouse

Reviewed by Baker, Univ. Kansas Publ., Mus. Nat. Hist., 5:251–258, 1 fig., April 10, 1952.

External measurements: 235–280; 122–163; 26–29; 18–21 (dry). Upper parts ochraceous, buffy, or tawny, finely sprinkled with dusky; underparts creamy white with much basal plumbeous showing through; feet white; tail bicolored, dusky above, white below. Skull with trenchant supraorbital ridge; smoothly arched dorsally in lateral view; temporal region swollen; nasals approx. parallel-sided; auditory bullae smaller than in *truei*-group but larger than in *mexicanus*-group.

This mouse is a representative of a rather striking group of large, notably long-tailed mice inhabiting the more arid regions of southern Mexico. It lives chiefly among rocks, and differs from nearly all other species [of *Peromyscus*] of Mexico in having a long tail and a slightly beaded skull (after Osgood, 1909:186).

Peromyscus melanophrys coahuilensis Baker

1952. *Peromyscus melanophrys coahuilensis* Baker, Univ. Kansas Publ., Mus. Nat. Hist., 5:257, April 10, type from 7 mi. S and 1 mi. E Gómez Farías, 6500 ft., Coahuila.

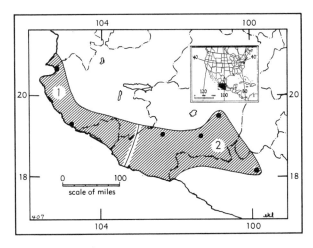

Map 407. *Peromyscus perfulvus.*

1. *P. p. chrysopus*   2. *P. p. perfulvus*

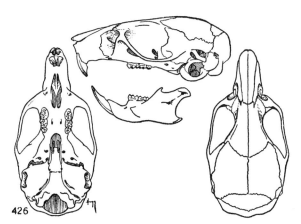

Fig. 426. *Peromyscus melanophrys micropus,* 4 mi. NE Ocotlán, Jalisco, No. 31768 K.U., ♀, X 1½.

MARGINAL RECORDS.—Coahuila: 17 mi. N, 8 mi. W Saltillo; type locality; ½ mi. S, 2 mi. E La Ventura.

Peromyscus melanophrys consobrinus Osgood

1904. *Peromyscus melanophrys consobrinus* Osgood, Proc. Biol. Soc. Washington, 17:66, March 21, type from Berriozábal, Zacatecas.

MARGINAL RECORDS.—San Luis Potosí: 1 mi. E La Paz. Tamaulipas (Alvarez, 1963:446): 14 mi. N, 6 mi. W Palmillas, 5500 ft.; 9 mi. SW Tula, 3900 ft. San Luis Potosí: 2 mi. NW Tepeyac, 3400 ft., 14 mi. N, 29 mi. W Ciudad del Maíz; 1½ mi. E Río Verde. Guanajuato: Silao. Zacatecas: 1⅗ km. N Santa Rosa (Baker, *et al.*, 1967:226); Monte Escobedo; 2 mi. ESE Troncosa, 7000 ft.

Peromyscus melanophrys melanophrys (Coues)

1874. *Hesperomys (Vesperimus) melanophrys* Coues, Proc. Acad. Nat. Sci. Philadelphia, 26:181, December 15, type from Santa Efigenia, Oaxaca.
1897. *P[eromyscus]. melanophrys*, J. A. Allen, Bull. Amer. Mus. Nat. Hist., 9:51, March 15.

1894. *Peromyscus leucurus* Thomas, Ann. Mag. Nat. Hist., ser. 6, 14:364, November, type from Tehuantepec, Oaxaca.
1903. *Peromyscus leucurus gadovii* Thomas, Ann. Mag. Nat. Hist., ser. 7, 11:484, May, type from San Carlos [= Yautepec], Oaxaca.

MARGINAL RECORDS.—Distrito Federal: Contreras, 2600 m. Puebla: Chalchicomula; Tehuacán, 1700 m. Oaxaca (Goodwin, 1969:185, 186): Teotitlán; Sierra de Juárez; Nizanda; Arroyo Encantado. Chiapas: San Bartolomé; San Vicente. Oaxaca (Goodwin, 1969:185, 186): Caja de Agua; Huamelula; Puerto Angel (62900 KU); Miahuatlán; Sola de Vega. Guerrero: 15 mi. S Chilpancingo, 4500 ft.; Cañón de Zopilote, 14½ mi. by road N Zumpango, *ca.* 2000 ft. (Winkelmann, 1962:108, as *P. melanophrys* only); *Los Sabinos, 1210 m.*; 14 mi. S, 1 mi. W Iguala, 2600 ft.

Peromyscus melanophrys micropus Baker

1952. *Peromyscus melanophrys micropus* Baker, Univ. Kansas Publ., Mus. Nat. Hist., 5:255, April 10, type from 3 mi. N Guadalajara, Jalisco.

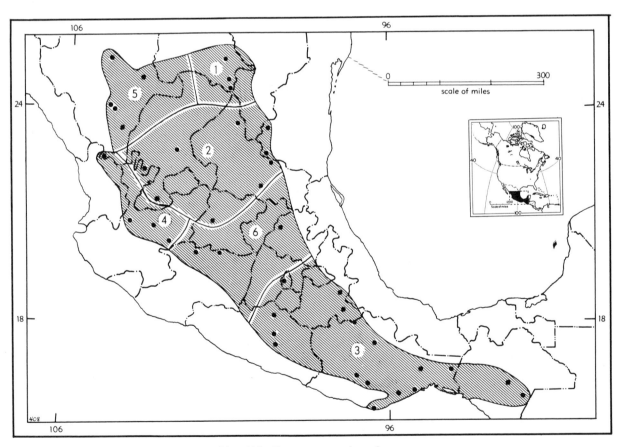

Map 408. *Peromyscus melanophrys.*

1. *P. m. coahuilensis*	3. *P. m. melanophrys*	5. *P. m. xenurus*
2. *P. m. consobrinus*	4. *P. m. micropus*	6. *P. m. zamorae*

MARGINAL RECORDS.—Durango: Paso de Sihuacori (Crossin, et al., 1973:199). Jalisco: 2 mi. NNW Magdalena; *2 mi. NW Magdalena, 4500 ft.*; type locality; 4 mi. NE Ocotlán, 5050 ft.

Peromyscus melanophrys xenurus Osgood

1904. *Peromyscus xenurus* Osgood, Proc. Biol. Soc. Washington, 17:67, March 21, type from City of Durango, Durango.
1952. *Peromyscus melanophrys xenurus*, Baker, Univ. Kansas Publ., Mus. Nat. Hist., 5:256, April 10.

MARGINAL RECORDS (Baker and Greer, 1962:117, unless otherwise noted).—Durango: 8½ mi. N Alamillo, 5900 ft.; 3 mi. NNW Cuencamé; 2 km. SW Mezquital (Crossin, et al., 1973:197); 5 mi. S Durango; 8 mi. NW Durango.

Peromyscus melanophrys zamorae Osgood

1904. *Peromyscus melanophrys zamorae* Osgood, Proc. Biol. Soc. Washington, 17:65, March 21, type from Zamora, Michoacán.

MARGINAL RECORDS.—Hidalgo: Zimapán. Michoacán: Queréndaro; type locality.

Peromyscus mekisturus Merriam
Puebla Deer Mouse

1898. *Peromyscus mekisturus* Merriam, Proc. Biol. Soc. Washington, 12:124, April 30, type from Chalchicomula, 8400 ft., Puebla.

External measurements of the type, an adult female, are 249; 155; 24; 18.4 (dry). Upper parts as in *P. melanophrys*; underparts creamy buff, wholly without white; hind feet dusky to toes, toes white; tail brownish above, mottled brownish and whitish below. Skull of the *melanophrys* type, but smaller; nasals short; frontals much constricted and without supraorbital beading. Tail strikingly long relative to length of head and body. See Map 405.

MARGINAL RECORDS.—Puebla: type locality; Tehuacán, 1700 m.

mexicanus-group
Peromyscus ochraventer Baker
El Carrizo Deer Mouse

1951. *Peromyscus ochraventer* Baker, Univ. Kansas Publ., Mus. Nat. Hist., 5:213, December 15, type from 70 km. by highway S Ciudad Victoria, 6 km. W Pan-American Highway at El Carrizo, Tamaulipas.

External measurements: 226–249; 116–128; 24–26; 20–21. Upper parts ochraceous tawny, brighter on sides, duller on back; underparts cinnamon buff; brighter on pectoral region; feet

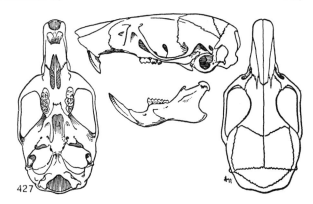

Fig. 427. *Peromyscus ochraventer*, 70 km. by highway S Victoria, 6 km. W highway, Tamaulipas, No. 37039 K.U., ♂, X 1½.

white; tail scaly, indistinctly bicolored with dark hairs above, light hairs below. Skull lacking beaded or ridged supraorbital border; rostrum almost parallel-sided; P1 and P2 with well-developed outer accessory cusps; anteriormost loph (parastyle-protoconule) of M1 almost as broad as greatest breadth of M1.

This species resembles *P. furvus* but has buffy instead of grayish underparts, and may be allied to the *P. boylii* species-group as much as to the *P. mexicanus* species-group. See Map 409.

MARGINAL RECORDS.—Tamaulipas: type locality; *Gómez Farías; Rancho del Cielo; La Joya de Salas.* San Luis Potosí: *vic. El Naranjo* (Brown and Welser, 1968:425); 8 mi. W El Naranjo (= 8 mi. W Naranjos?), 2400 ft. (Baker and Phillips, 1965:338).

Peromyscus yucatanicus
Yucatán Deer Mouse

External measurements: 208–232; 95–122; 23–26; 15.2–18.3 (dry). Upper parts essentially as in *P. banderanus*; underparts yellowish white, becoming nearly pure white on throat, pectoral spot usually absent; tail as in *P. banderanus*. "While apparently quite distinct, this species is little more than a miniature of *P. mexicanus*. In all general characters except size it shows no marked departure from *mexicanus*. Its slightly blotched tail, slightly beaded skull, small teeth, and small audital bullae readily distinguish it from any species approximating it in size." (Osgood, 1909:212.)

Lawlor (1965) considered this species to be monotypic, but on the basis of my examination of the specimens examined by him, I recognize two subspecies, the pale, northern *P. y. yucatanicus* and the dark, southern *P. y. badius*.

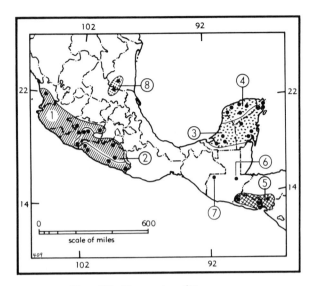

Map 409. Six species of *Peromyscus*.

1. *P. banderanus banderanus*   5. *P. stirtoni*
2. *P. banderanus vicinior*   6. *P. grandis*
3. *P. yucatanicus badius*   7. *P. mayensis*
4. *P. yucatanicus yucatanicus*   8. *P. ochraventer*

Peromyscus yucatanicus badius Osgood

1904. *Peromyscus yucatanicus badius* Osgood, Proc. Biol. Soc. Washington, 17:70, March 21, type from Apazote, Campeche.

MARGINAL RECORDS.—Quintana Roo: Pueblo Nuevo X-Can (92476 KU); 68 km. N, 16 km. E Chetumal (93709 KU); 27 km. NW Chetumal (93708 KU). Campeche: 7 km. N, 51 km. E Escárcega (93712 KU); 7½ km. W Escárcega (92435 KU); type locality. Quintana Roo: Esmeralda (Lawlor, 1965:432).

Peromyscus yucatanicus yucatanicus J. A. Allen and Chapman

1897. *Peromyscus yucatanicus* J. A. Allen and Chapman, Bull. Amer. Mus. Nat. Hist., 9:8, February 23, type from Chichén-Itzá, Yucatán.

MARGINAL RECORDS.—Yucatán: 66 km. NE Mérida (93730 KU); 6 km. N Tizimín (93726 KU). Quintana Roo: 5 km. WSW Puerto Juárez (92474 KU); La Vega (108418 USNM); Puerto Morelos (108408 USNM). Yucatán: type locality; Calcehtok (35198 KU); Mérida Airport, 6 km. S Mérida (92443 KU). Lawlor (1965:432) credited Gaumer (1917:117) with recording mice of this species from seven localities in Yucatán, including Yaxcach and Xbac. If correctly identified to species, they probably are *P. y. yucatanicus*.

Peromyscus mexicanus
Mexican Deer Mouse

External measurements: 191–277; 92–143; 23–29; 16–24. Upper parts Clay Color to blackish,

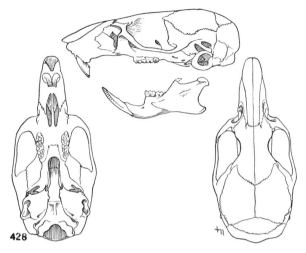

Fig. 428. *Peromyscus mexicanus saxatilis*, 5 km. SE Turrialba, 1950 ft., Prov. Cartago, Costa Rica, No. 26986 K.U., ♂, X 1½.

usually darkest mid-dorsally; underparts white to creamy, often with buffy pectoral region, certain subspecies tinted Pinkish Buff; feet white; carpus and tarsus white to dusky; tail almost naked to moderately haired, darker above, blotched below. Skull with long and slender rostrum, broader in certain subspecies; braincase broad, arched to rounded; supraorbital ridges moderately to well developed, forming shelf over orbit, faintly to strongly beaded; molariform teeth and auditory bullae relatively small.

Peromyscus mexicanus angelensis (Osgood)

1904. *Peromyscus banderanus angelensis* Osgood, Proc. Biol. Soc. Washington, 17:69, March 21, type from Puerto Angel, Oaxaca.
1969. *Peromyscus m[exicanus]. angelensis*, Musser, Amer. Mus. Novit., 2357:5, January 27.

MARGINAL RECORDS (Musser, 1969:3, unless otherwise noted).—Guerrero: near Ometepec, 200 ft. Oaxaca: Zarzamora, 15 mi. W Tequisistlán, 3000 ft.; *Santa Lucía, 17 mi. W Tehuantepec, 4000 ft.*; Arroyo San Juan, 15 mi. W Tehuantepec; *Cerro Tres Cruces, 8 mi. E Tenango, 4000 ft.*; type locality; mainland near Escondido Bay; Pinotepa de Don Luis (Goodwin, 1969:189, 190).

Peromyscus mexicanus mexicanus (Saussure)

1860. *H[esperomys]. mexicanus* Saussure, Revue et Mag. Zool., Paris, ser. 2, 12:103, México; assumed to be vic. Mirador, Veracruz; restricted to 10 km. E Mirador (Dalquest, Occas. Pap. Mus. Zool., Louisiana State Univ., 23:8, July 10, 1950).
1894. *P[eromyscus]. mexicanus*, Thomas, Ann. Mag. Nat. Hist., ser. 6, 14:364, November.
1898. *Peromyscus tehuantepecus* Merriam, Proc. Biol. Soc.

Washington, 12:122, April 30, type from Tehuantepec, Oaxaca (Goodwin, Amer. Mus. Novit., 1732:2, June 10, 1955, implies that *tehuantepecus* is a valid subspecies of *P. mexicanus*, and lists specimens from Ixcuintepec, Oaxaca).

1955. *Peromyscus banderanus sloeops* Goodwin, Amer. Mus. Novit., 1732:2, June 10, type from Río Mono Blanco, 25 km. NE Zanatepec, 3000 ft., District of Juchitán, Oaxaca, México. Musser (Amer. Mus. Novit., 2357:7, January 27, 1969) found that this name applied to a population of *P. mexicanus*, not *P. banderanus*, but did not decide whether it applied to an otherwise unnamed subspecies or to one already named.

MARGINAL RECORDS.—San Luis Potosí: 10 km. E Platanito. Veracruz (Hall and Dalquest, 1963:311): Piedras Clavadas; *35 km. NW Tuxpan; 25 km. NW Tuxpan;* 14 km. NW Tuxpan, thence along coast to Catemaco; Achotal. Oaxaca (Goodwin, 1969:187): Ubero; Monte Rico. Chiapas: 8 km. N Berriozábal, 1065 m. (Baker, *et al.,* 1973:78, 85); *10–11 km. W Tuxtla Gutiérrez, 800 m.;* mts. near Tonalá. Oaxaca (Goodwin, 1969:187, unless otherwise noted): 10 mi. W Tehuantepec; Cerro Arenal; Lachillo; Lagunas (Hall and Kelson, 1959:650). Veracruz: Otatitlán; *2 km. N Motzorongo* (Hall and Dalquest, 1963:311); 3 km. N Presidio (*ibid.*); 4 km. WNW Fortín, 3200 ft. (*ibid.*); Jico. Puebla: Rancho El Ajengibre, Kilometer 264, México–Tuxpan road (3617 U. Minn.). San Luis Potosí: Xilitla.

Peromyscus mexicanus putlaensis Goodwin

1964. *Peromyscus mexicanus putlaensis* Goodwin, Amer. Mus. Novit., 2183:5, June 4, type from San Vicente, District of Putla, 4000 ft., Oaxaca.

MARGINAL RECORDS (Goodwin, 1969:189).—Oaxaca: Santo Domingo Chicahuastla [= Chicahuaxtla]; type locality.

Peromyscus mexicanus saxatilis Merriam

1898. *Peromyscus mexicanus saxatilis* Merriam, Proc. Biol. Soc. Washington, 12:121, April 30, type from Jacaltenango, 5400 ft., Huehuetenango, Guatemala.
1908. *Peromyscus nicaraguae* J. A. Allen, Bull. Amer. Mus. Nat. Hist., 24:649, October 13, type from Matagalpa, Matagalpa, Nicaragua.
1928. *Peromyscus mexicanus philombrius* Dickey, Proc. Biol. Soc. Washington, 41:3, January 25, type from Los Esemiles, 8000 ft., Chalatenango, El Salvador. Regarded as a synonym of *P. m. saxatilis* by Burt and Stirton, Misc. Publ. Mus. Zool., Univ. Michigan, 117:57, September 22, 1961.
1928. *Peromyscus mexicanus salvadorensis* Dickey, Proc. Biol. Soc. Washington, 41:4, January 25, type from Mt. Cacaguatique, 3500 ft., San Miguel, El Salvador. Regarded as indistinguishable from *P. m. saxatilis* by Felten, Senckenbergiana Biol., 39:134, August 30, 1958.

MARGINAL RECORDS.—Chiapas: San Bartolomé; San Vicente. Guatemala: Chanquejelve. Honduras (Hall and Kelson, 1959:647, and Musser, 1969:9, regard specimens from Honduras formerly assigned to *P.*

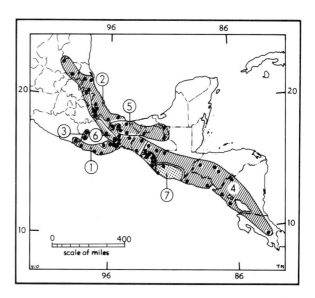

Map 410.   *Peromyscus mexicanus* and *Peromyscus gymnotis*

Guide to kinds	
1. *P. m. angelensis*	4. *P. m. saxatilis*
2. *P. m. mexicanus*	5. *P. m. teapensis*
3. *P. m. putlaensis*	6. *P. m. totontepecus*
	7. *P. gymnotis*

guatemalensis tropicalis as best allocated to *P. m. saxatilis*): Santa Bárbara; El Calíche Orica; Rancho Quemado. Nicaragua: 3½ km. S, 2 km. W Jalapa (Jones and Genoways, 1970:8, as *P. mexicanus* only). Costa Rica: 5 km. SE Turrialba. Nicaragua: 3 mi. NNW Diriamba (Jones and Genoways, 1970:14, as *P. mexicanus* only); San Antonio (Jones and Genoways, 1970:7, as *P. mexicanus* only). El Salvador (Burt and Stirton, 1961:57): Pine Peaks, 3 mi. W Volcán de Conchagua; Hda. Chilata. Chiapas: Chicharras; Catarina, 1300 m.

Peromyscus mexicanus teapensis Osgood

1904. *Peromyscus mexicanus teapensis* Osgood, Proc. Biol. Soc. Washington, 17:69, March 21, type from Teapa, 800 ft., Tabasco.

MARGINAL RECORDS.—Veracruz: 14 km. SW Coatzacoalcos, 100 ft. Tabasco: Montecristo. Chiapas: Laguna Ocotal; 10 km. S Solusuchiapa, 395 m. (Baker, *et al.,* 1973:78, 85). Veracruz (Hall and Dalquest, 1963:311): 63 km. ESE Jesús Carranza, 500 ft.; Jesús Carranza, 250 ft.; 15 km. SW Jimba, 750 ft.

Peromyscus mexicanus totontepecus Merriam

1898. *Peromyscus mexicanus totontepecus* Merriam, Proc. Biol. Soc. Washington, 12:120, April 30, type from Totontepec, 6500 ft., Oaxaca.
1898. *Peromyscus mexicanus orizabae* Merriam, Proc. Biol. Soc. Washington, 12:121, April 30, type from Orizaba, Veracruz.
1956. *Peromyscus banderanus coatlanensis* Goodwin, Amer. Mus. Novit., 1791:7, September 28, type from Agua Sarca [Agua Zarca], about 7 km. SW Coatlán, District of Tehuan-

tepec, Oaxaca, México. Musser (Amer. Mus. Novit., 2357:7, January 27, 1969) found that this name applied to a population of P. *mexicanus*, not P. *banderanus*, but did not decide whether it applied to an otherwise unnamed subspecies or to one already named.

MARGINAL RECORDS.—Veracruz: 7 km. W El Brinco, 800 ft.; *Cautlapán, 4000 ft.*; *Orizaba*; 3 km. SE Orizaba, 5500 ft. (Hall and Dalquest, 1963:311); *Motzorongo*. Oaxaca: *mts. near Santo Domingo; Santo Domingo; Nueva Raza* (Carleton, 1977:34); Agua Zarca (Goodwin, 1969:190); *Las Cuevas* (*op. cit.*:188); type locality.

Peromyscus gymnotis Thomas
Naked-eared Deer Mouse

1894. *Peromyscus gymnotis* Thomas, Ann. Mag. Nat. Hist., ser. 6, 14:365, November, type from Guatemala.
1904. *Peromyscus allophylus* Osgood, Proc. Biol. Soc. Washington, 17:71, March 21, type from Huehuetán, 200 ft., Chiapas. A synonym of P. *gymnotis* according to Musser, Amer. Mus. Novit., 2453:1, February 26, 1971, who (*op. cit.*:10) accords specific (instead of subspecific) rank to *Peromyscus gymnotis*.

External measurements: 193–242; 91–124; 22–27; 16–20. Coloration as described in P. *mexicanus*. Skull differing from that of P. *mexicanus saxatilis* and from Oaxacan specimens of P. *m. mexicanus* as follows: smaller; tooth-row shorter (3.8–4.5 vs. 4.1–4.7); tympanic bullae relatively (not actually) larger and anterior border of bulla directed anteromedially instead of more nearly at a right angle with longitudinal axis of skull; supraorbital ridges less conspicuous or wanting and not forming as much of a shelf over orbit; dorsal outline of skull viewed laterally more convex. See Map 410.

MARGINAL RECORDS.—Chiapas: Finca Liquidámbar, 1210 m.; Paval, 20 km. by road NE Mapastepec; Finca Esperanza, 150 m.; Huehuetán. Guatemala: Finca Ciprés; Hda. California

Peromyscus mayensis Carleton and Huckaby
Mayan Mouse

1975. *Peromyscus mayensis* Carleton and Huckaby, Jour. Mamm., 56:444, May 30, type from 7 km. NW Santa Eulalia, "Yaiquich," 2950 m., Guatemala. Known only from type locality.

External measurements of 15 specimens: 232(209–245); 111 (102–120); 26.3 (25–27); 21.2 (19–23). Length of skull, 32.0 (30.6–33.1); zygomatic breadth, 15.5 (15.0–16.1). Pelage drab; upper parts dark umber brown; underparts slate gray tipped with white imparting frosted effect;

no well-marked lateral line; ears dark brown, lined with fine black hairs; dorsal surface of metatarsus dusky brown, phalanges of hind foot paler; metacarpus dirty white; 6 plantar tubercles and plantar surface naked to heel; tail blackish, only slightly paler below. Rostrum of skull long, attenuated, no flaring of nasals distally; zygomatic arches slope gently anteriorly (not squared); interorbital region hourglass in outline, lacking shelf or bead; frontals anterolaterally expanded (bulbous) creating shallow depression immediately posterior to nasals; suture between frontals and parietals obtusely V-shaped; anterior border of mesopterygoid fossa U-shaped, lacking spine; in lateral view, border of upper diastema barely concave rather than straight. Tooth-rows parallel; mesolophostyles, mesolophostylids and ectolophostylids present; anterocone deeply bifurcated forming anterolabial and anterolingual conules and, on M1 labial and lingual cones appearing only slightly smaller than primary cones; anterolabial stylids but no anterolingual stylids on first lower molars; m3 having 2nd primary fold; enamel surface of incisors yellow-orange. Glans penis, 12.7 (12.0–13.5); bacular length, 18.6 (18.2–19.1). Three pairs of mammae—1 pair axillary and 2 pairs inguinal (adapted from original description). See Map 409.

Peromyscus stirtoni Dickey
Stirton's Deer Mouse

1928. *Peromyscus stirtoni* Dickey, Proc. Biol. Soc. Washington, 41:5, January 25, type from Río Goascorán, 100 ft., 13° 30′ N, La Unión, El Salvador.

Average external measurements of 17 (?) specimens are: 197; 95; 23.7 (Dickey, 1928:5). Upper parts approximating ochraceous buff moderately mixed with dusky hairs; pectoral spot usually absent; tail conspicuously furred, distinctly bicolored. Skull of average size; rostrum fairly heavy; nasals usually truncate posteriorly; supraorbital border markedly trenchant, not beaded, usually not upturned; braincase narrow but not especially produced posteriorly; auditory bullae small; molar tooth-row and especially M3, weak. See Map 409.

MARGINAL RECORDS.—El Salvador: El Tablón (Burt and Stirton, 1961:58). Honduras: La Piedra de Jesús, Sabana Grande. El Salvador: type locality; Pine Peaks, Volcán Conchagua.

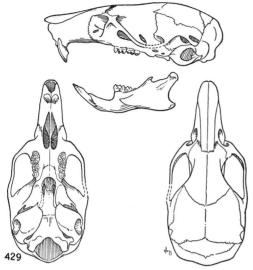

Fig. 429. *Peromyscus nudipes nudipes*, San Rafael, 30 km. S San José, Prov. San José, Costa Rica, No. 39254 K.U., ♂, X 1½.

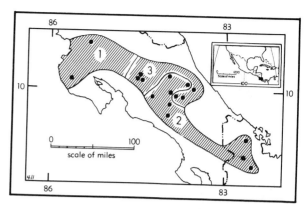

Map 411. *Peromyscus nudipes*.

1. *P. n. hesperus* 2. *P. n. nudipes*
3. *P. n. orientalis*

Peromyscus nudipes
Naked-footed Deer Mouse

External measurements: 240–280; 120–140; 26–30; 18.4–22 (dry). Upper parts dark brown to blackish mid-dorsally, sides lighter, richer brown; underparts varying from yellowish white with buffy pectoral suffusion to entirely buffy except for throat and chin; forefeet white; hind feet white above on distal half, dusky proximally; tail usually entirely brownish black but sometimes irregularly blotched below. Skull with narrow braincase; small auditory bullae; occasional faint beading supraorbitally; widely opened palatine slits; zygomata notched slightly anteriorly. Pelage full and soft; tail clothed with short hairs, prominently scaled.

Peromyscus nudipes hesperus Harris

1940. *Peromyscus nudipes hesperus* Harris, Occas. Pap. Mus. Zool., Univ. Michigan, 423:1, November 29, type from Hda. Santa María, 3200 ft., about 15 mi. NE Liberia, Guanacaste, Costa Rica.

MARGINAL RECORDS.—Costa Rica: type locality; [Cerro de] San Juan.

Peromyscus nudipes nudipes (J. A. Allen)

1891. *Hesperomys (Vesperimus?) nudipes* J. A. Allen, Bull. Amer. Mus. Nat. Hist., 3:213, April 17, type from La Carpintera, Cartago, Costa Rica.
1894. *Peromyscus nudipes*, Thomas, Ann. Mag. Nat. Hist., ser. 6, 14:365, November.
1902. *Peromyscus cacabatus* Bangs, Bull. Mus. Comp. Zool., 39:29, April, type from Boquete, Chiriquí, Panamá.

MARGINAL RECORDS.—Costa Rica: El Coronel de Carrillo; Rancho de Río Jiménez; Volcán Irazú; El Muñeco. Panamá: upper Río Changena, 2400–4800 ft. (Handley, 1966:783); Cerro Punta, 5000–7800 ft. (*ibid.*); Boquete. Costa Rica: San Joaquín; Los Higuerones Escazú.

Peromyscus nudipes orientalis Goodwin

1938. *Peromyscus nudipes orientalis* Goodwin, Amer. Mus. Novit., 987:3, May 13, type from El Sauce Peralta, about 1000 ft., Cartago, Costa Rica.

MARGINAL RECORDS.—Costa Rica: Western localities: Lajas Villa Quesada; Zarcero; Tapesco. Eastern localities: type locality; Juan Viñas; Cervantes.

Peromyscus furvus J. A. Allen and Chapman
Blackish Deer Mouse

1897. *Peromyscus furvus* J. A. Allen and Chapman, Bull. Amer. Mus. Nat. Hist., 9:201, June 16, type from Jalapa, Veracruz.
1950. *Peromyscus latirostris* Dalquest, Occas. Pap. Mus. Zool., Louisiana State Univ., 23:8, July 10, type from Apetsco, 2700 ft., near Xilitla, San Luis Potosí. Regarded as inseparable from *P. furvus* by Hall, Anal. Inst. Biol., 39:154, September 11, 1971.
1961. *Peromyscus angustirostris* Hall and Alvarez, Proc. Biol. Soc. Washington, 74:203, August 11, type from 3 km. W of Zacualpan, 6000 ft., Veracruz. Regarded as inseparable from *P. furvus* by Hall, Anal. Inst. Biol., 39:154, September 11, 1971.

External measurements: 229–300; 114–162; 26–33; 20–23. Upper parts dark brown to blackish, approximating russet on sides; underparts grayish (much of slaty basal parts of hairs showing through whitish tipping); feet white; blackish brown marking on tarsal joint; tail either entirely

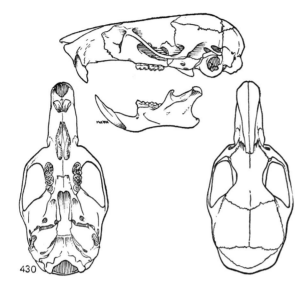

Fig. 430. *Peromyscus furvus*, 2 km. W Jico, 4200 ft., Veracruz, No. 19561 K.U., ♂, X 1½.

blackish or irregularly bicolored and blotched underneath; skull narrow interorbitally; supraorbital border of frontals not beaded but in some specimens trenchant; nasals usually broad anteriorly.

MARGINAL RECORDS (Hall, 1971:154, unless otherwise noted).—San Luis Potosí: Miramar Grande, 6000 ft. Hidalgo: 13 mi. NE Metepec (Hwy. 53), 660 ft.

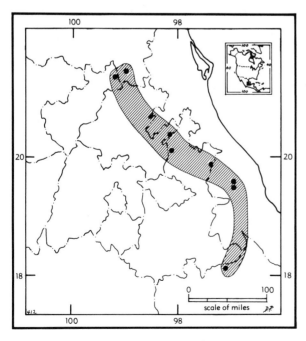

Map. 412. *Peromyscus furvus*.

Puebla: 19° 52′ N, 97° 20′ W, 12$\frac{1}{10}$ km. by road NE Tezuitlán (Heaney and Birney, 1977:543). Veracruz: 5 km. N Jalapa, 4500 ft.; 5 km. S Jalapa; *2 km. W Jico, 4200 ft.* Oaxaca: Huehuetlán (Goodwin, 1969:192, as *P. melanocarpus;* identified as *P. furvus* by D. G. Huckaby *in Litt.*, April 12, 1972). Puebla: 7$\frac{3}{10}$ mi. by road SW Huauchinango, 6800 ft. Querétero: 6 mi. W Ahuacatlan, 5800 and 5600 ft.

Peromyscus altilaneus Osgood
Todos Santos Deer Mouse

1904. *Peromyscus altilaneus* Osgood, Proc. Biol. Soc. Washington, 17:74, March 21, type from Todos Santos, 10,000 ft., Huehuetenango, Guatemala. Known only from type locality.

External measurements of the type, the only known specimen, are: 228; 115; 28; 20.6 (dry). ". . . Scarcely any character can be found distinguishing it from *guatemalensis* except that of size" (Osgood, 1909:198).

Peromyscus guatemalensis
Guatemalan Deer Mouse

External measurements: 252–290; 122–153; 28–34; 16–25. Upper parts approx. ferruginous heavily mixed with dusky, especially mid-dorsally, which imparts an overall appearance of dark brown to blackish; underparts buffy whitish but with slaty basal color showing through conspicuously; pectoral region washed with cinnamon·rufous; forefeet white; hind feet chiefly white but dusky of tarsal joint sometimes extending over as much as half of antiplantar surface of foot; tail usually somewhat bicolored, underside sometimes irregularly blotched with yellowish and dusky, tail sometimes entirely dusky. Skull much as in the *mexicanus*-group, but generally larger; frontals constricted, trenchant supraorbitally but not beaded; pelage of a wooly texture.

This species seems to be restricted to the highlands.

Peromyscus guatemalensis guatemalensis Merriam

1898. *Peromyscus guatemalensis* Merriam, Proc. Biol. Soc. Washington, 12:118, April 30, type from Todos Santos, 10,000 ft., Huehuetenango, Guatemala.

MARGINAL RECORDS.—Chiapas: Triunfo, 1950 m.; Pinabete. Guatemala: type locality; Volcán San Lucas; Volcán Santa María. Chiapas: Volcán Tacaná, 3000 m.

Peromyscus guatemalensis tropicalis Goodwin

1932. *Peromyscus guatemalensis tropicalis* Goodwin, Amer. Mus. Novit., 560:3, September 16, type from Chimoxan,

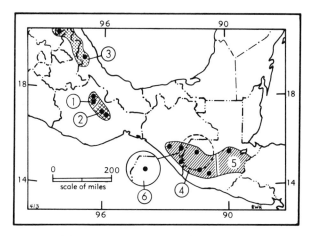

Map 413. Four species of *Peromyscus.*

1. *P. lepturus ixtlani*
2. *P. lepturus lepturus*
3. *P. simulatus*
4. *P. guatemalensis guatemalensis*
5. *P. guatemalensis tropicalis*
6. *P. altilaneus*

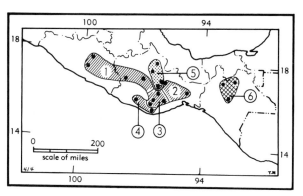

Map 414. Three species of *Peromyscus.*

1. *P. megalops auritus*
2. *P. megalops azulensis*
3. *P. megalops megalops*
4. *P. megalops melanurus*
5. *P. melanocarpus*
6. *P. zarhynchus*

about 40 mi. NE Cobán, Alta Verapaz, 1500 ft., Guatemala. Known only from type locality. The name *P. g. tropicalis* was based on material that Musser (Amer. Mus. Novit., 2357:9, 10, January 27, 1969) suggests is a subspecies of *Peromyscus mexicanus.*

Peromyscus megalops
Brown Deer Mouse

External measurements: 238–288; 127–150; 26–31; 15.8–23 (dry). Upper parts tawny and blackish brown, darkest mid-dorsally, lightest and clearest on sides; underparts varying from creamy to faintly buffy, pectoral and axillary region usually tawny; feet white; haired part of tail usually bicolored, scaly portion dusky above and blotched below. Skull large and broad; supraorbital border prominent and distinctly beaded; frontal region channeled; lachrymal region swollen; auditory bullae relatively small. Tail decidedly longer than head and body; pelage long, lax.

Peromyscus megalops auritus Merriam

1898. *Peromyscus auritus* Merriam, Proc. Biol. Soc. Washington, 12:119, April 30, type from mts. 15 mi. W City of Oaxaca, 9300 ft., Oaxaca.
1909. *Peromyscus megalops auritus*, Osgood, N. Amer. Fauna, 28:214, April 17.
1898. *Peromyscus comptus* Merriam, Proc. Biol. Soc. Washington, 12:120, April 30, type from mts. near Chilpancingo, Guerrero.

MARGINAL RECORDS.—Guerrero: Puerto Chico, 20 mi. N Chilpancingo (Brown and Welser, 1968:425, as *P. megalops* only). Oaxaca: type locality; San Andrés Chicahuaxtla (Goodwin, 1969:191). Guerrero: mts. near Chilpancingo; *Omilteme.*

Peromyscus megalops azulensis Goodwin

1956. *Peromyscus megalops azulensis* Goodwin, Amer. Mus. Novit., 1791:6, September 28, type from Cerro Azul, 7000 ft., District of Juchitán, Oaxaca.

MARGINAL RECORDS (Goodwin, 1969:191).— Oaxaca: Estancia; type locality.

Peromyscus megalops megalops Merriam

1898. *Peromyscus megalops* Merriam, Proc. Biol. Soc. Washington, 12:119, April 30, type from mts. near Ozolotepec, Oaxaca.

MARGINAL RECORDS.—Oaxaca (Goodwin, 1969:190, unless otherwise noted): Cerro Madreña; Santo Tomás Teipan, 7000 ft.; Río Guajolote (Schaldach, 1966:296); *Río Molino;* type locality.

Peromyscus megalops melanurus Osgood

1909. *Peromyscus megalops melanurus* Osgood, N. Amer. Fauna, 28:215, April 17, type from Pluma, 4600 ft., Oaxaca.

MARGINAL RECORDS.—Oaxaca: *ca.* 11 mi. E Juquila, 6100 ft. (Baker and Womochel, 1966:306); *Lacháo* (Goodwin, 1969:192); *4 mi. S Jalatengo, 5000 ft.* (Musser, 1964:12); type locality.

Peromyscus melanocarpus Osgood
Zempoaltepec Deer Mouse

1904. *Peromyscus melanocarpus* Osgood, Proc. Biol. Soc. Washington, 17:73, March 21, type from Mt. Zempoaltepec, 8000 ft., Oaxaca.

External measurements of the type, a subadult female, are: 241; 125; 27; 19.2 (dry). An adult male from Totontepec, Oaxaca, measures: 262; 132; 30. Upper parts blackish brown; underparts blackish slate washed with white; tail covered with short blackish hairs, little paler below than above; feet dusky brownish to base of toes. Skull much as in *P. megalops*.

The chief dignoastic feature of this species is the dusky color of the feet. See Map 414.

MARGINAL RECORDS.—Oaxaca (Goodwin, 1969:192, unless otherwise noted): Puerto Elijio; type locality; Totontepec (Hall and Kelson, 1959:654); Comaltepec.

Peromyscus zarhynchus Merriam
Chiapan Deer Mouse

1898 *Peromyscus zarhynchus* Merriam, Proc. Biol. Soc. Washington, 12:117, April 30, type from Tumbalá, 5500 ft., Chiapas.
1898. *Peromyscus zarhynchus cristobalensis* Merriam, Proc. Biol. Soc. Washington, 12:117, April 30, type from San Cristóbal, Chiapas.

External measurements: 303–327; 157–178; 33–38; 21.2–24 (dry). Upper parts dark brown, darkest medially, clearest along sides; underparts pale yellowish white, with or without pectoral spot; feet whitish, proximal part sometimes brownish; tail either bicolored or dusky above and blotched below. Skull large but not especially robust; nasals and rostrum markedly elongate; frontals constricted; supraorbital border trenchant but seldom beaded; lower part of infraorbital plate produced anteriorly. See Map 414.

MARGINAL RECORDS.—Chiapas: type locality; San Cristóbal; 6½ km. SE Rayón, 1675 m. (Baker, *et al.*, 1973:78, 85).

Peromyscus grandis Goodwin
Big Deer Mouse

1932. *Peromyscus grandis* Goodwin, Amer. Mus. Novit., 560:4, September 16, type from Finca Concepción, 35 mi. E Cobán, 3750 ft., Alta Verapaz, Guatemala. Known only from type locality.

External measurements of the type, an adult female, are: 315; 150; 32; 22. Upper parts rich hazel mixed with black, hazel everywhere predominating; middle of back from crown of head to base of tail darker than sides; underparts dull white, strongly overlaid with pinkish cinnamon, especially in pectoral region; fore- and hind-feet creamy white; tail unevenly bicolored, dusky above and irregularly blotched with white below. Skull large, resembling that of *P. zarhynchus*, but general form more massive; rostrum and nasals broader; incisors heavier and molar tooth-rows shorter (after Goodwin, 1932:4, 5). See Map 409.

Subgenus Megadontomys Merriam

Revised by Musser, Occas. Pap. Mus. Zool., Univ. Michigan, 636:13–19, 1 fig., June 17, 1964.

1898. *Megadontomys* Merriam, Proc. Biol. Soc. Washington, 12:115, April 30. Type, *Peromyscus thomasi* Merriam.

Size (large), cranial features, and coloration as described in the account of *P. thomasi*, the only species in the subgenus. Baculum long, approx. half length of hind foot and longer than glans penis; basal two-thirds of glands fluted and densely set with spines; urethral opening essentially terminal; midway in glans, urethra expanded and forming elongate crater provided with urethral flap—an arrangement unique in the genus. Glands accessory to male reproductive tract fundamentally as in subgenus *Peromyscus*.

Peromyscus thomasi
Thomas' Deer Mouse

External measurements: 300–350; 155–188; 31–34; 21–24.8. Selected cranial measurements are: greatest length of skull, 34.7–37.6; rostral breadth, 5.8–6.7; interorbital constriction, 5.2–5.8; alveolar length of upper molar tooth-row, 6.0–6.7. Upper parts rich tawny to darker; nose, orbital ring, region at base of vibrissae black. Pelage of venter creamy white with slate undercolor showing through (*P. t. cryophilus* darker, owing to black-tipped hairs); feet white to griz-

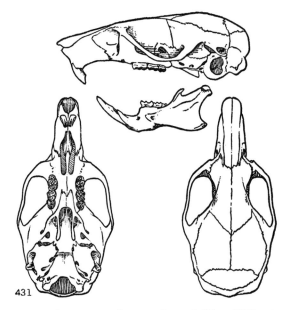

Fig. 431. *Peromyscus thomasi thomasi*, 2 km. W Puentecillas, 18 km. SSW Chichichualco, 2500 m., Guerrero, No. 35250 K.U., ♂, X 1½.

zled; hairs of feet numerous and long to few and short. Supraorbital ridge variably elevated and beaded; zygomata slightly to strongly notched by infraorbital foramen.

Peromyscus thomasi cryophilus Musser

1964. *Peromyscus thomasi cryophilus* Musser, Occas. Pap. Mus. Zool., Univ. Michigan, 636:13, June 17, type from Cerro Pelón, 9200 ft., 13 mi. NE Llano de las Flores, Distrito de Ixtlán, Oaxaca.

MARGINAL RECORDS.—Oaxaca: type locality; $6\frac{1}{2}$ mi. SSW *Vista Hermosa* (Jones and Genoways, 1967:321).

Peromyscus thomasi nelsoni Merriam

1898. *Peromyscus (Megadontomys) nelsoni* Merriam, Proc. Biol. Soc. Washington, 12:116, April 30, type from Jico, 6000 ft., Veracruz.
1964. *Peromyscus thomasi nelsoni*, Musser, Occas. Pap. Mus. Zool., Univ. Michigan, 636:18, June 17.

MARGINAL RECORDS.—Puebla: 19° 52′ N, 97° 20′ W, $12\frac{1}{10}$ km. by road NE Tezuitlán (Heaney and Birney, 1977:543, as *P. thomasi* only). Veracruz: type locality.

Peromyscus thomasi thomasi Merriam

1898. *Peromyscus (Megadontomys) thomasi* Merriam, Proc. Biol. Soc. Washington, 12:116, April 30, type from mts. near Chilpancingo, 9700 ft., Guerrero.

MARGINAL RECORDS.—Guerrero: Puerto Chico, 20 mi. N Chilpancingo (Brown and Welser, 1968:425); 15 km. S Chilpancingo, 9000 ft.; type locality.

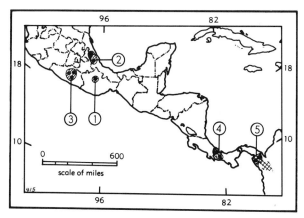

Map 415. Three species of *Peromyscus*.

1. *P. thomasi cryophilus*
2. *P. thomasi nelsoni*
3. *P. thomasi thomasi*
4. *P. flavidus*   5. *P. pirrensis*

Subgenus Isthmomys Hooper and Musser

1964. *Isthmomys* Hooper and Musser, Occas. Pap. Mus. Zool., Univ. Michigan, 635:12, April 22. Type, *Megadontomys flavidus* Bangs.

Size large (total length 320–376); M1 with anterior lamina divided (in *P. flavidus*), making 6 cusps instead of 5, to undivided in some specimens of *P. pirrensis;* supraorbital bead strongly developed. Baculum long, approx. half length of hind foot, and longer than glans penis; glans cone-shaped, flared distally (where diameter is two-thirds of length), marked by small groove ventrally; glans a third length of hind foot, armed with large, long, sharp, proximally directed spines; urethral opening ventral and subterminal. Glands accessory to male reproductive tract fundamentally as in subgenus *Peromyscus*.

KEY TO SPECIES OF SUBGENUS ISTHMOMYS

1. Length of hind foot less than 33; upper parts rich ochraceous; anterior lobe of M1 broad and distinctly divided, making 6 cusps instead of 5. . . . . . *P. flavidus,*   p. 717
1′. Length of hind foot more than 33; upper parts dark brownish cinnamon to cinnamon rufous lined with black; anterior lobe of M1 narrow and entire, or only slightly notched. . . . . *P. pirrensis,*   p. 718

Peromyscus flavidus (Bangs)
Yellow Deer Mouse

1902. *Megadontomys flavidus* Bangs, Bull. Mus. Comp. Zool., 39:27, April, type from Boquete, Volcán de Chiriquí, Panamá.
1909. *Peromyscus flavidus*, Osgood, N. Amer. Fauna, 28:221, April 17.

External measurements: 320–375; 155–205; 31–33; 20–24. Upper parts rich ochraceous, paler than in *P. pirrensis;* underparts yellowish white; forefeet white; hind feet white mixed with brownish to base of toes; tail thinly clothed with short hairs, faintly bicolored. Skull differs from that of *P. thomasi* in being larger, higher, and relatively narrower; rostrum much heavier; premaxillae much more swollen laterally; teeth and auditory bullae actually and relatively smaller; supraorbital beads strongly developed; teeth relatively short and broad; M1 with anterior lamina distinctly divided, making 6 cusps instead of 5 (after Osgood, 1909:222).

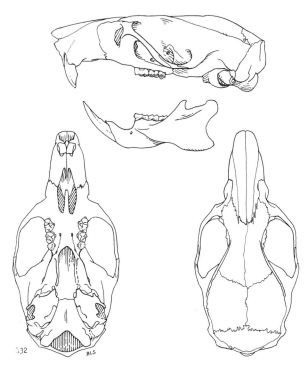

Fig. 432. *Peromyscus flavidus*, Boquete, Panamá, No. 156909 U.S.N.M., ♂, X 1½.

MARGINAL RECORDS.—Panamá: upper Río Changena, 4800 ft. (Handley, 1966:783); type locality.

Peromyscus pirrensis Goldman
Mount Pirri Deer Mouse

1912. *Peromyscus pirrensis* Goldman, Smiths. Miscl. Coll., 60(2):5, September 20, type from head Río Limón, Mt. Pirri, 4500 ft., Panamá.

External measurements of six adult topotypes are: 342–376; 185–204; 36–36.5. Upper parts dark brownish cinnamon to cinnamon rufous lined with black, becoming grayish brown on head and more rusty on rump; sides brighter, more rufescent; underparts dull buffy white, with plumbeous basal color of fur showing through everywhere; nose and upper sides of forearms to near base of toes dusky; toes of forefeet whitish; hind feet whitish more or less clouded with dusky over proximal half of metatarsus; tail brownish and nearly unicolored. Skull resembling that of *P. flavidus* but rostrum less swollen laterally; nasals more attenuate posteriorly; anterior lobe of 1st upper molar decidedly narrow, its longitudinal notch shown in *flavidus* being faint or absent (after Goldman, 1912b:5, 6).

MARGINAL RECORDS.—Panamá: Cerro Malí, 4900 ft. (Handley, 1966:783); *Tacarcuna Village, 1950 ft. (ibid.)*; type locality.

Subgenus **Habromys** Hooper and Musser

1964. *Habromys* Hooper and Musser, Occas. Pap. Mus. Zool., Univ. Michigan, 635:12, April 22. Type, *Peromyscus lepturus* Merriam.

Tail approx. as long as, or shorter than, head and body, and closely haired; superior outline of skull nearly flat or only slightly arched. Baculum short, approx. a quarter length of hind foot, but longer than glans penis; glans slightly rugose but completely nonspinous; urethral opening ventral and subterminal. Of glands accessory to male reproductive tract in subgenus *Peromyscus*, *Habromys* has vesiculars greatly reduced, anterior- and posterior prostates greatly reduced, ventral prostates greatly enlarged, and no trace of a preputial (which is vestigial in *Peromyscus*).

KEY TO NOMINAL SPECIES OF SUBGENUS HABROMYS

1. Hindfoot more than 25. . . . *P. lepturus*, p. 719
1'. Hindfoot less than 25.
  2. Total length less than 180. *P. simulatus*, p. 718
  2'. Total length more than 180.
    3. Zygomatic breadth less than 13.7; length of incisive foramina less than 5.3. . . . . . *P. chinanteco*, p. 719
    3'. Zygomatic breadth more than 13.7; length of incisive foramina more than 5.3. . . . . . . *P. lophurus*, p. 719

Peromyscus simulatus Osgood
Jico Deer Mouse

1904. *Peromyscus simulatus* Osgood, Proc. Biol. Soc. Washington, 17:72, March 21, type from near Jico, 6000 ft., Veracruz.

External measurements of the type, an adult, are: 169; 87; 21; 14.3 (dry). Color as in *P. lophurus* but "dark markings of feet and face slightly more intense; tail chiefly brown, but with a narrow line of white on under side" (Osgood, 1909:193). Skull essentially as in *P. lophurus* but smaller; braincase more inflated; rostrum more depressed; auditory bullae relatively larger; interorbital constriction wider; teeth small.

This species is distinguishable from its congeners by a combination of small size, crested tail, and dark brown feet. See Map 413.

MARGINAL RECORDS.—Veracruz: 3 km. W Zacualpan, 6000 ft. (83263 KU); type locality.

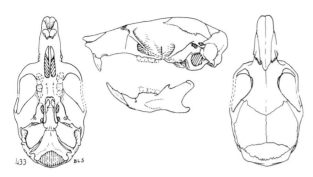

Fig. 433. *Peromyscus simulatus*, Jico, Veracruz, No. 55027 U.S.N.M., ♂, X 1½.

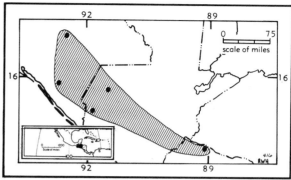

Map 416. *Peromyscus lophurus*.

Peromyscus chinanteco Robertson and Musser
Chinanteco Deer Mouse

1976. *Peromyscus chinanteco* Robertson and Musser, Univ. Kansas Mus. Nat. Hist., Occas. Pap., 47:1, March 1, type from N slope Cerro Pelón, 31⅜ km. S Vista Hermosa, 2650 m., Oaxaca.

External measurements of three males (one young, two subadults) are, respectively: 212, 194, 205; 121, 103, 113; 23, 24, 24; two females (one adult, one subadult): —, 192; —, 104; 23, 23. Upper parts grayish-brown; underparts grayish-white; tail unicolored; dorsal pelage of all four feet proximal to toes grayish-brown, toes white; eye-ring dark; patch of white fur from which mystacial vibrassae arise.

P. chinanteco differs from its nearest relative, *P. simulatus*, as follows: larger; tail both absolutely and relatively longer; nasals narrower and slightly longer; cranium slightly more inflated; incisive foramina anterior to upper molars instead of extending behind anterior borders of 1st molars.

MARGINAL RECORDS (Robertson and Musser, 1976:7).—Oaxaca: 21 km. S Vista Hermosa, 2080 m.; 28⅜ km. S Vista Hermosa, 2350 m.; type locality. These occurrences are covered by the northernmost dot on Map 413 representing the geographic occurrence of *Peromyscus lepturus ixtlani*. Not mapped.

Peromyscus lophurus Osgood
Crested-tailed Mouse

1904. *Peromyscus lophurus* Osgood, Proc. Biol. Soc. Washington, 17:72, March 21, type from Todos Santos, 8500 ft., Huehuetenango, Guatemala.

Average external measurements of two adult topotypes: 208; 105; 24.5; 16 (dry). Upper parts between wood brown and fawn color, with small dusky area in middle of back; underparts white; tail long, well haired, markedly penicillate, unicolored sepia brown; forefeet white, hind feet dusky or brownish to base of toes, which are white. Skull resembling that of *lepturus*, but smaller and having rostral part decidedly shorter; molar teeth actually about same size, but larger relative to length of skull; interparietal unusually large (after Osgood, 1909:192). This highland species is known from 6400 to 10,200 ft.

MARGINAL RECORDS.—Chiapas: San Cristóbal [= Ciudad de las Casas]. Guatemala: type locality. El Salvador: Los Esesmiles, 8000 ft. (Musser, 1969:20). Guatemala: S slope Volcán Tajumulco (Musser, 1969:20). Chiapas: Ambos de Triunfo, 1950 m.

Peromyscus lepturus Merriam
Slender-tailed Deer Mouse

External measurements: 210–285; 109–145; 26–30; 16.4–23 (dry). Upper parts brownish black lightly sprinkled with cinnamon; sides, shoulders, and head paler; underparts creamy white with slaty basal tones showing through clearly, pectoral region sometimes faintly washed with russet; forefeet white, dusky color of hind leg extending onto dorsum of foot, in some specimens to toes; tail either bicolored or uniformly dark. Skull: nasals and palatine slits long; zygomata slightly to strongly convergent anteriorly; frontals small, constricted, not beaded. Some specimens of *P. mexicanus totontepecus* and *P. melanocarpus*, also from Mt. Zempoaltepec, closely resemble *P. lepturus* in color but their skulls are larger and their teeth relatively smaller. See Map 413.

Peromyscus lepturus ixtlani Goodwin

1964. *Peromyscus ixtlani* Goodwin, Amer. Mus. Novit., 2183:2, June 4, type from Cerro Machín, 5 km. NE Macuiltianguis, 9000 ft., Oaxaca.

1969. *Peromyscus lepturus ixtlani*, Musser, Amer. Mus. Novit., 2357:17, January 27.

MARGINAL RECORDS.—Oaxaca (Musser, 1969: 12): type locality; *Cerro Pelón, 13 road mi. NE Llano de las Flores, 8900 ft.*; 4 mi. by road SW Llano de las Flores, 8700 ft.

Peromyscus lepturus lepturus Merriam

1898. *Peromyscus lepturus* Merriam, Proc. Biol. Soc. Washington, 12:118, April 30, type from Mt. Zempoaltepec, 8200 ft., Oaxaca.

MARGINAL RECORDS.—Oaxaca: Totontepec; type locality.

Subgenus **Podomys** Osgood

Revised by Osgood, N. Amer. Fauna, 28:226–228, April 17, 1909.

1909. *Podomys* Osgood, N. Amer. Fauna, 28:226, April 17. Type, *Hesperomys floridanus* Chapman.

Plantar tubercles of hind foot, 5, instead of 6 as in other subgenera of *Peromyscus;* digital tubercles, 3; phalangeal, 2, latter much reduced and subcircular; molars slightly more hypsodont than in *Peromyscus* but less so than in *Onychomys;* accessory tubercles in salient interangles of molars small, as seen in transverse section, and never forming loop extending to outer edge of tooth, as in *Peromyscus;* mammae, 6 (i. $\frac{2}{2}$, a. $\frac{0}{0}$, p. $\frac{1}{1}$) (after Osgood, 1909:226, 227).

Baculum only a quarter as long as hind foot; glans penis small, simple, distal half and extreme basal part superficially smooth, and urethral opening terminal when ventral process of glands is folded. Of glands accessory to male reproductive tract in subgenus *Peromyscus, Podomys* has greatly reduced vesiculars, greatly reduced anterior- and dorsal prostates, enlarged ampullaries, and a deferent duct of unique structure.

Peromyscus floridanus (Chapman)
Florida Mouse

1889. *Hesperomys floridanus* Chapman, Bull. Amer. Mus. Nat. Hist., 2:117, June 7, type from Gainesville, Alachua Co., Florida.
1896. *Peromyscus floridanus*, Bangs, Proc. Biol. Soc. Washington, 10:122, November 5.
1890. *Hesperomys macropus* Merriam, N. Amer. Fauna, 4:53, October 8, type from Lake Worth, Palm Beach Co., Florida.

External measurements: 186–221; 80–95; 24–29; 22–25. Pelage long, soft, silky. Upper parts bright ochraceous buff, finely mixed with dusky dorsally, clearest on sides; underparts

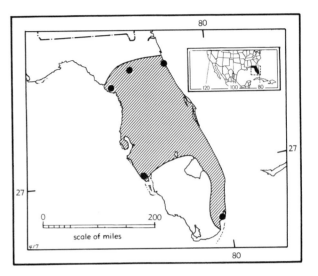

Map 417.   *Peromyscus floridanus.*

creamy, sometimes with buffy pectoral suffusion; feet white; tail brownish dusky above, white below. Skull large, relatively deep; supraorbital border developed into a shelf; palatine slits short and widely open; interpterygoid fossa wide and truncate anteriorly molars large and broad.

This species seems to occur only on high well-drained sandy ridges covered with pines and/or palmetto.

MARGINAL RECORDS.—Florida: type locality; Anastasia Island; Miami; Englewood (Layne, 1974:393); Levy County (Fertig and Layne, 1963:322).

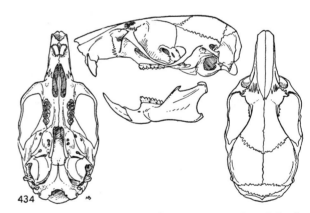

Fig. 434. *Peromyscus floridanus*, 20 mi. SW Orlando, Orange Co., Florida, No. 96832 M.V.Z., ♂, X 1½.

Subgenus **Osgoodomys** Hooper and Musser

1964. *Osgoodomys* Hooper and Musser, Occas. Pap. Mus. Zool., Univ. Michigan, 635:12, April 22. Type, *Peromyscus banderanus* J. A. Allen.

Skull narrow; posterior part of braincase so much elongated that more than half of interparietal lies behind tympanic bullae; supraorbital beads well developed, forming trenchant shelf above orbit and bounded on inner side by distinct groovelike channel extending from lachrymal region to or beyond parietofrontal suture; lachrymal region swollen. Baculum minute, only a fifth as long as hind foot; glans penis small, simple, and urethral opening terminal (no ventral process beyond opening). Of glands accessory to male reproductive tract in subgenus *Peromyscus*, *Osgoodomys* lacks vesicular and both anterior- and dorsal-prostates but has exceptionally large and distinctive ampullaries and well-developed preputials.

Peromyscus banderanus
Michoacán Deer Mouse

External measurements: 216–245; 107–132; 24–28; 14–18.5. Upper parts ochraceous buff with mixture of cinnamon to dusky; underparts

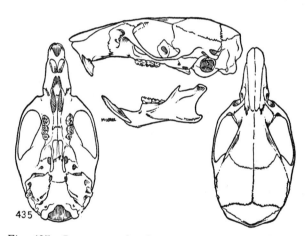

Fig. 435. *Peromyscus banderanus vicinior*, 3.2 km. SSE Iguala, 970 m., Guerrero, No. 28367 K.U., ♀, X 1½.

creamy, broad ochraceous buff pectoral patch may be absent; orbital ring and spot at base of whiskers Vandyke Brown to sooty; feet white; tarsals Prout's Brown to sooty; tail brownish to blackish above, uniform white to blotched with dusky or yellow below. Auditory bullae small; plantar surface of hind feet partially to totally naked.

This species is distinguishable from its congeners by the combination of large size, naked soles of hind feet, buffy color, and long, narrow, beaded skull. See Map 409.

Peromyscus banderanus banderanus J. A. Allen

1897. *Peromyscus banderanus* J. A. Allen, Bull. Amer. Mus. Nat. Hist., 9:51, March 15, type from Valle de Banderas, Nayarit.

MARGINAL RECORDS.—Nayarit: Navarrete. Michoacán: Los Reyes; Zitácuaro; 6 mi. from Apatzingán. Guerrero: ½ mi. E Coalcomán; El Limón; Río Aguacatillo, 30 km. N Acapulco, 1000 ft.; near Ometepec, thence northwestward along Pacific Coast to Michoacán: *3 km. S Melchor Ocampo* (Alvarez, 1968:32) and presumably to Nayarit.

Peromyscus banderanus vicinior Osgood

1904. *Peromyscus banderanus vicinior* Osgood, Proc. Biol. Soc. Washington, 17:68, March 21, type from La Salada, Michoacán.

MARGINAL RECORDS.—Michoacán: type locality; 2½ mi. S, 1 mi. E Tacámbaro, 4700 ft. Guerrero: 3⅕ km. SSE Iguala, 970 m.; Acahuizotla. Michoacán: 22 km. S Arteaga (Alvarez, 1968:33).

Genus Ochrotomys Osgood—Golden Mouse

Revised by Packard, Univ. Kansas Mus. Nat. Hist., Miscl. Publ., 51:373–406, July 11, 1969.

1909. *Ochrotomys* Osgood, N. Amer. Fauna, 28:222, April 17. Type, *Arvicola nuttalli* Harlan. (Proposed as subgenus of *Peromyscus*. Raised to generic rank by Hooper, Miscl. Publ. Mus. Zool., Univ. Michigan, No. 105:23, December 29.)

Plantar tubercles, 6, with rudimentary 7th adjacent to large tubercle at base of 5th digit. Mammae, 6 (i. $\frac{2}{2}$, p. $\frac{1}{1}$). Baculum small, shorter in proportion to width of base than in *Peromyscus* (Blair, 1942:201); glans short, urn-shaped, and with fleshy bilobed process (Hooper, 1958:20, but see Hooper and Musser, 1964:9). Skull: posterior palatine foramina farther back than in *Peromyscus*. Molariform teeth relatively wide, and with enamel folds much compressed; tubercles relatively low; tendency for development of a raised cingulum marked by subsidiary tubercles in inner salient angles of M1 and M2; enamel thicker than in *Peromyscus*, generally occupying more of occlusal surfaces of worn teeth; pattern of occlusal surfaces in partly worn upper and lower molars much compressed laterally and longitudinally, leaving 5 subtriangular islands of dentine in M1 and 4 in M2. Lacking entepicondylar foramen (Rinker, 1960:276, but see Manville, 1961:104). Young behaviorally and physically precocious as compared with various species of *Peromyscus* (Layne, 1960:52). (Description after Osgood, 1909:222, 223, except where otherwise noted.)

Ochrotomys nuttalli
Golden Mouse

External measurements: 150–190; 68–93; 17–20; 13.4–16.4 (dry). Pelage soft and thick; underparts heavily furred. Upper parts uniformly rich tawny ochraceous; feet and underparts creamy, often washed with ochraceous on venter; tail pale golden brown above, creamy below. Skull relatively broad, braincase inflated; nasals short, compressed posteriorly; interpterygoid fossa truncate anteriorly.

Ochrotomys nuttalli aureolus (Audubon and Bachman)

1841. *Mus (Calomys) aureolus* Audubon and Bachman, Proc. Acad. Nat. Sci. Philadelphia, 1:98, type from the oak forests of South Carolina. Neotype designated from Marshall, Madison Co., North Carolina, by Packard, Univ. Kansas Mus. Nat. Hist., Miscl. Publ., 51:395, July 11, 1969.

1963. *Ochrotomys nuttali* [sic] *aureolus*, Rippy and Harvey, Trans. Kentucky Acad. Sci., 24:5.

1909. *Peromyscus nuttalli aureolus*, Osgood, N. Amer. Fauna, 28:225, April 17.

MARGINAL RECORDS (Packard, 1969:396).— Kentucky: Salt Lick Creek, 2½ mi. W Charters; 10 mi. E Olive Hill. Virginia: vic. Blacksburg. North Carolina: Asheville. Georgia: Toccoa. North Carolina: Murphy. Georgia: Rome. Kentucky: 8 mi. NNE Golden Pond; 1.37 mi. NW jct. Kentucky Highway 109 and U.S. Highway 62. Not found: Kentucky: Crancreek.

Ochrotomys nuttalli flammeus Goldman

1941. *Peromyscus nuttalli flammeus* Goldman, Proc. Biol. Soc. Washington, 54:190, December 8, type from Delight, Pike Co., Arkansas.

1959. O[chrotomys]. n[uttalli]. *flammeus*, McCarley, Texas Jour. Sci., 11:410, December.

MARGINAL RECORDS.—Missouri: 7 mi. SE Cassville (Brown, 1963:425). Arkansas: Beebe (Packard, 1969:397); 10 mi. N Pine Bluff (*ibid.*). Oklahoma: 15 mi. SE Broken Bow; 5 mi. E Nashoba (McCarley,

1961:109); Redland (Packard, 1969:397). Arkansas: ½ mi. W Fayetteville.

Ochrotomys nuttalli floridanus Packard

1969. *Ochrotomys nuttalli floridanus* Packard, Univ. Kansas Mus. Nat. Hist., Miscl. Publ., 51:397, July 11, type from Welaka, Putnam Co., Florida.

MARGINAL RECORDS (Packard, 1969:398).— Florida: Chattahoochee; New Berlin, thence southward along Atlantic Coast to opposite Hicoria, thence northward along Gulf Coast to Spring Creek.

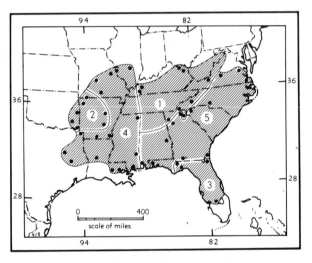

Map 418. *Ochrotomys nuttalli.*

Guide to subspecies	
1. *O. n. aureolus*	3. *O. n. floridanus*
2. *O. n. flammeus*	4. *O. n. lisae*
	5. *O. n. nuttalli*

Ochrotomys nuttalli lisae Packard

1969. *Ochrotomys nuttalli lisae* Packard, Univ. Kansas Mus. Nat. Hist., Miscl. Publ., 51:398, July 11, type from La Nana Creek bottoms, 1 mi. E Stephen F. Austin State College Campus, Nacogdoches, Nacogdoches Co., Texas.

MARGINAL RECORDS (Packard, 1969:400, unless otherwise noted).—Illinois: Salem (Hall and Kelson, 1959:657). Mississippi: 2½ mi. S, 2½ mi. E Tishomingo; Trigg Area; Harrison County (Kennedy, *et al.*, 1974:18). Louisiana (Lowery, 1974:249): 2 mi. S Ponchatoula; East Baton Rouge Parish; 5 mi. E Lena Station. Texas: Trinity County (Davis, 1966:194, as *O. nuttalli* only); 20 mi. NW Palestine; Bowie County (Davis, 1966:194, as *O. nuttalli* only). Arkansas: El Dorado; Island 80. Missouri: Willowspring; Hahatouka; Meramec State Park; St. Louis (Goldman, 1941:191, as *P. n. flammeus*).

Ochrotomys nuttalli nuttalli (Harlan)

1832. *Arvicola nuttalli* Harlan, Month. Amer. Jour. Geol. Nat. Sci., Philadelphia, p. 446, April. Type locality, Norfolk, Nor-

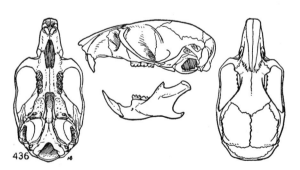

Fig. 436. *Ochrotomys nuttalli aureolus,* Chepanoke, Perquimaus Co., North Carolina, No. 54089 M.V.Z., ♂, X 1½.

folk Co., Virginia. Neotype designated from Lake Drummond, Dismal Swamp, Nansemond Co., Virginia, by Packard, Univ. Kansas Mus. Nat. Hist., Miscl. Publ., 51:401, July 11, 1969.

1958. *Ochrotomys nuttalli*, Hooper, Miscl. Publ. Mus. Zool., Univ. Michigan, No. 105:23, December 29.

1939. *Peromyscus nuttalli lewisi* A. H. Howell, Jour. Mamm., 20:498, November 14, type from Amelia Court House, Amelia Co., Virginia.

MARGINAL RECORDS (Packard, 1969:402, unless otherwise noted).—Virginia: Bosher's Dam, James River (Hall and Kelson, 1959:658); Lake Drummond, thence southward along Atlantic Coast to Georgia: St. Mary's; 4 mi. S Beachton. Florida: Fort Walton. Alabama: Mobile; Auburn. Georgia: Marietta; White County. South Carolina: Cliff Ridge Road. North Carolina: Charlotte. Virginia: Lynchburg.

Genus Baiomys True—Pygmy Mice

Revised by Packard, Univ. Kansas Publ., Mus. Nat. Hist. 9:579–670, 4 pls., 12 figs., June 16, 1960.

1894. *Baiomys* True, Proc. U.S. Nat. Mus., 16:758, February 7. Type, *Hesperomys (Vesperimus) taylori* Thomas.

External measurements: 93–135; 33–56; 12–17. Upper parts blackish sepia to ochraceous buff; underparts slaty gray to white or pale buff; 6 plantar pads; soles nearly naked except for some hairs on anterior parts of soles and anteriorly to base of toes and between toes; occipitonasal length of skull, 17.0–21.5; zygomatic breadth, 9.0–11.5; coronoid process of mandible well developed, strongly recurved; ascending ramus of mandible short and erect; posterior palatine foramina nearly opposite middle of M2; interorbital space wide relative to widest part of frontals; nasals projecting only slightly anteriorly to incisors; primary 1st fold of M3 obliterated at an early stage of wear; major cusps of upper and lower anteriormost 2 molars alternating, more so in m1–m2 than in M1–M2; dentition, i. $\frac{1}{1}$, c. $\frac{0}{0}$, p. $\frac{0}{0}$, m. $\frac{3}{3}$.

The two living species differ most in size and shape of body and length of skull in the one area where they occur together (are sympatric), which is west-central México. Elsewhere there is less difference in these morphological features. According to Packard (1960:664) "This is a documented instance of character displacement. . . . On the basis of internal morphological characters . . . *Baiomys* seems to be more closely related to a South American hesperom[y]ine, perhaps *Calomys*, than to any North American cricetine." Lesser size, relatively smaller and more rounded ears, and a longer coronoid process are three of several features differentiating *Baiomys* from *Peromyscus*.

KEY TO SPECIES OF BAIOMYS

1. Longitudinal, dorsal outline of skull (viewed from side) evenly convex; entoglossal process of basihyal pointed; baculum rounded at tip and 3.0–3.9 mm. long; short process of incus knob-shaped.
   Baiomys musculus, p. 723
1'. Longitudinal, dorsal outline of skull (viewed from side) not evenly convex (outline anteriorly deflected ventrally from frontoparietal suture); entoglossal process of basihyal absent or much reduced; baculum notched at tip and rarely longer than 2.9 mm.; short process of incus attenuated. . . . .*Baiomys taylori*, p. 725

Baiomys musculus
Southern Pygmy Mouse

External measurements: 100–135; 33–56; 14.1–17.0. Upper parts dark reddish brown or ochraceous-buff to nearly black; underparts pale pinkish buff to white or pale buffy; occipitonasal length of skull, 18.2–21.5; zygomatic breadth, 9.4–11.5; rostrum gradually curved toward anterior end of nasals; entoglossal process of basihyal pointed; baculum more than 2.9 mm. long; short process of incus knob-shaped.

Baiomys musculus brunneus (J. A. Allen and Chapman)

1897. *Peromyscus musculus brunneus* J. A. Allen and Chapman, Bull. Amer. Mus. Nat. Hist., 9:203, June 16, type from Jalapa, Veracruz.

1912. *Baiomys musculus brunneus*, Miller, Bull. U.S. Nat. Mus., 79:137, December 31.

MARGINAL RECORDS (Packard, 1960:614).—Veracruz: type locality; Santa María, near Mirador, 1800 ft.; *Chichicaxtla*; Boca del Río; Río Blanco, 20 km. WNW Piedras Negras, 400 ft.; Presidio; 3 km. SE Orizaba, 5500 ft.; *Teocelo*.

Baiomys musculus grisescens Goldman

1932. *Baiomys musculus grisescens* Goldman, Proc. Biol. Soc. Washington, 45:121, July 30, type from Comayabuela [= Comayaguela], 3100 ft., just S of Tegucigalpa, Honduras.

MARGINAL RECORDS (Packard, 1960:616).—Guatemala: ½ mi. N, 1 mi. E Salama, 3200 ft. Honduras: El Caliche, Cedros; Hatillo; La Piedra de Jesús, Sabana Grande; Cementerio. Guatemala: Lake Atescatempa; 1 mi. S Rabinal, 3450 ft.

Baiomys musculus handleyi Packard

1958. *Baiomys musculus handleyi* Packard, Univ. Kansas Publ., Mus. Nat. Hist., 9:399, December 19, type from Sacapulas, El Quiche, Guatemala. Known only from type locality.

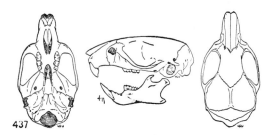

Fig. 437. *Baiomys musculus brunneus*, 7 km. NNW Cerro Gordo, Veracruz, No. 24320 K.U., ♀, X 1½.

Baiomys musculus infernatis Hooper

1952. *Baiomys musculus infernatis* Hooper, Jour. Mamm., 33:96, February 18, type from Teotitlán, Oaxaca.

MARGINAL RECORDS.—Puebla: Tepanco, 6000 ft.; Tehuacán, 5400 ft. Oaxaca: type locality.

Baiomys musculus musculus (Merriam)

1892. *Sitomys musculus* Merriam, Proc. Biol. Soc. Washington, 7:170, September 29, type from Colima, Colima.
1907. *Baiomys musculus*, Mearns, Bull. U.S. Nat. Mus., 56:381, April 13.

MARGINAL RECORDS (Packard, 1960:622, 623, unless otherwise noted).—Nayarit: 3 mi. NNW Las Varas, 150 ft. Jalisco: 13 mi. S, 15 mi. W Guadalajara. Michoacán: 12 mi. S Tzitzio; San José Prura [= Purúa]. Guerrero (Alvarez, 1968:34): *4 km. N Zacatula;* Zacatula. Michoacán: *3 km. S Melchor Ocampo (ibid.).* Colima: Armería. Jalisco: Chamela Bay.

Baiomys musculus nigrescens (Osgood)

1904. *Peromyscus musculus nigrescens* Osgood, Proc. Biol. Soc. Washington, 17:76, March 21, type from Valley of Comitán, Chiapas.
1912. *Baiomys musculus nigrescens*, Miller, Bull. U.S. Nat. Mus., 79:137, December 31.

MARGINAL RECORDS (Packard, 1960:625, unless otherwise noted).—Chiapas: Bochil; 25 mi. E Comitán, Las Margaritas, 1250 ft. Guatemala: Chanquejelve; La Primavera; Jacaltenango, 5400 ft.; 4 mi. S Guatemala City, 4700 ft.; El Progresso. El Salvador: Mt. Cacaguatique (Burt and Stirton, 1961:58); San Miguel (*ibid.*); 1 mi. S Los Planes; Chilata (Burt and Stirton, 1961:58). Guatemala: El Zapote. Chiapas: Mapastepec, 45 m.; Pijijiapan, 10 m.; 6 mi. NW Tonalá; 15 mi. SW Las Cruces; Cintalpa, 555 m.; Ocuilapa, 3500 ft.

Baiomys musculus pallidus Russell

1952. *Baiomys musculus pallidus* Russell, Proc. Biol. Soc. Washington, 65:21, January 29, type from 12 km. NW Axochiapan, 3500 ft., Morelos.
1959. *Baiomys musculus nebulosus* Goodwin, Amer. Mus. Novit., 1929:1, March 5, type from Xanangua [= Xanaguia], District of Miahuatlan, Oaxaca, México, 6000 ft.

MARGINAL RECORDS (Packard, 1960:627, 628, unless otherwise noted).—Morelos: 5 mi. W Tepoztlán. Puebla: 2 mi. S Atlixco, 5800 ft.; Acatlán, 4100 ft. Oaxaca: 2 mi. NW Tamazulapán, 6550 ft.; Tepantepec; Oaxaca, 500 ft.; Yalalag, 3000 ft.; El Campanario; Tapanatepec (Goodwin, 1969:194), thence up coast to Guerrero: Zihuatanejo Bay; El Limón; Texalzintla, 6 km. NNW Teloloapan (Hall and Kelson, 1959:662). Morelos: Tetacala (*ibid.*); Cuernavaca (*ibid.*).

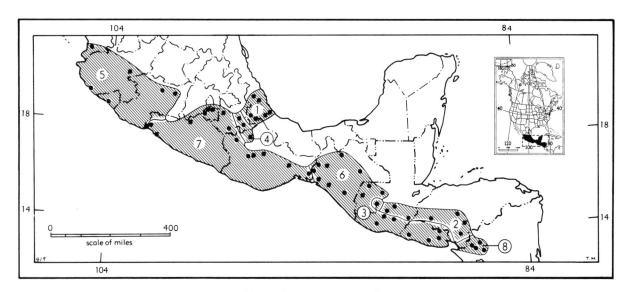

Map 419. *Baiomys musculus.*

1. *B. m. brunneus*	3. *B. m. handleyi*	5. *B. m. musculus*	7. *B. m. pallidus*
2. *B. m. grisescens*	4. *B. m. infernatis*	6. *B. m. nigrescens*	8. *B. m. pullus*

Baiomys musculus pullus Packard

1958. *Baiomys musculus pullus* Packard, Univ. Kansas Publ., Mus. Nat. Hist., 9:401, December 19, type from 8 mi. S Condega, Estelí, Nicaragua.

MARGINAL RECORDS (Packard, 1960:629, unless otherwise noted).—Nicaragua: San Rafael Del Norte; *Santa María de Ostuma* (Jones and Genoways, 1970:2, as *B. musculus* only); Matagalpa; 9 mi. NNW Estelí (Jones and Genoways, 1970:4, as *B. musculus* only); type locality.

Baiomys taylori
Northern Pygmy Mouse

External measurements: 87–123; 34–53; 12–15. Upper parts pale drab or reddish brown to almost black; underparts grayish to cream-buff; occipitonasal length, 16.8–19.0; zygomatic breadth, 8.7–10.2; rostrum deflected ventrally at fronto-nasal suture; entoglossal process of basihyal much reduced or absent; baculum less than 3 mm. long; short process of incus attenuated.

Baiomys taylori allex (Osgood)

1904. *Peromyscus allex* Osgood, Proc. Biol. Soc. Washington, 17:76, March 21, type from Colima, Colima.
1958. *Baiomys taylori allex*, Packard, Proc. Biol. Soc. Washington, 71:17, April 11.

MARGINAL RECORDS (Packard, 1960:636, unless otherwise noted).—Nayarit: 3 mi. SE Mirador; 1 mi. E Ixtlán, 4000 ft. Jalisco: Etzatlán; Ameca; 27 mi. S, 12 mi. W Guadalajara (33562 KU). Michoacán: 9 mi. S Lombardía; 10 mi. S, 1 mi. W Apatzingán. Colima: type locality. Nayarit: 2 mi. WNW Valle de Banderas, near sea level; 3 mi. NNW Las Varas, 150 ft. (Hall and Kelson, 1959:659).

Baiomys taylori analogus (Osgood)

1909. *Peromyscus taylori analogus* Osgood, N. Amer. Fauna, 28:256, April 17, type from Zamora, Michoacán.
1912. *Baiomys taylori analogus*, Miller, Bull. U.S. Nat. Mus., 79:137, December 31.

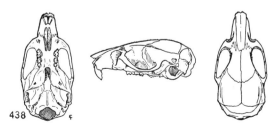

Fig. 438. *Baiomys taylori analogus*, 23 km. E Mexico City, México, No. 99873 M.V.Z., ♂, X 1½.

MARGINAL RECORDS (Packard, 1960:639, 640).—San Luis Potosí: Hda. Capulín; 3$\frac{3}{10}$ mi. by road N Tamazunchale. Hidalgo: Tula, 2050 m. México: Templo del Sol, Pyramides de San Juan, Teotihuacán. Veracruz: Acultzingo. México: 4 km. ENE Tlalmanalco. Distrito Federal: 200 m. N San Mateo Xalpa (Jalpa), 2390 m. Michoacán: 2 mi. SW Zitacuaro; 1 mi. E, 6 mi. S Tacámbaro; La Huacana; Uruapan. Jalisco: 2 mi. N Ciudad Guzmán; 27 mi. S, 12 mi. W Guadalajara; 13 mi. S, 15 mi. W Guadalajara; 7 mi. NW Tepatitlán; 1 mi. S Jalostotitlán, 5700 ft. Guanajuato: 4 mi. N, 5 mi. W León. Querétaro: 6 mi. E Querétaro, 6550 ft.; Tolimán.

Baiomys taylori ater Blossom and Burt

1942. *Baiomys taylori ater* Blossom and Burt, Occas. Pap. Mus. Zool., Univ. Michigan, 465:2, October 8, type from 7 mi. W Hereford, Cochise Co., Arizona.

MARGINAL RECORDS (Packard, 1960:642, unless otherwise noted).—Arizona: 1½ mi. SW Fort Grant, Graham Mts. New Mexico: 18 mi. S, 2 mi. W Animas; 25½ mi. S Animas (in Big Bill Canyon). Chihuahua: Casas Grandes; *Río Piedras Verdes, 5 mi. WNW Colonia Juárez* (Anderson, 1972:351); 35 mi. NW Dublán, 5300 ft. (ibid.); 2½ mi. N, 3 mi. W San Francisco, 5200 ft. Arizona: 9 mi. W Hereford; Patagonia; 1½ mi. ENE Greaterville, Thurber Ranch.

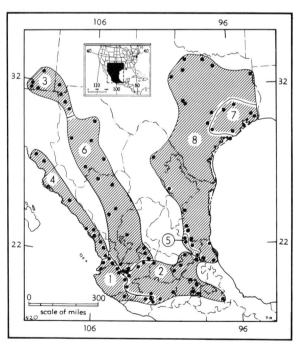

Map 420. *Baiomys taylori.*

1. *B. t. allex*	5. *B. t. fuliginatus*
2. *B. t. analogus*	6. *B. t. paulus*
3. *B. t. ater*	7. *B. t. subater*
4. *B. t. canutus*	8. *B. t. taylori*

Baiomys taylori canutus Packard

1960. *Baiomys taylori canutus* Packard, Univ. Kansas Publ., Mus. Nat. Hist., 9:643, June 16, type from 1 mi. S Pericos, Sinaloa.

MARGINAL RECORDS (Packard, 1960:645).— Sonora: [Ciudad] Obregón; *10³/5 mi. SE [Ciudad] Obregón;* 1 mi. NNW Navajoa [Navojoa]. Sinaloa: Culiacán, 175 ft.; 15 mi. N Rosario, Chelé, 300 ft. Nayarit: Acaponéta; 2 mi. SW Rosa Morada [Rosamorada]; 3 mi. SE Tepic; *2 mi. WNW Tepic, 3200 ft.* Sinaloa: Escuinapa; Mazatlán.

Baiomys taylori fuliginatus Packard

1960. *Baiomys taylori fuliginatus* Packard, Univ. Kansas Publ., Mus. Nat. Hist., 9:645, June 16, type from 2 mi. N, 10 mi. E Ciudad del Maíz, 4000 ft., San Luis Potosí.

MARGINAL RECORDS.—San Luis Potosí: El Salto (Packard, 1960:647); 8 mi. W El Naranjo, 2400 ft. (Baker and Phillips, 1965:338); type locality.

Baiomys taylori paulus (J. A. Allen)

1903. *Peromyscus paulus* J. A. Allen, Bull. Amer. Mus. Nat. Hist., 19:598, November 12, type from Río Sestín, northwestern Durango.
1912. *Baiomys taylori paulus,* Miller, Bull. U.S. Nat. Mus., 79:137, December 31.

MARGINAL RECORDS (Packard, 1960:649, unless otherwise noted).—Chihuahua (Anderson, 1972:351): 40 mi. E Gallego, 5000 ft.; 5 mi. SE Chihuahua; 15 mi. ESE Boquillas, 4700 ft. Durango: 14 mi. E Zarca (Hall and Kelson, 1959:659); San Gabriel. Zacatecas: Valparaíso, 6500 ft. Aguascalientes: 1 mi. N Chicalote; 16 mi. S Aguascalientes. Jalisco: 2 mi. WNW Lagos de Moreno; 4 mi. W Guadalajara, 5100 ft.; 3 mi. W Tala, 4300 ft.; 2 mi. NW Magdalena, 4500 ft. Durango: 5 mi. N Durango, 6400 ft.; type locality. Chihuahua (Anderson, 1972:351): 10 mi. W Balleza; *2 mi. W Miñaca, 6900 ft.;* Rancho San Ignacio, 4 mi. S, 1 mi. W Santo Tomás.

Baiomys taylori subater (V. Bailey)

1905. *Peromyscus taylori subater* V. Bailey, N. Amer. Fauna, 25:102, October 24, type from Bernard Creek, near Columbia, Brazoria Co., Texas.
1912. *Baiomys taylori subater,* Miller, Bull. U.S. Nat. Mus., 79:137, December 31.

MARGINAL RECORDS (Packard, 1960:651, unless otherwise noted).—Texas: Huntsville; Sour Lake; 7 mi. S La Belle; Virginia Point; *14 mi. SSE Alvin;* type locality; 4 mi. NNE Yoakum (Hall and Kelson, 1959:660); 3 mi. W College Station, 1 mi. W Easterwood Airport.

Baiomys taylori taylori (Thomas)

1887. *Hesperomys (Vesperimus) taylori* Thomas, Ann. Mag. Nat. Hist., ser. 5, 19:66, January, type from San Diego, Duval Co., Texas.

1907. *Baiomys taylori,* Mearns, Bull. U.S. Nat. Mus., 56:381, April 13.

MARGINAL RECORDS (Packard, 1960:655, unless otherwise noted).—Texas: 3 mi. N Bowie; 3 mi. S Denton (Baccus, *et al.,* 1971:148); Lotta (*op. cit.:* 149); 25 mi. E Austin; 7 mi. SE Luling (Hall and Kelson, 1959:660); Matagorda Peninsula (*ibid.*), thence southward along coast to Veracruz: Tampico Alto; *Cerro Azul;* Potrero Llano; Platón Sánchez, 800 ft. (Hall and Dalquest, 1963:311). San Luis Potosí: Ciudad Valles. Tamaulipas (Alvarez, 1963:447): Antiguo Morelos; Rancho Pano Ayuctle; Jaumave; Hidalgo. Nuevo León: Santa Catarina. Coahuila: 6 mi. SW San Gerónimo. Texas: 6 mi. S Mason (Packard and Garner, 1964:388); 15 mi. S Brady (Packard and Judd, 1968:537); 7 mi. E Roby (Hart, 1972:214, as *B. taylori* only); *16 mi. N Albany* (Packard and Judd, 1968:537); 20 mi. N Albany (*ibid.*); 5 mi. SW Bomarton (Baccus, 1971:182).

Genus **Onychomys** Baird—Grasshopper Mice

Revised by Hollister, Proc. U.S. Nat. Mus., 47:427–489, October 29, 1914.

1858. *Onychomys* Baird, Mammals, *in* Repts. Expl. Surv. . . . , 8(1):458, July 14. Type, *Hypudaeus leucogaster* Wied-Neuwied.

External measurements: 119–190; 29–62; 17–25; 11–24. Stout mice with short, relatively thick tails; forefeet with 5 plantar tubercles, hind feet with 4; sole of hind foot densely furred from heel to tubercles. Pelage sharply bicolored; underparts white, upper parts variable according to species. Skull with distinctly wedge-shaped nasals, which extend well beyond premaxillary tongues; interorbital constriction narrow; zygomatic plate narrow, straight anteriorly; molars more hypsodont than in *Peromyscus;* M3 reduced; coronoid process of mandible high.

Much confusion has arisen regarding the colors of *Onychomys* because of the complex nature of the molts. Briefly, the juvenal pelage is an even gray; the adult pelage strongly buffy or tawny; and old individuals often are uniformly gray above, thus closely resembling subadults in color.

KEY TO SPECIES OF ONYCHOMYS

1. Tail usually less than ½ length of head and body; M3 subcircular in cross section; M1 less than ⅓ length of tooth-row.
   O. leucogaster, p. 727
1'. Tail usually more than ½ length of head and body; M3 transversely ovoid in cross section; M1 more than ⅓ length of tooth-row.
   O. torridus, p. 730

Onychomys leucogaster
Northern Grasshopper Mouse

x ⅔

External measurements: 130–190; 29–61; 17–25; 13–24. Upper parts brownish to pinkish cinnamon or buffy, most intense along dorsal area; underparts pure white; tail like upper parts in basal two-thirds above, but whitish distally and below.

Onychomys leucogaster usually occurs in the Upper Sonoran Life-zone but occasionally is found in the Lower Sonoran or Transition life-zones.

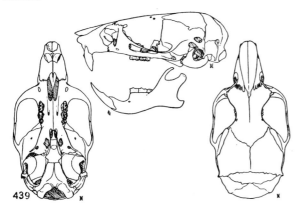

439

Fig. 439. *Onychomys leucogaster brevicaudus*, 16 mi. NE Iron Point, Nevada, No. 68276 M.V.Z., ♂, X 1½.

Onychomys leucogaster albescens Merriam

1904. *Onychomys leucogaster albescens* Merriam, Proc. Biol. Soc. Washington, 17:124, June 9, type from Samalayuca, Chihuahua.

MARGINAL RECORDS.—Chihuahua: Mexican boundary line, opposite El Paso, Texas. Texas: *near El Paso* (Anderson, 1972:351). Chihuahua (Anderson, 1972:352): *10 mi. SE Zaragosa, 3700 ft.; 13 mi. SE Zaragosa; 8 mi. NE Samalayuca, 3700 ft.; 5 mi. NE Samalayuca, 4300 ft.; 1 mi. E Samalayuca, 4500 ft.;*

type locality; *28 mi. S, 2 mi. W Ciudad Juárez*. **See** addenda.

Onychomys leucogaster arcticeps Rhoads

1898. *Onychomys arcticeps* Rhoads, Proc. Acad. Nat. Sci. Philadelphia, 50:194, May 3, type from Clapham, Union Co., New Mexico.
1914. *Onychomys leucogaster arcticeps*, Hollister, Proc. U.S. Nat. Mus., 47:439, October 29.

MARGINAL RECORDS.—Wyoming (Long, 1965a: 636, 637): 4 mi. N Garland; 6 mi. NW Greybull; head Bridgers Creek; 12 mi. N, 7 mi. W Bill; *23 mi. SW Newcastle;* Newcastle. South Dakota: Buffalo Gap; Smithville; Stanley County; Bonesteel. Nebraska (Jones, 1964c:210): Perch; Callaway; 5 mi. W Holbrook. Kansas: 4 mi. N, 3 mi. E Hays, 2000 ft.; Nekoma; Rezeau Ranch, 5 mi. N Belvidere. Texas: Lipscomb; *Hemphill County* (Davis, 1966:177, as *O. leucogaster* only); Mobeetie; 8 mi. NE Electra (Dalquest, 1968:18); 7 mi. NW Bomarton (Baccus, 1971:182); Howard County (Davis, 1966:177, as *O. leucogaster* only); Fort Lancaster; Pecos County (Davis, 1966:177, as *O. leucogaster* only); Monahans. New Mexico: Carlsbad; Roswell; Santa Rosa; near Cimarron. Colorado: Westcliffe; Salida; Colorado Springs; Golden; Longmont; 12 mi. NW Fort Collins (Armstrong, 1972:213). Wyoming: Laramie (Long, 1965a:636). Colorado (Armstrong, 1972:213): 5 mi. E Canadian Creek; *1 mi. W Walden.* Wyoming: 1 mi. N Encampment (Long, 1965a:636). Colorado (Armstrong, 1972:213): Three Forks, 30 mi. above Baggs, Wyoming; 8 mi. NE Craig; *Craig;* 1 mi. W Elk Springs. Wyoming (Long, 1965a:636, 637): 32 mi. S, 22 mi. W Rock Springs; Fort Bridger; *Mountainview; Cumberland;* Kemmerer; *Fontenelle;* Big Sandy; 2 mi. N, 6 mi. W Burris; 13 mi. N, 1 mi. E Cody.

Onychomys leucogaster breviauritus Hollister

1913. *Onychomys leucogaster breviauritus* Hollister, Proc. Biol. Soc. Washington, 26:216, December 20, type from Fort Reno, Canadian Co., Oklahoma.

MARGINAL RECORDS.—Nebraska: 2⅕ mi. S Niobrara (Jones, 1964c:212); *1–1½ mi. SE Niobrara (ibid.);* 4 mi. SE Carroll (*ibid.*); ½ mi. W Manley (Genoways and Choate, 1970:121); *⅖ mi. N, 2 mi. W Weeping Water (ibid.).* Kansas: Fort Riley; Neosho Falls. Oklahoma: type locality; 14 mi. S Olustee; area near Hollis (Ruffer, 1965:242, as *O. leucogaster* only); *11 mi. N, 4 mi. W Hollis* (Martin and Preston, 1970:53); Camp Supply. Kansas: 3½ mi. S, 1 mi. W Kinsley. Nebraska: 2 mi. S Franklin (Jones, 1964c:212).

Onychomys leucogaster brevicaudus Merriam

1891. *Onychomys leucogaster brevicaudus* Merriam, N. Amer. Fauna, 5:52, July 30, type from Blackfoot, Bingham Co., Idaho.

MARGINAL RECORDS.—Idaho: Double Springs, 16 mi. NE Dickey; Big Lost River; Blackfoot;

Montpelier Creek. Wyoming (Long, 1965a:637): Cokeville; *Bear River, 14 mi. N Evanston;* Evanston. Nevada: Goose Creek, 2 mi. W Utah boundary, 5000 ft.; Tecoma, 4900 ft.; 3 mi. E Smith Creek Cave, Mt. Moriah, 5500 ft.; 1 mi. N Baker; Coal Valley, 10 mi. N Seeman Pass, 4650 ft.; NW base Timber Mtn., 4200 ft.; 15 mi. NE Tonopah, Ralston Valley, 5800 ft.; Fingerrock Wash, Stewart Valley, 5400 ft. California: Benton; Farrington Ranch, near Mono Lake. Nevada: Gardnerville. California: Sierra Valley; Amedee. Nevada: 5 mi. N Summit Lake, 5900 ft.; Hot Spring, Thousand Creek, 4300 ft. Idaho: Murphy; Glenns Ferry.

Onychomys leucogaster fuliginosus Merriam

1890. *Onychomys fuliginosus* Merriam, N. Amer. Fauna, 3:59, September 11, type from Black Tank lava beds, NE of San Franciso Mtn., Coconino Co., Arizona.
1913. O[*nychomys*]. *l*[*eucogaster*]. *fuliginosus*, Hollister, Proc. Biol. Soc. Washington, 26:216, December 20.
1913. *Onychomys leucogaster capitulatus* Hollister, Proc. Biol. Soc. Washington, 26:215, December 20, type from lower end Prospect Valley [about 25 mi. WSW Supai, 4500 ft.], Hualpai Indian Reservation, Grand Canyon, Arizona. Regarded as a synonym of *O. l. fuliginosus* by Van Cura and Hoffmeister, Jour. Mamm., 47:622, December 2, 1966.

MARGINAL RECORDS (Van Cura and Hoffmeister, 1966:623, 624, unless otherwise noted).—Arizona: 1 mi. N Pipe Spring National Monument (Hoffmeister and Durham, 1971:39); ½ mi. S Dry Lake (*ibid.*); Aubrey Valley, 10 mi. S Pine Spring, 6000 ft.; 10 mi. N Red Lake (Cockrum, 1961:158); Pasture Wash, Grand Canyon National Park; Hermit Basin, Grand Canyon National Park; E side Cedar Mtn., Grand Canyon National Park; *3 mi. above mouth Cedar Ranch Wash, 4500 ft.;* Heiser Spring, Wupatki Monument; *Walnut Tank, 10 mi. N Angell; Angell;* ¾ *mi. S Angell;* [2 *mi. E*] Flagstaff (Hall and Kelson, 1959:665, as *O. l. pallescens*); *cedar belt E O'Leary Peak, 6000–7250 ft., San Francisco Mts.;* 2 mi. N Rimrock; 3½ mi. S, 2½ mi. E Camp Verde; Kirkland (Cockrum, 1961:158); ½ mi. S Valentine; *3 mi. N Valentine;* 6 mi. E Peach Spring; Gardner Ranch, Shivwits Plateau (Hoffmeister and Durham, 1971:39); Wolf Hole (Hall and Kelson, 1959:665, as *O. l. melanophrys*).

Onychomys leucogaster fuscogriseus Anthony

1913. *Onychomys leucogaster fuscogriseus* Anthony, Bull. Amer. Mus. Nat. Hist., 32:11, March 7, type from Ironside, 4000 ft., Malheur Co., Oregon.

MARGINAL RECORDS.—Washington: Baird; Coulee City; Asotin. Idaho: Weiser; Nampa. Oregon: Alvord Valley. Nevada: 8½ mi. E Vya, 5900 ft.; 10 mi. SE Hausen, 4675 ft. California: Box Springs, Madeline Plains, N base Observation Peak; near Tule Lake; Picard, near Lower Klamath Lake. Oregon: Klamath Falls; Swan Lake Valley; Buck Creek; Willows Junction. Washington: Yakima; Douglas.

Onychomys leucogaster leucogaster (Wied-Neuwied)

1841. *Hypudaeus leucogaster* Wied-Neuwied, Reise in das innere Nord-America . . . , 2:99, type from Mandan Indian village, near Fort Clark, Oliver Co., North Dakota.
1858. *Onychomys leucogaster,* Baird, Mammals, *in* Repts. Expl. Surv. . . . , 8(1):459, July 14.
1885. O[*nychomys*]. *leucogaster* var. *pallidus* Herrick, Geol. Nat. Hist. Surv. Minnesota, Ann. Rept. for 1884, p. 183, type from Lake Traverse, near sources of Minnesota and Boix de Siouz rivers, South Dakota.

MARGINAL RECORDS.—Manitoba: border Riding Mountain National Park; Aweme. North Dakota: Pembina. Minnesota: 4 mi. W Karlstad; Parkers Prairie; Willmar (Thoma and Gunderson, 1963:27); ½ mi. S Alpha (Heaney and Birney, 1975:32). Iowa (Bowles, 1975:94): West Okoboji Lake; *S ¹/₂ of SE ¹/₄ of sec. 2, Preston Twp., Plymouth Co.* South Dakota: 1 mi. W Vermillion. North Dakota: Linton; type locality; Fort Berthold; Minot; Bottineau.

Onychomys leucogaster longipes Merriam

1889. *Onychomys longipes* Merriam, N. Amer. Fauna, 2:1, October 30, type from Concho County, Texas.
1913. *Onychomys leucogaster longipes,* Hollister, Proc. Biol. Soc. Washington, 26:216, December 20.

MARGINAL RECORDS.—Texas: San Angelo; type locality; Refugio County (Davis, 1966:177, as *O. leucogaster* only); Rockport; Mustang Island [18 mi. S Port Aransas]; Brownsville. Tamaulipas: Soto la Marina (Alvarez, 1963:448); Victoria. Nuevo León: Linares. Coahuila: 1 mi. S Hermanas, 1300 ft.; 6 mi. N, 2 mi. E La Babia. Texas: Pecos River, 25 mi. S Sheffield.

Onychomys leucogaster melanophrys Merriam

1889. *Onychomys leucogaster melanophrys* Merriam, N. Amer. Fauna, 2:2, October 30, type from Kanab, Kane Co., Utah.

MARGINAL RECORDS.—Utah: 2 mi. E Price, thence southward along Green and Colorado rivers to Arizona (Van Cura and Hoffmeister, 1966:625, unless otherwise noted): 5 mi. N, 2 mi. W Page; *16 mi. SW Navajo Bridge; 10 mi. S Jacobs Pools* (Hall and Kelson, 1959:665); 5 mi. S, 1 mi. E Kane; 4½ mi. S Fredonia. Utah: Zion National Park; 2 mi. N Kanab, 5200 ft.; Bryce National Park; Thurber.

Onychomys leucogaster missouriensis (Aud. and Bach.)

1851. *Mus missouriensis* Audubon and Bachman, The viviparous quadrupeds of North America, 2:327, type from Fort Union near present town of Buford, Williams Co., North Dakota.
1914. *Onychomys leucogaster missouriensis,* Hollister, Proc. U.S. Nat. Mus., 47:438, October 29.

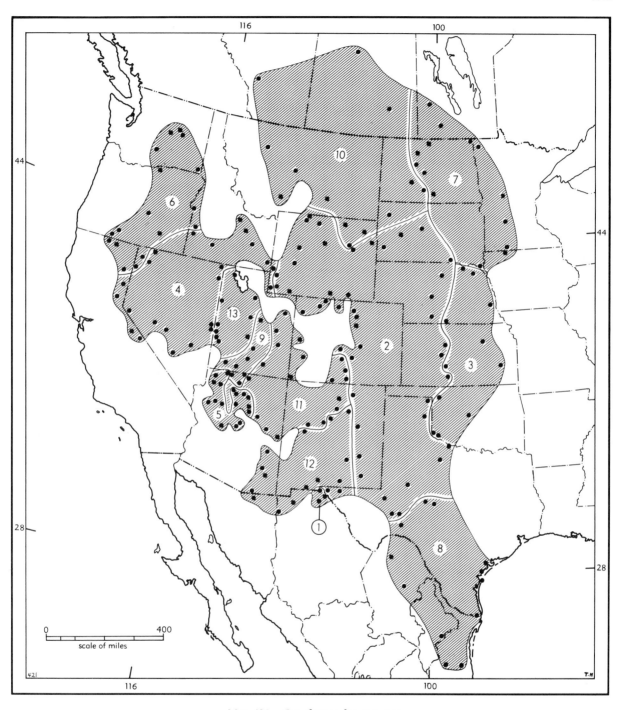

Map 421. *Onychomys leucogaster.*

1. *O. l. albescens*	5. *O. l. fuliginosus*	10. *O. l. missouriensis*
2. *O. l. arcticeps*	6. *O. l. fuscogriseus*	11. *O. l. pallescens*
3. *O. l. breviauritus*	7. *O. l. leucogaster*	12. *O. l. ruidosae*
4. *O. l. brevicaudus*	8. *O. l. longipes*	13. *O. l. utahensis*
	9. *O. l. melanophrys*	

MARGINAL RECORDS.—Saskatchewan: Carlton; Indian Head. North Dakota: Glen Ullin; Cannon Ball. South Dakota: 14 mi. S, 4 mi. W Reva (Andersen and Jones, 1971:380). Wyoming (Long, 1965a:638): Moorecroft; 42 mi. S, 13 mi. W Gillette; 5 mi. NE Clearmont. Montana: Fort Custer; Great Falls; Bozeman; Robare. Alberta: Calgary.

Onychomys leucogaster pallescens Merriam

1890. *Onychomys melanophrys pallescens* Merriam, N. Amer. Fauna, 3:61, September 11, type from Moki Pueblos, Navajo Co., Arizona. (Regarded as valid by Benson, Univ. California Publ. Zool., 40:451, December 31, 1935, and later authors.)

1895. *Onychomys leucogaster pallescens*, J. A. Allen, Bull. Amer. Mus. Nat. Hist., 7:225, June 29.

MARGINAL RECORDS.—Utah: Kennedys Hole, jct. White and Green rivers. Colorado (Armstrong, 1972:214): 2 mi. W Grand Junction; Norwood; Four Corners; Conejos River, 8300 ft.; *Moffat, 7568 ft.;* Crestone, 7871 ft.; 22 mi. E Mosca; 5 mi. SSE Fort Garland. New Mexico: Española; Santa Fe; Albuquerque; [15 mi. S] Acoma. Arizona (Van Cura and Hoffmeister, 1966:622, unless otherwise noted): 3½ mi. SE Springerville (Cockrum, 1961:158); Taylor (*ibid.*); 1 mi. S, 2½ mi. E Winslow; 4½ mi. S, 1½ mi. W Cameron; 5 *mi. SW Cameron* (Cockrum, 1961:157), thence along E side Colorado River to 5 mi. SW Page. Utah: River Mile 121, Glen Canyon, E bank (Durrant and Dean, 1959:92); brush flat, Hwy. 160, 6 mi. N Moab, 4200 ft.; 1 mi. E Green River, 4080 ft.

Onychomys leucogaster ruidosae Stone and Rehn

1903. *Onychomys ruidosae* Stone and Rehn, Proc. Acad. Nat. Sci. Philadelphia, 55:22, May 7, type from Hales Ranch, Ruidoso, Lincoln Co., New Mexico.

1913. *Onychomys leucogaster ruidosae*, Hollister, Proc. Biol. Soc. Washington, 26:216, December 20.

MARGINAL RECORDS.—New Mexico: Las Vegas; type locality; Jarilla. Texas: Hudspeth County (Davis, 1966:177, as *O. leucogaster* only); 7½ mi. E El Paso City Hall (Jones and Lee, 1962:78, as *O. leucogaster* only). New Mexico: Mesilla; Monument 15, Mexican boundary line. Chihuahua: Colonia Díaz; Llano de las Carretas, 27 mi. W Cuervo, 4300 ft. (Anderson, 1972:352). Sonora: Río Santa Cruz. Arizona: Locheil (Cockrum, 1961:159); Willcox (*ibid.*); 1½ mi. SW Fort Grant (Van Cura and Hoffmeister, 1966:625); *Ash Creek, San Carlos Indian Reservation* (Cockrum, 1961:159); Chiricahua Ranch, 20 mi. NE Calva (*ibid.*). New Mexico: Burleigh; San Pedro, Sandia Mts. **See** addenda.

Onychomys leucogaster utahensis Goldman

1939. *Onychomys leucogaster utahensis* Goldman, Jour. Mamm., 20:354, August 14, type from S end Stansbury Island, 4250 ft., Great Salt Lake, Tooele Co., Utah.

1942. *Onychomys leucogaster aldousi* Goldman, Proc. Biol. Soc. Washington, 55:77, June 25, type from Desert Range Experiment Station, 50 mi. W Milford, Beaver Co., Utah. (Regarded as indistinguishable from *O. l. utahensis* by Durrant, Univ. Kansas Publ., Mus. Nat. Hist., 6:324, August 10, 1952.)

MARGINAL RECORDS.—Utah: Kelton; 24 mi. E Salt Lake City, 6500 ft.; Nephi; foothills S of Salina; Desert Range Experiment Station, 50 mi. W Milford, 5252 ft.; Warm Cove, 55 mi. W Milford, 5500 ft.; 5 mi. S Garrison, 5400 ft.; 4 mi. S Gandy, 4000 ft.; 5 mi. N Ibapah, 5175 ft.

Onychomys torridus
Southern Grasshopper Mouse

X ²⁄₃

External measurements: 119–163; 33–62; 18–23; 11–18. Upper parts grayish or pinkish cinnamon; underparts white; tail above like upper parts in basal two-thirds, white below and distally. See key for selected cranial differences between *O. torridus* and *O. leucogaster*.

The southern grasshopper mouse occurs in low, hot valleys, principally in the Lower Sonoran Life-zone.

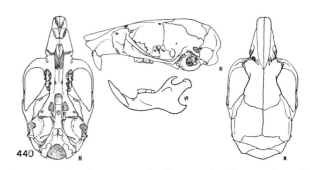

Fig. 440. *Onychomys torridus longicaudus,* ¼ mi. E Crystal Spring, Nevada, No. 52957 M.V.Z., ♂, X 1½.

Onychomys torridus ater Anderson

1972. *Onychomys torridus ater* Anderson, Bull. Amer. Mus. Nat. Hist., 148:353, September 8, type from 11 mi. E La Junta, Chihuahua.

MARGINAL RECORDS (Anderson, 1972:353, unless otherwise noted).—Chihuahua: 4 mi. S, 4 mi. W Santo Tomás; *Rancho San Ignacio, 4 mi. S, 1 mi. W Santo Tomás;* 2 mi. S, 2 mi. E Ciudad Guerrero; *El Rosario;* type locality; 2 *mi. W Miñaca* (Anderson and Long, 1961:2, as *O. torridus* only, in owl pellet).

Onychomys torridus canus Merriam

1904. *Onychomys torridus canus* Merriam, Proc. Biol. Soc. Washington, 17:124, June 9, type from San Juan Capistrano, Zacatecas.

MARGINAL RECORDS.—Durango: Hda. Atotonilco, 6680 ft. (Baker and Greer, 1962:121). San Luis Potosí: 3 km. S Santo Domingo; 10 mi. NE San Luis Potosí, 6000 ft.; Río Verde; Jesús María. Jalisco: 2 mi. SW Matanzas, 7550 ft. (Genoways and Jones, 1973:15). Aguascalientes: Chicalote. Zacatecas: type locality. Durango (Baker and Greer, 1962:121): 16 mi. S, 20 mi. W Vicente Guerrero, 6675 ft.; 4 mi. E Durango, 6400 ft.; *4 mi. S Morcillo, 6450 ft.;* 9 mi. N Durango, 6200 ft.

Onychomys torridus clarus Hollister

1913. *Onychomys torridus clarus* Hollister, Proc. Biol. Soc. Washington, 26:215, December 20, type from Keeler, E shore Owens Lake, Inyo Co., California.

MARGINAL RECORDS.—California: near Lone Pine; type locality; near Coso Mts.; Hot Springs Valley; neighborhood of Olancha.

Onychomys torridus longicaudus Merriam

1889. *Onychomys longicaudus* Merriam, N. Amer. Fauna, 2:2, October 30, type from St. George, Washington Co., Utah.
1904. *O[nychomys]. torridus longicaudus* Merriam, Proc. Biol. Soc. Washington, 17:123, June 9.

MARGINAL RECORDS.—Nevada: 3 mi. NNE Toulon, 3900 ft.; 5½ mi. NE San Antonio, 5700 ft.; 8 mi. NE Springdale, 4250 ft.; Big Creek, 5700 ft., Quinn Canyon Mts.; Railroad Valley, 2½ mi. S Lock's Ranch, 500 ft.; Panaca, 4700 ft. Utah: Beaverdam Wash, 2800 ft., 8 mi. N Utah–Arizona border; S boundary Zion National Park. Arizona (Van Cura and Hoffmeister, 1966:630, unless otherwise noted): Wolf Hole, 3500 ft. (Cockrum, 1961:162); *3 mi. S, 8 mi. E Pakoon Spring;* 3 mi. W Lower Pigeon Spring; ½ mi. E Vulcan[s] Throne, S end Toroweap Valley, Grand Canyon National Park; *Grand Wash, 8 mi. S Pakoon Spring, 1800 ft.* (Cockrum, 1961:162). Nevada: Cedar Basin, 3500 ft.; Jap Ranch, Colorado River, 14 mi. E Searchlight, 5000 ft.; *5½ mi. S, 9 mi. W Searchlight, 4300 ft.* California: Amargosa; Maturango Springs; Independence; Benton. Nevada: Cat Creek, 4500 ft., 4 mi. W Hawthorne; 1 mi. NW Fernley, 4090 ft. See addenda.

Onychomys torridus macrotis Elliot

1903. *Onychomys macrotis* Elliot, Field Columb. Mus., Publ. 74, Zool. Ser., 3(10):155, May 7, type from head San Antonio River, W slope Sierra San Pedro Mártir, Baja California.
1914. *Onychomys torridus macrotis*, Hollister, Proc. U.S. Nat. Mus., 47:469, October 29.

MARGINAL RECORDS.—Baja California: El Alamo; Trinidad; type locality; San Quintín.

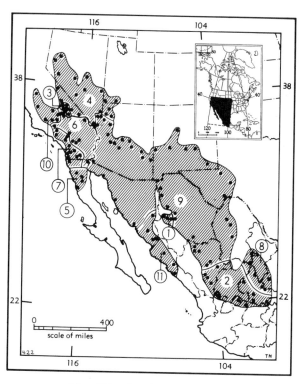

Map 422. *Onychomys torridus.*

Guide to subspecies
1. *O. t. ater*
2. *O. t. canus*
3. *O. t. clarus*
4. *O. t. longicaudus*
5. *O. t. macrotis*
6. *O. t. pulcher*
7. *O. t. ramona*
8. *O. t. surrufus*
9. *O. t. torridus*
10. *O. t. tularensis*
11. *O. t. yakiensis*

Onychomys torridus pulcher Elliot

1904. *Onychomys pulcher* Elliot, Field Columb. Mus., Publ. 87, Zool. Ser., 3(14):243, January 7, type from Morongo Pass, San Bernardino Mts., California.
1913. *O[nychomys]. t[orridus]. pulcher*, Hollister, Proc. Biol. Soc. Washington, 26:215, December 20.

MARGINAL RECORDS.—California: Little Lake; Walker Pass; SE side Clark Mtn., 5100 ft. Nevada: 8 mi. SSE Dead Mtn., 1900 ft., thence southward W of Colorado River to Baja California: Seven Wells. California: Palm Springs; Cabezon; type locality; Cushenbury Springs; Fairmont, Antelope Valley; Onyx.

Onychomys torridus ramona Rhoads

1893. *Onychomys ramona* Rhoads, Amer. Nat., 27:833, September, type from [Reche Canyon, 1250 ft., 4 mi. SE Colton] San Bernardino Valley, California.
1904. *Onychomys torridus ramona*, Merriam, Proc. Biol. Soc. Washington, 17:124, June 9.

MARGINAL RECORDS.—California: Mint Canyon, 6 mi. W Palmdale; type locality; Valle Vista; Warner Pass; La Puerta Valley; Jacumba. Baja California:

Tecarte Valley. California: mouth Tia Juana River; San Diego; Santee Mts.; Riverside; San Fernando.

Onychomys torridus surrufus Hollister

1914. *Onychomys torridus surrufus* Hollister, Proc. U.S. Nat. Mus., 47:472, October 29, type from Miquihuana, Tamaulipas.

MARGINAL RECORDS.—Coahuila: 7 mi. S, 4 mi. E Bella Unión. Tamaulipas: 4 mi. N Jaumave (Alvarez, 1963:448); *Jaumave.* San Luis Potosí: Tepeyac; 7 km. SE Presa de Guadalupe; *Presa de Guadalupe*; 6 km. S Matehuala. Coahuila: *La Ventura*; 8 mi. N La Ventura, 6000 ft.

Onychomys torridus torridus (Coues)

1874. *Hesperomys (Onychomys) torridus* Coues, Proc. Acad. Nat. Sci. Philadelphia, 26:183, December 15, type from Old Camp Grant . . . Arizona—see Cockrum, Jour. Mamm., 41:516, November 11, 1960.
1889. *Onychomys torridus*, Merriam, N. Amer. Fauna, 2:3, October 30.
1896. *Onychomys torridus arenicola* Mearns, Preliminary diagnoses of new mammals from the Mexican border of the United States, p. 3, May 25 (preprint of Proc. U.S. Nat. Mus., 19:139, December 21, 1896), type from Rio Grande, about 6 mi. above El Paso, El Paso Co., Texas. **See** addenda.
1896. *Onychomys torridus perpallidus* Mearns, Preliminary diagnoses of new mammals from the Mexican border of the United States, p. 4, May 25 (preprint of Proc. U.S. Nat. Mus., 19:140, December 21, 1896), type from left bank Colorado River at Monument 204, Mexican boundary line, Yuma Co., Arizona.

MARGINAL RECORDS.—Arizona (Van Cura and Hoffmeister, 1966:628, unless otherwise noted): Colorado River, Pierce Ferry (Cockrum, 1961:160); Gold Basin, 20 mi. S Gregg's Ferry (*ibid.*); Peach Springs, 4000 ft. (*ibid.*); Big Sandy Creek, 2800 ft. (*ibid.*); ½ mi. N Congress; *4½ mi. S Congress*; 5 mi. N Wickenburg (Hall and Kelson, 1959:667); New River (Cockrum, 1961:160); ½ mi. S, 1 mi. W Camp Verde; *4½ mi. S, ½ mi. E Camp Verde*; 10 mi. S Payson (Cockrum, 1961:159); N end Roosevelt Lake, mouth Bumblebee Creek, 2700 ft. (*ibid.*); Cazador Spring, 4000 ft. (*ibid.*); Calva, 2600 ft., San Carlos Indian Reservation (*ibid.*); 2½ mi. S, 1 mi. W Guthrie. New Mexico (Findley, *et al.*, 1975:231–233, as *O. torridus* only): Glenwood; San Mateo Mts., 5⅜ mi. E Springtime Canyon; 1½ mi. S U.S. [Hwy.] 66, Albuquerque, Sandia Mts. foothills; 2 mi. N, 5 mi. W Carrizozo; 44 mi. NW Roswell; 3 mi. N, 3½ mi. E Loving. Texas: Monahans; Fort Lancaster; Maxon Springs. Coahuila: 6 mi. NW Tanque Alvarez (Baker, 1956:276); 18 mi. S, 14 mi. E Tanque Alvarez, 4000 ft.; 3 mi. NW Cuatro Ciénegas. Durango (Baker and Greer, 1962:121): 34 mi. SSE La Zarca; 6½ mi. S Alamillo, 4600 ft.; Indé; Río Sestín; Rosario. Chihuahua (Anderson, 1972:353, 354): 5 mi. E Parral, 5700 ft.; 7 mi. WSW Cuauhtémoc; 20 mi. (by road) N Cuauhtémoc; 8 mi. SE Zaragosa; *San Diego*; Colonia Juárez; 35 mi. NW Dublán, 5300 ft.; *1½ mi. N San Francisco*; *2½ mi. N, ½ mi. W*

San Francisco. Sonora: San José de Guaymas; ½ mi. N Puerto Libertad; Cholla Bay, 5 mi. NW Punta Peñasco (Cockrum and Bradshaw, 1963:8). Arizona: left bank Colorado River, Monument 204; Harpers; 8 mi. SW Pierce Ferry (Cockrum, 1961:160). **See** addenda.

Onychomys torridus tularensis Merriam

1904. *Onychomys torridus tularensis* Merriam, Proc. Biol. Soc. Washington, 17:123, June 9, type from Bakersfield, Kern Co., California.

MARGINAL RECORDS.—California: 11 mi. E Llanda, 1250 ft.; Alila [= Earlimart]; Weldon; Kelso Valley; Caliente Creek Wash; Carrizo Plain; Alcalde.

Onychomys torridus yakiensis Merriam

1904. *Onychomys torridus yakiensis* Merriam, Proc. Biol. Soc. Washington, 17:124, June 9, type from Camoa, Río Mayo, Sonora.

MARGINAL RECORDS.—Sonora: Tecoripa; Alamos. Sinaloa: Sinaloa; 12 mi. N Culiacán (Jones, *et. al.*, 1962:157); 6 mi. N, 1½ mi. E El Dorado (*ibid.*). Sonora: Chinobampo; Obregón.

Genus Zygodontomys J. A. Allen—Cane Rats

Revised by Hershkovitz, Fieldiana: Zool., 46:196–207, December 20, 1962. In 1966 Handley, Ectoparasites of Panama, Field Mus. Nat. Hist., p. 783, November 22, applied the name *Z. microtinus* (instead of *Z. brevicauda*) to specimens from the mainland of Panamá.

1897. *Zygodontomys* J. A. Allen, Bull. Amer. Mus. Nat. Hist., 9:38, March 11. Type, *Oryzomys cherriei* J. A. Allen.

Total length, 217–320; tail, 75–140; hind foot, 26–34. Pelage full and soft, not unlike that of *Sigmodon*; tail shorter than head and body, scantily haired. Interparietal usually large; supraorbital ridges usually well marked; anterior margin of zygomatic plate cut back above; anterior palatine foramina broad, extending to level of tooth-rows; tympanic bullae of moderate size for an oryzomyine rodent; cross furrows between successive pairs of cusps of molar teeth cut off by longitudinal bar of enamel, yoking together pairs of cusps on median line of tooth; thus anterior cone of M1 connected with 2 succeeding pairs of cones by median longitudinal ridge, and 2 pairs of cones in M2 similarly connected (after J. A. Allen, 1897:38).

Zygodontomys brevicauda
Cane Rat

External measurements (based on North American specimens): 217–255; 75–110; 26–28. Upper parts grayish brown, sometimes with yellowish or ochraceous suffusion, darkest mid-dorsally, paler and grayer along flanks; underparts creamy white, hairs plumbeous basally; feet whitish; tail scantily haired, grayish above, whitish below. Skull with short, broad, tapered rostrum; braincase large and moderately inflated. This species occurs from sea level up to approx. 3500 ft.

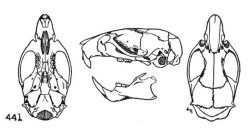

Fig. 441. *Zygodontomys brevicauda ventriosus*, Gatún, Canal Zone, Panamá, No. 1360 K.U., ♂, X 1.

Zygodontomys brevicauda cherriei (J. A. Allen)

1895. *Oryzomys cherriei* J. A. Allen, Bull. Amer. Mus. Nat. Hist., 7:329, November 8, type from Boruca, Costa Rica.
1962. *Zygodontomys brevicauda cherriei*, Hershkovitz, Fieldiana: Zool., 46:203, December 20.
1897. *Zygodontomys cherriei* J. A. Allen, Bull. Amer. Mus. Nat. Hist., 9:38, March 11.
1966. Z[*ygodontomys*]. m[*icrotinus*]. *cherriei*, Handley, Ectoparasites of Panama, Field Mus. Nat. Hist., p. 783, November 22.

MARGINAL RECORDS.—Costa Rica: type locality. Panamá (Handley, 1966:783): El Banco; *Boquerón*; Bugaba. Costa Rica: Palmar.

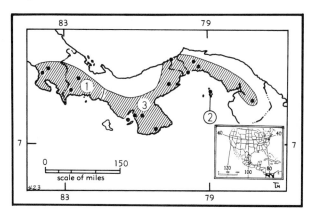

Map 423. *Zygodontomys brevicauda.*

1. Z. b. cherriei   2. Z. b. seorsus
3. Z. b. ventriosus

Zygodontomys brevicauda seorsus Bangs

1901. *Zygodontomys seorsus* Bangs, Amer. Nat., 35:642, August, type from San Miguel Island, Panamá. Known only from San Miguel Island [= Isla del Rey]. Handley, Ec-

toparasites of Panama, Field Mus. Nat. Hist., p. 783, November 22, 1966, recognized the population on Isla del Rey as a species, *Z. seorsus*, rather than a subspecies.
1962. *Zygodontomys brevicauda seorsus*, Hershkovitz, Fieldiana: Zool., 46:204, December 20.

Zygodontomys brevicauda ventriosus Goldman

1912. *Zygodontomys cherriei ventriosus* Goldman, Smiths. Miscl. Coll., 56(36):8, February 19, type from Tabernilla, Canal Zone, Panamá.
1962. *Zygodontomys brevicauda ventriosus*, Hershkovitz, Fieldiana: Zool., 46:204, December 20.
1966. Z[*ygodontomys*]. m[*icrotinus*]. *ventriosus*, Handley, Ectoparasites of Panama, Field Mus. Nat. Hist., p. 783, November 22.

MARGINAL RECORDS.—Panamá (Handley, 1966: 783): Fort Sherman; Cerro Azul; El Real; Pacora; Panamá Viejo; El Valle; Guánico; Altos Cacao; Paracoté; Isla Cébaco.

Genus Scotinomys Thomas—Brown Mice

Revised by Hooper, Occas. Pap. Mus. Zool., Univ. Michigan, 665:1–32, 12 figs., April 11, 1972.

1913. *Scotinomys* Thomas, Ann. Mag. Nat. Hist., ser. 8, 11:408, April. Type, *Hesperomys teguina* Alston.

External measurements: 115–164; 45–80; 15–20; 12–18. Small blackish-brown or dark cinnamon, almost unicolored, mice; tail sparsely haired, usually shorter than head and body; hind foot narrow; molars hypsodont and narrow (especially M1), cusps adapted for both piercing and crushing; 3 inner roots on m1, 2 on m2, 2 on m3 (*Zygodontomys* has 2, 1, and 1, respectively); 6 plantar- and 5 palmar-tubercles; glans penis and baculum closely resembling those of *Baiomys*; glans elongate, spinous, cylindrical; terminal crater containing meatus urinarius and mound-marking tip of baculum; dorsal papilla and urethral process lacking; slightly expanded subapically, twice as long as greatest diameter, and a fifth length of hind foot; baculum relatively simple, its minute form described by Hooper (1960:16).

KEY TO SPECIES OF SCOTINOMYS

1. Tail less than 64; transverse diameter of optic foramen more than that of occlusal area of M3. . . . . *S. teguina,* p. 734
1'. Tail more than 64; transverse diameter of optic foramen less than that of occlusal area of M3. . . . . . . *S. xerampelinus,* p. 734

Scotinomys xerampelinus (Bangs)
Chiriquí Brown Mouse

1902. *Akodon xerampelinus* Bangs, Bull. Mus. Comp. Zool., 39:41, April, type from Volcán de Chiriquí, 10,300 ft., Chiriquí, Panamá.
1913. *S[cotinomys]. xerampelinus,* Thomas, Ann. Mag. Nat. Hist., ser. 8, 11:409, April.
1945. *Scotinomys longipilosus* Goodwin, Amer. Mus. Novit., 1279:2, February 21, type from Volcán Irazú, 9400 ft., Cartago, Costa Rica.
1945. *Scotinomys harrisi* Goodwin, Amer. Mus. Novit., 1279:3, February 21, type from savanna at Las Vueltas, 8000 ft., Cartago, Costa Rica.

External measurements: 136–164; 65–80; 17–20; 12–17. Mammae, 6 (4 inguinal and 2 pectoral); optic foramen distinctly smaller than the more ventral orbital fissure; transverse diameter of optic foramen much less than that of occlusal surface of M3; masseteric process of maxilla not projecting below lower margin of maxilla when viewed laterally; in ventral view, base of baculum deeply concave.

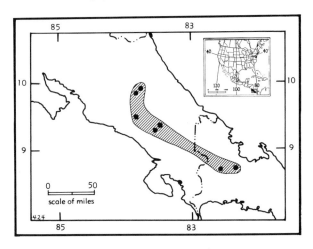

Map 424. *Scotinomys xerampelinus.*

MARGINAL RECORDS (Hooper, 1972:24, 27, 28, unless otherwise noted).—Costa Rica: Volcán Turrialba, 2590–2800 m.; Cerro Chirripo, headwaters Río Talari, 11,000 ft. Panamá: *Cerro Punta, 6800–7800 ft.* (Handley, 1966:784); Cylindro; *Casita Alta, 7800 ft.;* type locality. Costa Rica: 7½ km. E Canaan, Fila La Maquina, 8700 ft.; Las Vueltas, 8000 ft.; Volcán Irazú, 2850–3200 m. Not found: Panamá: Copeta; Hortigal; Sardinia; Potrero.

Scotinomys teguina
Alston's Brown Mouse

External measurements: 115–144; 45–63; 15–19; 12–18. Mammae, 6 or 4 (2 pectoral missing

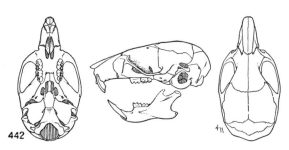

Fig. 442. *Scotinomys teguina irazu,* Volcán Irazú, Costa Rica, No. 116626 U.S.N.M., ♂, 1½.

in southern populations); optic foramen almost as large as more ventral orbital fissure; transverse diameter of optic foramen more than that of occlusal surface of M3; masseteric process of maxilla projecting below lower margin of maxilla when viewed laterally; in ventral view, base of baculum slightly cleft.

Scotinomys teguina apricus (Bangs)

1902. *Akodon teguina apricus* Bangs, Bull. Mus. Comp. Zool., 39:40, April, type from Boquete, 4000 ft., Chiriquí, Panamá.
1913. *S[cotinomys]. teguina apricus,* Thomas, Ann. Mag. Nat. Hist., ser. 8, 11:409, April.
1939. *Scotinomys teguina episcopi* Enders and Pearson, Not. Naturae, Acad. Nat. Sci. Philadelphia, 34:1, November 9, type from Siolo, Río Colorado, tributary Río Chiriquí Viejo, 10 mi. WNW El Volcán P.O., 4100 ft., Chiriquí, Panamá.
1939. *Scotinomys teguina garichensis* Enders and Pearson, Not. Naturae, Acad. Nat. Sci. Philadelphia, 34:2, November 9, type from Río Gariché, 3200 ft., 5 mi. SW El Volcán P.O., Chiriquí, Panamá.
1939. *Scotinomys teguina leridensis* Enders and Pearson, Not. Naturae, Acad. Nat. Sci. Philadelphia, 34:3, November 9, type from Casita Alta, 7000 ft., Finca Lerida, Boquete, Chiriquí, Panamá.
1946. *Scotinomys teguina endersi* Goodwin, Bull. Amer. Mus. Nat. Hist., 87:409, December 31, type from Agua Buena, about 3500 ft., locally known as Cañas Gordas, but W of the real Cañas Gordas, Savanna de Potrero Grande, Puntarenas, Costa Rica.

MARGINAL RECORDS.—Panamá (Hooper, 1972: 22, 24): Cerro Pando; *Cerro Punta;* type locality; *El Banco;* Río Gariché. Costa Rica: Agua Buena; Boruca. Not found: Panamá: Río Chebo.

Scotinomys teguina irazu (J. A. Allen)

1904. *Akodon irazu* J. A. Allen, Bull. Amer. Mus. Nat. Hist., 20:46, February 29, type from Volcán de Irazú, Cartago, Costa Rica.
1945. *Scotinomys t[eguina]. irazu,* Goodwin, Amer. Mus. Novit., 1279:2, February 21.
1945. *Scotinomys teguina cacabatus* Goodwin, Amer. Mus. Novit., 1279:1, February 21, type from rocky ravines above Villa Quesada, 10 mi. NW Volcán Poás, near Tapesco on main road to San Carlos, 5000 ft., Alajuela, Costa Rica.

1945. *Scotinomys teguina escazuensis* Goodwin, Amer. Mus. Novit., 1279:2, February 21, type from Los Higuerónes, 5000 ft., in humid tropical highlands above town of Escazú, San José, Costa Rica.

1967. *Scotinomys teguina stenopygius* Buchanan and Howell, Jour. Mamm., 48:414, August 21, type from Santa María de Ostuma, 12 km. N Matagalpa, 1250 m., Matagalpa, Nicaragua.

MARGINAL RECORDS (Hooper, 1972:19, 21).— Nicaragua: Río Coco, 5100 ft.; Hda. La Trampa. Costa Rica: 1⅓ mi. N Angel Falls, 5000 ft.; type locality; Moravia de Chirripó, 1116 m.; 3 km. E Canaan; *9–10¹/₂* [*sic*] *N San Isidro del General*; El Copey de Dota, 6000 ft.; *S of Escazú, Los Higuerónes;* San Ramón; Monte Verde. Nicaragua: *Santa María de Ostuma;* San Rafael del Norte.

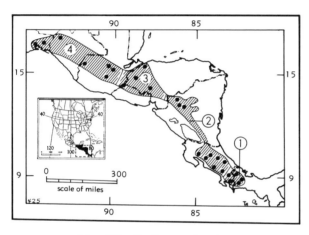

Map 425. *Scotinomys teguina.*

1. *S. t. apricus*   3. *S. t. rufoniger*
2. *S. t. irazu*   4. *S. t. teguina*

Scotinomys teguina rufoniger Sanborn

1935. *Scotinomys teguina rufoniger* Sanborn, Field Mus. Nat. Hist., Publ. 340, Zool. Ser., 20(11):84, May 15, type from mts. W of San Pedro, 4500 ft., Honduras.

MARGINAL RECORDS (Hooper, 1972:19, unless otherwise noted).—Honduras: type locality; *Humuya;* Cerro Linderos, 8700 ft.; *Muya* (Hall and Kelson, 1959:671, as *S. t. teguina*). El Salvador: *Los Esesmiles,*

7100–8000 ft.; Hda. Montecristo (Felten, 1958:137). Honduras: *Monte Verde* (Hall and Kelson, 1959:671, as *S. t. teguina); Cerro Pucca (ibid.);* Lepaera, 3300 ft.

Scotinomys teguina teguina (Alston)

1877. *Hesperomys teguina* Alston, Proc. Zool. Soc. London, p. 755, for 1876, April, type from Cobán, Alta Verapaz, Guatemala.

1913. *Scotinomys teguina,* Thomas, Ann. Mag. Nat. Hist., ser. 8, 11:409, April.

1935. *Scotinomys teguina subnubilus* Goldman, Proc. Biol. Soc. Washington, 48:141, August 22, type from Ocuilapa, 3500 ft., 10 mi. NW Ocozocoautla, and about 25 mi. W Tuxtla Gutiérrez, Chiapas.

MARGINAL RECORDS (Hooper, 1972:15, 16, unless otherwise noted).—Chiapas: Ocuilapa, 10 mi. NW Ocozocoautla, 3500 ft. Guatemala: type locality; near Tucurú, Finca Concepción, 1100 m.; 5 mi. N, 1 mi. W El Chol, 6000 ft.; Hda. El Injerto, 1600 m. Oaxaca: Juchitán, Zanatepec, 5000 ft.; Sierra Madre, N of Zanatepec (Goodwin, 1969:196, as *S. t. subnubilus*).

Genus **Sigmodon** Say and Ord—Cotton Rats

Revised by V. Bailey, Proc. Biol. Soc. Washington, 15:101–116, June 2, 1902.

1825. *Sigmodon* Say and Ord, Jour. Acad. Nat. Sci. Philadelphia, 4(2):352. Type, *Sigmodon hispidus* Say and Ord.

External measurements: 207–365; 75–166; 26–41; approx. 17–25. Body thickset; tail shorter than head and body; ear relatively small; fur short, hispid; 1st and 5th digits of hind foot considerably shorter than 2nd, 3rd, and 4th; 6 plantar tubercles. Mammae, 3 pectoral, 2 inguinal, total 10. Skull robust, rostrum well developed; pronounced supraorbital ridges extending posteriorly over temporal region; interparietal broad; zygomatic plate sharply cut back above, with forward projecting processes on upper border; incisive foramina extending to tooth-rows; palate broad reaching behind M3; lateral pits well developed; pterygoid fossae unusually deep. Upper molars flat-crowned with long, narrow folds surrounded by thick layer of enamel.

KEY TO SPECIES OF SIGMODON

(Modified from Baker, 1969:185, 186)

1. Tail sparsely haired and scaly in appearance, individual scales broad, .75 mm. wide; skull generally long and narrow, basioccipital long and broad, palatal pits shallow.
    2. Hind foot averages 32.5 or less; distance between temporal and occipital crests averages 3.2 or less; lateral border of nasal concave. . . . . *S. hispidus,* p. 736
    2'. Hind foot averages 34 or more; distance between temporal and occipital crests averages 3.9 or more; lateral border of nasal concave or straight.
        3. Keel on palate present; greatest length of skull averages about 40; foramen ovale large; lateral border of nasal not straight. . . . . *S. arizonae,* p. 741

3'. Keel on palate absent; greatest length of skull averages about 36; foramen ovale small; lateral border of nasal concave or straight. . . . . *S. mascotensis,* p. 740
1'. Tail heavily haired and not scaly in appearance, individual scales narrow, .50 mm. wide; skull generally short and broad, basioccipital either long and narrow or short and broad, palatal pits deep.
    4. Inside of pinna of ear whitish, in marked contrast to color of dorsum; interparietal generally less than 2.0 mm. long at midline; upper part of each premaxillary with pronounced rostral depression; mesopterygoid fossa generally parallel-sided at anterior end; lingual root of m1 small or absent. . . . . *S. leucotis,* p. 744
    4'. Inside of pinna of ear not conspicuously different from color of dorsum; interparietal usually 2.0 mm. or more long at midline; upper part of premaxillary with slight or no rostral depression; mesopterygoid fossa generally not parallel-sided anteriorly; lingual root of m1 large.
        5. Buff coloring on nose and around eye conspicuous; adult size small, length of head and body averaging less than 155 mm. and condylopremaxillary length averaging less than 33.2; auditory bullae small and elongate; lateral bulge of capsular projections of upper incisors pronounced; interparietal with slight to marked median posterior notch; paroccipital process (when viewed from below) curved and notched on anterior base. . . . . *S. ochrognathus,* p. 744
        5'. Buff color on nose and around eye usually not in marked contrast to rest of dorsum; adult size large, length of head and body averaging more than 155 mm. and condylopremaxillary length averaging more than 33.2; auditory bullae large and broad (relative to length of skull); lateral bulge of capsular projections of upper incisors slight to moderate; interparietal usually lacking any indication of a median posterior notch; paroccipital process (when viewed from below) generally straight or slightly hooked.
           6. Dorsum having brownish appearance; underparts washed with whitish or pale buff; adult size medium (length of head and body averaging 168 mm. and condylopremaxillary length averaging 34.5); skull flattened, long and narrow; incisive foramina not extending to line drawn between anterior surfaces of 1st upper molars; basioccipital short and wide; mesopterygoid fossa broad anteriorly; median keel on palate slightly developed; palatal pits moderately deep; incisors usually much recurved. . . . . *S. alleni,* p. 743
           6'. Dorsum having "pepper and salt" appearance; underparts washed with buff; adult size large (length of head and body averaging 179 mm. and condylopremaxillary length averaging 36.5); skull arched, short and broad; incisive foramina extending to or beyond a line drawn between anterior surfaces of 1st upper molars; basioccipital long and narrow; mesopterygoid fossa narrow anteriorly; median keel on palate well developed; palatal pits markedly deep; incisors not much recurved. . . . . *S. fulviventer,* p. 742

Sigmodon hispidus
Hispid Cotton Rat

External measurements: 224–365; 81–166; 28–41; 16–24. Upper parts coarsely grizzled; blackish or dark brownish hairs interspersed with buffy or grayish hairs, varying in overall tone, according to subspecies; sides usually only slightly paler; underparts usually pale to dark grayish, sometimes faintly washed with buff; tail coarsely annulated, sparsely haired, hairs not obscuring scaly annulations. Skull relatively long and narrow, mastoid breadth usually less than 46 per cent of basal length. Chromosomes, 2N = 52.— See addenda.

Sigmodon hispidus alfredi Goldman and Gardner

1947. *Sigmodon hispidus alfredi* Goldman and Gardner, Jour. Mamm., 28:57, February 17, type from I. N. Pruitt Farm, near William's Corner, 11 mi. N Springfield, Baca Co., Colorado.

MARGINAL RECORDS (Armstrong, 1972:215).— Colorado: 5 mi. E Kit Carson; Nee Noshe Reservoir, 18 mi. N Lamar; type locality; *Fred Gold Farm, 6 mi. W William's Corner;* 2 mi. S, 2 mi. E Hasty. See addenda.

Sigmodon hispidus berlandieri Baird

1855. *Sigmodon berlandieri* Baird, Proc. Acad. Nat. Sci. Philadelphia, 7:333, type from Río Nazas, Coahuila.
1902. *Sigmodon hispidus berlandieri,* V. Bailey, Proc. Biol. Soc. Washington, 15:106, June 2.
1897. *Sigmodon hispidus pallidus* Mearns, Preliminary diagnoses of new mammals of the genera *Sciurus, Castor,*

Neotoma and *Sigmodon*, from the Mexican border of the United States, p. 4, March 5 (preprint of Proc. U.S. Nat. Mus., 20:504, January 19, 1898), type from left bank of Rio Grande, about 6 mi. above El Paso, El Paso Co., Texas.

MARGINAL RECORDS.—New Mexico: 3 mi. W Kenton (Oklahoma), 4500 ft. (Mohlenrich, 1961:23). Texas: Randall County (Davis, 1966:197, as *S. hispidus* only); 2½ mi. W, 5 mi. S Old Mobeetie; Del Rio; Maverick County (Davis, 1966:197, as *S. hispidus* only); Zavala County (*ibid.*, as *S. hispidus* only); about 2 mi. N Dilley; Bee County (Davis, 1966:197, as *S. hispidus* only); Calhoun County (*ibid.*, as *S. hispidus* only); Mustang Island. Tamaulipas (Alvarez, 1963:450): 1 mi. E La Pesca; Soto la Marina; Ejido Santa Isabel. Querétaro: Tequisquiapam. Hidalgo: 97 km. N Mexico City (9 km. SW Pachuca), 8400 ft. Guanajuato: 2 mi. E Celaya, 5800 ft. (Zimmerman, 1970:447, as *S. hispidus* only). Jalisco (Zimmerman, 1970:447, as *S. hispidus* only): 4 mi. W Guadalajara; 2 mi. SW Ameca, 4000 ft.; 5 mi. NE H[u]ejuquilla, 6200 ft. Durango: 9 mi. N Alamillo, 6000 ft. (Baker and Greer, 1962:123, 124); Villa Ocampo, 4575 ft. (Zimmerman, 1970:447, as *S. hispidus* only). Chihuahua (Anderson, 1972:357): General Trías, 1797 m.; 6 mi. NW Galeana, 4350 ft.; 35 mi. NW Dublán, 5300 ft.; 1½ mi. N San Francisco, 5100 ft. New Mexico (Mohlenrich, 1961:23, unless otherwise noted): 11 mi. WSW Lordsburg (Zimmerman, 1970:447, as *S. hispidus* only); 2 mi. E Lordsburg, 4200 ft.; 2 mi. S, 1 mi. W Faywood, 4900 ft.; Monticello, 5300 ft.; *3 mi. N, 3 mi. E Monticello, 7000 ft.; 8 mi. N, 13 mi. E Monticello, 5400 ft.;* 2 mi. N, 1 mi. E San Antonio, 4500 ft.; *5 mi. N, 1 mi. E San Antonio, 4500 ft.; 1 mi. E Socorro, 4500 ft.; 1 mi. N, 6 mi. W Bernardo, 5100 ft.;* Pato Arroyo, 13 mi. W Bosque, 5200 ft.; 1 mi. S, 2 mi. E Belen, 4700 ft.; Scholle, 5800 ft.; 4 mi. S, 1 mi. E Carrizozo, 5500 ft.; 3 mi. N, 3 mi. W Mesa, 4700 ft.; 4 mi. E Cárdenas, 5200 ft.; 4 mi. SW Santa Rosa, 4700 ft. (Anderson and Berg, 1959:40); *Santa Rosa, 4600 ft.;* Variadero, 4500 ft.; Mosquero, 5500 ft.; 5 mi. S Moses, 4800 ft.

Sigmodon hispidus borucae J. A. Allen

1897. *Sigmodon borucae* J. A. Allen, Bull. Amer. Mus. Nat. Hist., 9:40, March 11, type from Boruca, near Río Diquís, 1600 ft., about 12 mi. from Pacific Coast, Puntarenas, Costa Rica.
1902. *Sigmodon hispidus borucae*, V. Bailey, Proc. Biol. Soc. Washington, 15:112, June 2.
1902. *Sigmodon austerulus* Bangs, Bull. Mus. Comp. Zool., 39:32, April, type from Volcán de Chiriquí, 10,000 ft., Chiriquí, Panamá. (Regarded as a synonym of *S. borucae* J. A. Allen, by Enders, Jour. Mamm., 34:509, November 13, 1953.)

MARGINAL RECORDS.—Costa Rica: Hda. Santa María; San José; Palmar; type locality; San Juanillo. [Panamá: Volcán de Chiriquí (Bangs, 1902:32)—the specimen from this locality is presumed by Enders (1953:508) to have come from near Boruca, Costa Rica.]

Sigmodon hispidus chiriquensis J. A. Allen

1904. *Sigmodon borucae chiriquensis* J. A. Allen, Bull. Amer. Mus. Nat. Hist., 20:68, February 29, type from Boquerón, Chiriquí, Panamá.
1912. *Sigmodon hispidus chiriquensis*, Miller, Bull. U.S. Nat. Mus., 79:184, December 31.

MARGINAL RECORDS.—Costa Rica: "Talamanca." Panamá (Handley, 1966:784, unless otherwise noted): Almirante; *7 km. SSW Changuinola;* Santa Fé; Gatún (Hall and Kelson, 1959:672); Cerro Azul; Cerro Campana; El Valle; Guánico; Cerro Hoya, 3000 ft.; Bugaba; *type locality.* Specimens from Almirante and 7 km. SSW Changuinola were listed as an undescribed subspecies by Handley (1966:784).

Sigmodon hispidus confinis Goldman

1918. *Sigmodon hispidus confinis* Goldman, Proc. Biol. Soc. Washington, 31:21, May 16, type from Safford, Graham Co., Arizona.

MARGINAL RECORDS (Zimmerman, 1970:447, as *S. hispidus* only).—Arizona: 11 mi. E Globe, 3400 ft.; *1 mi. S San Carlos;* 6 mi. SSW Pima; type locality; 1 mi. S, ½ mi. E Franklin; 17 mi. N, 10 mi. E Willcox.

Sigmodon hispidus eremicus Mearns

1897. *Sigmodon hispidus eremicus* Mearns, Preliminary diagnoses of new mammals of the genera *Sciurus, Castor, Neotoma* and *Sigmodon*, from the Mexican border of the United States, p. 4, March 5 (preprint of Proc. U.S. Nat. Mus., 20:504, January 19, 1898), type from Cienega Well, 30 mi. S Monument 204, Mexican boundary line, on E bank Colorado River, Sonora.

MARGINAL RECORDS.—California: a few miles below Palo Verde. Arizona: 4 mi. E Yuma (Zimmerman, 1970:447, as *S. hispidus* only). Sonora: opposite mouth Hardy River.

Sigmodon hispidus exsputus G. M. Allen

1920. *Sigmodon hispidus exsputus* G. M. Allen, Jour. Mamm., 1:236, December 4, type from Big Pine Key, one of the southern Florida Keys, Monroe Co., Florida.

MARGINAL RECORDS.—Florida: *Little Torch Key;* type locality.

Sigmodon hispidus floridanus A. H. Howell

1943. *Sigmodon hispidus floridanus* A. H. Howell, Proc. Biol. Soc. Washington, 56:73, June 16, type from Canal Point, Palm Beach Co., Florida.

MARGINAL RECORDS.—Florida (A. H. Howell, 1943:74, unless otherwise noted): Anastasia Island (Pournelle and Barrington, 1953:135); Ponce Park; Canaveral; *Sebastian;* Miami (Sherman, 1937:119); Everglade (*ibid.*); Hillsborough Co., Lake Mobley; Citronella (Elliot, 1907:245); Silver Springs; *4 mi. SW St. Augustine* (27089 KU).

Sigmodon hispidus furvus Bangs

1903. *Sigmodon hispidus furvus* Bangs, Bull. Mus. Comp. Zool., 39(6):158, July, type from La Ceiba, Atlántida, Honduras.
1904. *Sigmodon hispidus fervidus* Lydekker, Zool. Record, 40(Mammals):34, an accidental renaming of *furvus*.

MARGINAL RECORDS.—Belize: Grant's Works, thence along coast to Honduras: type locality. Guatemala: Puebla. Belize: Silkgrass; *Stann Creek Valley.*

Sigmodon hispidus griseus J. A. Allen

1908. *Sigmodon hispidus griseus* J. A. Allen, Bull. Amer. Mus. Nat. Hist., 24:657, October 13, type from lowlands E of Lake Nicaragua, Chontales, Nicaragua.

Map 426. *Sigmodon hispidus.*

Guide to subspecies
1. S. h. alfredi
2. S. h. berlandieri
3. S. h. borucae
4. S. h. chiriquensis
5. S. h. confinis
6. S. h. eremicus
7. S. h. exsputus
8. S. h. floridanus
9. S. h. furvus
10. S. h. griseus
11. S. h. hispidus
12. S. h. insulicola
13. S. h. komareki
14. S. h. littoralis
15. S. h. microdon
16. S. h. obvelatus
17. S. h. saturatus
18. S. h. solus
19. S. h. spadicipygus
20. S. h. texianus
21. S. h. toltecus
22. S. h. tonalensis
23. S. h. villae
24. S. h. virginianus
25. S. h. zanjonensis

MARGINAL RECORDS.—El Salvador (Burt and Stirton, 1961:62): Los Esesmiles; Mineral Montecristo. Honduras: Sabana Grande. Nicaragua: 9 mi. NNW Estelí (Jones and Genoways, 1970:4, as *S. hispidus* only); *type locality;* 3 mi. NNW Diriamba (Jones and Genoways, 1970:14, as *S. hispidus* only); San Antonio (Jones and Genoways, 1970:7, as *S. hispidus* only). El Salvador (Burt and Stirton, 1961:62): Pine Peaks; Puerto El Triunfo; Amate de Campo; *Chilata;* Hda. San Antonio; Laguna de Guija.

Sigmodon hispidus hispidus Say and Ord

1825. S[*igmodon*]. *hispidus* Say and Ord, Jour. Acad. Nat. Sci. Philadelphia, 4(2):354, type from St. Johns River, Florida.

MARGINAL RECORDS.—Missouri: Lamont: Wayne County. Tennessee: 8½ mi. NE Samburg. Kentucky: W side Cumberland River Valley, 9 mi. SSE Kuttawa (Robinson and Quick, 1965:100, as *S. hispidus* only). Tennessee: 1 mi. E Pulaski. Alabama: Leighton; Jackson. Florida: Milton; Chattahoochee. South Carolina: AEC Savannah River Plant (Wiegert and Mayenschein, 1966:119). North Carolina (Paul and Cordes, 1969:373, as *S. hispidus* only): Greenville; Camp Lejeune; Wilmington. South Carolina: Georgetown. Georgia: Riceboro. Florida: Welaka; Crystal River, thence westward along Gulf Coast to Louisiana (Lowery, 1974:250, Map 37): Cameron County; Calcasieu County; Caddo County. Arkansas: Fourche River bottom near Little Rock. Missouri: 5 mi. SE Branson.

Sigmodon hispidus insulicola A. H. Howell

1943. *Sigmodon hispidus insulicola* A. H. Howell, Proc. Biol. Soc. Washington, 56:74, June 16, type from Captiva Island, Lee Co., Florida.

MARGINAL RECORDS.—Florida: *Englewood; Pine Island; Sanibel Island;* type locality.

Sigmodon hispidus komareki Gardner

1948. *Sigmodon hispidus komareki* Gardner, Proc. Biol. Soc. Washington, 61:97, June 16, type from Woodville, 616 ft., Jackson Co., Alabama.

MARGINAL RECORDS.—North Carolina: Elkin; Raleigh. South Carolina: Society Hill. Georgia: Augusta; Nashville. Alabama: Ashford; Castleberry; Gallion; 4 mi. N Gainesville (Wolfe and Rogers, 1969:610, as *S. hispidus* only); Ardell; Huntsville. Tennessee: Rathburn; White Oak Lake, Oak Ridge National Laboratory (Dunnaway and Kaye, 1961:265); 1 mi. W Bearden.

Sigmodon hispidus littoralis Chapman

1889. *Sigmodon hispidus littoralis* Chapman, Bull. Amer. Mus. Nat. Hist., 2:118, June 7, type from East Peninsula, opposite Micco, Brevard Co., Florida. Known only from type locality, but probably occurs some distance north and south thereof on the coastal beach.

Sigmodon hispidus microdon V. Bailey

1902. *Sigmodon hispidus microdon* V. Bailey, Proc. Biol. Soc. Washington, 15:111, June 2, type from Puerto Morelos, Quintana Roo.

MARGINAL RECORDS.—Quintana Roo: type locality; Esmeralda. Campeche: Apazote, thence around coast to point of beginning.

Sigmodon hispidus obvelatus Russell

1952. *Sigmodon hispidus obvelatus* Russell, Proc. Biol. Soc. Washington, 65:81, April 25, type from 5 mi. S Alpuyeca, 3700 ft., Morelos.

MARGINAL RECORDS.—Morelos: Cuernavaca; *Yautepec;* 2 km. S Jonacatepec. Puebla: Tehuacán, 1700 m. (Hooper, 1947:54, as *S. h. mascotensis* but here referred to *S. h. obvelatus* on geographic grounds). Oaxaca (Goodwin, 1969:203): Teotitlán; Tamazulápam; 4 mi. W Mitla; Sacatepec. Guerrero: Cueva del Cañón del Zopilote, 13 km. S puente de Mexcala, 720 m. (Ramírez-P. and Sánchez-H., 1974:107, 110, 111, as *S. hispidus* only). Morelos: type locality.

Sigmodon hispidus saturatus V. Bailey

1902. *Sigmodon hispidus saturatus* V. Bailey, Proc. Biol. Soc. Washington, 15:111, June 2, type from Teapa, Tabasco.

MARGINAL RECORDS.—Tabasco: Frontera; Monte Cristo. Guatemala: Uaxactun. Belize: El Cayo; Mountain Pine Ridge. Guatemala: Chimoxan; Hda. Chancol; Zunil. Chiapas: Huehuetán; Prusia; Ocuilapa. Oaxaca (Goodwin, 1969:197, unless otherwise noted): Río Negro; La Gloria; mts. near Santo Domingo; 1 mi. S Tollosa (Baker and Greer, 1960:415); Soledad; Cuicatlán (Hall and Kelson, 1959:675); Tuxtepec. Veracruz: 3 km. E San Andrés Tuxtla, 1000 ft. (Hall and Dalquest, 1963:314); Coatzacoalcos.

Sigmodon hispidus solus Hall

1951. *Sigmodon hispidus solus* Hall, Univ. Kansas Publ., Mus. Nat. Hist., 5:42, October 1, type from island 88 mi. S, 10 mi. W Matamoros, Tamaulipas. Known only from type locality.

Sigmodon hispidus spadicipygus Bangs

1898. *Sigmodon hispidus spadicipygus* Bangs, Proc. Boston Soc. Nat. Hist., 28:192, March, type from Cape Sable, Monroe Co., Florida.

MARGINAL RECORDS.—Florida: 10 mi. E Pine Crest; Planter; Northwest Cape.

Sigmodon hispidus texianus (Audubon and Bachman)

1853. *Arvicola texiana* Audubon and Bachman, The viviparous quadrupeds of North America, 3:229, type from Brazos River, Texas.

1891. *Sigmodon hispidus texianus,* J. A. Allen, Bull. Amer. Mus. Nat. Hist., 3:287, June 30.

MARGINAL RECORDS.—Iowa: Waubonsie State Park, 6 mi. N, 2 mi. W Hamburg (Bowles, 1975:95, owl pellets). Kansas: 5 mi. S Hiawatha. Missouri: *Platte County* (Easterla, 1968c:364, as *S. hispidus* only); 1 mi. W Parkville (Genoways and Schlitter, 1967:357). Arkansas: Huntsville; Ozark. Texas: La Belle; Matagorda; Fort Clark; Fisher County (Davis, 1966:197, as *S. hispidus* only); Paducah; Hemphill County (Davis, 1966:197, as *S. hispidus* only); Stinnet. Kansas: Morton County; Greeley County; Logan County. Nebraska (Choate and Genoways, 1967:239): 3 mi. S Kearney, in Kearney Co. (Farney, 1975:12); 3½ mi. S, 1 mi. W Dawson.

Sigmodon hispidus toltecus (Saussure)

1860. [*Hesperomys*] *toltecus* Saussure, Revue et Mag. Zool., Paris, ser. 2, 12:98, type from mts. of Veracruz [probably near Mirador, Dalquest, Louisiana State Univ. Studies, Biol. Sci. Ser., 1:163, December 28, 1953].
1902. *Sigmodon hispidus toltecus,* V. Bailey, Proc. Biol. Soc. Washington, 15:110, June 2.

MARGINAL RECORDS.—Tamaulipas (Alvarez, 1963:450): Sierra de Tamaulipas, 2 mi. S, 10 mi. W Piedra; *16 km. N Tampico;* Tampico, thence southward along coast to Veracruz: Tlacotalpan (Hall and Dalquest, 1963:314). Oaxaca: 5 km. W Tuxtepec; Tierra Blanca (Goodwin, 1969:199). Veracruz: Orizaba; Mt. Orizaba. San Luis Potosí: Xilitla at Apetesco; 2 mi. N, 10 mi. E Ciudad del Maíz; El Salto. Tamaulipas (Alvarez, 1963:450): Rancho Pano Ayuctle; *10 km. N, 8 km. W El Encino; 2 km. W El Carrizo.*

Sigmodon hispidus tonalensis V. Bailey

1902. *Sigmodon hispidus tonalensis* V. Bailey, Proc. Biol. Soc. Washington, 15:109, June 2, type from Tonalá, Chiapas. Known from type locality only.

Sigmodon hispidus villae Goodwin

1958. *Sigmodon hispidus villae* Goodwin, Amer. Mus. Novit., 1871:2, February 26, type from Teopisca, 6000 ft., Chiapas. Known only from type locality.

Sigmodon hispidus virginianus Gardner

1946. *Sigmodon hispidus virginianus* Gardner, Proc. Biol. Soc. Washington, 59:137, October 25, type from Triplet, 160 ft., Brunswick Co., Virginia.

MARGINAL RECORDS.—Virginia: 8½ mi. S Richmond (Pagels and Adleman, 1971:195); 1⁷⁄₁₀ mi. NE Waverly (Peacock, 1967:243); type locality; Clarksville; Carroll County (Peacock, 1967:243); near Wilson (*ibid.*).

Sigmodon hispidus zanjonensis Goodwin

1932. *Sigmodon zanjonensis* Goodwin, Amer. Mus. Novit., 528:1, May 23, type from Zanjón, 9000 ft., Quezaltenango, Guatemala.
1934. *Sigmodon hispidus zanjonensis* Goodwin, Bull. Amer. Mus. Nat. Hist., 68:53, December 12.

MARGINAL RECORDS.—Honduras: Catacombas; Monte Redondo; Comayaguela. Guatemala: San Lucas; Huehuetenango; Sacapulas.

Sigmodon mascotensis
Jaliscan Cotton Rat

External measurements: 204–314; 93–153; 30–38; 17–23. In general resembles *S. hispidus.* More closely resembles *S. arizonae* in long hind foot and great distance between temporal and occipital crests. Differs from *S. arizonae* in smaller foramen ovale and in having lateral margins of nasals flared anteriorly rather than straight. Differs from *arizonae* and *hispidus* in longer skull and absence of keel on palate. Chromosomes, 2N = 28.—See addenda.

Sigmodon mascotensis inexoratus Elliot

1903. *Sigmodon hispidus inexoratus* Elliot, Field Columb. Mus., Publ. 71, Zool. Ser., 3(8):144, March 20, type from Ocotlán, Jalisco.
1949. *Sigmodon hispiduas* [*sic*] *atratus* Hall, Proc. Biol. Soc. Washington, 62:149, August 23, type from 6½ mi. W Zamora, 5950 ft., Michoacán. (Regarded as inseparable from *S. h. inexoratus* by Russell, Proc. Biol. Soc. Washington, 65:82, April 25, 1952.)

MARGINAL RECORDS.—Jalisco: 2 mi. SW Tepatitlan, 6150 ft. (63073 KU); *4 mi. NE Ocotlán* (31799 KU); *type locality;* 1 mi. S Ocotlán, 5000 ft. (Zimmerman, 1970:449, as *S. mascotensis* only). Michoacán: Zamora; *6¹⁄₂ mi. W Zamora, 5950 ft.;* 1 mi. E Jiquilpán (*ibid.*). *S. h. inexoratus* is here included in the species *S. mascotensis* owing to assignment of near topotypes of *inexoratus* to *mascotensis* by Zimmerman (1970:449). Nevertheless, these specimens should be compared with the holotype of *S. h. inexoratus* to determine their true affinities.

Sigmodon mascotensis ischyrus Goodwin

1956. *Sigmodon hispidus ischyrus* Goodwin, Amer. Mus. Novit., 1791:8, September 28, type from "El Arco" Gorge of Río Grande, Santo Domingo Chontecomatlán, District of Yautepec, 2600 ft., Oaxaca. Known only from type locality.

Sigmodon mascotensis mascotensis J. A. Allen

1897. *Sigmodon mascotensis* J. A. Allen, Bull. Amer. Mus. Nat. Hist., 9:54, March 15, type from Mineral San Sebastián, Mascota, Jalisco.

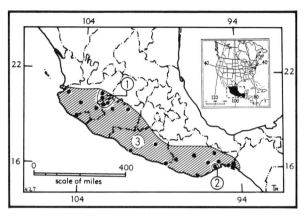

Map 427. *Sigmodon mascotensis*.

1. *S. m. inexoratus*   2. *S. m. ischyrus*
3. *S. m. mascotensis*

1902. *Sigmodon hispidus mascotensis*, V. Bailey, Proc. Biol. Soc. Washington, 15:108, June 2.
1897. *Sigmodon colimae* J. A. Allen, Bull. Amer. Mus. Nat. Hist., 9:55, March 15, type from plains of Colima, Colima.

MARGINAL RECORDS (Zimmerman, 1970:449, unless otherwise noted; as *S. mascotensis* only).— Jalisco: type locality; 11 km. SW Tamazula, 3800 ft.; ½ mi. NW Mazamitla (Hall and Kelson, 1959:449). Michoacán: 2 mi. SE Zacapu, 6600 ft.; Queréndaro (Hall and Kelson, 1959:675). Guerrero: Chilapa, 470 m. Oaxaca: 7 mi. S Putla, 2500 ft.; 3 mi. ESE Oaxaca; Santo Domingo (Goodwin, 1969:199); Tapanatepec (*ibid.*); Reforma (Hall and Kelson, 1959:675); Nizanda (Goodwin, 1969:199); La Reforma (*ibid.*); Puerto Angel (Hall and Kelson, 1959:675). Guerrero: 4½ mi. SW Cuajiniciulapa, 300 ft.; 5 mi. ESE Tecpan, −50 ft. Jalisco: ½ mi. W Barra de Navidad.

Sigmodon arizonae
Arizona Cotton Rat

External measurements: 202–317; 86–150; 29–38; 18–25. Resembles *S. hispidus*, but hind foot averaging longer; distance between temporal and occipital crests is greater, foramen ovale larger; lateral margins of nasals straight, rather than flared. Chromosomes, 2N = 22.—**See** addenda.

Sigmodon arizonae arizonae Mearns

1890. *Sigmodon hispidus arizonae* Mearns, Bull. Amer. Mus. Nat. Hist., 2:287, February 21, type from Fort Verde, Yavapai Co., Arizona.

MARGINAL RECORDS.—Arizona (Zimmerman, 1970:448, unless otherwise noted; as *S. arizonae* only): type locality; Granite Reef, 30 mi. NE Mesa; 10 mi. W Casa Grande; Sacaton (Hall and Kelson, 1959:448); Gila Bend Canal, Theba; Buckeye.

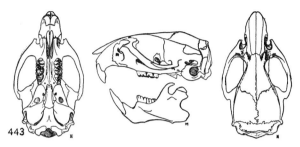

Fig. 443. *Sigmodon arizonae plenus*, Colorado River, ½ mi. above California–Nevada Monument, Nevada, No. 61836 M.V.Z., ♀, X 1.

Sigmodon arizonae cienegae A. B. Howell

1919. *Sigmodon hispidus cienegae* A. B. Howell, Proc. Biol. Soc. Washington, 32:161, September 30, type from Bullock's Ranch, 4 mi. E Fort Lowell, Pima Co., Arizona.

MARGINAL RECORDS.—Arizona: Aravaipa Creek, 5½ mi. S, 12 mi. E Winkleman, 2700 ft. (Zimmerman, 1970:448, as *S. arizonae* only); Fort Grant; 12½ mi. S, 4½ mi. E Willcox (Zimmerman, 1970:448, as *S. arizonae* only); 18 mi. E Douglas, Black Draw (*ibid.*). Sonora: Granados; Ures; Hermosillo; 2½ mi. W Caborca (Zimmerman, 1970:448, as *S. arizonae* only). Arizona: Continental, Santa Cruz River, 26 mi. S Tucson; type locality.

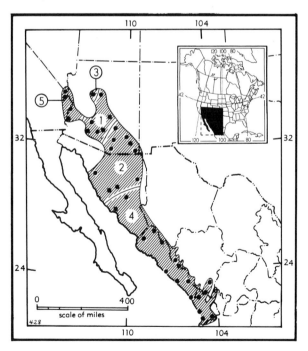

Map 428. *Sigmodon arizonae*.

1. *S. a. arizonae*   3. *S. a. jacksoni*
2. *S. a. cienegae*   4. *S. a. major*
5. *S. a. plenus*

Sigmodon arizonae jacksoni Goldman

1918. *Sigmodon hispidus jacksoni* Goldman, Proc. Biol. Soc. Washington, 31:22, May 16, type from 3 mi. N Fort Whipple, 5000 ft., near Prescott, Yavapai Co., Arizona. Known only from type locality.

Sigmodon arizonae major V. Bailey

1902. *Sigmodon hispidus major* V. Bailey, Proc. Biol. Soc. Washington, 15:109, June 2, type from Sierra de Choix, 50 mi. [probably 10 or 15 mi.] NE Choix, Sinaloa.

MARGINAL RECORDS (Zimmerman, 1970:448, 449, unless otherwise noted, as *S. arizonae* only).— Sonora: Tecoripa (Hall and Kelson, 1959:675). Sinaloa: type locality; 15 km. N, 65 km. E Sinaloa. Durango: Chacala (Baker and Greer, 1962:124); Santa Ana (Jones, 1964a:753). Sinaloa: 19⅕ km. (by road) NE Santa Lucía, 6200 ft. Nayarit: 4 mi. SE Tepic. Durango (as *S. arizonae* only): Paso de Sihuacori (Crossin, *et al.*, 1973:199); 2 km. SW Mezquital (*op cit.*:197). Nayarit: 2 mi. N Ahuacatlán; 3 mi. NNW Las Varas, 150 ft.; ½ mi. E San Blas. Sinaloa: 6 mi. NNW Teacapán; Mazatlán; El Dorado; El Fuerte. Sonora: 3 mi. S Maytorena (Cockrum and Bradshaw, 1963:9).

Sigmodon arizonae plenus Goldman

1928. *Sigmodon hispidus plenus* Goldman, Proc. Biol. Soc. Washington, 41:205, December 18, type from Parker, 350 ft., Yuma Co., Arizona.

MARGINAL RECORDS.—Nevada: Colorado River, ½ mi. N California–Nevada Monument, 500 ft. (but see Bradley, 1966:349, 350). California: Needles. Arizona: Parker; 15 mi. SW Ehrenberg.

fulviventer-group

Revised by Baker, Univ. Kansas Mus. Nat. Hist., Miscl. Publ., 51:177–232, July 11, 1969.

Sigmodon fulviventer
Zacatecan Cotton Rat

External measurements: 223–270; 94–109; 26–36. Upper parts varying according to subspecies from clear, coarsely grizzled gray to grizzled golden gray or yellowish brown lightened on sides by white bristles; underparts varying from buffy to clear, rich fulvous; tail brownish or brownish black, faintly bicolored; feet variable. Skull short, wide, and heavily ridged, not expanded interorbitally or much arched; zygomata widely spreading; interparietal narrow and not divided, median-posterior notch in interparietal absent; supraoccipital having slight median ridge; basioccipital elongate in comparison to width; nasals short, wide, and rounded at ends; posterior ends of incisive foramina extending to

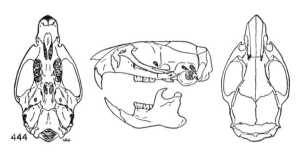

Fig. 444. *Sigmodon fulviventer melanotis*, 2 mi. W Pátzcuaro, 7700 ft., Michoacán, No. 100623 M.V.Z., ♀, X 1.

or beyond a line drawn between anterior surfaces of 1st upper molars; and paroccipital process, when viewed from below, straight instead of hooked; lingual root of 1st lower molar normal, not reduced or absent; mandible short and heavy with short, wide coronoid.

Sigmodon fulviventer fulviventer J. A. Allen

1889. *Sigmodon fulviventer* J. A. Allen, Bull. Amer. Mus. Nat. Hist., 2:180, October 21, type from Zacatecas, Zacatecas.

MARGINAL RECORDS (Baker, 1969:212).— Durango: 5 km. SE Tepehuanes, 1780 m.; 4 km. SE Atotonilco, 2037 m. Zacatecas: type locality. Guanajuato: 8 km. SW Ibarra. Zacatecas: 13 km. S Villanueva, 2090 m.; Laguna Valderama, 67 km. W Fresnillo, 2380 m. Durango: Hda. Coyotes, 2475 m.

Sigmodon fulviventer goldmani V. Bailey

1913. *Sigmodon minimus goldmani* V. Bailey, Proc. Biol. Soc. Washington, 26:132, May 21, type from 7 mi. N Las Palomas (at Hot Springs), 4200 ft., Sierra Co., New Mexico. Known only from type locality.
1962. *Sigmodon fulviventer goldmani*, Baker and Greer, Publ. Mus., Michigan State Univ., Biol. Ser., 2:123, August 27.

Sigmodon fulviventer melanotis V. Bailey

1902. *Sigmodon melanotis* V. Bailey, Proc. Biol. Soc. Washington, 15:114, June 2, type from Pátzcuaro, 7000 ft., Michoacán.
1969. *Sigmodon fulviventer melanotis*, Baker, Univ. Kansas Mus. Nat. Hist., Miscl. Publ., 51:212, July 11.

MARGINAL RECORDS.—Jalisco: 2 km. NW La Barca, 1525 m. (Baker, 1969:213). Michoacán: 18 km. E Zamora (Baker, 1969:213); *2 mi. W Pátzcuaro*; type locality; 3½ mi. S Pátzcuaro; Tancítaro. Jalisco: ½ mi. NW Mazamitla. Michoacán: 3 km. E La Palma, SE side Lago de Chapala (Baker, 1969:213).

Sigmodon fulviventer minimus Mearns

1894. *Sigmodon minima* Mearns, Proc. U.S. Nat. Mus., 17:130, July 19, type from near Monument 40, 1500 m., Hidalgo Co., New Mexico, on Mexican boundary line, 100 miles W of initial monument on W bank Rio Grande.

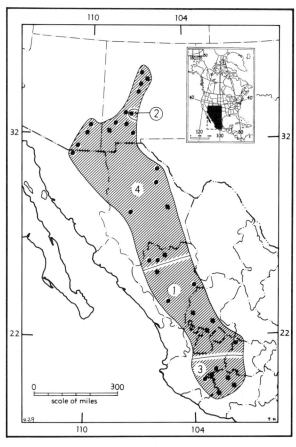

Map 429. *Sigmodon fulviventer.*

1. *S. f. fulviventer*   3. *S. f. melanotis*
2. *S. f. goldmani*   4. *S. f. minimus*

1962. *Sigmodon fulviventer minimus*, Baker and Greer, Publ. Mus., Michigan State Univ., Biol. Ser., 2:123, August 27.
1948. *Sigmodon minimus woodi* Gardner, Jour. Mamm., 29:65, February 13, type from E side Rio Grande, 51 mi. S Albuquerque, near Bernardo, 5000 ft., Socorro Co., New Mexico. Regarded as indistinguishable from "*S. m. minimus*" by Findley and Jones, Jour. Mamm., 44:315, August 22, 1963.

MARGINAL RECORDS.—New Mexico (Mohlhenrich, 1961:23, unless otherwise noted): 5 mi. N, 5 mi. E Algodones, 5200 ft.; 1$\frac{1}{10}$ mi. S, 6 mi. W Golden (Findley, *et al.*, 1975:237, as *S. fulviventer* only); 1 mi. S, 16 mi. W Las Cruces (*ibid.*). Chihuahua (Anderson, 1972:356): 3$\frac{1}{2}$ mi. ESE Los Lamentos, 1420 m. (owl pellets); 40 mi. E Gallego, 5000 ft.; 8 mi. N, 11 mi. E Charco de Peña, 4700 ft. Durango (Baker, 1969:211): 11 km. NNE Boquilla, 1952 m.; Rancho Bailón; Guanaceví. Chihuahua: 2 mi. W Miñaca (Anderson and Long, 1961:2, as *S. minimus*). Sonora: Río Santa Cruz; *Santa Cruz River, near Monument 111* (Baker, 1969:211). Arizona: Fort Huachuca; $\frac{1}{2}$ mi. SW Willcox (Lee and Zimmerman, 1969:333, as *S. fulviventer* only); Willcox (*ibid.*); $\frac{1}{2}$ mi. NE Willcox (*ibid.*); 20 mi.

N, 10 mi. E Willcox (Zimmerman, 1970:449, as *S. fulviventer* only). New Mexico (Mohlhenrich, 1961:23, unless otherwise noted): 1 mi. W Lordsburg (Lee and Zimmerman, 1969:334, as *S. fulviventer* only); 7 mi. N, 4$\frac{1}{2}$ mi. E Lordsburg, 4500 ft.; Silver City (Hall and Kelson, 1959:677); *Bayard, 6000 ft.*; 2 mi. N, 1$\frac{1}{2}$ mi. E Lake Valley, 5200 ft.; *Ladder Ranch, T. 14 S, R. 7 W, sec. 4, 6100 ft.*; 1 mi. S, 1 mi. W Winston, 6000 ft.; 6 mi. N, 1 mi. W Winston, 5700 ft.; E bank Rio Grande, opposite Bernardo (Hall and Kelson, 1959:677); 14 mi. S, 4 mi. W Albuquerque (Ivey, 1957:498); *1 mi. N Los Lunas; 5 mi. N, 1 mi. E Los Lunas, 4900 ft.; Isleta, 4900 ft.; 5 mi. S, 2 mi. W Bernalillo, 5100 ft.; Bernalillo, 5100 ft.; Algodones, 5100 ft.*

Sigmodon alleni
Allen's Cotton Rat

External measurements: 207–228; 88–128; 27–34. Brownish dorsum; underparts whitish or pale buff. Skull long and narrow, flattened appearance when viewed laterally; bulge of capsular projections for upper incisors slight; paroccipital processes, when viewed from below, slightly hooked; auditory bullae small in relation to breadth; basioccipital short and wide.

Sigmodon alleni alleni V. Bailey

1902. *Sigmodon alleni* V. Bailey, Proc. Biol. Soc. Washington, 15:112, June 2, type from San Sebastián, Mascota, Jalisco.

MARGINAL RECORDS (Baker, 1969:194).—Sinaloa: 2 km. E Santa Lucía, 1723 m. Nayarit: Tepic. Jalisco: type locality. Nayarit: Valle de Banderas; San Blas. Sinaloa: Copalá.

Sigmodon alleni planifrons Nelson and Goldman

1933. *Sigmodon planifrons* Nelson and Goldman, Proc. Biol. Soc. Washington, 46:197, October 26, type from Juquila, 5000 ft., Oaxaca.
1969. *Sigmodon alleni planifrons*, Baker, Univ. Kansas Mus. Nat. Hist., Miscl. Publ., 51:195, July 11.
1955. *Sigmodon planifrons minor* Goodwin, Amer. Mus. Novit., 1705:1, February 4, type from Santa Lucía, 4000 ft., 12 km. NE Tenango, Tehuantepec, Oaxaca. Not *Sigmodon minor* Gidley, 1922.
1955. *Sigmodon macdougalli* Goodwin, Amer. Mus. Novit., 1705:3, February 4, type from Santo Tomás Teipán (rain forest above village), 7000 ft., 12 km. S Yautepec, Oaxaca.
1955. *Sigmodon macrodon* Goodwin, Amer. Mus. Novit., 1705:4, February 4, type from Cerro San Pedro (rocky summit), 3600 ft., 20 km. W Mixtequilla, Tehuantepec, Oaxaca.
1959. *Sigmodon planifrons setzeri* Goodwin, Jour. Mamm., 40:447, August 20. Type, *Sigmodon planifrons minor* Goodwin.

MARGINAL RECORDS.—Oaxaca: Cerro San Pedro, 20 km. W Mixtequilla; Santa Lucía, 4000 ft.; 2 km. NNW Soledad, 1433 m. (Baker, 1969:196); 8 km.

ESE Río Grande, 30 m. (*ibid.*); *type locality*; 13 km. SSW Juchatengo, 1921 m. (Baker, 1969:196); Santo Tomás Teipán, 7000 ft., 12 km. S Yautepec.

Sigmodon alleni vulcani J. A. Allen

1906. *Sigmodon vulcani* J. A. Allen, Bull. Amer. Mus. Nat. Hist., 22:247, July 25, type from Volcán de Fuego, 10,000 ft., Jalisco.
1969. *Sigmodon alleni vulcani*, Baker, Univ. Kansas Mus. Nat. Hist., Miscl. Publ., 51:194, July 11.
1933. *Sigmodon guerrerensis* Nelson and Goldman, Proc. Biol. Soc. Washington, 46:196, October 26, type from Omilteme, 8000 ft., Guerrero.

MARGINAL RECORDS (Baker, 1969:195).—Jalisco: 10 km. SSW Autlán, 1372 m.; type locality. Michoacán: 14 km. E on road from Angahuan, 2300 m.; 10 km. W Capácuaro, 2059 m.; 3 km. W Pátzcuaro, 2380 m. Guerrero: Omilteme. Michoacán: 23 km. W Dos Aguas, 2135 m. Colima: 3 km. E Santiago. Jalisco: 9 km. [= 5 km. on label, KU 87630] NNW Barro de Navidad.

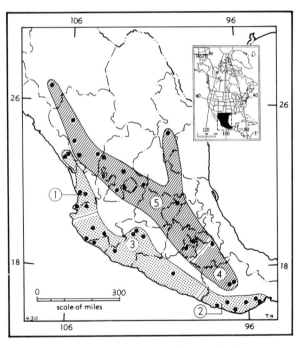

Map 430. *Sigmodon alleni* and *Sigmodon leucotis*.

1. *S. alleni alleni*   3. *S. alleni vulcani*
2. *S. alleni planifrons*   4. *S. leucotis alticola*
            5. *S. leucotis leucotis*

Sigmodon leucotis
White-eared Cotton Rat

External measurements: 230–252; 101–105; 25–31. Brownish-gray head and body; ears and underparts whitish; tail black becoming brown at

base below. Pronounced premaxillary depressions on each side of rostrum; lingual root on 1st lower molar greatly reduced or absent; interparietal short, length at midline less than 2 mm.; angular process of lower jaw slightly hooked; anterior portion of mesopterygoid fossa parallel-sided.

Sigmodon leucotis alticola V. Bailey

1902. *Sigmodon alticola* V. Bailey, Proc. Biol. Soc. Washington, 15:116, June 2, type from Cerro San Felipe, 10,000 ft., Oaxaca.
1969. *Sigmodon leucotis alticola*, Baker, Univ. Kansas Mus. Nat. Hist. Miscl. Publ., 51:222, July 11.

MARGINAL RECORDS.—Puebla: 15 km. NE Acatzingo (Baker, 1969:223). Oaxaca: type locality; 15 mi. W Oaxaca.

Sigmodon leucotis leucotis V. Bailey

1902. *Sigmodon leucotis* V. Bailey, Proc. Biol. Soc. Washington, 15:115, June 2, type from Sierra de Valparaíso, 8700 ft., Zacatecas.
1902. *Sigmodon alticola amoles* V. Bailey, Proc. Biol. Soc. Washington, 15:116, June 2, type from Pinal de Amoles, 7000 ft., Querétaro.

MARGINAL RECORDS (Baker, 1969:222, unless otherwise noted).—Chihuahua: La Unión, 10 km. N Guachochic, 8400 ft. (Anderson, 1972:358). Durango: 30 km. SSW Tepehuanes, 2500 m.; 28 mi. S, 17 mi. W Vicente Guerrero, 8350 ft. (Baker and Greer, 1962:125). Zacatecas: 13 km. S Chalchuites, 2623 m.; 15 km. W Zacatecas, 2135 m.; 17 km. S Pinos, 2165 m. Nuevo León: 20 km. SSW Galeana, 1891 m. Querétero: Pinal de Amoles. México: Monte Río Frío, 45 km. ESE México. Morelos: *3 km. W Huitzilac*; 7 km. W Huitzilac, 2806 m. Aguascalientes: 7½ km. NW Calvillo, 1830 m. Zacatecas: 9 mi. WNW Jalpa, 8250 ft. (Zimmerman, 1970:449, as *S. leucotis* only). Durango (Baker and Greer, 1962:124, 125): 2 mi. N Pueblo Nuevo, 6000 ft.; 1½ mi. W San Luis, 7550 ft.; *San Luis; ½ mi. E San Luis*.

Sigmodon ochrognathus V. Bailey
Yellow-nosed Cotton Rat

1902. *Sigmodon ochrognathus* V. Bailey, Proc. Biol. Soc. Washington, 15:115, June 2, type from Chisos Mts., 8000 ft., Brewster Co., Texas.
1903. *Sigmodon baileyi* J. A. Allen, Bull. Amer. Mus. Nat. Hist., 19:601, November 12, type from La Ciénega de las Vacas, 8500 ft., Durango.
1940. *Sigmodon ochrognathus montanus* Benson, Proc. Biol. Soc. Washington, 53:157, December 19, type from Peterson's Ranch ("Sylvania"), 6100 ft., 2 mi. N Sunnyside, Huachuca Mts., Cochise Co., Arizona.
1947. *Sigmodon ochrognathus madrensis* Goldman and Gardner, Jour. Mamm., 28:58, February 17, type from foothills of Sierra Madre, about 30 mi. NW Parral, 6200 ft., Chihuahua.

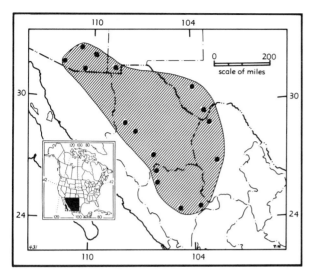

Map 431. *Sigmodon ochrognathus.*

External measurements: 223–260; 98–117; 25–30. Dorsum drab gray; nose and eye-ring ochraceous; underparts whitish; median-posterior notch on interparietal; tail scales narrow (0.50 mm. vs. 0.75).

MARGINAL RECORDS.—Arizona (Zimmerman, 1970:449): near Oracle; 11 mi. NE Willcox. New Mexico: Indian Creek Canyon, 6000 ft. (Findley and Jones, 1960:467). Texas: Davis Mts.; type locality. Coahuila: Tinaja del Telles, 4800 ft., El Jardin Ranch, Sierra de Carmen; 20 mi. S, 4 mi. W Ocampo, 5300 ft. Durango (Baker, 1969:230): 2 km. ESE Atotonilco, 2043 m.; 9 km. NNW Canatlán; Guanaceví; Rancho Santuario. Chihuahua (Baker, 1969:230, unless otherwise noted): *ca.* 30 mi. NW Parral, 6200 ft. (Hall and Kelson, 1959:677); 2 mi. W Miñaca (Anderson and Long, 1961:2); Cherry Ranch, 18 km. NW Cocomorachic. Arizona: Peterson's Ranch ("Sylvania"), 6100 ft., 2 mi. N Sunnyside; 11 mi. E Topawa (Hoffmeister, 1959:19).

Genus **Neotomodon** Merriam—Volcano Mouse

1898. *Neotomodon* Merriam, Proc. Biol. Soc. Washington, 12:127, April 30. Type, *Neotomodon alstoni* Merriam.

External measurements: 176–233; 78–105, 23–27; 19–23. Ears large and nearly naked; fur soft, dense; plantar tubercles, 6; mammae, 6. Skull broad; braincase strongly vaulted, short; zygomata widely spreading; zygomatic plate extended anteriorly; incisive foramina long, wide. Molars rooted, massive (almost disproportionately large); flat-crowned, heavily enameled. M1 and M2 similar, each having 3 external loops and 2 deep external re-entrant angles, and 2 internal loops and 1 shallow internal re-entrant angle; M3

peglike. Enamel pattern changes markedly with wear. With wear on M1 only the last inner and outer re-entrant angles persist. In M2 only external posterior angle persists.

Neotomodon alstoni
Volcano Mouse

For measurements and cranial details see account of the genus. Upper parts grayish to grayish buff, sometimes (seasonally) becoming fulvous brown; underparts white (underlying plumbeous showing conspicuously), often faintly washed with buff pectorally; tail sharply bicolored, dusky above, white below.

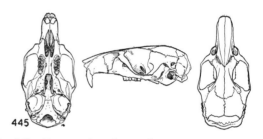

Fig. 445. *Neotomodon alstoni alstoni,* 5 mi. E Amecameca, 9600 ft., México, No. 91969 M.V.Z., ♂, X 1.

Neotomodon alstoni alstoni Merriam

1898. *Neotomodon alstoni* Merriam, Proc. Biol. Soc. Washington, 12:128, April 30, type from Nahuatzin, 8500 ft., Michoacán.

MARGINAL RECORDS.—Michoacán: type locality. Distrito Federal: 2½ km. SW Contreras, 2850 m. Puebla: Río Otlati, 15 km. NW San Martín, 8700 ft. [Texmeluacán]. México: Mt. Popocatépetl. Morelos: Huitzilac. Michoacán: Mt. Tancítaro.

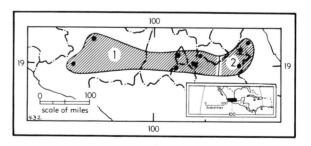

Map 432. *Neotomodon alstoni.*

1. *N. a. alstoni*   2. *N. a. perotensis*

Neotomodon alstoni perotensis Merriam

1898. *Netomodon perotensis* Merriam, Proc. Biol. Soc. Washington, 12:129, April 30, type from Cofre de Perote, 9500 ft., Veracruz.

1944. *Neotomodon alstoni perotensis*, Davis, Jour. Mamm., 25:398, December 12.
1898. *Neotomodon orizabae* Merriam, Proc. Biol. Soc. Washington, 12:129, April 30, type from Volcán de Orizaba, 9500 ft., Puebla.

MARGINAL RECORDS.—Veracruz: Las Vigas (Hall and Dalquest, 1963:315). Puebla: W slope Mt. Orizaba, 11,000 ft. Veracruz: type locality; *Perote* (Hall and Dalquest, 1963:315).

Genus **Neotoma** Say and Ord—Wood Rats

Revised by Goldman, N. Amer. Fauna, 31:1–124, 8 pls., 14 figs., October 19, 1910; partly rearranged by Goldman, Jour. Mamm., 13:59–67, February 10, 1932.

1825. *Neotoma* Say and Ord, Jour. Acad. Nat. Sci. Philadelphia, 4(2):345. Type, *Mus floridana* Ord.

External measurements: 225–470; 76–241; 28–52. Form ratlike; tail relatively long, usually clothed with short hairs, but in some species distichous and in others nearly naked. Upper parts varying from pale buff or gray to rich buff or ferruginous, more or less mixed with black; underparts usually white but sometimes pale buff. Skull narrow interorbitally; supraorbital ridges present but not markedly developed; rostrum relatively slender; incisive foramina long, extending to level of M1; bullae relatively large (but variable), obliquely situated, tapered anteriorly. Molars flat-crowned, prismatic, hypsodont, somewhat resembling those of microtines but simpler; M1 with two outer and two inner folds, the anterointernal fold not excessively deep; M2 with 2 outer folds and 1 inner fold, inner fold deep in subgenus *Hodomys*; M3 essentially as in M2; m1 and m2 with 2 outer folds, m3 with 1; m1 with 3 inner folds (usually), m2 with 2, m3 with 1; enamel pattern of m3 distinctly S-shaped in subgenus *Hodomys*.

KEY TO SUBGENERA OF NEOTOMA

1. Auditory bullae greatly enlarged (almost semicircular in profile but not greatly widened transversely), little tapered anteriorly, and situated only slightly obliquely to the long axis of the skull. . . . . *Teanopus*, p. 772

1′. Auditory bullae not greatly enlarged; bullae tapered anteriorly, and situated obliquely.
  2. Enamel pattern of chewing-surface of m3 distinctly S-shaped. . . *Hodomys*, p. 771
  2′. Enamel pattern of chewing-surface of m3 not S-shaped.
    3. Tail markedly distichous; basal part of baculum quadrate in cross section with slight concavity on dorsal and ventral surfaces.
              Teonoma, p. 767
    3′. Tail terete or almost naked; baculum not quadrate in basal cross section and lacking concavity on dorsal and ventral surfaces.
              Neotoma, p. 746

Subgenus **Neotoma** Say and Ord

1825. *Neotoma* Say and Ord, Jour. Acad. Nat. Sci. Philadelphia, 4(2):345. Type, *Mus floridana* Ord.
1910. *Homodontomys* Goldman, N. Amer. Fauna, 31:86, October 19. Type, *Neotoma fuscipes* Baird. (Arranged as a synonym by Burt and Barkalow, Jour. Mamm., 23:296, August 14, 1942.)

First upper molar with anterointernal angle of variable depth or obsolete; M3 with middle loop either divided or undivided by re-entrant angles; auditory bullae of variable size; "tail terete, tapering, and short-haired; hind foot naked below along outer side at least to tarsometatarsal joint" (Goldman, 1910:20). Length of baculum more than 5 mm.; lateral diameter of the base more than 1.7.

The following nominal species are known only from the islands in the Gulf of California and off the Pacific Cost of Baja California.

Neotoma varia . . . . .Turners Island, Sonora
Neotoma bryanti . . . . .Cedros Island, Baja California
Neotoma anthonyi . . . . .Todos Santos Island, Baja California
Neotoma martinensis . . . . .San Martín Island, Baja California
Neotoma bunkeri . . . . .Coronados Island, Baja California

These species are not considered in the following key.

KEY TO NOMINAL SPECIES OF SUBGENUS NEOTOMA

1. Occurring north of México.
  2. Occurring in California or Oregon.
    3. Tail markedly bicolored.
      4. Hairs on throat white to base. . . . . *N. albigula*, p. 751
      4′. Hairs on throat plumbeous basally. . . . . *N. lepida*, p. 755
    3′. Tail not markedly bicolored. . . . . *N. fuscipes*, p. 765

2'. Occurring other than in California or Oregon.
    5. Anterointernal re-entrant angle of M1 extending more than halfway across crown.
        6. Lateral diameter of base of baculum more than 2.80; total length of rat more than 375; occurring east of Mississippi River. . . . . *N. floridana* (subspecies *magister*), p. 749
        6'. Lateral diameter of base of baculum less than 2.80; total length of rat less than 375; occurring west of Mississippi River. . . . . *N. mexicana*, p. 761
    5'. Anterointernal re-entrant angle of M1 not extending more than halfway across crown.
        7. Occurring west of a line drawn through central Colorado and central New Mexico.
            8. Hairs on throat white to base.
                9. Upper parts steely gray. . . . . *N. micropus*, p. 750
                9'. Upper parts buffy or grayish more or less strongly washed with fulvous.
                        N. albigula, p. 751
            8'. Hairs on throat plumbeous basally.
                10. Upper parts usually pinkish buff; tail sharply bicolored; hairs on tail short.
                        N. lepida, p. 755
                10'. Upper parts ochraceous or yellowish; tail only faintly or not at all bicolored; hairs on tail moderately long. . . . . *N. stephensi*, p. 760
        7'. Occurring east of a line drawn through central Colorado and central New Mexico.
            11. Upper parts steely gray (except in an occasional specimen from southern Texas).
                        N. micropus, p. 750
            11'. Upper parts not steely gray.
                12. Tail sharply bicolored; baculum curved in profile and with rounded base.
                        N. albigula, p. 751
                12'. Tail not sharply bicolored; baculum straight in profile and with truncated base. . . . . *N. floridana*, p. 748
1'. Occurring south of United States.
    13. Occurring in western Sonora or Baja California.
        14. Tail sharply bicolored; base of baculum not dumbbell-shaped in cross section.
            15. Hairs of throat plumbeous basally; baculum a long, slender rod. . . . . *N. lepida*, p. 755
            15'. Hairs of throat white to base; baculum with expanded base that is shallowly U-shaped in cross section. . . . . *N. albigula*, p. 751
        14'. Tail not sharply bicolored; baculum short and with much expanded base that is dumbbell-shaped in cross section. . . . . *N. fuscipes*, p. 765
    13'. Not occurring in western Sonora or Baja California.
        16. Anterointernal re-entrant angle of M1 extending less than halfway across crown.
            17. Tail unicolored. . . . . *N. nelsoni*, p. 754
            17'. Tail sharply bicolored.
                18. Sphenopalatine vacuities present.
                    19. Palate with posterior median spine; posterior half of septum that divides anterior palatine foramina lacking a foramen; upper parts slaty gray (except in *N. m. littoralis*) . . . . *N. micropus*, p. 750
                    19'. Palate lacking posterior median spine; or if the spine is present, then posterior half of septum that divides anterior palatine foramina perforated by a foramen; upper parts not slate gray.
                      20. Hairs on throat white to base (except *N. a. melanura*—see 21' below).
                        N. albigula, p. 751
                    20'. Hairs on throat plumbeous basally.
                        21. Hind foot less than 35. . . . . *N. goldmani*, p. 760
                        21'. Hind foot more than 35. . . . . *N. a. melanura*, p. 753
                18'. Sphenopalatine vacuities absent. . . . . *N. palatina*, p. 754
    16'. Anterointernal re-entrant angle of M1 extending more than halfway across tooth.
        22. Occurring south of México.
            23. Zygomatic arches parallel or almost so; total length usually less than 370.
                        N. mexicana, p. 761
            23'. Zygomatic arches strongly convergent anteriorly; total length usually more than 370. . . . . *N. chrysomelas*, p. 765
        22 . Occurring in México.
            24. Total length more than 370, occurring in Tamaulipas. . . . . *N. angustapalata*, p. 765
            24'. Total length less than 370, not occurring in Tamaulipas. . . . . . *N. mexicana*, p. 761

Neotoma floridana
Eastern Wood Rat

External measurements: 310–441; 129–203; 35–46. Upper parts varying from pale cinnamon to buffy gray, sides fairly distinct from back; underparts white or grayish, the plumbeous of the hairs showing through clearly, especially on the belly; tail shorter than head and body, sparsely haired, not sharply bicolored, dusky or brownish above, whitish to gray below. Skull large, elongated; premaxillary tongues extending posteriorly beyond nasals; sphenopalatine vacuities relatively small; interpterygoid fossa broad; presphenoid constricted; auditory bullae relatively small, rounded; palate lacking posterior median spine; M1 with moderately developed anterointernal re-entrant angle.

The only place so far (1972) known to me where the geographic ranges of *Neotoma floridana* and *Neotoma micropus* meet is on the north side of the Cimarron River in Major County, Oklahoma, along the two sides of the north-to-south U.S. Highway 281. Here interchange of genes occurs. In this area, which is less than one mile in diameter, approx. half the individuals taken, it is verbally reported, are "intermediates," a fourth are *floridana*, and a fourth are *micropus*. Whether the intermediates are hybrids or intergrades is undetermined. If they be the latter, the several taxa now arranged as subspecies of the species *N. micropus* will be arranged as subspecies of the earlier-named species *Neotoma floridana*. Initially D. L. Spencer and later E. C. Birney sampled the population where the geographic ranges of the two alleged species meet, and it is hoped that in due time these authors will publish more details of their findings.

Neotoma floridana attwateri Mearns

1897. *Neotoma attwateri* Mearns, Proc. U.S. Nat. Mus., 19:721, July 30, type from Lacey's Ranch, Turtle Creek, Kerr Co., Texas.
1901. [*Neotoma floridana*] *attwateri*, Elliot, Field Columb. Mus., Publ. 45, Zool. Ser., 2:157, March 6.

MARGINAL RECORDS.—Texas: Tehuacana Creek, S of Waco; Robertson County (Davis, 1966:200, as *N.*

floridana only); Walker County (*ibid.*, as *N. floridana* only); Victoria; Rock Springs; Kerr County (Davis, 1966:200, as *N. floridana* only); Williamson County (*ibid.*, as *N. floridana* only).

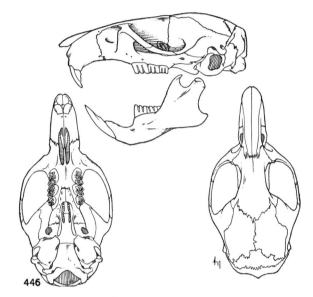

446

Fig. 446. *Neotoma floridana magister*, 2¼ mi. NW Lenhartsville, 800 ft., Berks Co., Pennsylvania, No. 38640 F.M.N.H., ♂, X 1.

Neotoma floridana baileyi Merriam

1894. *Neotoma baileyi* Merriam, Proc. Biol. Soc. Washington, 9:123, July 2, type from Valentine, Cherry Co., Nebraska.
1905. *Neotoma floridana baileyi*, V. Bailey, N. Amer. Fauna, 25:109, October 24.

MARGINAL RECORDS.—South Dakota: Spring Creek, 18 mi. SE Rapid City. Nebraska (Jones, 1964c:218): Long Pine; *3 mi. SSE Valentine*; 10 mi. S Cody.

Neotoma floridana campestris J. A. Allen

1894. *Neotoma campestris* J. A. Allen, Bull. Amer. Mus. Nat. Hist., 6:322, November 7, type from Pendennis, Lane Co., Kansas.
1914. *Neotoma floridana campestris*, R. Kellogg, Kansas Univ. Mus. Nat. Hist., Publ. 1, Zool. Ser., 1(1):5, January 30.

MARGINAL RECORDS.—Nebraska (Jones, 1964c:219): North Platte; *5 mi. S, 2½ mi. W Brady*; Quinn's Canyon, 10 mi. S Gothenburg. Kansas: 6 mi. SW Woodston; Hays; type locality; 9 mi. S, 4½ mi. E Wallace. Colorado (Finley, 1958:318): near Holly, Arkansas River bottom; Fort Lyon; Olney; Pueblo; *3 mi. S, 2 mi. W Fountain*; 10 mi. S Colorado Springs; 8 mi. NE Agate; Wray.

Neotoma floridana floridana (Ord)

1818. *Mus floridanus* Ord, Bull. Soc. Philom. Paris, p. 181, December, type from St. Johns River, Florida; probably near Jacksonville, Duval Co. (See Bangs, Proc. Boston Soc. Nat. Hist., 28:184, March, 1898.)

1825. *N[eotoma]. floridana,* Say and Ord, Jour. Acad. Nat. Sci. Philadelphia, 4(2):346.

MARGINAL RECORDS.—North Carolina: Smith Island, Brunswick Co., thence down coast to Florida: near Vero Beach (Layne, 1974:394); Murdock; Tallahassee. Georgia: Reynold's Refuge, Dougherty Co.; S of Macon. South Carolina: Savannah River Plant (Golley, *et al.,* 1965:4, as *N. floridana* only).

Neotoma floridana haematoreia A. H. Howell

1934. *Neotoma floridana haematoreia* A. H. Howell, Proc. Acad. Nat. Sci. Philadelphia, 86:403, October 15, type from near summit Blood Mtn., 4440 ft., Lumpkin Co., Georgia.

MARGINAL RECORDS.—Tennessee: 3 mi. above Townsend, on Little River. North Carolina: Horsepasture River, near exit from gorge, 1300 ft. (Adams, 1965:499). South Carolina: 1¼ mi. from Caesars Head Hotel, Geer Hwy., 3000 ft. Georgia: type locality.

Neotoma floridana illinoensis A. H. Howell

1910. *Neotoma floridana illinoensis* A. H. Howell, Proc. Biol. Soc. Washington, 23:28, March 23, type from Wolf Lake, Union Co., Illinois.

MARGINAL RECORDS.—Missouri: Black. Illinois: Aldridge. Tennessee: 5 mi. NE Samburg (Beasley and Severinghaus, 1973:111). Alabama: Sand Mtn., Jackson Co. Florida: Marianna. Mississippi (Kennedy, *et al.,* 1974:19, as *N. floridana* only): Lauderdale County; Holmes County. Louisiana (Lowery, 1974:260): 2 mi. N Tremont; Fishville; ½ mi. N Simpson. Texas: Texarkana. **See** addenda.

Neotoma floridana magister Baird

1858. *N[eotoma]. magister* Baird, Mammals, *in* Repts. Expl. Surv. . . . , 8(1):498, July 14, type from a cave near Carlisle, Cumberland Co., or near Harrisburg, Dauphin Co., Pennsylvania.

1957. *Neotoma floridana magister,* Schwartz and Odum, Jour. Mamm., 38:204, May 27.

1893. *Neotoma pennsylvanica* Stone, Proc. Acad. Nat. Sci. Philadelphia, 45:16, type from near top South Mtn., Cumberland Co., Pennsylvania, some 6 mi. from Pine Grove at a place known as Lewis' Cave.

MARGINAL RECORDS.—New York: Lake Mohonk. Connecticut: Scattergoat [= Schagticoke] Mtn., near Kent. New York: near Piermont. Pennsylvania: Lancaster County. Maryland: Plummer Island. Virginia: Bedford County; Wythe County. North Carolina: Grandfather Mtn., Avery Co., 4100 ft. Tennessee: Walden Ridge, near Soddy, 3 mi. SW Rathburn. Georgia: Dade County (Golley, 1962:142). Alabama: Woodville; N shore Muscle Shoals, 10 mi. SE Florence. Tennessee: Duck River, 2 mi. SW Waverly. Indiana: just W of Valeene; 1½ mi. W Springville (Bader and Hall, 1960:111, as *N. floridana* only). Ohio: Sugar

Grove. Pennsylvania: 1 mi. SW Utica; Lycoming County.

Neotoma floridana osagensis Blair

1939. *Neotoma floridana osagensis* Blair, Occas. Pap. Mus. Zool., Univ. Michigan, 403:5, June 16, type from Okesa, Osage Co., Oklahoma.

MARGINAL RECORDS.—Kansas: 2 mi. S Marysville; Leavenworth County. Missouri: Hahatonka. Arkansas: Sylamore Experimental Forest, Calico Rock (Ward and Leonard, 1968:530, as *N. floridana* only). Texas (Davis, 1966:200, as *N. floridana* only, unless otherwise noted): Nacogdoches County; Cherokee County; Henderson County; 4 mi. E Stoneburg (Dalquest, 1968:19). Oklahoma: Chattanooga; Wichita Mts. Wildlife Refuge (Glass and Halloran, 1961:238); North Canadian River, 3 mi. S Chester (116943 KU). Kansas: 3 mi. SE Arkansas City; Smoky Hill River, 3 mi. S Wilson.

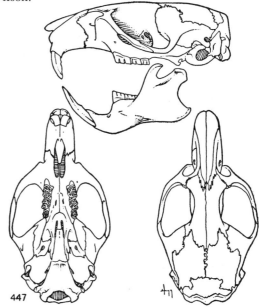

Fig. 447. *Neotoma floridana osagensis,* 5 mi. N, 12 mi. E Arkansas City, Cowley Co., Kansas, No. 39283 K.U., ♂, X 1.

Neotoma floridana rubida Bangs

1898. *Neotoma floridana rubida* Bangs, Proc. Boston Soc. Nat. Hist., 28:185, March, type from Gibson, Terrebonne Paris, Louisiana.

MARGINAL RECORDS.—Louisiana: Bonita (Lowery, 1974:260). Mississippi (Kennedy, *et al.,* 1974:19, as *N. floridana* only): Yazoo County; Covington County. Alabama: Mobile. Mississippi: Bay St. Louis (Kennedy, *et al.,* 1974:19, as *N. floridana* only). Louisiana (Lowery, 1974:259, 260): Biloxi Marsh; Golden Meadow; Weeks Island; Sulphur. Texas: Sour Lake; Trinity County (Davis, 1966:200, as *N. floridana* only); N of Camden along Neches River. Louisiana (Lowery,

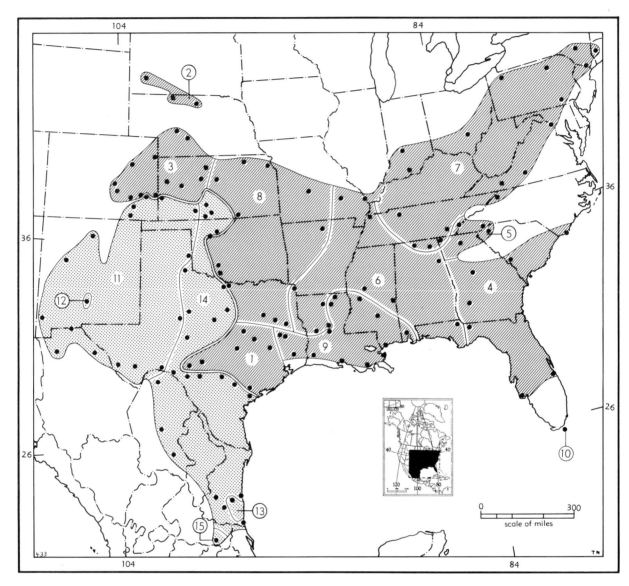

Map 433.   *Neotoma floridana* and *Neotoma micropus*.

1. *N. floridana attwateri*	6. *N. floridana illinoensis*	11. *N. micropus canescens*
2. *N. floridana baileyi*	7. *N. floridana magister*	12. *N. micropus leucophaea*
3. *N. floridana campestris*	8. *N. floridana osagensis*	13. *N. micropus littoralis*
4. *N. floridana floridana*	9. *N. floridana rubida*	14. *N. micropus micropus*
5. *N. floridana haematoreia*	10. *N. floridana smalli*	15. *N. micropus planiceps*

1974:260): 10 mi. SE Alexandria; N of Hwy. 80, Russell Sage Game Management Area.

Neotoma floridana smalli Sherman

1955. *Neotoma floridana smalli* Sherman, Jour. Mamm., 36:119, February 25, type from Key Largo, Monroe Co., Florida. Known only from type locality.

Neotoma micropus
Southern Plains Wood Rat

External measurements: 300–380; 120–185; 34–41. Upper parts steely to slaty gray, some-times washed faintly with buff (*N. m. littoralis* and occasional specimens of *N. m. micropus* from southern Texas are distinctly brownish); underparts gray on belly, white on gular and pectoral regions; tail dusky above, whitish below. Skull generally resembling that of *N. floridana* but more robust and sculptured; interpterygoid fossa broad; palate usually with posterior median spine; sphenopalatine vacuities large; palatal bridge relatively short; anterior palatine foramina relatively long; anterointernal re-entrant angle of M1 shallow.

Hybridization, or possibly intergradation, between *Neotoma micropus* and *Neotoma floridana* at one place in Oklahoma has been mentioned in the account of *N. floridana*. Finley (1958:301–303, 315, 546) describes hybridization between *Neotoma micropus* and *Neotoma albigula* in southeastern Colorado. Anderson (1969:43, 48) speculates about gene flow between *N. micropus* and *N. albigula* in eastern Coahuila, but his data may indicate only convergent adaptation there in certain morphological features of the two species.

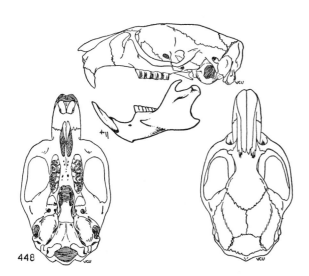

Fig. 448. *Neotoma micropus canescens*, State Park, Meade Co., Kansas, No. 12753 K.U., ♂, X 1.

Neotoma micropus canescens J. A. Allen

1891. *Neotoma micropus canescens* J. A. Allen, Bull. Amer. Mus. Nat. Hist., 3:285, June 30, type from North Beaver Creek [= North Canadian River], Cimarron Co., Oklahoma.

MARGINAL RECORDS.—Colorado: 2 mi. S, 2 mi. E Hasty (Armstrong, 1972:219). Kansas: Coolidge; Stephenson Ranch, 7 mi. S Kingsdown. Texas (Davis, 1966:202, as *N. micropus* only, unless otherwise noted): Wheeler County; Colorado (Hall and Kelson, 1959:685); Lozier (*ibid.*); Brewster County; Presidio County. Chihuahua: 3¼ mi. ESE Los Lamentos (76497 KU); Monument 15, Mexican–U.S. boundary (Anderson, 1969:29); 12 mi. N Nueva Casas Grandes (Bogan and Williams, 1975:278, as *N. micropus* only). New Mexico: 6 mi. SSE Lordsburg; 8 mi. SE Grants; Rinconada. Colorado: 11 mi. N, 8 mi. E Branson, 5600 ft. (Armstrong, 1972:219); 18 mi. S La Junta.

Neotoma micropus leucophaea Goldman

1933. *Neotoma micropus leucophaea* Goldman, Jour. Washington Acad. Sci., 23:472, October 15, type from White Sands, 10 mi. W Point of Sands, White Sands National Monument, 4100 ft., Otero Co., New Mexico. Known only from type locality.

Neotoma micropus littoralis Goldman

1905. *Neotoma micropus littoralis* Goldman, Proc. Biol. Soc. Washington, 18:31, February 2, type from Altamira, 1000 ft., Tamaulipas.

MARGINAL RECORDS.—Tamaulipas: Sierra de Tamaulipas, 2 mi. S, 10 mi. W Piedra (Alvarez, 1963:453); *6 mi. N, 6 mi. W Altamira* (*ibid.*); type locality.

Neotoma micropus micropus Baird

1855. *Neotoma micropus* Baird, Proc. Acad. Nat. Sci. Philadelphia, 7:333, April, type from Charco Escondido, Tamaulipas.
1899. *Neotoma macropus* [*sic*] *surberi* Elliot, Field Columb. Mus., Publ. 37, Zool. Ser., 1(14):279, May 15, type from 3 mi. W Alva, Woods Co., Oklahoma.

MARGINAL RECORDS.—Kansas: 5 mi. N Belvidere; 8 mi. SW Medicine Lodge. Oklahoma: 5 mi. W Orienta; North Canadian River, 3 mi. S Chester (120861 KU). Texas: Henrietta; Brazos; Eastland County (Davis, 1966:202, as *N. micropus* only); Kinney County (*ibid.*, as *N. micropus* only); Uvalde County (*ibid.*, as *N. micropus* only); San Antonio; Karnes County (Davis, 1966:202, as *N. micropus* only); Refugio County (*ibid.*, as *N. micropus* only), thence southward along coast to Tamaulipas (Alvarez, 1963:454): 1 mi. E La Pesca; *La Pesca*; Forlón; Ciudad Victoria. Coahuila: 3 mi. N, 5 mi. W La Rosa, 3600 ft.; 3 mi. NW Cuatro Ciénegas (Baker, 1956:287); Cañón del Cochino, 3200 ft., 16 mi. N, 21 mi. E Piedra Blanca. Texas: Comstock; Tom Green County (Davis, 1966:202, as *N. micropus* only); Fisher County (*ibid.*, as *N. micropus* only); Newlin. Kansas: Cave Creek.

Neotoma micropus planiceps Goldman

1905. *Neotoma micropus planiceps* Goldman, Proc. Biol. Soc. Washington, 18:32, February 2, type from Río Verde, 3000 ft., San Luis Potosí. Known only from type locality.

Neotoma albigula
White-throated Wood Rat

Revised by Hall and Genoways, Jour. Mamm., 51:504–516, August 28, 1970.

External measurements: 282–400; 76–185; 30–38.5. Upper parts grayish washed with fulvous to ochraceous mixed with dusky; underparts white or grayish, individual hairs plumbeous basally except on throat where they are white to the base; tail brownish or even dusky above, whitish below. Skull: rostrum relatively broad; auditory bullae of moderate size; sphenopalatine vacuities large; anterointernal re-entrant angle of M1 shallow.

The species *N. albigula* is placed after *N. micropus* and before *N. lepida* on the basis of characteristics of the baculum as described and figured by Burt and Barkalow (1942:290–292).

Instances of hybridism or intergradation between *Neotoma albigula* and *Neotoma micropus* in southeastern Colorado and possibly convergent adaptation in certain morphological features in *N. albigula* and *N. micropus* in eastern Coahuila are mentioned in the account of *N. micropus*.

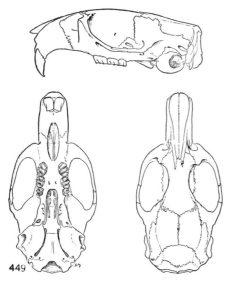

Fig. 449. *Neotoma albigula venusta*, Carrizo Creek, San Diego Co., California, No. 7559 M.V.Z., ♂, X 1.

Neotoma albigula albigula Hartley

1894. *Neotoma albigula* Hartley, Proc. California Acad. Sci., ser. 2, 4:157, May 9, type from vic. Fort Lowell, near Tucson, Pima Co., Arizona.
1894. *Neotoma intermedia angusticeps* Merriam, Proc. Biol. Soc. Washington, 9:127, July 2, type from NW corner Hidalgo Co., New Mexico.

MARGINAL RECORDS (Hall and Genoways, 1970:512, 513, unless otherwise noted).—New Mexico: Chama Canyon; Rinconado; 4 mi. SW Cimarron (Findley, *et al.*, 1975:243, as *N. albigula* only); Bell Ranch (*op. cit.*:245); 25 mi. SW Tucumcari. Texas: Washburn; Armstrong County; 6 mi. SSE Lazare; 16 mi. NW Seymour (Baccus, 1971:182); Llano; *Black Gap region, Brewster Co.*; The Basin, Chisos Mts., 5200 ft. Chihuahua: Santa Rosalía; 8 mi. NE Laguna, *ca.* 3 mi. from Laguna de Bustillos (Anderson and Nelson, 1960:100, as *N. albigula* only). Sonora: Hermosillo. Arizona: Nogales. Sonora: Santo Domingo. Arizona: *Papago Well;* 9 mi. E Papago Well; E base Crater Mtn., 13 mi. N Ajo, 1200 ft.; *10 mi. S Gila Bend* (Cockrum, 1961:192); Gila Bend; near Buckeye; *Wickenburg, 2500 ft.;* Congress Junction; Big Sandy Creek; *Hualpai Mts., 5800 ft.;* Kingman, 3300 ft.; Gold Basin, 3000 ft.; *Peach Spring, 4000 ft.;* Seligman; Montezuma Well, 3500 ft.; 7 mi. N Payson, 4500 ft.; Salt River, about 12 mi. N McMillenville, 3000 ft.; White

River; Chiricahua Ranch, 4700 ft., 20 mi. NE Calva; San Francisco River, 13 mi. above Clifton, 4000 ft. New Mexico: Glenwood; Datil Mts.; Riley; Grants; *San Rafael;* Gallup; Canyon de Chelly; 2 mi. S, 3 mi. E Estrella (Findley, *et al.*, 1975:244, as *N. albigula* only).

Neotoma albigula brevicauda Durrant

1934. *Neotoma albigula brevicauda* Durrant, Jour. Mamm., 15:65, February 16, type from Castle Valley, about 15 mi. NE Moab, Grand Co., Utah.

MARGINAL RECORDS.—Utah: type locality. Colorado (Hall and Genoways, 1970:514): 1 mi. SW Gateway, 4600 ft.; Coventry, 6800 ft.; Bedrock, 5150 ft.

Neotoma albigula durangae J. A. Allen

1903. *Neotoma intermedia durangae* J. A. Allen, Bull. Amer. Mus. Nat. Hist., 19:602, November 12, type from San Gabriel, Durango.
1910. *Neotoma albigula durangae,* Goldman, N. Amer. Fauna, 31:37, October 19.

MARGINAL RECORDS.—Chihuahua: *possibly jct. ríos Conchos and Grande* (Anderson, 1969:Fig. 1, p. 28); Consolación (82797 KU). Coahuila (Hall and Genoways, 1970:512, as *N. a. albigula*): Fortín, 3300 ft.; Monclova; Jaral; Jimulco. Durango (Hall and Genoways, 1970:514): Mt. San Gabriel; Rancho Santuario. Chihuahua (on authority of S. Anderson, Univ. Kansas Mus. Nat. Hist., Miscl. Publ., 51:25–50, July 11, 1969, and *in Litt.*): Parral, 6000 ft.; Las Trincheras, 4900 ft. (76961 MVZ); 2 mi. N, 6 mi. E Camargo, 4150 ft. (54812 KU); 1 mi. E Julimes (86067 KU). Not found: Durango: La Ciénega de las Vacas (Hall and Genoways, 1970:514).

Neotoma albigula laplataensis F. W. Miller

1933. *Neotoma albigula laplataensis* F. W. Miller, Proc. Colorado Mus. Nat. Hist., 12(1):2, July 22, type from near Bondad, La Plata Co., Colorado.

MARGINAL RECORDS.—Utah: Recapture Canyon, 12 mi. N Blanding, 6000 ft. (Hall and Genoways, 1970:514). Colorado (Armstrong, 1972:221): 2 mi. S, 4 mi. W Cortez, 5900 ft.; 2 mi. NE Bondad, 6100 ft.; Stollsteimer Creek, near mouth Deep Cañon; *1 mi. N Juanita.* New Mexico (Findley, *et al.*, 1975:244, as *N. albigula* only): ¼ mi. SE base Huerfanito. Arizona (Cockrum, 1961:194, 195): Canyon de Chelley; St. Michaels, 7000 ft.; *Zuñi River;* 8 mi. S St. Johns, 5800 ft.; *Springerville, 7000 ft.;* 3 mi. SE Springerville; Turkey Creek, 3400 ft.; Canyon Padre; *Winona, 6400 ft.;* Red Lake; Cataract Canyon, 12 mi. WSW Anita, 5400 ft.; Supai Canyon; *1 mi. N Bass Camp* (Hoffmeister, 1971:175), thence to and southeastward along S bank Colorado River to Palisades Creek, Mile 66 (*op. cit.:* 176); Rainbow Lodge, Navajo Mtn., 6400 ft.

Neotoma albigula latifrons Merriam

1894. *Neotoma latifrons* Merriam, Proc. Biol. Soc. Washington, 9:121, July 2, type from Queréndaro, Michoacán.
1970. *Neotoma albigula latifrons*, Hall and Genoways, Jour. Mamm., 51:506, August 28.

MARGINAL RECORDS.—Michoacán: *Isla Palmitas* (Hall and Genoways, 1970:514); type locality.

Neotoma albigula leucodon Merriam

1894. *Neotoma leucodon* Merriam, Proc. Biol. Soc. Washington, 9:120, July 2, type from San Luis Potosí, San Luis Potosí.
1910. *Neotoma albigula leucodon*, Goldman, N. Amer. Fauna, 31:36, October 19.
1905. *Neotoma montezumae* Goldman, Proc. Biol. Soc. Washington, 18:29, February 2, type from Zimapán, 7500 ft., Hidalgo.
1905. *Neotoma leucodon zacatecae* Goldman, Proc. Biol. Soc. Washington, 18:30, February 2, type from Plateado, 7600 ft., Zacatecas.

MARGINAL RECORDS (Hall and Genoways, 1970:514).—Durango: Hda. Atotonilco, 6680 ft. San Luis Potosí: 6 km. S Matehuala; Presa de Guadalupe; *3 mi. NW Tepeyac;* 10 mi. NW Ciudad del Maíz. Hidalgo: *Zimapán;* Ixmiquilpán; Marqués. Guanajunto: La Quemada. Jalisco: 3 mi. E Unión de San Antonio, 6100 ft.; 10 mi. NE Yahualica. Zacatecas; 3 mi. SW Jalpa. Jalisco: 3 mi. E Totatiche, 5600 ft.; La Mesa María de León, 7400 ft., 22° 25′ N, 103° 24′ W. Zacatecas: Valparaíso. Durango: 16 mi. S, 29 mi. W Vicente Guerrero, 6675 ft.; 9 mi. N Durango; 1 mi. N Chorro; 26 mi. SW Yerbanís, 6725 ft.

Neotoma albigula mearnsi Goldman

1915. *Neotoma albigula mearnsi* Goldman, Proc. Biol. Soc. Washington, 28:135, June 29, type from Tinajas Altas, Gila Mts., Yuma Co., Arizona.

MARGINAL RECORDS (Hall and Genoways, 1970:514, unless otherwise noted).—Arizona: S of Wellton; *type locality;* Tule Tank (Cockrum, 1961:194); 9 mi. E Papago Well, 1100 ft.; Alamo Canyon, Ajo Mts.; *26 mi. S Wellton, 500 ft.*

Neotoma albigula melanura Merriam

1894. *Neotoma intermedia melanura* Merriam, Proc. Biol. Soc. Washington, 9:126, July 2, type from Ortiz, Sonora.
1905. N[eotoma]. a[lbigula]. melanura, Goldman, Proc. Biol. Soc. Washington, 18:29, February 2.

MARGINAL RECORDS (Hall and Genoways, 1970:514).—Sonora: type locality. Chihuahua; Mojarachic; Batopilas. Sinaloa: 1 mi. N Topolobampo, 50 ft. (Birney and Jones, 1972:198), thence up coast to Sonora: Batamotal; Bahía San Pedro.

Neotoma albigula melas Dice

1929. *Neotoma albigula melas* Dice, Occas. Pap. Mus. Zool., Univ. Michigan, 203:3, June 19, type from Malpais Spring, malpais lava beds near Carrizozo, Lincoln Co., New Mexico.

MARGINAL RECORDS.—New Mexico: 13 mi. NW Carrizozo; type locality.

Neotoma albigula robusta Blair

1939. *Neotoma albigula robusta* Blair, Occas. Pap. Mus. Zool., Univ. Michigan, 403:3, June 16, type from Limpia Canyon, 4300 ft., 16 mi. N Fort Davis, Jeff Davis Co., Texas.

MARGINAL RECORDS.—Texas: type locality; Paisano; Marfa; La Mota Mtn. area; Limpia Canyon, 2 mi. NW Fort Davis.

Neotoma albigula seri Townsend

1912. *Neotoma albigula seri* Townsend, Bull. Amer. Mus. Nat. Hist., 31:125, June 14, type from Tiburón Island, Gulf of California, Sonora. Known only from type locality.

Neotoma albigula sheldoni Goldman

1915. *Neotoma albigula sheldoni* Goldman, Proc. Biol. Soc. Washington, 28:136, June 29, type from Pápago Tanks, Sierra Pinacate, Sonora.

MARGINAL RECORDS.—Sonora: type locality; Sáric.

Neotoma albigula subsolana Alvarez

1962. *Neotoma albigula subsolana* Alvarez, Univ. Kansas Publ., Mus. Nat. Hist., 14:141, April 30, type from Miquihuana, 6400 ft., Tamaulipas.

MARGINAL RECORDS (Hall and Genoways, 1970:514).—Coahuila: 9 mi. E Hermanas; Panuco, 3000 ft. Nuevo León: Santa Catarina; Ojo de Agua; Iturbide, Sierra Madre Oriental, 5000 ft.; 9 mi. S Aramberri, 3900 ft. Tamaulipas: Joya Verde, 35 km. SW Ciudad Victoria (on Jaumave Road), 3800 ft.; *Jaumave;* 9 mi. SW Tula, 3900 ft.; *Nicolás, 56 km. NW Tula, 5500 ft.* Nuevo León: Doctor Arroyo, 5800 ft. Coahuila: 8 mi. N La Ventura, 5500 ft.; N slope Sierra Guadalupe, 10 mi. S, 5 mi. W General Cepeda, 6500 ft.; *6 mi. E Hermanas.*

Neotoma albigula venusta True

1894. *Neotoma venusta* True, Diagnoses of some undescribed wood rats (genus *Neotoma*) in the National Museum, p. 2, June 27 (preprint of Proc. U.S. Nat. Mus., 17:354, November 15, 1894), type from Carrizo Creek, Imperial Co., California.
1910. *Neotoma albigula venusta*, Goldman, N. Amer. Fauna, 31:33, October 19.
1897. *Neotoma cumulator* Mearns, Preliminary diagnoses of new mammals of the genera *Sciurus, Castor, Neotoma,* and

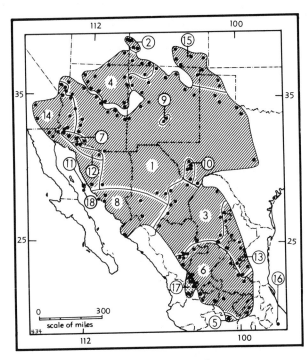

Map 434.   *Neotoma albigula*-group.

1. *N. albigula albigula*	10. *N. albigula robusta*
2. *N. albigula brevicauda*	11. *N. albigula seri*
3. *N. albigula durangae*	12. *N. albigula sheldoni*
4. *N. albigula laplataensis*	13. *N. albigula subsolana*
5. *N. albigula latifrons*	14. *N. albigula venusta*
6. *N. albigula leucodon*	15. *N. albigula warreni*
7. *N. albigula mearnsi*	16. *N. nelsoni*
8. *N. albigula melanura*	17. *N. palatina*
9. *N. albigula melas*	18. *N. varia*

Sigmodon, from the Mexican border of the United States, p. 3, March 5 (preprint of Proc. U.S. Nat. Mus., 20:503, January 19, 1898), type from old Fort Yuma, Imperial Co., California, on right bank of Colorado River opposite present town of Yuma, Arizona.

1904. *Neotoma desertorum grandis* Elliot, Field Columb. Mus., Publ. 87, Zool. Ser., 3(14):247, January 7, type from Cameron Lake, Kern Co., California (?; see Grinnell, 1933:178).

MARGINAL RECORDS (Hall and Genoways, 1970:514, 515, unless otherwise noted).—Arizona: Colorado River, 31 mi. N, 2½ mi. W Camp Mohave; Mineral Park; Kingman; Johnsons Ranch, Wickieup, 2000 ft. (Cockrum, 1961:195); Congress Junction; Harquahala Mts., 7½ mi. S Salome, 5000 ft.; Norton; Wellton; 4 mi. S Gadsden. Sonora: Costa Rica Ranch. Baja California: Colonia Lerdo; E base Cocopah Mts. California: Borrego Spring; Long Canyon, 29 mi. N, 19½ mi. W Mecca. Arizona: Fort Mohave.

Neotoma albigula warreni Merriam

1908. *Neotoma albigula warreni* Merriam, Proc. Biol. Soc. Washington, 21:143, June 9, type from Gaume Ranch, 4600 ft., NW corner Baca Co., Colorado.

MARGINAL RECORDS (Hall and Genoways, 1970:515, unless otherwise noted).—Colorado (Armstrong, 1972:222): 2 mi. E Wetmore, 5700 ft.; 6 mi. NW Higbee, 4550 ft.; Two Buttes Peak, 4500 ft.; *Two Buttes*. Oklahoma: Regnier, 4500 ft. Texas: 10 mi. S, 3 mi. W Gruver; 2 mi. S, 11 mi. E Pringle. New Mexico: Clayton. Colorado (Armstrong, 1972:222): 1 mi. S, 7 mi. E Trinidad; 9 mi. SW Walsenburg, 6600 ft.

Neotoma nelsoni Goldman
Nelson's Wood Rat

1905. *Neotoma nelsoni* Goldman, Proc. Biol. Soc. Washington, 18:29, February 2, type from Perote, 7800 ft., Veracruz. Known only from type locality.

Average external measurements of five topotypes are: 349; 154; 38. Upper parts pale cinnamon heavily overlaid with brown, especially mid-dorsally; underparts dull white over plumbeous except on small pectoral area where hairs are white to base; feet white; tail faintly bicolored, smoky brown above, paler below.

Neotoma palatina Goldman
Bolaños Wood Rat

1905. *Neotoma palatina* Goldman, Proc. Biol. Soc. Washington, 18:27, February 2, type from Bolaños, 2800 ft., Jalisco.

Extreme external measurements of 8 males (2 old and 6 adult) and 6 adult females are, respectively: 333–404, 326–378; 139–180, 144–171; 36–39, 34–37. Pelage less Ochraceous-Buff on

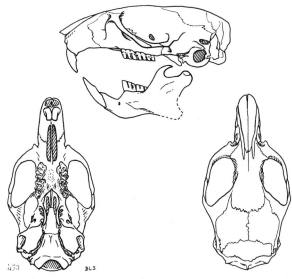

Fig. 450.   *Neotoma palatina*, 6 mi. ENE Bolaños, 5350 ft., Jalisco, No. 99037 K.U., ♂, X 1.

sides and in most specimens less Ochraceous-Buff on back than in *N. a. leucodon*. Skull large, robust, smoothly rounded, well arched at level of anterior zygomatic root; rostrum short, robust; frontals large, sides slightly upturned; interpterygoid fossa especially broad, rounded anteriorly; sphenopalatine vacuities absent; vomer markedly prolonged posteriorly, terminating at presphenoid-basisphenoid suture; auditory bullae small but robust. Proximal end of baculum more deeply notched dorsally than in *N. albigula*.

MARGINAL RECORDS (Hall and Genoways, 1970:515).—Jalisco: 10 mi. NE Huejuquilla, 6800 ft.; 1 mi. NW Mezquitic, 5000 ft.; *3 mi. N Villa Guerrero, 5600 ft.*; 3 mi. E Totatiche, 5600 ft.; *6 mi. ENE Bolaños, 5350 ft.; 4 mi. ENE Bolaños, 4400 ft.; 2 mi. E Bolaños, 3550 ft.*; type locality; *5 mi. NE Huejuquilla, 6200 ft.*

Neotoma varia Burt
Turner Island Wood Rat

1932. *Neotoma varia* Burt, Trans. San Diego Soc. Nat. Hist., 7:178, October 31, type from Turner Island (lat. 28° 43′ N, long. 112° 19′ W), Gulf of California, Sonora. Known only from type locality. See Map 434.

External measurements of the type, an adult female, the only known specimen, are: 305; 140; 33. Differs from contiguous subspecies of *Neotoma albigula* in minor details except those of dentition. In *varia* the molar tooth-row is convex laterally rather than straight and M3 has 2 rather than 3 lobes and consequently 1 rather than 2 external re-entrant angles. The constancy of these unusual deviations from the usual pattern found in the *albigula*-group will remain unknown until additional specimens of *varia* have been studied.

Neotoma lepida
Desert Wood Rat

See Goldman, Jour. Mamm., 13:59–67, February 9, 1932, for revised list of subspecies and closely related species.

External measurements: 225–383; 95–188; 28–41. Upper parts varying much according to subspecies but usually approx. buffy gray; underparts grayish or faintly buffy, all hairs plumbeous basally; tail bicolored, dark gray or dusky above, pale gray below. Skull robust, sculptured; auditory bullae well inflated; interpterygoid space moderately broad posteriorly, but variable in shape anteriorly; anterointernal angle of M1 shallow.

Neotoma lepida abbreviata Goldman

1909. *Neotoma abbreviata* Goldman, Proc. Biol. Soc. Washington, 22:140, June 25, type from San Francisco Island

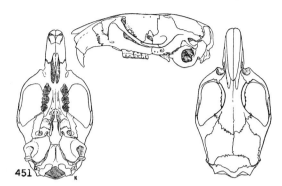

Fig. 451. *Neotoma lepida lepida*, Baker Creek, 7300 ft., Nevada, No. 42031 M.V.Z., ♀, X 1.

(near southern end of San José Island), Gulf of California, Baja California. Known only from type locality.
1932. *Neotoma lepida abbreviata*, Burt, Trans. San Diego Soc. Nat. Hist., 7:182, October 31.

Neotoma lepida arenacea J. A. Allen

1898. *Neotoma arenacea* J. A. Allen, Bull. Amer. Mus. Nat. Hist., 10:150, April 12, type from San José del Cabo, Baja California.
1932. *Neotoma lepida arenacea*, Goldman, Jour. Mamm., 13:65, February 9.

MARGINAL RECORDS.—Baja California: La Paz; Cape San Lucas.

Neotoma lepida aridicola Huey

1957. *Neotoma lepida aridicola* Huey, Trans. San Diego Soc. Nat. Hist., 12:287, 288, September 25, type from El Barril (near 28° 20′ N), Gulf of California, Baja California.

MARGINAL RECORDS.—Baja California (Huey, 1957:288): 7 mi. W San Francisquito Bay; type locality.

Neotoma lepida aureotunicata Huey

1937. *Neotoma lepida aureotunicata* Huey, Trans. San Diego Soc. Nat. Hist., 8:349, June 15, type from Punta Peñascosa, Sonora. Known only from type locality.

Neotoma lepida auripila Blossom

1933. *Neotoma auripila* Blossom, Occas. Pap. Mus. Zool., Univ. Michigan, 273:1, October 31, type from Agua Dulce Mts., 9 mi. E Papago Well, Pima Co., Arizona.
1935. *N[eotoma]. l[epida]. auripila* Blossom, Occas. Pap. Mus. Zool., Univ. Michigan, 315:1, May 29.

MARGINAL RECORDS.—Arizona: Crow Butte, W side Cabeza Prieta Mts., 9 mi. E Tinajas Altas; Tule Tank; type locality.

Neotoma lepida bensoni Blossom

1935. *Neotoma lepida bensoni* Blossom, Occas. Pap. Mus. Zool., Univ. Michigan, 315:1, May 29, type from Pápago Tanks, Sierra Pinacate, Sonora.

MARGINAL RECORDS.—Sonora: type locality; Elegante Crater, 41 mi. SW Sonoyta.

Neotoma lepida californica Price

1894. *Neotoma californica* Price, Proc. California Acad. Sci., ser. 2, 4:154, pl. 11, May 9, type from Bear Valley, San Benito Co., California.
1938. *Neotoma lepida californica*, von Bloeker, Proc. Biol. Soc. Washington, 51:201, December 23.

MARGINAL RECORDS.—California: Mt. Diablo; Herrero Canyon, 22 mi. S Los Banos; 1 mi. SE summit San Benito Mtn., Gabilan Mts.; Diablo Range, 4 mi. E King City; mouth Stonewall Creek, 4 mi. NE Soledad; Fremont Peak, Gabilan Range.

Neotoma lepida devia Goldman

1927. *Neotoma intermedia devia* Goldman, Proc. Biol. Soc. Washington, 40:205, December 2, type from Tanner Tank, 5200 ft., Painted Desert, Arizona.
1932. *Neotoma lepida devia* Goldman, Jour. Mamm., 13:62, February 9.

MARGINAL RECORDS.—Arizona: across from mouth Kanab Canyon (Hoffmeister, 1971:175), thence eastward along S bank Colorado River to McCormick Mine (*ibid.*); Cameron; *type locality*; New River Valley, 30 mi. N Phoenix; base Castle Dome Peak; 10 mi. below Cibola, Colorado River, thence northward E of Colorado River to Hoover Dam Ferry, Colorado River.

Neotoma lepida egressa Orr

1934. *Neotoma lepida egressa* Orr, Proc. Biol. Soc. Washington, 47:109, June 13, type from 1 mi. E El Rosario, 200 ft., Baja California.

MARGINAL RECORDS.—Baja California: San Telmo; Arroyo Nuevo York, 15 mi. S Santo Domingo; type locality; thence northward along coast to point of beginning.

Neotoma lepida felipensis Elliot

1903. *Neotoma bella felipensis* Elliot, Field Columb. Mus., Publ. 79, Zool. Ser., 3(12):217, August 15, type from San Felipe, Baja California.
1932. *Neotoma lepida felipensis*, Goldman, Jour. Mamm., 13:64, February 9.

MARGINAL RECORDS.—Baja California: Palomar; type locality; Parral; Cañón Viento.

Neotoma lepida flava Benson

1935. *Neotoma lepida flava* Benson, Occas. Pap. Mus. Zool., Univ. Michigan, 317:7, July 1, type from Tinajas Altas, 1150 ft., Yuma Co., Arizona. Known only from type locality.

Neotoma lepida gilva Rhoads

1894. *Neotoma intermedia gilva* Rhoads, Amer. Nat., 28:70, January, type from Banning, Riverside Co., California.
1932. *Neotoma lepida gilva*, Goldman, Jour. Mamm., 13:63, February 9.
1894. *Neotoma desertorum sola* Merriam, Proc. Biol. Soc. Washington, 9:126, July 2, type from San Emigdio, Kern Co., California. (See also Jones and Fisher, Proc. Biol. Soc. Washington, 86:435–437, December 14, 1973.)

MARGINAL RECORDS.—California: Paraiso Springs; Milpitas Ranch; Stanley; Kern River, 15 mi. N Bakersfield; S. Fork, Kern River; Tehachapi; Mt. Waterman; Cabezon; Jacumba. Baja California: Calamahue; Yubay; Pozo San Augustín, 20 mi. E San Fernando; Esperanza Canyon, E base Sierra San Pedro Mártir. California: Mt. Piños; Cuyama Valley.

Neotoma lepida grinnelli Hall

1942. *Neotoma lepida grinnelli* Hall, Univ. California Publ. Zool., 46(5):369, July 3, type from Colorado River 20 mi. above (by river; about 12½ airline mi. N) Picacho, Imperial Co., California.

MARGINAL RECORDS.—Nevada: 4½ mi. W Boulder City, 2600 ft. California: Pilot Knob; Beal Well, Chocolate Mts.

Neotoma lepida harteri Huey

1937. *Neotoma lepida harteri* Huey, Trans. San Diego Soc. Nat. Hist., 8:351, June 15, type from 10 mi. S Gila Bend (2 mi. N Black Gap), Maricopa Co., Arizona.

MARGINAL RECORDS.—Arizona: type locality; just N of Growler Mine.

Neotoma lepida insularis Townsend

1912. *Neotoma insularis* Townsend, Bull. Amer. Mus. Nat. Hist., 31:125, June 14, type from Angel de la Guarda Island, Gulf of California, Baja California. Known only from type locality.
1932. *Neotoma lepida insularis*, Burt, Trans. San Diego Soc. Nat. Hist., 7:182, October 31.

Neotoma lepida intermedia Rhoads

1894. *Neotoma intermedia* Rhoads, Amer. Nat., 28:69, January, type from Dulzura, San Diego Co., California.
1932. *Neotoma lepida intermedia*, Goldman, Jour. Mamm., 13:64, February 9.

MARGINAL RECORDS.—California: San Luis Obispo; San Fernando; San Bernardino Mts.; Redlands; Julian. Baja California: Tecate River; summit

San Matías Pass; San Fernando; Rancho San Antonio, W base Sierra San Pedro Mártir; 20 mi. E Ensenada, thence northward along coast to point of beginning. Also, discontinuously in California: 8 mi. E Porterville; Piute Mts.

Neotoma lepida latirostra Burt

1932. *Neotoma lepida latirostra* Burt, Trans. San Diego Soc. Nat. Hist., 7:180, October 31, type from Danzante Island, lat. 25° 47′ N, long. 111° 11′ W, Gulf of California, Baja California. Known only from type locality.

Neotoma lepida lepida Thomas

1893. *Neotoma lepida* Thomas, Ann. Mag. Nat. Hist., ser. 6, 12:235, September, type from somewhere on "Simpson's route" between Camp Floyd [= Fairfield], Utah, and Carson City, Nevada.
1894. *Neotoma desertorum* Merriam, Proc. Biol. Soc. Washington, 9:125, July 2, type from Furnace Creek, Death Valley, Inyo Co., California.
1899. *Neotoma bella* Bangs, Proc. New England Zool. Club, 1:66, July 31, type from Palm Springs, Riverside Co., California.

MARGINAL RECORDS.—Nevada: Cedar Creek, 10 mi. NE San Jacinto; Goose Creek, 2 mi. W Utah boundary, 5000 ft. Utah: Kelton, 4225 ft.; Promontory; 7 mi. SW Nephi, 6000 ft.; Salina; Cedar City; Castle Cliffs, 1 mi. N Beaverdam Wash, 2500 ft. Nevada: Cedar Basin, 3500 ft.; N side Potosi Mtn., 7000 ft. California: Purdy; Colton Well. Baja California: Cerro Prieto, near volcanic lake; Sierra Cocopah. California: N base San Jacinto Mtn., near Whitewater; Hesperia; Mohave; Owens Lake; Owens Valley near Benton. Nevada: 1 mi. SW Pyramid Lake, 3950 ft.; 2 mi. E Antelope, 4400 ft.

Neotoma lepida marcosensis Burt

1932. *Neotoma lepida marcosensis* Burt, Trans. San Diego Soc. Nat. Hist., 7:179, October 31, type from San Marcos Island, lat. 27° 13′ N, long. 112° 05′ W, Gulf of California, Baja California. Known only from type locality.

Neotoma lepida marshalli Goldman

1939. *Neotoma lepida marshalli* Goldman, Jour. Mamm., 20:357, August 14, type from Carrington Island, about 4250 ft., Great Salt Lake, Tooele Co., Utah.

MARGINAL RECORDS.—Utah: type locality; Stansbury Island.

Neotoma lepida molagrandis Huey

1945. *Neotoma lepida molagrandis* Huey, Trans. San Diego Soc. Nat. Hist., 10:307, August 31, type from site of old well near mesalike shelf, some 3 mi. inland from landing beach, Santo Domingo Landing, about 50 ft., lat. 28° 15′ N, Baja California.

MARGINAL RECORDS.—Baja California: Punta Prieta; Santa Gertrudis Mission; 12 mi. E El Arco, Rancho Miraflores; San Ignacio; Tinaja Santa Clara, Sierra Santa Clara; type locality.

Neotoma lepida monstrabilis Goldman

1932. *Neotoma lepida monstrabilis* Goldman, Jour. Mamm., 13:62, February 9, type from Ryan, 6000 ft., Kaibab National Forest, Coconino Co., Arizona.

MARGINAL RECORDS.—Utah: 7 mi. N Escalante, 6500 ft.; Mile 105, Glen Canyon, W bank (Durrant and Dean, 1959:94); Mile 41, Glen Canyon, W bank (*ibid.*). Arizona: Lee's Ferry; Kwagunt delta on Colorado River (Hoffmeister, 1971:175), thence southward and westward along N bank Colorado River to ½ mi. E Vulcans Throne (Hoffmeister and Durham, 1971:40); Green Spring (*ibid.*); Lake Mead, foot Grand Wash (*ibid.*); Littlefield. Utah: Zion National Park, 4300 ft.; Kaiparowits Plateau.

Neotoma lepida nevadensis Taylor

1910. *Neotoma nevadensis* Taylor, Univ. California Publ. Zool., 5(6):289, February 12, type from Virgin Valley, 4800 ft., Humboldt Co., Nevada.
1946. *Neotoma lepida nevadensis*, Hall, Mammals of Nevada, p. 530, July 1.

MARGINAL RECORDS.—Oregon: Vale. Idaho: 2 mi. S Melba; 8 mi. W Rogerson. Nevada: 36 mi. NE Paradise Valley, 5500 ft.; 2½ mi. E Flanigan, 4250 ft. California: Petes Valley, 12 mi. NE Susanville. Nevada: 12-Mile Creek, ½ mi. E California boundary, 5300 ft. Oregon: Voltage.

Neotoma lepida notia Nelson and Goldman

1931. *Neotoma intermedia notia* Nelson and Goldman, Proc. Biol. Soc. Washington, 44:108, October 17, type from La Laguna, 5500 ft., Sierra de la Victoria, Baja California.
1932. *Neotoma lepida notia*, Goldman, Jour. Mamm., 13:65, February 9.

MARGINAL RECORDS.—Baja California: type locality; Mt. Miraflores.

Neotoma lepida nudicauda Goldman

1905. *Neotoma nudicauda* Goldman, Proc. Biol. Soc. Washington, 18:28, February 2, type from Carmen Island, Gulf of California, Baja California. Known only from type locality.
1932. *Neotoma lepida nudicauda*, Burt, Trans. San Diego Soc. Nat. Hist., 7:182, October 31.

Neotoma lepida perpallida Goldman

1909. *Neotoma intermedia perpallida* Goldman, Proc. Biol. Soc. Washington, 22:139, June 25, type from San José Island, Gulf of California, Baja California. Known only from type locality.
1932. *Neotoma lepida perpallida* Goldman, Jour. Mamm., 13:65, February 9.

Neotoma lepida petricola von Bloeker

1938. *Neotoma lepida petricola* von Bloeker, Proc. Biol. Soc. Washington, 51:203, December 23, type from Abbotts Ranch, Arroyo Seco, 670 ft., Monterey Co., California.

MARGINAL RECORDS.—California: Mal Paso Creek, S of Point Lobos; Big Pines, 4000 ft., Santa Lucia Mts.; type locality; Mt. Mars.

Neotoma lepida pretiosa Goldman

1909. *Neotoma intermedia pretiosa* Goldman, Proc. Biol. Soc. Washington, 22:139, June 25, type from Matancita, 100 ft. (also called Soledad), 50 mi. N Magdalena Bay, Baja California.
1932. *Neotoma lepida pretiosa* Goldman, Jour. Mamm., 13:64, February 9.

MARGINAL RECORDS.—Baja California: San Jorge, SW of Comondú; type locality; [Santa] Margarita Island; [Santa] Magdalena Island.

Neotoma lepida ravida Nelson and Goldman

1931. *Neotoma intermedia ravida* Nelson and Goldman, Proc. Biol. Soc. Washington, 44:107, October 17, type from Comondú, 700 ft., Baja California.
1932. *Neotoma lepida ravida*, Goldman, Jour. Mamm., 13:64, February 9.

MARGINAL RECORDS.—Baja California: Aguaje de la Natividad, 25 mi. NW San Ignacio; Sierra de la Giganta; type locality; Paso Hondo, 16 mi. N La Purísima.

Neotoma lepida sanrafaeli Kelson

1950. *Neotoma lepida sanrafaeli* Kelson, Jour. Washington Acad. Sci., 39:418, January 9, type from Rock Canyon Corral, 5 mi. SE Valley City, 4500 ft., Grand Co., Utah.

MARGINAL RECORDS.—Colorado (Finley, 1958: 332): 5 mi. W Rangely, 5300 ft.; 18 mi. S, 2 mi. E Rangely; foot of Book Cliffs, 5 mi. W Palisade; 1½ mi. S Loma; *1 mi. SW Mack.* Utah: type locality; Mile 148, Glen Canyon, W bank (Durrant and Dean, 1959:94); King's Ranch, 5000 ft.; Notom, 6200 ft.; 7 mi. N Greenriver, 4100 ft.

Neotoma lepida vicina Goldman

1909. *Neotoma intermedia vicina* Goldman, Proc. Biol. Soc. Washington, 22:140, June 25, type from Espíritu Santo Island, Gulf of California, Baja California. Known only from type locality.
1932. *Neotoma lepida vicina* Goldman, Jour. Mamm., 13:65, February 9.

Neotoma bryanti Merriam
Bryant's Wood Rat

1887. *Neotoma bryanti* Merriam, Amer. Nat., 21:191, February, type from Cerros [= Cedros] Island, Baja California. Known only from type locality.

Average external measurements of eight topotypes are: 377; 168; 38.1. Upper parts rich creamy buff with strong admixture of black-tipped hairs on top of head and back; underparts creamy white sometimes suffused with pinkish buff; tail brownish above, grayish below. Skull large, angular; frontals broadened abruptly anteriorly, deeply channeled above, lateral margins strongly upturned in nearly straight posteriorly diverging lines; interpterygoid fossa narrow.

Neotoma anthonyi J. A. Allen
Anthony's Wood Rat

1898. *Neotoma anthonyi* J. A. Allen, Bull. Amer. Mus. Nat. Hist., 10:151, April 12, type from Todos Santos Island, Baja California. Known only from type locality.

External measurements: 330–345; 132–149; 34–36.3. Upper parts grayish brown becoming darker dorsally owing to admixture of dusky hairs; underparts white with suffusion of pale pinkish buff on throat and belly; feet white; tail bicolored brownish black above, dirty white below. Skull relatively large, robust, and angular; in general resembling that of *N. lepida* but maxillary arm of zygomata more robust, less abruptly spreading; frontals more elevated; anterior palatine foramina larger; teeth larger.

Neotoma martinensis Goldman
San Martín Island Wood Rat

1905. *Neotoma martinensis* Goldman, Proc. Biol. Soc. Washington, 18:28, February 2, type from San Martín Island, Baja California. Known only from type locality.

Average external measurements of four topotypes are: 352; 165; 37.6. Upper parts creamy buff darkened dorsally by dusky hairs; underparts creamy white, hairs plumbeous basally; tail brownish black above, grayish below. Skull relatively smoothly rounded, high and well arched in profile anteriorly; nasals especially long, reaching well beyond level of lachrymals, abruptly tapering posteriorly; premaxillary tongues extending posteriorly beyond nasals; interorbital region much constricted; zygomatic arches convergent anteriorly; interpterygoid fossa narrow and somewhat constricted; auditory bullae small but with especially large meatūs.

Neotoma bunkeri Burt
Bunker's Wood Rat

1932. *Neotoma bunkeri* Burt, Trans. San Diego Soc. Nat. Hist., 7:181, October 31, type from Coronados Island, lat. 26° 06′ N, long. 111° 18′ W, Gulf of California, Baja California. Known only from type locality.

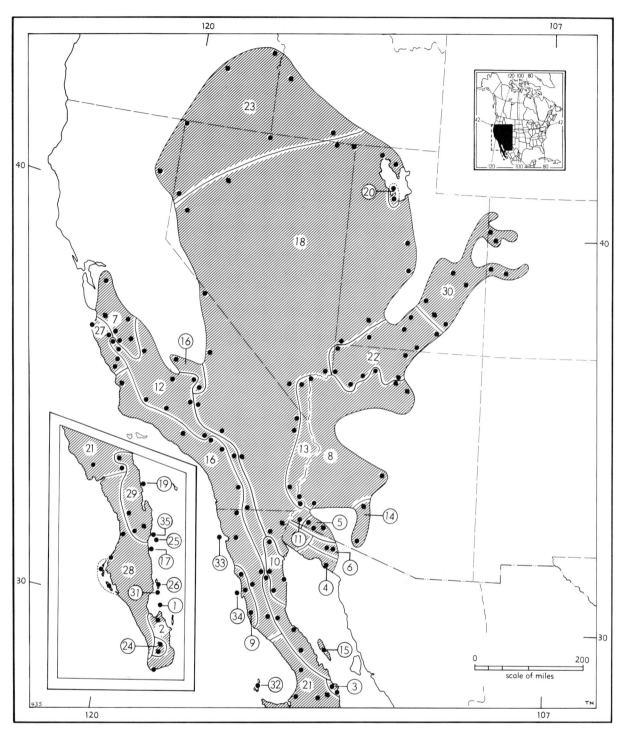

Map 435. *Neotoma lepida* and allied species.

Guide to kinds
1. *N. lepida abbreviata*
2. *N. lepida arenacea*
3. *N. lepida aridicola*
4. *N. lepida aureotunicata*
5. *N. lepida auripila*
6. *N. lepida bensoni*
7. *N. lepida californica*
8. *N. lepida devia*

9. *N. lepida egressa*
10. *N. lepida felipensis*
11. *N. lepida flava*
12. *N. lepida gilva*
13. *N. lepida grinnelli*
14. *N. lepida harteri*
15. *N. lepida insularis*
16. *N. lepida intermedia*
17. *N. lepida latirostra*

18. *N. lepida lepida*
19. *N. lepida marcosensis*
20. *N. lepida marshalli*
21. *N. lepida molagrandis*
22. *N. lepida monstrabilis*
23. *N. lepida nevadensis*
24. *N. lepida notia*
25. *N. lepida nudicauda*
26. *N. lepida perpallida*

27. *N. lepida petricola*
28. *N. lepida pretiosa*
29. *N. lepida ravida*
30. *N. lepida sanrafaeli*
31. *N. lepida vicina*
32. *N. bryanti*
33. *N. anthonyi*
34. *N. martinensis*
35. *N. bunkeri*

759

External measurements of the type are: 390; 168; 43. Upper parts dark gray; underparts deep mouse gray washed with white; tail brownish above, white below. Skull resembling that of *Neotoma fuscipes macrotis* Thomas but rostrum relatively shorter and broader, antorbital foramina wider, tympanic bullae less rounded, and braincase more flattened (after Burt, 1932:181).

Neotoma stephensi
Stephens' Wood Rat

Revised by Hoffmeister and de la Torre, Jour. Mamm., 41:476–491, November 11, 1960.

External measurements: 286–312; 135–141; 28–31. Upper parts yellowish to grayish buff, darkened dorsally by dusky hairs; underparts white or creamy, often washed with pinkish buff, fur plumbeous basally except on throat, inner sides of forelegs, pectoral and inguinal regions; feet usually white; tail pale gray to grayish brown above, paler below. Skull closely resembling that of *N. lepida* but differing in its most extreme development in being smaller, less angular; braincase more smoothly rounded; zygomata more squarely spreading; frontal region broader and more flattened.

The tail is notably hairier in *N. stephensi* than in *N. lepida*.

Neotoma stephensi relicta Goldman

1932. *Neotoma stephensi relicta* Goldman, Jour. Mamm., 13:66, February 9, type from Keams Canyon, Navajo Co., Arizona.

MARGINAL RECORDS (Hoffmeister and de la Torre, 1960:491, unless otherwise noted).—Utah: Rainbow Bridge, 4000 ft.; *near War God Spring, 8500 ft., Navajo Mtn.*; Navajo Mtn. Trading Post, SE side Navajo Mtn. Arizona: Long Canyon, 6450 ft.; type locality. New Mexico (Findley, *et al.*, 1975:246, as *N. stephensi* only): Pine River road, SW¼, sec. 25, T. 32 N, R. 7 W; 2 mi. S, 3 mi. E Estrella; 6½ mi. S, 9 mi. W San Ysidro; 12 mi. S Gallup. Arizona: Tuba [City]; Cedar Ridge, 6000 ft.

Neotoma stephensi stephensi Goldman

1905. *Neotoma stephensi* Goldman, Proc. Biol. Soc. Washington, 18:32, February 2, type from Hualpai Mts., 6300 ft., Mohave Co., Arizona.

MARGINAL RECORDS (Hoffmeister and de la Torre, 1960:488, 489).—Arizona: Hilltop, thence eastward along S side Colorado River to Little Colorado River and eastward along S side Little Colorado River to jct. Puerco River; Zuñi River. New Mexico: Grants; *4 mi. W McCartys*; Burley; Burro Mts.; Glenwood, San Francisco Valley, 5000 ft. Arizona: 3 mi. W Cosper

Ranch; McMillenville; 7 mi. N Payson; Mayer; Harquahala Mts., 5000 ft.; Lucky Star Mine, Chemehuevis Mts.; type locality; Hualapai Mtn. Park; Hackberry; 8 mi. N Pine Spring; lower end Prospect Valley, 5200 ft.

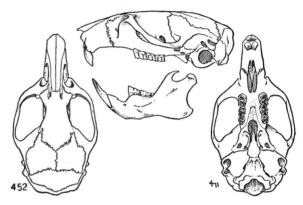

Fig. 452. *Neotoma stephensi stephensi*, Hualpai Mts., Arizona, No. 244156 U.S.N.M., ♂, X 1.

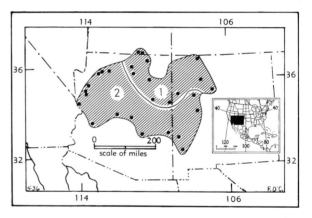

Map 436. *Neotoma stephensi.*

1. *N. s. relicta*   2. *N. s. stephensi*

Neotoma goldmani Merriam
Goldman's Wood Rat

1903. *Neotoma goldmani* Merriam, Proc. Biol. Soc. Washington, 16:48, March 19, type from Saltillo, 5000 ft., Coahuila.

Average measurements of four topotypes are: total length, 279; tail, 128; hind foot, 30. Upper parts creamy buff, paler on head, darkened on back by admixture of dusky hairs; underparts white; feet white; tail blackish above, white below. Skulls resembling those of *N. lepida* and *N. stephensi* but smaller; premaxillary tongues broader posteriorly; auditory bullae relatively small; molars relatively large.

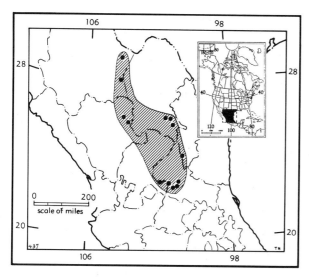

Map 437. *Neotoma goldmani.*

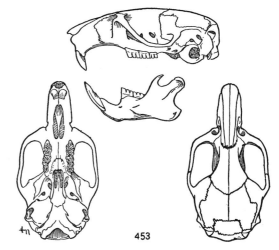

Fig. 453. *Neotoma mexicana fallax,* 1½ mi. NW Golden, Jefferson Co., Colorado, No. 29182 K.U., ♂, X 1.

MARGINAL RECORDS.—Chihuahua: 4 mi. NW Escobillas (Anderson, 1972:362); Sierra Almagre, 6000 ft., 12 mi. S Jaco. Coahuila: Jaral; 17 mi. N, 8 mi. W Saltillo, 5200 ft.; type locality. Nuevo León: approx. 12 mi. N Matehuala [12 mi. due N would be in San Luis Potosí], 5900 ft. (Musser, 1964:19). San Luis Potosí: Ventura; 10 mi. NE San Luis Potosí; San Luis Potosí; Santa Teresa; Cerro Peñón Blanco. Durango: 4 mi. WSW Lerdo, 3800 ft.; 1 mi. SSE Mapimí, 4100 ft. Chihuahua: *5 mi. NE Las Cruces* (Anderson, 1972:362).

Neotoma mexicana
Mexican Wood Rat

Arranged according to Hall, Jour. Washington Acad. Sci., 45:328–332, October 31, 1955.

External measurements: 290–417; 105–206; 31–41. Upper parts, depending on subspecies, grayish, grayish buff, dull brown, russet or bright rufous, usually darkened dorsally by admixture of blackish hairs; underparts, depending on subspecies, white or yellowish, hairs plumbeous basally except that inguinal and gular hair of many subspecies is white to base; tail black to dusky above, grayish to whitish below. Skull of average size for the genus; anterointernal re-entrant angle of M1 deep, extending more than halfway across crown; interorbital region much constricted.

The combination of size, color, and deep anterointernal angle of M1 serves to distinguish this large and variable species from its congeners.

Neotoma mexicana atrata Burt

1939. *Neotoma mexicana atrata* Burt, Occas. Pap. Mus. Zool., Univ. Michigan, 400:1, March 1, type from 4 mi. W Car-

rizozo, Lincoln Co., New Mexico. Known only from type locality.

Neotoma mexicana bullata Merriam

1894. *Neotoma mexicana bullata* Merriam, Proc. Biol. Soc. Washington, 9:122, July 2, type from Santa Catalina Mts., Pima Co., Arizona. Known only from type locality.

Neotoma mexicana chamula Goldman

1909. *Neotoma ferruginea chamula* Goldman, Proc. Biol. Soc. Washington, 22:141, June 25, type from mts. near San Cristóbal, 8400 ft., Chiapas.
1955. *Neotoma mexicana chamula*, Hall, Jour. Washington Acad. Sci., 45:329, October 31.

MARGINAL RECORDS.—Chiapas: type locality. Guatemala: Hda. Chancol, about 13 mi. N Huehuetanango; Volcán Santa María.

Neotoma mexicana distincta Bangs

1903. *Neotoma distincta* Bangs, Proc. Biol. Soc. Washington, 16:89, June 25, type from Texolo [= Teocelo, near Jalapa], Veracruz. Known only from type locality.
1955. *Neotoma mexicana distincta*, Hall, Jour. Washington Acad. Sci., 45:329, October 31.

Neotoma mexicana eremita Hall

1955. *Neotoma mexicana eremita* Hall, Jour. Washington Acad. Sci., 45:328, October 31, type from 1 mi. S San Francisco, 50 ft., Nayarit.

MARGINAL RECORDS.—Nayarit: type locality. Jalisco: 4 mi. SW Puerto Vallarta, 20 ft. (Genoways and Jones, 1973:18).

Neotoma mexicana fallax Merriam

1894. *Neotoma fallax* Merriam, Proc. Biol. Soc. Washington,
     9:123, July 2, type from Gold Hill, Boulder Co., Colorado.
1910. *Neotoma mexicana fallax*, Goldman, N. Amer. Fauna,
     31:56, October 19.

MARGINAL RECORDS (Armstrong, 1972:223,
224).—Colorado: 21 mi. NW Fort Collins; *1 mi. N Owl
Canyon*; Loveland; Boulder; *Green Mtn., 4 mi. W Denver, 5500 ft.*; Frankstown; 3 mi. N, 2 mi. W Peyton,
7400 ft. (den record only); Pueblo; Buelah [Beulah]; 7
mi. SW Salida, 8300 ft.; *Salida, 7300 ft.*; Garden Park, 6
mi. up Red Cañon, N of Cañon City; *5 mi. SW Colorado
Springs*; Forks Creek; Estes Park.

Neotoma mexicana ferruginea Tomes

1862. *Neotoma ferruginea* Tomes, Proc. Zool. Soc. London,
     for 1861, p. 282, April, type from Dueñas, Sacatepequez,
     Guatemala.
1955. *Neotoma mexicana ferruginea*, Hall, Jour. Washington
     Acad. Sci., 45:330, October 31.

MARGINAL RECORDS.—Guatemala: San Lucas;
type locality. El Salvador: Volcán de Santa Ana (Burt
and Stirton, 1961:60).

Neotoma mexicana griseoventer Dalquest

1951. *Neotoma ferruginea griseoventer* Dalquest, Jour. Wash-
     ington Acad. Sci., 41:363, November 14, type from Xilitla,
     San Luis Potosí.
1955. *Neotoma mexicana griseoventer*, Hall, Jour. Washing-
     ton Acad. Sci., 45:330, October 31.

MARGINAL RECORDS.—San Luis Potosí: El
Salto; type locality.

Neotoma mexicana inopinata Goldman

1933. *Neotoma mexicana inopinata* Goldman, Jour. Washing-
     ton Acad. Sci., 23:471, October 15, type from Chuska Mts.,
     8800 ft., San Juan Co., New Mexico.

MARGINAL RECORDS.—Colorado (Finley, 1958:
288): 2½ mi. S Fruita, 4800 ft., S side Colorado
River; *Grand Junction, S side Colorado River*; 8 mi.
NW Olathe; 7½ mi. W Montrose; 2 mi. S, 6½ mi. E
Cahone, 6800 ft.; 2 mi. E Durango; Pagosa Springs; *1
mi. S, 2 mi. W Chromo, 7200 ft.* New Mexico: Pine
River Canyon, sec. 20, T. 31 N, R. 7 W (Findley, *et al.*,
1975:249); 18 mi. N, 1 mi. E Farmington (Finley,
1958:288); Gallup; Zuñi Mts. Arizona: Canyon del
Muerto and head of Spruce Creek, 9000 ft., Tunitcha
Mts. Utah: Mile 69, Glen Canyon, E bank (Durrant and
Dean, 1959:93); Mile 142, Glen Canyon, E bank
(*ibid.*); mouth Nigger Bill Canyon, E side Colorado
River, 4 mi. above Moab Bridge.

Neotoma mexicana inornata Goldman

1938. *Neotoma mexicana inornata* Goldman, Proc. Biol. Soc.
     Washington, 51:60, March 18, type from Carmen Mts., 6100
     ft., Coahuila.

MARGINAL RECORDS.—Coahuila: type locality;
Sierra de la Madera, 22 mi. S, 5 mi. W Ocampo, 6200 ft.

Neotoma mexicana isthmica Goldman

1904. *Neotoma isthmica* Goldman, Proc. Biol. Soc. Washing-
     ton, 17:80, March 21, type from Huilotepec, 100 ft., 8 mi. S
     Tehuantepec, Oaxaca.
1955. *Neotoma mexicana isthmica*, Hall, Jour. Washington
     Acad. Sci., 45:330, October 31.

MARGINAL RECORDS.—Oaxaca: Coixtlahuaca.
Chiapas: Teopisca; Canjob. Oaxaca: Rincón Bamba
(Goodwin, 1969:205); Puerto Angel; Oaxaca.

Neotoma mexicana mexicana Baird

1855. *Neotoma mexicana* Baird, Proc. Acad. Nat. Sci.
     Philadelphia, 7:333, April, type from [mts.] near Chi-
     huahua, Chihuahua.
1905. *Neotoma mexicana madrensis* Goldman, Proc. Biol.
     Soc. Washington, 18:31, February 2, type from Sierra
     Madre, 7000 ft., near Guadalupe y Calvo, Chihuahua. Re-
     garded as inseparable from *N. m. mexicana* by Anderson,
     Bull. Amer. Mus. Nat. Hist., 148:363, September 8, 1972.

MARGINAL RECORDS.—Arizona: Rincon Mts.;
Cave Creek, Chiricahua Mts., 6200 ft. New Mexico:
Animas Mts.; Dark Canyon, 17 mi. S, 22 mi. W
Carlsbad (Findley, *et al.*, 1975:249, as *N. mexicana*
only). Texas: Fort Davis; Paisano; Brewster County
(Davis, 1966:205, as *N. mexicana* only). Chihuahua:
near Parral (Anderson, 1972:363). Zacatecas: 4 mi. E
Calabazal; Sierra de Valparaíso. Durango (Baker and
Greer, 1962:128): 2 mi. N Pueblo Nuevo, 6000 ft.; ½ mi.
W Revolcaderos, 6600 ft.; 1½ mi. W San Luis.
Chihuahua: Sierra Madre, near Guadalupe y Calvo,
7000 ft.; 3 mi. NE Temoris, 5600 ft. (Anderson,
1972:363); Carimechi. Sonora: Cuchita; Oposura. Not
found: Durango: La Ciénega de las Vacas (Hall and
Kelson, 1959:696).

Neotoma mexicana navus Merriam

1903. *Neotoma navus* Merriam, Proc. Biol. Soc. Washington,
     16:47, March 19, type from Sierra Guadalupe, Coahuila.
1955. *Neotoma mexicana navus*, Hall, Jour. Washington Acad.
     Sci., 45:330, October 31.

MARGINAL RECORDS.—Coahuila: type locality;
Diamante Pass, 8500 ft., Sierra Guadalupe, 3 mi. N, 18
mi. E Saltillo.

Neotoma mexicana ochracea Goldman

1905. *Neotoma ferruginea ochracea* Goldman, Proc. Biol. Soc.
     Washington, 18:30, February 2, type from Atemajac, 4000
     ft., near Guadalajara, Jalisco. Known only from type locality.
1955. *Neotoma mexicana ochracea*, Hall, Jour. Washington
     Acad. Sci., 45:330, October 31.

Neotoma mexicana parvidens Goldman

1904. *Neotoma parvidens* Goldman, Proc. Biol. Soc. Washing-
     ton, 17:81, March 21, type from Juquila, 5000 ft., Oaxaca.

1955. *Neotoma mexicana parvidens,* Hall, Jour. Washington Acad. Sci., 45:330, October 31.

MARGINAL RECORDS.—Oaxaca: type locality; *ca. 11 mi. E Juquila, 6100 ft.* (Baker and Womochel, 1966:306); 1 mi. NNW Soledad [approx. 30 km. N Punta Angel], 4700 ft.

Neotoma mexicana picta Goldman

1904. *Neotoma picta* Goldman, Proc. Biol. Soc. Washington, 17:79, March 21, type from mts. near Chilpancingo, 10,000 ft., Guerrero.
1955. *Neotoma mexicana picta,* Hall, Jour. Washington Acad. Sci., 45:331, October 31.

MARGINAL RECORDS.—Guerrero: Omilteme; type locality. Oaxaca: San Andrés Chicahuaxtla (Goodwin, 1969:206); Chontecomatlán (*ibid.*); Río Guajolote (Schaldach, 1966:296).

Neotoma mexicana pinetorum Merriam

1893. *Neotoma pinetorum* Merriam, Proc. Biol. Soc. Washington, 8:111, July 31, type from San Francisco Mtn., Coconino Co., Arizona. [= "Little Springs at north base of the San Francisco Mountains," Bailey, V., Grand Canyon Nat. Hist. Assoc., Nat. Hist. Bull., 1:16, June 1935.]
1910. *Neotoma mexicana pinetorum,* Goldman, N. Amer. Fauna, 31:58, October 19.

MARGINAL RECORDS.—Arizona: S rim Grand Canyon; type locality; Winona. New Mexico: 10 mi. SW Quemado; 10 mi. W Chloride; Kingston. Arizona: Prieto Plateau; Bradshaw Mts.; Simmons; Pine Spring.

Neotoma mexicana scopulorum Finley

1953. *Neotoma mexicana scopulorum* Finley, Univ. Kansas Publ., Mus. Nat. Hist., 5:529, August 15, type from 3 mi. NW Higbee, 4300 ft., Otero Co., Colorado.

MARGINAL RECORDS.—Colorado (Finley 1958:283): type locality; Two Buttes [the peak], 4600–4650 ft. Oklahoma: Tesquite Canyon. New Mexico (Findley, *et al.,* 1975:248–250, as *N. mexicana* only): Apache Canyon, Clayton; Bell Ranch; Santa Rosa; 9 mi. N, 12 mi. W Fort Sumner; 1 mi. S Ruidoso; 20 mi. S Cloudcroft; 1 mi. E Elephant Butte; San Mateo Peak; Datil Mts.; Point of Malpais, T. 6 N, R. 11 W; Grants; 12 mi. S Gallup. Colorado (Finley, 1958:283): 3 mi. W San Acacio; 5 mi. SSE Fort Garland; 20 mi. E Walsenburg. [Specimens from central New Mexico formerly assigned to *N. m. fallax* are here assigned to *N. m. scopulorum* because Finley (1953 and 1958) separated them geographically from the northern population of *fallax,* including its type locality, when he named *scopulorum.* Study of specimens may warrant a different assignment.]

Neotoma mexicana sinaloae J. A. Allen

1898. *Neotoma sinaloae* J. A. Allen, Bull. Amer. Mus. Nat. Hist., 10:149, April 12, type from Tatámeles, Sinaloa.

1910. *Neotoma mexicana sinaloae,* Goldman, N. Amer. Fauna, 31:60, October 19.

MARGINAL RECORDS.—Sonora: San Javier; Mira Sol. Sinaloa: 15 km. N, 65 km. E Sinaloa, 4700 ft. (Birney and Jones, 1972:202). Durango: Chacala. Sinaloa (*loc. cit.*): 10 km. NE Santa Lucía, 6400 ft.; type locality; Mazatlán, thence up coast to Huitacochi, 8 mi. S Navolato. Sonora: Camoa.

Neotoma mexicana solitaria Goldman

1905. *Neotoma ferruginea solitaria* Goldman, Proc. Biol. Soc. Washington, 18:31, February 2, type from Nentón, 3500 ft., Guatemala.
1955. *Neotoma mexicana solitaria,* Hall, Jour. Washington Acad. Sci., 45:331, October 31.

MARGINAL RECORDS.—Guatemala: type locality; Sacapulas. Honduras: Cerro Puca.

Neotoma mexicana tenuicauda Merriam

1892. *Neotoma tenuicauda* Merriam, Proc. Biol. Soc. Washington, 7:169, September 29, type from N slope Sierra Nevada de Colima, 12,000 ft., Jalisco.
1955. *Neotoma mexicana tenuicauda,* Hooper, Occas. Pap. Mus. Zool., Univ. Michigan, 565:22, March 31.

MARGINAL RECORDS.—Sinaloa: Plomosas, 2500 ft. (Birney and Jones, 1972:202). Jalisco: mts. *ca.* 10 mi. N Bolaños. Zacatecas: Plateado. Aguascalientes: Sierra Fría. Jalisco: 2 mi. NNW Magdalena. Michoacán: Zamora; 9 mi. SE Pátzcuaro, 8000 ft.; Tancítaro, 7850 ft. Jalisco: type locality; Talpa; San Sebastián. Sinaloa: *2 mi. SW Plomosas, 3050 ft.* (Birney and Jones, 1972:202).

Neotoma mexicana torquata Ward

1891. *Neotoma torquata* Ward, Amer. Nat., 25:160, February, type from abandoned mine between Tetela del Volcán and Zacualpan, Morelos.
1955. *Neotoma mexicana torquata,* Hall, Jour. Washington Acad. Sci., 45:331, October 31.
1894. *Neotoma fulviventer* Merriam, Proc. Biol. Soc. Washington, 9:121, July 2, type from Toluca Valley, México.
1894. *Neotoma orizabae* Merriam, Proc. Biol. Soc. Washington, 9:122, July 2, type from Volcán de Orizaba, Puebla.

MARGINAL RECORDS.—Hidalgo: Encarnación. Veracruz: 3 km. E Las Vigas, 8000 ft. Puebla: Tehuacán, 1700 m. Morelos: 2 km. S Jonacatepec. México: N slope Volcán Toluca.

Neotoma mexicana tropicalis Goldman

1904. *Neotoma tropicalis* Goldman, Proc. Biol. Soc. Washington, 17:81, March 21, type from Totontepec, 6500 ft., Oaxaca.
1955. *Neotoma mexicana tropicalis,* Hall, Jour. Washington Acad. Sci., 45:331, October 31.

MARGINAL RECORDS (Goodwin, 1969:208).—Oaxaca: type locality; Río Mono Blanco.

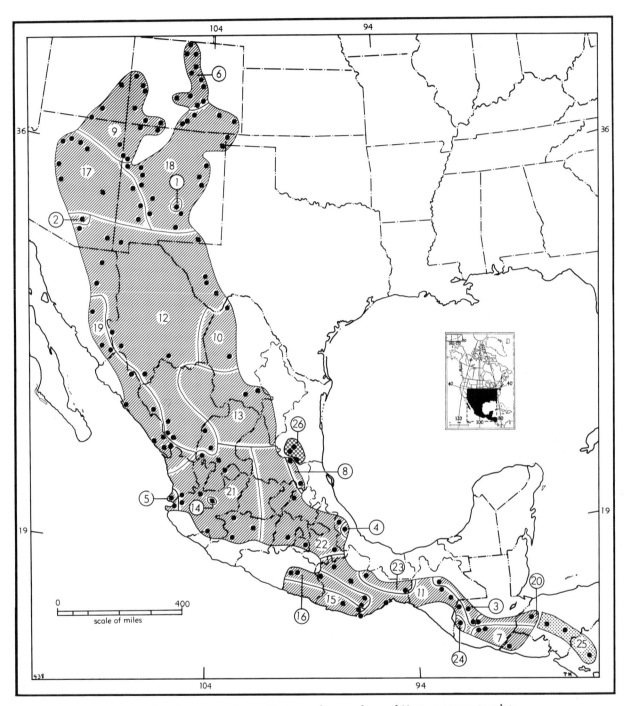

Map 438. *Neotoma mexicana, Neotoma chrysomelas,* and *Neotoma angustapalata.*

Guide to kinds
1. *N. mexicana atrata*
2. *N. mexicana bullata*
3. *N. mexicana chamula*
4. *N. mexicana distincta*
5. *N. mexicana eremita*
6. *N. mexicana fallax*
7. *N. mexicana ferruginea*
8. *N. mexicana griseoventer*

9. *N. mexicana inopinata*
10. *N. mexicana inornata*
11. *N. mexicana isthmica*
12. *N. mexicana mexicana*
13. *N. mexicana navus*
14. *N. mexicana ochracea*
15. *N. mexicana parvidens*
16. *N. mexicana picta*
17. *N. mexicana pinetorum*

18. *N. mexicana scopulorum*
19. *N. mexicana sinaloae*
20. *N. mexicana solitaria*
21. *N. mexicana tenuicauda*
22. *N. mexicana torquata*
23. *N. mexicana tropicalis*
24. *N. mexicana vulcani*
25. *N. chrysomelas*
26. *N. angustapalata*

Neotoma mexicana vulcani Sanborn

1935. *Neotoma ferruginea vulcani* Sanborn, Field Mus. Nat. Hist., Publ. 340, Zool. Ser., 20(11):84, May 15, type from S slope Volcán Tajumulco, San Marcos, 13,200 ft., Guatemala. Known only from type locality.
1955. *Neotoma mexicana vulcani,* Hall, Jour. Washington Acad. Sci., 45:332, October 31.

Neotoma chrysomelas J. A. Allen
Nicaraguan Wood Rat

1908. *Neotoma chrysomelas* J. A. Allen, Bull. Amer. Mus. Nat. Hist., 24:653, October 13, type from Matagalpa, Matagalpa, Nicaragua.

Average external measurements of 4 topotypes are: 375; 160; 36. "Upper parts light orange rufous . . . , purest along sides, darkened on top of head and back by overlying black-tipped hairs; . . . underparts white, the fur plumbeous basally except small areas, mainly on throat and on pectoral and inguinal regions, where it is pure white to roots; . . . tail brownish or blackish above, grayish below" (Goldman, 1910:75). Compared with the skull of *N. m. chamula,* that of *chrysomelas* "has zygomata more squarely spreading posteriorly, narrowing anteriorly, the sides less nearly parallel than in *chamula;* squamosal arm of zygoma heavier; frontals broader posteriorly, the lateral border slightly convex and overhanging orbits" (Goldman, *loc. cit.*).

Specimens are needed from Honduras in the area between Cerro Puca (where *Neotoma mexicana solitaria* has been recorded) and Montaña Vasquez (where *Neotoma chrysomelas* has been recorded), to learn whether *chrysomelas* is a subspecies of *Neotoma mexicana.*

MARGINAL RECORDS.—Honduras: Montaña Vasquez; Hatillo. Nicaragua: type locality.

Neotoma angustapalata Baker
Tamaulipan Wood Rat

1951. *Neotoma angustapalata* Baker, Univ. Kansas Publ. Mus. Nat. Hist., 5:217, December 15, type from 70 km. by highway S Ciudad Victoria, 6 km. W Pan-American Highway at El Carrizo, Tamaulipas.

External measurements of a male topotype are: 380; 195; 42. Upper parts dusky brown; head, especially cheeks, grayer; underparts plumbeous washed with white, entirely white on gular and inguinal regions; tail sparsely haired, blackish above, whitish below. Skull with large auditory bullae; external auditory meatus large; sides of interpterygoid fossa concave and broadly excavated near posterior end of tooth-row; anterointernal re-entrant angle of M1 deep.

Hooper suggests that the specimens on which this species was described, and assigned by its author to the *mexicana*-group, are instead "large, deeply pigmented examples of the species *N. micropus.* . . ." The form of the baculum of *N. angustapalata,* however, suggests no close affinities with *N. micropus* or *mexicana;* indeed, the baculum is indistinguishable from that of *N. albigula* figured by Burt and Barkalow.

MARGINAL RECORDS.—Tamaulipas: type locality; El Pachón; Rancho de Cielo; *10 km. N, 8 km. W El Encino* (Alvarez, 1963:453).

Neotoma fuscipes
Dusky-footed Wood Rat

External measurements: 335–468; 158–241; 32–47. Upper parts ochraceous buff, darkened dorsally by dusky hairs, usually much grayer on face; underparts white, sometimes washed with tawny or pale buff, hairs plumbeous basally except on throat, chest, and inguinal region where hairs are white to roots; feet and ankles usually dusky, sometimes white distally; tail dark above sometimes lighter below but not sharply bicolored. Skull large, long relatively narrow; premaxillary tongues long; faint to moderate supraorbital ridge present; interpterygoid fossa usually narrow, or variable in shape anteriorly; M3 broad, the middle enamel loop partially or wholly divided by deep inner re-entrant angle.

Neotoma fuscipes annectens Elliot

1898. *Neotoma fuscipes annectens* Elliot, Field Columb. Mus., Publ. 27, Zool. Ser., 1(10):201, April 16, type from Portola, San Mateo Co., California.
1898. *Neotoma fuscipes affinis* Elliot, Field Columb. Mus., Publ. 27, Zool. Ser., 1(10):202, April 16, type from Alum Rock Park, Santa Clara Co., California.

MARGINAL RECORDS.—California: W side Mt. Diablo, 1750 ft.; Strawberry Canyon, 200–500 ft., thence along coast to point of beginning.

Neotoma fuscipes bullatior Hooper

1938. *Neotoma fuscipes bullatior* Hooper, Univ. California Publ. Zool., 42:225, March 1, type from 2 mi. S San Miguel, 620 ft., San Luis Obispo Co., California.

MARGINAL RECORDS.—California: type locality; Carrizo Plains; San Diego Joes, 16 mi. SW McKittrick.

Neotoma fuscipes fuscipes Baird

1858. *Neotoma fuscipes* Baird, Mammals, *in* Repts. Expl. Surv. . . . , 8(1):495, July 14, type from Petaluma, Sonoma Co., California.

1894. *Neotoma splendens* True, Diagnoses of some undescribed wood rats (genus *Neotoma*) in the National Museum, p. 1, June 27 (preprint of Proc. U.S. Nat. Mus., 17:353, November 15, 1894), type from Nicasio, Marin Co., California.

MARGINAL RECORDS.—Northeastern segment (Murray and Barnes, 1969:43–45, unless otherwise noted): Oregon: upper Coyote Valley. California: Point Ranch; 8⅓ mi. S town of Davis Creek; Soldier Creek (Murray, *in Litt.*, Feburary 26, 1971); 9 mi. E Likely; 5 *mi. N, 1²/₅ mi. W Madeline;* 1⅓ mi. N, 6 mi. W Madeline; 7 mi. S, 4 mi. E Ravendale; Grasshopper Ranger Station; Hayden Hill (Goldman, 1910:89); ½ mi. N, 5 mi. E Cassel (Murray and Barnes, Fig. 1, and MS); Old Station; Burney (Goldman, 1910:89); Bald Mtn. (*ibid.*); Dana (*ibid.*); 2⅖ mi. NW Day (Murray and Barnes, Fig. 1, and MS); Crescent Butte (Hooper, 1938:221); 5⅓ mi. E Tennant (*ibid.*); Beswick (Goldman, 1910:89); caves, 11 mi. NE Weed (Hooper, 1938:221); *1¹/₂ mi. SW Edgewood* (*ibid.*); *Kangaroo Creek* (*ibid.*); Scott River, 6 mi. NW Callahan (*ibid.*); Scott Valley, near Fort Jones (Goldman, 1910:89); Clear Creek, 1400 ft., 3 mi. W Klamath River (Hooper, 1938:221); Seiad Valley (*ibid.*); Hornbrook (Goldman, 1910:89). Southwestern segment (Goldman, 1910:89, unless otherwise noted): California: McCloud River, near Baird Station (Hooper, 1938:222); 5 mi. W Fruto (Hooper, 1938:221); *3 mi. W Stoneyford* (*ibid.*); Sites (*ibid.*); Freshwater Creek, Colusa Co.; Rumsey (Hooper, 1938:222); 3 mi. W Vacaville, 700 ft. (Hooper, 1938:221); 1½ mi. S Petaluma (*ibid.*); Mt. St. Helena; Lower Lake, Lake Co.; K3 Lodge, Stubbs [= Clearlake Oaks] (Hooper, 1938:221); Eel River, near South Yolla Bolly Mtn.; 1 mi. NW Mad River Bridge, 2300 ft. (Hooper, 1938:222); Helena, 1405 ft. (*ibid.*); Bully Choop Mtn.

Neotoma fuscipes luciana Hooper

1938. *Neotoma fuscipes luciana* Hooper, Univ. California Publ. Zool., 42:229, March 1, type from Seaside, Monterey Co., California.

MARGINAL RECORDS.—California: Camp Ord, 3½ mi. E Marina, 100 ft.; San Lucas; Salinas Valley, San Ardo, 450 ft.; Paso Robles, 750 ft.; 1 mi. NE Morro, thence along coast to point of beginning.

Neotoma fuscipes macrotis Thomas

1893. *Neotoma macrotis* Thomas, Ann. Mag. Nat. Hist., ser. 6, 12:234, September, type from San Diego, San Diego Co., California.
1894. *Neotoma fuscipes macrotis*, Merriam, Proc. Acad. Nat. Sci., Philadelphia, 14:246, September 25.

MARGINAL RECORDS.—California: Santa Margarita, 996 ft.; Matilija; 7 mi. S Simi; Arroyo Seco, near Pasadena; Glendora; 2 mi. NE Grapeland, 2000 ft.; base San Jacinto Mts., 1700 ft., near Cabezon; Garnet Queen Mine, Santa Rosa Mts. Baja California: Sierra Juárez; 5 mi. S Monument 258, W coast, thence northward along coast to opposite point of beginning.

Neotoma fuscipes martirensis Orr

1934. *Neotoma fuscipes martirensis* Orr, Proc. Biol. Soc. Washington, 47:110, June 13, type from Valladares, 2700 ft., Sierra San Pedro Mártir, Baja California.

MARGINAL RECORDS.—Baja California: Ensenada; Las Cruces, 20 mi. E Ensenada, 2600 ft.; Laguna Hanson, 5200 ft.; Concepción, 6000 ft.; Vallecitos, 8500 ft., Sierra San Pedro Mártir; 1 mi. E El Rosario, 200 ft.

Neotoma fuscipes monochroura Rhoads

1894. *Neotoma monochroura* Rhoads, Amer. Nat., 28:67, January, type from Grants Pass, Josephine Co., Oregon.
1906. *Neotoma fuscipes monochroura*, Stephens, California mammals, p. 116.

MARGINAL RECORDS (Hooper, 1938:220, unless otherwise noted).—Oregon: Burlington (Young, 1962:6); Salem; Prospect; Belmont Orchard, 6 mi. S Medford. California: E. Fork Illinois River, ¼ mi. S Oregon line, 1900 ft.; Hoopa Valley (Goldman, 1910:89); Horse Mtn., 4700 ft.; Fair Oaks; 3 mi. S Covelo; Lierly's Ranch, 2340 ft., 4 mi. S Mt. Sanhedrin; Calpella (Goldman, 1910:89); Novato (*ibid.*); Mt. Tamalpais (*ibid.*); *Lagunitas; Mt. Wittenberg, 1000 ft., 2 mi. W Olema;* Point Reyes (Goldman, 1910:89), thence northward along coast to Oregon: Gold Beach (*ibid.*); *Elkhead* [= *Yoncalla*]; Drain; vic. Corvallis (Walters and Roth, 1950:290).

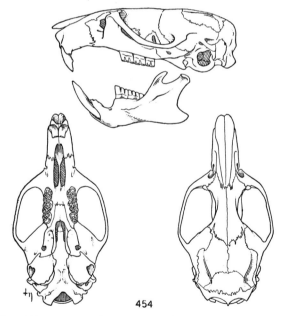

Fig. 454. *Neotoma fuscipes monochroura*, Van Duzen River, 5 mi. E Carlotta, Humboldt Co., California, No. 47261 K.U., ♂, X 1.

Neotoma fuscipes perplexa Hooper

1938. *Neotoma fuscipes perplexa* Hooper, Univ. California Publ. Zool., 42:224, March 1, type from Sweeney's Ranch, 22 mi. S Los Banos, Merced Co., California.

MARGINAL RECORDS.—California: Pacheco Pass; type locality; 2 mi. NNE New Idria, 1900 ft.; Arroyo Pasajero, 5 mi. E Coalinga; Priest Valley, 2400–2500 ft.; 1½ mi. S Soledad, 182 ft.

Neotoma fuscipes riparia Hooper

1938. *Neotoma fuscipes riparia* Hooper, Univ. California Publ. Zool., 42:223, March 1, type from Kincaids Ranch, 2 mi. NE Vernalis, Stanislaus Co., California.

MARGINAL RECORDS.—California: San Joaquin River, 3 mi. NE Vernalis; El Nido.

Neotoma fuscipes simplex True

1894. *Neotoma macrotis simplex* True, Diagnoses of some undescribed wood rats (genus *Neotoma*) in the National Museum, p. 2, June 27 (preprint of Proc. U.S. Nat. Mus., 17:354, November 15, 1894), type from Fort Tejon, Kern Co., California.
1901. *Neotoma fuscipes simplex,* Miller and Rehn, Proc. Boston Soc. Nat. Hist., 30(1):105, December 27.
1894. *Neotoma fuscipes dispar* Merriam, Proc. Biol. Soc. Washington, 9:124, July 2, type from Lone Pine, Inyo Co., California.
1904. *Neotoma fuscipes mohavensis* Elliot, Field Columb. Mus., Publ. 87, Zool. Ser., 3(14):246, January 7, type from Oro Grande, San Bernardino Co., California.
1904. *Neotoma fuscipes cnemophila* Elliot, Field Columb. Mus., Publ. 90, Zool. Ser., 3:267, March, type from Lockwood Valley, Mt. Piños, Ventura Co., California.

MARGINAL RECORDS.—California: Kearsarge Pass, 6000 ft., Sierra Nevada; Carl Walters Ranch, 2 mi. N Independence; Onyx, 2750 ft.; Oro Grande, 2700 ft.; Doble, 7000 ft.; Seven Oaks, 5000–5100 ft.; Fairmont; 10 mi. W Lebec; Mt. Piños, 5500–6500 ft.; Buena Vista Lake, N side; head Cuddy Valley, 5900 ft.

Neotoma fuscipes streatori Merriam

1894. *Neotoma fuscipes streatori* Merriam, Proc. Biol. Soc. Washington, 9:124, July 2, type from Carbondale, Amador Co., California.

MARGINAL RECORDS.—California: Manton, 2300 ft.; Lyman's, 3300 ft., 4 mi. NW Lyonsville; ⅒ mi. E Virgilia (K. F. Murray, *in Litt.*); Blue Canyon, 4600 ft.; Michigan Bluff, 3500 ft.; Cascades, Yosemite National Park; Kings River Canyon, 5000 ft.; Canyon Creek 700 ft., 7 mi. W Orosi; 3 mi. SE Friant; Raymond, 740 ft.; 3 mi. NE Coulterville, 3200 ft.; type locality; 1 mi. NE Red Bluff, 300 ft.; *Dale's, 600 ft.* (Grinnell, *et al.,* 1930:516).

Subgenus **Teonoma** Gray

1843. *Teonoma* Gray, List of the . . . Mammalia in the . . . British Museum, p. 117. Type, *Myoxus drummondii* Richardson.

"Tail large, bushy, and somewhat distichous; sole of hind foot normally densely furred from

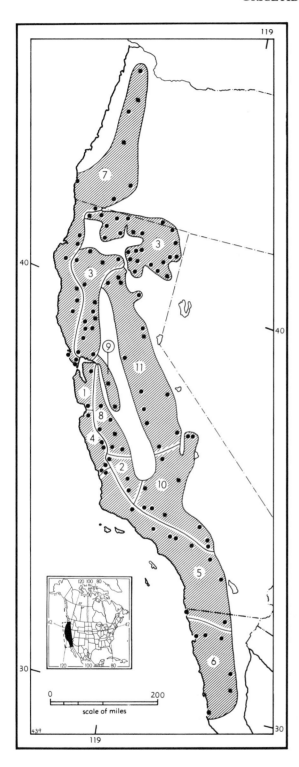

Map 439. *Neotoma fuscipes.*

Guide to subspecies
1. *N. f. annectens*
2. *N. f. bullatior*
3. *N. f. fuscipes*
4. *N. f. luciana*
5. *N. f. macrotis*
6. *N. f. martirensis*
7. *N. f. monochroura*
8. *N. f. perplexa*
9. *N. f. riparia*
10. *N. f. simplex*
11. *N. f. streatori*

heel to posterior tubercle. Skull large and angular; temporal ridges prominent, diverging posteriorly to near anterior border of interparietal, whence they turn abruptly inward and again outward in crossing interparietal to lamboidal crest; frontal region narrow, constricted near middle, somewhat depressed, and excavated above along median line; maxillary arms of zygomata broad and heavy; bullae large; interpterygoid fossa narrow." (Goldman, 1910:95.)

Neotoma cinerea
Bushy-tailed Wood Rat

External measurements: 273–470; 120–223; 30–52. Upper parts varying from pale gray lightly washed with buff to dark brownish black; underparts varying from white to pinkish or buff; tail dusky above, much paler (usually whitish) below. Skull with relatively short braincase; anterior palatine foramina longer than palatal bridge; sphenopalatine vacuities present or absent, but relatively constant within a series; M1 with deep anterointernal re-entrant angle; M3 with anterior closed triangle and two confluent posterior loops.

This is the only species of the genus that has a bushy, somewhat squirrellike tail.

Neotoma cinerea acraia (Elliot)

1904. *Teonoma cinerea acraia* Elliot, Field Columb. Mus., Publ. 87, Zool. Ser., 3(14):247, January 7, type from Jordan Hot Springs, near Kern River, Sierra Nevada, Tulare Co., California.
1940. *Neotoma cinerea acraia,* Hooper, Univ. California Publ. Zool., 42:413, May 17.

MARGINAL RECORDS.—Idaho: E side Bear Lake. Utah: Stockmore; Fish Lake Plateau; Henry Mts.; Mile 101, Glen Canyon, W bank (Durrant and Dean, 1959:95). Arizona (Hoffmeister, 1971:176): Point Impe-

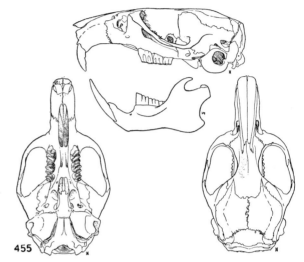

455

Fig. 455. *Neotoma cinerea acraia,* Baker Creek, 8500 ft., Nevada, No. 41994 M.V.Z., ♀, X 1.

rial; *Point Honan;* Shiva Temple. Utah: Cottonwood [Wash], 8 mi. N Kanab, 4800 ft. Nevada: Hidden Forest; 11 mi. E Panaca, 6700 ft.; Water Canyon, 8 mi. N Lund; Monitor Valley, 9 mi. E Toquima Peak, 7000 ft.; Lapon Canyon, Mt. Grant, 8900 ft. California: Hanaupah Canyon, 8900–9500 ft., Panamint Mts.; type locality; Fresno. Nevada: 2 mi. S Zephyr Cove; Carson City; 5 mi. W Fallon; Cherry Creek, 6800 ft.; Goose Creek, 2 mi. W Utah boundary, 5000 ft. Utah: Standrod, Raft River Mts., 5500 ft.

Neotoma cinerea alticola Hooper

1940. *Neotoma cinerea alticola* Hooper, Univ. California Publ. Zool., 42:409, May 17, type from Parker Creek [= Shields Creek, U.S. Forest Service map, 1932 ed.], 5500 ft., Warner Mts., Modoc Co., California.

MARGINAL RECORDS.—Washington: Stay-awhile Spring. Idaho: John Day Ranch, 10 mi. SW White Bird; Arco; N side, S fork, Snake River, 3 mi. W Swan Valley; vic. Pocatello [Schutt's Mine]. Utah: Grouse Creek, 6000 ft. Nevada: Steels Creek, 7000 ft.; Willow Creek, Ruby Mts., 2 mi. S Elko Co. line, 6500 ft.; Evans. California: ½ mi. N Tahoe City; Lake Helen, 8500 ft. Oregon: Tule Lake; Bend; near mouth John Day River.

Neotoma cinerea arizonae Merriam

1893. *Neotoma arizonae* Merriam, Proc. Biol. Soc. Washington, 8:110, July 31, type from Keams Canyon, Navajo Co., Arizona.
1910. *Neotoma cinerea arizonae,* Goldman, N. Amer. Fauna, 31:106, October 19.

MARGINAL RECORDS.—Colorado (Finley, 1958: 270, unless otherwise noted): 11 mi. N, 11 mi. W Rangley, 6000 ft.; White River, 20 mi. E Rangley; 6 mi.

NE Meeker; *Dry Fork, Ute Creek;* 8½ mi. W Rifle; New Castle (Armstrong, 1972:226); *12 mi. SE Rifle; 4 mi. S, 3 mi. E Collbran; ca.* 1 mi. W Cameo; Coventry; 1½ mi. W Dolores; *head Prater Canyon.* New Mexico: 18 mi. N, 1 mi. E Farmington (Finley, 1958:270); Rio Los Pinos, 2 mi. S Colorado [state line] (Findley, *et al.,* 1975:251, as *N. cinerea* only); Pueblo Bonito, Chaco Canyon. Arizona: Holbrook; Winslow; Walpai [*sic*]; Lees Ferry. Utah: Mile 83, Glen Canyon, E bank (Durrant and Dean, 1959:95); Castle Valley, 18 mi. NE Moab.

Neotoma cinerea cinerea (Ord)

1815. *Mus cinereus* Ord, *in* Guthrie, A new geog., hist. coml. grammar . . . , Philadelphia, 2nd Amer. ed., 2:292. Type locality, near Great Falls, Cascade Co., Montana.
1858. *Neotoma cinerea*, Baird, Mammals, *in* Repts. Expl. Surv. . . . , 8(1):499, July 14.

MARGINAL RECORDS.—Alberta: Banff National Park. Saskatchewan: Govenlock. Montana: Terry; Miles City; Billings; Red Lodge. Wyoming (Long, 1965a:639): Lamar River; mouth Grinnell Creek, Pahaska; 25 mi. S, 28 mi. W Cody; 5 mi. W Union; Bull Lake; *Halfmoon Lake;* Big Sandy; *jct. Green River and New Fork;* Cokeville; Salt River, 10 mi. N Afton; *Teton Pass.* Idaho: Lost River Mts., near Arco; Challis; Sawtooth; Riggins; Fiddle Creek. Montana: Rock Creek; Bass Creek; Ravalli; Columbia Falls. British Columbia: Newgate; Morrissey (Cowan and Guiguet, 1965:192).

Neotoma cinerea cinnamomea J. A. Allen

1895. *Neotoma cinnamomea* J. A. Allen, Bull. Amer. Mus. Nat. Hist., 7:331, November 8, type from Kinney Ranch, Bitter Creek, Sweetwater Co., Wyoming.
1944. *Neotoma cinerea cinnamomea*, Hooper, Jour. Mamm., 25:415, December 12.

MARGINAL RECORDS.—Wyoming (Long, 1965a:640): Fontenelle; Thayer Junction; *6 mi. S Point of Rocks;* type locality. Colorado (Finley, 1958:270, as *N. c. arizonae;* Long, 1965a:639, as *N. c. cinnamomea*): Little Snake River, 25 mi. down river from Baggs P.O.; 8 mi. NE Craig; *5 mi. W Craig;* 1 mi. S Cross Mtn.; Castle Park, Dinosaur National Monument. Wyoming (Long, 1965a:640): W side Green River, 1 mi. N Utah boundary; Beaver Creek, 4 mi. S Lonetree; Fort Bridger. For differing estimates of the relationship to one another of *N. c. arizonae, cinnamomea,* and *orolestes,* see Finley (1958:269), Long (1965a:640), and Armstrong (1972:226).

Neotoma cinerea drummondii (Richardson)

1828. *Myoxus drummondii* Richardson, Zool. Jour., 3:517. Type locality, probably near Jasper House, Alberta. (See Goldman, N. Amer. Fauna, 31:99, October 19, 1910.)
1892. *Neotoma cinerea drummondi*, Merriam, Proc. Biol. Soc. Washington, 7:25, April 13.

MARGINAL RECORDS.—Mackenzie (Scotter and Simmons, 1975:489, 490): N of Little Dal Lake (62° 40', 126° 45') in Silverberry (South Redstone) River area; *S of Little Dal Lake (62° 40', 126° 42' 30");* a cave, 61° 17', 124° 06'; Fort Liard. British Columbia: St. John Creek. Alberta: 2 mi. E Brownvale (Kelsall, 1971:326); Pipestone Creek and Wapiti River; type locality. British Columbia: *Kootenay Park;* Paradise Mine; Glacier House; Barkerville; Laurier Pass; *Robb Lake;* Fort Halkett.

Neotoma cinerea fusca True

1894. *Neotoma occidentalis fusca* True, Diagnoses of some undescribed wood rats (genus *Neotoma*) in the National Museum, p. 2, June 27 (preprint of Proc. U.S. Nat. Mus., 17:354, November 15, 1894), type from Fort Umpqua, Douglas Co., Oregon.
1897. [*Neotoma cinerea*] *fusca*, Trouessart, Catalogus mammalium . . . , fasc. 3, p. 544.
1903. *Neotoma fuscus apicalis* Elliot, Field Columb. Mus., Publ. 74, Zool. Ser., 3:160, May 7, type from Gardiner, Douglas Co., Oregon.

MARGINAL RECORDS.—Oregon: Portland; Wells; Eugene; type locality; Gardiner, thence northward along coast and Columbia River to point of beginning.

Neotoma cinerea lucida Goldman

1917. *Neotoma cinerea lucida* Goldman, Proc. Biol. Soc. Washington, 30:111, May 23, type from Charleston Peak, Charleston Mts., Clark Co., Nevada.

MARGINAL RECORDS.—Nevada: Clark Canyon, 8000 ft.; type locality; N side Potosi Mtn., 7000 ft.

Neotoma cinerea macrodon Kelson

1949. *Neotoma cinerea macrodon* Kelson, Jour. Washington Acad. Sci., 30:417, December 15, type from E side confluence Green and White rivers, 4700 ft., 1 mi. SE Ouray, Uintah Co., Utah.

MARGINAL RECORDS.—Utah: Willow Creek, 29 mi. S Ouray, 5400 ft.; type locality.

Neotoma cinerea occidentalis Baird

1855. *Neotoma occidentalis* Baird, Proc. Acad. Nat. Sci. Philadelphia, 7:335, April, type from Shoalwater [= Willapa] Bay, Pacific Co., Washington.
1891. *Neotoma cinerea occidentalis*, Merriam, N. Amer. Fauna, 5:58, July 30.
1899. *Neotoma c[inerea]. columbiana* Elliot, Field Columb. Mus., Publ. 32, Zool. Ser., 1(13):255, May 17, type from Ducks, British Columbia.
1900. *Neotoma saxamans* Osgood, N. Amer. Fauna, 19:33, October 6, type from Bennett City, head Lake Bennett, British Columbia. Regarded as inseparable from *N. c. occidentalis* by Cowan and Guiguet, The mammals of British Columbia, Handbook No. 11, British Columbia Provincial Mus., p. 195, October 1965.

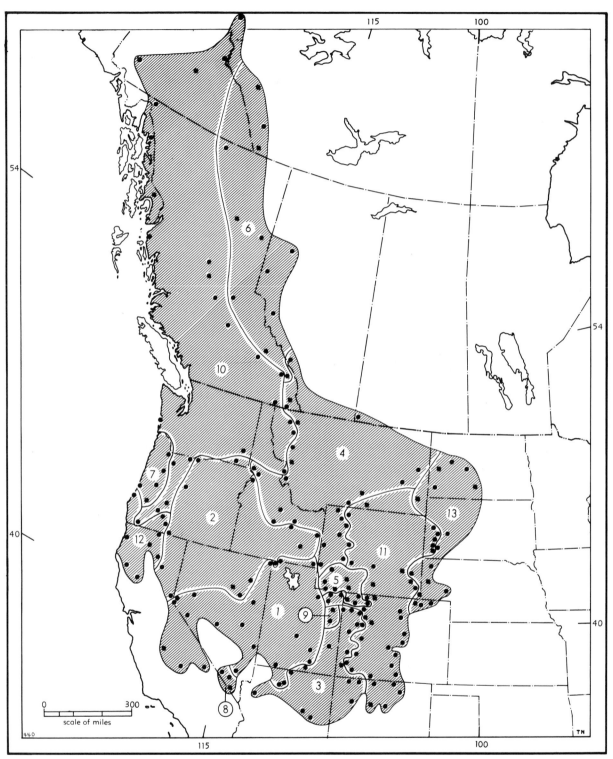

Map 440. *Neotoma cinerea.*

Guide to subspecies
1. *N. c. acraia*
2. *N. c. alticola*

3. *N. c. arizonae*
4. *N. c. cinerea*
5. *N. c. cinnamomea*
6. *N. c. drummondii*

7. *N. c. fusca*
8. *N. c. lucida*
9. *N. c. macrodon*
10. *N. c. occidentalis*

11. *N. c. orolestes*
12. *N. c. pulla*
13. *N. c. rupicola*

770

MARGINAL RECORDS.—Mackenzie: 65° 23' N, 131° 16' W (Martell, 1974a:348, as *N. cinerea* only). British Columbia: Stuart Lake; Nulki Lake; Tiltzarone. Lake; Lac la Hache; Revelstoke; Invermere; Yahk. Montana: Upper Stillwater Lake, 3500 ft.; Flathead Lake; Gird Creek. Washington: Almota. Oregon: The Dalles; Crater Lake; Prospect; Oregon Caves National Monument; Bissell. Washington: type locality, thence northward along coast to British Columbia: Inverness. Alaska (Shaw, 1962:431): Unuk River; mouth Lower Taku River. British Columbia: Bennett City. Yukon (Youngman, 1975:83): Keele Peak, Selway Range, 275 mi. NNE Whitehorse; Lapie River, Canol Road, Mile 132; N Cultus Bay, Kluane Lake.

Neotoma cinerea orolestes Merriam

1894. *Neotoma orolestes* Merriam, Proc. Biol. Soc. Washington, 9:128, July 2, type from Saguache Valley, 20 mi. W Saguache, Saguache Co., Colorado.
1910. *Neotoma cinerea orolestes*, Goldman, N. Amer. Fauna, 31:104, October 19.
1894. *Neotoma grangeri* J. A. Allen, Bull. Amer. Mus. Nat. Hist., 6:324, November 7, type from Custer, Custer Co., South Dakota.

MARGINAL RECORDS.—Montana: Pryor Mts. South Dakota (Turner, 1974:105): Davenport Cave, 3 mi. S, ½ mi. W Sturgis, 4400 ft.; Rapid City; Glendale; Housing Area, Wind Cave National Park, 4100 ft.; Angostura Dam. Wyoming (Long, 1965a:641): Laramie Peak; 2½ mi. S Chugwater; 3 mi. W Meriden. Colorado (Finley, 1958:259): Pinewood; Boulder; Palmer Lake; Colorado Springs; Salida; Querida; 24 mi. E Hooper; Culebra Canyon. New Mexico: Costilla Pass; near Cimarron; Willis; Jemez Mts.; Tierra Amarillo. Colorado (Finley, 1958:259, 260): 1 mi. S, 2 mi. W Chromo; Vallecito Camp; 15 mi. N, 10 mi. E Dolores; 2 mi. NE Ridgway; 1 mi. NE Bowie; *P. Slideler Ranch*; 12 mi. above Glenwood Springs, Grand River Canyon; E fork Rifle Creek (20 mi. NE Rifle); *9 mi. NE Buford, Lost Creek*. Utah: Myton; Granite Park; Hideout. Colorado (Finley, 1958:259, 260): 8 mi. S, 4 mi. W Craig; 16 mi. N Craig; Three Forks, 30 mi. above Baggs Crossing. Wyoming (Long, 1965a:641): *4 mi. N, 8 mi. E Savery;* Bridger Pass; Ferris Mts.; ½ mi. N, 3 mi. E South Pass City; *Lake Fork; Crowheart;* Black Mtn., head Pat O'Hara Creek.

Neotoma cinerea pulla Hooper

1940. *Neotoma cinerea pulla* Hooper, Univ. California Publ. Zool., 42:411, May 17, type from Kohnenberger's Ranch, South Fork Mtn., 3200 ft., Trinity Co., California.

MARGINAL RECORDS.—Oregon: Empire; Fort Klamath; Naylox. California: Brownell; Burney; Castle Lake, 5434 ft.; Elk Creek; near Blake Lookout, South Fork Mtn., 5500 ft.; Orick.

Neotoma cinerea rupicola J. A. Allen

1894. *Neotoma rupicola* J. A. Allen, Bull. Amer. Mus. Nat. Hist., 6:323, November 7, type from Corral Draw, 3700 ft., Pine River Indian Reservation, Black Hills, South Dakota.
1910. *Neotoma cinerea rupicola*, Goldman, N. Amer. Fauna, 31:107, October 19.

MARGINAL RECORDS.—North Dakota: Oakdale; Fort Clark; Cannonball. Nebraska (Jones, 1964c:215): Glen; *4 mi. E Agate;* Scottsbluff; *3 mi. NW Lisco;* Oshkosh. Colorado (Armstrong, 1972:229): sec. 3, T. 11 N, R. 54 W; 12 mi. NW New Raymer; 1 mi. N Geary Reservoir. Wyoming (Long, 1965a:642): 2 mi. S Pine Bluffs; *Uva;* Bordeaux. South Dakota: type locality; Quinns Draw, Cheyenne River; *2 mi. S, 5 mi. E Harding* (Andersen and Jones, 1971:380). Montana: 11 mi. S, 6½ mi. W Ekalaka (Lampe, *et al.,* 1974:20). North Dakota: 25 mi. S Medora; Mikkelson.

Subgenus Hodomys Merriam

1894. *Hodomys* Merriam, Proc. Acad. Nat. Sci. Philadelphia, 46:232, September 24. Type, *Neotoma alleni* Merriam.

Enamel pattern of chewing surface of last lower molar S-shaped rather than having 2 transverse loops as in other members of the genus; incisive foramina extending to level of toothrows; auditory bullae small.

Neotoma alleni
Allen's Wood Rat

Arranged according to Kelson, Univ. Kansas Publ., Mus. Nat. Hist., 5:240–242, April 10, 1952. See also Genoways and Birney, Mammalian Species, 41:1, 2, June 28, 1974, for comprehensive biological information.

External measurements: 368–446; 158–224; 37–45. Upper parts rich reddish brown to dusky brown; underparts plumbeous washed with white, sometimes faintly buffy; tail sparsely haired, more or less dusky above and below. Enamel pattern of m3 S-shaped instead of 2 transverse loops as in other species of *Neotoma.*

Neotoma alleni alleni Merriam

1892. *Neotoma alleni* Merriam, Proc. Biol. Soc. Washington, 7:168, September 29, type from Manzanillo, Colima.

MARGINAL RECORDS (Genoways and Birney, 1974:2).—Sinaloa: Isla Palmito de la Virgen, 15 ft.; Rosario. Nayarit: Acaponeta. Jalisco: 7 mi. N Guadalajara; 10 mi. SE Tuxpan, 4200 ft. Colima: Colima; Armería, thence up coast to point of beginning.

Neotoma alleni elattura (Osgood)

1938. *Hodomys vetulus elatturus* Osgood, Field Mus. Nat. Hist., Publ. 431, Zool. Ser., 20(35):475, December 31, type

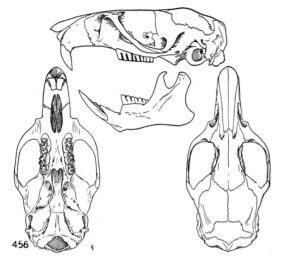

Fig. 456. *Neotoma alleni alleni*, Chamela Bay, Jalisco, No. 80946 U.M.M.Z., ♂, X 1.

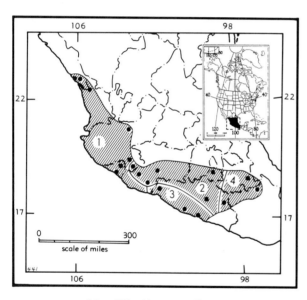

Map 441. *Neotoma alleni*.

1. *N. a. alleni*   3. *N. a. guerrerensis*
2. *N. a. elattura*   4. *N. a. vetula*

from Cuapongo, Guerrero [probably = Coapango, Guerrero, 5½ km. S, 14 km. W Chilpancingo]. *Not* Chilpancingo as stated by Osgood (*supra cit.*), see Kelson, *sub cit.*, 241, 242.
1952. *Neotoma alleni elattura*, Kelson, Univ. Kansas Publ., Mus. Nat. Hist., 5:241, April 10.

MARGINAL RECORDS.—Jalisco: 8 mi. E Jilotlán de los Dolores (Genoways and Birney, 1974:2). Michoacán: La Huacana. Morelos: Cañón del Lobo, 6 mi. W Yautepec. Guerrero: Chilpancingo; *type locality*. Michoacán: 7 mi. S Tumbiscatío, 2700 ft.; 11 mi. by

road E Dos Aguas, 4500 ft. (Winkelmann, 1962:109, as *N. alleni* only).

Neotoma alleni guerrerensis (Goldman)

1938. *Hodomys alleni guerrerensis* Goldman, Jour. Washington Acad. Sci., 28:498, November 15, type from Acapulco, Guerrero.
1952. *Neotoma alleni guerrerensis*, Kelson, Univ. Kansas Publ., Mus. Nat. Hist., 5:241, April 10.

MARGINAL RECORDS.—Guerrero: El Limón; 5 mi. SE Tecpan, 50 ft.; type locality.

Neotoma alleni vetula (Merriam)

1894. *Hodomys vetulus* Merriam, Proc. Acad. Nat. Sci. Philadelphia, 46:236, September 24, type from Tehuacán, Puebla.
1952. *Neotoma alleni vetula*, Kelson, Univ. Kansas Publ., Mus. Nat. Hist., 5:241, April 10.

MARGINAL RECORDS.—Puebla: type locality. Oaxaca: Teotitlán, 950 m. Guerrero: Tlalixtaquilla.

Subgenus **Teanopus** Merriam

Accorded subgeneric rank by Burt and Barkalow, Jour. Mamm., 23:296, August 14, 1942.

1903. *Teanopus* Merriam, Proc. Biol. Soc. Washington, 16:81, May 29. Type, *Teanopus phenax* Merriam.

Skull resembling that of *Neotoma* and *Teonoma*, but auditory bullae enormously inflated ventrally, blunt anteriorly, nearly parallel, almost exactly as in *Xenomys*; antorbital slits large and broadly open; sphenoidal vacuities open; braincase without temporal shield; lower jaw with distinct prominence over root of incisor; angle elongate, its lower border strongly inflected and upturned, forming a long shallow trough as in *Teonoma*, but less extreme; infracondylar notch deeper than in either *Neotoma* or *Teonoma;* m3 with re-entrant enamel loop on inner side passing obliquely forward in front of its mate on outer side, thus approaching the condition in *Hodomys* (after Merriam, 1903:81).

Neotoma phenax (Merriam)
Sonoran Wood Rat

1903. *Teanopus phenax* Merriam, Proc. Biol. Soc. Washington, 16:81, May 29, type from Camoa, Río Mayo, Sonora.
1942. [*Neotoma*] *phenax*, Burt and Barkalow, Jour. Mamm., 23:289, August 14.

External measurements of the type, an adult female, are: 352, 172; 37.5. Corresponding, average, measurements of three topotypes are: 365; 183; 37.7. Upper parts buffy gray; underparts yel-

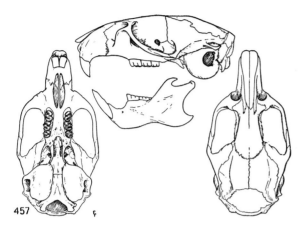

Fig. 457. *Neotoma phenax*, Camoa, Río Mayo, Sonora, No. 86067 M.V.Z., ♀, X 1.

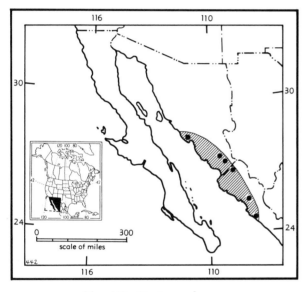

Map 442. *Neotoma phenax*.

lowish white anteriorly, plumbeous showing through posteriorly; tail unicolored dusky.

MARGINAL RECORDS.—Sonora: San José de Guaymas; type locality; Alamos, 54 km. E Navajoa, 1000 ft. (Birney and Jones, 1972:204). Sinaloa (*ibid.*): Laguna, 17 km. SW Choix, 500 ft.; Rancho Viejo; 2 mi. W El Dorado, 20 ft., thence up coast to point of beginning.

Genus **Xenomys** Merriam—Magdalena Rat

1892. *Xenomys* Merriam, Proc. Biol. Soc. Washington, 7:160, September 29. Type, *Xenomys nelsoni* Merriam.

Skull in general resembling that of *Neotoma;* supraorbital shelf much elevated and beaded;

lachrymals much enlarged; auditory bullae greatly enlarged and inflated, and parallel to long axis of skull; slight postorbital process present; paroccipital processes long and stout; interparietal much enlarged; enamel pattern of occlusal surface of m3 S-shaped much as in *Neotoma alleni.*

Xenomys nelsoni Merriam
Magdalena Rat

1892. *Xenomys nelsoni* Merriam, Proc. Biol. Soc. Washington, 7:161, September 29, type from Hda. Magdalena, between city of Colima and Manzanillo, Colima.

External measurements of three males are, respectively: 315, 300, 335; 155, 143, 170; 31, 30, 32. Upper parts fulvous or tawny-rufous, . . . back sparsely mixed with black-tipped hairs; an ill-defined dusky ring around each eye, above which is a whitish spot about as large as the eye itself; . . . tail concolored, dark umber-brown all around; . . . underparts creamy white to the very roots of the hairs except along the sides of the belly, where the basal part of the fur is plumbeous; line of demarkation between colors of upper and lower parts everywhere sharp and distinct. (After Merriam, 1892:162.) See Map 443.

MARGINAL RECORDS.—Jalisco: Chamela Bay. Colima: type locality; Armería.

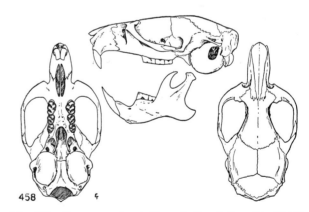

Fig. 458. *Xenomys nelsoni*, Chamela Bay, Jalisco, No. 80945 U.M.M.Z., ♀, X 1.

Genus **Nelsonia** Merriam—Diminutive Wood Rat

1897. *Nelsonia* Merriam, Proc. Biol. Soc. Washington, 11:277, December 17. Type, *Nelsonia neotomodon* Merriam.

Skull resembling that of a large *Peromyscus* but more nearly flat; zygomata heavier, less depressed, and more spreading anteriorly; inferior angle of antorbital slit thickened and protruding forward and outward as distinct process; auditory bullae subconical as in *Peromyscus* and *Hodomys*, not bullate as in *Neotoma* and *Xenomys*; braincase depressed as in *Peromyscus*, not elevated as in *Neotoma*, *Xenomys*, and *Hodomys*; incisive foramina large and open, broader anteriorly than in *Neotoma* or *Peromyscus*; coronoid process of mandible small, hardly larger than in *Peromyscus* (after Merriam, 1897:277,278); M3 with single external re-entrant angle, which nearly divides tooth into separate prisms; m2 with one internal and one external re-entrant angle; m3 with but one re-entrant angle and that internal.

Nelsonia neotomodon
Diminutive Wood Rat

Revised by Hooper, Occas. Pap. Mus. Zool., Univ. Michigan, 558:1–12, 1 fig., 1 map, September 17, 1954.

External measurements: 233–256; 106–129; 27–30; 22–24.5. Greatest length of skull, 31.9–33.9; zygomatic breadth, 16.5–18.6; alveolar length of cheek-teeth, 6.2–6.9. Upper parts grayish brown to dark slaty gray, washed with pale ochraceous to black; sides tending to be more ochraceous than back; underparts white, hairs plumbeous basally; tail dusky above and white below; tip of tail white in subspecies *neotomodon*; cheek-teeth prismatic as in *Neotoma*, but prisms alternate instead of opposite; lateral face of interior part of zygomatic root of zygomatic arch concave.

Nelsonia neotomodon cliftoni Genoways and Jones

1968. *Nelsonia neotomodon cliftoni* Genoways and Jones, Proc. Biol. Soc. Washington, 81:97, April 30, type from 2½ mi. ENE Jazmín, 6800 ft., Jalisco.

MARGINAL RECORDS (Genoways and Jones, 1968:100).—Jalisco: type locality; *4 mi. ENE Jazmín.*

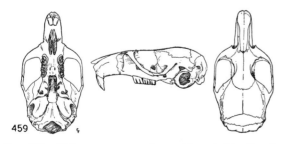

Fig. 459. *Nelsonia neotomodon goldmani*, Mt. Tancítaro, Michoacán, No. 125813 U.S.N.M., ♂, X 1.

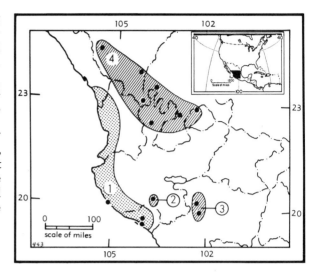

Map 443. *Xenomys nelsoni* and *Nelsonia neotomodon*.

1. *Xenomys nelsoni*
2. *Nelsonia n. cliftoni*
3. *Nelsonia n. goldmani*
4. *Nelsonia n. neotomodon*

Nelsonia neotomodon goldmani Merriam

1903. *Nelsonia goldmani* Merriam, Proc. Biol. Soc. Washington, 16:80, May 29, type from Mt. Tancítaro, Michoacán.
1954. *Nelsonia neotomodon goldmani*, Hooper, Occas. Pap. Mus. Zool., Univ. Michigan, 558:7, September 17.

MARGINAL RECORDS.—Michoacán: Cerro Patambán; type locality.

Nelsonia neotomodon neotomodon Merriam

1897. *Nelsonia neotomodon* Merriam, Proc. Biol. Soc. Washington, 11:278, December 17, type from mts. near Plateado, 8200 ft., Zacatecas.

MARGINAL RECORDS.—Durango: 1½ mi. W San Luis, 8000 ft.; *E slope Cerro Huehuento, 9500 ft.*; 28 mi. S, 17 mi. W Vincente Guerrero, 8350 ft. (Baker and Greer, 1962:129). Zacatecas: Sierra de Valparaíso, W of Valparaíso, 8700 ft. Aguascalientes: 15 mi. W Presa Calles, Sierra Fría, 8000–8200 ft. Zacatecas: type locality. Jalisco: Sierra Madre near Bolaños, 7600 ft. Zacatecas: Sierra Madre, 8500 ft., between Huazamata, Durango, and San Juan Capistrano, Zacatecas.

Genus **Rheomys** Thomas—Water Mice

1906. *Rheomys* Thomas, Ann. Mag. Nat. Hist., ser. 7, 17:421, April. Type, *Rheomys underwoodi* Thomas.

Small glossy blackish mice, often frosted with white-tipped hairs; underparts paler than, but not sharply distinct from, upper parts; hind feet fimbriated and slightly webbed. Skull relatively smooth, constricted interorbitally; anterior palatine slits long; upper incisors sometimes grooved, resulting in worn edge of each being deeply notched.

Possibly *Rheomys* should be treated as a subgenus of the South American fish mice, genus *Ichthyomys;* certainly the two taxa do not differ much morphologically.

Water mice inhabit the smaller jungle streams and most have been caught by placing traps in the water. According to Hooper (1968:552), six stomachs contained aquatic larvae of insects, insects and spiders, and fish; the last probably was carrion.

KEY TO SPECIES OF RHEOMYS

1. Tail less than 100; occurring in eastern Panamá. . . . . *R. raptor,*   p. 776
1'. Tail more than 100; not occurring in eastern Panamá.
  2. Occurring south of Nicaragua.
    3. Tail more than 125.
                  R. underwoodi,   p. 776
    3'. Tail less than 125. *R. hartmanni,*   p. 775
  2'. Occurring north of Nicaragua.
    4. Total length less than 265; cutting edge of upper incisors one straight line when viewed from front; occurring in Chiapas south into El Salvador. . . . . . .*R. thomasi,*   p. 775
    4'. Total length more than 265; cutting edge of upper incisors forming an inverted V when viewed from front; occurring in Oaxaca.
              R. mexicanus, p. 776

Subgenus **Rheomys** Thomas

1906. *Rheomys* Thomas, Ann. Mag. Nat. Hist., ser. 7, 17:421, April. Type, *Rheomys underwoodi.*

Size medium to small (total length 208–253); upper parts blackish mixed with brown in some species and with cinnamon in others; cutting edge of upper incisors one straight line when viewed from front.

Rheomys thomasi
Thomas' Water Mouse

External measurements of the types of *R. t. thomasi* (female), *R. t. stirtoni* (male), and *R. t. chiapensis* (subadult male) are, respectively: 253, 233, 208; 120, 125, 114; 32, 33, 30; 7, 6, 9. Color essentially as in *underwoodi.* Skull smaller than that of *underwoodi;* braincase lower, narrower; anterior palatine foramina longer. According to Hooper (1968:553) the examples from El

Salvador cranially and externally closely resemble Costa Rican specimens of *hartmanni,* but are larger, and the examples from Chiapas and Guatemala closely resemble those of *underwoodi* in cranial characters, diameter of tail, size of hind feet, and length of fringes on feet, but the pinnae of the ears are almost as large as those of *hartmanni.*

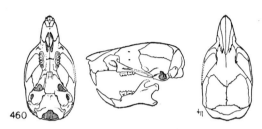

Fig. 460. *Rheomys thomasi stirtoni,* Los Esesmiles, Chalatenango, El Salvador, No. 12565 U.C.L.A., ♂, X 1.

Rheomys thomasi chiapensis Hooper

1947. *Rheomys thomasi chiapensis* Hooper, Jour. Mamm., 28:53, February 17, type from Prusia, 1100 m., Chiapas.

MARGINAL RECORDS.—Chiapas: type locality. Guatemala: Finca Injerto (Hooper, 1968:551, as *Rheomys* only).

Rheomys thomasi stirtoni Dickey

1928. *Rheomys thomasi stirtoni* Dickey, Proc. Biol. Soc. Washington, 41:12, February 1, type from Los Esemiles, 8000 ft., Chalatenango, El Salvador. Known only from type locality.

Rheomys thomasi thomasi Dickey

1928. *Rheomys thomasi* Dickey, Proc. Biol. Soc. Washington, 41:11, February 1, type from Finca San Felipe, Mt. Cacaguatique, 3500 ft., San Miguel, El Salvador. Known only from type locality.

Rheomys hartmanni Enders
Hartmann's Water Mouse

1939. *Rheomys hartmanni* Enders, Proc. Acad. Nat. Sci. Philadelphia, 90:295, February 3, type from hot springs on Río Cotito, Chiriquí, 4900 ft., Panamá.

External measurements of the type, an adult female, are: 212; 104; 25. Upper parts "uniformly and finely mixed black and cinnamon; underparts pale silvery gray; guard hairs not conspicuously longer than underfur, both Plumbeous basally; . . . fore and hind feet Sepia, the fringing bristles on hind feet grayish white. Skull small and flat, frontal region depressed; audital bullae heavy, narrow." (Goodwin, 1946:403.) For comparison with *underwoodi,* see account of that species.

MARGINAL RECORDS.—Costa Rica (Hooper, 1968:552): Monte Verde (sight record, as *Rheomys* only); 11 mi. by road N San Isidro; *9 mi. N San Isidro.* Panamá: type locality.

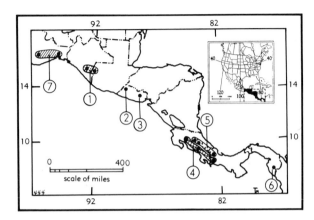

Map 444. *Rheomys.*

Guide to kinds
1. *R. thomasi chiapensis*
2. *R. thomasi stirtoni*
3. *R. thomasi thomasi*
4. *R. hartmanni*
5. *R. underwoodi*
6. *R. raptor*
7. *R. mexicanus*

Rheomys underwoodi Thomas
Underwood's Water Mouse

1906. *Rheomys underwoodi* Thomas, Ann. Mag. Nat. Hist., ser. 7, 17:422, April, type from Tres Ríos, Cartago, Costa Rica.

External measurements: 280–292; 148–156; 35; 5–7.8. Upper parts brown, darkened by black-tipped overhairs, tips of longer overhairs white; underparts grayish white; forefeet white, hind feet dark brown above, fringe-hairs glossy white; tail dark brown above, paler below. "Skull smoothly rounded, interorbital region broad, its edges rounded, palatal foramina not extending back to level of the front of first upper molars, incisors normal, narrow, and slightly rounded in front" (Goodwin, 1946:403). Hooper (1968:552) compares *underwoodi* with *hartmanni*, stating that the latter is smaller, with pelage less aquatic or muskratlike, underparts and muzzle darker, pinnae larger, hind feet shorter and narrower with shorter fringes, tail smaller in diameter and length, cranium smaller and more angular in dorsal view, with relatively smaller braincase, thicker incisors, and a longer masseteric process of the zygomatic plate.

MARGINAL RECORDS.—Costa Rica: Río Poasita (Hooper, 1968:552); type locality. Panamá: El Volcán de Chiriquí.

Rheomys raptor Goldman
Goldman's Water Mouse

1912. *Rheomys raptor* Goldman, Smiths. Miscl. Coll., 60(2):7, September 20, type from near head Río Limón, Mt. Pirri, 4500 ft., Panamá. Known only from type locality.

External measurements of the type, an adult male, are: 201; 94; 23.5; 7.7 (dry). Upper parts finely mixed black and cinnamon, black predominating on back, especially on rump; underparts pale smoky gray or slightly darker; sides of rump with a few projecting white hairs. Skull "smoothly rounded, the frontal region not depressed as in *Ichthyomys hydrobates;* audital bullae short and rounded; incisors of the ordinary murine type; molars about as in *Ichthyomys hydrobates*" (Goldman, 1912b:8). See Map 444.

Subgenus Neorheomys Goodwin

1959. *Neorheomys* Goodwin, Amer. Mus. Novit., 1967:3, October 29. Type, *Rheomys mexicanus.*

Size large (total length 280–320); upper parts Ochraceous Tawny; cutting edges of upper incisors forming inverted V when viewed from front.

Rheomys mexicanus Goodwin
Goodwin's Water Mouse

1959. *Rheomys (Neorheomys) mexicanus* Goodwin, Amer. Mus. Novit., 1967:4, October 29, type from San José Lachiguirí, District of Miahuatlán, Oaxaca.

External measurements of nine specimens are: 280–302 (320?); 140–168; 32–41. Upper parts Ochraceous Tawny, darkest on rump, paler on shoulders and sides; basal half of hairs Deep Mouse Gray; scattered blackish guard-hairs on back, longer whitish guard-hairs on lower rump; underparts and undersides of fore- and hind-limbs Light Buff; tail Mummy Brown above and white below; tip of tail white; anterior palatine foramina widest posteriorly and terminating posteriorly on line across anterior border of front molars. Differs from *R. thomasi* and *R. stirtoni* in being larger, and in having upper parts Ochraceous Tawny instead of Prout's Brown or Mummy Brown, pelage longer and more lax, tail longer and bicolored (instead of nearly unicolored), incisors narrower, and interorbital region shorter and more depressed. (After Goodwin, 1959b:6.) See Map 444.

MARGINAL RECORDS (Goodwin, 1969:208).—Oaxaca: type locality; Unión Hidalgo.

SUBFAMILY **MICROTINAE**

Revised by Miller, N. Amer. Fauna, 12:1–84, 3 pls., 40 figs., July 23, 1896. Synopsis of North American species by Hall and Cockrum, Univ. Kansas Publ., Mus. Nat. Hist., 5:373–498, 149 figs., January 15, 1953.

Thickset animals with bluntly rounded muzzles; most of each limb hidden in the general integument of the trunk, thus resulting in a short-legged appearance; 5 toes on each foot; tail never so long as head and body; postorbital processes absent; outer wall of infraorbital canal transverse or oblique to long axis of skull and with front edge emarginate or undercut and not projecting in front of anterior border of superior ramus of maxillary root of zygoma; palatine bones notably thick; ectopterygoid plates well developed; cheek-teeth 12 (3 on each side of upper- and lower-jaw), long-crowned and in most genera persistently growing; occlusal faces made up of triangles and loops.

KEY TO NORTH AMERICAN GENERA OF MICROTINAE

(After Ellerman, *et al.*, 1941:550–555)

1. Lower incisors wholly lingual to molars and terminating in horizontal ramus opposite, or in front of, alveolus of m3.
    2. Cheek-teeth longitudinally complex (many loops); inner and outer salient angles approx. equal in size; m1 with 7 closed triangles between terminal loops; supraorbital ridges strong but not fusing in interorbital region; foreclaws of D2 and D3 with periodic seasonal (winter) extra basal growths. . . . . *Dicrostonyx*, p. 835
    2'. Cheek-teeth longitudinally simplified (few loops); inner salient angles of upper molars and outer angles of lower molars smaller than those of the opposite sides; m1 with 3 closed triangles between termination loops (or with 2 transverse loops, if closed triangles absent); supraorbital ridges fusing in adults and forming median orbital crest; foreclaws not developing seasonal outgrowths.
        3. Posterior palate not terminating as simple transverse shelf; upper incisors strongly grooved; tooth-rows not, or less widely, divergent posteriorly; forefeet not much specialized in that claws are not notably thickened, soles almost hairless and ungual phalanges not notably lengthened. . . . . *Synaptomys*, p. 830
        3'. Posterior palate terminating as a simple transverse shelf; upper incisors not grooved; tooth-rows widely divergent posteriorly; forefeet specialized in that claws of some individuals are notably thickened, soles almost concealed by hairs, and ungual phalanges notably lengthened. . . . . *Lemmus*, p. 828
1'. Lower incisors passing from lingual to labial side of molars between bases of roots of m2 and m3 and ascending behind molars to termination within or near condylar process.
    4. Cheek-teeth rooted in adults.
        5. External form modified for aquatic life in that tail is laterally compressed, and swimming fringes on hind feet conspicuous; basal length of skull more than 50 mm. . . . . *Ondatra*, p. 824
        5'. External form not modified for aquatic life; basal length of skull less than 50 mm.
            6. Posterior palate terminating as a simple transverse shelf; lower molars with inner re-entrant angles little if any deeper than outer re-entrant angles. . . . . . *Clethrionomys*, p. 778
            6'. Posterior palate terminating with a median spinous process converted into a sloping septum between posterolateral pits; lower molars with inner re-entrant angles deeper than outer. . . . . *Phenacomys*, p. 785
    4'. Cheek-teeth not rooted (ever-growing) in adults.
        7. External form lemminglike in that ears are so short as to be concealed in fur and tail is so short as to extend scarcely beyond hind feet in prepared skins; tympanic bullae and mastoid bullae much enlarged (see Fig. 485); M1 and M2 with traces of extra complexities between main inner folds (m1 with 5 closed triangles). *Lagurus*, p. 821
        7'. External form volelike (not lemminglike) in that ears are not entirely concealed in fur and tail extends beyond hind feet in prepared skins; tympanic bullae and mastoid bullae of normal size (see Figs. 482 and 486); M1 and M2 usually without traces of extra complexities between the main inner folds (ml with 3, 4, or 5 closed triangles).
            8. Underfur muskratlike in that it obviously is adapted to aquatic existence; m3 with triangles (only 2 between terminal loops) tightly closed; alveolar length of upper molars more than 9.9 mm. . . . . *Neofiber*, p. 823
            8'. Underfur mouselike in that it is not adapted to aquatic existence; m3 without tightly closed triangles; alveolar length of upper molars less than 9.9 mm. . . . . . . *Microtus*, p. 789

Genus **Clethrionomys** Tilesius/Red-backed Mice

North American species revised (under name *Evotomys*) by V. Bailey, Proc. Biol. Soc. Washington, 11:113–138, May 13, 1897.

1850. *Clethrionomys* Tilesius, Isis, 2:28. Type, *Mus glareolus* Schreber.
1874. *Evotomys* Coues, Proc. Acad. Nat. Sci. Philadelphia, 26:186, December 15. Type, *Mus rutilus* Pallas.
1900. *Craseomys* Miller, Proc. Washington Acad. Sci., 2:87, July 26. Type, *Hypudaeus rufocanus* Sundevall.
1905. *Paulomys* Thomas, Ann. Mag. Nat. Hist., ser. 7, 15:493. Type, *Evotomys smithii* Thomas.

Ellerman and Morrison-Scott [Checklist of Palaearctic and Indian mammals, 1758–1946, British Museum (Natural History), November 19, 1951], on page 659, suggest that the three names listed below are synonyms of *Clethrionomys*.

1898. *Aschizomys* Miller, Proc. Acad. Nat. Sci. Philadelphia, p. 369, October 11. Type, *Aschizomys lemminus* Miller.
1911. *Caryomys* Thomas, Abstr. Proc. Zool. Soc. London, p. 4, February 14, Proc. Zool. Soc. London, p. 175. Type, *Microtus (Eothenomys) inez* Thomas.
1935. *Neoaschizomys* Tokuda, Mem. Coll. Sci. Kyoto, ser. B, 10(No. 3):241. Type, *Neoaschizomys sikotanensis* Tokuda.

External measurements: 120–165; 50–53; 16–21.0; 10–16. Weight (of *C. gapperi*), 16–42 grams; basal length of skull, 21.0–24.3; zygomatic breadth, 12.5–14.4. External form not much modified, back usually red; mammae, 2–2 = 8; plantar pads, 6. Skull "weak"; temporal ridges poorly developed and not fused anteriorly in interorbital region in adults; interorbital constriction medium; braincase broad; bullae relatively large, lacking spongy tissue (in living species); palate ends posteriorly as straight transverse shelf with no median septum; cheek-teeth rooted in adults.

KEY TO NORTH AMERICAN SPECIES OF CLETHRIONOMYS

1. Postpalatal bridge incomplete except in an occasional extremely old individual; pretympanic fenestra more than $\frac{2}{3}$ of a circle; tail short, thick, and with closely set bristly hairs. . . . .*C. rutilus*,   p. 778
1'. Postpalatal bridge complete even in half-grown young; pretympanic fenestra less than $\frac{2}{3}$ of a circle; tail slender, with short hairs except at the tip where hairs are longer.
   2. Occurring only in Oregon and California; described as paler (less red) and averaging larger than *C. gapperi* of Washington. . . . . *C. californicus*,   p. 784
   2'. Not occurring in Oregon or California; Washington populations described as darker (more red) and averaging smaller than *C. californicus*.
                   C. gapperi,   p. 779

Clethrionomys rutilus
Northern Red-backed Mouse

R.P.Grossenheider   X ½

American species revised (under name *Clethrionomys dawsoni*) by Orr, Jour. Mamm., 26:67–74, February 27, 1945. American species listed by Rausch, Jour. Washington Acad. Sci., 40:135, April 21, 1950. Canadian subspecies revised by Manning, Bull. Nat. Mus. Canada, 144:iv + 1–67, 21 figs., 1957.

External measurements: 130–158; 30–44; 18.5–21; 10–14. Upper parts bright; dorsal stripe well defined, varying from bright reddish to dark rufous; sides grayish; venter grayish with the tips of guard-hairs varying from whitish to deep buff; tail sharply bicolored, varying from reddish to blackish above. Tail short, thick and closely set with bristly hairs. Postpalatal bridge incomplete except occasionally in extremely old individuals; enamel patterns of M3 and m1 more complex than in *C. californicus*.

Clethrionomys rutilus albiventer Hall and Gilmore

1932. *Clethrionomys albiventer* Hall and Gilmore, Univ. California Publ. Zool., 38:398, September 17, type from Sevoonga, 2 mi. E of North Cape, St. Lawrence Island, Bering Sea, Alaska. Known only from type locality.
1952. *Clethrionomys rutilus albiventer*, Rausch, Jour. Parasit., 38:416, October.

Clethrionomys rutilus dawsoni (Merriam)

1888. *Evotomys dawsoni* Merriam, Amer. Nat., 22:650, July, type from Finlayson River, 3000 ft., a northern source of Liard River, lat. 61° 30′ N, long. 129° 30′ W, Yukon.
1950. *Clethrionomys rutilus dawsoni*, Rausch, Jour. Washington Acad. Sci., 40:134, April 21.

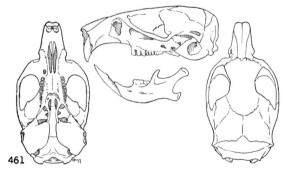

461

Fig. 461. *Clethrionomys rutilus dawsoni*, E side Deadman Lake, 1800 ft., 15 mi. SE Northway, Alaska, No. 21410 K.U., ♀, X 1½.

1898. *Evotomys alascensis* Miller, Proc. Acad. Nat. Sci. Philadelphia, 50:364, October 15, type from St. Michael, Norton Sound, Alaska. (See Osgood, N. Amer. Fauna, 24:34, November 23, 1904.)

MARGINAL RECORDS.—Alaska: 1½ mi. ESE Point Barrow, 10 ft.; ¾ mi. N, 1½ mi. W, Umiat, 69° 22′ 18″, 152° 08′ 10″, 370 ft. Yukon: 4 mi. WSW mouth Blow River (Youngman, 1975:86). Mackenzie: Fort McPherson; Bathurst Inlet. Keewatin: Baker Lake; Smoke Lake. Manitoba: Nonala, 80 mi. N Churchill. Mackenzie: Fort Reliance; Fort Rae; Fort Providence; Kakisa River (Murie and Dickinson, 1973:123). British Columbia: *N bank Tetsa River, 10 mi. S, 63 mi. W Muskwa* (64283 KU); Summit Pass, 4500 ft., 10 mi. S, 70 mi. W Fort Nelson; Junction. Alaska: Juneau; Yakutat; Mt. Logan area, Hubrick's Camp [N side Chitina River, 25 mi. W Alaska–Yukon line]; 5 mi. NNE Gulkana (Anderson, 1960:195); Moose Camp, Kenai Peninsula; Ugaguk River, near outlet Becharof Lake; Nushagak; Kuskokwim River; St. Michael; Teller, thence westward into Siberia and eastward to Alaska: Lava Lake; *Kuzitrin Lake; Mount Boyan; Trail Creek;* Cape Sabine (Childs, 1969:53).

Clethrionomys rutilus glacialis Orr

1945. *Clethrionomys dawsoni glacialis* Orr, Jour. Mamm., 26:71, February 27, type from Glacier Bay, Alaska.
1950. *Clethrionomys rutilus glacialis*, Rausch, Jour. Washington Acad. Sci., 40:135, April 21.

MARGINAL RECORDS.—Alaska (Glacier Bay area): Coppermine; *Bartlett Cove.*

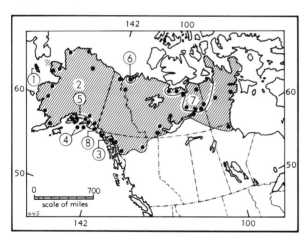

Map 445. *Clethrionomys rutilus.*

1. *C. r. albiventer*	5. *C. r. orca*
2. *C. r. dawsoni*	6. *C. r. platycephalus*
3. *C. r. glacialis*	7. *C. r. washburni*
4. *C. r. insularis*	8. *C. r. watsoni*

Clethrionomys rutilus insularis (Heller)

1910. *Evotomys dawsoni insularis* Heller, Univ. California Publ. Zool., 5:339, March 5, type from W side Canoe Passage, Hawkins Island, Prince William Sound, Alaska.

1950. *Clethrionomys rutilus insularis*, Rausch, Jour. Washington Acad. Sci., 40:135, April 21.

MARGINAL RECORDS.—Alaska (Prince William Sound): type locality; Northeast Bay, Hinchinbrook Island.

Clethrionomys rutilus orca (Merriam)

1900. *Evotomys orca* Merriam, Proc. Washington Acad. Sci., 2:24, March 14, type from Orca, Prince William Sound, Alaska.
1950. *Clethrionomys rutilus orca*, Rausch, Jour. Washington Acad. Sci., 40:135, April 21.

MARGINAL RECORDS.—Alaska (Prince William Sound area): Valdez Narrows; Cordova; La Touche, La Touche Island; head of Port Nellie Juan.

Clethrionomys rutilus platycephalus Manning

1957. *Clethrionomys rutilus platycephalus* Manning, Bull. Nat. Mus. Canada, 144 (for 1956–Biol. Series, No. 49):50, prior to March 15, type from Tuktoyaktuk, 100 ft., lat. 69° 22′, long. 133° 05′, Mackenzie.

MARGINAL RECORDS.—Mackenzie: type locality; *8 mi. S of type locality.*

Clethrionomys rutilus washburni Hanson

1952. *Clethrionomys rutilus washburni* Hanson, Jour. Mamm., 33:500, November 19, type from Perry River, lat. 67° 34′ N, long. 102° 07′ W, Mackenzie.

MARGINAL RECORDS.—Mackenzie: type locality; Deer Pass; Thelon Cabin; Artillery Lake; Aylmer Lake; Coppermine; Port Epworth.

Clethrionomys rutilus watsoni Orr

1945. *Clethrionomys dawsoni watsoni* Orr, Jour. Mamm., 26:73, February 27, type from Cape Yakataga, Alaska. Known only from type locality.
1950. *Clethrionomys rutilus watsoni*, Rausch, Jour. Washington Acad. Sci., 40:135, April 21.

Clethrionomys gapperi
Gapper's Red-backed Mouse

External measurements: 120–158; 30–50; 16–21; 12–16; weight, 16–42 grams. Dorsal stripe varies from bright chestnut through hazel, to yellowish brown (occasionally black); sides gray to buffy gray, sometimes with yellowish wash; venter silvery, some having buffy wash; tail bicolored, whitish to gray below, dark brown to black above. Tail slender, short-haired except at tip where hairs are longer. Postpalatal bridge complete even in half-grown young, with result that posterior margin of palate forms solid, truncate shelf; dentition relatively light; enamel patterns of M3 and m1 more complex than in *C. californicus.*

Clethrionomys gapperi arizonensis Cockrum and Fitch

1952. *Clethrionomys gapperi arizonensis* Cockrum and Fitch, Univ. Kansas Publ., Mus. Nat. Hist., 5:291, November 15, type from Little Colorado River, 8300 ft., White Mts., Apache Co., Arizona.

MARGINAL RECORDS.—Arizona: type locality; *Hannagan Creek, 8600 ft.*

Clethrionomys gapperi athabascae (Preble)

1908. *Evotomys gapperi athabascae* Preble, N. Amer. Fauna, 27:178, October 26, type from Fort Smith, Slave River, Mackenzie.
1932. *Clethrionomys gapperi athabascae,* Harper, Jour. Mamm., 13:28, February 9.

MARGINAL RECORDS.—Mackenzie: Fort Resolution. Keewatin: mouth Windy River (Harper, 1956:29). Manitoba (Soper, 1961:197): Nueltin Lake; Flin Flon; Atikameg Lake. Alberta: Tornado Pass. British Columbia: Invermere; *Assiniboine* (Cowan and Guiguet, 1965:203); *Thompson Pass.* Alberta: Jasper National Park. British Columbia: Moose Pass; Tupper Creek; Charlie Lake; Laurier Pass; N bank Tetsa River, 11 mi. S, 56 mi. W Muskwa (Anderson, 1960:195); Liard River. Mackenzie: Kakisa River (Murie and Dickinson, 1973:123).

Clethrionomys gapperi brevicaudus (Merriam)

1891. *Evotomys gapperi brevicaudus* Merriam, N. Amer. Fauna, 5:119, July 30, type from 3 mi. N Custer, 6000 ft., Black Hills, Custer Co., South Dakota.
1942. *Clethrionomys gapperi brevicaudus,* Bole and Moulthrop, Sci. Publs., Cleveland Mus. Nat. Hist., 5:153, September 11.

MARGINAL RECORDS.—Wyoming: 3 mi. NW Sundance, 5900 ft. South Dakota (Turner, 1974:108): Big Spearfish Canyon, 6 mi. S, 2 mi. W Spearfish, 4600 ft.; *Deadwood; Nemo, 4700 ft.;* Diamond S Ranch, near Rapid City; *3 mi. S, 1 mi. W Rockerville; 5¾ mi. N, 5¾ mi. E Custer, 5220 ft.;* type locality. Wyoming: 12 mi. SE Newcastle.

Clethrionomys gapperi carolinensis (Merriam)

1888. *Evotomys carolinensis* Merriam, Amer. Jour. Sci., ser. 3, 36:460, December, type from Roan Mtn., 6000 ft., Mitchell Co., North Carolina.
1939. *Clethrionomys gapperi carolinensis,* R. Kellogg, Proc. Biol. Soc. Washington, 52:38, March 11.

MARGINAL RECORDS.—West Virginia: Hardy County; *Franklin;* Cranberry Glades. Virginia: Whitetop Mtn. North Carolina: type locality; *Toxaway River Gorge* (Paul and Quay, 1963:124). Georgia (Wharton and White, 1967:670, as *C. gapperi* only):

Rabun Bald; *Beech Creek.* Tennessee: Spence Field, 1 mi. W Thunderhead Mtn. West Virginia: *Odd;* Cheat Mtn., 3 mi. E Cheat Bridge.

Clethrionomys gapperi cascadensis Booth

1945. *Clethrionomys gapperi cascadensis* Booth, Murrelet, 26:27, August 10, type from 2 mi. S Blewett Pass, 3000 ft., Kittitas Co., Washington.

MARGINAL RECORDS.—British Columbia (Cowan and Guiguet, 1965:205): Alta Lake; *Manning Park.* Washington: Pasayten River; Liberty; Simcoe Mts.; Mt. St. Helens; Nooksak River. British Columbia (Cowan and Guiguet, 1965:205): Lihumitson Mtn.; North Vancouver; *Mt. Seymour.*

Clethrionomys gapperi caurinus (V. Bailey)

1898. *Evotomys caurinus* V. Bailey, Proc. Biol. Soc. Washington, 12:21, January 27, type from Lund, E shore Malaspina Inlet, British Columbia.
1935. *Clethrionomys gapperi caurinus,* Racey and Cowan, Prov. British Columbia, Rept. Provincial Mus. Nat. Hist. for 1935, p. H25; also Sheppe, Canadian Field-Nat., 74:173, November 28, 1960, despite Hall and Cockrum, Univ. Kansas Publ., Mus. Nat. Hist., 5:300, November 17, 1952.

MARGINAL RECORDS.—British Columbia: Inverness, mouth Skeena River (Hall and Cockrum, 1952:300); Kimsquit (*ibid.*); Hagensborg (Cowan and Guiguet, 1965:205); Lewis Creek near Powell River (*ibid.*), thence northward along coast and on some coastal islands to point of beginning.

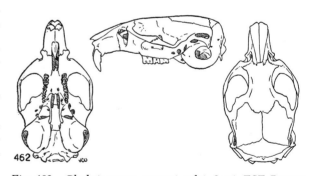

Fig. 462. *Clethrionomys gapperi galei,* 3 mi. ESE Browns Peak, 10,000 ft., Albany Co., Wyoming, No. 17288 K.U., ♂, X 1½.

Clethrionomys gapperi galei (Merriam)

1890. *Evotomys galei* Merriam, N. Amer. Fauna, 4:23, October 8, type from Ward, 9500 ft., Boulder Co., Colorado.
1931. *Clethrionomys gapperi galei,* Hall, Univ. California Publ. Zool., 37:6, April 10.
1941. *Clethrionomys gapperi uintaensis* Doutt, Proc. Biol. Soc. Washington, 54:161, December 8, type from Paradise Park, 10,050 ft., Uintah Co., Utah. (Arranged as a synonym of *C. g. galei* by Long, Univ. Kansas Publ., Mus. Nat. Hist., 14:644, July 6, 1965.)

MARGINAL RECORDS.—Alberta: "extreme southwestern Alberta." Montana: Big Snowy Mts. Wyoming (Long, 1965a:645): 38 mi. E Lovell; 2 mi. S, 6¼ mi. W Buffalo; *21¹/₂ to 22 mi. S, 24¹/₂ mi. W Douglas;* Laramie Peak; 10 mi. E Laramie. Colorado (Armstrong, 1972:232, 233): Gold Hill, 8400 ft.; *Minnehaha;* Lake Moraine, 10,250 ft.; 17 mi. W Salida, 11,000 ft.; Black Mesa, 9 mi. WNW Sapinero, 9500 ft.; Grand Mesa, 28 mi. E Grand Junction; Baxter Pass, 8500 ft.; 9½ mi. SW Pagoda Peak, 7700 ft. Wyoming (Long, 1965a:645): Bridgers Pass, 18 mi. SW Rawlins, 7500 ft.; 5 mi. SW Maxon, 9000 ft. Utah (formerly arranged as *C. g. uintaensis*): Beaver Dams; *15 mi. E Ephraim;* summit, 18 mi. E Mayfield; Bald Peak, Uinta Mts.; Silver Lake P.O.; *Emigration Canyon, 8 mi. above forks;* Monte Cristo, 18 mi. W Woodruff. Wyoming (Long, 1965a: 645): 31 mi. N Pinedale; *Jackeys Creek, 3 mi. S Dubois;* Needle Mtn.; *Harebell Creek;* Snow Pass, Mammoth Hot Springs. Montana: Deer Lodge County; St. Mary Lake. British Columbia: *Wall Lake; Akamina Pass* (Cowan and Guiguet, 1965:206).

Clethrionomys gapperi gapperi (Vigors)

1830. *Arvicola gapperi* Vigors, Zool. Jour., 5:204, type from between York [= Toronto] and Lake Simcoe, Ontario.
1928. *Clethrionomys gapperi gapperi*, Green, Jour. Mamm., 9:255, August 9.
1894. *Evotomys fuscodorsalis* J. A. Allen, Bull. Amer. Mus. Nat. Hist., 6:103, April 14, type from Trousers Lake, New Brunswick.

MARGINAL RECORDS.—Quebec: Mingan (Harper, 1961:61); Moisie Bay (*ibid.*); Godbout; *Mont Orford* (Wrigley, 1969:208); *South Bolton* (*ibid.*); Glen Sutton (*ibid.*). New York: Lake George. Massachusetts: 7 mi. N North Amherst (Muul and Carlson, 1963:416, as *C. gapperi* only); *7 mi. E Amherst* (Platt, 1968:332, as *C. gapperi* only). Connecticut: Woodstock; Glastonbury. New Jersey: Walkill Valley, near Long Lake. Virginia: Washington; Peaks of Otter; Mountain Lake; Elliot Knob. Maryland: 3 mi. E Grantsville (Paradiso, 1969:103). Pennsylvania: Renovo. Ontario: Middlesex County. Michigan: Alcona County; *Crawford County;* Leelanau County. Wisconsin (Jackson, 1961:228, 229): Oak Creek; Millston; Meenon Twp., Burnett Co. Minnesota: Carlton County; Lake of the Woods County. Manitoba (Soper, 1961:196): Marchand; Pinawa; Pine Falls. Ontario: Humbolt Bay, Lake Nipigon region. Quebec: Bark Lake; Camp de la Roche (Wrigley, 1969:208).

Clethrionomys gapperi gaspeanus Anderson

1943. *Clethrionomys gapperi gaspeanus* Anderson, Ann. Rept. Provancher Soc. Nat. Hist. of Canada, Quebec, for 1942, p. 57, September 7, type from Berry Mountain Camp, about 1500 ft., near jct. Berry Mountain Brook with Grand Cascapedia River, Matane Co., Quebec.

MARGINAL RECORDS.—Quebec: Ste. Anne de Monts; 2 mi. W Coin du Banc (Manville, 1961:108); Red Camp, Cascapedia Valley. New Brunswick:

Dalhousie; St. Leonard; Baker Lake. Quebec: Matapedia.

Clethrionomys gapperi gauti Cockrum and Fitch

1952. *Clethrionomys gapperi gauti* Cockrum and Fitch, Univ. Kansas Publ., Mus. Nat. Hist., 5:289, November 15, 1952, type from Twining, 10,700 ft., Taos Co., New Mexico.

MARGINAL RECORDS.—Colorado: 3 mi. N, 21 mi. W Saguache; San Isabel (Armstrong, 1972:233). New Mexico: 15 mi. SW Cimarron; Pecos Baldy, Pecos Mts.; Santa Fe ski basin (Findley, *et al.*, 1975:253, as *C. gapperi* only); Goat Peak; 1 mi. N, 6 mi. E Cuba, Jemez Mts., San Gregorio Lake, T. 21 N, R. 1 E (*loc. cit.*); 11½ mi. NE Chama (*loc. cit.*). Colorado: 8 mi. N, 2 mi. W Hesperus (Armstrong, 1972:233).

Clethrionomys gapperi hudsonius Anderson

1940. *Clethrionomys gapperi hudsonius* Anderson, Rapport Annuel 1939, Société Provancher d'Histoire Naturelle du Canada, Québec, p. 73, February 29, type from Kapuskasing, on Kapuskasing River, about 64 mi. W Cochrane, Ontario.

MARGINAL RECORDS.—Manitoba: Sandhill Lake (59° 21′ N, 98° 43′ W); [Fort] Churchill. Quebec: Richmond Gulf; Lake Albanel; Lake St. John area (Harper, 1961:71); 47° 50′ N, 75° 35′ W, S of Clova (MacLeod and Cameron, 1961:282). Ontario: Murdock Creek, S of Swastika; type locality. Manitoba: Ilford.

Clethrionomys gapperi idahoensis (Merriam)

1891. *Evotomys idahoensis* Merriam, N. Amer. Fauna, 5:66, July 30, type from Sawtooth (Alturas) Lake, 7200 ft., E base Sawtooth Mts., Blaine Co., Idaho.
1933. *Clethrionomys gapperi idahoensis*, Whitlow and Hall, Univ. California Publ. Zool., 40:265, September 30.

MARGINAL RECORDS.—Montana: Yellow Bay, Flathead Lake region. Idaho: Salmon River Mts. [= Timber Creek, Lemhi Mts.]. Wyoming (Long, 1965a:646): NW corner Teton National Forest, 18 mi. N, 9 mi. W Moran; *Upper Arizona Creek; Black Rock Meadows, 3 mi. S, 17 mi. E Moran;* Togwotee Pass; 10 mi. SE Afton. Idaho: 3 mi. S Victor; N rim Copenhagen Basin; Indian Creek; Edna. Oregon: 10 mi. N Harney; Kamela. Washington: Humpeg Falls. Idaho: ½ mi. E Black Lake.

Clethrionomys gapperi limitis (V. Bailey)

1913. *Evotomys limitis* V. Bailey, Proc. Biol. Soc. Washington, 26:133, May 21, type from Willow Creek, 8500 ft., a branch of the Gilita, Mogollon Mts., Catron Co., New Mexico.
1952. *Clethrionomys gapperi limitis*, Cockrum and Fitch, Univ. Kansas Publ., Mus. Nat. Hist., 5:290, November 10.

MARGINAL RECORDS.—New Mexico: Copper Canyon, Magdalena Mts.; San Mateo Peak, 10,000 ft., San Mateo Mts.; type locality.

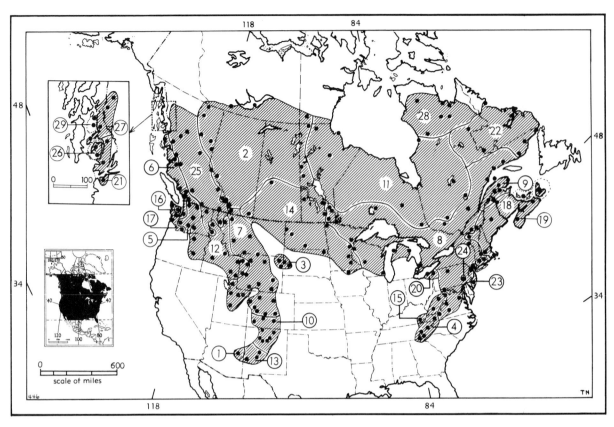

Map 446. *Clethrionomys gapperi.*

Guide	6. *C. g. caurinus*	14. *C. g. loringi*	22. *C. g. proteus*
to	7. *C. g. galei*	15. *C. g. maurus*	23. *C. g. rhoadsii*
subspecies	8. *C. g. gapperi*	16. *C. g. nivarius*	24. *C. g. rupicola*
1. *C. g. arizonensis*	9. *C. g. gaspeanus*	17. *C. g. occidentalis*	25. *C. g. saturatus*
2. *C. g. athabascae*	10. *C. g. gauti*	18. *C. g. ochaceus*	26. *C. g. solus*
3. *C. g. brevicaudus*	11. *C. g. hudsonius*	19. *C. g. pallescens*	27. *C. g. stikinensis*
4. *C. g. carolinensis*	12. *C. g. idahoensis*	20. *C. g. paludicola*	28. *C. g. ungava*
5. *C. g. cascadensis*	13. *C. g. limitis*	21. *C. g. phaeus*	29. *C. g. wrangeli*

Clethrionomys gapperi loringi (V. Bailey)

1897. *Evotomys gapperi loringi* V. Bailey, Proc. Biol. Soc. Washington, 11:125, May 13, type from Portland, Traill Co., North Dakota.

1929. *Clethrionomys gapperi loringi*, B. Bailey, Jour. Mamm., 10:162, May 9.

MARGINAL RECORDS.—Saskatchewan: Wingard. Manitoba (Soper, 1961:197, 198): Duck Mtn.; Icelandic River; Pine Ridge. Minnesota: Kittson County; Hinckley; Minneapolis; 2 mi. N, 4 mi. W Wanamingo (Heaney and Birney, 1975:32). Iowa: 5 mi. E [= ESE] Forest City. South Dakota: Fort Sisseton. North Dakota: Cannon Ball; 8 mi. N, 7 mi. W Killdeer (Genoways and Jones, 1972:28); Buford. Alberta: Cypress Hills, Eagle Butte.

Clethrionomys gapperi maurus R. Kellogg

1939. *Clethrionomys gapperi maurus* R. Kellogg, Proc. Biol. Soc. Washington, 52:37, March 11, type from Black Mts., 4½ mi. SE Lynch, 4100 ft., Harlan Co., Kentucky.

MARGINAL RECORDS.—Kentucky: type locality. *Virginia: Big Stone Gap.*

Clethrionomys gapperi nivarius (V. Bailey)

1897. *Evotomys nivarius* V. Bailey, Proc. Biol. Soc. Washington, 11:136, May 13, type from NW slope Mt. Ellinor, 4000 ft., Olympic Mts., Washington.

1929. *Evotomys gapperi nivarius*, Taylor and Shaw, Occas. Pap. Charles R. Conner Mus., State Coll. Washington, 2:23, December. See also Johnson and Ostenson, Jour. Mamm., 40:574, 575, November 20, 1959.

MARGINAL RECORDS.—Washington: *Lake Sutherland;* Staircase, on Lake Cushman; *Sol Duc Hot Springs.*

Clethrionomys gapperi occidentalis (Merriam)

1890. *Evotomys occidentalis* Merriam, N. Amer. Fauna, 4:25, October 8, type from Aberdeen, Grays Harbor Co., Washington.

1929. *Evotomys gapperi occidentalis,* Taylor and Shaw, Occas. Pap. Charles R. Conner Mus., State Coll. Washington, 2:23, December. See also Johnson and Ostenson, Jour. Mamm., 40:574, 575, November 20, 1959.
1894. *Evotomys pygmaeus* Rhoads, Proc. Acad. Nat. Sci. Philadelphia, 46:284, October 23, type from mouth Nisqually River, Pierce Co., Washington.

MARGINAL RECORDS.—British Columbia: *Port Moody* (Sheppe, 1960:173). Washington: *Lake Whatcom;* Cottage Lake; *Steilacoom;* 15 mi. N Carson; Ilwaco; Ozette Lake; near Shelton. British Columbia (Sheppe, 1960:173): *Point Grey; Stanley Park, Vancouver.*

Clethrionomys gapperi ochraceus (Miller)

1894. *Evotomys gapperi ochraceus* Miller, Proc. Boston Soc. Nat. Hist., 26:193, March 24, type from Alpine Garden, near head Tuckerman's Ravine, Mt. Washington, 5500 ft., Coos Co., New Hampshire.
1929. [*Clethrionomys gapperi*] *ochraceus,* Goodwin, Jour. Mamm., 10:243, August 10.

MARGINAL RECORDS.—Prince Edward Island: Kensington, thence S along Atlantic Coast to Massachusetts: Wareham; Petersham. New York: St. Huberts; *Archer and Anna Huntington Wildlife Forest Station, near Newcomb* (Patric, 1962:200, as *Clethrionomys* only). Quebec: Hatley. Maine: South Twin Lake. New Brunswick: *Andover;* Trousers Lake.

Clethrionomys gapperi pallescens Hall and Cockrum

1940. *Clethrionomys gapperi rufescens* R. W. Smith, Amer. Midland Nat., 24:233, July, type from Wolfville, Kings Co., Nova Scotia. Not *Arvicola rufescens* de Sélys-Longchamps, 1836, a synonym of *Clethrionomys glareolus glareolus* Schreber, 1780.
1952. *Clethrionomys gapperi pallescens* Hall and Cockrum, Univ. Kansas Publ., Mus. Nat. Hist., 5:302, November 17, a renaming of *C. g. rufescens* R. W. Smith.

MARGINAL RECORDS.—Nova Scotia: Frizzleton, down E coast Cape Breton Island and Nova Scotia and back up W coast to Albany; thence to James River.

Clethrionomys gapperi paludicola Doutt

1941. *Clethrionomys gapperi paludicola* Doutt, Proc. Biol. Soc. Washington, 54:162, December 8, type from Pymatuning Swamp, 1000 ft., 4 mi. W Linesville, Crawford Co., Pennsylvania.

MARGINAL RECORDS.—Ohio: Conneaut Creek [= Farnham]; *Padanaram.* Pennsylvania: type locality.

Clethrionomys gapperi phaeus (Swarth)

1911. *Evotomys phaeus* Swarth, Univ. California Publ. Zool., 7:127, January 12, type from Marten Arm, Boca de Quadra, Alaska.
1952. *Clethrionomys gapperi phaeus,* Hall and Cockrum, Univ. Kansas Publ., Mus. Nat. Hist., 5:302, November 17.

MARGINAL RECORDS.—Alaska: Chickamin River; type locality. British Columbia: Fort [= Port] Simpson.

Clethrionomys gapperi proteus (Bangs)

1897. *Evotomys proteus* Bangs, *in* V. Bailey, Proc. Biol. Soc. Washington, 11:137, May 13, type from Rigoulette, Labrador. (See Harper, Univ. Kansas Mus. Nat. Hist., Miscl. Publ., 27:61, August 11, 1961.)
1938. *Clethrionomys gapperi proteus,* C. F. Jackson, Jour. Mamm., 19:433, November 14.

MARGINAL RECORDS (Harper, 1961:61–63, 68, unless otherwise noted).—Labrador: Nain; type locality; Black Bay (Hall and Kelson, 1959:716). Quebec: Kecarpoui (Weaver, 1940:420); Washicoutai. Labrador: Ashuanipi Lake. Quebec: *Mollie T. Lake;* Leroy Lake. Not found: Carol Lake; Opiscoteo Lake area.

Clethrionomys gapperi rhoadsii (Stone)

1893. *Evotomys gapperi rhoadsii* Stone, Amer. Nat., 27:55, January, type from Mays Landing, Atlantic Co., New Jersey (about halfway between Mare Run and Mays Landing dam on Great Egg Harbor River—see Miller and Kellogg, Bull. U.S. Nat. Mus., 205:570, March 3, 1955).
1942. *Clethrionomys gapperi rhoadsii,* Bole and Moulthrop, Sci. Publs., Cleveland Mus. Nat. Hist., 5:153, September 11.

MARGINAL RECORDS.—New York: southern New York. New Jersey: *Lakehurst;* type locality; *Tuckahoe; Port Norris; Mauricetown; Ancora; Medford.*

Clethrionomys gapperi rupicola E. L. Poole

1949. *Clethrionomys gapperi rupicola* E. L. Poole, Not. Naturae, Acad. Nat. Sci. Philadelphia, 212:2, January 21, type from Pinnacle, Berks Co., Pennsylvania. Known only from Kittatinny Ridge of Berks and Schuylkill cos., Pennsylvania.

Clethrionomys gapperi saturatus (Rhoads)

1894. *Evotomys gapperi saturatus* Rhoads, Proc. Acad. Nat. Sci. Philadelphia, 46:284, October 23, type from Nelson, on the banks of a small stream flowing into Kootenai [sic] Lake, British Columbia.
1933. [*Clethrionomys gapperi*] *saturatus,* Whitlow and Hall, Univ. California Publ. Zool., 40:265, September 30.

MARGINAL RECORDS.—British Columbia: Bear Lake; *Tetana Lake* (Cowan and Guiguet, 1965:206); Summit Lake; Moose Pass (Cowan and Guiguet, 1965:206); *Yellowhead Lake;* Glacier; Paradise Mine (Cowan and Guiguet, 1965:207); Kitchener. Montana: Prospect Creek. Idaho: Craig Mtn. Washington: Sherman Creek Pass. British Columbia: Anarchist Mtn.; Botanie Lake; Stuie; W end Eutsuk Lake; Lakelse; Hazelton.

Clethrionomys gapperi solus Hall and Cockrum

1952. *Clethrionomys gapperi solus* Hall and Cockrum, Univ. Kansas Publ., Mus. Nat. Hist., 5:304, November 17, type from Loring, Revillagigedo Island, Alaska.

MARGINAL RECORDS.—Alaska (Revillagigedo Island): type locality; mouth Fish Creek, Ketchikan.

Clethrionomys gapperi stikinensis Hall and Cockrum

1952. *Clethrionomys gapperi stikinensis* Hall and Cockrum, Univ. Kansas Publ., Mus. Nat. Hist., 5:305, November 17, type from Stikine River at Great Glacier, British Columbia.

MARGINAL RECORDS.—British Columbia: type locality; Stikine River at Flood Glacier. Alaska: Bradfield Canal; Helm Bay.

Clethrionomys gapperi ungava (V. Bailey)

1897. *Evotomys ungava* V. Bailey, Proc. Biol. Soc. Washington, 11:130, May 13, type from Fort Chimo, Quebec.
1939. *Clethrionomys gapperi ungava*, Anderson, Ann. Rept. Provancher Soc. Nat. Hist. of Canada, Quebec, for 1938, p. 83, February 28.

MARGINAL RECORDS (Harper, 1961:70, 71).—Quebec: Payne Lake; Irony Lake S of Leaf Bay; type locality; Lac Aulneau; Lac Aigneau; Lower Seal Lake.

Clethrionomys gapperi wrangeli (V. Bailey)

1897. *Evotomys wrangeli* V. Bailey, Proc. Biol. Soc. Washington, 11:120, May 13, type from Wrangell, Wrangell Island, Alaska.
1952. *Clethrionomys gapperi wrangeli*, Hall and Cockrum, Univ. Kansas Publ., Mus. Nat. Hist., 5:303, November 17.

MARGINAL RECORDS.—Alaska: Sergief Island; type locality.

Clethrionomys californicus
California Red-backed Mouse

External measurements: 155–160; 47–54; 17–21. Compared with *C. gapperi* immediately to the north in Washington, *C. californicus* is described as larger, upper parts paler (less red), tail sharply instead of indistinctly bicolored, and underparts white, whitish, or buffy white, instead of salmon. Separation of *C. californicus* and *C. gapperi* as two species rests on the statement of Johnson and Ostenson (1959:574, 575) that "all *Clethrionomys* north of the Columbia River should be referred to the species *gapperi* and the populations in western Oregon and northern California should be retained in the species *californicus.*" Degree of difference between populations near each other but from opposite sides of the river remains to be ascertained.

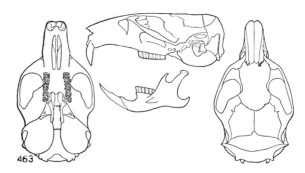

Fig. 463. *Clethrionomys californicus obscurus,* head Doggett Creek, 5800 ft., Siskiyou Mts., Siskiyou Co., California, No. 69490 M.V.Z., ♀, X 1½.

Clethrionomys californicus californicus (Merriam)

1890. *Evotomys californicus* Merriam, N. Amer. Fauna, 4:26, October 8, type from Eureka, Humboldt Co., California.

MARGINAL RECORDS.—Oregon: Astoria; Camp Millard, near Estacada; Wells. California: near Fair Oaks; *Willits;* 7 mi. W Cazadero, thence north along coast to point of beginning.

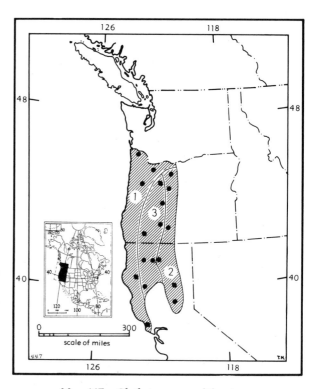

Map 447. *Clethrionomys californicus.*

Guide to subspecies   2. *C. c. mazama*
1. *C. c. californicus*   3. *C. c. obscurus*

Clethrionomys californicus mazama (Merriam)

1897. *Evotomys mazama* Merriam, Proc. Biol. Soc. Washington, 11:71, April 21, type from Crater Lake, 7000 ft., Mt. Mazama, Oregon.
1937. *Clethrionomys californicus mazama*, V. Bailey, N. Amer. Fauna, 55:192, August 29.

MARGINAL RECORDS.—Oregon: Mt. Hood; The Three Sisters; type locality. California: Mt. Shasta; vic. Lassen Peak; near Quincy.

Clethrionomys californicus obscurus (Merriam)

1897. *Evotomys obscurus* Merriam, Proc. Biol. Soc. Washington, 11:72, April 21, type from Prospect, 2600 ft., Upper Rogue River Valley, Oregon.
1937. *Clethrionomys californicus obscurus*, V. Bailey, N. Amer. Fauna, 55:192, August 29.

MARGINAL RECORDS.—Oregon: W base Mt. Jefferson; Lane County (Quay, 1962:303, as *C. occidentalis* only); W side Crater Lake. California: Castle Lake; 2 mi. S South Yolla Bolly Mtn.; Jackson Lake. Oregon: Grants Pass.

Genus Phenacomys Merriam
Heather Vole and Tree Mice

Revised by A. B. Howell, N. Amer. Fauna, 48:iv + 66, 7 pls., 11 figs., October 12, 1926.

1889. *Phenacomys* Merriam, N. Amer. Fauna, 2:28, October 30. Type, *Phenacomys intermedius* Merriam.

External measurements: 130–193; 26–87; 16–22. Basal length, 22.3–25.7; zygomatic breadth, 13.9–16.3. Upper parts gray to cinnamon. Mammae usually 8 (p., 4; i., 4); plantar pads, 6. Temporal ridges widely separated in adult skull; bullae of medium size, without internal spongy tissue; palate neither transversely continuous nor shelflike; cheek-teeth rooted in adults; enamel pattern of lower molars unique, chiefly in depth of re-entrant angles, as compared with the external angles.

KEY TO SPECIES OF PHENACOMYS

1. Tail less than 50 mm.   *P. intermedius,* p. 785
1'. Tail more than 50 mm.
    2. Incisors strongly recurved (see Fig. 467); tail thick, quite hairy; arboreal.
                     P. longicaudus, p. 788
    2'. Incisors not strongly recurved (see Fig. 465); tail slender, scantily haired; terrestrial. . . . . *P. albipes,* p. 787

Subgenus Phenacomys Merriam

1889. *Phenacomys* Merriam, N. Amer. Fauna, 2:28, October 30. Type, *Phenacomys intermedius* Merriam.

Phenacomys intermedius
Heather Vole

External measurements: 130–153; 26–41; 16–18.2; 11–17; weight, up to 40.8 grams. Upper parts agouti gray to brownish but face, in some subspecies, yellowish; underparts whitish; tail sharply bicolored.

The species *Phenacomys intermedius,* of which nine subspecies are listed below, poses an unsolved problem in that it may be a composite of two or three allopatric species. A. B. Howell's (1926) excellent revision of the genus was based on too few specimens from too few areas to enable him to solve the problem. He tentatively recognized three species; Crowe (1943:403) thought there was only one; and Cowan and Guiguet (1965:212) thought there might be at least two. Solving the problem will be worthwhile and likely will require obtaining specimens from a critical area in Alberta and from farther east where the geographic ranges of *P. i. mackenzii* and *P. i. celatus* would be expected to meet.

Phenacomys intermedius celatus Merriam

1889. *Phenacomys celatus* Merriam, N. Amer. Fauna, 2:33, October 30, type from Godbout, Quebec.
1953. *Phenacomys intermedius celatus,* Hall and Cockrum, Univ. Kansas Publ., Mus. Nat. Hist., 5:395, January 15.
1889. *Phenacomys latimanus* Merriam, N. Amer. Fauna, 2:34, October 30, type from Fort Chimo, Quebec.
1889. *Phenacomys ungava* Merriam, N. Amer. Fauna, 2:35, October 30, type from Fort Chimo, Quebec. (This name

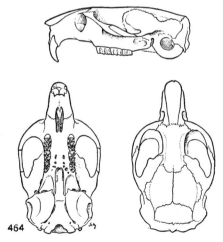

Fig. 464. *Phenacomys intermedius celsus,* Humphreys Basin, 10,800 ft., Sierra Nevada, Fresno Co., California, No. 41125 M.V.Z., ♂, X 1½.

selected to apply to the subspecies by Miller, Proc. Biol. Soc. Washington, 11:84, April 21, 1897, and was used by Howell (N. Amer. Fauna, 48:25, October 12, 1926), who revised the members of genus *Phenacomys*.)

MARGINAL RECORDS.—Quebec (Harper, 1961: 72): Irony Lake, S of Leaf Bay; Fort Chimo; Lac Aulneau; Harrington Harbor; St. Charles Point; type locality. Ontario: Nipissing; Algoma. Minnesota: Ely [= Lake States Experimental Forest, Timm, 1975:39]. Ontario: Kenora; Lake Abitibi. Quebec: Authier-Nord region (Foster, 1961:181, as *P. ungava* only); Merry Island (Edwards, 1963:7, as *P. ungava* only); Golfcourse Cove, Richmond Gulf (*ibid.*); Seal Lake (Harper, 1961:72).

Phenacomys intermedius celsus A. B. Howell

1923. *Phenacomys intermedius celsus* A. B. Howell, Proc. Biol. Soc. Washington, 36:158, May 1, type from Muir Meadow, 9300 ft., Tuolumne Meadows, Yosemite National Park, California.

MARGINAL RECORDS.—Nevada: ½ mi. S Grass Lake. California: Ten Lakes, Yosemite National Park; Mt. Lyell; Humphreys Basin; Fletcher Creek, Yosemite National Park; Pyramid Peak.

Phenacomys intermedius crassus Bangs

1900. *Phenacomys celatus crassus* Bangs, Proc. New England Zool. Club, 2:39, September 20, type from Rigolet, Hamilton Inlet, Labrador.
1943. *Phenacomys intermedius crassus*, Crowe, Bull. Amer. Mus. Nat. Hist., 80:404, February 4.

MARGINAL RECORDS (Harper, 1961:72).— Quebec: Leroy Lake; Attikamagen Lake. Labrador: type locality; L'Anse au Loup; Ashuanipi Lake.

Phenacomys intermedius intermedius Merriam

1889. *Phenacomys intermedius* Merriam, N. Amer. Fauna, 2:32, October 30, type from a basaltic plateau, 5500 ft., about 20 mi. NNW Kamloops, British Columbia.
1891. *Phenacomys orophilus* Merriam, N. Amer. Fauna, 5:65, July 30, type from near head Timber Creek, Lemhi Mts. [= "Salmon River Mountains"], Lemhi Co., Idaho.
1894. *Phenacomys truei* J. A. Allen, Bull. Amer. Mus. Nat. Hist., 6:331, November 7, type from Black Hills, now Laramie Mts., Wyoming.
1897. *Phenacomys preblei* Merriam, Proc. Biol. Soc. Washington, 11:45, March 16, type from side of Twin Peak, 9000 ft., near Longs Peak, Boulder Co., Colorado.
1899. *Phenacomys constablei* J. A. Allen, Bull. Amer. Mus. Nat. Hist., 12:4, March 4, type from Telegraph Creek, British Columbia.

MARGINAL RECORDS.—British Columbia (Cowan and Guiguet, 1965:211, unless otherwise noted): Telegraph Creek; mts. near head Chapa-atan River (Howell, 1926:19); Tetana Lake; Wells Gray Park; Glacier; Paradise Mine; Newgate. Montana: Big

Snowy Mts.; Bear Tooth Mts. Wyoming (Long, 1965a:647): Beartooth Lake; *4 mi. N, 19 mi. W Lander;* 3 mi. N, 18 mi. W Lander; *17 mi. S, 6½ mi. W Lander;* Merna; 11½ mi. S, 2 mi. E Robertson; Bridgers Pass, 18 mi. SW Rawlins, 7500 ft.; "Black Hills" (Laramie Mts.); 10 mi. E Laramie. Colorado: 9 mi. N Colorado Springs, 7200 ft. (Armstrong, 1972:235). New Mexico (Findley, *et al.*, 1975:253, as *P. intermedius* only): E slope Taos Mts.; Pecos Baldy; Santa Fe; 11½ mi. NE Chama. Colorado: Wolf Creek Pass, 10,850 ft. (Armstrong, 1972:235); 28 mi. E Grand Junction (Anderson, 1959:412). Utah: Provo River, 3 mi. N Soapstone Guard Station, Wasatch National Forest; Spectacle Lake; Mt. Timpanogos, 10,200 ft. Idaho: Sawtooth City. Oregon: Blue Mts. California: Mt. Shasta. Oregon: Diamond Lake. Washington: Loomis [Okanogan Co.]. British Columbia: Fairview (Cowan and Guiguet, 1965:211); Kamloops (*ibid.*); *type locality;* Wistaria, near Burns Lake; Hazelton (Howell, 1926:19).

Phenacomys intermedius laingi Anderson

1942. *Phenacomys intermedius laingi* Anderson, Canadian Field-Nat., 56:59, June 8, type from Kimsquit River, Cornice Creek, near head Dean Inlet, 52° 54' N and 127° W, 2500 ft., British Columbia.

MARGINAL RECORDS.—British Columbia: head Eutsuk Lake; Lac la Hache; *Atnarko River;* Cariboo Mtn. near Stuie (Cowan and Guiguet, 1965:211) *type locality.*

Phenacomys intermedius levis A. B. Howell

1923. *Phenacomys intermedius levis* A. B. Howell, Proc. Biol. Soc. Washington, 36:157, May 1, type from Saint Marys Lake, Glacier Co., Montana.

MARGINAL RECORDS.—Alberta: head Smoky River; Thoral Creek, 6000 ft.; Braggs Creek; Waterton Lakes National Park. Montana: Midvale; *Summit;* Kintla Lake. British Columbia (Cowan and Guiguet, 1965:211): Tornado Pass; Assiniboine; Field; Moose River.

Phenacomys intermedius mackenzii Preble

1902. *Phenacomys mackenzii* Preble, Proc. Biol. Soc. Washington, 15:182, August 6, type from Fort Smith, Slave River, Mackenzie.
1943. *Phenacomys intermedius mackenzii*, Crowe, Bull. Amer. Mus. Nat. Hist., 80:403, February 4.

MARGINAL RECORDS.—Mackenzie: Lake St. Croix; Fort Reliance. Manitoba: Churchill; *Sutton Lake* (Foster, 1961:197, as *P. ungava* only); Flin Flon (Soper, 1961:198). Alberta: Mud Lake near Macleod; *Maycroft, 4700 ft.;* Bearberry Creek, W of Olds; Entrance, 3000 ft.; Muskeg Creek. British Columbia: Tupper Creek; Charlie Lake; Atlin. Yukon (Youngman, 1975:88); SW end Dezadeash Lake; Christmas Bay, Kluane Lake; Sheldon Lake, Canol Road, Mile 222.

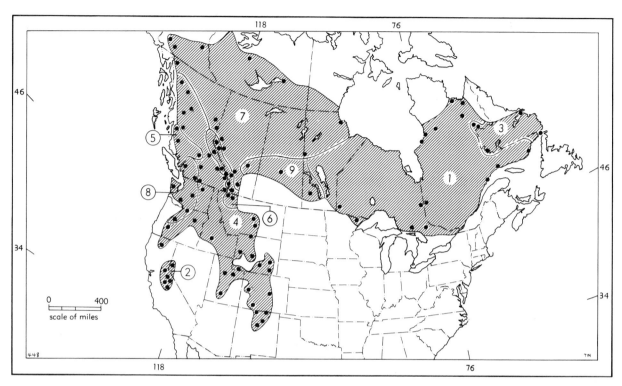

Map 448. *Phenacomys intermedius.*

1. *P. intermedius celatus*
2. *P. intermedius celsus*
3. *P. intermedius crassus*
4. *P. intermedius intermedius*
5. *P. intermedius laingi*
6. *P. intermedius levis*
7. *P. intermedius mackenzii*
8. *P. intermedius oramontis*
9. *P. intermedius soperi*

Phenacomys intermedius oramontis Rhoads

1895. *Phenacomys oramontis* Rhoads, Amer. Nat., 29:941, October, type from Church Mtn. (Lihumitson Mtn.), 6000 ft., Lihumitson Park, Mt. Baker Range, New Westminster District, British Columbia.
1942. *Phenacomys intermedius oramontis*, Anderson, Canadian Field-Nat., 56:59, June 8.
1899. *Phenacomys olympicus* Elliot, Field Columb. Mus., Publ. 30, Zool. Ser., 1:225, February 2, type from Happy Lake, 5000 ft., Olympic Mts., Clallam Co., Washington. Anderson (Canadian Field-Nat., 56:59, June 8, 1942) considers *P. olympicus* to be a synonym of *P. i. oramontis* Rhoads.
1899. *Microtus (Lagurus) pumilus* Elliot, Field Columb. Mus., Publ. 30, Zool. Ser., 1:226, February 2, type from Happy Lake, Olympic Mts., Clallam Co., Washington. Dalquest (Univ. Kansas Publ., Mus. Nat. Hist., 2:340, April 9, 1948) considers *M. pumilus* to be a synonym of *P. i. oramontis* Rhoads.

MARGINAL RECORDS.—British Columbia: Mt. Whistler (Cowan and Guiguet, 1965:212); *Avalanche Pass, 5500 ft.;* Manning Park. Washington: near Stephens Pass. Oregon: Deschutes River; Three Sisters; *Mt. Hood.* Washington: Canyon Creek; Boulder Lake.

Phenacomys intermedius soperi Anderson

1942. *Phenacomys ungava soperi* Anderson, Canadian Field-Nat., 56:58, June 8, type from near Swanson Creek, in middle of sec. 34, T. 19, R. 17, Riding Mountain National Park, Manitoba.
1943. *Phenacomys intermedius soperi*, Crowe, Bull. Amer. Mus. Nat. Hist., 80:404, February 4.

MARGINAL RECORDS.—Saskatchewan: Emma Lake, SE of Prince Albert National Park. Manitoba: type locality. Alberta: Battle Lake (head Battle River).

Phenacomys albipes Merriam
White-footed Vole

1901. *Phenacomys albipes* Merriam, Proc. Biol. Soc. Washington, 14:125, July 19, type from Redwoods, near Arcata, Humboldt Bay, Humboldt Co., California.

Average external measurements of six adult males: 171 (165–181); 63 (62–71); 19.5 (19–20). Upper parts close to Prout's Brown, with mixture of black-tipped hairs. Underparts clear gray, often washed with pinkish buff in autumn.

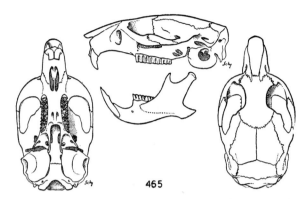

Fig. 465. *Phenacomys albipes*, Old Fort Clatsop, 100 ft., Clatsop Co., Oregon, No. 94499 M.V.Z., ♀, X 1½.

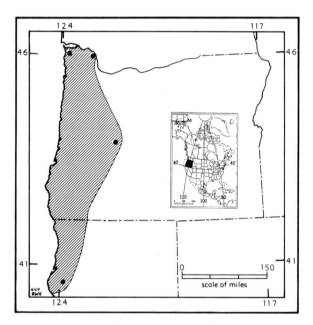

Map 449. *Phenacomys albipes*.

MARGINAL RECORDS (Maser and Johnson, 1968:25, unless otherwise noted).—Oregon: Old Fort Clatsop; 7 mi. SE Ranier; 23 mi. SE Vida. California: type locality, thence along coast to point of beginning.

Subgenus **Arborimus** Taylor

1915. *Arborimus* Taylor, Proc. California Acad. Sci., ser. 4, 5:119, December 30. Type, *Phenacomys longicaudus* True.

Taylor (1915:119, *supra*) proposed the name *Arborimus* as a subgenus; A. B. Howell (1926:5) placed the name as a synonym of *Phenacomys;* M. L. Johnson (1973:239–244) elevated the name to generic rank. Each of the three mammalogists clearly stated reasons for his handling of the

name *Arborimus*. Because the degree of difference between *Arborimus* and *Phenacomys* may be about the same as between some subgenera in the genus *Microtus, Arborimus* here is provisionally accorded subgeneric rank.

Phenacomys longicaudus
Reddish Tree Mouse

External measurements of six adult males and five adult females of *P. l. longicaudus:* 166 (158–176), 182 (170–187); 67 (60–72), 73 (66–83); 20 (19–21), 21 (21–22). Corresponding measurements of one male and one female of *P. l. silvicola:* 193, 191; 87, 81; 20, 22. Upper parts uniform cinnamon, near ochraceous tawny with many hairs sparingly tipped with black in *P. l. longicaudus;* upper parts near cinnamon brown in *P. l. silvicola;* underparts whitish and tail long and well haired in both subspecies.

Placing *P. silvicola* and *P. longicaudus* as subspecies of one species is a provisional arrangement. The existing uncertainty about the relationship of the two kinds probably could be removed by studying in one place most of the existing specimens of the two taxa. So doing also could help to clarify the worth of the subgenera proposed for recognition in the genus *Phenacomys*.

A. B. Howell (1926:35), on the basis of the few specimens then known, considered the possibility that *P. silvicola* was only a subspecies of *P. longicaudus* but decided on specific status in the absence of specimens that could be identified as intergrades. M. L. Johnson (1968:27), on the basis of degree of resemblance in serum proteins, suggested subspecific relationship of the two taxa but did not employ a name combination to implement his suggestion, something that Olterman and Verts (1972:29) failed to realize when they wrote "The species *Phenacomys longicaudus* and *Phenacomys silvicola* have been combined to form the single species *Arborimus longicaudus* [by] Johnson, 1968:27. . . ."

Phenacomys longicaudus longicaudus True

1890. *Phenacomys longicaudus* True, Proc. U.S. Nat. Mus., 13:303, November 15, type from Marshfield, Coos Co., Oregon.

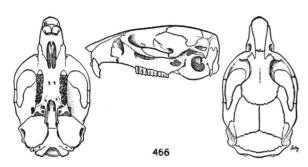

Fig. 466. *Phenacomys longicaudus longicaudus*, 1 mi. S Occidental, Sonoma Co., California, No. 94761 M.V.Z., ♀, X 1½.

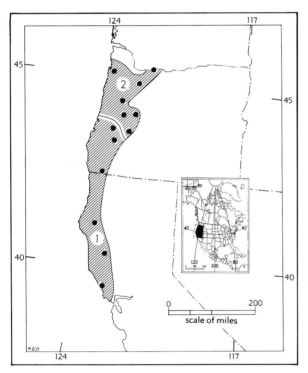

Map 450. *Phenacomys longicaudus.*

1. *P. l. longicaudus*   2. *P. l. silvicola*

MARGINAL RECORDS.—Oregon (Olterman and Verts, 1972:29): 3 mi. E Elkton; near Roseburg; 23 mi. E Gold Beach. California: 1 mi. W Bridgeville; Mt. Sanhedrin; 1 mi. S Occidental, thence north along coast to point of beginning.

Phenacomys longicaudus silvicola A. B. Howell

1921. *Phenacomys silvicolus* A. B. Howell, Jour. Mamm., 2:98, May 2, type from 5 mi. SE Tillamook, Tillamook Co., Oregon.

MARGINAL RECORDS.—Oregon: 1 mi. E Cascade Locks (Olterman and Verts, 1972:29, as *Arborimus lon-*

gicaudus); near Molalla (Brown, 1964:648); *8 mi. N Corvallis* (Brown, 1964:647); near Corvallis; *Mary's Peak, 15 mi. W Corvallis* (Brown, 1964:647); Eugene (Brown, 1964:648); Vida (Olterman and Verts, 1972:30, as *Arborimus longicaudus*); 17 mi. SE Cottage Grove (*ibid.*), thence west to coast and up coast to *Netarts*; 3 to 4 mi. W Tillamook; *type locality.*

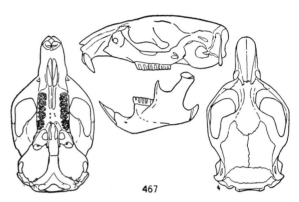

Fig. 467. *Phenacomys longicaudus silvicola*, Netarts, Tillamook Co., Oregon, No. 61411 M.V.Z., sex?, X 1½.

Genus **Microtus** Schrank—Meadow Voles

North American species revised by V. Bailey, N. Amer. Fauna, 17:1–88, 5 pls., 17 figs., June 6, 1900.

1798. *Microtus* Schrank, Fauna Boica . . . , 1(Abth. 1):72. Type, *Microtus terrestris* Schrank [= *Mus arvalis* Pallas].
1799. *Arvicola* Lacépède, Tableau des divisions, sous divisions, ordres, et genres des mannifères; 10. Type, *Mus amphibius* Linnaeus.
1817. *Mynomes* Rafinesque, Amer. Month. Mag., 2:45. Type, *Mynomes pratensis* Rafinesque [= *Mus pennsylvanica* Ord].
1830. *Psammomys* Le Conte, Ann. Lyc. Nat. Hist., New York, 3:132 (for 1829). Type, *Psammomys pinetorum* Le Conte. Not *Psammomys* Cretzschmar, 1828.
1831. *Pitymys* McMurtrie, The animal kingdom . . . by the Baron Cuvier . . . Amer. ed., 1:434. Type, *Psammomys pinetorum* Le Conte.
1831. *Ammomys* Bonaparte, Saggio di una distribuzione methodica degli animali vertebrati, p. 20. Type, *Psammomys pinetorum* Le Conte.
1836. *Hemiotomys* de Sélys-Longchamps, Essai monog. sur les campagnols des environs de Liége, p. 7. Included *fulvus* [= *Mus arvalis* Pallas] and *amphibius* [= *Mus terrestris* Linnaeus (= *Arvicola terrestris* of authors)].
1849. *Neodon* Hodgson, Ann. Mag. Nat. Hist., ser. 1, 3:203. Type, *Neodon sikimensis* Hodgson.
1857. *Agricola* Blasius, Fauna der Wirbelthiere Deutschlands, Säugethiere, p. 334. Type, *Mus agrestis* Linnaeus.
1858. *Chilotus* Baird, Mammals, *in* Repts. Expl. Surv. . . . , 8(1):516, July 14. Type, *Arvicola oregoni* Bachman.
1858. *Pedomys* Baird, Mammals, *in* Repts. Expl. Surv. . . . , 8(1):517, July 14. Type, *Arvicola austerus* Le Conte [= *Hypudaeus ochrogaster* Wagner].

1863. *Phaiomys* Blyth, Jour. Asiatic Soc. Bengal, 32, 1:89. Type, *Phaiomys leucurus* Blyth.

1887. *Lasiopodomys* Lataste, Ann. Mus. Civ. Storia Nat. Genova, 2a, 4:268. Type, *Arvicola brandti* Radde.

1890. *Campicola* Schulze, Schrift. Naturw. Vereins der Harzes in Wernigerode, 5:24. Contained *arvalis* and *agrestis* according to Ellerman and Morrison-Scott, Checklist of Palaearctic and Indian mammals. . . . British Museum (Nat. History), p. 690, November 19, 1951.

1894. *Tetramerodon* Rhoads, Proc. Acad. Nat. Sci. Philadelphia, p. 282, Oct. 23. Type, *Arvicola tetramerus* Rhoads.

1894. *Aulacomys* Rhoads, Amer. Nat., 28:182, February. Type, *Aulacomys arviculoides* Rhoads.

1898. *Orthriomys* Merriam, Proc. Biol. Soc. Washington, 12:106, April 30. Type, *Microtus umbrosus* Merriam.

1898. *Herpetomys* Merriam, Proc. Biol. Soc. Washington, 12:107, April 30. Type, *Microtus guatemalensis* Merriam.

1899. *Euarvicola* Aeloque, Faune de France, Mammals, p. 49. Type, *Mus agrestis* Linnaeus.

1901. *Stenocranius* Kastschenko, Ann. Imp. Mus. St. Pétersbourg, 6:167. Type, *Microtus slowzowi* Poljakoff.

1908. *Chinomys* Miller, Ann. Mag. Nat. Hist., ser. 8, 1:97. Type, *Arvicola nivalis* Martins.

1911. *Proedromys* Thomas, Proc. Zool. Soc. London, p. 177, March 22. Type *Proedromys bedfordi* Thomas.

1914. *Alexandromys* Ognev, Moskva Dnev. Zool. otd. obsc. liub. jest., 2:109. Type, *Microtus pelliceus* Thomas.

1919. *Abusticola* Shidlovsky, Tiflis Bull. Terr. Exper. Stat., 2:21. Type, *Microtus rubelianus* Shidlovsky [= *Microtus (Pitymys) majori* Thomas].

1933. *Sumeriomys* Argyropulo, Zeitschr. f. Säugetierkunde, 8:180. Type, *Mus socialis* Pallas.

1941. *Lemmimicrotus* Tokuda, Trans. Biogeog. Soc. Japan, 4(1):68, December. Type, *Arvicola mandarinus* Milne-Edwards.

External measurements: 101–261; 15–93; 13–30. Basal length, 19.3–36.0; zygomatic breadth, 10.7–23.0. Pelage usually long and loose; tail usually less than half as long as head and body, but not excessively reduced; ears short and rounded and nearly concealed by pelage. Mammae usually 2–2 = 8 (exceptions: 1–1 = 4, *mexicanus*-group; 1–2 = 6, *ochrogaster*-group; 2–0 = 4, *umbrosus;* and 2–1 = 6, *guatemalensis*-group). Lower incisors with roots extending far behind, and on outer side of, molar series; upper incisors not grooved; molars rootless, with outer and inner re-entrant angles approx. equal; the pattern, a series of loops and triangles of enamel surrounding areas of dentine, varies somewhat with the species.

Arrangement of the living species of voles according to subgenera so as to show natural relationships will be easier when more is known about the fossil relatives. Hinton's (1926) monograph helped immensely. Since 1926 much new information (see Camp, *et al.*, 1964, and previous volumes) has appeared in print, but gaps remain in the fossil record. Since 1926 neozoologists also have contributed new information about living microtines that is useful for anyone who attempts to make a natural classification—see the "Literature cited" in Hooper and Hart (1962) as well as their account of the glans penis and synopsis of cranial and external features.

KEY TO NORTH AMERICAN SPECIES OF MICROTUS

1. Last lower molar (m3) with 2 transverse loops and 1 or 2 median triangles; plantar tubercles, 5.
  2. M3 with 3 or 5 triangles; mammae 6 (? in *oaxacensis*); tail *ca.* 25% of total length.
    3. Last upper molar (M3) with 3 triangles; last lower molar (m3) with 2 median triangles; mammae 6. . . . . *M. guatemalensis,* p. 803
    3'. Last upper molar (M3) with 5 triangles; last lower molar (m3) with 1 median triangle; mammae possibly 6. . . . . *M. oaxacensis,* p. 800
  2'. Last upper molar (M3) with 2 triangles; mammae 4; tail *ca.* 33% of total length. *M. umbrosus,* p. 800
1'. Last lower molar (m3) with 3 transverse loops and no closed triangles; plantar tubercles, 5 or 6.
  4. M3 with 3 closed triangles (except: *M. breweri* with 2 usually confluent; *M. chrotorrhinus* with 5 closed); mammae 8 (except in *M. mexicanus*, which has 4).
    5. Plantar tubercles 5; side glands either on flanks or inconspicuous.
      6. Side glands conspicuous on flanks of adult males; external measurements more than 176, 55, 22. . . . . (Subgenus *Arvicola*) *M. richardsoni,* p. 820
      6'. Side glands obscure or wanting; external measurements less than 176, 55, 22. *M. oregoni,* p. 809
    5'. Plantar tubercles, 6; side glands on hips in adult males (on flanks in *M. xanthognathus*).
      7. Skull narrow as in Fig. 483 and with pronounced median crest; claws enlarged.
                               (Subgenus *Stenocranius*).
        8. Hind foot more than 21.5; confined to Hall and St. Matthew islands in Bering Sea.
                                   M. abbreviatus, p. 819
        8'. Hind foot less than 21.5; not on Hall and St. Matthew islands but instead on mainland of North America. . . . . *M. miurus,* p. 818
    7'. Skull not so narrow as in Fig. 483 and without pronounced median crest; claws not enlarged.
      9. M2 with 4 closed angular sections and a rounded posterior loop.
        10. M3 with 2 of the 3 triangles usually confluent; insular species.

11. Interparietal approx. as wide as long; pale (upper parts buffy gray); confined to Muskeget Island, Massachusetts. . . . .*M. breweri,* p. 796
11′. Interparietal notably wider than long; dark (upper parts dark yellowish bister); confined to Great Gull and Little Gull islands, New York.
M. nesophilus, p. 796
10′. M3 with 3 closed triangles. . . . .*M. pennsylvanicus,* p. 792
9′. M2 with 4 closed angular sections and no posterior loop (except irregularly in *M. californicus*).
12. Mammae, 4 (i., 1–1; p., 1–1); skull short and wide with incisive foramina not constricted (see Fig. 480). . . . .*M. mexicanus,* p. 814
12′. Mammae, 8 (i., 2–2; p., 2–2). Skull short or wide with foramina constricted or unconstricted (see figures of skulls).
13. m1 with 4 closed triangles (5 in some *M. o. sitkensis*) and rounded anterior loop. . . . .*M. oeconomus,* p. 805
13′. m1 with 5 or 6 closed triangles.
14. A pair of glands on flanks of males; nose yellowish.
15. Hind foot more than 23.5; side glands conspicuous in adult males; M3 with 3 closed triangles. . . . .*M. xanthognathus,* p. 811
15′. Hind foot less than 23.5; glands obscure or wanting; M3 with 5 closed triangles. . . . .*M. chrotorrhinus,* p. 810
14′. A pair of glands on hips of males; nose not yellowish.
16. Incisive foramina gradually tapered posteriorly (not abruptly constricted) or not constricted posteriorly (see Figs. 471 and 475).
17. Tail averaging $\frac{1}{3}$ or more of total length; skull without prominent ridging and with incisive foramina open and not constricted posteriorly, or if constricted only gradually tapered (see Fig. 475); not confined to Oregon and California and not in Baja California.
18. On Forester, Coronation, and Warren islands of SE Alaska. . . . .*M. coronarius,* p. 809
18′. Not on Forester, Coronation, and Warren islands of SE Alaska. . . . .*M. longicaudus,* p. 806
17′. Tail averaging less than $\frac{1}{3}$ of total length; skull prominently ridged and with incisive foramina wide open and not constricted posteriorly (see Fig. 471); confined to Oregon, California and Baja California. . . . .*M. californicus,* p. 801
16′. Incisive foramina abruptly constricted and narrower posteriorly than anteriorly (see Figs. 469 and 473).
19. Confined to Pacific coastal area west of Cascade Mts.
20. Fraser River of extreme SW British Columbia south to 40° lat. in California; upper parts dark brown to blackish; hind foot more than 23.7.   *M. townsendii,* p. 804
20′. Clark County, Washington, S in Oregon at least to Eugene; upper parts yellowish; hind foot less than 23.7. . . . .*M. canicaudus,* p. 800
19′. West of Cascade–Sierra Nevada mtn. chain; upper parts some shade of brownish, with buffy or grayish wash.
M. montanus, p. 797
4′. M3 with 2 closed triangles; mammae, 4 or 6.
21. Mammae, 4; fur short, fine, molelike, reddish on upper parts; tail less than 26 mm. (Subgenus *Pitymys*).
22. Hind foot of adults less than 16; upper parts pale (near tawny); geographical range Florida and extreme southern Georgia, *M. p. parvulus* a subspecies of *M. pinetorum*.
22′. Hind foot of adults more than 16; upper parts dark (russet brown or darker); geographical range not including Florida and extreme southern Georgia.
23. Ear from notch usually more than 12.5; upper parts dark (near dark umber); occurs in mts. of eastern México. . . . .*M. quasiater,* p 817
23′. Ear from notch usually less than 12.5; upper parts pale (bright russet brown to brownish chestnut); occurs in eastern and central North America north of México. . . . .*M. pinetorum,* p. 816
21′. Mammae, 6; fur long, coarse, mouselike, grayish as opposed to reddish on upper parts; tail more than 26 mm. in areas where subgenera *Pitymys* and *Pedomys* occur together (Subgenus *Pedomys*). . . . .*M. ochrogaster,* p. 812

Subgenus **Microtus** Schrank

1798. *Microtus* Schrank, Fauna Boica, 1(Abth. 1):72. Type, *Microtus terrestris* Schrank [= *Mus arvalis* Pallas].

Plantar tubercles, 6; lateral glands on hips in adult males; m1 with 5 closed triangles; m3 with 3 transverse loops and no triangles; M2 with 4 closed sections, and in most eastern species an additional posterior inner loop; M3 with 3 closed triangles (except in *M. chrotorrhinus*).

The 17 species of this subgenus in non-forested areas of North America, along with a corresponding number of species in the Old World, are a large part of the fauna of small mammals in the grasslands of the boreal and temperate life zones of the World.

These rodents subsist on the grasses and forbs and in turn themselves are the principal food of the carnivorous mammals and raptorial birds.

Short gestation periods characteristic of these mice enable most of the species to maintain large populations despite the pressures from native predators.

Microtus pennsylvanicus
Meadow Vole

Total length, 140–195; tail, 33–64; hind foot, 18–24; ear, 12–16. Body 2.0 to 3.1 times as long as tail; tail 1.9–2.7 times as long as hind foot. Color of pelage varies according to subspecies; in general, upper parts varying from bright yellowish chestnut to dull bister that is much obscured by black-tipped hairs; northern subspecies generally reddish and southern subspecies generally more blackish or grayish; underparts usually some shade of gray washed with whitish or buffy. Middle upper molar with 5th posterior loop; incisive foramina long and not constricted posteriorly; Ellerman (1941:590) indicated that this species and the Old World *M. agrestis* are closely related; in fact Ellerman and Morrison-Scott (1951:702) suggest that *M. agrestis* and *M. pennsylvanicus* are conspecific when they state that *M. agrestis* occurs "Doubtless also in much of North America."

Microtus pennsylvanicus has the largest geographic range of any American species in the genus *Microtus*.

Microtus pennsylvanicus acadicus Bangs

1897. *Microtus pennsylvanicus acadicus* Bangs, Amer. Nat., 31:239, March, type from Digby, Nova Scotia.

MARGINAL RECORDS.—Nova Scotia: Victoria County; Guysborough County; Shelburne County; Kings County. Prince Edward Island: Mt. Herbert (Cameron, 1959:47).

Microtus pennsylvanicus admiraltiae Heller

1909. *Microtus admiraltiae* Heller, Univ. California Publ. Zool., 5:256, February 18, type from Windfall Harbor, Admiralty Island, Alaska.
1933. *Microtus pennsylvanicus admiraltiae*, Swarth, Proc. Biol. Soc. Washington, 46:208, October 26.

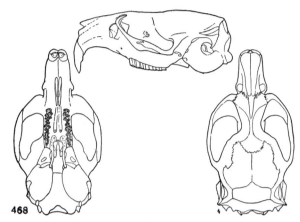

Fig. 468. *Microtus pennsylvanicus pullatus*, 17 mi. E, 4 mi. N Ashton, 6275 ft., Fremont Co., Idaho, No. 89191 M.V.Z., ♂, X 1½.

MARGINAL RECORDS.—Alaska: *Hawk Inlet;* type locality; *Mole Harbor; lake W of Mole Harbor.*

Microtus pennsylvanicus alcorni Baker

1951. *Microtus pennsylvanicus alcorni* Baker, Univ. Kansas Publ., Mus. Nat. Hist., 5:105, November 28, type from 6 mi. SW Kluane, 2550 ft., Yukon.

MARGINAL RECORDS.—Alaska: E side Deadman Lake, 1800 ft., 15 mi. SE Northway. Yukon: type locality. Alaska: E side Chilkat River, 100 ft., 4 mi. N, 9 mi. W Haines; 20 mi. NE Anchorage (Anderson, 1960:206); Tyonek, Cook Inlet; Katmai Bay (Cahalane, 1959:212, 213); N base Mt. Dumpling (*ibid.*); Nushagak; Skwentna [= Skuventna River?] (Schiller and Rausch, 1956:198).

Microtus pennsylvanicus aphorodemus Preble

1902. *Microtus aphorodemus* Preble, N. Amer. Fauna, 22:52, October 31, type from barren grounds about 50 mi. S Cape Eskimo, near mouth Thlewiaza River, Keewatin.
1937. *Microtus pennsylvanicus aphorodemus*, Anderson, *in* Canada's Western Northland (Mammals), Dept. Mines and Resources, Ottawa, p. 112, July 9.

MARGINAL RECORDS.—Keewatin: type locality. Manitoba: Churchill.

Microtus pennsylvanicus chihuahuensis Bradley and Cockrum

1968. *Microtus pennsylvanicus chihuahuensis* Bradley and Cockrum, Amer. Mus. Novit., 2325:3, June 19, type from 3 mi. SE Galeana, Chihuahua. Known only from type locality.

Microtus pennsylvanicus copelandi Youngman

1967. *Microtus pennsylvanicus copelandi* Youngman, Jour. Mamm., 48:581, November 20, type from near North Head, Grand Manan (Island), New Brunswick.

MARGINAL RECORDS (Youngman, 1967:581).—New Brunswick: Grand Manan Island: North Head; *type locality; Grand Harbour.*

Microtus pennsylvanicus drummondii (Audubon and Bachman)

1853. *Arvicola drummondii* Audubon and Bachman, The viviparous quadrupeds of North America, 3:166, type from "Valleys of the Rocky Mountains"; probably vic. Jasper House, Alberta.
1913. *Microtus pennsylvanicus drummondii,* Hollister, Canadian Alpine Jour., Special number, p. 23. February 17.
1899. *Microtus stonei* J. A. Allen, Bull. Amer. Mus. Nat. Hist., 12:5, March 4, type from Liard River, British Columbia. See J. A. Allen, Bull. Amer. Mus. Nat. Hist., 19:550, who points out valid characters of this subspecies.
1951. *Microtus pennsylvanicus arcticus* Cowan, Jour. Mamm., 32:353, August 23, type from Kidluit Bay, NE corner Richards Island, lat. 69° 31′ N, long. 133° 49′ W, Mackenzie. Indistinguishable from *M. p. drummondii* according to Martell (Jour. Mamm., 56:256, February 20, 1975).

MARGINAL RECORDS.—Mackenzie: *Fish Island, 69° 21′ N, 134° 54′ W* (Martell, 1975:255); Kidluit Bay, NE corner Richards Island, 69° 31′ N, 133° 49′ W (*ibid.*); Fort Anderson; Clinton-Colden Lake. Manitoba: Herchmer. Ontario: Fort Albany; Lake Attawapiskat; Thunder Bay; Quetico Park; Rainey River. North Dakota: Pembina (Bailey, 1927:93); Stump Lake (*ibid.*); Portland; *Lisbon;* Ludden; Napoleon; Lostwood. Saskatchewan: Big Muddy Lake; Lost Mountain. Montana: St. Mary Lake; Blackfoot. Idaho: Coeur d'Alene; 5 mi. W Cocolalla; *Priest Lake.* British Columbia: Crows Nest Station, on Canadian Pacific Railroad, 4444 ft.; Mt. Assiniboine (Cowan and Guiguet, 1965:219); *Vermilion Crossing* (*ibid.*); Thompson Pass (*ibid.*). Alberta: 50 mi. N Jasper House. British Columbia: *Ootsa Lake* (Cowan and Guiguet, 1965:219); Wistaria. Yukon: Caribou Crossing; Fort Selkirk; Sixtymile Creek; Yukon–Alaska boundary, Yukon River (Youngman, 1975:92); Old Crow River, at Timber Creek (*ibid.*).

Microtus pennsylvanicus enixus Bangs

1896. *Microtus enixus* Bangs, Amer. Nat., 30:1051, December, type from Hamilton Inlet, Labrador.
1936. *M[icrotus]. p[ennsylvanicus]. enixus,* D. L. Davis, Jour. Mamm., 17:290, August 17.

MARGINAL RECORDS (Harper, 1961:76, 80).—Labrador: Hebron, down coast to Black Bay; Ashuanipi Lake. Quebec: Charlton Island; Cape Jones; Lac Aulneau.

Microtus pennsylvanicus finitus S. Anderson

1956. *Microtus pennsylvanicus finitus* S. Anderson, Univ. Kansas Publ., Mus. Nat. Hist., 9:96, May 10, type from 5 mi. N, 2 mi. W Parks, Dundy Co., Nebraska.

MARGINAL RECORDS.—Nebraska (Jones, 1964c: 229): type locality; *Haigler.* Colorado: Wray.

Microtus pennsylvanicus fontigenus Bangs

1896. *Microtus fontigenus* Bangs, Proc. Biol. Soc. Washington, 10:48, March 9, type from Lake Edward, Quebec. (Regarded as inseparable from *M. p. pennsylvanicus* by Weaver, Jour. Mamm., 21:420, November 14, 1940, but admitted as valid by Rand, Canadian Field-Nat., 57:119, January 24, 1944, who studied actual specimens. Harper, 1961:74, 75, regards *fontigenus* as inseparable from *M. p. pennsylvanicus* without studying specimens.)
1897. *Microtus pennsylvanicus fontigenus,* Miller, Proc. Boston Soc. Nat. Hist., 28:14, April 30.

MARGINAL RECORDS.—Quebec: Mutton Bay (Harper, 1961:75); Natashkwan, down NW shore St. Lawrence River to near Quebec City; Labelle County. Ontario: Rossport; Nipigon; Macdiarmid; Kapuskasing.

Microtus pennsylvanicus funebris Dale

1940. *Microtus pennsylvanicus funebris* Dale, Jour. Mamm., 21:338, August 13, type from Coldstream, 1450 ft., 3½ mi. SE Vernon, British Columbia. (Regarded by Cowan and Guiguet, Handbook No. 11, British Columbia Provincial Mus., p. 219, October 1965, as inseparable from *M. p. drummondii.*)

MARGINAL RECORDS.—British Columbia: island in Anahim Lake; Okanagan (Dale, 1940:338); *Goatfell* (Cowan and Guiguet, 1965:220); Yahk (*ibid.*). Washington: Newport; Conconully. British Columbia: Princeton.

Microtus pennsylvanicus insperatus (J. A. Allen)

1894. *Arvicola insperatus* J. A. Allen, Bull. Amer. Mus. Nat. Hist., 6:347, December 7, type from Custer, Custer Co., South Dakota.
1943. *Microtus pennsylvanicus insperatus,* Anderson, Canadian Field-Nat., 57:92, October 17.
1920. *Microtus pennsylvanicus wahema* V. Bailey, Jour. Mamm., 1:72, March 2, type from Glendive, Dawson Co., Montana.

MARGINAL RECORDS.—Saskatchewan: Crane Lake; Swift Current. Montana: 3 mi. S Medicine Lake City. North Dakota: *10 mi. S Williston; Bismarck;* mouth Cannonball River. South Dakota: type locality. Wyoming (Long, 1965a:648): *1¹/₂ mi. E Buckhorn;* Newcastle; 10 mi. N, 6 mi. W Bill; 1½ mi. S, 5½ mi. W Buffalo; *4 mi. NNE Banner;* 3 mi. WNW Monarch. Alberta: Milk River.

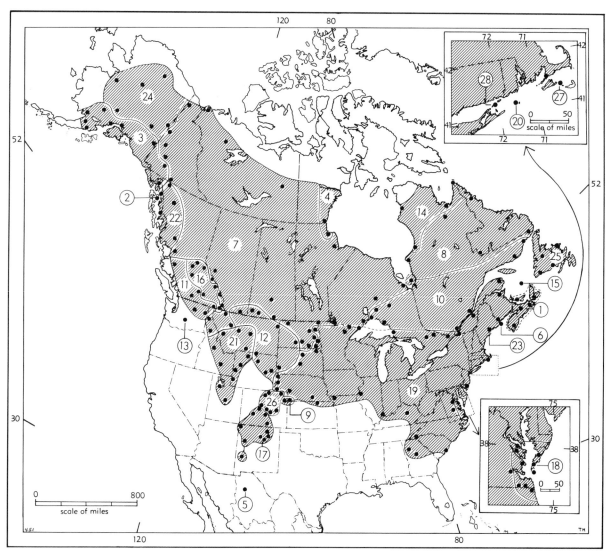

Map 451. *Microtus pennsylvanicus* and two allied species.

Guide to kinds

1. *M. p. acadicus*
2. *M. p. admiraltiae*
3. *M. p. alcorni*
4. *M. p. aphorodemus*
5. *M. p. chihuahuensis*

6. *M. p. copelandi*
7. *M. p. drummondii*
8. *M. p. enixus*
9. *M. p. finitus*
10. *M. p. fontigenus*
11. *M. p. funebris*
12. *M. p. insperatus*
13. *M. p. kincaidi*

14. *M. p. labradorius*
15. *M. p. magdalenensis*
16. *M. p. microcephalus*
17. *M. p. modestus*
18. *M. p. nigrans*
19. *M. p. pennsylvanicus*
20. *M. p. provectus*
21. *M. p. pullatus*

22. *M. p. rubidus*
23. *M. p. shattucki*
24. *M. p. tananaensis*
25. *M. p. terraenovae*
26. *M. p. uligocola*
27. *M. breweri*
28. *M. nesophilus*

Microtus pennsylvanicus kincaidi Dalquest

1941. *Microtus montanus kincaidi* Dalquest, Proc. Biol. Soc. Washington, 54:145, September 30, type from The Potholes, 10 mi. S Moses Lake, Grant Co., Washington. Known only from type locality and *Moses Lake*.

1948. *Microtus pennsylvanicus kincaidi* Dalquest, Univ. Kansas Publ., Mus. Nat. Hist., 2:347, April 9.

Microtus pennsylvanicus labradorius V. Bailey

1898. *Microtus pennsylvanicus labradorius* V. Bailey, Proc. Biol. Soc. Washington, 12:88, April 30, type from Fort Chimo, Quebec.

MARGINAL RECORDS (Harper, 1961:81).—Quebec: Port Burwell; Fort Chimo; Great Whale River, up coast to Povungnituk Bay.

Microtus pennsylvanicus magdalenensis Youngman

1967. *Microtus pennsylvanicus magdalenensis* Youngman, Jour. Mamm., 48:579, November 20, type from Grindstone (Island), Magdalen Islands, Quebec.

MARGINAL RECORDS (Youngman, 1967:581).— Quebec: Magdalen Islands: Grosse Islands; *type locality; Amherst.*

Microtus pennsylvanicus microcephalus (Rhoads)

1894. *Arvicola (Mynomes) microcephalus* Rhoads, Proc. Acad. Nat. Sci. Philadelphia, 46:286, October 23, type from Lac LaHache, British Columbia. (Regarded as inseparable from *M. p. drummondii* by Cowan and Guiguet, Handbook No. 11, British Columbia Provincial Mus., p. 219, October 1965.)
1940. *Microtus pennsylvanicus microcephalus*, Dale, Jour. Mamm., 21:337, August 14.

MARGINAL RECORDS.—British Columbia: Indianpoint Lake, 15 mi. NE Barkerville (Dale, 1940:338); Mt. Robson, Mt. Robson Park; Field; *Morrissey* (Cowan and Guiguet, 1965:220, as *M. p. modestus*); Newgate (*ibid.*, as *M. p. modestus*); Monashee Pass (Cowan and Guiguet, 1965:219); *Salmon Arm* (*ibid.*); type locality; Cottonwood P.O.

Microtus pennsylvanicus modestus (Baird)

1858. *Arvicola modesta* Baird, Mammals, *in* Repts. Expl. Surv. . . . , 8(1):535, July 14, type from Cochetopa [= "Sawatch"] Pass, Saguache Co., Colorado.
1900. *Microtus pennsylvanicus modestus*, V. Bailey, N. Amer. Fauna, 17:20, June 6.
1893. *Arvicola (Mynomes) aztecus* J. A. Allen, Bull. Amer. Mus. Nat. Hist., 5:73, April 28, type from Aztec, 5900 ft., San Juan Co., New Mexico. Considered to be a synonym of *M. p. modestus* by Anderson and Hubbard (Amer. Mus. Novit., 2460:8, April 21, 1971).

MARGINAL RECORDS.—Colorado (Armstrong, 1972:236): Twin Lakes; Westcliffe, 7800 ft.; *ca.* 20 mi. NW Weston. New Mexico (Findley, *et al.*, 1975:256, as *M. pennsylvanicus* only): 15 mi. S Cimarron; Wagon Mound; 1 mi. N Las Vegas; Santa Fe Canyon; San Rafael; Fruitland. Colorado: East Elk Creek, 1½ mi. N Gunnison River, 7429 ft. (Armstrong, 1972:236). Isolated segment: New Mexico: 7 mi. SW Aragon, 6400 ft. (S. Anderson, 1961:1); *8 mi. SW Aragon, 7000 ft.* (Anderson and Hubbard, 1971:2).

Microtus pennsylvanicus nigrans Rhoads

1897. *Microtus pennsylvanicus nigrans* Rhoads, Proc. Acad. Nat. Sci. Philadelphia, 49:307, June 18, type from Currituck, Currituck Co., North Carolina.

MARGINAL RECORDS (Handley and Patton, 1947:173, unless otherwise noted).—Virginia: Richmond County; Accomack County; Northampton County; Smiths Island (Bailey, 1900:19). North

Carolina: type locality. Virginia: Norfolk County; Nansemond County; Gloucester County.

Microtus pennsylvanicus pennsylvanicus (Ord)

1815. *Mus Pennsylvanica* Ord, *in* Guthrie, A new geog., hist. coml. grammar. . . . Philadelphia, 2nd Amer. ed., 2:292, type from meadows below Philadelphia, Pennsylvania.
1895. *M[icrotus]. pennsylvanicus*, Rhoads, Amer. Nat., 29:940, October.
1817. *Mynomes pratensis* Rafinesque, Amer. Month. Mag., 2:45, based on Wilson's (Amer. Ornith., 6:Pl. 50, Fig. 3) description of the meadow vole from meadows below Philadelphia and along the seashore.
1820. *Lemmus noveboracensis* Rafinesque, Annals of Nature, p. 3, type from New York or New Jersey.
1825. *Arvicola riparius* Ord, Jour. Acad. Nat. Sci. Philadelphia, 4(2):305, 306, type locality not given.
1825. *Arvicola palustris* Harlan, Fauna Americana, p. 136, type from swamps along shores of Delaware River.
1840. *Arvicola hirsutus* Emmons, Report on the quadrupeds of Massachusetts, p. 60, type from Massachusetts.
1840. *Arvicola albo-rufescens* Emmons, Report on the quadrupeds of Massachusetts, p. 60, type from Williamstown, Massachusetts.
1841. *Arvicola fulva* Audubon and Bachman, Proc. Acad. Nat. Sci. Philadelphia, 1:96, type from "One of the Western States; we believe Illinois."
1841. *Arvicola nasuta* Audubon and Bachman, Proc. Acad. Nat. Sci. Philadelphia, 1:96, 97, type from near Boston, Massachusetts.
1842. *Arvicola rufescens* DeKay, Zoology of New-York . . . Mammals, 1:88, 89, type from Oneida Lake, New York.
1842. *Arvicola oneida* DeKay, Zoology of New-York . . . Mammals, 1:88, 89, type from Oneida Lake, New York.
1854. *Arvicola dekayi* Audubon and Bachman, The viviparous quadrupeds of North America, 3:287, 288, type from New York or Illinois.
1858. [*Arvicola riparia*] var. *longipilis* Baird, Mammals, *in* Repts. Expl. Surv. . . . , 8(1):524, July 14, type from West Northfield, Illinois, or Racine, Wisconsin.
1858. *Arvicola rufidorsum* Baird, Mammals, *in* Repts. Expl. Surv. . . . , 8(1):526, July 14, type from Holmes Hole, Marthas Vineyard, Massachusetts.

MARGINAL RECORDS.—Quebec: Gaspé Peninsula. New Brunswick: near Bathhurst. Maine: Calais. Massachusetts: Woods Hole. New Jersey: Tuckerton. Virginia: Westmoreland County; Prince George County. South Carolina: Charleston. Georgia (Golley, 1962:144): Newton County; Polk County. North Carolina: Highlands; *near Smokemount Campground, Great Smoky Mountains National Park* (Linzey and Linzey, 1967:310). Kentucky: Lexington; *Midway.* Indiana (Mumford, 1969:74): Gibson County. Illinois: Good Hope. Missouri: 2 mi. N, 6 mi. E Maryville (Easterla and Damman, 1977:10, as *M. pennsylvanicus* only). Kansas: SW¼ sec. 4, T. 2 S, R. 6 W (Fleharty and Anderson, 1964:129). Nebraska (Jones, 1964c:231, unless otherwise noted): 9 mi. S Hastings (Choate and Genoways, 1967:240); 4 mi. WNW Keystone; Smeed; Scotts Bluff; Monroe Creek. South Dakota: Pierre;

Swan Creek, 1600 ft., 13 mi. S Selby; Fort Sisseton. North Dakota (Bailey, 1927:90): Lidgerwood; Fargo; *Larimore*; Drayton. Minnesota: Tower; Pigeon River, SW ¼ sec. 20, T. 64 N, R. 6 E (Timm, 1975:24). Ontario: Pancake Bay; Nipissing; Algonquin Park; Ottawa. Quebec (Wrigley, 1969:208, as *M. pennsylvanicus* only): Dundee; *St. Anne de Bellevue*; St. Lambert.

Microtus pennsylvanicus provectus Bangs

1908. *Microtus provectus* Bangs, Proc. New England Zool. Club, 4:20, March 6, type from Block Island, Newport Co., Rhode Island. Known only from Block Island. Chamberlain (Jour. Mamm., 35:589, November 29, 1954) considers *M. provectus* to be "merely a subspecies of *M. pennsylvanicus*."

Microtus pennsylvanicus pullatus S. Anderson

1956. *Microtus pennsylvanicus pullatus* S. Anderson, Univ. Kansas Publ., Mus. Nat. Hist., 9:97, May 10, type from 12 mi. N and 2 mi. E Sage, 6100 ft., Lincoln Co., Wyoming.

MARGINAL RECORDS.—Montana: Highwood Mts.; 7 mi. NE Higler; 10 mi. NW Park City. Wyoming (Long, 1965a:649): ⅗ mi. S, 3⅕ mi. E Cody; 34 mi. N, 4 mi. W Pinedale; *Kendall;* Salt River, 10 mi. W Afton; *Cokeville;* 6 mi. N, 2 mi. E Sage. Utah: 2 mi. S Provo; *½ mi. W Salt Lake Airport, U.S. Hwy. 40, 4200 ft.;* small marsh between Hwy. 40 and Western Pacific Railroad Tracks, 3½ mi. N Mills Junction (Egoscue, 1965:685). Idaho: Bannock Creek, 4 mi. S Portneuf River; Challis. Montana: Florence.

Microtus pennsylvanicus rubidus Dale

1940. *Microtus pennsylvanicus rubidus* Dale, Jour. Mamm., 21:339, August 13, type from Sawmill Lake, near Telegraph Creek, British Columbia. (Regarded as inseparable from *M. p. drummondii* by Cowan and Guiguet, Handbook No. 11, British Columbia Provincial Mus., p. 219, October 1965.

MARGINAL RECORDS.—British Columbia: Atlin; type locality; Hazelton. Alaska: Fort Wrangell; Taku River.

Microtus pennsylvanicus shattucki Howe

1901. *Microtus pennsilvanicus* [sic] *shattucki* Howe, Proc. Portland Soc. Nat. Hist., 2:201, December 31, type from Tumble Down Dick Island, near Long Island, Penobscot Bay, Maine.

MARGINAL RECORDS (Youngman, 1967:583).— Maine: Dark Harbor, Islesboro Island (=Long Id.); *type locality; North Haven Island.*

Microtus pennsylvanicus tananaensis Baker

1951. *Microtus pennsylvanicus tananaensis* Baker, Univ. Kansas Publ., Mus. Nat. Hist., 5:107, November 28, type from Yerrick Creek, 21 mi. W, 4 mi. N Tok Junction, Alaska.

MARGINAL RECORDS.—Alaska: Old John Lake (Libby, 1959:609); Eagle; type locality; head Glacier Creek, Mt. McKinley; Nulato; Bettles.

Microtus pennsylvanicus terraenovae (Bangs)

1894. *Arvicola terraenovae* Bangs, Proc. Biol. Soc. Washington, 9:129, July 27, type from Codroy, Newfoundland.
1936. *M[icrotus]. p[ennsylvanicus]. terraenovae*, D. L. Davis, Jour. Mamm., 17:290, August 17.

MARGINAL RECORDS (Cameron, 1959:85, unless otherwise noted).—Newfoundland: Penguin Island (Hall and Cockrum, 1953:411); type locality; Tompkins; South Brook.

Microtus pennsylvanicus uligocola S. Anderson

1956. *Microtus pennsylvanicus uligocola* S. Anderson, Univ. Kansas Publ., Mus. Nat. Hist., 9:94, May 10, type from 6 mi. W and ½ mi. S Loveland, 5200 ft., Larimer Co., Colorado.

MARGINAL RECORDS.—Colorado (Armstrong, 1972:237): 14⅘ mi. NNE Fort Collins; 4 mi. W Proctor; 2 mi. W Ramah, 5000 ft.; 12 mi. S Colorado Springs; 3 mi. SW Florissant; Fairplay; 7 mi. NW Fort Collins.

Microtus breweri (Baird)
Beach Vole

1858. *Arvicola breweri* Baird, Mammals, *in* Repts. Expl. Surv. . . . , 8(1):525, July 14, type from Muskeget Island, off Nantucket, Massachusetts. Known from type locality only.
1896. *Microtus breweri*, Miller, Proc. Boston Soc. Nat. Hist., 27:83, June.

Average and extreme external measurements of five males and eight females: 200.8 (191–211), 186 (176–203); 49.8 (45–58), 47.9 (41–56); 23.7 (22.4–25), 23.1 (22–24). Upper parts buffy gray, with scattered brown and black-tipped hairs; sides paler; venter tinged with sulphur yellow; tail bicolored, rusty brown or blackish above, soiled whitish below. This insular species is closely related to *M. pennsylvanicus* and is said to differ as follows: paler, longer, and coarser pelage; anteriorly wider nasals; and longer interparietal. See Map 451.

Whether reported hybrids with *M. pennsylvanicus* can back-cross or are even fertile was unknown to Fivush, *et al.* (1975:272), who questioned whether "true species differentiation" has yet been achieved by *M. breweri*.

Microtus nesophilus V. Bailey
Gull Island Vole

1898. *Microtus insularis* V. Bailey, Proc. Biol. Soc. Washington, 12:86, April 30, type from Great Gull Island, off eastern extremity of Long Island, Suffolk Co., New York. Not *Lemmus insularis* [= *Microtus agrestis*] Nilsson, 1844. Known only from type locality and adjacent *Little Gull Island* (Miller, Bull. New York State Mus., 6:324, November 18, 1899).
1898. *Microtus nesophilus* V. Bailey, Science, n.s., 8:783, December 2, a renaming of *Microtus insularis* V. Bailey.

External measurements: 185; 41; 20–21. Upper parts dark yellowish bister heavily mixed with black hairs, darkest on nose and face; venter dusky, washed with cinnamon; tail bicolored, blackish above, dark brown below. This insular species, closely related to *M. pennsylvanicus*, is said to differ as follows: darker pelage; shorter and wider skull; more widely spreading zygomata; deeper prezygomatic notches. This taxon probably is extinct (see V. Bailey, 1900:27 and Miller, 1899:324). See Map 451.

Microtus montanus
Montane Vole

Revised by S. Anderson, Univ. Kansas Publ., Mus. Nat. Hist., 9:415–511, 12 figs., 2 tables, August 1, 1959.

External measurements: 140–220; 24–69; 14–27. Weight, 37.3–85.0 grams. Upper parts some shade of brownish, often with buffy or grayish wash, and with mixture of black-tipped hairs; sides paler and more buffy; venter white to gray, sometimes washed with buffy; tail bicolored, blackish brown to black above, grayish below. Body 2.4–2.8 times as long as tail; tail 2.2–2.4 times as long as hind foot. Middle upper molar with 4 closed triangles; incisive foramina constricted posteriorly.

Microtus montanus amosus Hall and Hayward

1941. *Microtus montanus amosus* Hall and Hayward, The Great Basin Naturalist, 2:105, July 20, type from Torrey, Wayne Co., Utah.

MARGINAL RECORDS (S. Anderson, 1959:477, 478).—Utah: Vernal; Jensen; *Steep Creek, 15 mi. N Boulder;* Hall Ranch, Salt Gulch, 8 mi. W Boulder; Round Willow Bottom Reservoir, 9800 ft., 10 mi. N, 13

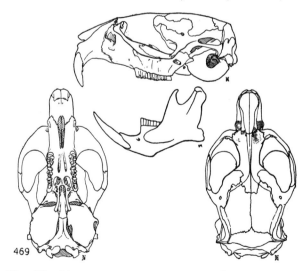

Fig. 469. *Microtus montanus micropus*, Cleveland Ranch, Nevada, No. 46288 M.V.Z., ♂, X 1½.

mi. W Escalante; Pine Lake, 6 mi. SE Widsoe; Panguich; Griffin Spring, 8 mi. NE Widsoe; type locality; Mammoth Ranger Station, Manti National Forest; ½ mi. E Soldier Summit.

Microtus montanus arizonensis V. Bailey

1898. *Microtus montanus arizonensis* V. Bailey, Proc. Biol. Soc. Washington, 12:88, April 30, type from Springerville, Apache Co., Arizona.

MARGINAL RECORDS (S. Anderson, 1959:484, unless otherwise noted).—Arizona: type locality; Alpine, 8000 ft.; Prieto Plateau, S end Blue Range, 9000 ft.; N fork White River; *4 mi. S, 16 mi. W Springerville* (Findley and Jones, 1962:156, as *M. montanus* only).

Microtus montanus canescens V. Bailey

1898. *Microtus nanus canescens* V. Bailey, Proc. Biol. Soc. Washington, 12:87, April 30, type from Conconully, Okanogan Co., Washington.
1938. *Microtus montanus canescens*, Hall, Proc. Biol. Soc. Washington, 51:133, August 23.

MARGINAL RECORDS (S. Anderson, 1959:451, unless otherwise noted).—British Columbia: Cariboo, Frazer River, 40 mi. above mouth Chilcotin River; Lac du Bois, 11 mi. NW Kamloops; *Kamloops;* Ducks; *Vernon;* Coldstream; Midway, Kettle River Valley. Washington: Steamboat Rock, Grand Coulee; *1 mi. S Devil's Punchbowl, Grand Coulee;* Dave Lewis Ranch, 10 mi. N Coulee City; Benton City (Oberg Ranch); *Sunnyside;* Zillah; Wiley City; Bumping Lake (Hall and Kelson, 1959:728); Blewett Pass; *Wenatchee;* Chelan; type locality; 30 Mile Lake, Loomis, E base Windy Mts. British Columbia: Oliver; Trout Creek (1 mi. up), Summerland, Okanagan Lake; Glimpse Lake, 20 mi. NE Nicola; Westwick Lake, 12 mi. W Williams Lake.

Microtus montanus codiensis S. Anderson

1954. *Microtus montanus codiensis* S. Anderson, Univ. Kansas Publ., Mus. Nat. Hist., 7:497, July 23, type from 3⅛ mi. E and ⅜ mi. S Cody, 5020 ft., Park Co., Wyoming.

MARGINAL RECORDS (S. Anderson, 1959:464).— Montana: Beartooth Mts. Wyoming: 13 mi. N, 1 mi. E Cody, 6300 ft.; type locality; 3 mi. N, 10 mi. W Thermopolis; Needle Mt., 10,500 ft.; SW slope Whirlwind Peak, 9000 ft.; Black Mt., head Pat O'Hara Creek.

Microtus montanus dutcheri V. Bailey

1898. *Microtus dutcheri* V. Bailey, Proc. Biol. Soc. Washington, 12:85, April 30, type from Big Cottonwood Meadows, 10,100 ft., SE of Mt. Whitney, Inyo Co., California.
1913. *Microtus montanus dutcheri*, Grinnell, Proc. California Acad. Sci., ser. 4, 3:317, August 28.

MARGINAL RECORDS (S. Anderson, 1959:473, unless otherwise noted).—California: head N fork Kern River, 9600 ft.; *Cottonwood Lakes, 11,100 ft.* (Grinnell, 1933:186); Little Cottonwood Creek, 9500 ft.; Olancha Peak; *Monache Meadows, 7500 ft.;* Jackass Meadows,

7750 ft.; *Jordan Hot Springs, 6700 ft.; Red Rock Meadows, 9000 ft.; Whitney Creek.*

Microtus montanus fucosus Hall

1935. *Microtus montanus fucosus* Hall, Univ. California Publ. Zool., 40:421, October 25, type from Hiko, 4000 ft., Pahranagat Valley, Lincoln Co., Nevada.

MARGINAL RECORDS (S. Anderson, 1959:481).— Nevada: 3 mi. N Crystal Spring, 4000 ft.; *type locality;* 4 mi. S Alamo (skulls from owl pellets); *Ash Spring, 3800 ft.; Crystal Spring, 4000 ft.*

Microtus montanus fusus Hall

1938. *Microtus montanus fusus* Hall, Proc. Biol. Soc. Washington, 51:133, August 23, type from 2½ mi. E Cochetopa Pass, Saguache Co., Colorado.

MARGINAL RECORDS (S. Anderson, 1959:483, unless otherwise noted).—Colorado: 4 mi. S, 3 mi. E Collbran, 6800 ft.; *5 mi. W Independence Pass, 11,000 ft.;* Independence Pass, 12,095 ft.; Twin Lakes (Hall and Kelson, 1959:730); 10 mi. SW Salida, 8500 ft. (Armstrong, 1972:239); Monshower Meadows, 3 mi. E Cochetopa Pass (*ibid.*); 8 mi. S, 11 mi. W Monte Vista (*ibid.*); 12 mi. E Cumbres (*ibid.*). New Mexico: Raton Pass, 7800 ft. (Dalquest, 1975:138); Capulin Mountain National Monument, 6500 ft. (*ibid.*); 3 mi. SW Tres Piedras, 9000 ft.; Jemez Creek, 6 mi. NW Bland; *Fenton Lake, 10 mi. N Jemez Springs;* American Creek; sec. 10, T. 32 N, R. 6 W, San Juan Co. (Findley and Jones, 1962:156, as *M. montanus* only); *sec. 15, T. 32 N, R. 6 W, San Juan Co.* (*ibid.*, as *M. montanus* only). Colorado: Florida; ¼ mi. N Middle Well, Prater Canyon, Mesa Verde National Park, 7500 ft. Utah: *Lockerby;* 10 mi. E Monticello. Colorado: 9 mi. S, 1 mi. W Glade Park P.O., 8800 ft. (Armstrong, 1972:238); 2½ mi. S, 8 mi. E Skyway, Grand Mesa, 9600 ft.

Microtus montanus micropus Hall

1935. *Microtus montanus micropus* Hall, Univ. California Publ. Zool., 40:417, October 25, type from Cleveland Ranch, 6000 ft., Spring Valley, White Pine Co., Nevada.

MARGINAL RECORDS (S. Anderson, 1959:475, 476).—Idaho: Grasmere. Nevada: summit between heads of Copper and Coon creeks, Jarbidge Mts. Utah: Salt Springs (Egoscue, 1961:124); Fish Springs, 4400 ft.; 5 mi. S Garrison, 5400 ft. Nevada: 3½ mi. N Ursine, Eagle Valley, 5600 ft.; Steptoe Creek, 5½ mi. SE Ely, 6400 ft.; Newark Valley; Fish Spring Valley, 6½ mi. N Fish Lake, 6600 ft.; Monitor Valley, 9–9¼ mi. E Toquima Peak, 7000 ft.; 1 mi. NE Rogers Ranch, Big Smoky Valley, 5500 ft.; Bells Ranch, Reese River, 6890 ft.; Campbell Creek Ranch, 5500 ft.; Iron Point, 4300 ft.; Deep Hole, S end Granite Range; Soldier Meadows, 4600 ft.; Pine Forest Range; McDermitt.

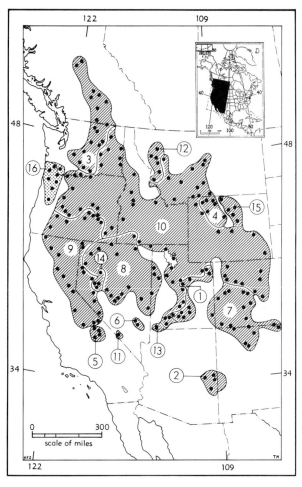

Map 452. *Microtus montanus* and *Microtus canicaudus.*

1. *M. m. amosus*
2. *M. m. arizonensis*
3. *M. m. canescens*
4. *M. m. codiensis*
5. *M. m. dutcheri*
6. *M. m. fucosus*
7. *M. m. fusus*
8. *M. m. micropus*
9. *M. m. montanus*
10. *M. m. nanus*
11. *M. m. nevadensis*
12. *M. m. pratincola*
13. *M. m. rivularis*
14. *M. m. undosus*
15. *M. m. zygomaticus*
16. *M. canicaudus*

Microtus montanus montanus (Peale)

1848. *Arvicola montana* Peale, Mammalia and ornithology, *in* U.S. Expl. Exped. . . . , 8:44, type from headwaters Sacramento River, near Mt. Shasta, Siskiyou Co., California.
1897. *[Microtus] montanus,* Trouessart, Catalogus mammalium . . . , p. 563.
1858. *Arvicola longirostris* Baird, Mammals, *in* Repts. Expl. Surv. . . . , 8(1):531, July 14, type perhaps from upper Pit River, California (see Kellogg, Univ. California Publ. Zool., 21:263, 264, April 18, 1922).
1914. *Microtus montanus yosemite* Grinnell, Proc. Biol. Soc. Washington, 27:207, October 31, type from Yosemite Valley, 4000 ft., Mariposa Co., California.

MARGINAL RECORDS (S. Anderson, 1959:468, 469, unless otherwise noted).—Oregon: 28 mi. NW Burns; 5 mi. NNE Burns; *4 mi. E Burns;* 2 mi. E Crane; summit NE Steens Mts.; *Steens Mts.* Nevada: Barrel Spring, 3½ mi. N, 9½ mi. E Fort Bidwell, California, 5700 ft.; 4 mi. SW Diessner, 5800 ft.; Round Hole, 3900 ft.; Pyramid Lake, 2 mi. W Sutcliffe; *2 mi. E Fernley;* 2½ mi. W Hazen, Truckee Canal; Schurtz Indian Dam, Walker River, 4300 ft.; N fork Cat Creek, Mt. Grant, 8800 ft.; *Indian Creek, White Mts., 7400 ft.;* Fish Lake, 4800 ft. California: *Cottonwood Creek, White Mts., 9500 ft.; Robert's Ranch, Wyman Creek, White Mts., 8200 ft.;* Deep Spring Lake, 5000 ft.; Big Pine Creek; Shaver Ranger Station, 5300 ft.; Gentry's, Big Oak Flat Road, 5800 ft.; Tallac; Cisco; Buck's Ranch, 5100 ft.; Lyonsville P.O., 3500 ft.; Fort Crook, 8 mi. SSW Dana, 3000 ft.; *Sisson (city of Mount Shasta);* Dale Meadows, Mt. Eddy (Hall and Kelson, 1959:730); Stud Horse Canyon, Siskiyou Mts., 6500 ft. Oregon: Brownsboro; Diamond Lake; *Yamsay Mts.;* West Silver Creek, 10 mi. SW Silver Lake; Squaw Butte.

Microtus montanus nanus (Merriam)

1891. *Arvicola (Mynomes) nanus* Merriam, N. Amer. Fauna, 5:63, July 30, type from Pahsimeroi Mts., head Pahsimeroi River, 9350 ft., Custer Co., Idaho.
1938. *Microtus montanus nanus,* Hall, Proc. Biol. Soc. Washington, 51:133, August 23.
1917. *Microtus montanus caryi* V. Bailey, Proc. Biol. Soc. Washington, 30:29, February 21, type from Milford, Fremont Co., Wyoming. Regarded as inseparable from *nanus* by S. Anderson, Univ. Kansas Publ., Mus. Nat. Hist., 7:494, July 23, 1954.
1941. *Microtus montanus nexus* Hall and Hayward, The Great Basin Naturalist, 2:106, July 20, type from West Canyon, Oquirrh Range, Utah Co., Utah.

MARGINAL RECORDS (S. Anderson, 1959:460–462, unless otherwise noted).—Washington: Gifford; Liberty Lake; Oakesdale; Pullman. Idaho: Nez Perce; New Meadows; Van Wyck [possibly now submerged]; Smiths Ferry, W side Payette River; Challis. Montana: E fork Blacktail Creek; Brown's Lake, Powell Co. (Hoffmann, *et al.,* 1969:590, as *M. montanus* only); 10 mi. W Butte (*ibid.*); 6 mi. N Manhattan (*ibid.*); Yogo Peak, Judith Basin Co. (*ibid.*); *Trask Gulch, Judith Basin Co.* (*ibid.*); Judith Mts., Fergus Co. (*ibid.*); Big Snowy Mts.; *22 mi. S, 12 mi. E Lewistown, Fergus Co. line;* Cremer's Ranch, 13 mi. ENE Melville; *Cayuse Hills, 11 mi. ENE Melville;* 14 mi. S Big Timber; Gardiner. Wyoming (Long, 1965a:652): *Canyon Camp;* N end Lake [Yellowstone?]; *Pacific Creek;* Togwotee Pass; within 1 mi. Milford; 16 mi. S, 11 mi. W Waltman; *6 mi. S, 2 mi. W Casper;* Beaver; 5½ mi. E Cheyenne. Colorado (Armstrong, 1972:239, 240): ½ mi. S, 6 mi. W Loveland, 5200 ft.; Boulder; 1 mi. N Gillett; Trout Creek Ranch, 2 mi. N Garo; near Eagle, *ca.* 10,000 ft.; Deep Lake, 16 mi. N Glenwood Springs; between Flag Creek and Grand Hogback, 9 mi. S Meeker; 18 mi. NW Meeker; 14 mi. NW Craig. Utah: La Point; Pines Picnic Area SE Mt. Nebo; *Nebo Mts.;* Manti; *10 mi. N*

Fishlake, Fishlake Mts., 10,000 ft.; 9 mi. N Fishlake; Fishlake; Marysvale; Panguich Lake; Blue Springs, Zion National Park (retained here as *M. m. nanus* of Anderson, 1959:462, although Stock, 1970:432 lists as *M. m. rivularis*); Brian Head, Parawan Mts.; Little Valley, Sheeprock Mts.; 6 mi. S Timpie Junction (Egoscue, 1961:124, as *M. m. nexus*); 3⅓ mi. N Mills Junction (Egoscue, 1965:686); Fremont Island, Great Salt Lake; Kelton. Nevada: Goose Creek, 2 mi. W Utah boundary, 5000 ft. Idaho: Three Creek; Bowmont. Oregon: 16 mi. W Jordan Valley, Arcuenaga Ranch; head Crooked Creek, Owyhee Desert, 4000 ft.; ridge N of Crane (Hall and Kelson, 1959:730); Strawberry Mts.; Schauffler Ranch, 5 mi. W Mt. Vernon; 2 mi. SW Barnes; 25 mi. SE Lapine, Deschutes Co., in Lake Co., sec. 17, T. 24 S, R. 14 E; Lapine; Little Lava Lake; Camp Sherman; Wapinitia; 6 mi. E Cascade Locks. Washington: 8 mi. E Bingen; Smithville–Centerville Road, *ca.* 1500 ft. above Columbia River. Oregon: 1½ mi. S Heppner Junction. Washington: Connell; Ruff; 1 mi. E Bluestem.

Microtus montanus nevadensis V. Bailey

1898. *Microtus nevadensis* V. Bailey, Proc. Biol. Soc. Washington, 12:86, April 30, type from Ash Meadows, Nye Co., Nevada.
1935. *Microtus montanus nevadensis,* Hall, Univ. California Publ. Zool., 40:423, October 25.

MARGINAL RECORDS (S. Anderson, 1959:480).—Nevada: Ash Meadows, 4¾ mi. NW Devils Hole, 2200 ft.; *type locality.*

Microtus montanus pratincola Hall and Kelson

1951. *Microtus montanus pratincolus* Hall and Kelson, Univ. Kansas Publ., Mus. Nat. Hist., 5:75, October 1, type from 6 mi. E Hamilton, 3700 ft., Ravalli Co., Montana.

MARGINAL RECORDS (S. Anderson, 1959:452).—Montana: W arm Flathead Lake; [*National*] *Bison Range;* Ravalli; Missoula; 8 mi. NE Stevensville; type locality; *Hamilton;* Hot Springs Creek, Flathead Reservation.

Microtus montanus rivularis V. Bailey

1898. *Microtus nevadensis rivularis* V. Bailey, Proc. Biol. Soc. Washington, 12:87, April 30, type from St. George, Washington Co., Utah. Known only from type locality (S. Anderson, 1959:479).
1900. *Microtus montanus rivularis* V. Bailey, N. Amer. Fauna, 17:29, June 6.

Microtus montanus undosus Hall

1935. *Microtus montanus undosus* Hall, Univ. California Publ. Zool., 40:420, October 25, type from Lovelock, Pershing Co., Nevada.

MARGINAL RECORDS (S. Anderson, 1959:474).—Nevada: type locality; 2 mi. N, 2 mi. E Stillwater; *Stillwater;* 10 mi. SSE Fallon; 4 mi. W Fallon.

Microtus montanus zygomaticus S. Anderson

1954. *Microtus montanus zygomaticus* S. Anderson, Univ. Kansas Publ., Mus. Nat. Hist., 7:500, July 23, type from Medicine Wheel Ranch, 9000 ft., 28 mi. E Lovell, Big Horn Co., Wyoming.

MARGINAL RECORDS (S. Anderson, 1959:464).—Wyoming: type locality; 1 mi. S, 7½ mi. W Buffalo, 6500 ft.; Buffalo Creek, 27 mi. N, 1 mi. E Powder River, 6075 ft.; *Big Horn Mts., head Powder River;* 5 mi. N, 9 mi. E Tensleep, 7400 ft.; W slope, head Trappers Creek, 9500 ft.

Microtus canicaudus Miller
Gray-tailed Vole

1897. *Microtus canicaudus* Miller, Proc. Biol Soc. Washington, 11:67, April 21, type from McCoy, Willamette Valley, Polk Co., Oregon.

External measurements: 140–168; 20–45; 15–21. Upper parts yellowish; auditory bullae much inflated; auditory meatus large; lateral pits at posterior edge of bony palate unusually shallow. These features differentiate *M. canicaudus* from *M. m. nanus*. Additional differences in *canicaudus* are: interpterygoid space more acuminate anteriorly; distance between 1st upper molars less; incisive foramina wider; braincase broader relative to its length.

According to T. C. Hsu and M. L. Johnson (1970:824–826) the X chromosome is nearly metacentric in *M. canicaudus* instead of acrocentric, serum proteins and hemoglobin when "examined by electrophoresis" differ in *M. canicaudus* and *M. montanus,* and the two taxa are allopatric. See Map 452.

MARGINAL RECORDS (S. Anderson, 1959:455, unless otherwise noted).—Oregon: Old Fort Warren. Washington: S [*of*] *Ridgefield;* 7 mi. NE Vancouver. Oregon: Gresham; 7 mi. SE Moladda; 5 mi. SE Albany; Eugene (Hall and Kelson, 1959:729); Sheridan; Banks.

Microtus umbrosus Merriam
Zempoaltepec Vole

1898. *Microtus umbrosus* Merriam, Proc. Biol. Soc. Washington, 12:107, April 30, type from Mt. Zempoaltepec, 8200 ft., Oaxaca.

Average external measurements of seven specimens: 184; 65; 23. Upper parts uniform dusky, with brown-tipped hairs; underparts dark plumbeous thinly washed with fulvous. Plantar tubercles, 5 (a rudiment of 6th); side glands rudimentary or absent; mammae, 2–0 = 4; ears large and almost naked; feet large; tail long, approx. half as long as head and body, and scantily

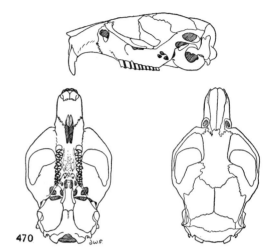

Fig. 470.   *Microtus umbrosus,* Mt. Zempoaltepec, 8200 ft., Oaxaca, No. 68474 U.S.N.M., ♀, X 1½.

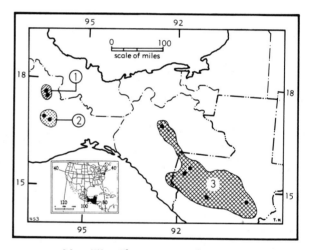

Map 453.   Three species of *Microtus.*

1. *M. oaxacensis*   2. *M. umbrosus*
        3. *M. guatemalensis*

haired. Skull long and narrow; bullae small; M3 with 2 closed, rounded triangles and 1 open triangle; m2 with anterior pair of triangles confluent; m3 with posterior loop, 2 closed triangles, and anterior loop.

MARGINAL RECORDS (Goodwin, 1969:213).—Oaxaca: Totontepec; type locality.

Microtus oaxacensis Goodwin
Tarabundí Vole

1966. *Microtus oaxacensis* Goodwin, Amer. Mus. Novit., 2243:1, May 19, type from an evergreen rain forest, about 5000 ft., at Tarabundí Ranch near Vista Hermosa, 135 km. N Oaxaca City, District of Ixtlan, Oaxaca, México.

External measurements: 159–163; 31–38; 21–22.5; 12–14.5. Selected measurements are: condylobasal length, 27.8–28.6; zygomatic breadth, 15.8–16.2; interorbital constriction, 3.6–4.4; length of maxillary tooth-row, 7.5–7.8. Pelage long, soft, wooly. Upper parts blackish brown, base of hairs Sooty Black, tipped with Ochraceous-Tawny, mixed with black hairs; underparts slightly paler. Mammae, 6; 2 pairs pectoral, 1 pair inguinal. M3 with 4 or 5 closed triangles; incisive foramina short, narrow. In external characters superficially resembles *M. (Pitymys) quasiater* and certain individuals of *M. (Herpetomys) guatemalensis*. See Map 453.

MARGINAL RECORDS (Goodwin, 1969:211).—Oaxaca: type locality; 6¹/₂ mi. SSW *Vista Hermosa;* 10 mi. SSW Vista Hermosa.

Microtus californicus
California Vole

Revised by R. Kellogg, Univ. California Publ. Zool., 21:1–42, 1 fig., December 28, 1918.

External measurements: 157–211; 39–68; 20–25. Upper parts vary from tawny olive through cinnamon brown, with mixture of long, dark overhairs that vary from light seal brown to black; sides lighter, with fewer long overhairs; venter neutral gray, often with white-tipped hairs; tail bicolored, clove brown to black above, grayish below. Body 2.0–2.5 times as long as tail; tail 2.3–2.6 times as long as hind foot; 4 closed triangles in M2; incisive foramina are not con-

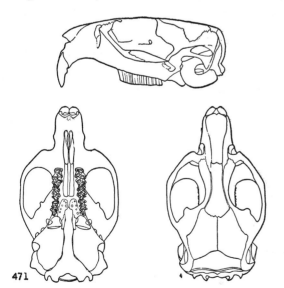

Fig. 471. *Microtus californicus kernensis,* Thompson Canyon, 3900 ft., Walker Basin, Kern Co., California, No. 60281 M.V.Z., ♂, X 1½.

stricted posteriorly but rounded at both ends and widest in middle.

Donald M. Hatfield's (1935:261–271) study revealed that this species has three distinct pelages: juvenal, by the 5th day; postjuvenal, by the 3rd week; and adult by the 8th or 9th week. He was the first to learn that there is a regular pattern in the postjuvenal and juvenal molts but that the progress of molt in the adults is spotty and irregular—a phenomenon now known to characterize most, or all, microtine rodents.

Microtus californicus aequivocatus Osgood

1928. *Microtus californicus aequivocatus* Osgood, Jour. Mamm., 9:56, February 9, type from San Quintín, Baja California.

MARGINAL RECORDS.—Baja California: San José; San Antonio; 1 mi. E El Rosario; type locality; San Telmo.

Microtus californicus aestuarinus R. Kellogg

1918. *Microtus californicus aestuarinus* R. Kellogg, Univ. California Publ. Zool., 21:15, December 28, type from Grizzly Island, Solano Co., California.

MARGINAL RECORDS.—California: Mill Creek, 2 mi. NE Tehama; Chico; Chambers Ravine, 4 mi. N Oroville; Tracy Lake, 6 mi. SW Galt; La Grange; Tulare Lake basin; Tracy; type locality; Bolinas; Petaluma; Cordelia.

Microtus californicus californicus (Peale)

1848. *Arvicola californica* Peale, Mammalia and ornithology, *in* U.S. Expl. Exped. . . . , 8:46, type from vicinity of San Francisco Bay, California, probably at San Francisquito Creek, near Palo Alto, Santa Clara Co.
1897. [*Microtus*] *californicus,* Trouessart, Catalogus mammalium . . . , p. 563.
1853. *Arvicola edax* Le Conte, Proc. Acad. Nat. Sci. Philadelphia, 6:405, type from California, S of San Francisco. Probably Monterey according to R. Kellogg, Univ. California Publ. Zool., 21:18, December 28, 1918.
1858. *Arvicola trowbridgii* Baird, Mammals, *in* Repts. Expl. Surv. . . . , 8(1):529, July 14, type from Monterey, California.

MARGINAL RECORDS.—California: Walnut Creek; slopes of Mt. Diablo; Sweeney's Ranch, 22 mi. S Las Baños; Pozo; Morro; Monterey; Salinas Valley; Boulder Creek; Pescadero; Black Mtn.

Microtus californicus constrictus V. Bailey

1900. *Microtus californicus constrictus* V. Bailey, N. Amer. Fauna, 17:36, June 6, type from Cape Mendocino, Humboldt Co., California.

MARGINAL RECORDS.—California: Eureka; Fair Oaks; Cuddeback; Capetown.

Microtus californicus eximius R. Kellogg

1918. *Microtus californicus eximius* R. Kellogg, Univ. California Publ. Zool., 21:12, December 28, type from Lierly's Ranch, 2340 ft., 4 mi. S Mt. Sanhedrin, Mendocino Co., California.

MARGINAL RECORDS.—Oregon: *ca.* 13 mi. NE Drain [E ½, NW ¼, sec. 29, T. 21 S, R. 3 W] (Borrecco and Hooven, 1972:32, as *M. californicus* only); Siskiyou. California: Mayten; Cassel; Turner's; 2 mi. S South Yolla Bolly Mtn.; Rumsey; Olema; Mendocino City; Helena. Oregon: Rogue River Valley near Grants Pass; Drain.

Microtus californicus grinnelli Huey

1931. *Microtus californicus grinnelli* Huey, Trans. San Diego Soc. Nat. Hist., 7:47, December 19, type from Sangre de Cristo in Valle San Rafael on W base Sierra Juárez, Baja California, lat. 31° 52′ N, long. 116° 06′ W. Known only from type locality.

Microtus californicus halophilus von Bloeker

1937. *Microtus californicus halophilus* von Bloeker, Proc. Biol. Soc. Washington, 50:156, September 10, type from Moss Landing, Monterey Co., California.

MARGINAL RECORDS.—California: mouth Elkhorn Slough; Seaside.

Microtus californicus huperuthrus Elliot

1903. *Microtus californicus huperuthrus* Elliot, Field Columb. Mus., Publ. 74, Zool. Ser., 3:161, May 7, type from La Grulla, San Pedro Mártir Mts., Baja California (see Elliot, Field Columb. Mus., Publ. 115, Zool. Ser., 8:292, March 4, 1907).
1903. *Microtus californicus hyperythrus* Elliot, Field Columb. Mus., Publ. 79, Zool. Ser., 3:218, August 15. This was a change of transliteration.
1926. *Microtus californicus perplexabilis* Grinnell, Jour. Mamm., 7:223, August 9, type from La Grulla, 7000 ft., Sierra San Pedro Mártir, Baja California.

MARGINAL RECORDS.—Baja California: Concepción; Vallecitos; San Ramón; type locality.

Microtus californicus kernensis R. Kellogg

1918. *Microtus californicus kernensis* R. Kellogg, Univ. California Publ. Zool., 21:26, December 28, type from Fay Creek, 4100 ft., Kern Co., California.

MARGINAL RECORDS.—California: Taylor Meadow; Onyx; Fort Tejon; Buena Vista Lake; Bakersfield.

Microtus californicus mariposae R. Kellogg

1918. *Microtus californicus mariposae* R. Kellogg, Univ. California Publ. Zool., 21:19, December 28, type from 1¾ mi. W El Portal, 1800 ft., Mariposa Co., California.

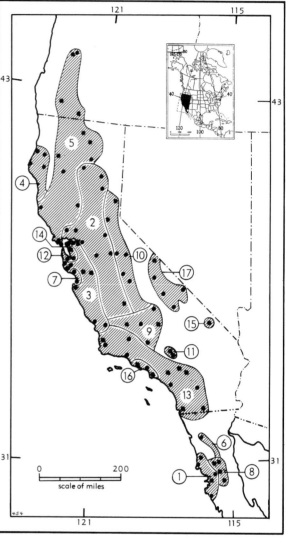

Map 454. *Microtus californicus.*

Guide to subspecies	
1. *M. c. aequivocatus*	9. *M. c. kernensis*
2. *M. c. aestuarinus*	10. *M. c. mariposae*
3. *M. c. californicus*	11. *M. c. mohavensis*
4. *M. c. constrictus*	12. *M. c. paludicola*
5. *M. c. eximius*	13. *M. c. sanctidiegi*
6. *M. c. grinnelli*	14. *M. c. sanpabloensis*
7. *M. c. halophilus*	15. *M. c. scirpensis*
8. *M. c. huperuthrus*	16. *M. c. stephensi*
	17. *M. c. vallicola*

MARGINAL RECORDS.—California: Dutch Flat; type locality; Minkler; Clovis; Pleasant Valley; Carbondale.

Microtus californicus mohavensis R. Kellogg

1918. *Microtus californicus mohavensis* R. Kellogg, Univ. California Publ. Zool., 21:29, December 28, type from Victorville, 2700 ft., San Bernardino Co., California.

MARGINAL RECORDS.—California: Oro Grande; type locality.

Microtus californicus paludicola Hatfield

1935. *Microtus californicus paludicola* Hatfield, Jour. Mamm., 16:316, November 15, type from Melrose Marsh, Alameda Co., California.

MARGINAL RECORDS.—California: *W base El Cerrito (Knoll);* ½ mi. E Bay Farm Island; Alvarado; Redwood City.—Thaeler (1961:81) regards the shorter tail and hind foot of this subspecies as insufficient grounds for distinguishing it from *M. c. californicus.*

Microtus californicus sanctidiegi R. Kellogg

1918. *Microtus californicus neglectus* R. Kellogg, Univ. California Publ. Zool., 21:31, December 28, type from Escondido, 640 ft., San Diego Co., California. Not *Arvicola neglectus* Jenyns, 1841 [= *Microtus agrestis neglectus*].
1922. *Microtus californicus sanctidiegi* R. Kellogg, Proc. Biol. Soc. Washington, 35:78, March 20, a renaming of *M. c. neglectus* R. Kellogg.

MARGINAL RECORDS.—California: Gaviota Pass; Mt. Piños; Bluff Lake; Fish Creek; Garnet Queen Mine, Santa Rosa Mts.; Mountain Spring; mouth Tiajuana River; San Onofre; Santa Ana Cañon; Matilija; Santa Barbara. Introduced on *San Clemente Island, California* (von Bloeker, 1967:257).

Microtus californicus sanpabloensis Thaeler

1961. *Microtus californicus sanpabloensis* Thaeler, Univ. California Publ. Zool., 60:81, November 28, type from San Pablo Creek (salt marsh), Contra Costa Co., California.

MARGINAL RECORDS.—California (Thaeler, 1961:82): Giant (salt marsh); *1 mi. E mouth of San Pablo Creek;* thence W to bay and NE to point of beginning.

Microtus californicus scirpensis V. Bailey

1900. *Microtus scirpensis* V. Bailey, N. Amer. Fauna, 17:38, June 6, type from spring near Shoshone, 1560 ft., on Amargosa River, eastern Inyo Co., California. Known only from type locality.
1918. *Microtus californicus scirpensis,* R. Kellogg, Univ. California Publ. Zool., 21:24, December 28.

Microtus californicus stephensi von Bloeker

1932. *Microtus californicus stephensi* von Bloeker, Proc. Biol. Soc. Washington, 45:134, September 9, type from Playa del Rey, Los Angeles Co., California.

MARGINAL RECORDS.—California: Point Mugu; type locality; Sunset Beach.

Microtus californicus vallicola V. Bailey

1898. *Microtus californicus vallicola* V. Bailey, Proc. Biol. Soc. Washington, 12:89, April 30, type from Lone Pine

Creek, 4500 ft., where it cuts through Alhambra Hills near Lone Pine, Inyo Co., California. (See A. B. Howell, Jour. Mamm., 4:266, November 1, 1923.)

MARGINAL RECORDS.—California: Benton; head Willow Creek at N end Panamint Mts.; Olancha; Kearsarge Pass (Grays); Bishop Creek.

Microtus guatemalensis Merriam
Guatemalan Vole

1898. *Microtus guatemalensis* Merriam, Proc. Biol. Soc. Washington, 12:108, April 30, type from Todos Santos, 10,000 ft., Huehuetenango, Guatemala.

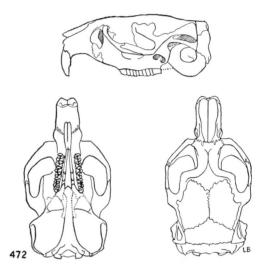

472

Fig. 472. *Microtus guatemalensis,* Todos Santos, Huehuetenango, Guatemala, No. 76767 U.S.N.M., ♂, X 1¼.

Average external measurements of 20 specimens: 150; 37; 21. Upper parts dark umber brown; nose blackish; lips white; underparts clear plumbeous or lightly washed with dull ochraceous. Plantar tubercles, 5; side glands on flanks of males small and obscure or absent; mammae, 2–1 = 6, inguinal pair functionless; ears large; pelage long and soft; tail approx. one-third as long as head and body; auditory bullae not reduced; M3 with 3 closed triangles; m1 with 3 closed triangles and an interior confluent pair of triangles opening into terminal loop; m3 with 4 closed sections including a pair of subequal median triangles. See Map 453.

MARGINAL RECORDS (Smith and Jones, 1967: 189).—Chiapas: Tzontehuitz Mtn., 7 mi. N San Cristóbal. Guatemala: 2 mi. S San Juna Ixcoy, 9340 ft.; 7⅕ mi. E Mataquescuintla, 8400 ft.; 4 mi. S, 10 mi. E Tononicapán; type locality.

Microtus townsendii
Townsend's Vole

External measurements: 169–225; 48–70; 20–26; 15–17. Fur thin and harsh; ears prominent above fur. Upper parts dark brownish with heavy mixture of black-tipped guard-hairs; venter grayish to grayish brown or smoky; tail slightly bicolored, blackish above; body 2.2–2.4 times as long as tail; tail 2.2–2.6 times as long as hind foot; hip glands conspicuous in adult males; M2 with 4 closed triangles; incisive foramina long, narrow, and constricted posteriorly.

Microtus townsendii cowani Guiguet

1955. *Microtus townsendii cowani* Guiguet, Rept. Provincial Mus. Nat. Hist. and Anthro., British Columbia, for 1954, p. B67, August 1, type from Triangle Island, lat. 129° 05′ W, long. 50° 55′ N, British Columbia. Known only from type locality.

Microtus townsendii cummingi Hall

1936. *Microtus townsendii cummingi* Hall, Murrelet, 17:15, March 28, type from Bowen Island, Howe Sound, British Columbia.

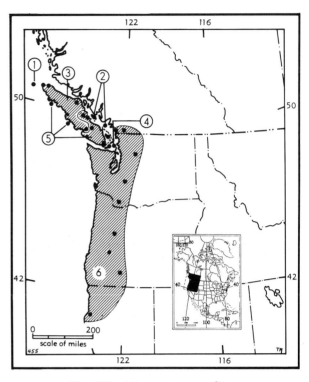

Map 455. *Microtus townsendii.*

1. *M. t. cowani*	4. *M. t. pugeti*
2. *M. t. cummingi*	5. *M. t. tetramerus*
3. *M. t. laingi*	6. *M. t. townsendii*

MARGINAL RECORDS.—British Columbia: Cape Scott; *Hope Island;* Nigei Island; *Hurst Island;* Sayward; Beaver Creek, W of Port Alberni, thence southwestward along *Alberni Inlet* and up coast to point of beginning.

Microtus townsendii pugeti Dalquest

1940. *Microtus townsendii pugeti* Dalquest, Murrelet, 21:7, April 30, type from Neck Point, NW part Shaw Island, San Juan Co., Washington.

MARGINAL RECORDS.—Washington: Friday Harbor, San Juan Island; Strawberry Bay, Cypress Island.

Microtus townsendii tetramerus (Rhoads)

1894. *Arvicola (Tetramerodon) tetramerus* Rhoads, Proc. Acad. Nat. Sci. Philadelphia, 46:283, October 23, type from Beacon Hill Park, Victoria, Vancouver Island, British Columbia.
1936. *Microtus townsendii tetramerus,* Hall, Murrelet, 17:15, March 28.

MARGINAL RECORDS.—British Columbia: Forbidden Plateau; Comox, thence southeastward along coast to type locality, thence northwestward up coast of Vancouver Island and northeast along *Alberni Inlet* to

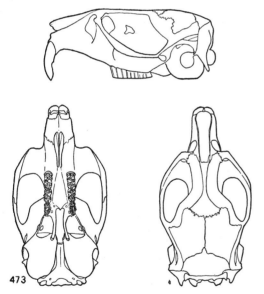

Fig. 473. *Microtus townsendii townsendii,* 1 mi. N Oregon City, Clackamas Co., Oregon, No. 32484 M.V.Z., ♂, X 1½.

MARGINAL RECORDS.—British Columbia: Texada Island; type locality.

Microtus townsendii laingi Anderson and Rand

1943. *Microtus townsendi laingi* Anderson and Rand, Canadian Field-Nat., 57:74, October 17, type from Port Hardy, Vancouver Island, British Columbia.

Alberni, thence to point of beginning. Also on Bunsby Island, Vargas Island (Cowan and Guiguet, 1965:224), *small island 2 mi. NE Tofino* (*ibid.*), Saltspring Island, and *Pender Island* (Cowan and Guiguet, 1965:224).

Microtus townsendii townsendii (Bachman)

1839. *Arvicola townsendii* Bachman, Jour. Acad. Nat. Sci. Philadelphia, 8:60, type from Columbia River; according to V. Bailey (N. Amer. Fauna, 17:46, June 6, 1900), near mouth Willamette [River], on or near Wappatoo (or Sauvie) Island, Oregon.

1896. *M*[*icrotus*]. *townsendi*, Miller, N. Amer. Fauna, 12:66, July 23.

1848. *Arvicola occidentalis* Peale, Mammalia and ornithology, *in* U.S. Expl. Exped. . . . , 8:45, type from Puget Sound.

MARGINAL RECORDS.—British Columbia: Port Moody; Chilliwack. Washington: Sauk; Nisqually Flats; Clark County. Oregon: *1 mi. N Oregon City*; Salem; Eugene; Prospect. California: Humboldt Bay, near Eureka, thence up coast, including *Lulu Island* (Cowan and Guiguet, 1965:224), to point of beginning.

Microtus oeconomus
Tundra Vole

American forms revised by Paradiso and Manville (Proc. Biol. Soc. Washington, 74:77–92, May 19, 1961).

External measurements: 152–225; 30–54; 19–25. Upper parts vary from dusky gray, with traces of brownish, through rich buff, tawny or cinnamon brown and rusty brown. In all there is a mixture of black-tipped hairs; sides paler; venter white and in some subspecies washed with dark buff; tail varies from slightly to sharply bicolored, pale dusky to black above, whitish to pale buff below; body 2.8–3.6 times as long as tail; tail 1.8 to 2.4 times as long as hind foot; anterior lower

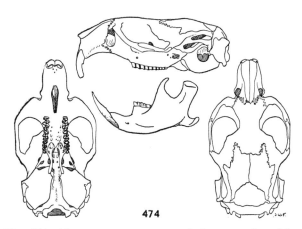

474

Fig. 474. *Microtus oeconomus macfarlani*, N side Salcha River, 600 ft., 25 mi. S, 20 mi. E Fairbanks, Alaska, No. 21488 K.U., ♂, X 1½.

molar (m1) has only 4 closed triangles, 5th triangle being open and confluent with short terminal loop (except in *M. o. sitkensis*, where 5th triangle usually closed); incisive foramina short and constricted posteriorly.

The specific name *oeconomus* almost certainly is not the correct name for the Tundra Vole. Possibly *ratticeps* of Keyserling and Blasius 1841 is the correct one, for reasons advanced by Ognev (1950:258, 281 [= 215, 235 of English translation of 1956], and on preceding and succeeding pages). Probably *kamtschaticus* Pallas 1776 is the correct name if *ratticeps* does not occur in Kamtchatka (see Ellerman and Morrison-Scott, 1951:705). Pending knowledge of whether or not *ratticeps* occurs there, the not unfamiliar name *oeconomus* is used here.

Microtus oeconomus amakensis O. J. Murie

1930. *Microtus amakensis* O. J. Murie, Jour. Mamm., 11:74, February 11, type from Amak Island, Bering Sea, Alaska. Known only from type locality.

1952. *Microtus oeconomus amakensis*, Hall and Cockrum, Univ. Kansas Publ., Mus. Nat. Hist., 5:309, November 17.

Microtus oeconomus elymocetes Osgood

1906. *Microtus elymocetes* Osgood, Proc. Biol. Soc. Washington, 19:71, May 1, type from E side of Montague Island, Prince William Sound, Alaska. Restricted to Montague Island (see Heller, Univ. California Publ. Zool., 5:342, March 5, 1910).

1942. *M*[*icrotus*]. *oec*[*onomus*]. *elymocetes*, Zimmermann, Wiegmann's Arch. für Naturgesch., 11:188, September 10.

Microtus oeconomus innuitus Merriam

1900. *Microtus innuitus* Merriam, Proc. Washington Acad. Sci., 2:21, March 14, type from Northeast Cape, St. Lawrence Island, Bering Sea, Alaska.

1942. *M*[*icrotus*]. *oec*[*onomus*]. *innuitus*, Zimmermann, Wiegmann's Arch. für Naturgesch., 11:188, September 10.

MARGINAL RECORDS.—Alaska (St. Lawrence Island): type locality; Sevoonga.

Microtus oeconomus macfarlani Merriam

1900. *Microtus macfarlani* Merriam, Proc. Washington Acad. Sci., 2:24, March 14, type from Fort Anderson, Anderson River, Mackenzie.

1942. *M*[*icrotus*]. *oec*[*onomus*]. *macfarlani*, Zimmermann, Wiegmann's Arch. für Naturgesch., 11:187, September 10.

1909. *Microtus operarius endoecus* Osgood, N. Amer. Fauna, 30:23, October 7, type from mouth Charlie Creek, Yukon River, about 50 mi. above Circle, Alaska.

MARGINAL RECORDS (Paradiso and Manville, 1961:82, 83, unless otherwise noted).—Alaska: Meade River, 50½ mi. S, 9 mi. W Point Barrow. Mackenzie: Coppermine; Bathurst Inlet (Hall and Kelson,

1959:735); *Glacier Lake, Iron Mtn., 4500 ft.* (Youngman, 1968:78). Yukon (Youngman, 1975:96, 97): Little Hyland River, 128 mi. N Watson Lake, 4000 ft.; *Ida Lake (= McPherson Lake)*; Canol Road, Mile 11; SW end Dezadeash Lake; Burwash Landing. Alaska: Chitine; *5 mi. NNE Gulkana* (Hall and Kelson, 1959:735); Mt. McKinley; Huslia River (N fork); Utukok River, 200 mi. SW Point Barrow; Kaolak River (Hall and Kelson, 1959:735).

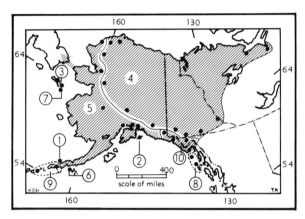

Map 456. *Microtus oeconomus.*

1. *M. o. amakensis*
2. *M. o. elymocetes*
3. *M. o. innuitus*
4. *M. o. macfarlani*
5. *M. o. operarius*
6. *M. o. popofensis*
7. *M. o. punukensis*
8. *M. o. sitkensis*
9. *M. o. unalascensis*
10. *M. o. yakutatensis*

Microtus oeconomus operarius (Nelson)

1893. *Arvicola operarius* Nelson, Proc. Biol. Soc. Washington, 8:139, December 28, type from St. Michael, Norton Sound, Alaska.
1942. *M[icrotus]. oec[onomus]. operarius,* Zimmermann, Wiegmann's Arch. für Naturgesch., 11:187, September 10.
1897. *Microtus kadiacensis* Merriam, Proc. Biol. Soc. Washington, 11:222, July 15, type from Kodiak Island, Alaska.
1952. *Microtus oeconomus gilmorei* Setzer, Proc. Biol. Soc. Washington, 65:75, April 25, type from Point Lay, lat. 69° 46′ N, long. 163° 04′ W, Alaska.

MARGINAL RECORDS (Paradiso and Manville, 1961:81, unless otherwise noted).—Alaska: Point Lay; Kowak [= Kobuk] River (Hall and Kelson, 1959:736); Ophir; 20 mi. NE Anchorage (43044 KU); Prince William Sound; Tyonek (Hall and Kelson, 1959:736), thence southwest along Alaska Peninsula to *Itzamek Bay* (Hall and Kelson, 1959:735), thence north along coast to point of beginning.

Microtus oeconomus popofensis Merriam

1900. *Microtus unalascensis popofensis* Merriam, Proc. Washington Acad. Sci., 2:22, March 14, type from Popof Island, Shumagin Islands, Alaska.
1942. *M[icrotus]. oec[onomus]. popofensis,* Zimmermann, Wiegmann's Arch. für Naturgesch., 11:187, September 10.

MARGINAL RECORDS.—Alaska: type locality; *Unga Island* (Paradiso and Manville, 1961:88).

Microtus oeconomus punukensis Hall and Gilmore

1932. *Microtus innuitus punukensis* Hall and Gilmore, Univ. California Publ. Zool., 38:399, September 17, type from Big Punuk Island, near E end St. Lawrence Island, Alaska. Known only from type locality.
1942. *M[icrotus]. oec[onomus]. punukensis,* Zimmermann, Wiegmann's Arch. für Naturgesch., 11:188, September 10.

Microtus oeconomus sitkensis Merriam

1897. *Microtus sitkensis* Merriam, Proc. Biol. Soc. Washington 11:221, July 15, type from Sitka, Alaska.
1942. *M[icrotus]. oec[onomus]. sitkensis,* Zimmermann, Wiegmann's Arch. für Naturgesch., 11:187, September 10.

MARGINAL RECORDS.—Alaska: Mud Bay, Chichagof Island (Paradiso and Manville, 1961:90); type locality.

Microtus oeconomus unalascensis Merriam

1897. *Microtus unalascensis* Merriam, Proc. Biol. Soc. Washington, 11:222, July 15, type from Unalaska, Alaska.
1942. *M[icrotus]. oec[onomus]. unalascensis,* Zimmermann, Wiegmann's Arch. für Naturgesch., 11:187, September 10.

MARGINAL RECORDS (Paradiso and Manville, 1961:87).—Alaska: *Urilia Bay*; False Pass; *Ikatan Peninsula; Sanak Island;* Unalaska Island.

Microtus oeconomus yakutatensis Merriam

1900. *Microtus yakutatensis* Merriam, Proc. Washington Acad. Sci., 2:22, March 14, type from N shore Yakutat Bay, Alaska.
1942. *M[icrotus]. oec[onomus]. yakutatensis,* Zimmermann, Wiegmann's Arch. für Naturgesch., 11:187, September 10.

MARGINAL RECORDS.—Alaska: Chitina River Glacier, 4500 ft. British Columbia: Stonehouse Creek, 5½ mi. W of jct. with Kelsall River (Cowan and Guiguet, 1965:229). Alaska: Glacier Bay, Point Gustavus (Paradiso and Manville, 1961:84); type locality.

Microtus longicaudus
Long-tailed Vole

External measurements: 155–221; 50–93; 20–25. Upper parts vary from dull grayish bister through brownish gray to dark sepia brown, all with mixture of numerous black-tipped hairs; sides slightly paler causing more reddish back to stand out as a wide band; venter plumbeous with light to heavy wash of whitish to dull buffy; tail long, indistinctly to distinctly bicolored, brownish to black above, soiled whitish to gray below. Body 1.6–1.9 times as long as tail; tail 2.8–3.5 times as long as hind foot; middle upper molar

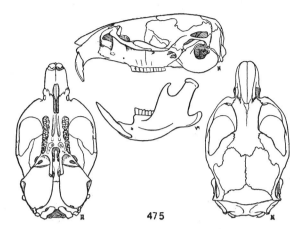

Fig. 475. *Microtus longicaudus latus,* Wisconsin Creek, 8500 ft., Nevada, No. 45846 M.V.Z., ♀, X 1½.

with 4 closed triangles; incisive foramina not abruptly constricted posteriorly, but gradually tapered or as wide posteriorly as anteriorly; skull relatively smooth, not heavily ridged even in adults.

Microtus longicaudus abditus A. B. Howell

1923. *Microtus mordax abditus* A. B. Howell, Jour. Mamm., 4:36, February 9, type from Walker's Ranch, Pleasant Valley, 8 mi. S Tillamook, Tillamook Co., Oregon.
1938. *Microtus longicaudus abditus,* Goldman, Jour. Mamm., 19:491, November 14.

MARGINAL RECORDS.—Oregon: Netarts; type locality.

Microtus longicaudus alticola (Merriam)

1890. *Arvicola (Mynomes) alticolus* Merriam, N. Amer. Fauna, 3:67, September 11, type from Little Spring, 8200 ft., San Francisco Mtn., Coconino Co., Arizona.
1938. *Microtus longicaudus alticola,* Goldman, Jour. Mamm., 19:491, November 14.

MARGINAL RECORDS.—Utah: P.R. Springs, 43 mi. S Ouray, 7950 ft.; Warner Ranger Station, La Sal Mts., 9750 ft.; 5 mi. W Monticello. Arizona: type locality.

Microtus longicaudus angusticeps V. Bailey

1898. *Microtus angusticeps* V. Bailey, Proc. Biol. Soc. Washington, 12:86, April 30, type from Crescent City, Del Norte Co., California.
1938. *Microtus longicaudus angusticeps,* Goldman, Jour. Mamm., 19:491, November 14.

MARGINAL RECORDS.—Oregon: mouth Rogue River. California: Mendocino.

Microtus longicaudus baileyi Goldman

1938. *Microtus longicaudus baileyi* Goldman, Jour. Mamm., 19:492, November 14, type from Greenland Spring, head Bright Angel Creek, 8000 ft., Grand Canyon National Park, Coconino Co., Arizona.

MARGINAL RECORDS (Hoffmeister, 1971:176, unless otherwise noted).—Arizona: Big Springs (6300–6700 ft.), Kaibab Plateau (Hall and Kelson, 1959:737); Point Imperial; *Snowshoe Cabin, 8400 ft., Walhalla Plateau; Bright Angel Spring, Kaibab Plateau* (Hall and Kelson, 1959:737); *Marble Flats;* Kanabownits Spring; *Swamp Lake.*

Microtus longicaudus bernardinus Merriam

1908. *Microtus mordax bernardinus* Merriam, Proc. Biol. Soc. Washington, 21:145, June 9, type from Dry Lake, 9000 ft., San Bernardino Mts., San Bernardino Co., California.
1938. *Microtus longicaudus bernardinus,* Goldman, Jour. Mamm., 19:492, November 14.

MARGINAL RECORDS.—California: Bluff Lake, 7500 ft.; type locality; S fork Santa Ana River, 8500 ft.

Microtus longicaudus halli Hayman and Holt

1931. *Microtus mordax angustus* Hall, Univ. California Publ. Zool., 37:13, April 10, type from Godman [= Goodman] Springs, 5700 ft., Blue Mts., Columbia Co., Washington. Not *Microtus angustus* Thomas, 1908.
1941. *Microtus mordax halli* Hayman and Holt, *in* Ellerman, The families and genera of living rodents, British Mus., 2:603, March 21, a renaming of *Microtus mordax angustus* Hall.
1948. *Microtus longicaudus halli,* Dalquest, Univ. Kansas Publ., Mus. Nat. Hist., 2:353, April 9.

MARGINAL RECORDS.—Washington: Humpeg [= Hornpegg] Falls; *Butte Creek; type locality.*

Microtus longicaudus incanus Lee and Durrant

1960. *Microtus longicaudus incanus* Lee and Durrant, Proc. Biol. Soc. Washington, 73:168, December 30, type from ¼ mi. SE Burned Ridge, Mount Ellen, Henry Mts., 10,300 ft., Garfield Co., Utah.

MARGINAL RECORDS (Lee and Durrant, 1960:170).—Utah: Sawmill Basin, Henry Mts., 9100 ft.; *type locality;* Straight Creek, E Slope Mt. Pennell, 9000 ft.

Microtus longicaudus latus Hall

1931. *Microtus mordax latus* Hall, Univ. California Publ. Zool., 37:12, April 10, type from Wisconsin Creek, 8500 ft., Toyabe Mts., Nye Co., Nevada.
1938. *Microtus longicaudus latus,* Goldman, Jour. Mamm., 19:491, November 14.

MARGINAL RECORDS.—Utah: Standrod, Raft River Mts., 5500 ft.; Butterfield Canyon, 3 mi. SW But-

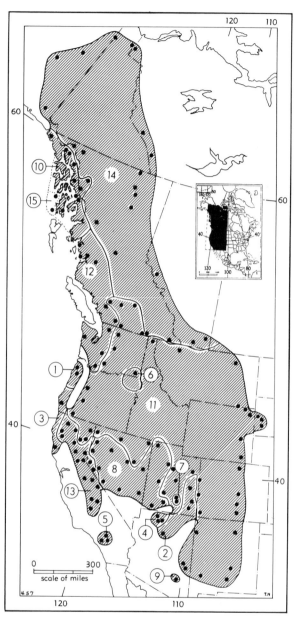

Map 457. *Microtus longicaudus* and *Microtus coronarius*.

Guide to kinds
1. *M. l. abditus*
2. *M. l. alticola*
3. *M. l. angusticeps*
4. *M. l. baileyi*
5. *M. l. bernardinus*
6. *M. l. halli*
7. *M. l. incanus*
8. *M. l. latus*
9. *M. l. leucophaeus*
10. *M. l. littoralis*
11. *M. l. longicaudus*
12. *M. l. macrurus*
13. *M. l. sierrae*
14. *M. l. vellerosus*
15. *M. l. coronarius*

terfield Tunnel, 8000 ft.; Britts Meadow, near Delano Ranger Station, 14 mi. E Beaver; Bryce National Park; Blue Springs, Zion National Park. Nevada: 3½ mi. N Ursine, Eagle Valley, 5900 ft.; Breen Creek, Kawich Range, 7000 ft.; Chiatovich Creek, 8200 ft.; Lapon Canyon, Mt. Grant, 8800 ft.; Hardscrabble Canyon; Horse

Canyon, Pahrum Peak, 5800–6000 ft.; El Dorado Canyon, Humboldt Range, 7800 ft.; Union; E side Schellbourne Pass, 6800 ft.

Microtus longicaudus leucophaeus (J. A. Allen)

1894. *Arvicola leucophaeus* J. A. Allen, Bull. Amer. Mus. Nat. Hist., 6:320, November 7, type from Graham Mts., Graham Co., Arizona. Known only from type locality.
1938. *Microtus longicaudus leucophaeus*, Goldman, Jour. Mamm., 19:491, November 14.

Microtus longicaudus littoralis Swarth

1933. *Microtus mordax littoralis* Swarth, Proc. Biol. Soc. Washington, 46:209, October 26, type from Shakan, Prince of Wales Island, Alaska [Shakan is on Kosciusko Island, and Swarth probably took his specimen on Prince of Wales Island, opposite Shakan].
1938. *Microtus longicaudus littoralis*, Goldman, Jour. Mamm., 19:491, November 14.

MARGINAL RECORDS.—Alaska: Yakutat; [Point Gustavus] Glacier Bay; E side Chilkat River, 100 ft., 4 mi. N, 9 mi. W Haines; Juneau; Sundum. British Columbia: Telegraph Creek; *Glenora* (Cowan and Guiguet, 1965:227); Flood Glacier. Alaska: Thomas Bay; Bradfield Canal; Dall Island.

Microtus longicaudus longicaudus (Merriam)

1888. *Arvicola (Mynomes) longicaudus* Merriam, Amer. Nat., 22:934, October, type from Custer, 5500 ft., Black Hills, Custer Co., South Dakota.
1895. *Microtus (Mynomes) longicaudus*, J. A. Allen, Bull. Amer. Mus. Nat. Hist., 7:266, August 21.
1891. *Arvicola (Mynomes) mordax* Merriam, N. Amer. Fauna, 5:61, July 30, type from Sawtooth (or Alturas) Lake, 7200 ft., E base Sawtooth Mts., Blaine Co., Idaho. (Arranged as a synonym of *M. l. longicaudus* by Long, Univ. Kansas Publ., Mus. Nat. Hist., 14:653, July 6, 1965.)

MARGINAL RECORDS.—British Columbia: Shuswap; Cascade. Idaho: 4 mi. W Meadow Creek. Montana: Blackfoot; Piney Buttes, S of Glasgow (Hoffmann, *et al.*, 1969:591, as *M. longicaudus* only); Moorehead (*ibid.*). Wyoming (Long, 1965a:655, 656): 1¼ mi. N, ½ mi. E Rockypoint; Warren Peak, Bear Lodge Mts.; *3 mi. NW Sundance; 1½ mi. NW Sundance; Sundance*. South Dakota (Turner, 1974:110): Big Spearfish Canyon, 3 mi. S Spearfish, 4200 ft.; *Boxelder Creek, Steamboat Rock Campgrounds*; 7 mi. S, 1 mi. W Sturgis, 4700 ft.; *Palmer Gulch, 5 mi. SE Hill City; Palmer Gulch, 8 mi. SE Hill City, 5300 ft.*; 4 mi. E Custer, Harney National Forest; *type locality*. Wyoming (Long, 1965a:656): *1½ mi. E Buckhorn*; 3 mi. N, 5 mi. E Orin; 3½ mi. W Horse Creek. Colorado (Armstrong, 1972:241, 242): 6 mi. W Fort Collins, 6500 ft.; Bear Creek, 6500 ft.; 1½ mi. SW San Isabel City, 9000 ft.; 4 mi. S La Veta, 7000 ft. New Mexico: Martinez; 5 mi. N, 9 mi. E Capitan, Capitan Mts. (Findley and Jones, 1962:157, as *M. longicaudus* only); 8 mi. E Cloudcroft, James Canyon, Sacramento Mts. (*ibid.*); *10 mi. S Cloudcroft, Sac-*

ramento Mts. (ibid.); [Kingston] Mimbres Mts. Arizona: Rose Peak; Reservation Creek. New Mexico: Chuska Mts. Colorado (Armstrong, 1972:241, 242): Ute Peak; West Paradox Valley; 28 mi. N, 5½ mi. W Mack, 7250 ft. Utah: Vernal; Lake Creek, 11 mi. E Mount Pleasant; Wildcat Ranger Station, Boulder Mtn.; 3½ mi. NE Pleasant Grove. Nevada: Goose Creek, 2 mi. W Utah boundary; W side Ruby Lake, 3 mi. S Elko Co. line; Hansen Canyon, 5 mi. W Paradise Valley; 2 mi. W Vya. California: *5 mi. N Fredonyer Peak;* Butte Lake; Parker Creek; Goose Lake. Oregon: Siskiyou; Upper Deschutes River, Little Meadow (near head Deschutes River); 15 mi. W Bend. Washington: 8 mi. S Glenwood, base Mt. Adams; 2 mi. S Blewett Pass; Hart Lake, Railroad Creek. British Columbia: Manning Park; Spences Bridge; Blackwater Lake; *McGillivary Creek.*

Microtus longicaudus macrurus Merriam

1898. *Microtus macrurus* Merriam, Proc. Acad. Nat. Sci. Philadelphia, 50:353, October 4, type from Lake Cushman, Olympic Mts., Mason Co., Washington.
1938. *Microtus longicaudus macrurus,* Goldman, Jour. Mamm., 19:491, November 14.

MARGINAL RECORDS.—British Columbia: Princess Royal Island; Kimsquit; Rainbow Mts.; Alta Lake; Silver Creek near Hope. Washington: Sauk; Naches River; Wind River; Granville [= Taholah], thence along coast to point of beginning, including in British Columbia: *Stuart, Goose, King,* and Swindle islands.

Microtus longicaudus sierrae R. Kellogg

1922. *Microtus mordax sierrae* R. Kellogg, Univ. California Publ. Zool., 21:288, April 18, type from Tuolumne Meadows, 8600 ft., Yosemite National Park, California.
1938. *Microtus longicaudus sierrae,* Goldman, Jour. Mamm., 19:491, November 14.

MARGINAL RECORDS.—California: Etna; Bear Creek; Lassen Peak; Sierra Valley, 5100 ft.; Independence Lake, 7000 ft. Nevada: ½ mi. S Mt. Rose, 9500 ft.; *½ mi. S Marlette Lake, 8150 ft.;* Desert Creek, Sweetwater Range, 6250 ft. California: Farringtons Ranch near Mono Lake, 6800 ft.; McCloud Camp, Cottonwood Creek, White Mts., 9200 ft.; *Roberts Ranch, Wyman Creek, White Mts., 8250 ft.;* Monache Meadow, 8000 ft.; Taylor Meadow; Hume, 5300 ft.; El Portal, 2000 ft.; Blue Cañon (4700–5000 ft.); 20 mi. SW Quincy, 5000 ft.; head Bear Creek, 6400 ft.; head Grizzly Creek, 6000 ft.

Microtus longicaudus vellerosus J. A. Allen

1899. *Microtus vellerosus* J. A. Allen, Bull. Amer. Mus. Nat. Hist., 12:7, March 4, type from upper Liard River, British Columbia.
1944. *Microtus longicaudus vellerosus,* Anderson and Rand, Canadian Field-Nat., 58:20, April 1.
1899. *Microtus cautus* J. A. Allen, Bull. Amer. Mus. Nat. Hist., 12:7, March 4, type from Hells Gate, Liard River, British Columbia.

MARGINAL RECORDS.—Yukon: Summit Lake, 67° 43′, 136° 29′ (Youngman, 1975:97). Northwest Territories: Horn Lake, 37 mi NW Fort McPherson, 1900 ft. (Youngman, 1964:3); Glacier Lake; Nahanni Gates. British Columbia (Cowan and Guiguet, 1965:228): 104 mi. NW Fort Nelson; Summit Pass; Tuchodi Lake; head Sikanni Chief River. Alberta: Jasper Park; Eagle Butte (Soper, 1965:223); Sweet Grass Hills. British Columbia (Cowan and Guiguet, 1965:228): Newgate; *Pend d'Oreille;* Rossland; Sinkut Mtn.; Hazelton; *Nine Mile Mtn.;* Atlin. Alaska: Skagway. British Columbia: *near jct. Stonehouse Creek and Kelsall River* (Cowan and Guiguet, 1965:228). Alaska: Chitina River Glacier; N side Salcha River, 600 ft., 25 mi. S, 20 mi. E Fairbanks; Circle. Yukon Territory: Rampart House (Youngman, 1964:3).

Microtus coronarius Swarth
Coronation Island Vole

1911. *Microtus coronarius* Swarth, Univ. California Publ. Zool., 7:131, January 12, type from Egg Harbor, Coronation Island, Alaska.

External measurements: 203–232; 70–92; 24–26. Upper parts Vandyke brown obscured by black-tipped hairs; sides paler, near broccoli brown; venter dark gray; tail bicolored, blackish above, whitish below. This species is closely related to *M. longicaudus.* See Map 457.

MARGINAL RECORDS.—Alaska: type locality; *Warren Island;* Forrester Island.

Microtus oregoni
Creeping Vole

Reviewed by Hatfield and Hooper, Murrelet, 16:33, 34, May, 1935.

External measurements: 129–154; 32–42; 16–19; 9–10. Average weights of 10 males and 10 females, respectively, of *M. o. oregoni,* 19.3, 19.1 grams. Upper parts varying from sooty gray through mixed bister to yellowish bister; underparts dusky with wash varying from dull buff to whitish; tail indistinctly bicolored, sooty to blackish above, dark gray to silvery gray below; ears blackish, scantily haired, and protruding from fur. Plantar tubercles, 5; side glands obscure or wanting; ears small; fur dense and without stiff hairs; molars small; M3 with 2 or 3 closed triangles; m1 with 5 closed triangles; m2 with anterior pair of triangles usually confluent; m3 with 3 transverse loops.

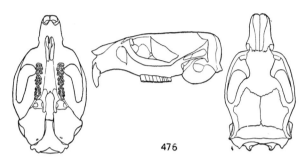

Fig. 476. *Microtus oregoni oregoni*, 1½ mi. S, ½ mi. E Willow Creek, 600 ft., Humboldt Co., California, No. 97516 M.V.Z., ♂, X 1½.

Microtus oregoni adocetus Merriam

1908. *Microtus oregoni adocetus* Merriam, Proc. Biol Soc. Washington, 21:145, June 9, type from 2 mi. S South Yolla Bolly Mtn., 7500 ft., Tehama Co., California.

MARGINAL RECORDS.—Oregon: Belmont Orchard, 6 mi. S Medford. California: type locality; Blake Lookout on South Fork Mtn.

Microtus oregoni bairdi Merriam

1897. *Microtus bairdi* Merriam, Proc. Biol. Soc. Washington, 11:74, April 21, type from Glacier Peak, 7800 ft., Crater Lake, Klamath Co., Oregon.
1920. *Microtus oregoni bairdi*, Taylor, Jour. Mamm., 1:180, August 24.

MARGINAL RECORDS.—Oregon: type locality. California: Beswick.

Microtus oregoni oregoni (Bachman)

1839. *Arvicola oregoni* Bachman, Jour. Acad. Nat. Sci. Philadelphia, 8(1):60, type from Astoria, Clatsop Co., Oregon.
1896. *Microtus oregoni*, Miller, N. Amer. Fauna, 12:9, July 23.
1899. *Microtus morosus* Elliot, Field Columb. Mus., Publ. 30, Zool. Ser., 1:227, February 2, type from Boulder Lake, 5000 ft., Olympic Mts., Clallam Co., Washington.
1920. *Microtus oregoni cantwelli* Taylor, Jour. Mamm., 1:180, August 24, type from Glacier Basin, 5935 ft., Mt. Rainier, Pierce Co., Washington (see Dalquest, Univ. Kansas Publ., Mus. Nat. Hist., 2:357, April 9, 1948, who regards *M. o. cantwelli* as inseparable from *M. o. oregoni*).

MARGINAL RECORDS.—Washington: Mount Vernon; Stehekin, head Lake Chelan; Entiat River, 20 mi. from mouth; Mt. Aix, head Hindoo Creek; Signal Peak. Oregon: N base Three Sisters, 5000 ft.; Elk

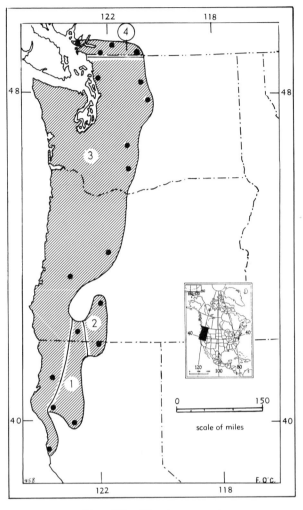

Map 458. *Microtus oregoni*.

1. *M. o. adocetus*   3. *M. o. oregoni*
2. *M. o. bairdi*   4. *M. o. serpens*

Head. California: Hoopa Valley; Mendocino City, thence north along coast to point of beginning.

Microtus oregoni serpens Merriam

1897. *Microtus serpens* Merriam, Proc. Biol. Soc. Washington, 11:75, April 21, type from Agassiz, British Columbia.
1920. *M[icrotus]. o[regoni]. serpens*, Taylor, Jour. Mamm., 1:180, August 24.

MARGINAL RECORDS.—British Columbia: Vancouver (Cowan and Guiguet, 1965:215); *Port Moody*; type locality; Allison Pass; Huntingdon.

Microtus chrotorrhinus
Rock Vole

Reviewed by Komarek, Jour. Mamm., 13:155–158, May 11, 1932.

External measurements: 137–170; 45–50; 19.4–22. Upper parts grayish bister to bright glossy bister, lined with black-tipped hairs; face, especially from eyes to nose, yellowish to dull orange rufous; venter plumbeous, sometimes with thin whitish wash; tail slightly bicolored, grayish brown to brownish above, paler below, body 2.4–2.7 times as long as tail; tail 2.0–2.5 times as long as hind foot; M3 with 5 closed triangles; incisive foramina short and wide.

Microtus chrotorrhinus carolinensis Komarek

1932. *Microtus chrotorrhinus carolinensis* Komarek, Jour. Mamm., 13:158, May 11, type from Great Smoky Mts. of North Carolina, about 5 mi. (airline distance) N of Smokemont, Swain Co., on a tributary of Bradley Fork, a small branch of Oconalufty River, 3200 ft.

MARGINAL RECORDS.—West Virginia: Randolph County. North Carolina: type locality. Tennessee: Dry Sluice Trail, 4300 ft., near divide (Mt. Collins).

Microtus chrotorrhinus chrotorrhinus (Miller)

1894. *Arvicola chrotorrhinus* Miller, Proc. Boston Soc. Nat. Hist., 26:190, March 24, type from head of Tuckerman's Ravine, 5300 ft., Mt. Washington, Coos Co., New Hampshire.
1896. *Microtus chrotorrhinus*, Bangs, Proc. Biol. Soc. Washington, 10:49, March 9.

MARGINAL RECORDS.—Quebec: Seal River; Mt. Albert, 3500 ft.; New Derreen. New Brunswick: *ca. 3 ½ mi. SW Summit Mt. Carleton* (Peterson and Symansky, 1963:278, as *M. chrotorrhinus* only); Trousers Lake. Maine: Mt. Coburn, about 12 mi. N The Forks. New Hampshire: type locality. New York: Hunter Mtn. Pennsylvania: near Sonestown. New York: Beedes [= Breed's]. Ontario: Timagami; Pancake Bay. Minnesota: Burntside Lake. Ontario: near Lowbush, Lake Abitibi. Quebec: Mistassini Lake area. **See** addenda.

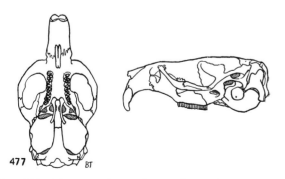

Fig. 477. *Microtus chrotorrhinus chrotorrhinus*, summit Mt. Washington, New Hampshire, No. 150181 U.S.N.M., ♀, X 1½.

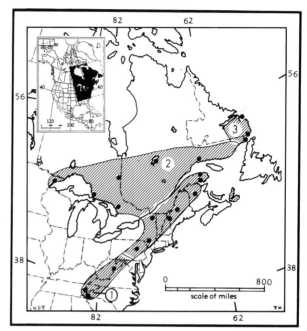

Map 459. *Microtus chrotorrhinus*.

1. *M. c. carolinensis*   2. *M. c. chrotorrhinus*
3. *M. c. ravus*

Microtus chrotorrhinus ravus Bangs

1898. *Microtus chrotorrhinus ravus* Bangs, Proc. Biol. Soc. Washington, 12:188, November 16, type from Black Bay, Strait of Belle Isle, Labrador.

MARGINAL RECORDS (Peterson, 1962:420).—Labrador: Hare Harbour; type locality; *Red Bay;* L'Anse au Loup.

Microtus xanthognathus (Leach)
Yellow-cheeked Vole

1815. *Arvicola xanthognatha* Leach, Zool. Miscl., 1:60, type from Hudson Bay.
1896. *M[icrotus]. xanthognathus*, Miller, N. Amer. Fauna, 12:66, July 23.

External measurements: 186–226; 45–53; 24–27. Large adults weigh as much as 170 grams. Upper parts dark sepia to bister, heavily lined with coarse black hairs on back; sides of nose and ear patch bright rusty yellowish; venter dusky gray; tail indistinctly bicolored, blackish above, dusky gray below; body 3.2–3.8 times as long as tail; tail 1.8–1.9 times as long as hind foot; M3 with only 3 closed triangles; middle section of m3 frequently divided into 2 nearly closed triangles; incisive foramina long and narrow.

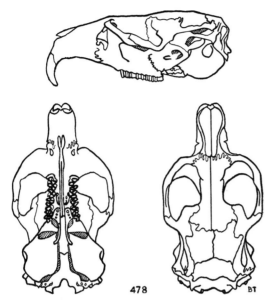

478

Fig. 478. *Microtus xanthognathus*, Fort Smith, Mackenzie, No. 109370 U.S.N.M., ♂, X 1½.

MARGINAL RECORDS.—Mackenzie: Arctic Coast [Franklin Bay] E of Fort Anderson; Fort Franklin; Fort Rae; Fort Smith. Manitoba: Fort Churchill; Nelson River. Alberta: Athabaska River, 30 mi. above Pelican Portage; Cache Pecotte, 40 mi. E Jasper House. Mackenzie: Liard River. Yukon: Dominion Creek, head Indian River (Youngman, 1975:101). Alaska: Eagle; S of Fairbanks; Toklat River, McKinley National Park; near mouth Takotna River; Yukon River 200 mi. SW mouth Porcupine; mouth Porcupine River. Yukon: Yukon–Alaska boundary at 69°20' (*loc. cit.*).

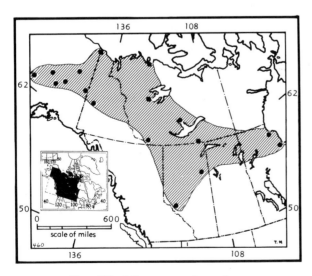

Map 460. *Microtus xanthognathus.*

Microtus ochrogaster
Prairie Vole

x½

Reviewed by Bole and Moulthrop, Sci. Publ., Cleveland Mus. Nat. Hist., 5:155–161, September 11, 1942.

External measurements: 130–172; 24–41; 17–22; 11–15. Weight, 37.0–48.0 grams. Pelage long, lax, and coarse; upper parts light gray to dark bister, with peppery appearance resulting from mixture of black and whitish, pale fulvous or hazel-tipped hairs; sides paler; venter neutral gray or washed with whitish or pale cinnamon; tail sharply bicolored, dusky to dark bister above, whitish to buffy below; plantar tubercles, 5; side glands obscure or wanting, rarely discernible; mammae, 2–1 = 6; ears of medium size; molars having wide re-entrant angles; M3 with 2 closed triangles; m1 with 3 closed and 2 open triangles; m2 with anterior pair of triangles confluent; m3 with 3 transverse loops, middle loop sometimes constricted, or even divided into 2 triangles.

Microtus ochrogaster haydenii (Baird)

1858. *Arvicola (Pedomys) haydenii* Baird, Mammals, *in* Repts. Expl. Surv. . . . , 8(1):543, July 14, type from Fort Pierre, Stanley Co., South Dakota.
1907. *Microtus ochrogaster haydeni*, Osgood, Proc. Biol. Soc. Washington, 20:48, April 18.

MARGINAL RECORDS.—Montana: 3 mi. E, 5 mi. S Culbertson, 1860 ft. North Dakota: Cannon Ball. South Dakota: 1 mi. W Oacoma. Nebraska (Jones, 1964c:225, 226): *Verdigre;* Oakdale; 10 mi. S Ord; *1³⁄₄ mi. S. Kearney;* 1 mi. S Alma. Kansas: 3 mi. N, 2 mi. W Hoisington. Oklahoma: Garnett; Fort Reno; Alva. Kansas: Cairo; Scott County State Park. New Mexico: 4 mi. S, 2½ mi. W Cimarron (Rowlett, 1972:640, as *M. ochrogaster* only). Colorado: Cañon City; 2½ mi. S, 16 mi. W Loveland, 7280 ft. (Armstrong, 1972:245). Wyoming: 26 mi. N, 4½ mi. E Laramie (Long, 1965a:659). Nebraska (Jones, 1964c:226): Mitchell; *Glen.* South Dakota: Bald Hills, S Pactola (Turner, 1974:114).

Microtus ochrogaster ludovicianus V. Bailey

1900. *Microtus ludovicianus* V. Bailey, N. Amer. Fauna, 17:74, June 6, type from Iowa, Calcasieu Parish, Louisiana.
1972. *Microtus ochrogaster ludovicianus*, Ruan and Laughlin, Southwestern Nat., 16:439, February 18; also Lowery, The mammals of Louisiana . . . , p. 264, June 7, 1974.

MARGINAL RECORDS.—Louisiana: type locality. Texas: Sour Lake.

Microtus ochrogaster minor (Merriam)

1888. *Arvicola austerus minor* Merriam, Amer. Nat., 22:600, July, type from Bottineau, at base Turtle Mts., Bottineau Co., North Dakota.
1942. *Microtus ochrogaster minor*, Bole and Moulthrop, Sci. Publs., Cleveland Mus. Nat. Hist., 5:160, September 11.

MARGINAL RECORDS—Alberta: SE of Astotin Lake. Saskatchewan: Wingard. Manitoba: Riding Mountain National Park; Winnipeg; Ridgeville. Minnesota: 5 mi. SE Nimrod (Heaney and Birney, 1975:32, as *M. ochrogaster* only); Elk River. Wisconsin (Long, 1976:1): Juneau County; Waushara County; *Portage County*; Clark County. Minnesota: Nicollet County; 4½ mi. N, 8 mi. W Holland (Heaney and Birney, 1975:32, as *M. ochrogaster* only); Ortonville. North Dakota: Oakes; Goodall. Saskatchewan: Big Muddy Lake. Alberta: "A short distance east of Red River"; Edmonton.

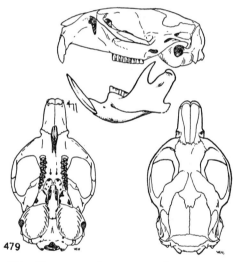

Fig. 479. *Microtus ochrogaster ochrogaster*, Lawrence, Douglas Co., Kansas, No. 1075 K.U., ♀, X 1½.

Microtus ochrogaster ochrogaster (Wagner)

1842. *Hypudaeus ochrogaster* Wagner, *in* Schreber, Die Säugthiere . . . , Suppl., 3:592, type from America, probably from New Harmony, Indiana (see Bole and Moulthrop, Sci. Publs., Cleveland Mus. Nat. Hist., 5:157, September 11, 1942).
1898. *Microtus (Pedomys) ochrogaster*, J. A. Allen, Bull. Amer. Mus. Nat. Hist., 10:459, November 10.
1853. *Arvicola austerus* Le Conte, Proc. Acad. Nat. Sci. Philadelphia, 6:405, type from Racine, Wisconsin.
1858. *Arvicola (Pedomys) cinnamonea* Baird, Mammals, *in* Repts. Expl. Surv. . . . , 8(1):541, July 14, type supposed by Baird to have come from Pembina, Minnesota [= Pembina, North Dakota?], but see V. Bailey (N. Amer. Fauna, 17:74, June 6, 1900), who thinks that the type came from somewhere within the geographic range of *M. o. ochrogaster*.

MARGINAL RECORDS.—Minnesota: Winona. Wisconsin: Sauk County (Long, 1976:1); Dodge County (*ibid.*); Racine (Jackson, 1961:239). Michigan: Cass County. Ohio: Union Twp., Clermont Co. Kentucky: 2 1/10 km. SE Somerset (Fassler, 1974:42, as *M. ochrogaster* only). Tennessee: ¾ mi. S Allardt (Dimmick, 1969:126); 8 mi. N Pikeville (*ibid.*). Alabama: N side Tennessee River, 12 mi. S Huntsville (Whitaker and Zimmerman, 1968:328, as *M. ochrogaster* only). Tennessee: S of Lawrenceburg (Dimmick, 1969:126); 4 mi. SW Bathsprings (Severinghaus and Beasley, 1973:131); 5½ mi. W Brownsville (*ibid.*); 6 mi. SW Miston (*ibid.*). Arkansas (Kee and Enright, 1970:358, unless otherwise noted): Jonesboro; 1 mi. W El Paso; 1 mi. E Brasfield; 2 mi. SW Slovak (Sealander, *et al.*, 1975:423); Gateway (Brown, 1964:471). Kansas: 18 mi. SW Columbus; *Cherryvale*; Smoky Hill River, 1 mi. S, ½ mi. W Lindsborg; ½ mi. S, 3½ mi. W Beloit. Nebraska (Jones, 1964c:226, 227): *1½ mi. S Franklin;* 1 mi. E Ravenna; Norfolk; 8 mi. SW Niobrara. Iowa: *Lyon Twp., Lyon Co.* (Bowles, 1975:100); Melvin.

Microtus ochrogaster ohionensis Bole and Moulthrop

1942. *Microtus ochrogaster ohionensis* Bole and Moulthrop, Sci. Publs., Cleveland Mus. Nat. Hist., 5:155, September 11, type from Symmes Creek, 2 mi. N Chesapeake, Lawrence Co., Ohio.

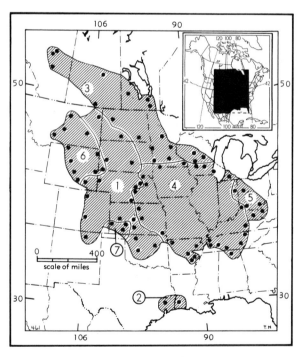

Map 461. *Microtus ochrogaster.*

1. *M. o. haydenii*   4. *M. o. ochrogaster*
2. *M. o. ludovicianus*   5. *M. o. ohionensis*
3. *M. o. minor*   6. *M. o. similis*
                    7. *M. o. taylori*

MARGINAL RECORDS.—Indiana: Salamonia. Ohio: Shelby County; type locality. West Virginia: Spring Valley Golf Course, near Huntington. Kentucky: Rowan County. Ohio: Ripley.

Microtus ochrogaster similis Severinghaus

1977. *Microtus ochrogaster similis* Severinghaus, Proc. Biol. Soc. Washington, 90:49, June 16, type from KULR-TV Tower, Billings, Yellowstone Co., Montana.

MARGINAL RECORDS.—Montana (Severinghaus, 1977:53): 4 mi. N James Kipp State Park; Tongue River, 3 mi. S Miles City. Wyoming: 15 mi. ENE Sundance (Severinghaus, 1977:53). South Dakota: 9 mi. S, 3 mi. W Oelrichs (*ibid.*). Nebraska: *Agate* (*op. cit.*:54). Wyoming: 2 mi. S, ½ mi. E Lusk (*ibid.*); Pass, Carbon Co. (*ibid.*); Sun (Long, 1965a:660); 12 mi. N, 3 mi. W Shoshoni (*ibid.*); 4 mi. N Garland (*ibid.*). Montana (Hoffmann, *et al.*, 1969:591, as *M. ochrogaster* only): *3 mi. E Bear Creek, Carbon Co.*; Bozeman; *13 mi. ENE Melville;* 4 mi. SE Shawmut.

Microtus ochrogaster taylori Hibbard and Rinker

1943. *Microtus ochrogaster taylori* Hibbard and Rinker, Univ. Kansas Sci. Bull., 29(pt. 2, no. 4):256, October 15, type from farm of H. H. Hildebrand, ½ mi. N Fowler, Meade Co., Kansas.

MARGINAL RECORDS.—Kansas: Coolidge; ½ mi. NW Bellfont; type locality.

Microtus mexicanus
Mexican Vole

External measurements: 121–152; 24–35; 17–21; 12–15. Weight, 26–43.4 grams. Pelage coarse and lax; upper parts cinnamon buff to dark cinnamon brown, with mixture of black hairs resulting in grizzled-brownish color; sides paler; venter washed with grayish buff to cinnamon, sometimes whitish; tail slightly bicolored, dusky to dark brown above, paler below; incisive foramina short, wide, and truncate posteriorly. Body 3.3–3.8 times as long as tail; tail only 1.5–1.7 times as long as hind foot; mammae 1–1 = 4; anterior lower molar (m1) with extra anterior trefoil. In several morphological features *M. mexicanus* is intermediate between *M. ochrogaster* and *M. pinetorum.* Hooper and Hart (1962:66) would place *M. mexicanus* in subgenus *Pitymys.*

Microtus mexicanus fulviventer Merriam

1898. *Microtus fulviventer* Merriam, Proc. Biol. Soc. Washington, 12:106, April 20, type from Cerro San Felipe, 10,200 ft., Oaxaca.
1964. *Microtus mexicanus fulviventer,* Musser, Occas. Pap. Mus. Zool., Univ. Michigan, 636:20, June 17.

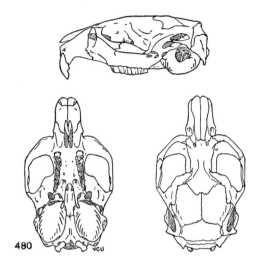

Fig. 480. *Microtus mexicanus fundatus,* 3½ mi. S Pátzcuaro, 7900 ft., Michoacán, No. 100637 M.V.Z., ♂, X 1½.

MARGINAL RECORDS.—Oaxaca: Reyes [Pápalo Santos Reyes]; near Cerro Pelón, 9200 ft. (Musser, 1964:20); Mt. Zempoaltepec; mts. near Ozolotepec; *near Campemento Río Molino, 7300 ft.* (*ibid.*); 15 mi. W Oaxaca; San Andrés Chicahuaxtla (Goodwin, 1969:209); 16 km. SW Cuquila, 2380 m. (Webb and Baker, 1971:144).

Microtus mexicanus fundatus Hall

1948. *Microtus mexicanus fundatus* Hall, Univ. Kansas Publ., Mus. Nat. Hist., 1:425, December 24, type from 3½ mi. S Pátzcuaro, 7900 ft., Michoacán.

MARGINAL RECORDS.—Michoacán: Nahuatzen; 10 mi. SE Pátzcuaro, 9200 ft.; *6 mi. S Pátzcuaro, 8000 ft.*

Microtus mexicanus guadalupensis V. Bailey

1902. *Microtus mexicanus guadalupensis* V. Bailey, Proc. Biol. Soc. Washington, 15:118, June 2, type from Guadalupe Mts., 7800 ft., El Paso Co., Texas.

MARGINAL RECORDS.—New Mexico (Findley, *et al.*, 1975:259, as *M. mexicanus* only): Capulin Springs, Sandia Mts.; 5 mi. N, 9 mi. E Capitan, Capitan Mts. Texas: The Bowl, 8000 ft., Guadalupe Mts. New Mexico: Lightning Lake, sec. 9, T. 18 S, R. 12 E.

Microtus mexicanus hualpaiensis Goldman

1938. *Microtus mexicanus hualpaiensis* Goldman, Jour. Mamm., 19:493, November 14, type from Hualpai Peak, 8400 ft., Hualpai Mts., Mohave Co., Arizona. Known only from type locality.

Microtus mexicanus madrensis Goldman

1938. *Microtus mexicanus madrensis* Goldman, Jour. Mamm., 19:493, November 14, type from Río Gavilán, 6700 ft., 5 mi.

W Colonia García, about 60 mi. SW Casas Grandes, Chihuahua.

MARGINAL RECORDS.—Chihuahua: type locality; *Colonia García* (Anderson, 1972:365); *Meadow Valley* (*ibid.*); 2 mi. W Miñaca, 6900 ft. (*ibid.*, in owl pellet). Durango (Baker and Greer, 1962:130): Ciudad; 10 mi. SW El Salto; 1½ mi. W San Luis. Chihuahua (Anderson, 1972:365): Sierra Madre, near Guadalupe y Calvo, 7000 ft.; Chuhuichupa.

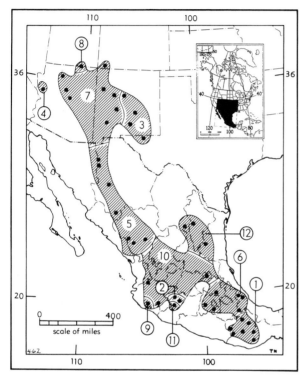

Map 462. *Microtus mexicanus.*

1. *M. m. fulviventer*	7. *M. m. mogollonensis*
2. *M. m. fundatus*	8. *M. m. navaho*
3. *M. m. guadalupensis*	9. *M. m. neveriae*
4. *M. m. hualpaiensis*	10. *M. m. phaeus*
5. *M. m. madrensis*	11. *M. m. salvus*
6. *M. m. mexicanus*	12. *M. m. subsimus*

Microtus mexicanus mexicanus (Saussure)

1861. *Arvicola* (*Hemiotomys*) *mexicanus* Saussure, Revue et Mag. Zool., Paris, ser. 2, 13:3, January, type from Volcán de Orizaba, Puebla.
1897. [*Microtus*] *mexicanus*, Trouessart, Catalogus mammalium . . . , p. 564.

MARGINAL RECORDS.—Hidalgo: Sierra de Pachuca. Veracruz (Hall and Dalquest, 1963:318): 6 km. SSE Altotonga, 9000 ft.; 3 km. E Las Vigas, 8000 ft.; 4 mi. SW Acultzingo, 7000 ft. Morelos: Huitzilac (Ramírez-P., 1971:279). México: N slope Volcán de Toluca.

Microtus mexicanus mogollonensis (Mearns)

1890. *Arvicola mogollonensis* Mearns, Bull. Amer. Mus. Nat. Hist., 2:283, February 21, type from Bakers Butte, Mogollon Mts., Coconino Co., Arizona.
1932. *Microtus mexicanus mogollonensis,* V. Bailey, N. Amer. Fauna, 53:204, March 1.

MARGINAL RECORDS.—Colorado: Prater Canyon, Mesa Verde National Park, 7600 ft. (Armstrong, 1972:243). New Mexico (Findley, *et al.,* 1975:259, 260, as *M. mexicanus* only): 1 mi. E McGaffey; *15 mi. NE San Mateo;* 4 mi. WSW Cebolleta; head Water Canyon, Magdalena Mts.; 23 mi. S Wall Lake, Black Range. Arizona: type locality; 8 mi. NW Flagstaff (Brown, 1968:159, as *M. mexicanus* only); Pasture Wash Ranger Station, 6300 ft. (Hoffmeister, 1971:176); *Pasture Wash, jct. of roads W-9A and W-9* (*ibid.*).

Microtus mexicanus navaho Benson

1934. *Microtus mexicanus navaho* Benson, Proc. Biol. Soc. Washington, 47:49, February 9, type from Soldier Spring, 8800 ft., E slope Navajo Mtn., San Juan Co., Utah.

MARGINAL RECORDS.—Utah: type locality. Arizona: *5 mi. E Rainbow Lodge, Navajo Mtn.*

Microtus mexicanus neveriae Hooper

1955. *Microtus mexicanus neveriae* Hooper, Occas. Pap. Mus. Zool., Univ. Michigan, 565:23, March 31, type from mts. about 20 mi. SE Autlán, near lumber mill of La Nevería, 8200 ft., Jalisco.

MARGINAL RECORDS.—Jalisco: *mts. S Autlán, 9000 ft.;* type locality; *20 mi. SSE Autlán, 6500 ft.*

Microtus mexicanus phaeus (Merriam)

1892. *Arvicola phaeus* Merriam, Proc. Biol. Soc. Washington, 7:171, September 29, type from N slope El Nevado de Colima, 10,000 ft., Jalisco.
1900. *Microtus mexicanus phaeus,* V. Bailey, N. Amer. Fauna, 17:54, June 6.

MARGINAL RECORDS.—Querétaro: Pinal de Amoles. Jalisco: SE slope El Nevado de Colima, 9100 ft. (Baker and Phillips, 1965:692); *NW slope El Nevado de Colima, 8000 ft.;* 9 mi. N, 10 mi. W Magdalena.

Microtus mexicanus salvus Hall

1948. *Microtus mexicanus salvus* Hall, Univ. Kansas Publ., Mus. Nat. Hist., 1:426, December 24, type from Mt. Tancítaro, 11,400 ft., Michoacán. Known only from Mt. Tancítaro at elevations of 7800–11,400 ft.

Microtus mexicanus subsimus Goldman

1938. *Microtus mexicanus subsimus* Goldman, Jour. Mamm., 19:494, November 14, type from Sierra Guadalupe, southeastern Coahuila.

MARGINAL RECORDS.—Coahuila: 12 mi. E Antonio de las Alazanas. Tamaulipas: mountains near Miquihuana. Coahuila: type locality.

Subgenus Pitymys McMurtrie

1831. *Pitymys* McMurtrie, The animal kingdom . . . by the Baron Cuvier . . . , Amer. ed., 1:434. Type, *Psammomys pinetorum* Le Conte.

Skull weak, with relatively small squamosal crests, and supraorbital ridges widely separated in interorbital region of adults; tympanic bullae small; molars narrow; M3 with 2 closed triangles; m1 with 3 closed and 2 open triangles; m2 with anterior pair of triangles confluent; m3 with 3 transverse loops; plantar pads, 5. External form modified for semifossorial life; fur soft and dense; ear reduced. Sole not fully haired; tail short; mammae, 0–2 = 4. Cranially this subgenus closely resembles *M. ochrogaster*, which has been thought by several authors to merit subgeneric rank under the subgeneric name *Pedomys*. In areas where both subgenera occur, *Pitymys* can be distinguished by reddish (not grayish or blackish) upper parts, and molelike (not volelike) fur.

Microtus pinetorum
Pine Vole

External measurements: 105–145; 17–25; 13–20; 8–12. Weight, 25.2–38.8 grams. Upper parts bright russet brown to brownish chestnut, sometimes lined with blackish-tipped hairs on back and rump; venter plumbeous with wash of dull buff to bright cinnamon; tail indistinctly bicolored or even unicolored, usually same color as back.

Microtus pinetorum auricularis (V. Bailey)

1898. *Microtus pinetorum auricularis* V. Bailey, Proc. Biol. Soc. Washington, 12:90, April 30, type from Washington, Adams Co., Mississippi.

MARGINAL RECORDS.—Ohio: Union Twp., Clermont Co.; Scioto County (Gottschang, 1965:48, as *Pitymys pinetorum* only). Kentucky: Monticello. Tennessee: LaFollette. North Carolina: Cherokee County. Georgia (Golley, 1962:147): White County; Polk

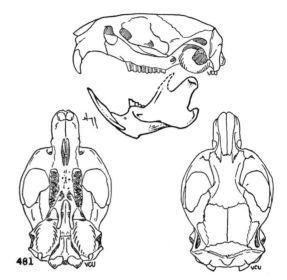

Fig. 481. *Microtus pinetorum nemoralis*, 8½ mi. SW Toronto, Greenwood Co., Kansas, No. 8038 K.U., ♀, X 1½.

County. Mississippi (Kennedy, *et al.*, 1974:20): Wayne County; Pearl River County. Louisiana: Varnado (Crain and Packard, 1966:324); Baton Rouge; Mansfield (Lowery, 1974:268). Texas: Newton County (Packard, 1961:195); 3 mi. S Kerrville; Gillespie County (Davis, 1966:207, as *Pitymys pinetorum* only); Wood County (*ibid.*, as *Pitymys pinetorum* only); Jefferson. Illinois: Olive Branch; Reevesville. Indiana: Worthington; *Ripley County* (Lindsay, 1960:259).

Microtus pinetorum carbonarius (Handley)

1952. *Pitymys pinetorum carbonarius* Handley, Jour. Washington Acad. Sci., 42:152, May 28, type from Eubank, Pulaski Co., Kentucky.
1953. *Microtus pinetorum carbonarius*, Hall and Cockrum, Univ. Kansas Publ., Mus. Nat. Hist., 5:449, January 15.

MARGINAL RECORDS.—Ohio: Symmes Creek. Virginia: *Cleveland*. Tennessee: Watauga Valley; Jefferson County; *10 mi. SW Knoxville* (Tuttle, 1964:146, as *Pitymys pinetorum* only); environs Oak Ridge National Laboratory, Roane Co. (Cosgrove and O'Farrell, 1965:510, as *M. pinetorum* only); High Cliff. Kentucky: Quicksand.

Microtus pinetorum nemoralis V. Bailey

1898. *Microtus pinetorum nemoralis* V. Bailey, Proc. Biol. Soc. Washington, 12:89, April 30, type from Stilwell, Adair Co., Oklahoma.

MARGINAL RECORDS.—Minnesota: La Crescent (Heaney and Birney, 1975:33, as *M. pinetorum* only). Wisconsin: Lynxville. Iowa: Iowa City. Missouri: Kimmswick; Williamsville. Arkansas: Beebe; Fourche Bayou, near Little Rock. Texas: Bowie County (Davis, 1966:207, as *Pitymys pinetorum* only); S of Woodbine;

4 mi. E Stoneburg (Dalquest, 1968:19). Oklahoma: *Indiahoma;* Wichita Mountains Wildlife Refuge (Glass and Halloran, 1961:238); Logan County. Kansas: 3 mi. SE Arkansas City; Manhattan. Nebraska (Jones, 1964c:234): 2 mi. S, ½ mi. E Barnston; *Nebraska City.* Iowa: Council Bluffs; Floyd County (Bowles, 1975:103). Minnesota: *Caledonia.*

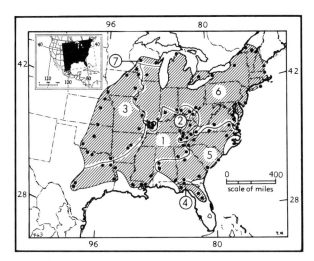

Map 463. *Microtus pinetorum.*

Guide to subspecies
1. *M. p. auricularis*
2. *M. p. carbonarius*
3. *M. p. nemoralis*
4. *M. p. parvulus*
5. *M. p. pinetorum*
6. *M. p. scalopsoides*
7. *M. p. schmidti*

Microtus pinetorum parvulus (A. H. Howell)

1916. *Pitymys parvulus* A. H. Howell, Proc. Biol. Soc. Washington, 29:83, April 4, type from "near the town of Lynne, Marion County, Florida" (see A. H. Howell, Jour. Mamm., 15:72, February 15, 1934).
1953. *Microtus parvulus,* Hall and Cockrum, Univ. Kansas Publ., Mus. Nat. Hist., 5:451, January 15.
1952. *Pitmys* [sic] *pinetorum parvulus,* Sherman, Quart. Jour. Florida Acad. Sci., 15:93, June; also, *P[itymys]. p[inetorum]. parvulus,* Arata, Jour. Mamm., 46:87, February 20, 1965.

MARGINAL RECORDS.—Georgia: 10 mi. SSW Thomasville. Florida: Gainesville; type locality; Quincy.

Microtus pinetorum pinetorum (Le Conte)

1830. *Psammomys pinetorum* Le Conte, Ann. Lyc. Nat. Hist., New York, 3:133, type from pine forests of Georgia, probably on the Le Conte plantation, near Riceboro, Liberty Co.
1896. *Microtus pinetorum,* Miller, N. Amer. Fauna, 12:9, July 23.

MARGINAL RECORDS.—Virginia: Prince George County. North Carolina: Currituck; Tarboro. South Carolina: Georgetown; Beaufort. Georgia (Golley, 1962:147): Appling County; Brady County. Alabama:

Ashford; Greensboro; 2 mi. W Tuscaloosa (Wolfe and Rogers, 1969:610, as *Pitymys pinetorum* only). Georgia: De Kalb County (Golley, 1962:147). North Carolina: Highlands; Marshall. Virginia: Patrick County.

Microtus pinetorum scalopsoides (Audubon and Bachman)

1841. *Arvicola scalopsoides* Audubon and Bachman, Proc. Acad. Nat. Sci. Philadelphia, 1:97, October, type from Long Island, New York.
1896. *Microtus pinetorum scalopsoides,* Batchelder, Proc. Boston Soc. Nat. Hist., 27:187, October.
1853. *Arvicola apella* Le Conte, Proc. Acad. Nat. Sci. Philadelphia, 6:405, type from Pennsylvania.
1858. *Arvicola kennicottii* Baird, Mammals, *in* Repts. Expl. Surv. . . . , 8(1):547, July 14, type from Illinois.

MARGINAL RECORDS.—Michigan: Emmet County. Ontario: near London; Point Abino. New York: Lewis County. Quebec (Wrigley, 1969:209, as *M. pinetorum* only): Mont St. Hilaire; South Bolton. Vermont: S side Bluff Mtn., Island Pont (Miller, 1964:627, as *M. pinetorum* only). New Hampshire: Durham. Massachusetts: Saugus, about 9 mi. N Boston. New York: Montauk Point, Long Island. New Jersey: Tuckerton. Virginia: Cape Charles; Wallaceton; Campbell County; Washington County. Ohio: Hocking. Indiana: Brookville. Illinois: Metropolis; Alto Pass; Warsaw. Wisconsin: near Prairie du Sac.

Microtus pinetorum schmidti (Jackson)

1941. *Pitymys pinetorum schmidti* Jackson, Proc. Biol. Soc. Washington, 54:201, December 8, type from Worden Twp., Clark Co., Wisconsin.
1953. *Microtus pinetorum schmidti,* Hall and Cockrum, Univ. Kansas Publ., Mus. Nat. Hist., 5:451, January 15.

MARGINAL RECORDS.—Wisconsin: type locality; Bairds Creek (Jackson, 1961:245).

Microtus quasiater (Coues)
Jalapan Pine Vole

1874. *Arvicola (Pitymys) pinetorum* var. *quasiater* Coues, Proc. Acad. Nat. Sci. Philadelphia, 26:191, December 15, type from Jalapa, Veracruz.
1896. *M[icrotus]. quasiater,* Miller, N. Amer. Fauna, 12:60, July 23.

External measurements: 112–137; 17–25; 17–19; 11–18. Uniform dark umber or seal brown above, slightly paler on venter; tail slightly paler below than above.

MARGINAL RECORDS.—San Luis Potosí: Apetsco. Veracruz (Hall and Dalquest, 1963:320): 4 km. W Tlapacoyan, 1700 ft.; *5 km. N Jalapa;* Jalapa; *Huatusco, 5000 ft.;* Tuxpango. Oaxaca: Huautla (Goodwin, 1969:213). Puebla: 7½ mi. by road NE from

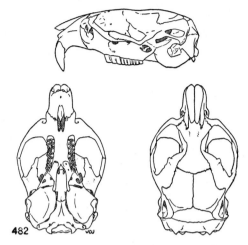

Fig. 482. *Microtus quasiater*, Teocelo, 4500 ft., Veracruz, No. 30706 K.U., ♀, X 1½.

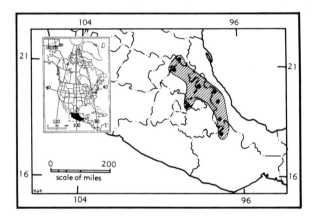

Map 464. *Microtus quasiater*.

Tezuitlan (*sic*) (5965 U. Minn.); Huauchinango. Hidalgo: 16 km. NW Jacala, 1550 m.

Subgenus **Stenocranius** Kastschenko

1901. *Stenocranius* Kastschenko, Ann. Mus. Zool., Acad. Imp. Sci., St. Pétersbourg, 6:167. Type, *Microtus slowzowi* Poljakoff [= *Mus gregalis* Pallas].

Dentition normal but skull unusually narrow; median crest sharp; claws tend to be enlarged; ears small.

Some authors since 1960 have arranged the American taxa of the subgenus *Stenocranius* as subspecies of the Old World species *Microtus gregalis*. The important literature bearing on this matter is cited by Fedyk (1970:143–152). His systematic arrangement, which was based on microscopic and macroscopic resemblances and dif-ferences to be seen between all named taxa of the subgenus, is followed here.

Microtus miurus
Singing Vole

External measurements: 101–155; 19–41; 19–21; 10–14. Weight, 22–52.1 grams. Upper parts vary from pale buffy gray through pale ochraceous to pale tawny, lightly mixed with buffy-tipped to black-tipped guard-hairs; venter whitish gray to buffy; tail dusky above, buffy to ochraceous below.

Microtus miurus andersoni Rand

1945. *Microtus andersoni* Rand, Bull. Nat. Mus. Canada, 99:42, prior to June 20, type from near headwaters Little Keele River, 5500 ft., 82 mi. W Mackenzie River on Canol Road, Mackenzie. Known only from type locality. According to Youngman, Nat. Mus. Canada Publ. Zool., 10:102, September 4, 1975, the holotype and 15 topotypes "are barely recognizable as a local deme. Cranially, and in pelage colour, these specimens should be referred to *M. m. muriei.*"
1952. *Microtus miurus andersoni*, Hall and Cockrum, Univ. Kansas Publ., Mus. Nat. Hist., 5:312, November 17.

Microtus miurus cantator Anderson

1947. *Microtus cantator* Anderson, Bull. Nat. Mus. Canada, 102:161, January 24, type from mtn. top near Tepee Lake, lat. 61° 35′ N, long. 140° 22′ W, N slope Saint Elias Range, Yukon.
1952. *Microtus miurus cantator*, Hall and Cockrum, Univ. Kansas Publ., Mus. Nat. Hist., 5:312, November 17.

MARGINAL RECORDS.—Alaska: 25 km. W Paxcon (Rausch, 1964:348); *Wrangell Mountains* (*ibid.*). Yukon: type locality; head Kluane Lake (Banfield, 1960:2); 25 mi. SSE Destruction Bay, 6400 ft. (Youngman, 1975:101); *Sheep Mtn., Alaska Hwy., Mile 1061, 5500 ft.* (Youngman, 1964:4, as *M. miurus* only).

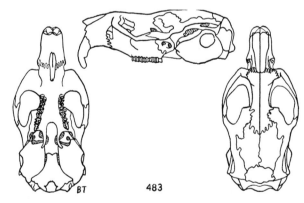

Fig. 483. *Microtus miurus muriei*, Chandler Lake, lat. 68° 12′ N, long. 152° 45′ W, 2900 ft., Alaska, No. 43758 K.U., ♀, X 1½.

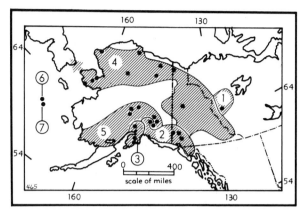

Map 465. *Microtus miurus* and *Microtus abbreviatus*.

Guide to kinds
1. *M. m. andersoni*
2. *M. m. cantator*
3. *M. m. miurus*
4. *M. m. muriei*
5. *M. m. oreas*
6. *M. a. abbreviatus*
7. *M. a. fisheri*

Microtus miurus miurus Osgood

1901. *Microtus miurus* Osgood, N. Amer. Fauna, 21:64, September 26, type from head Bear Creek, in mts. near Hope City, Turnagain Arm, Cook Inlet, Alaska.

MARGINAL RECORDS (Rausch, 1964:348, 349).—Alaska: Willow Pass, 90 km. N Anchorage; type locality; *Palmer Creek*; N fork Indian Creek.

Microtus miurus muriei Nelson

1931. *Microtus muriei* Nelson, Jour. Mamm., 12:311, August 24, type from Kutuk River (tributary of Alatna River), Endicott Mts., Alaska.
1952. *Microtus miurus muriei*, Hall and Cockrum, Univ. Kansas Publ., Mus. Nat. Hist., 5:312, November 17.
1950. *Microtus miurus paneaki* Rausch, Jour. Washington Acad. Sci., 40:135, April 21, type from Tolugak Lake (lat. 68° 24′ N, long. 152° 10′ W), Brooks Range, Alaska (regarded as inseparable from *M. m. muriei* by Hall and Cockrum, Univ. Kansas Publ., Mus. Nat. Hist., 5:311, November 17, 1952).

MARGINAL RECORDS.—Alaska: Fish Creek, SE of Teshekpuk Lake; Lake Schrader. Yukon (Youngman, 1975:104): Firth River, 15 mi. S mouth Joe Creek; *British Mountains, 20 mi. SE mouth Joe Creek*; 20 mi. S Chapman Lake. Alaska: Arctic Village; type locality; *Kuzitrin Lake*; Anvil Peak, Cooper Gulch; Lava Lake; Magnet Creek; Cape Sabine (Childs, 1969:54).

Microtus miurus oreas Osgood

1907. *Microtus miurus oreas* Osgood, Proc. Biol. Soc. Washington, 20:61, April 18, type from Toklat River, Alaskan Range, Alaska.

MARGINAL RECORDS.—Alaska: Savage River; Jarvis Creek; Gakona Glacier; Swede Lake; 25 km. W Paxson (Rausch, 1964:348); N side Mt. McKinley; Glacier Creek.

Microtus abbreviatus
Insular Vole

External measurements: 160–178; 25–32; 22–24. Fur long, concealing ears; upper parts yellowish brown to rich dark buff, sprinkled with black-tipped hairs; underparts creamy white to strong, clear buff; tail sharply or slightly bicolored, a narrow line of dusky to dark brownish above, creamy to buff below. For placement of this species at the end of subgenus *Stenocranius*, see Rausch and Rausch (1968:96).

Microtus abbreviatus abbreviatus Miller

1899. *Microtus abbreviatus* Miller, Proc. Biol. Soc. Washington, 13:13, January 31, type from Hall Island, Bering Sea, Alaska. Known only from type locality.

Microtus abbreviatus fisheri Merriam

1900. *Microtus abbreviatus fisheri* Merriam, Proc. Washington Acad. Sci., 2:23, March 14, type from St. Matthew Island, Bering Sea, Alaska. Known only from type locality.

Subgenus Arvicola Lacépède

1799. *Arvicola* Lacépède, Tableau des divisions, sous-divisions, ordres, et genres des mammifères, 10. Type, *Mus amphibius* Linnaeus.
1894. *Aulacomys* Rhoads, Amer. Nat., 28:182, February 17. Type, *Aulacomys arvicoloides* Rhoads. Inseparable from *Arvicola* according to Zimmermann, Säugetierkund. Mitteil., 3:110–112, July 1, 1955.

Plantar tubercles, 5; side glands on flanks of males conspicuous; musk-bearing anal gland; feet large; tail long; fur full and long; bullae small; incisors projecting far beyond premaxillae; molars with constricted and tightly closed sections; M2 with 4 closed triangles; M3 with 2 or 3 closed triangles; m1 with 5 closed triangles; m3 with 3 transverse loops.

Remarks under the genus *Microtus* on the need for additional material to fill gaps in the fossil record apply especially well to *Arvicola*. It is difficult to find any one feature setting *M. terrestris* and *M. richardsoni* apart from other microtines in the genus *Microtus*. If *Arvicola* be recognized at all (Heptner, 1952, would not do so), recognition would rest only on a combination of features not duplicated in other species. Miller (1912: 723), Ognev (1950:588), Zimmerman (1955:112), Hooper and Hart (1962:65), and Jannett and Jannett (1974:232), among others, accord *Arvicola* generic rank. *Arvicola* is provisionally accorded only subgeneric rank in the present work.

Microtus richardsoni
Water Vole

X 1

External measurements: 198–261; 69–92; 25–30; 15–20. Weight of two adult males, 112, 123.3 grams. Pelage long; upper parts grayish sepia to dark sepia or dark reddish brown, often darkened with black-tipped hairs; underparts plumbeous, with white or silvery white wash; tail bicolored, dusky above, grayish below; other characters as given above for the subgenus. Number of embryos varies from 4 to 7.

Microtus richardsoni arvicoloides (Rhoads)

1894. *Aulacomys arvicoloides* Rhoads, Amer. Nat., 28:182, February, type from Lake Keechelus, 8000 ft., Kittitas Co., Washington. (Inferentially regarded as indistinguishable from *M. r. richardsoni* by Cowan and Guiguet, Handbook No. 11, British Columbia Provincial Mus., pp. 215–217, 222, July 15, 1956.)

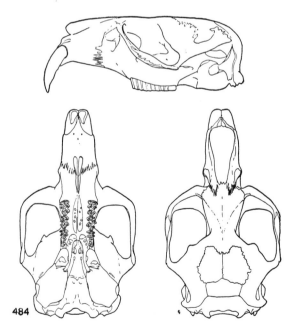

484

Fig. 484. *Microtus richardsoni arvicoloides*, Tumalo Creek, 6100 ft., 15 mi. W Bend, Deschutes Co., Oregon, No. 83968 M.V.Z., ♂, X 1½.

1900. *Microtus richardsoni arvicoloides,* V. Bailey, N. Amer. Fauna, 17:62, June 6.
1895. *Microtus principalis* Rhoads, Amer. Nat., 29:940, October, type from Mt. Baker Range, British Columbia.

MARGINAL RECORDS.—British Columbia: Gold Bridge; Hope–Princeton Summit. Washington: Wenatchee. Oregon: Mt. Hood; Mt. Jefferson; Crater Lake; *Prospect;* Detroit. Washington: Mt. St. Helens. British Columbia: Lihumitson Mtn.; Avalanche Pass; *London Mtn.* (Cowan and Guiguet, 1965:217); Tenquille Lake (*ibid.*).

Microtus richardsoni macropus (Merriam)

1891. *Arvicola (Mynomes) macropus* Merriam, N. Amer. Fauna, 5:60, July 30, type from Pahsimeroi Mts., 9700 ft., Custer Co., Idaho. (Regarded as indistinguishable from *richardsoni* by Anderson and Rand, Canadian Field-Nat., 57:106, December 10, 1943; but regarded as distinguishable by Dalquest, Univ. Kansas Publ., Mus. Nat. Hist., 2:356, April 9, 1948.)
1900. *Microtus richardsoni macropus,* V. Bailey, N. Amer. Fauna, 17:61, June 6.

MARGINAL RECORDS.—Montana: 1 mi. W, 2 mi. S Summit; Palace Butte Camp, Hyalite Creek (Hoffmann, *et al.,* 1969:592, as *Arvicola richardsoni* only); Beartooth Mts. Wyoming (Long, 1965a:658, 659): Medicine Wheel Ranch, 28 mi. E Lovell; *23½ mi. S, 5 mi. W Lander;* South Pass City; 12 mi. N Pinedale; LaBarge Creek; 9 mi. S Robertson; *14 mi. S, 2 mi. E Robertson;* 10 mi. SE Afton. Idaho: Camp Tendoy; Mt. Jefferson, 3 mi. S Montana line; Alturas Lake; SW slope Cuddy Mtn. Oregon: Strawberry Butte; Wallowa Mts. Washington: Stay-a-while Spring. Idaho: Priest Lake.

Microtus richardsoni myllodontus Rasmussen and Chamberlain

1959. *Microtus richardsoni myllodontus* Rasmussen and Chamberlain, Jour. Mamm., 40:54, February 20, type from head Boulger Canyon, 2 mi. NE Huntington Reservoir, 10,000 ft., Wasatch Plateau, Sanpete Co., Utah.

MARGINAL RECORDS (Rasmussen and Chamberlain, 1959:56, unless otherwise noted).—Idaho: head Crow Creek, Caribou Co. (Long, 1965a:658). Utah: 12 mi. W Garden City (Hall and Cockrum, 1953:439); N fork Ashley Creek (Egoscue, 1965:686); Huntington Canyon, 9 mi. NW Huntington, 8000 ft.; Elkhorn Guard Station, 14 mi. N Torrey, 9400 ft.; Fish Lake Mts. (Hall and Cockrum, 1953:439); Lambs Canyon, 13 mi. SE Salt Lake City (*ibid.*). (See also Long, 1965a:658, 659.)

Microtus richardsoni richardsoni (DeKay)

1842. *A[rvicola]. richardsoni* DeKay, Zoology of New-York, Mammals, p. 91, type from "near the foot of the Rocky Mountains." According to V. Bailey (N. Amer. Fauna, 17:60, June 6, 1900) the type was obtained by Drummond in vic. Jasper House, Alberta.

1897. [*Microtus*] *richardsoni*, Trouessart, Catalogus mammalium . . . , p. 565.

MARGINAL RECORDS.—Alberta: 5 mi. W Henry House; Waterton Lake Park. British Columbia: Rossland; Mt. Revelstoke; *latitude of Mt. Robson.*

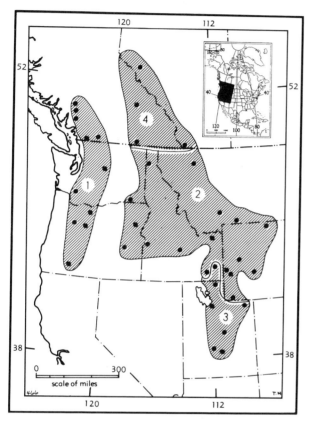

Map 466. *Microtus richardsoni.*

1. *M. r. arvicoloides*   3. *M. r. myllodontus*
2. *M. r. macropus*   4. *M. r. richardsoni*

Genus Lagurus Gloger—Sagebrush Vole

1841. *Lagurus* Gloger, Gemeinnütziges Hand- und Hilfs-buch der Naturgeschichte, 1:97. Type, *Lagurus Migratorius* Gloger [= *Mus lagurus* Pallas].
1881. *Eremiomys* Polyakoff, Mém. Acad. Imp. Sci., St. Pétersbourg, 39, supplement, p. 34. Type, *Mus lagurus* Pallas.
1912. *Lemmiscus* Thomas, Ann. Mag. Nat. Hist., ser. 8, 9:401, April. Type, *Arvicola curtata* Cope.

External measurements given below for species; basal length, 20.0–23.2; zygomatic breadth, 13.2–15.0. Body thickset and lemminglike; tail short, approx. same length as pes; sole of pes densely haired, having 5 concealed pads; ears short and supporting pale gray fur; mammae, 2-2 = 8; check-teeth ever-growing,

having widely open folds and lacking cement in folds; skull angular, flattened; peglike squamosal crests prominent; supraorbital ridges, although strong, not fused in interorbital region; auditory bullae large, projecting posteriorly beyond plane of occiput, cancellous and foamlike in structure.

The American species have been separated, under the subgeneric name *Lemmiscus* Thomas, from the Old World representatives, on the basis of the following characters: antitragus present (rather than absent); ear more than half as long as hind foot; cement present in re-entrant angles of molars; 4 (not 5) closed triangles in m3.

Lagurus curtatus
Sagebrush Vole

External measurements: 108–142; 16–28; 14–18; 9–13. Weight, 17.5–24.8 grams. Pelage long and lax; upper parts pale buffy gray to ashy gray; ears and nose tinged with buff; sides paler; venter silvery or soiled whitish to buffy; tail indistinctly bicolored, with dusky line above but silvery white to buffy below.

Lagurus curtatus curtatus (Cope)

1868. *Arvicola curtata* Cope, Proc. Acad. Nat. Sci. Philadelphia, 20:2, type from Pigeon Spring, Mt. Magruder, Nevada, near boundary between Inyo Co., California, and Esmeralda Co., Nevada.
1912. *Lagurus* (*Lemmiscus*) *curtatus*, Thomas, Ann. Mag. Nat. Hist., ser. 8, 9:401, April.
1877. *Arvicola decurtata* Coues, in Coues and Allen, Monog. N. Amer. Rodentia, p. 215 (in text), August, a *nomen nudum.*

MARGINAL RECORDS.—Nevada: 11 mi. SE Reno; Breen Creek, Kawich Mts. Utah: ½ mi. E Pine Valley, 6500 ft. (Stock, 1970:432). Nevada: S end Belted Range, 5½ mi. NW Whiterock Spring; Mt. Magruder, Indian Spring. California: Inyo Mts.; Mono Mills. Nevada: Lapon Canyon, Mt. Grant.—Additional study of specimens probably will alter our knowledge of the ranges of subspecies in Utah.

Lagurus curtatus intermedius (Taylor)

1911. *Microtus* (*Lagurus*) *intermedius* Taylor, Univ. California Publ. Zool., 7:253, June 24, type from head Big Creek, 8000 ft., Pine Forest Mts., Humboldt Co., Nevada.
1934. *Lagurus curtatus intermedius*, Borell and Ellis, Jour. Mamm., 15:35, February 15.

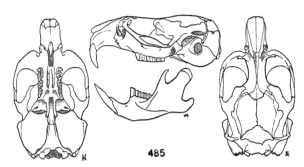

485

Fig. 485. *Lagurus curtatus curtatus*, Chiatovich Creek, 8200 ft., Esmeralda Co., Nevada, No. 38731 M.V.Z., ♂, X 1½.

MARGINAL RECORDS.—Nevada: 4 mi. S Diessner; 5 mi. N Summit Lake; 1 mi. SE Tuscarora, 5900 ft. Utah: Grouse Creek, Raft River Mts., 6500 ft.; Butterfield Canyon, 3 mi. SW Butterfield Tunnel, 7000 ft.; Daves Hollow, Bryce Canyon National Park. Nevada: Dutch Flat Schoolhouse, Reese River. California: Madeline Plains, 6 mi. E Ravendale. Nevada: 15 mi. S Vya.

Lagurus curtatus levidensis (Goldman)

1941. *Lemmiscus curtatus levidensis* Goldman, Proc. Biol. Soc. Washington, 54:70, July 31, type from 5 mi. E Canadian River, W base Medicine Bow Range, 8000 ft., E of Walden, North Park, Jackson Co., Colorado.
1951. *Lagurus curtatus levidensis*, Kelson, Jour. Mamm., 32:114, February 15.

MARGINAL RECORDS.—Montana (Hoffmann, *et al.*, 1969:592, as *L. curtatus* only): 8 mi. W Butte; 2 mi. NE Cooke. Wyoming (Long, 1965a:661): 40½ mi. S, 17½ mi. W Gillette; *42 mi. S, 13 mi. W Gillette; 10 mi. N, 6 mi. W Bill;* 5 mi. W Horse Creek P.O. Colorado (Armstrong, 1972:246): Fort Collins; Egeria Park, near Toponas; 6 mi. NE Meeker. Utah: Browns Corral, 20 mi. S Ouray, 6250 ft.; Blacks Fork, 10,000 ft., Uinta Mts. Idaho: American Falls; 45 mi. SE Silver City; Salmon River Mts. Montana: near Wisdom (Hoffmann, *et al.*, 1969:592, as *L. curtatus* only).

Lagurus curtatus orbitus Dearden and Lee

1955. *Lagurus curtatus orbitus* Dearden and Lee, Jour. Mamm., 36:271, May 26, type from Steep Creek, 15 mi. N Boulder, 8500 ft., Garfield Co., Utah.

MARGINAL RECORDS.—Utah: W of Skougaard's Resort, Fishlake, 9000 ft.; Spectacle Lake, 10,500 ft., 13 mi. S Teasdale; *type locality;* Cowpuncher Ranger Station, 18 mi. N Escalante, 9500 ft.

Lagurus curtatus pallidus (Merriam)

1888. *Arvicola (Chilotus) pallidus* Merriam, Amer. Nat., 22:704, August, type from Fort Buford, Williams Co., North Dakota.
1953. *Lagurus curtatus pallidus*, Hall and Cockrum, Univ. Kansas Publ., Mus. Nat. Hist., 5:455, January 15.

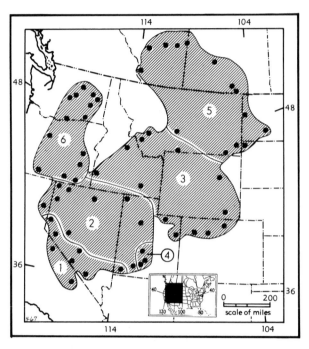

Map 467. *Lagurus curtatus.*

1. *L. c. curtatus*   4. *L. c. orbitus*
2. *L. c. intermedius*   5. *L. c. pallidus*
3. *L. c. levidensis*   6. *L. c. pauperrimus*

MARGINAL RECORDS.—Alberta: Compeer (Soper, 1965:231). Saskatchewan: 10 mi. SE Beechy (Nero, 1964:119, as *L. curtatus* only); Big Muddy Coulee, 10 mi. SE Willowbunch Lake; Big Muddy Lake. North Dakota: type locality; Glen Ullin. South Dakota: 18 mi. N, ½ mi. E Camp Crook (Birney and Lampe, 1972:466). Montana (Hoffmann, *et al.*, 1969:592, as *L. curtatus* only, unless otherwise noted): 5 mi. SE Ekalaka; 2 mi. E Biddle; Philbrook (Hall and Kelson, 1959:753). Alberta (Soper, 1965:231, unless otherwise noted): 49° 50′ N, 114° 11′ W (Salt and Wershler, 1975:184); Calgary; 9 mi. NW Drumheller; Youngstown.

Lagurus curtatus pauperrimus (Cooper)

1868. *Arvicola pauperrima* Cooper, Amer. Nat., 2:535, December, type from plains of the Columbia, near Snake River, southwestern Washington.
1946. [*Lagurus curtatus*] *pauperrimus*, Hall, Mammals of Nevada, p. 560, July 1.
1913. *Microtus (Lagurus) curtatus artemisiae* Anthony, Bull. Amer. Mus. Nat. Hist., 32:14, March 7, type from Ironside, 4000 ft., Malheur Co., Oregon (V. Bailey, N. Amer. Fauna, 55:214, August 29, 1936, regards *artemisiae* as inseparable from *L. c. pauperrimus*).

MARGINAL RECORDS.—Washington: 3 mi. W Delrio; 11 mi. E Creston; 5 mi. N Edwall; 2 mi. SW Ritzville; Great Plain of Columbia near Snake River. Oregon: Ironside; Creston; Steens Mts.; Rock Creek, N side Hart Mtn.; sec. 20, T. 38 S, R. 12 E, Klamath Co. (Hammer, 1971:26, as *Lagurus curtatus* only); "Crooked River National Grassland in the vicinity east of the town of Culver, Jefferson County" (Maser, 1974:195, as *Lagurus curtatus* only). Washington: 10 mi. SE Bickleton; 8 mi. N Kittitas; 8 mi. SW Waterville. Not found: Washington: 2 mi. E Vomer.

Genus Neofiber True—Round-tailed Muskrat

Revised by Schwartz, Occas. Pap. Mus. Zool., Univ. Michigan, 547:1–27, 3 pls., July 29, 1953.

1884. *Neofiber* True, Science, 4:34, July 11. Type, *Neofiber alleni* True.

External measurements: 285–381; 99–168; 40–50; 15–22. Weight, 155–330 grams; condylobasal length, 41.5–50.1; zygomatic breadth, 25.3–30.4. Externally modified for aquatic life, but less so than *Ondatra*; tail round, with long hairs partly covering it; hind foot larger than forefoot; swimming-fringes on feet and tail not highly developed. Mammae, 1–2 = 6. Upper parts sorghum brown to blackish brown; venter grayish-white with buffy to avellaneous wash. Skull much as in *Ondatra* but smaller; m1 with 5 [not 6] triangles between anterior loop and posterior loop; anterior loop rounded and with shallow re-entrant angles; m3 reduced, has only 1 outer fold [not 2] and 2 [instead of 3] salient angles.

Neofiber alleni
Round-tailed Muskrat

See characters of the genus.

Neofiber alleni alleni True

1884. *Neofiber alleni* True, Science, 4:34, July 11, type from Georgiana, Brevard Co., Florida.

MARGINAL RECORDS.—Florida: Lake City; ⅘ mi. E Grandin; 1⅘ mi. W Flagler Beach; Canaveral; 1 mi.

NE Stuart; Winterhaven; *Sumter County;* Belleview; 3½ mi. S Gainesville.

Neofiber alleni apalachicolae Schwartz

1953. *Neofiber alleni apalachicolae* Schwartz, Occas. Pap. Mus. Zool., Univ. Michigan, 547:14, July 29, type from Apalachicola, E side of river, Franklin Co., Florida.

MARGINAL RECORDS.—Florida: Lake Miccosukee; 2 mi. W Madison; St. Marks River, Wakulla Co.; Carabelle; type locality.

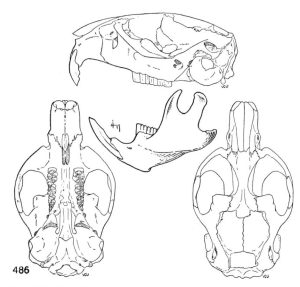

Fig. 486. *Neofiber alleni alleni*, 1 mi. E Courtenay, Merritt Island, Brevard Co., Florida, No. 27124 K.U., ♀, X 1.

Neofiber alleni exoristus Schwartz

1953. *Neofiber alleni exoristus* Schwartz, Occas. Pap. Mus. Zool., Univ. Michigan, 547:12, July 29, type from 12⅒ mi. SE Waycross, Ware Co., Georgia.

MARGINAL RECORDS.—Georgia: type locality; Woodbine; ¼ mi. W Chessers Island, Okefinokee Swamp; Billy's Island, Okefinokee Swamp.

Neofiber alleni nigrescens Howell

1920. *Neofiber alleni nigrescens* Howell, Jour. Mamm., 1:79, March 2, type from Ritta, S shore Lake Okeechobee, Palm Beach Co., Florida.

MARGINAL RECORDS.—Florida: 2⅘ mi. E Frostproof; 14 mi. NW Delray Beach, Loxahatchee Wildlife Refuge; 4 mi. W Clewiston; S of Sarasota.

Neofiber alleni struix Schwartz

1952. *Neofiber alleni struix* Schwartz, Chicago Acad. Sci., Nat. Hist. Miscl., 101:1, February 15, type from 21 mi. W Miami, Dade Co., Florida.

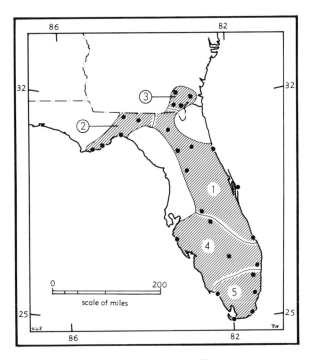

Map 468. *Neofiber alleni.*

1. *N. a. alleni*   3. *N. a. exoristus*
2. *N. a. apalachicolae* 4. *N. a. nigrescens*
            5. *N. a. struix*

MARGINAL RECORDS.—Florida: about 22 mi. NW Fort Lauderdale; 2 mi. NW Miami Springs; 7½ mi. SE Florida City; Cape Sable; Everglades.

Genus **Ondatra** Link—Muskrat

Revised (under name *Fiber*) by N. Hollister, N. Amer. Fauna, 32:1–47, 6 pls., April 29, 1911.

1795. *Ondatra* Link, Beyträge zur Naturgeschichte, 1(2):76. Type by tautonymy, *Castor zibethicus* Linnaeus (International Commission on Zoological Nomenclature, Opinion 55, Smiths. Inst. Spec. Publ., 2169:126, 127, May 12, 1913).
1800. *Fiber* Cuvier, Leçons d'anatomie comparée . . . , 1:Tabl. 1. Type, *Castor zibethicus* Linnaeus.
1817. *Simotes* Fischer, Mém. Soc. Imp. Nat. Moscow, 5:444. Type, *Mus zibethicus* [= *Castor zibethicus* Linnaeus].
1827. *Moschomys* Billberg, Synopsis fauna Scandinaviae, ed. 1:1, Mamm., Conspectus A (before p. 1). Type, *Castor zibethicus* Linnaeus.

External measurements: 409–620; 180–295; 64–88. Weight, 541–1575 grams; basal length of skull, 52.4–65.1; zygomatic breadth, 34.6–44.0. Externally highly modified for aquatic life; tail laterally compressed, relatively long, and more or less naked but with well-developed swimming-fringe below; hind foot much larger than forefoot, with conspicuous swimming-fringes; ears short; fur thick and soft. Mammae, 1–2 = 6. Upper parts

vary from a bright rusty red through several shades of brown to almost black; sides from bright rusty red to Prout's Brown; venter from whitish to Broccoli Brown with washes of varying intensities of rufous, cinnamon and brownish; feet pinkish gray to dark brown.

Skull resembling those of *Microtus* and *Neofiber* but larger and more massive. Differs from *Neofiber* in structure of lower molars (m1 with 6 [not 5] triangles between anterior loop and posterior loop; 1st of 6 triangles not closed; anterior loop bilobed and with deep re-entrant angles; m3 with 3 [not 2] outer salient angles).

Ondatra zibethicus
Muskrat

Measurements and description are as given in the account of the genus.

The muskrat has been introduced into many areas where it did not occur naturally. Some of these introductions have been deliberate; others have been accidental—escapes from fur farms. Storer (1937:443–460) and Bleich (1974:7, 8) have listed several known introductions in California and on several islands off the coast of British Columbia. Hansen (1965:669, 670) reported introductions in south-central Oregon and northern California, as did Wood (1974:2–4) for some coastal areas of those two states. Newsom (1937: 437) reported that muskrats have been introduced on Anticosti Island in the Gulf of St. Lawrence. Lowery (1974:269, 272) thinks an occurrence in extreme northwestern Louisiana represents an introduction by man. So far as known, this is the only American microtine that has become established in Eurasia; the muskrat now occurs at many places in Europe and in northern Asia.

In North America trappers receive more dollars for muskrat pelts than they do for the pelts of any other one kind of furbearer.

Ondatra zibethicus albus (Sabine)

1823. *Fiber zibethicus–albus* Sabine, *in* Franklin, Narrative of a journey to the shores of the Polar Sea in . . . 1819–22, p. 660, type from Cumberland House, Saskatchewan.

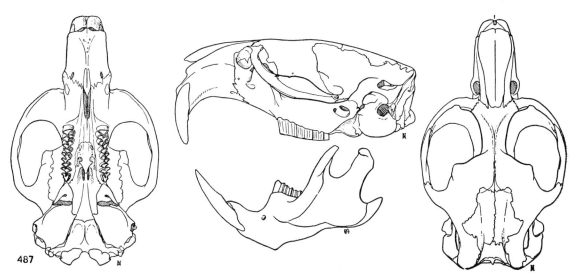

Fig. 487. *Ondatra zibethicus mergens*, 10 mi. SE Fallon, Nevada, No. 90544 M.V.Z., ♂, X 1.

1912. *Ondatra zibethica alba*, Miller, Bull. U.S. Nat. Mus., 79:231, December 31.

1902. *Fiber zibethicus hudsonius* Preble, N. Amer. Fauna, 22:53, October 31, type from Fort Churchill, Keewatin.

MARGINAL RECORDS.—Keewatin: 16 km. N Eskimo Point (Stewart, *et al.*, 1975:507). Manitoba: Fort Churchill; York Factory; Echimamish River; 6 mi. SE Delta (Tamsitt, 1962:75, as *O. zibethicus* only); Whitewater Lake, Riding Mountain National Park; Whitewater Lake (Soper, 1961:200); Oak Lake (*ibid.*). Saskatchewan: Prince Albert National Park. Keewatin: Windy River, SW of Simon's Lake (Harper, 1956:44).

Ondatra zibethicus aquilonius (Bangs)

1899. *Fiber zibethicus aquilonius* Bangs, Proc. New England Zool. Club, 1:11, February 28, type from Rigolet, Hamilton Inlet, Labrador.

1912. *Ondatra zibethica aquilonia*, Miller, U.S. Nat. Mus., 79:230, December 31.

MARGINAL RECORDS (Harper, 1961:84).—Labrador: Killinek area. Quebec: head George River. Labrador: Nain, down E coast, and W along Strait of Belle Isle, to Quebec: Kecarpoui. Thence inland to Labrador: Kenamu River. Quebec: Nichicum Lake; Cape Jones, thence north along coast to Kikkerteluk River; Payne Lake, thence southeast along coast to area of Fort Chimo.

Ondatra zibethicus bernardi Goldman

1932. *Ondatra zibethica bernardi* Goldman, Proc. Biol. Soc. Washington, 45:93, June 21, type from 4 mi. S Gadsden, Yuma Co., Arizona.

MARGINAL RECORDS.—Nevada: Colorado River, Durban Ranch, 14 mi. E Searchlight. Arizona (Cockrum, 1961:210): *William Roberts Ranch, 500 ft., op-*

posite Needles; Topock; *Colorado River, 6 mi. S, 2¹/₂ mi. E Topock; Yuma;* type locality. Baja California: 15 mi. S Volcano Lake. California: Calipatria.

Ondatra zibethicus cinnamominus (Hollister)

1910. *Fiber zibethicus cinnamominus* Hollister, Proc. Biol. Soc. Washington, 23:125, September 2, type from Wakeeney, Trego Co., Kansas.

1912. *Ondatra zibethica cinnamomina*, Miller, Bull. U.S. Nat, Mus., 79:232, December 31.

MARGINAL RECORDS.—Manitoba: Treesbank District (Soper, 1961:201). North Dakota: Oakes. Kansas: 6 mi. SW Garnett; Halstead. Oklahoma: 2 mi. N Dawson. Texas (Davis, 1966:210, as *O. zibethicus* only, unless otherwise noted): Lake of the Pines; Trinity County; Wise County; Archer County; Baylor County; Hardeman County; Canadian (Hall and Kelson, 1959:757); Hutchinson County. Colorado: Arkansas River, 10 mi. W Pueblo (Armstrong, 1972:247); Ward. Wyoming (Long, 1965a:662): 7½ mi. N, 8½ mi. E Savery; Sun; 28 mi. N, 12 mi. W Casper. Montana: Little Dry Creek. Alberta: Lodge Creek. Manitoba: Whitewater Lake (Soper, 1961:200).

Ondatra zibethicus goldmani Huey

1938. *Ondatra zibethica goldmani* Huey, Trans. San Diego Soc. Nat. Hist., 8:409, January 18, type from Saint George, Washington Co., Utah.

MARGINAL RECORDS.—Utah: Orderville. Nevada: Muddy Creek near St. Thomas.

Ondatra zibethicus macrodon (Merriam)

1897. *Fiber macrodon* Merriam, Proc. Biol. Soc. Washington, 11:143, May 13, type from Lake Drummond, Dismal Swamp, Norfolk Co., Virginia.

1912. *Ondatra zibethica macrodon*, Miller, Bull. U.S. Nat. Mus., 79:230, December 31.
1911. *F[iber]. niger* Brass, Aus dem Reiche der Pelze, p. 604, type locality in New Jersey or Delaware.

MARGINAL RECORDS.—Pennsylvania: Chester County, thence east to coast and down coast into North Carolina: *Pender County* (Funderburg, 1961:268); *Onslow County* (*ibid.*); *New Hanover County* (*ibid.*); Brunswick County (*ibid.*); Raleigh. Virginia: Nelson County; Washington. Maryland: *Jefferson.*

Ondatra zibethicus mergens (Hollister)

1910. *Fiber zibethicus mergens* Hollister, Proc. Biol. Soc. Washington, 23:1, February 2, type from Fallon, Churchill Co., Nevada.
1912. *Ondatra zibethica mergens*, Miller, Bull. U.S. Nat. Mus., 79:231, December 31.

MARGINAL RECORDS.—Nevada: Paradise; Marys River, 22 mi. N Deeth, 5000 ft.; W side Ruby Lake, 3 mi. N White Pine Co. line, 6200 ft.; 20 mi. S Schurz, W

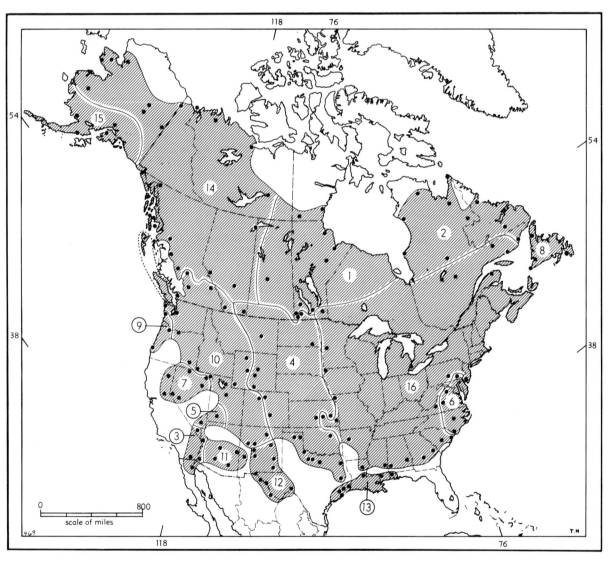

Map 469. *Ondatra zibethicus.*

1. *O. z. albus*	5. *O. z. goldmani*	9. *O. z. occipitalis*	13. *O. z. rivalicius*
2. *O. z. aquilonius*	6. *O. z. macrodon*	10. *O. z. osoyoosensis*	14. *O. z. spatulatus*
3. *O. z. bernardi*	7. *O. z. mergens*	11. *O. z. pallidus*	15. *O. z. zalophus*
4. *O. z. cinnamominus*	8. *O. z. obscurus*	12. *O. z. ripensis*	16. *O. z. zibethicus*

side Walker Lake; Carson River, 5 mi. SE Minden, 4900 ft. California: *Lake Tahoe;* Lake Arthur, sec. 19, T. 13 N, R. 9 E, 1600 ft. (Goertz, 1964:480); *Lake Theodore, sec. 8, T. 13 N, R. 9 E, 1700 ft. (ibid.);* Eagle Lake.

Ondatra zibethicus obscurus (Bangs)

1894. *Fiber obscurus* Bangs, Proc. Biol. Soc. Washington, 9:133, September 15, type from Codroy, Newfoundland.
1959. *Ondatra zibethicus obscurus,* Cameron, Nat. Mus. Canada Bull., 154:85.

MARGINAL RECORDS (Cameron, 1959:89, unless otherwise noted).—Newfoundland: Port Saunders; Salmonier; type locality; Bay St. George (Hall and Cockrum, 1953:470).

Ondatra zibethicus occipitalis (Elliot)

1903. *Fiber occipitalis* Elliot, Field Columb. Mus., Publ. 74, Zool. Ser., 3:162, May 2, type from Florence, Lane Co., Oregon.
1912. *Ondatra zibethica occipitalis,* Miller, Bull. U.S. Nat. Mus., 79:231, December 31.

MARGINAL RECORDS.—Washington: Aberdeen. Oregon: Portland; Coquille.

Ondatra zibethicus osoyoosensis (Lord)

1863. *Fiber osoyoosensis* Lord, Proc. Zool. Soc. London, p. 97, October, type from Lake Osoyoos, British Columbia.
1912. *Ondatra zibethica osoyoosensis,* Miller, Bull. U.S. Nat. Mus., 79:231, December 31.

MARGINAL RECORDS.—British Columbia: Williams Lake; Golden. Alberta: Stoney and Crooked Highwood Mts. Wyoming (Long, 1965a:663): 12 mi. S, 15 mi. W Cody; 4 mi. NW Milford; 1 mi. N Fort Bridger. Colorado: Hot Sulphur Springs. New Mexico (Findley, *et al.,* 1975:265, as *O. zibethicus* only): 3 mi. SE Cimarron; *15 mi. S Cimarron;* Albuquerque; San Rafael. Utah: Utah Lake; Ogden. Nevada: Goose Creek, 2 mi. W Utah boundary, 5000 ft. Idaho: S fork Owyhee River, 12 mi. N Nevada line. Oregon: Shirk, W of Steens Mts.; The Dalles. Washington: Tenino; N part Olympic Peninsula (Dalquest, 1948:Fig. 120). British Columbia: Vancouver (Cowan and Guiguet, 1965:230); *Port Moody;* Ashcroft.

Ondatra zibethicus pallidus (Mearns)

1890. *Fiber zibethicus pallidus* Mearns, Bull. Amer. Mus. Nat. Hist., 2:280, February 21, type from Fort Verde, Yavapai Co., Arizona.
1912. *Ondatra zibethica pallida,* Miller, Bull. U.S. Nat. Mus., 79:232, December 31.

MARGINAL RECORDS.—Arizona (Cockrum, 1961:212): Montezuma Well, 3500 ft.; Beckers Lake, 2 mi. NW Springerville. New Mexico: Upper Tularosa River. Arizona (Cockrum, 1961:212): *Safford, 2900 ft.;* Camp Grant; 20 mi. SW Phoenix; *type locality.*

Ondatra zibethicus ripensis (V. Bailey)

1902. *Fiber zibethicus ripensis* V. Bailey, Proc. Biol. Soc. Washington, 15:119, June 2, type from Eddy, near Carlsbad, Eddy Co., New Mexico.
1912. *Ondatra zibethica ripensis,* Miller, Bull. U.S. Nat. Mus., 79:232, December 31.

MARGINAL RECORDS.—New Mexico (Findley, *et al.,* 1975:265, as *O. zibethicus* only): Santa Rosa; Bitter Lakes National Wildlife Refuge; Carlsbad. Texas: Val Verde County (Davis, 1966:210, as *O. zibethicus* only). Chihuahua: 1 mi. NW Ojinaga, 2400 ft. (Anderson, 1972:366). Texas (*ibid.*): Little Box, Rio Grande, *ca.* 45 mi. SE Fort Hancock; within 4 or 5 mi. of Fort Hancock, on Rio Grande; N bank Rio Grande, 4⅘ mi. NW City Hall, El Paso, 3750 ft. New Mexico (Findley, *et al.,* 1975:265, as *O. zibethicus* only); San Antonio; Los Lunas.

Ondatra zibethicus rivalicius (Bangs)

1895. *Fiber zibethicus rivalicius* Bangs, Proc. Boston Soc. Nat. Hist., 26:541, July 31, type from Burbridge, Plaquemines Parish, Louisiana.
1940. *Ondatra z[ibethicus]. rivalicius,* Davis and Lowery, Jour. Mamm., 21:212, May 16.

MARGINAL RECORDS.—Alabama: Three-mile Creek, near Mobile, thence down coast to Texas: Galveston County (Davis, 1966:210, as *O. zibethicus* only); 2 mi. SW Pasadena; Chambers County (Davis, 1966:210, as *O. zibethicus* only); Orange County (*ibid.,* as *O. zibethicus* only). Louisiana: Avoyelles Parish. Mississippi: Pearl River County (Kennedy, *et al.,* 1974:21, as *O. zibethicus* only).

Ondatra zibethicus spatulatus (Osgood)

1900. *Fiber spatulatus* Osgood, N. Amer. Fauna, 19:36, October 6, type from Lake Marsh, Yukon.
1912. *Ondatra zibethica spatulata,* Miller, Bull. U.S. Nat. Mus., 79:231, December 31.

MARGINAL RECORDS.—Mackenzie: Richards Island; Fort Anderson; 6 mi. SW Coppermine (Ellis, 1957:6, sight record as *O. zibethicus* only); Fort Reliance. Alberta: Athabaska Lake; Blindman River; Henry House. British Columbia: Chezacut Lake; Wistaria; Kispiox (Cowan and Guiguet, 1965:232). Alaska: Portage Cove, Revillagigedo Island; Sergief Island. British Columbia: Bennett. Alaska: Anklin River, Yakutat Bay; Eagle; Ladd; *Fairbanks area* (Johansen, 1962:65, as *O. zibethicus* only); Russian Mission, Yukon River; Nome; Lava Lake; Kobuk Valley; Fort Hamlin. Yukon: Old Crow Flats, international boundary, 65 mi. N Porcupine River (Youngman, 1975:106).

Ondatra zibethicus zalophus (Hollister)

1910. *Fiber zibethicus zalophus* Hollister, Proc. Biol. Soc. Washington, 23:1, February 2, type from Becharof Lake, Alaska Peninsula, Alaska.

1912. *Ondatra zibethica zalophus,* Miller, Bull. U.S. Nat. Mus., 79:231, December 31.

MARGINAL RECORDS.—Alaska: Knik Valley; Fort Kenai; Ugashik; Nushagak.

Ondatra zibethicus zibethicus (Linnaeus)

1766. [*Castor*] *zibethicus* Linnaeus, Syst. nat., ed. 12, 1:79, type from eastern Canada.
1795. [*Ondatra*] *zibethicus* Link, Beyträge zur Naturgeschichte, 1(2):76.
1808. *Ondatra americana* Tiedemann, Zoologie . . . , 1:481, a renaming of *Mus zibethicus* Schreber and *Castor zibethicus* Linnaeus.
1829. *Fiber zibethicus,* var. B, *nigra* Richardson, Fauna Boreali-Americana, 1:119 (no definite locality).
1829. *Fiber zibethicus,* var. C, *maculosa* Richardson, Fauna Boreali-Americana, 1:119 (no definite locality).
1867. *Fiber zibethicus varius* Fitzinger, Sitzungsb. k. Akad. Wiss., Berlin, 56:47, a renaming of *F. z. maculosa* Richardson.
1867. *Fiber zibethicus niger* Fitzinger, Sitzungsb. k. Akad. Wiss., Berlin, 56:47, a renaming of *F. z. nigra* Richardson.

MARGINAL RECORDS.—Quebec: 75 mi. above mouth Romaine River (Harper, 1961:82); lower Natashquan River (*ibid.*); Ste. Anne de Monts, thence along coast to New Jersey: Mullica and Wading River meadows. Pennsylvania: Carlisle. Virginia: Blacksburg. South Carolina (Golley, 1966:113): Georgetown County; Calhoun County. Georgia (Golley, 1962:150): Schreven County; Pulaski County. Alabama: Seale; Coffeeville. Mississippi: Lauderdale County (Kennedy, *et al.,* 1974:21). Louisiana (Lowery, 1974:272, Map 41): *Tensas Parish;* Caldwell Parish.—Lowery (*op. cit.:* 269, 272) comments on the establishment of the species in the eight northeasternmost parishes since the mid-1960's, probaby by natural range-extension from SE Arkansas.—Arkansas: Ozark. Kansas: *Arkansas City;* Hamilton; *Neosho Falls;* vic. Frontenac (Long and Hays, 1962:104, as *O. zibethicus* only). Missouri: Squaw Creek National Wildlife Refuge, Mound City (Elder and Shanks, 1962:144, as *O. zibethicus* only). Iowa: Granite (Bowles, 1975:105). Minnesota: S end Ten Mile Lake. Manitoba: Whiteshell River (Soper, 1961:199). Quebec: headwaters Peribonca River (Harper, 1961:81).

Genus **Lemmus** Link—Lemmings

American forms revised by W. B. Davis, Murrelet, 25:19–25, September 19, 1944; see also Rausch, R., Arctic, 6:126, 127, July, 1953, and Rausch, R. L. and Rausch, V. R., Zeitschr. für Säugetierkunde, 40:8–34, February 1975.

1795. *Lemmus* Link, Beyträge zur Naturgeschichte, 1(2):75. Type, *Mus lemmus* Linnaeus.

External measurements: 120–160; 18–26; 18–23. Weight up to 113 grams; basal length, 28.8–31.4; zygomatic breadth, 19.7–22.0. Pelage long, with much reddish; form heavy, thickset; plantar pads vestigial; sole heavily haired; legs short; tail not projecting beyond hind feet in study skins; ear much reduced; mammae, 2–2 = 8. Skull exceptionally massive, with extreme of interorbital constriction; temporal ridges fused into sharp median interorbital crest; zygomata widely divergent; tooth-rows widely divergent posteriorly; tympanic bullae large; outer salient angles and infolds deeper on outer than inner side of upper molars and deeper on inner, than on outer, side of lower molars.

Lemmus sibiricus
Brown Lemming

External measurements: 122–160; 18–23; 18–23; 8–11. Weight of adult males, of *L. s. alascensis* from northern Alaska, up to 113 grams. Upper parts anteriorly grizzled grayish (summer) to grizzled tawny (winter); rump varying from dark chestnut to rusty or burnt umber (summer) and from tawny to fulvous (winter); sides varying from light ochraceous-buff to cinnamon (summer) and from fuscous to cinnamon (winter); underparts washed with buff.

Lemmus sibiricus alascensis Merriam

1900. *Lemmus alascensis* Merriam, Proc. Washington Acad. Sci., 2:26, March 14, type from Point Barrow, Alaska.
1953. *Lemmus sibiricus alascensis,* Rausch, Arctic, 6:127, July.

MARGINAL RECORDS.—Alaska: type locality; 30 mi. N Thetis Island; Colville Delta; Topagaruk; Wainwright, then along coast to point of beginning.

Lemmus sibiricus harroldi Swarth

1931. *Lemmus harroldi* Swarth, Proc. Biol. Soc. Washington, 44:101, October 17, type from Nunivak Island, Alaska. Known only from type locality.
1953. *Lemmus sibiricus harroldi,* Rausch, Arctic, 6:127, July.

Lemmus sibiricus helvolus (Richardson)

1828. *Arvicola (Lemmus) helvolus* Richardson, Zool. Jour., 3:517, type from near headwaters of one of southern tributaries of Peace River, or between there and Jasper

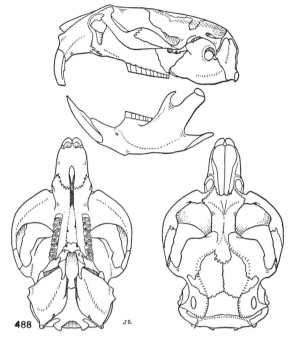

488

Fig. 488. *Lemmus sibiricus alascensis,* 1⅖ mi. S, ⅗ mi. E Barrow Village, long. 156° 45′ 25″ W, lat. 71° 16′ 20″ N, 20 ft., Alaska, No. 50838 K.U., ♂, X 1½.

House region, Alberta (Preble, N. Amer. Fauna, 27:182, October 26, 1908).

1953. *Lemmus sibiricus helvolus,* Rausch, Arctic, 6:127, July.

MARGINAL RECORDS.—Yukon (Youngman, 1975:110): Dominion Creek, head Indian River; forks Macmillan River; 138 mi. N Watson Lake, 5 mi. E Little Hyland River (*ibid.*). British Columbia (Cowan and Guiguet, 1965:202, unless otherwise noted): Sikanni Chief River headwaters; Laurier Pass; *Mt. Brilliant;* Stuie (Hall and Kelson, 1959:760); Wistaria; Kiskagus Creek; Telegraph Creek (Hall and Kelson, 1959:760); Atlin. Yukon: Haeckel Hill, 8 mi. NW Whitehorse. Also possibly *Alberta: ? Rocky Mountain region of northwestern Alberta.*

Lemmus sibiricus minusculus Osgood

1904. *Lemmus minusculus* Osgood, N. Amer. Fauna, 24:36, November 23, type from Kakhtul River, near jct. with Malchatna River, Alaska.

MARGINAL RECORDS.—Alaska: upper forks Chalitna River; type locality; Nushagak River near Kakwok; Port Heiden (Rausch and Rausch, 1975:25).

Lemmus sibiricus nigripes (True)

1894. *Myodes nigripes* True, Diagnoses of new North American mammals, p. 2, April 24 (preprint of Proc. U.S. Nat.

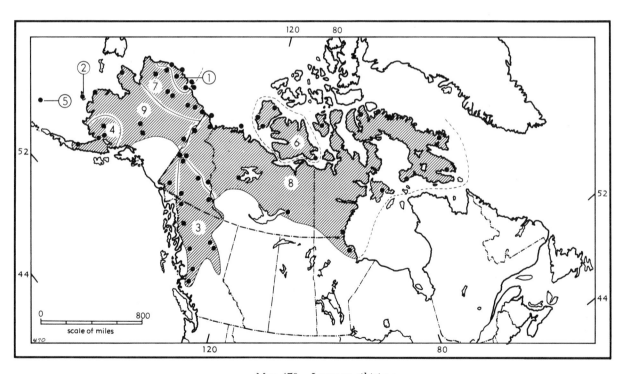

Map 470. *Lemmus sibiricus.*

1. *L. s. alascensis*	4. *L. s. minusculus*	7. *L. s. subarcticus*
2. *L. s. harroldi*	5. *L. s. nigripes*	8. *L. s. trimucronatus*
3. *L. s. helvolus*	6. *L. s. phaiocephalus*	9. *L. s. yukonensis*

Mus., 17:242, November 15, 1894), type from St. George Island, Pribilof Islands, Alaska. Known only from type locality except as introduced on St. Paul Island (see O'Malley, Science, n.s., 63:596, June 11, 1926).

1953. *Lemmus sibiricus nigripes*, Rausch, Arctic, 6:127, July.

Lemmus sibiricus phaiocephalus Manning and Macpherson

1958. *Lemmus trimucronatus phaiocephalus* Manning and Macpherson, Arctic Inst. of N. Amer., Tech. Pap. 2, p. 27, type locality at Mahogany Point, Castel Bay, N coast Banks Island.

MARGINAL RECORDS (Manning and Macpherson, 1958:34).—Franklin (Banks Island): type locality; *Providence Point, Banks Island; Back Point, Banks Island;* Cape Adelaide, Victoria Island; *Taylor Island; De Haven Point, Victoria Island;* De Salis Bay, Banks Island; *Masik River Valley, Banks Island; Sachs Harbour, Banks Island;* Cape Kellet, Banks Island.

Lemmus sibiricus subarcticus Bee and Hall

1956. *Lemmus trimucronatus subarcticus* Bee and Hall, Univ. Kansas Mus. Nat. Hist., Miscl. Publ., 8:109, March 10, type from Lake Schrader, long. 145° 09' 50" W, lat. 69° 24' 28" N, 2900 ft., Brooks Range, Alaska.

MARGINAL RECORDS.—Alaska: 10 mi. N Pingok Island; S to coast and E along it to Demarcation Point; Mt. Mary, S end Lake Peters; Wahoo Lake; Tulugak Lake; Chandler Lake; 2 mi. W Utukok River, long. 161° 15' 30" W, lat. 68° 54' 50" N, 1275 ft.; Kaolak; *Kaolak River, long. 159° 47' 40" W, lat. 70° 11' 15" N, 30 ft.*

Lemmus sibiricus trimucronatus (Richardson)

1825. *Arvicola trimucronata* Richardson, *in* Parry, Journal of a second voyage . . . , App., p. 309, type from Point Lake, Mackenzie.

1953. *Lemmus sibiricus trimucronatus*, Rausch, Arctic, 6:127, July.

MARGINAL RECORDS.—Franklin: Prince of Wales Island (Manning and Macpherson, 1961:118); Strathcona Sound; Pond Inlet; Pangnirtung Fjord; Bishop Island, Frobisher Bay; Cape Dorset. Keewatin: Coral Harbor, Southampton Island; 50 mi. S Cape Eskimo. Manitoba: Churchill. Mackenzie: Artillery Lake; Fort Franklin. Yukon (Youngman, 1975:111): Keele Lake; 13 mi. S Chapman Lake; Christmas Bay, Kluane Lake; Swede Dome, 34 mi. E Dawson; Rampart House; Herschel Island, Paulina Cove; 4 mi. WSW mouth Blow River. Mackenzie: Baillie Island, Cape Bathurst.

Lemmus sibiricus yukonensis Merriam

1900. *Lemmus yukonensis* Merriam, Proc. Washington Acad. Sci., 2:27, March 14, type from Charlie Creek, Yukon River, Alaska.

1843. ? *Myodes albigularis* Wagner, *in* Schreber, Die Säugthiere . . . , Suppl., p. 602, according to Hinton (Monograph of the voles and lemmings . . . , 1926, p. 207) may be from NW coast North America. If identifiable, this name might apply to one of the kinds of *Lemmus* currently bearing a name proposed at a later date.

MARGINAL RECORDS.—Alaska: type locality; head Toklat River, McKinley Range; Glacier Creek, McKinley Range; Hooper Bay; Grantley Harbor.

Genus **Synaptomys** Baird—Bog Lemmings

Revised by A. B. Howell, N. Amer. Fauna, 50:ii + 38, August 5, 1927.

1858. *Synaptomys* Baird, Mammals, *in* Repts. Expl. Surv. . . . , 8(1):558, July 14. Type, *Synaptomys cooperi* Baird.

External measurements: 118–154; 16–27; 16–24. Basal length of skull, 22–26; zygomatic breadth, 15.0–17.7. Pelage long and loose; tail averaging slightly longer than hind foot; soles of feet not heavily haired; plantar pads, 6; mammae, 6 or 8. Supraorbital ridges forming median interorbital crest in adults; squamosal crests well developed; rostrum thick and short (less than 25% of total length of skull); tympanic bullae large; upper incisors broad and prominently grooved; molars rootless, with persistent pulp; roots of mandibular incisors on lingual sides of molars; re-entrant angles excessively deep on buccal sides of upper molars and on lingual sides of lower molars; M3 with 4 simple, transverse loops, peculiarly arranged; external form not highly specialized.

KEY TO SPECIES OF SYNAPTOMYS

1. None of the hairs at base of ears appreciably brighter than remainder of pelage; mammae, 6; mandibular molars with triangles on outer side; palate with broad, blunt median projection. . . . *S. cooperi*, p. 831
1'. A few hairs at base of ears distinctly brighter (with bright rusty tinge) than remainder of pelage; mammae, 8; mandibular molars without triangles on outer side; palate with sharply pointed median projection pointed backward. . . *S. borealis*, p. 833

Subgenus **Synaptomys** Baird

Revised by Wetzel, Jour. Mamm., 36:1–20, February 25, 1955.

1858. *Synaptomys* Baird, Mammals, *in* Repts. Expl. Surv. . . . , 8(1):558, July 14. Type, *Synaptomys cooperi* Baird.

Mandibular molars with closed triangles on outer sides; outer re-entrant angles well developed; mandibular incisors heavy, outer edges never noticeably unworn or prolonged into sharp splinters of enamel; palate with poorly developed posterior spinous process; no hairs at base of ears appreciably brighter than remainder of pelage; mammae, 1–2 = 6.

Synaptomys cooperi
Southern Bog Lemming

External measurements: 118–154; 13–24; 16–24; 8–14. Weight, 21.4–50 grams. Upper parts varying from bright cinnamon to sayal brown, often heavily suffused with black; underparts plumbeous; for additional characters see account of subgenus.

Synaptomys cooperi cooperi Baird

1858. *Synaptomys cooperi* Baird, Mammals, *in* Repts. Expl. Surv. . . . , 8(1):558, July 14, type locality fixed by Bole and Moulthrop, Sci. Publs., Cleveland Mus. Nat. Hist., 5:146, September 11, 1942, at Jackson, Carroll Co., New Hampshire.
1896. *Synaptomys fatuus* Bangs, Proc. Biol. Soc. Washington, 10:47, March 9, type from Lake Edward, Quebec.

MARGINAL RECORDS.—Quebec: Lake Mistassini; St. Margaret River; Mingan; 2 mi. W Coin du Banc (Manville, 1961:108). New Brunswick: Miramichi

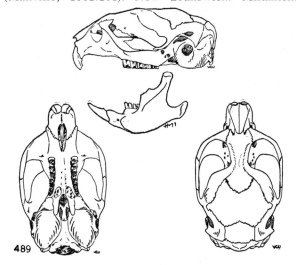

Fig. 489. *Synaptomys cooperi gossii*, 1 mi. W Lawrence, Douglas Co., Kansas, No. 4626 K.U., ♂, X 1½.

Road, 15 mi. from Bathurst. Nova Scotia: Cape Breton Island; Digby. Maine: Mount Desert Island (Manville, 1960:416, as *S. cooperi*). New Hampshire: Durham. Connecticut: Macedonia Park. Pennsylvania: Bushkill Creek; Warfordsburg; Somerset; Markleysburg; Waynesburg. Ohio: Powhatan Point; Athens; Meags; *Ripley;* Owensville. Indiana: Bascom; Mitchell; Bicknell; Warren County; Jasper County (Quay, 1962:303, as *S. cooperi* only). Michigan: Warren Woods. Wisconsin: Kilbourn; Camp Douglas; *Millston* (Jackson, 1961:224); Withee. Minnesota: Elk River; *10 mi. N Nimrod* (Heaney and Birney, 1975:33, as *S. cooperi* only); Itasca State Park, in Clearwater Co.; Roseau County. Manitoba: near Dawson, Sandilands Forest Reserve; Pine Falls. Ontario: Lake Nipigon; Franz; Chapleau; Ridout; Biscatasing; Timagami.

Synaptomys cooperi gossii (Coues)

1877. *Arvicola (Synaptomys) gossii* Coues, *in* Coues and Allen, Monog. N. Amer. Rodentia, p. 235 (published as synonym of *Synaptomys cooperi*, but name stated to apply to Kansas specimens, description and measurements of which are printed on p. 236), type from Neosho Falls, Woodson Co., Kansas.—*Synaptomys helaletes gossii* Merriam, Proc. Biol. Soc. Washington, 10:60, March 19, 1896, was proposed as a new name. Merriam (*loc. cit.*) regarded the *Arvicola (Synaptomys) gossii* as a *nomen nudum.*
1897. *Synaptomys cooperi gossi*, Rhoads and Young, Proc. Acad. Nat. Sci. Philadelphia, 49:307, June.
1942. *Synaptomys cooperi saturatus* Bole and Moulthrop, Sci. Publs., Cleveland Mus. Nat. Hist., 5:149, September 11, type from Bloomington, McLean Co., Illinois.

MARGINAL RECORDS.—Iowa: Lansing. Wisconsin: Lynxville. Indiana: *Hebron;* Roselawn. Illinois: Bloomington; *Le Roy; Muncie;* Kickapoo State Park; Flat Rock. Indiana: Winslow; Henryville. Illinois: Rosiclare; Alto Pass. Arkansas: Lake City; 1 mi. E Saffel (Sealander, *et al.,* 1975:424); *2 mi. N Helena* [S of probable range] (Easterla, 1968b:364, as *S. cooperi* only; no specimen was saved). Missouri: Horseshoe Lake; Ashland; Bellair. Kansas: type locality; 3 mi. N, 1 mi. E Sharon; Little Salt Marsh. Nebraska: 1 mi. NW Callaway (Jones, 1964c:221); Valentine. Iowa: sec. 22, Waterford Twp., Clay Co. (Bowles, 1975:98, owl pellets).

Synaptomys cooperi helaletes Merriam

1896. *Synaptomys helaletes* Merriam, Proc. Biol. Soc. Washington, 10:59, March 19, type from Dismal Swamp, Norfolk Co., Virginia.
1927. *Synaptomys cooperi helaletes*, A. B. Howell, N. Amer. Fauna, 50:9, August 5.

MARGINAL RECORDS.—Virginia: Lake Drummond; *type locality.* North Carolina: Chapanoke.

Synaptomys cooperi kentucki Barbour

1956. *Synaptomys cooperi kentucki* Barbour, Jour. Mamm., 37:414, September 11, type from Sadieville, Scott Co., Kentucky.

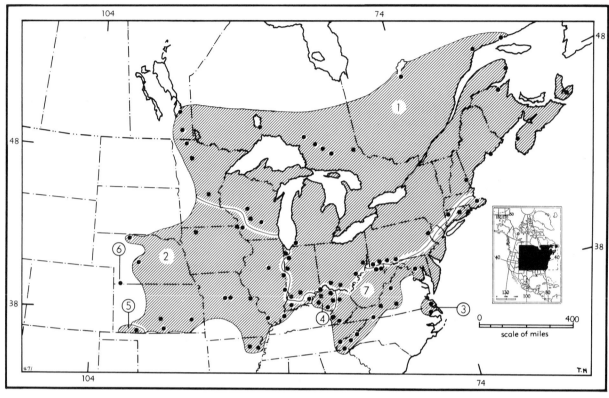

Map 471. *Synaptomys cooperi.*

1. *S. c. cooperi*	3. *S. c. helaletes*	6. *S. c. relictus*
2. *S. c. gossii*	4. *S. c. kentucki*	7. *S. c. stonei*
	5. *S. c. paludis*	

MARGINAL RECORDS.—Kentucky: Maysville; *2 mi. W Flemingsburg*; 3 mi. E Owingsville; *4 mi. E Waco*; Richmond; 1 mi. E Lawrenceburg; Kleber Sanctuary, 2 mi. N Franklin Co. line on Kentucky Hwy. 368; *1 mi. N Milford.*

Synaptomys cooperi paludis Hibbard and Rinker

1942. *Synaptomys cooperi paludis* Hibbard and Rinker, Univ. Kansas Sci. Bull., 28:26, May 15, type from Meade County State Park, Meade Co., Kansas. Known only from type locality.

Synaptomys cooperi relictus Jones

1958. *Synaptomys cooperi relictus* Jones, Univ. Kansas Publ., Mus. Nat. Hist., 9:387, May 12, type from Rock Creek Fish Hatchery, 5 mi. N, 2 mi. W Parks, Dundy Co., Nebraska. Known only from type locality.

Synaptomys cooperi stonei Rhoads

1893. *Synaptomys stonei* Rhoads, Amer. Nat., 27:53, January, type from May's Landing (on Egg River), Atlantic Co., New Jersey.
1897. *Synaptomys cooperi stonei* Rhoads, Proc. Acad. Nat. Sci. Philadelphia, 49:305, June.

MARGINAL RECORDS.—Massachusetts: Plymouth. Rhode Island: Kingston Fairgrounds, West Kingston (West and Hutchison, 1964:479), thence south along Atlantic Ocean and Chesapeake Bay and up E side Potomac River to Maryland: Hyattsville. Virginia: Elliot Knob; Campbell County; Blacksburg; White Top Mtn. North Carolina: Mt. Mitchell; Weaverville; Highlands; *Great Smoky Mountains National Park at Indian Gap.* Tennessee: Great Smoky Mountains National Park at Upper Greenbriar Cove. Kentucky: Goldbug; 1¾ km. E Science Hill (Fassler, 1974:42, as *S. cooperi* only); Morehead. West Virgina: Morgantown; Terra Alta. Connecticut: Mansfield.

Subgenus **Mictomys** True

1894. *Mictomys* True, Diagnoses of new North American mammals, p. 2, April 24 (preprint of Proc. U.S. Nat. Mus., 17:242, November 15, 1894). Type, *Mictomys innuitus* True.

Mandibular molars without closed triangles on outer sides, and almost no outer re-entrant angles; mandibular incisors relatively slender and sharply pointed; palate with well-developed spinous process behind; few hairs at bases of ears appreciably brighter than remainder of pelage. Mammae, 2–2 = 8.

Synaptomys borealis
Northern Bog Lemming

External measurements: 118–135; 19–27; 16–22; 12–13. Weight, 32–34 grams (two individuals). Upper parts varying from grizzled gray to Brussels Brown, with admixture of black hairs; underparts plumbeous; tail bicolored; feet grayish to almost black; for additional characters see account of subgenus.

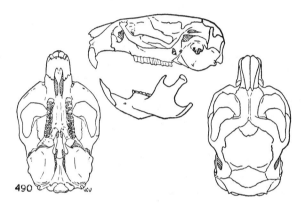

Fig. 490. *Synaptomys borealis dalli*, McIntyre Creek, 2250 ft., 3 mi. NW Whitehorse, Yukon, No. 21332 K.U., ♀, X 1½.

Synaptomys borealis artemisiae Anderson

1932. *Synaptomys borealis artemisiae* Anderson, Bull. [Ann. Rept. for 1931] Nat. Mus. Canada, 70:104, November 24, type from Stevenson Creek, SW Princeton, British Columbia, on Hope–Princeton Trail, E slope Cascade Range, 2400 ft., lat. 49° 23′ N, long. 120° 35′ W.

MARGINAL RECORDS.—British Columbia: the Okanagan (Cowan and Guiguet, 1965:196); type locality; *Whipsaw Creek, 5600 ft.*

Synaptomys borealis borealis (Richardson)

1828. *Arvicola borealis* Richardson, Zool. Jour., 3:517, type from Fort Franklin, Great Bear Lake, Mackenzie.
1907. *Synaptomys borealis*, Osgood, Proc. Biol. Soc. Washington, 20:49, April 18.
1902. *Synaptomys (Mictomys) bullatus* Preble, Proc. Biol. Soc. Washington, 15:181, August 6, type from Trout Rock, near Fort Rae, Great Slave Lake, Mackenzie.

MARGINAL RECORDS.—Mackenzie: 68° 20′ N, 133° 38′ W (Martell, 1974b:348); Fort Franklin; Sarahk Lake. Alberta: Smith Landing. Saskatchewan: Brightsand Lake, about 65 mi. NNW North Battleford. Alberta: "Edmonton district"; Peace River Landing.

British Columbia: near Tupper Creek School House; "*Peace River block in northeastern British Columbia.*" Mackenzie: Fort Providence.

Synaptomys borealis chapmani J. A. Allen

1903. *Synaptomys (Mictomys) chapmani* J. A. Allen, Bull. Amer. Mus. Nat. Hist., 19:555, October 10, type from Glacier, Selkirk Range, British Columbia.
1927. *Synaptomys borealis chapmani*, A. B. Howell, N. Amer. Fauna, 50:9, August 5.

MARGINAL RECORDS.—Alberta: Ray Lake; Mt. Forgetmenot. Montana: Camas Creek in Glacier National Park. Washington: Sema Meadows, 3000 ft. British Columbia: Rossland; Mt. Revelstoke; Indianpoint Lake.

Synaptomys borealis dalli Merriam

1896. *Synaptomys (Mictomys) dalli* Merriam, Proc. Biol. Soc. Washington, 10:62, March 19, type from Nulato, Alaska.
1927. *Synaptomys borealis dalli*, A. B. Howell, N. Amer. Fauna, 50:9, August 5.
1903. *Synaptomys (Mictomys) andersoni* J. A. Allen, Bull. Amer. Mus. Nat. Hist., 19:554; October 10, type from Level Mts., northern British Columbia.

MARGINAL RECORDS.—Yukon (Youngman, 1975:112, 114, as *S. b. borealis*, because he undervalued paler color of *S. b. dalli*): Summit Lake, 67° 43′, 136°29′; Canol Road, Mile 249; North Toobally Lake. British Columbia: Sixteen Mile Lake; Chezacut Lake; Caribou Mtn., near Stuie; Telegraph Creek; Atlin. Yukon: Burwash Landing. Alaska: Prince William Sound; Seldovia; Cape Ugyak (Schiller and Rausch, 1956:197); Brooks River (*ibid.*); Lake Aleknagik; type locality; Bettles. Yukon: Old Crow (Youngman, 1975:112, 114, as *S. b. borealis* because he undervalued paler pelage in *S. b. dalli*).

Synaptomys borealis innuitus (True)

1894. *Mictomys innuitus* True, Diagnoses of new North American mammals, p. 3, April 24 (preprint of Proc. U.S. Nat. Mus., 17:243, November 15), type from Fort Chimo, Quebec.
1927. *Synaptomys borealis innuitus*, A. B. Howell, N. Amer. Fauna, 50:9, August 5.

MARGINAL RECORDS (Harper, 1961:57, unless otherwise noted).—Quebec: Esker Lake (61° 37′ N, 74° 40′ W) (Anderson, 1962:421, listed as *S. borealis*); Payne Bay (*ca.* 60° N, 70° W) (*ibid.*); Irony Lake, S of Leaf Bay; type locality; Swampy Bay River [at Rattogobas Lake]; [upper part] Ste. Marguerite River [*ca.* 100 mi. N Gulf of St. Lawrence]; Godbout; Merry Island (Edwards, 1963:10, recorded only as *S. borealis*); Golfcourse Cove, Richmond Gulf (*ibid.*); Lower Seal Lake; Payne Lake [59° 30′ N, 74° W].

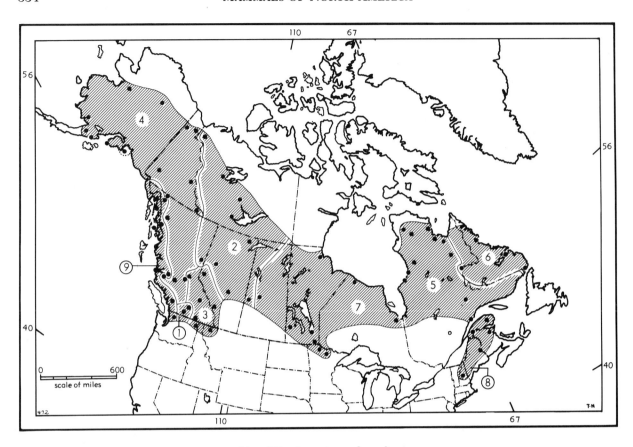

Map 472. *Synaptomys borealis.*

1. *S. b. artemisiae*	4. *S. b. dalli*	7. *S. b. smithi*
2. *S. b. borealis*	5. *S. b. innuitus*	8. *S. b. sphagnicola*
3. *S. b. chapmani*	6. *S. b. medioximus*	9. *S. b. truei*

Synaptomys borealis medioximus Bangs

1900. *Synaptomys (Mictomys) innuitus medioximus* Bangs, Proc. New England Zool. Club, 2:40, September 20, type from L'Anse au Loup, Strait of Belle Isle, Labrador.
1927. *Synaptomys borealis medioximus*, A. B. Howell, N. Amer. Fauna, 50:9, August 5.

MARGINAL RECORDS (Harper, 1961:57–59, unless otherwise noted).—Labrador: Nain; Hamilton Inlet (Hall and Kelson, 1959:764); *Rigolet; Red Bay;* type locality; Little Slimy Lake ($\frac{5}{8}$ mi. W Knob Lake, Quebec); *Camp Brook; brook entering Northwest Bay of Attikamagen Lake.*

Synaptomys borealis smithi Anderson and Rand

1943. *Synaptomys borealis smithi* Anderson and Rand, Canadian Field-Nat., 57:101, December 10, type from Thicket Portage, Mile 165, H. B. Railroad, Manitoba.

MARGINAL RECORDS.—Manitoba: Churchill. Ontario: Port Severn; Moosonee. Minnesota (Heaney and Birney, 1975:33, as *S. borealis* only): 10 mi. S Big Falls; Williams; *Warroad.* Manitoba: near Dawson

Cabin, N part Sandilands Forest Reserve; near Pine Falls, Winnipeg River; Clear Lake, Riding Mtn. National Park. Saskatchewan: E of Nesslin Lake, Prince Albert National Park.

Synaptomys borealis sphagnicola Preble

1899. *Synaptomys (Mictomys) sphagnicola* Preble, Proc. Biol. Soc. Washington, 13:43, May 29, type from Fabyans, Coos Co., New Hampshire.
1927. *Synaptomys borealis sphagnicola*, A. B. Howell, N. Amer. Fauna, 50:9, August 5.

MARGINAL RECORDS.—Quebec: Tabletop Mtn. New Brunswick: Miramichi Road, 15 mi. from Bathurst. Maine: Mt. Katahdin. New Hampshire: type locality. Quebec: Ste. Rose.

Synaptomys borealis truei Merriam

1896. *Synaptomys (Mictomys) truei* Merriam, Proc. Biol. Soc. Washington, 10:62, March 19, type from Skagit Valley, Skagit Co., Washington.
1896. *Synaptomys (Mictomys) wrangeli* Merriam, Proc. Biol. Soc. Washington, 10:63, March 19, type from Wrangell, Alexander Archipelago, Alaska.

MARGINAL RECORDS.—British Columbia: Atlin Lake; Great Glacier; Anahim Lake (Cowan and Guiguet, 1965:199); Alta Lake; *Allison Pass*. Washington: Mt. Baker. British Columbia (Cowan and Guiguet, 1965:199): Campbell Island; *Yeo Island; Neckis River;* Metlakatla; *Port Simpson.* Alaska: *Anan Creek;* Wrangell, Alexander Archipelago; Port Snettisham.

Genus **Dicrostonyx** Gloger—Collared Lemmings

Revised by G. M. Allen, Bull. Mus. Comp. Zool., 62:509–540, February, 1919. Canadian forms treated by Anderson and Rand, Jour. Mamm., 26:301–306, November 14, 1945.

1841. *Dicrostonyx* Gloger, Gemeinnütziges Hand- und Hilfs-buch der Naturgeschichte, 1:97. Type, an American species, probably *Mus hudsonius* Pallas.
1854. *Myolemmus* Pomel, Catalogue méthodíque Vert. Foss. Bassin de la Loire, p. 27. Type, *Arvicola ambiguus* from the Pleistocene of France. According to Hinton (Monograph of the voles and lemmings . . . , 1926, p. 137), *Arvicola ambiguus* is an unidentifiable species.
1855. *Misothermus* Hensel, Zeitschr. der Deutsch. geol. Gesell., 7:492. Type, *Mus torquatus* Pallas.
1881. *Borioikon* Poljakov, Mém. Acad. Imp. Sci., St. Pétersbourg, 39: Supplement, p. 34. Type, *Mus torquatus* Pallas.
1897. *Tylonyx* Schulze, Mamm. Europaea in Helios, Abh. und Vorträge Gesammtgebiete Naturwiss., Berlin, 14:83. Type, *Mus torquatus* Pallas.

External measurements: 132–160; 10–22; 15–24; 6. Weight, 45–75 grams; basal length of skull, 25.0–30.5; zygomatic breadth, 17.6–23.1. Ears entirely concealed in fur; foot-soles densely haired. In summer, upper parts brownish gray, buffy gray or variegated tawny and gray with black mid-dorsal stripe in some specimens; underparts (*D. g. rubricatus*) whitish but hairs plumbeous basally and lightly tipped with tawny. In winter all white (*D. g. unalescensis* and *D. g. stevensoni* excepted); white animals with foreclaws of digits 3 and 4 each with bulbous ventral part as large as dorsal part and separated from dorsal part by lateral groove and notch at end; ventral part absent in summer. Skull with squamosal pegs projecting anteriolaterally behind orbits; tympanic bullae cancellous; lower incisor not extending behind m3; m1 with 7 closed triangles between terminal loops.

This is the only rodent known to turn white in winter.

KEY TO SPECIES OF DICROSTONYX

1. First and 2nd upper cheek-teeth without an accessory fold at the posterointernal corner; last lower molar normally lacking an accessory anterointernal fold and an anteroexternal enamel fold. *D. hudsonius*, p. 837

1'. First and 2nd upper cheek-teeth with small accessory enamel fold at posterointernal corner; last lower molar provided with accessory anterointernal enamel fold, and, in some forms, with anteroexternal accessory fold.
2. Restricted to St. Lawrence Island
D. exsul, p. 837
2'. Not on St. Lawrence Island; on Alaska Peninsula and North American mainland north of 58°.
D. groenlandicus, p. 835

Dicrostonyx groenlandicus
Collared Lemming

External measurements: 132–157; 10–20; 15–22. In summer, upper parts grizzled gray to grayish brown with more or less distinct dorsal blackish stripe; underparts grayish white, usually tinged with buff. All white in winter. M1 and M2 usually each with well-developed posterointernal vestigial angle and hind walls of posterointernal triangles retaining thick enamel and concave form; m3 usually with pair of anterior vestigial angles.

Ognev (1948:479, 508) regards *D. groenlandicus* as conspecific with the earlier named *D. torquatus*, but Gileva and Lobanova (1974:191, 192) report differences of specific rank between chromosomes of Palaearctic and Nearctic populations. Youngman (1975:114) adds some data on numbers of chromosomes. He (*op. cit.*: 115) tentatively proposes a geographic range for *D. g. lentus* more extensive than is shown on Map 473 herewith. His proposal merits examination of specimens, from critical areas, not reported on as of 1975.

Dicrostonyx groenlandicus clarus Handley

1953. *Dicrostonyx groenlandicus clarus* Handley, Jour. Washington Acad. Sci., 43:197, June 24, type from near sea level at Cherie Bay, 5⅝ mi. ENE Mould Bay Station (lat. 76° 19′ N, long. 119° 02′ W), Prince Patrick Island, Franklin.

MARGINAL RECORDS.—Franklin: S coast, South Borden Island; Winter Harbor, Melville Island; Liddon Gulf, Melville Island; type locality.

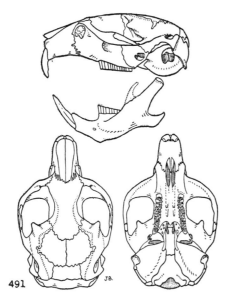

Fig. 491. *Dicrostonyx groenlandicus rubricatus*, Kaolak River, long. 159° 47′ 40″ W, lat. 70° 11′ 15″ N, 30 ft., Alaska, No. 43324 K.U., ♂, X 1½.

Dicrostonyx groenlandicus groenlandicus (Traill)

1823. *Mus groenlandicus* Traill, *in* Scoresby, Journal of voyage to the northern whale-fisher . . . , p. 416, type from Jamesons Land, Greenland.
1919. *Dicrostonyx groenlandicus*, G. M. Allen, Bull. Mus. Comp. Zool., 62:533, February.

MARGINAL RECORDS.—Greenland: "northeast Greenland, N to 83° 24′"; type locality. Franklin (Baffin Island): Pangnirtung Fjord. Keewatin: Coral Harbour, Southhampton Island; Baker Lake. Franklin: Iglulik Island, Fury and Hecla Strait; Axel Heiberg Island.

Dicrostonyx groenlandicus kilangmiutak Anderson and Rand

1945. *Dicrostonyx groenlandicus kilangmiutak* Anderson and Rand, Jour. Mamm., 26:305, November 14, type from De-Haven Point, southeastern point of Victoria Island, W side Victoria Strait, Franklin, approx. lat. 69° N, long. 101° 30′ W.

MARGINAL RECORDS.—Franklin: Cape Kellett, Banks Island; Taylor Island. Keewatin: Queen Maud Sea area. Mackenzie: Sifton Lake; Aklavik. Yukon (Youngman, 1975:116): British Mts., 20 mi. SE mouth Joe Creek; Herschel Island, Pauline Cove.

Dicrostonyx groenlandicus lentus Handley

1953. *Dicrostonyx groenlandicus lentus* Handley, Jour. Washington Acad. Sci., 43:198, June 24, type from Lake Harbor (lat. 62° 43′ N, long. 69° 41′ W), Baffin Island, Franklin.

MARGINAL RECORDS.—Franklin (Baffin Island): Netilling Lake; Cape Mercy; type locality; Bowdoin Harbor (Schooner Harbor); Bowman Bay (Camp Kungovik).

Dicrostonyx groenlandicus nelsoni Merriam

1900. *Dicrostonyx nelsoni* Merriam, Proc. Washington Acad. Sci., 2:25, March 14, type from St. Michael, Norton Sound, Alaska.
1953. *Dicrostonyx groenlandicus nelsoni*, Handley, Jour. Washington Acad. Sci., 43:200, June 24.

MARGINAL RECORDS.—Alaska: near Nome; Russian Mission; type locality.

Dicrostonyx groenlandicus nunatakensis Youngman

1967. *Dicrostonyx torquatus nunatakensis* Youngman, Proc. Biol. Soc. Washington, 80:31, March 24, type from 20 mi. S Chapman Lake, 5500 ft. (lat. 64° 35′, long. 138° 13′), Yukon.

MARGINAL RECORDS (Youngman, 1967:31).—Yukon: Ogilvie Mts., 52 mi. NE Dawson, 14 mi. S Lomand [= Lomond] Lake, 5400 ft.; *type locality*.

Dicrostonyx groenlandicus peninsulae Handley

1953. *Dicrostonyx unalascensis peninsulae* Handley, Jour. Washington Acad. Sci., 43:199, June 24, type from near sea level at Urilla Bay, Unimak Island, Alaska.

MARGINAL RECORDS.—Alaska: Nushagak; Chignik; type locality; Muller [= Moller] Bay.

Dicrostonyx groenlandicus richardsoni Merriam

1900. *Dicrostonyx richardsoni* Merriam, Proc. Washington Acad. Sci., 2:26, March 14, type from Fort Churchill, Manitoba.
1935. *Dicrostonyx groenlandicus richardsoni*, Degerbøl, Rept. Fifth Thule Exped., Mammals, 2:9.

MARGINAL RECORDS.—Mackenzie: Artillery Lake. Keewatin: Kazan River; Tavani. Manitoba: type locality. Keewatin: 1 mi. NW mouth Windy River (Harper, 1956:25).

Dicrostonyx groenlandicus rubricatus (Richardson)

1839. *Arvicola rubricatus* Richardson, The zoology of Captain Beechey's voyage . . . , p. 7, type from American side of Bering Strait, Alaska.
1937. *Dicrostonyx groenlandicus rubricatus*, Anderson, Mammals and birds of the western Arctic district, *in* Canada's Western Northland, Dept. of Mines and Resources, Ottawa, p. 110, July 9.
1900. *Dicrostonyx hudsonius alascensis* Stone, Proc. Acad. Nat. Sci. Philadelphia, 52:37, March 24, type from Point Barrow, Alaska.

MARGINAL RECORDS.—Alaska: Point Barrow; Barter Island; Porcupine River, 200 mi. above Yukon; Fort Yukon; Driftwood.

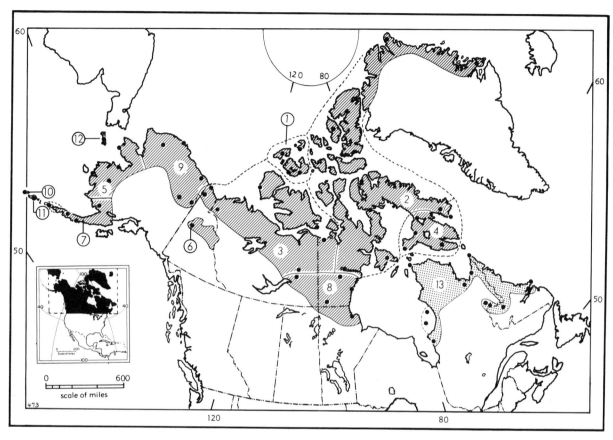

Map 473.   American *Dicrostonyx*.

1. *D. groenlandicus clarus*
2. *D. groenlandicus groenlandicus*
3. *D. groenlandicus kilangmiutak*
4. *D. groenlandicus lentus*

5. *D. groenlandicus nelsoni*
6. *D. groenlandicus nunatakensis*
7. *D. groenlandicus peninsulae*
8. *D. groenlandicus richardsoni*
9. *D. groenlandicus rubricatus*

10. *D. groenlandicus stevensoni*
11. *D. groenlandicus unalascensis*
12. *D. exsul*
13. *D. hudsonius*

Dicrostonyx groenlandicus stevensoni Nelson

1929. *Dicrostonyx unalascensis stevensoni* Nelson, Proc. Biol. Soc. Washington, 30:145, pl. 2, March 30, type from Umnak Island, Alaska. Known only from type locality.

1953. *Dicrostonyx groenlandicus stevensoni*, Hall and Cockrum, Univ. Kansas Publ., Mus. Nat. Hist., 5:484, January 15.

Dicrostonyx groenlandicus unalascensis Merriam

1900. *Dicrostonyx unalascensis* Merriam, Proc. Washington Acad. Sci., 2:25, March 14, type from Unalaska, Unalaska. Known only from type locality.

1953. *Dicrostonyx groenlandicus unalascensis*, Hall and Cockrum, Univ. Kansas Publ., Mus. Nat. Hist., 5:484, January 15.

Dicrostonyx exsul G. M. Allen
St. Lawrence Island Collared Lemming

1919. *Dicrostonyx exsul* G. M. Allen, Bull. Mus. Comp. Zool., 62:532, February, type from St. Lawrence Island, Bering Sea, Alaska. Known only from type locality.

External measurements: 146; 17; 20. Upper parts in summer pinkish gray, darker on rump; ear-patch mixture of ochraceous-buff and tawny; blackish median line indistinct; feet pale buff above in central part with whitish dusky tinge peripherally. In winter presumably all white.

Dicrostonyx hudsonius (Pallas)
Labrador Collared Lemming

1778. *Mus hudsonius* Pallas, Novae species quadrupedum e glirium ordine . . . , p. 208, type from Labrador.

1897. *Dicrostonyx hudsonius*, Bangs, Proc. Biol. Soc. Washington, 11:237, September 17.

Average external measurements of six adults: 148; 20.3; 22.1. Maximum recorded: 160; 22; 24. In summer, adults nearly uniform buffy gray above, with indistinct blackish spinal stripe; ear tufts, flanks, and throat buffy or rusty; underparts gray, usually strongly tinged with buffy; M1 and

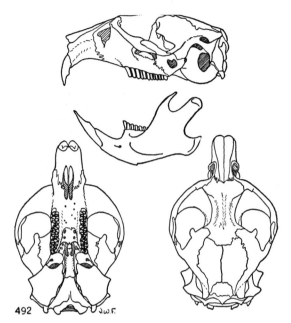

Fig. 492. *Dicrostonyx hudsonius*, George River, Hudson Bay Post, 800 ft., Quebec, No. 29248 K.U., ♀, X 1½.

M2 each lacks posterointernal vestigial angle and with wall of hinder inner triangle reduced; m3 without anterior vestigial angles.

MARGINAL RECORDS (Harper, 1961:54, 56, unless otherwise noted).—Quebec: Digges Islands; Port Burwell. Labrador: Hamilton Inlet; Ptarmigan Lake; Michikamau Lake. Quebec: Knob Lake; Burnt Creek; George River Post; Fort Chimo; Kakachischuan Point; Belcher Islands; Christie Islands, Nastapoka Bay (Anderson and Rand, 1945:303).

SUBFAMILY MURINAE

A characterization that applies to the particular species of Old World rats and mice that have been introduced into North America is as follows: upper parts uniform grayish-brown to black; underparts paler. Tail long, sparsely haired, and scaly. Molars provided with small, rounded cusps (tubercles) arranged in 3 longitudinal rows. Palate extending posteriorly to beyond plane of last molars; palatine slits extending posteriorly to plane of anterior borders of last upper molars. Dentition, i. $\frac{1}{1}$, c. $\frac{0}{0}$, p. $\frac{0}{0}$, m. $\frac{3}{3}$.

KEY TO SPECIES OF INTRODUCED MURINAE

1. Total length less than 250; tail less than 110; occipitonasal length less than 35; 1st upper molar longer than combined lengths of 2nd and 3rd upper molars. . . . . *Mus musculus*, p. 840
1'. Total length more than 250; tail more than 110; occipitonasal length more than 35; 1st upper molar shorter than combined lengths of 2nd and 3rd upper molars.
2. Tail longer than head and body; M1 with distinct outer notches on 1st row of cusps; length of parietal measured along temporal ridge less than distance between temporal ridges. . . . . *Rattus rattus*, p. 838
2'. Tail not longer than head and body; M1 without distinct outer notches on first row of cusps; length of parietal measured along temporal ridge not less than distance between temporal ridges. . . . . . . *Rattus norvegicus*, p. 839

Genus **Rattus** Fischer—Old World Rats

1803. *Ruttus* [*sic*] Fischer, Das Nationalmuseum der Naturgeschichte zu Paris . . . , 2:128. Type, *Mus decumanus* Pallas.
1881. *Epimys* Trouessart, Bull. Soc. d'Études Sci. d'Angers, 10:117. Type, *Mus rattus* Linnaeus.
1917. *Epinomys* Elliot, A check-list of mammals of the North American continent . . . , Suppl., p. 41, March 27, an emendation of *Epimys* Trouessart.

External measurements: 320–480; 153–255; 30–44. Weight, 150–540 grams. Occipitonasal length, 39.0–51.5; zygomatic breadth, 18.1–25.2; mastoid breadth, 15.8–19.2; alveolar length of maxillary tooth-row, 6.3–7.7. Upper parts tawny to black; underparts gray to white; tail sparsely haired and scaled.

Rattus rattus
Black Rat

External measurements: 325–455; 160–255; 30–40. Weight, 115–350 grams. Occipitonasal length, 39.0–45.4; zygomatic breadth, 18.1–21.6; alveolar length of maxillary tooth-row, 6.3–6.6. M1 with distinct outer notches on first row of cusps. Mammae, 2–3 = 10. Upper parts black to tawny; underparts slate colored to white; tail not bicolored and usually longer than head and body; ears more than 20 and naked.

At least three subspecies of black rat have been introduced in North America: *Rattus rattus rattus* (Linnaeus), *Rattus rattus alexandrinus* (Desmarest), and *Rattus rattus frugivorus* (Rafinesque). When first named, each of the three kinds was regarded as a species. Citations to the original descriptions are given immediately below.

1758. [*Mus*] *rattus* Linnaeus, Syst. nat., ed. 10, 1:61. Type locality, Uppsala, Sweden.
1803. *Mus alexandrinus* É. Geoffroy Saint-Hilaire, Catalogue des Mammifères du Muséum National d'Histoire Naturelle, Paris, p. 192. Type locality, Alexandria, Egypt. For availability of this name proposed in 1803 see remarks on page 72 concerning *Scalopus*.
1814. *Musculus frugivorus* Rafinesque, Précis des decuverts et travaux somiologiques . . . , p. 13. Type locality, Sicily.

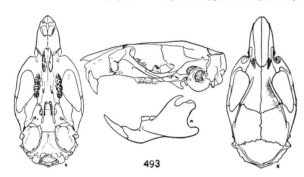

Fig. 493. *Rattus rattus alexandrinus*, Reno, Nevada, No. 94922 M.V.Z., ♀, X 1.

Because interbreeding occurs in North America between the three subspecies and because stocks from other continents have repeatedly been introduced, it is often impossible to allocate specimens to a definite subspecies. Not mapped.

[Although synonyms for the subspecific names of the nonnative black rat are in the literature, we have not listed the synonyms here because none was based on a North American specimen. For the Norway rat and house mouse, also nonnative species, treated beyond, several synonyms are listed for the subspecies that occur in North America because some of the synonyms were based on North American specimens.]

Rattus norvegicus
Norway Rat

A characterization that applies to the one subspecies that has been introduced into North America is as follows: external measurements: 316–460; 122–215; 30–45. Weight, 195–485 grams. Occipitonasal length, 41.0–51.5; zygomatic breadth, 21.0–25.2; mastoid breadth, 16.0–19.2. Upper parts reddish or grayish brown to black; underparts paler, grayish to yellow-white. White, black, or mottled specimens are found occasionally. Tail bicolored, being paler below and darker above; ears usually smaller than 20 and covered with short, fine hair. M1 without distinct notches on first row of cusps. Skull narrower across braincase than in *R. rattus*. Not mapped.

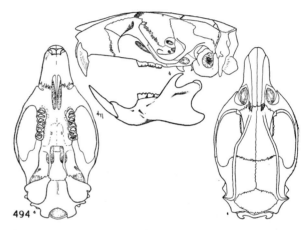

Fig. 494. *Rattus norvegicus norvegicus*, Redlands, San Bernardino Co., California, No. 77258 M.V.Z., ♂, X 1.

Rattus norvegicus norvegicus (Berkenhout)

1769. *Mus norvegicus* Berkenhout, Outlines of the natural history of Great Britain and Ireland, 1:5. Type locality, England.
1916. *Rattus norvegicus*, Hollister, Proc. Biol. Soc. Washington, 29:126, June 6.
1778. *Mus decumanus* Pallas, Novae species quadrupedum e glirium ordine . . . , p. 91. Type locality, Europe.
1779. *M[us]. surmolottus* Severinus, Tentamen Zool. Hungaricae, p. 73. Type locality, central Europe.
1800. *M[us]. d[ecumanus]. hybridus* Bechstein, Thomas Pennant's allgemeine Uebersicht der vierfüssigen Thiere, 2:713. Type locality, Waltershausen, Germany.
1816. *Mus capius* Oken, Lehrbuch der Naturgeschichte, 3:895, a renaming of *M. decumanus* Pallas.
1837. *Mus hibernicus* Thompson, Proc. Zool. Soc. London, p. 52, December 20. Type locality, Rathfriland, Co. Down, Ireland.
1839. *Mus maurus* Waterhouse, The zoology of H.M.S. Beagle, Mammalia, p. 31. Type locality, America.
1841. *Mus decumanoides* Hodgson, Jour. Asiatic Soc. Bengal, 10:915, a *nomen nudum*.
1842. *Mus leucosternum* Rüppell, Mus. Senckenbgiana, 3:108, 116. Type locality, Eritrea.
1848. *Mus maniculatus* Wagner, Wiegmann's Arch. für Naturgesch., 14:186. Type locality, Egypt.

Genus **Mus** Linnaeus—House Mouse

1758. *Mus* Linnaeus, Syst. nat., ed. 10, 1:59. Type, *Mus musculus* Linnaeus (by tautonomy).

Small, head and body usually less than 100; hind foot narrow; 5th digit longer than hallux. Skull resembling that of *Rattus*, but light, flattened, with supraorbital ridges faint or absent, and short muzzle. M1 elongated, 3-rooted; crown length of M1 more than length of M2 and M3 combined; M1 with 8 cusps; M2 usually with 6 cusps; M3 small; upper incisors compressed and with subapical notch. Mammae, 3–2 = 10.

Mus musculus
House Mouse

Reviewed by Schwarz and Schwarz, Jour. Mamm., 24:59–72, February 20, 1943.

External measurements: 130–198; 63–102; 14–21; 11–18. Weight, 18–23 grams. Occipitonasal length, 20.1–22.9; zygomatic breadth, 10.6–12.1; mastoid breadth, 9.3–10.2; alveolar length of maxillary tooth-row, 3.1–3.7. Upper parts light brown to black; underparts whitish, often with buffy wash. Tail not sharply bicolored, but paler below than above. Skull as described for the family and genus.

According to Schwarz and Schwarz (1943:59–72), this mouse rapidly evolves taxonomically distinguishable forms which are commensals of man. All wild forms are characterized by medium size, tail shorter than head and body, and underparts white. The commensal stocks tend to have longer tails, to be darker, and to have smaller molars and a shorter molar tooth-row.

Schwarz and Schwarz (*op. cit.*) suggest that two subspecies have been introduced into North America: *Mus musculus brevirostris* Waterhouse and *M. m. domesticus* Rutty. The former ranges from South America to the southern United States, where *brevirostris* overlaps the range of *M. m. domesticus* that ranges from Alaska to the northern part of the central states of the United States. *Mus musculus musculus* Linnaeus has not been introduced onto this continent. The two subspecies and the names (synonyms) for their commensals stand as given here. Not mapped.

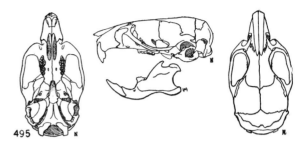

Fig. 495. *Mus musculus* subsp. ?, Baker, 5800 ft., White Pine Co., Nevada, No. 41874 M.V.Z., ♂, X 1½.

Mus musculus brevirostris Waterhouse

1837. *Mus brevirostris* Waterhouse, Proc. Zool. Soc. London, p. 19, November 21. Type locality, Maldonado, Uruguay.
1943. *Mus musculus brevirostris*, Schwarz and Schwarz, Jour. Mamm., 24:64, February 20.
1897. *Mus musculus jalapae* J. A. Allen and Chapman, Bull. Amer. Mus. Nat. Hist., 9:198, June 16, type from Jalapa Veracruz.
1942. *Mus musculus percnonotus* Moulthrop, Sci. Publs., Cleveland Mus. Nat. Hist., 5:79, June 1, type from Mina Carlota, Barrio de Cumanayagua, Trinidad Mts., Santa Clara, Cuba.

Mus musculus domesticus Rutty

1772. *Mus domesticus* Rutty, Essay Nat. Hist. County Dublin, 1:281. Type locality, Dublin, Ireland.
1943. *Mus musculus domesticus*, Schwarz and Schwarz, Jour. Mamm., 24:65, February 20.
1862. *H[esperomys]. indianus* Wied-Neuwied, Wiegmann's Arch. für Naturgesch., Jahrg. 28, 1:111. Type locality, New Harmony, Posey Co., Indiana.

FAMILY **ZAPODIDAE**—Jumping Mice

Hind legs longer than forelegs; hind feet with 5 functional digits; tail longer than head and body; antorbital foramen large and rounded; auditory bullae not inflated; zygoma depressed; jugal prolonged and forming suture with lachrymal; molars semihypsodont, rooted; P4 small or absent; lower premolars absent; M2 with 4 or more narrow labial re-entrant folds; anterior and posterior cingula of M3 large; back with broad dorsal band of yellow-brown, flecked with black; sides paler than back; underparts white or suffused with color of sides; tail more or less bicolored and sparsely haired.

Genus **Zapus** Coues—Jumping Mice

Revised by Krutzsch, Univ. Kansas Publ., Mus. Nat. Hist., 7:349–472, April 21, 1954.

1876. *Zapus* Coues, Bull. U.S. Geol. and Geog. Surv. Territories, ser. 2, 1:253, January 8. Type, *Dipus hudsonius* Zimmermann.

External measurements: 188–255; 112–155; 28–35. Back with a broad dorsal band of some shade of brown or brownish-yellow darkened with brownish-black; sides pale and slightly streaked with brownish-black; underparts snow-white, in some animals suffused with color of sides by sharp line of clear brownish-yellow; tail bicolored, dark brown above, yellowish-white below. Upper premolar small; M2 smaller than M1; 2 labial re-entrant folds of equal length on M3; anterior cingulum of M1 small; frontal region moderately swollen.

KEY TO SPECIES OF ZAPUS AND NAPAEOZAPUS

(*Zapus*-part after Krutzsch, 1954:384)

1. Fourth upper premolar absent (3 cheek-teeth present); tip of tail white. . . *Napaeozapus insignis*,   p. 847
1'. Fourth upper premolar present (4 cheek-teeth present); tip of tail not white.
    2. Baculum with tip spade-shaped and tip wider than .43; underfur with medullary pattern rectangular, cuticular scales small; coronoid process of mandible long and slender, angle of divergence from condyle broad; angle of mandible turned in and wide; pterygoid fossae wide; skull broad in relation to length; premolars with crescentine fold on occlusal surface. . . . *Zapus trinotatus*,   p. 846
    2'. Baculum with tip lanceolate (not spade-shaped) and tip less than .43 wide; underfur with medullary pattern square or rectangular; but, if rectangular, cuticular scales large; coronoid process short and broad, angle of divergence from condyle narrow; angle of mandible turned inward and small to medium; pterygoid fossae usually narrow; skull not broad in relation to length; premolars without crescentine fold on occlusal surface.
      3. Baculum less than 5.1 in total length; guard-hair averaging 115 μ in diameter (range, 96–140); medullary pattern of underfur rectangular and its cuticular scales large; skull small; incisive foramina shorter than 4.6; condylobasal length averaging less than 20; length of maxillary tooth-row averaging less than 3.7; palatal breadth at M3 less than 4.2.   *Zapus hudsonius*,   p. 841
      3'. Baculum more than 5.1 in total length; guard-hair more than 140 μ in diameter (range, 130–168); medullary pattern of underfur squarish and its cuticular scales moderately large; skull large; incisive foramina longer than 4.7; condylobasal length averaging more than 21; maxillary tooth-row averaging more than 3.8; palatal breadth at M3 more than 4.4.
                                                 Zapus princeps,   p. 844

Zapus hudsonius
Meadow Jumping Mouse

×⅔

External measurements: 188–216; 112–134; 28–31. Back ochraceous to dark brown; sides pale; lateral line ochraceous-buff, indistinct, or absent; belly white or suffused with ochraceous; tail brown to brownish-black above, white to yellowish-white below. Skull narrow in proportion to length; upper premolar small (.30 long and .35 wide), with shallow re-entrant fold; tooth-row short. *Zapus hudsonius* differs from the other species of the genus in smaller size, narrower skull, smaller premolars, and shorter tooth-row.

Zapus hudsonius acadicus (Dawson)

1856. *Meriones acadicus* Dawson, Edinburgh New Philos. Jour., n.s., 3:2. Type from Nova Scotia, Canada.
1942. *Zapus hudsonius acadicus*, Anderson, Ann. Rept. Provancher Soc. Nat. Hist. Canada, Quebec, p. 38 (for 1941), July 14.
1899. *Zapus hudsonius hardyi* Batchelder, Proc. New England Zool. Club, 1:5, February 8, type from Mount Desert Island, Hancock Co., Maine.

MARGINAL RECORDS.—Quebec: Ste. Anne des Monts. New Brunswick: Miramichi Road, 15 mi. from

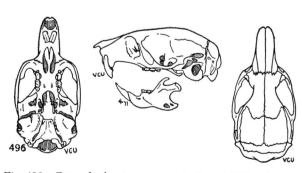

Fig. 496. *Zapus hudsonius campestris*, 3 mi. NW Sundance, Crook Co., Wyoming, No. 20844 K.U., ♀, X 1½.

Bathurst. Prince Edward Island. Nova Scotia: Big Intervale (Cameron, 1959:28); Doctors Cove, N of Barrington Passage. Maine: Columbia Falls; Small Point Beach. Massachusetts: *Marthas Vineyard* (Choate, 1972:215, as Z. *hudsonius* only); Wilmington; 2 mi. N Gilbertville. New York: Berlin; *Glenville;* North Wood; Keene Valley. Quebec: South Bolton (Wrigley, 1969:209, as Z. *hudsonius* only). Maine: E. Branch Penobscot River. New Brunswick: Baker Lake.

Zapus hudsonius alascensis Merriam

1897. *Zapus hudsonius alascensis* Merriam, Proc. Biol. Soc. Washington, 2:223, July 15, type from Yakutat Bay, Alaska.

MARGINAL RECORDS.—Alaska: Lake Aleknagik; head Chalitna River; E side Chilkat River, 100 ft., 9 mi. W, 4 mi. N Haines; Portage Cove, Revillagigedo Island; Yakutat; Cook Inlet, Tyonek; Chignik Bay; Frosty Peak.

Zapus hudsonius americanus (Barton)

1799. *Dipus americanus* Barton, Trans. Amer. Philos. Soc., 4:115. Type locality, near Schuykill River, a few miles from Philadelphia, Pennsylvania.
1899. *Zapus hudsonius americanus*, Batchelder, Proc. New England Zool. Club, 1:6, February 8.
1839. *Meriones microcephalus* Harlan, Proc. Zool. Soc. London, p. 1, July. Type locality, Beck's Farm, a few miles NE Philadelphia, Philadelphia Co., Pennsylvania.
1942. *Zapus hudsonius brevipes* Bole and Moulthrop, Sci. Publs., Cleveland Mus. Nat. Hist., 5:168, September 11, type from Bettsville, Seneca Co., Ohio.
1942. *Zapus hudsonius rafinesquei* Bole and Moulthrop, Sci. Publs., Cleveland Mus. Nat. Hist., 5:169, September 11, type from Cat Run, Bellmont Co., Ohio.

MARGINAL   RECORDS.—Michigan: Douglas Lake; Bloomfield. Pennsylvania: *East Springfield; 4¹/₂ mi. SW* [town of] *North East;* Bensons Swamp, 5 mi. E Columbus. New York: E Aurora; Peterboro; Catskills. Connecticut: Sharon; South Woodstock. Massachusetts: Middleboro; *Nantucket Island* (Waters and Stockley, 1965:67, as Z. *hudsonius* only). New York: Locust Grove. New Jersey: Tuckerton. Maryland: Assateagus, 5 mi. S Ocean City. Virginia: near Hampton. North Carolina: Raleigh. Georgia (Golley, 1962:161): Oconee County; Meriwether County. Alabama: 4 mi. N Auburn. Kentucky: Nelson Valley, 3⅛ km. N Somerset on State Hwy. 39 (Fassler, 1974:42, as Z. *hudsonius* only). Indiana: Mineral Springs. Michigan: 9 mi. N Ludington.

Zapus hudsonius campestris Preble

1899. *Zapus hudsonius campestris* Preble, N. Amer. Fauna, 15:20, August 8, type from Bear Lodge Mts., Crook Co., Wyoming.

MARGINAL RECORDS.—Montana: 3 mi. S, 1¾ mi. E Ekalaka (Lampe, *et al.*, 1974:23). South Dakota: 10 mi. S, 5 mi. W Reva (Andersen and Jones, 1971:382); *1*

mi. S, 8¹/₂ mi. W Rapid City (Turner, 1974:121); 4 mi. SE Hill City (*op. cit.:* 122); Wind Cave National Park Game Ranch, Gold Spring Creek. Wyoming (Long, 1965a:665): 1½ mi. E Buckhorn; Devils Tower, floodplain Belle Fourche River. Montana: Rotten Grass Creek, N base Big Horn Mts.; *Long Pine Hills* (Andersen and Jones, 1971:382).

Zapus hudsonius canadensis (Davies)

1798. *Dipus canadensis* Davies, Trans. Linnean Soc. London, 4:157. Type locality, near city of Quebec, Canada.
1899. *Zapus hudsonius canadensis*, Batchelder, Proc. New England Zool. Club, 1:4, February 8.
1943. *Zapus hudsonius ontarioensis* Anderson, Ann. Rept. Provancher Soc. Nat. Hist. Canada, Quebec, p. 59 (for 1942), type from Pancake Bay (Batchawana Bay) SE end of Lake Superior, Algoma Dist., about 40 mi. NE Sault Ste. Marie, Ontario.

MARGINAL RECORDS.—Quebec: ½ mi. N Mistassini Post; Camp de la Roche (Wrigley, 1969:209, as Z. *hudsonius* only); Valcartier. New York: Spectacle Pond, Brighton Twp.; Elba. Ontario: St. Thomas; Pancake Bay; Schreiber. Quebec: Liamika Lake.

Zapus hudsonius hudsonius (Zimmermann)

1780. *Dipus hudsonius* Zimmermann, Geographische Geschichte . . . , 2:358. Type locality, Hudson Bay, Canada (considered by Anderson, Ann. Rept. Provancher Soc. Nat. Hist. Canada, Quebec, p. 37 (for 1941), July 14, 1942, to be Fort Severn, Ontario).
1876. *Zapus hudsonius*, Coues, Bull. U.S. Geol. and Geog. Surv. Territories, ser. 2, 1:253, January 8.
1792. *Dipus labradorius* Kerr, The animal kingdom . . . , p. 276, based on "Labradore Jerboid Rat" of Pennant's History of Quadrupeds, 1781.

MARGINAL RECORDS.—Alaska: Fairbanks. Yukon: 14 mi. E Dawson City, 1300 ft. (Youngman, 1964:5); forks MacMillan River. Mackenzie: Fort Resolution. Manitoba: York Factory. Ontario: Fort Severn; Silver Islet. Michigan: Marquette National Forest; 5 mi. N Menominee. Wisconsin (Jackson, 1961:266): *Cravitz* [= *Crivitz*]; Antigo; North Twin Lake; Solon Springs. Manitoba: Sandilands Forest Reserve (Soper, 1961:202); Delta. Alberta (Soper, 1965:242): Stettler; Lobstick Creek; Demmitt. British Columbia: Charlie Lake; Driftwood River (Cowan and Guiguet, 1965:241); 1 mi. NW jct. Irons Creek and Liard River. Yukon: SW end Dezadeash Lake (Youngman, 1975:117).—See addenda.

Zapus hudsonius intermedius Krutzsch

1954. *Zapus hudsonius intermedius* Krutzsch, Univ. Kansas Publ., Mus. Nat. Hist., 7:447, April 21, type from Ridgeway, Winneshiek Co., Iowa.

MARGINAL RECORDS.—Manitoba: Pine Ridge (Soper, 1961:202, as Z. *h. campestris*). North Dakota: Pembina. Wisconsin (Jackson, 1961:266): St. Croix

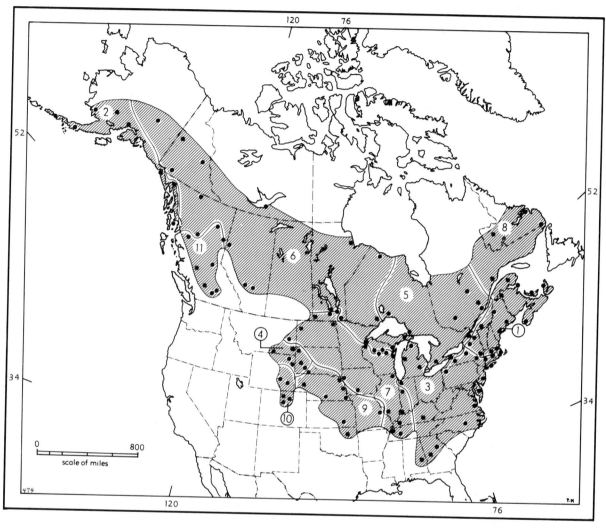

Map 474. *Zapus hudsonius.*

Guide to subspecies		
1. *Z. h. acadicus*	3. *Z. h. americanus*	6. *Z. h. hudsonius*	9. *Z. h. pallidus*
2. *Z. h. alascensis*	4. *Z. h. campestris*	7. *Z. h. intermedius*	10. *Z. h. preblei*
	5. *Z. h. canadensis*	8. *Z. h. ladas*	11. *Z. h. tenellus*

River, above mouth Namakagon River; Rib Hill; Keshena; *Green Bay;* Point Beach State Forest. Illinois: Fox Lake. Indiana: Lafayette; New Harmony. Kentucky: Lyon County. Tennessee: S part Benton County (Severinghaus and Beasley, 1973:130, Fig. 1); *SE part Carroll County (ibid.);* ½ mi. ESE Lasiter's Corner. Illinois: 6 mi. S Pinckneyville (near Pyatt). Iowa (Bowles, 1975:107): 3 mi. E Afton; Sioux City. Montana: Yellostone River, 7 mi. NE Glendive, 2000 ft. North Dakota: Grinnell. Manitoba (Soper, 1961:202, as *Z. h. campestris*): Oak Lake; Portage la Prairie.

Zapus hudsonius ladas Bangs

1899. *Zapus hudsonius ladas* Bangs, Proc. New England Zool. Club, 1:10, February 28, type from Rigoulette, Hamilton Inlet, Labrador.

MARGINAL RECORDS.—Quebec: northwestern Ungava. Labador: Makkovik; Red Bay, thence along N shore Gulf of St. Lawrence to Quebec: Godbout (Harper, 1961:85). Labador: Flour Lake (Harper, 1961:86).

Zapus hudsonius pallidus Cockrum and Baker

1950. *Zapus hudsonius pallidus* Cockrum and Baker, Proc. Biol. Soc. Washington, 63:1, April 26, type from NW corner sec. 4, T. 12 S, R. 20 E, 5½ mi. N, 1¾ mi. E Lawrence, Douglas Co., Kansas.

MARGINAL RECORDS.—South Dakota: Batesland. Nebraska (Jones, 1964c:240, 241): 3 mi. NE Ponca; Engineer Cantonment; 4 mi. E Barada; *5 mi. SE Rulo.* Missouri: Pevely. Oklahoma: Mohawk Park.

Kansas: ½ mi. S Hamilton. Nebraska (Jones, 1964c:241): Platte Meadows, Kearney; Crescent Lake National Wildlife Refuge.

Zapus hudsonius preblei Krutzsch

1954. *Zapus hudsonius preblei* Krutzsch, Univ. Kansas Publ., Mus. Nat. Hist., 7:452, April 21, type from Loveland, Larimer Co., Colorado.

MARGINAL RECORDS.—Wyoming: Springhill, 12 mi. N Laramie Peak, 6300 ft.; Chugwater; *Cheyenne.* Colorado (Armstrong, 1972:249):6 mi. E Hudson, 5000 ft.; Semper; *Boulder;* type locality.

Zapus hudsonius tenellus Merriam

1897. *Zapus tenellus* Merriam, Proc. Biol. Soc. Washington, 11:103, April 26, type from Kamloops, British Columbia.
1934. *Zapus hudsonius tenellus,* Hall, Univ. California Publ. Zool., 40:377, November 5.

MARGINAL RECORDS.—British Columbia: E side Minaker River, 1 mi. W Trutch (Krutzsch, 1954:454); 3 mi. N, 5 mi. W Fort St. John; Indianpoint Lake; Enderby; *Coldstream near Vernon;* Douglas Lake (Cowan and Guiguet, 1965:241); *Nicola Lake (ibid.);* Lillooet *(ibid.);* Chezacut *(ibid.);* Hazelton (Krutzsch, 1954:454, but Cowan and Guiguet, 1965:241, place Hazelton and 1 mi. W Trutch in the geographic range of *hudsonius,* apparently without examining actual specimens from those places).

Zapus princeps
Western Jumping Mouse

External measurements: 216–247; 129–148; 31–34. Back yellowish-gray to salmon-brown and ochraceous; sides paler than back; lateral line ochraceous-buff or indistinct or absent; belly white, usually suffused with ochraceous; tail pale brown to grayish brown above, white to yellowish white below. Skull large; upper premolars medium (.55 long and .50 wide), with shallow re-entrant fold. *Zapus princeps* differs from *Z. trinotatus* in having smaller premolars, a narrower and shallower skull, and a shorter baculum; from *Z. hudsonius, Z. princeps* differs in having larger premolars, a longer rostrum, and a longer baculum.

Zapus princeps chrysogenys Lee and Durrant

1960. *Zapus princeps chrysogenys* Lee and Durrant, Proc. Biol. Soc. Washington, 73:171, December 30, type from 2½ mi. NE La Sal Peak, La Sal Mountains, 8500 ft., Grand Co., Utah.

MARGINAL RECORDS (Lee and Durrant, 1960:174).—Utah: type locality; *Beaver Creek, 1¹/₂ mi. E La Sal Peak, La Sal Mts., 9000 ft.; Beaver Creek, 2 mi. NE Mt. Waas, La Sal Mts., 8720 ft.*

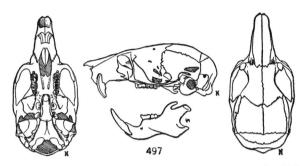

Fig. 497. *Zapus princeps oregonus,* Wisconsin Creek, 7800 ft., Nevada, No. 45871 M.V.Z., ♂, X 1½.

Zapus princeps cinereus Hall

1931. *Zapus princeps cinereus* Hall, Univ. California Publ. Zool., 37:7, April 10, type from Pine Canyon, 6600 ft., Raft River Mts., 17 mi. NW Kelton, Boxelder Co., Utah.

MARGINAL RECORDS.—Idaho: Camp Tendoy, Pocatello. Utah: Pine Creek, 3 mi. N Rosette, Raft River Mts., 6100 ft. Idaho: Mt. Harrison, 10 mi. S Albion.

Zapus princeps curtatus Hall

1931. *Zapus princeps curtatus* Hall, Univ. California Publ. Zool., 37:7, April 10, type from head Big Creek, 8000 ft., Pine Forest Mts., Humboldt Co., Nevada.

MARGINAL RECORDS.—Nevada: Alder Creek, 6000 ft.; *type locality;* Leonard Creek, 6500 ft.

Zapus princeps idahoensis Davis

1934. *Zapus princeps idahoensis* Davis, Jour. Mamm., 15:221, August 10, type from 5 mi. E Warm Lake, 7000 ft., Valley Co., Idaho.

MARGINAL RECORDS.—Alberta: Boom Creek, 5600 ft., 27 mi. W Banff; *Crows Nest Pass;* Waterton Lakes Park. Montana: Highwood Mts.; 15 mi. S Heath, N fork Flat Willow Creek; 2 mi. E Shriver, 6500 ft. Wyoming (Long, 1965a:667): 16¼ mi. N, 17 mi. W Cody; 23½ mi. S, 5 mi. W Lander; LaBarge Creek; *2 mi. S, 19 mi. W Big Piney; 12 mi. N Kendall;* 31 mi. N Pinedale; Valley; *25 mi. S, 28 mi. W Cody;* Glen Creek. Idaho: 4 mi. N, 17 mi. E Ashton, 6275 ft.; Prairie Creek, 12 mi. NW Ketchum, 2400 ft.; 5 mi. W Warm Lake, 7000 ft.; 1 mi. N Bear Creek R. S., SW slope Smith Mtn., 5400 ft.; N fork Potlatch River. Washington: Kamiak Butte. Idaho: 5 mi. N, 13 mi. E Coeur d'Alene. British Columbia: Newgate; Invermere (Cowan and Guiguet, 1965:242); *Kootenay Crossing (ibid.); Vermilion Crossing, Kootenay.*

Zapus princeps kootenayensis Anderson

1932. *Zapus princeps kootenayensis* Anderson, Ann. Rept. Nat. Mus. Canada, for 1931, p. 108, November 24, type from near summit Green Mtn., head Murphy Creek, about 10 mi. N Rossland, West Kootenay Dist., British Columbia.

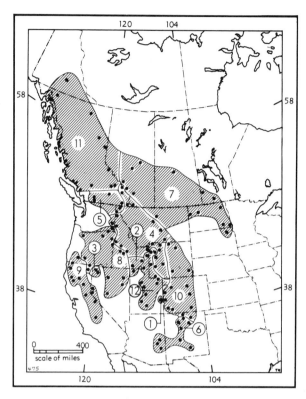

Map 475. *Zapus princeps.*

1. *Z. p. chrysogenys*	7. *Z. p. minor*
2. *Z. p. cinereus*	8. *Z. p. oregonus*
3. *Z. p. curtatus*	9. *Z. p. pacificus*
4. *Z. p. idahoensis*	10. *Z. p. princeps*
5. *Z. p. kootenayensis*	11. *Z. p. saltator*
6. *Z. p. luteus*	12. *Z. p. utahensis*

MARGINAL RECORDS.—British Columbia: Monashee Pass, 4000 ft.; Moyie; *Camp 6, Meadow Creek, 7 mi. SE Yahk.* Idaho: 4 mi. W Meadow Creek, 3000 ft.; 5 mi. W Cocololla, 3500 ft. Washington: *Sullivan Lake.* British Columbia: Manning Park; Hope–Princeton Summit, 5500 ft.

Zapus princeps luteus Miller

1911. *Zapus luteus* Miller, Proc. Biol. Soc. Washington, 24:253, December 23, type from Española, 5000 ft., Rio Arriba Co., New Mexico.
1954. *Zapus princeps luteus,* Krutzsch, Univ. Kansas Publ., Mus. Nat. Hist., 7:406, April 21.
1913. *Zapus luteus australis* Bailey, Proc. Biol. Soc. Washington, 26:132, May 21, type from Socorro, Socorro Co., New Mexico.

MARGINAL RECORDS (Findley, *et al.*, 1975:272, as *Z. princeps* only, unless otherwise noted).—New Mexico: *3 mi. E Eagle Nest;* 1 mi. W Cowles; Socorro; Penasco Creek, 12 mi. E Cloudcroft. Arizona (Cockrum, 1961:214): Hannagan Creek, 8200 ft.; W fork Black River, 7700 ft.; N fork White River, 8200 ft.; *Alpine, 8000–8500 ft.* New Mexico: 3 mi. NW Albuquer-

que; 1 mi. S Seven Springs fish hatchery; Arroyo Hondo.

Zapus princeps minor Preble

1899. *Zapus princeps minor* Preble, N. Amer. Fauna, 15:23, August 8, type from Wingard, near Carleton House, Saskatchewan.

MARGINAL RECORDS.—Alberta: 4 mi. N Marinville [= Morinville]. Saskatchewan: type locality. Manitoba: Shoal Lake; Aweme. North Dakota: Grafton; 4 mi. S Blackner. South Dakota: Webster. North Dakota: Oakes; Grinnell; Buford. Montana: Eagle Creek, 25 mi. SE Big Sandy. Alberta: High River; Kananaskis Valley, 2000 ft.; Red Deer River; Blindman River.

Zapus princeps oregonus Preble

1899. *Zapus princeps oregonus* Preble, N. Amer. Fauna, 15:24, August 8, type from Elgin, Blue Mts., Union Co., Oregon.
1899. *Zapus major* Preble, N. Amer. Fauna, 15:24, August 8, type from Warner Mts., Lake Co., Oregon.
1899. *Zapus nevadensis* Preble, N. Amer. Fauna, 15:25, August 8, type from Ruby Mts., Elko Co., Nevada.
1931. *Zapus princeps palatinus* Hall, Univ. California Publ. Zool., 37:8, April 10, type from Wisconsin Creek, 7800 ft., Toyabe Mts., Nye Co., Nevada.

MARGINAL RECORDS.—Washington: Anatone, 3300 ft. Oregon: East Pine Creek, 2½ mi. NE Cornucopia. Idaho: 1 mi. NE Heath, SW slope Cuddy Mtn., 4000 ft.; Bald Mtn., Ranger Station, 10 mi. S Idaho City, 7400 ft.; Trail Creek, Boise National Forest. Nevada: Harrison Pass Ranger Station, Ruby Mts.; Steel Creek, 7000 ft.; Wisconsin Creek, 7000 ft.; 13 mi. N Paradise Valley, 6700 ft. California: Lassen Creek; Buck Creek Ranger Station. Oregon: Fort Klamath; Marks Creek, 12 mi. N Howard; Meacham. Washington: Twin Buttes, 25 mi. SE Dayton.

Zapus princeps pacificus Merriam

1897. *Zapus pacificus* Merriam, Proc. Biol. Soc. Washington, 11:104, April 26, type from Prospect, Rogue River Valley, Jackson Co., Oregon.
1954. *Zapus princeps pacificus,* Krutzsch, Univ. Kansas Publ., Mus. Nat. Hist., 7:412, April 21.
1898. *Zapus alleni* Elliot, Field Columb. Mus., Publ. 27, Zool. Ser., 1:212, April 19, type from Pyramid Peak, Lake Tahoe, Eldorado Co., California.

MARGINAL RECORDS.—Oregon: type locality. Nevada: ½ mi. S Mt. Rose, 9500 ft.; *¹/₂ mi. S Marlette Lake, 8150 ft.;* ½ mi. E Zephyr Cove, Lake Tahoe, 6400 ft. California: Mono Lake P. O., 6500 ft.; 2 mi. E Kern Peak, 9300 ft., Sierra Nevada Mts.; 1 mi. W Fyffe; Rich Gulch, 3850 ft., 11 mi. W, 8 mi. N Quincy; Warner Creek, 8000 ft., Lassen Peak; 2 mi. S Yolla Bolly Mtn.; 8 mi. NE Hyampon, 2900 ft.; Siskiyou Mts., 6000 ft.; Poker Flat, 5000 ft., 12 mi. NW Happy Camp.

Zapus princeps princeps J. A. Allen

1893. *Zapus princeps* J. A. Allen, Bull. Amer. Mus. Nat. Hist., 5:71, April 28, type from Florida, La Plata Co., Colorado.

MARGINAL RECORDS.—Wyoming (Long, 1965a: 667, 668): Medicine Wheel Ranch, 28 mi. E Lovell; *Granite Creek Campground, Bighorn Mts.; 4¹/₂ mi. S, 17¹/₂ mi. E Shell;* ⅛ mi. S, 1 mi. W Buffalo; Meadow. Colorado: Gold Hill; Minnehaha; 7 mi. SE Russell. New Mexico: E slope Taos Mts.; Tierra Amarilla. Colorado: type locality; 5 mi. N, 3 mi. E Stoner, 8800 ft. (Armstrong, 1972:251); 15 mi. E Gateway (*ibid.*); 35 mi. S Rangely, 8000 ft. (*ibid.*); Meeker. Wyoming (Long, 1965a:667, 668): *4 mi. N, 8 mi. E Savery;* Bridger Pass.

Zapus princeps saltator J. A. Allen

1899. *Zapus saltator* J. A. Allen, Bull. Amer. Mus. Nat. Hist., 12:3, March 4, type from Telegraph Creek, British Columbia.
1931. *Zapus princeps saltator,* Hall, Univ. California Publ. Zool., 37:10, April 10.

MARGINAL RECORDS.—Yukon: Rose River, Mile 95 on Canol Road. British Columbia: Mile 114 on Alaska Highway N of Summit Pass; Charlie Lake; Tupper Creek; Moose Pass; Glacier; Mt. Revelstoke; Lillooet; Knight Inlet; Kynoch Inlet; Inverness; Port Simpson (Cowan and Guiguet, 1965:244). Alaska: Taku River. British Columbia: Atlin.

Zapus princeps utahensis Hall

1934. *Zapus princeps utahensis* Hall, Occas. Pap. Mus. Zool., Univ. Michigan, 296:3, November 2, type from Beaver Creek, 19 mi. S Manila, Daggett Co., Utah.

MARGINAL RECORDS.—Wyoming (Long, 1965a: 668): Upper Arizona Creek; Whetstone Creek; 10 mi. SE Afton; 5 mi. E Lonetree. Utah: jct. Deep and Carter creeks, 7900 ft.; Paradise Park, 15 mi. N, 21 mi. W Vernal, 10,500 ft.; Ephraim, 8500 ft.; E fork Boulder Creek, 10 mi. N Boulder; Puffer Lake; Payson Lake, 8300 ft., 12 mi. SE Payson, Mt. Nebo; *Ophir Canyon, southwestern Oquirrh Mts., ca. 7500 ft.* (Egoscue, 1961:124); Butterfield Canyon, approx. 5 mi. above Butterfield Tunnel. Idaho: Strawberry Creek, 20 mi. NE Preston, 6700 ft.; 3 mi. SW Victor.

Zapus trinotatus
Pacific Jumping Mouse

External measurements: 221–238; 131–149; 31–34. Back of various shades of ochraceous and tawny; sides paler than back; lateral line distinct and bright; belly white, usually suffused with ochraceous; ears dark; tail brown above, white to yellowish-white below. Skull broad and deep in proportion to length; upper premolars large (.70 long and .75 wide), with labial re-entrant fold

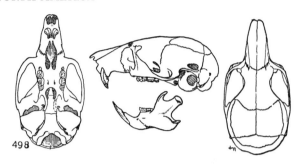

Fig. 498. *Zapus trinotatus montanus,* E slope Cascade Divide, 6400 ft., Crater Lake National Park, Oregon, No. 22165 K.U., ♀, X 1½.

forming crescentic loop. *Zapus trinotatus* differs from *Z. princeps* and *Z. hudsonius* in having larger premolars, broader skull, and longer baculum.

Zapus trinotatus eureka A. B. Howell

1920. *Zapus trinotatus eureka* A. B. Howell, Univ. California Publ. Zool., 21:229, May 20, type from Fair Oaks, Humboldt Co., California.

MARGINAL RECORDS.—California: Trinidad; Russian Gulch State Park; Albion River, ⅛ mi. E MacDonalds Ranch; Mendocino City; Carlotta.

Zapus trinotatus montanus Merriam

1897. *Zapus trinotatus montanus* Merriam, Proc. Biol. Soc. Washington, 11:104, April 26, type from Crater Lake, Klamath Co., Oregon.

MARGINAL RECORDS.—Oregon: Brooks Meadow, 4300 ft., 9 mi. ENE Mt. Hood; Tumalo Creek, 15 mi. W Bend, 6100 ft.; Anna Creek, Mt. Mazama, 6000 ft.; E slope Cascade Divide, 6400 ft., Crater Lake National Park; Diamond Lake; North Santiam River, 3400 ft.

Zapus trinotatus orarius Preble

1899. *Zapus orarius* Preble, N. Amer. Fauna, 15:29, August 8, type from Point Reyes, Marin Co., California.
1944. *Zapus trinotatus orarius,* Hooper, Miscl. Publ. Mus. Zool., Univ. Michigan, 59:67, January 12.

MARGINAL RECORDS.—California: 3 mi. W Inverness, 300 ft.; West Portal, Fort Barry.

Zapus trinotatus trinotatus Rhoads

1895. *Zapus trinotatus* Rhoads, Proc. Acad. Nat. Sci. Philadelphia, 47:421, January 15, type from Lulu Island, mouth Frazer River, British Columbia.
1899. *Zapus imperator* Elliot, Field Columb. Mus., Publ. 30, Zool. Ser., 1:228, February 2, type from Siegs Ranch, Elwah River, Clallam Co., Washington.

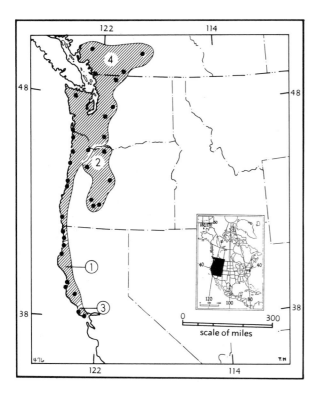

Map 476. *Zapus trinotatus.*

1. *Z. t. eureka*   3. *Z. t. orarius*
2. *Z. t. montanus*   4. *Z. t. trinotatus*

MARGINAL RECORDS.—British Columbia: Alta Lake, 2600 ft.; Okanagan; Manning Park. Washington: Baker Lake; Snoqualmie Pass; Mt. Rainier; Ice Caves, 2800 ft., 5 mi. WSW Gulch. Oregon: Portland, Council Crest, 950 ft. California: Requa; Crescent City. Oregon: Gold Beach; Gardiner; Sutton Lake, 6 mi. N Florence; Newport; Netarts; Old Fort Clatsop, 100 ft. Washington: 3½ mi. SE Chinook, 10 ft.; Deer Lake, 3800 ft. British Columbia: Stanley Park, Vancouver; *Brackendale.*

Genus **Napaeozapus** Preble
Woodland Jumping Mouse

Revised by Wrigley, Illinois Biol. Monog., 47:8 + 118, 22 figs., 8 tables, December, 1972.

1899. *Napaeozapus* Preble, N. Amer. Fauna, 15:33, August 8. Type, *Zapus insignis* Miller.

External measurements: 220–255; 131–160; 29–33. Back with a broad band of brown, much darkened with black; sides buff-yellow, tinged with clay color; underparts white, bordered by yellow lateral line; tail sharply bicolored, brown above, paler below; tip of tail normally white. Premolars absent; M2 about same size as M1; 3 narrow labial re-entrant folds of unequal length

on M3; anterior cingulum of M1 large; frontal region somewhat swollen.

Napaeozapus insignis
Woodland Jumping Mouse

See account of the genus.

Napaeozapus insignis abietorum (Preble)

1899. *Zapus (Napaeozapus) insignis abietorum* Preble, N. Amer. Fauna, 15:36, August 8, type from Peninsula Harbor, N shore Lake Superior, Ontario.
1900. *Napaeozapus insignis abietorum*, Miller, Bull. New York State Mus. Nat. Hist., 8:114, November 21.

MARGINAL RECORDS (Wrigley, 1972:37, 40, unless otherwise noted).—Ontario: Albany River, 20 mi. E Kenogami River; Moosonee. Quebec: Lake Albanel; Mistassini Post (Hall and Kelson, 1959:778); 47° 50′ N, 75° 35′ W, S of Clova (MacLeod and Cameron, 1961:282). Ontario: Balsam Creek, 25 mi. NE North Bay; Trout Lake, East Ferris Twp.; Frank Bay, Lake Nipissing; *Bigwood, Bigwood Twp.;* French River, S of Sudbury; MacLennan, Tarbutt Adl. Twp.; Pancake Bay, Lake Superior, 46° 57′ N, 84° 41′ W, thence northwestward along N shore Lake Superior to Murillo, Oliver Twp.; 90 mi. N Port Arthur; Nakina.

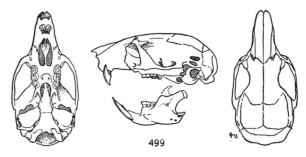

Fig. 499. *Napaeozapus insignis insignis,* 6½ mi. NE Amherst, 600 ft., Franklin Co., Massachusetts, No. 41109 K.U., ♂, X 1½.

Napaeozapus insignis frutectanus Jackson

1919. *Napaeozapus insignis frutectanus* Jackson, Proc. Biol. Soc. Washington, 32:9, February 14, type from Crescent Lake, Oneida Co., Wisconsin.

MARGINAL RECORDS (Wrigley, 1972:42, 43, unless otherwise noted).—Manitoba: Forbes Lake (Iverson and Turner, 1973:16); Caddy Lake, Whiteshell Forest Reserve. Ontario: Rainy River. Minnesota: 2½ mi. N, 1 mi. E Mineral Center (Timm, 1975:27), thence southwestward and then eastward along S shore Lake Superior to Ontario: Batchawana Bay (Hall and Kelson, 1959:779). Michigan: Sugar Island; Alcona County; S branch Au Sable River; Crawford County; 7 mi. E Kalkaska. Wisconsin: Jordan Pond Cedar Swamp, 1½ mi. E Jordan Pond; Worden Township; Danbury. Minnesota:

Cloverdale; Upper Long Lake; Itasca State Park. Manitoba: Cedar Lake, 2 mi. N Vivian.

Napaeozapus insignis insignis (Miller)

1891. *Zapus insignis* Miller, Amer. Nat., 25:742, August, type from Restigouche River, New Brunswick.
1899. *Napaeozapus insignis,* Miller, Bull. New York State Mus. Nat. Hist., 6:330, November 18.
1941. *Napaeozapus insignis algonquinensis* Prince, Occas. Pap. Royal Ontario Mus. Zool., 7:1, December 12, type from Smoke Lake, Algonquin Park, Ontario.
1942. *Napaeozapus insignis gaspensis* Anderson, Ann. Rept. Provancher Soc. Nat. Hist. Canada, Quebec, for 1941, p. 39, July 14, type from near Federal Zinc and Lead Mine, about 1500 ft., on upper waters Berry Mtn. Brook, a tributary of

Grand Cascepedia River, Shickshock Mts., about halfway between Gulf of St. Lawrence and Baie de Chaleur, Gaspé Co., Quebec.

MARGINAL RECORDS (Wrigley, 1972:45–52, unless otherwise noted).—Quebec: 2 mi. W Cap de la Madeleine; 2 mi. W Coin du Banc (Manville, 1961:108, as *N. i. gaspensis*). New Brunswick: Rexton. Prince Edward Island: 1 mi. NE Alma; 5 mi. S Hunter River; Fortune. Nova Scotia: *2 mi. W Cape North;* Cape North; Boularderie; West Bay, Cemetery Brook; Moser River; Halifax; Barrington Passage; 16 mi. S Annapolis Royal; *2 mi. S South Milford,* thence to and along coast to New Brunswick: Fundy National Park. Thence southwestward along coast of Atlantic Ocean to Maine: Mount Desert Island; Brunswick. New Hampshire: 2

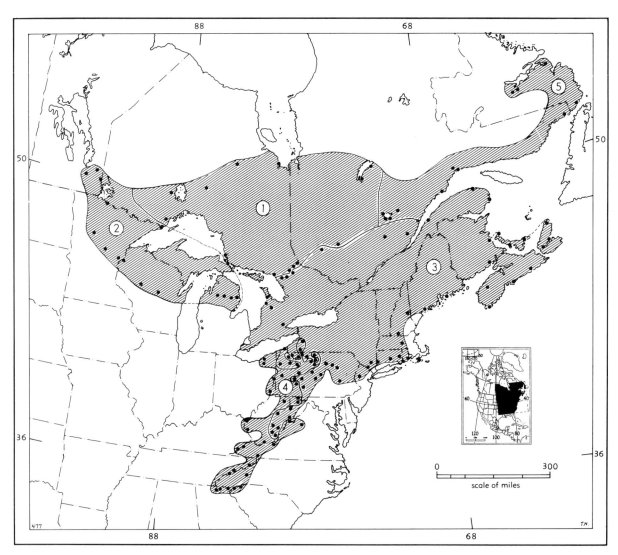

Map 477. *Napaeozapus insignis.*

Guide to subspecies   2. *N. i. frutectanus*   4. *N. i. roanensis*
1. *N. i. abietorum*   3. *N. i. insignis*   5. *N. i. saguenayensis*

mi. S Ossipee Center; Jaffrey. Massachusetts: $1\frac{1}{5}$ mi. S, 5 mi. E Holden; Martha's Vineyard (Keith, 1969:73, as *N. insignis* only). Rhode Island (Cronan and Brooks, 1962:70): South Kingstown; Westerly. Connecticut: Plainfield; Portland; New Fairfield. New York: $5\frac{1}{2}$ mi. E Cold Spring; Tuxedo. New Jersey: *Chester (Blackwater River)*; Fairmount. Pennsylvania: $2\frac{1}{2}$ mi. NNW Shartlesville; 8 mi. WNW Sweet Valley; *Ganoga Lake;* Eagles Mere; *Ogdonia Creek, 4 1/4 mi. SSE Hillsgrove;* Slacks Run, 1 mi. E Bodines; Penns Creek Mtn., 1 mi. S Glen Iron; *3 1/2 mi. WNW Milroy;* 3 mi. WSW Milroy. West Virginia: $3\frac{1}{10}$ mi. NW Circleville; S slope Cranberry Mtn., 9 mi. WSW Marlinton; White Sulphur Springs; $1\frac{1}{5}$ mi. N Kieffer; 2 mi. N Richwood; French Creek; Davis. Pennsylvania: *1 1/2 mi. W Imler;* 7 mi. NW Imler; *9 mi. SSE Indiana;* $2\frac{1}{2}$ mi. SSE Indiana; $1\frac{3}{4}$ mi. SW Patton; Gum Stump, $3\frac{1}{2}$ mi. NNW Wingate; Cherry Springs; Red Mill Brook, 4 mi. NE Clermont; *4 mi. N Clermont;* Sugar Run, 10 mi. SW Bradford; 10 mi. SE Warren; 1 mi. SE Garland; *Benson Swamp, 5 mi. E Columbus; 1 mi. SW President;* McCrea Run, 9 mi. E Oil City; $5\frac{1}{2}$ mi. NE Siegel; 2 mi. N Clarks Mills; Burgess Gulf, 4 mi. SW North East. Thence northeast along shore of Lake Erie, north along Niagara River, and westward along shore of Lake Ontario to Ontario: 3 mi. W Freelton; Bentinck Township; Southhampton; Red Bay, thence along Georgian Bay and inland to Noganosh Lake; Arnstein; South Bay, Lake Nipissing. Quebec: La Vérendrye Park; Lake Edward; Camp de la Roche, Laurentides Park, 47° 41′ N, 70° 50′ W; Trois Pistoles, thence along coast to point of beginning. Wrigley (1972:42) considers a record of *Napaeozapus* from Turkey Run, Parke Co., Indiana, to be invalid. Not found: New York: Tamarack Swamp. Pennsylvania: Howard Station.

Napaeozapus insignis roanensis (Preble)

1899. *Zapus (Napaeozapus) insignis roanensis* Preble, N. Amer. Fauna, 15:35, August 8, type from Roan Mountain, Mitchell Co., North Carolina.

1900. *Napaeozapus insignis roanensis,* Miller, Bull. New York State Mus. Nat. Hist., 8:114, November 21.

MARGINAL RECORDS (Wrigley, 1972:54–56).—New York: 2 mi. N Springville. Pennsylvania: *Mile Run, 8 mi. NNW Emporium;* Costello; *Carter Camp; Rapplee Hollow, 5 1/2 mi. SW Ulysses;* Woodcock Run, $7\frac{1}{2}$ mi. WSW Ulysses; Tamarack, 9 mi. NNW Renovo; McGees Mills; $1\frac{1}{2}$ mi. SE New Florence. Maryland: Finzel; Dans Mtn. Virginia: Laurel Fork, 9 mi. NNW Monterey; Shenandoah National Park; Eggleston; Mt. Rogers; *Whitetop Mtn.* North Carolina: $4\frac{7}{10}$ mi. SW Busick; *5 2/5 mi. SW Busick.* South Carolina: Caesars Head; 15 mi. N Walhalla. Georgia: near North Carolina line; Union County; Fannin County. Tennessee: Indian Gap; *Eagle Rocks Creek;* $4\frac{1}{2}$ mi. SE Cosby; NE slope Road Mtn., $6\frac{1}{2}$ mi. SW town of Roan Mountain. Kentucky: Black Mtn. Virginia: 6 mi. N Wise. West Virginia: $1\frac{1}{2}$ mi. NE Cooper Rock; Cabwaylingo State Forest, Arkansas branch of Twelvepole Creek; Huntington; 8 mi. S Morgantown. Ohio: Cat Run, Powhatan Point. West Virginia: 5 mi. SSW Chester. Pennsylvania: 2 mi. E Middle Lancaster. Ohio: North Chagrin Metropolitan Park; *Gates Mills;* Willoughby; *Mentor.* Pennsylvania: $4\frac{1}{2}$ mi. SE Spartansburg. New York: *Quaker Run, Allegany Indian Res.;* Red House, Allegany Park. Not found: New York: Buffalo Camp, Allegany Park. Ohio: Holden Arboretum; Mechanicsville. Virginia: Little Meadows.

Napaeozapus insignis saguenayensis Anderson

1942. *Napaeozapus insignis saguenayensis* Anderson, Ann. Rept. Provancher Soc. Nat. Hist. Canada, Quebec, for 1941, p. 40, July 14, type from Trout Lake, near Moisie Bay, N shore Gulf of St. Lawrence, Saguenay Co., Quebec.

MARGINAL RECORDS (Wrigley, 1972:57, 58, unless otherwise noted).—Labrador: Hamilton Inlet; Black Bay (Harper, 1961:87). Quebec: Strait of Belle Isle (Hall and Kelson, 1959:779); type locality; Godbout; Bell River; Val Jalbert; St. Félicien; *Ashuapmuchuan River* (Harper, 1961:87, as *N. i. algonquinensis*); Lake Chitigama; Third Portage, St. Margaret River. Labrador: Hamilton River; Hamilton River, Muskrat Falls.

SUBORDER HYSTRICOMORPHA
Porcupines, Capybaras and Allies

This suborder includes a diverse assemblage of rodents principally of South American distribution. Members of this suborder appeared in South America possibly in the Oligocene.

Infraorbital canal greatly enlarged for transmission of part of masseter muscle; superficial lateral masseter does not extend forward on rostrum; angular process of mandible usually arises from side of ramus (exception: Cavioidea); postorbital process usually absent; dental formula, i. $\frac{1}{1}$, c. $\frac{0}{0}$, pm. $\frac{1}{1}$, m. $\frac{3}{3}$; premolars large and usually strongly molariform; molariform teeth flat-crowned, hypsodont, sometimes persistently growing, and of highly variable pattern; tibia and fibula never fully fused; tendency toward zygomatic complexity.

In attempting to diagnose or prepare a key to the groups of this suborder, one must be constantly aware of the limitations of our present knowledge. Many fossil genera are of uncertain position. The same is true of a number of modern genera. Therefore the composition of each of several of the higher categories is unknown. Consequently the following keys are to be used with caution.

KEY TO NORTH AMERICAN FAMILIES OF HYSTRICOMORPHA

1. M3 greatly enlarged, exceeding in alveolar length combined length of other upper molariform teeth; molariform teeth strongly prismatic and multilaminar. . . . . Hydrochaeridae, p. 854
1'. M3 not greatly enlarged; molariform teeth not strongly prismatic and, if laminate, not excessively so (fewer than 8 laminae in M3).
   2. Upper cheek-teeth laminate, each lamina bearing enamel only on anterior surface; cheek-teeth reduced (apparently) to $\frac{2}{2}$. . . . . Heptaxodontidae, p. 855
   2'. Upper cheek-teeth not laminate or, if somewhat laminate, then with enamel not confined to anterior surface of laminae; cheek-teeth more than $\frac{2}{2}$.
      3. Inferior border of angular process of mandible strongly inrolled; spiny pelage with conspicuous quills. . . . . Erethizontidae, p. 850
      3'. Inferior border of angular process of mandible not strongly inrolled; pelage sometimes spiny but lacking large quills.
         4. External form much modified for cursorial life; clavicles reduced or wanting; hind foot functionally tridactyl or, if pentadactyl, then jugal enormously expanded into a broad, curved, externally rugose plate. . . . . Dasyproctidae, p. 856
         4'. External form not specially modified for cursorial life; clavicles well developed; hind foot not tridactyl nor is jugal excessively expanded.
            5. Paroccipital processes long, relatively straight, and not appressed to auditory bullae. . . . . Capromyidae, p. 862
            5'. Paroccipital processes long, curved forward under, but not appressed to, auditory bullae. . . . . Echimyidae, p. 871

SUPERFAMILY ERETHIZONTOIDEA

FAMILY ERETHIZONTIDAE
New World Porcupines

Form robust; body covered above with thick, sharp spines distributed among the hairs; molariform teeth flat-crowned, rooted, subhypsodont; lower border of angular process inflected; upper zygomatic root over anterior part of tooth-row; feet noticeably modified for arboreal life.

KEY TO NORTH AMERICAN GENERA OF ERETHIZONTIDAE

1. Hallux well developed; tail short, non-prehensile. . . . .*Erethizon*, p. 850
1'. Hallux vestigial or absent; tail long, prehensile. . . . .*Coendou*, p. 853

Genus **Erethizon** F. Cuvier—Porcupine

Subspecies reviewed by Anderson and Rand, Canadian Jour. Res., sec. D, 21:292–309, September 6, 1943.

1822. *E*[*rethizon*]. F. Cuvier, Mém. Mus. Hist. Nat., Paris, 9:432. Type, *Hystrix dorsata* Linnaeus, 1758.

Size large; body stout; tail short, thick, non-prehensile; 4 digits on forefeet, 5 on hind feet. Skull compact, broad, heavily constructed; nasals broad; frontals broad, heavily ridged, the ridges convergent posteriorly and forming sagittal crest; zygoma uncomplicated, jugal deeper anteriorly;

mandible with low coronoid and angular processes; upper cheek-teeth with 1 persistent internal and 1 persistent external fold, 2 additional outer folds tend to become isolated with wear.

Erethizon dorsatum
Porcupine

x ¹⁄₁₀

External measurements: 648–1030; 148–300; 86–124. Upper parts blackish or dark brownish; wooly underfur and long, stiff guard-hairs clearly distinct.

Erethizon dorsatum bruneri Swenk

1916. *Erethizon epixanthum bruneri* Swenk, Univ. Nebraska Studies, 16:117, November 21, type from 3 mi. E Mitchell, Scotts Bluff Co., Nebraska.
1947. *Erethizon dorsatum bruneri*, Anderson, Bull. Nat. Mus. Canada, 102:173, January 24.

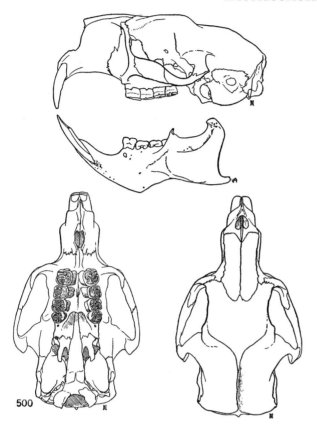

Fig. 500. *Erethizon dorsatum epixanthum*, Wilson Creek, 7200 ft., Nevada, No. 88223 M.V.Z., ♂, X ½.

MARGINAL RECORDS.—North Dakota: McKenzie County (Hewston, 1962:270, as "porcupine" only); 7 *mi. N, 5½ mi. W Killdeer* (Genoways and Jones, 1972:30). Nebraska (Jones, 1964c:243, 244): Fort Niobrara National Wildlife Refuge; North Fork, 14 mi. NW Norfolk; *Cuming County*. Kansas: 4 mi. S, 2 mi. E Gardner; 6 mi. W Arkansas City. Oklahoma: Wichita Mountains Wildlife Refuge. Texas: *Henrietta* (Kaspar and Parrish, 1974:214); *Wilbarger County* (Dalquest, 1968:19); near Seymour along Brazos River (Baccus, 1971:183); *near Quanah* (Dalquest, 1968:19): Stinnett. Oklahoma: E tip Black Mesa. Colorado: Geary Reservoir, 9 mi. W Grover. Wyoming (Long, 1965a:670): *Woods Creek*; Encampment; 24 mi. N, 12 mi. E Sinclair; 1 mi. S, 4½ mi. W Buffalo; Arvada. Montana: 7 mi. N, 3 mi. W Camp Crook, 3400 ft. (Lampe, *et al.*, 1974:23).

Erethizon dorsatum couesi Mearns

1897. *Erethizon epixanthus couesi* Mearns, Proc. U.S. Nat. Mus., 19:723, July 30, type from Fort Whipple, Yavapai Co., Arizona.
1946. *Erethizon dorsatum couesi*, Hall, Mammals of Nevada, p. 649, July 1.

MARGINAL RECORDS.—Colorado (Armstrong, 1972:253): Grandview Overlook, Colorado National Monument; near Chromo, 8500 ft. New Mexico: Santa

Fe. Texas (Davis, 1966:214, unless otherwise noted, and as *E. dorsatum* only): 2³⁄₁₀ mi SW Bovina (Best and Kennedy, 1972:351, as *E. dorsatum* only); Foard County; 12 mi. N Albany, Taylor County (Kasper and Parrish, 1974:214); Mason County; Kerr County. Coahuila (Jones and Genoways, 1968:709): Hda. Las Margaritas; Cuatro Ciénegas. Sinaloa: 16 km. NNE Choix, 1700 ft. (*ibid.*). Sonora (*ibid.*): 6 mi. N Puerto Kino; Rancho Santa Ana, *ca.* 45 mi. W Hermosillo. Arizona: 13⅝ mi. by road N Cabeza Prieta Game Range boundary (sec. 20, T. 10 S, R. 14 W) (Simmons, 1966:122); near Baldy Tanks, Castle Dome Mts. (Cockrum, 1961:216); Bill Williams Fork. Nevada: *ca.* 2½ mi. NE California–Nevada boundary at tip of state; Kyle Canyon, Charleston Peak; Sheep Mts. Arizona (Hoffmeister, 1971:177): S park boundary, 1 mi. S Pasture Wash Ranger Station; *Desert View*. Utah: head of Nigger Bill Canyon, E side Colorado River, 4 mi. above Moab bridge, 3995 ft. Also, once reported from San Bernardino Mts., California (Stephens, 1906:179, as *E. epixanthus*).

Erethizon dorsatum dorsatum (Linnaeus)

1758. [*Hystrix*] *dorsata* Linnaeus, Syst. nat., ed. 10, 1:57. Type locality, eastern Canada.
1822. *E*[*rethizon*]. *dorsatum*, F. Cuvier, Mém. Mus. Hist. Nat., Paris, 9:432.
1900. *Erethizon dorsatus picinus* Bangs, Proc. New England Zool. Club, 2:37, September 20, type from L'Anse au Loup, Strait of Belle Isle, Labrador. Regarded as inseparable from *E. d. dorsatum* by Harper, Univ. Kansas Mus. Nat. Hist., Miscl. Publ., 27:87, 89, 90, August 11, 1961.
1937. *Erethizon epixanthum doani* H. H. Bailey, Bailey Mus. and Library Nat. Hist., Bull. 12, p. [1], January 15, type from timber, 15 mi. NW Red Bay, Labrador.

MARGINAL RECORDS.—Mackenzie: northern part of Anderson River; E of Lynx Lake. Keewatin: 18 mi. N Windy River (Harper, 1956:45). Manitoba: Herchmer. Quebec: Richmond Gulf (Harper, 1961:91); Fort Chimo (*ibid.*); L'Anse au Loup; Gaspé. Nova Scotia: East Roman Valley; Barrington Passage. Maine: Mount Desert Island (Manville, 1960:416, as *E. dorsatum* only). West Virginia: Spruce Knob. Tennessee: S bank Tennessee River, 85° 33' W, 35° 1' N (Parmalee and Guilday, 1966:81). Illinois: Riverton, 2 mi. NE Palestine (Parmalee, 1962:90). Iowa: Grant City (Bowles, 1975:108). Minnesota: near Elk River; Warren. Alberta: near Rochester (Keith and Meslow, 1966:541); Ray Lake.

Erethizon dorsatum epixanthum Brandt

1835. *Erethizon epixanthus* Brandt, Mém. Acad. Imp. Sci., St. Pétersbourg, ser. 6, Sci. Math. Phys. et Nat., 3:390. Type locality, California. (See Hollister, Canadian Alpine Jour., Special Number, p. 27, February 17, 1913.)
1884. *Erethrizon* [*sic*] *dorsatus epixanthus*, True, Proc. U.S. Nat. Mus., 7(App., Circ. 29):600, November 29.

MARGINAL RECORDS.—Alberta: Lodge Creek, near W end Cypress Hills. Saskatchewan: Lonesome

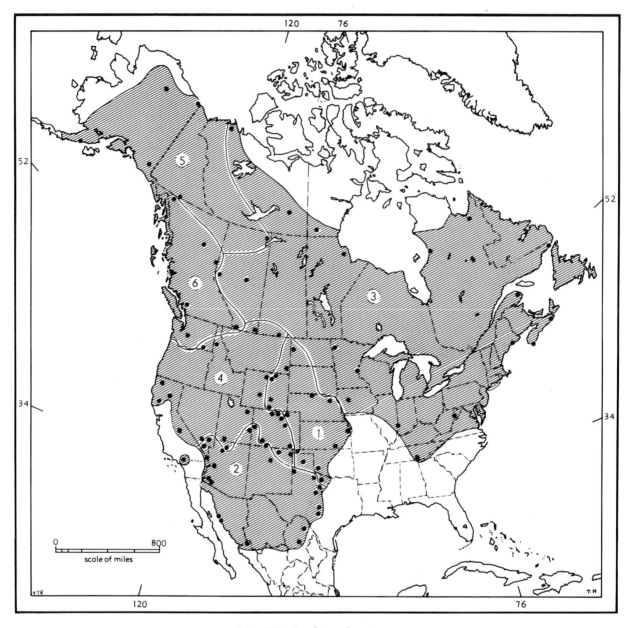

Map 478. *Erethizon dorsatum.*

| 1. *E. d. bruneri* | 3. *E. d. dorsatum* | 5. *E. d. myops* |
| 2. *E. d. couesi* | 4. *E. d. epixanthum* | 6. *E. d. nigrescens* |

Butte, SW of Wood Mtn. Wyoming (Long, 1965a:671): Shell; Hailey. Colorado (Armstrong, 1972:253, 254): Medicine Bow Range; 18 mi. N Fort Collins; Greeley; Deer Trail; 3½ mi. S, 5 mi. W Kim. New Mexico: Clayton (Best and Kennedy, 1972:351); near Cimma-ron. Colorado (Armstrong, 1972:254): *Osier*; jct. Gordon Creek and Piedra River. Utah: Carter Creek. Arizona (Hoffmeister, 1971:177, sight records only): *Long Road in Thompson Canyon; Bright Angel Point; Shiva Temple; Point Sublime Road.* Utah: 1 mi. W Kanab, 5000 ft. Nevada: Oak Spring, 6000 ft. California: Whitney Creek, 10,850 ft. (Grinnell, 1933:196);

Battle Creek Meadows, 4800 ft. (Grinnell, *et al.,* 1930:531); vic. North Yolla Bolly Mtn.; Forest House Mtn., 4000 ft., 8 mi. W Yreka. Oregon: *"almost the whole state . . . to the coast."* Washington: Walla Walla River, E of Wallula. Idaho: Coeur d'Alene.

Erethizon dorsatum myops Merriam

1900. *Erethizon epixanthus myops* Merriam, Proc. Washington Acad. Sci., 2:27, March 14, type from Portage Bay, Alaska Peninsula, Alaska.

1943. *Erethizon dorsatum myops*, Anderson and Rand, Canadian Jour. Res., sec. D, 21:293, September 6.

MARGINAL RECORDS.—Alaska: Ice Reef. Alberta: Wood Buffalo Park. Yukon: Teslin Lake. Alaska: Chitina River; type locality; Chandler Lake.

Erethizon dorsatum nigrescens J. A. Allen

1903. *Erethizon epizanthus* [*sic*] *nigrescens* J. A. Allen, Bull. Amer. Mus. Nat. Hist., 19:558, October 10, type from Shesley River, British Columbia.
1943. *Erethizon dorsatum nigrescens*, Anderson and Rand, Canadian Jour. Res., sec. D, 21:293, September 6.

MARGINAL RECORDS.—British Columbia: 1 mi. N Atlin; Robb Lake; Tupper Creek. Alberta: Waterton Lakes. Washington: Tye. British Columbia (Cowan and Guiguet, 1965:246): Alta Lake; Neckis River; *Atlin*.

Genus **Coendou** Lacépède
Prehensile-tailed Porcupines

1799. *Coendou* Lacépède, Tableau des divisions, sous-divisions, ordres et genres des mammifères, p. 11. (Published as supplement to Discours d'ouverture et de clôture du cours d'histoire naturelle et tableaux méthodiques des mammifères et des oiseaux.) Type, by subsequent designation (Palmer, N. Amer. Fauna, 23:633, January 23, 1904), *Coendou prehensilis* [probably equals *Hystrix prehensilis* Linnaeus, 1758].
1825?. *Sphiggurus* F. Cuvier, Des dents des mammifères. . ., p. 256. Type, *Sphiggurus spinosus* F. Cuvier. (*Sphiggure* F. Cuvier, Mém. Mus. Hist. Nat., Paris, 9:427, 1822, possibly not intended as a technical name, since F. Cuvier formalized the name 2 years later.)
1825?. *Sinoetherus* F. Cuvier, Des dents des mammifères. . ., p. 256. Type, *Hystrix prehensilis* Linnaeus, 1758. (For *Synéthère* F. Cuvier, 1822, the same remarks as above, pertaining to *Sphiggure*, apply.)
1835. *Cercolabes* Brandt, Mém. Acad. Imp. Sci., St. Pétersbourg, ser. 6, Sci. Math. Phys. et Nat., 3:391, a renaming of *Coendou* Lacépède, 1799. [See also Krumbiegel, Zool. Anz., 137:18–25, January 15, 1942, for discussion of taxonomy and geographic distribution of *Coendou*.]

KEY TO NOMINAL NORTH AMERICAN SPECIES OF COENDOU

1. Upper parts "pale fulvous or rufous-yellow, with a slight admixture of brown" (Waterhouse, 1848:434); occurring in West Indies [?]. . . . .*C. pallidus*, p. 854
1'. Upper parts darker; not occurring in West Indies.
    2. Tips of spine extending well beyond hairs. . . . .*C. rothschildi*, p. 854
    2'. Tips of spines not extending well beyond hairs. . . . . . .*C. mexicanus*, p. 853

Skull broad, robust; frontal region much broadened, and strongly arched in dorsal profile; nasals short; palate relatively broad; pterygoid hamulae thick; tympanics well inflated; molariform teeth essentially as in *Erethizon*. Body clothed above with short, thick spines or quills; tail prehensile, modified for contact on distal dorsal surface, quill-bearing above and below proximally; both forefeet and hind feet with 4 digits.

Coendou mexicanus
Mexican Porcupine

Skull moderately (for genus) inflated and arched in infraorbital region (but varying much in this respect according to subspecies); somewhat elongated and narrow; bullae relatively long and narrow; upper premolar larger than molars. Upper parts blackish to dark brown; underparts grizzled.

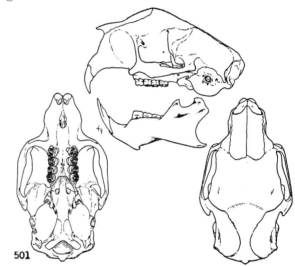

Fig. 501. *Coendou mexicanus mexicanus*, Río Blanco, 20 km. WNW Piedras Negras, Veracruz, No. 18191 K.U., ♀, X ½.

Coendou mexicanus laenatus Thomas

1903. *Coendou laenatus* Thomas, Ann. Mag. Nat. Hist., ser. 7, 11:381, April, type from Boquete, Chiriquí, Panamá.
1920. *Coendou mexicanum laenatum*, Goldman, Smiths. Miscl. Coll., 69(5):133, April 24.

MARGINAL RECORDS.—Nicaragua: San Antonio (Jones and Genoways, 1970:7, as *C. mexicanus* only). Costa Rica: Esparta; Escazú; Pozo Ancho. Panamá: Sibube (Handley, 1966:785); type locality; *Cerro Punta* (Handley, 1966:785).

Coendou mexicanus mexicanus (Kerr)

1792. *Hystrix mexicana* Kerr, The animal kingdom . . ., 1:214. Type locality, mts. of México [probably in Veracruz,

Dalquest, Occas. Pap. Mus. Zool., Louisiana State Univ., 23:12, July 10, 1950].

1901. *Coendou mexicanum*, Miller and Rehn, Proc. Boston Soc. Nat. Hist., 30(1):173, December 27.

1844. *Cercolabes liebmani* Reinhardt, Wiegmann's Arch. für Naturgesch., p. 241. Type locality, México.

MARGINAL RECORDS.—San Luis Potosí: above Xilitla at Miramar Grande. Veracruz (Hall and Dalquest, 1963:322): Jalapa; 3 km. E San Andrés Tuxtla, 1000 ft.; *Catemaco; 10 km. NW Minatitlán, 100 ft.;* Minatitlán. Honduras: Guaymaca; Laguna Archaga. El Salvador (Burt and Stirton, 1961:64): Lake Olomega; Puerto del Triunfo; Amate de Campo. Guatemala: Finca Primavera. Oaxaca: La Concepción (Goodwin, 1969:217); 4 mi. S Soledad (63133 KU). Veracruz: Río Blanco, 20 km. WNW Piedras Negras, 300 ft.

Coendou mexicanus yucataniae Thomas

1902. *Coendou mexicanus yucataniae* Thomas, Ann. Mag. Nat. Hist., ser. 7, 10:249, September, type from Yucatán (probably near Izamal).

MARGINAL RECORDS.—Yucatán: Loché. Quintana Roo: Pueblo Nuevo X-Can (92557 KU); Chetumal (Birney, *et al.*, 1974:16). Campeche: 46 km. S Champotón (93792 KU). Yucatán: Izamal.

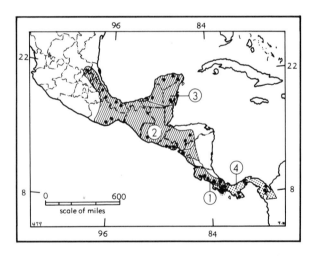

Map 479. *Coendou mexicanus* and *Coendou rothschildi*.

1. *C. mexicanus laenatus*
2. *C. mexicanus mexicanus*
3. *C. mexicanus yucataniae*
4. *C. rothschildi*

Coendou pallidus (Waterhouse)
Antillean Porcupine

1848. *Cercolabes pallidus* Waterhouse, A natural history of the Mammalia, 2:434, type purported to be from West Indies. Known only by the two immature specimens originally described.

1897. [*Coendu*] *pallidus*, Trouessart, Catalogus mammalium . . . , fasc. 3, p. 622.

"General hue of an extremely pale fulvous brown; spines pale sulphur yellow, or white, with a short black point, and almost entirely hidden by the long fur; muzzle dusky brown; feet suffused with brown; tail brown-black beneath" (Waterhouse, 1848:434). Not mapped.

Coendou rothschildi Thomas
Rothschild's Porcupine

1902. *Coendou rothschildi* Thomas, Ann. Mag. Nat. Hist., ser. 7, 10:169, August, type from Sevilla Island, Chiriquí, Panamá.

"Rothschild's porcupine is readily distinguished from its Panama congener, *Coendou mexicanum laenatum*, by the exposed spiny covering, the spines in the latter species being mainly concealed by the long overlapping fur" (Goldman, 1920:134). In view of remarks by Thomas (1902:169), Tate (1935:306), and Cabrera (1958 [pt. 2, 1961]:599), it would not be surprising if examination of specimens from critical localities reveals that *Coendou rothschildi* is a subspecies of *Coendou bicolor* (Tschudi). See Map 479.

MARGINAL RECORDS.—Panamá (Handley, 1966:785): Fort Sherman; Armila; Boca de Río Paya, thence into South America and northwestward along Pacific coast to Panamá (*ibid.*): Fort Kobbe; Parita; Guánico; type locality; Boquerón.

SUPERFAMILY CAVIOIDEA

FAMILY HYDROCHAERIDAE—Capybaras

Large sized semiaquatic rodents; M3 greatly enlarged, exceeding in alveolar length the other molariform teeth combined; molariform teeth strongly prismatic, especially M3, which has a series of approx. 10 narrow transverse laminae; paroccipital processes greatly enlarged.

The capybaras are the largest living rodents. They possess partially webbed feet. There is but one known genus, *Hydrochaeris*, extant.

SUBFAMILY HYDROCHAERINAE

Genus Hydrochaeris Brünnich—Capybaras

1772. *Hydrochaeris* Brünnich, Zoologiae Fundamenta . . . , p. 44. Type, *Sus hydrochoeris* Linnaeus, 1766.

Hydrochaeris hydrochaeris
Capybara

External measurements of the type of *H. h. isthmius*, an adult male, are: 1025, 200. Upper parts "varying from very dark rusty reddish to dull pale clay color . . . becoming more or less blackish in some specimens on middle of face, cheeks, lower part of rump, and outer sides of hind legs . . . ears and feet brownish, thinly haired" (Goldman, 1912b:11); underparts slightly paler than upper parts. Other characters as for the family.

Hydrochaeris hydrochaeris isthmius (Goldman)

1912. *Hydrochoerus isthmius* Goldman, Smiths. Miscl. Coll., 60(2):11, September 20, type from Marragantí, near head tidewater, Río Tuyra, Darién, Panamá.
1961. *Hydrochaeris hydrochaeris isthmius*, Cabrera, Rev. Mus. Argentino de Cienc. Nat., 4:584, August 25.

MARGINAL RECORDS.—Panamá (Handley, 1966: 785): Juan Mina; *type locality;* El Real; Río Seteganti, thence into South America and northward to Panamá (*ibid.*): 15 mi. E Panamá.

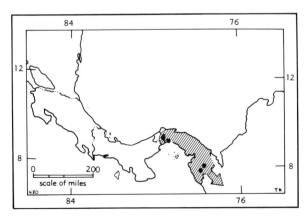

Map 480. *Hydrochaeris hydrochaeris isthmius.*

FAMILY HEPTAXODONTIDAE

First tooth of maxillary series mechanically dominant; cheek-teeth composed of 4–7 vertical laminae, the crests of which are approx. parallel and are arranged obliquely to long axis of skull, resulting in anterior surface of each lamina of a maxillary tooth facing anteromedially; each lamina bearing enamel on anterior surface only.

Owing to the fragmentary nature of the known remains, it seems inadvisable to attempt to describe this family more precisely. This family is thought to be extinct and is known only from osseous material from cave deposits on islands in the Caribbean Sea. Two monotypic genera are described below because they may have persisted long enough to have coexisted with early man. The four other known genera are not described here because it is believed they vanished before early man reached the areas where these rodents lived. Three of these genera, *Clidomys*, *Spirodontomys*, and *Speoxenus* are known only from Jamaica, and the fourth, *Amblyrhiza*, is known from the islands of Anguilla and St. Martin, which are near the northern end of the Lesser Antillean chain. A fifth related genus, *Alterodon* from Jamaica, of uncertain family placement, also may have disappeared before early man reached Jamaica. The literature essential to learning as much as is known about these five genera can be traced through the bibliographies in the three papers: Ray (1964b:126, 127; 1965:11, 12), and Varona (1974:121–139).

Genus Elasmodontomys Anthony

1916. *Elasmodontomys* Anthony, Ann. New York Acad. Sci., 27:199, August 9. Type, *Elasmodontomys obliquus* Anthony, 1916.
1917. *Heptaxodon* Anthony, Bull. Amer. Mus. Nat. Hist., 37:183, January 29.

Cheek-teeth, $\frac{4}{4}$, "the crown patterns of which are made up of five enamel plates and connecting dentine" (Anthony, 1918:380). Mandibular symphysis not extending posteriorly beyond level of middle of P4.

Elasmodontomys obliquus Anthony

1916. *Elasmodontomys obliquus* Anthony, Ann. New York Acad. Sci., 27:199, August 9, type from cave deposits near Utuado, Puerto Rico. Known also from Morovis and Ciales, Puerto Rico (Anthony, Mem. Amer. Mus. Nat. Hist., n.s., 2:380, June, 1918).
1917. *Heptaxodon bidens* Anthony, Bull. Amer. Mus. Nat. Hist., 37:183, January 29, type from cave deposits near Utuado, Puerto Rico. Known from type locality and from

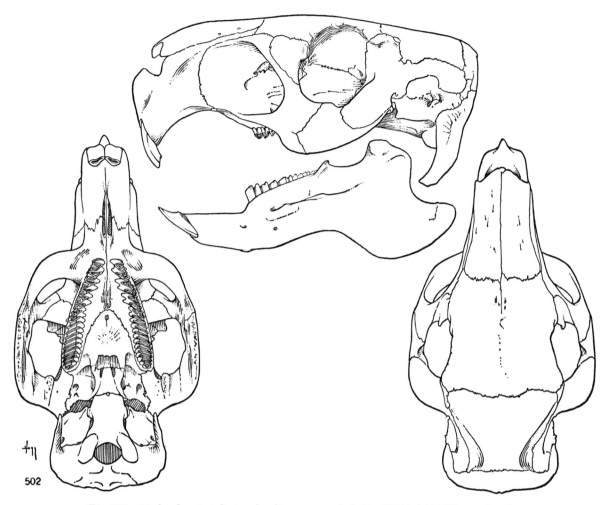

Fig. 502. *Hydrochaeris isthmius,* locality not recorded, No. 38194 A.M.N.H., sex ?, X ½.

Ciales, Puerto Rico (Anthony, Mem. Amer. Mus. Nat. Hist., n.s., 2:398, June, 1918). Found to be a juvenile of *E. obliquus* by Ray, Bull. Mus. Comp. Zool., Harvard Univ., 131:107–127, May 1964.

Not mapped. Extinct.

Genus **Quemisia** Miller

1929. *Quemisia* Miller, Smiths. Miscl. Coll., 81(9):22, March 30. Type, *Quemisia gravis* Miller, 1929.

Enamel pattern of mandibular cheek-teeth much as in *Elasmodontomys,* but transverse axis of laminae slanting forward at an angle of about 21° instead of about 50°; mandibular symphysis much elongated, extending posteriorly to or beyond level of m1; lower incisors short. Until more nearly complete osteological remains of *Quemisia* become available, it is here provisionally retained in the family Heptaxodontidae. In his important paper on the relationships of *Quemisia gravis,* Ray (1965:10) speculates that it

may be "an evolutionary bridge between Antillean capromyids and heptaxodontines."

Quemisia gravis Miller

1929. *Quemisia gravis* Miller, Smiths. Miscl. Coll., 81(9):23, March 30, type from cave near Atalaye Plantation, 4 mi. E St. Michel, Haiti. Known also from Boca de Infierno [cave, Samaná Bay], Dominican Republic [Miller, Smiths. Miscl. Coll., 82(5):10, December 11, 1929]. Not mapped. Extinct.

Family **DASYPROCTIDAE**
Agoutis and Pacas

Cheek-teeth hypsodont, semirooted, having deep re-entrant enamel folds that with wear form isolated narrow lakes; hind feet functionally perissodactylous, forefeet functionally artiodactylous; digits with strong, thick, blunt, almost hooflike claws.

This family contains two relatively diverse groups, the agoutis (*Dasyprocta*) and the pacas (*Agouti*). Some authors (Miller and Gidley, 1918;

Ellerman, *et al.,* 1940) place them in separate families, principally on the basis of the singular zygomatic structure of the paca.

KEY TO NORTH AMERICAN GENERA OF DASYPROCTIDAE

1. Zygoma much expanded to form broad cheek-plate, which is deeply hollowed beneath and, in maturity, extremely rugose above. . . . .*Agouti,*   p. 857
1'. Zygoma lacking broad cheek-plate.
                                    Dasyprocta,   p. 858

SUBFAMILY AGOUTINAE

Genus **Agouti** Lacépède—Pacas

1799. *Agouti* Lacépède, Tableau des divisions, sous-divisions, ordres et genres des mammifères . . . , p. 9. Type by monotypy, *Agouti paca* Lacépède, 1799 [= *Mus paca* Linnaeus, 1766, p. 81].
1807. *Coelogenus* F. Cuvier, Ann. Mus. Hist. Nat. Paris, 10:203. [*Coelogenys auct.*] Included species, *Coelogenus subniger* F. Cuvier, 1807, and *C. fulvus* F. Cuvier, 1807 [both = *Mus paca* Linnaeus, 1766, p. 81].

Jugal and part of maxillary expanded into conspicuous cheek-plate, the surface of this becoming excessively rugose in adult; cheek-teeth strongly hypsodont, but enamel structures not completely multilaminar; external form robust; ears large; tail vestigial; 4 toes on forefoot; 5 toes on hind foot; pelage stiff and appressed. Terrestrial but enters water readily; lives in burrows.

Only one species, *Agouti paca,* occurs in Central America. From there the subspecies *A. p. nelsoni* was introduced into Cuba in the past century, according to Varona (1974:77).

Agouti paca
Paca or Spotted Cavy

Upper parts seal brown, darkest mid-dorsally; sides having 4 lines or rows of spots of creamy white; upper line obscure, tiny, narrow spots; 2nd line consists of distinct, narrow spots; 3rd

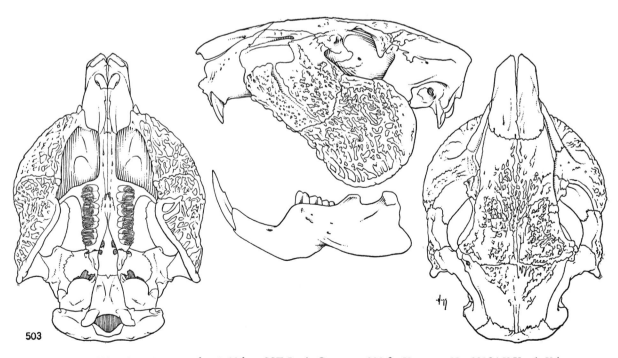

Fig. 503.   *Agouti paca nelsoni,* 30 km. SSE Jesús Carranza, 300 ft., Veracruz, No. 32184 K.U., ♂, X ½.

line a stripe changing to spots on shoulder and hip; 4th line obscure and incomplete; claws pink.

Agouti paca nelsoni Goldman

1913. *Agouti paca nelsoni* Goldman, Smiths. Miscl. Coll., 60(22):9, February 28, type from Catemaco, Veracruz.

MARGINAL RECORDS.—San Luis Potosí: 9 km. NNE Xilitla. Veracruz: Chichicaxtle; type locality. Campeche: 7½ km. W Escárcega (92558 KU). Yucatán: 13 km. W Peto (93797 KU); 8 km. N, 10 km. W Tizimín (93795 KU). Quintana Roo: Cozumel Island (Jones and Lawlor, 1965:416, as *A. paca* only). Belize: Bokowina. Honduras: Guaymaca. El Salvador: Colinas de Jacuarán (Burt and Stirton, 1961:64). Chiapas: 20 km. SE Teopisca. Oaxaca (Goodwin, 1969:219): Río Mono Blanco; Río Grande. Veracruz: 30 km. SSE Jesús Carranza, 300 ft. (Hall and Dalquest, 1963:324); Río Blanco, 15 km. W Piedras Negras, 300 ft.

Agouti paca virgatus Bangs

1902. *Agouti paca virgatus* Bangs, Bull. Mus. Comp. Zool., 39:47, April, type from Divalá, Chiriquí, Panamá.

MARGINAL RECORDS.—Costa Rica: San Gerónimo. Panamá (Handley, 1966:786): Almirante; Gatún; Armila; Cana, thence into South America and northwestward along coast to Panamá (*ibid.*): Cerro Hoya; type locality.

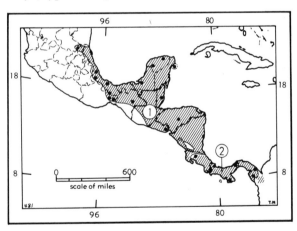

Map 481. *Agouti paca.*

1. *A. p. nelsoni*   2. *A. p. virgatus*

SUBFAMILY DASYPROCTINAE

Genus **Dasyprocta** Illiger—Agoutis

1811. *Dasyprocta* Illiger, Prodromus systematis mammalium et avium . . . , p. 93. Type by subsequent designation by Thomas (Ann. Mag. Nat. Hist., ser. 7, 12:464, October, 1903) who, by removing to a new genus one of the two species on which *Dasyprocta* was based, left *Mus aguti* Linnaeus, 1766, as the type.

Body form slender, cursorial; hind foot long, narrow, tridactyl, bearing hooflike claws; front foot tetradactyl; tail obsolete; fur on posterior part of body long and thick; length of head and body, 420–610. Skull slender, somewhat rectangular in dorsal aspect; nasals shorter than frontals; frontals broad, relatively flat; weak sagittal crest present in adults; auditory bullae relatively large; paroccipital processes prominent; palatine foramina short, placed well anteriorly; lachrymal much enlarged; angular process of mandible reflected strongly outward; molariform teeth, $\frac{4}{4}$, hypsodont.

There is need for a revisionary study of the agoutis. At the present time it is not possible clearly to characterize the species; ranges of individual variation are not known. The following remarks, then, must be regarded as only the merest guide to a complex problem. Indeed, it seems unwise to construct a key based on much more than known geographic distributions.

KEY TO NOMINAL NORTH AMERICAN SPECIES OF DASYPROCTA

1. Occurring on West Indian islands.
                              D. aguti,   p. 858
1′. Not occurring on West Indian islands.
   2. Occurring on Coiba Island, Panamá.
                              D. coibae,   p. 862
   2′. Not occurring on Coiba Island, Panamá.
     3. Occurring in Veracruz.
                              D. mexicana,   p. 859
     3′. Not occurring in Veracruz.
       4. Occurring on Ruatan Island, Honduras. . . .*D. ruatanica*,   p. 860
       4′. Not occurring on Ruatan Island, Honduras. *D. punctata*,   p. 860

Dasyprocta aguti
Brazilian agouti

See characters of the genus, and remarks in the terminal paragraph there. Size approx. as in *D. punctata*. According to Varona (1974:75) the kinds of agoutis named from the Lesser Antilles very probably were introduced from the mainland in pre-Columbian times. In reference to the Virgin Islands, Miller (1918:508) noted that the golden-rumped Brazilian agouti was obtained in the winter of 1916–1917 at St. Thomas, that the animal was reported from there as early as 1852, but that no bones of the genus were found on the Virgin Islands in kitchen middens of pre-Columbian Indians, and that there is now little doubt that introductions of agoutis have been made from both Brazil and the Lesser Antilles.

The relationship of *D. aguti* to the other South American species has never been satisfactorily determined.

Dasyprocta aguti aguti (Linnaeus)

1766. [*Mus*] *aguti* Linnaeus, Syst. nat., ed. 12, 1:80. Type locality, Brazil. Introduced on St. Thomas, Virgin Islands, West Indies (Miller, Proc. U.S. Nat. Mus., 54:508, October 15, 1918).
1829. *D*[*asyprocta*]. *aguti*, Fischer, Synopsis mammalium, p. 379.

MARGINAL RECORDS.—Brazil: *type locality*. Virgin Islands: St. Thomas. The animals referred to by Varona (1974:76) as *"Dasyprocta aguti* ssp." on *St. Croix, Virgin Islands*, and the animals, if any, on *Barbados, Lesser Antilles*, may be this subspecies.

Dasyprocta aguti albida Gray

1842. *Dasyprocta albida* Gray, Ann. Mag. Nat. Hist., ser. 1, 10:264, December, type from St. Vincent, Lesser Antilles.
1974. *Dasyprocta aguti albida*, Varona, Catálogo mam. viv. y exting. de las Antillas, Acad. Cienc. Cuba, p. 75.

MARGINAL RECORDS.—Lesser Antilles: type locality; Grenada.

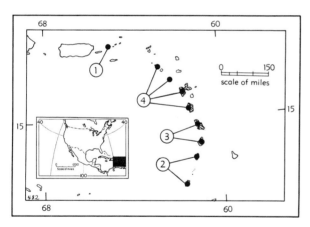

Map 482. *Dasyprocta aguti.*

| 1. *D. a. aguti* | 3. *D. a. fulvus* |
| 2. *D. a. albida* | 4. *D. a. noblei* |

Dasyprocta aguti fulvus (Kerr)

1792. *M*[*us*]. *Pilorides fulvus* Kerr, The animal kingdom . . . , p. 227—based primarily on Pennant's *"musk cavy."*
1959. *Dasyprocta aguti fulvus*, Hershkovitz, Jour. Mamm., 40:349, August 20, who points out that Pennant, 1771, Syn. Quadr., p. 247, specifies the island of Martinique as the type locality.
1815. [*Dasyprocta*] *moschata* Illiger, Abh. k. Akad. Wiss., Berlin, 1804–1811:108, 114—name based by implication on Pennant's *"musk cavy."*
1875. *Dasyprocta antillensis* Sclater, Proc. Zool. Soc. London, for 1874. Type locality, Santa Lucia, Lesser Antilles.

MARGINAL RECORDS.—Lesser Antilles: type locality; Santa Lucia.

Dasyprocta aguti noblei G. M. Allen

1914. *Dasyprocta noblei* G. M. Allen, Proc. New England Zool. Club, 5:69, October 7, type from Goyave, Guadeloupe, Lesser Antilles.
1974. *Dasyprocta aguti noblei*, Varona, Catálogo mam. viv. y exting. de las Antillas, Acad. Cienc. Cuba, p. 75.

MARGINAL RECORDS.—Lesser Antilles: St. Kitts; Dominica; Montserrat; type locality.

Dasyprocta mexicana Saussure
Mexican Agouti

1860. *Dasyprocta mexicana* Saussure, Revue et Mag. Zool., Paris, ser. 2, 12:53, type from "hot zone of Mexico"; probably in the State of Veracruz.

The following description is based on specimens in the Museum of Natural History, University of Kansas, that are provisionally referred to this species. Upper parts black mid-dorsally and over rump, fading to grizzled dark brown over shoulders and sides, this coloration resulting from a fading of the black to brown and the addition of pale tips on the hairs; underparts paler and light tips of hairs more prominent, hairs on belly sometimes all dirty white; ears nearly naked, hairs on outer side blackish to brownish, buffy on inside; feet blackish. With wear, enamel loops of

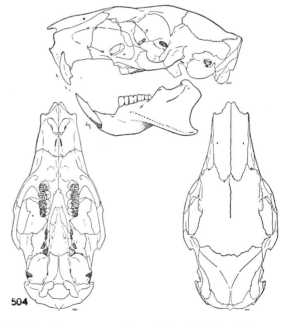

Fig. 504. *Dasyprocta mexicana*, 20 km. ENE Jesús Carranza, 200 ft., Veracruz, No. 32188 K.U., ♀, X ½.

molariform teeth become isolated as separate islets. See Map 483.

MARGINAL RECORDS.—Veracruz: 2 km. N Paraje Nuevo, 1750 ft.; 60 km. SE Jesús Carranza, 450 ft. Oaxaca: bridge on National Route 185 over Río Sarabia (Schaldach, 1965:137). Introduced into western and eastern *Cuba* (Varona, 1974:76).

Dasyprocta ruatanica Thomas
Ruatan Island Agouti

1901. *Dasyprocta ruatanica* Thomas, Ann. Mag. Nat. Hist., ser. 7, 8:272, October, type from Ruatan Island, Bay Islands, Honduras. Known only from type locality.

"Closely allied to *D. punctata*, but much smaller. . . . Size markedly less than in *D. punctata*. Fur . . . everywhere annulated to the roots with black and ochraceous yellow. . . . Under surface like back, but more olivaceous, and there is a white spot on the chin and a yellow patch on the middle of the lower part of the belly. . . . Ears nearly naked, their few hairs blackish. Limbs like body proximally, darkening terminally on hands and feet to grizzled or deep brown. . . . Skull in general shape closely agreeing with that of the type of *D. punctata*, but conspicuously smaller in all dimensions." (Thomas, 1901:272.) See Map 483.

Dasyprocta punctata
Agouti

× ⅒

External measurements: 490–620; 10–35; 110–133. Upper parts variable both geographically and individually; blackish, especially over head and nape, mixed with buff or russet, or distinctly reddish, usually brightest on posterior part of body; underparts usually paler, sometimes even tawny; hairs on rump usually much longer than elsewhere on upper parts.

Varona (1974:77) reports the introduction of an unknown subspecies of this species into the *Caimán Islands* at the close of the past century.

Dasyprocta punctata bellula R. Kellogg

1946. *Dasyprocta punctata bellula* R. Kellogg, Proc. Biol. Soc. Washington, 59:59, March 11, type from San José Island, Golfo de Panamá, Panamá. Known only from type locality.

Dasyprocta punctata callida Bangs

1901. *Dasyprocta callida* Bangs, Amer. Nat., 35:635, August, type from San Miguel Island, Golfo de Panamá, Panamá.
1946. *Dasyprocta punctata callida*, R. Kellogg, Proc. Biol. Soc. Washington, 59:59, March 11.

MARGINAL RECORDS.—Panamá: Isla Pedro Gonzáles (Handley, 1966:786); type locality.

Dasyprocta punctata chiapensis Goldman

1913. *Dasyprocta punctata chiapensis* Goldman, Smiths. Miscl. Coll., 60(22):13, February 28, type from Huehuetán, Chiapas.

MARGINAL RECORDS.—Guatemala: *probably in highlands*. Chiapas: *Chicharras;* type locality.

Dasyprocta punctata dariensis Goldman

1913. *Dasyprocta punctata dariensis* Goldman, Smiths. Miscl. Coll., 60(22):11, February 28, type from near head Río Limón, 5200 ft., Mt. Pirri, Panamá.

MARGINAL RECORDS.—Panamá (Handley, 1966: 786): Mandinga; Armila, thence into South America and northward to Panamá (*ibid.*): Chepigana; El Real; type locality; Ancón Hill.

Dasyprocta punctata isthmica Alston

1876. *Dasyprocta isthmica* Alston, Proc. Zool. Soc. London, p. 347, August, type from Colón, Canal Zone, Panamá.
1913. *D[asyprocta]. punctata isthmica,* Goldman, Smiths. Miscl. Coll., 60(22):11, February 28.

MARGINAL RECORDS.—Panamá (Handley, 1966:786): Cerro Azul; Paraíso; Río Trinidad; Boquete; Gatún.

Dasyprocta punctata nuchalis Goldman

1917. *Dasyprocta punctata nuchalis* Goldman, Proc. Biol. Soc. Washington, 30:113, May 23, type from Divalá, Chiriquí, Panamá.

MARGINAL RECORDS.—Panamá: Bugaba; type locality.

Dasyprocta punctata pallidiventris Bole

1937. *Dasyprocta punctata pallidiventris* Bole, Sci. Publs., Cleveland Mus. Nat. Hist., 7:182, August 31, type from Paracoté, 1½ mi. S mouth Angulo River, Mariato-Suay lands, Veraguas, Panamá.

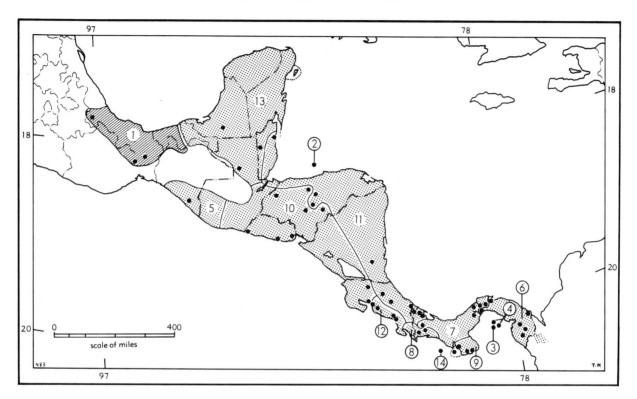

Map 483.   Four species of *Dasyprocta.*

Guide to kinds		
1. *D. mexicana*	5. *D. punctata chiapensis*	10. *D. punctata punctata*
2. *D. ruatanica*	6. *D. punctata dariensis*	11. *D. punctata richmondi*
3. *D. punctata bellula*	7. *D. punctata isthmica*	12. *D. punctata underwoodi*
4. *D. punctata callida*	8. *D. punctata nuchalis*	13. *D. punctata yucatanica*
	9. *D. punctata pallidiventris*	14. *D. coibae*

MARGINAL RECORDS.—Panamá (Handley, 1966: 786): type locality; Guánico; Cerro Hoya; Isla Cébaco.

Dasyprocta punctata punctata Gray

1842. *Dasyprocta punctata* Gray, Ann. Mag. Nat. Hist., ser. 1, 10:264, December. Type locality restricted (Goodwin, Bull. Amer. Mus. Nat. Hist., 87:417, December 31, 1946) to Realejo, Nicaragua.

MARGINAL RECORDS.—Honduras: La Florida; Roman River; El Guayabal; El Horno. Costa Rica: Vijagua; Esparta. El Salvador (Burt and Stirton, 1961:65): Lake Olomega; Puerto del Triunfo; Barra de Santiago, thence to Guatemala: *probably Pacific lowlands.*

Dasyprocta punctata richmondi Goldman

1917. *Dasyprocta punctata richmondi* Goldman, Proc. Biol. Soc. Washington, 30:114, May 23, type from Escondido River, 50 mi. above Bluefields, Nicaragua.

MARGINAL RECORDS.—Belize: Belize, thence along Caribbean Coast to Costa Rica: Cuabre. Panamá (Handley, 1966:786): Sibube; 7 km. SSW Changuinola; Almirante. Costa Rica: San José; Vijagual. Nicaragua:

type locality. Honduras: Monte Redondo; El Caliche Cedros.

Dasyprocta punctata underwoodi Goldman

1931. *Dasyprocta punctata underwoodi* Goldman, Jour. Washington Acad. Sci., 21:481, November 19, type from San Gerónimo, Pirrís, San José, Costa Rica.

MARGINAL RECORDS.—Costa Rica: Pozo Pital; type locality; Agua Buena; Palmar.

Dasyprocta punctata yucatanica Goldman

1913. *Dasyprocta punctata yucatanica* Goldman, Smiths. Miscl. Coll., 60(22):12, February 28, type from Apazote, near Yohaltún, Campeche.

MARGINAL RECORDS.—Quintana Roo: *Cozumel Island* (Jones and Lawlor, 1965:416, as reported by natives only). Belize: El Cayo. Guatemala: Sayaxché (Ryan, 1960:15, as *D. p. chiapensis*). Campeche: La Tuxpana. Introduced into western and eastern *Cuba* (Varona, 1974:76).

Dasyprocta coibae Thomas
Coiba Island Agouti

1902. *Dasyprocta coibae* Thomas, Nov. Zool., 9:136, April 10, type from Coiba Island, Panamá. Known only from type locality.

"Size about as in *D. isthmica*. Fur coarse and sparse; rump hairs less conspicuously lengthened than usual—about 3 in. in length. General colour above coarsely grizzled brown . . . the hairs ringed with brown and orange; where they lengthen on the rump they are still ringed with the same two colours, but the tips are broadly orange, almost as in *D. isthmica*. Undersurface soiled yellowish, the hairs both of throat and belly brown basally and yellow terminally. Crown blackish brown. Ears nearly naked, the sparse hairs brown. Arms and legs like body, uppersurface of hands and feet black. Tail naked, about an inch in length, and therefore decidedly longer than usual. Skull broad and thickly built. Nasals parallel-sided, not tapering." (Thomas, 1902:136.) See Map 483.

FAMILY CAPROMYIDAE
HUTIAS AND COYPUS

Taxa here arranged essentially as by L. S. Varona, Catálogo mam. viv. y exting. de las Antillas, Acad. Cienc. Cuba, 1974: viii + 139, 1 map.

Upper molars ever-growing (Plagiodontinae and Capromyinae) or not (Myocastorinae), but when not ever-growing then extremely hypsodont; paroccipital processes long, usually straight, and not closely appressed to auditory bullae; size varying from that of a guinea pig to that of a small dog.

The family—as here constituted—is made up of a diverse assemblage of genera, some of which are known from specimens of living species and others from only bones of extinct species recovered from cave deposits and Indian middens. Students disagree on the classification of the Antillean hystricomorph genera and especially on those here included in this family. The classification below is essentially that of Varona (1974), which overall seems to be superior to earlier classifications.

KEY TO SUBFAMILIES AND GENERA OF CAPROMYIDAE

1. External body form strongly modified for aquatic life; upper molars strongly hypsodont.
   subfamily Myocastorinae, p. 871
   Myocastor, p. 871
1'. External body form not modified for aquatic life; upper molars not strongly hypsodont.
   2. Enamel pattern of upper molars tetramerous, folds long, deep, and obliquely arranged.
      subfamily Plagiodontinae, p. 867
      3. Root of upper incisor occurring at anterior margin of zygomatic process of maxillary; P4 with one external re-entrant angle.
         4. Enamel pattern conspicuously compressed anteroposteriorly and folds conspicuous and sharp; length of mandibular molar series more than 25 and diastema more than 16.7. . . . .*Hyperplagiodontia*, p. 870
         4'. Enamel pattern not conspicuously compressed anteroposteriorly and folds not conspicuous or sharp; length of mandibular molar series less than 25 and diastema less than 16.7. . . . .*Plagiodontia*, p. 867
      3'. Root of upper incisor occurring in infraorbital foramen; P4 with 2 external re-entrant angles, its enamel pattern markedly different from that of molars. . . . .*Isolobodon*, p. 869
   2'. Enamel pattern of upper molars pentamerous or hexamerous. . . . . subfamily Capromyinae, p. 862
      5. Enamel pattern of upper molars pentamerous; roots of lower cheek-teeth evenly spaced in mandible.
         6. Humerus shorter than 82 mm. . . . .*Capromys*, p. 862
         6'. Humerus longer than 82 mm. . . . .*Macrocapromys*, p. 866
      5'. Enamel pattern of upper molars hexamerous; root of p4 diverging sharply forward away from that of m1. . . . .*Hexolobodon*, p. 867

SUBFAMILY CAPROMYINAE

Genus Capromys Desmarest—Hutias

Revised by Mohr, Mitt. Zool. Mus. Hamburg, 48:48–118, May 1939. The *Capromys elegans* of Cabrera (Bol. Soc. Española Hist. Nat. Madrid, 1:372, December 1901) based on

specimen supposed to have been collected in Cuba, is a member of the Philippine genus *Phloeomys*. (See Cabrera, Trab. Mus. Cienc. Nat., Madrid, No. 3, p. 30, 1912.)

1822. *Capromys* Desmarest, Bull. Soc. Philom. Paris, p. 185, December. Type, *Capromys fournieri* Desmarest [= *Isodon pilorides* Say].

Skull long and rather flat, a postorbitallike ridge can be present, jugal distinct. Bullae prominent. Paroccipital process usually slightly lengthened. Infraorbital foramen lacking canal for nerve transmission. Palate slightly constricted anteriorly; palatal foramina medium. Mandible with angular process drawn backward, and strongly lifted outward; condyle high; coronoid process low. Fur harsh; feet broad, of arboreal type. Claws prominent; D. 5 relatively long, hallux medium.

Remains of *Capromys* (subgenus unknown) from St. Croix Island are mentioned by Varona (1974:62).—**See addenda.**

KEY TO SUBGENERA OF CAPROMYS

1. Tail less than 36% of total length.
   2. Tail more than ¼ of total length; bases of alveoli of right and left P4 not in contact; root capsule of I1 ending anterior to P4. . . . . *Capromys*, p. 863
   2'. Tail less than ¼ of total length; bases of alveoli of right and left P4 in contact; root capsule of I1 ending over middle of P4. . . . . *Geocapromys*, p. 865
1'. Tail more than 36% of total length.
   3. Tail *ca.* 48% of total length; horizontal part of zygomatic arch vertically shallow. . . . . *Mysateles*, p. 863
   3'. Tail *ca.* 40% of total length; horizontal part of zygomatic arch vertically deep. . . . . *Mesocapromys*, p. 865

Subgenus Capromys Desmarest—Hutias

1822. *Capromys* Desmarest, Bull. Soc. Philom. Paris, p. 185, December. Type, *Capromys fournieri* Desmarest [= *Isodon pilorides* Say].

Tail *ca.* ⅓ of total length; tail weak, hair thereon sparse and short; preorbital bar of maxillary bone from top to bottom slightly (less than in *Mysateles*) inclined posteriorly; horizontal part of zygomatic arch vertically deep with posterior process on lower border; bases of alveoli of right and left P4 not in contact; root capsule of I1 ending anterior to P4. Not agile climbers (after Mohr, 1939:65).

Capromys garridoi Varona
Garrido's Hutia

1970. *Capromys garridoi* Varona, Poeyana, ser. A, No. 74:2, July 30, type from Cayo Majá, next to NW extremity of Cayo Largo (21° 35' N, 81° 30' W), Cuba. Known only from the Cayos Majá, in some of which *C. garridoi* and *C. pilorides* are sympatric.

Externally resembling *C. pilorides* but smaller; tail relatively longer; tail shorter than in subgenera *Mysateles* and *Mesocapromys;* braincase longer and more arched posteriorly than in *C. pilorides.* Not mapped.

Capromys pilorides
Desmarest's Hutia

External measurements of six adults, including both sexes, are: 710–810; 170–255; 95–100; 27, 28 (two specimens only). Upper parts varying from black and pale buff to black and deep rusty; underparts varying from buffy whitish to that of the back, but usually somewhat intermediate; tail unicolored, varying from buffy to dark reddish brown. Skull large and heavily built.

Varona (1974:63) refers to remains of this species, subsp. ?, from caves on Cayman Brac where the animal is extinct.

Capromys pilorides pilorides (Say)

1822. *I[sodon]. pilorides* Say, Jour. Acad. Nat. Sci. Philadelphia, 2:333, November. Type locality, "South America or one of the West Indian islands." Known from various localities throughout Cuba (see Mohr, Mitt. Zool. Mus. Hamburg, 48:73, May, 1939).
1848. *Capromys pilorides*, Waterhouse, A natural history of the Mammalia, 2:287.
1822. *Capromys fournieri* Desmarest, Mém. Soc. Hist. Nat. Paris, 1:43, pl. Type locality, Cuba.
1958. *Capromys intermedius* Arredondo, El Cartero Cubano, Ano. 17 (No. 12):11, 48, found in caves and Indian middens at different localities in Cuba. A synonym of *C. p. pilorides* according to Varona, Catálogo mam. viv. y exting. de las Antillas, Acad. Cienc. Cuba, 1974:62.

MARGINAL RECORDS.—Known from various localities throughout Cuba. Not mapped.

Capromys pilorides relictus G. M. Allen

1911. *Capromys pilorides relictus* G. M. Allen, Bull. Mus. Comp. Zool., 54:207, July, 1911, type from Casas Mts., Nueva Gerona, Isle of Pines, Cuba. Known only from type locality. Not mapped.

Subgenus Mysateles Lesson—Hutias

1842. *Mysateles* Lesson, Nouveau tableau du règne animal . . . Mammifères, p. 124. Type, *Mysateles poeppingii* Lesson [= *Capromys prehensilis* Poeppig].

Tail almost half of total length; hair thereon long, thick, to bushy; preorbital bar of maxillary bone from top to bottom inclined posteriorly; horizontal part of zygomatic arch vertically shallow with posterior process on lower border; bases of alveoli of right and left P4 not in contact; root capsule of I1 ending at middle of anterior border of alveolus of P4. Agile climbers (after Mohr, 1939:54).

Capromys melanurus Poey *in* Peters
Bushy-tailed Hutia

1864. *Capromys melanurus* Poey *in* Peters, Monatsb. preuss. Akad. Wiss., Berlin, p. 384, type from Manzanillo, Cuba.
1939. *Mysateles melanurus rufescens* Mohr, Mitt. Zool. Mus. Hamburg, 48:62, May, 1939, type from Manzanillo, Cuba.

Closely resembling *C. prehensilis* except as follows: *C. melanurus* has slightly more inflated frontals; less developed postorbital processes; markedly narrower basioccipital region; tail not used as a prehensile organ and is easily detached basally in life; tail bushier and much darker (black to rusty).

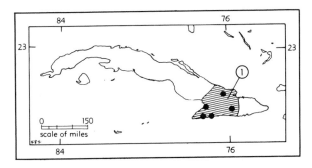

Map 484. *Capromys melanurus.*

MARGINAL RECORDS.—Cuba: Holguin; San Luis; Turquino; Portillo; type locality. Localities not found: Yao Arriba; Cuabitas.

Capromys nana G. M. Allen
Dwarf Hutia

1917. *Capromys nana* G. M. Allen, Proc. New England Zool. Club, 6:54, March 29, type from Sierra de Hato Nuevo, Matanzas, Cuba.—See addenda.

External measurements of an adult female are: 395; 176; 45; 19. Size about that of a Norway rat. Upper parts a mixture of black hairs having ochraceous tips; tips becoming whitish on muzzle and in front of ear and darker over rump; underparts pale grayish on chin, middle of belly, and inner surface of legs, elsewhere like upper parts but paler; tail sharply bicolored, like rump in proximal 2 cm., distally hairs becoming much shorter and black above, pale ochraceous buff below; ears sparsely clothed with short, black hairs. Braincase shorter and more rounded than in other species; orbit especially large (for the genus), larger than antorbital foramen; supraorbital ridge beaded.

MARGINAL RECORDS.—Cuba: near Limones; Daiquirí; Ciénega de Zapata. Not mapped.

Capromys prehensilis
Prehensile-tailed Hutia

External measurements of an adult male from San Diego de los Baños, Cuba, are: 710; 305; 82; 30. Upper parts a mixture of blackish, buffy, and rufous shades; underparts whitish to brownish, often white anteriorly and brown posteriorly; tail bright rufous to brownish, sometimes partly or almost completely denuded by abrasion. Skull less than half the bulk of that of *C. pilorides.*

Capromys prehensilis gundlachi Chapman

1901. *Capromys prehensilis gundlachi* Chapman, Bull. Amer. Mus. Nat. Hist., 14:317, November 12, type from Nueva Gerona, Isle of Pines, Cuba. Known only from type locality.

Capromys prehensilis prehensilis Poeppig

1824. *Capromys prehensilis* Poeppig, Jour. Acad. Nat. Sci. Philadelphia, 4:11, type from wooded parts of southern Cuba. Known from the six provinces of Cuba (Varona, Catálogo mam. viv. y exting. de las Antillas, Acad. Cienc. Cuba, 1974:65).

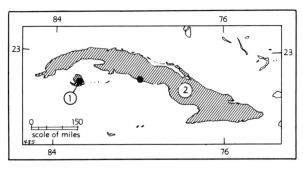

Map 485. *Capromys prehensilis.*

1. *C. p. gundlachi*   2. *C. p. prehensilis*

1834. *Capromys poeyi* Guérin, Mag. Zool., Paris, Ann. 4, Cl. I, Pl. 15, type from Cuba. Trinidad mentioned as one locality.

1865. *C[apromys]. pallidus* Poey *in* Peters, Monatsb. preuss. Akad. Wiss., Berlin, p. 384, type from Cuba. Vic. Trinidad mentioned as one locality.

Subgenus **Mesocapromys** Varona—Hutias

1970. *Mesocapromys* Varona, Poeyana, ser. A, No. 73:8, July 30. Type, *Capromys auritus* Varona.

According to Varona (1970a:ser. A, No. 73:8) characterized by a combination of characters seen in other subgenera plus some features peculiar to the species. Tail *ca.* two-fifths of total length; preorbital bar of maxillary bone from top to bottom inclined slightly (less than in *Mysateles* and even less than in *Capromys*) posteriorly; horizontal part of zygomatic arch vertically deep with short posterior process on lower border; position of root capsule not recorded. Frontals depressed in interorbital region; supraorbital crests conspicuous; auditory bullae large, approx. same length as molariform tooth-row; first lingual fold of p4 short, wide, and anterior to major labial fold. Size of animal less than in any other Cuban species of *Capromys* except for *C. nana*, which is smaller (Varona and Garrido, 1970:ser. A, No. 75:6).

Capromys auritus Varona
Hutia Rat

1970. *Capromys auritus* Varona, Poeyana, ser. A, No. 73:1, July 30, type from canal zone in central part of Cayo Fragoso (79° 27′ W, 22° 41′ N), Archipelago de Sabana, Las Villas, Cuba. Known only from type locality.

External measurements of holotype (adult male), a second male, and a "young adult" female are: 462, 478, 478; 177, 181, 210; 57, 57, 60; 30, 29, 30. Upper parts dark brown; tail same color. Total length of skull, 65.5, 65.0, 61.0; zygomatic breadth, 33.0, 33.7, 30.0; 22 caudal vertebrae. Nasals convex laterally and not extending so far posteriorly as do premaxillae. Not mapped.

Capromys sanfelipensis Varona and Garrido
Land Hutia

1970. *Capromys sanfelipensis* Varona and Garrido, Poeyana, ser. A, No. 75:3, August 12, type from Cayo Juan García (83° 31′ W, 21° 59′ N), of Cayos de San Felipe, Archipelago de los Canarreos, Pinar del Río, Cuba. Known only from type locality.

External measurements of holotype, a male, a second male, and a female (all adults) are: 480, 475, —; 205, 203, —; 60, 55, 60; 25, 26, 24. Upper

parts dark brown; tail same at base but otherwise reddish. Total length of skull, 61.5, 63.5, 63.3; zygomatic breadth, 32.0, 31.0, 33.4; 27 caudal vertebrae. Nasals parallel-sided and extending posteriorly as far as do premaxillae. Not mapped.

Subgenus **Geocapromys** Chapman—Hutias

Reviewed by Lawrence, Occas. Pap. Boston Soc. Nat. Hist., 8:189–193, November 7, 1934. *Geocapromys* regarded as a subgenus of *Capromys* and revised by Mohr (Mitt. Zool. Mus. Hamburg, 48:75–80, May, 1939), who uses *Geocapromys* in name combinations as a generic name. Varona (Catálogo mam. viv. y exting. de las Antillas, Acad. Cienc. Cuba, 1974, p. 67) uses *Geocapromys* only as a subgeneric name.

1901. *Geocapromys* Chapman, Bull. Amer. Mus. Nat. Hist., 14:314, November 12. Type, *Capromys brownii* Fischer.

1917. *Synodontomys* G. M. Allen, Bull. Mus. Comp. Zool., 61:5, January. Type, *Capromys columbianus* Chapman.

Tail a tenth to an eighth of total length; tail short-haired; hair thereon weak, fine and short; preorbital bar of maxillary bone vertical (*C. ingrahami*) or from top to bottom inclined anteriorly (*C. brownii*); horizontal part of zygomatic arch vertically deeper than in *Capromys* and with or without posterior process on lower border; bases of alveoli of right and left P4 in contact, encroaching on floor of narial passage; root capsule of I1 ending above and lateral to anterior half of outer border of alveolus of P4; P4 with a small 3rd re-entrant angle on lingual side (after Miller, 1929b:2); nonclimbing (after Mohr, 1939:76).

KEY TO SPECIES OF SUBGENUS GEOCAPROMYS

(Modified from Lawrence, 1934:193)

1. Alveolar length of lower cheek-teeth less than 17.5 (rarely more).
  2. Lower posterior margin of jugal forming nearly a right angle with a more or less prominent posteriorly directed spine. . . . .*C. brownii*, p. 866
  2′. Lower posterior margin of jugal sloping strongly forward forming an obtuse angle; no posterior projection.
      C. ingrahami, p. 866
1′. Alveolar length of lower cheek-teeth usually more than 18.0.
  3. Frontals inflated anteriorly; postorbital region markedly constricted; postorbital processes short and bluntly triangular. . . . .*C. brownii*, p. 866
  3′. Frontals not inflated; no postorbital constriction or postorbital processes.
      C. columbianus, p. 866
      When better known, *C. pleistocenicus* may or may not key out under 3′. . . p. 866

Capromys brownii
Brown's Hutia

External measurements (dry) and description of an adult male from Jamaica are: 450, 35, 60, 6 (from crown). Upper parts blackish, underparts uniform dusky brown. Skull differs from those of other species of *Geocapromys* in "absence of supraorbital processes, the relatively greater width at the interorbital constriction, the narrowness and relative greater height of the condyle of the lower jaw" (Chapman, 1901:321).

Capromys brownii brownii Fischer

1830. *Capromys brownii* Fischer, Synopsis mammalium, Addenda, p. 389 (= 589). Type locality, Jamaica. Known only from Jamaica.
1851. *Capromys brachyurus* Gosse, A naturalist's sojourn in Jamaica, p. 471. Type locality, Jamaica. Not mapped.

Capromys brownii thoracatus True

1889. *Capromys brachyurus thoracatus* True, Proc. U.S. Nat. Mus., 11:469, September 3, type from Little Swan Island, Gulf of Honduras. Known only from Little Swan Island. Not mapped.

Capromys columbianus (Chapman)
Cuban Hutia

1892. *Capromys columbianus* Chapman, Bull. Amer. Mus. Nat. Hist., 4:314, December 29, type from cave near Trinidad, Cuba.
1917. *Geocapromys cubanus* G. M. Allen, Bull. Mus. Comp. Zool., 61(1):9, January, type from Sierra de Hato-Nuevo, Matanzas, Cuba.

Known only from subfossil fragments. Space between inner borders of alveoli at anterior margin of upper premolars exceedingly narrow because maxillary tooth-rows converge anteriorly and nearly touch each other; median palatal ridge disappears posteriorly; posterior nares much narrowed. Alveolar length of upper cheek-teeth of an adult, 21.

MARGINAL RECORDS.—Cuba: Sierra de Hato-Nuevo; type locality; Daiquirí; Isla de Pinos. Not mapped. Patton (1966:181) records remains of the subgenus *Geocapromys* sp. from Cayman Brac, British West Indies.

Capromys ingrahami
Ingraham's Hutia

External measurements: 370–389; 45; 50–65; mean weight (Clough, 1974:671), for *C. i. ingrahami*, 17 males, 613 grams; seven females, 676 grams. Pelage rough and harsh on upper parts and finer below; yellowish-brown, mixed with reddish- or yellowish-brown, sprinkled with gray and black. Characterized by lower posterior margin of jugal sloping strongly forward and forming an obtuse angle. Jugal lacking a posterior projection. *C. i. abaconis* and *C. i. irrectus* known only from fragments of bone. Not mapped.

Capromys ingrahami abaconis (Lawrence)

1934. *Geocapromys ingrahami abaconis* Lawrence, Occas. Pap. Boston Soc. Nat. Hist., 8:191, November 7, type from Imperial Lighthouse Caves, Hole in the Wall, Great Abaco Island, Bahama Islands. Known only from Great Abaco Island. Extinct.
1974. *Capromys ingrahami*, Varona, Catálogo mam. viv. y exting. de las Antillas, Acad. Cienc. Cuba, 1974:67.

Capromys ingrahami ingrahami J. A. Allen

1891. *Capromys ingrahami* J. A. Allen, Bull. Amer. Mus. Nat. Hist., 3:329, August 31, type from East Plana Cay, Plana Keys, Bahama Islands. Known only from type locality, until January, 1973, when introduced "on a small cay in Exumas National Sea and Land Park" (Clough, Jour. Mamm., 55:670, August 20).

Capromys ingrahami irrectus (Lawrence)

1934. *Geocapromys ingrahami irrectus* Lawrence, Occas. Pap. Boston Soc. Nat. Hist., 8:190, November 7, type from "Burial Cave No. 1," Gordon Hill Caves, Crooked Island, Bahama Islands.
1974. *Capromys ingrahami irrectus*, Varona, Catálogo mam. viv. y exting. de las Antillas, Acad. Cienc. Cuba, 1974:68. Said to be extinct.

MARGINAL RECORDS.—Bahama Islands: type locality; Long Island; Eleuthera Island; Exuma Island.

Capromys pleistocenicus (Arredondo)

1958. *Geocapromys columbianus pleistocenica* [sic] Arredondo, El Cartero Cubano, 17 (12):48. No type or exact type locality designated but name proposed for remains found in caves in different localities in Cuba and Isle of Pines according to Varona, Catálogo mam. viv. y exting. de las Antillas, Acad. Cienc. Cuba, 1974:68. Varona (*ibid.*) lists this species as extinct under the subgenus *Geocapromys*, where he also lists *Capromys* sp. from Cayman Brac as extinct. Not mapped. Extinct.

Genus Macrocapromys Arredondo
Arredondo's Hutia

1958. *Macrocapromys* Arredondo, El Cartero Cubano, 17 (12):10. Type, *Macrocapromys acevedo* [sic] Arredondo.

Humerus a fifth longer than in the largest species of *Capromys* (= *C. pilorides*) according to Arredondo (1958:10).

Macrocapromys acevedoi Arredondo
Arredondo's Hutia

1958. *Macrocapromys acevedo* [*sic*] Arredondo, El Cartero Cubano, 17 (12):10, type from Lamas Cave, Santa Fe, Havana, Cuba.

Characters as given for the genus.

MARGINAL RECORDS.—Cuba (Varona, 1974:69): type locality; Paredones Cave; Ceiba del Agua. Not mapped. Extinct.

Genus Hexolobodon Miller—Haitian Hutias

1929. *Hexolobodon* Miller, Smiths. Miscl. Coll., 81(9):19, March 30. Type, *Hexolobodon phenax* Miller.

So far as known resembling *Geocapromys*, but differing as follows: cheek-teeth with roots becoming closed at, or soon after, the stage when crowns are worn flat; root of lower incisor passing beneath root of m3 and terminating, in fully adult individuals, on outer side of tooth-row beneath floor of groove that separates alveolus of m3 from base of coronoid process; p4 with only 2 re-entrant angles on inner side (as in *Capromys*); all maxillary teeth having 2 about equally developed re-entrant angles on each side, these imparting to crowns an evenly 6-lobed structure (after Miller, 1929a:19, 20).

KEY TO SPECIES OF GENUS HEXOLOBODON

1. Frontoparietal constriction accentuated; one external re-entrant fold on M2.
   H. poolei, p. 867
1'. Frontoparietal constriction not accentuated; two external re-entrant folds on M2. . . . . *H. phenax,* p. 867

Hexolobodon poolei Rímoli
Poole's Hutia

1977. *Hexolobodon poolei* Rímoli, Roedores Fósiles de la Hispaniola. Univ. Central del Este, República Dominicana. Ser. Cientifica III, page 21, January 10. Type from Cueva Grande (Large Cave), near St. Michel, Haiti. Extinct.

Other than the characters noted in the key, *H. poolei* differs from *H. phenax* in anterolabial lobule of M2 presenting an island of enamel completely separated (surrounded by cement) and sagittal crest long and high rather than short and low.

Hexolobodon phenax Miller
Haitian Hutia

1929. *Hexolobodon phenax* Miller, Smiths. Miscl. Coll., 81 (9):20, March 30, type from small cave near St. Michel, Haiti.

"An animal about the size of *Capromys pilorides,* but skull probably differing from that of all species of *Capromys* and *Geocapromys* in shorter rostrum and generally more robust form." See also generic description (Miller, 1929a:20).

MARGINAL RECORDS.—Haiti: type locality; caves near l'Atalaye. Not mapped. Extinct.

SUBFAMILY PLAGIODONTINAE

Genus Plagiodontia F. Cuvier

1836. *Plagiodontia* F. Cuvier, Ann. Sci. Nat., ser. 2, 6:347. Type, *Plagiodontia aedium* F. Cuvier.

Length of head and body of known males, 312–405. Form stout, feet robust, tail naked and "of moderate length" (Ellerman, *et al.,* 1940:134). Enamel pattern of upper molars tetramerous, folds long, deep, obliquely arranged; molars not strongly hypsodont; zygomatic arches slender; paroccipital processes especially long and slender; little interorbital and palatal constriction.

The systematic arrangement below of the seven taxa of the genus *Plagiodontia* is based on the findings of D. H. Johnson (1948), C. E. Ray (1964a), S. Anderson (1965), authors cited by them, and, in part, on Rímoli (1977).

KEY TO SPECIES OF GENUS PLAGIODONTIA

(After Johnson, 1948:73, 74, and Ray, 1964a:1–4)

1. Maximum diameter of occlusal surface of P4 less than 9.6 mm.
   2. Skull smaller, alveolar length of mandibular tooth-row less than 18; postorbital processes (at least in *P. aedium*) almost absent.
      3. M3 with posterior fold and accessory process; length of occlusal surface of M3 2.7; alveolar length of mandibular tooth-row 16.3.
         P. spelaeum, p. 868
      3'. M3 lacking posterior fold and accessory process; length of occlusal surface of M3 3.8; alveolar length of mandibular tooth-row 16.5.
         P. aedium, p. 868

2'. Skull larger, alveolar length of mandibular tooth-row more than 18; postorbital processes prominent.
   4. Labial and posterior lingual folds of lower cheek-teeth overlapping; postorbital processes relatively long and slender; alveolar length of mandibular tooth-row approx. 20. . . . .*P. hylaeum*,   p. 868
   4'. Labial and posterior lingual folds of lower cheek-teeth meeting without overlapping; postorbital processes blunter; alveolar length of mandibular tooth-row approx. 24. . . . .*P. ipnaeum*,   p. 868
1'. Maximum diameter of occlusal surface of P4 more than 9.6 mm. . . . .*P. araeum*,   p. 869

[See also *P. caletensis* and *P. velozi* beyond, named by Rímoli (1977), who was unable (*in Litt.*) to construct a key. Rímoli (1977:24, 25), unlike Anderson (1965:95–97), regarded *P. aedium* and *P. hylaeum* as separate species.]

Plagiodontia aedium F. Cuvier
Cuvier's Hutia

1836. *Plagiodontia aedium* F. Cuvier, Ann. Sci. Nat., ser. 2, 6:347. Type locality, Dominican Republic.

External measurements of a specimen in alcohol with skull removed are: 465; 153; 67; 18. Pelage of the same specimen is as follows: upper parts chiefly "gray-based and whitish-tipped hairs between 30 and 40 mm. long, with a sparse intermixture of gray underfur and a few longer, wholly dark bristles; fur of underparts shorter, whitish basally and tipped with a buffy shade" (Johnson, 1948:70, 71). Skull relatively small (greatest length approx. 69); postorbital processes of frontal barely discernible; a single well-defined pair of palatal pits; incisors and cheek-teeth small; labial and lingual folds of lower cheek-teeth moderately overlapping.

MARGINAL RECORDS.—Haiti: near Miragoane. Dominican Republic: Anadel; [mouth of] Río San Juan; Kilometer 2 site; Kilometer 4 site. Not mapped.

Plagiodontia hylaeum Miller
Dominican Hutia

1927. *Plagiodontia hylaeum* Miller, Proc. U.S. Nat. Mus., 72(16):4, September 30, type from Guarabo, 10 mi. E Jovero, Samaná Province, Dominican Republic.

Measurements of males are as follows: length of head and body, 348–405; tail, 127–146. As compared with *P. aedium*, "smaller scales on tail and feet; narrower feet with longer and sharper claws; pinna of ear less hairy, its margin more thickened and more evenly rounded; color seemingly darker in general, contour hairs with less contrasting tips, cheeks and underparts essentially similar in color to back rather than distinctly paler, a whitish pectoral spot occasionally present" (Johnson, 1948:72).

MARGINAL RECORDS.—Dominican Republic: type locality; near Constanza, La Vega Province; San Gabriel [Island]. Haiti: Trujin, Massif de La Selle; Atalaye Plantation, near St. Michel. Not mapped.

Plagiodontia ipnaeum Johnson
Johnson's Hutia

1948. *Plagiodontia ipnaeum* Johnson, Proc. Biol. Soc. Washington, 61:72, June 16, type from kitchen-midden deposit at a Ciguayan village site at Anadel, 2 km. E. Samaná, Samaná Prov., Dominican Republic.

Judging from the skull, *P. ipnaeum* is the largest species of the genus; postorbital processes prominent, but projecting less than in *P. hylaeum*; skull deep; palate constricted anteriorly; palatal pits opposite middle of M1; pterygoid vacuity broad, its anterior margin obtusely pointed; upper tooth-rows closely approximated anteriorly, divergent posteriorly; cheek-teeth progressively smaller from P4 to M3, with greatest difference between P4 and M1; mandible . . . more massive than in other species; angular process greatly expanded horizontally (after Johnson, 1948:72).

This species was incorrectly identified as *Plagiodontia aedium* by Miller, for example, in 1929 (1929a:18).

MARGINAL RECORDS.—Dominican Republic: type locality; mouth San Juan River, 10 km. N Samaná; near Monte Cristi; near Constanza, La Vega Prov.; San Pedro de Macoris. Haiti: Atalaye Plantation, near St. Michel. Not mapped. Extinct.

Plagiodontia spelaeum Miller
Little Hutia

1929. *Plagiodontia spelaeum* Miller, Smiths. Miscl. Coll., 81(9):18, March 30, type from Crooked Cave near Atalaye Plantation, 4 mi. E St. Michel, Haiti. Known only from type locality. Not mapped.

Smallest species of *Plagiodontia*; in such characters as can be distinguished in fragmentary material resembling *P. hylaeum* rather than any other kind; palate moderately constricted anteriorly; upper cheek-teeth as in *P. hylaeum* except that M3 relatively smaller; secondary fold on posterior margin of M3 well developed; labial and lingual folds of lower cheek-teeth overlapping about as in *P. a. aedium* and less so than in *P. hylaeum* (modified from Johnson, 1948:73). Not mapped. Extinct.

Plagiodontia araeum Ray
Ray's Hutia

1964. *Plagiodontia araeum* Ray, Breviora, 203:2, April 10, type from Late Pleistocene or Recent deposit in first chamber of unnamed cave 2 km. SE Rancho de La Guardia, Municipio de Hondo Valle, Prov. San Rafael, República Dominicana. Coordinates of Rancho de La Guardia are 18° 43′ N, 71° 39′ W. Known only from type locality.

Known only from one left upper cheek-tooth thought to be P4. Differing from any upper cheek-tooth of other species of genus in greater maximum diameter of occlusal surface (11.0 mm.) and its extreme anterolingual-posterolabial compression. Maximum diameter of crown of P4 perpendicular to its longitudinal axis 10.7 mm. Lingual re-entrant with slight posterad flexure at internal extremity, not observed in other species of the genus; labial re-entrant with concomitant flexure near its mouth; posterolabial concavity in enamel wall shallow in comparison with that in other species of the genus (after Ray, 1964a:2–4). Not mapped. Extinct.

Plagiodontia caletensis Rímoli
La Caleta Hutia

1977. *Plagiodontia caletensis* Rímoli, Roedores Fósiles de la Hispaniola. Univ. Central del Este, República Dominicana, Ser. Científica III, p. 30, January 10, type from pre-Columbian deposits of the native cemetery of La Caleta, Dominican Republic. Known only from type locality.

Resembles *P. ipnaeum* but frontonasal region more prominent; postorbital processes larger; angular processes more wide-spreading ventrally with a greater development of the internal fold; mandibular condyle larger and wider (Rímoli, 1977:31).

Plagiodontia velozi Rímoli
Veloz's Hutia

1977. *Plagiodontia velozi* Rímoli, Roedores Fósiles de la Hispaniola. Univ. Central del Este, República Dominicana. Ser. Científica III, p. 32, January 10, type from a small cave (or Cave No. 1 according to the notes of Miller) near the "Plantación L'Atalaye, St. Michel, Department de L'Artibonite, Haiti." Known only from type locality. Extinct.

Resembles *P. ipnaeum* and *P. caletensis* but larger; molar series longer and molars and premolars more robust (Rímoli, 1977:33).

Genus Isolobodon J. A. Allen—Hutias

KEY TO SUBGENERA AND SPECIES OF GENUS ISOLOBODON

1. Crowns and alveoli of upper and lower molars markedly wider than long (species *I. montanus*). . . . subgenus *Aphaetreus*, p. 870
1′. Crowns and alveoli of upper and lower molars approx. as long as wide.
                subgenus *Isolobodon*, p. 869
  2. Mandibular tooth-row 17.4 or less.
                    I. levir, p. 870
  2′. Mandibular tooth-row more than 17.4. . . . .*I. portoricensis*, p. 869

Subgenus Isolobodon J. A. Allen

1916. *Isolobodon* J. A. Allen, Ann. New York Acad. Sci., 27:19, January 25. Type, *Isolobodon portoricensis* J. A. Allen.
1922. *Ithydontia* Miller, Smiths. Miscl. Coll., 74(3):4, October 16. Type, *Ithydontia levir* Miller. Shown to be a synonym of *Isolobodon*, by Miller, Smiths. Miscl. Coll., 81(9):14, March 30, 1929.

Curve of upper incisor long, its root lying in antorbital foramen; lower incisor terminating beneath M3; P4 with 2 outer re-entrant angles, its enamel pattern obviously different from that of molars; re-entrant folds in upper teeth not very oblique, their slant more than 45° as referred to corresponding alveolar line; re-entrant folds on inner side of lower teeth extending more than halfway across crowns; frontal sinus not sufficiently inflated to produce obvious swelling over anterior zygomatic root or to encroach on area of antorbital foramen or of orbit; posterior margin of zygomatic process of maxillary lying at or behind level of middle of alveolus of P4.

Opposed inner and outer re-entrant angles of all teeth remaining distinct throughout life, the enamel pattern of each tooth entire; crowns and alveoli of both upper and lower molars nearly as long as wide (after Miller, 1929a:17).

Isolobodon portoricensis J. A. Allen
Allen's Hutia

1916. *Isolobodon portoricensis* J. A. Allen, Ann. New York Acad. Sci., 27:19, January 25, type from Cueva de la Ceiba, near Utuado, Puerto Rico.

Owing to the unsatisfactory nature of known material, it is not yet possible adequately to describe the species nor to compare them in as much detail as is desirable. Seemingly, *I. portoricensis* differs from *I. levir* in longer mandibular tooth-row (more than 17.4 vs. less); greater length of frontal along median suture (more than

21.0 vs. less); wider rostrum (more than 12.0 at premaxillary suture vs. less); greater interorbital breadth (more than 19.0 vs. less).

This species may have been introduced by pre-Columbian man onto several of the islands where its remains have been found. See Reynolds, *et al.* (1953), and Varona (1974:71).

MARGINAL RECORDS.—Puerto Rico: various localities, probably throughout the island. Dominican Republic: various sites vic. Samaná Bay. Virgin Islands: St. Thomas; St. Croix. Mona Island (Varona, 1974:71). Not mapped. Extinct.

Isolobodon levir (Miller)
Miller's Hutia

1922. *Ithydontia levir* Miller, Smiths. Miscl. Coll., 74(3):5, October 16, type from larger of 2 caves NE St. Michel de l'Atalye (*sic*), Haiti.
1929. *Isolobodon levir* Miller, Smiths. Miscl. Coll., 81(9):14, March 30.

Reynolds, *et al.* (1953) examined the material then available of this Hispaniolian "species" and concluded that it was indistinguishable from the earlier named *I. portoricensis* J. A. Allen, known from both Puerto Rico and Hispaniola. Varona (1974:71) placed *I. levir* as a synonym of *I. portoricensis* because he relied on the "opinion" of Reynolds, *et al.* Rímoli (1977:38–40), on the basis of additional crania and lower jaws, concluded that *I. levir* was a valid species, differing from *I. portoricensis* in shorter maxillary tooth-row and smaller skull evident in six measurements (*op. cit.*:38). Possibly in this subgenus or in a related taxon the secondary sexual differences, and especially the changes in proportions of the skull with increasing ontogenetic age, need to be ascertained to decide whether *I. levir* is a valid species. Although Rímoli (*op. cit.*:39) records no overlap in length of the maxillary tooth-row in 14 *I. portoricensis* and two *I. levir*, he records extensive overlap in measurements of the mandibular tooth-row in 34 *I. portoricensis* and 38 *I. levir*. The lack of overlap in length of the maxillary tooth-row may be due to small size of sample. Because Rímoli is the latest author to evaluate the status of *I. levir*, it is here listed as a species.

MARGINAL RECORDS.—Haiti: type locality. Dominican Republic: sites in vic. Samaná Bay. Not mapped.

Subgenus Aphaetreus Miller—Montane Hutia

1922. *Aphaetreus* Miller, Smiths. Miscl. Coll., 74(3):3, October 16. Type, *Aphaetreus montanus* Miller.

Mandible and its teeth resembling those of the subgenera *Isolobodon* and *Plagiodontia;* mandibular cheek-teeth prismatic, growing from persistent pulps, their essential structure as in the two related subgenera but entire tooth-row appearing as if compressed anteroposteriorly with result that spaces occupied by dentine and those by cement are narrowed, owing to enamel plates being closer together; crown of each tooth obviously wider than long instead of seemingly longer than wide; inner re-entrant angles confluent with posteroexternal angle, causing enamel pattern to consist of an anterior Y and a posterior I completely isolated from each other by band of cement; m3 nearly as large as m2 (after Miller, 1922:3).

Isolobodon montanus (Miller)
Montane Hutia

1922. *Aphaetreus montanus* Miller, Smiths. Miscl. Coll., 74(3):4, October 16, type from larger of 2 caves NE St. Michel de l'Atalye (*sic*), Haiti.
1977. *Isolobodon montanus*, Rímoli, Roedores Fósiles de la Hispaniola. Univ. Central del Este, República Dominicana. Ser. Cientifica III, p. 40, January 10.

See characters of the genus.

MARGINAL RECORDS.—Haiti: type locality. Dominican Republic: San Gabriel Island. Not mapped. Extinct.

Genus Hyperplagiodontia Rímoli
Rímoli's Hutia

1977. *Hyperplagiodontia* Rímoli, Roedores Fósiles de la Hispaniola. Univ. Central del Este, República Dominicana. Ser. Cientifica III, page 33, January 10. Type, *Hyperplagiodontia stenocoronalis* Rímoli.

Larger than *Plagiodontia*, its nearest generic relative; molariform teeth slightly hypsodont and roots short; folds in molariform teeth more compressed anteroposteriorly and sharper and more conspicuous than in *Plagiodontia* (adapted from Rímoli, 1977:33).

Hyperplagiodontia stenocoronalis Rímoli
Rímoli's Hutia

1977. *Hyperplagiodontia stenocoronalis* Rímoli, Roedores Fósiles de la Hispaniola. Univ. Central del Este, República Dominicana. Ser. Cientifica III, page 34, January 10. Type from St. Michel, cerca de l'Atalaye, Département de l'Artibonite, Haiti. Known only from type locality. Extinct.

Characters as described for the genus.

SUBFAMILY MYOCASTORINAE

Genus Myocastor Kerr—Coypus

1792. *Myocastor* Kerr, The animal kingdom . . . , p. 225. Type, *Mus coypus* Molina.
1805. *Myopotamus* É. Geoffroy St.-Hilaire, Ann. Mus. Hist. Nat. Paris, 6:82. Type, *Myopotamus bonariensis* É. Geoffroy St.-Hilaire.

Large (length of head and body approx. 500), robust, compact; strongly specialized for aquatic life; skull heavy, well-ridged; paroccipital process greatly elongated and curved slightly anteriorly at distal end; frontals broad and flattened; small postorbital process present; palate constricted anteriorly; no accessory canal in infraorbital foramen; upper cheek-teeth decreasing markedly in size anteriorly, markedly hypsodont, semirooted, with strong and persistent inner and outer re-entrant folds surrounded by wide enamel. Feet short, pentadactyl, and webbed posteriorly.

The genus seems to be monotypic.

Myocastor coypus
Nutria

Pelage consisting of two types of hair, a thick, dense underfur and the overlying harsh guardhairs. Upper parts rich glossy brown or chestnut; underparts paler; tail nude and scaly.

The habitat of this animal is much the same as that of the native muskrat (*Ondatra*). The nutria has been introduced into North America and has become established and is spreading, to some extent at the expense of the muskrat. The introduced subspecies of the nutria probably is:

Myocastor coypus bonariensis (É. Geoffroy St.-Hilaire)

1805. *Myopotamus bonariensis* É. Geoffroy St.-Hilaire, Ann. Mus. Hist. Nat. Paris, 6:82, type from Argentina.
1917. *Myocastor coypus bonariensis*, Thomas, Ann. Mag. Nat. Hist., ser. 8, 20:100, July.

SOME RECORDS.—Louisiana; Ohio; Montana; Washington; Oregon; British Columbia; California; Nebraska (Jones, 1964c:335, 336); Kansas; Oklahoma (Glass and Halloran, 1961:239); Texas (Davis, 1966:218); New Mexico; Florida; Minnesota (Timm, 1975:39); Ontario (*ibid.*). Not mapped.

FAMILY ECHIMYIDAE—Spiny Rats

Upper molars rooted, flat-crowned; inner and outer re-entrant folds becoming isolated on crown surface with wear; or, heavier, more complex dentition that tends toward a series of transverse laminae. Paroccipital processes curved forward under prominent auditory bullae; frontals broad, flattened. Externally [so far as known] more or less ratlike; tendency present for development of spiny covering, which is in rare cases strongly developed. Forefeet tetradactyl, hind feet pentadactyl.—See addenda.

KEY TO NORTH AMERICAN GENERA OF ECHIMYIDAE

1. Occurring on Antillean islands.[1]
                    Heteropsomys,   p. 874
1'. Not occurring on Antillean islands.
    2. Both upper and lower cheek-teeth with strong tendency to formation of series of transverse laminae; pelage harsh but not spiny.   . . .*Diplomys*,   p. 874
    2'. Upper and lower cheek-teeth lacking tendency toward formation of series of transverse laminae; pelage spiny.
        3. Pelage excessively spiny; upper cheek-teeth with 4 clearly marked outer re-entrant folds. *Hoplomys*,   p. 871
        3'. Pelage moderately spiny; upper cheek-teeth usually with 3 or fewer outer re-entrant folds. *Proechimys*,   p. 872

[1] See account on p. 874 of *Proechimys corozalus* from Puerto Rico, one of the Antillean islands.

SUBFAMILY ECHIMYINAE

Genus Hoplomys J. A. Allen—Armored Rat

Revised by Handley, Smiths. Miscl. Coll., 139(4):1–10, 1 fig., July 3, 1959.

1908. *Hoplomys* J. A. Allen, Bull. Amer. Mus. Nat. Hist., 24:649, October 13. Type, *Hoplomys truei* J. A. Allen.

Resembling *Proechimys*, especially externally. Cheek-teeth with 4 transverse furrows, the maxillary series being more obliquely arranged than in *Proechimys*. Spines well developed, possibly maximum for the family; fur coarse; tail shorter than head and body, nearly naked. Attains a head-body length of approx. 400.

The genus is monotypic.

Hoplomys gymnurus
Armored Rat

The type, an adult male, of *H. g. truei*, measures as follows: length of head and body, 380; tail, 170; hind foot (dry), 53. Upper parts blackish underlaid with pale to grayish brown, palest on sides; underparts white or nearly so; ears nearly naked, dark brown to black; tail naked, scaly, brownish above, whitish below.

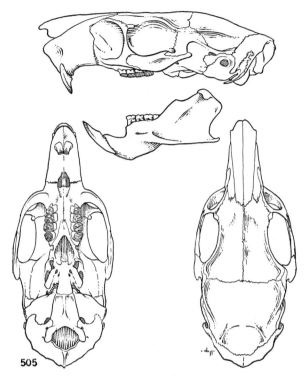

Fig. 505. *Hoplomys gymnurus goethalsi*, Cana, Panamá, No. 179035 U.S.N.M., ♂, X 1.

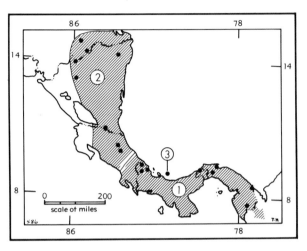

Map 486. *Hoplomys gymnurus.*

1. *H. g. goethalsi*   2. *H. g. truei*
3. *H. g. wetmorei*

Hoplomys gymnurus wetmorei Handley

1959. *Hoplomys gymnurus wetmorei* Handley, Smiths. Miscl. Coll., 139(4): 9, July 3, type from Isla Escudo de Veraguas, Prov. Bocas del Toro, Panamá. Known only from type locality.

Hoplomys gymnurus goethalsi Goldman

1912. *Hoplomys goethalsi* Goldman, Smiths. Miscl. Coll., 56(36):10, February 19, type from Río Indio, near Gatún, Canal Zone, Panamá.
1920. *Hoplomys gymnurus goethalsi* Goldman, Smiths. Miscl. Coll., 69(5):123, April 24.

MARGINAL RECORDS.—Panamá (Handley, 1966:787): Sibube; upper Río Changena; Almirante; Fort Sherman; Mandinga; Armila, thence into South America, and northward to Panamá (*ibid.*): Cana; Cerro Azul.

Hoplomys gymnurus truei J. A. Allen

1908. *Hoplomys truei* J. A. Allen, Bull. Amer. Mus. Nat. Hist., 24:650, October 13, type from Lavala [= Savala], Matagalpa, Nicaragua.
1920. *Hoplomys gymnurus truei*, Goldman, Smiths. Miscl. Coll., 69(5):124, April 24.

MARGINAL RECORDS.—Honduras: 40 km. E Catacamas (Pine and Carter, 1970:804). Nicaragua: Leicus Creek (Buchanan and Howell, 1965:550). Costa Rica: Suerre; Santa Teresa Perálta. Nicaragua: El Castillo (Pine and Carter, 1970:804); Vijagua (Buchanan and Howell, 1965:549); Río Coco (*ibid.*). Honduras: 78 mi. ENE Danlí (Pine and Carter, 1970:804).

Genus **Proechimys** J. A. Allen—Spiny Rats

1899. *Proechimys* J. A. Allen, Bull. Amer. Mus. Nat. Hist., 12:264, December 26. Type, *Echimys trinitatis* J. A. Allen and Chapman.

Length of head and body, 165–290; tail, 121–242; hind foot, 35–57; ear, 17–29. "Muriform echimyids of medium size; pelage with flattened and lanceolate and sometimes clavate aristiforms, varying greatly in width and distributed over most of the dorsal surface from shoulders to hips or base of tail . . . [skull usually] elongate and strongly built, with supraorbital ridges well developed, frequently extending across parietals toward occipital region; zygomatic arches . . . always with postorbital process; . . . mesopterygoid fossa extending forward at least to plane of third molars; . . . upper molariform teeth with a main internal fold and one to five external counterfolds which usually appear as enamel islands in worn teeth, these counterfolds barely implicating the lateral wall; lower molariform teeth with folds as in the upper molariform teeth except that they are reversed and the number of internal counterfolds is usually fewer in the molars. (Moojen, 1948:333.)

Proechimys semispinosus
Tomes' Spiny Rat

Size large; tail short and hairy; aristiforms wide and stiff, especially well developed on back; general color on upper parts some shade of ochraceous, usually much darker on back and forming a conspicuous dorsal band; feet dark; ventral surfaces and inner sides of legs white; skull elongate and strong with ridges well developed; incisive foramina long and narrow; bullae large; usually 4 counterfolds in M3 and M2; usually 3 but sometimes 4 counterfolds in M1 and even P4; lower premolar with 4, and lower molars with 3, counterfolds.

Proechimys semispinosus burrus Bangs

1901. *Proechimys burrus* Bangs, Amer. Nat., 35:640, August, type from San Miguel Island, Golfo de Panamá, Panamá. Known only from type locality.
1920. *Proechimys semispinosus burrus*, Goldman, Smiths. Miscl. Coll., 69(5):120, April 24.

Proechimys semispinosus centralis (Thomas)

1889. *Echinomys semispinosus* True, Proc. U.S. Nat. Mus., 11:467, September 3, type from San Emilio, Lake Nicaragua, Nicaragua. Not *Echimys semispinosus* Tomes, 1860, from Gualaquiza, Ecuador.
1896. *Echinomys centralis* Thomas, Ann. Mag. Nat. Hist., ser. 6, 18:312, October, a renaming of *E. semispinosus* True.
1920. *Proechimys semispinosus centralis*, Goldman, Smiths. Miscl. Coll., 69(5):120, April 24.

MARGINAL RECORDS.—Honduras: Río Coco, within 1 mi. Waspam [= Huaspan] (Pine, 1969:643). Nicaragua: Leicus Creek, 65 m. (Buchanan and How-

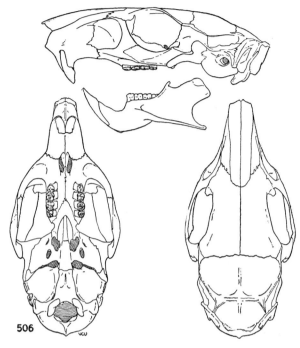

Fig. 506. *Proechimys semispinosus panamensis*, Vijagual, 37 km. N Quepos, 600 m., Costa Rica, No. 39251 K.U., ♂, X 1.

ell, 1965:554). Costa Rica: Vijagual San Carlos. Nicaragua: type locality.

Proechimys semispinosus goldmani Bole

1937. *Proechimys semispinosus goldmani* Bole, Sci. Publs., Cleveland Mus. Nat. Hist., 7:178, August 31, type from Altos Cacao, on ridge between Río Mariato and Río Negro, Mariato-Suay Lands, Veraguas, Panamá.

MARGINAL RECORDS.—Panamá: Mariato Rubber Camp; type locality; *Cerro Viejo; Paracoté.*

Proechimys semispinosus ignotus R. Kellogg

1946. *Proechimys semispinosus ignotus* R. Kellogg, Proc. Biol. Soc. Washington, 59:61, March 11, type from San José Island, Golfo de Panamá, Panamá. Known only from type locality.

Proechimys semispinosus panamensis Thomas

1900. *Proechimys centralis panamensis* Thomas, Ann. Mag. Nat. Hist., ser. 7, 5:220, February, type from savanna of Panamá, near City of Panamá, Panamá.
1920. *Proechimys semispinosus panamensis*, Goldman, Smiths. Miscl. Coll., 69(5):120, April 24.
1900. *Proechimys centralis chiriquinus* Thomas, Ann. Mag. Nat. Hist., ser. 7, 5:220, February, type from Bogava [= Bugaba], 800 ft., Chiriquí, Panamá.

MARGINAL RECORDS.—Costa Rica: Puerto Viejo (88249 KU). Panamá (Handley, 1966:786): Almirante; Mandinga; Armila, thence into South America, and

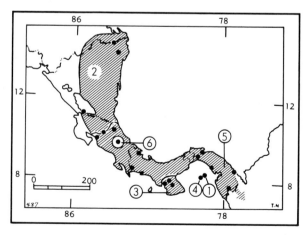

Map 487. *Proechimys semispinosus.*

1. *P. s. burrus*   4. *P. s. ignotus*
2. *P. s. centralis*   5. *P. s. panamensis*
3. *P. s. goldmani*   6. *P. s. rubellus*

northward to Panamá (*ibid.*): Cana; Chimán; Cerro Azul; Bugaba. Costa Rica: San Gerónimo Pirrís; Vijagual.

Proechimys semispinosus rubellus Hollister

1914. *Proechimys rubellus* Hollister, Proc. Biol. Soc. Washington, 27:57, March 20, type from Angostura Valley, Cartago, Costa Rica. Known only from type locality.
1920. *Proechimys semispinosus rubellus*, Goldman, Smiths. Miscl. Coll., 69(5):120, April 24.

Proechimys corozalus Williams and Koopman
Puerto Rican Spiny Rat

1951. *Proechimys corozalus* Williams and Koopman, Amer. Mus. Novit., 1515:4, June 4, type from Corozal Limestone Quarry, Puerto Rico. The type and only known specimen, perhaps not a *Proechimys*, is part of a left mandibular ramus, lacking two teeth, a fossil of unknown geologic age, and possibly is an unnamed genus.

Resembles *Proechimys* (*Trinomys*) *iheringi* of southeastern Brazil but: larger (p4–m3 *ca.* 14.1 instead of 9.2 mm.); m2 larger than m1; mandibular masseteric ridge less well developed (after Williams and Koopman, 1951:4). Not mapped. Extinct.

Genus Diplomys Thomas—Spiny Rats

1916. *Diplomys* Thomas, Ann. Mag. Nat. Hist., ser. 8, 18:240, August. Type, *Loncheres caniceps* Günther, 1876.

In general like *Echimys*, but lower cheek-teeth in a series of transverse plates; same is true of upper teeth; 4 laminae in each upper tooth; 4 in p4, 3 in each lower molar (after Ellerman, 1940:115). Goldman (1920:124) states that this genus, as represented in Panamá, has short and conspicuously tufted ears. The pelage is spiny but much less so than in *Hoplomys* or *Proechimys*. Tail hairy; hind feet markedly broadened, adapted to arboreal habits.

Diplomys labilis
Gliding Spiny Rat

External measurements of the type, an adult male, are: 540; 240; 47; 17. Pelage long, stiff, but without spines. Upper parts bright ferruginous, darkened mid-dorsally (variable individually) by black-tipped hairs; yellowish-white patch at base of vibrissae, above eye, and behind ear; underparts varying from strong buff to clear ferruginous, sometimes both tones present in irregular patches; tail like upper parts basally, distally blackish, sometimes with thin whitish pencil; feet yellowish-brown.

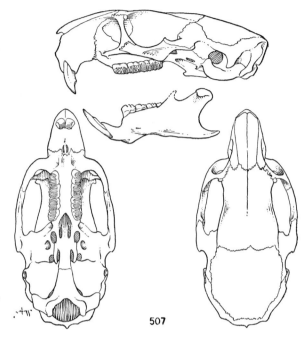

507

Fig. 507. *Diplomys labilis labilis*, San Miguel Island, Panamá, No. 116667 U.S.N.M., ♂, X 1.

Diplomys labilis darlingi (Goldman)

1912. *Isothrix darlingi* Goldman, Smiths. Miscl. Coll., 60(2):12, September 20, type from Marrangantí, near Real de Santa María, on the Río Tuyra, Panamá.
1966. *D[iplomys]. l[abilis]. darlingi*, Handley, Ectoparasites of Panama, Field Mus. Nat. Hist., p. 787, November 22.

MARGINAL RECORDS.—Panamá (Handley, 1966:787): Mandinga; Armila; Cerro Malí, 5000 ft.; Tapalisa, 400 ft.; type locality; *Ancón*; Fort Kobbe; Juan Mina. Almost certainly occurs in adjacent Colombia.

Diplomys labilis labilis (Bangs)

1901. *Loncheres labilis* Bangs, Amer. Nat., 35:638, August 22, type from San Miguel Island, Golfo de Panamá, Panamá. Known only from San Miguel Island.

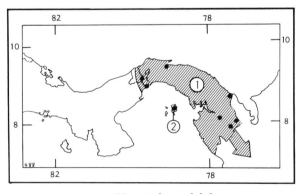

Map 488. *Diplomys labilis*.

1. *D. l. darlingi*   2. *D. l. labilis*

1916. *D[iplomys]. labilis,* Thomas, Ann. Mag. Nat. Hist., ser. 8, 18:296, September.

Genus **Heteropsomys** Anthony

1916. *Heteropsomys* Anthony, Ann. New York Acad. Sci., 27:202, August 9. Type, *Heteropsomys insulans* Anthony.
1916. *Neopsomys* Miller, Smiths. Miscl. Coll., 16(12):2, December 7, an accidental renaming of *Heteropsomys* Anthony in such a way that the type species and holotype (specimen) is the same as for *Heteropsomys*.
1917. *Homopsomys* Anthony, Bull Amer. Mus. Nat. Hist., 37:187, January 29. Type, *Homopsomys antillensis* Anthony. *Homopsomys* included in synonymy of *Heteropsomys* by Varona, Catálogo mam. viv. y exting. de las Antillas, Acad. Cienc. Cuba, 1974:73.

"Dorsal outline [of skull] gently curving from tip of nasals to parietal region, then abruptly curving downward; nasals unknown; ascending portion of premaxilla not reaching to nasal termination; lacrymal indicated as large; frontal with well developed postorbital process, interorbital constriction slight; parietals strongly convex; supra-occipitals with low, sharp, sagittal crest. . . . Infraorbital foramen of medium size. Palate short. . . ." (Anthony, 1918:407, 408.) Upper incisors weak but long; upper cheek-teeth having 1 conspicuous internal fold and 3 separate transverse enamel-surrounded lakes; lower cheek-teeth having pattern of uppers reversed and but 2 lakes.

KEY TO SUBGENERA OF GENUS HETEROPSOMYS

1. Upper cheek-teeth with 1 internal re-entrant fold and 3 isolated enamel lakes. . . . . .*Heteropsomys,* p. 875
1'. Upper cheek-teeth having outer as well as inner re-entrant fold(s) and lacking large, transverse, isolated enamel lakes.
  2. Course of upper incisors readily traced on surface of rostrum; M3 with 2 inner and 2 outer re-entrant folds; infraorbital canal with conspicuous accessory canal for transmission of nerve. . . . .*Boromys,* p. 876
  2'. Course of upper incisors not readily traced on surface of rostrum; M3 with 1 inner and 1 outer re-entrant fold; infraorbital canal lacking accessory canal for transmission of nerve. *Brotomys,* p. 875

Subgenus **Heteropsomys** Anthony

1916. *Heteropsomys* Anthony, Ann. New York Acad. Sci., 27:202, August 9. Type, *Heteropsomys insulans* Anthony. For description see account of the genus.

Heteropsomys insulans Anthony

1916. *Heteropsomys insulans* Anthony, Ann. New York Acad. Sci., 27:202, August 9, type from Cueva de la Ceiba, Hda. Jobo, near Utuado, Puerto Rico. Known only from type locality.

In addition to characters listed for the genus, of which this species is the type, see account of *H. antillensis* below. Not mapped. Extinct.

Heteropsomys antillensis Anthony

1917. *Homopsomys antillensis* Anthony, Bull. Amer. Mus. Nat. Hist., 37:187, January 29, type from Utuado, Puerto Rico.
1974. *Heteropsomys antillensis,* Varona, Catálogo mam. viv. y exting. de las Antillas, Acad. Cienc. Cuba, p. 73.

Differs from *H. insulans* in heavier incisors, longer sagittal crest, greater zygomatic breadth, longer palate, and other features indicative of greater ontogenetic age; possibly the name *H. insulans,* based on subadult or young animal re-

mains, is the senior synonym for *H. antillensis,* based on remains of an adult or old animal. Anthony (1918:407), himself, recognized this possibility, and his excellent illustrations (*loc. cit.,* Fig. 40) support the possibility.

MARGINAL RECORDS.—Puerto Rico: type locality; Ciales. Not mapped. Extinct.

Subgenus **Brotomys** Miller

1916. *Brotomys* Miller, Smiths. Miscl. Coll., 66(12):6, December 7. Type, *Brotomys voratus* Miller. Reduced to subgeneric rank by Varona, Catálogo mam. viv. y exting. de las Antillas, Acad. Cienc. Cuba., 1974:73, 74.

Rostrum short, wide, deep; infraorbital canal especially large, lacking accessory canal for transmission of nerve; teeth weakly developed; toothrows approx. parallel, each tooth with 3 poorly developed roots; enamel pattern of P4 and M1 (much worn) with single re-entrant fold on each side (a tiny lake near inner tip of outer fold); tip of outer fold posterior to tip of inner fold. Distinguishable from subgenus *Heteropsomys* by relatively large antorbital foramina, horizontal anterior zygomatic roots, slightly emarginate palate, and perhaps more simplified enamel pattern (after Miller, 1916:6, 7).

KEY TO SPECIES OF SUBGENUS BROTOMYS

1. Frontopalatal depth less than 88% of interorbital breadth. . . . .*H. contractus,*   p. 876
1'. Frontopalatal depth more than 88% of interorbital breadth. . . . . .*H. voratus,*   p. 876

Heteropsomys voratus (Miller)

1916. *Brotomys voratus* Miller, Smiths. Miscl. Coll., 66(12):7, December 7, type from San Pedro de Macoris, Dominican Republic.
1974. *Heteropsomys voratus,* Varona, Catálogo mam. viv. y exting. de las Antillas, Acad. Cienc. Cuba, p. 74.

Selected measurements of the type are: palatal length, 25.4; palatilar length, 23.0; interorbital breadth, 14.6; least breadth of palate between premolars, 2.0; maxillary tooth-row (alveoli), 10.0.

MARGINAL RECORDS.—Dominican Republic, all in immediate vic. Samaná Bay: type locality; Boca del Infierno; San Gabriel [Island]; [Río] Naranjo Abajo[= a key about ½ mi. E mouth]; Anadel; foothills SE Monte Cristi (Kilometer 2 site; Kilometer 4 site) Not mapped. Probably extinct.

Heteropsomys contractus (Miller)

1929. *Brotomys ? contractus* Miller, Smiths. Miscl. Coll., 81(9):13, March 30, type from small cave near St. Michel, Haiti. Known only from type locality.
1974. *Heteropsomys contractus,* Varona, Catálogo mam. viv. y exting. de las Antillas, Acad. Cienc. Cuba, p. 74.

Resembling *Heteropsomys voratus* but size slightly less, rostrum relatively shorter, interorbital region narrower in proportion to frontopalatal depth and more arched transversely; teeth broader than in *H. voratus,* and palate noticeably constricted, its interalveolar width at middle conspicuously less than transverse diameter of the adjoining alveoli (after Miller, 1929a:13).
Not mapped. Extinct.

Subgenus Boromys Miller

1916. *Boromys* Miller, Smiths. Miscl. Coll., 66(12):7, December 7. Type, *Boromys offella* Miller.—Regarded as a synonym of subgenus *Brotomys* (therefore of genus *Heteropsomys*) by Varona, Catálogo mam. viv. y exting. de las Antillas, Acad. Cienc. Cuba, 1974:73, but here retained as a subgenus because of characters enumerated immediately below.

Resembling subgenus *Brotomys* but antorbital foramen with well-developed secondary groove for nerve; root of incisor producing evident swelling on side of maxillary above neural groove; cheek-teeth relatively larger, but of same form and root structure; enamel pattern of M3 (and probably of all other maxillary teeth) consisting of 2 narrow re-entrant folds from each side; extremities of all folds reaching about to median line of crown (after Miller, 1916:7).

KEY TO SPECIES OF SUBGENUS BOROMYS

1. Alveolar length upper cheek-teeth more than 9.0. . . . .*H. offella,*   p. 876
1'. Alveolar length upper cheek-teeth less than 9.0. . . . .*H. torrei,*   p. 876

Heteropsomys offella (Miller)

1916. *Boromys offella* Miller, Smiths. Miscl. Coll., 66(12):8, December 7, type from Maisí, Baracoa, Cuba.
1974. *Heteropsomys offella,* Varona, Catálogo mam. viv. y exting. de las Antillas, Acad. Cienc. Cuba, p. 74.

Selected measurements of type are: length from in front of incisor to posterior edge of alveolus of M3, 29.2; palatal breadth including alveoli of 2nd molars, 9.0; least breadth of palate between premolars, 3.0; alveolar length maxillary cheek-teeth, 10.8.

MARGINAL RECORDS.—Cuba: "State of Pinar Del Rio"; "State of" Las Villas; "State of" Camaguey (7 km. W Banao); "State of" Oriente (type locality); also Isla de Pinos. Not mapped.

Heteropsomys torrei (G. M. Allen)

1917. *Boromys torrei* G. M. Allen, Bull. Mus. Comp. Zool., 61:6, January, type from cavern in Sierra de Hato Nuevo, Matanzas, Cuba.
1974. *Heteropsomys torrei,* Varona, Catálogo mam. viv. y exting. de las Antillas, Acad. Cienc. Cuba, p. 74.

Resembles *H. offella* but differs in lesser size and deeper indentation of posterior margin of palate, which extends forward to level of center of M2. Selected measurements of type are: palatal breadth including alveoli of M1, 6.4; palatal breadth including alveoli of M3, 6.5; alveolar length maxillary cheek-teeth, 7.6.

MARGINAL RECORDS.—"States" as follows, all in Cuba: Pinar del Río; Matanzas (type locality); Las Villas; Camaguey (7 km. W Banao); Oriente; also Isla de Pinos. Not mapped.

ORDER **CETACEA**—Whales and Porpoises

Fusiform; exclusively aquatic; thick, subcutaneous layer of blubber; sebaceous glands absent, sudiferous glands absent or reduced; anterior limbs paddle-shaped, not externally divided into segments; external hind limbs absent; pelvic vestigial, not in contact with vertebral column; tail flattened dorsoventrally, laterally expanded into 2 pointed flukes; skin smooth, hairless except for bristles in region of mouth in some genera; pinnae absent, external auditory meatus vestigial and all but closed; external nares (single or paired) opening through aperture near crown of head; mammae a single pair, situated in depressions on either side of vulva; cervical vertebrae flattened and frequently fused; odontoid process of axis short and blunt; clavicles absent; sacrum absent, sacral vertebrae similar to adjacent lumbars and caudals; skull elongate anteriorly; braincase telescoped (except in fossil suborder Archaeoceti); teeth simplified or absent.

Characters in the keys and diagnoses of taxa have been drawn mainly from the literature instead of from firsthand examination of specimens, and therefore are to be used with caution. Synonymies with a few exceptions are built on those of Hershkovitz (1966). The aims in the present account are to list all synonyms but to omit many of the misspellings, misidentifications, and most homonyms. Whoever works with names of cetaceans owes a great debt of gratitude to Hershkovitz for preparing and publishing his "Catalog of living whales" (1966:viii + 259).

KEY TO SUBORDERS OF CETACEA

1. Whalebone (baleen) absent; teeth present; single narial opening. . . . . Odontoceti, p. 877
1'. Whalebone (baleen) present; teeth absent after birth; paired narial openings.
    Mysticeti, p. 912

SUBORDER **ODONTOCETI**

Whale bone absent; teeth present; single narial opening; maxillae expanded and covering greater part of orbital processes of frontal bones; maxillae lacking orbital plates; nasal bones forming no part of root of narial passage, passage directed upward and backward.

KEY TO NORTH AMERICAN FAMILIES OF ODONTOCETI

1. Tubercula of posterior ribs articulating with metapophyses; throat having external grooves.
    2. Narial opening median; functional teeth in mandible never more than 4. . . . . Ziphiidae, p. 904
    2'. Narial opening left of median line of skull; functional teeth in mandible 18 or more. . . . . . . Physeteridae, p. 901
1'. Tubercula of posterior ribs articulating with diapophyses; throat lacking external grooves.
    3. Atlas and axis not fused; dorsal fin absent or rudimentary (see illustrations). . . . . Monodontidae, p. 898
    3'. Atlas and axis fused; dorsal fin well developed (except absent in *Lissodelphis*). . . . . Delphinidae, p. 877

FAMILY **DELPHINIDAE**
Porpoises and Dolphins

Total length to 30 ft. Tail notched posteromedially; pectoral appendage sickle-shaped; dorsal fin well developed (except absent in *Lissodelphis*). Skull with rostrum approx. as long as, or even longer than, cranial portion of skull; rostrum not deepened or solidified; lachrymal bone indistinctly separated from jugal bone; pterygoid bones meeting medially or separate; anterior 2 cervical vertebrae fused; fewer than 8 doubleheaded ribs; tubercula of posterior ribs articulating with diapophyses; functional teeth numerous in both jaws.

KEY TO NORTH AMERICAN GENERA OF DELPHINIDAE

1. Distinctly beaked.
    2. Mandibular symphysis more than ¼ length of ramus; crowns of teeth rugose. . . . .*Steno*, p. 878
    2'. Mandibular symphysis approx. ⅛ or less of length of ramus; crowns of teeth smooth or conical.
        3. Rostrum approx. same length as cranial portion of skull; usually 80 (74–92) or more vertebrae. . . . . *Lagenorhynchus*, p. 888

3'. Rostrum longer than cranial portion of skull; never more than 76 vertebrae.
    4. Black ring around eye (in *Delphinus* but no distinct ring reported in *Lagenodelphis*); a deep lateral groove on each side (right and left) of palate.
        5. Length of fusion of premaxillae on midline of skull less than 70% (*ca.* 40%) of width of base of rostrum measured in antorbital notches. . . . .*Lagenodelphis,*   p. 885
        5'. Length of fusion of premaxillae on midline of skull more than 70% (*ca.* 95%) of width of base of rostrum measured in antorbital notches. . . . .*Delphinus,*   p. 883
    4'. Black ring around eye absent; palate lacking lateral grooves.
        6. Dorsal fin absent; pterygoid bones separate. . . . .*Lissodelphis,*   p. 887
        6'. Dorsal fin present; pterygoid bones in contact.
            7. Teeth large, smooth, 19–26 in each tooth-row; 61–64 vertebrae. . . .*Tursiops,*   p. 885
            7'. Teeth small, acute, 37–52 in each tooth-row; never less than 69 vertebrae. . . . .*Stenella,*   p. 879
1'. Beak absent.
    8. Dorsal fin straight on posterior border; pectoral appendages large, ovate, rounded at distal end. . . . .*Orcinus,*   p. 890
    8'. Dorsal fin falcate or sickle-shaped on posterior border; pectoral appendages long, pointed.
        9. Pterygoid bones separate; most anterior point of lower jaw exceeding that of upper jaw.
            10. Approx. 95 vertebrae; white coloration limited to sides and venter posterior to pectoral appendages. . . . .*Phocoenoides,*   p. 898
            10'. Approx. 65 vertebrae; venter white from region of chin to caudal fin. . . .*Phocoena,*   p. 896
        9'. Pterygoid bones in contact; most anterior point of lower jaw not exceeding that of upper jaw.
            11. Anterior end of head rounded, not blunt; teeth in upper jaw not confined to anterior half of rostrum.
                12. Vertebrae fewer than 60 (*ca.* 50); color blackish throughout. . . . . .*Pseudorca,*   p. 889
                12'. Vertebrae more than 60; color with some white markings.
                    13. Vertebrae approx. 67; color dark gray with white lip bands and white bands in anal region. . . . .*Feresa,*   p. 895
                    13'. Vertebrae approx. 80; color dark gray with fine white spots; underparts paler gray; white around anus, genital aperture, patch anterior to navel, and between flippers. . . . .*Peponocephala,*   p. 895
            11'. Anterior end of head blunt with high forehead; teeth in upper jaw, if present, confined to anterior half of rostrum.
                14. Approx. 60 vertebrae; anterior part of head, excluding lower jaw, black. . . . .*Globicephala,*   p. 893
                14'. Approx. 68 vertebrae; anterior part of head, including lower jaw, white. *Grampus,*   p. 891

Genus **Steno** Gray—Rough-toothed Dolphin

1846. *Steno* Gray, *in* The zoology of the voyage of H.M.S. Erebus and Terror. . . , 1(Mamm.):43. Type, *Delphinus rostratus* Desmarest [= *Delphinus bredanensis* Lesson].

1859. *Glyphidelphis* Gervais, Zool. Paleont. Française, p. 301. Type species, *Delphinus rostratus* Cuvier (= *Delphinus bredanensis* Lesson).

Steno bredanensis (Lesson)
Rough–toothed Dolphin

Total length to 8 ft. Color purplish-black dorsally, sides with yellowish-white spots; beak and ventral surface white, tinged with purple and rose color. Beaked, but beak not distinctly set off from "forehead"; rostrum long, narrow and compressed; mandibular symphysis one-fourth length of ramus; pterygoid bones in contact; postorbital processes of frontals slender; teeth large, crowns rugose, 20–27 in each tooth-row; 65 or 66 vertebrae.

The genus *Steno* is monotypic.

1817. *Delphinus rostratus* Desmarest, *in* Nouv. Dict. Hist. Nat., ed. 2, 9:160. Type locality, Paimpol, France. Not *Delphinus rostratus* Shaw, 1801.

1823. [*Delphinus*] *frontatus* G. Cuvier, Recherches sur les ossemens fossiles . . . , ed. 2, 5:278, 400. Name restricted by Hershkovitz, U.S. Natl. Mus. Bull., 246:11, 1966, to skin of *Inia geoffrensis* Blainville, accompanied by mismatched skull of *Steno bredanensis* Lesson.

1828. *Delphinus bredanensis* Lesson, Histoire naturelle . . . des mammifères et des oiseaux découverts depuis 1788, cétacés, p. 206. Type locality, European seas.

1955. *Steno bredanensis,* Ellerman and Morrison-Scott, Checklist of Palaearctic and Indian mammals, 1758–1946, p. 734, November 19.

1829. *Delphinus planiceps* Van Breda, Verhandl. Nederlands Inst., p. 235, Pl. 1, 2—*Delphinus planiceps* Schlegel, 1841, Abh. Gebiete Zool., 1:27, Pl. 4, Fig. 8. Type locality, Dutch Coast.

1836. *Delphinorhynchus santonicus* Lesson, Histoire naturelle des mammifères . . . , p. 330. Type locality, Ile d'Aix, River Charente, France.

1841. *Delphinus reinwardtii* Schlegel, Abh. Gebiete Zool., 1:27, Pl. 3, Figs. 2, 3. Type locality, Indian Archipelago.

1841. *Delphinus chamissonis* Wiegmann (or earlier, *fide* Hershkovitz, U.S. Nat. Mus. Bull., 246:17, 1966), Schreber, Die Säugethiere, Pl. 359.

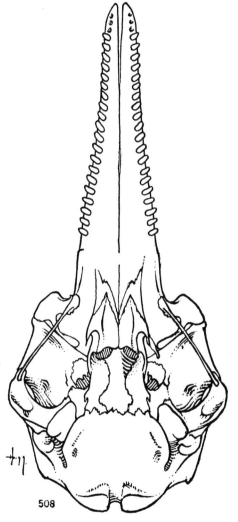

508

Fig. 508. *Steno bredanensis,* outline from a specimen received at Aburatsu, Kyushu (probably a whaling station), Japan, presumably from the North Pacific. After Ogawa (Studien über die Zahnwale in Japan. 4 Mitteilung. Botany and Zoology, theoretical and applied, Tokyo, vol. 4, No. 7 to vol. 5, No. 3, July 1936–March 1937. Fig. b, p. 1686). Details from No. 49628 U.S.N.M., sex unknown, from Tampa, Florida. Ventral view, X ¼.

1846. *Steno compressus* Gray, *in* The zoology of the voyage of H.M.S. Erebus and Terror . . . , 1 (Mamm.):43, Pl. 27. Type locality, South Sea, Antarctic expedition.

1876. *Delphinus (Steno) perspicillatus* Peters, Monatsb. k. Akad. Wiss., Berlin, Pls. 2. Type from S. Atlantic, 32° 29′ S, 2° 1′ W.

See characters of the genus.

RANGE.—Atlantic. Virginia: Norfolk. Florida: *ca.* 1 mi. NNW Rock Island (Layne, 1965:144). Texas: San Luis Pass at W end Galveston Island (Schmidly and Melcher, 1974:457), thence southward to South American waters. Pacific (Perrin and Walker, 1975:905–907). California: Marin County, Stinson Beach. Southward in ocean: 14° 36′ N, 104° 19′ W; 14° 22′ N, 102° 40′ W; 10° 13′ N, 91° 30′ W; 7° 12′ N, 90° 11′ W.

Genus **Stenella** Gray—Dolphins

1864. *Clymene* Gray, Proc. Zool. Soc. London, p. 237, November. Type, *Delphinus euphrosyne* Gray [= *Delphinus styx* Gray]. Not *Clymene* Oken, 1815, a mollusk, nor Lamarck, 1818, a polychaete, nor Savigny, 1822, a polychaete.

1866. *Stenella* Gray, Proc. Zool. Soc. London, p. 213, September. Type, *Steno attenuatus* Gray.

1868. *Clymenia* Gray, Synopsis of the . . . whales and dolphins in the . . . British Museum, p. 6. No type species designated. Not *Clymenia* Munster, 1839, a mollusk.

1868. *Micropia* Gray, Synopsis of the . . . whales and dolphins in the . . . British Museum, p. 6, as a subgenus of *Clymenia.* Type species, *Clymenia stenorhyncha* Gray (= *Delphinus microps* Gray) by virtual tautonomy and monotypy.

1880. *Prodelphinus* Gervais, *in* Van Bénéden and Gervais, Ostéographie des cétacés . . . , p. 604. Type, *Delphinus marginatus* Duvernoy [= *Delphinus styx* Gray].

1934. *Fretidelphis* Iredale and Troughton, Mem. Australian Mus., 6:65. Type species, *Delphinus roseiventris* Wagner (= *Delphinus longirostris* Gray) by original designation.

Total length to 8 ft. Beaked; rostrum long and narrow; rostral portion of intermaxillae convex; pterygoid bones in contact; lateral grooves on palate absent or rudimentary; mandibular symphysis less than one-fifth length of ramus; teeth small, acute, numerous (34–56 in each tooth-row); 68–81 vertebrae.

Some authors including Mitchell (1975:967) have grouped the species of this genus in three categories: (1) Spinner Dolphin, (2) Striped Dolphin, and (3) Spotted Dolphin. These vernacular names are of more than ordinary use owing to the confusion about the technical names properly to be used for species and subspecies. This confusion was noted in 1951 by J. R. Ellerman and T. C. S. Morrison-Scott in their "Checklist of Palaearctic and Indian Mammals (1758–1946)" (British Museum, Natural History) where they remarked (p. 732): "This genus is in chaos. . . ." In 1966 in his indispensable "Catalog of Living Whales (U.S. Nat. Mus. Bull. 246)" Philip Hershkovitz took a necessary step toward an orderly arrangement of kinds and correct application of technical names by listing 43 competing names proposed for species and/or subspecies of *Stenella*. This number of synonyms, junior and senior, was exclusive of the several names in his list of *incertae sedis*, and a number of homonyms. See also the remarks on the spotted dolphin(s) beyond (p. 882).

KEY TO NORTH AMERICAN SPECIES, AND TWO SUBSPECIES, OF THE GENUS STENELLA

1. Rostrum more than 60% of total length of skull. . . . . *S. longirostris*,   p. 880
1′. Rostrum less than 60% of total length of skull.
    2. Narrow black bands from eye to anus and from eye to base of pectoral appendage.
                                                    S. coeruleoalba,   p. 880
    2′. Narrow black bands from eye to anus and from eye to base of pectoral appendage lacking.
        3. Maxillary teeth small, less than 3 mm. in width at alveoli; vertebrae, 79–81. *S. a. attenuata*,   p. 882
        3′. Maxillary teeth relatively large, more than 5 mm. in width at alveoli; vertebrae usually 70 or fewer.
            4. General color-effect purplish-gray dorsally; greatest breadth of skull more than 192.
                5. From Atlantic. . . . . *S. pernettensis*,   p. 883
                5′. Sea ?; color ?; skull breadth? . . . . *S. dubia*,   p. 883
            4′. General color-effect blackish dorsally; greatest breadth of skull less than 192.
                6. General color-effect blackish throughout; known from the Pacific.   *S. a. graffmani*,   p. 882
                6′. Ventral surface whitish; known from the Atlantic. . . . .*S. frontalis*,   p. 882

SPINNER DOLPHIN
Stenella longirostris (Gray)—Spinner Dolphin

1828. *Delphinus longirostris* Gray, Spicilegia zoologica . . . , 1:1. Type locality, unknown (not Cape of Good Hope as ordinarily cited).
1934. *Stenella longirostris*, Iredale and Troughton, Mem. Australian Mus., 6:66, May 4.
1846. *Delphinus microps* Gray, *in* The zoology of the voyage of H.M.S. Erebus and Terror . . . , 1(Mamm.): 42, Pl. 25. Type locality, unknown.
1846. *Delphinus alope* Gray, *in* The zoology of the voyage of H.M.S. Erebus and Terror . . . , 1(Mamm.):Pl. 32. Type locality, unknown.
1846. *Delphinus roseiventris* Wagner, *in* Schreber, Die Säugthiere, 7:Pl. 360, Type locality, unknown.
1866. *Delphinus stenorhynchus* Gray, Catalogue of seals and whales in the British Museum, pp. 240 (*b* under *Delphinus microps*), 396 (name). Type locality, unknown.
1871. *Steno consimilis* Malm, Kongl. Svenska Vetensk. Akad. Handl., 9:104, Pl. 6, Figs. 53*a*, *b* (skull), type from Madagascar.

Total length to approx. 7 ft. Color dark gray dorsally, mottled with light gray; white ventrally, blotched irregularly with grayish. Pectoral fins small; rostrum twice length of cranial portion of skull; palate with traces of lateral grooves; 46–65 (usually 50–52) teeth in each tooth-row; 72 or 73 vertebrae.—**See** addenda.

Perrin (1975:1059–1067, and in earlier papers cited in his terminal listing of literature) has significantly increased our knowledge of *Stenella* in the eastern Pacific. He (*op. cit.*:1066) speculates about the basis of *Delphinus longirostris* Gray 1828 but uses it [*Stenella longirostris* (Gray)] as the specific name of the spinner dolphin. He recognizes four "unnamed subspecies" in the eastern Pacific and diagnoses (*op. cit.*:1063–1065) each as to color, size, shape, and cranial features. Also he describes and maps (*op. cit.*:Fig. 5) the geographic ranges and lists specimens examined. He does not designate type specimens. He does not provide subspecific names because "assignment of trinomials must wait until the question of how many species . . . exist in the world's oceans has been settled," apparently not realizing that subspecific names could remain unchanged should the specific name(s) have to be changed.

RANGE.—PACIFIC (Miller and Kellogg, 1955:656). Known from: María Madre, Tres Marías Islands, off Nayarit (88976 USNM, as *S. microps*); between Panamá and Galapagos Islands. See also paragraph immediately above. ATLANTIC. South Carolina (Leatherwood, *et al.*, 1976:110). Florida: 1 mi. N St. Augustine Inlet (Caldwell and Caldwell, 1975:3); Bunces Pass, near St. Petersburg Beach (*ibid.*); 4 mi. off mainland at Carabelle, between St. George Island on the west and Alligator Point on the east (Lowery, 1974:332). Also west of Bahamas: Turtle Rocks (Varona, 1974:83).

STRIPED DOLPHIN
Stenella coeruleoalba (Meyen)—Striped Dolphin

X ¹/₃₂

1833. *Delphinus coeruleo-albus* Meyen, Nova Acta Acad. Cesar. Leop.-Carol. Nat. Curios, 16(2):610, Pl. 43, Fig. 2. Type from E coast of South America vic. Río de La Plata.

1957. *Stenella coeruleo-albus,* Tomilin, Mammals of eastern Europe and northern Asia, 9(Cetacea): 554, Figs. 96, 97, and 98.

1846. *Delphinus styx* Gray, *in* The zoology of the voyage of H.M.S. Erebus and Terror . . . , 1(Mamm.):39, Pl. 21. Type locality, W coast Africa.

1846. *Delphinus euphrosyne* Gray, *in* The zoology of the voyage of H.M.S. Erebus and Terror . . . , 1(Mamm.):40, Pls. 22, 31 (skull). Type locality, unknown.

1847. *Delphinus holbollii* Nilsson, Skand. Fauna I—Däggdjuren, 596, Lund (*fide* van Bree, Beaufortia, 20 (267):130, February 6, 1973).

1848. *Delphinus lateralis* Peale, U.S. Expl. Exped., Mammalia, p. 35, Pl. 8, Fig. 1. Type locality, 13° 58′ N, 161° 22′ W (*fide* Hershkovitz, Bull. U.S. Nat. Mus., 246:29, February 28, 1966).

1853. *Delphinus tethyos* Gervais, Bull. Soc. Agric. Hérault, 40:150, Pl. 1, Figs. 1–4. Type from Valréas, mouth Orb River, France.

1856. *Delphinus marginatus* Duvernoy, *in* Pucheran, Revue et Mag. Zool., Paris, ser. 2, 8:545, pl. 25. Type locality, near Dieppe, France.

1866. *Tursio dorcides* Gray, Catalogue of seals and whales in the British Museum, p. 400. Type locality, unknown.

1866. *Clymene dorides* Gray, Proc. Zool. Soc. London, p. 214, September. Type locality, unknown.

1868. *Clymene similis* Gray, Proc. Zool. Soc. London, p. 146, Fig. 2, May. Type locality, Cape of Good Hope, S. Atlantic, Republic of South Africa.

1868. *Clymenia euphrosynoides* Gray, Synopsis of the . . . whales and dolphins in the . . . British Museum, p. 6. Type locality, unknown. New name for *Delphinus euphrosyne* Gray 1846.

1871. *Clymenia burmeisteri* Malm, Kongl. Svenska Vetensk. Akad. Handl., 9(2):63, Pl. 6, Figs. 54a, b, type from S. Atlantic off Brazil.

1889. *Prodelphinus petersii* Lütken, Vidensk. Selsk. Skrift. Copenhagen, (6), 5(1):40, 43, type from Indian Ocean.

1893. *Delphinus amphitriteus* Philippi, Anal. Mus. Nac. Chile, (1), Zool., 6:7, Pl. 1, Fig. 3, type from S. Atlantic, 29° 15′, captured in 1851 by R. A. Philippi on his voyage from Hamburg, Germany, to Valparaíso, Chile.

Striped dolphins occur in all major oceans. When studies, yet to be made, reveal how these dolphins should be divided into kinds, possibly into two or more species and almost certainly into subspecies, most of the names needed likely are among the 13 listed immediately above as junior synonyms of *Stenella coeruleoalba* (Meyen, 1833).

Total length to 9 ft. Black dorsally; white ventrally; narrow black band from eye to anus and another from eye to base of flipper; rostrum less than twice length of remainder of skull; 44–50 teeth in each tooth-row; *ca.* 76 vertebrae.

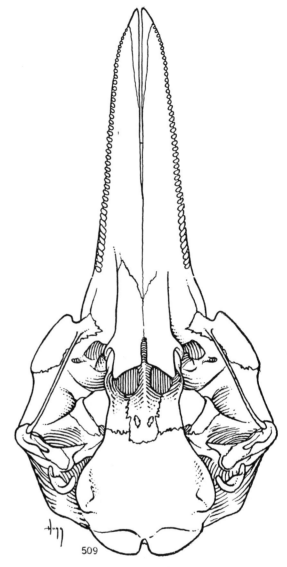

Fig. 509. *Stenella coeruleoalba*, 5 mi. S Westport, Washington, No. 33133 U.S.N.M., sex unknown, after Kenyon and Scheffer (Jour. Mamm., 30:Pl. 1, p. 268, August 17, 1949). Some details from No. 276066 U.S.N.M., sex unknown, from same locality. Ventral view, X ⅔.

RANGE.—Atlantic. Nova Scotia: Halifax (Leatherwood, *et al.,* 1976:113), and possibly S Greenland S to Jamaica, and westward to Florida: Indian Rocks Beach (Caldwell and Caldwell, 1969:612). Louisiana: 4 mi. W Holly Beach (Lowery, 1974:330). Pacific. Bering Sea, thence southward to Oregon: 10 mi. S mouth Columbia River. See also Hubbs, Perrin, and Balcomb (1973:549–552) for sight records in Pacific down coast off México and off Central America (9° 36′ N, 89° 29′ W).

SPOTTED DOLPHIN(S)

In correspondence about the nomenclature for species and subspecies of *Stenella* in North American waters, a junior contemporary pointed out that the names *S. dubia* (G. Cuvier, 1812), *S. frontalis* (G. Cuvier, 1829), *S. attenuata* (Gray, 1846), and *S. plagiodon* (Cope, 1866) apply to the Spotted Dolphin(s) of one, two, or three species.

Both of us suspect there are two species, but the correct names could be, as he put it, "almost any permutation" of the four names listed above, three of which are based on specimens from unknown localities. He added (*in Litt.*), "General usage today, although it may be incorrect, is to apply the name *S. plagiodon* to the large, robust spotted dolphin occurring in the northwestern tropical and subtropical Atlantic, *S. frontalis* to a smaller spotted dolphin found on both sides of the warm N. Atlantic, and *S. attenuata* to spotted dolphins in the South Atlantic, Indian, and Pacific Oceans. Complicating factors are that, if there are two species, each probably occurs on both sides of the Atlantic, and each probably has large, robust and small, slender racial forms. Because postcranial and color pattern characters may be diagnostic of whatever species truly exist, allocation of some of the junior synonyms based on skulls may prove impossible."

Of the 23 synonyms listed below for the four following taxa of spotted dolphins, it is certain that some or most of those names will not apply to dolphins in North American waters. However, considering the opportunity for one or more species to range around the coast of South Africa, westward across the Atlantic, and from the coast of South Africa eastward through the Indian Ocean and into the Pacific Ocean, a trinomen for a North American subspecies conceivably *could* contain any one of the 23 names.

More recently Perrin (1975:1066, Fig. 9) has tentatively identified simply as *S. coeruleoalba* two "stocks" in the eastern Pacific, possibly with a break between the two in waters from 10° to 17° N lat. He remarks that enough specimens now are available from the eastern Tropical Pacific to permit meaningful comparisons with specimens from the eastern Pacific.

Stenella frontalis (G. Cuvier)
Bridled Spotted Dolphin

1829. *D[elphinus]. frontalis* G. Cuvier, Le règne animal . . . , 1:288, April. Type locality, off Cape Verde Islands, North Atlantic Ocean.
1950. *Stenella frontalis*, Fraser, Atlantide Rept. No. 1 (Scientific results of the Danish expedition to the coasts of tropical West Africa (1945–1946), p. 61.

1836. *Delphinus fraenatus* F. Cuvier, Histoire naturelle des cétacés . . . , p. 155, Pl. 10, Fig. 1. Type locality, 30 leagues S Cape Verde Islands, North Atlantic Ocean.
1846. *Delphinus doris* Gray, in The zoology of the voyage of H.M.S. Erebus and Terror . . . , 1(Mamm.):39, Pl. 20. Type locality, unknown.
1850. *Delphinus clymene* Gray, Catalogue of the . . . Mammalia in the . . . British Museum, Cetacea, p. 115. Type locality, unknown.—See addenda.
1860. *Delphinus mediterraneus* Loche, Revue et Mag. Zool., Paris, ser. 2, 12:475, Pl. 22, Fig. 2. Type locality, coast of Algeria.
1866. *Delphinus normalis* Gray, Proc. Zool. Soc. London, p. 214, September. Type locality, unknown.

Total length to at least 7 ft. Color blackish dorsally, shading to grayish laterally; whitish ventrally; sides mottled with blackish. Pectoral appendages small; rostrum less than twice length of cranial portion of skull; 29–34 teeth in each upper jaw and 33–36 in each lower jaw (according to Leatherwood, *et al.*, 1976:108); *ca.* 70 vertebrae.

RANGE.—Atlantic. North Carolina: Fort Macon. Texas: Yarbrough Pass (Schmidly, *et al.*, 1972:625).

Stenella attenuata (Gray)
Gray's Spotted Dolphin

Total length 6–8 ft. Color blackish dorsally, ash-gray ventrally including pectoral appendages, mottled dorsally with grayish spots. Rostrum less than twice length of cranial portion of skull; 35–47 teeth in each tooth-row; 70–81 vertebrae.

Stenella attenuata attenuata (Gray)

1846. *Steno attenuatus* Gray, in The zoology of the voyage of H.M.S. Erebus and Terror . . . , 1(Mamm.):44. Type locality, unknown. Regarded as a synonym of *Stenella dubia* (G. Cuvier) by Hershkovitz, U.S. Nat. Mus. Bull., 246:32, February 28, 1966.
1934. *Stenella attenuata*, Iredale and Troughton, Mem. Australian Mus., 6:66, May 4.

RANGE.—Known from Honolulu, Hawaii, may inhabit waters farther east.

Stenella attenuata graffmani (Lönnberg)

1934. *Prodelphinus graffmani* Lönnberg, Arkiv för Zool., 26A(19):1, April 24, type from off beach approx. 20 mi. N Acapulco, Guerrero.
1940. *Stenella graffmani*, Kellogg, Nat. Geog. Mag., 77(1):85, January.

RANGE.—Nayarit: María Magdalena Island (Van Gelder, 1960:15). Guerrero: type locality. Panamá (Miller and Kellogg, 1955:656): Islas Secas, Golfo de Chiriquí; Golfo de Panamá.

Stenella pernettensis
Blainville's Spotted Dolphin

Total length to 7½ ft. Color purplish-gray dorsally, mottled with white or light gray, shading to grayish laterally; whitish ventrally, mottled with dark gray. Pectoral appendages large; rostrum less than twice length of cranial portion of skull; 34–37 teeth in each tooth-row; 68–69 vertebrae.

As of June 30, 1977, I am aware [Bull. Zool. Nomencl., vol. 31 (pt. 1):44–48, 1974] (1) that the International Commission on Zoological Nomenclature was requested to suppress the names *Delphinus pernettensis* de Blainville 1817, and *Delphinus pernettyi* Desmarest 1820, and (2) that the Commission (*ser. cit.*:33:157, 158, opinion 1067, March 1977) granted the request. *S. pernettensis* is retained in this volume because the law of priority was adopted (see p. 1 of preface in this work) for competing names and because suppression of *D. pernettensis* could result in a multiplication of names if and when study of actual specimens reveals the existence of two or more subspecies of a single species. Apparently the Commission did not consider the needless multiplication of names, and may wish to revise the ruling in opinion 1067.

Stenella pernettensis plagiodon (Cope)

1866. *Delphinus plagiodon* Cope, Proc. Acad. Nat. Sci. Philadelphia, 18:296. Type locality, unknown.
1940. *Stenella plagiodon*, Kellogg, Nat. Geog. Mag., 77(1):83, January.

RANGE.—Atlantic. North Carolina: Cape Hatteras and Texas: channel in front of Port Aransas, thence southward to South American waters (Miller and Kellogg, 1955:657).

Stenella dubia G. Cuvier
Cuvier's Spotted Dolphin

1812. *D[elphinus]. dubius* G. Cuvier, Ann. Mus. Hist. Nat., Paris, 19:9–10.
1966. *Stenella dubia*, Hershkovitz, U.S. Nat. Mus. Bull., 246:31, February 28.
1826. *Delphinus malayanus* Lesson, Voyage autour du monde sur la Corvette . . . La Coquille, Zoologie, 1(1):184, Atlas, Pl. 9, Fig. 5. Type locality, Karimata Strait, between Java and Borneo.
1829. *D[elphinus]. velox* Cuvier, Le règne animal . . . , ed. 2, 1:288. Type locality, Ceylon.
1840. *Delphinus pseudodelphis* Wiegmann (1840 or earlier), *in* Schreber, Die Säugthiere, Pl. 358 (no text; no type locality; possibly a *nomen nudum*).
1841. *Delphinus loriger* Wiegmann (1841?), *in* Schreber, Die Säugthiere, Pl. 362 (no text; no type locality; possibly a *nomen nudum*).
1845. *Delphinus rappii* Reichenbach, Synopsis Mammalium,

Leipzig, 1:12, Pl. 18, Fig. 57. Name based on *D. capensis* Rapp (not Gray), type being from Cape of Good Hope, Republic of South Africa.
1846. *Delphinus brevimanus* Wagner, *in* Schreber, Die Säugthiere, 7:Pl. 361, Fig. 2. Type locality, uncertain.
1850. *"Delphinus? microbrachium"* Gray, Catalogue of the . . . Mammalia in the British Museum, Cetacea, p. 119, type from unknown locality.
1865. *Delphinus asthenops* Cope, Proc. Acad. Nat. Sci. Philadelphia, pp. 200, 201. Types, two skulls (Nos. 595, 596) in Acad. Nat. Sci., from an unknown locality.
1865. *Delphinus crotaphiscus* Cope, Proc. Acad. Nat. Sci. Philadelphia, pp. 200, 203. Type, a skull, Mus. Peabody Acad., Salem, Massachusetts; presented by F. W. Putnam. Type locality, unknown.
1866. *Clymene punctata* Gray, Proc. Zool. Soc. London, 1865:738. Regarded as "a synonym of *attenuatus*" by Hershkovitz, U.S. Nat. Mus. Bull., 246:34, February 28, 1966.
1871. *Clymenia esthenops* Gray, Suppl. Catalogue of seals and whales in the British Museum, p. 72. Possibly only a misspelling of *asthenops*.
1874. *Clymenia aesthenops* Dall, *in* Scammon, Marine mammals of the north-western coast of North America . . . , p. 267, type "skull only, Mus. Peabody Academy of Science, Salem, Mass." Probably only a misspelling of *asthenops*.

Length up to 7 ft. Dorsum dark, marked by pale spots; underparts whitish. Skull recorded (Hershkovitz, 1966:31) as in the "Paris Museum." When the skull is identified as, or arbitrarily assigned to, a particular kind of dolphin, the name *Stenella dubia* as a specific name will logically serve as a base from which to start in arranging subsequently proposed names (specific, subspecific, and mere junior synonyms) in a zoologically meaningful order for the spotted dolphin(s) and so clear away the existing "chaos" referred to by Ellerman and Morrison-Scott (1951:732) and mentioned above in the account of the genus *Stenella*.

RANGE.—Not certainly known at present, but when a natural classification is worked out for taxa within the genus, the name *dubia* may appear as part of the trinomen or binomen for one or more kinds of *Stenella* that occur in North American waters.

Genus Delphinus Linnaeus—Dolphins

1758. *Delphinus* Linnaeus, Syst. nat., ed. 10, 1:77. Type, *Delphinus delphis* Linnaeus.
1846. *Rhinodelphis* Wagner, *in* Schreber, Die Säugthiere, 7:viii, 281, 316 [included many species], type, *Delphinus delphis* Linnaeus, designated by G. M. Allen, Bull. Mus. Comp. Zool., 83:250, 1939.
1880. *Eudelphinus* Van Bénéden and Gervais, Ostéographie des cétacés, p. 600. Type, *Delphinus delphis* Linnaeus by original designation on p. 601.

Total length to 8½ ft. Beaked; rostrum long and narrow; palate with 2 deep lateral grooves;

pterygoid bones in contact; mandibular symphysis less than one-fifth length of ramus; black ring encircling eye; teeth small, acute, numerous (47–65 in each tooth-row); 73–76 vertebrae.

Delphinus delphis
Common Dolphin

X ¹/₂₈

Total length to 8 ft. Color blackish dorsally, including dorsal fin, flukes and dorsal surface of pectoral appendages; sides shading to grayish-green; mixed with elongated elliptical bands of whitish on flanks; whitish ventrally; whitish band over "forehead" with narrow black band in center that connects black eye-rings; black band from snout to leading edge of pectoral appendage frequently present.

Delphinus delphis delphis Linnaeus

1758. [*Delphinus*] *delphis* Linnaeus, Syst. nat., ed. 10, 1:77. Type locality, European seas.
1804. *Delphinus vulgaris* Lacépède, Histoire naturelle des cétacées . . . , pp. xlii, 250, Pl. 13, Fig. 1, Pl. 14, Fig. 1. Type locality, all seas.
1828. *Delphinus capensis* Gray, Spicilegia zoologica, 1:2, Pl. 2, Fig. 1, type from Cape of Good Hope, Republic of South Africa.
1830. *Delphinus novae-zelandiae* Quoy and Gaimard, Voyage . . . de l'Astrolabe, Zoologie, 1:149, Pl. 28. Type locality, near Cape Gable, not far from Tolaga Bay, New Zealand.
1846. *Delphinus fulvifasciatus* Wagner, *in* Schreber, Die Säugthiere, 7:Pl. 361, Fig. 1, name based on the "dauphin à bande fauve" of Hombron and Jaquinot, 1842–1853, Voyage au Pôle Sud . . . de l'Astrolabe et la Zélée, Zoologie, Pl. 21, Fig. 1. According to Pucheran, 1853, Voyage au Pôle Sud . . . de l'Astrolabe et la Zélée, Zoologie, 3:37 (Atlas, Pl. 21, Fig. 1), the type is from Hobart Town, Tasmania.
1846. *Delphinus janira* Gray, *in* The zoology of the voyage of H.M.S. Erebus and Terror . . . , 1(Mamm.):41, Pl. 23. Type locality, Newfoundland.
1846. *Delphinus forsteri* Gray, *in* The zoology of the voyage of H.M.S. Erebus and Terror . . . , 1(Mamm.):42, Pl. 24. Type locality, between New Caledonia and Norfolk Island.
1848. *Delphinus albimanus* Peale, Mammalia and ornithology, *in* U.S. Expl. Exped. . . . , 1838–1842 . . . , Philadelphia, 8:33, Pl. 7, Fig. 1. Type locality, coast of Chile, lat. 27° 16′ S, long. 75° 30′ W.
1850. *Delphinus sao* Gray, Catalogue of the . . . Mammalia in the . . . British Museum, Cetacea, p. 125. Type locality, off Madagascar.

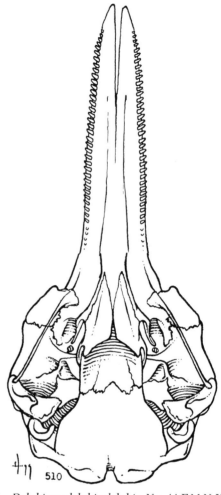

510

Fig. 510. *Delphinus delphis delphis*, No. 44 F.M.N.H., after Elliot (Field Columb. Mus., Publ. 95, Zool. Ser., 4(1):Pl. 24, 1904). Details from No. 21524 U.S.N.M., ♀, from Woods Hole, Massachusetts. Ventral view, X approx. ⅓.

1859. *Delphinus frithii* Blyth, Jour. Asiatic Soc. Bengal, 28:492, type from between England and India.
1860. *Delphinus algeriensis* Loche, Revue et Mag. Zool., Paris, 12:474, Pl. 22, Fig. 1. Type locality, coast of Algeria.
1866. *Delphinus moorii* Gray, Proc. Zool. Soc. London, for 1865, p. 736, Fig. 1, April. Type locality, South Atlantic, lat. 34° S, long. 7° 3′ W.
1866. *Delphinus walkeri* Gray, Proc. Zool. Soc. London, for 1865, p. 737, Fig. 2, April. Type locality, South Atlantic, lat. 35° 38′ S, long. 0° 10′ E.
1866. *Delphinus pomeegra* Owen, Trans. Zool. Soc. London, 6:23, Pl. 6, Fig. 3, August 15. Type locality, off coast of Madras, India.
1866. *Delphinus major* Gray, Catalogue of seals and whales in the British Museum, p. 396. Type locality, unknown.
1868. *Delphinus marginatus* Lafont, Actes Soc. Linnéene Bordeaux, 26:518. Type locality, Arcachon, France. Not Duvernoy, 1856.
1873. *Delphinus bairdii* Dall, Proc. California Acad. Sci., 5:12, January 29, type from Cape [now Point] Arguello,

Santa Barbara Co., California. Here arranged as a junior synonym on basis of information provided by van Bree and Purves, Jour. Mamm., 53:372–374, June 23, 1972.

1880. *Eudelphinus tasmaniensis* Van Bénéden and Gervais, Ostéographie des cétacés, p. 604, Pl. 39, Fig. 9, type from Hobart Town, Tasmania.

1881. *Delphinus* [*delphis* var.] *fusus* Fischer, Actes Soc. Linnéenne Bordeaux, (4), 35:127, 128, 218, Pl. 4, Fig. 1, type from Arcachon, Bay of Biscay.

1881. *Delphinus delphis souverbianus* Fischer, Actes Soc. Linnéenne Bordeaux, 35:127. Type locality, Arcachon, France.

1881. *Delphinus delphis variegatus* Fischer, Actes Soc. Linnéenne Bordeaux, 35:127. Type locality, Arcachon, France.

1881. *Delphinus delphis balteatus* Fischer, Actes Soc. Linnéenne Bordeaux, 35:127. Type locality, Arcachon, France.

1881. *Delphinus delphis moschatus* Fischer, Actes Soc. Linnéenne Bordeaux, 35:127. Type locality, Arcachon, France.

1883. *Delphinus delphis* var. *curvirostris* Riggio, Naturalista Siciliano, 2:158. Type locality, possibly Mediterranean Sea.

RANGE.—Atlantic segment. "Small gams . . . appear off the edge of the Grand Banks and Flemish Cap in July to September. . . . They frequent the warmed 'slope water' between the Labrador and Gulf currents at temperatures between 11° and 17°C. This is the northern limit of their range in the western Atlantic Ocean. . . . A single atypical specimen was taken in Trinity Bay, Newfoundland." (Banfield, 1974:257, 258.) Nova Scotia (Banks and Brownell, 1969:269). Massachusetts: Woods Hole, thence southward to the Caribbean Sea and South American waters. Pacific segment. British Columbia: Victoria (Guiguet, 1955:136). California: La Jolla (Banks and Brownell, 1969:265, 266). Costa Rica: 8° 14′ N, 84° 17′ W (*op. cit.:* 270), thence southward to waters off Perú (*ibid.*).

Genus **Lagenodelphis** Fraser
Fraser's Dolphin

1956. *Lagenodelphis* Fraser, Sarawak Mus. Jour., 7(8, n.s.):496, type *Lagenodelphis hosei* Fraser.

Length to 8 ft. 10½ in. Tympanoperiotic bones as in *Lagenorhynchus;* premaxillae fused on midline dorsally in the manner seen in *Delphinus* but not in same extent (after Perrin, *et al.*, 1973:346). The skull possesses features of both *Delphinus* and *Lagenorhynchus* and is intermediate also in some respects in shape of body. The genus is monotypic.

Lagenodelphis hosei Fraser
Fraser's Dolphin

1956. *Lagenodelphis hosei* Fraser, Sarawak Mus. Jour., 7(8, n.s.):496, type from mouth Lutong River, Baram, Sarawak.

Length to 8 ft. 10½ in. Basal width of rostrum *ca.* 53 per cent of its length; 78–80 vertebrae; post-temporal fossae suboval; prominent dark gray lateral stripe from near eye to anus; dorsum grayish-blue with brownish hue; dorsal fin weakly falcate (see Perrin, *et al.*, 1973:345–350, who also describe variation according to age). The species is monotypic.

RANGE (Perrin, *et al.*, 1973:345–350).—No specimen yet (1975) saved from North American waters. The specimen nearest thereto is from *ca.* 800 km. W Cocos Island (5° N, 95° 45′ W). Other specimens are from farther west in tropical waters of the Pacific and another is from the Indian Ocean side of the Cape region of Africa. Sight records off the Pacific Coast of Central America warrant listing the species here.

Genus **Tursiops** Gervais—Bottle-nosed Dolphins

Reviewed by True, Ann. Durban Mus., 1(1):10–24, June 1, 1914.

1843. *Tursio* Gray, List of the . . . Mammalia in the . . . British Museum, pp. xxiii, 105, May 13. Type, *Delphinus truncatus* Montague. Not *Tursio* Wagler, 1830, nor *Tursio* Fleming, 1822.

1855. *Tursiops* Gervais, Histoire naturelle des mammifères . . . , 2:323. Type, *Delphinus truncatus* Montague, a renaming of *Tursio* Gray.

Total length to 12 ft. Rostrum long, tapering; vomer visible in median line of palate; pterygoid bones in contact; mandibular symphysis less than one-fifth length of ramus; postorbital process of frontal triangular; teeth large, smooth, 19–26 in each tooth-row; 61–64 vertebrae.

Study of adequate material of *Tursiops* from the several parts of its range probably would reveal a number of species different from the number now recognized.

KEY TO NORTH AMERICAN SPECIES OF TURSIOPS

1. Portion of parietal bone forming border of temporal fossa broad throughout; mandibular condyle less than half as long as height of mandible at coronoid process; known from the Atlantic.   *T. nesarnack,*   p. 886
1′. Portion of parietal bone forming border of temporal fossa distinctly narrowed ventrally; mandibular condyle half as long as height of mandible at coronoid process; known from the Pacific. . . . . . .*T. gillii,*   p. 887

Tursiops nesarnack
Lacépède's Bottle-nosed Dolphin

X 1/48

Total length to 12 ft. Color purplish-gray to clear gray dorsally, whitish ventrally; sides light gray. Posterior end of vomer narrow and rectangular; posterior margins of orbital plates of maxillae rounded; portion of parietal bone forming border of temporal fossa broad throughout; mandibular condyle less than half as long as height of mandible at coronoid process; 20–26 teeth in each upper jaw and 18–24 in each lower jaw.

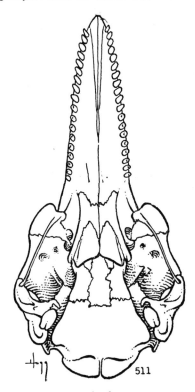

Fig. 511. *Tursiops nesarnack aduncus*, approximate locality, 12° N, 120° W, Pacific, No. 35045 A.M.N.H., after Andrews (Bull. Amer. Mus. Nat. Hist., 30:Pl. 10, August 26, 1911). Details from No. 258642 U.S.N.M., ♂, from Charles Island, Post Office Bay, Galapagos Islands. Ventral view, X ⅕.

Tursiops nesarnack aduncus (Ehrenberg)

1832. *Delphinus aduncus* Ehrenberg, *in* Hemprich and Ehrenberg, Symbolae Phisicae Mammalia, 2:ftn. 1 (last page of fascicle headed "*Herpestes leucurus* H. E.").

1841. *Delphinus hamatus* Wiegmann, 1841 (or earlier), *in* Schreber, Die Säugthiere, 7:Pl. 369. Type from Red Sea of Indian Ocean.

1842. *Delphinus abusalam* Rüppell, Mus. Senckenbergianum (1845), 3:140, Pl. 12. Type from Red Sea, Indian Ocean.

1848. *Delphinus perniger* Blyth, Jour. Asiatic Soc. Bengal, 17:249, 250. Type from Bay of Bengal, Indian Ocean.

1862. *Tursiops catalania* Gray, Proc. Zool. Soc. London, 1862:143. Type locality, off Cape Melville, northeastern Queensland.

1866. *Steno? gadamu* Gray, Catalogue of the seals and whales in the British Museum, p. 394, a prior publication of Owen's MS. See Owen, Trans. Zool. Soc. London, 6:17, Pls. 3, 4. Type from Vizagapatam, Madras.

1874. *Delphinus caerulescens* Giglioli, Zoologia Magenta cetacei osservati durante il viaggio intorno al globo della R. pirocorvetta Magenta, 1865–1868, p. 88. Type locality, entrance to Van Diemen (= Osumi) Strait between Kyushu and Tanega islands, Japan. Possibly belongs here.

1886. *Delphinus salam* [*sic*] Van Bénéden, Ann. Mus. Hist. Nat. Belgique, Paléontologie, 13:111. Probably a *lapsus* for *Delphinus abusalam* Rüppell.

1903. *Tursiops fergusoni* Lydekker, Jour. Bombay Nat. Hist. Soc., 15:41, Pl. 3. Type (an immature of *T. catalania*) from Trivandrum, Travancore, India.

1908. *Tursiops gephyres* Lahille, Anal. Mus. Nac. Buenos Aires, 16:347. Type from Punta Lara, Río de La Plata, Argentina.

1909. *Tursiops dawsoni* Lydekker, Proc. Zool. Soc. London, 1908:802–806. Type from off Trivandrum, southern India. Possibly belongs here.

1911. *Tursiops nuuanu* Andrews, Bull. Amer. Mus. Nat. Hist., 30:233, Pl. 10. Type from 12° N, 120° W, Santa Catalina Island, Gulf of California, Mexico.

1934. *Tursiops maugeanus* Iredale and Troughton, Mem. Australian Mus., 6:68. Type from Tamar River, Tasmania. According to Hershkovitz, U.S. Nat. Mus. Bull., 246:50, 1966, based on *Tursiops tursio*, Scott and Lord (not Fabricius), 1919.

In addition to the above names thought to be junior synonyms, Hershkovitz, U.S. Nat. Mus. Bull., 246:48–51, 1966, lists other names applied to this subspecies. These other names, as he indicates, appear to be misidentifications or homonyms.

RANGE.—Baja California (Hershkovitz, 1966:51), thence southward into South America.

Tursiops nesarnack nesernack Lacépède

1789. *Delphinus tursio* Bonnaterre, Tab[leau] Encyclopédique et Méthodique des trois Régnes de la Nature, . . . Cétologie, Paris, p. 21, Pl. 11. A homonym of *Delphinus tursio* Fabricius 1780 [= *Phocoena phocoena* (Linnaeus) 1758, see beyond].

1804. *Delphinus nesarnack* Lacépède, Histoire naturelle des cétacées. . . , pp. xliii, 307, Pl. 15, Fig. 2, type from N. Atlantic. The basis clearly is the account by Bonnaterre, 1789:21 (*supra*).

1961. *Tursiops nesarnack*, Hershkovitz, Fieldiana: Zool., 39:550, July 25.

1821. *Delphinus truncatus* Montagu, Mem. Wernerian Nat. Hist. Soc., 3:75, Pl. 3. Type locality, Duncannon Pool, near Stoke Gabriel, approx. 5 mi. up River Dart, Devonshire, England.

1828. *Delphinus compressicauda* Lesson, Histoire naturelle . . . des mammifères et des oiseaux découverts depuis 1788, cétacés, p. 272. Type locality, Atlantic Ocean, near equator, lat. 04° S, long. 26° W.

1846. *Delphinus eurynome* Gray, *in* The zoology of the voyage of H.M.S. Erebus and Terror . . . , 1(Mamm.):38, Pl. 17. Type locality, unknown.

1846. *Delphinus metis* Gray, *in* The zoology of the voyage of H.M.S. Erebus and Terror . . . , 1(Mamm.):38, Pl. 18. Type locality, unknown.

1846. *Delphinus cymodoce* Gray, *in* The zoology of the voyage of H.M.S. Erebus and Terror . . . , 1(Mamm.):38, Pl. 19. Type locality, unknown.

1862. *Delphinus tursio obtusus* Schlegel, Nat. Hist. van Nederland, Zoogdieren, Pl. 13. Type from N. Atlantic, Netherlands.

1865. *Delphinus erebennus* Cope, Proc. Acad. Nat. Sci. Philadelphia, p. 281. Type from Red Bank, opposite Philadelphia, Pennsylvania.

1884. *Tursio subridens* Flower, Proc. Zool. Soc. London for 1883:481, based on papier-mâché casts. Type locality, unknown.

1886. *Tursiops parvimanus* Van Bénéden, Ann. Mus. Hist. Nat. Belge, Paléontologie, 13:100. Type, a mounted skin in Copenhagen Museum; type locality, unknown.

1940. *Tursiops truncatus ponticus* Barabash-Nikiforov, Fauna kitsobraznykh chernogo Moria, Izd. Voronezh. Gas. Un-ta, p. 56, Figs. 14, 15. Type from Novorosisk, Black Sea.

RANGE.—Atlantic. Newfoundland: Trinity Bay (Banfield, 1974:258). Massachusetts: Cape Cod. Texas: Aransas Pass. Cuba (Varona, 1974:85). Puerto Rico (*ibid.*). Lesser Antilles: St. Vincent (Caldwell, *et al.*, 1971:305); Barbados (Varona, 1974:85), thence southward to South American waters.

Tursiops gillii Dall
Gill's Bottle-nosed Dolphin

1873. *Tursiops gillii* Dall, Proc. California Acad. Sci., 5:13, January 29, type from Monterey, California.

Total length to 12 ft. Color grayish-black dorsally, whitish ventrally. Posterior end of vomer narrow and rectangular; posterior margins of orbital plates of maxillae rounded; portion of parietal bone forming border of temporal fossa narrowed ventrally; mandibular condyle half as long as height of mandible at coronoid process.

RANGE.—Pacific. California: Possibly San Francisco Bay (Orr, 1963:424); type locality, thence southward to Baja California: San Bartolomé Bay. Known also from Golfo de California from Sonora: Puerto Peñasco.

Genus **Lissodelphis** Gloger
Right-whale Dolphins

1841. *Lissodelphis* Gloger, Gemeinnütziges Hand- und Hilfs-buch der Naturgeschichte, 1:169. Type, *Delphinus peronii* Lacépède.

1830. *Tursio* Wagler, Natürliches System des Amphibien . . . , p. 34. Type, *Delphinus peronii* Lacépède. Not *Tursio* Flemming, 1822.

1861. *Leucorhamphus* Lilljeborg, Upsala Univ. Arsskrift, 4:5. Type, *Delphinus peronii* Lacépède.

Total length to approx. 8 ft. Color black dorsally, whitish ventrally, well-defined border between dorsal and ventral colors. Beaked; rostrum long, broad, and tapering; rostral portion of intermaxillae flattened; pterygoid bones not in contact, their inner margins parallel; mandibular symphysis less than one-fifth length of ramus; teeth small, acute, numerous (43–45 in each tooth-row); dorsal fin absent; prominent caudal ridge.

A single species, *L. borealis*, is known from North American waters.

Lissodelphis borealis (Peale)
Northern Right-whale Dolphin

X ¹/₃₂

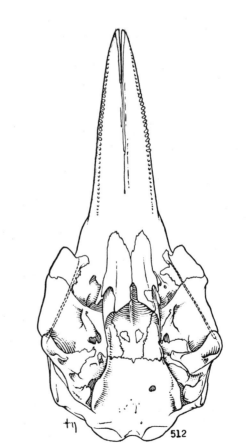

Fig. 512. *Lissodelphis borealis*, 100 mi. off Cape Mendocino, California, No. 8160 U.S.N.M., ♂. Ventral view, X ¼.

1848. *Delphinapterus borealis* Peale, Mammalia and ornithology, *in* U.S. Expl. Exped. . . . , 1838–1842 . . . ,
Philadelphia, 8:35. Type locality, North Pacific Ocean, lat.
46° 06′ 50″ N, long. 134° 05′ W, approx. 500 mi. off mouth
Columbia River, Oregon.
1901. [*Lissodelphis*] *borealis*, Elliot, Field Columb. Mus.,
Publ. 45, Zool. Ser., 2:30, March 6.

See characters of the genus. A specimen from
Washington weighed 124 pounds.

Possibly *L. borealis* is a subspecies of *L.
peronii* Lacépède, named in 1804 from Tasman-
ian waters. Hershkovitz (1966:59) notes that
"Nishiwaki (1957, Coll. Rep. Fish. Sci. Univ. Tokyo) records both *Lissodelphis peronii* and *L. p.
borealis* Peale as occurring in Japanese waters."

RANGE.—Pacific. Bering Sea, thence southward to
California: San Diego Bay.

Genus **Lagenorhynchus** Gray
White-sided Dolphins

1846. *Lagenorhynchus* Gray, Ann. Mag. Nat. Hist., ser. 1,
17:84, February. Type, *Lagenorhynchus albirostris* Gray.
1866. *Electra* Gray, Catalogue of seals and whales in the
British Museum, p. 268. Type, *Lagenorhynchus electra*
Gray.
1866. *Leucopleurus* Gray, Proc. Zool. Soc. London, p. 216,
September. Type, *Delphinus leucopleurus* Rasch [= *Delphinus acutus* Gray].
1866. *Sagmatias* Cope, Proc. Acad. Nat. Sci. Philadelphia,
18:294. Type, *Sagmatias amblodon* Cope.

Total length to 10 ft. Beak short but distinct;
rostrum broad, approx. as long as cranial portion
of skull; rostral portion of premaxillae flat or
slightly convex; pterygoid bones in contact or
separate; mandibular symphysis less than one-
fifth length of ramus; spines and transverse processes of vertebrae long and slender; teeth variable in size, 22–45 in each tooth-row; 73–92
vertebrae.

KEY TO NORTH AMERICAN SPECIES OF
LAGENORHYNCHUS

1. Interorbital breadth less than 200; 74 vertebrae; known from the Pacific.
                         L. obliquidens,   p. 889
1′. Interorbital breadth more than 200; 78 or
   more vertebrae; known from the Atlantic.
   2. Black circle around eye; 88–92 verte-
      brae. . . . .*L. albirostris*,   p. 889
   2′. Black circle around eye lacking; 78–82
      vertebrae. . . . .*L. acutus*,   p. 888

Lagenorhynchus acutus (Gray)
Atlantic White-sided Dolphin

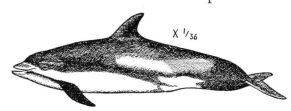

1828. *Delphinus* (*Grampus*) *acutus* Gray, Spicilegia zoologica
. . . , 1:2. Type locality, unknown.
1846. *Lagenorhynchus acutus* Gray, *in* The zoology of the
voyage of H.M.S. Erebus and Terror . . . , 1(Mamm.):36.
1841. *Delphinus eschrichtii* Schlegel, Abh. Gebiete Zool.,
1:23, Pls. 1 and 4. Type locality, Faeroe Islands, Great Britain.
1843. *Delphinus leucopleurus* Rasch, Nytt. Mag. Naturv.,
4:100. Type locality, Gulf of Christiana, Norway.
1868. *Leucopleurus arcticus* Gray, Synopsis of the . . .
whales and dolphins in the . . . British Museum, p. 7. Type
locality, North Sea.
1876. *Lagenorhynchus perspicillatus* Cope, Proc. Acad. Nat.
Sci. Philadelphia, 28:136, type from Woods Hole or Cape
Cod, Massachusetts.
1876. *Lagenorhynchus gubernator* Cope, Proc. Acad. Nat. Sci.
Philadelphia, 28:138, type from Casco Bay, near Portland,
Maine.

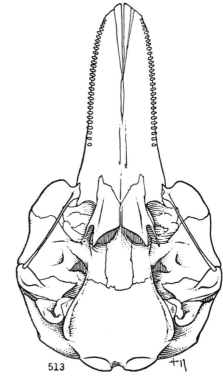

Fig. 513. *Lagenorhynchus acutus*, No. 45 F.M.N.H., after
Elliot (Field Columb. Mus., Zool. Ser., 2:Pl. 8, 1901). Ventral
view, X 1/2.7.

Total length to 9 ft. Color grayish dorsally, whitish ventrally, yellowish on flanks; frequently a narrow black band from base of flukes along side to level of dorsal fin; black circle around eye. Premaxillae flat, their outer margins sinuate; temporal fossae elongate; pterygoid bones frequently in contact; 30–37 teeth in each tooth-row; 78–82 vertebrae.

RANGE.—Atlantic. Southern Greenland, thence southward to Massachusetts: Monomoy Island (Waters and Rivard, 1962:118).

Lagenorhynchus albirostris Gray
White-beaked Dolphin

1846. *Lagenorhynchus albirostris* Gray, Ann. Mag. Nat. Hist., ser. 1, 17:84, February. Type locality, Great Yarmouth, England.
1846. *Delphinus pseudotursio* Reichenbach, Naturg. Ins.-und Auslandes, Cetaceen, Pl. 24, Fig. 7 (animal, *D. tursio* of Brightwell, 1846) [name based on the *D. tursio* of Brightwell (not Fabricius)].
1846. *Delphinus ibseni* Eschricht, Kongl. Danske Vidensk. Selsk. Naturv. Math., Afd., 12:297, type from N. Atlantic, Denmark.

Total length to 10 ft. Color grayish-black dorsally, whitish ventrally; base of forehead, beak and mandible light gray; irregular whitish patches frequently on sides; whitish areas on dorsal median line posterior to nasal aperture and anterior to flukes. Premaxillae broad and flat, their outer margins convex; temporal fossae elliptical; pterygoid bones frequently in contact; 26 or 27 teeth in each tooth-row (usually only 22–25 erupted past gumline); 88–92 vertebrae.

RANGE.—Atlantic. Franklin: Davis Strait, and waters off western and southern Greenland, and to Newfoundland.

Lagenorhynchus obliquidens Gill
Pacific White-sided Dolphin

1865. *Lagenorhynchus obliquidens* Gill, Proc. Acad. Nat. Sci. Philadelphia, 17:177. Type locality, Pacific Ocean, near San Francisco, California.
1866. *Delphinus longidens* Cope, Proc. Acad. Nat. Sci. Philadelphia, 18:295. Type from unknown locality (No. 3886 in U.S. Nat. Mus., obtained by U.S. Expl. Exped. 1838–1842).
1955. *Lagenorhynchus ognevi* Slensov, Trans. Inst. Oceanology, Acad. Sci. U.S.S.R., Moscow, 18:60. Type from Nemoro Sea, 15–20 mi. E Kunashir Island, South Kuriles.

Total length to approx. 7 ft. Color blackish dorsally, whitish ventrally, the two colors distinctly separated on sides. Premaxillae rounded, their outer margins weakly sinuate; temporal fossae

large, rounded; pterygoid bones frequently separate; 29–31 teeth in each tooth-row; 74 vertebrae.

L. obliquidens Gill 1865 regarded by Scheffer and Rice (1963:6) as possibly identical with *L. thicolea* Gray 1849, and conspecific *L. cruciger* (Quoy and Gaimard 1824). Consequently, the correct name for this dolphin may be *Lagenorhynchus cruciger obliquidens* or *L. c. thicolea*. At present (1975) the skull of *thicolea* cannot be located; also it was originally reported to be from the "west coast of America" but subsequently as "west coast of N. America."

RANGE.—Pacific. Alaska: Valdez (Swan Point), thence southward to Baja California: Ballenas Bay.

Genus Pseudorca Reinhardt—False Killer Whale

1862. *Pseudorca* Reinhardt, Overs. Kongl. Danske Vidensk. Selsk. Förh., p. 151. Type, *Phocaena crassidens* Owen.

Total length to 18 ft. Blackish throughout. Dorsal fin relatively small, recurved; snout rounded, head flattened. Skull with short, broad rostrum; rostral portion of premaxillae equal to, or greater than, breadth of proximal portion; rostrum truncate distally; pterygoid bones in contact; palatines elongated laterally across optic canal; 8–11 teeth in each tooth-row, circular in cross section; approx. 50 vertebrae. This genus is monotypic.

Pseudorca crassidens (Owen)
False Killer Whale

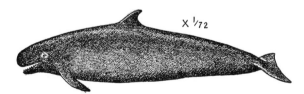

x ¹/₇₂

1846. *Phocaena crassidens* Owen, A history of British fossil mammals and birds, p. 516. Type locality, Lincolnshire Fens, near Stamford, England (subfossil).
1862. *Pseudorca crassidens*, Reinhardt (on title page of reprint of article from Overs. Kongl. Danske Vidensk. Selsk. Förh., pp. 103–152).
1865. *Orca meridionalis* Flower, Proc. Zool. Soc. London, for 1864, p. 420, Figs. 1, 2, May. Type locality, Tasmania.
1866. *Orca destructor* Cope, Proc. Acad. Nat. Sci. Philadelphia, p. 293, October. Type locality, Paita, Perú.
1868. *Globicephalus grayi* Burmeister, Anal. Mus. Pub. Buenos Aires, 1:373, Pl. 21. Type locality, coast of Argentina S LaPlata River.
1882. *Pseudorca? mediterranea* Giglioli, Zoöl. Anzeiger, 5:268–289, type locality, Mediterranean Sea.

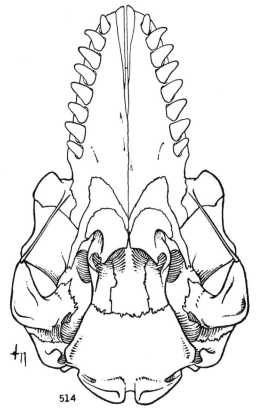

514

Fig. 514. *Pseudorca crassidens*, near Princeton, Florida, No. 218360 U.S.N.M., sex unknown. Ventral view, X ⅛.

See characters of the genus.

RANGE.—Atlantic. North Carolina: Hatteras Village. Texas: 120 mi. SSE Galveston (Schmidly and Melcher, 1974:458). Cuba: Cojímar. Lesser Antilles: St. Vincent (Caldwell, *et al.*, 1971:308), thence into South American waters. Pacific. Washington: Puget Sound, thence southward to Baja California Sur: San José Island (Van Gelder, 1960:19). Guerrero: Acapulco.

Genus **Orcinus** Fitzinger—Killer Whale

1846. *Orca* Gray, Ann. Mag. Nat. Hist., ser. 1, 17:84, February. Type by tautonomy, *Delphinus orca* Linnaeus. Not *Orca* Wagler, 1830.
1860. *Orcinus* Fitzinger, Wiss.-pop. Naturgesch. Säugethiere, 6:204. Type, *Delphinus orca* Linnaeus.
1868. *Ophysia* Gray, Synopsis of the . . . whales and dolphins in the . . . British Museum, p. 8. Type *Orca capensis* Gray.
1870. *Gladiator* Gray, Proc. Zool. Soc. London, 1870:71. Type, *Orca stenorhyncha* Gray (= *Orca gladiator* Gray = *Delphinus orca* Linnaeus); name preoccupied by *Gladiator* Gistl, a genus of Coleoptera.

Total length to 31 ft. (male), largest of delphinids. Black dorsally, patches of white poste-

rior to eye and ventral to dorsal fin; ventral surface white forming trident region posterolaterally. Dorsal fin high, prominent, straight on posterior border; pectoral appendages large, ovate, rounded at distal ends. Rostrum of skull broad, elongate; premaxillae narrowed and flattened; pterygoid bones not in contact; roots of teeth flattened, crowns recurved, 10–13 in each tooth-row; approx. 52 vertebrae, anteriormost 2 or 3 cervicals fused, 7 anterior ribs doubleheaded. The genus is monotypic.

Orcinus orca (Linnaeus)
Killer Whale

X ¹⁄₁₂₀

1758. [*Delphinus*] *orca* Linnaeus, Syst. nat., ed. 10, 1:77. Type locality, European seas.
1899. *Orcinus orca*, Palmer, Proc. Biol. Soc. Washington, 13:24, January 31.
1780. *Delphinus serra* Borowski, Gemeinnüzzige Naturgeschichte des Thierreichs, 2:38. Type locality, Spitzbergen, Davis Strait, etc.; "épée de mer" of Brisson, 1756, Le règne animal, p. 372.
1780. *Physeter microps* Fabricius (not Linnaeus), Fauna Groenlandica, p. 44. Type locality, Greenland seas, N. Atlantic.
1789. *Delphinus gladiator* Bonnaterre, Tableau . . . des trois règnes de la nature, cétologie, p. 23. Type locality, Spitzbergen or Davis Strait or New England Coast.
1792. *D*[*elphinus*] *orca ensidorsatus* Kerr, The animal kingdom, p. 364. Type, the "sword grampus" of the Atlantic, Antarctic, and European seas.
1804. *Delphinus duhamelii* Lacépède, Histoire naturelle des cétacées. . . . , p. xliii, 314. Type locality, mouth Loire, France.
1817. *Delphinus grampus* Blainville, *in* Desmarest, Nouv. Dict. Hist. Nat., 9:168. Name based on the North Atlantic killer whale of authors but primarily on a "grampus" of Hunter.
1846. *Orca capensis* Gray, *in* The zoology of the voyage of H.M.S. Erebus and Terror . . . , 1(Mamm.):39, Pl. 9. Type locality, Cape of Good Hope, Republic of South Africa.
1858. *Delphinus victorini* Grill, Kongl. Svenska Vetensk.-Akad. Handl. Stockholm, 2(10):21, Pl. 1. Type locality, Knysua River, W Cape Town, Republic of South Africa.
1866. *Orcinus eschrichtii* Steenstrup, *in* note by J. Reinhardt, Rec. Memoirs on Cetacea, Ray Soc., p. 188. Type locality, Kollefjord of Strömö, Faeroe Islands, Great Britain.
1866. *Orcinus schlegelii* Lilljeborg, Rec. Memoirs on Cetacea, Ray Soc., p. 235. Type locality, W coast Norway.

1866. *Orca magellanica* Burmeister, Ann. Mag. Nat. Hist., ser. 3, 18:101, Pl. 9, Fig. 5, August. Type locality, S Cape Corrientes, Argentina, mouth Río del Cristiano, lat. 38° 50′ S.

1869. *Orca rectipinna* Cope, Proc. Acad. Nat. Sci. Philadelphia, 21:22. Type locality, coast of California.

1869. *Orca ater* Cope, Proc. Acad. Nat. Sci. Philadelphia, 21:22. Type locality, northwest coast from Oregon to Aleutian Islands.

1869–1871. [*Orca gladiator*] *australis* Gervais, Van Bénéden and Gervais, Ostéographie des cétacés, Atlas, Pl. 47, Fig. 2, p. 540. Type locality, Algoa Bay, Republic of South Africa.

1869–1871. *Orca gladiator* var. *arcticus* Gervais, Van Bénéden and Gervais, Ostéographie des cétacés, Atlas, Pl. 47, Fig. 3, type from Faeroe Islands, N. Atlantic.

1869–1871. *Orca gladiator europaeus* Gervais, Van Bénéden et Gervais, Ostéographie des cétacés, Atlas, Pl. 47, Figs. 4, 5, type (skull in Paris Museum), from (according to Hershkovitz, U.S. Nat. Mus. Bull., 246:85, February 28, 1966) an unspecified locality in N. Atlantic.

1870. *Orca stenorhyncha* Gray, Proc. Zool. Soc. London, p. 71, Figs. 1, 3, June. Type locality, North Sea, Weymouth, England.

1870. *Orca latirostris* Gray, Proc. Zool. Soc. London, p. 76, June. Type locality, North Sea.

1871. *Orca africana* Gray, Supplement . . . catalogue of seals and whales in the British Museum, p. 91. Type locality, Algoa Bay, Republic of South Africa.

1871. *Orca tasmanica* Gray, Supplement . . . catalogue of seals and whales in the British Museum, p. 92. Type locality, Tasmania, Indian Ocean. Name based on "*Orca gladiator* var. *australis*, Gervais [Ostéogr. Cet. t. 47, fig. 1″ (skull)].

1871. *Orca minor* Malm, Kongl. Svenska Vetensk. Akad. Handl., 9:81. Type (a skeleton in Göteborg Museum), from Sweden (restricted to Varberg by Hershkovitz, U.S. Nat. Mus. Bull., 246:85, February 28, 1966), N. Atlantic.

1874. *Orca ater* var. *fusca* Dall, *in* Scammon, Marine mammals of the north-western coast of North America . . . , p. 298, pl. 17, fig. 3. Type locality, coast of California or Oregon.

1875. *O*[*rca*] *pacifica* Gray, Proc. Zool. Soc. London, p. 76, June. Type locality, North Pacific.

1876. *Orca antarctica* P. Fischer, Jour. de Zool., 5:146. Type locality, Powell and South Shetland Islands, Antarctic Ocean.

1904. [*Orcinus orca*] *magellanicus*, Trouessart, Catalogus Mammalium . . . , Suppl., p. 771.

See characters of the genus. The species is here regarded as monotypic.

RANGE.—Atlantic. Coast of Greenland, thence southward to Bahamas: Great Abaco Island (Backus, 1961:418). Lesser Antilles: St. Vincent (Caldwell and Caldwell, 1969:636), westward to Florida: Marineland (Layne, 1965:147). Texas: Port Aransas (Caldwell and Caldwell, 1969:636, sight record only). Pacific. Alaska: Cape Sabine (Childs, 1969:54); St. Matthew and Pribilof islands, thence southward to Baja California.

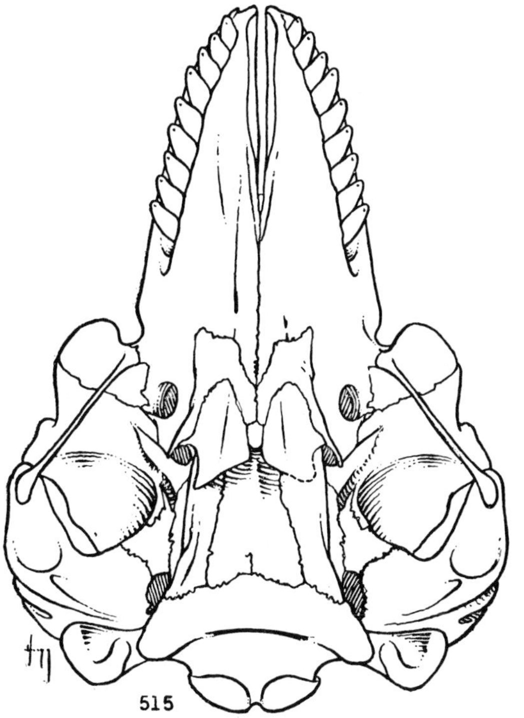

Genus **Grampus** Gray
Grampus

1828. *Grampus* Gray, Spicilegia zoologica, 1:2. Type species [according to Hershkovitz, U.S. Nat. Mus. Bull., 246:87, February 28, 1966], "*Delphinus griseus* Cuvier [1812], by virtual selection (Gray, 1846, The zoology of the voyage of H.M.S. Erebus and Terror . . . , 1(Mamm.):30) and by formal subsequent selection (Reinhardt, 1862, Overs. Kongl. Danske Vidensk. Selsk. Förh., pp. 107, 148)."

1873. *Grayius* Scott, Mammalia, Recent and extinct . . . , p. 104, a renaming of *Grampus* Gray, 1846, but not of Gray, 1828. Not *Grayius* Bonaparte, 1856.

1933. *Grampidelphis* Iredale and Troughton, Rec. Australian Mus., 19(1):31, August 2. Type, *Grampidelphis exilis* Iredale and Troughton.

Total length to 13 ft. Body grayish, paler on ventral surface, pectoral appendages and flukes black. Dorsal fin high, strongly recurved; forehead prominent and bulging. Skull with rostrum expanded anterior to antorbital notches, tapering to obtuse termination anteriorly; area of rostrum anterior to external nares raised, triangular, convex; rostral portion of premaxillae broad and slightly rounded; mandibular symphysis short; pterygoid bones in contact; 2–7 teeth in each mandibular ramus, clustered in region of symphysis; teeth lacking in upper jaws, or if present small and confined to anterior half of ros-

Fig. 515. *Orcinus orca*, coast of California, No. 16487 U.S.N.M., sex unknown. Ventral view, X approx. ⅓.

trum; approx. 68 vertebrae. This genus is monotypic.

Grampus griseus (G. Cuvier)
Risso's Dolphin or Grampus

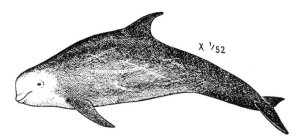

X 1/52

1812. *Delphinus griseus* G. Cuvier, Ann. Mus. Hist. Nat. Paris, 19:14, also uppermost fig. on Pl. 1, at the end of the 17-page article. Type locality, Brest, France. The binomen "*Delphinus griseus*" appears on p. 14 and as a caption immediately above the uppermost of four figs. on Pl. 1 at the end of the 17-page article. Consequently, *Delphinus griseus* has page priority, as well as line priority on Pl. 1, over "*Delphinus aries?*."

1837. *Grampus griseus*, Hamilton, Jardine's Naturalist's Library, Mammalia, 6(Whales):233, Pl. 21.

1812. *Delphinus aries* G. Cuvier, Ann. Mus. Hist. Nat. Paris, 19:12, also lowermost fig. on Pl. 1, at the end of the 17-page article. "*Delphinus aries?*" is a caption immediately above the lowermost of four figs. on Pl. 1, and appears nowhere else in the article. The animal depicted in the lowermost figure is referred to on p. 12 as "1 *aries marinus* de Pliny et Elien."

1822. *Delphinus rissoanus* Desmarest, Encyclopédie méthodique, mammalogie, Suppl., p. 519. Type locality, Nice, Mediterranean Coast of France.

1837. *Globicephalus rissii*, Hamilton, Jardine's Naturalist's Library, Mammalia, 6(Whales):219, Pl. 18. Type, according to Hershkovitz, U.S. Nat. Mus. Bull., 246:88, February 28, 1966, "[name based on Cuvier's 'dauphin de Risso'].—Anonymous, 1838, Chinese repository, Canton 6:411 [S Pacific: China (Lewchew, or Lu-chu, 110° E, 21° N, South China Sea); specimen taken on board the Morrison on its voyage to Lu-chu and Japan]."—According to David H. Johnson, *in verbis* 1965, Lu-chu = Loo Choo = Riu Kiu.

1846. *Grampus cuvieri* Gray, Ann. Mag. Nat. Hist., ser. 1, 17:85. Type locality, Isle of Wight, England.

1846. *Grampus sakamata* Gray, *in* The zoology of the voyage of H.M.S. Erebus and Terror . . . , 1(Mamm.):31. Type locality, Japan.

1850. *Grampus richardsoni* Gray, Catalogue of the . . . Mammalia in the . . . British Museum, Cetacea, p. 85. Type locality, Kalks Bay, Simons Bay, Cape of Good Hope, Republic of South Africa.

1859. *Globiocephalus rissoi* Blyth, Jour. Asiatic Soc. Bengal, 28:481. Type locality, unknown.

1866. *Globiocephalus chinensis* Gray, Catalogue of seals and whales in the British Museum, p. 323. Type locality, near Leuchen, South China Seas.

1873. *Grampus stearnsii* Dall, Proc. California Acad. Sci., 5:13, January, type from Monterey, California.

1881. *Grampus sowerbianus* Fischer, Actes Soc. Linnéenne Bordeaux, 35:210. Type locality, unknown.

1933. *Grampidelphis exilis* Iredale and Troughton, Rec. Australian Mus., 19:32, Pl. 10, Figs. 1–5. Type from Ocean Beach, Manley, Sydney, New South Wales, Australia.

1933. *Grampidelphis kuzira* Iredale and Troughton, Rec. Australian Mus., 19:34. New name for *Grampus sakamata* Van Bénéden and Gervais, 1880, allegedly not of Gray 1846.

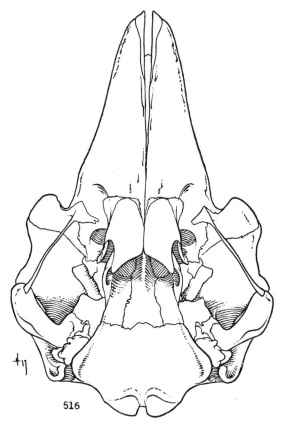

Fig. 516. *Grampus griseus*, outline from photograph of skull No. 27.68 Charleston Museum, sex unknown, from Folly Island, near Charleston, South Carolina. Details from No. 49347 U.S.N.M., sex unknown, from California. Ventral view, X approx. 1/4.4.

See characters of the genus.

RANGE.—Atlantic (Leatherwood, *et al.*, 1976:97, and Paul, 1968:746–748). Eastern Newfoundland; Massachusetts; New Jersey; South Carolina (Charleston); Florida (Tarpon Springs); Bahamas; Virgin Islands. Cuba: Bahía de Matanzas (Varona, 1974:85). Lesser Antilles: St. Vincent (Caldwell, *et al.*, 1971:305). Pacific. British Columbia (southern coast, Orr, 1966:342). Washington: Makkaw Bay, 5 mi. SW Neah Bay (Stroud, 1968:347). California (near Princeton and at Monterey, Orr, 1966:341); Baja California (Paul, 1968:746).

Genus **Globicephala** Lesson
Blackfishes or Pilot Whales

1828. *Globicephala* Lesson, Histoire naturelle . . . des mammifères et des oiseaux découverts depuis 1788, cétacés, p. 441. Type, *Delphinus deductor* Scoresby [= *Delphinus melas* Traill].

1830. *Cetus* Wagler, Natürliches System der Amphibien, p. 33 [new name for *Globicephala* Lesson; preoccupied by *Cetus* Billberg, 1828 (= *Physeter* Linnaeus, 1758)].

1864. *Sphaerocephalus* Gray, Proc. Zool. Soc. London, 1864:244 [subgenus of *Globiocephalus* Gray (= *Globicephala* Lesson); type species: *Globiocephalus incrassatus* Gray by monotypy; name preoccupied by *Sphaerocephalus* Eschscholtz, a beetle].—Flower, 1884, Proc. Zool. Soc. London, 1883:510 [type species: *Sphaerocephalus incrassatus* Gray = *Globiceps* (= *Globicephala*) *melas* Traill].

1884. *Globiceps* Flower, Proc. Zool. Soc. London, 1883:508 [new name for *Globicephala* Lesson; taxonomy; *Sphaerocephalus* Gray 1864 a synonym; name preoccupied by *Globiceps* Lepeletier and Serville, 1825, an insect, and Ayres, 1852, a coelenterate; *Globiceps* Van Bénéden, 1868 a *nomen nudum*].

Total length to 28 ft. Black, some individuals with narrow white stripe from throat to belly; gray area on back immediately posterior to dorsal fin in some individuals of *G. sieboldii.* Head globular, lip with rounded protuberance; mouth oblique; dorsal fin long, low, thick; pectoral appendages narrow and long. Skull with short, broad rostrum; premaxillae covering anterior half of rostrum, or nearly so; symphysis of mandible short; pterygoid bones in contact; teeth confined to anterior half of rostrum, large, 7–12 in each tooth-row; 57–60 vertebrae.

If only two species, each without subspecies, of *Globicephala* are to be recognized, some of the names that appear to be junior synonyms are listed below. Van Bree (1971:85), in abbreviated form, lists additional names that may be junior synonyms.

KEY TO NORTH AMERICAN SPECIES OF
GLOBICEPHALA

1. Pectoral appendages long, approx. $\frac{1}{5}$ of total length of body. `` . . . . .G. melaena, p. 893
1′. Pectoral appendages short, approx. $\frac{1}{6}$ or less of total length of body. G. sieboldii, p. 894

Globicephala melaena (Traill)
Long-finned Pilot Whale

X ¹/₁₁₂

1809. *Delphinus melas* Traill, Jour. Nat. Philos. Chem. Arts, 22:81, Pl. 3, February. Type locality, Scapay Bay, Pomona, Orkney Islands, Scotland.

1898. *Globicephala melaena*, Thomas, The zoologist, ser. 4, 2:99, March. (Feminine of *melas.*)

1812. *Delphinus globiceps* G. Cuvier. Ann. Mus. Hist. Nat. Paris, 19:14, Pl. 1, August. Type locality, Sainte Brieux, France.

1820. *Delphinus deductor* Scoresby, Account . . . Arctic regions . . . , 1:496, a renaming of *Delphinus melas* Traill.

1825. *Delphinus grinda* Lyngbye, Tidsskr. Naturv., 4:232. Type locality, Faeroe Islands, Great Britain.

1827. *Delphinus intermedius* Harlan, Jour. Acad. Nat. Sci. Philadelphia, 6:51, pl. 1, fig. 3, June. Type locality, Salem Harbor, Massachusetts.

1829. *Delphinus harlani* J. B. Fischer, Synopsis mammalium, p. (456) 656. Type locality, coast of New England, a renaming of *Delphinus intermedius* Harlan, owing to the wrong assumption that it was preoccupied by *Delphinus intermedius* Gray, 1827.

1834. *Phocaena edwardii* A. Smith, African Zoology, South African Quart. Jour., (2), 3(2):239, type from Slang-kop, near Cape of Good Hope, Republic of South Africa.

1837. *Globicephalus conductor* Rapp, Die Cetaceen, p. 34 [technical name for *Globicéphale conducteur* of Lesson (1828, Hist. Nat. Cétacées, p. 216) in synonymy of *Delphinus globiceps* Cuvier].

1846. *Globiocephalus affinis* Gray, *in* The zoology of the voyage of H.M.S. Erebus and Terror . . . , 1(Mamm.):32. Type locality, unknown.

1846. *Globiocephalus svineval* Gray, *in* The zoology of the voyage of H.M.S. Erebus and Terror . . . , 1(Mamm.):32. Type locality, E coast North America.

1846. *D[elphinus]. carbonarius* Wagner, *in* Schreber, Die Säugthiere, 7:305, Pl. 352, Fig. 1. Based on *Phocaena* sp. of Bennett, 1840, recorded from the Atlantic, Pacific, and Indian oceans from 50 ° N to 35 ° S, including coast of California and Indian Archipelago.

1862. *Globiocephalus incrassatus* Gray, Proc. Zool. Soc. London, for 1861, p. 309, 2 figs., April, type from Bridgeport, coast of Dorsetshire, England.

1896. *Globicephalus chilensis* Philippi, Anal. Mus. Nac. Chile, Zool. Entr., 12a:7, Pl. 1, Figs. 3, 4. Type from Los Vilos, Coquimbo, and Chiloé Island, Chile.

1933. *Globicephalus ventricosus* Iredale and Troughton, Rec. Australian Mus., 19(1):35, August 2. Not Lacépède, 1804. (See Fraser, Ann. Mag. Nat. Hist., ser. 12, 4:942–944, September, 1951, also Miller and Kellogg, Bull. U.S. Nat. Mus., 205:63, March 3, 1955.)

1939. *Globicephala leucosagmaphora* Rayner, Ann. Mag. Nat. Hist., (11), 4:543. Type from S Cape of Good Hope, Republic of South Africa.

Total length to 28 ft. Atlantic males average "at least 20 feet" (Leatherwood, *et al.*, 1976:91). Color black; in some a narrow white stripe from throat to belly. Dorsal fin recurved, situated anterior to' mid-point of body; pectoral appendages elongate, slender, and pointed, approx. one-fifth length of body. Skull with rostrum longer than its breadth at base; premaxillae large, flat, not projecting over lateral and anterior margin of maxillae as viewed from above; external nares broad, transversely bordered by narrow plates of premaxillae; approx. 10 teeth in each tooth-row; 58–60 vertebrae.

RANGE.—Atlantic. Coast of Greenland, thence southward to North Carolina: Nags Head (485823 USNM).

Globicephala sieboldii Gray
Short-finned Pilot Whale

1846. *Globiocephalus sieboldii* Gray, *in* The zoology of the voyage of H.M.S. Erebus and Terror . . . , 1(Mamm.):32.
1841. *Delphinus globiceps* Schlegel, Abh. Gebiete Zool., p. 33, a *nomen dubium*. Type locality, Japan. Not G. Cuvier, 1812.
1846. *Globiocephalus macrorhynchus* Gray, *in* The zoology of the voyage of H.M.S. Erebus and Terror . . . , 1(Mamm.):33. Type locality, South seas.
1852. *Globiocephalus indicus* Blythe, Jour. Asiatic Soc. Bengal, 21:358. Type locality, Serampore, Hooghly River, Bengal, India.
1869. *Globiocephalus scammonii* Cope, Proc. Acad. Nat. Sci. Philadelphia, 21:21, type from 10 mi. off Pacific Coast Baja California, in lat. 31° N.
1871. *Globiocephalus guadaloupensis* Gray, Supplement . . . catalogue of seals and whales in the British Museum, p. 84 [N Atlantic: French West Indies (type locality: Guadeloupe Island); name based on the *Globiocephalus intermedius* of Gervais and Van Bénéden, 1868–1880 (not Harlan); type skull in the Paris Museum].
1871. *Globiocephalus australis* Gray, Supplement . . . catalogue of seals and whales in the British Museum, p. 85. Type locality, "Coast of Australia"—a *nomen nudum*.
1871. *Globiocephalus sibo* Gray, Supplement . . . catalogue of seals and whales in the British Museum, p. 85. Type locality, N Pacific, Japan. Name based on the Japanese "sibo golo" described by Gray (1846, The zoology of the voyage of H.M.S. Erebus and Terror . . . , 1(Mamm.)) under his *Globiocephalus sieboldii*.
1871. *Globiocephalus propinquus* Malm, Kongl. Svenska Vetensk.- Akad. Handl., 9:85. Type locality, N. Atlantic, near equator.
1876. *Globiocephalus brachypterus* Cope, Proc. Acad. Nat. Sci. Philadelphia, 28:129, type from E coast Delaware Bay, at mouth Maurice River. (Regarded as identical with *G. macrorhyncha* by Fraser, Atlantide Rept. No. 1, pp. 50, 52, 58, 59, 1950.)

Total length to 20 ft.; grayish area in some individuals immediately posterior to dorsal fin; pectoral appendages long, slender, pointed, one-sixth length of body. Skull large and massive in proportion to body; breadth of rostrum approx. four-fifths of rostral length; 7–12 teeth in each tooth-row; approx. 57 vertebrae.

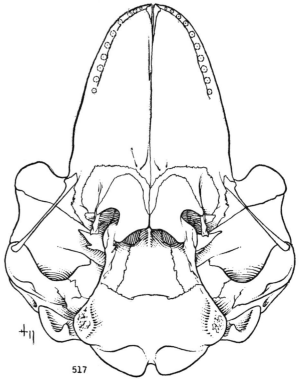

Fig. 517. *Globicephala sieboldii*, 10 mi. off Pacific Coast of Baja California, lat. 31° N, No. 9074 U.S.N.M., holotype. Ventral view, X ⅓.

Lowery (1974:361) regards this species as monotypic, notes that Hershkovitz (1966:94) "goes so far as to synonymize *macrorhyncha* under the larger, distinctively marked species of the North Atlantic, *G. melaena*," but emphasized "that Sergeant (1962) and others have demonstrated ample justification for recognizing two well-defined species."

RANGE.—Atlantic. Coast of New Jersey, thence southwestward to Texas: Aransas Pass. Gulf of Mexico. Lesser Antilles: St. Vincent (Caldwell, *et al.*, 1971:308). Pacific (Miller and Kellogg, 1955:664, as *G. scammonii* [= *G. sieboldii* Gray]). Alaska: Kanatak, western shore of Alaska Peninsula, thence southward to Guatemala: Gulf of Duce. But, van Bree (1971:86) notes that specimens of pilot whales from the Pacific Ocean north of a line drawn from California westward to Nagasaki need to be studied to determine the species occurring there.

Because of that need I inquired about the characters of the skull (California Acad. Sci. No. 8055) from Kanatak, Alaska, and on January 28, 1976, Dr. Robert T. Orr and Ms. Jacqueline

Schonewald provided the following information: length of rostrum (from tip of rostrum to center of line joining bases of maxillary notches), 36 cm.; width of rostrum at base of notches, 33.7 cm.; viewed dorsally the premaxillae completely cover the maxillae anteriorly; and obvious alveoli for teeth number 7 on each side of the upper jaw and 7 in each ramus of the lower jaw. Consequently, the specimen is here recorded as *Globicephala sieboldii.*

Genus **Peponocephala** Nishiwaki and Norris
Melon-headed Whale

1966. *Peponocephala* Nishiwaki and Norris, Sci. Rept. Whales Research Inst. Tokyo, 20:95, type *Lagenorhynchus electra* Gray 1864.

Length 6½–9 ft. Number of teeth (or alveoli) in each jaw 19–24 (= total number of 76–96); total length of skull, 417–459; rostrum, 54–56 per cent of length of skull; basal width of rostrum, 29–31 per cent of length of skull (after van Bree and Cadenat, 1968:193–202). These authors (*op. cit.*:200) place *Peponocephala* in the subfamily Globicephalinae containing also *Globicephala* Lesson 1828, *Orcinus* Fitzinger 1860, *Pseudorca* Reinhardt 1862, *Orcaella* Gray 1866, and *Feresa* Gray 1870.

Peponocephala electra (Gray)
Melon-headed Whale

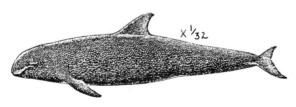

1846. *Lagenorhynchus electra* Gray, The zoology of the voyage of H.M.S. Erebus and Terror . . . 1(Mamm.):35, Pl. 13. Type locality, unknown.
1966. *Peponocephala electra*, Nishiwaki and Norris, Sci. Rept. Whales Research Inst. Tokyo, 20:95.
1846. *Lagenorhynchus asia* Gray, The zoology of the voyage of H.M.S. Erebus and Terror . . . , 1(Mamm.):35, Pl. 14. Type locality, unknown.
1848. *Delphinus pectoralis* Peale, U.S. Expl. Exped., 8(Mammalogy and Ornithology):32, Pl. 6, Fig. 1, type from Hilo Bay, Hawaii.
1866. *Delphinus (Lagenorhynchus) fusiformis* Owen, Trans. Zool. Soc. London, 6:22, Pl. 5, Fig. 1, type from Madras, India.
1868. *Electra obtusa* Gray, Synopsis of the . . . whales and dolphins in the . . . British Museum, p. 7, Pl. 13, new name for *Lagenorhynchus electra* Gray.

Upper parts including sides dark gray, having fine white spots; underparts paler gray; skin of both lips, area around anus and genital aperture, patches anterior to navel, and anchor-shaped area between flippers white. Total of 82 vertebrae. Other characters noted above in account of the genus. Dawbin, Noble, and Fraser (1970) graphically recorded changes in skull with increasing age.

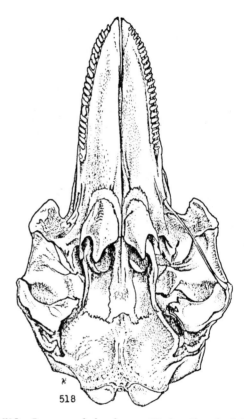

518

Fig. 518. *Peponocephala electra*, Niodor, Iles de Saloum, Sénégal, after van Bree and Cadenat (Beaufortia, Fig. 1, p. 198, July 10, 1968). Adult, sex unknown. Ventral view, X ¼.

RANGE.—All oceans between 37° N and 37° S. Eastern Pacific: 14° 20' N, 91° 52' W, approx. 90 nautical mi. due W Champerico, Guatemala (504087 USNM).

Genus **Feresa** Gray—Pygmy Killer Whale

1870. *Feresa* Gray, Proc. Zool. Soc. London, 1870:77, type *Orca intermedia* Gray (= *Feresa attenuata* Gray), by monotypy. See also Gray, 1871, Supplement . . . catalogue of seals and whales in the British Museum, p. 78.
1884. *Feresia* Flower, Proc. Zool. Soc. London, 1883:510, is an emendation of *Feresa* Gray.

Total length 7–9 ft. in males; females slightly smaller. Dark gray, almost black, except for white margins of lips and small white anal area. Snout overhangs tip of lower jaw. A ventral groove on midline of abdomen, from anus to beyond navel. Rostrum slightly shorter than post-rostral part of skull. Nine to 11 large conical teeth in each maxillary row, and 11–13 in each mandibular row; diameter of largest tooth *ca.* 8 mm. The genus is monotypic.

Feresa attenuata Gray
Pygmy Killer Whale

1875. *Feresa attenuata* Gray, Jour. Mus. Godeffroy (Hamburg), 8:184, type locality unknown, but type specimen was skull only, No. 1672a—74.11.25.1 British Museum (Natural History), according to Hershkovitz, Bull. U.S. Nat. Mus., 246:99, 1966.
1956. *Feresa occulta* Jones and Packard, Proc. Biol. Soc. Washington, 69:167, December 31, a new name for *Delphinus intermedius* Gray, November 1827, which is a junior primary homonym of *Delphinus intermedius* Harlan, June 1827 (proposed for the Atlantic Blackfish earlier named *Delphinus* [= *Globicephla* Lesson, 1828] *melas* by Traill in 1809). Hershkovitz, Bull. U.S. Nat. Mus., 246:98, 99, gives an extended synonymy.

See description above of the genus, which is monotypic.

RANGE.—East Pacific. "approximately 300 to 400 nautical miles off Costa Rica" (Perrin and Hubbs, 1969:297). West Atlantic. Texas: Padre Island, *ca.* 26° 04′ N, 97° 09′ W (Lowery, 1974:348). Florida: Singer Island (Caldwell and Caldwell, 1975:1). Lesser Antilles: off St. Vincent (Lowery, 1974:348).

Genus Phocoena G. Cuvier—Harbor Porpoises

1816. *Phocoena* G. Cuvier, Le règne animal . . . , 1:279, December. Type, *Delphinus phocoena* Linnaeus. (*Phocoena* G. Cuvier, 1816, considered unavailable by Sherborn, Index animalium, sec. 2 (1801–1805), pt. 1, p. xli, 1922; if so, *Phocoena* G. Cuvier, 1817, *in* Nouv. Dict. Hist. Nat., 9:163, is available.)
1866. *Acanthodelphis* Gray, Catalogue of seals and whales in the British Museum, p. 304. Type, *Phocaena* (sic) *spinipinnis* Burmeister.

Total length to 6 ft. Upper parts slate gray to black or brown. Dorsal fin triangular. Skull small;

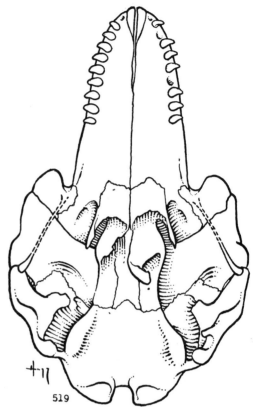

Fig. 519. *Feresa attenuata*, "near Taiji, Wakayama-Ken, Japan, after Yamada (Sci. Repts. Whales Research Institute, Tokyo, 9:64, Fig. 5a, 1954), adult ♀. Some details added from a cast of the holotype. Ventral view, X ¼.

rostrum short and broad; pterygoid bones small and not in contact; proximal ends of premaxillae with raised irregular bosses in front of external nares; symphysis of mandible short; 16–28 teeth in each tooth-row, small, compressed, spadelike; 64 or 65 vertebrae.

KEY TO NORTH AMERICAN SPECIES OF PHOCOENA

1. Posterior edge of palate with medial U-shaped indentation; maxillary bone not entering orbit. . . . .*P. sinus*, p. 896
1′. Posterior edge of palate with W-shaped indentation; maxillary bone entering orbit. . . . .*P. phocoena*, p. 897

Phocoena sinus Norris and McFarland
Vaquita or Cochita

1958. *Phocoena sinus* Norris and McFarland, Jour. Mamm., 39:24, February 20, type from NE short Punta San Felipe, Baja California Norte, Gulf of California.

Total length less than 5 ft. Uniform brown dorsally. Maxillary bone excluded from orbit by lateral margin of frontal; anteroventral extension of nasals covered by mesethmoid; posterior edge of palate having medial U-shaped indentation, formed by medial edges of rounded, roughly triangular pterygoids, and ventral extension of vomer that enters palate immediately posterior to palatines. Teeth (of type)$\frac{20-21}{18-18}$.

RANGE (Norris and McFarland, 1958:22, 23).—Baja California Norte: 8 mi. NE Punta San Felipe (sight record only); type locality, thence southward to Jalisco: Banderas Bay. Not found: Jalisco: Piginto River.

Phocoena phocoena (Linnaeus)
Harbor Porpoise

1758. [*Delphinus*] *phocoena* Linnaeus, Syst. nat., ed. 10, 1:77. Type locality, Swedish seas (see Thomas, Proc. Zool. Soc. London, p. 158, March 22, 1911).
1947. *Phocoena phocoena*, Anderson, Bull. Nat. Mus. Canada, 102 (1946):85, January 24.
1780. *Delphinus tursio* Fabricius, Fauna Groenlandica, p. 49. Type locality, North Greenland, on the high seas. On basis of "all blackish, with the abdomen a little whitish" and maxillary teeth "spaced apart" while ignoring other alleged characters, here assigned to synonymy of [*Delphinus*] *phocaena* Linnaeus 1758.
1792. *D*[*elphinus*]. *phocaena fuscus* Kerr, The animal kingdom. . . , p. 363. Type, the "poursille" of Bomare, 1768, Dict. Univ. Hist. Nat., article: "Baleine." Possibly belongs here.
1804. *Delphinus ventricosus* Lacépède, Histoire naturelle des cétacées . . . , p. xliii, 311. Type locality, River Thames, England.
1827. *Phocaena communis* Lesson, Manuel de mammalogie . . . , p. 413. Type locality, Atlantic Ocean, and French estuaries of la Loire, la Charente, la Seine.
1865. *Phocaena vomerina* Gill, Proc. Acad. Nat. Sci. Philadelphia, 17:178, type from Puget Sound, Washington.
1865. *Phocaena brachycium* Cope, Proc. Acad. Nat. Sci. Philadelphia, p. 279. Type locality, Salem Harbor, Massachusetts.
1865. *Phocaena tuberculifera* Gray, Proc. Zool. Soc. London, 1865:320. Type from Margate, England.
1870. *Phocoena rondeletii* Giglioli, Note interno alla distribuzione della fauna vertebrata nell' oceane, R. Magenta, 1865–68, p. 78. Type locality, N. Atlantic Coast Europe. Name borrowed from Willoughby, 1686, Hist. Pisc., 4:31.
1876. *Phocaena lineata* Cope, Proc. Acad. Nat. Sci. Philadelphia, pp. 134, 135. Type locality (in N. Atlantic of United States), "Harbour of New York" (p. 134), Eastport, Maine (p. 136).

1905. *Phocaena relicta* Abel, Jahrb. k. k. Geol. Reichsanstalt, Wien, 55:387, 388, Figs. 3, 4. Type locality, Crimean Coast, Black Sea, U.S.S.R.
1935. *Phocaena phocaena acuminata* van Deinse, Lev. Nat., 40:115. Type locality, Dishoeck Zoutelande, Walcheren, Holland.
1942. *Phocaena vomerina*, Scheffer, Murrelet, 23:45, August 14.
1946. *Phocaena phocaena acuminata* var. *conidens* van Deinse, Zool. Med. Rijks museum Nat. Hist. Leiden, 26:158, 159. Type locality, Zoutelande, Walcheren, Holland.

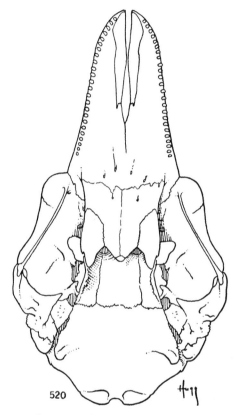

Fig. 520. *Phocoena phocoena*, Bolinas Bay, Marin Co., California, No. 47936 K.U., ♀. Ventral view, X ⅓.

Total length to 5 ft. Slate gray to black above, flanks grayish, fading to white on venter. Maxillary bone completely covering lateral margin of frontal, and entering orbit; posterior edge of palate having W-shaped indentation, formed by pointed, usually acutely triangular, pterygoids and medial pointed extension of palatines; palate covering ventral extension of vomer but in most skulls revealing vomer as small point of bone at apex of palatines. Teeth $\frac{27-28}{26-26}$.

RANGE.—Atlantic. Greenland: Davis Strait. New Jersey–Delaware border: Delaware River. South Carolina (Leatherwood, *et al.*, 1976:150). Pacific.

Alaska: Point Barrow, thence southward through Bering Strait to California: San Pedro Channel (Norris and McFarland, 1958:35).

Genus **Phocoenoides** Andrews—Dall's Porpoise

1911. *Phocoenoides* Andrews, Bull. Amer. Mus. Nat. Hist., 30:31, May 16. Type, *Phocoenoides truei* Andrews [= *Phocaena dalli* True].

Total length to 7 ft. Color blackish except area of white on sides posterior to pectoral appendages and region surrounding vent; dorsal fin tipped with white. Color variants carrying "grayish-brown" areas in normally black areas have been described and depicted in drawings by Morejohn, *et al.* (1973:977–982). They show also a secondary sexual difference in the anal region where black pigment extends farther anteriorly in females than in males. Posterior margin of dorsal fin and posterior margin of each pectoral appendage slightly concave. Skull having relatively short, flat beak; temporal fossae small; maxillae short proximally; external nares large; pterygoids separate; lower jaw projecting slightly anterior to upper jaw; 23–27 teeth in each tooth-row; 97 or 98 vertebrae; all cervicals united; last 4 caudals united.

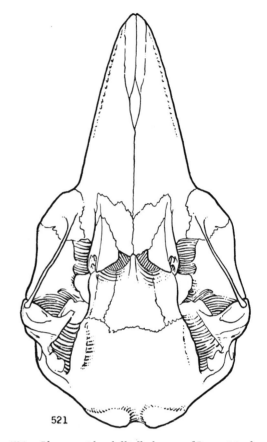

521

Fig. 521. *Phocoenoides dalli*, [holotype of *P. truei* Andrews], Aikawahama, Japan, No. 31425 A.M.N.H., ♂ after Andrews (Bull. Amer. Mus. Nat. Hist., 30(Art. 4): Fig. 5, 1911). Ventral view, X ⅓.

Phocoenoides dalli (True)
Dall's Porpoise

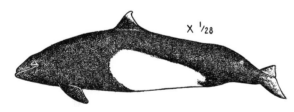

X ¹/₂₈

1885. *Phocaena dalli* True, Proc. U.S. Nat. Mus., 8:95, June 19, type from strait west of Adakh [= Adak], Aleutian Islands, Alaska.
1911. *P[hocoenoides]. dalli*, Andrews, Bull. Amer. Mus. Nat. Hist., 30:34, May 16.
1911. *Phocoenoides truei* Andrews, Bull. Amer. Mus. Nat. Hist., 30:32, May 16, type from Aikawahama, Rikuzen Prov., Japan. (Regarded as inseparable from *P. dalli* by Benson, Jour. Mamm., 27:373, November 25, 1946.)

See characters of the genus.

RANGE.—Pacific. Alaska: Western Aleutian Islands, thence southward to Baja California: cool waters of NW part (Hubbs, 1960:144).

FAMILY **MONODONTIDAE**
White Whale and Narwhal

Total length to 18 ft. Dorsal fin absent; pectoral appendages short and broad; snout blunt; throat lacking external grooves; external nares posterior to orbits; palatine bones participating in formation of posterior foramina of infraorbital canals; foramina of antorbital canals distinctly posterior to level of anterior border of orbits; cervical vertebrae unfused; 8 double-headed ribs, tubercula of posterior ribs articulating with diapophyses; teeth tusklike in manner of growth.

KEY TO GENERA OF MONODONTIDAE

1. Teeth never less than 32; color whitish throughout. . . . . . .*Delphinapterus*, p. 899
1'. Teeth never more than 2; color marbled grayish above, whitish below. .*Monodon*, p. 900

Genus **Delphinapterus** Lacépède—White Whale

1804. *Delphinapterus* Lacépède, Histoire naturelle des cétacées . . . , p. xli. Type, *Delphinapterus beluga* Lacépède [= *Delphinus leucas* Pallas].
1815. *Beluga* Rafinesque, Anal. Nat., p. 60, a renaming of *Delphinapterus* Lacépède.

Total length to 18 ft. Teeth $\frac{8}{8}$ to $\frac{10}{10}$, irregularly set in jaw; teeth implanted obliquely in alveoli and directed forwards in jaw; 9 lumbar vertebrae, 23 caudal vertebrae; premaxillae flat or slightly concave anterior to external nares; skull more or less symmetrical; posterior portion of pterygoid bones spreading laterally over alisphenoid bones; dorsal fin reduced to indistinct ridge; tail strongly furcate; external constriction between head and body; skin white in adults but having some pigment in young animals.

White whales have a circumpolar distribution in Arctic seas, seemingly prefer coastal waters and bays, have been known to travel considerable distances up freshwater rivers, and are gregarious (usually traveling in gams of 5–10). The genus, as here considered, is monotypic.

Delphinapterus leucas (Pallas)
White Whale

X $^{1}/_{72}$

1773. *Physeter Katodon* Müller, Des Ritters Carl von Linné . . . vollständiges Natursystem nach der zwölften Latinischen Aufgabe, 1:497. [Homonym of *Physeter Catodon* Linnaeus 1758.]
1776. *Delphinus leucas* Pallas, . . . Reise durch verschiedene Provinzen des Russischen Reichs, 3(book 1):85, footnote. Type locality, mouth Obi [Ob] River, northeastern Siberia, U.S.S.R.
1812. *D[elphinapterus]. leucas*, G. Cuvier, Ann. Mus. Hist., Nat. Paris, 19:13.
1776. *B[alaena]. albicans* Müller, Zoologiae Danicae prodromus, p. 7.
1777. *Delphinus canadensis* Erxleben, Systema Regni animalis . . . Mammalia, pp. 601–628 (Canada)—*fide* Tomilin, 1957, Zveri S.S.S.R. i Prilezhashchikh Stran, 9:705, but not found on pp. 601–628 of 1777 edition of Erxleben examined by E. R. Hall.
1792. *D[elphinus]. Phocaena albus* Kerr, The animal kingdom . . . , p. 363 [N. Atlantic: Canada (type locality: St. Lawrence River); name based on the "moine de mer" of Bomare

(Dict. Univ. Hist. Nat., entry: "Baleine")]—*fide* Hershkovitz, 1966, U.S. Nat. Mus. Bull., 246:109, February 28.
1804. *Delphinapterus beluga* Lacépède, Histoire naturelle des cétacées . . . , pp. xli, 243 (type locality, p. 247, Arctic Ocean and N. Atlantic Ocean, particularly Davis Strait).
1827. *Delphinus (Delphinapterus?) kingii* Gray, Philos. Mag. or Ann., (2), 2(2):375 [type a skull only, "brought to this country [England] and presented to the [British] Museum by Capt. P. P. King, R. N., when he returned from his survey of the coast of New Holland" [Australia]].—True, 1889, Bull. U.S. Nat. Mus., 36:146, 147 [New Holland locality regarded as erroneous; type skull agrees with Alaskan beluga]. *Fide* Hershkovitz, 1966, U.S. Nat. Mus. Bull., 246:110, February 28.
1828. *C[atodon]. Sibbaldi* Fleming, A History of British Animals . . . , p. 39 [N. Atlantic: Great Britain (type locality: Kairston, Orkney). According to Hershkovitz, 1966, U.S. Nat. Mus. Bull., 246:110, February 28, name based on a gam of belugas recorded by Sibbald (1692, Phalainologia, p. 24).
1828. *Beluga glacialis* Lesson, Complément des oeuvres de Buffon, Cétacés, p. 194, Pl. 3, Fig. 2 (animal with name in caption) [N. Atlantic: Scotland (type locality: Firth of Forth)]—*fide* Hershkovitz, 1966, U.S. Nat. Mus. Bull., 246:110, February 28.

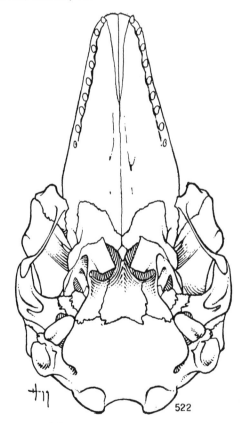

522

Fig. 522. *Delphinapterus leucas*, outline from No. 46 F.M.N.H., after Elliot (Field Columb. Mus., Zool. Ser., 2:Pl. 5, 1901). Details from No. 238104 U.S.N.M., juvenile ♀, from St. Lawrence River. Ventral view, X $\frac{1}{4}$.

1828. *"beluga borealis"* Lesson, Complément des oeuvres de Buffon, Cétacés, pp. 440, 192 [new name for *Delphinus leucas* Pallas; characters; habits; distribution].

1846. *Beluga Catodon* Gray, The zoology of the voyage of H.M.S. Erebus and Terror. . . , 1(Mamm.):29 [N. Atlantic: Greenland (type locality); type a skull, British Museum (Natural History)].

1865. *Beluga rhinodon* Cope, Proc. Acad. Nat. Sci. Philadelphia, 17:278, type from Upernavik, Greenland.

1865. *Beluga declivis* Cope, Proc. Acad. Nat. Sci. Philadelphia, 17:278. Type locality unknown.

1865. *Beluga concreta* Cope, Proc. Acad. Nat. Sci. Philadelphia, 17:278. Type locality, Melville Island (see Koopman, Proc. Acad. Nat. Sci. Philadelphia, 128:20, November 15, 1976).

1866. *Beluga angustata* Cope, Proc. Acad. Nat. Sci. Philadelphia, 18:293, type from Upernavik, Greenland.

1935. *Delphinapterus freimani* Klumov, Biulleten' rybnogo Khoziaistvo S.S.S.R. [Sea Fishery Economics in U.S.S.R.], Moscou, No. 7, pp. 26–28, Fig. 2, type from White Sea, northeastern U.S.S.R.

1935. *Delphinapterus dorofeevi* Barabash and Klumov, Biulleten' rybnogo Khoziaistvo S.S.S.R. [Sea Fishery Economics in U.S.S.R.], Moscou, No. 11, p. 24, type from Okhotsk Sea. (Regarded as a subspecies of *D. leucas* by Ellerman and Morrison-Scott, Checklist of Palaearctic and Indian mammals, 1758–1946. British Museum, London, p. 727, November 19, 1951. Miller and Kellogg, Bull. U.S. Nat. Mus., 205:654, March 3, 1955, list *D. dorofeevi* as a species and state, "Presumably enters Bering Sea, and may represent forms recorded at Point Barrow. . . .")

1935. *Delphinapterus leucas maris-albi* Ostroumov, Fish Industry of the North, U.S.S.R., no. 11 [Arctic: U.S.S.R. (type locality: Gulf of Onega, White Sea)]—*fide* Hershkovitz, 1966, U.S. Nat. Mus. Bull., 246:111, February 28.

See characters of the genus. Banfield (1974:249–251) uses the name combination *D. leucus leucus* for Canadian animals, thereby implying that *D. dorofeevi* Barabash and Klumov 1935 is a recognizable subspecies. He may be correct, although the name *D. dorofeevi* is here listed (*supra*) as a junior synonym of *D. leucas*.

According to Sergeant and Brodie (1969:2561), "Measurements of length, girth, and weight show that male white whales grow larger than females. The smallest white whales come from western Hudson Bay, the White Sea, and Bristol Bay, Alaska. Animals of intermediate size inhabit all other arctic Canadian localities sampled and also the St. Lawrence River and the Kara and Barents seas. The largest white whales inhabit West Greenland waters, the Okhotsk Sea, and the coast of Sakhalin. Extreme differences in body weight of adult males are about threefold. Nonoverlapping differences in size indicate isolation of some adjacent populations of white whales; equal or overlapping sizes suggest, but cannot prove, mixing of other populations. Size can be positively correlated with marine productivity, being lowest in the arctic and in estuaries and highest in subarctic seas."

To decide on the taxonomic validity of names proposed for different sizes of white whales (and possibly shapes of the skull) an estimate is desirable as to how much of the observed difference is genetic (inherited from preceding generations) and how much is environmental (caused by living conditions for only the living generation). As of 1977 no such estimate has been found, and therefore the white whale is here provisionally treated as a monotypic species.

RANGE.—Arctic and subarctic seas N from Cook Inlet of Alaska, James Bay of Canada (Edwards, 1963:12), and Gulf of St. Lawrence (Anderson, 1947:86).—Four southern, extralimital occurrences are: in Washington (Scheffer and Slipp, 1948:303, 304), Tacoma in 1940 and Steilacoom in 1939; Connecticut (Leatherwood, *et al.*, 1976:101); Massachusetts, Cape Cod (J. A. Allen, 1869:206).

Genus **Monodon** Linnaeus—Narwhal

1758. *Monodon* Linnaeus, Syst. nat., ed. 10, 1:75. Type, *Monodon monoceros* Linnaeus.

1804. *Narwalus* Lacépède, Histoire naturelle des cétacées . . . , p. xxxvii. Type, *Narwalus vulgaris* Lacépède [= *Monodon monoceras* Linnaeus].

Total length to 15½ ft. excluding tusk; only 2 teeth; in adult males one of these spiraling in sinistral direction and projecting anteriorly for as much as 9 ft.; other tooth usually not exposed; females rarely showing tusks externally; 6 lumbar and 26 caudal vertebrae; premaxillae convex anterior to external nares; skull asymmetrical; pterygoid bones small, not spreading laterally over alisphenoid bones; dorsal fin absent; tail only slightly furcate; color marbled grayish above, whitish below.

Narwhals have a circumpolar distribution in Arctic seas.

Monodon monoceros Linnaeus
Narwhal

1758. [*Monodon*] *monoceros* Linnaeus, Syst. nat., ed. 10, 1:75. Type locality, Arctic seas.

1804. *Narwalus vulgaris* Lacépède, Histoire naturelle des cétacées . . . , p. xxxvii, 142. Type locality, unknown.

1804. *Narwalus microcephalus* Lacépède, Histoire naturelle

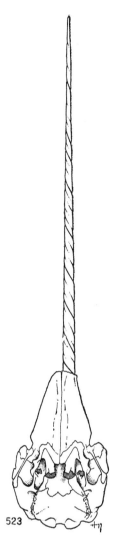

Fig. 523. *Monodon monoceros*, Holstenberg, Greenland, No. 2378 K.U., ♂. Ventral view, X 1/14.

des cétacées. . . , p. xxxviii, 159, Pl. 9, Fig. 1. Type locality, unknown.

1804. *Narwalus andersonianus* Lacépède, Histoire naturelle des cétacées. . . , p. xxxviii, 163. Type locality, unknown.

See characters of the genus.

RANGE.—Arctic seas northward from Alaska: mouth Caribou River, in Nelson Lagoon (Geist, *et al.*, 1960:252), and northward from Labrador Coast.

FAMILY PHYSETERIDAE—Sperm Whales

Mandible small and narrow, ending short of snout; narial opening to left of median line of head; pectoral appendages small (see drawings of whole animals, *Kogia* and *Physeter* below).

KEY TO GENERA OF PHYSETERIDAE

1. Adults more than 18 ft. long; dorsal fin absent, replaced by median longitudinal row of bumps on posterior half of back; snout high and blunt; more than 17 teeth in each ramus of lower jaw; on ventral side of skull anterior ends of palatal bones nearer posterior than anterior end of skull. *Physeter*,   p. 902
1'. Adults less than 18 ft. long; dorsal fin falcate; snout conical; fewer than 17 teeth in each ramus of lower jaw; on ventral side of skull anterior ends of palatal bones nearer anterior than posterior end of skull. *Kogia*,   p. 901

Genus **Kogia** Gray
Pygmy and Dwarf Sperm Whales

1846. *Kogia* Gray, *in* The zoology of the voyage of H.M.S. Erebus and Terror . . . , 1(Mamm.):22. Type, *Physeter breviceps* Blainville.
1851. *Euphysetes* Wall, Mem. Australian Mus., 1:46. Type, *Euphysetes grayii* Wall [= *Physeter breviceps* Blainville].
1871. *Callignathus* Gill, Amer. Nat., 4:737, 738, 740 (footnote), February. Type, *Physeter simus* Owen [= *Physeter breviceps* Blainville]. Not *Callignathus* Costa, 1853, a fish.
1876. *Cogia* Wallace, The geographical distribution of animals, 2:208, an emendation of *Kogia* Gray.

Total length to 13 ft. Dorsal fin falcate; color blackish above, grayish to whitish on sides and below; rostrum short; sagittal crest well defined; anterior border of lateral crest of supracranial basin elevated outside of antorbital notches; lachrymal bones massive, fused with jugals; zygoma incomplete; palatine bones small; pterygoid bones expanded; facial depression moderately developed, distorting but not obliterating longitudinal ridge behind nares; brain relatively large, extending forward to level of orbit; mandibular symphysis short; each lower jaw having 9–15 needlelike teeth.

KEY TO SPECIES OF KOGIA

1. Total length more than 9 ft.; dorsal fin low, posterior to center of back; condylobasal length more than 350. . . . *K. breviceps*,   p. 901
1'. Total length less than 9 ft.; dorsal fin high, near center of back; condylobasal length less than 325. . . . . *K. simus*,   p. 902

Kogia breviceps (Blainville)
Pygmy Sperm Whale

1838. *Physeter breviceps* Blainville, Ann. Anat. Phys., 2:337, pl. 10. Type locality, region of Cape of Good Hope, Republic of South Africa.

1846. *Kogia breviceps*, Gray, in The zoology of the voyage of H.M.S. Erebus and Terror . . . , 1(Mamm.):22.

1851. *Euphysetes grayii* Wall, Mem. Australian Mus., 1:37, pl. 2. Type locality, Maroubra Beach, New South Wales, Australia.

1866. *Euphysetes macleayi* Krefft, Proc. Zool. Soc. London, for 1865, p. 713, Figs. 1–6, April. Type locality, Manly Beach, New South Wales, Australia.

1871. *Kogia floweri* Gill, Amer. Nat., 4:738, February, type from near Mazatlán, Sinaloa.

See characters of the genus. Total length 9–11½ ft.; condylobasal length 391–469 mm.; dorsal fin low, located posterior to center of back; maxillary teeth absent; mandibular teeth 12–16 pairs; mandibular symphysis long (86–120 mm.) and ventrally keeled; dorsal cranial fossae not cupped posteriorly; left fossa conspicuously longer and narrower than right; least breadth of dorsal sagittal septum 20–38 mm.; pterygoid-basioccipital wings long; foramen magnum near midpoint of skull height (After Handley, 1966:67).

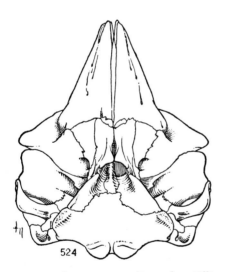

Fig. 524. *Kogia breviceps*, outline after Elliot (Field Columb. Mus., Publ. 95, Zool. Ser., 4(1):Pl. 17, 1904). Details from No. 283625 U.S.N.M., ♀, from Crescent Beach, St. Johns Co., Florida. Ventral view, X approx. 1/7.5.

RANGE.—Atlantic. Nova Scotia: Halifax Harbor, thence southward and westward to Florida: St. Petersburg (Lowery, 1974:321). Texas (*ibid.*): Mustang Island; Padre Island. Cuba: Bahía de Nuevas Grandes, N coast, at boundary of provinces of Camagüey and Oriente (Varona, 1974:82). Pacific. California: Morro Bay (Roest, 1970:415).

Kogia simus (Owen)
Dwarf Sperm Whale

1866. *Physeter (Euphysetes) simus* Owen, Trans. Zool. Soc. London, 6:30, Pls. 10–14, August 15, type from Waltair, Madras Presidency, India.

Resembles *K. breviceps* except smaller (total length 7–9 ft.; condylobasal length 262–302 mm.); dorsal fin high, located near center of back; 1–3 pairs maxillary teeth present; mandibular teeth, 8–11 pairs; mandibular symphysis short (37–46 mm.) and ventrally plane; dorsal cranial fossae cupped posteriorly and subsymmetrical; least breadth of dorsal sagittal septum, 6–14 mm.; pterygoid-basioccipital wings short; foramen magnum well below midpoint of skull height (After Handley, 1966:67).

RANGE.—Atlantic. Central eastern United States (Handley, 1966:68). Florida: 3 mi. E Dustin, Okaloosa Co. (Lowery, 1974:323). Texas: Galveston Island (*ibid.*). Cuba (Caldwell, *et al.*, 1973:516). Lesser Antilles: St. Vincent (*ibid.*). Pacific. California: 3 mi. S Oceano (Roest, 1970:415) and Baja California (Rice and Scheffer, 1968:10).

Genus **Physeter** Linnaeus—Sperm Whale

1758. *Physeter* Linnaeus, Syst. nat., ed. 10, 1:76. Type, *Physeter macrocephalus* Linnaeus (selected by Desmarest, 1855–1856, in J. C. Chenu, Encyclopédie d'Histoire naturelle . . . , [5]:296–298.

1761. *Catodon* Linnaeus, Fauna Suecia, ed. 2, p. 18. Type, *Physeter macrocephalus* Linnaeus.

1804. *Physalus* Lacépède, Histoire naturelle des cétacées. . . . , pp. xi, 219. Type, *Physalus cylindricus* Bonnaterre.

1828. *Megistosaurus* Harlan (or anonymous in Harlan), Amer. Jour. Sci. Arts, 14:186. Type, bones exhumed at mouth Mississippi River, later identified as *Physeter macrocephalus*.

1828. *Cetus* Billberg, Synop. faunae scandinaviae, p. 39. Type, *Cetus cylindricus* Lacépède.

1865. *Meganeuron* Gray, Proc. Zool. Soc. London, 1865:439, as subgenus of *Catodon*. Type, *Catodon (Meganeuron) kreffti* (= *Physeter catodon* Linnaeus).

Total length to 69 ft. (male), 39 ft. (female), largest of toothed whales. Snout high and blunt; head square in lateral view, approx. one-third total length of animal; dorsal fin absent, replaced by median, longitudinal row of bumps on posterior half of back; color dark bluish-gray, sometimes with whitish on venter and lower jaw; skull massive in proportion to postcranial skeleton; facial depression greatly developed, obliterating longitudinal ridge behind nares; zygoma complete; brain small, situated far behind orbital region; each lower jaw with 20–25 teeth, some of which may be 8 in. long; vestigial teeth in upper jaw.

For the sperm whale the specific name *catodon* is used here because it has line priority over *macrocephalus* and *microps* and page priority over *tursio*. See synonymy below. Cogent reasons for using the specific name *macrocephalus* are given by Husson and Holthius, Zool. Med. Rijksmuseum Nat. Hist. Leiden, 48:205–217, 2 pls., 1974. Many of the junior synonyms listed immediately below are from Hershkovitz, Bull. U.S. Nat. Mus., 246:115–122, 1966, instead of from the original sources.

Physeter catodon Linnaeus
Sperm Whale

X ¹/₂₄₀

1758. *Physeter catodon* Linnaeus, Syst. nat., ed. 10, 1:76, line 22. Type locality, Oceano septentrionali, but subsequently restricted by Thomas (1911, Proc. Zool. Soc. London, p. 157) to Kairston, Orkney Islands, Scotland.

1758. [*Physeter*] *macrocephalus* Linnaeus, Syst. nat., ed. 10, 1:76, line 28. Type locality, Oceano Europaeo, but restricted to Berkhey, province of Zuid-Holland, The Netherlands, by Husson and Holthius, Zool. Med. Rijksmuseum Nat. Hist. Leiden, 48(19):209, 1974, who reproduce a description in 1605 by Clusius and an original drawing (by Goltzius or Matham), both relating to a sperm whale stranded in 1598 at their type locality. Husson and Holthius (*op. cit.*:210) select the Berkhey specimen as lectotype of *Physeter macrocephalus*. So far as is known, no part of that specimen is preserved in 1975.

1758. [*Physeter*] *microps* Linnaeus, Syst. nat., ed. 10, 1:76, line 38. Type locality, Oceano septentrionali.

1758. [*Physeter*] *tursio* Linnaeus, Syst. nat., ed. 10, 1:77, line 5. Type locality, Oceano septentrionali.

1780. *Physeter novae angliae* Borowski, Gemeinnützige Hand-und Hilfs-buch der Naturgeschichte des Thierreichs, 2:32. Type locality, New England. Name based on the "cachalot de la Nouvelle Angleterre" of Brisson, Le règne animal . . . , p. 360, 1756.

1780. *Physeter andersonii* Borowski, *op. cit.*: 33. Type locality, Iceland and Greenland. Name based on the "cachalot à dents pointues" of Brisson, Le règne animal . . . , p. 362, 1756.

1789. *P[hiseter (sic)]. trumpo* Bonnaterre, Tableau encyclopédique et méthodique de cétologie, p. 14, Pl. 8, Fig. 1. Type locality, Bayonne, France.

1789. *P[hiseter (sic)]. cylindricus* Bonnaterre, *op. cit.*: 16, Pl. 7, Fig. 1. Type locality, Greenland, Atlantic.

1789. *P[hiseter (sic)]. mular* Bonnaterre, *op. cit.*: 17, Pl. 8, Fig. 5. Type locality, Greenland, Atlantic.

1792. *Ph[yseter] macrocephalus niger* Kerr, The animal kingdom. . . , p. 360. Type, black sperm whale with two dorsal fins.

1792. *Ph[yseter] macrocephalus cinereus* Kerr, *op. cit.*:361. Type locality, New England, North Atlantic.

1792. *Ph[yseter] micr[ops] falcidentatus* Kerr, *ibid.* Name based on the "cachalot à dents en faucilles" of Brisson, Le règne animal . . . , p. 363, 1756.

1792. *Ph[yseter] micr[ops] rectidentatus* Kerr, *op. cit.*:362. Name based on the straight-toothed sperm whale of Brisson, Le règne animal . . . , p. 362, 1756.

1798. *Physeter maximus* G. Cuvier, Tableau elémentaire de l'histoire naturelle, p. 176. Type, one of 31 sperm whales stranded near Andierne, France.

1804. *Physeter orthodon* Lacépède, Histoire naturelle des cétacées. . . , pp. xli, 236. Type locality, Greenland, Atlantic. Name based on the Arctic sperm whale described by Anderson, Hist. Groenland, 1747.

1818. *Physeterus [sic] sulcatus* Lacépède, Mém. Mus. Hist. Nat. Paris, 4:474. Type locality, Japan, North Pacific.

1822. *Physeter australasianus* Desmoulins, Dict. Class. Hist. Nat., 2:618. Type locality, Indian Ocean in Moluccas or near New Zealand.

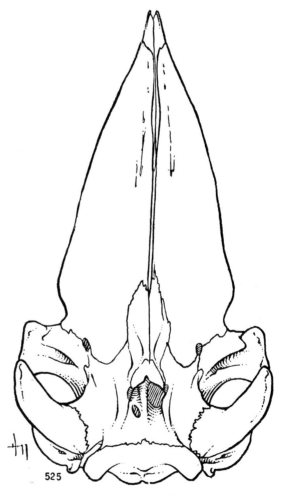

525

Fig. 525. *Physeter catodon*, outline from photograph by Stejneger of a skull from Bering Island. Details from skull from Kinkasan, Japan, No. 253051 U.S.N.M. Sex unknown. Ventral view, X approx. ¹/₂₅.

1822. *Tursio vulgaris* Fleming, Philos. Zool., 2:211. A new name for *Physeter tursio* Linnaeus.

1824. *Physeter polycephus* Quoy and Gaimard, *in* Freycinet, Voyage autour du monde . . . Zoologie, p. 77; Atlas, Pl. 12. Name based on animal observed at sea.

1826. *Delphinus bayeri* Risso, Histoire naturelle de l'Europe Mériodional, 3:22. Type locality, Nice, Mediterranean, France.

1846. *P[hyseter]. australis* Gray, *in* The zoology of the voyage of H.M.S. Erebus and Terror. . . , 1(Mamm.):22. A *lapsus* for *Physeter australasianus* Desmoulins.

1850. *Catodon colneti* Gray, Catalogue of the . . . Mammalia in the . . . British Museum, Cetacea, p. 52. Type locality, Point Angles, México (*fide* Hershkovitz, Bull. U.S. Nat. Mus., 246:115–122, 1966).

1865. *Catodon (Meganeuron) krefftii* Gray, Proc. Zool. Soc. London, 1865:440, Figs. 1–4. Type locality, New South Wales, Australia.

1866. *Physeter australis asiaticus* Gray, Catalogue of seals and whales in the British Museum, p. 210. A *lapsus* for *Physeter australasianus* Desmoulins 1822.

1898. *[Physeter] pterodon,* Trouessart, Catalogus Mammalium, p. 1056. Listed in synonymy of *?macrocephalus,* attributed to "Lesson, Descr. Mamm., p. 167."

See characters of the genus.

RANGE.—Atlantic. Franklin: Davis Strait. Florida: Delray Beach (Layne, 1965:143). Texas: Padre Island, N Mansfield Channel (Schmidly and Melcher, 1974:454), thence southward to the Lesser Antilles and South American waters. Pacific. Alaska: Pribilof Islands, thence southward to Panamá: Golfo de Panamá and South American waters.

FAMILY ZIPHIIDAE—Beaked Whales

Revised by True, Bull. U.S. Nat. Mus., 73:1–89, 42 pls., September 28, 1910. See also Moore, Fieldiana: Zool., 53(4): iv + 298, 24 figs., December 13, 1968.

Total length to 42 ft.; narial aperture median, half-moon-shaped, directed forward; external longitudinal grooves on throat divergent posteriorly, nearly meeting anteriorly; pectoral fins small, ovate, having 5 moderately well-developed digits; dorsal fin obtuse, usually falcate, situated posterior to middle of back. Skull with slender rostrum or beak; lachrymal bone distinct from jugal bone; bones of skull asymmetrical except in *Berardius*; tubercula of posterior ribs articulate with metapophyses; numerous vestigial teeth imbedded in gums of both upper and lower jaws; functional teeth confined to lower jaw in all North American species, never less than 2 nor more than 4; in females, teeth rarely above gum-line.

KEY TO GENERA OF ZIPHIIDAE

1. Four functional mandibular teeth; dorsal fin straight on posterior border.*Berardius*,   p. 904
1'. Functional mandibular teeth never more than 2; dorsal fin concave on posterior margin.
  2. All cervical vertebrae fused; prominent longitudinal maxillary crests anterior to external nares causing pronounced forehead in lateral external view. . . . .*Hyperoodon*,   p. 910
  2'. Fusion of cervical vertebrae confined to no more than first 4; maxillary crests absent or, if present, small and lateral to external nares.
    3. Premaxillae ridged posteriorly, forming distinct cup around external nares; mouth small, reaching posteriorly less than halfway to level of eye. . . . . . .*Ziphius*,   p. 908
    3'. Premaxillae flattened posteriorly, not forming cup around external nares; mouth large, reaching posteriorly nearly to level of eye.
             Micropteron,   p. 905

Genus **Berardius** Duvernoy—Beaked Whales

1851. *Berardius* Duvernoy, Ann. Sci. Nat., Paris, ser. 3 (Zool.), 15:52. Type, *Berardius arnouxii* Duvernoy.

Total length to 42 ft., largest of beaked whales. Blackish dorsally, grayish to whitish ventrally. Four functional teeth in lower jaw, anterior pair the larger; skull symmetrical (asymmetrical in other genera of the family Ziphiidae); maxillae crested anterior to external nares but not so much as in *Hyperoodon*; nasal bones broad, massive and rounded; antorbital notch distinct; mesethmoid only partly ossified; 3 anterior cervical vertebrae fused; dorsal fin small, straight on posterior border.

Berardius bairdii Stejneger
Baird's Beaked Whale

X ¹⁄₁₆₈

1883. *Berardius bairdii* Stejneger, Proc. U.S. Nat. Mus., 6:75, June 30, type from Stare Gavan, Bering Island, Commander Islands, Bering Sea.

1884. *Berardius vegae* Malm, Bihang Kongl. Svenska Vetensk.-Akad. Handl., 8(4):109. Type locality, Komandorskye Islands, Bering Sea.—Wilhelm Odelberg of the Library of the Royal Swedish Acad. Sci., in a letter of October 10, 1975, to P. J. H. van Bree, states that A. W. Malm's paper (heretofore frequently cited as 1883) was not distributed (= published in the zoological sense) before an unknown date in the first half of 1884.

See characters of the genus.

The other species of the genus, *B. arnuxii*, is known only from the waters of the Southern Hemisphere.

RANGE.—Pacific. Alaska: St. Matthew Islands, thence southward to California: Monterey Bay.

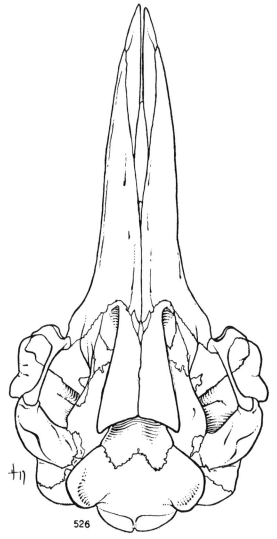

Fig. 526. *Berardius bairdii*, about 12½ mi. N Cape Mendocino, Humboldt Co., California, No. 49725 U.S.N.M., ♂?. Ventral view, X 1/10.

Genus **Micropteron** Eschricht—Beaked Whales

1828. *Aodon* Lesson, Histoire naturelle . . . des mammifères et des oiseaux découverts depuis 1788, cétacés, p. 155, 440, Pl. 3, Fig. 1. Type, *Aodon dalei* Lesson [= *Physeter bidens* Sowerby]. Not *Aodon* Lacépède, 1798, a fish.

1846. *Micropterus* Wagner, *in* Schreber, Die Säugthiere . . . , 7:281, 352, Pl. 348. Type, *Delphinus micropterus* Cuvier [= *Physeter bidens* Sowerby]. Not *Micropterus* Lacépède, 1802, a fish.

1849. *Micropteron* Eschricht, Kongl. Danske Vidensk. Selsk. Skrift. Nat. Math. Afd. Kjöbenhavn, (5), 1:97. Type, *Delphinus micropterus* G. Cuvier, 1829 [= *Physeter bidens* Sowerby].

1850. *Mesoplodon* Gervais, Ann. Sci. Nat., Paris, ser. 3 (Zool.), 14:16, July. Type, *Delphinus sowerbensis* Blainville [= *Physeter bidens* Sowerby].

1850. *Dioplodon* Gervais, Compt. Rend. Acad. Sci. Paris, 31:512. Type, *Delphinus densirostris* Blainville. Valid as a subgenus according to Ellerman and Morrison-Scott, Checklist of Palaearctic and Indian mammals, 1758–1946, p. 724, November 19, 1951.

1851. *Mesodiodon* Duvernoy, Ann. Sci. Nat., Paris, ser. 3 (Zool.), 15:41, Pl. 2, Fig. 2, 2'. Type, *Dioplodon sowerbyi* Gervais [= *Physeter bidens* Sowerby].

1866. *Mesiodon* Gray, Catalogue of seals and whales in the British Museum, p. 349, misspelling of *Mesodiodon* Duvernoy, 1851.

1866. *Diplodon* Gray, Catalogue of seals and whales in the British Museum, p. 349, misspelling of *Dioplodon* Gervais, 1850.

1866. *Dolichodon* Gray, Catalogue of seals and whales in the British Museum, p. 353, Fig. 72. Type, *Ziphius Layardii* Gray.

1871. *Callidon* Gray, Ann. Mag. Nat. Hist., ser. 4, 7:368, May. Type, *Mesoplodon güntheri* Krefft [= *Ziphius Layardii* Gray].

1871. *Neoziphius* Gray, Supplement . . . catalogue of seals and whales in the British Museum, p. 101. Type, *Dioplodon europaeus* Gervais [= *D. gervaisi* Deslongchamps].

1876. *Oulodon* von Haast, Proc. Zool. Soc. London, p. 457, October. Type, *Mesoplodon grayi* von Haast.

1922. *Paikea* Oliver, Proc. Zool. Soc. London, p. 574, September 28. Type, *Berardius hectori* Gray.

Total length to 22 ft. At most, a single pair of teeth in lower jaw project above gum-line [in some males additional teeth are buried in the gum tissue and in females all teeth are buried in gum tissue. The single pair of functional (visible) teeth in some species is anterior to the posterior margin of the mandibular symphysis and in other species is posterior to the symphysis]; skull asymmetrical; maxillary tuberosities present or absent; nasal bones narrow; mesethmoid ossified in adults; rostrum long and narrow; anterior 2 or 3 cervical vertebrae fused; dorsal fin concave on posterior surface.

On the basis of Moore's (1966:35, 36) account and Figs. 1–9 (*op. cit.*) it may be added that: premaxillae curved upward and expanded beside, above, and behind superior nares; each nasal laterally almost as high as premaxilla but depressed mesially.

For diagnoses of several species, see Moore (1966:32–61). North American species here arranged according to Moore (1968:248, 277, 278).

KEY TO SKULLS OF SPECIES OF GENUS MICROPTERON IN NORTH AMERICAN WATERS

(After Moore, 1966)

1. Premaxillary foramina situated on or below horizontal plane of beak.
  2. Exposed part of vomer on ventral side of rostrum less than 95 mm. long. . . . .*M. europaeus*, p. 906
  2'. Exposed part of vomer on ventral side of rostrum more than 95 mm. long.
    3. In side view of antorbital tubercle, lachrymal wrapped around anterior end of frontal.
      M. mirus, p. 906
    3'. In side view of antorbital tubercle, lachrymal not wrapped around anterior end of frontal.
      M. ginkgodens, p. 907
1'. Premaxillary foramina situated above horizontal plane of beak.
  4. On vertex of skull, right premaxilla extended posteriorly beyond right nasal for distance equal to ¾ or more of length of nasal as exposed on vertex.
    5. Both lateral and mesial margins of spiracular plate, in lateral view, continuous with plane of dorsal profile of beak. . . . .*M. densirostris*, p. 908
    5'. Only lateral margin of spiracular plate, in lateral view, continuous (aligned) with plane of dorsal profile of beak, whereas mesial margin raised posteriorly. . . . . . .*M. stejnegeri*, p. 907
  4'. On vertex of skull, right premaxilla extended posteriorly beyond right nasal for distance of less than ¾ of length of nasal as exposed on vertex.
    6. Anterior part of jugal bent (not curled) up in front of edge of maxillary in antorbital notch; length of mandibular symphysis more than 200 mm. . . . .*M. bidens*, p. 908
    6'. Anterior part (tongue) of jugal curled up in front of edge of maxillary in antorbital notch; length of mandibular symphysis less than 200 mm. . . . .*M. carlhubbsi*, p. 907

Micropteron mirus (True)
True's Beaked Whale

1913. *Mesoplodon mirum* True, Smiths. Miscl. Coll., 60(25):1, March 14, type from Beauford Harbor, Carteret Co., North Carolina.
1919. *Mesoplodon mirus*, Harmer, Report on Cetacea stranded on the British coasts during 1918, British Mus., p. 18, March 22.

Total length to 17 ft. Type specimen, adult female (according to True, 1913:651): back, slate-black; lower sides, yellow-purple, flecked with black; median line of belly somewhat darker; a grayish area in front of vent; fins same color as back. With skull upright and dorsal profile of beak horizontal, premaxillary foramina (viewed from anterior end of skull) lower than horizontal plane transecting centers of maxillary foramina or straddling that plane; lachrymal extending 10 mm. or more beyond maxilla and forming apex of antorbital tubercle. Greatest length of vomer seen on palate, 137–187 mm.; tooth in mandible, 140–167 mm. anterior to mandibular symphysis and at apex of jaw; mandibular symphysis, 181–210 mm. long. (Cranial characters after Moore, 1966:44, 45.)

RANGE (Moore, 1966:54).—Nova Scotia: South Gut, St. Annes Bay, Cape Breton Island. Maine: Wells Beach. Connecticut: Mason Island, off Mystic. New York: Edgemere, Rockaway Beach, Long Island. New Jersey: 4 mi. below Seaside Park, Ocean Co. North Carolina: between Gull Island and Oregon Inlet, Dare Co.; type locality. Florida: Flagler Beach.

Micropteron europaeus (Gervais)
Gervais' Beaked Whale

1848–52. *Dioplodon europaeus* Gervais, Zoologie et paléontologie Françaises. . . , ed. 1, p. 4, a *nomen nudum*. Type locality, English Channel.
1855. *Dioplodon europaeus* Gervais, Histoire naturelle des Mammifères (Paris), 2:320, type locality, English Channel.
1866. *Dioplodon gervaisi* Deslongchamps, Bull. Soc. Linnéenne, Normandie, 10:176, type from English Channel.

Total length to 22 ft. Adult female from Rockaway Beach, New York (according to Raven, 1937:3), slaty black above, and paler on sides and underparts; pure white area, irregular in shape, *ca.* 10 in. in diameter, enclosing genital and mammary grooves. With skull upright and long axis of beak horizontal, premaxillary foramina lower than, or straddling, horizontal plane transecting centers of maxillary foramina. Greatest length of right nasal on vertex of skull 40–50 mm. long; visible part of vomer on palate less than 80 mm. long; mandibular symphysis less than 158 mm. long.

Total length to 16 ft. in the American specimen. With skull upright and long axis of beak horizontal, an anterior view reveals at least one of the pair of premaxillary foramina on or below a horizontal plane (paralleling long axis of beak) that transects centres of maxillary foramina. Visible part of vomer on palate more than 100 mm. long; in lateral view of antorbital tubercle, lachrymal relatively thick and not curling up around anterior end of frontal bone; greatest length (69, 76, and 81 mm.) of right nasal bone on vertex of each of the three known skulls more than twice the least distance between premaxillary foramina (30, 35, and 36 mm. respectively); greatest transverse span of combined premaxillary bones at midpoint in length of beak between 40 and 60 mm. (45–49 mm.); greatest width of temporal fossa measured approx. perpendicular to long axis of temporal fossa less than 60 mm. (After Moore and Gilmore, 1965:1239, 1240.)

RANGE.—California: Delmar, ca. 15 mi. N San Diego.

Micropteron carlhubbsi (Moore)
Hubbs' Beaked Whale

1963. *Mesoplodon carlhubbsi* Moore, Amer. Midland Nat., 70:422, October 18, type from La Jolla, California.

Total length to 18 ft. Blackish throughout except for whitish on anterior part of head and much of anterior part of each flipper. Maxillary ridge never lower than maxillary prominence when skull is upright and long axis of beak is horizontal; greatest breadth of nasals 59–82 mm.; least distance between right and left parts of premaxillary crests more than 27 mm.

RANGE (Moore, 1966:55, 56).—Pacific. Washington: Oyhut, Grays Harbor. California: Drakes Bay; San Simeon Bay; type locality.

Micropteron stejnegeri (True)
Stejneger's Beaked Whale

1885. *Mesoplodon stejnegeri* True, Proc. U.S. Nat. Mus., 8:585, November 21, type from Bering Island, Commander Islands, U.S.S.R.

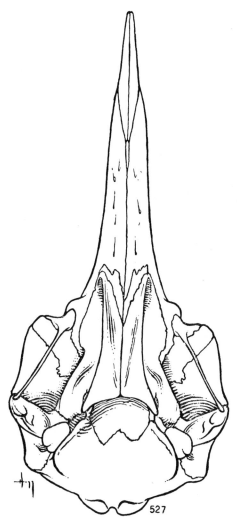

Fig. 527. *Micropteron europaeus*, Atlantic City, New Jersey, No. 23346 U.S.N.M., young ♂. Ventral view, X ⅕.

RANGE (Moore, 1966:51, 52).—New York: Rockaway Beach, Long Island. New Jersey: North Long Branch; Atlantic City. North Carolina: Kitty Hawk; Oregon Inlet. Florida: St. Augustine; Melbourne Beach; 8 mi. N Vero Beach; Key Largo; Boca Grande. Texas: Padre Island. Cuba, Pinar del Río: Cayo Alacranes. Jamaica, Parish of St. Thomas: Bull Bay.

Micropteron ginkgodens (Nishiwaki and Kamiya)
Ginkgo-toothed Beaked Whale

1958. *Mesoplodon ginkgodens* Nishiwaki and Kamiya, Bull. Japanese Soc. of Scientific Fisheries, 24(6, 7):445. Type from Oiso Beach, Japan.
1963. *Mesoplodon hotaula* Deraniyagala, Ceylon Today, 12(3):13, Pl. 1. Type from Ratmalana, Ceylon. See also Spolia Zeylanica, 30:79, 84.

Total length to 20 ft. Blackish throughout. With skull upright, premaxilla in lateral view rising posteriorly above nares in a curve abruptly turning (*ca.* 90° angle) anteriorly; with skull upright and with upper profile of beak horizontal, only mesial margin of spiracular plate rising posteriorly; pterygoid sinus not extending anterior to transverse plane intersecting base of beak; greatest span of premaxillary crest less than 140 mm.; least distance between premaxillary foramina, 35–48 mm. (after Moore, 1966:43, 44).

RANGE (Moore, 1966:53).—North Pacific. Alaska: St. Paul Island, Pribilof Islands, Bering Sea; Nushagak Peninsula, 30 mi. S Dillingham; 4 mi. S Kasilof River, W coast Kenai Peninsula; Egg Island, near Copper River. British Columbia: Long Beach near Trofino, Vancouver Island. Washington: Waatch River, Clallam Co. Oregon: Yaquina Bay, South Beach, near Newport.

Micropteron bidens (Sowerby)
Sowerby's Beaked Whale

1804. *Physeter bidens* Sowerby, Trans. Linnaean Soc. London, 7:310, The British Miscellany, p. 1, Pl. 1, type from Brodie House, coast of Elginshire, Scotland.
1871. *Micropteron bidens*, Malm, Kongl. Svenska Vetensk.-Akad. Handl., n.s., 9(1):96.
1817. *Delphinus sowerbensis* Blainville, Nouv. Dict. Hist. Nat., 9:177. New name for *Physeter bidens* Sowerby.
1827. *Heterodon dalei* Lesson, Manuel de Mammologie, p. 419, type from Le Hâvre, France.
1829. *Delphinus micropterus* G. Cuvier, Le règne animal . . ., ed. 2, 1:288, type from Le Hâvre, France since "name based solely on the incorrectly identified 'dauphin de Dale' of Blainville and F. Cuvier" according to Hershkovitz, U.S. Nat. Mus. Bull., 246:128, 129, February 28, 1966.
1850. *Nodus dalei*, Gray, Catalogue of the . . . Mammalia in the . . . British Museum, Cetacea, p. 74, where name is attributed to Wagler, 1830:34, and in synonymy of *Delphinorhynchus micropterus* [= *Delphinus micropterus* G. Cuvier, 1829].

Total length to 16½ ft. Blackish to bluish-black dorsally and usually grayish or whitish ventrally. With skull upright and long axis of beak horizontal, maxillary prominence higher than maxillary ridge; lachrymal extends no more than 10 mm. beyond maxilla (and in some not extended beyond maxilla); posterior lip of alveolus of mandibular tooth no more than 25 mm. anterior to mandibular symphysis; least distance between right and left parts of premaxillary crest less than 25 mm.

RANGE (Moore, 1966:48, 49).—North Atlantic. Newfoundland: Notre Dame Bay, Wild Bight; Trinity Bay, Chapel Arm. Massachusetts: Nantucket Island.

Micropteron densirostris (Blainville)
Blainville's Beaked Whale

1817. *Delphinus densirostris* Blainville, Nouv. Dict. Hist. Nat., ed. 2, 9:178. Type locality, unknown.
1846. *Ziphius sechellensis* Gray, *in* The zoology of the voyage of H.M.S. Erebus and Terror. . . , 1 (Mamm.):28, Pl. 6, Figs. 1, 2 (skull), type from Seychelles Islands, Indian Ocean.

Length to *ca.* 17 ft. Adult male from Peggy's Cove, Nova Scotia, nearly completely black; some reddish patches on underside between flippers and about preputial opening and some irregular spots on belly may have been pale gray, blue gray, or white in life. Weight 1783 pounds (after Raven, 1942:25). With dorsal profile of beak horizontal, mesial and lateral margins of left spiracular plate horizontal and on same plane as dorsal profile of beak; dorsal profile of proximal part of premaxilla in lateral view describing smooth curve behind and above nares. Maxillary foramina lacking sharp margins and exact distance between them therefore difficult to measure, differing therein from other North American species.

RANGE (Moore, 1966:50, unless otherwise noted).—Atlantic. Nova Scotia: Peggy's Cove. Massachusetts: Annisquam. New York: Long Island. New Jersey: Carsons Inlet. North Carolina: Bogue Banks, near Beaufort. South Carolina: vic. Charleston. Florida: near Crescent Beach (Caldwell and Caldwell, 1971:4). Bahama Islands: Abaco. Pacific. Midway Island.—This rare species has been recorded from all major oceans on both sides of the equator, but not from waters near the polar ice (see Hershkovitz, 1966:135, and Moore, 1966:51, 52).

Genus Ziphius G. Cuvier
Cuvier's Beaked Whale

1814. *Epiodon* Rafinesque, Précis des découvertes et travaux somiologiques . . ., p. 13. Type, *Epiodon urganantus* Rafinesque, a *nomen dubium*.
1823. *Ziphius* G. Cuvier, Recherches sur les ossemens fossiles. . . . Type, *Ziphius cavirostris* G. Cuvier.
1828. *Diodon* Lesson, Complément des oeuvres de Buffon, Histoire naturelle des animaux rares . . ., 1(Cétacés):123, 440 (name). Included species: *Diodon desmaresti* Risso type (= *Ziphius cavirostris* Cuvier) by original designation (p. 123), *D. sowerbyi* (sic) Blainville (= *Mesoplodon bidens* Sowerby); generic name preoccupied by *Diodon* Linnaeus 1758, a genus of fish and *Diodon* Storr, 1780, for the narwal (= *Monodon* Linnaeus).
1841. *Hypodon* Haldeman, 1841, Proc. Acad. Nat. Sci. Philadelphia, 1:127. Included species: *H. desmarestii* (type = *Ziphius cavirostris* Cuvier), *H. sowerbyi* (= *Mesoplodon bidens* Sowerby); proposed as a new name for *Diodon* Lesson, preoccupied.

1846. *Xiphius* Agassiz, Nomenclator zoologici index universalis, p. 389, an emendation of *Ziphius*.

1864. *Aliama* Gray, Proc. Zool. Soc. London, p. 242, November. Type, *Delphinus Desmarestii* Risso [= *Ziphius cavirostris* G. Cuvier].

1865. *Petrorhynchus* Gray, Proc. Zool. Soc. London, p. 524, October. Type, *Hyperoodon capensis* Gray [= *Ziphius cavirostris* G. Cuvier].

1865. *Ziphiorrhynchus* Burmeister, Revista Farmacéutica, October (also Ann. Mag. Nat. Hist., ser. 3, 17:94, pl. 3, April, 1866). Type, *Ziphiorrhynchus cryptodon* Burmeister [= *Ziphius cavirostris* G. Cuvier].

Total length to 23 ft. Color variable, usually grayish throughout, darker above than below. Two functional teeth in lower jaw; 1st 3 or 4 cervical vertebrae fused; mesethmoid ossified; premaxillae massive posteriorly, forming distinct cup around external nares; skull asymmetrical; nasal bones obscuring part of external nares when skull viewed from above; antorbital notch indistinct; dorsal fin convex on posterior surface; mouth reaching posteriorly less than halfway to level of eye.

RANGE.—Seas of both the Northern and Southern hemispheres. The genus is monotypic.

Ziphius cavirostris G. Cuvier
Cuvier's Beaked Whale

X ¹/₁₁₂

1814. *Epiodon urganantus* Rafinesque, Précis des découvertes et travaux somiologiques . . . , p. 13, a *nomen dubium*. Type locality, Sicily.

1823. *Ziphius cavirostris* G. Cuvier, Recherches sur les ossemens fossiles . . . , ed. 2, 5:352. Type locality, near Fos, Bouches-du-Rhone, France.

1826. *Delphinus Desmarestii* Risso, Histoire naturelle des animaux de l'Europe méridionale, 3:24, Pl. 3, Fig. 3. Type locality, Mediterranean Sea.

1846. *Delphinus philippii* Cocco, Wiegmann's Arch. für Naturgesch., sect. 1, 12:104, Pl. 4, Fig. C. Type locality, Straits of Messina, Mediterranean Sea.

1850. *Hyperoodon doumetii* Gray, Catalogue of the . . . Mammalia in the . . . British Museum, Cetacea, p. 68. Type locality, Corsica.

1851. *Hyperoodon gervaisi* Duvernoy, Ann. Sci. Nat., Paris, ser. 3 (Zool.), 15:49, 67. Type locality, coast of Aresquiers, near Frontignan, Dept. Hérault, France.

1864. *Ziphius indicus* Van Bénéden, Mém. Couronnés . . . Acad. Roy. Sci. Lettres et Beaux-Arts Belgique, 16:23, Pl. 1, June. Type locality, Cape of Good Hope, Republic of South Africa.

1865. *Delphinorhynchus australis* Burmeister, Allgem.

Zeitschr. Gesell. Naturw., Dresden, 26:262, August. Type locality, near Buenos Aires, Argentina.

1865. *Hyperoodon capensis* Gray, Proc. Zool. Soc. London, p. 359, October, type from Cape of Good Hope, Republic of South Africa.

1865. *Ziphiorrhynchus cryptodon* Burmeister, Revista Farmacéutica, October (also Ann. Mag. Nat. Hist., ser. 3, 17:94, pl. 3, April, 1866). Type locality, Buenos Aires, Argentina.

1865. *Hyperodon* [sic] *semijunctus* Cope, Proc. Acad. Nat. Sci. Philadelphia, 17:280 (also 274), type from Charleston, South Carolina. A misspelling of the generic name (*Hyperoodon*). A second misspelling thereof, incorrectly ascribed to Cope (*supra*) is *Hyperondon* . . . , Hall and Kelson, Mammals of North America, p. 810, March 31, 1959.

1867. *Epiodon patachonicum* Burmeister, Allgem. Zeitschr. gesammt. Naturw. (Dresden), 29:5, new name for *Ziphiorrhynchus cryptodon* Burmeister.

1871. *Ziphius aresques* Gray, Supplement . . . catalogue of seals and whales in the British Museum, p. 98—name in synonymy of *Epiodon desmarestii*, attributed to Gervais (1869–187?, Atlas ostéographie des cétacés, Pl. 21, Figs. 1–4), who figured the Aresquiès beaked whale (but did not name it in text).

1871. *Petrorhynchus mediterraneus* Gray, Supplement . . . catalogue of seals and whales in the British Museum, p. 98, type from Mediterranean Sea.

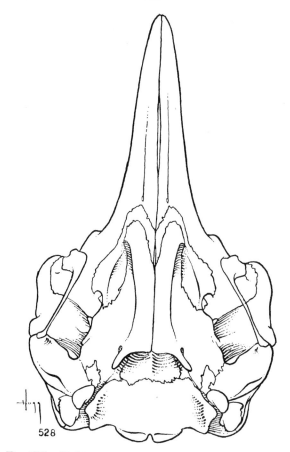

528

Fig. 528. *Ziphius cavirostris*, Newport, Rhode Island, No. 49599 U.S.N.M., ♂. Ventral view, X ⅛.

1872. *Epiodon Heraultii* Gray, Ann. Mag. Nat. Hist., (4), 10:469, type from Atlantic (France). Name based on the *Ziphius* of Aresquiès (Hérault), figured in Van Bénéden and Gervais, 1869–1871, Atlas ostéographie des cétacés, Pl. 21, Figs. 1–4 (skull).
1873. *Ziphius savii* Richiardi, Arch. per la Zool. (2), 3, Pls. 7–8 "[work not seen].—Van Bénéden and Gervais, 1880, Ostéographie des cétacés, p. 377 [N. Atlantic: Italy (type locality: coast of Pisa, Ligurian Sea)" . . .]. After Hershkovitz, U.S. Nat. Mus. Bull., 246:141, February 28, 1966.
1873. *Epiodon chathamiensis* Hector, Trans. Proc. New Zealand Inst., 5:164, Pls. 4–5, Figs. a–b, May. Type locality, Chatham Islands, E of New Zealand.
1880. *Ziphius (Epiodon) novae-zealandiae* von Haast, Trans. Proc. New Zealand Inst., 12:241, Pl. 8, May. Type locality, Lyttelton Harbor, New Zealand.
1883. *Ziphius grebnitzkii* Stejneger, Proc. U.S. Nat. Mus., 6:77, June 30, type from Bering Island, Commander Islands, Bering Sea.

See characters of the genus.

RANGE.—Atlantic. Rhode Island: Newport, thence southward to Florida: Conch Key, Monroe Co. Louisiana: outer beach 5 mi. S of N end of main island of Chandeleur Island Chain (Lowery, 1974:312). Texas: Galveston's West Beach (Schmidly and Melcher, 1974:456). Bahamas: *ca.* 10 mi. N Norman's Cay (Caldwell and Caldwell, 1971:158). Cuba (Erdman, 1962:277): Bahía de Matanzas; Caibarien. Barbados: Bathsheba, St. Joseph (Caldwell, *et al.*, 1971:52), thence into South American waters. Pacific. Alaska: Trident Bay, Akun Island (Fiscus, *et al.*, 1969:127), thence eastward to Kiska Island, and southward to Baja California: 4 mi. N San Luis Gonzaga Bay (Orr, 1966:339); Smith Island, N of Bahía de los Angeles (Orr, 1967:328).

Genus **Hyperoodon** Lacépède
Bottle-nosed Whales

1804. *Hyperoodon* Lacépède, Histoire naturelle des cétacées . . . , p. xliv, 319. Type, *Hyperoodon butskopf* Lacépède [= *Balaena ampullata* Forster].
1804. *Anarnak* Lacépède, Histoire naturelle des cétacées . . . , p. xxxviii, 164. Type, *Anarnak Groenlandicus* Lacépède, a *nomen dubium*.
1806. *Hyprodon* Duméril, Zool. Analytique, p. 28. Emendation of *Hyperoodon* Lacépède.
1811. *Uranodon* Illiger, Prodromus systematis mammalium et avium . . . , p. 143. Emendation of *Hyperoodon* Lacépède, 1804. Type, *Delphinus butzkopf* (sic) Bonnaterre, 1789.
1811. *Ancylodon* Illiger, Prodromus systematis mammalium et avium . . . , p. 143. Type, *Monodon Spurius* Fabricius [= *Balaena ampullata* Forster].
1811. *Uranodon* Illiger, Prodromus systematis mammalium et avium . . . , p. 143. Type, *Delphinus Butskopf* Bonnaterre [= *Balaena ampullata* Forster].
1814. *Bidens* Fischer, Zoognosia, Tabulis synopticis illustrata, 3:686. Type, *Delphinus diodon* "Hunter" (1787, Philos. Trans.) (= *Delphinus diodon* Lacépède, 1804, by monotypy).
1821. *Hyperdordon* Gray, London Med. Repos., 15:210. Typographical error for *Hyperoodon* Lacépède. Included species, *Delphinus edentus* (sic) Schreber (= *D. edentulus* Schreber = *Balaena ampullata* Forster).
1825. *Ceto-diodon* Jacob, Dublin Philos. Jour. Sci. Review, 1:72, Pl. [3], Fig. 8, March. Type, *Ceto-diodon hunteri* Jacob [= *Balaena ampullata* Forster].
1830. *Anodon* Wagler, Natürliches System der Amphibien, ftn. 3, p. 34. Perhaps a misspelling of *Aodon* Lesson. From included species, Hershkovitz, U.S. Nat. Mus. Bull., 246:143, February 28, 1966, selected *Balaena ampullata* Forster as type. Generic name preoccupied by *Anodon* Smith, a reptile.
1830. *Nodus* Wagler, Natürliches System der Amphibien, p. 34. Type, *Delphinus edentulus* Schreber, *in* Cuvier, Mammifères liv. 53. [*Nodus* Wagler 1830 is here placed as a synonym of *Hyperoodon* because of his *designation* of *Delphinus edentulus* Schreber as the type species. Some authors (*e.g.*, Hershkovitz, 1961:556) have chosen instead to make the *concept* of Wagler the basis of the generic name *Nodus* and point out that Wagler's concept differed from Schreber's. Basing the generic name *Nodus* 1830 on the alleged *concept* instead of on the designated *type* would make *Nodus* the senior generic synonym for the several species now listed under the generic name *Micropteron* 1849.]
1830. *Orca* Wagler, Natürliches System der Amphibien, p. 34 [included species: *Delphinus bidentatus* "Hunter" (type by subsequent designation (Iredale and Troughton, 1934, Mem. Australian Mus., 6:60) = *Delphinus bidentatus* Bonnaterre = *Balaena ampullata* Forster), *D. demarestii* (= *Ziphius cavirostris* Cuvier)].
1843. *Uperoodon* Gray, List of the . . . Mammalia in the . . . British Museum, p. xxiii. Misprint for *Hyperoodon* Lacépède or *Uranodon* Illiger.
1843. *Chaenodelphinus* Eschricht, Förh. Skand. Naturf. Ottende Møde, Kjøbenhavn, p. 651. Type, *Balaena rostrata* Müller [= *Balaena ampullata* Forster].
1846. *Chaenocetus* Eschricht, Overs. Kongl. Dansk. Vidensk. Selsk. Forh., for 1845, p. 17. Type, The "Naebhvae" or *Balaena ampullata* Forster.
1863. *Lagenocetus* Gray, Proc. Zool. Soc. London, p. 200, October. Type, *Lagenocetus latifrons* Gray [= *Balaena ampullata* Forster].
1866. *Lagocetus* Gray, Catalogue of seals and whales in the British Museum, ed. 2, p. 82, an emendation of *Lagenocetus* Gray.

Total length to 32 ft. Grayish-black dorsally, light gray to yellowish-gray ventrally. Single pair of small, functional teeth in lower jaw, situated at anterior tip of mandible; beak of skull with high crests formed by elevation of maxillary bones on either side of premaxillae anterior to external nares; mesethmoid not fully ossified; all cervical vertebrae fused; dorsal fin concave on posterior border; forehead high and beak distinct owing to high maxillary crest.

Hyperoodon ampullatus (Forster)
Northern Bottle-nosed Whale

1770. *Balaena ampullata* Forster, *in* Kalm, Travels into North America . . . , 1:18, footnote. (Based on "beaked whale" of Pennant, The British Zoology, pt. 3, p. 43 [= p. 59, Pl. 5, Fig. 1], 1766.) Type locality, Maldon, Essex, England.

1902. *Hyperoodon ampullatus*, Rhoads, Science, n.s., 15:756, May 9.

1776. *B[alaena]. rostrata* Müller, Zoologiae Danicae prodromus . . . , p. 7. Type locality, unknown.

1780. *Monodon Spurius* Fabricius, Fauna Groenlandica . . . , p. 31. Type locality, Greenland Seas. Probably a junior synonym of *Hyperoodon ampullatus* Forster.

1789. *D[elphinus]. bidentatus* Bonnaterre, Tableau . . . des trois règnes de la nature, cétologie, p. 25, Pl. 11, Fig. 3. Type locality, River Thames, England.

1789. *D[elphinus]. Butskopf* Bonnaterre, Tableau . . . des trois règnes de la nature, cétologie, p. 25. Type locality, Honfleur, France.

1801. *Delphinus bidens* Shaw, General Zoology, 2(2):514. Type, the "bottle-nosed whale of Dale."

1802[?]. *Delphinus? edentulus* Schreber, Die Säugthiere, Pl. 347 (Honfleur whale *ex* Baussard, 1789, with technical name as caption). Type, the animal figured; no text.

1804. *Anarnak Groenlandicus* Lacépède, Histoire naturelle des cétacées. . . , p. 164. Type locality, Greenland Seas. Possibly a junior synonym of *Hyperoodon ampullatus* Forster.

1804. *Delphinus diodon* Lacépède, Histoire naturelle des cétacées . . . , p. 309, Pl. 13, Fig. 3. Type locality, near London, England.

1812. *Delphinus coronatus* Fremenvillé, Bull. Soc. Philom. Paris, 3:71. Type locality, Spitzbergen. Possibly a junior synonym of *Hyperoodon ampullatus* Forster.

1817. *Delphinus chemnitzianus* Blainville, *in* Desmarest, Nouv. Dict. Hist. Nat., 9:175, name based on *Balaena rostrata* of Chemnitz, 1779. Type locality, Spitzbergen, North Atlantic.

1820. *Hyperoodon borealis* Nilsson, Skand. Fauna, 1:404, substitute for the "bottle-head whale" of authors.

1822. *Delphinus Hunteri* Desmarest, Encyclopédie méthodique . . . mammalogie, 2:520. Type locality, River Thames, England.

1822. *Delphinus hyperoodon* Desmarest, Encyclopédie méthodique . . . mammalogie, 2:521. Type locality, near Honfleur, France.

1828. *Hyperoodon honfloriensis* Lesson, Histoire naturelle . . . des mammifères et des oiseaux découverts depuis 1788, cétacés, p. 137, 440. Type locality, Honfleur, France.

1843. *Delphinus quadridens* Burguet, Actes Soc. Linnéenne Bordeaux, 13:304. Type from Arcachon, France.

1846. *Hyperoodon latifrons* Gray, *in* The zoology of the voy-

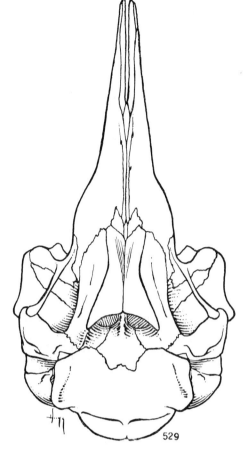

Fig. 529. *Hyperoodon ampullatus*, Norway, No. 14449 U.S.N.M., sex unknown. Ventral view, X approx. ⅑.

age of H.M.S. Erebus and Terror . . . , 1 (Mamm.):27, Pl. 4, type from Orkney Islands, Scotland.

1851. *H[yperoodon]. baussardi* Duvernoy, Ann. Sci. Nat., Paris (3), 15:67. Name based primarily on the "butzkopf" or "duaphin de Honfleur" of Baussard, according to Hershkovitz, U.S. Nat. Mus. Bull., 246:148, February 28, 1966.

See characters of the genus. The only other species of the genus, *H. planifrons*, inhabits waters of the Southern Hemisphere.

RANGE (Mitchill and Kozicki, 1975:1030, 1031, unless otherwise noted).—"reported from Davis Straits and the entrance to Hudson Strait" (Leatherwood, *et al.*, 1976:69). Quebec: Cap Martin, Gulf of St. Lawrence (49° 00′ N). Newfoundland: Deldo Arm, Trinity Bay. Nova Scotia: Cobequid Bay; South Beach, Wallace Lake, Sable Island (1 near W end and 1 from *ca.* 2 mi. E of east end of Lake). Massachusetts: Beverly Farms; North Dennis. Rhode Island: Newport.

SUBORDER **MYSTICETI**

North American species revised by True, Smiths. Contrib. Knowl., 33:1–332, 97 figs., 50 pls., August 29, 1904.

Whale bone present; teeth absent after birth; paired narial openings; maxillae ending in front of orbital processes of frontal bones; maxillae having orbital plates; nasal bones forming anterior part of roof of nasal passages; passage of nares directed upward and forward.

KEY TO FAMILIES OF MYSTICETI

1. External grooves on throat absent; cervical vertebrae fused; rostral and cranial elements of skull interdigitate only slightly or not at all; nasals and nasal branches of premaxillae situated entirely anterior to level of orbital wings of frontals.

Balaenidae, p. 919
1'. External grooves on throat present; cervical vertebrae separate; rostral and cranial elements of skull interdigitate; nasals and nasal branches of premaxillae not situated entirely anterior to level of orbital wings of frontals.
    2. Dorsal fin absent; external throat grooves 2–4; parietal bones entirely behind nasals and nasal branches of maxillae and premaxillae [occipital shield not extending forward over or beyond articular portion of squamosal]; frontal bones clearly exposed on vertex. . . . . . .Eschrichtidae, p. 912
    2'. Dorsal fin present; external throat grooves numerous; parietal bones extending forward laterally beyond level attained by nasals and nasal branches of maxillae and premaxillae [occipital shield extending forward over orbit and beyond articular portion of squamosal]; frontal bones barely or not at all exposed on vertex.

Balaenopteridae, p. 913

FAMILY **ESCHRICHTIDAE**—Gray Whale

Total length to 50 ft. Mottled gray throughout. Dorsal fin absent, small humps on posterior mid-dorsal region; 2–4 short throat grooves; baleen yellowish, less than 20 in. long; snout high and rigid; mouth bisects head. Skull having deep rather than broad rostrum; supraorbital processes of frontal bones broad at base, nasal bones greatly enlarged; occipital shield separating parietal bones at vertex of skull, not extending forward over level of orbit or beyond anterior level attained by articular part of squamosal; parietal bone transversely united by narrow band of in-terparietal; surface of occipital shield tuberculate; premaxillae forming rim to external nares; mandible only slightly bowed laterally; cervical vertebrae separate.

Gray whales are migratory, moving from Arctic waters southward along the California Coast in December and January to their breeding lagoons along the coasts of Baja California, Sonora, and Sinaloa. This family contains a single genus.

Genus **Eschrichtius** Gray—Gray Whale

1864. *Eschrichtius* Gray, Ann. Mag. Nat. Hist., ser. 3, 14:350, November. Type, *Balaenoptera robusta* Lilljeborg [= *Balaena gibbosa* Erxleben].
1869. *Rhachianectes* Cope, Proc. Acad. Nat. Sci. Philadelphia, 21:15. Type, *Agaphelus glaucus* Cope [= *Balaena gibbosa* Erxleben].

See characters of the family. The genus, as here considered, is monotypic.

Eschrichtius gibbosus (Erxleben)
Gray Whale

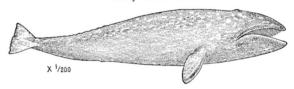

X ¹⁄₂₀₀

1777. [*Balaena*] *gibbosa* Erxleben, Systema regni animalis . . . , p. 610. Type locality, coast of New England. (Based on "scrag whale" of Dudley, Phil. Trans. Roy. Soc. London, 33:258, 1725.)
1937. *Eschrichtius gibbosus,* van Deinse and Junge, Temminckia, 2:181.
1861. *Balaenoptera robusta* Lilljeborg, Förh. Skand. Naturf. Ottende Møde, Kjöbenhavn, for 1860, 8:602. Type locality, Sweden.
1868. *Agaphelus glaucus* Cope, Proc. Acad. Nat. Sci. Philadelphia, 20:160, type from Monterey Bay, California.

The law of priority requires use of the name *E. gibbosus* (Erxleben 1777) instead of *E. glaucus* (Cope 1868) for the gray whale. See Hall and Kelson (1959:834). The correct name for the north Pacific population would be *Eschrichtius gibbosus glaucus* Cope should the existing population be found to differ subspecifically from the north Atlantic population, which was exterminated "about 1725" (Morzer Bruyns, 1971:163) and now is known only from osseous remains.

See characters for the family.

RANGE.—Pacific. Alaska (Maher, 1960:262, as *E. glaucus,* and as sight records, unless otherwise noted):

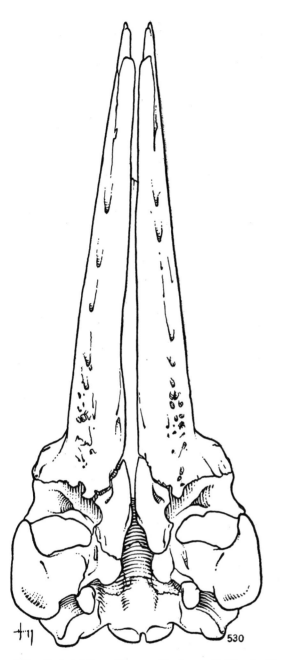

Fig. 530. *Eschrichtius gibbosus*, Monterey, California, No. 13803 U.S.N.M., sex unknown. Ventral view, X $\frac{1}{15}$.

Point Barrow (Maher and Wilimovsky, 1963:17, caught by Eskimos), thence eastward along coast to Beechey Point; Foggy Island; mouth Shaviovik River; Flaxman Island; Barter Island, and westward along coast to Wainwright (Maher, 1960:259, caught by Eskimos); Icy Cape; Cape Sabine (Maher, 1960:260); Bering Strait (Hall and Kelson, 1959:834), thence southward to latitude of Jalisco. Range formerly included North Atlantic.

FAMILY BALAENOPTERIDAE
Fin-backed Whales

Total length to 105 ft. Dorsal fin present; throat grooves numerous; sheets of whalebone numerous, only moderately long (shorter and broader than in Balaenidae). Skull having rostrum flat and broad; lateral bowing at lower jaws increases buccal capacity; parietal bones extending anterolaterally beyond posterior margin of nasal bones; frontal bones not exposed on vertex of skull; occipital shield extending anteriorly over level of orbit and beyond anterior level attained by articular part of squamosal; ventral surface of squamosal bones deeply concave; ribs articulating with transverse processes of vertebrae; cervical vertebrae separate.

KEY TO NORTH AMERICAN GENERA OF BALAENOPTERIDAE

1. Pectoral appendages long, approx. $\frac{1}{3}$ total length of animal, serrate on anterior border; acromion and coracoid processes of scapula rudimentary. . . . .*Megaptera*, p. 918
1'. Pectoral appendages short, $\frac{1}{6}$–$\frac{1}{7}$ of total length of animal, smooth on anterior border; acromion and coracoid processes of scapula well developed. *Balaenoptera*, p. 913

Genus **Balaenoptera** Lacépède
Fin-backed Whales

1804. *Balaenoptera* Lacépède, Histoire naturelle des cétacées . . . , pp. xxxvi, 114. Type *Balaenoptera gibbar* Lacépède (= *Balaenoptera physalus* Linnaeus) by original designation and monotypy (*i.e.*, "première espèce" of the "premier sous-genre," which included only the first species; cf. Hershkovitz, Fieldiana: Zool., 39:559, 1961).

1831. *Rorqual* Voigt, *in* Cuvier, Das Thierreich, 1:342 [subgenus of *Balaena* Linnaeus; included species: *B. boops* ("die Jubarte" = *Balaenoptera musculus*), *B. musculus* ("der Rorqual" = *B. physalus* Linnaeus, type by virtual tautonomy)].—Palmer, 1904, Index Gen. Mamm., p. 612 [name attributed to G. Cuvier, 1829, Le règne animal . . . , ed. 2, 1:298, wherein only the French vernacular plural "rorquals" is used for a subdivision of the genus *Balaena* Linnaeus].

1836. *Rorqualus* Cuvier, Histoire naturelle des cétacées . . . , p. 303 [included species: *R. boops* (= *Balaenoptera musculus* Linnaeus), *R. musculus* (= *B. physalus* Linnaeus, type by virtual tautonomy), *R. antarctica* (= *Megaptera novaeangliae* Borowski)].

1842. *Ptychocetus* Gloger, Gemeinnützige Hand- und Hilfs-buch der Naturgeschichte, 1:xxxiv, 174 [included species: *Balaenoptera borealis*, *B. boops*, *B. rostrata*, *B. musculus* Linnaeus designated as type by Hershkovitz, U.S. Nat. Mus. Bull., 246:152, February 28, 1966].

1849. *Pterobalaena* Eschricht, Kongl. Danske Vidensk. Selsk. Skrift. Nat. Math. Afd., (5), 1:108 [type species: the "finhval" of the northern seas (= *Balaena physalus* Linnaeus)].

1849. *Ogmobalaena* Eschricht, Kongl. Danske Vidensk. Selsk. Skrift. Nat. Math. Afd., (5), 1:108 [type species: the "furehvaler eller Rörhvaler" of the northern sea (= *Balaena physalus* Linnaeus)].

1864. *Benedenia* Gray, Proc. Zool. Soc. London, p. 211, November. Type, *Benedenia knoxii* Gray [= *Balaena physalus* Linnaeus].

1864. *Sibbaldus* Gray, Proc. Zool. Soc. London, p. 222, November. Type, by tautonomy, *Sibbaldus borealis* Gray [= *Balaena musculus* Linnaeus].

1865. *Sibbaldius* Flower, Proc. Zool. Soc. London, for 1864, p. 391, May, an emendation of *Sibbaldus* Gray.

1866. *Cuvierius* Gray, Catalogue of seals and whales in the British Museum, p. 114. Type, *Physalus latirostris* Flower [= *Balaena musculus* Linnaeus].

1866. *Rudolphius* Gray, Catalogue of seals and whales in the British Museum, p. 170. Type, *Balaenoptera laticeps* Gray [= *Balaenoptera borealis* Lesson].

1866. *Swinhoia* Gray, Catalogue of seals and whales in the British Museum, p. 382 [subgenus of *Balaenoptera* Lacépède; type species: *Balaenoptera swinhoei* Gray (= *Balaena physalus* Linnaeus), by tautonomy and monotypy].

1867. *Flowerius* Lilljeborg, Nova Acta Reg. Soc. Sci. Upsala, (3), 6:6, 11 [type species: *Flowerius gigas* Eschricht, (= *Balaena musculus* Linnaeus), by monotypy].

1868. *Agaphelus* Cope, Proc. Acad. Nat. Sci. Philadelphia, 20:159 [included species: "*Balaena gibbosa* of Erxleben" (= *Balaenoptera acutorostrata* Lacépède), type by original designation; *Agaphelus glaucus* Cope (= *Balaena gibbosa* Erxleben)].

1874. *Stenobalaena* Gray, Ann. Mag. Nat. Hist., (4), 14:305 [type species: *Stenobalaena xanthogaster* Gray (= *Balaena physalus* Linnaeus), by monotypy].

1874. *Dactylaena* Gray, Ann. Mag. Nat. Hist., (4), 13:449 [subgenus of *Balaenoptera* Lacépède; type species: *Balaenoptera huttoni* Gray (= *Balaenoptera acutorostrata* Lacépède), by monotypy].

Total length to 105 ft. Black or bluish dorsally; white to yellowish on underparts or at least on flanks. Narrow but evident part of nasal bones exposed on vertex of skull; acromion and coracoid processes of scapula well developed. Pectoral appendages short, smooth on anterior border.

KEY TO NORTH AMERICAN SPECIES OF BALAENOPTERA

1. Rostrum triangular as viewed from above; occipital bone separate from nasal bones.
 2. Length of baleen, exclusive of bristles, approx. 10 in.; mid-dorsal surface of pectoral appendage with white band. . . . .*B. acutorostrata*, p. 914
 2'. Length of baleen, exclusive of bristles, 28–36 in.; dorsal surface of pectoral appendage dark gray, bluish or brown.
  3. Major baleen plates 350–400; plates purple and white; ventral surface of tail flukes white. *B. physalus*, p. 916
  3'. Major baleen plates fewer than 350; plates black with white fringe, or white in front of mouth, gray-black in rear; ventral surface of tail flukes usually dark blue to brown.
   4. Major baleen plates 320–340, and longer than 21½ in.; plates black with white fringe; chin, throat, and chest white; lateral ridges on snout poorly developed. . . . . . .*B. borealis*, p. 915
   4'. Major baleen plates approx. 270, and shorter than 21½ in.; plates white in front, gray-black in rear; chin, throat, and chest dark; lateral ridges on snout well developed. *B. edeni*, p. 915
1'. Rostrum parallel-sided posteriorly and curved inward at tip as viewed from above; occipital bone touching posterior margin at nasal bones. . . .*B. musculus*, p. 917

Balaenoptera acutorostrata Lacépède
Minke Whale

Total length to 33 ft. Grayish-black dorsally, white ventrally including ventral aspects of pectoral appendages and flukes; dorsal surface of pectoral appendage with conspicuous white band; baleen yellowish-white to pure white. External throat grooves reaching to chin; dorsal fin prominent, falcate; greatest length of baleen, exclusive of bristles, approx. 10 in. Calves 9 ft. long at birth.

Balaenoptera acutorostrata acutorostrata Lacépède

1780. *Balaena rostrata* O. Fabricius, Fauna Groenlandica . . . , p. 40. Type locality, Greenland seas. Not Müller, 1776.

1804. *Balaenoptera acuto-rostrata* Lacépède, Histoire naturelle des cétacées . . . , p. xxxvii. Type locality, European seas.

1837. *Rorqualus minor* Hamilton, Jardine's Naturalist's Library, Mammalia, 6(Whales):142, Pl. 7. Type locality, Firth of Forth, Scotland. Name based on the *Balaena rostrata* of Knox, 1834 (Edinburgh New Philos. Jour., 18:197).

1838. *Balaena minimus borealis* Knox, Catalogue of anatomical preparations illustrative of the whale, p. 14, type from Firth of Forth.

1845. *Balaenoptera eschrichtii* Rasch, Nytt. Mag. Naturv., 4:123. Type locality, Swedish coasts.

1849. *Pterobalaena minor groenlandica* Eschricht, Kongl. Dansk Vidensk. Selsk. Skrift. Nat. Math. Afd., 1:109. Type locality, Greenland. New name for Greenland variety of *B. rostrata* Fabricius.

1849. *Pterobalaena minor bergensis* Eschricht, Kongl. Dansk. Vidensk. Selsk. Skrift., Nat. Math. Afd., 1:109. Type locality, Norway.

1850. *Balaenoptera microcephala* Gray, Catalogue of the . . . Mammalia in the . . . British Museum, Cetacea, p. 32. Brandt MS. name in synonymy of *Balaenoptera rostrata*.

1862. *P[terobalaena]. n[ana]. pentadactyla* Barkow, Das Leben der Walle . . . , pp. 5–17 (name p. 17). Type from unknown locality, possibly from North Atlantic.

1862. *P[terobalaena]. n[ana]. tetradactyla* Barkow, Das Leben der Walle, p. 17. New name for *P. minor* Eschricht.

Balaenoptera acutorostrata bonaerensis Burmeister, 1867, a Southern Hemisphere subspecies, and *B. acutorostrata thalmaha* Deranyagala, 1963, from Ceylon are extralimital.

RANGE.—Atlantic. Franklin: Baffin Bay, thence southward to Florida: Spring Creek, Wakulla Co. (Lowery, 1974:374); Little Duck Key. Louisiana: 4 mi. W Holly Beach (*loc. cit.*). Bahamas: Little Bahama Bank, *ca.* 15 mi. E Great Abaco Island, 20 mi. N en-

trance Northeast Providence Channel (Struhsaker, 1967:483).

Balaenoptera acutorostrata davidsoni Scammon

1872. *Balaenoptera davidsoni* Scammon, Proc. California Acad. Sci., 4:269, October 4, type from Admiralty Inlet, Puget Sound, Washington.

1977. *Balaenoptera a[cutorostrata]. davidsoni*, Rice, U.S. Dept. of Commerce, National Oceanic and Atmospheric Administration, Tech. Rept., National Marine Fisheries Serv., SSRF-711:6, April.

RANGE.—Pacific. Bering Sea, thence southward to Baja California.

Balaenoptera edeni Anderson
Bryde's Whale

1879. *Balaenoptera edeni* Anderson, Anatomical and zoological researches . . . expeditions to western Yunnan in 1868 and 1875 . . . , p. 551, Pl. 44. Type locality, Thaybyoo Choung, Gulf of Martaban, between Sittang and Beeling rivers, Burma.

1912. *Balaenoptera brydei* Olsen, Tidens Tegn. [a Norwegian newspaper], November 12, type from Saldanha Bay, western Cape Province, Republic of South Africa.

Total length to 50 ft. Prominent lateral ridges on each side of median ridge of snout; chin, throat, and chest dark; gray band crosses white belly; greatest length of baleen 19 in.; otherwise closely resembles sei whale.

RANGE.—Tropical and subtropical oceans. Atlantic. Virginia: Walnut Point, Northumberland Co. (Miller, 1927:111, as "Pollack whale. . . . *Balaenoptera borealis*"). Florida: Pablo Beach, *ca.* 18 mi. E Jacksonville (Miller, 1924b:1, as "Pollack whale. . . . *Balaenoptera borealis*"); near Panacea (Rice, 1965:114), thence southward to Grenada (Hershkovitz, 1966:159). Pacific. California: *ca.* 1 km. from shore, *ca.* 32° 47′, 118° 44′ (Morejohn and Rice, 1973:313), thence southward to Baja California (Hershkovitz, 1966:159).

Balaenoptera borealis Lesson
Sei Whale

Total length to 62 ft. Gray to bluish-gray dorsally, laterally and on venter posterior to anus; anterior to anus, venter white; baleen blackish with curly white "hairs" on proximal surface. External throat grooves not reaching chin; dorsal fin high and falcate, approx. two-thirds of the way back along mid-dorsal line; major baleen plates 320–340; greatest length of baleen, exclusive of bristles, approx. 26 in. Calves are reported to be 15–16 ft. long at birth. The subspecies *Balaenoptera borealis schlegelli* Flower, 1865, of the Southern Hemisphere (of which *Balaenop-*

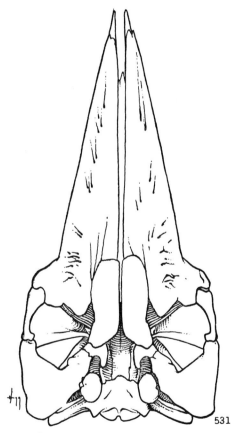

Fig. 531. *Balaenoptera acutorostrata acutorostrata*, Monomoy Point Lighthouse, Harwich Port, Massachusetts, No. 20931 U.S.N.M., sex unknown. Ventral view, X $\frac{1}{10}$.

tera alba Giglioli, 1870, is a synonym) is a larger animal than *B. b. borealis* of the Northern Hemisphere.

Balaenoptera borealis borealis Lesson

1828. *Balaenoptera borealis* Lesson, Histoire naturelle . . . des mammifères et des oiseaux découverts depuis 1788, cétacés, p. 342. Type locality, Gromitz, Lubeck Bay, Schleswig-Holstein, Germany.

1822. *Balaena rostrata* Rudolphi, Abh. k. Akad. Wiss., Berlin, for 1820–21, pp. 27–40, 5 pl. Type locality, North Sea. Not Müller, 1776, nor O. Fabricius, 1780.

1844. *Balaenoptera arctica* Temminck and Schlegel, *in* Siebold, Fauna Japonica, mamms. marins, p. 26. Type locality, Japan. Not Schlegel, 1841.

1846. *Balaenoptera laticeps* Gray, *in* The zoology of the voyage of H.M.S. Erebus and Terror . . . , 1(Mamm.):20. Type locality, North Sea, coast of Holstein.

1846. *Balaenoptera iwasi* Gray, *in* The zoology of the voyage of H.M.S. Erebus and Terror . . . , 1(Mamm.):20, name based primarily on the *Balaenoptera arctica* or *"iwasi"* of Temminck, 1841.

RANGE.—Atlantic. Coast of Labrador, thence southward to Mississippi: Mississippi Sound side of Ship Island (Gunter and Overstreet, 1974:479). Louisiana: near mouth Fort Bayou on Breton Sound (Lowery, 1974:370). Campeche: Gulf of Campeche (Miller, 1928:171). Cuba: Ensenada de Mora (Varona, 1965:1). Pacific. Bering Sea, thence southward to Baja California.

Balaenoptera physalus (Linnaeus)
Fin Whale or Fin-backed Whale

X ¹⁄₃₂₄

Total length to 81 ft. Grayish-black dorsally, white ventrally including tail flukes; left ramus grayish externally, whereas right ramus white externally; baleen purple and white, sometimes lead-colored. Posterior mid-dorsal region distinctly and acutely ridged; dorsal fin small, situated posterior to middle of body; major baleen plates 350–400, their greatest length, exclusive of bristles, approx. 36 in. Calves 21–22 ft. long at birth. The subspecies *Balaenoptera physalus quoyi* (Fischer, 1829), of the Southern Hemisphere (of which *Balaenoptera patachonicus* Burmeister, 1865, is a synonym) is said to be a larger animal than *B. p. physalus* of the Northern Hemisphere.

Balaenoptera physalus physalus (Linnaeus)

1758. [*Balaena*] *physalus* Linnaeus, Syst. nat., ed. 10, 1:75. Type locality, Spitzbergen seas (see Thomas, Proc. Zool. Soc. London, p. 156, March 22, 1911).

1862. *Balaenoptera physalus*, Schlegel, De Dieren van Nederland . . . , Zoogdieren, p. 101.

1758. [*Balaena*] *boops* Linnaeus, Syst. nat., ed. 10, 1:76. Type locality, "In Oceano septentrionali."

1792. *B*[*alaena*]. *m*[*ysticetus*]. *major* Kerr, The animal kingdom . . . , p. 357. Name based on "Raj. pisc. 16" (= Ray, 1713, Synopsis methodica piscium, p. 16), a composite, but restricted by Hershkovitz, U.S. Nat. Mus. Bull., 246:165, February 28, 1966, who designated "Arctic seas" as type locality.

1804. *Balaenoptera gibbar* Lacépède, Histoire naturelle des cétacées . . . , pp. xxxvi and 114. Type locality, Arctic Ocean near Greenland.

1804. *Balaenoptera rorqual* Lacépède, Histoire naturelle des cétacées . . . , p. xxxvi and 126. Type locality, Eastern North Atlantic, Scotland to Mediterranean.

1811. *Balaena sulcata* Neill, Mem. Wernerian Nat. Hist. Soc., 1:212. Type locality, near Alloa, Scotland.

1828. *Balaenoptera mediterraneensis* Lesson, Histoire naturelle . . . des mammifères et des oiseaux découverts depuis 1788, cétacés, p. 442, a renaming of *Balaenoptera rorqual* Lacépède. Type locality, Mediterranean.

1828. *Physalis vulgaris* Flemming, A history of British animals . . . , p. 32. Type locality, unknown.

1828. [*Physalus*] *verus* Bilberg, Synop. faunae Scandinaviae, p. 41. New name for *Balaena physalus* Linnaeus.

1829. *Balaena antiquorum* J. B. Fischer, Synopsis mammalium, p. 525. Type locality, Mediterranean Sea.

1829. *Balaenoptera aragous* Farines and Carcassonne, Mém. sur un Cétacé échoué 27 Nov. 1928 [*sic*] sur la Côte de Saint-Cyprien, p. 6. Type locality, Coast of St. Cyprien, Corsica.

1840. *Balaenoptera tenuirostris* Sweeting, Charlesworth's Mag. Nat. Hist., 4:343, July. Type locality, Charmouth Beach, England.

1841. *Balaena sulcata arctica* Schlegel, Abh. Gebiete Zool., 1:38, 6 pls., 5 figs. Type locality, coast of Holland.

1856. *Physalus duguidii* Heddle, Proc. Zool. Soc. London, p. 187, Pls. 44–45, December 11. Type locality, Island of Laman, Orkney Islands, Scotland.

1857. *Pterobalaena communis* Van Bénéden, Bull. Acad. Roy. Sci., Lettres, Beaux-Arts, Belgique, (26), (2), 1:390–403, type from near Vlieland Island, Netherlands.

1861. *Balaena robusta* Lilljeborg, Förh. Skand. Naturf. Ottende Møde, Kjöbenhavn, for 1860, 8:602. Type locality, Sweden.

1862. *Pterobalaena Gigantea michrochira* Barkow, Das Leben der Walle, p. 17. New name for *Balaena physalus* authors and *B. arctica* Schlegel.

1864. *Benedenia knoxii* Gray, Proc. Zool. Soc. London, p. 212, 2 figs., November. Type locality, coast of North Wales.

1866. *Balaenoptera swinhoii* Gray, Proc. Zool. Soc. London, for 1865, p. 725, 6 figs., April. Type locality, Takow, Formosa.

1868. *Swinhoai chinensis* Gray, Synopsis of the . . . whales and dolphins in the . . . British Museum, p. 3, a renaming of *Balaenoptera swinhoii* Gray.

1869. *Balaenoptera velifera* Cope, Proc. Acad. Nat. Sci. Philadelphia, p. 16, July 20. Type locality, "Oregon coasts."

1869. *S[ibbaldius]. tuberosus* Cope, Proc. Acad. Nat. Sci. Philadelphia, p. 16, July 20. Type locality, coast of eastern North America.

1869. *Sibbaldius tectirostris* Cope, Proc. Acad. Nat. Sci. Philadelphia, p. 17, July 20. Type locality, Sinepuxent Bay, Maryland.

1879. *Balaenoptera blythii* Anderson, Anatomical and zoological researches . . . expeditions to western Yunnan in 1868 and 1875 . . . , p. 564. Type locality, Indian Coast.

1884. *Dubertus rhodinsulensis* Trumbull, *in* G. B. Goode, Fisheries and fishery industries of the United States, sec. 1, 1:29, a *nomen nudum*. Type locality, Rhode Island.

1901. [*Balaenoptera velifera*] *copei* Elliot, Field Columb. Mus., Publ. 45, Zool. Ser., 2:13, March 16, type from Shumagin Islands, Alaska.

RANGE.—Atlantic. Greenland: Melville Island, thence southward to the Caribbean Sea. Pacific. Alaska: St. Paul Island, Pribilof Islands, thence southward to Baja California.

Balaenoptera musculus (Linnaeus)
Blue Whale or Sulphur-bottomed Whale

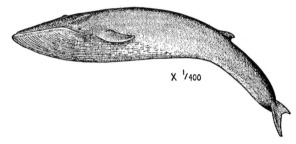

X 1/400

Total length to 105 ft. Bluish dorsally with patches of gray on back and sides; yellowish to white ventrally; pectoral appendages white below; baleen black. Rostrum parallel-sided posteriorly but sides curved inward at tip as viewed from above instead of triangular as in other species of *Balaenoptera;* occipital shield touching nasal bones, instead of not touching nasal bones as in other species of the genus; calves 23–26 ft. long at birth. Blue whales, the largest known animals, occur in summer near polar pack ice in both hemispheres.

Balaenoptera musculus intermedia Burmeister, 1871 (of which *Sibbaldius* [*sic*] *antarcticus* Burmeister, 1866 and *Balaenoptera miramaris* Lahille, 1898, are synonyms), of the Southern Hemisphere, is a larger animal than *B. m. musculus* of the Northern Hemisphere, and *B. m. brevicauda* Zemsky and Boronin, 1964, possibly a *nomen nudum*, validly named by Ichihara, 1966, of the Indian Ocean, is a smaller animal than *B. m. musculus*.

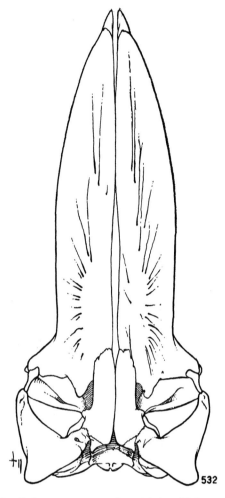

Fig. 532. *Balaenoptera musculus*, vicinity of Balaena Station, Hermitage Bay, Newfoundland, No. 49757 U.S.N.M., ♂. Ventral view, X 1/16.

Balaenoptera musculus musculus (Linnaeus)

1758. [*Balaena*] *musculus* Linnaeus, Syst. nat., ed. 10, 1:76. Type locality, Firth of Forth, Scotland (see Thomas, Proc. Zool. Soc. London, p. 156, March 22, 1911).

1903. *Balaenoptera musculus*, Racovitza, Expédition Antarctique Belgique, Cetacea, pp. 33, 54, Figs. 12–13.

1804. *Balaenoptera jubartes* Lacépède, Histoire naturelle des cétacées . . . , p. xxxvii and 120. Type locality, Greenland seas to Iceland.

1832. *Balaenoptera rorqual* Dewhurst, Mag. Nat. Hist., 5:214, type, the Ostend skeleton (described while exhibited in Charing Cross, London) from a whale found dead floating in the North Sea between England and Belgium.

1837. *Rorqualus borealis* Hamilton, Jardine's Naturalist's Library, Mammalia, 6(Whales):125, 138, Pl. 6, type from North Berwick, Firth of Forth, Scotland. Not *Balaenoptera borealis* Lesson, 1828.

1847. *Physalus (Rorqualus) sibbaldi* Gray, Proc. Zool. Soc. London, p. 92, July 20. Type locality, coast of Yorkshire, England.

918 MAMMALS OF NORTH AMERICA

1857. *Balaenoptera gigas* Reinhardt, *in* Rink, Grönland, geografisk og statistisk beskrevet, Kjöbenhavn, 1(2):10. Type locality, Greenland seas.

1859. *Balaenoptera indica* Blyth, Jour. Asiatic Soc. Bengal, 29:488. Type locality, Sondip, Bay of Bengal, India.

1865. *Physalus latirostris* Flower, Proc. Zool. Soc. London, for 1864, p. 414, May. Type locality, coast of Holland.

1866. *Balaenoptera carolinae* Malm, Nagra blad om Hvaldjur i Allmaenhet og Balaenoptera carolinae i Synnerhet, Goeteborg (pagination unknown). Type locality, unknown.

1869. *Sibbaldius sulfureus* Cope, Proc. Acad. Nat. Sci. Philadelphia, p. 20, July 20. Type locality, Northwest Coast of United States.

1870. *Rorqualus major* Knox, Trans. New Zealand Inst., (1869), 2:21, 23, Pl. 2a, Fig. 1. New name for the holotype of *Rorqualus borealis* Hamilton, 1837.

1877. *Pterobalaena gryphus* Munter, Mitth. Naturw. Verein von Neu-Vorpommern und Rügen, 9:1, 2 pls. Type locality Wieck bei Greifswald, Germany.

RANGE.—Atlantic. Franklin: Baffin Bay, thence southward to Panamá: Canal Zone. Pacific. Alaska: St. Matthew Islands, thence southward to waters off Nicaragua.

Genus **Megaptera** Gray—Humpback or Hump-backed Whale

1846. *Megaptera* Gray, Ann. Mag. Nat. Hist., ser. 1, 17:83, February. Type, *Megaptera longipinna* Gray [= *Balaena novae angliae* Borowski].

1846. *Perqualus* Gray, *in* The zoology of the voyage of H.M.S. Erebus and Terror. . . , 1(Mamm.):Pl. 32. A subgenus of *Balaenoptera*. Type, *B. boops* (= of Fabricius not Linnaeus = *Balaena novaeangliae* Borowski).

1846. *Megapteron* Gray, *in* The zoology of the voyage of H.M.S. Erebus and Terror . . . , 1(Mamm.):51. Apparently a *lapsus* for *Megaptera*. Type, *Balaena longimana* Rudolphi as designated by Hershkovitz, U.S. Nat. Mus. Bull., 246:176, February 28, 1966.

1849. *Kyphobalaena* Eschricht, Kongl. Dansk. Vidensk. Selsk. Skrift., Nat. Math. Afd., 1:108. Type, *Balaena boops* O. Fabricius [= *Balaena novae angliae* Borowski].

1864. *Poescopia* Gray, Proc. Zool. Soc. London, p. 207, Fig. 3, November. Type, *Balaena lalandii* Fischer [= *Balaena novae angliae* Borowski].

1873. *Cyphobalaena* Marschall, Nomen. Zool. Mamm., p. 5. An emendation of *Kyphobalaena* Eschricht, 1849.

1957. *Megapterina* Tomilin, Mammals of eastern Europe and northern Asia, 9(Cetacea):274. A supergeneric category proposed by Gray (1864, Proc. Zool. Soc. London, 1864:205) erroneously listed as a generic name in synonymy of *Megaptera* Gray.

Total length to 53 ft. Dorsum and sides black; venter white, most posterior part mottled black and white; pectoral appendages black above, white below; approx. 300, nearly black baleen plates. Knoblike protuberances on anterodorsal part of snout and on lower jaw; each pectoral appendage serrate on anterior border and approx. a third of total length of animal; dorsal fin small,

falcate, situated two-thirds length of body posterior to snout; acromion and coracoid processes of scapula rudimentary. Calves 15–16 ft. long at birth.

Megaptera novaeangliae (Borowski) Humpback or Hump-backed Whale

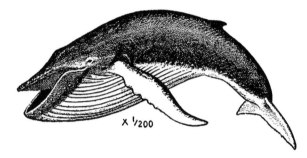

X 1/200

1780. *Balaena boops* O. Fabricius, Fauna Groenlandica . . . , p. 36. Type locality, Greenland seas, especially Pamiut and Pissukbik. Not Linnaeus, 1758.

1781. *Balaena novae angliae* Borowski, Gemeinnüztige Naturgeschichte des Thierreichs . . . , 2(1):21. Type locality, coast of New England.

1932. *Maegaptera novaeangliae*, Kellogg, Proc. Biol. Soc. Washington, 45:148, September 9.

1789. *B[alaena]. nodosa* Bonnaterre, Tableau . . . des trois règnes de la nature, cétologie, p. 5. Type locality, New England Coast.

1828. *Balaenoptera australis* Lesson, Histoire naturelle . . . des mammifères et des oiseaux découverts depuis 1788, cétacés, p. 372. Type locality, Cape of Good Hope, Republic of South Africa.

1829. *B[alaena]. lalandii* J. B. Fischer, Synopsis mammalium, p. 525. Type locality, Cape of Good Hope, Republic of South Africa.

1832. *Balaena longimana* Rudolphi, Abh. k. Akad. Wiss., Berlin, p. 133, Pls. 1–5, type from mouth Elbe River, Germany, N. Atlantic.

1834. *Balaenoptera capensis* A. Smith, South African Quart. Jour., 2:130. Type locality, Cape of Good Hope, Republic of South Africa.

1836. *R[orqualus]. antarcticus* F. Cuvier, Cétacés, p. 347, Pl. 20, Figs. 2–4. Name based on the "rorqual du Cap" of G. Cuvier, 1823 [*fide* Hershkovitz, U.S. Nat. Mus. Bull., 246:180, February 28, 1966].

1842. *Balaenoptera leucopteron* Lesson, Nouveau tableau du règne animal . . . mammifères, p. 202. Type locality, high southern latitudes.

1846. *Megaptera longipinna* [*sic*] Gray, Ann. Mag. Nat. Hist., (1), 17:83. A *lapsus* for *Balaena longimana* Rudolphi.

1846. *Megaptera poeskop* Gray, *in* The zoology of the voyage of H.M.S. Erebus and Terror . . . , 1(Mamm.):17, type from Cape of Good Hope, South Africa, S. Atlantic.

1846. *Megaptera americana* Gray, *in* The zoology of the voyage of H.M.S. Erebus and Terror . . . , 1(Mamm.):17. Type locality, Bermuda Islands.

1850. *Megaptera kuzira* Gray, Catalogue of the . . . Mammalia in the . . . British Museum, Cetacea, p. 30. Name based on *Balaenoptera antarctica* of Temminck (1842, Fauna Japonica, p. 21, Pl. 30), type from Japan, N. Pacific.

1863. *Balaenoptera syncondylus* A. Müller, Schrift. Phys.-oekonom. Ges. Konigsberg, 4:38, Pls. 1–3. Type locality, "Ostsee an die kurische Nehrung," Germany.

1865. *M[egaptera]. gigas* Cope, Proc. Acad. Nat. Sci. Philadelphia, p. 179, December. Type locality, unknown ["North Atlantic species; characters; comparisons; name erroneously attributed to Rudolphi or Gray without bibliographic references—*fide* Hershkovitz, U.S. Nat. Mus. Bull., 246:182, February 28, 1966].

1865. *Megaptera osphyia* Cope, Proc. Acad. Nat. Sci. Philadelphia, p. 180, December, type from 40 mi. off Petit Manan Lighthouse, Maine.

1866. *Megaptera longimana* var. *moorei* Gray, Catalogue of seals and whales in the British Museum, p. 122. Type locality, Estuary of the Dee, Scotland.

1866. *Megaptera? burmeisteri* Gray, Catalogue of seals and whales in the British Museum, p. 129, type from coast of Buenos Aires, Argentina, S. Atlantic.

1866. *Megaptera kuzira* Gray, Catalogue of seals and whales in the British Museum, p. 130, a renaming of *Balaena sulcata antarctica* Schlegel.

1867. *Megaptera braziliensis* Cope, Proc. Acad. Nat. Sci. Philadelphia, p. 32, type from Bahia, Brazil, S. Atlantic.

1868. *Kyphobalaena képorkak* Van Bénéden, Bull. Acad. Roy. Sci., Lettres et Beaux-Arts Belgique, 25:116. Type locality, Davis Stait, Greenland, N. Atlantic.

1869. *Megaptera versabilis* Cope, Proc. Acad. Nat. Sci. Philadelphia, p. 15, Figs. 5, 6. Type locality not specified; description based on the "North Pacific humpback" of Scammon.

1871. *Megaptera bellicosa* Cope, Proc. Amer. Phil. Soc., 12:103, 107, 7 figs. Type locality, St. Bartholomew's, West Indies.

1883. *Megaptera indica* Gervais, Comptes Rendus Acad. Sci. Paris, 97:1566. Type locality, Persian Gulf.

1897. *Balaena atlanticus* Hurdis, Natural history of the Bermudas, pp. 330, 333. Name probably a *lapsus* for *Megaptera americana* Gray.

See characters of the genus.

RANGE.—Atlantic. Greenland: Disco Bay, thence southward to Florida (Layne, 1965:137, 138): Delray Beach; mouth Tampa Bay. Cuba: waters vic. Cayo Bahía de Cádiz, to the north of Las Villas (Varona, 1974:89). Virgin Islands: off Camanoe Island, 18° 33′ N, 64° 32′ W (Erdman, 1970:637). Lesser Antilles (Varona, 1974:89): San Bartolomé; Grenadines. Thence southward to South American waters. Pacific. Alaska: Pribilof Islands, thence southward to Panamá: Golfo de Panamá.

FAMILY **BALAENIDAE**
Right Whales and Bowhead Whale

Total length to 70 ft. Head large, approx. one-third length of body; baleen plates numerous, long; dorsal fin absent; pectoral appendages quadradactylous or pentadactylous; cleft of mouth a curved line; external throat grooves absent. Skull with rostrum convex in lateral view, small, attenuated in general outline when viewed from above; supraorbital wings of frontal bones narrow; nasal bones subequal; cervical vertebrae fused; most ribs attached to vertebrae; lumbar vertebrae, 10 or more.

Genus **Balaena** Linnaeus
Bowhead Whale and Northern Right Whale

1758. *Balaena* Linnaeus, Syst. nat., ed. 10, 1:75. Type, *Balaena mysticetus* Linnaeus.

1777. *Balena* Scopoli, Introductio ad historium naturalem, p. 486, an emendation of *Balaena* Linnaeus.

1842. *Baloena* Lesson, Nouveau tableau du règne animal, p. 202, an emendation of *Balaena* Linnaeus.

1849. *Leiobalaena* Eschricht, Kongl. Danske Vidensk. Selsk. Skrift. Nat. Math., Afd., (5), 1:108. Type, the "glathvaler" or "rithvaler" of the northern seas (= *Balaena mysticetus* Linnaeus).

1864. *Eubalaena* Gray, Proc. Zool. Soc. London, p. 201, November. Type, *Balaena australis* Desmoulins.

1864. *Hunterus* Gray, Ann. Mag. Nat. Hist., ser. 3, 14:349, November. Type, *Hunterus Temminckii* Gray [= *Balaena glacialis* Borowski].

1865. *Macleayius* Gray, Proc. Zool. Soc. London, 1864:588,

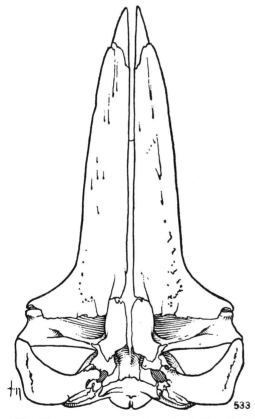

Fig. 533. *Megaptera novaeangliae*, Acad. Nat. Sci. Philadelphia, after Elliot (Field Columb. Mus., Publ. 95, Zool. Ser., 4(1):Pl. 11, 1904), sex unknown. Ventral view, X approx. $\frac{1}{26}$.

589. Type, *Macleayius australiensis* Gray (= *Eubalaena glacialis australis* Desmoulins, 1822) by original designation.

1866. *Hunterius* Gray, Catalogue of seals and whales in the British Museum, p. 78, an emendation of *Hunterus* Gray.

1873. *Macleayanus* Marschall, Nomen Zool. Mamm., p. 8, an emendation of *Macleayius* Gray.

1873. *Halibalaena* Gray, Proc. Zool. Soc. London, 1873:140. Type, *Balaena britannica* Gray (= *Balaena glacialis* Müller) by monotypy.

Total length to 65 ft. Blackish throughout except for ash-colored area at tip of mandible, cream-colored throat and chin, and occasionally white underparts. For other characters see accounts of the species.

KEY TO NORTH AMERICAN SPECIES OF BALAENA

1. Length of skull approx. 25% of total length of animal; throat and chin black.
   B. glacialis,   p. 920
1'. Length of skull approx. 40% of total length of animal; throat and chin cream-colored.
   B. mysticetus,   p. 921

Balaena glacialis
Northern Right Whale

X ¹/220

See characters of the genus. It is monotypic and inhabits subpolar and temperate waters.

Balaena glacialis glacialis Müller

1758. *B[alaena]. glacialis* Müller, Zoologiae Danicae prodromus, p. 7.

1788. *Balaena islandica* Gmelin, Syst. nat., ed. 13:1. Name based on "la baleine d'Islande" of Brisson.

1804. *Balaena nordcaper* Lacépède, Histoire naturelle des cétacées. . . , pp. xxxvi, 103, Pls. 2, 3. Type locality, seas between Spitzbergen, Norway, Iceland; also in the Greenland seas.

1860. *Balaena biscayensis* Eschricht, Revue et Mag. Zool., Paris, (2), 12:229, type from San Sebastián, Spain.

1865. *Balaena cisarctica* Cope, Proc. Acad. Nat. Sci. Philadelphia, type from Delaware River, opposite Philadelphia, on coast of New Jersey.

1866. *Balaena mysticetus* var. *angulata* Gray, Catalogue of seals and whales in the British Museum, p. 86, Fig. 5, type from Orkneys? (*fide* Hershkovitz, 1966, U.S. Nat. Mus. Bull., 246:190, February 28), N. Atlantic.

1867. *Hunterius svedenborgii* Lilljeborg, Nova Acta Reg. Soc. Upsala, (3), 6(2):35, Pls. 9–11, type from Wanga, west Gotland, Sweden.

1870. *Balaena mediterranea* Gray, Ann. Mag. Nat. Hist., (4),

6:198, type supposedly from the Ile de Saint Marguerite, Var, Mediterranean Sea, France.

1870. *Macleayius britannicus* Gray, Ann. Mag. Nat. Hist., (4), 6:200, type from Lyme Regis, Dorsetshire, England, N. Atlantic.

1871. *Balaena eubalaena* Gray, Supplement . . . catalogue of seals and whales in the British Museum, p. 44. Name in synonymy of *Hunterius biscayensis*; attributed to Flower (1864, Proc. Zool. Soc. London, p. 391), who used the combination *Eubalaena australis* where cited—*fide* Hershkovitz, U.S. Nat. Mus. Bull., 246:191, February 28, 1966.

1873. *Balaena Van Benediana* Capellini, Mem. Accad. Sci. Inst. Bologna (3), 3:12, ftn. (separate). Name based on fused cervical vertebrae figured by Lacépède (1804, Histoire naturelle des cétacées, Pl. 7, Fig. 1), and erroneously believed to be of the Ile de Saint Marguerite whale represented by a skeleton in the Paris museum.

1881. *Balaena Van Benedeniana* Fischer, Actes Soc. Linnéenne Bordeaux, (4), 35:46. An emendation of *benediana* Capellini.

1877. *Balaena tarentina* Capellini, Mem. Acead. Sci. Inst. Bologna (3), 7:11 (separate), Pls. 1–3, type from Gulf of Taranto, Italy, Mediterranean Sea.

1890. *Balaena euskariensis* Real, La ballena euskara, Memoria . . . San Sebastián, p. 3—listed but not seen by Hershkovitz, U.S. Nat. Mus. Bull., 246, p. 191, February 28, 1966. Rice (1977:7) notes that *B. g. australis* Desmoulins, 1822, is available as the subspecific name for the Right Whale of the Southern Hemisphere.

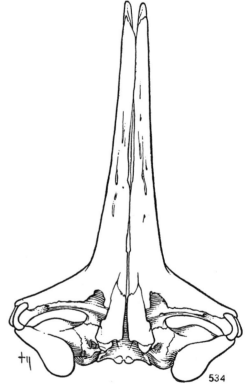

534

Fig. 534. *Balaena glacialis*, No. 23077 U.S.N.M., sex unknown. Ventral view, X ¹/₃₂.

RANGE.—Atlantic. Gulf of St. Lawrence and coast of Newfoundland, thence southward to coast of New England and, in winter, to Florida: Hillsboro Lighthouse, Pompano (Moore, 1953:122); Sarasota (Moore and Clark, 1963:269). Texas: near Freeport (Schmidly, et al., 1972:215, as *Balaena glacialis*).

Balaena glacialis japonica Lacépède

1818. *Balaena japonica* Lacépède, Mém. Mus. Hist. Nat., Paris, 4:469, 473. Type, description based on a colored figure by a Japanese artist. Type locality, Japan.
1958. *Eubalaena glacialis japonica*, Imaizumi, Nat. Sci. and Mus. Tokyo, 25(11–12):13.
1818. *Balaena lunulata* Lacépède, Mém. Mus. Hist. Nat., Paris, 4:473. Name based on a Japanese drawing of a right whale.
1825. *Balaena kuliomoch* Chamisso, Nova Acta Acad. Cesar. Leop.-Carol., Nat. Curios, 12:254, Pl. 17, Fig. 1. Name based on an Aleut wood-carving. Type locality, Bering Sea.
1864. *Balaena sieboldi* Gray, Ann. Mag. Nat. Hist., (3), 14:349. Type locality, Japan, N. Pacific.
1865. *Balaena aleoutiensis* Van Bénéden, Bull. Acad. Roy. Sci., Lettres, Beaux-Arts, Belgique, 20:854. Type locality, 40–60° in N. Pacific.
1868. *Balaena cullamach* Cope, Proc. Acad. Nat. Sci. Philadelphia, 20:225. An emendation or a misspelling of *kuliomach* Chamisso.

Omura, *et al.* (1969:64) found no difference, of specific worth, between a Right Whale from the North Pacific and another from the North Atlantic. This influenced Rice (1977:7) to place *B. japonica*, and its junior synonyms, as synonyms of *Balaena glacialis* which he treats as a monotypic species.

RANGE.—Pacific. Alaska: Cape Prince of Wales (Miller and Kellogg, 1955:669); Aleutian Islands, thence southward to Baja California.

Balaena mysticetus Linnaeus
Bowhead Whale

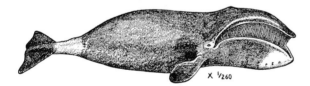

X ¹/₂₆₀

1758. [*Balaena*] *mysticetus* Linnaeus, Syst. nat., ed. 10, 1:75. Type locality, Greenland Seas.
1792. *Balaena mysticetus groenlandica* Kerr, The animal kingdom . . . , p. 356. Type locality, Greenland, N. Atlantic.
1828. *Balaena borealis* Lesson, Complément des oeuvres de Buffon, Cétacées, pp. 394, 442. New name for *Balaena mysticetus* Linnaeus.
1841. *Balaena mysticetus arctica* Schlegel, Abh. Gebiete Zool., 1:36, type from ice cap south to 75° N.
1874. *Balaena mysticetus roysii* Dall, *in* Scammon, Marine mammals of the north-western coast of North America . . . , p. 305. Type locality, Okhotsk Sea.
1883. *Balaena mysticetus* forma *pitlekajensis* Malm, Bihang

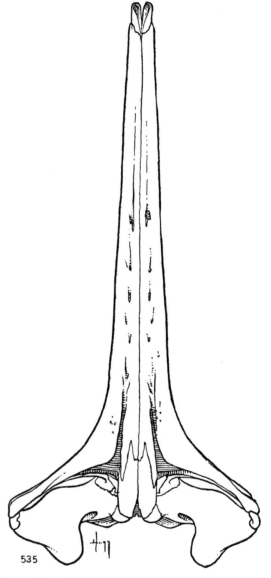

535

Fig. 535. *Balaena mysticetus*, 5 mi. off Iglurak Island, 20 mi. E Point Barrow, Alaska, No. 257313 U.S.N.M., ♀. Ventral view, X approx. ¹/₃₅.

Kongl. Svenska Vetensk.-Akad. Handl., 8(4):37. Type locality, Pitlek, northeastern Siberia.

Depth of skull from dorsalmost point on rostrum to ventralmost point on mandible more than greatest depth of thorax; anterodorsal part of head lacking "bonnet" seen in *Eubalaena*; approx. 360 plates of baleen, central plates longest (up to 12 ft.); see other characters under account of genus.

RANGE.—Arctic and subarctic seas northward from the Pribilof Islands (Alaska) and the St. Lawrence River. Also reported from off coast of Massachusetts (Hershkovitz, 1966:193).

Order CARNIVORA—Carnivores

Size small (weasels) to large (bears); no fewer than 4 toes on each foot; toes having claws; P4 and m1, referred to as carnassial teeth, functioning much as do the two blades of scissors in cutting fibers of flesh that constitute the principal food of most kinds of carnivores (teeth of bears reflect their omnivorous food habits and carnassial teeth as such are not well developed in bears and in several genera of procyonids); upper tooth-row bent at union of P4 and M1 (straight in bears and in most Creodonta); condyle of lower jaw transverse; clavicle reduced or absent; stomach simple. The combination above of characters instead of any one character has to be used to distinguish carnivores from some kinds of mammals that are members of other orders.

Other characters include the following: canine teeth large, long, and recurved; 3 incisor teeth in each side of each jaw; outer incisor largest and inner one smallest; disparity in size of the 3 incisors greater in upper jaw than in lower jaw; diphodont (possessing milk teeth and permanent teeth); roots on both the milk teeth and permanent teeth; radius and ulna distinct; fibula slender and distinct from tibia; scaphoid and lunar bones of the wrist united into one; brain large and having deep convolutions.

Each major continent, even Australia, has one or more representatives of the order. Several of the genera have exceptionally extensive geographic ranges, for example *Canis*, *Felis*, and *Mustela*. The same can be said of several of the species of the mentioned genera. As is generally known, the more primitive living members of the order mostly are found in the southern parts of the major land areas. Examples include the mustelid genus *Lynchodon* and the ursid genus *Tremarctos* in South America, and perhaps the canid genus *Lycaon* of Africa.

The Order Carnivora apparently arose independently from ancestral insectivore stock, and was initially represented by the family Miacidae of the late Paleocene and Eocene. The Miacidae gave rise to two superfamilies, (1) the Aeluroidea, which include the Viverridae, Hyaenidae, and Felidae, and (2) the Arctoidea, which include the Mustelidae, Canidae, Ursidae, and Procyonidae. Possibly the Order Pinnipedia should be included within the arctoid group of carnivores.

The Order Carnivora has provided modern man with three domestic, or at least semidomestic, mammals—the ferret, the cat, and the dog. By selective breeding, many color varieties of the mink (*Mustela vison*) and the red fox (*Vulpes fulva*) have been produced.

For man the Carnivora have provided more trouble in his pastoral enterprises than have any other mammals. Wolves in the early days and coyotes in later times levied on cattle and sheep, respectively, in the grazing areas of North America. In Australia the dingo is said to compete with man for mutton. In Africa the lion is always under pressure from the Masai in their area, and his hope for survival lies in reserves set aside to preserve wildlife. North American hunters who take deer dislike the puma (mountain lion) and have helped to extirpate him in many regions. The outlook for the large carnivores is not good.

Key to North American Families of Carnivora

1. Digitigrade; caecum present;[1] longitudinal septa in tympanic bullae.
    2. Four digits on forefoot; entepicondylar foramen of humerus lacking; 3[2] lower molars. . . . . .Canidae, p. 923
    2′. Five[3] digits on forefoot; entepicondylar foramen of humerus present; 1 or 2 lower molars.
        3. Molariform teeth $\frac{3-4}{3}$; trigonid of 1st lower molar lacking metaconid and therefore consisting of only 2 cuspids; posterior palatine foramina on or posterior to maxillopalatine suture. . . . . Felidae, p. 1035
        3′. Molariform teeth $\frac{5-6}{4-6}$; trigonid of 1st lower molar having distinct metaconid and, therefore, a trigonid in fact; posterior palatine foramina anterior to maxillopalatine suture (nonnative). Viverridae, p. 1034
1′. Plantigrade or subplantigrade; caecum absent; no longitudinal septa in tympanic bullae.
    4. Alisphenoid canal present; 3 lower molars; entepicondylar foramen in

---

[1] The African genus *Nandinia* of the Viverridae is said to lack a caecum.

[2] The American genus *Speothus* and Old World genus *Cuon* have only 2 lower molars. The African genus *Otocyon* has 4.

[3] *Bdeogale* and *Suricata*, African genera of the Viverridae, are said to have only 4 toes on the forefoot.

humerus absent; length of head and body more than 41 in. in adults; length of tail-vertebrae less than 14% of total length. . . . . .Ursidae, p. 946

4'. Alisphenoid canal absent; 2 lower molars; entepicondylar foramen in humerus present;[4] length of head and body less than 41 in. in adults; length of tail-vertebrae more than 14% of total length.

    5. Molars $\frac{1}{2}$;[5] cutting carnassial (except in *Enhydra*). . .Mustelidae, p. 980

    5'. Molars $\frac{2}{2}$; crushing carnassial. Procyonidae, p. 960

[4] The entepicondylar foramen in the Mustelidae is absent in some genera and present in others.

[5] Molars are $\frac{1}{1}$ in the Old World genus *Mellivora*.

FAMILY CANIDAE
Coyote, Wolves, Foxes, and Dogs

Digitigrade, claws nonretractile, 4 or 5 front- and 4 hind-toes; long-legged and cursorial; rostrum of skull elongated; carnassials highly sectorial; tympanic bullae low, rarely much inflated; alisphenoid canal well developed; dental formula usually i. $\frac{3}{3}$, c. $\frac{1}{1}$, p. $\frac{4}{4}$, m. $\frac{2}{3}$, rarely with m. $\frac{2}{2}$ or m. $\frac{3-4}{4}$.

These cursorial carnivores occur on all continents, except Antarctica. Even Australia has the dingo, which is doglike and supposedly arrived in the canoes of prehistoric man. No fossil remains of an ancestral dingo have been found in Australian Tertiary beds.

KEY TO NORTH AMERICAN GENERA OF CANIDAE

1. Molars $\frac{2}{3}$. . . . .*Speothos*, p. 946

1'. Molars $\frac{2}{3}$.

    2. Ventral border of dentary with prominent "step"; tail with mane of stiff black-tipped hairs on dorsal surface. . . . .*Urocyon*, p. 941

    2'. Ventral border of dentary without a prominent "step"; tail without mane.

        3. Postorbital processes thickened and convex dorsally; upper incisors prominently lobed; basilar length usually more than 147. . .*Canis*, p. 923

        3'. Postorbital processes thin and concave dorsally; upper incisors usually not prominently lobed; basilar length usually less than 147.

            4. Ears long and pointed; rostrum relatively narrow, its width, measured at a point opposite the cone of P2 less than 18% of condylobasal length; teeth of

rami relatively widely spaced, anteroposterior diameter of il at base usually less than distance between alveoli of il and c. . . . .*Vulpes*, p. 936

    4'. Ears short and rounded; rostrum relatively broad, its width measured at a point opposite the cone of P2 more than 18% of condylobasal length; teeth of rami relatively closely spaced, anteroposterior diameter of il greater than distance between alveoli of il and c. . . . .*Alopex*, p. 935

Genus Canis Linnaeus
Coyote, Wolves, and Dogs

1758. *Canis* Linnaeus, Syst. nat., ed. 10, 1:38. Type, *Canis familiaris* Linnaeus.

1837. *Vulpecanis* Blainville, Ann. Sci. Nat., Paris, Zool., 8(2):279. Type, *Canis aureus* Linnaeus.

1839. *Lyciscus* Hamilton-Smith, *in* The naturalist's library (edit. Jardine), 25:160. Type, *Canis latrans* Say.

1839. *Sacalius* Hamilton-Smith, The naturalist's library (edit. Jardine), 25:214. Type, *Canis aureus* Linnaeus.

1841. *Oxygoüs* Hodgson, Calcutta Jour. Nat. Hist., 2:213. Type, *Oxygoüs indicus* Hodgson.

1855. *Lupulus* Gervais, Histoire naturelle des mammifères, 2:60–62. Based on *Canis aureus* Linnaeus and others. Not *Lupulus* Blainville, 1843.

1869. *Dieba* Gray, Catalogue of carnivorous, pachydermatous, and edentate Mammalia in the British Museum, p. 180. Type, *Canis anthus* Cuvier, 1820 [= *Canis aureus* Linnaeus].

1906. *Lupulella* Hilzheimer, Zool. Beobachter, 47:363. Type, *Canis mesomelas* Schreber, 1778.

1906. *Schaeffia* Hilzheimer, Zool. Beobachter, 47:364. Type, *Canis adustus* Sundevall, 1846.

1906. *Alopedon* Hilzheimer, Zool. Beobachter, 47:365. Type, *Canis thooides* Hilzheimer [= *Canis anthus* Cretzschmar, not *C. anthus* Cuvier, 1820].

1974. *Cubacyon* Arredondo and Varona, Poeyana, 131:2. Type, *Cubacyon transversidens* [= *Canis familiaris* L.], Poeyana, 131:3, September 10, from Pleistocene, Cueva del Túnel, La Salud, Habana, Cuba.

Size medium to large, adults usually more than 18 pounds; tail bushy, cylindrical, and of greatest diameter at the middle; postorbital processes convex above, evenly continuous with inflated frontal area; temporal ridges uniting to form a sagittal crest; inferior margin of lower jaw without prominent step.

KEY TO NORTH AMERICAN SPECIES OF CANIS

1. Nose pad 1 in. or less in diameter; heel pad less than 1¼ in. in diameter; anteroposterior diameter of canine less than 11 mm.; greatest length of skull of males less than 215 mm., and of females less than 205 mm.; rostrum relatively narrow, braincase relatively broad dorsoposteriorly, postorbital constriction relatively broad, and short anteroposteriorly (see Fig. 536); tail usually held low when animal is running. . . . .*C. latrans,*   p. 924

1'. Nose pad 1 in. or more in diameter; heel pad more than 1¼ in. in diameter; anteroposterior diameter of canine more than 11 mm.; greatest length of skull of males more than 215 mm., and of females more than 205 mm.; rostrum relatively broad, braincase relatively narrow dorsoposteriorly, postorbital constriction relatively narrow, and elongated anteroposteriorly (see Figs. 537 and 538); tail usually held high when animal is running.

     2. Larger; condylobasal length of males more than 213 mm., condylobasal length of females more than 203 mm.; premolars relatively broad; M2 and inner lobe of M1 relatively small; cingulum on outer edge of M1 not conspicuous (see Fig. 537); not recorded from Arkansas, eastern Oklahoma, eastern Texas, or other states bordering on the Gulf of Mexico. . . . .*C. lupus,*   p. 928

     2'. Smaller; condylobasal length of males less than 235 mm.; condylobasal length of females less than 215 mm.; premolars relatively narrow; M2 and inner lobe of M1 relatively large; cingulum on outer edge of M1 usually conspicuous (see Fig. 538); recorded from Indiana, Illinois, Missouri, Georgia, southern Tennessee, eastern Oklahoma, Arkansas, and states of the United States bordering on the Gulf of Mexico. . . . .*C. rufus,*   p. 933

Canis latrans
Coyote

×¹⁄₁₆

C.G.T.Pritchard

Revised by Jackson, Classification of the races of the coyote, pt. 2, pp. 227–341, Pls. 58–81, Figs. 20–28, *in* Young and Jackson, The clever coyote, Amer. Wildlife Manag. Inst., Washington, D.C., pp. xv + 411, 81 pls., 28 figs., November 29, 1951.

External measurements: 1052–1320; 300–394; 177–220. Condylobasal length of skull, 160.2–203.5. Nose pad (rhinarium) averaging 25 mm. or less in diameter; canines relatively long and narrow, tips of C1 extending below level of anterior mental foramina when lower jaw is in place; anteroposterior diameter of C1 at base less than 11 mm.; premolars relatively widely spaced; premolars and m1 relatively narrow; M2, inner lobe of M1, and talonid of m1 relatively large with trenchant compressed cusps; distance from posterior border of alveolus of M2 to anterior border of alveolus of P1 3.1 or more times distance between medial borders of alveoli of P1's. Tail directed downward when animal runs.

Map 489 does not show record stations of occurrence in several parts of the eastern United States where the presence of the coyote is thought to have resulted from introductions of the animals by man. Schultz (1955:44–46) mentions such occurrences in West Virginia, Virginia, North Carolina, South Carolina, Georgia, Mississippi, and Alabama, and lists in detail occurrences in Tennessee.

Coyotelike animals that have recently appeared in the northeastern and south-central United States apparently represent range extensions of *C. latrans.* These populations, however, seem to have been influenced in part by interbreeding with *C. lupus* or *C. rufus,* and some individuals may approach wolves in certain characters.

Canis latrans cagottis (Hamilton-Smith)

1839. *Lyciscus cagottis* Hamilton-Smith, *in* The naturalist's library (edit. Jardine), 25:164, type from Río Frío, W slope Mt. Iztaccíhuatl, México.
1932. *Canis latrans cagottis,* Nelson, Proc. Biol. Soc. Washington, 45:224, November 26.

MARGINAL RECORDS.—San Luis Potosí: Hda. La Parada. Veracruz: 15 km. W Piedras Negras, 300 ft. (Hall and Dalquest, 1963:328). Oaxaca (Goodwin, 1969:221): Tequisistlán; *San Felipe del Agua;* Cerro San Felipe. México: *type locality.* Michoacán: 15 mi. NE Morelia.

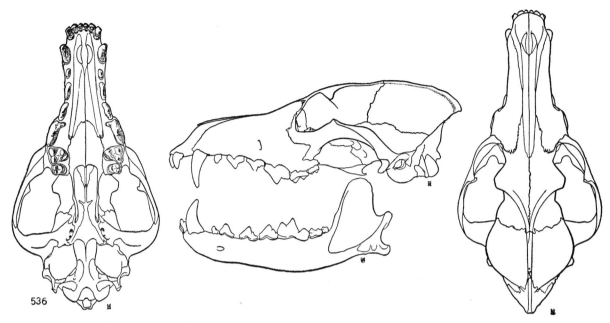

Fig. 536. *Canis latrans lestes*, 5 mi. SE Fallon, Nevada, No. 89940 M.V.Z., ♂, X ⅔.

Canis latrans clepticus Elliot

1903. *Canis clepticus* Elliot, Field Columb. Mus., Publ. 79, Zool. Ser., 3:225, August 15, type from Vallecitos, Sierra San Pedro Mártir, 8500 ft., Baja California.
1932. *Canis latrans clepticus*, Nelson, Proc. Biol. Soc. Washington, 45:224, November 26.

MARGINAL RECORDS.—California: San Marcos; Julian; Jacumba. Baja California: San Felipe; La Grulla; San Quintín.

Canis latrans dickeyi Nelson

1932. *Canis latrans dickeyi* Nelson, Proc. Biol. Soc. Washington, 45:224, November 26, type from Cerro Magote, 2 mi. W Río Gaoscorán, District of La Unión, El Salvador.

MARGINAL RECORDS.—El Salvador: type locality. Costa Rica: Miravalles; Bebedero; Nicoya.

Canis latrans frustror Woodhouse

1851. *Canis frustror* Woodhouse, Proc. Acad. Nat. Sci. Philadelphia, 5(1850–1851):147, June 30, type from Red Fork of Arkansas River, probably near 97° W long., near present town of Perkins, Payne Co., now Cimarron River, Oklahoma, about 100 mi. W Fort Gibson, Oklahoma.
1951. *Canis latrans frustror*, Jackson, *in* Young and Jackson, The clever coyote, p. 271.

MARGINAL RECORDS.—Missouri: Madisonville; Barren. Mississippi (Kennedy, *et al.*, 1974:23, 24): 3 mi. W Enid; Webster County; 5 mi. E of N end Eagle Lake, *in* Issaquena Co.; Adams County. Louisiana (Lowery, 1974:387, as *C. latrans* only): 5 mi. N Baton Rouge; ca.

2 mi. W Addis. Texas: Jefferson County (Davis, 1966:110, as *C. latrans* only); Matagorda County (*ibid.*, as *C. latrans* only); Aransas National Wildlife Refuge; Baby Head; *Cherokee*; Murray; Archer County (Davis, 1966:110, as *C. latrans* only). Oklahoma: Wichita Mountains Wildlife Refuge; Calumet; type locality. Kansas: Hamilton; Eudora. Missouri: Randolph County.

Canis latrans goldmani Merriam

1904. *Canis goldmani* Merriam, Proc. Biol. Soc. Washington, 17:157, October 6, type from San Vicente, Chiapas.
1932. *Canis latrans goldmani*, Nelson, Proc. Biol. Soc. Washington, 45:224, November 26.

MARGINAL RECORDS.—Oaxaca: San Antonio (Goodwin, 1969:221). Chiapas: type locality.

Canis latrans hondurensis Goldman

1936. *Canis hondurensis* Goldman, Jour. Washington Acad. Sci., 26:33, January 15, type from Cerro Quinote, NE Archaga, on Talanga Road, N Tegucigalpa, Honduras.
1951. *Canis latrans hondurensis*, Jackson, *in* Young and Jackson, The clever coyote, Amer. Wildlife Manag. Inst., Washington, D.C., p. 313.

MARGINAL RECORDS.—Honduras: type locality; *La Cueva (Archaga, Department of Tegucigalpa).*

Canis latrans impavidus J. A. Allen

1903. *Canis impavidus* J. A. Allen, Bull. Amer. Mus. Nat. Hist., 19:609, November 12, type from Río de las Bocas, northwestern Durango.

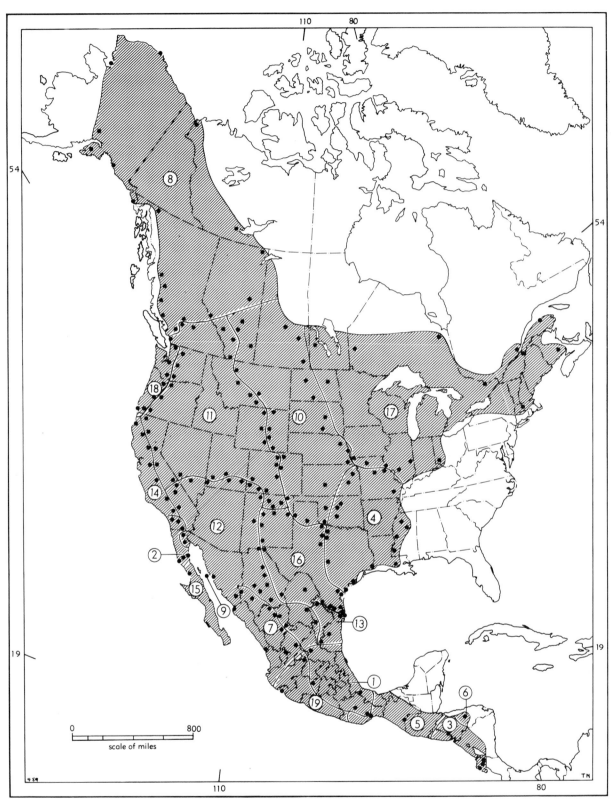

Map 489. *Canis latrans.*

Guide to subspecies	5. *C. l. goldmani*	10. *C. l. latrans*	15. *C. l. peninsulae*
1. *C. l. cagottis*	6. *C. l. hondurensis*	11. *C. l. lestes*	16. *C. l. texensis*
2. *C. l. clepticus*	7. *C. l. impavidus*	12. *C. l. mearnsi*	17. *C. l. thamnos*
3. *C. l. dickeyi*	8. *C. l. incolatus*	13. *C. l. microdon*	18. *C. l. umpquensis*
4. *C. l. frustror*	9. *C. l. jamesi*	14. *C. l. ochropus*	19. *C. l. vigilis*

1932. *Canis latrans impavidus*, Nelson, Proc. Biol. Soc. Washington, 45:224, November 26.

MARGINAL RECORDS.—Chihuahua (Anderson, 1972:373):Carimeche; Samachique. Durango (Baker and Greer, 1962:131, unless otherwise noted): Villa Ocampo (Hall and Kelson, 1959:844); *11 mi. SE Las Nieves, 5800 ft.;* 25 mi. SE Las Nieves; 5 mi. NE La Pila. Zacatecas: Fresnillo. Sinaloa (Armstrong, *et al.,* 1972:50): 6 mi. NNW Teacapán; 12 mi. N Higueras de Zaragosa. Sonora: Camoa (Río Mayo).

Canis latrans incolatus Hall

1934. *Canis latrans incolatus* Hall, Univ. California Publ. Zool., 40:369, November 5, type from Isaacs Lake, 3000 ft., Bowron Lake region, British Columbia.

MARGINAL RECORDS.—Alaska: Point Barrow. Mackenzie: Mackenzie River Delta; near Providence. Alberta: Fort Smith; *near Murdock Creek;* 20 mi. S Athabaska Landing; 15 mi. S Henry House. British Columbia: Horse Lake (Cowan and Guiguet, 1965:280); Pemberton; west branch Homalko [= Homathko?] River; Stuie; Ootsa Lake; *Quick* (Cowan and Guiguet, 1965:280); Telkwa (*ibid.*); Spruce Mtn. Alaska: near Whitehorse, Alsek River; Copper River Flat, 20 mi. from Cordova; Kenai Lake; Wonder Lake, Mt. McKinley region; Kotzebue. Not located: Alaska: Big Delta River; Eagle River; Healy. Alberta: Whitemud.

Canis latrans jamesi Townsend

1912. *Canis jamesi* Townsend, Bull. Amer. Mus. Nat. Hist., 31:130, June 14, type from Tiburón Island, Sonora. Known only from type locality.
1932. *Canis latrans jamesi*, Nelson, Proc. Biol. Soc. Washington, 45:224, November 26.

Canis latrans latrans Say

1823. *Canis latrans* Say, *in* Long, Account of an exped. . . . to the Rocky Mts. . . . , 1:168, type from Engineer Cantonment, about 12 mi. SE of present town of Blair, Washington Co., Nebraska, on W bank Missouri River.
1897. *Canis pallidus* Merriam, Proc. Biol. Soc. Washington, 11:24, March 15, type from Johnstown, Brown Co., Nebraska. Not *Canis pallidus* Cretzschmar, 1826, type from Kardofan, North Africa, *qui Vulpes pallida.*
1898. *Canis nebracensis* Merriam, Science, n.s., 8:782, December 2, a renaming of *Canis pallidus* Merriam, 1897.

MARGINAL RECORDS.—Saskatchewan: Prince Albert National Park; Melfort (specimens from here and Prince Albert National Park referred to subspecies *C. l. thamnos* by Beck, 1958:16); Windhorst. North Dakota: Shell Creek; Palace Buttes, 6 mi. N Cannonball. South Dakota: Crow Creek Reservation; Fort Randall. Nebraska (Jones, 1964c:249): Beemer; *type locality;* Fontanelle Forest; Pawnee County. Kansas: Fort Larned. Oklahoma: Anthon; Frederick. Texas: Armstrong County (Davis, 1966:110, as *C. latrans* only); 70 mi. W Amarillo. New Mexico: Des Moines. Colorado (Armstrong, 1972:257): Denver; 1 mi. S Nunn. Wyoming (Long, 1965a:674): Jelm; Shirley;

Casper; Arvada. Montana:*Lodge Grass;* Delphia; Conrad. Alberta: Buffalo; Red Deer; Lacombe.

Canis latrans lestes Merriam

1897. *Canis lestes* Merriam, Proc. Biol. Soc. Washington, 11:25, March 15, type from Toyabe Mts., near Cloverdale, Nye Co., Nevada.
1913. *Canis latrans lestes*, Grinnell, Proc. California Acad. Sci., ser. 4, 3:285, August 28.

MARGINAL RECORDS.—Alberta: Eagle Pass, head James River; Calgary; *New Oxley;* Waterton Lakes National Park. Montana: *Camas Creek, Glacier Park;* Little Belt Mts.; Wallis; Bridger. Wyoming (Long, 1965a:675): Shell; 8 mi. N, 2½ mi. W Waltman; 16 mi. N, 2 mi. W Saratoga. Colorado (Armstrong, 1972:257, 258): 18 mi. N Fort Collins; *Loveland;* Boulder; 20 mi. S Colorado Springs. New Mexico: Halls Peak; Santa Fe Forest; Abiquiu. Colorado (Armstrong, 1972:258): Jim Creek, W of La Jara Reservoir; Coventry. Utah: Dark Valley, 5 mi. SW Teasdale; *Panguitch;* Parowan. Nevada: Silver Peak Mts. California: Deep Spring Lake; Wawona; Clipper Gap; Glade; *Winthrop;* Gottville. Oregon: Harriman; Bend; *Hay Creek;* Sink. Washington: Goldendale; Ellensberg; *Cashmere;* Lake Chelan; *Conconully.* British Columbia (Cowan and Guiguet, 1965:280): *Ladner; Lulu Island;* North Vancouver; Hat Creek; *Tranquille Lake;* Shuswap. Cowan and Guiguet (1965:280) separate the geographic range of *lestes* along the 52nd parallel from that of *incolatus* to the north, whereas Jackson (1951:263) indicates that the separation in western British Columbia is farther to the south.

Canis latrans mearnsi Merriam

1897. *Canis mearnsi* Merriam, Proc. Biol. Soc. Washington, 11:30, March 15, type from Quitobaquito, Pima Co., Arizona.
1932. *Canis latrans mearnsi*, Nelson, Proc. Biol. Soc. Washington, 45:224, November 26.
1897. *Canis estor* Merriam, Proc. Biol. Soc. Washington, 11:31, March 15, type from Noland's Ranch, San Juan River Valley, San Juan Co., Utah.

MARGINAL RECORDS.—Utah: La Sal. Colorado (Armstrong, 1972:258): Dove Creek; SW base Pagosa Peak, 8000 ft., Archuleta Co. New Mexico: El Vado; Magdalena; 15 mi. W El Paso. Chihuahua (Anderson, 1972:373): 16 km. NNW Moctezuma; *2 mi. W El Sueco;* 6 mi. SE Station Arados; 7 mi. SE Ciudad Chihuahua. Coahuila (Baker, 1956:293): 3 mi. S, 1 mi. W Hda. Guadalupe; 3 mi. S, 12½ mi. E San Antonio de las Alazanas, 8900 ft. Durango (Baker and Greer, 1962:131): 3 mi. S Hda. Atotonilco; 7 mi. NNW La Zarca, 6000 ft. Chihuahua (Anderson, 1972:373): 9 mi. (by road) S Boquillas de Conchos; 4 mi. NW San Francisco de Borja, 5700 ft.; *8 mi. NW San Francisco de Borja, 5700 ft.;* 2 mi. W Miñaca, 6900 ft. Sonora: Costa Rica Ranch. Baja California: 80 mi. S Mexicali; E base Laguna Hanson Mts. California: Hesperia; Coso; Independence. Nevada: Oak Springs. Utah: New Castle; Mt. Ellen, Henry Mts.

Canis latrans microdon Merriam

1897. *Canis microdon* Merriam, Proc. Biol. Soc. Washington, 11:29, March 15, type from Mier, on Río Grande, Tamaulipas.
1932. *Canis latrans microdon*, Nelson, Proc. Biol. Soc. Washington, 45:224, November 26.

MARGINAL RECORDS.—Nuevo León: Rodríguez [= Anahuac]. Texas: *Roma*. Tamaulipas: Matamoros; Bagdad; 9½ mi. SW Padilla (Alvarez, 1963:455); Nicolás, 53 km. N Tula (*ibid.*).

Canis latrans ochropus Eschscholtz

1829. *Canis ochropus* Eschscholtz, Zoologischer Atlas . . . , 3:1, type from Sacramento River Valley, not far south of Sacramento, Sacramento Co., California.
1932. *Canis latrans ochropus*, Nelson, Proc. Biol. Soc. Washington, 45:224, November 26.

MARGINAL RECORDS.—California: Maple Creek; Chico; Wheatland; Mariposa County; Orosi; Onyx; San Jacinto.

Canis latrans peninsulae Merriam

1897. *Canis peninsulae* Merriam, Proc. Biol. Soc. Washington, 11:28, March 15, type from Santa Anita, Cape Saint Lucas, Baja California.
1932. *Canis latrans peninsulae*, Nelson, Proc. Biol. Soc. Washington, 45:224, November 26.

MARGINAL RECORDS.—Baja California: Playa María Bay, thence south throughout the rest of the peninsula.

Canis latrans texensis V. Bailey

1905. *Canis nebrascensis texensis* V. Bailey, N. Amer. Fauna, 25:175, October 24, type from 45 mi. SW Corpus Christi at Santa Gertrudis, Kleberg Co., Texas.
1932. *Canis latrans texensis* V. Bailey, N. Amer. Fauna, 53:312, March 11.

MARGINAL RECORDS.—New Mexico: Cowles; Tucumcari. Texas (Davis, 1966:110, as *C. latrans* only, unless otherwise noted): Vernon (Hall and Kelson, 1959:846); Throckmorton County; Shackelford County; San Patricio County; Nueces Bay (Hall and Kelson, 1959:846); 50 mi. N Brownsville (*ibid.*); Cameron County; Hidalgo County; Starr County. Tamaulipas: Nuevo Laredo. Coahuila: 25 mi. NW Muzquiz. Chihuahua: 12 mi. S Jaco, 5500 ft. (Anderson, 1972:373). Texas: El Paso. New Mexico: Organ Mts.; Carthage; Isleta.

Canis latrans thamnos Jackson

1949. *Canis latrans thamnos* Jackson, Proc. Biol. Soc. Washington, 62:31, March 17, type from Basswood Island, Apostle Islands, Ashland Co., Wisconsin.

MARGINAL RECORDS.—Manitoba: Kenville. Ontario: Ghost Narrows, Lac Seul; Smoky Falls; Dunrobin, 25 mi. W Ottawa. Quebec: near Baie St.-Paul (78 Quebec Wildlife Serv.); near St.-Aubert (77 Quebec Wildlife Serv.); near Tourelle, 3 mi. E Ste. Anne des Monts (Georges, 1976:78, as *Canis latrans* only). New Brunswick: near Sussex (Wolfram, 1964:113, as *C. latrans* only). Massachusetts: Prescott Peninsula [in Quabbin Reservoir] (Pringle, 1960:278). Ohio: Stonelick Lake, Clermont Co. (Goodpaster and Hoffmeister, 1968:117, as *C. latrans* only). Illinois: Douglas County; Macoupin County. Missouri: Spencerberg; Carollton; Tarkio. North Dakota: Sullys Hill National Park (?) [in Benson County]. Lawrence and Bossert (1969:12) suggested that the coyote population of New England has been modified through the introduction of genes from *C. lupus* and/or *C. familiaris*, and designated this population as *C. latrans* var.

Canis latrans umpquensis Jackson

1949. *Canis latrans umpquensis* Jackson, Proc. Biol. Soc. Washington, 62:31, March 17, type from 5 mi. SE Drew, Douglas Co., Oregon.

MARGINAL RECORDS.—Washington: Nooksack River; Renton; Trout Lake. Oregon: Marmot; *Blue River;* 20 mi. SE Oakridge; Ashland; Pyramid Rock, Pistol River, 12 mi. from mouth.

Canis latrans vigilis Merriam

1897. *Canis vigilis* Merriam, Proc. Biol. Soc. Washington, 11:33, March 15, type from Manzanillo, Colima.
1932. *Canis latrans vigilis*, Nelson, Proc. Biol. Soc. Washington, 45:224, November 26.

MARGINAL RECORDS.—Jalisco: Ojo de Agua, Tepic. Colima: type locality.

Canis lupus
Gray Wolf

X ¹⁄₁₆   C G Pritchard

Revised by Goldman (Classification of wolves, pt. 2, pp. 389–507, in part, Pls. 88–125, 131, 1 fig. in text), *in* Young and Goldman, The wolves of North America, Amer. Wildlife Manag. Inst., Washington, D.C., pp. xx + 636, 131 pls., 15 figs. in text, 1944.

External measurements: 1003–2046; 356–500; 220–310. Condylobasal length of skull, 203.3–269.3. Nose pad (rhinarium) averaging 31 mm. or more in diameter; C1 more than 12 mm. in anteroposterior diameter at base, C1 not extending below level of anterior mental foramen when lower jaw is in place; premolars relatively broad and closely spaced; M2 relatively small; color varying from nearly black to white, some shade of gray in most subspecies. In life, tail usually carried horizontally or directed upward when animal runs.

Canis lupus alces Goldman

1941. *Canis lupus alces* Goldman, Proc. Biol. Soc. Washington, 54:109, September 30, type from Kachemak Bay, Kenai Peninsula, Alaska. Known only from type locality.

Canis lupus arctos Pocock

1935. *Canis lupus arctos* Pocock, Proc. Zool. Soc. London, p. 682, September 12, type from Melville Island, Franklin.

MARGINAL RECORDS.—Franklin: Discovery Bay, Eureka Sound, and Bear Peninsula, all on Ellesmere Island; type locality; Thomsen River, Banks Island (Manning and Macpherson, 1958:55; these two

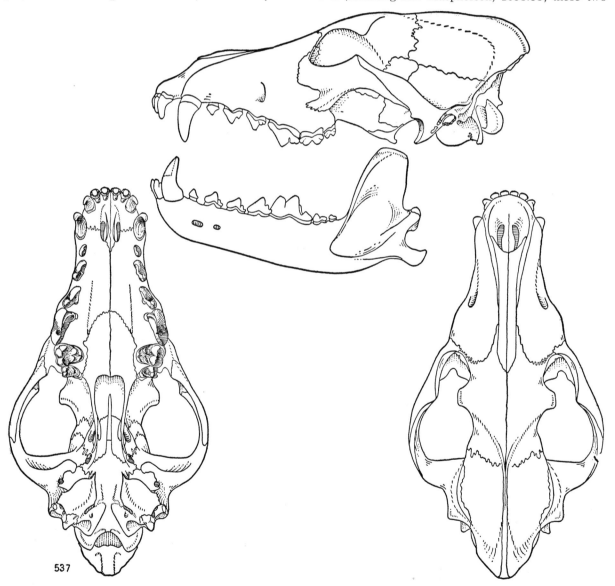

Fig. 537. *Canis lupus tundrarum*, Wahoo Lake long. 146° 58′ W, lat. 69° 08′ N, Alaska, No. 51561 K.U., ♀, X ⅔.

authors, *op. cit.*: 53, 54, suggest that *C. l. arctos* in the past 45 years has replaced *C. l. bernardi*); Sverdrup Islands.

Canis lupus baileyi Nelson and Goldman

1929. *Canis nubilus baileyi* Nelson and Goldman, Jour. Mamm., 10:165, May 9, type from Colonia García (about 60 mi. SW Casas Grandes), Chihuahua.
1937. *Canis lupus baileyi*, Goldman, Jour. Mamm., 18:45, February 14.

MARGINAL RECORDS.—New Mexico: Hatch. Texas: 16 mi. NW Fort Davis. San Luis Potosí: Bledos. Oaxaca (Goodwin, 1969:224): Tequisistlán. Durango: El Salto. Chihuahua: Sierra Madre near Guadalupe y Calvo. Arizona: Arivaca; Helvetia; *Santa Rita Range, 20 mi. SE Tucson* (Cockrum, 1961:218, 219); Fort Bowie.

Canis lupus beothucus G. M. Allen and Barbour

1937. *Canis lupus beothucus* G. M. Allen and Barbour, Jour. Mamm., 18:230, May 12, type from Newfoundland. Known only from Newfoundland; now extinct.

Canis lupus bernardi Anderson

1943. *Canis lupus bernardi* Anderson, Jour. Mamm., 24:389, August 17, type from Cape Kellett, southwestern part Banks Island, Franklin, lat. about 72° N, long. 125° W. Known only from type locality.
1943. [*Canis lupus*] *banksianus* Anderson, Jour. Mamm., 24:390, August 17, an accidental renaming of *C. l. bernardi* Anderson.

Canis lupus columbianus Goldman

1941. *Canis lupus columbianus* Goldman, Proc. Biol. Soc. Washington, 54:110, September 30, type from Wistaria, N side Ootsa Lake, British Columbia.

MARGINAL RECORDS.—Yukon: Lapie Lake, Mile 105, Canol Road. British Columbia: 94 mi. S Fort Nelson; Little Prairie (Cowan and Guiguet, 1965:284); Barkerville. Alberta: Jasper National Park. British Columbia: Avola; Kettle River; Roche River, tributary of Similkameen River; near Hope; Atnarko; Wistaria (Cowan and Guiguet, 1965:284); Quick (*ibid.*); 60 mi. S Telegraph Creek; Telegraph Creek; Atlin.

Canis lupus crassodon Hall

1932. *Canis occidentalis crassodon* Hall, Univ. California Publ. Zool., 38:420, November 8, type from Tahsis Canal, Nootka Sound, Vancouver Island, British Columbia.
1937. *Canis lupus crassodon*, Goldman, Jour. Mamm., 18:45, February 11.

MARGINAL RECORDS.—Vancouver Island: Quatsino; Saanich Peninsula (Cowan and Guiguet, 1965:284).

Canis lupus fuscus Richardson

1839. *Canis lupus* var. *fusca* Richardson, *in* Beechey, The zoology of Captain Beechey's voyage . . . , p. 5. Type locality, banks of Columbia River, below The Dalles, between Oregon and Washington.
1850. *Lupus gigas* Townsend, Acad. Nat. Sci. Philadelphia, 2:75, November, type from near Vancouver, Clark Co., Washington.

MARGINAL RECORDS.—British Columbia: Kimsquit, at head Dean Channel; Pemberton (Cowan and Guiguet, 1965:285). Washington: Twin Peaks; type locality. Oregon: South Ice Cave, about 40 mi. S Bend; Sycan, 30 mi. S Silver Lake. Nevada: Little High Rock Canyon. California: Litchfield. Oregon: Crane Prairie, 25 mi. NE Ashland, thence up coast to British Columbia: Calvert Island (Cowan and Guiguet, 1965:285); Swindle Island.

Canis lupus griseoalbus Baird

1823. *Canis Lupus-Griseus* Sabine, *in* Franklin, Narrative of a journey to the shores of the Polar Sea in . . . 1819–22, p. 654 (App.). Not *Canis Griseus* Boddaert, 1794.
1858. *Canis occidentalis* var. *griseo-albus* Baird, Mammals, *in* Repts. Expl. Surv. . . . , 8(1):104, July 14, type from Cumberland House, Saskatchewan.
1952. *Canis lupus griseoalbus*, Hall and Kelson, Univ. Kansas Publ., Mus. Nat. Hist., 5:338, December 5.
1947. *Canis lupus knightii* Anderson, Bull. Nat. Mus. Canada, 102:54, January 24, type from Cumberland House, Saskatchewan.

MARGINAL RECORDS.—Saskatchewan: type locality. Manitoba: Norway House; Dacotah, 30 mi. SE Delta (Tamsitt, 1962:76, as *C. lupus*); Riding Mountain National Park. Saskatchewan: *ca.* Carlton; Prince Albert National Park.

Canis lupus hudsonicus Goldman

1941. *Canis lupus hudsonicus* Goldman, Proc. Biol. Soc. Washington, 54:112, September 30, type from head Shultz Lake, Keewatin.

MARGINAL RECORDS.—Keewatin: Perry River District, Queen Maud Sea; Wager River; Cape Fullerton; *Hudson Bay*; 1 mi. S mouth Windy River (Harper, 1956:48). Mackenzie: Artillery Lake; Back River, 20 mi. below Lake Beechey.

Canis lupus irremotus Goldman

1937. *Canis lupus irremotus* Goldman, Jour. Mamm., 18:41, February 14, type from Red Lodge, Carbon Co., Montana.

MARGINAL RECORDS.—Alberta: Calgary; Gleichen. Montana: Fort Conrad; Ingomar; Powderville. South Dakota: Belle Fourche; *3 mi. S, 1 1/2 mi. W Sturgis* (Turner, 1974:126); *18 mi. NE Dewey* (*ibid.*). Wyoming (Long, 1965a:677): 15 mi. NW Manville; *10 mi. N Lost Springs*; Glenrock; 8 mi. NW Splitrock; Cokeville. Idaho: Castleford; Hammett; Boise National

Forest. Oregon: Logan Valley. Idaho: Priest River. Alberta: Waterton Lakes National Park.

Canis lupus labradorius Goldman

1937. *Canis lupus labradorius* Goldman, Jour. Mamm., 18:38, February 11, type from vic. Fort Chimo (now Chimo), Quebec.

MARGINAL RECORDS (Harper, 1961:94).—Quebec: Stupart Bay; Fort Chimo. Labrador: Nain, down coast to Adlavik; Michikamau Lake. Quebec: Lower Seal Lake area.

Canis lupus ligoni Goldman

1937. *Canis lupus ligoni* Goldman, Jour. Mamm., 18:39, February 11, type from head Duncan Canal, Kupreanof Island, Alaska.

MARGINAL RECORDS.—Alaska: Eleanor Cove, Yakutat Bay; Juneau. British Columbia: Port Simpson. Alaska: Dall Island; Kuiu Island.

Canis lupus lycaon Schreber

1775. *Canis lycaon* Schreber, Die Säugthiere . . . , theil 2, heft 13, Pl. 89. Type locality, Quebec, Quebec (fixed by Goldman, Jour. Mamm., 18:38, February 11, 1937).
1937. *Canis lupus lycaon*, Goldman, Jour. Mamm., 18:45, February 11.
1843. *Canis lupus canadensis* Blainville, Ostéographie ou description iconographique . . . , vol. 2, fasc. 13 (Atlas), Genre Canis, p. 45, Pl. 7, based on a specimen presumably from Canada.
1940. *Canis tundrarum ungavensis* Comeau, Ann. de l'Acfas [= Association Canadienne—Francoise pour l' avancement des Sciences], 6:121, Montreal, June 15, type from 35 mi. N Godbout, Quebec.

MARGINAL RECORDS.—Labrador (Harper, 1961:97): Sandwich Bay area; Porcupine; St. Charles River, down N shore St. Lawrence River to Quebec: 50 mi. N Quebec. Rhode Island: Aquidneck Island (Cronan and Brooks, 1962:90), thence westward to Pennsylvania: *no precise locality* (Goldman, 1944:441). Ohio: *no precise locality* (*ibid.*). Wisconsin: Grant County (Jackson, 1961:293, Map 55, pre-1900). Minnesota: Elk River. Ontario: Rainy Lake; Hurkett, Thunder Bay; 40 mi. NE Mattawa. Labrador: area E of Menihek Lakes (Harper, 1961:95).

According to R. M. Nowak (pers. comm.), a specimen from an unknown locality in Florida, assigned by Goldman (1944:440) to *C. l. lycaon*, is referable to *C. rufus*. The tooth-bearing part of a mandibular ramus from Citico Mound in southeastern Tennessee was identified as *C. rufus floridanus* by Kellogg (1939:267), but was identified as *C. lupus lycaon* by Goldman (1944:441). If Goldman (*loc. cit.*) was correct, *C. lupus* occurred farther south in post-Pleistocene time than is shown on Map 490.

Canis lupus mackenzii Anderson

1823. *Canis Lupus-Albus* Sabine, *in* Franklin, Narrative of a journey to the shores of the Polar Sea in . . . 1819–22, p. 655 (App.), and pl. opposite p. 312. Type from Fort Enterprise, Mackenzie. Not *C[anis]. Lupus albus* Kerr, 1792.
1943. *Canis lupus mackenzii* Anderson, Jour. Mamm., 24:388, August 17, type from Imnanuit, W of Kater Point, Bathurst Inlet, lat. 67° 44' 20" N, long. 109° 04' 03" W, Mackenzie.

MARGINAL RECORDS.—Mackenzie: type locality; upper Back River; Reliance, Great Slave Lake; Fort Good Hope; Peel River.

Canis lupus manningi Anderson

1943. *Canis lupus manningi* Anderson, Jour. Mamm., 24:392, August 17, type from Hantzch River, E side Foxe Basin, W side Baffin Island, lat. about 67° N, long. 24° W, Franklin.

MARGINAL RECORDS.—Baffin Island: Salmon River, Pond Inlet, about 72° 30' N; Ashe Inlet, Hudson Strait.

Canis lupus mogollonensis Goldman

1937. *Canis lupus mogollonensis* Goldman, Jour. Mamm., 18:43, February 11, type from S. A. Creek, 10 mi. NW Luna, Catron Co., New Mexico.

MARGINAL RECORDS.—Arizona: Kendrick Peak. New Mexico: type locality; Magdalena; Monticello; Capitan Mts.; Hurley. Arizona: Galiuro Mts.; Aguila; *6 mi. E Williams* (Cockrum, 1961:218).

Canis lupus monstrabilis Goldman

1937. *Canis lupus monstrabilis* Goldman, Jour. Mamm., 18:42, February 11, type from 10 mi. S Rankin, Upton Co., Texas.

MARGINAL RECORDS.—Texas: Fort Richardson; Llano. Tamaulipas: Matamoros. Texas: Guadalupe Mts. New Mexico: Mescallero; Capitan Mts.

Canis lupus nubilus Say

1823. *Canis nubilus* Say, *in* Long, Account of an exped. . . . to the Rocky Mts. . . . , 1:169, type from Engineer Cantonment, near present town of Blair, Washington Co., Nebraska.
1829. *Canis lupus* var. *nubilus*, Richardson, Fauna Boreali-Americana, p. 69.
1841. *Canis variabilis* Wied-Neuwied, Reise in das innere Nord-Amerika in . . . 1832–34, 2:95. Type locality, Fort Clark, near Stanton, Mercer Co., North Dakota.

MARGINAL RECORDS.—Manitoba: Duck Mtn. Minnesota: Crookston. Iowa (Bowles, 1975:111, 112, historical records predating 1884): Turkey River; Clinton County; Missouri River, Fremont Co. Nebraska: ½ mi. NE Falls City (Jones, 1964c:252). Oklahoma: Beaver Creek. New Mexico: Gallo Canyon, 40 mi. SE Corona; Carthage; *Mountainair*; 18 mi. N Santa Rosa.

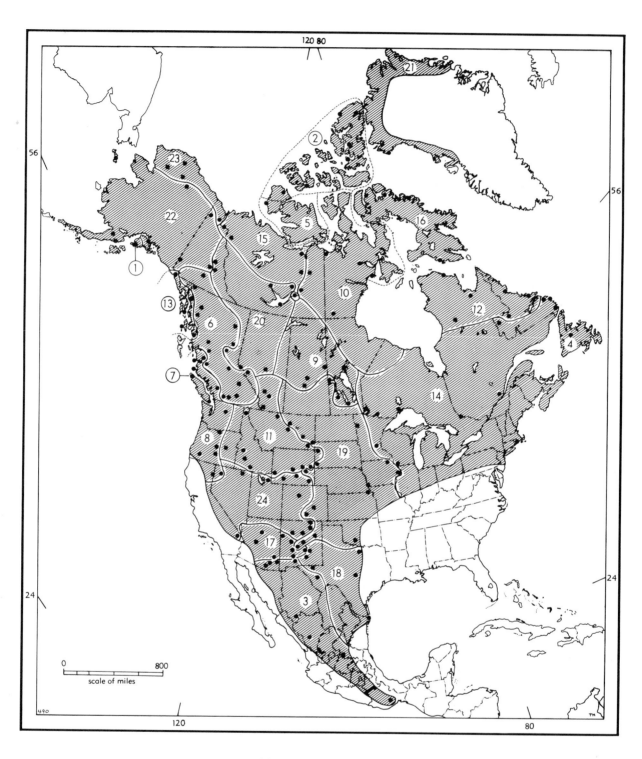

Map 490. *Canis lupus.*

1. *C. l. alces*
2. *C. l. arctos*
3. *C. l. baileyi*
4. *C. l. beothucus*
5. *C. l. bernardi*
6. *C. l. columbianus*
7. *C. l. crassodon*
8. *C. l. fuscus*
9. *C. l. griseoalbus*
10. *C. l. hudsonicus*
11. *C. l. irremotus*
12. *C. l. labradorius*
13. *C. l. ligoni*
14. *C. l. lycaon*
15. *C. l. mackenzii*
16. *C. l. manningi*
17. *C. l. mogollonensis*
18. *C. l. monstrabilis*
19. *C. l. nubilus*
20. *C. l. occidentalis*
21. *C. l. orion*
22. *C. l. pambasileus*
23. *C. l. tundrarum*
24. *C. l. youngi*

932

Colorado: Fort Garland; Pueblo (Armstrong, 1972:259, 260). Wyoming (Long, 1965a:678): *Natrona County;* Douglas. South Dakota: *Dewey;* Folsom. Montana: Ekalaka; Miles City; upper Missouri River. Saskatchewan: *central Saskatchewan.*

Canis lupus occidentalis Richardson

1829. *Canis lupus occidentalis* Richardson, Fauna Boreali-Americana, 1:60. Type locality, by restriction (Miller, Smiths. Miscl. Coll., 59(15):4, June 8, 1912), Fort Simpson, near mouth Liard River, Mackenzie.
1829. [*Canis lupus occidentalis*] var. C, *Lupus sticte* Richardson, *loc. cit.*, p. 68. Type locality, banks of Mackenzie River. Applied to a color phase of *C. l. occidentalis.*
1829. [*Canis lupus occidentalis*] var. E, *Lupus ater* Richardson, *op. cit.*, p. 70. Type locality, banks of Mackenzie and Saskatchewan rivers. Applied to a color phase of *C. l. occidentalis.*

MARGINAL RECORDS.—Mackenzie: Aylmer Lake; Fort Resolution. Alberta: Edmonton. British Columbia: Yellowhead Pass Area; Fort McLeod, Cariboo District. Yukon: Pelly Lakes; N fork Macmillan River. Mackenzie: Fort Rae. Inferentially, Cowan and Guiguet (1965:284, 285) refer the British Columbian specimens to *C. l. columbianus.*

Canis lupus orion Pocock

1935. *Canis lupus orion* Pocock, Proc. Zool. Soc. London, p. 683, September 12, type from Cape York, on Baffin Bay, northwestern Greenland. Known only from type locality.

Canis lupus pambasileus Elliot

1905. *Canis pambasileus* Elliot, Proc. Biol. Soc. Washington, 18:79, February 21, type from Susitna River, region of Mt. McKinley, Alaska.
1937. *Canis lupus pambasileus*, Goldman, Jour. Mamm., 18:45, February 11.

MARGINAL RECORDS.—Alaska: head Salmon Fork, Black River, 120 mi. NW Fort Yukon. Yukon: 6 mi. N New Rampart House; Hoole Canyon, Pelly River; White River. Alaska: Kenai Peninsula, probably near base; delta of Katmai River (Schiller and Rausch, 1956:194); S shores Naknek Lake (Cahalane, 1959:193); St. Lawrence Island.

Canis lupus tundrarum Miller

1912. *Canis tundrarum* Miller, Smiths. Miscl. Coll., 59(15):1, June 8, type from Point Barrow, Alaska.
1935. *Canis lupus tundrarum*, Pocock, Proc. Zool. Soc. London, p. 681, September 12.

MARGINAL RECORDS.—Alaska: Meade River, near Point Barrow. Yukon: Porcupine River. Alaska: John River Valley; near head Noatak River.

Canis lupus youngi Goldman

1937. *Canis lupus youngi* Goldman, Jour. Mamm., 18:40, February 11, type from Harts Draw, N slope Blue Mts., 20 mi. NW Monticello, San Juan Co., Utah.

MARGINAL RECORDS.—Wyoming: Dry Lake, 15 mi. N Rawlins; Federal. Colorado: Castle Peak, 15 mi. NE Eagle (Armstrong, 1972:260). New Mexico: Tusas; Haynes; Lamy; San Mateo Mts. California: Old Barnett Mine, 12 mi. E Lanfair, Providence Mts. Nevada: Gold Creek. Utah: South Eden. Wyoming: Rock Springs.

Canis rufus
Red Wolf

Revised by Goldman (Classification of wolves, pt. 2, pp. 389–507, in part, Pls. 126–129, 131, 1 fig. in text), *in* Young and Goldman, The wolves of North America, Amer. Wildlife Manag. Inst., Washington, D.C., pp. xx + 636, 131 pls., 15 figs. in text, 1944. Nowak (Defenders of Wildlife News, 42:60, 61, April 1967) pointed out that according to opinion 447 of the International Commission on Zoological Nomenclature, *Lupus niger* Bartram 1791 was unavailable, as were all Bartram's names, because he had not been consistently binomial.

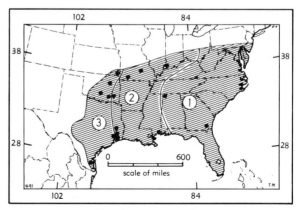

Map 491. *Canis rufus.*

1. *C. r. floridanus*   2. *C. r. gregoryi*
3. *C. r. rufus*

External measurements: 1403–1650; 343–420; 210–254. Condylobasal length of skull, 187.5–233.7. Intermediate in size between *C. latrans* and *C. lupus.* Color tending toward tawny, especially on muzzle, ears, and outer surfaces of limbs; a black phase occurs; differs from *C. latrans* in: size larger; build more robust, pelage somewhat coarser; rhinarium and feet larger; cranium higher; rostrum deeper; zygomata more widely spreading; jugal heavier and more deeply inserted on maxilla; occipital condyles usually extending farther transversely; bullae larger; M1 with posteroexternal cusp more nearly equaling anteroexternal cusp. Differs from *C. lupus* in: smaller size and more slender build; rhinarium and foot pads smaller; general coloration more tawny; molariform teeth with crowns more deeply cleft, cusps more compressed laterally and points and shearing edges more trenchant; P4 with protocone more prominent but sometimes absent; M1 with more prominent cingulum and posteroexternal cusp distinctly smaller compared to anteroexternal cusp, M2 relatively larger.

Canis rufus floridanus Miller

1912. *Canis floridanus* Miller, Proc. Biol. Soc. Washington, 25:95, May 4, type from Horse Landing, St. Johns River, about 12 mi. S Palatka, Putnam Co., Florida.

MARGINAL RECORDS.—Alabama: 12 mi. S Cherokee. Georgia: Okefinokee Swamp, thence around coast of Florida to point of beginning.

Canis rufus gregoryi Goldman

1937. *Canis rufus gregoryi* Goldman, Jour. Mamm., 18:44, February 11, type from Mack's Bayou, 3 mi. E Tensas River, 18 mi. SW Tallulah, Madison Parish, Louisiana.

MARGINAL RECORDS.—Indiana: Wabash River (Mumford, 1969:86). Mississippi: Biloxi. Texas: Anahuac National Wildlife Refuge (Paradiso, 1965:319, as *C. niger* only); *Mount Belview* [*Mont Belvieu*]; Segno; Wakefield. Oklahoma: Nashoba (Garland Creek); *Talahina* [*Talihina*] (Goldman, 1944:486). Missouri: Reeds Spring (*ibid.*); Cook's Station; *Westover.*

Canis rufus rufus Audubon and Bachman

1851. *Canis lupus* var. *rufus* Audubon and Bachman, The viviparous quadrupeds of North America, 2:240. Type locality, designated by Goldman (Jour. Mamm., 18:45, February 11, 1937), 15 mi. W Austin, Texas.

MARGINAL RECORDS.—Kansas: "south of Columbus." Oklahoma: Atoka. Texas: 6 mi. N Dayton; Armstrong (Davis, 1966:113, as *C. niger* only). Oklahoma: Cherokee Town, 40 mi. N Ardmore (8098 USNM); Red Fork.

A specimen from 22 mi. N Sheffield, Texas, referred

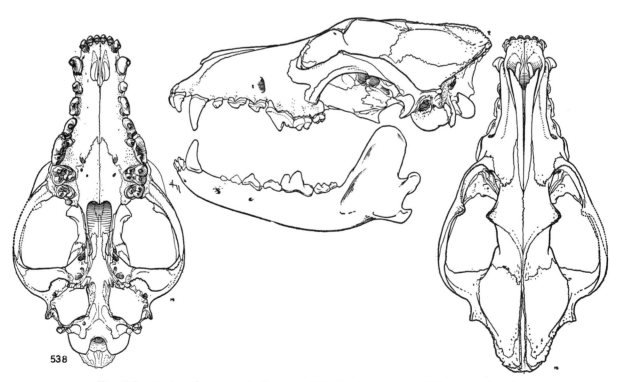

Fig. 538. *Canis rufus gregoryi*, Cremona, Polk Co., Texas, No. 95809 M.V.Z., ♀, X ⅔.

by Goldman (1944:489) to *C. r. rufus*, was considered by Paradiso and Nowak (1972:14) to be *C. latrans*. All specimens from the Edwards Plateau area of central Texas, referred by Goldman (1944:488, 489) to *C. r. rufus*, were considered by Paradiso and Nowak (1972:14) to be hybrids between *C. rufus* and *C. latrans*. According to R. M. Nowak (pers. comm.), a specimen from the Wichita Mountains Refuge, Oklahoma, assigned by Goldman (1944:488) to *C. r. rufus*, is referable to *C. latrans*; all Arkansas specimens assigned by Goldman (1944:488) to *C. r. rufus* are either referable to *C. r. gregoryi* or are hybrids of *C. rufus* and *C. latrans*; specimens from Reeds Spring, Missouri, referred by Hall and Kelson (1952:340) to *C. r. rufus*, are referable to *C. r. gregoryi*; and specimens from 3 mi. N Thomasville and 5 mi. N Gainesville, Missouri, referred by Leopold and Hall (1945:143) to *C. r. rufus*, are hybrids of *C. rufus* and *C. latrans*.

Genus **Alopex** Kaup—Arctic Fox

1829. *Alopex* Kaup, Skizzerte Entwickelungs-Geschichte und Natürliches System der europäischen Thierwelt, pt. 1, p. 85. Type, *Canis lagopus* Linnaeus.
1869. *Leucocyon* Gray, Proc. Zool. Soc. London, for 1868, p. 521. Type, *Canis lagopus* Linnaeus.

Resembles *Vulpes* (of which it is regarded as a subgenus by Bobrinski, 1944:146) but differs in: short rounded ears, thickly haired feet, and, usually, seasonal dichromatism, being brownish-gray in summer and white or "blue" in winter.

Whoever assesses geographic variation in this circumboreal species needs to consult Vibe (1967:101–150) and his extensive bibliography (*op. cit.*:207–227).

Alopex lagopus
Arctic Fox

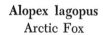

External measurements of an adult female from Point Barrow, Alaska, are: 1097; 422; 161 (ear from crown); 56. Characters those of genus.

Alopex lagopus groenlandicus (Bechstein)

1799. *Canis groenlandicus* Bechstein, Thomas Pennant's allgemeine Uebersicht der vierfüssigen Theire, 1:270. Type locality, Greenland.
1947. *Alopex lagopus groenlandicus*, Anderson, Bull. Nat. Mus. Canada, 102:50, January 24.

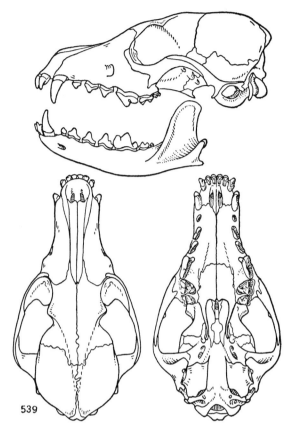

539

Fig. 539. *Alopex lagopus lagopus*, Point Barrow, Alaska, No. 52366 K.U., ♂, X ½.

1898. *?Canis lagopus spitzbergenensis* Barrett-Hamilton and Bonhote, Ann. Mag. Nat. Hist., ser. 7, 1:287, April, type from Spitzbergen.

MARGINAL RECORDS.—*Ellesmere Island; Greenland*.

Alopex lagopus hallensis (Merriam)

1900. *Vulpes hallensis* Merriam, Proc. Washington Acad. Sci., 2:15, March 14, type from Hall Island, Bering Sea, Alaska. Known also from St. Matthew Island.
1953. *Alopex lagopus hallensis*, Rausch, Murrelet, 34:20, October 5.

Alopex lagopus lagopus Linnaeus

1758. *Canis lagopus* Linnaeus, Syst. nat., 10th ed., 1:40, type locality Lapland.
1912. *Alopex lagopus*, Miller, Catalogue of the mammals of western Europe, British Museum, p. 319, November 23.
1820. *Canis vulpes caerulea* Nilsson, Skand. fauna, 1:88, type locality Lapland.
1827. [*Canis lagopus*] *argenteus* Billberg, Synop. Faunae Scandinaviae, p. 14, type locality Lapland.
1902. *Vulpes lagopus innuitis* Merriam, Proc. Biol. Soc. Washington, 15:170, August 6, type from Karogar River, Point

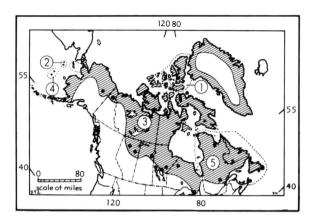

Map 492. *Alopex lagopus.*

1. *A. l. groenlandicus*   3. *A. l. lagopus*
2. *A. l. hallensis*   4. *A. l. pribilofensis*
              5. *A. l. ungava*

Barrow, Alaska (see Rausch, Murrelet, 34(2):20–22, October 5, 1953, who regards *Alopex lagopus lagopus* as "a single, highly variable form of arctic fox [with a] . . . circumpolar distribution").

1911. *V[ulpes]. lagopus kenaiensis* Brass, Aus dem Reiche der Pelze, p. 466, April, type locality mainland of Alaska. Not *V. kenaiensis* Merriam, 1900.

MARGINAL RECORDS.—Franklin: Bylot Island; *Baffin Island.* Ontario: *ca.* 75 mi. NE Sault Ste. Marie (Banfield, 1974:297, Map 131); Severn River. Manitoba: Oxford Lake; Norway House. Saskatchewan: Reindeer Lake (Banfield, 1974:298). Alberta (Soper, 1965:267): *ca.* 60 mi. N McMurray; *ca.* 140 mi. W Athabasca. Yukon: Porcupine River, mouth Berry Creek (Youngman, 1975:129). Alaska: Yukon Flats; Igagnik, thence northward along coast to Seward Peninsula, westward into Siberia, eastward to Alaska, N and E along coast to Karogar River, Point Barrow.

Alopex lagopus pribilofensis (Merriam)

1902. *Vulpes pribilofensis* Merriam, Proc. Biol. Soc. Washington, 15:171, August 6, type from St. George Island, Pribilof Islands, Alaska. Known only from type locality and St. Paul Island.
1953. *Alopex lagopus pribilofensis*, Rausch, Murrelet, 34:20, October 5.

Alopex lagopus ungava (Merriam)

1902. *Vulpes lagopus ungava* Merriam, Proc. Biol. Soc. Washington, 15:170, August 6, type from Fort Chimo, S of Ungava Bay, Quebec.
1912. *Alopex lagopus ungava,* Miller, Bull. U.S. Nat. Mus., 79:82, December 31.

MARGINAL RECORDS.—Quebec: Cape Wolstenholme; type locality. Newfoundland: *northern and southern Newfoundland* (Cameron, 1959:29, as *Alopex lagopus* only). Nova Scotia: Lingan Bay (*ibid.*). Quebec:

Anticosti Island (*ibid.*); Lake St. John area (Harper, 1961:98).

Genus **Vulpes** Bowdich
Kit Fox and Red Fox

1821. *Vulpes* Bowdich, An analysis of the natural classification of Mammalia, p. 40. Type, *Canis vulpes* Linnaeus. (*Vulpes* Oken, 1816, is not available because his names are non-Linnaean.)
1839. *Cynalopex* Hamilton-Smith, The naturalist's library (edit. Jardine), 25:222. Based on *Canis corsac* Linnaeus [= *Vulpes corsac*].

Medium to small canids with long, pointed ears, muzzle elongate and attenuated, tail round and bushy, usually as long as length of head and body; width of rostrum measured at a point opposite cone of P2 usually less than 18 per cent of condylobasal length; teeth relatively widely spaced and relatively narrow.

KEY TO NORTH AMERICAN SPECIES OF VULPES

1. Tail with white tip; feet black. *V. vulpes,*   p. 936
1'. Tail with black tip; feet not black. *V. velox,*   p. 939

Vulpes vulpes
Red Fox

Several authors, including Rausch (1953:107), who used the name combination *Vulpes vulpes alascensis,* have stated that the red fox of North America and Eurasia is a single species, and accordingly *Vulpes vulpes* is here used as the specific name. According to Vogt and Arakaki (1971:319), "The karyotypes of the American red fox and the European red fox are very similar, the only apparent difference being that the European animal has two pairs of microchromosomes with a 2n number of 38, while the American fox has only one pair of microchromosomes with a 2n number of 36." The subspecies and provenience of the one American red fox examined are unknown. It would be interesting to ascertain the 2n number of chromosomes in additional red foxes of several subspecies in Eurasia and North America.

External measurements: 900–1025; 350–400; 145–171. Color normally reddish-yellow, feet and dorsal surface of ears black, tail tip white; black, silver, and cross color-phases occur in nature.

ts range
'lains re-
had been

Vulpes vulpes alascensis Merriam

1900. *Vulpes alascensis* Merriam, Proc. Washington Acad. Sci., 2:668, December 28, type from Andreafski, *ca.* 70 mi. above delta of Yukon River, Alaska.
1953. *Vulpes vulpes alascensis,* Rausch, Arctic, 6:107, July.

MARGINAL RECORDS.—Alaska: Barrow. Mackenzie: Perry River District, Queen Maud Sea; Great Slave Lake. Yukon: Ross River (Mile 90), Canol Road. Alaska: Knik Valley; Cold Bay; Alaska Peninsula, opposite Popof Island; Dolgoi and Sanak islands (Murie, 1959:287); Unimak Island (Peterson, 1967:125); Akun Island (Murie, 1959:287); Akutan, Unalaska, and Umnak islands (Peterson, 1967:125); Chuginadak, Amlia, Adak, and Kanaga islands (Murie, 1959:287); type locality, thence northeastward along coast to *Seward Peninsula,* westward into Siberia, eastward to Alaska and along coast to point of beginning. Peterson (1967:124, 125, Table 1) questions whether the original range of *V. vulpes* in the Aleutian Islands extended west of Umnak Island.

Vulpes vulpes cascadensis Merriam

1900. *Vulpes cascadensis* Merriam, Proc. Washington Acad. Sci., 2:665, December 28, type from Trout Lake, S base Mt. Adams, Skamania Co., Washington.

MARGINAL RECORDS.—British Columbia: Anahim Lake; Chezacut; Baldy Mtn., near Bridesville. Washington: Loomis. Oregon: The Dalles; 40 mi. S Bend; *Yamsay Mtn.;* Swan Lake Valley; Mt. McLoughlin; Mt. Hebo; Tillamook. Washington: type locality; Mount Rainier National Park. British Columbia: Boundary Bay; Homathko River.

Vulpes vulpes fulva (Desmarest)

1820. *Canis fulvus* Desmarest, Mammalogie . . . pt. 1, p. 203, *in* Encyclopédie méthodique . . . . Type locality, Virginia.
1960. *V[ulpes]. v[ulpes]. fulva,* Churcher, Jour. Mamm., 41:359, August 15.
1842. *Vulpes fulvus,* De Kay, The zoology of New-York, Mammalia, p. 44.
1894. *Vulpus [sic] pennsylvanicus [sic]* Rhoads, Amer. Nat., 28:524, June. Not *[Vulpes] Pensylvanicus* Boddaert, 1784 [= *Urocyon cinereoargenteus* (Schreber), 1775].

MARGINAL RECORDS.—Michigan: Houghton County; Drummond Island (Manville, 1950:359, as *V. fulva*); Huron County. Ontario: King Twp., York Co. Maine: Mount Desert Island (Manville, 1960:416, as *V. fulva*); islands of Casco Bay. Virginia: Northampton County; Brunswick County. South Carolina (Golley, 1966:142): Georgetown County; Charleston County. Georgia (Golley, 1962:176): Brantley County; Thomas County. Alabama: Orange Beach. Louisiana: 6 mi. S Abbeville (Lowery, 1974:398). Texas: Sour Lake; along Howard Creek and Pecos River; 5⅜ mi. E Big Spring (Packard and Bowers, 1970:450); 2 mi. W Lubbock (*ibid.*); 5¹/₁₀ mi. W Plainview (*ibid.*); 3⅛ mi. W Kress (*ibid.*); Wichita County (Dalquest, 1968:19, as *V. vul-*

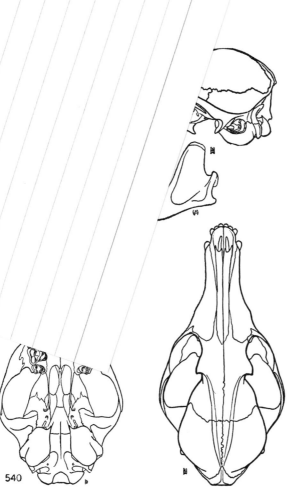

540

Fig. 540. *Vulpes vulpes necator,* Marlette Lake, 8000 ft., Ormsby Co., Nevada, No. 69636 M.V.Z., ♂, X ½.

Vulpes vulpes abietorum Merriam

1900. *Vulpes alascensis abietorum* Merriam, Proc. Washington Acad. Sci., 2:669, December 28, type from Stuart Lake, British Columbia. Indistinguishable from *V. v. alascensis* according to Youngman, Nat. Mus. Canada, Publ. Zool., 10:132, September 4, 1975.
1911. ? *V[ulpes]. argenteus sitkaensis* Brass, Aus dem Reiche dem Pelze, p. 448, April. Type locality, West Coast, especially the Aleutian Islands.

MARGINAL RECORDS.—Mackenzie: Fort Reliance, Great Slave Lake; mouth Salt River. Alberta: 20 mi. S Athabaska Landing. British Columbia: Little Prairie; Isaacs Lake; Akolkolex River; Lac La Hache; Quick; Telegraph Creek; Atlin. Yukon: 6 mi. SW Kluane, 2559 ft.

pes only, possibly introduced). Oklahoma (Glass and Halloran, 1960:71, 72): near Bray; vic. Stillwater. Kansas: 7 mi. N Fowler (Anderson and Nelson, 1958:307); 5 mi. S, 7 mi. W Friend (*ibid.*); Waconda Springs; *1½ mi. SW Glasco* (Getz, 1961:283). Wisconsin (Jackson, 1961:301, Map 56): Pierce County; Douglas County.—A tempting speculation is that on the Midgrass and Shortgrass prairies of the High Plains, wolves were abundant as long as natural mortality in the bison herds provided most of the wolves' food. When the wolf population declined, the coyote population increased. When the coyote population decreased in the 1900s the red fox moved westward and now lives where few or none did two or three decades ago. Janes and Gier (1966:23–31) thought that the red fox by that time occurred in every Kansas county and added:

"There are numerous reliable reports of red foxes in the watersheds of the South Platte, Republican, and Arkansas rivers in eastern Colorado, adjacent to Kansas." See the shaded area for subspecies 4 on Map 493.

Vulpes vulpes harrimani Merriam

1900. *Vulpes harrimani* Merriam, Proc. Washington Acad. Sci., 2:14, March 14, type from Kodiak Island, Alaska. Known only from Kodiak Island.

Vulpes vulpes kenaiensis Merriam

1900. *Vulpes kenaiensis* Merriam, Proc. Washington Acad. Sci., 2:670, December 28, type from Kenai Peninsula, Alaska. Known only from Kenai Peninsula.

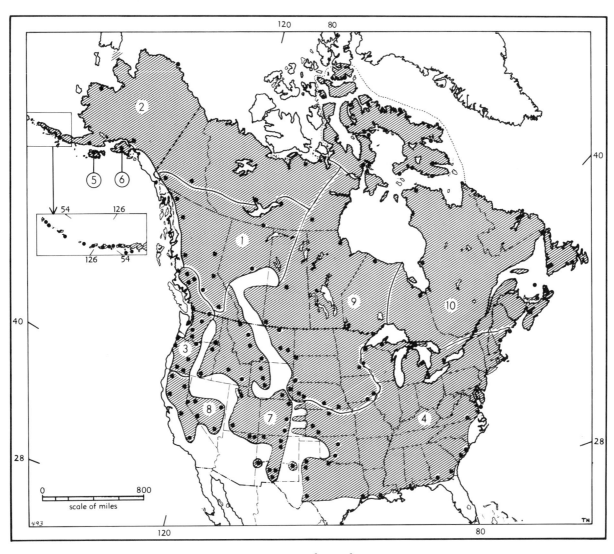

Map 493. *Vulpes vulpes.*

1. V. v. abietorum	4. V. v. fulva	8. V. v. necator
2. V. v. alascensis	5. V. v. harrimani	9. V. v. regalis
3. V. v. cascadensis	6. V. v. kenaiensis	10. V. v. rubricosa
	7. V. v. macroura	

Vulpes vulpes macroura Baird

1852. *Vulpes macrourus* Baird, *in* Stansbury, Exploration and survey of the Valley of the Great Salt Lake of Utah . . . , (Spec. Sess., U.S. Senate, Exec. No. 3), App. C, p. 309, June, type from Wasatch Mts., bordering Great Salt Lake, Utah.
1962. *Vulpes vulpes macroura*, Halloran, Jour. Mamm., 43:432, August 20.

MARGINAL RECORDS.—Canada: *Rocky Mts.* Montana (Hoffmann, *et al.*, 1969:594, as *V. vulpes*, unless otherwise noted): Swiftcurrent Creek (Hall and Kelson, 1959:857); Deep Creek, 10 mi. SW Choteau; East Helena; Mystic Lake. Wyoming: 3½ mi. S, 27 mi. W Cody; Lake Fork, Wind River Mts. (Long, 1965a:680); 17 mi. S, 6½ mi. W Lander, 8450 ft.; Laramie (Long, 1965a:680). Colorado (Armstrong, 1972:261): 7 mi. N Greely; 3 mi. S Colorado Springs; 12 mi. W La Veta. New Mexico: W slope Wheeler Peak, 12,400 ft.; Pecos Baldy; Point of Sands, Tularosa Basin; Organ Mts.; Ash Canyon, San Andres Mts.; Farmington; *Shiprock* [Also shown in New Mexico on Map 493 is a possible occurrence near head Willow Creek less than 30 mi. E of Arizona border (V. Bailey, 1932:297), and an occurrence 1 mi. S Railroad Mtn. (Findley, *et al.*, 1975:287) less than 50 mi. W Texas border]. Arizona (Halloran and Taber, 1965:139): Round Rock; near Kayenta. Utah: 15 mi. S Henrieville; type locality. Idaho: near Pocatello; Sawtooth Mts. Oregon: Steens Mtn.; about Elgin. Washington: Blue Mts.; *NE Washington.*

Vulpes vulpes necator Merriam

1900. *Vulpes necator* Merriam, Proc. Washington Acad. Sci., 2:664, December 28, type from Whitney Meadow, near Mt. Whitney, Tulare Co., California.

MARGINAL RECORDS.—California: W side Mt. Shasta; *Mt. Shasta, 7000 ft.*; near Eagle Lake. Nevada: Marlette Lake, 8000 ft.; Fairview; Shell Creek Range; 1 mi. W Hidden Forest, 9000 ft.; Silver Peak Range. California: *S side Cirque Peak*; summit Piute Mts.; *Alwell's Mill, E. Fork Kaweah River*; Buck Meadows; 7 mi. NE Maxwell.—Considerable information on probable origin and increase in geographic range of the population of red fox provisionally identified as *V. v. necator* in the Sacramento Valley of California has been summarized by Gray (1977:58).

Vulpes vulpes regalis Merriam

1900. *Vulpes regalis* Merriam, Proc. Washington Acad. Sci., 2:672, December 28, type from Elk River, Sherburne Co., Minnesota.
1963. *Vulpes vulpes regalis*, Jones and Henderson, Jour. Mamm., 44:283, May 20.

MARGINAL RECORDS.—Keewatin: Upper Kazan River (Harper, 1956:59). Manitoba: Churchill. Ontario: Severn. Minnesota: 10 mi. N, 1 mi. E Hovland (Timm, 1975:29); Dakota County, thence southward along W bank Mississippi River, to Iowa (Bowles, 1975:114,

115): Cedar Twp., Muscatine Co.; near Bloomfield. Nebraska (Jones, 1964c:257): 6½ mi. NW Odell; Hastings; 5 mi. W Lewellen; Bridgeport. Wyoming (Long, 1965a:680): Fort Laramie. South Dakota (Jones and Henderson, 1963:283): 10 mi. N Rumford; Harding County. Montana (Hoffmann, *et al.*, 1969:594, as *V. vulpes*): near Homestead; NW corner Prairie County; 35 mi. N Nashua. Saskatchewan (Soper, 1961:37): Cypress Hills; Harper Lake.

Vulpes vulpes rubricosa Bangs

1897. *Vulpes pennsylvanica vafra* Bangs, Proc. Biol. Soc. Washington, 11:53, March 16, type from Digby, Nova Scotia. Not *Vulpes vafer* Leidy, 1869, a fossil species.
1898. *Vulpes pennsylvania rubricosa* Bangs, Science, n.s., 7:272, February 25, a renaming of *V. p. vafra* Bangs.
1960. *V[ulpes]. v[ulpes]. rubricosa*, Churcher, Jour. Mamm., 41:359, August 15.
1898. *Vulpes deletrix* Bangs, Proc. Biol. Soc. Washington, 12:36, March 24, type from Bay St. George, Newfoundland. Regarded as inseparable from *V. v. rubricosa* by Churcher, Jour. Mamm., 41:359, August 15.
1900. *Vulpes rubricosa bangsi* Merriam, Proc. Washington Acad. Sci., 2:667, December 28, type from L'Anse au Loup, Strait of Belle Isle, Labrador. Regarded as inseparable from *V. v. rubricosa* by Churcher, Jour. Mamm., 41:359, August 15.

MARGINAL RECORDS.—Franklin (Macpherson, 1964:140, unless otherwise noted): Grise Fiord; Arctic Bay; Pond Inlet; Clyde River; Pangnirtung; Frobisher Bay; Lake Harbour (Macpherson, 1964:138). Labrador: L'Anse au Loup, Strait of Belle Isle. Newfoundland: Bay St. George. Quebec: *Anticosti Island* (Cameron, 1959:61); Magdalen Islands (Cameron, 1962:511, 512). Prince Edward Island: Rollo Bay (Cameron, 1959:49). Nova Scotia: Frizzleton, Cape Breton Island (Cameron, 1959:31). Quebec: Mont St. Hilaire (Wrigley, 1969:210, as *V. vulpes*). Ontario: Moose Factory. Quebec: N end James Bay; Hudson Strait. Franklin (Macpherson, 1964:138, 140): Amadjuak; Cape Dorset; Repulse Bay; Igloolik; Resolute Bay.

Vulpes velox
Kit Fox

X ⅟₁₅

The eight taxa heretofore regarded as constituting the species *Vulpes macrotis* are here listed in the earlier-named species *Vulpes velox* because of the geneflow in eastern New Mexico and western Texas reported by Rohwer and Kilgore (1973) between *Vulpes m. neomexicanus* and *V. v. velox.*

External measurements: 600–800; 225–300; 111–132. Color buffy yellow, tail black-tipped, feet no darker than rest of body.

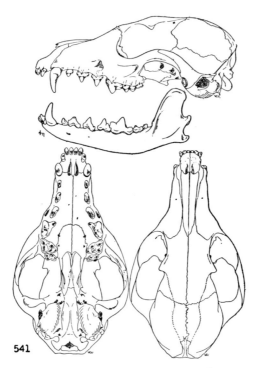

Fig. 541. *Vulpes velox velox*, First View, Cheyenne Co., Colorado, No. 1601 K.U., ♂, X ½.

Vulpes velox arsipus Elliot

1904. *Vulpes arsipus* Elliot, Field Columb. Mus., Publ. 87, Zool. Ser., 3(14):256, January 7, type from Daggett, San Bernardino Co., California.
1931. *Vulpes macrotis arizonensis* Goldman, Jour. Washington Acad. Sci., 21:249, June 4, type from 2 mi. S Tule Tanks, Yuma Co., Arizona.

MARGINAL RECORDS.—Utah (Stock, 1970:432): 1 mi. N St. George; *7 mi. SE St. George.* Arizona: Coolidge (Cockrum, 1961:221); Tucson; Fort Huachuca (Cockrum, 1961:221). Sonora: near Punta Peñasca. California: Brawley; Palm Springs; Victorville; Lovejoy Buttes, near Lancaster; Borax Flat; Triangle Spring, Death Valley; *Daylight Pass Road, 1000 ft.* (Wauer, 1961:109, as species only). Nevada: Indian Springs; 4 mi. N Mesquite.

Vulpes velox devia Nelson and Goldman

1909. *Vulpes macrotis devius* Nelson and Goldman, Proc. Biol. Soc. Washington, 22:25, March 10, type from Llano de Yrais, opposite Magdalena Island, Baja California. Known only from type locality.

Vulpes velox hebes Merriam

1902. *Vulpes velox hebes* Merriam, Proc. Biol. Soc. Washington, 15:73, March 22, type from Calgary, Alberta.

MARGINAL RECORDS.—Alberta: type locality. *Southwestern Saskatchewan.* North Dakota: Pembina Mts.; Souris River. *Western Iowa.* North Dakota: 9 mi. N, 2 mi. E Scranton (Pfeifer and Hibbard, 1970:835). *Southeastern British Columbia.* Alberta: Buffalo Lake.

Vulpes velox macrotis Merriam

1888. *Vulpes macrotis* Merriam, Proc. Biol. Soc. Washington, 4:136, February 18, type from Riverside, Riverside Co., California.

MARGINAL RECORDS.—California: type locality; Morino, San Jacinto Plain.

Vulpes velox mutica Merriam

1902. *Vulpes muticus* Merriam, Proc. Biol. Soc. Washington, 15:74, March 22, type from Tracy, San Joaquin Co., California.

MARGINAL RECORDS.—California: Tracy; vic. Lagrange; Goshen; Walker Basin; 1 mi. E Rose Station, 1200 ft.; Coalinga; 6 mi. E Panoche.

Vulpes velox neomexicana Merriam

1902. *Vulpes macrotis neomexicanus* Merriam, Proc. Biol. Soc. Washington, 15:74, March 22, type from Baird's Ranch, eastern side San Andres Mts., Dona Ana Co., New Mexico.

MARGINAL RECORDS.—Colorado: 3 mi. S Stateline Store, *ca.* 12 mi. SW Mack (Miller and McCoy, 1965:342, as *V. m. arsipus;* but conceivably an unnamed subspecies); McElmo Canyon (Egoscue, 1964:40). New Mexico: Albuquerque; 8 mi. W Vaughn (Findley, *et al.,* 1975:289); *ca.* Vaughn (Rohwer and Kilgore, 1973:Fig. 1); *ca.* 17 mi. S Cuervo (*op. cit.:*Fig. 1); *ca.* 19 mi. ENE Mesa (*op. cit.:*G of Fig. 1); *ca.* 42 mi. E Roswell (*op. cit.:*B of Fig. 1); Carlsbad. Texas: *ca.* Garden City (Rohwer and Kilgore, 1973:Fig. 1); *ca.* 14 mi. W Sterling City (*op. cit.:*J of Fig. 1); mouth Pecos River; Brewster County (Davis, 1966:104, as *V. macrotis* only). Chihuahua (Anderson, 1972:374); 6 mi. NNW Moctezuma, 4470 ft.; Colonia Díaz. New Mexico: S end Animas Valley. Arizona: Fort Grant. New Mexico: Cliff. Arizona: Wide Ruin (Halloran and Taber, 1965:140).

Vulpes velox nevadensis Goldman

1931. *Vulpes macrotis nevadensis* Goldman, Jour. Washington Acad. Sci., 21:250, June 4, type from near Willow Creek Ranch, near Jungo, Humboldt Co., Nevada.

MARGINAL RECORDS.—Oregon: Owyhee Valley between Rome and Pollock. Idaho: 20 mi. S Grandview. Nevada: Adelaide; 35 mi. SW Argenta; White Sage Valley; Spring Valley near Osceola. Utah: 15 mi.

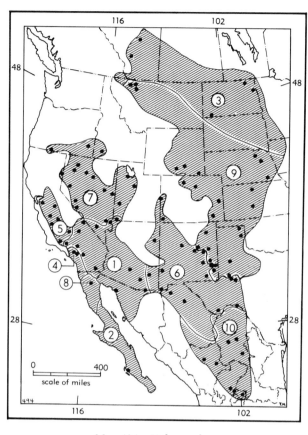

Map 494. *Vulpes velox.*

1. *V. v. arsipus*	6. *V. v. neomexicana*
2. *V. v. devia*	7. *V. v. nevadensis*
3. *V. v. hebes*	8. *V. v. tenuirostris*
4. *V. v. macrotis*	9. *V. v. velox*
5. *V. v. mutica*	10. *V. v. zinseri*

N Knolls; 2 mi. S Topliff; 30 mi. W Milford, 6500 ft. Nevada: 1 mi. N Rox; *Coyote Spring; 2½ mi. N Oasis, 5300 ft.;* Smith Ranch; 5 mi. SE Sodaville; *17 mi. S Fallon, 4100 ft.;* 5 mi. NE Hazen, 4000 ft. California: near Calneva. Nevada: Sulphur; type locality. Oregon: 6 mi. S Klamath Falls (Laughlin and Cooper, 1973:23).

Vulpes velox tenuirostris Nelson and Goldman

1931. *Vulpes macrotis tenuirostris* Nelson and Goldman, Jour. Mamm., 12:302, August 24, type from Trinidad Valley, NW base Sierra San Pedro Mártir, Baja California. Known only from type locality.

Vulpes velox velox (Say)

1823. *Canis velox* Say, in Long, Account of an exped. . . . to the Rocky Mts. . . . , 1:487, type from South Platte River (in Logan Co. ?), Colorado.
1851. *Vulpes velox*, Audubon and Bachman, The viviparous quadrupeds of North America, 2:13.

MARGINAL RECORDS.—Montana: Kipp (Long and Long, 1964a:108). Nebraska (Jones, 1964c:254): northern part, Antelope County; Madison County; Wilber. Kansas (Janes and Gier, 1966:25): Osborn County (sight record); Stafford County. Texas: Lipscomb County (Cutter, 1959:31); Hemphill County (*ibid.*); 4 mi. S Plainview (Packard and Garner, 1964:390); Martin County (Davis, 1966:105); *Midland.* New Mexico: 8 mi. S, 19 mi. E Maljamar (Packard and Bowers, 1970:450); 7³⁄₁₆ mi. W Tatum (Rohwer and Kilgore, 1973:D of Fig. 1 and p. 163); 106 mi. N Mexican boundary and 41 mi. W Texas border (*op. cit.*: Fig. 1); 175 mi. N Mexican boundary and 38 mi. W Texas border (*op. cit.*: Fig. 1); 180 mi. N Mexican boundary and 135 mi. W Texas border (*op. cit.*: A of Figs. 1 and 3); 280 mi. N Mexican boundary and 65 mi. W Texas border (*op. cit.*: Fig. 1). Colorado (Armstrong, 1972:264): 25 mi. NW Estelene; near Pueblo; W of Colorado Springs, El Paso Co.; 3 mi. NE Boulder. Wyoming (Long, 1965a:681): 2 mi. S, 5 mi. E Archer; Aurora; Bridgers Pass. Montana (Long and Long, 1964a:108, unless otherwise noted): Robare; Browning (Hoffmann, *et al.*, 1969:594, as *V. velox* only); *Blackfoot.*

Vulpes velox zinseri Benson

1938. *Vulpes macrotis zinseri* Benson, Proc. Biol. Soc. Washington, 51:21, February 18, type from San Antonio de Jaral, Coahuila.

MARGINAL RECORDS.—Coahuila: 3 mi. S, 16 mi. E Cuatro Ciénegas; type locality. Zacatecas: Real de Pinos (Anderson and Hadary, 1965:343); Mexican Hwy. 49, 20 mi. ESE Guadalupe (Baker, 1968:318, as *V. macrotis* only); Mexican Hwy. 45, 5 mi. SSE Victor Rosales (*ibid.*). Durango: *2 mi. SSE Chocolate;* 8 mi. S, 11 mi. W Lerdo; 7 mi. NNW LaZarca, 6000 ft. (Baker and Greer, 1962:133). Coahuila: 4 mi. S, 22 mi. W Cuatro Ciénegas.

Genus **Urocyon** Baird—Gray Foxes

1858. *Urocyon* Baird, Mammals, in Repts. Expl. Surv. . . . , 8(1):121, July 14. Type, *Canis virginianus* Erxleben [= *Canis cinereoargenteus* Schreber].

Median dorsal line of stiff black hair on tail; temporal ridges of skull widely separated but so approaching each other posteriorly as to form a lyrate or U-shaped figure; ventral border of ramus of lower jaw having a distinct "step" midway between tip of angular process and anterior border of coronoid process.

KEY TO SPECIES OF UROCYON

1. Total length less than 800, tail less than 300, condylobasal length less than 106; found only on certain islands off coast of southern California. . . . . .*U. littoralis,* p. 945

1'. Total length more than 800, tail more than 300, condylobasal length more than 106; not found on islands off coast of southern California. . . . . . *U. cinereoargenteus*, p. 942

Urocyon cinereoargenteus
Gray Fox

External measurements: 800–1125; 275–443; 100–150. Condylobasal length of skull, 110–130. Throat white, face gray, sides of neck, lower flanks, and underparts of tail rusty, tail with median dorsal streak of black hair. Ventral border of mandible with a distinct "step." Also occurs in South America.

Urocyon cinereoargenteus borealis Merriam

1903. *Urocyon cinereoargenteus borealis* Merriam, Proc. Biol. Soc. Washington, 16:74, May 29, type from Marlboro, 7 mi. from Monadnock, Cheshire Co., New Hampshire.

MARGINAL RECORDS.—Quebec: Compton County. Maine: Pennellville. New Hampshire: near Exeter; type locality. Massachusetts: Easthampton. Vermont: *Springfield; Rutland;* Whiting. Ontario: 6 mi. E Kaladar; near Glengarry County.

Urocyon cinereoargenteus californicus Mearns

1897. *Urocyon cinereoargenteus californicus* Mearns, Preliminary diagnoses of new mammals of the genera *Lynx*, *Urocyon*, *Spilogale* and *Mephitis*, from the Mexican boundary line, p. 3, January 12 (preprint of Proc. U.S. Nat. Mus., 20:459, December 24, 1897), type from Tahquitz Valley, San Jacinto Mts., Riverside Co., California.

MARGINAL RECORDS.—California: Rip Rap Mine, Piute Mts.; Cactus Flat; *Whitewater;* Santa Rosa Mtn. Baja California: Laguna Hanson; Rancho San Pablo, 10 mi. SE Alamo. California: Mt. Piños.

Urocyon cinereoargenteus cinereoargenteus (Schreber)

1775. *Canis cinereo argenteus* Schreber, Die Säugthiere . . . , theil 2, heft 13, Pl. 92. Type locality, eastern North America.
1894. *Urocyon cinereo-argenteus*, Rhoads, Amer. Nat., 28:524, June.
1775. *Urocyon virginianus virginianus* Schreber, Die Säugthiere . . . , theil 3, p. 585 (vernacular name and descrip-

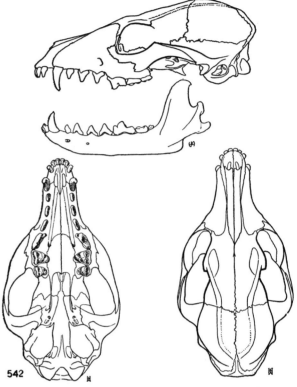

Fig. 542. *Urocyon cinereoargenteus scottii,* Colorado River, 2 mi. N California–Nevada Monument, Nevada, No. 61465 M.V.Z., ♂, X ½.

tion on p. 361), Pl. 92B. Type locality, "Carolina und die wärmer Gegenden von Nordamerica. . . ." Based primarily on the gray fox of Catesby.
1784. [*Vulpes*] *Pensylvanicus* Boddaert, Elenchus animalium . . . , p. 97, based on the "Brant fox" from Pennsylvania of Pennant (History of quadrupeds, 1781, p. 235).

MARGINAL RECORDS.—Michigan: Cheboygan County. Ontario: Oxford County. New York: Canajoharie; Nassau. Rhode Island: mainland portion (Cronan and Brooks, 1962:93). South Carolina: upper South Carolina. Tennessee: Walden Ridge near Soddy. Mississippi: 8½ mi. N Burnsville (Kennedy, *et al.,* 1974:24); Sunflower County (*op. cit.:* 25). Illinois: Union County; near Petersburg; Deerfield.

Urocyon cinereoargenteus costaricensis Goodwin

1938. *Urocyon cinereoargenteus costaricensis* Goodwin, Amer. Mus. Novit., 987:2, May 13, type from Sabanilla de Pirrís, Prov. San José, Costa Rica.

MARGINAL RECORDS.—Costa Rica: *Escazú; Pozo Azul;* type locality.

Urocyon cinereoargenteus floridanus Rhoads

1895. *Urocyon cinereo-argenteus floridanus* Rhoads, Proc. Acad. Nat. Sci. Philadelphia, 47:42, April 9, type from Tarpon Springs, Hillsboro Co., Florida.

MARGINAL RECORDS.—Mississippi: Mississippi State College. South Carolina: Savannah River Plant (Wood and Odum, 1964:545, as *U. cinereoargenteus* only). Georgia: Okefinokee Swamp, thence through peninsular Florida and westward along Gulf Coast to Mississippi: Harrison County (Kennedy, *et al.*, 1974:25). Louisiana (Lowery, 1974:402): 2 mi. N Pearl River; ½ mi. E Bayou Sorrel; 3 mi. SE Egan. Texas (Davis, 1966:107, as *U. cinereoargenteus* only, unless otherwise noted): Lavaca County; Comal County; Travis County; Cooke County; Harrison County (Buechner, 1944:185).

Urocyon cinereoargenteus fraterculus Elliot

1896. *Urocyon cinereo-argentatus* [*sic*] *fraterculus* Elliot, Field Columb. Mus., Publ. 11, Zool. Ser., 1(3):80, June 11, type from San Felipe, Yucatán.
1899. *Urocyon parvidens* Miller, Proc. Acad. Nat. Sci. Philadelphia, 51:276, July 26, type from Mérida, Yucatán.

MARGINAL RECORDS.—Yucatán: type locality. Quintana Roo: Cozumel Island (Jones and Lawlor, 1965:416, by native report only). Belize: Stann Creek Valley. Guatemala: 1–8 mi. S La Libertad. Campeche: 7½ km. W Escárcega (92561 KU). Yucatán: San Anselmo.

Urocyon cinereoargenteus furvus G. M. Allen and Barbour

1923. *Urocyon cinereoargenteus furvus* G. M. Allen and Barbour, Bull. Mus. Comp. Zool., 65:266, February, type from 3 mi. W Balboa, Canal Zone. Known only from type locality.

Urocyon cinereoargenteus guatemalae Miller

1899. *Urocyon guatemalae* Miller, Proc. Acad. Nat. Sci. Philadelphia, 51:278, July 26, type from Nentón, Huehuetenango, Guatemala.
1938. *Urocyon cinereoargenteus guatemalae*, Goldman, Jour. Washington Acad. Sci., 28(11):495, November 15.

MARGINAL RECORDS.—Chiapas: Tumbalá. Guatemala: type locality. Honduras: Belén Guacho; *Las Flores;* La Flor Archaga. Nicaragua: Matagalpa; San Rafael. Honduras: *Tegucigalpa;* La Cruz Grande. Guatemala: Volcán San Lucas. Chiapas: vic. Suchiate; Prusia, 1200 m. Felten (1958:221) applied the name *U. c. fraterculus* to specimens from El Salvador.

Urocyon cinereoargenteus madrensis Burt and Hooper

1941. *Urocyon cinereoargenteus madrensis* Burt and Hooper, Occas. Pap. Mus. Zool., Univ. Michigan, 430:4, May 27, type from Carimechi, Río Mayo, Chihuahua.

MARGINAL RECORDS.—Sonora: Guaymas. Chihuahua (Anderson, 1972:375): Cherry Ranch, 11 mi. WNW Cocomorachic; 10 km. N Guachochic, 8400 ft. Durango: Ciudad (Baker and Greer, 1962:134); 28 mi. S, 17 mi. W Vicente Guerrero, 8350 ft. (*ibid.*); San Luis (*ibid.*). Chihuahua: Cusáraga (Anderson, 1972:375).

Urocyon cinereoargenteus nigrirostris (Lichtenstein)

1850. *Canis nigrirostris* Lichtenstein, Abh. k. Akad. Wiss., Berlin, for 1827, p. 106, type from Real Arriba [= Temascaltepec], Estado México.
1959. *Urocyon cinereoargenteus nigrirostris*, Hall and Kelson, Mammals of North America, Ronald Press, p. 861, March 31.
1938. *Urocyon cinereoargenteus colimensis* Goldman, Jour. Washington Acad. Sci., 28:495, November 15, type from 3 mi. W Colima, Colima.

MARGINAL RECORDS.—Sinaloa: Sierra de Choix. Durango: Chacala; Pueblo Nuevo, 6000 ft. (Baker and Greer, 1962:134); La Pila (*ibid.*). San Luis Potosí: Presa de Guadalupe. Hidalgo: Tula (Goldman, 1938:496, as *U. c. colimensis*). Puebla: Piaxtla. Oaxaca: Juquila. Nayarit: Acaponeta. Sinaloa: 16 km. SE Topolobampo, 20 ft. (Armstrong, *et al.*, 1972:50).

Urocyon cinereoargenteus ocythous Bangs

1899. *Urocyon cinereoargenteus ocythous* Bangs, Proc. New England Zool. Club, 1:43, June 5, type from Platteville, Grant Co., Wisconsin.

MARGINAL RECORDS.—Manitoba: St. Adolphe (Sutton, 1958:440). Ontario: Wild Potatoe Lake, Rainy River Dist. Minnesota: Gunflint Lake (Timm, 1975:29). Michigan: *Gogebic County;* Ontonagan County; Delta County. Wisconsin (Jackson, 1961:308, Map 57): Door County; Kenosha County. Missouri: 5 mi. N Gainesville. Arkansas: Mississippi County; Chicot County; El Dorado. Oklahoma: 25 mi. S Smithville (Long and Long, 1964b:108); Wichita Mts. (*ibid.*). Kansas: 14 mi. SW Meade (Getz, 1961:283, as *U. cinereoargenteus* only); Lincoln (Andersen and Fleharty, 1964:193); Phillips County (Janes and Gier, 1966:25, as *U. cinereoargenteus* only). Nebraska (Jones, 1964c:260, 261): Hall County; NW corner Sioux Ordinance Depot. Wyoming: Owen [Owens] (Long, 1965a:682). Nebraska (Jones, 1964c:260, 261): Almeria; S of Center. South Dakota: *Sanborn County* (Jones and Henderson, 1963:283); Wessington Springs; Deer Ears Butte (Jones and Henderson, 1963:283). North Dakota: near Wilton; near Minot.

An unexplained extralimital occurrence is Old Fort Point, Lake Athabasca, Alberta.

Urocyon cinereoargenteus orinomus Goldman

1938. *Urocyon cinereoargenteus orinomus* Goldman, Jour. Washington Acad. Sci., 28:497, November 15, type from Orizaba, Veracruz.

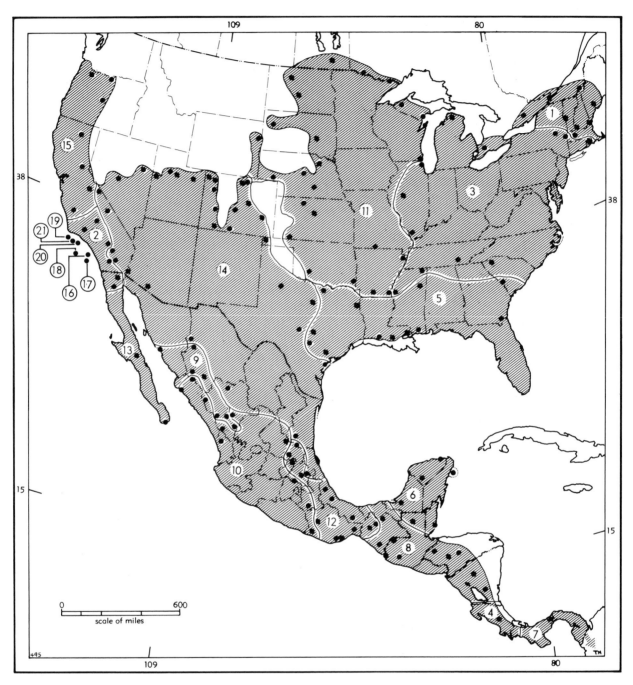

Map 495. *Urocyon cinereoargenteus* and *Urocyon littoralis*.

1. *U. c. borealis*
2. *U. c. californicus*
3. *U. c. cinereoargenteus*
4. *U. c. costaricensis*
5. *U. c. floridanus*
6. *U. c. fraterculus*
7. *U. c. furvus*

8. *U. c. guatemalae*
9. *U. c. madrensis*
10. *U. c. nigrirostris*
11. *U. c. ocythous*
12. *U. c. orinomus*
13. *U. c. peninsularis*
14. *U. c. scottii*

15. *U. c. townsendi*
16. *U. l. catalinae*
17. *U. l. clementae*
18. *U. l. dickeyi*
19. *U. l. littoralis*
20. *U. l. santacruzae*
21. *U. l. santarosae*

MARGINAL RECORDS.—San Luis Potosí: El Salto. Hidalgo: El Chico (Goldman, 1938:496, as *U. c. colimensis*). Veracruz: *3 km. E Las Vigas, 8000 ft.;* Jalapa; 15 km. W Piedras Negras, 300 ft.; 24 km. E and 7 km. S Jesús Carranza. Chiapas: Pueblo Nuevo Solisthuacán; 8 km. N Berriozábal, 1065 m. (Baker, *et al.,* 1973:78, 85). Oaxaca: 2 mi. E Juchitán; Salinera Garrapatera (Goodwin, 1969:224); Astata (*ibid.*); Totontepec. Querétero: Jalpan. San Luis Potosí: Río Verde.

Urocyon cinereoargenteus peninsularis Huey

1928. *Urocyon cinereoargenteus peninsularis* Huey, Trans. San Diego Soc. Nat. Hist., 5:203, September 1, type from San Ignacio, Cape Region, Baja California.

MARGINAL RECORDS.—Baja California: type locality; Cape San Lucas.

Urocyon cinereoargenteus scottii Mearns

1891. *Urocyon virginianus scottii* Mearns, Bull. Amer. Mus. Nat. Hist., 3:236, June 5, type from Pinal Co., Arizona.
1895. *Urocyon cinereo-argenteus scottii*, J. A. Allen, Bull. Amer. Mus. Nat. Hist., 7:253, June 29.
1897. *Urocyon cinereoargenteus texensis* Mearns, Preliminary diagnoses of new mammals of the genera *Lynx, Urocyon, Spilogale* and *Mephitis*, from the Mexican boundary line, p. 2, January 12 (preprint of Proc. U.S. Nat. Mus., 20:459, December 24, 1897), type from San Pedro, near Eagle Pass, Maverick Co., Texas.
1904. *Urocyon cinereo-argenteus inyoensis* Elliot, Field Columb. Mus., Publ. 90, Zool. Ser., 3:268, March 7, type from Beveridge Canyon, Inyo Mts., Inyo Co., California.

MARGINAL RECORDS.—Colorado (Armstrong, 1972:265, unless otherwise noted): Castle Park, Dinosaur National Monument; Lily; Little Hills Experiment Station, 35 mi. W Meeker; within 3 mi. of Montrose; Animas River; E of Antonito; Salida; Estes Park; Loveland (Hall and Kelson, 1959:863); 1 mi. W Larkspur; 8 mi. SE Haswell. Oklahoma: 6 mi. N Kenton (Urban and Wimmer, 1959:450). Texas: 3$\frac{3}{10}$ mi. E Crosbyton (Packard and Bowers, 1970:451); *15 mi. NW Spur* (*ibid.*); Brazos; Mason County; Aransas National Wildlife Refuge (Halloran, 1961:23); *Rockport.* Veracruz: 3 km. W Zacualpan, 6000 ft. (Hall and Dalquest, 1963:330). Tamaulipas: La Joya de Salas. Durango: Río Ocampo. Chihuahua: Sierra Madre Mts., 150 mi. S international boundary, 10 mi. E Sonoran boundary (Anderson, 1972:376). Arizona: Organ Pipe Cactus National Monument. California: Bard; Santa Rosa Mts.; Palm Springs; Mazourka Canyon, Inyo Mts.; *12 mi. W Big Pine;* Cottonwood Creek, White Mts. Nevada: *Silver Peak Range;* Millet P.O.; *foothills White Pine Mts.;* Cherry Creek Canyon. Utah: mouth Birch Creek; *between Callao and Trout Creek, E foothills Deep Creek Mts.* (Egoscue, 1966:143); Johnson Pass (*op. cit.*:144); Provo (Long and Long, 1964a:108); 12 mi. W Roosevelt.

Urocyon cinereoargenteus townsendi Merriam

1899. *Urocyon californicus townsendi* Merriam, N. Amer. Fauna, 16:103, October 28, type from Baird, Shasta Co., California.
1933. *Urocyon cinereoargenteus townsendi*, Grinnell, Univ. California Publ. Zool., 40(2):110, September 26.
1910. *Urocyon californicus sequoiensis* Dixon, Univ. California Publ. Zool., 5(7):303, February 12, type from Lagunitas, Marin Co., California.

MARGINAL RECORDS.—Oregon: Portland; The Dalles; Skeleton Cave, near Bend. California: type locality; Goldrun; Yosemite Valley; Nacimiento, thence up coast to mouth Columbia River.

Urocyon littoralis
Island Gray Fox

External measurements: 590–787; 110–290; 98–157. Condylobasal length of skull, 92.1–105.6.

Urocyon littoralis catalinae Merriam

1903. *Urocyon catalinae* Merriam, Proc. Biol. Soc. Washington, 16:74, May 29, type from Santa Catalina Island, California. Known only from type locality.
1937. *Urocyon littoralis catalinae*, Grinnell, Dixon, and Linsdale, The fur-bearing mammals of California, 2:467, August 10.

Urocyon littoralis clementae Merriam

1903. *Urocyon clementae* Merriam, Proc. Biol. Soc. Washington, 16:75, May 29, type from San Clemente Island, California. Known only from type locality.
1930. [*Urocyon littoralis*] *clementae*, Grinnell and Linsdale, Proc. Biol. Soc. Washington, 43:154, September 26.

Urocyon littoralis dickeyi Grinnell and Linsdale

1930. *Urocyon littoralis dickeyi* Grinnell and Linsdale, Proc. Biol. Soc. Washington, 43:154, September 26, type from San Nicolas Island, California. Known only from type locality.

Urocyon littoralis littoralis (Baird)

1858. *Vulpes (Urocyon) littoralis* Baird, Mammals, *in* Repts. Expl. Surv. . . . , 8(1):143, July 14, type from San Miguel Island, California. Known only from type locality.
1888. *Urocyon littoralis*, Merriam, Proc. Biol. Soc. Washington, 4:135, February 18.

Urocyon littoralis santacruzae Merriam

1903. *Urocyon littoralis santacruzae* Merriam, Proc. Biol. Soc. Washington, 16:75, May 29, type from Santa Cruz Island, California. Known only from type locality.

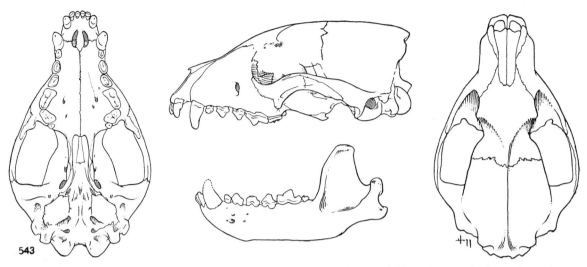

Fig. 543. *Speothos panamensis*, Mt. Pirre, near head Río Limón, Panamá, No. 179046 U.S.N.M., ♀, drawn from photograph, X approx. ½.

Urocyon littoralis santarosae Grinnell and Linsdale

1930. *Urocyon littoralis santarosae* Grinnell and Linsdale, Proc. Biol. Soc. Washington, 43:154, September 26, type from Santa Rosa Island, California. Known only from type locality.

Genus **Speothos** Lund—Bush Dogs

1839. *Speothos* Lund, Ann. Sci. Nat. Paris, ser. 2 (Zool.), 11:223, 224, 232, April. Type, *Speothos pacivorus* Lund (from bone caves in Minas Gerais, Brazil).
1843. *Icticyon* Lund, Overs. Kongl. Danske Vidensk. Selsk. Forh., p. 80. Type, *Icticyon venaticus* Lund.
1846. *Cynalicus* Gray, Ann. Mag. Nat. Hist., 17:293. Type, *Cynalicus melanogaster* Gray [= *Icticyon venaticus* Lund].
1848. *Melictis* Schinz, Revue zoologique, 176–178, June. Type, *Melictis beskii* Schinz [= *Icticyon venaticus* Lund].

Legs and tail short; hair long; molars $\frac{2}{2}$; in m1 talonid trenchant and inner cusp (metaconid) absent.

Speothos venaticus
Bush Dog

Characters those of the genus.

Speothos venaticus panamensis (Goldman)

1912. *Icticyon panamensis* Goldman, Smiths. Miscl. Coll., 60(2):14, September 20, type from head of Río Limón, 5000 ft., Mt. Pirre, eastern Panamá. Known only from type locality, but almost certainly occurs in most of Panamá and adjacent Colombia.
1957. *Speothos venaticus panamensis*, Hershkovitz, Novedades Colombianas, Mus. Hist. Nat. Univ. del Cauca, 3:161, December 1.

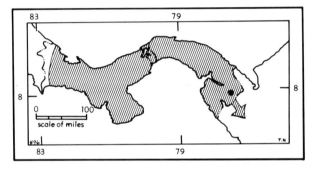

Map 496. *Speothos venaticus panamensis*.

FAMILY **URSIDAE**—Bears

Large, heavily built, plantigrade, tail short; pentadactyl, claws not retractile; tympanic bullae not inflated; alisphenoid canal present (except in *Ailuropoda*); no entepicondylar foramen in humerus; kidneys conglomerate; molars with broad flat tubercular crowns; 4th upper premolar lacking 3rd or inner root; first 3 premolars of each jaw usually rudimentary or lost. Dentition, i. $\frac{3}{3}$, c. $\frac{1}{1}$, p. $\frac{4}{4}$, m. $\frac{2}{3}$, except that 1st upper incisor missing in the sloth bear, *Ursus (Melursus) ursinus*, of India and Ceylon.

The family can conveniently be divided into four subfamilies: Ursinae (Old and New World, Pleistocene–Recent); Agriotheriinae (Eurasia, Miocene–Pliocene) having less complex cuspidation on the molars than do Ursinae; Tremarctinae (North and South America, Pliocene–Recent, the only living species is in South America) having the masseteric fossa on the lower jaw divided by a bony septum into 2 fossae, and; Ailuropodinae (western China, Pliocene–Recent) lacking an alisphenoid canal and having wider molars and more widely flaring zygomatic arches than do other bears. The recent publications by Erdbrink (1953), Davis (1964), and Kurtén (1966, 1967) and earlier works cited therein are indispensable to whoever undertakes to trace the evolution of the bears.

Genus Ursus Linnaeus
Bears

1758. *Ursus* Linnaeus, Syst. nat., ed. 10, 1:47. Type, *Ursus arctos* Linnaeus (brown bear).

1793. *Melursus* Meyer, Zool. Entdeckung, p. 155. Type, *Bradypus ursinus* Shaw, Nat. Misc. ii (unpaged), Pls. 58, 59 (sloth bear).

1809. *Arceus* Goldfuss, Vergleich. Naturbeschreib. Säugeth., 301, 302. Type, *Arceus niger* (sloth bear).

1811. *Prochilus* Illiger, Prodromus systematis mammalium et avium, 109. Type, *Bradypus ursinus* Shaw (sloth bear).

1814. *Chondrorhynchus* G. Fischer, Zoognosia, III, 142, 143. Type, none given but evidently the sloth bear from India.

1825. *Helarctos* Horsfield, Jour. Zool., ii, p. 221. Type, *Helarctos euryspilus* Horsfield [= *Ursus malayanus* Horsfield] (sun bear).

1825. *Danis* Gray, Ann. Philos., 26:60, July. Type, *Ursus ferox* Desmarest [= *Ursus horribilis* Ord, 1815] (brown bear).

1825. *Thalarctos* Gray, Ann. Philos., n.s., 10:62, July. Type, *Thalarctos polaris* Gray [= *Ursus maritimus* Phipps] (polar bear).

1825. *Thalassarctos* Gray, Ann. Philos., n.s., 10:339, November, an emendation of *Thalarctos* Gray (polar bear).

1864. *Myrmarctos* Gray, Proc. Zool. Soc. London, 694. Type, *Myrmarctos eversmanni* Gray [= *Ursus arctos* Linnaeus] (brown bear).

1865. *Euarctos* Gray, Proc. Zool. Soc. London, for 1864, p. 692, May. Type, *Ursus americanus* Pallas, 1780 (American black bear).

1896. *Thalassiarchus* Kobelt, Bericht Senckenberg. naturf. Gesell., Frankfurt am Main, p. 93, an emendation of *Thalarctos* Gray (polar bear).

1898. *Ursarctos* Heude, Mém. Nat. Hist. Empire Chinois, 1:17. Type, *Ursus arctus* [sic] *yesoensis* Lydekker, 1897 [= *Ursus arctos lasiotus* Gray, 1867] (brown bear).

1898. *Melanarctos* Heude, Mém. Nat. Hist. Empire Chinois, 4, 1:18. Type, *Melanarctos cavifrons* Heude [= *Ursus lasiotus* Gray] (brown bear).

1901. *Selenarctos* Heude, Mém. Nat. Hist. Empire Chinois, 5:2. Type, *Ursus thibetanus* Cuvier (Asiatic black bear).

1917. *Arcticonus* Pocock, Ann. Mag. Nat. Hist., 20:129. Type, *Ursus thibetanus* Cuvier (Asiatic black bear).

1918. *Vetularctos* Merriam, N. Amer. Fauna, 41:131, February 9. Type, *Vetularctos inopinatus* Merriam, 1918 (brown bear).

1923. *Mylarctos* Lönnberg, Proc. Zool. Soc. London, p. 91, April 12. Type, *Ursus pruinosus* Blyth, 1854 (brown bear).

Characters as listed above for the family apply also to the subfamily Ursinae, and to the genus *Ursus*, the only genus of that subfamily.

KEY TO NORTH AMERICAN SPECIES OF URSUS

1. Always entirely white; combined length of M1 and M2 less than palatal width; confined to Arctic region; primarily maritime. . . . . *U. maritimus,* p. 958
1'. Never entirely white (*Ursus americanus kermodei* nearly so but having brownish muzzle, *Ursus arctos* in local northern populations superficially so but having brownish feet and lower legs and having pelage of body basally brownish); combined length of M1 and M2 never less than palatal width; confined to Arctic region and never maritime.
   2. Claws on forefeet only a little longer than on hind feet; hump at shoulders lacking or only slightly developed; M2 broadest midway of its length; length of M2 less than 29.5 mm. in part of American where *U. arctos* occurs.
   U. americanus, p. 947
   2'. Claws on forefeet approx. twice as long as on hind feet; hump on shoulders prominent; M2 broadest anteriorly; length of M2 more than 29.5 mm. (all taxa listed beyond under subgenus *Ursus*). . . . .*U. arctos,* p. 951

Ursus americanus
Black Bear

Canadian forms reviewed by Anderson, Ann. Rept. Provancher Soc. Nat. Hist. Canada, Quebec, for 1944, pp. 17–33, November 2, 1945.

Pelage black or cinnamon in much of western North America, bluish on St. Elias Range of southern Alaska, and white or yellowish white along Pacific Coast of central British Columbia; claws on forefeet only a little longer than on hind feet. Compared with *Ursus arctos*, maximum size less, hump at shoulders lacking or only slightly developed instead of prominent. M2 broadest approx. halfway between anterior and posterior margins, and in part of America where *U. americanus* and *U. arctos* occur together, length of 2nd upper molar less than 29.5 mm. instead of more. First lower molar (m1) having broad, open, cuspless valley medially between metaconid and entoconid; posterior part of p4 ordinarily lacking medial accessory cusps or median anteroposterior sulcus.

Ursus americanus altifrontalis Elliot

1903. *Ursus altifrontalis* Elliot, Field Columb. Mus., Publ. 80, Zool. Ser., 3:234, August 27, type from Lake Crescent, Clallam Co., Washington.
1913. *Ursus americanus altifrontalis*, Grinnell, Proc. California Acad. Sci., ser. 4, 3:248, August 28.

MARGINAL RECORDS.—British Columbia: Atnarko (Cowan and Guiguet, 1965:291); *Redstone* (*ibid.*); Lillooet (*ibid.*); Glacier National Park; Yahk. Washington: Chelan; Signal Peak. Oregon: Dufur. California: divide between Grizzly Creek and Salmon River; vic. Yolla Bolly Mtn.; summit Sanhedrin Mtn.; vic. Bodega, thence northward along coast to British Columbia: Namu (Cowan and Guiguet, 1965:291); Bella Coola (*ibid.*); *Hagensborg*.

Ursus americanus amblyceps Baird

1859. *Ursus amblyceps* Baird, Rept. U.S. and Mexican boundary Surv., 2(2):29, January, type from Fort Webster (copper mines), on the Gila River, long. 108° 04′ W, lat. 32° 47′ N, Grant Co., New Mexico.

1905. *Ursus americanus amblyceps*, V. Bailey, N. Amer. Fauna, 25:187, October 24.

MARGINAL RECORDS.—Colorado: 4 mi. NW Columbine (Armstrong, 1972:270). Kansas: Kiowa County; Comanche County; small canyon just S of Meade. New Mexico: Capitan; Guadalupe Mts. Texas: 15 mi. N Fort Davis; Chisos Mts. New Mexico: Sacramento Mts.; type locality. Arizona: near Wakefield Mine, Huachuca Mts.; Sierra Ancha Mts. (Johnson and Johnson, 1964:323); within 15 mi. of Flagstaff (Hoffmeister and Carothers, 1969:188, as *U. americanus* only); South Rim of Grand Canyon. Utah: 10 mi. W Monticello, Indian Creek, Abajo Mts.; Castle Creek.

Ursus americanus americanus Pallas

1780. *Ursus americanus* Pallas, . . . Spicilegia zoologica, . . . , fasc. 14:5. Type locality, eastern North America.
1898. *Ursus (Euarctos) americanus sornborgeri* Bangs, Amer. Nat., 32:500, July, type from Okak, Labrador. (See Bangs, *in* Grenfell's Labrador, the country and the people, p. 467, 1909, and J. A. Allen, Bull. Amer. Mus. Nat. Hist., 28:1–5, January 5, 1910.)
1913. *Ursus arctos schwenki*, Shoemaker, Stories of great Pennsylvania hunters, p. 25, type from Union Co., Pennsylvania. Trimmings from the skin on which this name was based are in the U.S. National Museum.
1945. *Euarctos randi* Anderson, Ann. Rept. Provancher Soc. Nat. Hist. Canada, Quebec, for 1944, p. 19, November 2, type from Sheldon Mtn., Canol Road, Mile 222, Yukon. Inseparable from *U. a. americanus* according to Youngman, Nat. Mus. Canada Publ. Zool., 10:133, September 4, 1975.
1945. *Euarctos hunteri* Anderson, Ann. Rept. Provancher Soc. Nat. Hist., Canada, Quebec, for 1944, p. 22, November 2, type from near mouth Prairie Creek, South Nahanni River,

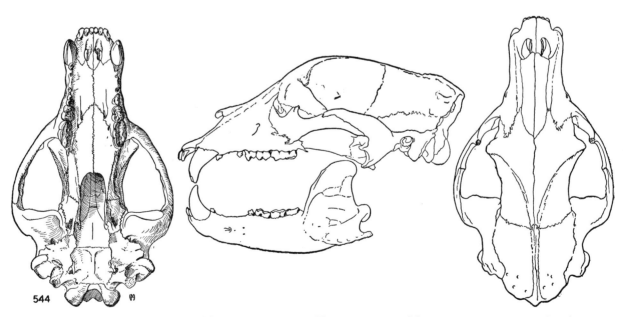

Fig. 544. *Ursus americanus californiensis*, 6 mi. N Weldon, Kern Co., California, No. 16269 M.V.Z., ♂, X ¼.

Mackenzie. Inseparable from *U. a. americanus* according to Youngman, Nat. Mus. Canada Publ. Zool., 10:133, September 4, 1975.

MARGINAL RECORDS.—Alaska: Carnivore Lakes. Yukon:[1] Old Crow River at 140° 00' (Youngman, 1975:135). Keewatin: Kazan River, 20 mi. E north end Ennadai Lake (Harper, 1956:61). Labrador: Okak. Quebec: Anticosti Island, thence down Atlantic Coast to Georgia and W to Texas: 7 mi. from Kerrville; Prairie Dog Fork near Washburn. Nebraska (Jones, 1964c:262): Sun Island, Missouri River, between Peru and Brownville; 40 mi. N Loup River on Cedar River; *between North Loup River and Bush Lake;* Bear Creek. North Dakota: Turtle Mts. Alaska: Cape Thompson; Kupowruk River; near mouth Inukpasukruk Creek.

Ursus americanus californiensis J. Miller

1900. *Ursus californiensis* J. Miller, True bear stories, Rand McNally and Co., Chicago and New York, p. 250. Type locality, "California." Restricted to Sierra Nevada (Grinnell, Univ. California Publ. Zool., 32:401, July 19, 1929).
1929. *Ursus americanus californiensis*, Grinnell, Univ. California Publ. Zool., 32:396, July 19.

MARGINAL RECORDS.—California: Callahan; McCloud; *McCloud River near Baird Station.* Nevada: W side Washoe Valley, 1 mi. S Bowers Mansion. California: S. Fork, Kern River; 6 mi. N Weldon; Tehachapi Mts.; *ca.* 7 mi. SE Tejon Ranch; near Wawona; Crocker's [= Sequoia]; Soda Springs; S. Fork Salmon River, 8000 ft. Following extirpation of *Ursus arctos* in the southern part of the Sierra Nevada, the Tehachapi Mts., and the Southern Coast Range of California, *Ursus americanus* (probably subspecies *U. a. californiensis*) has invaded and now occurs in these three mountainous areas.

Ursus americanus carlottae Osgood

1901. *Ursus (Euarctos) carlottae* Osgood, N. Amer. Fauna, 21:30, September 26, type from Massett, Graham Island, Queen Charlotte Islands, British Columbia. Known only from Queen Charlotte Islands.
1928. *Ursus americanus carlottae*, Hall, Univ. California Publ. Zool., 30:235, March 2.

[1] Youngman (1975:133–135), followed here, refers all black bears (*U. americanus*) from the Yukon Territory to the subspecies *U. a. americanus*, although animals from south-central Yukon seem to average smaller than black bears anywhere else. The same can be said of grizzly bears (*U. arctos*) except possibly the extirpated Mexican population. Study of specimens of *U. americanus* from all parts of its geographic range therefore may warrant recognition of the subspecies *U. a. randi* (see Hall and Kelson, 1959:866 and 869 for its provisional geographic range) in Yukon even if *U. a. hunteri* from southeastern Yukon and southwestern Mackenzie is not recognizable.

Ursus americanus cinnamomum Audubon and Bachman

1854. *Ursus americanus* var. *cinnamomum* Audubon and Bachman, The viviparous quadrupeds of North America, 3:125, type from Lower Clearwater River, Camp Chopunnish, near mouth Jim Ford Creek, Clearwater Co., western Idaho (see V. Bailey, N. Amer. Fauna, 55:319, August 29, 1936).

MARGINAL RECORDS.—British Columbia (Cowan and Guiguet, 1965:293): 10 mi. W Fort Nelson; Buckinghorse River, 94 mi. S Fort Nelson. Alberta: Jasper National Park; Waterton Lakes National Park. Montana: 7 mi. S, 14 mi. E Livingston. Wyoming (Long, 1965a:684, 685): northern Bighorn Mts.; 16 mi. N Tensleep; Riverton; *6 mi. S, ¹/₂ mi. E Lander;* Laramie Peak; Laramie; S *Jelm.* Utah: Elk Park; 6 mi. above mouth Nine Mile Canyon; North Ogden Canyon. Oregon: Silver Lake; La Pine. Washington: Republic. British Columbia (Cowan and Guiguet, 1965:293, 294, unless otherwise noted): Morrissey (Hall and Kelson, 1959:868); Glacier National Park (*ibid.*); *Mt. Revelstoke;* Kamloops; Chezacut; Eutsuk Lake; Quick; Bear Lake; 122 mi. S Telegraph Creek (Hall and Kelson, 1959:867, as *U. a. americanus*); 31 mi. S Telegraph Creek (*ibid.*, as *U. a. americanus*).

Ursus americanus emmonsii Dall

1895. [*Ursus americanus*] var. *emmonsii* Dall, Science, n.s., 2:87, July 26, based on specimens from Saint Elias Alps, near Yakutat Bay, Alaska.
1897. *Ursus Glacilis* [*sic*] Kells, Canadian Nat. Sci. News, 1:12, April. (Name was applied to the kind of bear living on Mt. Saint Elias, Alaska.)

MARGINAL RECORDS.—Alaska: Camp Rubrick, Chitina River; type locality; base Mt. Cook; Lituya Bay, near Mt. Fairweather, thence northward along coast to point of beginning.

Ursus americanus eremicus Merriam

1904. *Ursus americanus eremicus* Merriam, Proc. Biol. Soc. Washington, 17:154, October 6, type from Sierra Guadalupe, Coahuila.

MARGINAL RECORDS.—Coahuila (Baker, 1956:300): Santa Rosa Mts., 160 mi. NW Muzquiz; 2 mi. N, 6 mi. W Piedra Blanca, 4950 ft.; 2 mi. N Ocampo; type locality. Tamaulipas: Agua Linda. San Luis Potosí: near Matehuala.

Ursus americanus floridanus Merriam

1896. *Ursus floridanus* Merriam, Proc. Biol. Soc. Washington, 10:81, April 13, type from Key Biscayne, Dade Co., Florida.
1959. *Ursus americanus floridanus*, Hall and Kelson, Mammals of North America, Ronald Press, p. 868, March 31.

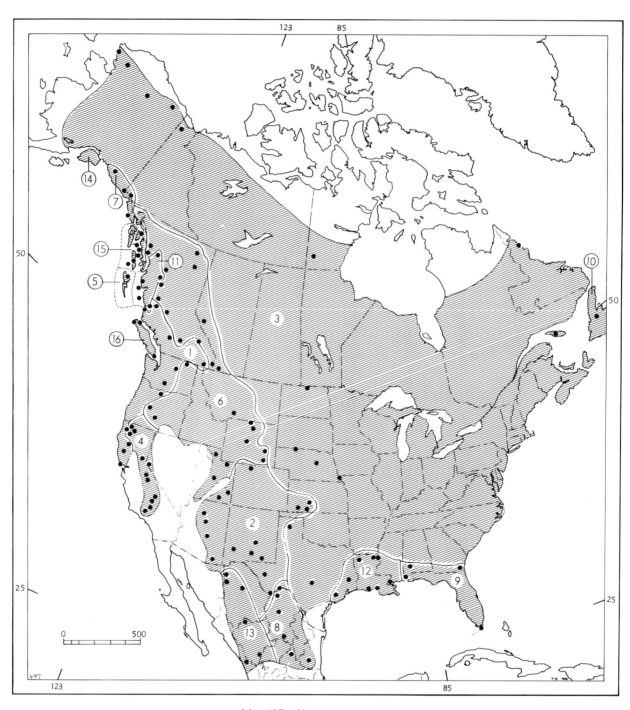

Map 497. *Ursus americanus.*

1. *U. a. altifrontalis*	5. *U. a. carlottae*	9. *U. a. floridanus*	13. *U. a. machetes*
2. *U. a. amblyceps*	6. *U. a. cinnamomum*	10. *U. a. hamiltoni*	14. *U. a. perniger*
3. *U. a. americanus*	7. *U. a. emmonsii*	11. *U. a. kermodei*	15. *U. a. pugnax*
4. *U. a. californiensis*	8. *U. a. eremicus*	12. *U. a. luteolus*	16. *U. a. vancouveri*

MARGINAL RECORDS.—Alabama: Carlton. Georgia: Okefinokee Swamp, thence throughout Florida: including Keys as far SW as Matecumbes (Layne, 1974:395). Alabama: near Bayou Labatre.

Ursus americanus hamiltoni Cameron

1957. *Ursus americanus hamiltoni* Cameron, Jour. Mamm., 37:538, January 9, type from Big Falls, Humber River, Newfoundland. Known only from Island of Newfoundland.

Ursus americanus kermodei Hornaday

1905. *Ursus kermodei* Hornaday, Ninth Ann. Rep. N.Y. Zool. Soc., for 1904, p. 82, January, type from Gribble Island, British Columbia.
1928. *Ursus americanus kermodei,* Hall, Univ. California Publ. Zool., 30:235, March 2.

MARGINAL RECORDS.—British Columbia: headwaters Nass River (Cowan and Guiguet, 1965:294); Hazelton, Skeena River; South Bentinck Arm; *Burke Channel, Kwatna Inlet;* Aristazable Island; Pitt Island; Port Essington (Cowan and Guiguet, 1965: 294).

Ursus americanus luteolus Griffith

1821. *Ursus luteolus* Griffith, Class Quadrimembria, order Carnivora, general and particular description of carnivorous animals . . . , p. 236. Type locality, Louisiana.
1959. *Ursus americanus luteolus,* Hall and Kelson, The mammals of North America, Ronald Press, p. 868, March 31.

MARGINAL RECORDS.—Louisiana (Lowery, 1974:411): Mer Rouge; Monticello; Mandeville; Morgan City; 25 mi. SW Abbeville. Texas: Wharton; Tarkington; *Kountze.* Louisiana: Bayou Bodcau near Sarepta.

Ursus americanus machetes Elliot

1903. *Ursus machetes* Elliot, Field Columb. Mus., Publ. 80, Zool. Ser., 3:235, September 3, type from Casas Grandes, Sierra Madre, Chihuahua.
1959. *Ursus americanus machetes,* Hall and Kelson, The mammals of North America, Ronald Press, p. 868, March 31.

MARGINAL RECORDS.—Chihuahua (Anderson, 1972:377): San Luis Mts., 55 mi. S U.S. boundary; 18 mi. SW Gallego, 6000 ft.; *Arroyo del Nido, 25 mi. SW Gallego, 6000 ft.* Durango: Río Ocampo; 28 mi. S, 17 mi. W Vicente Guerrero, 8350 ft. (Baker and Greer, 1962:135). Sinaloa: vic. Escuinapa. Chihuahua: within 15 mi. of Colonia García (Anderson, 1972:377).

Ursus americanus perniger J. A. Allen

1910. *Ursus americanus kenaiensis* J. A. Allen, Bull. Amer. Mus. Nat. Hist., 28:6, January 5, type from Homer, Kenai Peninsula, Alaska. Not *Ursus kenaiensis* Merriam, 1904.
1910. *Ursus americanus perniger* J. A. Allen, Bull. Amer. Mus. Nat. Hist., 28:115, April 30, a renaming of *Ursus americanus kenaiensis* J. A. Allen. Known only from Kenai Peninsula.

Ursus americanus pugnax Swarth

1911. *Ursus americanus pugnax* Swarth, Univ. California Publ. Zool., 7:141, January 12, type from Rocky Bay, now Bobs Bay, Dall Island, Alaska.

MARGINAL RECORDS.—Alaska: Taku River; Mitkof Island; Fools Inlet, Wrangell Island; San Alberto Bay, Prince of Wales Island; type locality; Kupreanof Island.

Ursus americanus vancouveri Hall

1928. *Ursus americanus vancouveri* Hall, Univ. California Publ. Zool., 30:231, March 2, type from King Solomon's (*sic*) Basin, Vancouver Island, British Columbia.

MARGINAL RECORDS.—British Columbia (Cowan and Guiguet, 1965:295): Port Hardy, thence southeastward along coast to Victoria, thence northwestward along coast to Sea Otter Cove. Confined to Vancouver Island.

Ursus arctos–group
Big Brown and Grizzly Bears

x ⅕₅

North American forms revised by Merriam, N. Amer. Fauna, 41:1–136, 16 pls., February 9, 1918. Eurasian forms revised by Couturier, "L'Ours Brun," printed in Grenoble, France, xiv + 905 pp., frontispiece, 83 pls., 43 figs., 1954.

1758. *Ursus* Linnaeus, Syst. nat., ed. 10, 1:47. Type, *Ursus arctos* Linnaeus.

Front claws longer than hind claws; m1 with one or more cusplets medially in valley between metaconid and entoconid; p4 with median accessory cusps and a median anteroposterior sulcus on posterior part; M2 broadest at anterior end.

In the species-subspecies category 94 or so names have been applied to North American specimens of the Big Brown and Grizzly Bear(s). These names are listed below. Most of them are in the name-combinations as left by C. Hart Merriam (1941) who was unable later to complete for publication his intended final systematic classification.

In the following accounts all the named kinds are listed together with their type localities. Map 498 shows North American type localities and also some marginal occurrences. For additional eastern occurrences not shown on that map— **See** addenda.

MARGINAL RECORDS.—Alaska: Alaktak River [154°58', 70°47'] (Rausch, 1951:165), eastward along (or to within a few miles of) Arctic Coast to Keewatin:

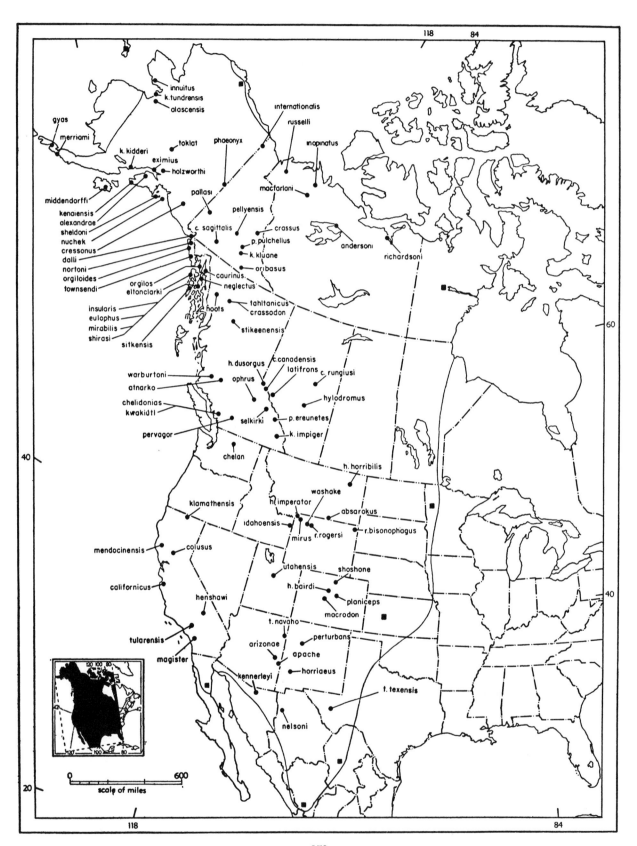

innuitus, k.tundrensis, alascensis, internationalis, russelli, gyas, merriami, toklat, phaeonyx, inopinatus, k.kidderi, eximius, holzworthi, pallasi, macfarlani, middendorffi, pellyensis, kenaiensis, alexandrae, sheldoni, nuchek, cressonus, dalli, nortoni, orgiloides, townsendi, orgilos, eltonclarki, c. sagittalis, crassus, p.pulchellus, k.kluane, andersoni, richardsoni, oribasus, caurinus, neglectus, insularis, eulophus, mirabilis, shirasi, sitkensis, hoots, tahltanicus, crassodon, stikeenensis, warburtoni, atnarko, h.dusorgus, ophrus, c.canadensis, latifrons, c.rungiusi, chelidonias, kwrakiutl, hylodromus, pervagor, selkirki, p.ereunetes, k.impiger, chelan, h.horribilis, klamathensis, h.imperator, washake, absarokus, r.bisonophagus, idahoensis, mirus, r.rogersi, mendocinensis, colusus, utahensis, shoshone, californicus, h.bairdi, planiceps, henshawi, macrodon, t.navaho, tularensis, arizonae, perturbans, magister, apache, horriaeus, kennerleyi, t.texensis, nelsoni

scale of miles 0 ... 600

952

Baker Lake (Anderson, 1947:43). Minnesota: Sandhill River (Swanson, 1945:62). Kansas: Trego County (Cockrum, 1952:239). Texas: Davis Mts. (Merriam, 1918:35). Durango: "southern Durango" (G. M. Allen, 1942a:153). Baja California: Sierra Juárez near Santa Catarina Mission (G. M. Allen, 1942a:145) thence northward along Pacific Coast of North America to point of beginning.

Ursus absarokus Merriam

1914. *Ursus absarokus* Merriam, Proc. Biol. Soc. Washington, 27:181, August 13, type from head Little Bighorn River, northern part Bighorn Mts., Carbon Co., Montana.

Ursus alascensis Merriam

1896. *Ursus horribilis alascensis* Merriam, Proc. Biol. Soc. Washington, 10:74, April 13, type from Unalaklik River, Alaska.
1918. *Ursus alascensis* Merriam, N. Amer. Fauna, 41:94, February 9.

Ursus alexandrae Merriam

1914. *Ursus alexandrae* Merriam, Proc. Biol. Soc. Washington, 27:174, August 13, type from Kasilof Lake, Kenai Peninsula, Alaska.

Ursus andersoni Merriam

1918. *Ursus andersoni* Merriam, N. Amer. Fauna, 41:83, February 9, type from E branch Dease River, near Great Bear Lake, Mackenzie.

Ursus apache Merriam

1916. *Ursus apache* Merriam, Proc. Biol. Soc. Washington, 29:134, September 6, type from Whorton Creek, S slope White Mts., a few miles W Blue, Greenlee Co., Arizona.

Ursus arctos beringianus Middendorff

1853. *Ur[sus]. arctos* var. *beringiana* Middendorff, Reise in den äussersten Norden und Osten Sibiriens . . . , 2(2):Pl. 1, Figs. 1–6. Type locality, Great Shantar Island, Sea of Okhotsk, Siberia.
1855. *Ursus piscator* Pucheran, Revue et Mag. Zool., Paris, ser. 2, 7:392, August. Type locality, Petropaulovski, southern Kamtchatka.
1924. *U[rsus]. a[rctos]. kolymnensis* Ognev, Priroda i Okhota na Ukraine (Nature and sport in Ukraine), Kharkov, 1(2):112, type from Saborzevo, Kolyma River, northwest of Sea of Okhotsk, Anadirsk Dist., Siberia.

MARGINAL RECORDS.—Alaska: St. Lawrence Island (A. H. Howell, 1940:216; Rausch, 1953:19).

Ursus arizonae Merriam

1916. *Ursus arizonae* Merriam, Proc. Biol. Soc. Washington, 29:135, September 6, type from east side Escudilla Mts., Apache Co., Arizona.

Ursus atnarko Merriam

1918. *Ursus atnarko* Merriam, N. Amer. Fauna, 41:22, February 9, type from Lonesome Lake, Atnarko River, one of upper forks of Bella Coola, British Columbia.

Ursus californicus Merriam

1896. [*Ursus horribilis*] subspecies *californicus* Merriam, Proc. Biol. Soc. Washington, 10:76, April 13, type from Monterey, Monterey Co., California.
1914. *Ursus californicus* Merriam, Proc. Biol. Soc. Washington, 27:188, August 13.

Ursus canadensis canadensis Merriam

1914. *Ursus shoshone canadensis* Merriam, Proc. Biol. Soc. Washington, 27:184, August 13, type from Moose Pass, near Mt. Robson, British Columbia.
1947. *Ursus canadensis canadensis*, Anderson, Bull. Nat. Mus. Canada, 102:41, January 24.
1916. *Ursus ophrus* Merriam, Proc. Biol. Soc. Washington, 29:148, September 6, type from eastern British Columbia; exact locality unknown (arranged as a synonym of *canadensis* by Anderson, *supra*).

Ursus canadensis rungiusi Merriam

1918. *Ursus rungiusi rungiusi* Merriam, N. Amer. Fauna, 41:49, February 9, type from Rocky Mts. on headwaters on Athabaska River, Alberta.
1947. *Ursus canadensis rungiusi*, Anderson, Bull. Nat. Mus. Canada, 102:41, January 24.

Ursus canadensis sagittalis Merriam

1918. *Ursus rungiusi sagittalis* Merriam, N. Amer. Fauna, 41:50, February 9, type from Champagne Landing, southwestern Yukon.
1947. *Ursus canadensis sagittalis*, Anderson, Bull. Nat. Mus. Canada, 102:41, January 24.

Map 498.   Solid circles denote type localities of nominal species and subspecies in America of the Big Brown and Grizzly Bears.

Solid squares denote marginal record stations of occurrence that are not type localities. These record stations, beginning in Alaska and proceeding in clockwise order, are as follows: Alaska: Alaktak River near Admiralty Bay (*Ursus horribilis richardsoni*—Bee and Hall, 1956:196). Keewatin: Baker Lake (*Ursus andersoni*—Anderson, 1947:43). Minnesota: Sandhill River in southern Polk Co. (*Ursus horribilis*—Swanson, 1945:62). Kansas: Trego Co. (*Ursus horribilis horribilis*—Cockrum, 1952:239). Coahuila: vicinity of Cuatro Ciénegas (*Ursus sp.*—Baker, 1956:301). Durango: southern Durango (*Ursus nelsoni*—G. M. Allen, 1942:153). Baja California: Sierra Juárez near Santa Catarina Mission (grizzly bear—G. M. Allen, 1942:145). Alaska: St. Lawrence Island (*Ursus arctos beringianus*—Rausch, 1953:19).

Ursus caurinus Merriam

1914. *Ursus caurinus* Merriam, Proc. Biol. Soc. Washington, 27:187, August 13, type from Berners Bay, E side Lynn Canal, southeastern Alaska.

Ursus chelan Merriam

1916. *Ursus chelan* Merriam, Proc. Biol. Soc. Washington, 29:136, September 6, type from T. 30 N, R. 16 E, Willamette Meridian, Wenatchee National Forest, E slope Cascade Mts., northern Chelan Co., Washington.

Ursus chelidonias Merriam

1918. *Ursus chelidonias* Merriam, N. Amer. Fauna, 41:21, February 9, type from head Jervis Inlet, British Columbia.

Ursus colusus Merriam

1914. *Ursus colusus* Merriam, Proc. Biol. Soc. Washington, 27:187, August 13, type from Sacramento River, probably between Colusa and Sacramento, California.

Ursus crassodon Merriam

1918. *Ursus crassodon* Merriam, N. Amer. Fauna, 41:90, February 9, type from Klappan Creek (Third South Fork Stikine River), British Columbia.

Ursus crassus Merriam

1918. *Ursus crassus* Merriam, N. Amer. Fauna, 41:90, February 9, type from upper Macmillan River, Yukon.

Ursus cressonus Merriam

1916. *Ursus cressonus* Merriam, Proc. Biol. Soc. Washington, 29:137, September 6, type from Lakina River, S slope Wrangell Range, Alaska.

Ursus dalli Merriam

1896. *Ursus dalli* Merriam, Proc. Biol. Soc. Washington, 10:71, April 13, type from Yakutat Bay (NW side), Alaska.

Ursus eltonclarki Merriam

1914. *Ursus eltonclarki* Merriam, Proc. Biol. Soc. Washington, 27:175, August 13, type from Freshwater Bay, Chichagof Island, Alaska.

Ursus eulophus Merriam

1904. *Ursus eulophus* Merriam, Proc. Biol. Soc. Washington, 17:153, October 6, type from Admiralty Island, Alaska.

Ursus eximius Merriam

1916. *Ursus eximius* Merriam, Proc. Biol. Soc. Washington, 29:139, September 6, type from head of Knik, Cook Inlet, Alaska.

Ursus gyas Merriam

1902. *Ursus dalli gyas* Merriam, Proc. Biol. Soc. Washington, 15:78, March 22, type from Pavlof Bay, Alaska Peninsula, Alaska.

1918. *Ursus gyas* Merriam, N. Amer. Fauna, 41:124, February 9.
1902. *Ursus merriami* J. A. Allen, Bull. Amer. Mus. Nat. Hist., 16:141, April 12, type from Portage Bay, opposite Port Muller, Alaska Peninsula, Alaska.

Ursus henshawi Merriam

1914. *Ursus henshawi* Merriam, Proc. Biol. Soc. Washington, 27:190, August 13, type from southern Sierra Nevada, near Havilah, Kern Co., California.

Ursus holzworthi Merriam

1929. *Ursus holzworthi* Merriam, Proc. Biol. Soc. Washington, 42:173, June 15, type from last slope of Talkeetna Mts., near headwaters of Oshetna or Black River, Alaska.

Ursus hoots Merriam

1916. *Ursus hoots* Merriam, Proc. Biol. Soc. Washington, 29:140, September 6, type from Clearwater Creek, a N branch Stikine River, British Columbia.

Ursus horriaeus Baird

1858. *U[rsus]. horribilis* var. *horriaeus* Baird, Mammals, *in* Repts. Expl. Surv. . . . , 8(1):224, July 14, type from old copper mines near present town of Santa Rita, Grant Co., New Mexico (see V. Bailey, N. Amer. Fauna, 53:357, March 1, 1932).
1914. [*Ursus*] *horriaeus*, Merriam, Proc. Biol. Soc. Washington, 27:191, August 13.

Ursus horribilis bairdi Merriam

1914. *Ursus bairdi* Merriam, Proc. Biol. Soc. Washington, 27:192, August 13, type from Blue River, Summit Co., Colorado.
1918. *Ursus horribilis bairdi* Merriam, N. Amer. Fauna, 41:19, February 9.

Ursus horribilis dusorgus Merriam

1918. *Ursus dusorgus* Merriam, N. Amer. Fauna, 41:33, February 9, type from head Jackpine River, near Mt. Bess, close to British Columbian boundary, Alberta.
1947. *Ursus horribilis dusorgus*, Anderson, Bull. Nat. Mus. Canada, 102:39, January 24.

Ursus horribilis horribilis Ord

1815. *Ursus horribilis* Ord, *in* Guthrie, A new geog., hist., coml. grammar . . . , Philadelphia, 2nd Amer. ed., 2:291 (described on p. 299). Type locality, Missouri River, a little above mouth Poplar River, northeastern Montana.
1820. *Ursus cinereus* Desmarest, Mammalogie . . . , p. 164, *in* Encyclopédie méthodique. . . . Type locality, not designated.
1822. *Ursus griseus* Choris, Voyage pittoresque autour du monde, Paris (unpaged). Name applied by Choris to the bear of the interior of North America, but Choris identifies his animal with *Ursus griseus* Cuv.
1827. *Ursus candescens* Hamilton-Smith, *in* Griffith's Cuvier, The Animal Kingdom . . . , 2:229 (*fide* Griffith, *loc. cit.*).

Ursus horribilis imperator Merriam

1914. *Ursus imperator* Merriam, Proc. Biol. Soc. Washington, 27:180, August 13, type from Yellowstone National Park, Wyoming.

1918. *Ursus horribilis imperator* Merriam, N. Amer. Fauna, 41:20, February 9.

Ursus hylodromus Elliot

1904. *Ursus hylodromus* Elliot, Field Columb. Mus., Publ. 87, Zool. Ser., 3:257, January 7, type from Alberta.

1916. *Ursus selkirki* Merriam, Proc. Biol. Soc. Washington, 29:150, September 6, type from Selkirk Mts., Upper Columbia River, British Columbia. (Regarded as indistinguishable from *U. hylodromus* Elliot by Merriam, N. Amer. Fauna, 41:77, February 9, 1918.)

Ursus idahoensis Merriam

1918. *Ursus idahoensis* Merriam, N. Amer. Fauna, 41:54, February 9, type from N fork Teton River, Fremont Co., Idaho.

Ursus innuitus Merriam

1914. *Ursus innuitus* Merriam, Proc. Biol. Soc. Washington, 27:177, August 13, type from Golofnin Bay, S side Seward Peninsula, western Alaska.

Ursus inopinatus (Merriam)

1918. *Vetularctos inopinatus* Merriam, N. Amer. Fauna, 41:132, February 9, type from Rendezvous Lake, NE of Fort Anderson, Mackenzie.

1924. *Ursus inopinatus*, Miller, Bull. U.S. Nat. Mus., 128:106, April 29.

Ursus insularis Merriam

1916. *Ursus eltonclarki insularis* Merriam, Proc. Biol. Soc. Washington, 29:141, September 6, type from Admiralty Island, southeastern Alaska.

1918. *Ursus insularis* Merriam, N. Amer. Fauna, 41:44, February 9.

Ursus internationalis Merriam

1914. *Ursus internationalis* Merriam, Proc. Biol. Soc. Washington, 27:177, August 13, type from Alaska–Yukon boundary, about 50 mi. S of Arctic Coast. (Arranged as a synonym of *richardsoni* by Bee and Hall, Univ. Kansas Mus. Nat. Hist., Miscl. Publ., 8:189, March 10, 1956.)

Ursus kenaiensis Merriam

1904. *Ursus kenaiensis* Merriam, Proc. Biol. Soc. Washington, 17:154, October 6, type from Cape Elizabeth, extreme W end Kenai Peninsula, Alaska.

Ursus kennerleyi Merriam

1914. *Ursus kennerleyi* Merriam, Proc. Biol. Soc. Washington, 27:194, August 13, type from mts. near Los Nogales, Sonora.

Ursus kidderi kidderi Merriam

1902. *Ursus kidderi* Merriam, Proc. Biol. Soc. Washington, 15:78, March 22, type from Chinitna Bay, Cook Inlet, Alaska.

Ursus kidderi tundrensis Merriam

1914. *Ursus kidderi tundrensis* Merriam, Proc. Biol. Soc. Washington, 27:196, August 13, type from Shaktolik River, Norton Sound, Alaska.

Ursus klamathensis Merriam

1914. *Ursus klamathensis* Merriam, Proc. Biol. Soc. Washington, 27:185, August 13, type from Beswick, near mouth Shovel Creek, Klamath River, Siskiyou Co., California.

Ursus kluane impiger Merriam

1918. *Ursus kluane impiger* Merriam, N. Amer. Fauna, 41:81, February 9, type from Columbia Valley, British Columbia.

Ursus kluane kluane Merriam

1916. *Ursus kluane* Merriam, Proc. Biol. Soc. Washington, 29:141, September 6, type from McConnell River, Yukon.

Ursus kwakiutl Merriam

1916. *Ursus kwakiutl* Merriam, Proc. Biol. Soc. Washington, 29:143, September 6, type from Jervis Inlet, coast of southern British Columbia.

Ursus latifrons Merriam

1914. *Ursus phaeonyx latifrons* Merriam, Proc. Biol. Soc. Washington, 27:183, August 13, type from Jasper House, Alberta.

1918. *Ursus latifrons* Merriam, N. Amer. Fauna, 41:97, February 9.

Ursus macfarlani Merriam

1918. *Ursus macfarlani* Merriam, N. Amer. Fauna, 41:51, February 9, type from Anderson River, 50 mi. below Fort Anderson, Mackenzie.

Ursus macrodon Merriam

1918. *Ursus macrodon* Merriam, N. Amer. Fauna, 41:38, February 9, type from Twin Lakes, Lake Co., Colorado.

Ursus magister Merriam

1914. *Ursus magister* Merriam, Proc. Biol. Soc. Washington, 27:189, August 13, type from Los Biacitos, head San Onofre Canyon, Santa Ana Mts., San Diego Co., California.

Ursus mendocinensis Merriam

1916. *Ursus mendocinensis* Merriam, Proc. Biol. Soc. Washington, 29:145, September 6, type from Long Valley, N of Sherwood, Mendocino Co., California.

Ursus middendorffi Merriam

1896. *Ursus middendorffi* Merriam, Proc. Biol. Soc. Washington, 10:69, April 13, type from Kodiak Island, Alaska.

1911. *Ursus kadiaki* Kleinschmidt, Outdoor Life, 27:3, January. Name applied to the big brown bear of "Kadiak Island, Alaska Peninsula, Montague Island, and Yacutat," all in Alaska.

Ursus mirabilis Merriam

1916. *Ursus mirabilis* Merriam, Proc. Biol. Soc. Washington, 29:146, September 6, type from Admiralty Island, Alaska.

Ursus mirus Merriam

1918. *Ursus mirus* Merriam, N. Amer. Fauna, 41:40, February 9, type from Slough Creek, Yellowstone National Park, Wyoming.

Ursus neglectus Merriam

1916. *Ursus kwakiutl neglectus* Merriam, Proc. Biol. Soc. Washington, 29:144, September 6, type from near Hawk Inlet, Admiralty Island, Alaska.

1918. *Ursus neglectus* Merriam, N. Amer. Fauna, 41:28, February 9.

Ursus nelsoni Merriam

1914. *Ursus nelsoni* Merriam, Proc. Biol. Soc. Washington, 27:190, August 13, type from Colonia García, Chihuahua.

Ursus nortoni Merriam

1914. *Ursus nortoni* Merriam, Proc. Biol. Soc. Washington, 27:179, August 13, type from Yakutat, Alaska.

Ursus nuchek Merriam

1916. *Ursus nuchek* Merriam, Proc. Biol. Soc. Washington, 29:146, September 6, type from head Nuchek Bay, Hinchinbrook Island, Prince William Sound, Alaska.

Ursus orgiloides Merriam

1918. *Ursus orgiloides* Merriam, N. Amer. Fauna, 41:46, February 9, type from Italio River, Alaska.

Ursus orgilos Merriam

1914. *Ursus orgilos* Merriam, Proc. Biol. Soc. Washington, 27:176, August 13, type from Bartlett Bay, E side Glacier Bay, southeastern Alaska.

Ursus oribasus Merriam

1918. *Ursus oribasus* Merriam, N. Amer. Fauna, 41:56, February 9, type from upper Liard River, Yukon, near British Columbian boundary.

Ursus pallasi Merriam

1916. *Ursus pallasi* Merriam, Proc. Biol. Soc. Washington, 29:149, September 6, type from Donjek River, southwestern Yukon.

Ursus pellyensis Merriam

1918. *Ursus pellyensis* Merriam, N. Amer. Fauna, 41:82, February 9, type from Ketza Divide, Pelly Mts., Yukon.

Ursus perturbans Merriam

1918. *Ursus perturbans* Merriam, N. Amer. Fauna, 41:64, February 9, type from a canyon on Mt. Taylor, 12 mi. E San Mateo, Valencia Co., New Mexico.

Ursus pervagor Merriam

1914. *Ursus pervagor* Merriam, Proc. Biol. Soc. Washington, 27:186, August 13, type from Pemberton [= Lillooet] Lake, British Columbia.

Ursus phaeonyx Merriam

1904. *Ursus horribilis phaeonyx* Merriam, Proc. Biol. Soc. Washington, 17:154, October 6, type from Glacier Mtn., Tanana Mts., Alaska (about 2 mi. below source Comet Creek, near Fortymile Creek, between Yukon and Tanana rivers).

1914. [*Ursus*] *phaeonyx* Merriam, Proc. Biol. Soc. Washington, 27:183, August 13.

Ursus planiceps Merriam

1918. *Ursus planiceps* Merriam, N. Amer. Fauna, 41:37, February 9, type from Colorado, exact locality unknown, but probably in foothills or on western edge of plains.

Ursus pulchellus ereunetes Merriam

1918. *Ursus pulchellus ereunetes* Merriam, N. Amer. Fauna, 41:56, February 9, type from Beaverfoot Range, Kootenai Dist., British Columbia.

Ursus pulchellus pulchellus Merriam

1918. *Ursus pulchellus pulchellus* Merriam, N. Amer. Fauna, 41:55, February 9, type from Ross River, Yukon.

Ursus richardsoni Swainson

1838. *Ursus richardsoni* Swainson, Animals in menageries . . . , p. 54. Type locality assumed to be shore of the Arctic Ocean, on W side Bathurst Inlet about 8 mi. from mouth Hood River, Mackenzie. (Arranged as a subspecies of *Ursus horribilis* by Bee and Hall, Univ. Kansas Mus. Nat. Hist., Miscl. Publ., 8:188, March 10, 1956.)

Ursus rogersi bisonophagus Merriam

1918. *Ursus rogersi bisonophagus* Merriam, N. Amer. Fauna, 41:66, February 9, type from Bear Lodge, Sundance National Forest, Black Hills, Crook Co., Wyoming.

Ursus rogersi rogersi Merriam

1918. *Ursus rogersi rogersi* Merriam, N. Amer. Fauna, 41:65, February 9, type from high up on Greybull River, Absaroka Mts., Yellowstone National Park, Wyoming.

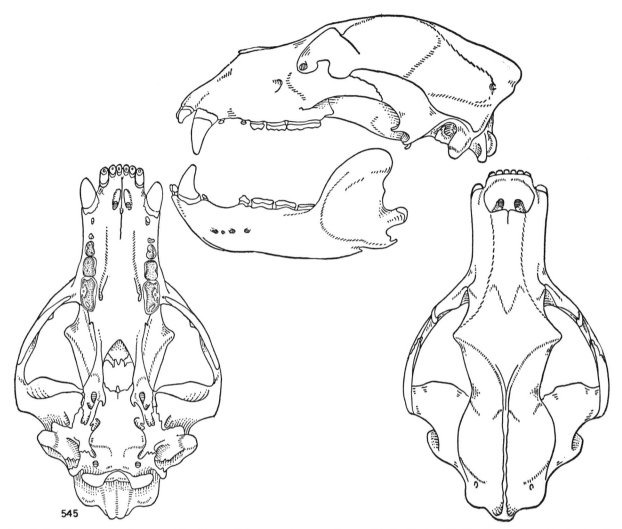

Fig. 545. *Ursus richardsoni*, jt. Chandler and Sic-sic-puk rivers, long. 152° 00′ W, lat. 68° 51′ N, approx. 1800 ft., Brooks Range, Alaska, No. 51562 K.U., probably ♂, X ¼.

Ursus russelli Merriam

1914. *Ursus russelli* Merriam, Proc. Biol. Soc. Washington, 27:178, August 13, type from Mackenzie Delta, Mackenzie. (Arranged as a subspecies of *Ursus internationalis* by Anderson, *in* Porsild, Canadian Field-Nat., 59:8, August 16, 1945.)

Ursus sheldoni Merriam

1910. *Ursus sheldoni* Merriam, Proc. Biol. Soc. Washington, 23:127, September 2, type from Montague Island, Prince William Sound, Alaska.

Ursus shirasi Merriam

1914. *Ursus shirasi* Merriam, Proc. Biol. Soc. Washington, 27:195, August 13, type from Bybus Bay, Admiralty Island, Alaska.

Ursus shoshone Merriam

1914. *Ursus shoshone* Merriam, Proc. Biol. Soc. Washington, 27:184, August 13, type from Estes Park, Larimer Co., Colorado.

Ursus sitkensis Merriam

1896. *Ursus sitkensis* Merriam, Proc. Biol. Soc. Washington, 10:73, April 13, type from near Sitka, Alaska.

Ursus stikeenensis Merriam

1914. *Ursus stikeenensis* Merriam, Proc. Biol. Soc. Washington, 27:178, August 13, type from Tatletuey Lake, near head Skeena River, northern British Columbia.

Ursus tahltanicus Merriam

1914. *Ursus tahltanicus* Merriam, Proc. Biol. Soc. Washington, 27:181, August 13, type from Klappan Creek ("= 3d So. Fk. Stikine River"), British Columbia.

Ursus texensis navaho Merriam

1914. *Ursus navaho* Merriam, Proc. Biol. Soc. Washington, 27:191, August 13, type from Navajo country near Fort Defiance (Mollhausen), Arizona, type probably killed in 1856 in Chuska Mountains, on boundary between northeastern Arizona and northwestern New Mexico.
1918. *Ursus texensis navaho* Merriam, N. Amer. Fauna, 41:37, February 9.

Ursus texensis texensis Merriam

1914. *Ursus horriaeus texensis* Merriam, Proc. Biol. Soc. Washington, 27:191, August 13, type from Davis Mts., Jeff Davis Co., Texas.
1918. *Ursus texensis texensis* Merriam, N. Amer. Fauna, 41:35, February 9.

Ursus toklat Merriam

1914. *Ursus toklat* Merriam, Proc. Biol. Soc. Washington, 27:182, August 13, type from head Toklat River, N base Alaska Range, near Mt. McKinley, Alaska.

Ursus townsendi Merriam

1916. *Ursus townsendi* Merriam, Proc. Biol. Soc. Washington, 29:151, September 6, type from mainland of southeastern Alaska, probably between Cross Sound and Alsek River Delta, but exact locality unknown.

Ursus tularensis Merriam

1914. *Ursus californicus tularensis* Merriam, Proc. Biol. Soc. Washington, 27:188, August 13, type from Fort Tejon, Tehachapi Mts., Kern Co., California.
1918. *Ursus tularensis* Merriam, N. Amer. Fauna, 41:30, February 9.

Ursus utahensis Merriam

1914. *Ursus utahensis* Merriam, Proc. Biol. Soc. Washington, 27:193, August 13, type from N fork Salina Creek, about 10 mi. SE Mayfield, Sanpete Co., Utah.

Ursus warburtoni Merriam

1916. *Ursus kwakiutl warburtoni* Merriam, Proc. Biol. Soc. Washington, 29:145, September 6, type from Atnarko River, British Columbia.
1918. *Ursus warburtoni* Merriam, N. Amer. Fauna, 41:27, February 9.

Ursus washake Merriam

1916. *Ursus washake* Merriam, Proc. Biol. Soc. Washington, 29:152, September 6, type from N. Fork Shoshone River, Absaroka Mts., between Bighorn Basin and Yellowstone National Park, Wyoming.

Ursus maritimus
Polar Bear

Revised by Manning, Canadian Wildlife Serv. Rept. Ser., 13:1–27, received Univ. Kansas, May 24, 1971. See addenda.

Size large; head relatively small; neck relatively long; claws not strongly curved but with well-developed cutting edges; fur dense; color uniformly white or buffy; canines and incisors relatively large and recurved; molars relatively small, the combined length of M1 and M2 less than palatal width; P4 relatively high; braincase relatively long; frontal shield not markedly elevated. All 5 digital pads, transverse plantar pad, external carpal pad, and small area of metatarsal pad (terminology of Pocock, 1932:102) hairless but *ca.* three-quarters of area of soles between these pads covered by hair (Seton, 1929, 2:209, Pl. 31), whereas entire soles, fore- and hind-feet, essentially hairless in black bear (Seton, 1929, 2:141) and brown bear (Seton, 1929, 2:29, Pl. 5, and p. 60, Pl. 8).

Polar bears live on the Arctic coasts and ice floes. They spend much time in the water and are excellent swimmers.

Manning (1971:25, 26) recognizes geographic variation in size in the Holocene species; individuals are smallest in eastern Greenland and largest in "Alaska south" (Saint Paul and Saint Lawrence islands). He considers the amount of variation in the Holocene populations too slight to divide them into subspecies, but he does tentatively recognize a Late Pleistocene subspecies, *Ursus maritimus tyrannus* Kurtén (1964:9) on the basis of greater size than he found in Holocene specimens.

The Holocene subspecies is circumboreal. The range map indicates the southern limits of distribution in only North America.

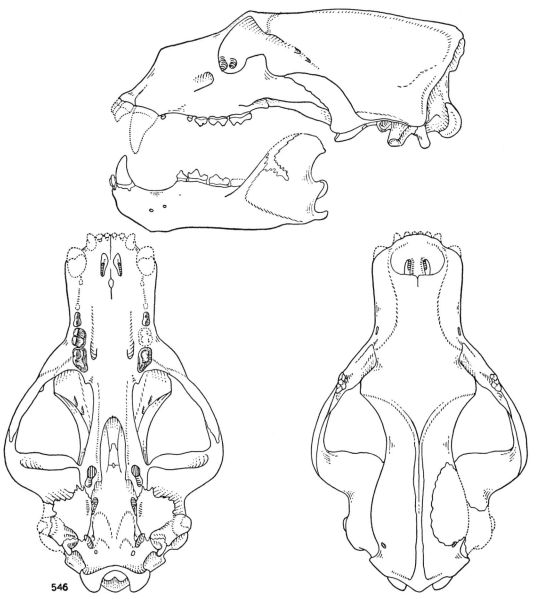

Fig. 546. *Ursus maritimus maritimus*, Birnirk Mounds, long. 156° 36′ 02″ W, lat. 71° 20′ 40″ N, 8 ft., Point Barrow, Alaska, No. 51982 K.U., probably ♂, lower jaw No. 51986 K.U., both X ¼.

Ursus maritimus maritimus Phipps

1774. *Ursus maritimus* Phipps, A voyage toward the North Pole. . . , p. 185. Type locality, Spitzbergen.

1776. *Ursus marinus* Pallas, . . . Reise durch verschiedene Provinzen des Russischan Reichs, 3:691. Type from Arctic Ocean. (Original publication not seen.)

1792. *Ursus polaris* Shaw, Museum Leverianum, 1:7. A renaming of *marinus*. (Original publication not seen.)

1908. [*Thalassarctos maritimus*] var. *ungavensis* Knottnerus-Meyer, Sitzungsb. Gesell. Naturforsch. Freunde, Berlin, p. 181, July, type from near Killinek (see Anderson, R. M., Nat. Mus. Canada, 102, Biol. Ser. No. 31, p. 47, January 24, 1947), Ungava Bay, [Quebec].

1908. *Thalassarctos eogroenlandicus* Knottnerus-Meyer, Sitzungsb. Gesell. Naturforsch. Freunde, Berlin, p. 182, July, type from pack ice off coast of eastern Greenland.

1908. *Th*[*alassarctos*]. *labradorensis* Knottnerus-Meyer, Sitzungsb. Gesell. Naturforsch. Freunde, Berlin, p. 183, July, type from Okak, Labrador.

1908. *Thalassarctos jenaensis* Knottnerus-Meyer, Sitzungsb. Gesell. Naturforsch. Freunde, Berlin, p. 184, July, type from Jena Island, Spitzbergen.

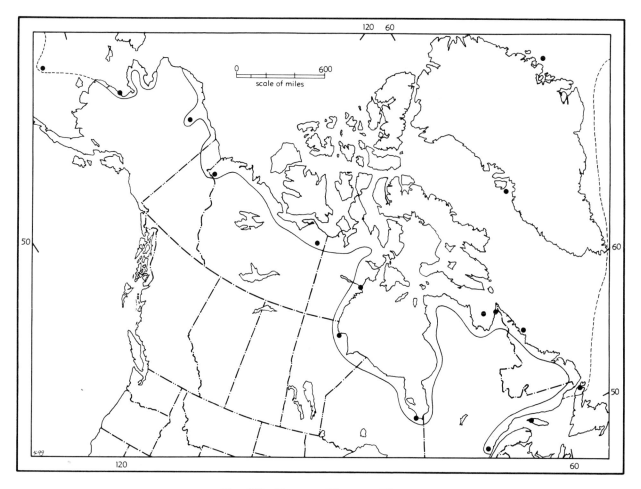

Map 499. *Ursus maritimus maritimus.*

Showing several southern occurrences of the polar bear (*Ursus maritimus*) in North America.

1908. *Th[alassarctos]. spitzbergensis* Knottnerus-Meyer, Sitzungsb. Gesell. Naturforsch. Freunde, Berlin, p. 184, July, type from Seven Islands, Spitzbergen.
1932. *Thalassarctos maritimus groenlandicus* Birula, Trud. Inst. Zool. Acad. Sci. U.S.S.R., vol. 1, pp. 11, 132. Type locality, western Greenland.

RECORDS marking the southern limits of distribution in North America are, from west to east, Alaska: Pribilof Islands; near St. Michaels; W. Fork Sagavanirktok River, long. 148° 51′ 00″ W, lat. 68° 55′ 30″ N. Mackenzie: Fort MacPherson, Mackenzie River; Teshierpi River, NE Great Slave Lake, *ca.* 75 mi. from salt water. Keewatin: Chesterfield Inlet. Manitoba: Churchill. Ontario: Moose Factory. Quebec: Ungava Bay (type locality of *T. m. ungavensis*); near Killinek, Ungava Bay. Labrador: Okak (type locality of *T. labradorensis*). Quebec: Lake St. John, Saguenay River; Cap de Rabast, Anticosti Island (Lewis and Doutt, 1942:370). Newfoundland. Western Greenland

(type locality of *T. m. groenlandicus*). Eastern Greenland (type locality of *T. m. eogroenlandicus*).

FAMILY PROCYONIDAE
Raccoons and Allies

Genera revised by Hollister, Proc. U.S. Nat. Mus., 49:143–150, August 13, 1915; and by Pocock, Proc. Zool. Soc. London, pp. 389–422, June 28, 1921.

Small to medium sized carnivores, tail long; feet plantigrade to semi-plantigrade and pentadactyl; alisphenoid canal absent (except *Ailurus*); claws non- or semiretractile; carnassial teeth not sectorial (except *Bassariscus*); molars $\frac{2}{2}$ or $\frac{2}{3}$, low-crowned, broad, and multituberculate.

At one time or another each of the genera has been made the type of a monotypic family or subfamily. The following classification is conservative.

SUBFAMILY **PROCYONINAE**

Small carnivores; pentadactyl; plantigrade or semiplantigrade; tail moderate to long, prehensile in *Potos*, annulated color pattern in some others; alisphenoid canal absent; digital, plantar, and carpal pads naked and well developed; baculum long, not upcurled apically, clavate at base, and not strongly keeled; no glandular area around anus; penis long, prepuce abdominal; anterior edge of coronoid inclined backward.

KEY TO NORTH AMERICAN GENERA OF PROCYONINAE

1. Tail not annulated; digits webbed ⅓ length beyond plantar pad; baculum tipped with 4 condylelike branches; premolars ⅗; mandible deep; palate parallel-sided. . . . . *Potos*, p. 977
1'. Tail annulated (except in *Bassaricyon*); digits not webbed or webbed more than ⅓ length beyond plantar pad; baculum not tipped with 4 condylelike branches; premolars ⁴⁄₄; mandible slender; palate arcuate.
   2. Molars and premolars with pointed or bladelike cusps; incisors with secondary lobes; canines rounded; tail long, long-haired, and annulated. . . . . *Bassariscus*, p. 961
   2'. Molars and premolars with heavy conical cusps; incisors without secondary lobes, or only I3 with secondary lobes; canines laterally compressed; tail short or long, short-haired or long-haired, and annulated or not (but no member combined long tail, long-haired tail and annulated tail).
      3. Rostrum greatly elongated and laterally compressed (see Fig. 551); P4 slightly longer than broad; M1 subquadrate, length and breadth about equal; claws long; snout greatly elongated. *Nasua*, p. 974
      3'. Rostrum not greatly elongated and not laterally compressed (see Figs. 548–550); P4 as broad as long or broader; M1 variable; claws short; snout not greatly elongated.
         4. Tail not annulated, longer than head and body; mastoid small; claws short, sharp, and much curved; proximal ⅔ of digits webbed; P4 has 3 tubercules; malar ending at M2. *Bassaricyon*, p. 978
         4'. Tail annulated, shorter than head and body; mastoid large; claws long and blunt, not much curved; less than proximal ½ of digits webbed; P4 has 5 tubercules; malar ending at M1. . . . . *Procyon*, p. 966

Genus **Bassariscus** Coues—Ringtails

1830. *Bassaris* Lichtenstein, Abh. preuss. Akad. Wiss., Berlin, p. 119. Type, *Bassaris astuta* Lichtenstein. Not *Bassaris* Hubner, 1816.
1887. *Bassariscus* Coues, Science, 9:516, May 27, a renaming of *Bassaris* Lichtenstein. Type, *Bassaris astuta* Lichtenstein.
1886. *Wagneria* Jentink, Notes Leiden Mus., 8:129. Type, *Bassaris sumichrasti* Saussure. Not *Wagneria* Robineau-Desvoidy, 1830, or Alenitzin, 1873.
1904. *Jentinkia* Trouessart, Catalogus mammalium . . . , Suppl., fasc. 1, p. 184, a renaming of *Wagneria* Jentink.

Body elongate; tail longer than body, distinctly annulated; muzzle moderately elongate; ears relatively long, with well-developed bursa; feet with 5 toes, claws semiretractile; baculum with rounded apex.

Skull elongate; postorbital processes well developed from frontal, moderately from zygomata; temporal crests uniting, if at all, only posteriorly to form sagittal crest; palate ending about posterior plane of M2; auditory bullae moderately inflated; P1–P3 unicusped; P4 sectorial with well-developed blade; deuterocone present posterior to protocone of P4; M1 variable, subquadrate to triangular, broader than long; canines doglike, smooth, slightly curved, not sculptured.

KEY TO SPECIES OF BASSARISCUS

1. Ears pointed; muzzle and feet blackish; 2nd to 5th digits of fore- and hind-limbs naked on lower surface behind digital pads; claws long, curved, compressed, and nonretractile; low ridges connecting cusps in molariform teeth; upper carnassial triangular; deuterocone lacking; I1 and I2 trifid. . . . . *B. sumichrasti*, p. 965
1'. Ears rounded; muzzle and feet grayish; 2nd to 5th digits of fore- and hind-limbs densely hairy on lower surface behind digital pads; claws short, straight, not much compressed, retractile; high, sharp ridges connecting cusps in molariform teeth; upper carnassial irregular in outline, deuterocone present; I1 and I2 not lobed. . . . . *B. astutus*, p. 962

Bassariscus astutus
Ringtail

x ¹⁄₁₀

External measurements: 616–811; 310–438; 57–78; 44–50. Weight, 870–1100 grams. Basilar length, 68.5–75.0; zygomatic breadth, 48.0–52.0; mastoid breadth, 33.4–35.4. Upper parts light buff to pinkish buff, overcast with black or dark brownish overhairs; underparts white or white washed with pale buff; eye ringed black or dark brown; supraorbital, suborbital, and subauricular patches white to pinkish buff; annulations distinct for full length of tail; margins of ears somewhat produced anteroexternally; feet with 5 toes, 2nd, 3rd, 4th, and 5th digits of fore- and hindlimbs densely haired on lower surface behind and around digital pads; claws short, straight, semiretractile; posterior border of maxillary portion of zygoma lying at plane of M1; foramen ovale opening somewhat downward; cusps in larger molariform teeth with high connecting ridges; upper carnassial irregular in outline, with prominent deuterocone; cutting edges of I1 and I2 normally smooth.

Bassariscus astutus arizonensis Goldman

1932. *Bassariscus astutus arizonensis* Goldman, Proc. Biol. Soc. Washington, 45:87, June 21, type from Cosper Ranch, about 12 mi. S Blue, 5000 ft., Greenlee Co., Arizona.

MARGINAL RECORDS.—Wyoming: 3 mi. N, 24 mi. W Kemmerer (Long, 1965a:687). Colorado (Armstrong, 1972:266): Hayden; 25 mi. S Hot Sulphur Springs, 8800 ft.; near Taylor Park Dam, *ca.* 9200 ft.; near Delta; San Miguel Cañon, 5 mi. above Cottonwood Creek; vic. Durango. Arizona: type locality. New Mexico: Redrock. Arizona: Santa Rita Mts.; 5 mi. N Wickenburg; mouth Diamond Creek, Colorado River; North Rim Ranger Station (Hoffmeister, 1971:166). Utah: Rainbow Lodge; Willow Creek, 42 mi. S Ouray; Dry Fork, 12 mi. NW Vernal.

Bassariscus astutus astutus (Lichtenstein)

1830. *B[assaris]. astuta* Lichtenstein, Abh. preuss. Akad. Wiss., Berlin, 1827, p. 119, type from near city of Mexico.
1887. *Bassariscus astutus*, Coues, Science, 9:516, May 27.
1904. *Bassariscus albipes* Elliot, Field Columb. Mus., Publ. 87, Zool. Ser., 3(14):258, January 7, type from Xico, near Jalapa, Veracruz.

MARGINAL RECORDS.—San Luis Potosí: 7 km. SE Presa de Guadalupe; 6 km. NE Ciudad Del Maíz; Hda. Capulín. Veracruz: *Acatlán, 4100 ft.*; 2 km. E Las Vigas, 8000 ft. (Hall and Dalquest, 1963:330); *Xico*; Orizaba; Maltrata (Hall and Dalquest, 1963:330). Oaxaca (Goodwin, 1969:225): Reyes; Mt. Zempoaltepec.

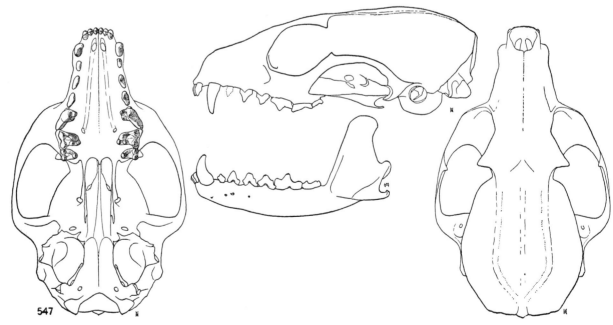

Fig. 547. *Bassariscus astutus nevadensis*, 10 mi. S Caliente, Nevada, No. 51665 M.V.Z., ♂, X 1.

Puebla: Pan-American Hwy. between Tepenene and Los Amates, SE Izúcar de Matamoros (Van Gelder, 1960:519, sight record only). Morelos (Ramírez-P., 1971:280): *vic. Huitzilac;* Lagunas de Zempoala, 3000 m. Distrito Federal: ½ mi. E San Gregorio Atlapulco, 2260 m. San Luis Potosí: Bledos.

Bassariscus astutus bolei Goldman

1945. *Bassariscus astutus bolei* Goldman, Proc. Biol. Soc. Washington, 58:105, July 18, type from Chilpancingo, Guerrero.

MARGINAL RECORDS.—Guerrero (Ramírez-P., 1971:280): Yerbabuena; 5 km. S Huitzuco. Oaxaca: Tenango. Guerrero: *type locality;* 18 km. SW Chichihualco (Ramírez-P., 1971:280).

Bassariscus astutus consitus Nelson and Goldman

1932. *Bassariscus astutus consitus* Nelson and Goldman, Jour. Washington Acad. Sci., 22:497, October 19, type from La Salada, 40 mi. S Uruapan, Michoacán.

MARGINAL RECORDS.—Sinaloa: 25 km. N, 3 km. E Choix, 1800 ft. (Armstrong, *et al.,* 1972:51). Durango (Baker and Greer, 1962:136): 2 mi. N Pueblo Nuevo, 6000 ft.; *Pueblo Nuevo, 5000 ft.* Jalisco: Bolaños; 3 mi. N Guadalajara; Ocotlán. Michoacán: 3 mi. NW Pátzcuaro, 6700 ft.; type locality. Sinaloa (Armstrong, *et al.,* 1972:51): Plomosas; 10 km. S, 38 km. E Sinaloa, 800 ft.

Bassariscus astutus flavus Rhoads

1894. *Bassariscus astutus flavus* Rhoads, Proc. Acad. Nat. Sci. Philadelphia, 45:417, January 30, type from Texas, exact locality unknown.

MARGINAL RECORDS.—Colorado: 3 mi. W Golden (Armstrong, 1972:267). Kansas: NE ¼ sec. 9, T. 25 S, R. 1 W, Sedgwick Co. (119392 KU); Badger Creek, 7 mi. E Emporia (68328 KU). Arkansas: Warren (Sealander and Gipson, 1972:458). Louisiana: 3 mi. S Sicily Island (Lowery, 1974:414). Texas: *ca.* 7 mi. E Beaumont; Matagordo County (V. Bailey, 1905:182); Aransas National Wildlife Refuge (Halloran, 1961:23, as *B. astutus* only, sight record); Corpus Christi (V. Bailey, 1905:182). Nuevo León: Potrero Prieto. Tamaulipas: Joya Verde, 35 km. SW Victoria (Alvarez, 1963:457); La Joya de Salas. Nuevo León: 7 mi. S Aramberri. Coahuila: Sierra Encarnación. Durango: 1½ mi. NW Nazas, 4100 ft. (Baker and Greer, 1962:136). Chihuahua (Anderson, 1972:379): Baborigame; Mojarachic; near Ocampo. Arizona: Frank Hands Ranch, Pinery Canyon (Cockrum, 1961:228). New Mexico: Black Range; Mogollon Mts.; near point of Malpais, T. 6 N, R. 11 W (Findley, *et al.,* 1975:298, as *B. astutus* only); 4 mi. N, 9 mi. W Abiquiu (*ibid.*); mouth Cimarroncito Canyon (*ibid.*). Colorado: W of Cañon City (Armstrong, 1972:267).—Seton (1929 (2):275) mentions records from Louisiana and other places including Matthews, Montgomery County,

Alabama. Bole and Moulthrop (1942:120, 121) and Goodpaster and Hoffmeister (1968:117) present information indicating the presence of *Bassariscus astutus* ssp. in Ohio since the mid-nineteenth century.

Bassariscus astutus insulicola Nelson and Goldman

1909. *Bassariscus astutus insulicola* Nelson and Goldman, Proc. Biol. Soc. Washington, 22:26, March 10, type from San José Island, Gulf of California, Baja California. Known only from type locality.

Bassariscus astutus macdougalli Goodwin

1956. *Bassariscus astutus macdougalli* Goodwin, Amer. Mus. Novit., 1757:10, March 8, type from La Ventosa, Salina Cruz, coastal lowlands, 20 km. S Tehuantepec, Oaxaca.

MARGINAL RECORDS.—Oaxaca (Goodwin, 1969:229): San Pedro Jilotepec; *Tehuantepec;* type locality; *Tenango.*

Bassariscus astutus nevadensis Miller

1913. *Bassariscus astutus nevadensis* Miller, Proc. Biol. Soc. Washington, 26:159, June 30, type from El Dorado Canyon, Clark Co., Nevada.

MARGINAL RECORDS.—Nevada: Elko (Yoakum, 1966:351, as *B. astutus* only). Utah: Birch Creek Canyon, Deep Creek Mts., 6000 ft. (*ibid.*); near Mountain Home; Sinbad, approx. 50 mi. SW Green River; 12 mi. SW Orderville. Arizona: *Grand Canyon.* Nevada (Bradley and Hansen, 1965:310, as *B. astutus* only): near Boulder City; type locality; Camp Bonanza, Desert Game Range. California: Union Mine, 4000 ft., Inyo Mts. Nevada: near Mercury (Bradley and Hansen, 1965:310, as *B. astutus* only); Meadow Valley Wash, 7 mi. S Caliente. Not found: Nevada: Basin Spring, Spring Range (Bradley and Hansen, 1965:310).

Bassariscus astutus octavus Hall

1926. *Bassariscus astutus octavus* Hall, Univ. California Publ. Zool., 30:39, September 8, type from San Luis River, 1700 ft., near Escondido, San Diego Co., California.

MARGINAL RECORDS.—California: Matilija; Mill Creek Cañon; Jacumba; type locality.

Bassariscus astutus palmarius Nelson and Goldman

1909. *Bassariscus astutus palmarius* Nelson and Goldman, Proc. Biol. Soc. Washington, 22:26, March 10, type from Comondú, Baja California.

MARGINAL RECORDS.—Baja California: Concepción, lat. 31°, 6000 ft., Sierra San Pedro Mártir; type locality.

Bassariscus astutus raptor (Baird)

1859. *Bassaris raptor* Baird, Mammals of the boundary, *in* Emory, Rept. U.S. and Mexican boundary survey . . . ,

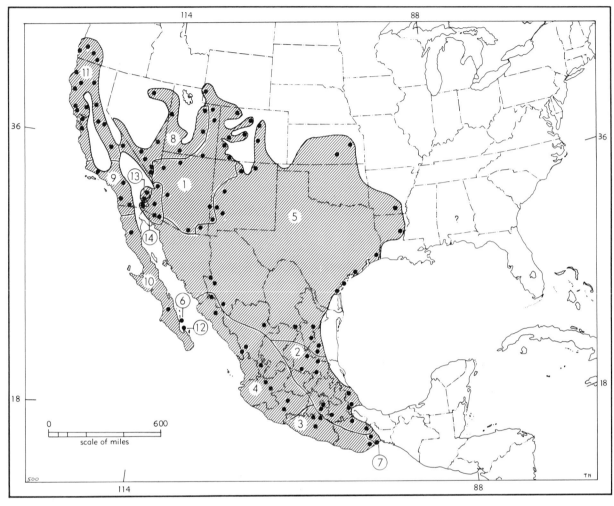

Map 500. *Bassariscus astutus.*

Guide to subspecies	5. *B. a. flavus*	10. *B. a. palmarius*
1. *B. a. arizonensis*	6. *B. a. insulicola*	11. *B. a. raptor*
2. *B. a. astutus*	7. *B. a. macdougalli*	12. *B. a. saxicola*
3. *B. a. bolei*	8. *B. a. nevadensis*	13. *B. a. willetti*
4. *B. a. consitus*	9. *B. a. octavus*	14. *B. a. yumanensis*

2(2):19, January, type from Glen Ellen, Sonoma Co., California (by restriction, Hall, Univ. California Publ. Zool., 30:44, September 8, 1926). The type specimen was caught in the city of Washington, D.C., where it had escaped from captivity.

1913. *Bassariscus astutus raptor*, Grinnell, Proc. California Acad. Sci., ser. 4, 3:289, August 28.

1894. *Bassariscus flavus oregonus* Rhoads, Proc. Acad. Nat. Sci. Philadelphia, 45:416, January 30, type from Grant's Pass, Josephine Co., Oregon.

MARGINAL RECORDS.—Oregon: Riddle, Umpqua Valley; Prospect; W side Upper Klamath Lake. California: Battle Creek; near Colfax; Yosemite; Tulare County; 6 mi. S Hetch Hetchy; Low Gap; 3 mi. NE Colusa; Salina; Eldridge; Diamond Canyon near

Fruitvale; Santa Rosa; Laytonville; 10 mi. NE Carlotta. Oregon: Gold Beach.

Bassariscus astutus saxicola Merriam

1897. *Bassariscus saxicola* Merriam, Proc. Biol. Soc. Washington, 11:185, July 1, type from Espíritu Santo Island, Gulf of California, Baja California. Known only from type locality.

1926. *Bassariscus astutus saxicola*, Hall, Univ. California Publ. Zool., 30:46, September 8.

Bassariscus astutus willetti Stager

1950. *Bassariscus astutus willetti* Stager, Proc. Biol. Soc. Washington, 63:203, December 29, type from Riverside Mts., Riverside Co., California.

MARGINAL RECORDS.—California: type locality; Palo Verde Mts.

Bassariscus astutus yumanensis Huey

1937. *Bassariscus astutus yumanensis* Huey, Trans. San Diego Soc. Nat. Hist., 8:357, June 15, type from Tinajas Altas, Gila Mts., Yuma Co., Arizona.

MARGINAL RECORDS.—Arizona: Parker (Cockrum, 1961:228); Sombrero Seep, Castle Dome Mts. (*ibid.*); Cabeza Prieta Game Range; type locality. California: Laguna Dam, Imperial Co.

Bassariscus sumichrasti
Cacomistle

External measurements: 790–1003; 396–508; 82–90; 44–45. Greatest length of skull, 82.9–90; basilar length, 74.2–79.5; zygomatic breadth, 49–57.5. General appearance as described for *B. astutus*, but muzzle and feet blackish instead of grayish; annulations of tail becoming obsolescent distally; pelage soft and lax; ears evenly rounded; 2nd, 3rd, 4th, and 5th digits of fore- and hind-limbs naked on lower surface behind digital pads; claws long, strongly curved, nonretractile; posterior border of maxillary part of zygoma at plane of M2; foramen ovale opening forward; cusps of larger molariform teeth with low connecting ridges; upper carnassial triangular; deuterocone lacking; cutting edges of I1 and I2 trifid.

Bassariscus sumichrasti campechensis (Nelson and Goldman)

1932. *Jentinkia sumichrasti campechensis* Nelson and Goldman, Jour. Washington Acad. Sci., 22:486, October 19, type from Apazote, near Yohaltún, Campeche.
1959. *Bassariscus sumichrasti campechensis*, Hall and Kelson, The mammals of North America, Ronald Press, p. 882, March 31.

MARGINAL RECORDS.—Quintana Roo: Nabalam. Campeche: La Tuxpeña; type locality.

Bassariscus sumichrasti latrans (Davis and Lukens)

1958. *Jentinkia sumichrasti latrans* Davis and Lukens, Jour. Mamm., 39:353, August 20, type from 2 mi. W Omiltemi, 7900 ft., Guerrero.
1969. *Bassariscus s[umichrasti]. latrans*, Goodwin, Bull. Amer. Mus. Nat. Hist., 141:229, April 30.

MARGINAL RECORDS (Davis and Lukens, 1958:353).—Guerrero: near Puentecilla [18 km. SSW Chichihualco]; type locality.

Bassariscus sumichrasti notinus Thomas

1903. *Bassariscus sumichrasti notinus* Thomas, Ann. Mag. Nat. Hist., ser. 7, 11:379, April, type from Boquete, 6000 ft., Chiriquí, Panamá.

MARGINAL RECORDS.—Costa Rica: Escazú. Panamá: type locality.

Bassariscus sumichrasti oaxacensis (Goodwin)

1956. *Jentinkia sumichrasti oaxacensis* Goodwin, Amer. Mus. Novit., 1757:11, March 8, type from La Gloria, 10 km. SE Santa María, Oaxaca.
1959. *Bassariscus sumichrasti oaxacensis*, Hall and Kelson, The mammals of North America, Ronald Press, p. 882, March 31.

MARGINAL RECORDS (Goodwin, 1969:229).—Chiapas: Pueblo Nuevo Solistahuacan. Oaxaca: type locality.

Bassariscus sumichrasti sumichrasti (Saussure)

1860. *Bassaris sumichrasti* Saussure, Revue et Mag. Zool., Paris, ser. 2, 12:7, January. Type locality, Mirador, Veracruz.
1903. *Bassariscus sumichrasti*, Thomas, Ann. Mag. Nat. Hist., ser. 7, 11:379, April.
1875. *Bassaris monticola* Cordero, La Naturaleza, 3:270. Type probably from vic. Jalapa, Veracruz.

MARGINAL RECORDS.—Veracruz: Jalapa; type locality. Oaxaca: Tehuantepec (J. A. Allen, 1879:339, as *B. sumichrasti* only); Tres Cruces (Goodwin, 1969:229); Río Sal, Lachao (*ibid.*).

Bassariscus sumichrasti variabilis (Peters)

1874. *Bassaris variabilis* Peters, Monatsb. preuss. Akad. Wiss., Berlin, p. 704, type from Cobán, Alta Verapaz, Guatemala.
1903. *Bassariscus sumichrasti variabilis*, Thomas, Ann. Mag. Nat. Hist., ser. 7, 11:379, April.

MARGINAL RECORDS.—Belize: Silkgrass. El Salvador (Burt and Stirton, 1961:45, as *Jentinkia sumichrasti*): Los Esesmiles; Chilata. Guatemala: near Joyabaj; Salache. Chiapas: W slope Volcán de Tacaná, 2800 m.

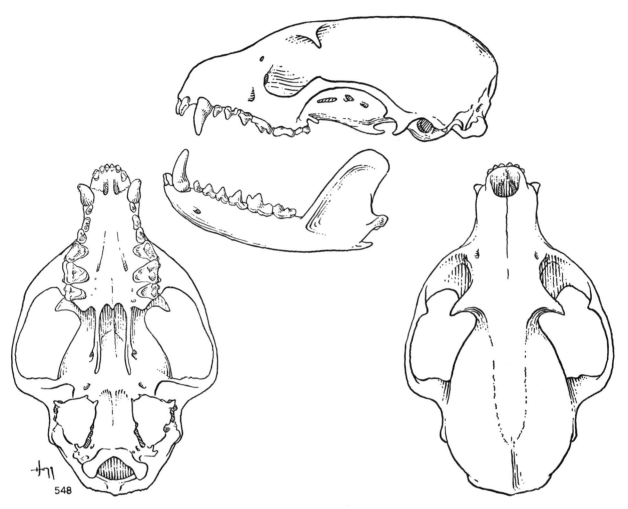

Fig. 548. *Bassariscus sumichrasti sumichrasti*, Mirador, Veracruz, No. 7082/8632 U.S.N.M., ♀, X 1.

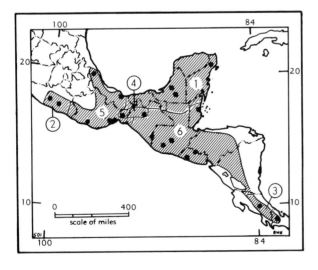

Map 501. *Bassariscus sumichrasti.*

1. *B. s. campechensis*	4. *B. s. oaxacensis*
2. *B. s. latrans*	5. *B. s. sumichrasti*
3. *B. s. notinus*	6. *B. s. variabilis*

Genus **Procyon** Storr—Raccoons

North American forms revised by Goldman, N. Amer. Fauna, 60:vi + 153, 22 pls., 2 figs., 1950.

1780. *Procyon* Storr, Prodromus methodi mammalium . . . , p. 35. Type, *Ursus lotor* Linnaeus.

Head broad behind, with pointed muzzle; ears medium-sized, pointed; toes not webbed; claws nonretractile; soles of feet naked; tail shorter than body, annulated, and not prehensile; baculum long and slender, bilobed anteriorly; mammae: p. 2, a. 2, i. 2.

Skull elongate; braincase broad and flat with medium interorbital constriction; zygomata with distinct postorbital processes; frontal processes greatly reduced; sagittal crest moderate or absent; palate flat or slightly concave, extending considerably beyond plane of last molars; vomer attached to palatal bones; foramen ovale small; postglenoid process small; auditory bullae large, much inflated on inner side, and not rising at sharp angle above tube of external auditory meatus; mandible long; ascending ramus low and without highly arched curve; condyles small.

Teeth large, maxillary tooth-row about a third greatest length of skull; molariform teeth with heavy, conical cusps; P4 and M1 subquadrate, only slightly broader than long or as broad as long; P4 subsectorial; incisors heavy; canines ovate at alveolus, slightly grooved on edges of outer surface and flattened on inner surface; P1 with a single root; P2 and P3 with 2 roots; P4 with 3 roots; P3 unicuspid with well-developed cingulum; P4 with 5th tubercle; m1 large, with 5 heavy conical cusps. Dentition, i. $\frac{3}{3}$, c. $\frac{1}{1}$, p. $\frac{4}{4}$, m. $\frac{2}{2}$.

KEY TO SUBGENERA AND SPECIES OF PROCYON

(After Goldman, 1950)

1. Pelage coarse and wiry, without underfur; hair on nape directed forward; palate extending behind posterior molars a distance of less than ¼ length of palate. . . (subgenus *Euprocyon*) *P. cancrivorus*, p. 974
1'. Pelage of 2 kinds, long guard-hairs and short, soft underfur; hair on nape normal, not directed forward; palate extending behind posterior molars a distance of more than ¼ length of palate. (subgenus *Procyon*)
   2. Geographic range on continental North America. *P. lotor*, p. 967
   2'. Geographic range insular (at least in part).
      3. Geographic range off Pacific Coast.
         4. Found on Vancouver Island. . . . . *P. lotor*, p. 967
         4'. Found on Tres Marías Islands. . . . . *P. insularis*, p. 972
      3'. Geographic range off Atlantic Coast (at least in part).
         5. Geographic range far offshore islands (Bahamas and Lesser Antilles).
            6. Found on Bahama Islands (New Providence Island). . . . .*P. maynardi*, p. 973
            6'. Found on Lesser Antilles.
               7. Color darker; upper carnassial longer than broad; found on Barbados.
                                   P. gloveralleni, p. 973
               7'. Color paler; upper carnassial shorter than broad; found on Guadeloupe. *P. minor*, p. 973
         5'. Geographic range coastal islands (at least in part).
            8. Size small; hind foot 90 or less; found on Cozumel Island off coast of Yucatán.
                                   P. pygmaeus, p. 973
            8'. Size larger; hind foot more than 90; found on southern Florida Keys and islands and coasts of Georgia and South Carolina. . . . .*P. lotor*, p. 967

Subgenus **Procyon** Storr

1780. *Procyon* Storr, Prodromus methodi mammalium . . . , p. 35. Type, *Ursus lotor* Linnaeus.
1795. *Campsiurus* Link, Beyträge zur Naturgeschichte, 1(2):87. Based on three species, including *Ursus lotor* Linnaeus.
1795. *Lotor* É. Geoffroy St.-Hilaire and G. Cuvier, Mag. Encyclopédique, 2:187. Based on the Raton [= *Ursus lotor* Linnaeus] of North America.

Pelage long, hair of neck not directed forward; claws sharp and curved; baculum strongly bowed and recurved at distal end; palate extending backward beyond plane of last molars a distance greater than a quarter of entire length of palate, usually nearly a third; posterior part of I3 usually with an independent cusp; upper molariform teeth with sharp, coniform cusps; M1 quadrituberculate and no accessory cusplet in cingulum.

Procyon lotor
Raccoon

External measurements: (males) 634–950; 200–405; 96–138; (females) 603–909; 192–340; 83–129. Weights range from 1.8 to 22.2 kg. Greatest length of skull: (males) 99.9–136.5; (females) 93.6–122.5; condylobasal length, 94.3–125.8, 89.4–115.9; zygomatic breadth, 60.2–89.1,

58.3–81.2; alveolar length of maxillary tooth-row, 36.4–47.4, 35.0–45.7. Upper parts iron gray to blackish, more or less suffused with buff, rusty, or "orange rufous," especially on nape; top of head grizzled; face with sharply delimited black mask usually reaching from cheeks across eyes and muzzle, with median extensions downward to rhinarium and upward on forehead, but more or less interrupted between eyes, in some subspecies; facial mask bordered by conspicuous white lines extending from near middle of forehead backward under ears or to sides of neck; sides of muzzle, lips and chin white; underparts thinly overlaid with long grayish or buffy overhairs, only partly concealing dense brownish underfur; throat crossed by distinct blackish or brownish area, separated from facial mask by narrow white lines extending posteriorly from muzzle; ears clothed with short grayish or buffy hairs, with black areas varying in size and distinctness at posterior base; forearms and thighs similar to underparts, but hind legs blackish near heels; forefeet whitish; hind feet usually whitish, but dusky of ankles sometimes extending down on metatarsus; tail above with 5–7 conspicuous black rings and a black tip, alternating with broader grayish or buffy rings, the black rings less sharply defined and sometimes interrupted below (after Goldman, 1950:32).

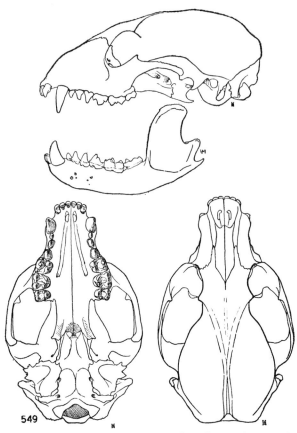

Fig. 549. *Procyon lotor psora*, S end Marlette Lake, Ormsby Co., Nevada, No. 70003 M.V.Z., ♀, X ⅓.

Procyon lotor auspicatus Nelson

1930. *Procyon lotor auspicatus* Nelson, Smiths. Miscl. Coll., 82(8):9, July 10, type from Marathon, Key Vaca, Monroe Co., Florida. Known only from type locality, and closely adjacent keys (Layne, 1974:396).

Procyon lotor crassidens Hollister

1914. *Procyon lotor crassidens* Hollister, Proc. Biol. Soc. Washington, 27:142, July 10, type from Talamanca, Limón, Costa Rica.

MARGINAL RECORDS.—Honduras: La Flor Archaga. Nicaragua: Jalapa; San Rafael del Norte. Costa Rica: type locality. Panamá: Almirante (Handley, 1966:789). El Salvador: Puerto del Triunfo; Lake Guija. Burt and Stirton (1961:43) imply that all specimens from El Salvador are subspecies *P. l. dickeyi*.

Procyon lotor dickeyi Nelson and Goldman

1931. *Procyon lotor dickeyi* Nelson and Goldman, Proc. Biol. Soc. Washington, 44:18, February 21, type from Barra de Santiago, Ahuachapán, El Salvador.

MARGINAL RECORDS.—*Guatemala*. El Salvador: type locality.

Procyon lotor elucus Bangs

1898. *Procyon lotor elucus* Bangs, Proc. Boston Soc. Nat. Hist., 28:219, March, type from Oak Lodge, East Peninsula opposite Micco, Brevard Co., Florida.

MARGINAL RECORDS.—Georgia: Okefinokee Swamp, thence along E coast of Florida to: Black Point; 15 mi. W Royal Palm State Park, 10 mi. E Pine Crest; Royal Palm Hammock, thence northward along coast to Aucilla River. Introduced on Grand Bahama, Bahama Islands.

Procyon lotor excelsus Nelson and Goldman

1930. *Procyon lotor excelsus* Nelson and Goldman, Jour. Mamm., 11:458, November 11, type from Owyhee River (near mouth N. Fork), Malheur Co., Oregon, 10 mi. W Fairylawn, Owyhee Co., Idaho.

MARGINAL RECORDS.—Montana (Hoffmann, *et al.*, 1969:595, 596, as *P. lotor* only): 4 mi. E Eureka; Abbott Creek and Middle Fork Flathead River at Martin City; Sunset Hill on Blackfoot River; Rock Creek; Beaverhead River at Dell. Idaho: Preuss Mts.; jct. Ross Fork Creek and Portneuf River, Bannock Co.; Goose Creek, 25 mi. SW Oakley. Nevada: Montello; near

Golconda; 14½ mi. E Hausen. California: Parker Creek. Oregon: Tupper. Washington: Wallula; Washtucna; Little Spokane River, 15 mi. W Idaho state line.

Procyon lotor fuscipes Mearns

1914. *Procyon lotor fuscipes* Mearns, Proc. Biol. Soc. Washington, 27:63, March 20, type from Las Moras Creek, 1011 ft., Fort Clark, Kinney Co., Texas.
1837. *Procyon nivea* Gray, Charlesworth's Mag. Nat. Hist., 1:580. Type locality, "Texas," may be referable to this subspecies.

MARGINAL RECORDS.—Texas: Canyon; *ca.* 4 mi. S, 17 mi. W Hollis (Martin and Preston, 1970:55, as *P. lotor* only, sight record); Cooke County (Davis, 1966:81, as *P. lotor* only); 10 mi. NW Texarkana; Panola County (Davis, 1966:81, as *P. lotor* only); Kountze; Jefferson County (Davis, 1966:81, as *P. lotor* only), thence southward along coast to Tamaulipas: Alta Mira; Gómez Farías. Coahuila: ½ mi. W Chorro, 3 mi. S, 11 mi. E Saltillo; Múzquiz. Texas: *1 mi. N Fort Davis;* near Limpia Creek, 5 mi. N Fort Davis, 4700 ft. Not found: Louisiana: Lake Ridge (Goldman, 1950:51).

Procyon lotor grinnelli Nelson and Goldman

1930. *Procyon lotor grinnelli* Nelson and Goldman, Jour. Washington Acad. Sci., 20:82, March 4, type from La Paz, Baja California.

MARGINAL RECORDS.—Baja California: San Ignacio; Mt. Miraflores.

Procyon lotor hernandezii Wagler

1831. *Pr[ocyon]. hernandezii* Wagler, Isis von Oken, 24:514, type from Tlalpam, Valley of Mexico.
1890. *Procyon lotor hernandezi,* J. A. Allen, Bull. Amer. Mus. Nat. Hist., 3:176, December 10.
1910. *Pr[ocyon]. hernandezi castaneus* de Beaux, Zoöl. Anzeiger, 35:624, April 26, type from "Mexico."

MARGINAL RECORDS.—Tamaulipas: Rancho Santa Rosa, 25 km. N, 13 km. W Ciudad Victoria (Alvarez, 1963:458). Querétaro: Jalpan. Veracruz (Hall and Dalquest, 1963:333): 4 km. W Tlapacoyan, 1700 ft.; Río Blanco, 20 km. WNW Piedras Negras, 300 ft.; 20 km. E Jesús Carranza, 300 ft. Chiapas: near Comitán. Oaxaca: Sola de Vega (Goodwin, 1969:230), thence northward along Pacific Coast to Nayarit: San Blas.

Procyon lotor hirtus Nelson and Goldman

1930. *Procyon lotor hirtus* Nelson and Goldman, Jour. Mamm., 11:455, November 11, type from Elk River, Sherburne Co., Minnesota.
1911. *Pr[ocyon]. hudsonicus* Brass, Aus dem Reiche der Pelze, p. 564, April. No type designated; may be referable to this subspecies.

MARGINAL RECORDS.—Alberta: Wood Buffalo Park. Saskatchewan: Meadow Lake (Beck, 1958:15); Christopher Lake (Houston and Houston, 1973:104);

Nipawin (Francis, 1959:30). Manitoba: N shore, Granville Lake (Lynch, 1971:622); Oxford House (Sutton, 1964:311, as raccoon only). Ontario (Simkin, 1966:144, as species only): jct. Severn and Rocksand rivers; Cordingly Lake (trapper's report). Minnesota: Roseau County; NE of Duluth, thence along S shore Lake Superior and W shore Lake Michigan to Illinois: Jackson Park, Chicago; Champaign County. Arkansas: Clark County. Oklahoma: 20 mi. from Frederick; 1 mi. N, 5 mi. W Hollis (Martin and Preston, 1970:55, as *P. lotor* only, sight record). Texas: Canadian; 5 mi. S, 10 mi. E Pringle (Cutter, 1959:31). New Mexico: Bear Canyon, Raton Range. Colorado (Armstrong, 1972:268): near Cañon City; Big Thompson Canyon. Wyoming (Long, 1965a:688): Sybelle Creek, 30 mi. N, 15 mi. E Laramie; *Douglas; 3 mi. E Bill; Little Missouri River, 2 mi. N, 16 mi. W Hulett;* New Haven. Montana (Hoffmann, *et al.,* 1969:594–596, as *P. lotor* only): Tongue River; 7 mi. S Belfry; Emigrant Creek, 3 mi. SW Emigrant; Jefferson River, 5 mi. E Cardwell; S. Fork Sun River, 1 mi. E Augusta; 10 mi. NW Fort Benton. Alberta: Fox (Hoffmann, *et al.,* 1969:595, as *P. lotor* only); Banff; Peace River country (precisely located on map, Soper, 1965:291).

Procyon lotor incautus Nelson

1930. *Procyon lotor incautus* Nelson, Smiths. Miscl. Coll., 82(8):10, July 10, type from Torch Key, Big Pine Key Group, Monroe Co., Florida.

MARGINAL RECORDS (Layne, 1974:396).—Florida: No Name Key; Key West; *and other keys intervening between the two just listed.*

Procyon lotor inesperatus Nelson

1930. *Procyon lotor inesperatus* Nelson, Smiths. Miscl. Coll., 82(8):8, July 10, type from Upper Matecumbe Key, Monroe Co., Florida.

MARGINAL RECORDS.—Florida: Virginia Key; Lower Matecumbe Key; *and other keys intervening between the two just listed* (Layne, 1974:396).

Procyon lotor litoreus Nelson and Goldman

1930. *Procyon lotor litoreus* Nelson and Goldman, Jour. Mamm., 11:457, November 11, type from Saint Simon Island, Glynn Co., Georgia.

MARGINAL RECORDS.—Georgia: Thunderbolt Creek, Chatham Co.; Ossabaw Island; type locality; mouth Altamaha River, McIntosh Co.

Procyon lotor lotor (Linnaeus)

1758. [*Ursus*] *lotor* Linnaeus, Syst. nat., ed. 10, 1:48. Type locality, Pennsylvania (fixed by Thomas, Proc. Zool. Soc. London, p. 140, March, 1911).
1815. *Procyon lotor,* Illiger, Abh. preuss. Akad. Wiss., Berlin, 1804–1811, pp. 70, 74.
1808. *L[otor]. vulgaris* Tiedemann, Zoologie . . . , 1:280 (original not seen).

1814. *Procyon annulatus* G. Fischer, Zoognosia, 3:177. Type locality, "Americae maritimis."
1837. *Procyon obscurus* Wiegmann, Arch. für Naturgesch., 3(1):370. Type locality, "Patria ignota," may be referable to this subspecies.
1848. *Procyon gularis* Hamilton-Smith, The naturalist's library (edit. Jardine), 15:222. Type locality, state of New York.
1850. [*Procyon brachyurus*] var. *fusca* Burmeister, Verzeich. Zool. Mus. Univ. Halle-Wittenberg Säugth. . . . , p. 13, may be referable to this subspecies.
1864. [*Procyon lotor*] var. *melanus* Gray, Proc. Zool. Soc. London, p. 704, may be referable to this subspecies.
1910. Pr[ocyon]. *lotor rufescens* de Beaux, Zoöl. Anzeiger, 35:625, April 26. Type locality, unknown. *P. l. rufescens* may be a synonym of *P. l. lotor*.

MARGINAL RECORDS.—Quebec: Madeline River, Gaspé. Nova Scotia: *no specific locality*. Maine: Mount Desert Island (Manville, 1960:416, as *P. lotor* only), thence southward to North Carolina: Coinjack; Hamstead (Funderburg, 1961:271); *Castle Hayne* (*ibid.*); Brunswick County (*ibid.*); *Old Dock* (*ibid.*); Moore County; Highlands. Kentucky: Mammoth Cave; Paducah. Illinois: Olive Branch; Belleville. Michigan: New Richmond. Ontario: Nipigon. Quebec: Meach Lake; Mt. Tremblant Provincial Park (Pirlot, 1962:132). *P. lotor*, subspecies unknown, introduced on Prince Edward Island (Cameron, 1959:50).

Procyon lotor marinus Nelson

1930. *Procyon lotor marinus* Nelson, Smiths. Miscl. Coll., 82(8):7, July 10, type from near Chokoloskee, Collier Co., Florida.

MARGINAL RECORDS.—Florida: Ritta; Flamingo; Coon Key, Ten Thousand Islands; type locality.

Procyon lotor maritimus Dozier

1948. *Procyon lotor maritimus* Dozier, Jour. Mamm., 29:286, August 31, type from Blackwater National Wildlife Refuge, Dorchester Co., Maryland.

MARGINAL RECORDS.—Delaware: Rehoboth Bay. Virginia: Saxis Island. Maryland: Crocheron; *type locality;* Vienna.

Procyon lotor megalodous Lowery

1943. *Procyon lotor megalodous* Lowery, Occas. Pap. Mus. Zool., Louisiana State Univ., 13:225, November 22, type from March Island, Iberia Parish, Louisiana.

MARGINAL RECORDS (Lowery, 1974:421, 422).—Louisiana: Toca; Chandeleur Islands opposite North Island; mouth Mississippi River; Grand Terre [Island]; Timbalier Island; Marsh Island; Johnsons Bayou; *Sabine National Wildlife Refuge;* Lacassine Refuge; 1½ mi. NW Esther; 3 mi. NE Schriever; *Tulane Univ. Riverside Campus, Belle Chasse.*

Procyon lotor mexicanus Baird

1858. *Procyon hernandezii* var. *mexicana* Baird, Mammals, *in* Repts. Expl. Surv. . . . , 8(1):215, July 14, type from Espía, Chihuahua.
1914. *Procyon lotor mexicanus,* Mearns, Proc. Biol. Soc. Washington, 27:65, March 20.
1855. *Procyon lotor,* variété mexicaine I. Geoffroy St.-Hilaire, Voyage autour du monde sur la frégate la Vénus . . . , Zoologie, p. 125. Type locality, Mazatlán, Sinaloa. (Geoffroy St.-Hilaire's account not seen by me; Goldman, N. Amer. Fauna, 60:53, November 7, 1950, states that Geoffroy St.-Hilaire did not formally propose a name.)

MARGINAL RECORDS.—Colorado: ½ mi. N Crestone, 8050 ft. (Armstrong, 1972:269, sight record only). New Mexico: 3 mi. SE Cimarron (Findley, *et al.,* 1975:299, as *P. lotor* only); 2 mi. S Fort Sumner (*ibid.*); Carlsbad. Texas: Pecos. Durango: Rancho Santuario; Ciudad (Baker and Greer, 1962:137); 28 mi. S, 17 mi. W Vicente Guerrero, 8350 ft. (*ibid.*). Nayarit: Acaponeta. Sinaloa: Isla Tachichilte (Armstrong, *et al.,* 1972:52). Sonora: Tecoripa; La Noria. Arizona: McCleary's Ranch, Santa Rita Mts.; Fort Lowell; Santa Catalina Mts. New Mexico: Gila National Forest; Tularosa Creek just above Apache Creek (Findley, *et al.,* 1975:298, as *P. lotor* only); 10 mi. N Grants (*ibid.*); Velarde.

Procyon lotor pacificus Merriam

1899. *Procyon psora pacifica* Merriam, N. Amer. Fauna, 16:107, October 28, type from Lake Keechelus, 3000 ft., Kittitas Co., Washington.
1923. *Procyon lotor pacifica,* Grinnell, Univ. California Publ. Zool., 21:316, January 27.
1911. *Procyon proteus* Brass, Aus dem Reiche der Pelze, p. 564, April, type from West coast from Puget Sound to Cascade Mts. Not J. A. Allen, 1904.

MARGINAL RECORDS.—British Columbia: Brackendale; Chilliwack Valley; Okanagan Falls; Wardner (Cowan and Guiguet, 1965:300). Washington: Toppenish. Oregon: Hardman; Big Summit Prairie, Ochoco National Forest; Riverside; 32 mi. S Grants Pass. California: Crescent City, thence northward along coast to point of beginning.

Procyon lotor pallidus Merriam

1900. *Procyon pallidus* Merriam, Proc. Biol. Soc. Washington, 13:151, June 13, type from New River, about 6 mi. W Imperial, Imperial Co., California.
1923. *Procyon lotor pallidus,* Grinnell, Univ. California Publ. Zool., 21:316, January 27.
1914. *Procyon lotor ochraceus* Mearns, Proc. Biol. Soc. Washington, 27:64, March 20, type from Monument 172, Río Sonoyta, 1181 ft., Sonora, near Quitobaquito, Arizona.

MARGINAL RECORDS.—Colorado: 7 mi. W Hayden (Armstrong, 1972:269); 3 mi. N, 3 mi. W Pea Green, 5000 ft. (*ibid.*); Navajo River, Archuleta Co. New Mexico: Whiskey Creek, *ca.* 9000 ft. (Halloran

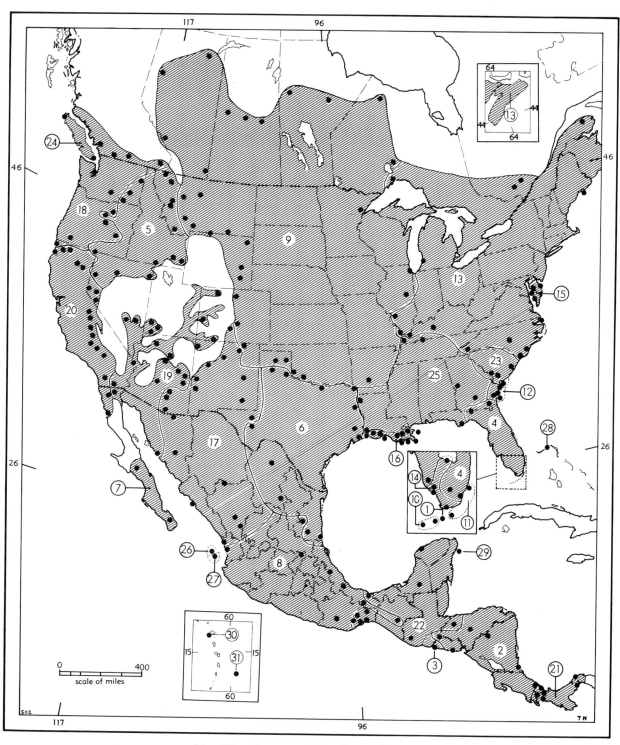

Map 502. *Raccoons of the subgenus Procyon.*

Guide to kinds

1. *P. lotor auspicatus*
2. *P. lotor crassidens*
3. *P. lotor dickeyi*
4. *P. lotor elucus*
5. *P. lotor excelsus*
6. *P. lotor fuscipes*
7. *P. lotor grinnelli*
8. *P. lotor hernandezii*
9. *P. lotor hirtus*
10. *P. lotor incautus*
11. *P. lotor inesperatus*
12. *P. lotor litoreus*
13. *P. lotor lotor*
14. *P. lotor marinus*
15. *P. lotor maritimus*
16. *P. lotor megalodous*
17. *P. lotor mexicanus*
18. *P. lotor pacificus*
19. *P. lotor pallidus*
20. *P. lotor psora*
21. *P. lotor pumilus*
22. *P. lotor shufeldti*
23. *P. lotor solutus*
24. *P. lotor vancouverensis*
25. *P. lotor varius*
26. *P. insularis insularis*
27. *P. insularis vicinus*
28. *P. maynardi*
29. *P. pygmaeus*
30. *P. minor*
31. *P. gloveralleni*

and Taber, 1965:140). Arizona: Wupatki National Monument; Phoenix; Rio de Flag (Hoffmeister and Carothers, 1969:188, as *P. lotor* only); Lakeside; 3 mi. NE Springerville; 25 mi. SE Fort Apache. Sonora: Bahía San Pedro. Baja California: Cocopah Mts. California: New River, Colorado Desert; Colorado River, 5 mi. below Needles. Nevada: Pahranagat Valley, 2 mi. N Crystal Spring; 3 mi. S Tennille Ranch, 10 mi. S Caliente. Utah: Pine Valley; Zion National Park; St. George. Arizona (Hoffmeister, 1971:67, 68): Havasu Canyon; *Tapeats Creek*.

Procyon lotor psora Gray

1842. *Procyon psora* Gray, Ann. Mag. Nat. Hist., ser. 1, 10:261, December, type from Sacramento, Sacramento Co., California.
1923. *Procyon lotor psora*, Grinnell, Univ. California Publ. Zool., 21:316, January 27.
1914. *Procyon lotor californicus* Mearns, Proc. Biol. Soc. Washington, 27:66, March 20, type from ocean beach near Mexican boundary Monument 258, San Diego Co., California.

MARGINAL RECORDS.—California: 5 mi. E Gazelle; Cassel; Spalding, Eagle Lake. Nevada: S end Marlette Lake; Walker River, 14 mi. N Yerington; E side Lake Tahoe. California: vic. Yosemite Valley; Soquel Mill, 40 mi. E Raymond; Three Rivers; Kern River, 25 mi. above Kernville; Isabella; Mohave River; Oro Grande; Julian. Baja California: Laguna Hanson; San Quintín, thence northward along coast to California: Trinidad; Klamath River, Siskiyou Co.

Procyon lotor pumilus Miller

1911. *Procyon pumilus* Miller, Proc. Biol. Soc. Washington, 24:3, January 28, type from Ancón, Canal Zone, Panamá.
1920. *Procyon lotor pumilus*, Goldman, Smiths. Miscl. Coll., 69(5):151, April 26.

MARGINAL RECORDS.—Panamá: Porto Bello; Balboa; Pedregal; Cerro Punta, 5300 ft. (Handley, 1966:789); Boquerón.

Procyon lotor shufeldti Nelson and Goldman

1931. *Procyon lotor shufeldti* Nelson and Goldman, Proc. Biol. Soc. Washington, 44:17, February 21, type from La Tuxpeña, Champotón, Campeche.

MARGINAL RECORDS.—Yucatán: 13 km. WSW Sisal (92563 KU). Honduras: Santa Bárbara. Guatemala: El Espino. Oaxaca (Goodwin, 1969:230): Santa María del Mar; Salina Cruz; *San Antonio*; Tequisistlán; Las Cuevas; Cerro Atravesado. Veracruz: Minatitlán. Campeche: type locality.

Procyon lotor solutus Nelson and Goldman

1931. *Procyon lotor solutus* Nelson and Goldman, Jour. Mamm., 12:308, August 24, type from Hilton Head Island, Beaufort Co., South Carolina.

MARGINAL RECORDS.—South Carolina: Santee; type locality; Yemassee; Savannah River Plant (Wood and Odum, 1964:545, as *P. lotor* only).

Procyon lotor vancouverensis Nelson and Goldman

1930. *Procyon lotor vancouverensis* Nelson and Goldman, Jour. Mamm., 11:458, November 11, type from Quatsino Sound, Vancouver Island, British Columbia.

MARGINAL RECORDS.—British Columbia (Cowan and Guiguet, 1965:300, unless otherwise noted): San Josef River Valley (Hall and Kelson, 1959:889), thence to and along coast, including *Saltspring, Pender*, and *Saturna* islands, to Cadboro Bay, thence northwestward along coast to point of beginning. Introduced on *Cox Island, Scott Island Group*, and *Graham Island* (Cowan and Guiguet, 1965:300), British Columbia.

Procyon lotor varius Nelson and Goldman

1930. *Procyon lotor varius* Nelson and Goldman, Jour. Mamm., 11:456, November 11, type from Castleberry, Conecuh Co., Alabama.
1910. *Pr[ocyon]. l[otor]. flavidus* de Beaux, Zool. Anzeiger, 35:626, April 26, type from "Southern United States," may be referable to this subspecies.

MARGINAL RECORDS.—Kentucky: Hickman. Tennessee: Clarksville. Georgia: Geneva; Nashville. Florida: Apalachicola. Louisiana (Lowery, 1974:422): 3 mi. WNW Paradis; *7 mi. SW Napoleonville; 4 mi. S Abbeville*; Choupique; *5 mi. W Sulphur; 1½ mi. N Cross Lake; 7¾ mi. NW Shreveport*.

Procyon insularis
Tres Marías Raccoon

External measurements (males): 840–904; 264–313; 128–135. Greatest length of skull, 119–121.8; condylobasal length, 114–115.2; zygomatic breadth, 82.5–86.4; alveolar length of maxillary tooth-row, 42.2–43.6. Pelage short and coarse, upper parts in general light cream buff, dorsal area thinly overlaid with black; underparts thinly overlaid with pale creamy buff, light brown underfur showing through; ears grayish, black patches at posterior base, usual in the group, obsolescent; skull large, angular, and massive, with remarkably heavy zygomata; squamosal arm broad anteriorly and extended vertically; crown of 2nd upper molar subquadrate, instead of subtriangular (after Goldman, 1950:72).

P. insularis is closely related to *P. lotor* of the mainland, but is regarded by Goldman (1950:20) as a separate species.

Procyon insularis insularis Merriam

1898. *Procyon lotor insularis* Merriam, Proc. Biol. Soc. Washington, 12:17, January 27, type from María Madre Island, Tres Marías Islands, México. Known only from María Madre Island.
1931. *Procyon insularis insularis*, Nelson and Goldman, Proc. Biol. Soc. Washington, 44:20, February 21.

Procyon insularis vicinus Nelson and Goldman

1931. *Procyon insularis vicinus* Nelson and Goldman, Proc. Biol. Soc. Washington, 44:20, February 21, type from María Magdalena, Tres Marías Islands, México. Known only from María Magdalena Island.

Procyon maynardi Bangs
Bahama Raccoon

1898. *Procyon maynardi* Bangs, Proc. Biol. Soc. Washington, 12:92, April 30, type from New Providence Island, Bahamas. Known only from type locality.

External measurements of a male topotype: 713, 240, 100 (dry). Cranial measurements of a male and a female topotype, respectively: greatest length, 105.9, 103.5; condylobasal length, 101.7, 101; zygomatic breadth, 79.3, 64.6; alveolar length of maxillary tooth-row, 38.7, 38.9. Upper parts grayish, becoming ochraceous buffy on nape and over shoulders, moderately overlaid with black, thinning out along sides; underparts thinly overlaid with grayish (after Goldman, 1950:75).

The Bahama raccoon is closely related to those of the Florida Keys in form and proportions of skull. In external size resembles *P. minor*, but differs in paler color and narrower palatal shelf.

Procyon pygmaeus Merriam
Cozumel Island Raccoon

1901. *Procyon pygmaeus* Merriam, Proc. Biol. Soc. Washington, 14:101, July 19, type from Cozumel Island, Yucatán. Known only from type locality.

External measurements of the type, a subadult male, and an adult female topotype, respectively, are: 667, 665; 230, 250; 90, 97. Cranial measurements of these same specimens: greatest length of skull, 100, 96.7; condylobasal length, 93.7, 91.9; zygomatic breadth, 58.8, 60.8; alveolar length of maxillary tooth-row, 35.3, 35.5. Upper parts light buffy gray; median dorsal area suffused with pale buff; underparts thinly overlaid with light buffy hairs but light brownish underfur showing through (after Goldman, 1950:77).

P. pygmaeus is the smallest species of the genus, and differs from its nearest relative, *P.*

lotor of the mainland, in small size, small teeth, and presence of a diastema between the 2nd and 3rd upper premolars.

Procyon minor Miller
Guadeloupe Raccoon

1911. *Procyon minor* Miller, Proc. Biol. Soc. Washington, 24:4, January 28, type from Pointe-à-Pitre, Guadeloupe, Lesser Antilles, West Indies. Known only from Guadeloupe Island.
1837. *Procyon brachyurus* Wiegmann, Arch. für Naturgesch., 3(1):369. Type locality, "Antillae?", may be referable to this species, Hershkovitz, Bull. Zool. Nomenclature, 22:338, January, 1966, notwithstanding.

Cranial measurements of the type, a young male, and a subadult topotype, respectively: greatest length of skull, 101.6, 104.5; condylobasal length, 94.5, 98.5; zygomatic breadth, 55.3, 62; alveolar length of maxillary tooth-row, 38.5, 37.3. Upper parts grayish, becoming "ochraceous buff" on nape and shoulders; dorsum heavily overlaid with black, with usual grizzled effect; underparts thinly overlaid with grayish but light brown underfur showing through (after Goldman, 1950:78).

Procyon gloveralleni Nelson and Goldman
Barbados Raccoon

1930. *Procyon gloveralleni* Nelson and Goldman, Jour. Mamm., 11:453, November 11, type from Barbados, Lesser Antilles, West Indies. Known only from type locality.

Cranial measurements of the type, a young male, and an adult female topotype, respectively, are: greatest length of skull, 94, 113.2; condylobasal length, 89.2,—; zygomatic breadth, 53.5, —; alveolar length of maxillary tooth-row, 37, 40. Upper parts near light ochraceous buff (most intense on nape and shoulders) rather heavily overlaid with black; underparts thinly overlaid with buffy grayish (after Goldman, 1950:79).

The Barbados raccoon resembles *P. minor* and *P. maynardi*, but in some features shows relationships to the raccoons of Central America. Formerly abundant on the south side of Barbados, *P. gloveralleni* is now rare, and may be extinct.

Subgenus Euprocyon Gray

1865. *Euprocyon* Gray, Proc. Zool. Soc. London, for 1864, p. 704, May. Type, *Ursus cancrivorus* Cuvier.

Pelage short, underfur absent; hair on nape directed forward; claws broad and blunt; baculum slightly bowed and not sharply recurved at distal end; palate extending behind molars a distance less than a quarter of total length of palate; upper molariform teeth (except 1st premolars) large and massive, with broad, round, blunt cusps, and with low connecting ridges; I3 with slightly developed cusp posteriorly; M1 5-tuberculate, with an additional accessory cusplet on posterior border of cingulum.

Procyon cancrivorus
Crab-eating Raccoon

External measurements of the type of *P. c. panamensis*, a male, are: 950, 350, 142. Greatest length of skull, 130; condylobasal length, 125.8; zygomatic breadth, 83.3; alveolar length of maxillary tooth-row, 48.3. Back ashy gray to ochraceous buff to yellowish ochraceous, becoming grayish white on throat, chin, and lips; tail with 7 or 8 black and grayish or yellowish rings and a black tip; proximal rings on tail more or less interrupted along ventral median line. For account of skull and teeth, see subgenus *Euprocyon*.

Procyon cancrivorus panamensis (Goldman)

1913. *Euprocyon cancrivorus panamensis* Goldman, Smiths. Miscl. Coll., 60(22):15, February 28, type from Gatún, Canal Zone, Panamá.
1920. *Procyon cancrivorus panamensis* Goldman, Smiths. Miscl. Coll., 69(5):152, April 24.

MARGINAL RECORDS.—Panamá (Handley, 1966: 788): $3\frac{7}{10}$ mi. SE Almirante; type locality; Portobelo; Mandinga; Armila; Tacarcuna Village; Boca de Río Paya, thence into South America, and northward to Panamá (*ibid.*): Cana; Panamá. Costa Rica: Cañas Gordas [= Agua Buena].

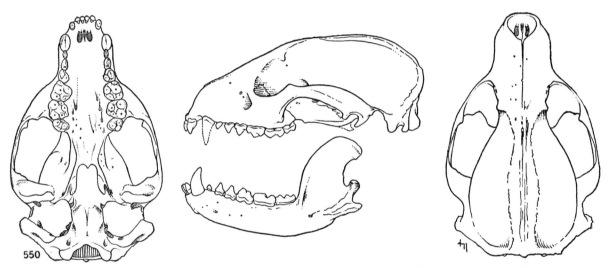

Fig. 550. *Procyon cancrivorus panamensis*, Gatún, Panamá, No. 171669 U.S.N.M., holotype, ♂, drawn from photographs, X approx. ½.

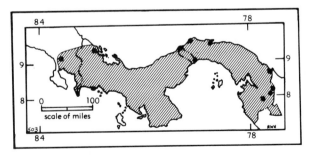

Map 503. *Procyon cancrivorus panamensis*.

Genus **Nasua** Storr—Coatis

1780. *Nasua* Storr, Prodromus methodi mammalium . . . , p. 35, tab. A. Type, *Viverra nasua* Linnaeus.

1799. *Coati* Lacépède, Tableau des divisions . . . des mammifères, des cétacés et des oiseaux, p. 7. Based on several species, including *Viverra nasua* Linnaeus.
1872. *Cuati* Liais, Climats, Géol., Faune et Géog. Botanique, Brésil, p. 427, an accidental renaming of *Coati* Lacépède.

Head long, rostral portion elongated; snout upturned, mobile; ears short, rounded; tail longer than body, nonprehensile; soles of feet naked; toes united for some distance from bases; claws of forefeet long, slightly curved; claws of hind feet strong, sharp, much curved.

Skull long and narrow; rostrum elongated and compressed laterally; interorbital constriction slight; sagittal crest well developed; palate flat or hollowed posteriorly, with distinct lateral grooves, and extending far beyond plane of last molars; palatine bones deeply notched on each side, with long processes extending laterally to the maxillary tuberosities; vomer usually attached to palatal bones; foramen ovale small; auditory bullae small, smoothly rounded, and rising at sharp angle from tube of external auditory meatus; ascending ramus of mandible relatively small and low.

Teeth of medium size; crowns comparatively high, with sharp, strong cusps; P4 longer than broad, nonsectorial; M1 subquadrate, length and breadth about equal, or longer than broad; I3 separated from I2 by diastema; lower incisors protruding forward; canines compressed laterally; premolars 2- or 3-rooted; P1–P3 each with single cusp; P4 with well-developed hypocone.

KEY TO NORTH AMERICAN SPECIES OF NASUA

1. Total length less than 800; length of tail less than 400; hind foot less than 90; found only on Cozumel Island, Quintana Roo. *N. nelsoni,* p. 976
1'. Total length more than 800; length of tail more than 400; hind foot more than 90; found on mainland. . . . . . *N. nasua,* p. 975

Nasua nasua
Coati

Use of *Nasua nasua* instead of *Nasua narica* follows Cabrera (Rev. Mus. Argentino de Cienc. Nat., 4:245, March 27, 1958) and Handley (Ectoparasites of Panama, Field Mus. Nat. Hist., Chicago, p. 789, November 22, 1966). *Viverra narica* Linnaeus, Syst. nat., 12th ed., 1:64, 1766, was described as *cauda concolore* (whereas *V. nasua* Linnaeus was described as *cauda albo annulata*) and may be applied incorrectly to animals of this species.

External measurements (males): 850–1340; 420–680; 95–122. Greatest length of skull, 107.2–136.2; condylobasal length, 100.8–128.9; zygomatic breadth, 50.8–74.6. Upper parts pale brown to reddish, often overlaid with yellow; neck and shoulders yellowish; eyes masked with pale umber to brown; muzzle, chin, and throat whitish; thin whitish streaks extending from muzzle between eyes and over eyes; ears white-tipped; postauricular patches whitish to yellowish; feet blackish; underparts yellowish to dark brown; tail colored about like dorsum, often annulated with yellow and brown.

Nasua nasua molaris Merriam

1902. *Nasua narica molaris* Merriam, Proc. Biol. Soc. Washington, 15:68, March 22, type from Manzanillo, Colima.
1904. *Nasua narica pallida* J. A. Allen, Bull. Amer. Mus. Nat. Hist., 20:53, February 29, type from Sierra Madre, near Guadalupe y Calvo, Chihuahua.
1942. *Nasua narica tamaulipensis* Goldman, Proc. Biol. Soc. Washington, 55:80, June 25, type from Cerro de la Silla, 2600 ft., near Monterrey, Nuevo León.

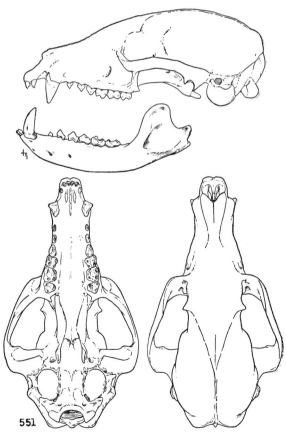

Fig. 551. *Nasua nasua narica,* 35 km. SE Jesús Carranza, 400 ft., Veracruz, No. 24531 K.U., ♂, X ½.

MARGINAL RECORDS.—Arizona: Petrified Forest National Monument (Cockrum, 1961:231); 40 mi. N Clifton. New Mexico: W fork Gila River, near Heart Bar Ranch, Gila National Forest; 25 mi. SE Animas. Texas: 20 mi. below Boquillas, Dead Horse Mts.; Rio Frio, about 40 mi. N Uvalde; Aransas National Wildlife Refuge (Halloran, 1961:23); *Aransas County* (Davis, 1966:82); Brownsville. Tamaulipas: 10 mi. NE Zamorina (Alvarez, 1963:458). Veracruz: Hda. Tamiahua, Cabo Rojo (Hall and Dalquest, 1963:335). Oaxaca: southwestern Oaxaca, thence northward along coast to Sonora: Chinobampo; 2 mi. E Ures. Arizona: Organ Pipe Cactus National Monument; W of Hillsdale, Santa Maria Mts.; T. 11 N, R. 12 E, edge Anderson Mesa.

INNER MARGINAL RECORDS.—Coahuila: foothills of Carmen Mts. Nuevo León: 12 km. W Allende, 3000 ft. Tamaulipas: Ciudad Victoria; 19 km. SW Mante. San Luis Potosí: 3 mi. NE Pujal; Xilitla. Morelos: 12 km. NW "Axochipan." Jalisco: Guadalajara. Chihuahua: Sierra Madre, near Guadalupe y Calvo.—See Kaufmann, Lanning, and Poole (1976:621–637) for a careful estimate of the natural versus apparent range of this subspecies in Arizona and southwestern New Mexico. The apparent range results from occasional wanderers and released or escaped captives. Map 504 herewith shows the apparent range, which is more extensive than the natural range.

Nasua nasua narica (Linnaeus)

1766. [*Viverra*] *narica* Linnaeus, Syst. nat., ed. 12, 1:64. Type locality, Achotal, Isthmus of Tehuantepec, Veracruz (by restriction, Hershkovitz, Fieldiana-Zool., Chicago Nat. Hist. Mus., 31:560, July 10, 1951).
1966. N[*asua*]. n[*asua*]. *narica*, Handley, Ectoparasites of Panama, Field Mus. Nat. Hist., p. 789, November 22.
1884. *Nasua narica*, True, Proc. U.S. Nat. Mus., 7(App., Circ. 29):608, November 29.
1860. *Nasua solitaria* var. *mexicana* Weinland, Zool. Garten, 1:191. Type locality, near Hda. El Mirador, about 4000 ft., 20 leagues from Veracruz, Veracruz.
1904. *Nasua narica bullata* J. A. Allen, Bull. Amer. Mus. Nat. Hist., 20:48, February 29, type from Pozo Azul, San José, Costa Rica. (Regarded as inseparable from *Nasua narica narica* by Hershkovitz, Fieldiana-Zool., Chicago Nat. Hist. Mus., 31:559, July 10, 1951.)
1904. *Nasua narica panamensis* J. A. Allen, Bull. Amer. Mus. Nat. Hist., 20:51, February 29, type from Boquerón, Chiriquí, Panamá. (Regarded as inseparable from *Nasua narica narica* by Hershkovitz, Fieldiana-Zool., Chicago Nat. Hist. Mus., 31:559, July 10, 1951.)
1932. *Nasua narica richmondi* Goldman, Jour. Washington Acad. Sci., 22:312, June 4, type from Río Escondido, 50 mi. above Bluefields, Zelaya, Nicaragua. (Regarded as inseparable from *Nasua narica narica* by Hershkovitz, Fieldiana-Zool., Chicago Nat. Hist. Mus., 31:559, July 10, 1951.)
1942. *Nasua narica isthmica* Goldman, Proc. Biol. Soc. Washington, 55:81, June 15, type from Santa Efigenia, about 8 mi. NW Tapanatepec, Oaxaca. (Regarded as inseparable from *Nasua narica narica* by Hershkovitz, Fieldiana-Zool., Chicago Nat. Hist. Mus., 31:559, July 10, 1951.)

MARGINAL RECORDS.—Veracruz (Hall and Dalquest, 1963:335): Jalapa; Carrizal; 14 km. SW Coatzacoalcos, 100 ft. Chiapas: 6 mi. SE Palenque. Belize: Kate's Lagoon. Honduras: Santa Bárbara, thence along Caribbean Coast to South America, and northward along Pacific Coast to El Salvador: Lake Olomega (Burt and Stirton, 1961:44). Oaxaca: Salinera Garrapatera (Goodwin, 1969:231). Veracruz: Orizaba (Hall and Dalquest, 1963:335).

Nasua nasua yucatanica J. A. Allen

1904. *Nasua narica yucatanica* J. A. Allen, Bull. Amer. Mus. Nat. Hist., 20:52, February 29, type from Chichén-Itzá, Yucatán.

MARGINAL RECORDS.—Quintana Roo: Pueblo Nuevo X-Can (92676 KU). Belize: Orange Walk. Guatemala: *Petén*. Campeche: 103 km. SE Escárcega (93821 KU); 5 km. S Champotón (92564 KU). Yucatán: Chablé.

Nasua nelsoni Merriam
Cozumel Island Coati

1901. *Nasua nelsoni* Merriam, Proc. Biol. Soc. Washington, 14:100, July 19, type from Cozumel Island, Quintana Roo. Known only from Cozumel Island.
1901. *Nasua thersites* Thomas, Ann. Mag. Nat. Hist., ser. 7, 8:271, October, type from Cozumel Island, Quintana Roo.

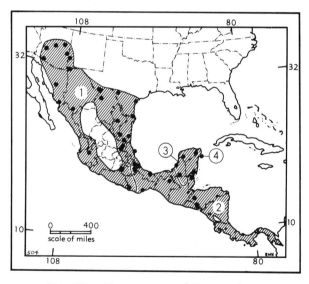

Map 504. *Nasua nasua* and *Nasua nelsoni*.

1. *N. nasua molaris*   3. *N. nasua yucatanica*
2. *N. nasua narica*   4. *N. nelsoni*

Average external measurements of three adult males and four adult females, respectively, are: 785, 744; 348, 328; 84, 79. Basal length of skull (type), 95; occipitonasal length, 95; zygomatic breadth, 61. Upper parts, belly, legs, and tail uniform dark seal brown; head and shoulders grizzled with golden fulvous; sides of neck and outer sides of arms grizzled with buffy whitish.

N. nelsoni differs from *N. nasua* of the mainland in being smaller, and in having relatively smaller teeth.

Genus **Potos** É. Geoffroy St.-Hilaire and G. Cuvier—Kinkajou

1795. *Potos* É. Geoffroy St.-Hilaire and G. Cuvier, Mag. Encyclopédique, n.s., 2:187. Type, *Viverra caudivolvula* Schreber.
1799. *Kinkajou* Lacépède, Tableau . . . des mammifères, des cétacés et des oiseaux, p. 7. Type, *Viverra caudivolvula* Schreber.
1806. *Caudivolvulus* Dumeril, Zoologie analytique . . . , p. 14. Type, "Le Kinkajou."
1811. *Cercoleptes* Illiger, Prodromus systematis mammalium et avium . . . , p. 127. Type, *Viverra caudivovula* Schreber.
1813. *Kinkashu* Fischer, Zoognosia, ed. 3, 1:14, a renaming of *Kinkajou* Lacépède.
1815. *Aesurus* Rafinesque, Analyse de la Nature, p. 59, a renaming of *Kinkajou* Lacépède.

Form elongated; head rounded, muzzle pointed; tail long, prehensile; ears short; posterior half of soles densely furred; claws short, sharply pointed; digits partially webbed; limbs short.

Skull short, rounded, highly arched; rostrum short and broad; braincase large, sharply constricted in postorbital region; postorbital processes well developed from frontals, slightly from zygomata; sagittal crest usually developed on forward half of braincase; palate flat, short, and extending only short distance posterior to plane of maxillary tuberosities; auditory bullae much flattend; mandible heavy, ramus high, straight, and little produced backward. Baculum unique in Procyonidae in terminating in 4 rounded processes, 2 directed laterodorsally and 2 directed anteriorly.

Teeth large, low-crowned, weakly tuberculate; incisor-canine diastema short; incisors heavy, broad; I3 large and deeply emarginate over posterior fold of cingulum; lower incisors with chisel-shaped crowns; canines large, flattened, emarginate lingually, deeply grooved labially, and nearly straight; premolars large; P2 and P3 simple, with bladelike crested crowns; P4 broad, with long shelf lateral to single internal, low, rounded cusp; M1 twice size of M2, subquadrate, and flat with raised outer rim; M2, m1, and m2 slightly sculptured and almost without tubercles. Dentition, i. $\frac{3}{3}$, c. $\frac{1}{1}$, p. $\frac{3}{3}$, m. $\frac{2}{2}$. (After Hollister, 1915:149, 150.)

Three names are omitted from the synonymies below because they may not relate to subspecies occurring in North America. They are *Lemur*

Simia-Sciurus [*sic*] Schreber 1774, *Cercoleptes brachyotus* Martin 1836, and *Viverra caudivolvula* Cuvier 1839. These names are discussed by Thomas (1902:267) and Cabrera (1958:249–252), and two of them by Kortlucke (1973:3–5). No one has learned the geographic origins of the specimens on which the names were based. Mostly on that account, they have been neither precisely applied nor appropriately disposed of otherwise.

Potos flavus
Kinkajou

North American taxa revised by Kortlucke, Univ. Kansas Mus. Nat. Hist., Occas. Pap., 17:1–36, October 26, 1973.

External measurements: 845–1300; 392–570; 75–140; 30–55. Condylobasal length, 76.8–95.1; zygomatic breadth, 52.9–69.3. Upper parts tawny olive, yellowish tawny, clay color, or wood brown; black mid-dorsal stripe present in some animals; underparts deep fulvous yellow, buff or brownish yellow; muzzle dark brown to blackish; tail, in general, colored above like upper parts, below like underparts, becoming dark brown distally in some subspecies.

Potos flavus chiriquensis J. A. Allen

1904. *Potos flavus chiriquensis* J. A. Allen, Bull. Amer. Mus. Nat. Hist., 20:72, February 29, type from Boquerón, Chiriquí, Panamá.
1931. *Potos flavus campechensis* Nelson and Goldman, Jour. Washington Acad. Sci., 21:482, November 19, type from La Tuxpeña, Champotón, Campeche.
1938. *Potos flavus arborensis* Goodwin, Amer. Mus. Novit., 987:1, May 13, type from El Sauce Peralta, 1000 ft., less than halfway from San José to Limón, Cartago, Costa Rica.
1944. *Potos flavus dugesii* Villa, Anal. Inst. Biol. Univ. Nac. Autó. México, 15:323, prior to October 25, type from SE La Esperanza coffee plantation, 45 km. NW Huixtla, 760 m., Chiapas.
1957. *Potos flavus boothi* Goodwin, Amer. Mus. Novit., 1830:1, May 3, type from Pueblo Nuevo, Solistahuacán, approx. 60 mi. N Tuxtla Gutiérrez, Chiapas. Known only from type locality.

MARGINAL RECORDS (Kortlucke, 1973:33, 34, unless otherwise noted).—Yucatán: 8 km. N, 10 km. W

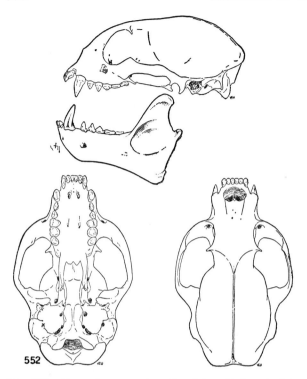

Fig. 552. *Potos flavus prehensilis*, 3 km. E San Andrés, Tuxtla, 1000 ft., Veracruz, No. 24532 K.U., ♂, X ½.

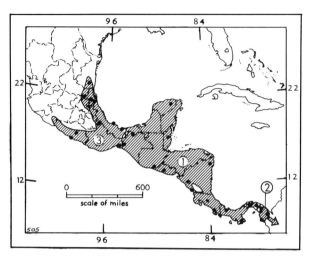

Map 505. *Potos flavus.*

1. *P. f. chiriquensis*   2. *P. f. megalotus*
3. *P. f. prehensilis*

Tizimín; to and down coast to Belize: Silkgrass, thence down coast to Nicaragua: Kukallaya [= Cuculaya] River, thence down coast to Panamá: Almirante; Madinga; jct. Río Bayano and Río Cañito; Guanico Arriba, thence up Pacific Coast to Costa Rica: Palmar. Nicaragua: 11 km. S, 3 km. E Rivas, thence up coast to El Salvador: La Libertad (Burt and Stirton, 1961:44). Guatemala: Astillero. Chiapas: Gineta Mts. Tabasco: 5 mi. SW Teapa, thence to and along coast to point of beginning.

Potos flavus megalotus (Martin)

1836. *Cercoleptes megalotus* Martin, Proc. Zool. Soc. London, p. 82, September, type locality subsequently designated as Santa Marta, Colombia, by Kortlucke, Univ. Kansas Mus. Nat. Hist., Occas. Pap., 17:29, October 26, 1973.
1902. *Potos flavus megalotus*, Thomas, Ann. Mag. Nat. Hist., ser. 7, 9:266, April.
1904. *Potos flavus caucensis* J. A. Allen, Bull. Amer. Mus. Nat. Hist., 20:75, February 29, type from Castilla Mts., upper Cauca region, 6000 ft., Colombia.
1913. *Potos flavus isthmicus* Goldman, Smiths. Miscl. Coll., 60(22):14, February 28, type from near head Río Limón, 5200 ft., Mt. Pirri, Darién, Panamá.
1913. *Potos flavus tolimensis* J. A. Allen, Bull. Amer. Mus. Nat. Hist., 32:481, September 25, type from Giradot, *ca.* 1500 ft., Magdalena Valley, Tolima, Colombia.
1914. *Potos flavus mansuetus* Thomas, Ann. Mag. Nat. Hist., ser. 8, 13:362, March, type from S. Domingo, W of Quito, 1760 ft., Eucador.

MARGINAL RECORDS (Kortlucke, 1973:34).—Panamá: Río Chucunaque; Cerro Tacarcuna, into South America and up southern coast of Panamá: Jaque.

Potos flavus prehensilis (Kerr)

1792. *Viverra prehensilis* Kerr, The animal kingdom, Class I, Mammalia, p. 169.
1959. *Potos flavus prehensilis*, Hershkovitz, Jour. Mamm., 40:341, August 20, type locality restricted to Atoyac, Veracruz.
1902. *Potos flavus aztecus* Thomas, Ann. Mag. Nat. Hist., ser. 7, 9:268, April, type from Atoyac, Veracruz.
1915. *Potos flavus guerrerensis* Goldman, Proc. Biol. Soc. Washington, 28:133, June 29, type from Ometepec, Guerrero.

MARGINAL RECORDS (Kortlucke, 1973:32, 33).—Tamaulipas: near Acuna. Veracruz: 75 mi. S Rivera; 12½ mi. N Tihuatlan; 13 mi. N San Andrés; Coatzocoalcos region. Oaxaca: Tapanatepec, thence up coast to Guerrero: Papayo; 1 mi. SW Omilteme. Veracruz: 3 km. N Presidio. San Luis Potosí: 9 km. NNE Xilitla.

Genus Bassaricyon J. A. Allen—Olingos

1876. *Bassaricyon* J. A. Allen, Proc. Acad. Nat. Sci. Philadelphia, 28:20, April 18. Type, *Bassaricyon gabbii* J. A. Allen.

Form elongated, head flattened, muzzle pointed; ears small, rounded; tail longer than head and body; posterior half of soles of feet furred; toes semipalmate; claws short, curved, sharply pointed; baculum small, slightly bowed, indistinctly bilobed anteriorly.

Skull short and broad; rostrum depressed, short; postorbital processes well developed from frontals and zygomata; orbits large; sagittal crest absent; palate flat, extending backward, having shallow lateral grooves; vomer not attached to palatal bones; posterior half of auditory bullae greatly inflated, anterior half low, flattened; tube of external auditory meatus short; mandible light, ascending ramus long and high.

Cheek-teeth low-crowned, only slightly cusped; canines large, not much flattened; upper canines nearly straight, not bowed outward; P1 and P2 compressed cones; P3 with posterointernal shelf; P4 with 3 cusps, hypocone lacking; M1 larger than M2; upper molars with 3 low, rounded cusps, and with well-developed outer cingulum. Dentition, i. $\frac{3}{3}$, c. $\frac{1}{1}$, p. $\frac{4}{4}$, m. $\frac{2}{2}$.

All named kinds of Central American *Bassaricyon* probably will prove to be a single species (*gabbii*) when sufficient material from that area is available.

KEY TO NORTH AMERICAN SPECIES OF BASSARICYON

1. Upper parts brownish; found in Nicaragua, Costa Rica, and Panamá, but not at Estrella de Cartago, Costa Rica, or at Cerro Pando, Panamá. . . . . *B. gabbii,* p. 979
1'. Upper parts grayish; occurring at Estrella de Cartago, Costa Rica, and at Cerro Pando, Panamá.
  2. Tail edged with hairs of a silvery sheen; ears gray, edged with white; found only at Estrella de Cartago, Costa Rica. . . . . *B. lasius,* p. 980
  2'. Tail not edged with hairs of a silvery sheen; ears brown, edged with white; found only at Cerro Pando, Panamá.
  B. pauli, p. 980

Bassaricyon gabbii
Bushy-tailed Olingo

External measurements of a male and a female, respectively, are: 950, 850; 480, 432; 80, 90 [*sic*]. Condylobasal length, 80.5, 78.7; zygomatic breadth, 53, 53; length of maxillary tooth-row, 28, 28.4. Upper parts pinkish buff to golden finely mixed with black; top of head and face grayish; ears edged with buffy white; underparts light yel-

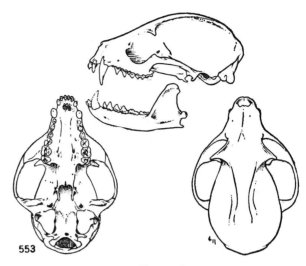

Fig. 553. *Bassaricyon gabbii medius,* Cana, Panamá, No. 179157 U.S.N.M., holotype, ♂, drawn from photographs, approx. X ½.

lowish buff. Tail above like dorsum, grayer below.

Bassaricyon gabbii gabbii J. A. Allen

1876. *Bassaricyon gabbii* J. A. Allen, Proc. Acad. Nat. Sci. Philadelphia, 28:21, April 18, type from Talamanca, Limón, Costa Rica.

MARGINAL RECORDS.—Costa Rica: *type locality.* Panamá (Handley, 1966:789, as *B. gabbii* only): Almirante; Gatún; Mandinga; Pacora; Corozal; La Chorrera; Cerro Punta, 5300 ft.

Bassaricyon gabbii medius Thomas

1909. *Bassaricyon medius* Thomas, Ann. Mag. Nat. Hist., ser. 8, 4:232, 233, September, type from Jiménez, 2400 ft., mts. inland of Chocó, Colombia.
1921. *Bassaricyon gabbii medius,* Lönnberg, Arkiv Zool., 14(4):29, June 7.
1912. *Bassariscyon* [*sic*] *gabbii orinomus* Goldman, Smiths. Miscl. Coll., 60(2):16, September 20, type from Cana (Santa Cruz de Cana), 1800 ft., upper Río Tuyra [= Tuira], Darién, Panamá. Arranged as a synonym of *B. medius* by Handley (Ectoparasites of Panama, Field Mus. Nat. Hist., p. 789, November 22, 1966).

MARGINAL RECORDS.—Panamá (Handley, 1966: 789, as *B. gabbii* only): Armila; Tacarcuna Casita; *Boca de Río Paya;* Cana; *Cerro Pirre, 5000 ft.* Also occurs in South America.

Bassaricyon gabbii richardsoni J. A. Allen

1908. *Bassaricyon richardsoni* J. A. Allen, Bull. Amer. Mus. Nat. Hist., 24:662, October 13, type from Río Grande, below 1000 ft., Atlantic slope, Nicaragua.

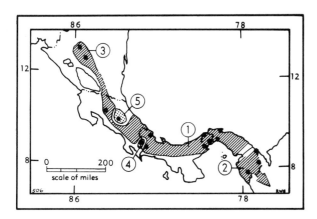

Map 506. Three species of *Bassaricyon*.

Guide to kinds	
1. *B. gabbii gabbii*	3. *B. gabbii richardsoni*
2. *B. gabbii medius*	4. *B. pauli*
	5. *B. lasius*

1946. *Bassaricyon gabbii richardsoni*, Goodwin, Bull. Amer. Mus. Nat. Hist., 87:426, December 31.

MARGINAL RECORDS.—Nicaragua: Hda. Tepeyac (104639 KU); type locality. Costa Rica: Lajas Villa Quesada. Nicaragua: *Matagalpa*.

Bassaricyon pauli Enders
Chiriquí Olingo

1936. *Bassaricyon pauli* Enders, Proc. Acad. Nat. Sci. Philadelphia, 88:365, September 29, type from Cerro Pando, 4800 ft., between Río Chiriquí Viejo and Río Colorado (about 10 mi. from El Volcán), Chiriquí, Panamá. Known only from type locality.

External measurements of the type, an adult male, are: 862, 434, 84, 38 (crown). Weight, 1.58 kg. Condylopremaxillary length, 80; zygomatic breadth, 57; breadth of braincase, 36.7. Upper parts grizzled, head gray and black, body gray-fulvous; mid-dorsal stripe of black-tipped hairs; underparts light gray tinged with yellow; black

tail-rings faint, ventral surface of tail darker than underparts of body.

Skull massive, wide rostrally, and deep; braincase large and uniformly expanded; zygomatic arch heavy; tympanic bullae large, heavy; glenoid fossa wide and deep, with well-developed processes.

B. pauli is probably only subspecifically distinct from the other nominal species from Central America of this genus, but sufficient material to demonstrate this is not yet available.

Bassaricyon lasius Harris
Harris' Olingo

1932. *Bassaricyon lasius* Harris, Occas. Pap. Mus. Zool., Univ. Michigan, 248:3, August 4, type from Estrella de Cartago, about 4500 ft., near source Río Estrella, 6 or 8 mi. S Cartago, Cartago, Costa Rica. Known only from type locality.

External measurements of the type, an adult male, are: 820, 437, 77, 35 (crown). Condylopremaxillary [= condylobasal ?] length of skull, 78.3; zygomatic breadth, 55; breadth of braincase, 36.8. Upper parts grizzled blackish gray, median line of back slightly more blackish; ears gray, edged with white; underparts light yellowish buff; tail above like back, light gray below. (After Harris, 1932:3.)

B. lasius closely resembles *B. g. richardsoni* and may be only subspecifically distinct from the other Central American olingos.

FAMILY MUSTELIDAE—Mustelids

Small (*Mustela*) to medium (*Gulo*) size; short-legged; feet short; 5 toes on each foot; plantigrade, semiplantigrade, or digitigrade; anal scent glands paired and large; rostrum of skull short; alisphenoid canal absent; carnassial teeth (P4 and m1) sectorial in varying degree; molars ½ or ¼.

KEY TO NORTH AMERICAN GENERA OF MUSTELIDAE

1. Premolars ⁴⁄₄. . . . . *Martes*, p. 981
1'. Premolars fewer than ⁴⁄₄.
    2. Fleshy part of tail so thickened at base that tail merges gradually with body; toes of four feet webbed at least as far as terminal phalanx of each toe.
        3. Premolars ⁴⁄₄. . . . . *Lutra*, p. 1028
        3'. Premolars ³⁄₃. . . . . *Enhydra*, p. 1033
    2'. Fleshy part of tail not so thickened at base as to cause tail to merge gradually with body; toes of four feet not webbed so far distally as terminal phalanx of each toe.
        4. Premolars ⁴⁄₄. . . . . *Gulo*, p. 1006

4'. Premolars $\frac{3}{3}$ or $\frac{3}{4}$.
    5. Premolars $\frac{3}{4}$. . . . .*Conepatus*, p. 1024
    5'. Premolars $\frac{3}{3}$.
        6. Talonid of m1 trenchant.
            7. Venter white or brown. . . . .*Mustela*, p. 987
            7'. Venter black. . . . .*Galictis*, p. 1006
        6'. Talonid of m1 basined or semibasined.
            8. Sides of head not solid black (gray or brown in *Eira*); palate projecting behind molars for a distance of more than length of M1; basilar length more than 80.
                9. Tail less than $\frac{1}{3}$ of total length; longitudinal white stripe on top of neck. . . . .*Taxidea*, p. 1009
                9'. Tail more than $\frac{1}{3}$ of total length; no longitudinal white stripe on top of neck. . . . .*Eira*, p. 1004
            8'. Side of head all black; palate not projecting behind molars for a distance of more than length of M1; basilar length less than 80.
                10. Upper parts having 4 or more lines of broken white stripes or spots. *Spilogale*, p. 1013
                10'. Upper parts having unbroken white stripes or upper parts all black except white top of head. . . . .*Mephitis*, p. 1018

Genus **Martes** Pinel—Martens and Fisher

1792. *Martes* Pinel, Actes Soc. Hist. Nat. Paris, 1:55. Type, *Martes domestica* Pinel [= *Mustela foina* Erxleben].
1829. *Zibellina* Kaup, Skizzirte Entwickelungs-Geschichte und natürliches System der europäischen Thierwelt, 1:31, 34. Type, *Mustela zibellina* Linnaeus.
1865. *Pekania* Gray, Proc. Zool. Soc. London, p. 107, June. Type, *Mustela pennanti* Erxleben.
1865. *Charronia* Gray, Proc. Zool. Soc. London, p. 108, June. Type, *Mustela flavigula* Boddaert.
1928. *Lamprogale* Ognev, Mém. Soc. Amis Sci. Nat. Moscow, No. 2, Zool., p. 26, 30. Proposed to replace *Charronia* Gray thought by Ognev to be preoccupied by *Charonia* Gistel, 1848, a genus of Mollusca.

Total length, 570–1033; weight, $1\frac{1}{2}$–12 and perhaps 18 pounds; basilar length, 60–115. Facial angle slight; tympanic bullae moderately inflated but not in close contact with paroccipital processes; palate behind last upper molars; inner moiety of M1 larger than outer; P4 with simple deuterocone; in m1, trigonid longer than talonid, metaconid small and appressed to protoconid, hypoconid large, talonid semibasined. Dentition, i. $\frac{3}{3}$, c. $\frac{1}{1}$, p. $\frac{4}{4}$, m. $\frac{1}{2}$.

KEY TO NORTH AMERICAN SPECIES OF MARTES

1. Tail more than 290; outside length of P4 more than 9.5; length of m1 more than 11. *M. pennanti*, p. 985
1'. Tail less than 290; outside length of P4 less than 9.5; length of ml less than 11. *M. americana*, p. 981

Martes americana
Marten

X $\frac{1}{8}$

External measurements: males 570–682, 170–240, 79–90, weight up to $2\frac{3}{4}$ pounds; females, 513–560, 165–195, 70–78, weight up to .85 kg. Basilar length, 60–75. Golden brown, deepening to blackish brown on feet and tip of tail, orange or yellowish on throat or chest, head grayish in some subspecies.

Martes americana abieticola (Preble)

1902. *Mustela americana abieticola* Preble, N. Amer. Fauna, 22:68, October 31, type from Cumberland House, Saskatchewan.

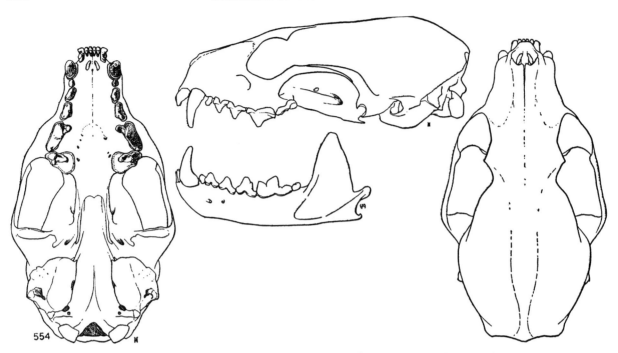

Fig. 554. *Martes americana sierrae*, Marlette Lake, Nevada, No. 69634 M.V.Z., subadult ♂, X 1.

1912. *Martes americana abieticola*, Miller, Bull. U.S. Nat. Mus., 79:92, December 31.

MARGINAL RECORDS.—Keewatin: 12 mi. N Windy River (Harper, 1956:65). Manitoba: Fort Churchill; Oxford House; Norway House. Saskatchewan: type locality; *northern part*.

Martes americana abietinoides Gray

1865. [*Martes americana*] var. 1., *abietinoides* Gray, Proc. Zool. Soc. London, p. 106, June. Type locality, "Edge of the humid western slope of the Rocky Mountains, somewhere between Kicking Horse Pass and the Columbia River." (See Rhoads, Proc. Acad. Nat. Sci. Philadelphia, 54:451, 1902.)
1912. *Martes americana abietinoides*, Miller, Bull. U.S. Nat. Mus., 79:93, December 31.

MARGINAL RECORDS.—British Columbia: Fort Grahame; Little Prairie; Moose Pass (Cowan and Guiguet, 1965:303). Alberta: Jasper National Park; Banff National Park. Montana: Cataract Creek (Manville, 1961:112, sight record only, as *M. americana*); Whitefish Range. Idaho: NE of Coeur d'Alene; Coolin. British Columbia: Manning Park; Tahtsa Lake (Cowan and Guiguet, 1965:303); Sustut River.

Martes americana actuosa (Osgood)

1900. *Mustela americana actuosa* Osgood, N. Amer. Fauna, 19:43, October 6, type from Fort Yukon, Alaska.
1912. *Martes americana actuosa*, Miller, Bull. U.S. Nat. Mus., 79:93, December 31.

1905. *Mustela boria* Elliot, Proc. Biol. Soc. Washington, 18:139, April 18, type from Lower Mackenzie River District, toward Arctic Ocean; exact locality unknown. Regarded by Preble (N. Amer. Fauna, 27:236, 237, October 26, 1908) as indistinguishable from *M. americana actuosa*.

MARGINAL RECORDS.—Mackenzie: Mackenzie Delta; Fort Anderson; 75 mi. W Contwoyto Lake. Alberta: Fort Chipewyan; Fort McMurray; near Henry House. British Columbia: vic. Laurier Pass; Bear Lake (Cowan and Guiguet, 1965:304); lower Iskut River; Atlin (Cowan and Guiguet, 1965:304). Yukon: Quiet Lake. Alaska: Georgetown; Andreafski; Hunt Fork, John River; 30 mi. NE Arctic Village.

Martes americana americana (Turton)

1806. [*Mustela*] *americanus* Turton, A general system of nature . . . , 1:60. Type locality, eastern North America.
1912. *Martes americana americana*, Miller, Bull. U.S. Nat. Mus., 79:92, December 31.
1820. *Mustela leucopus* Kuhl, Beiträge zur Zoologie und vergleichenden Anatomie, p. 74. Type locality, Canada.
1823. *Mustela Huro* F. Cuvier, Dictionaire des sciences naturelles . . . , 29:256. Type locality, upper Canada.
1874. *M*[*ustela*]. *martinus* Ames, Bull. Minnesota Acad. Nat. Sci., 1(No. 2):69. (Listed as "M. martinus *Sabine* Martin" two lines below "Mustela americana *Turton* Sable" under the title of "Mammalia of Minnesota.")

MARGINAL RECORDS.—Ontario: Trout Lake. Quebec: Lake Mistassini; Eric Lake (Harper,

1961:112); Musquaro Lake (*ibid.*). Massachusetts: Berkshire County. Pennsylvania: mts. N of Benton; Shippen Township. Ohio: vic. Chillicothe; Ashtabula County. Michigan: Sanilac County; Allegan County. Illinois: Cook County. Wisconsin (Jackson, 1961:330, Map 60): Vernon County; Dunn County. North Dakota: Grand Forks; Pembina River.

Martes americana atrata (Bangs)

1897. *Mustela atrata* Bangs, Amer. Nat., 31:162, February, type from Bay St. George, Newfoundland.

1942. *Martes americana atrata*, G. M. Allen, Extinct and vanishing mammals . . . , special publ. No. 11, Amer. Comm. for International Wild Life Protection, p. 166.

MARGINAL RECORDS (Cameron, 1959:98, 99).—Newfoundland: Gander Lake; *Gambo;* type locality; Grand Lake.

Martes americana brumalis (Bangs)

1898. *Mustela brumalis* Bangs, Amer. Nat., 32:502, July, type from Okkak, Labrador.

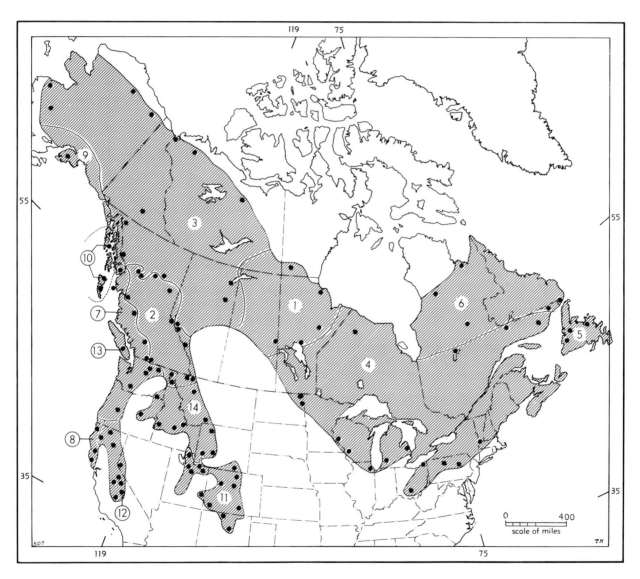

Map 507.   *Martes americana.*

Guide
to subspecies
1. *M. a. abieticola*
2. *M. a. abietinoides*

3. *M. a. actuosa*
4. *M. a. americana*
5. *M. a. atrata*
6. *M. a. brumalis*

7. *M. a. caurina*
8. *M. a. humboldtensis*
9. *M. a. kenaiensis*
10. *M. a. nesophila*

11. *M. a. origenes*
12. *M. a. sierrae*
13. *M. a. vancouverensis*
14. *M. a. vulpina*

1902. *Mustela americana brumalis*, Rhoads, Proc. Acad. Nat. Sci. Philadelphia, 54:448, May.
1934. *Martes americana brumalis*, Anderson, Mammals of the eastern Arctic and Hudson Bay, *in* Canada's Eastern Arctic, Dept. Interior, Ottawa, p. 95.

MARGINAL RECORDS (Harper, 1961:113).—Quebec: Fort Chimo. Labrador: L'Anse au Loup. Quebec: Strait of Belle Isle; Nichicun; Lower Seal Lake.

Martes americana caurina (Merriam)

1890. *Mustela caurina* Merriam, N. Amer. Fauna, 4:27, October 8, type from near Grays Harbor, Grays Harbor Co., Washington.
1953. *Martes americana caurina*, Wright, Jour. Mamm., 34:84, February 9.

MARGINAL RECORDS.—Alaska: Alaskan Panhandle. British Columbia: Anahim Lake; Lillooet; Lightning Creek. Washington: mts. near Republic; Riverside; Chelan; Trout Lake. Oregon: Three Sisters Peaks, thence to coast to southern Oregon and up coast, including *Hawkesbury*, Pitt, and *McCauley* islands, British Columbia (Cowan and Guiguet, 1965:305), to point of beginning.

Martes americana humboldtensis Grinnell and Dixon

1926. *Martes caurina humboldtensis* Grinnell and Dixon, Univ. California Publ. Zool., 21:411, March 17, type from ridge about 5 mi. NE Cuddeback [Carlotta], Humboldt Co., California.
1953. *Martes americana humboldtensis*, Wright, Jour. Mamm., 34:85, February 9.

MARGINAL RECORDS.—California: 12 mi. NE Requa; Longvale; neighborhood of Fort Ross, thence up coast to point of beginning.

Martes americana kenaiensis (Elliot)

1903. *Mustela americana kenaiensis* Elliot, Field Columb. Mus., Publ. 72, Zool. Ser., 3:151, March 20, type from Kenai Peninsula, Alaska.
1912. *Martes americana kenaiensis*, Miller, Bull. U.S. Nat. Mus., 79:93, December 31.

Limits of range unknown. See Schiller and Rausch (1956:195) for *Katmai National Monument*, Alaska.

Martes americana nesophila (Osgood)

1901. *Mustela nesophila* Osgood, N. Amer. Fauna, 21:33, September 26, type from Massett, Graham Island, Queen Charlotte Islands, British Columbia.
1953. *Martes americana nesophila*, Wright, Jour. Mamm., 34:85, February 9.

MARGINAL RECORDS.—Alaska: Three-mile Arm, Kuiu Island. British Columbia: type locality; Sewell Inlet (Cowan and Guiguet, 1965:305).

Martes americana origenes (Rhoads)

1902. *Mustela caurina origenes* Rhoads, Proc. Acad. Nat. Sci. Philadelphia, 54:458, September 30, type from Marvine Mtn., Garfield Co., Colorado.
1953. *Martes americana origenes*, Wright, Jour. Mamm., 34:84, February 9.

MARGINAL RECORDS.—Utah: Cache County; Henry's Fork. Wyoming (Long, 1965a:690): *Little Brooklyn Lake, Medicine Bow Mts.*; 3 mi. ESE Browns Peak. Colorado (Armstrong, 1972:275): near Arkins; summit Mt. Evans; East Spanish Peak. New Mexico (Findley, *et al.*, 1975:302, *M. americana*): Las Vegas Mts.; Chama. Colorado: La Plata Mts. Utah: San Juan Mts. Colorado: Divide Creek. Utah: Paradise Park, 10,000 ft.; Wasatch County; Salt Lake County.

Martes americana sierrae Grinnell and Storer

1916. *Martes caurina sierrae* Grinnell and Storer, Univ. California Publ. Zool., 17:2, August 23, type from head Lyell Canyon, Yosemite National Park, California.
1953. *Martes americana sierrae*, Wright, Jour. Mamm., 34:85, February 9.

MARGINAL RECORDS.—California: near Weed; *Mt. Shasta;* Warner Creek, 6600 ft. Nevada: 2 mi. S, 1 mi. E Incline, 6300 ft. California: Virginia Lakes, 10,000 ft.; vic. Mammoth; *¹/₂ mi. W Bishop;* Whitney Meadows, 9800 ft.; Jordan Hot Springs; Mt. Raymond; Canyon Creek; *Jackson Lake, 5900 ft.*

Martes americana vancouverensis Grinnell and Dixon

1926. *Martes caurina vancouverensis* Grinnell and Dixon, Univ. California Publ. Zool., 21:414, March 17, type from Golden Eagle Mine, 20 mi. S Alberni, Vancouver Island, British Columbia. Restricted to Vancouver Island. (Not recognized as subspecifically distinct from *M. a. caurina* by Cowan and Guiguet, Handbook No. 11 of British Columbia Provincial Mus., p. 305, October 1965.)
1953. *Martes americana vancouverensis*, Wright, Jour. Mamm., 34:85, February 9.

Martes americana vulpina (Rafinesque)

1819. *Mustela vulpina* Rafinesque, Amer. Jour. Sci., p. 82. Type locality, "regions watered by the Missouri." Characterized by tail white at end—a variation that occurs in American as well as Old World *Martes* (see de Vos, Ecology and management of fisher and marten in Ontario, Tech. Bull., Ontario Dept. Lands and Forests, pp. 22, 23, 1952).
1959. *Martes americana vulpina*, Hall and Kelson, The mammals of North America, Ronald Press, p. 901, March 31.

MARGINAL RECORDS.—Montana: Sun River region; Red Lodge region. Wyoming (Long, 1965a:691): Ishawooa Creek, 19 mi. S, 19 mi. W Cody; *Lake Fork;*

Middle Lake, 2 mi. S, 20¹/₂ mi. W Lander; 17 mi. S, 6½ mi. W Lander; LaBarge Creek. Idaho: Salmon River (Lemhi) Mts.; Baker Creek, 12 mi. N Ketchum; near Fort Boise. Oregon: Strawberry Butte. Washington: Blue Mts.

Martes pennanti
Fisher

x ¹/₁₂

External measurements: males 990–1033, 381–422, 113–128, weight up to 15 pounds (possibly up to 18 pounds); females 830–900, 340–380, 89–115; weight up to 6 pounds. Basilar length, 87–108. Dark brown, grayish on foreparts, blackish on rump, legs and tail; tail entirely black.

Martes pennanti columbiana Goldman

1935. *Martes pennanti columbiana* Goldman, Proc. Biol. Soc. Washington, 48:176, November 15, type from Stuart Lake, near headwaters Fraser River, British Columbia.

MARGINAL RECORDS.—Mackenzie: Simpson; Great Slave Lake. Alberta: Fort Smith. Manitoba: Cross Lake; Oxford House. Alberta: Prairie Creek, 250 mi. W Edmonton. British Columbia (Cowan and Guiguet, 1965:309): Cottonwood Creek; 40 mi. N Revelstoke. Idaho: jct. Haystack and Big creeks. Utah: Trial Lake. Idaho: Alturas Lake. British Columbia: Cache Creek (Cowan and Guiguet, 1965:309); Wistaria. Yukon: Morley Lake, 25 mi. SE Teslin (Youngman, 1968:81).

Martes pennanti pacifica (Rhoads)

1898. *Mustela canadensis pacifica* Rhoads, Trans. Amer. Philos. Soc., n.s., 19:435, September, type from Lake Keechelus, Kittitas Co., Washington.
1912. *Martes pennanti pacifica*, Miller, Bull. U.S. Nat. Mus., 79:94, December 31. [Regarded as inseparable from *M. p. pennanti* by Grinnell, Dixon, and Linsdale, The fur-bearing

mammals of California, 1:217, July 22, 1937 and by Cowan and Guiguet (on the authority of Hagmeier, a 1955 MS), The mammals of British Columbia, p. 309, October 1965, but see Goldman, Proc. Biol. Soc. Washington, 48:177, November 15, 1935.]

MARGINAL RECORDS.—Alaska: southern Alaskan Panhandle. British Columbia: Manning Park (Cowan and Guiguet, 1965:309). Washington: head Lake Chelan; Easton. Oregon: Blue Mts.; Lane County. California: Rock Creek Mts. near Cassel; *near Hetch Hetchy Valley;* Yosemite Valley; *Atwell's Mill;* Greenhorn Mts.; Northfolk; Buck Meadows; *Stonyford;* near Lakeport; Cahto; Denny, thence northward along coast to point of beginning.

Martes pennanti pennanti (Erxleben)

1777. [*Mustela*] *pennanti* Erxleben, Systema regni animalis . . . , p. 470. Type locality, Eastern Canada (= Quebec).
1912. *Martes pennanti pennanti,* Miller, Bull. U.S. Nat. Mus., 79:94, December 31.
1777. *Mustela canadensis* Schreber, Die Säugthiere . . . , 3:492, Pl. 126 (see Rhoads, Trans. Amer. Philos. Soc., n.s., 19:426, 434, September, 1898). Type: based on the Pekan of Buffon.
1784. *Mustela melanorhyncha* Boddaert, Elenchus animalium, p. 88. Type: based on Fisher of Pennant.
1800. *Viverra piscator* Shaw, General zoology . . . , 1:414. Type: based on Fisher of Pennant.
1802. *Mustela nigra* Turton, A general system of nature . . . , p. 60. Type: based on *Fisher Weasel* (of Pennant?).
1829. Mustela Canadensis *varietas alba,* Richardson, Fauna Boreali-Americana, p. 54. Type: based on an albinistic Pekan in Hudson's Bay Museum.
1829. *Mustela godmani* Fischer, Synopsis mammalium, p. 217. Based on *M. pennanti* Goldman, Amer. Nat. Hist., p. 203.

MARGINAL RECORDS.—Manitoba: 60 mi. E York Factory. Quebec: Mistassini; Eric Lake (Harper, 1961:113); Mingan. Maine: Mount Desert Island. New Hampshire: Monadnock. Massachusetts: *vic. Williamstown.* Pennsylvania: Mill Creek, about 3 mi. E Lancaster. Virginia: Peter's Mtn., thence southward in mts. to North Carolina. Tennessee: near Flat Rock. Indiana: Posey County; Hamilton County. Illinois: Cook County. Iowa: Elk Lake [?ca. 10 mi. E Spencer, Clay Co.] (Kearny, 1912:348, merely as "Fisher"). Wisconsin: Mississippi bottomlands near La Crosse. Minnesota (Balser and Longley, 1966:549, as *M. pennanti* only): Aitken County; Cass County. North Dakota: Grand Forks. Saskatchewan (Beck, 1958:17): Hazelcliffe; Prince Albert. Wyoming: Yellowstone National Park (Long, 1965a:692, as *M. p. columbiana*); near Ucross, on Piney Creek (Brown, 1965b:143). Montana: Glacier National Park. Saskatchewan (Beck, 1958:17): Prince Albert National Park; Red Earth Indian Reserve. Manitoba: N coast Lake Winnipegosis.

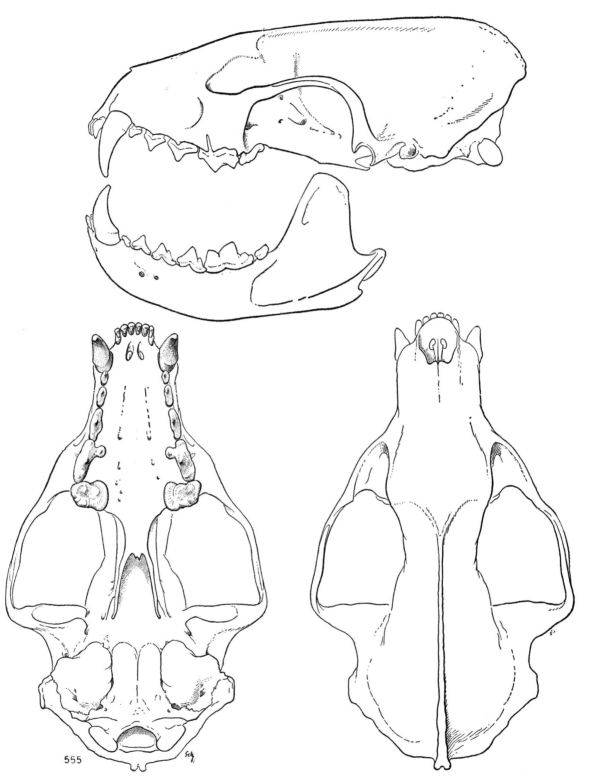

Fig. 555. *Martes pennanti pacifica*, Chinquapin, Yosemite National Park, Mariposa Co., California, No. 23668 M.V.Z., ♂, X 1.

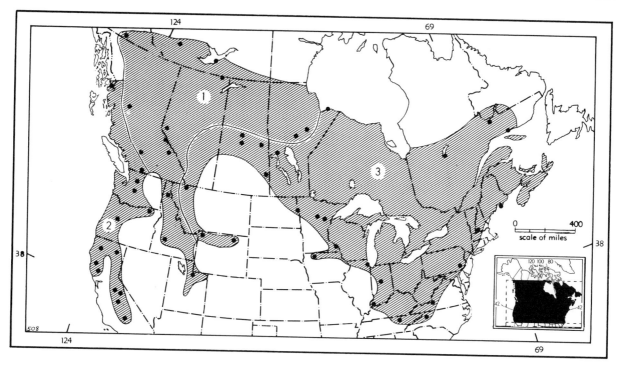

Map 508. *Martes pennanti.*

1. *M. p. columbiana*   2. *M. p. pacifica*   3. *M. p. pennanti*

Genus **Mustela** Linnaeus
Weasels, Ferret and Minks

1758. *Mustela* Linnaeus, Syst. nat., ed. 10, 1:45. Type, *Mustela erminea* Linnaeus (see Thomas, Proc. Zool. Soc. London, p. 138, March 22, 1911).

Total length, 190–700; legs short; body long; basilar length, 16–70; facial angle slight; tympanic bullae greatly inflated (moderately in subgenus *Lutreola*), cancellous, and with paroccipital processes closely appressed to bullae; palate behind upper molars; inner moiety of M1 larger than outer; P4 with simple deuterocone; in m1 trigonid longer than talonid, metaconid absent (incipiently developed in subgenus *Lutreola*), and talonid trenchant. Dentition, i. $\frac{3}{3}$, c. $\frac{1}{1}$, p. $\frac{2-3}{3-2}$, m. $\frac{1}{2}$.

KEY TO NORTH AMERICAN SPECIES OF MUSTELA

1. Length of upper tooth-rows less than 20 in males and 17.8 in females.
   2. Postglenoidal length of skull more than 47% of condylobasal length.
      3. Tail without black pencil and with at most few black hairs at extreme tip; in both sexes mastoid breadth ordinarily more than breadth of braincase. . . . . . .*M. nivalis*, p. 992
      3'. Tail with black pencil; in females mastoid breadth ordinarily less than breadth of braincase.
      M. erminea, p. 988
   2'. Postglenoid length of skull less than 47% of condylobasal length.
   M. frenata, p. 993
1'. Length of upper tooth-rows more than 20 in males and 17.8 in females.
   4. Abdomen white; face with blackish mask; m1 lacking even trace of metaconid; distance between upper canines more than width of basioccipital as measured between foramina situated midway along medial sides of tympanic bullae. . . . .*M. nigripes*, p. 999
   4'. Abdomen dark brown, like back; face uniformly brown without blackish mask; m1 with incipient metaconid; distance between upper canines less than width of basioccipital as measured between foramina situated midway along medial sides of tympanic bullae.
      5. Length of upper molar-premolar tooth-row more than 20.
      M. macrodon, p. 1004
      5'. Length of upper molar-premolar tooth-row less than 20. .*M. vison*, p. 1000

Subgenus **Mustela** Linnaeus—Weasels

Revised by Hall, Univ. Kansas Publ., Mus. Nat. Hist., 4:1–466, 41 pls., 31 figs., December 27, 1951.

1758. *Mustela* Linnaeus, Syst. nat., ed. 10, 1:45. Type, *Mustela erminea* Linnaeus (see Thomas, Proc. Zool. Soc. London, p. 138, March 22, 1911).
1829. *Arctogale* Kaup, Skizzirte Entwickelungs-Geschichte und natürliches System der europäischen Thierwelt, 1:30. Type, *Mustela erminea* Linnaeus, at least in part.
1829. *Ictis* Kaup, Skizzirte Entwickelungs-Geschichte und natürliches System der europäischen Thierwelt, 1:30. Type, *Mustela vulgaris* Erxleben [= *Mustela nivalis* Linnaeus]. Not Schinz, 1824–1828.
1841. *Gale* Wagner, *in* Schreber, Die Säugthiere . . ., Suppl., 2:234. Type, *Mustela vulgaris* Erxleben [= *Mustela nivalis* Linnaeus].
1865. *Neogale* Gray, Proc. Zool. Soc. London, p. 114, June. Included *Mustela brasiliensis* Sevastianoff, *Mustela aureoventris* Gray, and *Mustela xanthogenys* Gray [all are subspecies of *Mustela frenata* Lichtenstein].
1871. *Mustelina* Bogdanow, Trudy Obshtch. yestestvoispyateley Imp. Kazan. Univ. 1, otd. 1, p. 167 (not seen). Based on *Mustela erminea* Linnaeus and *Mustela vulgaris* Erxleben [= *Mustela nivalis* Linnaeus].
1899. *Eumustela* Acloque, Faune de France, mammifères, p. 62. Based on *Mustela vulgaris* Erxleben [= *Mustela nivalis* Linnaeus] and *Mustela erminea* Linnaeus.

Mustela erminea
Ermine

X ⅛

Total length in males, 225–340; in females, 190–290; tail-vertebrae, 16–19 in number and combined length 30–45 per cent of length of head and body; tail with distinct black pencil. Braincase long, precranial portion of skull short; postglenoidal length, when expressed as a percentage of the condylobasal length, more than 48 in females and ordinarily more than 46 in males. Upper parts brown; underparts whitish; whitish color ordinarily continuous from chin to inguinal region but in subspecies in humid region along Pacific Coast interrupted in some individuals by brown of upper parts encircling body in abdominal region. Differs from *Mustela nivalis* in presence of black pencil on tail, tail-vertebrae more than a quarter of length of head and body, and, in regions where both species occur, basilar length of skull more than 32.5 in males and more than 31.0 in females; from *Mustela frenata*, in regions where the two species occur together, by tail less than 44 per cent of length of head and body and

by postglenoidal length of skull more than 46 per cent of condylobasal length in males and more than 48 per cent in females.

The ermine is white in winter (tip of tail always black) except in the humid area along the Pacific Coast from southern British Columbia southward, where the winter coat is slightly paler brown than the summer coat.

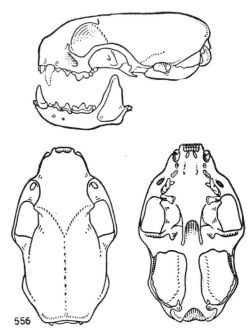

Fig. 556. *Mustela erminea arctica*, SE end Lake Peters, long. 145° 09′ 26″ W, lat. 69° 20′ 50″ N, Romanzof Mts., Alaska, No. 51551 K.U., ♂, X 1.

Mustela erminea alascensis (Merriam)

1896. *Putorius richardsoni alascensis* Merriam, N. Amer. Fauna, 11:12, June 30, type from Juneau, Alaska.
1944. *Mustela erminea alascensis*, Hall, Proc. Biol. Soc. Washington, 57:36, June 28.

MARGINAL RECORDS.—Alaska: Taku River; type locality; Caroll Inlet, Revillagigedo Island; Loring, thence northward along coast including Mitkof, Zarembo, and Wrangell islands, to point of beginning.

Mustela erminea anguinae Hall

1932. *Mustela cicognanii anguinae* Hall, Univ. California Publ. Zool., 38:417, November 8, type from French Creek, Vancouver Island, British Columbia. Occurring naturally, so far as is known, only on Vancouver and Salt Spring islands, but introduced, according to Cowan and Guiguet (Handbook No. 11, British Columbia Provincial Mus., p. 311, October 1965), on North Pender Island, British Columbia.
1945. *Mustela erminea anguinae* Hall, Jour. Mamm., 26:79, February 27.

Mustela erminea arctica (Merriam)

1896. *Putorius arcticus* Merriam, N. Amer. Fauna, 11:15, June 30, type from Point Barrow, Alaska.
1935. *Mustela erminea arctica*, Ognev, The mammals of the U.S.S.R. and adjacent countries, 3:31.
1904. *Putorius audax* Barrett-Hamilton, Ann. Mag. Nat. Hist., ser. 7, 13:392, May, type from Discovery Bay, Ellsemere Island.

MARGINAL RECORDS.—Franklin: Cape Sheridan; Bedford Pims Island; Craig Harbor. Keewatin: Franklin Isthmus. Mackenzie: Clinton Colden; Arctic Red River. Yukon (Youngman, 1975:143): forks Macmillan River; Selkirk. Alaska: Glacier Bay, thence along coast to Unimak Island, thence northward along coast westward into Siberia, eastward to Alaska and along coast (including Arctic Islands) to point of beginning.

Mustela erminea bangsi Hall

1945. *Mustela erminea bangsi* Hall, Jour. Mamm., 26:176, July 19, type from Elk River, Sherburne Co., Minnesota.

MARGINAL RECORDS.—Manitoba: Delta (Tamsitt, 1962:77, as *M. erminea* only); Red River Settlement. Minnesota: ½ mi. S, ¼ mi. W Pigeon River, NW ¼ sec. 29, T. 64, N, R. 6 E (Timm, 1975:32). Michigan: Tobin Harbor, Isle Royale; Sault Ste. Marie; Ann Arbor. Wisconsin: Pewaukee (Jackson, 1961:343). Iowa (Bowles, 1975:123): vic. Stockton; sec. 24, Marcy Twp., Boone Co.; Ireton. North Dakota: Brantford. Manitoba: Aweme.

Mustela erminea celenda Hall

1944. *Mustela erminea celenda* Hall, Proc. Biol. Soc. Washington, 57:38, June 28, type from Kasaan Bay, Prince of Wales Island, Alaska.

MARGINAL RECORDS.—Alaska: Craig, Prince of Wales Island; type locality; Howkan, Long Island; Otter Harbor, Dall Island.

Mustela erminea cicognanii Bonaparte

1838. *Mustela cigognanii* [sic] Bonaparte, Charlesworth's Mag. Nat. Hist., 2:37. Type locality, eastern United States.
1945. *M[ustela]. e[rminea]. cicognanii*, Hall, Jour. Mamm., 26:77, February 27.
1842. *Mustela pusilla* DeKay, Zoology of New-York . . . , pt. 1, Mammalia, p. 34, Pl. 14, Fig. 1, type from New York State.

MARGINAL RECORDS.—Ontario: Grand Bay, Lake Nipigon; Franz; Smoky Falls near Kapuskasing. Quebec: Kiamika Lake; Black Lake. Maine: Quimby; Lincoln; Naskeag, thence S along coast to Rhode Island: Middletown. Connecticut: Liberty Hill. New York: Babylon, Long Island (Connor, 1971:51). Maryland: 4 mi. NW Bethesda. Pennsylvania: North Shenango Township, Pymatuning Swamp, thence NE along shore Lake Erie, down NW shore of same to Lake

Huron and along northern shores of lakes Huron and Superior to Ontario: 20 mi. SW Fort Williams; Oscar.

Mustela erminea fallenda Hall

1945. *Mustela erminea fallenda* Hall, Jour. Mamm., 26:79, February 27, type from Huntingdon, British Columbia.

MARGINAL RECORDS.—British Columbia: Horseshoe Lake; Chilliwack (Cowan and Guiguet, 1965:311); Thurstons Ranch; Tami Hy Creek. Washington: Morovitz Guard Station, 831 ft.; Lake Whatcom.

Mustela erminea gulosa Hall

1945. *Mustela erminea gulosa* Hall, Jour. Mamm., 26:84, February 27, type from Trout Lake, Klickitat Co., Washington.

MARGINAL RECORDS.—Washington: 2 mi. E Skykomish; Easton; type locality; Mt. St. Helens, 6000 ft.; Meslers Ranch, 2000 ft., 1 mi. W Rainier Park.

Mustela erminea haidarum (Preble)

1898. *Putorius haidarum* Preble, Proc. Biol. Soc. Washington, 12:169, August 10, type from Massett, Graham Island, Queen Charlotte Islands, British Columbia.
1944. *Mustela erminea haidarum*, Hall, Proc. Biol. Soc. Washington, 57:38, June 28.

MARGINAL RECORDS.—British Columbia: type locality; Skidgate, Graham Island; Cumsheva Inlet, Moresby Island.

Mustela erminea initis Hall

1944. *Mustela erminea initis* Hall, Proc. Biol. Soc. Washington, 57:37, June 28, type from Saook Bay, Baranof Island, Alaska.

MARGINAL RECORDS.—Alaska: Freshwater Bay, Chichagof Island; type locality.

Mustela erminea invicta Hall

1945. *Mustela erminea invicta* Hall, Jour. Mamm., 26:75, February 27, type from Benewah, Benewah Co., Idaho.

MARGINAL RECORDS.—Alberta: Jasper House; forks of Blindman and Red Deer rivers; Rosebud (Soper, 1965:304). Montana: Duck Lake, 6 mi. NE St. Marys Lake; Darby; Tin Cup District. Idaho: Sawmill Creek, 1¼ mi. W Newsome; 14 mi. W Iron Mtn.; type locality. Washington: Lake Chelan; Twin Lakes, Winchester Mts. British Columbia: mts. in Chilliwack (Cowan and Guiguet, 1965:313); Hope (ibid.); Wentworth Lake. Alberta: Shovel Pass.

Mustela erminea kadiacensis (Merriam)

1896. [*Putorius arcticus*] subspecies *kadiacensis* Merriam, N. Amer. Fauna, 11:16, June 30, type from Kodiak Island, Alaska.

Map 509. *Mustela erminea.*

1. *M. e. alascensis*	6. *M. e. cicognanii*	11. *M. e. invicta*	16. *M. e. richardsonii*
2. *M. e. anguinae*	7. *M. e. fallenda*	12. *M. e. kadiacensis*	17. *M. e. salva*
3. *M. e. arctica*	8. *M. e. gulosa*	13. *M. e. muricus*	18. *M. e. seclusa*
4. *M. e. bangsi*	9. *M. e. haidarum*	14. *M. e. olympica*	19. *M. e. semplei*
5. *M. e. celenda*	10. *M. e. initis*	15. *M. e. polaris*	20. *M. e. streatori*

1945. *Mustela erminea kadiacensis*, Hall, Jour. Mamm., 26:179, July 19.

MARGINAL RECORDS.—Alaska (Kodiak Island): Kodiak; Karluk, and presumably over entire island.

Mustela erminea muricus (Bangs)

1899. *Putorius (Arctogale) muricus* Bangs, Proc. New England Zool. Club, 1:71, July 31, type from Echo, 7500 ft., El Dorado Co., California.

1945. *Mustela erminea murica*, Hall, Jour. Mamm., 26:77, February 27.
1903. *Putorius streatori leptus* Merriam, Proc. Biol. Soc. Washington, 16:76, May 29, type from Silverton, San Juan Co., Colorado.

MARGINAL RECORDS.—Montana: Camas Creek, Big Belt Mts., 4 mi. S Fort Logan. Wyoming: 3 mi. NW Sundance (Long, 1965a:692). South Dakota: Spring Creek, S Rapid City (Turner, 1974:129); 1/2 mi. E Sylvan Lake, 6250 *ft.* Wyoming: 30 mi. N, 10 mi. E

Laramie (Long, 1965a:692). Colorado: Denver; Turkey Creek, SW Colorado Springs, 6000 ft. New Mexico (Findley, *et al.*, 1975: 302, 303, as *M. erminea*): 15 mi. NW Cimarron; Santa Fe Baldy; Sandia Mts.; San Gregorio Lake, Jemez Mts. Utah: Wildcat Ranger Station, Boulder Mtn., 8700 ft.; Cedar Breaks National Monument, 10,500 ft. (Stock, 1970:433). Nevada: Baker Creek, 8500–11,100 ft.; South Twin River, Toyabe Mts.; ½ mi. S Marlette Lake, 8150 ft. California: Mammoth; *Vogelsang Lake, 10,350 ft., Yosemite National Park; Echo; Fallen Leaf Lake, 6500 ft.; 2 mi. W Black Butte, 6800 ft.; Castle Lake, 5434 ft.;* head of Rush Creek, 6400 ft. Oregon: Fort Klamath; Mill Creek, 20 mi. W Warmsprings. Washington: Stayawhile Spring, 5150 ft.; *Butte Creek.* Idaho: W. Fork Rapid Creek, 9½ mi. E Pocatello. Montana: Donovan.

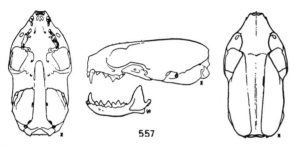

Fig. 557. *Mustela erminea muricus*, Baker Creek, White Pine Co., Nevada, No. 41501 M.V.Z., ♂, X 1.

Mustela erminea olympica Hall

1945. *Mustela erminea olympica* Hall, Jour. Mamm., 26:81, February 27, type from near head Soleduc [=Solduc] River, 4500 ft., Olympic Mts., Clallam Co., Washington.

MARGINAL RECORDS.—Washington: Clallam Bay, thence east and south along coast of Puget Sound to 4 mi. S Olympia, thence west to Pacific Coast and north to point of beginning.

Mustela erminea polaris (Barrett-Hamilton)

1904. *Putorius arcticus polaris* Barrett-Hamilton, Ann. Mag. Nat. Hist., ser. 7, 13:393, May, type from Gap Valley, 7¼ mi. NE Cape Brevoort, lat. 82° N, long. 59° 20′ W, northwestern Greenland.
1945. *Mustela erminea polaris,* Hall, Jour. Mamm., 26:179, July 19.

MARGINAL RECORDS.—Greenland: Dragon Point, thence S on E coast to Turner Sound and on W coast to type locality.

Mustela erminea richardsonii Bonaparte

1838. *Mustela Richardsonii* Bonaparte, Charlesworth's Mag. Nat. Hist., 2:38, January. Type locality, probably Fort Franklin, Canada.
1945. *Mustela erminea richardsonii,* Hall, Jour. Mamm., 26:77, February 27.
1903. *Putorius microtis* J. A. Allen, Bull. Amer. Mus. Nat.

Hist., 19:563, October 10, type from Shesley, British Columbia.
1904. *Putorius arcticus imperii* Barrett-Hamilton, Ann. Mag. Nat. Hist., ser. 7, 13:392, May, type from Fort Simpson, Mackenzie.
1913. *Mustela cicognanii mortigena* Bangs, Bull. Mus. Comp. Zool., 54:511, July, type from Bay St. George, Newfoundland.

MARGINAL RECORDS.—Mackenzie: Fort Goodhope. Keewatin: foot Baker Lake. Manitoba: Egg Island, Rabbit Point, south, thence east and thence north along the shores of Hudson Bay, including in Quebec: Charlton Island; Belcher Islands; Mistake Bay (Harper, 1961:115); Chubb Crater area (*ibid.*); Fort Chimo. Labrador: Killinek area (Harper, 1961:115); Okak, thence southward along coast including Newfoundland, to New Brunswick: Scotch Lake; Red Brook, Tobique River. Quebec: Godbout; Waswonaby Post. Ontario: Abitibi. Manitoba: Lake St. Martin; Whitewater Lake, Riding Mtn. (Soper, 1961:206). Saskatchewan: Indianhead; Cypress Hills (Beck, 1958:18); Livelong. Alberta: near Rochester (Keith and Meslow, 1966:541, as *M. erminea* only); Entrance. British Columbia: Isaacs Lake; Horse Lake; McGillivray Creek; Mons; Bute Inlet (Cowan and Guiguet, 1965:314), thence up coast to Port Simpson; Iskoot [= Iskut] River (Cowan and Guiguet, 1965:314); Sheslay. Yukon (Youngman, 1975:145): Kluane Lake; Macmillan Pass, Canol Road, Mile 282.

Mustela erminea salva Hall

1944. *Mustela erminea salva* Hall, Proc. Biol. Soc. Washington, 57:35, June 28, type from Mole Harbor, Admiralty Island, Alaska.

MARGINAL RECORDS.—Alaska (Admiralty Island): Hawk Inlet; Seymour Canal; *type locality; Gambier Bay.*

Mustela erminea seclusa Hall

1944. *Mustela erminea seclusa* Hall, Proc. Biol. Soc. Washington, 57:39, June 28, type from Port Santa Cruz, Suemez Island, Alaska. Known only from type locality.

Mustela erminea semplei Sutton and Hamilton

1932. *Mustela arctica semplei* Sutton and Hamilton, Ann. Carnegie Mus., 21:79, February 13, type from Coral Inlet, South Bay, Southampton Island, Keewatin.
1945. *Mustela erminea semplei,* Hall, Jour. Mamm., 26:179, July 19.
1935. *Mustela arctica labiata* Degerbøl, Rept. 5th Thule Exped., 2(no. 4):25, type from Malugsitaq, Melville Peninsula, Franklin.

MARGINAL RECORDS.—Franklin: Bylot Island; Ponds Inlet; Kingnait Fiord; SW coast Baffinland; Cape Dorset. Keewatin: Ranger Rim, Southampton Island; Eskimo Point; Rae Isthmus.

Mustela erminea streatori (Merriam)

1896. *Putorius streatori* Merriam, N. Amer. Fauna, 11:13, June 30, type from Mount Vernon, Skagit Valley, Skagit Co., Washington.
1945. *Mustela erminea streatori*, Hall, Jour. Mamm., 26:77, February 27.

MARGINAL RECORDS.—Washington: N end Whidbey Island, opposite Deception Pass; Oso, 550 ft.; *Hamilton*; 15 mi. N Government Springs, 1300 ft. Oregon: Deschutes River, 6 mi. E Crescent Lake. California: 10 mi. NE Carlotta; mouth Gualala River, thence northward along coast to Washington: Wallicut, 2 mi. E Ilwaco, thence to S end Puget Sound and north along E shore Puget Sound to point of beginning.

Mustela nivalis
Least Weasel

X ⅛

Total length less than 250 in males and 225 in females; tail a fourth or less of length of head and body and without black pencil (or at most with few black hairs at extreme tip); caudal vertebrae 11–16; basilar length (30.7) of longest skull less than in shortest skull of *M. frenata*, and same is true in comparison with *M. erminea* in areas where *M. nivalis* and *M. erminea* occur together. (SW of range of *M. nivalis*, basilar length is less than 30.7 in some *M. erminea*.)

In skulls of equal size, the mastoid breadth in *M. nivalis* equals or exceeds the greatest breadth of the braincase, whereas the reverse is ordinarily true in subspecies of *M. erminea* that are as small as *M. nivalis*. In summer, *nivalis* is brown above and white below; in the southeastern part of its range some individuals are brown in winter and are brown on most of the underparts as well as on the upper parts, but elsewhere in North America are white in winter.

For use of the name *nivalis* in place of *rixosa* for the species see Reichstein (1958:169).

Mustela nivalis allegheniensis (Rhoads)

1901. *Putorius alleghaniensis* Rhoads, Proc. Acad. Nat. Sci. Philadelphia, 52:751, March 25, type from near Beallsville, Washington Co., Pennsylvania.
1959. *Mustela nivalis allegheniensis*, Hall and Kelson, The mammals of North America, Ronald Press, p. 1082, March 31.

MARGINAL RECORDS.—Wisconsin (Jackson, 1961:345): Wausau; De Pere. Michigan: Osceola

Fig. 558. *Mustela nivalis rixosa*, Jadis, Roseau Co., Minnesota, No. 48005 K.U., subadult, probably ♂, X 1.

County; 8 mi. N Caro; *Deckerville*, thence along southern shore Lake Erie to Pennsylvania: McKeen Township; Bradford County (Richmond and McDowell, 1952:252, Fig. 1); Middle Paxton Township. Virginia: near Dayton; Blacksburg. North Carolina: 4 mi. NE Hendersonville, 2200 ft. (Barkalow, 1967:488); Balsam Gap, 3400 ft. (Stupka, 1960:519). Tennessee: Roan Mtn. (Tuttle, 1968:133, as *M. nivalis* only). Ohio: Vinton; *5 mi. SW Mt. Orab*; 6 mi. NNW New Richmond. Indiana: near Terre Haute (Whitaker and Zimmerman, 1965:516, as *M. nivalis* only). Illinois: Urbana; Henkel. Wisconsin: Prescott (Jackson, 1961:345); *Dunn County* (*op. cit.*:344, Map 63).

Mustela nivalis campestris Jackson

1913. *Mustela campestris* Jackson, Proc. Biol. Soc. Washington, 26:124, May 21, type from Beemer, Cuming Co., Nebraska.
1959. *Mustela nivalis campestris*, Hall and Kelson, The mammals of North America, Ronald Press, p. 1082, March 31.

MARGINAL RECORDS.—South Dakota: shore of Sand Lake, sec. 15, T. 126 N, R. 62 W; Waubay Migratory Waterfowl Refuge. Iowa: Chester; *National*; near McCausland (Bowles, 1975:126); *Davenport* (*ibid.*); Mount Pleasant. Missouri: 1 mi. N, 3½ mi. E Guilford (Easterla, 1970:333). Kansas: 3½ mi. NW Valley Falls (127983 KU); *NW ¼ sec. 22, T. 4 S, R. 9 W, Jewell Co.* (119962 KU); 8 mi. N, 4 mi. E Lincoln (Choate and Fleharty, 1975:64); SE ¼ sec. 11, T. 2 S, R. 12 W, Smith Co. (105067 KU). Nebraska (Jones, 1964c:279): 1¾ mi. N, 3 mi. W Holrege; Oshkosh; 4½ mi. N, 5 mi. W Rushville.

Mustela nivalis eskimo (Stone)

1900. *Putorius rixosus eskimo* Stone, Proc. Acad. Nat. Sci. Philadelphia, 52:44, March 24, type from Point Barrow, Alaska.
1958. *Mustela nivalis eskimo*, Reichstein, Zeitschr. für Säugetierkunde, 22:169.

MARGINAL RECORDS.—Alaska: Point Barrow, thence E along coast for undetermined distance and S to Yukon: La Pierre House; Ross River, near Sheldon Lake (Youngman, 1975:147); Klotassin River, tributary

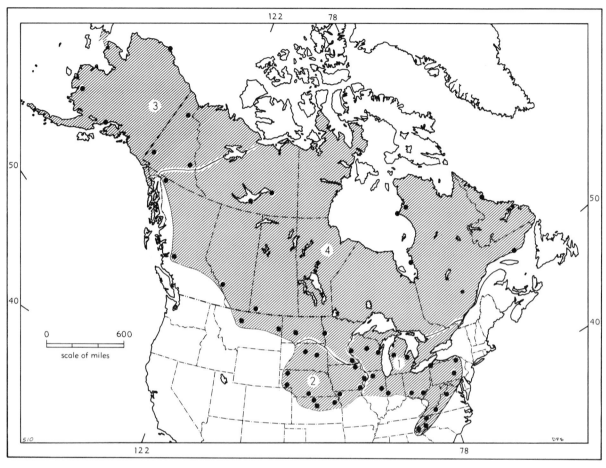

Map 510. *Mustela nivalis.*

| 1. *M. n. allegheniensis* | 2. *M. n. campestris* | 3. *M. n. eskimo* | 4. *M. n. rixosa* |

of White River. Alaska: Tyoonok [= Tyonek]; Bethel, and along coast to point of beginning.

Mustela nivalis rixosa (Bangs)

1896. *Putorius rixosus* Bangs, Proc. Biol. Soc. Washington, 10:21, February 25, type from Osler, Saskatchewan.
1958. *Mustela nivalis rixosa*, Reichstein, Zeitschr. für Säugetierkunde, 22:169.

MARGINAL RECORDS.—Mackenzie: Old Fort Reliance. Quebec (Harper, 1961:116): island S of Comb Hills; Elsie Island; Kogaluk River. Labrador: Davis Inlet, S along coast to Quebec: Natashkwan, thence along coast, St. Lawrence River, Ottawa River, and N shores of lakes Huron and Superior to Minnesota: Crystal Springs; Moorhead. North Dakota: Mandan. Montana: Wibaux. Saskatchewan: Eastend. Montana: Sun River Valley. Alberta: Miette River. British Columbia: Wistaria P.O.; 15 mi. E Atlin. Mackenzie: Fort Resolution.

Mustela frenata
Long-tailed Weasel

Total length, 300–550; tail two-fifths to seven-tenths of length of head and body, with distinct black pencil at end; caudal vertebrae 19–23; postglenoidal length, expressed as percentage of condylobasal length, less than 47 in females and ordinarily less than 46 in males; upper parts brown; light-colored underparts, in summer pelage, tinged with buffy or yellowish and continuous from chin to inguinal region; some subspecies (southwestern United States,

México, Central America, and Florida) with white or yellowish facial markings, which do not occur in any other American species of the genus *Mustela*. Northern subspecies are white in winter; the other subspecies are only a lighter shade of brown than in summer; tip of the tail at all times black. The black tip on the tail is shorter in relation to the length of the tail than in most subspecies of *M. erminea;* other differences from *M. erminea* and *M. nivalis* are given in the accounts of those species.

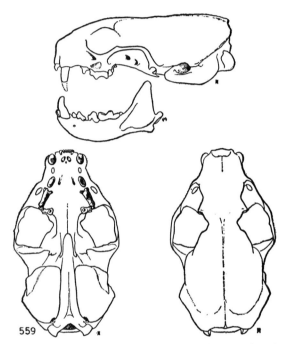

Fig. 559. *Mustela frenata nevadensis*, Baker Creek, White Pine Co., 8400 ft., Nevada, No. 41508 M.V.Z., ♂, X 1.

Mustela frenata alleni (Merriam)

1896. *Putorius alleni* Merriam, N. Amer. Fauna, 11:24, June 30, type from Custer, Custer Co., South Dakota.
1936. *Mustela frenata alleni*, Hall, Carnegie Inst. Washington Publ. 473:106, November 20.

MARGINAL RECORDS.—Wyoming: Sundance. South Dakota: Hill City; *Palmer Gulch, 8 mi. SE Hill City, 5300 ft.* (Turner, 1974:130); type locality; *20 mi. N Elk Mtn.*

Mustela frenata altifrontalis Hall

1936. *Mustela frenata altifrontalis* Hall, Carnegie Inst. Washington Publ. 473:94, November 20, type from Tillamook, Tillamook Co., Oregon.

MARGINAL RECORDS.—British Columbia: Chilliwack; *Lihumption Park, 4750 ft.* Washington: *Swamp Creek, 2050 ft., Nooksack River;* Tye; Auburn. Ore-

gon: Blaine; Reed; Langlois, thence up coast to British Columbia: *mouth Fraser River.*

Mustela frenata arizonensis (Mearns)

1891. *Putorius arizonensis* Mearns, Bull. Amer. Mus. Nat. Hist., 3:234, June 5, type from San Francisco Forest [then (1886?) Yavapai Co.], Arizona.
1936. *Mustela frenata arizonensis*, Hall, Carnegie Inst. Washington Publ. 473:106, November 20.

MARGINAL RECORDS.—Arizona: V. T. Park; Deadman Flat, 6400 ft.; Springerville. New Mexico: Willow Creek, 8500 ft., Mogollon Mts. Arizona: Prieto Plateau, 9000 ft., S end Blue Range; Quaking Asp Settlement; Government Prairie, near Parks.

Mustela frenata arthuri Hall

1927. *Mustela noveboracensis arthuri* Hall, Proc. Biol. Soc. Washington, 40:193, December 2, type from Remy, St. James Parish, Louisiana.
1936. *Mustela frenata arthuri* Hall, Carnegie Inst. Washington Publ. 473:105, November 20.

MARGINAL RECORDS.—Texas: 2⅘ mi. S Mt. Enterprise (Packard, 1961:195). Louisiana (Lowery, 1974:426): 4 mi. W Woodworth; 4¼ mi. S Angie. Mississippi: Saucier. Louisiana (*loc. cit.*): *Remy;* Convent. Texas: Chambers County (Davis, 1966:87, as *M. frenata* only); *3 mi. S Garwood;* 5 mi. W Eagle Lake; *12 mi. N Eagle Lake.*

Mustela frenata costaricensis Goldman

1912. *Mustela costaricensis* Goldman, Proc. Biol. Soc. Washington, 25:9, January 23, type from San José, Costa Rica.
1936. *Mustela frenata costaricensis*, Hall, Carnegie Inst. Washington Publ. 473:109, November 20.

MARGINAL RECORDS.—Costa Rica: [Volcán] Irazú, 3000 m.; *Tucurrique; Muñeco; Candelaria Mts.; type locality.*

Mustela frenata effera Hall

1936. *Mustela frenata effera* Hall, Carnegie Inst. Washington Publ. 473:93, November 20, type from Ironside, 4000 ft., Malheur Co., Oregon.

MARGINAL RECORDS.—Washington: Asotin. Oregon: 1½ mi. S Vale; 25 mi. NW Burns; Fort Rock; Sisters; 4 mi. S The Dalles, thence eastward along Columbia River to Washington: Wallula, thence *Prescott.*

Mustela frenata frenata Lichtenstein

1831. *Mustela frenata* Lichtenstein, Darstellung neuer oder wenig bekannter Säugethier . . . , Pl. 42 and corresponding text, unpaged, type from Ciudad México, México.
1813. *Mustela brasiliensis* Sevastianoff, Mém. Acad. Imp. Sci., St. Pétersbourg, 4:356–363, Tab. 4, name on plate only, the description being in the text. Not *Mustela brasiliensis* Gme-

lin, 1788, a name applied to an otter. Type locality, unknown.
1877. *Putorius mexicanus* Coues, Fur-bearing animals . . . , U.S. Geol. Surv. Territories, Miscl. Publ., 8:42, a *nomen nudum* [cited by Coues in synonymy as "*Putorius mexicanus*, Berlandier, MMS. ic. ined. 4 (Tamaulipas and Matamoras)"].

MARGINAL RECORDS.—Texas: San Antonio; Charco; Corpus Christi; Brownsville. Tamaulipas (Alvarez, 1963:459): *Matamoros;* 6 mi. N, 6 mi. W Altamira; Jaumave. Distrito Federal: Tlalpam. México: Nevada Toluca, 3200 m. San Luis Potosí: Alvarez. Coahuila: mts. near Saltillo.

Mustela frenata goldmani (Merriam)

1896. *Putorius frenatus goldmani* Merriam, N. Amer. Fauna, 11:28, June 30, type from Pinabete, Chiapas.
1912. *Mustela frenata goldmani*, Miller, Bull. U.S. Nat. Mus., 79:100, December 31.

MARGINAL RECORDS.—Chiapas: 20 mi. SE Teopisca. El Salvador: Los Esesmiles, 8000 ft.; *La Cebia, 2150 ft.* Guatemala: Finca San Isidro, San Sebastián. Chiapas: type locality.

Mustela frenata inyoensis Hall

1936. *Mustela frenata inyoensis* Hall, Carnegie Inst. Washington Publ. 473:99, November 20, type from Carl Walter's Ranch, 2 mi. N Independence, Inyo Co., California.

MARGINAL RECORDS.—California: Alvord, 4000 ft.; *Lone Pine; type locality.*

Mustela frenata latirostra Hall

1936. *Mustela frenata latirostra* Hall, Carnegie Inst. Washington Publ. 473:96, November 20, type from San Diego, San Diego Co., California.

MARGINAL RECORDS.—California: Cuyama Valley, 2200 ft.; Bluff Lake; *Cabazon;* Julian. Baja California: 32° 3′ N, 116° 54′ W (Ralston and Clark, 1971:209). California: Chula Vista; Rincon Point.

Mustela frenata leucoparia (Merriam)

1896. *Putorius frenatus leucoparia* Merriam, N. Amer. Fauna, 11:29, June 30, type from Pátzcuaro, Michoacán.
1912. *Mustela frenata leucoparia*, Miller, Bull. U.S. Nat. Mus., 79:100, December 31.

MARGINAL RECORDS.—Nayarit: Tepic. Jalisco: Río Santiago; 4½ mi. NE Comanja de Corona, 8000 ft. (Genoways and Jones, 1973:19); 4 mi. SE Degollado, 5600 ft. (*ibid.*). Michoacán: type locality. Morelos (Ramírez-P., 1971:281): *Cerro Cuautépetl, 3200 m.;* Huitzilac. Oaxaca (Goodwin, 1969:233): *San Juan Mixtepec;* San Pedro Mixtepec. Guerrero: Acahuizotla, 2800 ft. (Davis and Lukens, 1958:358). Jalisco: Los Masos; 2 mi. N Milpillas, 3000 ft. (Genoways and Jones, 1973:19).

Mustela frenata longicauda Bonaparte

1838. *Mustela longicauda* Bonaparte, Charlesworth's Mag. Nat. Hist., 2:38, January. Type locality, possibly Carlton House, Saskatchewan.
1936. *Mustela frenata longicauda*, Hall, Carnegie Inst. Washington Publ. 473:105, November 20.

MARGINAL RECORDS.—Alberta: Baptiste Lake, near Athabasca (Soper, 1965:309); Islay. Saskatchewan: Wingard. Manitoba: Riding Mtn. (Soper, 1961:207); *Delta* (Tamsitt, 1962:77, as *M. frenata* only); 1 mi. E Delta (*ibid.*); Winnipeg (Soper, 1961:207). North Dakota: Grafton; Larimore; Valley City. Nebraska (Jones, 1964c:270): Neligh; Inland. Kansas: Glasco; Trego County. Colorado (Armstrong, 1972:277): 3 mi. S, 4 mi. W Kit Carson; Deer Trail; 3 mi. NE Windsor. Wyoming: Fort Laramie. Nebraska (Jones, 1964c:271): Warbonnet Canyon; *21 mi. NW Crawford.* South Dakota: Rapid City. Montana: 7 mi. N, 4 mi. W Camp Crook (Lampe, *et al.*, 1974:25); ¾ mi. N Ingomar; 5 mi. NW Hilger; St. Marys Lake. Alberta: Waterton Lakes Park; Canmore; Moose Pass.

Mustela frenata macrophonius (Elliot)

1905. *Putorius macrophonius* Elliot, Proc. Biol. Soc. Washington, 18:235, December 9, type from Achotal, Veracruz.
1936. *Mustela frenata macrophonius*, Hall, Carnegie Inst. Washington Publ. 473:109, November 20.

MARGINAL RECORDS.—Veracruz: type locality; Pérez. Oaxaca (Goodwin, 1969:234): Limón; Ayutla.

Mustela frenata munda (Bangs)

1899. *Putorius xanthogenys mundus* Bangs, Proc. New England Zool. Club, 1:56, June 9, type from Point Reyes, Marin Co., California.
1936. *Mustela frenata munda*, Hall, Carnegie Inst. Washington Publ. 473:107, November 20.

MARGINAL RECORDS.—California: South Yolla Bolly Mt.; Mt. Sanhedrin; 4 mi. N Vallejo, thence W along Carquinez Straits and San Francisco Bay, thence N along coast at least to Fort Bragg; probably up coast to 40° N, thence to point of beginning.

Mustela frenata neomexicana (Barber and Cockerell)

1898. *Putorius frenatus neomexicanus* Barber and Cockerell, Proc. Acad. Nat. Sci. Philadelphia, 50:188, May 3, type from Armstrongs Lake, Mesilla, Dona Ana Co., New Mexico.
1912. *Mustela frenata neomexicana*, Miller, Bull. U.S. Nat. Mus., 79:100, December 31.

MARGINAL RECORDS.—Colorado: 2½ mi. S Two Buttes (Armstrong, 1972:277). Kansas: Liberal. Texas: Lozier. Durango: 1½ mi. W Gregio García (Baker and Greer, 1962:138); Durango. Chihuahua: Guachochic (Anderson, 1972:381). Arizona: Tombstone; *2 mi. W Willcox* (Cockrum, 1961:233); Safford. New Mexico: 3 mi. NW Albuquerque; 11 mi. SW Santa Fe; 5 mi. W

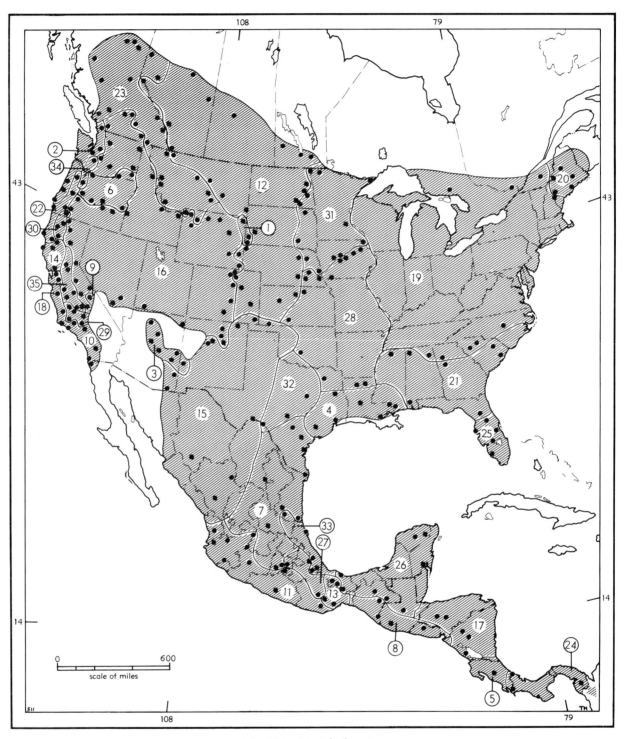

Map 511. *Mustela frenata*.

Guide to subspecies
1. *M. f. alleni*
2. *M. f. altifrontalis*
3. *M. f. arizonensis*
4. *M. f. arthuri*
5. *M. f. costaricensis*
6. *M. f. effera*
7. *M. f. frenata*
8. *M. f. goldmani*

9. *M. f. inyoensis*
10. *M. f. latirostra*
11. *M. f. leucoparia*
12. *M. f. longicauda*
13. *M. f. macrophonius*
14. *M. f. munda*
15. *M. f. neomexicana*
16. *M. f. nevadensis*
17. *M. f. nicaraguae*

18. *M. f. nigriauris*
19. *M. f. noveboracensis*
20. *M. f. occisor*
21. *M. f. olivacea*
22. *M. f. oregonensis*
23. *M. f. oribasus*
24. *M. f. panamensis*
25. *M. f. peninsulae*
26. *M. f. perda*

27. *M. f. perotae*
28. *M. f. primulina*
29. *M. f. pulchra*
30. *M. f. saturata*
31. *M. f. spadix*
32. *M. f. texensis*
33. *M. f. tropicalis*
34. *M. f. washingtoni*
35. *M. f. xanthogenys*

996

Capulin (Best, 1971:210, as *M. frenata* only). Colorado: *near Springfield* (Armstrong, 1972:277).

Mustela frenata nevadensis Hall

1936. *Mustela frenata nevadensis* Hall, Carnegie Inst. Washington Publ. 473:91, November 20, type from 3 mi. E Baker, White Pine Co., Nevada.

MARGINAL RECORDS.—British Columbia: Sicamous; *Monashee Pass;* Nelson. Washington: Ione. Idaho: Lochsa River [Locksaw Fork]; Leadore; Dry Creek, Targee National Forest. Wyoming (Long, 1965a: 694, 695, unless otherwise noted): *Yellowstone Lake;* Lamar River; 20 mi. NW Dubois; *1 mi. S, 3 mi. E Dubois;* Greybull River; Story; *Buffalo;* 42 mi. S, 13 mi. W Gillette; *10 mi. N, 6 mi. W Bill;* Garrett; *Hawk Springs Reservoir;* 15 mi. E Cheyenne. Colorado (Armstrong, 1972:278, 279): 10⅛ mi. NE Nunn; Fort Collins; Barr; 5 mi. E Colorado Springs; 1 mi. NW Blanca. New Mexico: 2 mi. N Twining, 10,500 ft.; Pecos Baldy, 11,600 ft.; Ribera. Arizona: 15 mi. E Lukachukai Navajo School, 8000 ft. Utah: St. George. Nevada: 3 mi. S Crystal Spring, 3900 ft., Pahranagat Valley; Nevada Test Site (Jorgensen and Hayward, 1965:10). California: near Big Prospector Meadow, 10,300 ft.; *Little Onion Valley, 7500 ft.;* Quaking Aspen Meadow, 7500 ft.; *E fork Kaweah River, 9800 ft.;* Hume; Merced Grove, Big Trees, 5400 ft.; Blue Canyon, 5000 ft.; Dale's, 600 ft., on Paines Creek; Mt. Shasta; Hornbrook. Oregon: Rustler Peak, Crater National Forest; *Anna Creek, Mt. Mazama;* 20 mi. W Crescent; Burns; 2 mi. NW Riverside. Idaho: 2 mi. S Payette, thence along Snake River to Washington: Snake River; 1 mi. W Moxee; Easton; Barron, 5000 ft. British Columbia: Hope (Cowan and Guiguet, 1965:319); Monte Creek, 20 mi. E Kamloops.

Mustela frenata nicaraguae J. A. Allen

1916. *Mustela tropicalis nicaraguae* J. A. Allen, Bull. Amer. Mus. Nat. Hist., 35:100, April 28, type from Matagalpa, Nicaragua.
1936. *Mustela frenata nicaraguae,* Hall, Carnegie Inst. Washington Publ. 473:109, November 20.

MARGINAL RECORDS.—Honduras: Cerro Grande La Paz; La Flor Archaga. Nicaragua: San Rafael del Norte; type locality; Mombacho.

Mustela frenata nigriauris Hall

1936. *Mustela frenata nigriauris* Hall, Carnegie Inst. Washington Publ. 473:95, November 20, type from 2½ mi. E Santa Cruz, Santa Cruz Co., California.

MARGINAL RECORDS.—California: *Glen Frazer Station; Concord;* 7 mi. E Clayton; ¼ mi. N Milpitas; Gonzales; 5 mi. SE Santa Margarita; 5 mi. N Las Cruces, thence N along coast and S shores of San Francisco Bay to point of beginning.

Mustela frenata noveboracensis (Emmons)

1840. *Putorius Noveboracensis* Emmons, A report on the quadrupeds of Massachusetts, p. 45. Type locality, Williamstown, Berkshire Co., Massachusetts.
1936. *Mustela frenata noveboracensis,* Hall, Carnegie Inst. Washington Publ. 473:104, November 20.
1842. *Mustela fusca* Audubon and Bachman, Jour. Acad. Nat. Sci. Philadelphia, 8(Pt. 2):288, type from state of New York.
1853. *Putorius agilis* Audubon and Bachman, The viviparous quadrupeds of North America, 3:184, Pl. 140, type from state of New York. Not *Mustela agilis* Tschudi, 1844, a name applied to the weasel of Perú.
1899. *Putorius noveboracensis notius* Bangs, Proc. New England Zool. Club, 1:53, June 9, type from Weaverville, Buncombe Co., North Carolina.

MARGINAL RECORDS.—Ontario: Metagama; Ottawa. Quebec: Black Lake. Maine: Bethel, thence along coast from southern Maine to North Carolina: Raleigh. South Carolina: 5 mi. E York; Laurens. Georgia: Roswell. Alabama: Fort Payne, west to Mississippi River, north up E bank to St. Croix River and north along Wisconsin–Minnesota boundary to Lake Superior, thence east along S shore to point of beginning.

Mustela frenata occisor (Bangs)

1899. *Putorius occisor* Bangs, Proc. New England Zool. Club, 1:54, June 9, type from Bucksport, Hancock Co., Maine.
1936. *Mustela frenata occisor,* Hall, Carnegie Inst. Washington Publ. 473:104, November 20.

MARGINAL RECORDS.—Maine: Moosehead Lake; 3rd Mopang Lake; type locality; *South Andover;* Umbagog Lake; Seven Pond Township.

Mustela frenata olivacea Howell

1913. *Mustela peninsulae olivacea* Howell, Proc. Biol. Soc. Washington, 26:139, May 21, type from Autaugaville, Autauga Co., Alabama.
1936. *Mustela frenata olivacea,* Hall, Carnegie Inst. Washington Publ. 473:104, November 20.

MARGINAL RECORDS.—Alabama: 9 mi. N Tennessee River near Leighton (in Lauderdale Co.); *White Creek.* Georgia: Spalding County. South Carolina: Society Hill; Sampit, thence south along coast to Florida: Silver Springs, thence west to coast and along coast to Alabama: 12 mi. NE Mobile. Mississippi: 7 mi. SW Holly Springs (Kennedy, et al., 1974:26).

Mustela frenata oregonensis (Merriam)

1896. *Putorius xanthogenys oregonensis* Merriam, N. Amer. Fauna, 11:25, June 30, type from Grants Pass, Rogue River Valley, Josephine Co., Oregon.
1936. *Mustela frenata oregonensis,* Hall, Carnegie Inst. Washington Publ. 473:107, November 20.

MARGINAL RECORDS.—Oregon: Forest Grove; McKenzie Bridge; 6 mi. S Medford. California: 12 mi. E Bridgeville; *2 mi. W Bridgeville;* Ferndale. Oregon: Goldbeach; *Buchanan.*

Mustela frenata oribasus (Bangs)

1899. *Putorius (Arctogale) longicauda oribasus* Bangs, Proc. New England Zool. Club, 1:81, December 27, type from source of Kettle River, 7500 ft., on the summit between middle fork Kettle River and Cherry Creek at Pinnacles, British Columbia.
1936. *Mustela frenata oribasa*, Hall, Carnegie Inst. Washington Publ. 473:105, November 20.

MARGINAL RECORDS.—British Columbia: W of Hudson Hope; Little Prairie. Alberta: Blueberry Mtn. (Soper, 1965:309, as *M. f. longicauda*); Thoral Creek, 7000 ft., 50 mi. NE Jasper. British Columbia: Isaacs Lake, 3200 ft.; E side Beaverfoot Range, 4000–4500 ft., between Fraser Creek and 6 mi. SE Fraser Creek; *head Cross River 10 mi. below Assiniboine Pass;* Paradise Mine (Cowan and Guiguet, 1965:319). Montana: Chief Mountain Lake; Highwood Mts.; Big Snowy Mts.; Tacy. Wyoming: Four Bears; Glen Creek, Mammoth Hot Springs. Montana: Donovan; Tin Cup District; Florence. British Columbia: camp E of Kootanie River; type locality; Lillooet; Alta Lake; Chezacut; Wistaria; Clearwater River, tributary to Peace River.

Mustela frenata panamensis Hall

1932. *Mustela frenata panamensis* Hall, Proc. Biol. Soc. Washington, 45:139, September 9, type from Río Indio, Canal Zone, near Gatún, Panamá.

MARGINAL RECORDS.—Panamá: Siola and Río Gariché, 5300 ft., eastward at least to Mt. Pierre.

Mustela frenata peninsulae (Rhoads)

1894. *Putorius peninsulae* Rhoads, Proc. Acad. Nat. Sci. Philadelphia, 46:152, June 19, type from Hudson's [14 mi. N Tarpon Springs], Pasco Co., Florida.
1936. *Mustela frenata peninsulae*, Hall, Carnegie Inst. Washington Publ. 473:105, November 20.

MARGINAL RECORDS.—Florida: Enterprise; *Osceola;* Melbourne (Moore, 1945:248); Hicoria (*ibid.*); 16 mi. SE Naples (1395 Larry N. Brown); Tarpon Springs; *Hernando County.*

Mustela frenata perda (Merriam)

1902. *Putorius tropicalis perdus* Merriam, Proc. Biol. Soc. Washington, 15:67, March 22, type from Teapa, Tabasco.
1936. *Mustela frenata perda*, Hall, Carnegie Inst. Washington Publ. 473:109, November 20.

MARGINAL RECORDS.—Yucatán: 14 km. SW X-Can (Birney, et al., 1974:17). Quintana Roo: 15 km. NNE Chetumal (Ramírez-P. and Phillips, 1971:145). Guatemala: *Cobán;* Finca la Providencia, San Cristóbal Verapaz. Chiapas: San Vicente; San Cristóbal. Veracruz: 35 km. SE Jesús Carranza, 400 ft. (Hall and Dalquest, 1963:338); Catemaco. Yucatán: Chichén-Itzá.

Mustela frenata perotae Hall

1936. *Mustela frenata perotae* Hall, Carnegie Inst. Washington Publ. 473:100, November 20, type from 12,500 ft., Cofre de Perote, Veracruz.

MARGINAL RECORDS.—Veracruz: Perote (Hall and Dalquest, 1963:338). Oaxaca (Goodwin, 1969:233): San Felipe del Agua; *Cerro San Felipe.* México: Monte Río Frio, 45 km. ESE Mexico City.

Mustela frenata primulina Jackson

1913. *Mustela primulina* Jackson, Proc. Biol. Soc. Washington, 26:123, May 21, type from 5 mi. NE Avilla, Jasper Co., Missouri.
1936. *Mustela frenata primulina*, Hall, Carnegie Inst. Washington Publ. 473:104, November 20.

MARGINAL RECORDS.—Iowa: Fayette, thence east to Mississippi River and down W bank to Louisiana (Lowery, 1974:427): Swartz; *Choudrant;* W of Ruston. Texas: 10 mi. S Sulphur Springs. Oklahoma: Norman. Kansas: Pratt; near Winkler. Nebraska (Jones, 1964c:272): *2 mi. N Geneva;* 1 mi. S Fairmont; Garrison; Nebraska City. Iowa: Taylor County; *sec. 21, Worth Twp. in Boone Co.;* 2 mi. E Ledges State Park; Union; New Hartford.

Mustela frenata pulchra Hall

1936. *Mustela frenata pulchra* Hall, Carnegie Inst. Washington Publ. 473:98, November 20, type from Buttonwillow, Kern Co., California.

MARGINAL RECORDS.—California: Coalinga; *4 mi. E Coalinga;* Alila [= Earlimart]; Isabella; Willow Spring; 2 mi. SW Simmler; *3 mi. S Coalinga.*

Mustela frenata saturata (Merriam)

1896. *Putorius saturatus* Merriam, N. Amer. Fauna, 11:21, June 30, type from Siskiyou, Jackson Co., Oregon.
1936. *Mustela frenata saturata*, Hall, Carnegie Inst. Washington Publ. 473:106, November 20.

MARGINAL RECORDS.—Oregon: Mt. Ashland; *Siskiyou.* California: Jackson Lake, 5900 ft.; South Fork Mtn., 5500 ft.; summit E of Hoopa, 5800 ft.

Mustela frenata spadix (Bangs)

1896. *Putorius longicauda spadix* Bangs, Proc. Biol. Soc. Washington, 10:8, February 25, type from Fort Snelling, Hennepin Co., Minnesota.
1936. *Mustela frenata spadiz*, Hall, Carnegie Inst. Washington Publ. 473:105, November 20.

MARGINAL RECORDS.—Minnesota: Kittson County; 2½ mi. SW Roseau; Lake Saganaga area (Timm, 1975:39); type locality. Iowa: Lansing; Ames; *sec. 19, Amaqua Twp., Boone Co.;* Pilot Mound. Nebraska (Jones, 1964c:273): Omaha; Norfolk. South Dakota: Armour; Fort Sisseton. North Dakota: Oakes; Casselton.

Mustela frenata texensis Hall

1936. *Mustela frenata texensis* Hall, Carnegie Inst. Washington Publ. 473:99, November 20, type from Kerr Co., Texas.

MARGINAL RECORDS.—Texas: Erath; *5 mi. N Waco;* Fredericksburg; *Kerr County;* Del Rio (Davis, 1961:273).

Mustela frenata tropicalis (Merriam)

1896. *Putorius tropicalis* Merriam, N. Amer. Fauna, 11:30, June 30, type from Jico, Veracruz.
1936. *Mustela frenata tropicalis,* Hall, Carnegie Inst. Washington Publ. 473:109, November 20.

MARGINAL RECORDS.—Tamaulipas: 50 mi. S Victoria. Veracruz (Hall and Dalquest, 1963:338): Hda. Tamiahua, Cabo Rojo; 3 km. N Paraje Nuevo, 1700 ft.; Orizaba; Las Vigas, 8500 ft.

Mustela frenata washingtoni (Merriam)

1896. *Putorius washingtoni* Merriam, N. Amer. Fauna, 11:18, June 30, type from Trout Lake, Mt. Adams, Washington.
1936. *Mustela frenata washingtoni,* Hall, Carnegie Inst. Washington Publ. 473:106, November 20.

MARGINAL RECORDS.—Washington: Spray Park; *Owyhigh Lakes; Gotchen Creek, 3500 ft.* Oregon: Permilia Lake, Mt. Jefferson. Washington: *type locality; Nisqually entrance.*

Mustela frenata xanthogenys Gray

1843. *Mustela xanthogenys* Gray, Ann. Mag. Nat. Hist., ser. 1, 11:118, type probably from bank of Sacramento River below mouth Feather River, or possibly from N shore San Francisco Bay, California.
1936. *Mustela frenata xanthogenys,* Hall, Carnegie Inst. Washington Publ. 473:107, November 20.

MARGINAL RECORDS.—California: type locality; 5 mi. S Fair Oaks; Clovis; Milo; *Poplar;* 2 mi. N Tipton; Mendota; Los Baños.

Subgenus Putorius Cuvier—Ferrets and Polecats

1817. *Putorius* Cuvier, Le règne animal . . . , 1:147. Type, *Mustela putorius* Linnaeus.
1840. *Foetorius* Keyserling and Blasius, Wirbelthiere Europas, p. 66. Type, *Mustela putorius* Linnaeus.
1877. *Cynomyonax* Coues, Fur-bearing animals . . . , U.S. Geol. Surv. Territories, Miscl. Publ., 8:99. Type, *Putorius nigripes* Audubon and Bachman.

Fur not modified for aquatic life; mastoid processes of skull remarkably angular and projecting.

This subgenus includes the Old World polecats and the ferrets of the Old and New World. The characters of some of the American weasels and some Old World polecats so nearly grade into each other that the subgeneric name *Putorius* could be discarded.

Mustela nigripes (Audubon and Bachman)
Black-footed Ferret

1851. *Putorius nigripes* Audubon and Bachman, The viviparous quadrupeds of North America, 2:297, type from Fort Laramie, Goshen Co., Wyoming.
1912. *Mustela nigripes,* Miller, Bull. U.S. Nat. Mus., 79:102, December 31.

External measurements of males: 500–533; 114–127; 60–73; ear from notch, 29–31. Basilar length of skulls of two males from Coolidge, Kansas, 62.2, 66.8; zygomatic breadth, 38.8 and 43.0. Tail vertebrae, 17 in number; their combined length 22 to 25 per cent of length of head and body. Females approx. 10 per cent smaller in linear measurements. Pelage pale yellowish buff becoming nearly white on face, throat, ventral half of body; top of head and middle of back brown; black mask across eyes; feet, legs, and terminal fourth of tail black to blackish brown; remainder of tail whitish or with top side like back. Pelage in winter only slightly paler than in summer.

The geographic range of the black-footed ferret nearly coincides with the range of the prairie dogs (genus *Cynomys*). The ferret may be dependent on the burrows of the prairie dog for shelter and on the occupants of burrows for food.

MARGINAL RECORDS.—Alberta: Rosebud (Soper, 1965:314). Saskatchewan: 3 mi. W Grayburn; 4 mi. SE Regina. North Dakota: near Stanton; near Fort Rice. South Dakota: Pierre; Wessington Springs. Nebraska: Fremont; Lincoln. Kansas: E of Lucas; Kingman. Oklahoma: 1 mi. E Norman. Texas: Gainesville; Seymour; Fort Stockton. New Mexico: Center-fire Creek, 10 mi. NE Luna. Arizona: *12 mi. W Winona* (Hoffmeister and Carothers, 1969:188); Government Prairie, near Parks (Cockrum, 1961:234); Red Lake

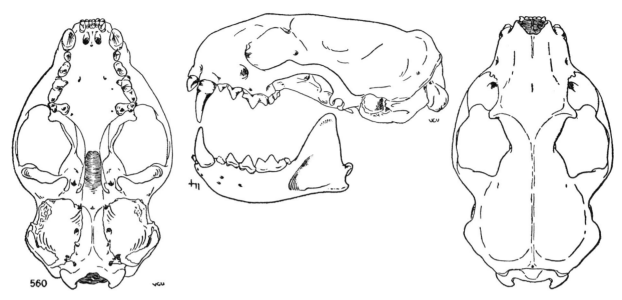

Fig. 560. *Mustela nigripes,* near Coolidge, Hamilton Co., Kansas, No. 11077 K.U., ♂, X 1.

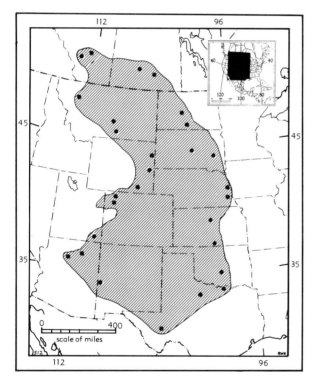

Map 512. *Mustela nigripes.*

Farming district N of Williams; Howell Mesa (Cockrum, 1961:234). Utah: 2 mi. S Blanding. Colorado: within a mile of Meeker; 22 mi. N Craig. Wyoming (Long, 1965a:697): 5 mi. W Laramie; Manville; Newcastle. Montana: 10 mi. SE Billings; 5 mi. S Lavina; Chouteau. Alberta: Calgary (Soper, 1965:314).

Subgenus **Lutreola** Wagner—Minks

Revised by Hollister, Proc. U.S. Nat. Mus., 44:471–480, April 18, 1913.

1841. *Lutreola* Wagner, *in* Schreber, Die Säugthiere . . . , Suppl., 2:239. Type, *Mustela lutreola* Linnaeus.
1843. *Vison* Gray, List of the . . . Mammalia in the . . . British Museum, p. 64. Type, *Mustela lutreola* Linnaeus.
1871. *Hydromustela* Bogdanow, Trudy Obshtch. yestestvoispytateley Imp. Kazan. Univ. 1, otd. 1, p. 167 (not seen). Type, *Mustela lutreola* Linnaeus.

Fur modified for aquatic life; m1 with suggestion of metaconid; tympanic bullae less flattened than in most members of the subgenera *Mustela* and *Putorius.* The characters of some members of the subgenus *Mustela* so nearly intergrade with those of the subgenus *Lutreola* that recognition of the latter may serve no useful purpose. The subgenus *Lutreola* includes the minks of Eurasia and North America.

Mustela vison
Mink

External measurements of males: 491–720; 158–194; 57–75. Basilar length of skull, 58–69; zygomatic breadth, 36–47. Tail vertebrae 18–20 in number and averaging 41–51 per cent of length of head and body. Females averaging 10 per cent or more smaller than males and half as heavy. Dark brown except for white spots somewhere on underparts (throat, chest or belly) of some individuals; brown most intense near tip of tail and along middle of back.

Mustela vison aestuarina Grinnell

1916. *Mustela vison aestuarina* Grinnell, Proc. Biol. Soc. Washington, 29:213, September 22, type from Grizzly Island, Solano Co., California.

MARGINAL RECORDS.—California: Jackson Lake; Eagle Lake. Nevada: 7 mi. S Reno; Wadsworth; 1 mi. NNW Fallon, 4000 ft.; Walker River, 14 mi. N Yerington; E. Fork Carson River, 5 mi. SE Minden. California: Fish Springs; Summit Lake, 12 mi. NW Lemoore; Firebaugh; type locality; Carlotta.

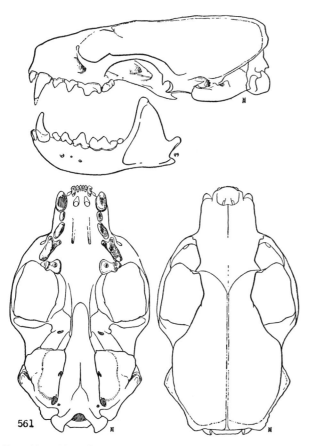

Fig. 561. *Mustela vison energumenos*, Goose Creek, 5000 ft., Elko Co., Nevada, No. 74391 M.V.Z., ♂, X 1.

Mustela vison aniakensis Burns

1964. *Mustela vison aniakensis* Burns, Canadian Jour. Zool., 42:1073, November, type from vic. Aniak, along Salmon River, Alaska.

MARGINAL RECORDS.—Alaska: type locality; Kakhtul River (reported occurrence by Osgood, 1904:45, as *L. v. melampeplus*); Nushagak River (*ibid.*); Platinum (Burns, 1964:1073).

Mustela vison energumenos (Bangs)

1896. *Putorius vison energumenos* Bangs, Proc. Boston Soc. Nat. Hist., 27:5, March, type from Sumas, British Columbia.
1912. *Mustela vison energumenos*, Miller, Bull. U.S. Nat. Mus., 79:101, December 31.

MARGINAL RECORDS.—Yukon: Old Crow (Youngman, 1975:149). Mackenzie: Fort Norman; Fort Simpson. Alberta: Athabaska River near Boiler Rapids; near Rochester (Keith and Meslow, 1966:541, as *M. vison* only). Montana: Glacier National Park. Wyoming (Long, 1965a:696): 12 mi. S, 15 mi. W Cody; *15¹/₂ mi. S, 13 mi. W Meeteetse*; Wind River, 3 mi. N, 6 mi. W Burris; Woods P.O. Colorado: Loveland; Colorado Springs; La Veta. New Mexico: vic. Cimarron; 3¹/₂ mi. E Watrous (Yarbough and Studier, 1968:105); Las Vegas (*ibid.*); Los Lunas, Rio Grande (Findley, *et al.*, 1975:308, as *M. vison*); Farmington (*ibid.*). Utah: La Sal National Forest. Colorado: Rangely; Green River near Ladore. Utah: Sheep Creek; below Murdock Power Plant, Provo River; Mt. Timpanogos; small cave, S slope, mouth Emigration Canyon, 4800 ft.; Logan River. Idaho: Snake River, 3 mi. SE Acequia. Nevada: Goose Creek, 5000 ft., 100 yds. W Utah boundary; Deeth; Ruby Lake. California: Sisson, Cold Creek, thence west to Pacific Coast at approx. California–Oregon boundary and up the coast, including *Broughton, Price*, and *Princess Royal islands*, British Columbia (Cowan and Guiguet, 1965:322), to Alaska: Wrangell; lower Taku River. British Columbia: Atlin. Yukon: Champagne, N side Dezadeash River; Forty Mile (Youngman, 1975:148).

Mustela vison evagor Hall

1932. *Mustela vison evagor* Hall, Univ. California Publ. Zool., 38:418, November 8, type from Little Qualicum River [8–9 mi. W Parkaville], Vancouver Island, British Columbia. Geographic range Vancouver Island; possibly also San Juan Island (Long, The Murrelet, 44:25, December 1963); introduced (Cowan and Guiguet, 1965:322) on Lanz Island and possibly on Cox Island.

Mustela vison evergladensis Hamilton

1948. *Mustela vison evergladensis* Hamilton, Proc. Biol. Soc. Washington, 61:139, September 3, type from Tamiami Trail (U.S. Route 94), 5 mi. SE Royal Palm Hammock, Collier Co., Florida.

MARGINAL RECORDS.—Florida: type locality. Layne (1974:412, Fig. 1) plotted a cluster of 20 record stations, tentatively assigned to this subspecies, delineating the area shown on Map 513 herewith.

Mustela vison ingens (Osgood)

1900. *Lutreola vison ingens* Osgood, N. Amer. Fauna, 19:42, October 6, type from Fort Yukon, Alaska.

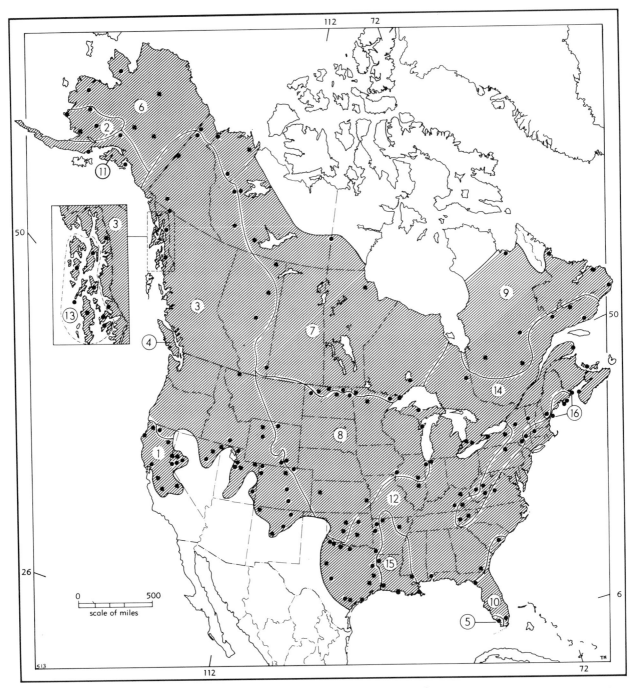

Map 513. *Mustela vison* and *Mustela macrodon.*

1. *M. v. aestuarina*	5. *M. v. evergladensis*	9. *M. v. lowii*	13. *M. v. nesolestes*
2. *M. v. aniakensis*	6. *M. v. ingens*	10. *M. v. lutensis*	14. *M. v. vison*
3. *M. v. energumenos*	7. *M. v. lacustris*	11. *M. v. melampeplus*	15. *M. v. vulgivaga*
4. *M. v. evagor*	8. *M. v. letifera*	12. *M. v. mink*	16. *M. macrodon*

1912. *Mustela vison ingens*, Miller, Bull. U.S. Nat. Mus., 79:101, December 31.

MARGINAL RECORDS.—Mackenzie: Mackenzie Delta; Anderson River; Fort Good Hope. Yukon: 40 mi. SE Crow Base, 68° 13', 141° (Youngman, 1975:150). Alaska: near Fairbanks; Wonder Lake, Mount McKinley Dist.; Andreafski; Imuruk Lake; Chandler Lake.

Mustela vison lacustris (Preble)

1902. *Lutreola vison lacustris* Preble, N. Amer. Fauna, 22:66, October 31, type from Echimamish River (near Painted Stone portage), Manitoba.
1912. *Mustela vison lacustris*, Miller, Bull. U.S. Nat. Mus., 79:101, December 31.

MARGINAL RECORDS.—Mackenzie: Great Bear Lake. Keewatin: Little Dubawnt River (Harper, 1956:68). Manitoba: up Churchill River. Ontario: Lake Nipigon. Minnesota: Grand Portage (Timm, 1975:34); *within 14 mi. radius N and W of Grand Marais (ibid.)*; Thief Lake Refuge. North Dakota: Pembina River; Turtle Mtn.; Souris River. Alberta: Lodge and North Fork creeks; Wood Buffalo Park. Mackenzie: Great Slave Lake.

Mustela vison letifera Hollister

1913. *Mustela vison letifera* Hollister, Proc. U.S. Nat. Mus., 44:475, April 18, type from Elk River, Sherburne Co., Minnesota.

MARGINAL RECORDS.—North Dakota: Lake Irwin; Drayton. Minnesota: Clearwater County. Michigan: Big Huron River. Wisconsin: Fish Creek (Jackson, 1961:357). Indiana: northwestern corner. Illinois: Urbana. Iowa: Burlington (Bowles, 1975:128). Kansas: Linn County. Oklahoma: Stillwater Creek near Stillwater; 2 mi. S Norman. Kansas: Logan County. Colorado: 3 mi. N Pawnee Butte. Wyoming (Long, 1965a:696): 6½ mi. SW Laramie; Story. North Dakota: Goodall.

Mustela vison lowii Anderson

1945. *Mustela vison lowii* Anderson, Ann. Rept. Provancher Soc. Nat. Hist. Canada, Quebec, for 1944, p. 57, November 2, type from Mistassini Post, Mistassini Lake, Mistassini Dist., Quebec.

MARGINAL RECORDS (Harper, 1961:118).— Quebec: Fort Chimo. Labrador: Nain; Sandwich Bay; Black Bay. Quebec: headwaters Peribonca River; Lake St. John; Lake Waswanipi.

Mustela vison lutensis (Bangs)

1898. *Putorius (Lutreola) lutensis* Bangs, Proc. Boston Soc. Nat. Hist., 28:229, March, salt marsh off Matanzas Inlet, St. Johns Co., Florida.
1913. *Mustela vison lutensis*, Hollister, Proc. U.S. Nat. Mus., 44:474, April 18.

MARGINAL RECORDS.—South Carolina: Mount Pleasant, down coast to Florida: Miami, thence westward to Gulf Coast and up coast to Cedar Keys.

Mustela vison melampeplus (Elliot)

1903. *Putorius vison melampeplus* Elliot, Field Columb. Mus., Publ. 74, Zool. Ser., 3:170, May 2, type from Kenai Peninsula, Alaska.
1912. *Mustela vison melampeplus*, Miller, Bull. U.S. Nat. Mus., 79:102, December 31.

MARGINAL RECORDS.—Alaska: Chulitna River; Disc Island, Prince William Sound; *Cold Bay;* Kukak Bay (Cahalane, 1959:186, as *Mustela vison*).

Mustela vison mink Peale and Palisot de Beauvois

1796. *Mustela mink* Peale and Palisot de Beauvois, A scientific and descriptive catalogue of Peale's museum, Philadelphia, p. 39. Type locality, Maryland.
1914. *Mustela vison mink*, Hollister, Proc. Biol. Soc. Washington, 27:215, October 31.
1825. *Mustela lutreocephala* Harlan, Fauna Americana, p. 63. Type locality, Maryland.
1858. *M[ustela]. rufa* Hamilton-Smith, The naturalist's library (edit. Jardine), 15:189. Type locality, New York.

MARGINAL RECORDS.—Maine: Flatt Island; Mount Desert Island (Manville, 1960:416, as *M. vison* only), thence down coast to South Carolina and Georgia: Hull's Branch, *ca.* 2 mi. above Camp Pinckney, Florida. Alabama: Bayou Labatre. Arkansas: near Augusta; around Huntsville; Alix. Texas: Texarkana; Panola County (Davis, 1966:89, as *M. vison* only); Rockland; Liberty County (Davis, 1966:89, as *M. vison* only); Matagorda (V. Bailey, 1905:196); lower Guadalupe River; DeWitt County (Davis, 1966:89, as *M. vison* only); Mason; Taylor County (Davis, 1966:89, as *M. vison* only); Wanderers Creek, 4 mi. SW Chillicothe; Vernon (Dalquest, 1968:20); Wichita Falls (*ibid.*); Cook County (Davis, 1966:89, as *M. vison* only). Oklahoma: Pittsburgh County; Mohawk Park. Illinois: near Ashmore. Indiana: South Bend. New York: Erie County; Peterboro. West Virginia: Peters Creek. Kentucky: Rowan County; N of Somerset (Fassler, 1974:43, as *M. vison* only). Virginia: Nelson County. New York: North Chatham. Massachusetts: Hampshire County.

Mustela vison nesolestes (Heller)

1909. *Lutreola vison nesolestes* Heller, Univ. California Publ. Zool., 5:259, February 18, type from Windfall Harbor, Admiralty Island, Alaska.
1912. *Mustela vison nesolestes*, Miller, Bull. U.S. Nat. Mus., 79:102, December 31.

MARGINAL RECORDS.—Alaska: type locality; Kupreanof Island; Cleveland Peninsula; Revillagigedo Island; Prince of Wales Island; Egg Harbor, Coronation Island; Sitka.

Mustela vison vison Schreber

1777. *Mustela vison* Schreber, Die Säugthiere . . . , Pl. 127b. Type locality, eastern Canada (= Quebec).

1854. *Putorius nigrescens* Audubon and Bachman, The viviparous quadrupeds of North America, 3:104. Name applied to animals from mts. of Pennsylvania, northern New York, Vermont, and Canada.

1858. *Mustela winingus* Baird, Mammals, *in* Repts. Expl. Surv. . . . , 8(1):177, July 14 (in synonymy).

1911. *Lutreola vison borealis* Brass, Aus dem Reiche der Pelze, p. 504, April. Type locality, northeastern North America.

MARGINAL RECORDS.—Quebec: Lac de Morhiban (Harper, 1961:116); lower Natashquan River (*ibid.*); 2 mi. W Coin du Banc (Manville, 1961:108, 109, sight record as *M. vison* only). Prince Edward Island (Anderson, 1945:58): *no specific locality*. Nova Scotia: *Cape Breton Island* (Cameron, 1959:33); Digby. Maine: Westbrook. Vermont: Rutland. New York: Catskill Mts. West Virginia: White Sulphur Springs. Tennessee: Roan Mtn.; Greenbrier. Kentucky: near Lynch. West Virginia: Shavers River. Pennsylvania: near Lopez. New York: Saranac Lake. Ontario: vic. Long Point; London; Mace, Lake Abitibi. Quebec: Eric Lake (Harper, 1961:116). Introduced in 1935 on Newfoundland and established there (Cameron, 1959:99).

Mustela vison vulgivaga (Bangs)

1895. *Putorius* (*Lutreola*) *vulgivagus* Bangs, Proc. Boston Soc. Nat. Hist., 26:539, July 31, type from Burbridge, Plaquemines Parish, Louisiana.

1912. *Mustela vison vulgivaga*, Miller, Bull. U.S. Nat. Mus., 79:102, December 31.

MARGINAL RECORDS.—Arkansas: Boone County. Louisiana (Lowery, 1974:431): 2 mi. NW Angie; *1 1/5 mi. S, 1 1/10 mi. E Pearl River;* 1 mi. NW Port Sulphur; *12 mi. SSE Golden Meadow;* Marsh Island; *Pecan Island; 7 mi. N, 1 mi. E Creole;* Carlyss; Interstate 20, 13½ mi. W Red River.

Mustela macrodon (Prentiss)
Sea Mink

1903. *Lutreola macrodon* Prentiss, Proc. U.S. Nat. Mus., 26:887, July 6, type from shellheaps at Brookline, Hancock Co., Maine.

1912. *Mustela macrodon*, Miller, Bull. U.S. Nat. Mus., 79:101, December 31.

1911. *Lutreola vison antiquus* Loomis, Amer. Jour. Sci., 31:228, March, type from Flagg Island, Casco Bay, Maine.

Differs from *M. vison* in larger size; for example, in males, the length of the upper molar–premolar tooth-row measures 21 (in holotype) instead of less than 20 mm. Known from skeletal remains only; according to Hardy (Forest and Stream, 61:125, August 15, 1903) the animal became extinct about the year 1860.

MARGINAL RECORDS (Waters and Ray, 1961:380).—New Brunswick: Campobello Island. Maine: Great Diamond Island, Casco Bay. Massachusetts: *Middleboro* (fossil remains).

Genus Eira Hamilton-Smith—Tayra

Reviewed (under the name *Tayra*) by Krumbiegel, Zoöl. Anzeiger, 139:81–96, August 15, 1942.

1842. *Eira* Hamilton-Smith, The naturalist's library (edit. Jardine), 35 (Mammalia, vol. 13, introduction):201, December. Type, *Eira barbara* [= *Mustela barbara* Linnaeus].

1816. *Tayra* Oken, Lehrbuch der Naturgeschichte, pt. 3 (Zoöl.), sect. 2, p. 1001. Type, *Mustela barbara* Linnaeus. Oken's names are not available (see Hershkovitz, Jour. Mamm., 30:295, August 17, 1949).

1839. *Eiraria* Lund, Ann. Sci. Nat., Paris, ser. 2, 11:232, April. Included species, *Mustela vittata* Linnaeus and *Mustela barbara* Linnaeus.

1843. *Galera* Gray, List of the . . . Mammalia in the . . . British Museum, pp. xx, 67. Type, *Mustela barbara* Linnaeus. Not *Galera* Browne, 1789 (see J. A. Allen, Bull. Amer. Mus. Nat. Hist., 16:377, October 11, 1902, and 24:586–589, September 11, 1908; and Hershkovitz, Jour. Mamm., 30:295, August 17, 1949).

Up to 1.1 m. long; body long, slender; tail long; neck long; head broad; legs short. Head and neck gray or brown; body black or dark brown; yellow or white spot on chest. Teeth sectorial, deuterocone of upper carnassial anterior in position and narrow-necked, occupying less than half inner surface of tooth; upper molar wider than long; in m1, talonid semibasined. Carpal pads fused, long and wide, and in contact with plantar pad; bursa of ear shallow and remote from posterior edge; subcaudal pouch shallow; baculum unbranched apically. Dentition, i. $\frac{3}{3}$, c. $\frac{1}{1}$, p. $\frac{3}{3}$, m. $\frac{1}{2}$.

Eira barbara
Tayra

External measurements of some Central American specimens: in males, 1075–1125, 425–455;

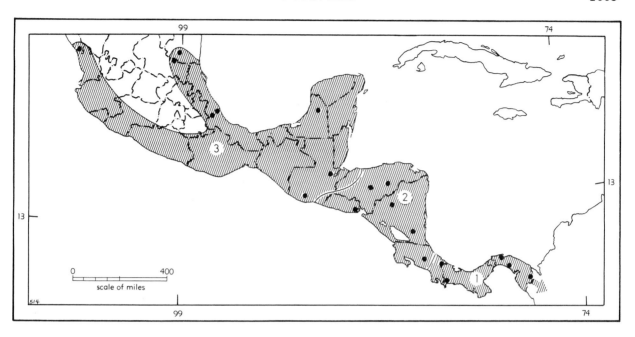

Map 514. *Eira barbara.*

1. *E. b. biologiae*   2. *E. b. inserta*   3. *E. b. senex*

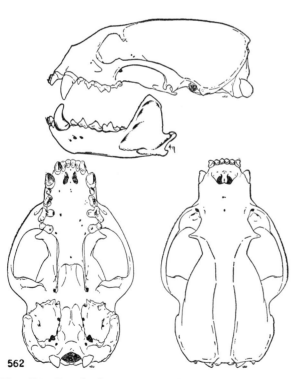

562

Fig. 562. *Eira barbara senex,* 38 km. SE Jesús Carranza, 500 ft., Veracruz, No. 24545 K.U., ♀, X ½.

in females 999–1095, 390–440. Other dimensions in males and females, respectively: condylobasal length, 101–120.5, 98–108.5; zygomatic breadth, 66.5–80.5, 60–68; length of maxillary tooth-row, 33–35, 33.9–38.7.

Eira barbara biologiae (Thomas)

1900. *Galictis barbara biologiae* Thomas, Ann. Mag. Nat. Hist., ser. 7, 5:146, January, type from Calovévora [= Calovébora], Veraguas, Panamá.
1959. *Eira barbara biologiae,* Hall and Kelson, The mammals of North America, Ronald Press, p. 920, March 31.

MARGINAL RECORDS.—Costa Rica: Cuabre. Panamá (Handley, 1966:790): Mandinga; Boca de Rio Paya, thence into South America, and northwestward along coast to Panamá (*ibid.*): Charco del Toro, Río Majé; Guánico; Bugaba.

Eira barbara inserta (J. A. Allen)

1908. *Tayra barbara inserta* J. A. Allen, Bull. Amer. Mus. Nat. Hist., 24:662, October 13, type from Uluse, Matagalpa, Nicaragua.
1959. *Eira barbara inserta,* Hall and Kelson, The mammals of North America, Ronald Press, p. 920, March 31.

MARGINAL RECORDS.—Honduras: Catacamas. Nicaragua: Peña Blanca. Costa Rica: Zarcéro. Nicaragua: type locality. El Salvador: Lake Olomega (Burt and Stirton, 1961:46). Honduras: Cantoral.

Eira barbara senex (Thomas)

1900. *Galictis barbara senex* Thomas, Ann. Mag. Nat. Hist., ser. 7, 5:146, January, type from Hda. Tortugas, approx. 600 ft., Jalapa, Veracruz.

1951. *Eira barbara senex*, Hershkovitz, Fieldiana-Zool., 31:561, July 10.

MARGINAL RECORDS.—Tamaulipas: Pano Ayuctle. Veracruz: type locality. Campeche: Dzibalchén (95113 KU). Guatemala: Finca Concepción; Finca El Espino, thence northward along coast to Sinaloa: Escuinapa. Veracruz: Mirador (Hall and Dalquest, 1963:339). San Luis Potosí: 8 mi. E Santa Barbarita.

Genus **Galictis** Bell—Grisóns

Revised (under the name *Grison*) by Krumbiegel, Zoöl. Anzeiger, 139:97–108, August 15, 1942, but see remarks below under species *Galictis vittata*.

1826. *Galictis* Bell, Zool. Jour., 2:552, April. Type, *Viverra vittata* Schreber.
1816. *Grison* Oken, Lehrbuch der Naturgeschichte, pt. 3 (Zoöl.), sect. 2, p. 1000. Type, *Viverra vittata* Schreber. Oken's names regarded as unavailable by Hershkovitz, Jour. Mamm., 30:289–301, August 17, 1949.
1825. *Grisonia* Gray, Ann. Philos., 26:339, a *nomen nudum*.
1829. *Grisonia* Fischer, Synopsis mammalium, p. 154, a *nomen nudum*.
1843. *Grisonia* Gray, List of the . . . Mammalia in the . . . British Museum, p. 68. Based on *Viverra vittata* of Schreber.
1912. *Grisonella* Thomas, Ann. Mag. Nat. Hist., ser. 8, 10:46. Type, *Mustela cuja* Molina.

Up to 96 cm. long; tail short; body slightly flattened dorsoventrally. Upper parts black or brown, grizzled with white- or yellow-tipped hairs; underparts black or brown, with few white- or yellow-tipped hairs. Teeth sectorial, deuterocone of upper carnassial occupying nearly half of inner surface of tooth; upper molar wider than long; in m1, talonid trenchant. Carpal pads fused, as long and wide as plantar pad, but separated from it. Pinna of ear well developed with small bursa in front of posterior margin. Subcudal pouch absent. Dentition, i. $\frac{3}{3}$, c. $\frac{1}{1}$, p. $\frac{3}{3}$, m. $\frac{1}{2}$.

Galictis vittata
Grisón

The specific name *G. vittata* Schreber, 1776, Die Säugthiere . . . , 3:447, is provisionally used here on the

assumption that it was based on a specimen of the larger of the two living species. The smaller species is currently designated as *G. cuja* and seems not to occur in Panamá or north thereof. The first "reviser" and most subsequent "revisers" applied the name *vittata* to only the larger species. Nevertheless, Krumbiegel, 1942, Zoöl. Anzeiger, 139:104, used the name *Grison allamandi* for the larger species on the assumption that the earlier name, *G. vittata*, was based on the smaller species. Cabrera, 1958, Rev. Mus. Argentino de Cienc. Nat., 4:258, 259, did not comment on Krumbiegel's publication probably because he did not see it, but Cabrera's arrangement of names for the two species of *Galictis* (*vittata* for the larger and *cuja* for the smaller) might not have differed if he had seen Krumbiegel's paper.

Characters essentially as for the genus. *G. vittatus* differs from *G. cuja*, the only other species in being larger (length of head and body to 800 mm., basal length to 90 (in males, 85 in females), with shorter tail (up to 20% of length of head and body), having fewer (17 or 18) tail vertebrae, having a metaconid on m1 and a stronger cingulum on the inner side of P3, and having the underparts extensively flecked with white.

Galictis vittata canaster Nelson

1901. *Galictis canaster* Nelson, Proc. Biol. Soc. Washington, 14:129, August 9, type from near Tunkás, Yucatán.
1958. *Galictis vittata canaster*, Cabrera, Rev. Mus. Argentino de Cienc. Nat., 4:259, March 27.

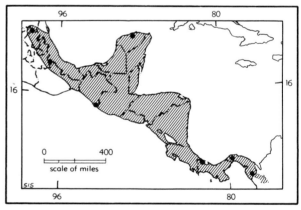

Map 515. *Galictis vittata canaster.*

MARGINAL RECORDS.—San Luis Potosí: Xilitla, thence eastward to Gulf Coast and along coast to Yucatán: Buctzatz, thence along Caribbean Coast to Panamá (Handley, 1966:790): Almirante; Buena Vista; Cana, and into South America; northward along Pacific Coast to Chiapas: Río Coatán. Veracruz: Orizaba.

Genus **Gulo** Pallas—Wolverines

1780. *Gulo* Pallas, Spicilegia zoologica . . . , fasc. 14, p. 25. Type, *Gulo sibiricus* Pallas [= *Mustela gulo* Linnaeus].

External measurements of male: 900–1125; 190–260; 180–192. Basal length of skull, 127–140.

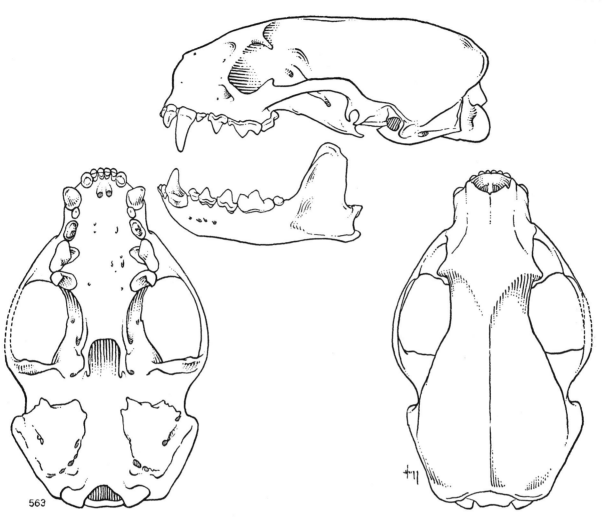

Fig. 563. *Galictis vittata canaster,* 20 km. ENE Jesús Carranza, 200 ft., Veracruz, No. 32222 K.U., Sex?, X 1.

Weight, 24–40 pounds. Females average 10 per cent less in linear measurements and 30 per cent less in weight. General color, blackish brown; paler and grayer on crown and cheeks; broad light brown stripe begins on each side of neck and extends back along side of body to join its opposite over and across base of tail; throat and chest more or less spotted with yellowish white; whitish on middle of abdomen over area of dermal gland there. Facial angle of skull steep; tympanic bullae moderately inflated and separate from paroccipital processes; palate behind upper molars; inner moiety of M1 larger than outer; P4 with simple deuterocone; in m1, talonid longer than trigonid, metaconid absent, talonid barely semitrenchant and with single, median longitudinal ridge. Dentition, i. $\frac{3}{3}$, c. $\frac{1}{1}$, p. $\frac{4}{4}$, m. $\frac{1}{2}$. Foot-soles naked in summer and densely haired in winter; largest living North American member of its family.

Gulo luscus
Wolverine

Characters as given for the genus. Some writers including Degerbøl (1935:35) regarded *Gulo lus-*

cus luscus of North America as only subspecifically different from *Gulo gulo* of Eurasia, as did also Kurtén and Rausch (1959:19). The two authors last named based their case on carefully prepared data obtained from a comparison of skulls from Fennoscandia and Alaska. Comparison of Alaskan skulls with skulls from northeastern Asia (instead of from Fennoscandia) would yield information more useful in deciding on subspecific versus specific relationship of Recent wolverines of the Old- and New-World. Until that information becomes available, the conservative arrangement of recognizing two Recent species is adopted here.

Gulo luscus katschemakensis Matschie

1918. *Gulo katschemakensis* Matschie, Sitzungsb. Gesell. Naturforsch. Freunde, Berlin, p. 151, July 30, type from Katschemak Bay, Kenai Peninsula, Alaska. Known only from the Kenai Peninsula. Examination of the holotype by Hall (in 1937), and examination of another individual from the Kenai Peninsula (Tustumena Lake, 2977 KU), indicate that *G. katschemakensis* is at most subspecifically different from *Gulo luscus*.

Gulo luscus luscus (Linnaeus)

1758. [*Ursus*] *luscus* Linnaeus, Syst. nat., ed. 10, 1:47. Type locality, Hudson Bay.

1823. *Gulo luscus*, Sabine, *in* Franklin, Narrative of a journey to the shores of the Polar Sea in . . . 1819–22, p. 650. (Regarded as a subspecies of *Gulo gulo* by Degerbøl, Rept. 5th Thule Exped., 2(No. 4):35, 1935, and by some later writers, including Kurtén and Rausch, Acta Arctica, 11:19, 1959.)

1905. *Gulo hylaeus* Elliot, Proc. Biol. Soc. Washington, 18:81, February 21, type from Susitna River, in region of Mt. McKinley, Alaska. (Regarded as inseparable from the subspecies *luscus* by Rausch, Arctic, 6:114, 1953.)

1918. *Gulo niediecki* Matschie, Sitzungsb. Gesell. Naturforsch. Freunde, Berlin, p. 148, July 30, type from Dease Lake, British Columbia. (Regarded by Anderson, Bull. Nat. Mus. Canada, 102:69, January 24, 1947, as indistinguishable from subspecies *luscus*.)

1918. *Gulo auduboni* Matschie, Sitzungsb. Gesell. Naturforsch. Freunde, Berlin, p. 153, July 30. Based on specimens from Newfoundland and New York. (Regarded by Anderson, Bull. Nat. Mus. Canada, 102:68, January 24, 1947, as indistinguishable from subspecies *luscus*.)

1918. *Gulo bairdi* Matschie, Sitzungsb. Gesell. Naturforsch. Freunde, Berlin, p. 153, July 30, type from NW of Fort Union, situated near present town of Buford, Williams Co., North Dakota. (Regarded by Anderson, Bull. Nat. Mus. Canada, 102:69, January 24, 1947, as indistinguishable from subspecies *luscus*.)

MARGINAL RECORDS.—Franklin: Ellesmere Island; Baffin Land, Pond Inlet; Cumberland Sound on SE coast. Labrador: Okak. Newfoundland. New Brunswick: Canterbury Station. New Hampshire:

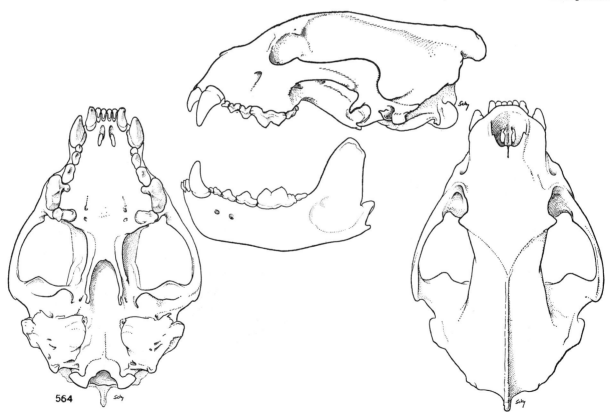

Fig. 564. *Gulo luscus luteus,* head Lyell Canyon, Tuolumne Co., California, No. 22121 M.V.Z., ♂, X ⅓.

Diamond region E of Connecticut Lakes. New York: Rensselaer County. Pennsylvania: near Great Salt Lick, Portage Twp. Indiana: Washington Twp., Noble Co.; near Edwardsport, Knox Co. Iowa: sec. 3, T. 68 N [86 N, according to Bowles, 1975:129], R. 15 W, Tama Co. (Haugen, 1961:546). Wisconsin: 2½ mi. W Gotham; La Crosse County. Minnesota: *ca.* 12 mi. N Bemidji. North Dakota: mouth Cherry Creek near Killdeer Mts. South Dakota: Timber Lake (Jones, 1964c:283). Nebraska (Jones, 1964c:284): near Chimney Rock; *near Gering.* Montana: Bear Paw Mts. (Newby and Wright, 1955:251); Dry Wolf Creek (Newby and McDougal, 1964:486); Flathead Pass (*ibid.*). Wyoming: *Yellowstone National Park;* National Timber Preserve just E of Yellowstone National Park (Wilson, 1908:351). Utah: Red Canyon country of Green River. Colorado: Elk River between mouth Mad and Big creeks; moraine S of Moraine Park; *near Boulder;* mts. of Clear Creek County; head Huerfano River. New Mexico: *northern New Mexico.* Colorado: La Plata Mts.; Alexander Mesa, Grand Mesa; 6 mi. E Meeker. Utah: Boulder Mts.; Mt. Baldy; Fillmore; near Brighton; Bear River. Nevada: 11⅜ km. S Baker (Barker and Best, 1976:133, skull from cave deposit). Idaho: Sawtooth Mts. British Columbia: Nelson; Horsefly River; Bella Coola, to Pacific Coast and along coast to Alaska: Polk Point on Wrangell Island; Yakutat, thence northwestward, excluding Kenai Peninsula, to Stepovak Bay; Imuruk Lake; 2 mi. up Avalik River. Mackenzie: Mackenzie Delta. Franklin: *Victoria Island; Melville Island,* and northeastward on mainland, and possibly on other Arctic islands, to point of beginning. Not found: near Montgomery, Colorado; Oizenoy Bay on Alaska Peninsula; Felix Harbor.

Birney's (1974:78–81) evidence suggests that the two animals listed above, one each from Minnesota and Iowa, were transported by man to the places where they were found dead. Bowles (1975:129) thinks the Iowa specimen "likely reached the state under natural circumstances."

Gulo luscus luteus Elliot

1904. *Gulo luteus* Elliot, Field Columb. Mus., Publ. 87, Zool. Ser., 3:260, January 7, type from Crater Meadows [= Groundhog Meadows], Whitney Creek [= Golden Trout Creek], 27 mi. S Mt. Whitney, 8700 ft., Tulare Co., California.
1913. *Gulo luscus luteus,* Grinnell, Proc. California Acad. Sci., ser. 4, 3:291, August 28.

MARGINAL RECORDS.—British Columbia: *western British Columbia.* Washington: 3 mi. S Riverside; *White River just outside [Park];* near Muddy Fork of Cowlitz River. Oregon: Mt. Hood section; upper Mackenzie Valley W of Three Sisters Peak. California: near Carberry Ranch; Virginia Lakes at 10,000 ft.; 8700 ft. at Pine City near Mammoth Pass; 8000 ft. in Monache Meadows; Piute Mts.; near Camp Nelson; 8000 ft. at Chiquito Lake; Mt. Linn; Mt. Sanhedrin; Mill Creek near Healdsburg; near San Rafael, thence up coast to area of beginning. (For California, Jones, 1955:569, cites and records 26 occurrences, some of which pro-

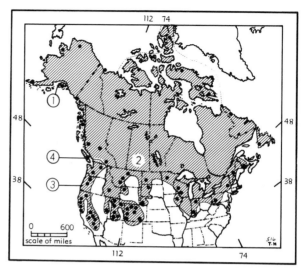

Map 516. *Gulo luscus.*

1. *G. l. katschemakensis*   3. *G. l. luteus*
2. *G. l. luscus*   4. *G. l. vancouverensis*

vide marginal records for the Sierra Nevada, but these are not shown on our small-scale map.)

Gulo luscus vancouverensis Goldman

1935. *Gulo luscus vancouverensis* Goldman, Proc. Biol. Soc. Washington, 48:177, November 15, type from Great Central Lake, Vancouver Island, British Columbia. See Cowan (Canadian Field-Nat., 50:145, December 3, 1936) for diagnostic characters additional to those pointed out in the original description.

MARGINAL RECORDS (Cowan and Guiguet, 1965:325).—British Columbia (Vancouver Island): *Zeballos;* Tsable River; *Fanny Bay; Shawnigan Lake; Jordan Meadows; Nootka Sound.*

Genus **Taxidea** Waterhouse—Badger

1839. *Taxidea* Waterhouse, Proc. Zool. Soc. London, p. 153, for 1838, May. Type, *Meles labradorius* Gmelin [= *Ursus taxus* Schreber].

Skull large, rugose; occiput depressed, facial angle steep; tympanic bullae highly inflated, not in contact with paroccipital processes; palate extending posteriorly beyond plane of upper molars; M1 triangular, often with cusps arranged in rows that are transverse to long axis of skull; P4 with accessory cusp behind deuterocone; trigonid of m1 longer than talonid; metaconid, hypoconid, and entoconid large; talonid often with 1–3 additional cusps.

The genus *Taxidea* has but one Recent species.

Taxidea taxus
Badger

X ⅑

Revised by C. A. Long, Jour. Mamm., 53:725–759, 11 figs., November 30, 1972.

External measurements: in males, 629–870; 108–155; 89–125; in females, 521–790; 98–155; 89–125. Weight, 3.6–10 kg. Other characteristics for males and females, respectively, are: greatest length of skull, 109.1–138.2, 108.5–130.5; zygomatic breadth, 68.9–92, 64.6–82.2; alveolar length of maxillary tooth-row, 36.7–44.7, 35.8–41.8. Upper parts grayish to reddish, guard-hairs long, white-tipped with subterminal black band, longest on sides; white stripe extending from nose at least to shoulders; nose, top of head, preauricular and postauricular spots brownish to black; chin, throat, and mid-ventral region whitish; underparts buffy; feet dark brown to black.

Long (1972) recognized two major groups of *T. taxus*: (1) the northern badger, consisting of the three subspecies *T. t. jacksoni*, *T. t. jeffersonii*, and *T. t. taxus*, and characterized by large size, usually, and short dorsal white stripe (extending, usually, only to neck or shoulders); (2) the southern badger or Mexican badger, consisting of the subspecies *T. t. berlandieri*, and characterized by small size, long dorsal white stripe (extending to mid-dorsum or, as is usual, to rump), low sagittal and lambdoidal crests, and reddish pelage.

Taxidea taxus berlandieri Baird

1858. *Taxidea berlandieri* Baird, Mammals, *in* Repts. Expl. Surv. . . . , 8(1):205, July 14, type from Llano Estacado, Texas, near New Mexico boundary.
1895. *Taxidea taxus berlandieri*, J. A. Allen, Bull. Amer. Mus. Nat. Hist., 7:256, June 29.
1865. *Taxidea americana* var. *Californica* Gray, Proc. Zool. Soc. London, p. 141, June, type probably from part of California that in 1833 adjoined México.
1898. *Taxidea taxus infusca* Thomas, Proc. Zool. Soc. London, for 1897, p. 899, April, type from Santa Anita, Baja California.

1918. *Taxidea taxus phippsi* Figgins, Proc. Colorado Mus. Nat. Hist., 2(2):1, April, type from NE of Chromo, Archuleta Co., Colorado. Regarded by Hall, Carnegie Inst. Washington Publ. 473:78, November 20, 1936, as a synonym of *T. t. taxus*; Schantz, Jour. Mamm., 30:91, February 21, 1950, regards it as a synonym of *T. t. berlandieri*.
1921. *Taxidea robusta* Hay, Proc. U.S. Nat. Mus., 59:535, October 13, type from Anita, Coconino Co., Arizona (Pleistocene).
1939. *Taxidea taxus sonoriensis* Goldman, Jour. Washington Acad. Sci., 29:300, July 15, type from Camoa, Río Mayo, about 15 mi. above Navojoa, Sonora.
1943. *Taxidea taxus papagoensis* Skinner, Bull. Amer. Mus. Nat. Hist., 80:166, November 6, type from Papago Springs Cave, Santa Cruz Co., Arizona (Pleistocene).
1948. *Taxidea taxus apache* Schantz, Proc. Biol. Soc. Washington, 61:175, November 12, type from San Pedro River, Arizona and Sonora, U.S.–Mexican boundary. Regarded by Hoffmeister and Goodpaster, Illinois Biol. Monog., 24(1):79, December 31, 1954, as a synonym of *T. t. sonoriensis*.
1949. *Taxidea taxus littoralis* Schantz, Jour. Mamm., 30:301, August 17, type from Corpus Christi, Nueces Co., Texas.
1949. *Taxidea taxus hallorani* Schantz, Jour. Mamm., 30:302, August 17, type from Castle Dome Mts., Kofa Game Refuge, Yuma Co., Arizona.
1949. *Taxidea taxus nevadensis* Schantz, Jour. Mamm., 30:304, August 17, type from White Sage Flat, between Desert Range and Sheep Range, Desert Game Range, Clark Co., Nevada. Not *Taxidea nevadensis* Butterworth, Univ. California Publ., Bull. Dept. Geol., 10:21, October 3, 1916.
1951. *Taxidea taxus halli* Schantz, Jour. Mamm., 32:127, February 15, a renaming of *T. t. nevadensis* Schantz.

MARGINAL RECORDS (Long, 1972:750, 751, unless otherwise noted).—California: Hayden Hill [Haydenhill]; Wawona; 7 mi. E Laws (Hall and Kelson, 1959:926). Nevada: Rhyolite; Pahranagat Valley. Utah: Days Spring; Kaiparowits Plateau (*ibid.*); Spring Creek, San Juan Co. Colorado: Paradox Valley (Armstrong, 1972:284); Pine [Los Piños] River; 8 mi. S La Jara, Rio Grande River. Oklahoma: E end Black Mesa (Hall and Kelson, 1959:926); Anthon; Alva (*ibid.*). Arkansas: Hwy. N Knob, *ca.* 10 mi. NE Fayetteville (Sealander and Forsyth, 1966:134, as *T. taxus* only). Oklahoma: 5 mi. S, 5 mi. E Grady. Texas: Montague County (Dalquest, 1968:20); Mason; San Antonio; Aransas National Wildlife Refuge; Mustang Island, 18 mi. S Port Aransas. Tamaulipas: Matamoros; 33 mi. S Washington Beach; 12 mi. S San Carlos, 1300 ft. San Luis Potosí: Bledos. Hidalgo: Tepeapulco. Puebla: 10 km. W Acatzingo, 2105 m. (Baker and Petersen, 1969:252). Durango: 3 km. N Atotonilco, 2037 m. (*ibid.*). Sinaloa: 16 km. NNE Choix, 1700 ft. Sonora: Camoa, Río Mayo, *ca.* 15 mi. above Navojoa; Guaymas; 8½ mi. N Quitovac (Hall and Kelson, 1959:928, as *T. t. sonoriensis*). Baja California: Volcano Lake; Calmalli; Santa Anita; Cape San Lucas, thence northward along Pacific Coast and inland to California: La Puerta Valley (*op. cit.*:926); San Jacinto (*ibid.*); Fairmont (*ibid.*); Orsi [Orosi]; Tracy; Burney.

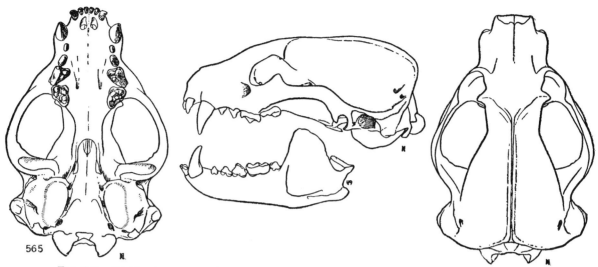

Fig. 565. *Taxidea taxus taxus*, Spring Valley, White Pine Co., Nevada, No. 41467 M.V.Z., ♂, X ½.

Taxidea taxus jacksoni Schantz

1946. *Taxidea taxus jacksoni* Schantz, Jour. Mamm., 26:431, February 12, type from 4 mi. E Milton, Rock Co., Wisconsin.

MARGINAL RECORDS (Long, 1972:745, unless otherwise noted).—Ontario: N shore Knife Lake, Quetico Provincial Park, thence along S shore Lake Superior and Lake Huron to Ontario: near Port Dover. Ohio: W part Pike Twp., Knox Co. (Hall and Kelson, 1959:928, as *T. t. taxus*); near Stoutsville; Findlay. Indiana: St. Joseph County. Wisconsin: Hurricane[e] (Jackson, 1961:369). Minnesota: Madison.—Occurrences from as far eastward as New York not listed above or by Long (*op. cit.*) are difficult to explain. Some may reflect an extension of range where the forest has been destroyed, and some other occurrences reflect releases of captive animals (see Nugent and Choate, 1970:626, 627, who cite a number of far eastern records). Conceivably some of the eastern records result from a repopulation of the natural geographic range when man began to trap fewer fur-bearing mammals than he had trapped at some earlier periods.

Taxidea taxus jeffersonii (Harlan)

1825. *Meles jeffersonii* Harlan, Fauna Americana, p. 309. On p. 310 it is stated that badgers of this kind "inhabit the open plains of Columbia, sometimes those of Missouri, and are sometimes found in the woods."
1972. *Taxidea taxus jeffersonii*, Long, Jour. Mamm., 53:732, November 30.
1878. *Taxidea sulcata* Cope, Proc. Amer. Philos. Soc., 17:227, type from Oregon Desert, Oregon (Pleistocene age).
1891. *Taxidea americana neglecta* Mearns, Bull. Amer. Mus. Nat. Hist., 3:250, June 5, type from Fort Crook, Shasta Co., California.
1950. *Taxidea taxus montana* Schantz, Jour. Mamm., 31:90,

February 21, type from 35 mi. S Dillon, Beaverhead Co., Montana.

MARGINAL RECORDS (Long, 1972:736–738, unless otherwise noted).—British Columbia: Narcosli Creek (Cowan and Guiguet, 1965:326); Invermere (*ibid.*); Tobacco Plains, E of Newgate. Montana: Blackfoot; Sun River; Clone's Gulch. Wyoming: 18 mi. NW Casper; 16–17 mi. E Laramie. Colorado (Armstrong, 1972:285, as *T. t. montana*): N of Turkey Creek, S of Colorado Springs, El Paso Co.; 4 mi. W Antonito; confluence Red Creek and Gunnison River. Utah: Dry Valley, San Juan Co.; 8 mi. S Panguitch; Garrison. Nevada (Hall and Kelson, 1959:928, as *T. t. taxus*): Cave Valley, 3 mi. W Cave, 6500 ft.; Kawich Valley, 3 mi. SE Kawich, 6000 ft.; 9½ mi. E Goldfield, 5500 ft.; Cave Spring, 6200–6248 ft. California: Mammoth (*ibid.*, as *T. t. neglecta*); Tuolumne Meadows, Yosemite National Park (*ibid.*, as *T. t. neglecta*); Quincy; Merriville; Cedarvale [Cedarville]; Fort Crook, Shasta Co.; South Yello Bally [Yolla Bolly] Mtn.; Huron; Twin Oaks; Mt. Piños (*ibid.*, as *T. t. neglecta*); Colton (*ibid.*, as *T. t. neglecta*); Santa Ysabel. Baja California: *lat. 31° 31' N* (*ibid.*, as *T. t. neglecta*), thence northward along Pacific Coast to California: Eden Valley Ranch, Mendocino Co. Oregon: Cottonwood, below Flynn. Washington: Divide above Trout Lake; Lake Chelan, Fish Creek. British Columbia: E of Riske Creek (Cowan and Guiguet, 1965:326).

Taxidea taxus taxus (Schreber)

1778. *Ursus taxus* Schreber, Die Säugthiere . . . , 3:520. Type locality, Labrador and Hudson Bay (probably SW of Hudson Bay).
1894. *Taxidea taxus*, Rhoads, Amer. Nat., 28:524, June.
1784. *Meles Taxus β. americanus* Boddaert, Elenchus animalium, 1:80, based on skin, lacking legs, from North America.

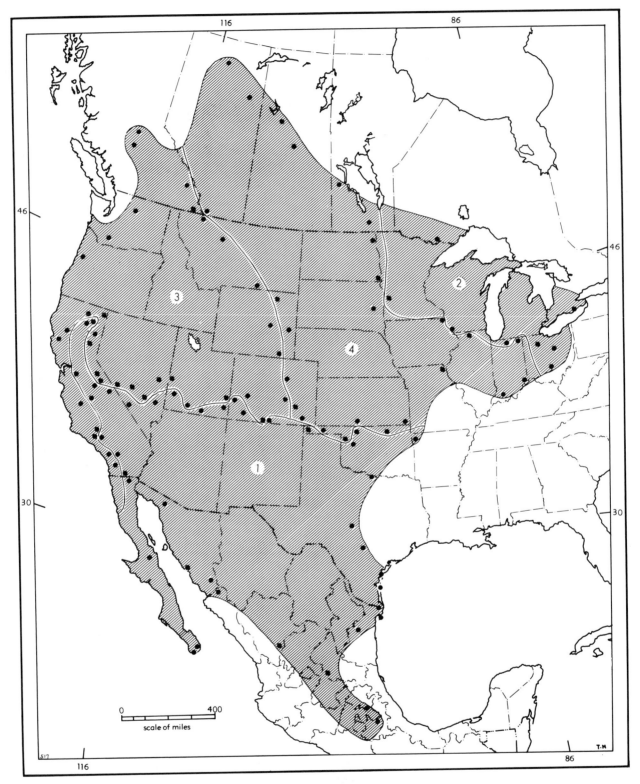

Map 517. *Taxidea taxus.*

1. *T. t. berlandieri*   2. *T. t. jacksoni*   3. *T. t. jeffersonii*   4. *T. t. taxus*

1788. *Ursus labradorius* Gmelin, Syst. nat., ed. 13, 1:102, a renaming of *Ursus taxus* Schreber.
1946. *Taxidea taxus dacotensis* Schantz, Proc. Biol. Soc. Washington, 59:81, June 19, type from Folsom, Custer Co., South Dakota.
1947. *Taxidea taxus iowae* Schantz, Jour. Mamm., 28:287, September 3, type from near Clarion, Wright Co., Iowa.
1950. *Taxidea taxus merriami* Schantz, Jour. Washington Acad. Sci., 40:92, March 21, type from Banner, Trego Co., Kansas.
1950. *Taxidea taxus kansensis* Schantz, Jour. Mamm., 31:346, August 21, type from 4 mi. SE McLouth, Leavenworth Co., Kansas.

MARGINAL RECORDS (Long, 1972:743, 744, unless otherwise noted).—Alberta: Peace River, lat. 58°; Athabaska [River] below mouth of Clearwater. Saskatchewan: Isla [Ile] à la Crosse; between Sugar Lake and Amyot Lake, Prince Albert National Park. Manitoba: Duck Mtn. (Soper, 1961:208); *Delta Marsh Region* (Tamsitt, 1962:77); Carman. North Dakota: Grafton; Lidgerwood. South Dakota: 3 mi. E Iroquois. Iowa: 3 mi. radius of New Albin (Bowles, 1975:130). Illinois: 3 mi. E Davis. Indiana: ½ mi. N, 1½ mi. E Mongo; Dearborn County (Mumford, 1969:97, sight record only); *Crawford County* (*ibid.*, sight record only); Dubois County (*ibid.*, sight record only). Illinois: *near Kaskasia River* (Hall and Kelson, 1959:928). Missouri: 4 mi. NE Wayland. Kansas: Cherokee County. Oklahoma: 10 mi. SE Fairfax. Kansas: 2½ mi. W Hardtner. Oklahoma: 4 mi. NW Peek; 3 mi. N, 4 mi. E Goodwell. Colorado (Armstrong, 1972:286): 13 mi. NW Estelene; Olney Springs; Coal Creek. Wyoming: Manville; Arvada. Alberta: Waterton Lakes National Park (Hall and Kelson, 1959:928).

Genus **Spilogale** Gray—Spotted Skunks

Revised by R. G. Van Gelder, Bull. Amer. Mus. Nat. Hist., 117:229–392, 47 figs., 32 tables, June 15, 1959 (but see also Mead, Jour. Mamm., 49:373–390, August 20, 1968, which bears on the question of whether *Spilogale* from the eastern United States are a different species from those farther west).

1865. *Spilogale* Gray, Proc. Zool. Soc. London, p. 150, June. Type, *Mephitis interrupta* Rafinesque.

External measurements: males, 240–610, 69–280, 32–59; females, 250–544, 68–210, 30–59. Basilar length, males, 35.9–58.0; females, 36.4–53.3. Pelage black with 4–6 white stripes, in some subspecies broken into spots on upper parts; tail-vertebrae shorter than head and body; skull flattened in longitudinal dorsal outline, of nearly equal height in frontal and parietal regions; mastoid bullae highly inflated; occiput widened and shallow; posterior margin of palate nearly on a line with posterior borders of upper molars; inferior margin of mandible relatively straight; "step" of *Mephitis* absent, or less developed in *S. putorius lucasana;* anterior and posterior borders

of M1 each less than outside length of P4; in m1 trigonid longer than talonid, metaconid high and distinct from protoconid, hypoconid low, entoconid low and separated from metaconid by wide notch. Dentition, i. $\frac{3}{3}$, c. $\frac{1}{1}$, p. $\frac{3}{3}$, m. $\frac{1}{2}$.

KEY TO SPECIES OF SPILOGALE

1. Skull lacks sagittal crest; dorsal stripes continuous to rump; forefeet and hind feet wholly white dorsally. . . . *S. pygmaea*, p. 1013
1'. Skull with sagittal crest; dorsal stripes discontinuous to rump; forefeet and hind feet not wholly white dorsally. . . *S. putorius*, p. 1014

Spilogale pygmaea
Pygmy Spotted Skunk

External measurements: males, 240–271, 69–84, 33–35; female, 250, 68, 34. Basilar length, males, 35.9–37.8; female, 36.4. Skull: sagittal crest lacking; cranium highly arched; mastoid bullae only slightly inflated. Two white dorsal stripes continuous to rump and usually to tail; nasal patch connected to preauricular patches; forefeet and hind feet wholly white dorsally (after Van Gelder, 1959:379). See Fig. 566. See Map 518.

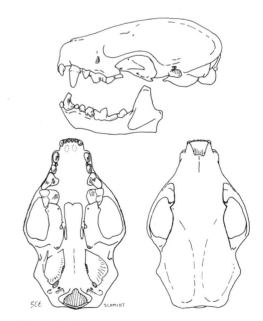

Fig. 566. *Spilogale pygmaea pygmaea*, 8 km. N Villa Unión, 450 ft., Sinaloa, No. 96185 K.U., ♂, X 1.

Spilogale pygmaea australis Hall

1938. *Spilogale pygmaea australis* Hall, Ann. Mag. Nat. Hist., ser. 11, 1:514, May, type from Acapulco, Guerrero.
1956. *Spilogale pygmaea albipes* Goodwin, Amer. Mus. Novit., 1757:13, March 8, type from Las Cuevas, in valley of and on eastern side of Río Tehuantepec, 17 km. NW city of Tehuantepec, Oaxaca. Regarded by Genoways and Jones (1971:124, 127) as inseparable from *S. p. australis.*

MARGINAL RECORDS (Genoways and Jones, 1971:127).—Guerrero: type locality. Oaxaca: 10 mi. S Juchatengo, 5350 ft.; Las Cuevas; Tehuantepec.

Spilogale pygmaea pygmaea Thomas

1898. *Spilogale pygmaea* Thomas, Proc. Zool. Soc. London, p 898, for 1897, April, type from Rosario, Sinaloa.

MARGINAL RECORDS.—Sinaloa (Genoways and Jones, 1971:127): 5 mi. NW Mazatlán; *8 km. N Villa Unión, 450 ft.*; type locality. Jalisco: vic. Chamela Bay (López-F., *et al.,* 1973:104). Colima: 17 km. S Colima (Greer and Greer, 1970:629).

Spilogale putorius
Spotted Skunk

External measurements: males, 310–610, 80–280, 32–59; females, 270–544, 85–210, 30–59. Basilar length, males, 39.2–58.0; females, 37.9–53.3. Sagittal crest present in males of subspecies of larger size and in some females; profile of cranium less convex than in S. *pygmaea;* mastoid bullae notably inflated. Dorsal stripes interrupted posteriorly; nasal patch not connected to shoulder stripes; forefeet and hind feet not completely white dorsally (after Van Gelder, 1959:255).

Spilogale putorius ambarvalis Bangs

1898. *Spilogale ambarvalis* Bangs, Proc. Boston Soc. Nat. Hist., 28:222, March, type from Oak Lodge, opposite Micco, Brevard Co., Florida.
1953. *Spilogale putorius ambarvalis,* Van Gelder, Jour. Mamm., 34:255, May 14.

MARGINAL RECORDS.—Florida: Silver Springs (Van Gelder, 1959:265); New Smyrna, thence down Atlantic Coast to Coconut Grove; *Perrine* (Layne, 1974:398); 1 mi. E Park entrance on State Route 27

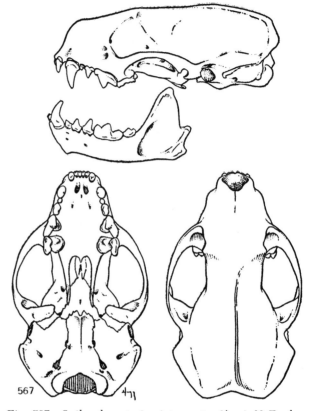

Fig. 567. *Spilogale putorius interrupta,* 1½ mi. N Fowler, Meade Co., Kansas, No. 14237 K.U., ♂, X 1.

(*ibid.*); 25 mi. SE Immokalee; Immokalee (Van Gelder, 1959:265); Chadwick Beach (*ibid.*); Palma Sola.

Spilogale putorius amphiala Dickey

1929. *Spilogale phenax amphialus* Dickey, Proc. Biol. Soc. Washington, 42:158, April 4, type from 2½ mi. N ranch house near coast, Santa Rosa Island, Santa Barbara Islands, Santa Barbara Co., California.
1959. *Spilogale putorius amphiala,* Van Gelder, Bull. Amer. Mus. Nat. Hist., 117:359, June 15.

MARGINAL RECORDS.—California, Santa Barbara Islands: type locality; Sur Ranch, 3 mi. E of main ranch, Santa Cruz Island (Van Gelder, 1959:369).

Spilogale putorius angustifrons A. H. Howell

1902. *Spilogale angustifrons* A. H. Howell, Proc. Biol. Soc Washington, 15:242, December 16, type from Tlalpam [= Tlalpan], Distrito Federal.
1959. *Spilogale putorius angustifrons,* Van Gelder, Bull. Amer. Mus. Nat. Hist., 117:309, June 15.

MARGINAL RECORDS.—Jalisco: 4 mi. W Villa Guerrero, 5500 ft. (Genoways and Jones, 1971:130); 10 mi. NW Matanzas, 8000 ft. (107868 KU); 4½ mi. NE Comanja de Corona, 8000 ft. (109457 KU). Distrito

Federal: type locality. Michoacán: Pátzcuaro (Van Gelder, 1959:312). Jalisco: 7½ mi. SE Tecomate (111976 KU); San Sebastián.

Spilogale putorius celeris Hall

1938. *Spilogale angustifrons celeris* Hall, Ann. Mag. Nat. Hist., ser. 11, 1:511, May, type from San Isidro, Costa Rica.
1959. *Spilogale putorius celeris,* Van Gelder, Bull. Amer. Mus. Nat. Hist., 117:325, June 15.

MARGINAL RECORDS.—Nicaragua: Jalapa (Van Gelder, 1959:327). Costa Rica: vic. Alajuela; *type locality;* Monte Rey, 22 km. S San José, 1100 m. Nicaragua: Matagalpa.

Spilogale putorius elata A. H. Howell

1906. *Spilogale angustifrons elata* A. H. Howell, N. Amer. Fauna, 26:27, November 24, type from San Bartolomé, Chiapas.
1959. *Spilogale putorius elata,* Van Gelder, Bull. Amer. Mus. Nat. Hist., 117:321, June 15.

MARGINAL RECORDS.—Chiapas: type locality. Honduras: Cantoral; La Flor Archaga (Van Gelder, 1959:325). El Salvador: San Salvador (*ibid.*). Guatemala: Capetillo (*ibid.*); Dueñas. Chiapas: Pinabete.

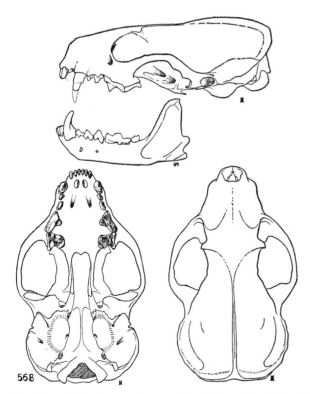

Fig. 568. *Spilogale putorius gracilis,* W side Ruby Lake, 6 mi. Elko Co. line, Nevada, No. 48058 K.U., ♂, X 1.

Spilogale putorius gracilis Merriam

1890. *Spilogale gracilis* Merriam, N. Amer. Fauna, 3:83, September 11, type from Grand Canyon of Colorado [River], 3500 ft., N of San Francisco Mtn., Arizona.
1959. *Spilogale putorius gracilis,* Van Gelder, Bull. Amer. Mus. Nat. Hist., 117:279, June 15.
1890. *Spilogale saxatilis* Merriam, N. Amer. Fauna, 4:13, October 8, type from Provo, Utah Co., Utah.
1902. *Spilogale tenuis* A. H. Howell, Proc. Biol. Soc. Washington, 15:241, December 16, type from Arkins, Larimer Co., Colorado.

MARGINAL RECORDS.—Washington: Kamiak Butte. Idaho: 4 mi. SW Selway Falls, 5800 ft. Montana (Hoffmann, *et al.,* 1969: 597, unless otherwise noted): Tin Cup Canyon, Ravalli Co. (Hall and Kelson, 1959:932); *Medicine Springs;* Ruby Range, Madison Co.; 50 mi. S Bozeman. Wyoming (Long, 1965a:701, unless otherwise noted): Story (63190 KU); Splitrock; Fort Steele; Iron Mtn. Colorado (Armstrong, 1972:287): 2 mi. N Dixon Dam; *Arkins;* Boulder; Cheyenne Mtn.; Furnace Canyon. Oklahoma: 3½ mi. N, 1⅛ mi. W Kenton (Geluso, 1972:457). New Mexico (Van Gelder, 1959:288, unless otherwise noted): Oak Canyon, near Folsom (Hall and Kelson, 1959:932); *Mouth of Cimarroncito Canyon;* Bear Cañon, Raton Range; 12 mi. E Los Lunas (Findley, *et al.,* 1975:310, as *S. gracilis* only); Thoreau. Arizona: Keams Canyon (Van Gelder, 1959:288); Shiva Temple (Hall, J. G., 1968:99); Democrat Mine, Hualpai Mts. (Van Gelder, 1959:288). California: Panamint Mts. (Hall and Kelson, 1959:931); Carroll Creek, Sierra Nevada Mts.; Wawona; *El Portal;* 6 mi. S Hetch Hetchy Valley; Rock Creek; Auburn; Susanville; Adin. Oregon (Van Gelder, 1959:292, unless otherwise noted): Fremont; Warm Springs River, 9 mi. NW Agency; The Dalles (Hall and Kelson, 1959:932); *mouth Deschutes River.* Washington: Almota; *Pullman* (Van Gelder, 1959:299). Not found: California: Swansea (Van Gelder, 1959:288).

Spilogale putorius interrupta (Rafinesque)

1820. *Mephitis interrupta* Rafinesque, Ann. Nat. . . . , 1:3. Type locality, Upper Missouri River (by restriction).
1952. *Spilogale putorius interrupta,* McCarley, Texas Jour. Sci., 4:108, March 30.
1859. *Mephitis quaterlinearis* Winans, in a [Kansas?] newspaper, see Coues, Fur-bearing animals . . . , U.S. Geol. Surv. Territories, Miscl. Publ. 8, pp. 239, 240, 1877.
1890. *Spilogale indianola* Merriam, N. Amer. Fauna, 4:10, October 8, type from Indianola, Matagorda Bay, Texas.

MARGINAL RECORDS.—Minnesota: sec. 6, Jadis Twp., Roseau Co.; St. Louis County. Wisconsin: sec. 25, T. 30 N, R. 19 W, St. Croix Co. Minnesota: Lanesboro (Van Gelder, 1959:278). Iowa: Lansing (Bowles, 1975:132). Wisconsin: Fort Atkinson (Jackson, 1961:374). Iowa: 4 mi. N Burlington (Bowles, 1975:132). Missouri: 7 mi. S Perry (Van Gelder, 1959:278); St. Francois County; Wayne County. Arkansas: Hot Springs. Texas (Van Gelder, 1959:279): 6 mi. E

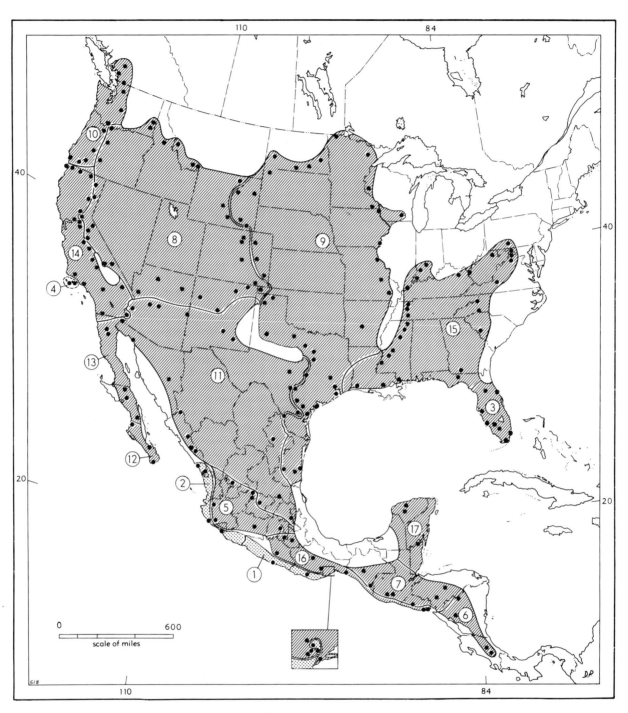

Map 518. *Spilogale.*

Guide to kinds

1. *S. pygmaea australis*
2. *S. pygmaea pygmaea*
3. *S. putorius ambarvalis*
4. *S. putorius amphiala*
5. *S. putorius angustifrons*

6. *S. putorius celeris*
7. *S. putorius elata*
8. *S. putorius gracilis*
9. *S. putorius interrupta*
10. *S. putorius latifrons*
11. *S. putorius leucoparia*

12. *S. putorius lucasana*
13. *S. putorius martirensis*
14. *S. putorius phenax*
15. *S. putorius putorius*
16. *S. putorius tropicalis*
17. *S. putorius yucatanensis*

Huntsville; 2½ mi. N Hackley; 1 mi. N Texas City; Aransas National Wildlife Refuge (Halloran, 1961:24); Corpus Christi. Tamaulipas (Alvarez, 1963:461): *3 mi. N La Pesca;* La Pesca; Sierra de Tamaulipas, 2 mi. S, 10 mi. W Piedra; Ciudad Victoria; *Rancho Santa Rosa, 2 km. N, 13 km. W Ciudad Victoria.* Texas: Laredo (Van Gelder, 1959:279); Beeville (*ibid.*); Atacosa County (*ibid.*); *Somerset* (*ibid.*); San Antonio; 8 mi. E Austin (Van Gelder, 1959:279); Waco; 4 mi. N Blum; Brazos; 6 mi. N, 12 mi. W Archer City (Van Gelder, 1959:279); 5 mi. W, ¼ mi. N Lubbock on U.S. Hwy. 84 (Packard and Garner, 1964:389); 4 mi. W Dumas; Bernstein (Van Gelder; 1959:279). Colorado (Armstrong, 1972:288): 10 mi. S Lamar; Hugo; near Fort Morgan. Wyoming (Long, 1965a:702): Iron Mtn.; North Laramie River, N of Garrett; Powder River, between Kaycee and Sussex; near Newcastle. South Dakota: Harding County (Andersen and Jones, 1971:385). North Dakota (Van Gelder, 1959:278): Dickenson; 3 mi. SW Linton; 1½ mi. S, 1 mi. W Jud; near Tower City. Recorded also from archaeological sites in *Illinois* (recorded only to species): *4 mi. SW Columbia* (Parmalee, *et al.*, 1961:119); *approx. 2 mi. NW Modoc.*

Spilogale putorius latifrons Merriam

1890. *Spilogale phenax latifrons* Merriam, N. Amer. Fauna, 4:15, October 8, type from Roseburg, Douglas Co., Oregon.
1959. *Spilogale putorius latifrons,* Van Gelder, Bull. Amer. Mus. Nat. Hist., 117:327, June 15.
1899. *Spilogale olympica* Elliot, Field Columb. Mus., Publ 32, Zool. Ser., 1:270, May 19, type from Lake Sutherland Olympic Mts., Clallam Co., Wasington.

MARGINAL RECORDS (Van Gelder, 1959:331, 334, unless otherwise noted).—British Columbia: Alta Lake, 70 mi. N Vancouver (Hall and Kelson, 1959:931); Sumas. Washington: Shafit Valley, Hamilton; Lake Keechelus (Hall and Kelson, 1959:931); 2 mi., N Carson Shepard Spring. Oregon: mouth Laurel Creek, N base Mt. Hood; Hall Creek, Oakridge; 43 mi. NE Grants Pass; Evans Creek; Goldbeach, thence N along coast into British Columbia: Howe Sound (Hall and Kelson, 1959:932).

Spilogale putorius leucoparia Merriam

1890. *Spilogale leucoparia* Merriam, N. Amer. Fauna, 4:11, October 8, type from Mason, Mason Co., Texas.
1959. *Spilogale putorius leucoparia,* Van Gelder, Bull. Amer. Mus. Nat. Hist., 117:299, June 15.
1890. *S* [*pilogale*]. *texensis* Merriam, N. Amer. Fauna, 4: table following p. 15, October 8. This name heads the column of measurements of *S. leucoparia* from Mason, Texas; *S. texensis* is a *lapsus* and a synonym of *S. leucoparia.*
1891. *Spilogale phenax arizonae* Mearns, Bull. Amer. Mus. Nat. Hist., 3:256, June 5, type from Fort Verde, Yavapai Co., Arizona.
1897. *Spilogale ambigua* Mearns, Preliminary diagnoses of new mammals . . . from the Mexican boundary line, p. 3, January 12 (preprint of Proc. U.S. Nat. Mus., 20:460, December 24, 1897), type from Eagle Mtn. (*ca.* 4 mi. S Monument 15, Mexican boundary line), Chihuahua.

MARGINAL RECORDS (Van Gelder, 1959:305, 309, unless otherwise noted).—Arizona: Whipple Barracks; Fort Verde; *Fossil Creek;* Springerville. New Mexico: Malpais Spring, Tularosa Basin (Hall and Kelson, 1959:931); Hope. Texas: *5 mi. N Fort McKavett;* 10 mi. N Mason; *Waring* (Hall and Kelson, 1959:931); Camp Bullis; San Diego (Hall and Kelson, 1959:931); Laredo. Nuevo León: Monterrey (Hall and Kelson, 1959:931). San Luis Potosí: Hda. Capulín (*ibid.*). Hidalgo: Tulancingo. Guanajuato: Santa Rosa (Hall and Kelson, 1959:933). Sinaloa: 6 km. E Cosalá (90981 KU); 7 mi. ENE Presa Sanalona, 600 ft. (Genoways and Jones, 1971:130); 1½ mi. N Badiraguato (96720 KU); 16 km. NNE Choix (89919 KU). Sonora: Sierra Seri, 9 mi. W San Javier; Punta Peñascosa. Arizona: 4 mi. S Gadsden; 10 mi. E Quartzite, in Plumosa Mts. Not found: Sonora: *Presidio.*

Spilogale putorius lucasana Merriam

1890. *Spilogale lucasana* Merriam, N. Amer. Fauna, 4:11, October 8, type from Cape St. Lucas, Baja California.
1959. *Spilogale putorius lucasana,* Van Gelder, Bull. Amer. Mus. Nat. Hist., 117:374, June 15.

MARGINAL RECORDS (Van Gelder, 1959:377, 379).—Baja California: Santo Domingo, lat. 25° 30′ N; 2 mi. SW La Paz; *Santa Anita;* type locality.

Spilogale putorius martirensis Elliot

1903. *Spilogale arizonae martirensis* Elliot, Field Columb. Mus., Publ. 74, Zool. Ser., 3:170, May 7, type from Vallecitos, San Pedro Mártir Mts., Baja California.
1959. *Spilogale putorius martirensis,* Van Gelder, Bull. Amer. Mus. Nat. Hist., 117:369, June 15.
1906. *Spilogale microdon* A. H. Howell, N. Amer. Fauna, 26:34, November 24, type from Comondú, Baja California.

MARGINAL RECORDS.—Baja California: 10 mi. SE Alamos (Van Gelder, 1959:374); type locality; San Pablo; San Ignacio; Comondú.

Spilogale putorius phenax Merriam

1838. *Mephitis Zorrilla* Lichtenstein, Abh. k. Akad. Wiss., Berlin, 1836, p. 281. Preoccupied by *M. zorilla* Lichtenstein, Darstellung . . . Säug. Abb. Besch., Berlin, vol. 2, 1827–1834, p. 105, Pl. 48.
1890. *Spilogale phenax* Merriam, N. Amer. Fauna, 4:13, October 8, type from Nicasio, Marin Co., California.
1959. *Spilogale putorius phenax,* Van Gelder, Bull. Amer. Mus. Nat. Hist., 117:335, June 15.
1926. *Spilogale phenax microrhina* Hall, Jour. Mamm., 7:53, February 12, type from Julian, San Diego Co., California.

MARGINAL RECORDS (Van Gelder, 1959:344, 345, 359, unless otherwise noted).—California: E. Fork Illinois River, between East Fork and Dunn Creek; Hornbrook; Sheepy Peak, 5 mi. W Tule Lake; Rich Gulch, 11 mi. N and 8 mi. W Quincy; S Marysville; 1¾ mi. SE Davis; Carbondale; Valley Springs; Raymond; Kings River; *Kaweah;* 2 mi. W Camp Nelson; Kern

River Lakes (Hall and Kelson, 1959:932); Converse Flats, San Bernardino Mts.; 5 mi. NE Granite Well, Providence Mtn.; Pilot Knob, Colorado River; La Puerta; Ventura; Santa Ynez Mission (Hall and Kelson, 1959:932), thence N along Pacific Coast to *Creek Station, Shelley Creek, Siskiyou Mts.* [Del Norte Co.].

Spilogale putorius putorius (Linnaeus)

1758. [*Viverra*] *Putorius* Linnaeus, Syst. nat., ed. 10, 1:44. Type locality, South Carolina. (See A. H. Howell, N. Amer. Fauna, 26:16, 17, November 24, 1906.

1875. *Spilogale putorius*, Coues, Bull. U.S. Geol. and Geog. Surv. Territories, ser. 2, 1:12.

1776. *Viverra Mapurita* Müller, Natursystems, Supplement, p. 32. Based on *"Le Zorille"* of Buffon (Histoire naturelle. . . , Paris, vol. 13, pl. 41, 1765).

1776. *Viverra Zorilla* Schreber, Die Säugthiere, Pl. 123; description, 1777, p. 445. Based in part on *"Le Zorille"* of Buffon.

1798. *Must*[*ela*]. *putida* Cuvier, Tableau elementaire de l'histoire naturelle des animaux, p. 116. A substitute name for *Viverra Putorius* Linnaeus.

1800. *Viverra Striata* Shaw, General zoology . . . , 1:387. Based in part on *Viverra Putorius* Linnaeus.

1822. [*mustela americana*] var. B Desmarest, Mammalogie . . . , p. 186. Based on *Putorius americanus striatus* of Catesby.

1837. *Mephitis bicolor* Gray, Mag. Nat. Hist., n. s., 1:581. Based on a spotted skunk for North America.

1890. *Spilogale ringens* Merriam, N. Amer. Fauna, 4:9, October 8, type from Greensborough, Hale Co., Alabama.

MARGINAL RECORDS.—Pennsylvania: Sidling Hill Mtn. West Virginia: 2 mi. S Gerrardstown (Van Gelder, 1959:261). Virginia: Winchester (Van Gelder, 1959:261); Rappahannock County; Roanoke County. North Carolina: Roan Mtn.; Bat Cave, Rutherford Co. (Van Gelder, 1959:261). South Carolina: Cleora. Georgia: Thomasville. Florida: 2³⁄₁₀ mi. E Tallahassee (Van Gelder, 1959:261). Mississippi: Bay St. Louis (Kennedy, et al., 1974:27). Louisiana (Lowery, 1974:438): *2 mi. SE Burnside; 7 mi. SW Sorrento; 16 mi. S, 1½ mi. W Lake Charles; 7½ mi. SE Lake Charles; Iowa.* Mississippi: Claiborne County; Hinds County; Madison County; Webster County; Pontotoc County; Corinth. Tennessee: eastern half of Henderson County; Carroll County; Henry County. Illinois: Golconda. Indiana: Posey County; Knox County; 1½ mi. W Springville (Bader and Hall, 1960:111, as *Spilogale putorius* only). Kentucky: Rowan County. West Virginia: near Huntington; Franklin.

Spilogale putorius tropicalis A. H. Howell

1902. *Spilogale angustifrons tropicalis* A. H. Howell, Proc. Biol. Soc. Washington, 15:242, December 16, type from San Mateo del Mar, Oaxaca.

1959. *Spilogale putorius tropicalis*, Van Gelder, Bull. Amer. Mus. Nat. Hist., 117:312, June 15.

MARGINAL RECORDS (Van Gelder, 1959:319, unless otherwise noted).—Morelos: Yautepec. Puebla: 5

mi. E Izúcar de Matamoros. Oaxaca (Goodwin, 1969:240, unless otherwise noted); Cerro San Felipe; *San Gerónimo* (Hall and Kelson, 1959:933); Santa María del Mar. El Salvador: SW edge Lake Olomega; Puerto de Triunfo. Oaxaca (Goodwin, 1969:240): Salina Cruz; Jalapa; Tehuantepec; Las Tejas; Tres Cruces; San Bartolo Yautepec. Guerrero: Chilpancingo.

Spilogale putorius yucatanensis Burt

1938. *Spilogale angustifrons yucatanensis* Burt, Occas. Pap. Mus. Zool., Univ. Michigan, 384:2, August 27, type from Chichén-Itzá, Yucatán.

1959. *Spilogale putorius yucatanensis*, Van Gelder, Bull. Amer. Mus. Nat. Hist., 117:319, June 15.

MARGINAL RECORDS.—Yucatán: Mérida; thence along coast to Belize: Orange Walk. Yucatán: Uxmal.

Genus Mephitis É. Geoffroy Saint-Hilaire and G. Cuvier—Striped and Hooded Skunks

Revised under the name *Chincha* by A. H. Howell, N. Amer. Fauna, 20:1–62, 8 pls., August 15, 1901; see also listing of named kinds by Hall, Carnegie Inst. Washington Publ. 473:63–74, November 20, 1936. For discussion of the nomenclature of this genus see A. H. Howell, N. Amer. Fauna, 20:14, August 31, 1901; Proc. Biol. Soc. Washington, 15:1–9, February 18, 1902; N. Amer. Fauna, 26:10, 11, November 24, 1906; J. A. Allen, Bull. Amer. Mus. Nat. Hist., 14:325–334, November 12, 1901; Proc. Biol. Soc. Washington, 15:59–66, March 22, 1902; J. A. Allen and others, Science, n.s., 16:114, 115, July 18, 1902.

1795. *Mephitis* É. Geoffroy Saint-Hilaire and G. Cuvier, Mag. Encyclopédique, n. s., 2:187. Type, *Viverra mephitis* Schreber.

1842. *Chincha* Lesson, Nouveau tableau du règne animal . . . mammifères, p. 67. Type, *Chincha americana* Lesson [= *Viverra mephitis* Schreber], from North America.

1901. *Leucomitra* A. H. Howell, N. Amer. Fauna, 20:39, August 31, 1901. Type, *Mephitis macroura* Lichtenstein. *Leucomitra* was proposed, and so far as known has been used only, as a subgenus.

External measurements of males: 558–800; 184–435; 58–90. Basal length of skull, 56–76. Females smaller, by as much as 15 per cent in several linear measurements. Pelage black with white markings on upper parts. Mastoid bullae relatively uninflated; posterior margin of palate nearly on a line with posterior border of upper molars; anteroposterior and transverse diameters of M1 each about equal to (usually more than) outside length of P4; metaconule of M1 not distinct; in m1, trigonid longer than talonid; metaconid high and distinct from protoconid, hypoconid high, entoconid high and separated from metaconid by deep, wide notch. Dentition, i. $\frac{3}{3}$, c. $\frac{1}{1}$, p. $\frac{3}{3}$, m. $\frac{1}{2}$.

KEY TO SPECIES OF MEPHITIS

1. Tympanic bullae small (see Fig. 569); back black or with white areas containing no black hairs . . . . *M. mephitis*, p. 1019
1'. Tympanic bullae large (see Fig. 570); back black or with white areas containing black hairs mixed with the white. *M. macroura*, p. 1022

Mephitis mephitis
Striped Skunk

External measurements of males, 575–800, 184–393, 60–90. Basal length, 60.5–76.0. Females smaller, by as much as 15 per cent in some linear measurements. Pelage mostly black; white on top of head and neck, in most individuals white extending posteriorly (usually separating into 2 stripes), in some individuals over all of top and sides of tail; white areas on body are entirely of white hair (not mixed with black hairs). Tympanic bullae smaller than in *Mephitis macroura* (see Figs. 569 and 570).

Mephitis mephitis avia Bangs

1898. *Mephitis avia* Bangs, Proc. Biol. Soc. Washington, 12:32, March 24, type from San Jose, Mason Co., Illinois.
1936. *Mephitis mephitis avia*, Hall, Carnegie Inst. Washington Publ. 473:65, November 20.
1908. *Mephitis mephitica newtonensis* Brown, Mem. Amer. Mus. Nat. Hist., 9(4): 175, Pl. 16, type from Conrad Fissure, Newton Co., Arkansas.

MARGINAL RECORDS.—Indiana: Kankakee Valley; Fowler. Illinois: Charleston. Kansas: 6 mi. S Garnett; Harvey County; Riley County.

Mephitis mephitis elongata Bangs

1895. *Mephitis mephitica elongata* Bangs, Proc. Boston Soc. Nat. Hist., 26:531, July 31, type from Micco, Brevard Co., Florida.
1921. *Mephitis mephitis elongata*, Howell, N. Amer. Fauna, 45:39, October 28.

MARGINAL RECORDS.—Virginia: Highland County; Buckingham County; Brunswick County. North Carolina: Raleigh, thence southward along Atlantic Coast and coast of Gulf of Mexico west into Mis-

sissippi: Bay St. Louis (Kennedy, *et al.*, 1974:27, as *M. mephitis* only). Louisiana: 4 mi. S University (Lowery, 1974:445). Mississippi: Pearl River County (Kennedy, *et al.*, 1974:27, as *M. mephitis* only). Alabama: Castleberry. Georgia: Nashville; Pinetucky. Tennessee: Greenbrier. Virginia: Lee County. West Virginia: Travellers Repose.

Mephitis mephitis estor Merriam

1890. *Mephitis estor* Merriam, N. Amer. Fauna, 3:81, September 11, type from Little Spring, 8200 ft., N base San Francisco Mt., Coconino Co., Arizona.
1931. *Mephitis mephitis estor*, Hall, Univ. California Publ. Zool., 37:1, April 10.

MARGINAL RECORDS.—Utah: Vernal, 5200 ft. Colorado (Armstrong, 1972:289): Headquarters, Colorado National Monument; ½ mi. S, 8 mi. E Skyway; Coventry; Mesa Verde National Park (sight record); Ute Creek, near Durango; *10 mi. N Arboles;* mouth Sandoval Cañon. New Mexico: Blanco; San Rafael; Datil. Chihuahua (Anderson, 1972:384): Ojo Palomo Viejo, 4000 ft.; San Diego. Durango: Río Sestín. Chihuahua: Sierra Madre near Guadalupe y Calvo. Sonora: La Noria. Baja California: Poso Vicente. California: 5 mi. below Needles. Nevada: 1 mi. N Rox. Utah: Zion National Park.

Mephitis mephitis holzneri Mearns

1897. *Mephitis occidentalis holzneri* Mearns, Preliminary diagnoses of new mammals . . . from the Mexican boundary line, p. 4, January 12 (preprint of Proc. U.S. Nat. Mus., 20:461, December 24, 1897), type from San Isidro Ranch, Baja California, within 2 mi. of border of San Diego Co., California.
1933. *Mephitis mephitis holzneri*, Grinnell, Univ. California Publ. Zool., 40:107, September 26.

MARGINAL RECORDS.—California: Monterey; Three Rivers; Weldon; Frazier Mtn.; Oro Grande; Seven Oaks; Whitewater; old stage station on Vallecito Creek; Campo. Baja California: type locality.

Mephitis mephitis hudsonica Richardson

1829. *Mephitis americana* var. *hudsonica* Richardson, Fauna Boreali-Americana, 1:55. Type locality, plains of Saskatchewan, Canada.
1934. *Mephitis mephitis hudsonica*, Hall, Univ. California Publ. Zool., 40:368, November 5.
1911. *Mephitis minnesotoe* Brass, Aus dem Reiche der Pelze, p. 532, April. Type locality, forested region of Minnesota.

MARGINAL RECORDS.—Mackenzie: Simpson; Great Slave Lake; Fort Smith. Manitoba: central Manitoba. Michigan: Chippewa County. Wisconsin (Jackson, 1961:377, Map 69): Door County; Kenosha County. Iowa (Bowles, 1975:134): near Maysville; *Louisa County;* Bloomfield; Hamburg. Nebraska (Jones, 1964c:295, 296): 3 mi. S Rulo; *2 mi. S, ½ mi. E Barnston;* 1 mi. W Endicott; *S of Parks;* Haigler. Col-

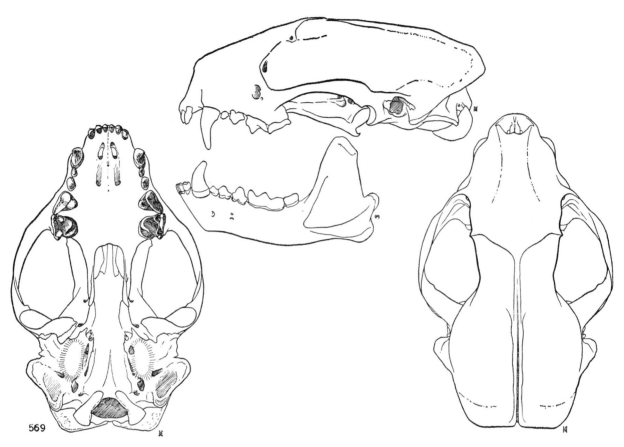

Fig. 569. *Mephitis mephitis major*, 1 mi. SE Lovelock, Pershing Co., Nevada, No. 90562 M.V.Z., ♂, X 1.

orado (Armstrong, 1972:291): Burlington; Cañon City; Salida; Chromo. New Mexico: Moreno Valley on E slope Taos Pass at 7800 ft.; Capitan Mts.; Lake Burford. Colorado (Armstrong, 1972:291): Gordon Creek, near Piedra River; Crawford; S bank Yampa River, 5 mi. NW Cross Mtn. Utah: Sheep Creek. Idaho: E side Bear Lake. Wyoming: Lower Geyser Basin; Shoshone Lake. Idaho: Coeur d'Alene. Washington: Spangle; Fort Spokane; Timentwa; Oroville. British Columbia: Ashcroft; Lonesome Lake; Bear Lake; Tuchodi Lake; jct. Liard and Nelson rivers.

Mephitis mephitis major (A. H. Howell)

1901. *Chincha occidentalis major* A. H. Howell, N. Amer. Fauna, 20:37, August 31, type from Fort Klamath, Klamath Co., Oregon.
1931. *Mephitis mephitis major,* Hall, Univ. California Publ. Zool., 37:1, April 10.

MARGINAL RECORDS.—Washington: Prescott. Idaho: Salmon River Mts.; jct. Ross Fork Creek and Portneuf River. Utah: 3 mi. W Woods Cross, 4343 ft.; Ephraim. Nevada: Hamilton; 1 mi. NW Wichman. California: Tahoe City; Spaldings' on Eagle Lake; Dry Creek. Oregon: Tule Lake; type locality; Harney. Washington: *Wallula;* Burbank.

Mephitis mephitis mephitis (Schreber)

1776. *Viverra mephitis* Schreber, Die Säugthiere . . . , theil 3, heft 17, Pl. 121. Type locality, eastern Canada [= Quebec].
1902. *Mephitis mephitis,* J. A. Allen, *et al.,* Science, n.s., 16:115, July 18.
1792. *Viverra mephitica* Shaw, Museum Leverianum . . . , p. 171. Type locality, unknown.
1818. *Mephitis americana* Desmarest, Nouv. Dict. Hist. Nat., Paris, 21:514, a composite species.

MARGINAL RECORDS.—Manitoba: Oxford House. Quebec: Rupert House (Harper, 1961:120); Mistassini Post. *N shore Gulf of St. Lawrence. New Brunswick.* Quebec: Lake Edward.

Mephitis mephitis mesomelas Lichtenstein

1832. *Mephitis mesomelas* Lichtenstein, Darstellung neuer oder wenig bekannter Säugethiere . . . , Pl. 45, Fig. 2. Type locality, Louisiana.
1936. *Mephitis mephitis mesomelas,* Hall, Carnegie Inst. Washington Publ. 473:66, November 20.
1896. *Mephitis mephitica scrutator* Bangs, Proc. Biol. Soc. Washington, 10:141, December 28, type from Cartville, Acadia Parish, Louisiana.

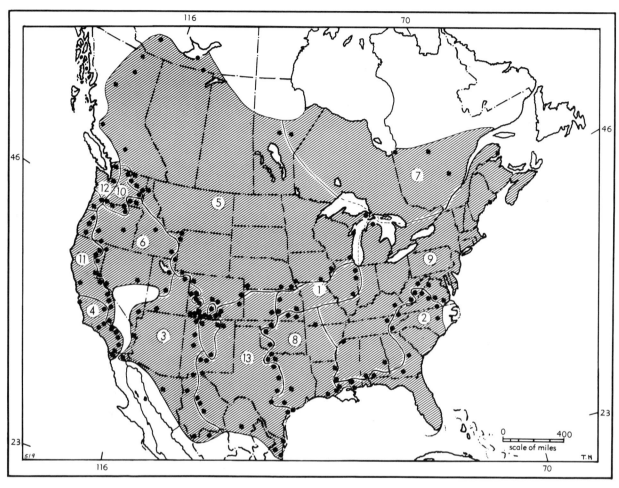

Map 519. *Mephitis mephitis.*

1. *M. m. avia*	5. *M. m. hudsonica*	10. *M. m. notata*
2. *M. m. elongata*	6. *M. m. major*	11. *M. m. occidentalis*
3. *M. m. estor*	7. *M. m. mephitis*	12. *M. m. spissigrada*
4. *M. m. holzneri*	8. *M. m. mesomelas*	13. *M. m. varians*
	9. *M. m. nigra*	

MARGINAL RECORDS.—Kansas: Hamilton; Neosho Falls. Missouri: Marble Cave. Louisiana (Lowery, 1974:445): Tallula; *16 mi. S Delta Point,* thence down W bank Mississippi River to 3 mi. S Plaquemine; *1½ mi. N Bayou Sorrel;* 22 mi. SW Abbeville, thence west along coast to Texas: O'Connorport [= Port O'Connor?]; DeWitt County (Davis, 1966:96, as *M. mephitis* only); San Antonio; Brazos County; Clay County (Davis, 1966:96, as *M. mephitis* only); Baylor County (Baccus, 1971:183); Wichita Falls. Oklahoma: Dewey County.

Mephitis mephitis nigra (Peale and Palisot de Beauvois)

1796. *Viverra nigra* Peale and Palisot de Beauvois, A scientific and descriptive catalogue of Peale's museum, Philadelphia, p. 37 Type locality, Maryland.

1921. *Mephitis mephitis nigra,* A. H. Howell, N. Amer. Fauna, 45:39, October 28.

1839. *Mephitis americana bivirgata* Hamilton-Smith, The naturalist's library (edit. Jardine), Mamm., 1:196. Type locality, unknown.

1842. *Mephitis putida* Boitard, Jardin des Plantes, Mamm., p. 147. Type locality, New Jersey.

`842. *Mephitis olida* Boitard, Jardin des Plantes, Mamm., p. 147. Type locality, unknown.

1842. *Mephitis fetidissima* Boitard, Jardin des Plantes, Mamm., p. 147. Type locality, unknown.

1875. *Mephitis frontata* Coues, Bull. No. 1, U.S. Geol. and Geog. Surv. Territories, p. 7, Fig. 1. Type locality, Pennsylvania.

1911. *Mephitis dentata* Brass, Aus dem Reiche der Pelze, p. 533, April. Type locality, from the Alleghenies to Connecticut.

MARGINAL RECORDS.—Quebec: near Anse aux Gascons (Manville, 1961:109, sight record as *M. mephitis* only). Maine: Mount Desert Island (Manville, 1960:416, as *M. mephitis* only), thence southward along Atlantic Coast of United States to Virginia: Prince George County; Amelia County; Shenandoah County. West Virginia: vic. Philippi; Odd, 2900 ft. Kentucky: Quicksand. Tennessee: Highcliff. Alabama: Auburn; Jackson. Mississippi (Kennedy, *et al.*, 1974:27); Wilkinson County; De Soto County. Michigan: Cheboygan County. *M. mephitis*, subspecies unknown, introduced on Prince Edward Island (Cameron, 1959:51).

Mephitis mephitis notata (A. H. Howell)

1901. *Chincha occidentalis notata* A. H. Howell, N. Amer. Fauna, 20:36, August 31, type from Trout Lake, Mt. Adams, Washington.
1936. *Mephitis mephitis notata*, Hall, Carnegie Inst. Washington Publ. 473:67, November 20.

MARGINAL RECORDS.—Washington: Loomis; Rock Lake. Oregon: Umatilla; Lena; Millers, near mouth Deschutes River. Washington: Skamania.

Mephitis mephitis occidentalis Baird

1858. *Mephitis occidentalis* Baird, Mammals, *in* Repts. Expl. Surv. . . . , 8(1):194, July 14, type from Petaluma, Sonoma Co., California.
1933. *Mephitis mephitis occidentalis*, Grinnell, Univ. California Publ. Zool., 40:106, September 26.
1901. *Chincha platyrhina* A. H. Howell, N. Amer. Fauna, 20:39, August 31, type from S. Fork Kern River, 3 mi. above Onyx, California.

MARGINAL RECORDS.—Oregon: Eugene; Roseburg; Grants Pass. California: Pitt River; Big Valley Mts.; Blue Canyon; *Lake Tahoe*; Hope Valley; Farrington's Ranch; near Laws; near Independence; S. Fork Kern River, 25 mi. E Kernville; Walnut Creek.

Mephitis mephitis spissigrada Bangs

1898. *Mephitis spissigrada* Bangs, Proc. Biol. Soc. Washington, 12:31, March 24, type from Sumas, British Columbia.
1936. *Mephitis mephitis spissigrada*, Hall, Carnegie Inst. Washington Publ. 473:67, November 20.
1899. *Mephitis foetulenta* Elliot, Field Columb. Mus., Publ. 32, Zool. Ser., 1:269, May 17, type from Lagune, near Port Angeles, Clallam Co., Washington.

MARGINAL RECORDS.—British Columbia: Nicomen Island (Cowan and Guiguet, 1965:329). Washington: Skyomish; Washougal River. Oregon: McCoy, thence to and up coast to opposite point of beginning.

Mephitis mephitis varians Gray

1837. *Mephitis varians* Gray, Charlesworth's Mag. Nat. Hist., 1:581. Type locality, Texas.
1936. *Mephitis mephitis varians*, Hall, Carnegie Inst. Washington Publ. 473:66, November 20.

MARGINAL RECORDS.—Colorado: Chivington (Armstrong, 1972:292). Kansas: 2 mi. N, 2 mi. E Hutchinson. Oklahoma: Alva; Wichita National Forest [Wichita Mountains National Wildlife Refuge?]. Texas: *Jack County* (Davis, 1966:96, as *M. mephitis* only); Brazos; McLennan County (Davis, 1966:96, as *M. mephitis* only); vic. Austin; San Antonio; Indianola; *Aransas National Wildlife Refuge* (Halloran, 1961:24, as *M. m. mesomelas*). Tamaulipas: Matamoros; 7 mi. W La Pesca (Baker and Webb, 1967:188); 2 mi. upstream from Marmolejo. Coahuila: 3 mi. S, 16 mi. E Quatro Ciénegas (Baker, 1956:310). Chihuahua (Anderson, 1972:385): 20 mi. SE Ciudad Chihuahua, 5100 ft.; *5 mi. N Chihuahua, 4700 ft.*; 11 mi. S, 1 mi. E Encinillas; *Rancho La Campana*; 18 mi. SW Gallego, 6000 ft. Texas: El Paso. New Mexico: Mountain (Mud) Tank; Tularosa; *Hall Peak*; vic. Cimarron. Colorado (Armstrong, 1972:292): Antonito; *La Jara*; 5 mi. NW Hooper; Costilla County; Huerfano Butte.

Mephitis macroura
Hooded Skunk

External measurements of males: 558–790; 275–435; 58–73. Basilar length of skull, 56.1–60.3. Females smaller, by as much as 15 per cent in some linear measurements. Pelage black usually with white on upper parts. This species can be distinguished from *M. mephitis* by larger tympanic bullae (see Figs. 569 and 570); in the white-backed color phase by having some black hairs mixed with the white hairs of the back, and in the black-backed phase by having the 2 white stripes widely separated and on the sides of the animal instead of narrowly separated and on the back of the animal.

Mephitis macroura eximius Hall and Dalquest

1950. *Mephitis macroura eximius* Hall and Dalquest, Univ. Kansas Publ., Mus. Nat. Hist., 1:579, January 20, type from 15 km. W Piedras Negras, 300 ft., Veracruz.

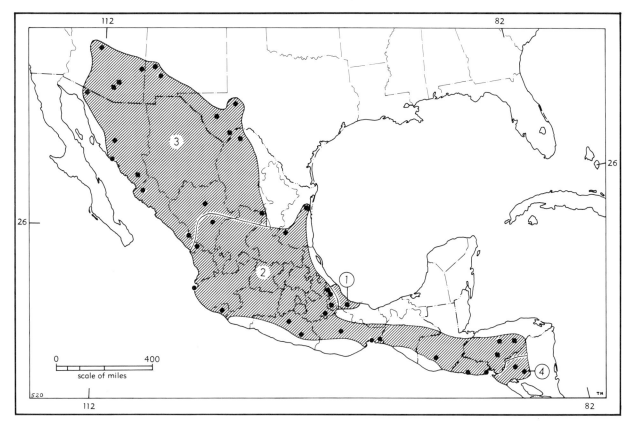

Map 520. *Mephitis macroura.*

1. *M. m. eximius*   2. *M. m. macroura*   3. *M. m. milleri*   4. *M. m. richardsoni*

MARGINAL RECORDS.—Veracruz: Río Blanco, 20 km. WNW Piedras Negras; *type locality.*

Mephitis macroura macroura Lichtenstein

1832. *Mephitis macroura* Lichtenstein, Darstellung neuer oder wenig bekannter Säugethiere . . . , Pl. 46, type from mts. NW of City of Mexico.
1832. *Mephitis vittata* Lichtenstein, Darstellung neuer oder wenig bekannter Säugethiere . . . , Pl. 47, type from San Mateo del Mar, Oaxaca.
1837. *Mephitis mexicana* Gray, Charlesworth's Mag. Nat. Hist., 1:581. Type locality, unknown.
1862. *Mephitis longicaudata* Tomes, Proc. Zool. Soc. London, for 1861, p. 280, April, type from Dueñas, Guatemala.
1865. *Mephitis concolor* Gray, Verraux MSS., Proc. Zool. Soc. London, p. 149, June, based on a variety of *M. vittata* Lichtenstein.
1869. *Mephitis vittata* var. *intermedia* Gray, Catalogue of the carnivorous, pachydermatous, and edentate Mammalia in the British Museum, p. 138, based on a variety of *M. vittata* Lichenstein.
1877. *Mephitis edulis* Coues, Berlandier MSS, Fur-bearing Animals . . . , U.S. Geol. Surv. Territories, Miscl. Publ., 8:236. Type locality, "Inhabits most of Mexico. I have found it about San Fernando de Bexar. . . ."

MARGINAL RECORDS.—Tamaulipas: San Fernando (Alvarez, 1963:462). Veracruz: Las Vigas; Jico (Hall and Dalquest, 1963:341); Orizaba. Puebla: Tehuacán. Honduras: El Caliche Orica; Catacamas; Tegucigalpa. El Salvador (Burt and Stirton, 1961:47): Volcán Conchagua; Rosario de Mora. Guatemala: vic. Dueñas. Oaxaca (Goodwin, 1969:241): Tapanatepec; Santa María del Mar; *Salina Cruz;* 15 mi. W Oaxaca de Juárez. Guerrero: 1 mi. SW Colotlipa, 2700 ft. (Davis and Lukens, 1958:357); 22 km. S Taxco, 400 ft. Colima: Hda. Magdalena. Jalisco: San Sebastián. Nayarit: Santa Teresa. Durango: 6 mi. NW La Pila, 6150 ft. (Baker and Greer, 1962:139). Nuevo León: *Monterrey* (Baird, 1858:193, as *Mephitis varians*, 709 USNM. According to C. O. Handley, Jr., *in Litt.*, February 21, 1966, No. 709 was in catalog as *Mephitis macroura* but was destroyed in 1886 and no part of the specimen remains in the museum). Tamaulipas: *Nicolás, 56 km. NW Tula* (Alvarez, 1963:462); Jaumave.

Mephitis macroura milleri Mearns

1897. *Mephitis milleri* Mearns, Preliminary diagnoses of new mammals . . . from the Mexican boundary of the United States, p. 1, February 11 (preprint of Proc. U.S. Nat. Mus., 20:467, December 24, 1897), type from Fort Lowell, near Tucson, Pima Co., Arizona.

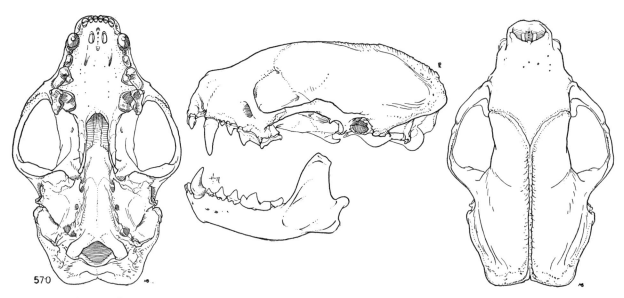

Fig. 570. *Mephitis macroura macroura*, 2 mi. W Pátzcuaro, 7600 ft., Michoacán, No. 100115 M.V.Z., ♂, X 1.

1901. *Mephitis macroura milleri*, J. A. Allen, Bull. Amer. Mus. Nat. Hist., 14:334, November 12.

MARGINAL RECORDS.—Arizona (Cockrum, 1961: 239, 240): Camp Verde; Cosper Ranch, Blue River, 30 mi. above Clifton. New Mexico: Lilly Mtn. (Hubbard, 1972:458); *ca.* 7 mi. above Mimbres P.O. Texas: 3 mi. N Mt. Livermore; $4\frac{7}{10}$ mi. S Royalty (Packard, 1965:102); Black Gap Wildlife Management Area. Coahuila: Juarez Cañón in Sierra del Carmen; La Ventura. Durango: 7 mi. NNW La Zarca (Baker and Greer, 1962:139). Sinaloa (Armstrong, *et al.*, 1972:54): 5 km. NE Santa Lucía, 5000 ft.; 2 mi. N San Blas, 50 ft. Sonora: Camoa; 10 mi. NW Guaymas (Packard, 1965:102); Hermosillo; Sierra Cuababi. Arizona: approx. 15 mi. W Tucson; Santa Catalina Mts.

Mephitis macroura richardsoni Goodwin

1957. *Mephitis macroura richardsoni* Goodwin, Amer. Mus. Novit., 1830:3, May 3, type from Matagalpa, Nicaragua, 2000 ft.

MARGINAL RECORDS.—Nicaragua: San Rafael del Norte; type locality.

Genus **Conepatus** Gray—Hog-nosed Skunks

1837. *Conepatus* Gray, Charlesworth's Mag. Nat. Hist., 1:581, November. Type, *Conepatus humboldtii* Gray.
1837. *Thiosmus* Lichtenstein, Abh. preuss. Akad. Wiss., Berlin, for 1836, p. 272. Included 10 species [now in genus *Conepatus* Gray].
1837. *Marputius* Gray, Charlesworth's Mag. Nat. Hist., 1:581, November. Type, *Mephitis chilensis* É. Geoffroy Saint-Hilaire [= *Viverra chinga* Molina].

1902. *Oryctogale* Merriam, Proc. Biol. Soc. Washington, 15:161, August 6. Type, *Mephitis leuconata* Lichtenstein.

External measurements: males 513–900; 174–410; 65–90. Condylobasal length of skull, 65.0–83.3; zygomatic breadth, 45.0–76.8. Females are smaller, perhaps by 10 per cent. Pelage black except for white back and tail; depending on the species the back is all white or instead marked with 2 white stripes; the proximal part of the ventral side of the tail is black. Claws on forefeet and bare area around nose pad much larger than in *Mephitis;* 3 or 4 pairs of mammae. Skull deepest in temporal region; mastoid bullae hardly inflated; posterior margin of palate behind upper molars; anteroposterior and transverse diameters of M1 each more than outside length of P4; metaconule of M1 not distinct; inferior margin of lower mandible longitudinally convex; in m1 trigonid shorter than talonid; metaconid high and distinct from protoconid; hypoconid low, entoconid high and separated from metaconid by distinct notch. Dentition, i. $\frac{3}{3}$, c. $\frac{1}{1}$, p. $\frac{2}{3}$, m. $\frac{1}{2}$.

KEY TO NORTH AMERICAN SPECIES OF CONEPATUS

1. Tail more than $\frac{1}{3}$ of total length; back with one broad white stripe; anterior opening of infraorbital canal single.
   2. White stripe as broad on rump as on shoulders; in M1, outside length less than 7.9 in male and less than 7.2 in female. . . . . *C. mesoleucus*, p. 1025
   2'. White stripe narrower (or even absent) on rump than on shoulders; in M1, outside length more than 7.9 in male

and more than 7.2 in female.

C. leuconotus, p. 1027

1'. Tail less than ⅓ of total length; back with 2 white stripes united at nape and separated by a black area narrower than either white stripe; each white stripe tapers to a tip on rump; anterior opening of infraorbital canal multiple (double or triple).

C. semistriatus, p. 1027

Conepatus mesoleucus
Hog-nosed Skunk

× ⅛

Maximum external measurements for males are near: 633; 275; 75. Condylobasal length of skull, 72.8; zygomatic breadth, 51.3. Females are smaller, by as much as 10 per cent in some linear measurements. Pelage black except that back and all but proximal part of underside of tail white.

In most populations of this species the white area of the back is continued broadly onto the tail whereas in Conepatus leuconotus the white area is only narrowly continuous from the back to the tail. The teeth are smaller in C. mesoleucus than in C. leuconotus. Actually there is intergradation in the color pattern, and I suppose that when specimens are obtained from areas intermediate between the geographic ranges of the two nominal species, they will prove to be only subspecifically different.

Conepatus mesoleucus figginsi F. W. Miller

1925. *Conepatus mesoleucus figginsi* F. W. Miller, Jour. Mamm., 6:50, January 9, type from Furnace Canyon, western Baca Co., Colorado.

MARGINAL RECORDS (Armstrong, 1972:293).— Colorado: *western Baca County;* type locality.

Conepatus mesoleucus filipensis Merriam

1902. *Conepatus filipensis* Merriam, Proc. Biol. Soc. Washington, 15:163, August 6, type from Cerro San Felipe, Oaxaca. Known only from type locality.
1952. *Conepatus mesoleucus filipensis*, Hall and Kelson, Univ. Kansas Publ., Mus. Nat. Hist., 5:335, December 5.

Conepatus mesoleucus fremonti F. W. Miller

1933. *Conepatus mesoleucus fremonti* F. W. Miller, Proc. Colorado Mus. Nat. Hist., 12(1):1, July 22, type from Garden Park, near Canon City, Fremont Co., Colorado.

MARGINAL RECORDS (Armstrong, 1972:294).— Colorado: Keaton Ranch, Little Fountain Creek; type locality.

Conepatus mesoleucus mearnsi Merriam

1902. *Conepatus mesoleucus mearnsi* Merriam, Proc. Biol. Soc. Washington, 15:163, August 6, type from Mason, Mason Co., Texas.
1902. *Conepatus pediculus* Merriam, Proc. Biol. Soc. Washington, 15:164, August 6, type from Sierra Guadalupe, Coahuila. Considered to be inseparable from *C. m. mearnsi* by Baker, Univ. Kansas Publ., Mus. Nat. Hist., 9:310, June 15, 1956.

MARGINAL RECORDS.—New Mexico (Findley, *et al.*, 1975:315, as *C. mesoleucus* only): Domingo Baca Canyon, Sandia Mts.; Jacarilla Mts.; Dexter. Texas (Davis, 1966:99, unless otherwise noted, as *C. mesoleucus* only): 11 mi. SE Lubbock (Packard and Garner, 1964:389, sight record only); 22 mi. SW Gail (*ibid.*, sight record only); Texas State Hwy. 156 [158], 17 mi. W Ballinger (near Bronte) (*ibid.*); Brazos bottom S of Waco (Hall and Kelson, 1959:941); Harris County; Waller County; De Witt County; 8 mi. NW Jourdanton (Raun and Wilks, 1961:205, as *C. mesoleucus* only); Dimmit County. Tamaulipas: Nicolás, 56 km. NW Tula (Alvarez, 1963:462). San Luis Potosí: Hda. La Parda. Jalisco: Zapotlán; Guadalajara. New Mexico (Findley, *et al.*, 1975:315, as *C. mesoleucus* only): Soledad Canyon, 17 mi. E Las Cruces, Organ Mts.; Big Rose Canyon; 8 mi. SE Paxton.

Conepatus mesoleucus mesoleucus (Lichtenstein)

1832. *Mephitis mesoleuca* Lichtenstein, Darstellung neuer oder wenig bekannter Säugethiere . . . , Pl. 44, Fig. 2, type from near Chico, Hidalgo.
1902. [*Conepatus*] *mesoleucus*, Merriam, Proc. Biol. Soc. Washington, 15:163, August 6.

MARGINAL RECORDS.—Hidalgo: type locality. Puebla: Río Otlati, 15 km. NW San Martín, 8700 ft. Chiapas: 17 mi. W Bochil. Morelos: 2 mi. W Huitzilac; *Cerro Cuautépetl, 3200 m.* (Ramírez-P., 1971:282). México: *Laguna Tonatiahua (Cerro Ocuilan), 3000 m.* (*ibid.*).

Conepatus mesoleucus nelsoni Goldman

1922. *Conepatus mesoleucus nelsoni* Goldman, Jour. Mamm., 3:41, February 8, type from Armería (near Manzanillo), 200 ft., Colima.

MARGINAL RECORDS.—Colima: Colima. Michoacán: Tancítaro, 6000 ft. Guerrero (Davis and Lukens, 1958:358): Acahuizotla, 2800 ft.; Ometepec,

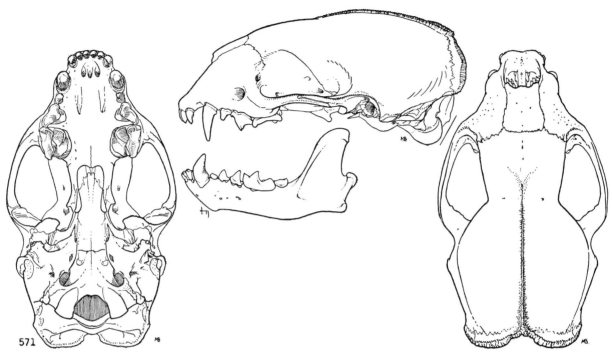

Fig. 571. *Conepatus mesoleucus mearnsi*, E base Burro Mesa, Chisos Mts., Brewster Co., Texas, No. 80336 M.V.Z., ♂, X 1.

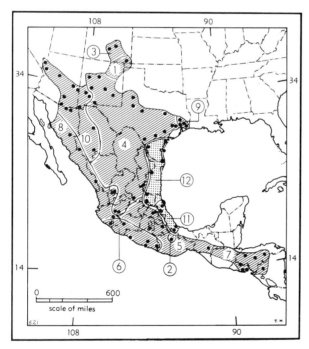

Map 521. *Conepatus mesoleucus* and *Conepatus leuconotus*.

1. C. m. figginsi	7. C. m. nicaraguae
2. C. m. filipensis	8. C. m. sonoriensis
3. C. m. fremonti	9. C. m. telmalestes
4. C. m. mearnsi	10. C. m. venaticus
5. C. m. mesoleucus	11. C. l. leuconotus
6. C. m. nelsoni	12. C. l. texensis

1000 ft.; Acapulco; Zihuatanejo Bay. Colima: *type locality; Hda. Magdalena.*

Conepatus mesoleucus nicaraguae J. A. Allen

1910. *Conepatus nicaraguae* J. A. Allen, Bull. Amer. Mus. Nat. Hist., 28:106, April 30, type from San Rafael del Norte, Nicaragua.
1946. *Conepatus mesoleucus nicaraguus*, Goodwin, Bull. Amer. Mus. Nat. Hist., 87:437, December 31.

MARGINAL RECORDS.—Honduras: Cementerio; El Caliche Cedros; Catacamas. Nicaragua: type locality. El Salvador (Burt and Stirton, 1961:47, as *Conepatus leuconotus nicaraguae*): Lake Olomega; Puerto del Triunfo; Los Esesmiles.

Conepatus mesoleucus sonoriensis Merriam

1902. *Conepatus sonoriensis* Merriam, Proc. Biol. Soc. Washington, 15:162, August 6, type from Camoa, Río Mayo, Sonora.
1952. *Conepatus mesoleucus sonoriensis*, Hall and Kelson, Univ. Kansas Publ., Mus. Nat. Hist., 5:335, December 5.

MARGINAL RECORDS.—Sonora: type locality. Sinaloa (Armstrong, et al., 1972:55): Rancho Rosalita, 26 mi. NE Choix, 1300 ft.; 19½ km. (by road) NE Santa Lucía, 6200 ft. Zacatecas: Valparaíso Mts. Jalisco: Los Masos.

Conepatus mesoleucus telmalestes V. Bailey

1905. *Conepatus mesoleucus telmalestes* V. Bailey, N. Amer. Fauna, 25:203, October 24, type from Big Thicket, 7 mi. NE Sour Lake, Hardin Co., Texas.

MARGINAL RECORDS.—Texas: San Jacinto County (Davis, 1966:99, as *C. mesoleucus* only); Tarkington Prairie; Big Thicket, 7–10 mi. NE Sour Lake; *type locality;* Montgomery County (Davis, 1966:99, as *C. mesoleucus* only).

Conepatus mesoleucus venaticus Goldman

1922. *Conepatus mesoleucus venaticus* Goldman, Jour. Mamm., 3:40, February 8, type from Cosper Ranch, Blue River, 12 mi. S Blue, 5000 ft., Greenlee Co., Arizona.

MARGINAL RECORDS.—Arizona: highway, ½ mi. S Temple Bar Junction (Huey, 1961:421); *ca.* 16 mi. SE Kingman; Camp Verde (Cockrum, 1961:241); Black River; type locality. New Mexico (Findley, *et al.*, 1975:315, as *C. mesoleucus* only): Diamond Creek, Gila National Forest; Hillsboro; *Lake Valley;* sec. 29, T. 33 S, R. 19 W in Hidalgo Co. Chihuahua (Anderson, 1972:386): 18 mi. SW Gallego, 6000 ft.; Sauceda, 27 mi. W Parral, 7100 ft. Sonora: San José Mts.; Santa Cruz. Arizona: La Osa; Baboquivari Mts. (Cockrum, 1961:242); Tucson.

Conepatus leuconotus
Eastern Hog-nosed Skunk

Maximum external measurements for males are near: 900; 410; 90. Condylobasal length of skull, 83.3; zygomatic breadth, 76.8. Females are smaller, by as much as 10 per cent in some linear measurements. The possibility that *C. leuconotus* and *C. mesoleucus* are only subspecifically different is commented on in the account of the latter species. See Map 521.

Conepatus leuconotus leuconotus (Lichtenstein)

1832. *Mephitis leuconata* Lichtenstein, Darstellung neuer oder wenig bekannter Säugethiere . . . , Pl. 44, Fig. 1, type from Río Alvarado, Veracruz.
1902. *Conepatus leuconotus*, Merriam, Proc. Biol. Soc. Washington, 15:161, August 6.

MARGINAL RECORDS.—Veracruz: Potrero Llano, 350 ft.; type locality; Río Blanco, 20 km. WNW Piedras Negras. Puebla: Rancho El Ajenjibre (Warner and Beer, 1957:20).

Conepatus leuconotus texensis Merriam

1902. *Conepatus leuconotus texensis* Merriam, Proc. Biol. Soc. Washington, 15:162, August 6, type from Brownsville, Cameron Co., Texas.

MARGINAL RECORDS.—Texas: Rockport; Padre Island; Brownsville, thence south along Gulf Coast probably into northern Veracruz and San Luis Potosí: Hda. Capulín; El Salto. Tamaulipas: near El Mulato. Texas: Laredo.

Conepatus semistriatus
Striped Hog-nosed Skunk

External measurements of males: 533–688; 160–193; 70–82. Condylobasal length of skull, up to at least 78; zygomatic breadth, up to at least 51. Females are smaller by as much as 10 per cent in certain linear measurements. Pelage black except for 2 white stripes on back and white tail. In specimens seen by me the white stripes are absent on the rump or only faintly indicated there. A few black hairs are mixed with the white hairs on the tail. The proximal part of the ventral surface of the tail is black. The white stripes are united on the nape, are separated on the body by a black area narrower than either white stripe, and each white stripe tapers to a point on the hip. From *Conepatus leuconotus* and *C. mesoleucus, Conepatus semistriatus* differs cranially as follows: dorsal profile longitudinally uneven (instead of evenly convex) owing to upturned tips of nasals, upward bulge of frontals in interorbital region, and a posterior prominence behind the postorbital constriction; margin of opening of anterior nares more nearly vertical (less oblique); postorbital constriction more pronounced; tympanic bullae less inflated; anterior opening of infraorbital canal double or triple instead of single. The species occurs also in South America.

Conepatus semistriatus conepatl (Gmelin)

1788. [*Viverra*] *conepatl* Gmelin, Syst. nat., ed. 13, 1:88. Type locality, "Nova Hispania" [= México]; restricted to Motzorongo, Veracruz, by Hershkovitz (Fieldiana-Zool., Chicago Nat. Hist. Mus., 31:562, July 10, 1951).
1951. *Conepatus semistriatus conepatl*, Hershkovitz, Fieldiana-Zool., Chicago Nat. Hist. Mus., 31:562, July 10.
1902. *Conepatus tropicalis* Merriam, Proc. Biol. Soc. Washington, 15:164, August 6, type from Motzorongo, Veracruz.

MARGINAL RECORDS.—Veracruz: type locality, presumably eastward to coast and along coast to Campeche: *no specific locality.* Veracruz: 35 km. SE Jesús Carranza, 400 ft.; *20 km. ENE Jesús Carranza, 200 ft.;* Pérez (Hall and Dalquest, 1963:342).

Conepatus semistriatus trichurus Thomas

1902. *Conepatus mapurito* Bangs, Bull. Mus. Comp. Zool., 39:48, April, type from Boquete, Volcán de Chiriquí, Chiriquí, Panamá. Not Gmelin, 1788.

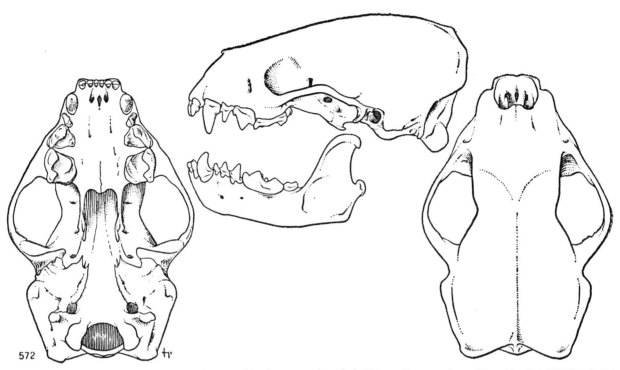

Fig. 572. *Conepatus semistriatus trichurus*, Pilón de Azucar (Peralta), 500 m., Cartago, Costa Rica, No. 32245 K.U., ♂, X 1.

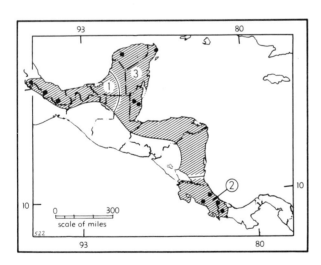

Map 522. *Conepatus semistriatus.*

1. *C. s. conepatl*   2. *C. s. trichurus*
3. *C. s. yucatanicus*

1905. *Conepatus tropicalis trichurus* Thomas, Ann. Mag. Nat. Hist., ser. 7, 15:585, June, a renaming of *Conepatus mapurito* Bangs.
1951. *Conepatus semistriatus trichurus*, Hershkovitz, Fieldiana-Zool., Chicago Nat. Hist. Mus., 31:562, July 10.

MARGINAL RECORDS.—Costa Rica: Jiménez; *Santa Teresa Peralta*. Panamá: Sibube (Handley,

1966:790); type locality. Costa Rica: Pozo Azul Pirrís; *Escazú*.

Conepatus semistriatus yucatanicus Goldman

1943. *Conepatus tropicalis yucatanicus* Goldman, Proc. Biol. Soc. Washington, 56:89, October 1, type from La Vega, on mainland opposite Isla Cancun, Quintana Roo.
1955. *Conepatus semistriatus yucatanicus*, Miller and Kellogg, Bull. U.S. Nat. Mus., 205:761, March 3.

MARGINAL RECORDS.—Quintana Roo: type locality. Belize: Stann Creek Valley; El Cayo. Guatemala: *no precise locality*. Yucatán: Mérida.

Genus **Lutra** Brünnich—River Otters

Revised by van Zyll de Jong, C. G., Royal Ontario Mus. Life Sci. Contrib., No. 80:1–104, 3 apps., 38 figs., 3 tables, June 15, 1972. (See also Pohle, Arch. für Naturgesch., 85(Abt. A, No. 9):1–247, 10 tables, 19 figs. in text, December, 1920, and Harris, C. J., Otters, a study of the Recent Lutrinae, pp. xiv + 397, illustrated, Weidenfeld and Nicolson, London, 1968.)

1762. *Lutra* Brisson, Regnum animale . . . , ed. 2, p. 13. Type, *Mustela lutra* Linnaeus. Brisson's names are regarded as non-Linnaean and as not consistently binomial. See Hopwood, Proc. Zool. Soc. London, 117(2–3):534, 535, October 30, 1947, and Hershkovitz, Proc. Biol. Soc. Washington, 62:12, March 17, 1949.
1772. *Lutra* Brünnich, Zoologiae Fundamenta . . . , p. 34, 42. Type, *Mustela lutra* Linnaeus.

1806. *Lutris* Duméril, Zoologie analytique . . . , p. 12, an emendation of *Lutra*.

1815. *Lutrix* Rafinesque, Anal. Nat., p. 59, a renaming of *Lutra*.

1842. *Leptonyx* Lesson, Nouveau tableau du règne animal . . . mammifères, p. 72. Type, *Leptonyx barang* Lesson. Not *Leptonyx* Swainson, 1821, a bird.

1865. *Barangia* Gray, Proc. Zool. Soc. London, p. 123, June. Type, *Lutra sumatrana* Gray.

1865. *Lutrogale* Gray, Proc. Soc. Zool. London, p. 127, June. Type, *Lutra perspicillata* I. Geoffroy Saint-Hilaire (misidentified by Gray as *L. monticola* Hodgson).

1865. *Hydrogale* Gray, Proc. Zool. Soc. London, p. 131, June. Type, *Lutra maculicollis* Lichtenstein. Not *Hydrogale* Kaup, 1829, a shrew.

1867. *Lutronectes* Gray, Proc. Zool. Soc. London, p. 180, May. Type, *Lutronectes whiteley* Gray [= *Mustela lutra* Linnaeus].

1921. *Hydrictis* Pocock, Proc. Zool. Soc. London, p. 543, September 9. Type, *Lutra maculicollis* Lichtenstein.

Trunk cylindrical; neck about as wide as skull, short; head flattened, rounded; legs short; toes broadly spread, webbed; claws laterally compressed, blunt; tail thick at base, oval in cross section; ears small, capable of being closed. Skull strongly flattened; muzzle short; braincase broad, long, arched; interorbital breadth less than width of muzzle; infraorbital foramen as large as, or larger than, alveolus of canine; orbit open, postorbital processes varying from large to rudimentary; alisphenoid canal absent; opening of external auditory meatus large; paroccipital process not in contact with bulla. M1 rhomboid, large; P4 with hollowed deuterocone; talonid and trigonid of m1 about equal in length; metaconid large; talonid basined. Dentition, i. $\frac{3}{3}$, c. $\frac{1}{1}$, p. $\frac{4}{3}$, m. $\frac{1}{2}$.

Subgenus **Lontra** Gray—American River Otters

1843. *Lontra* Gray, Ann. Mag. Nat. Hist., ser. 1, 11:118, February. Type, *Lutra canadensis* Schreber.

1843. *Latax* Gray, Ann. Mag. Nat. Hist., ser. 1, 11:119, February. Type, *Lutra lataxina* F. Cuvier. Not *Latax* Gloger, 1827.

1843. *Lataxina* Gray, List of the . . . Mammalia in the . . . British Museum, p. 70. Type, *Lataxina mollis* Gray [= *Lutra lataxina* F. Cuvier].

1855. *Lataxia* Gervais, Histoire naturelle des mammifères . . . , 2:118, a renaming of *Latax* Gray.

1865. *Nutria* Gray, Proc. Zool. Soc. London, p. 128, June. Type, *Lutra felina* Molina.

Dentition adapted for both cutting and crushing food; tip of baculum curved dorsad. All New World river otters of the genus *Lutra* belong to this subgenus. In the subgenus *Lutra*, the dentition is less adapted for crushing and the tip of the baculum is curved ventrad. Van Zyll de Jong (1972:80) arranged *Lontra* as a genus, but in

doing so he accords generic rank to a group of species that closely corresponds to a subgenus in other families of mammals. Consequently *Lontra* here is recognized as a subgenus.

KEY TO NORTH AMERICAN SPECIES OF LUTRA

1. Rhinarium furred laterally; sole of foot naked; skull convex dorsally; occurring south of 28° latitude. *L. longicaudis*, p. 1031
1'. Rhinarium naked; sole of foot with tufts of hair under toes; skull flat dorsally; occurring north of 28° latitude. *L. canadensis*, p. 1029

Lutra canadensis
River Otter

External measurements: 889–1300; 300–507; 100–146. Weight, 11–23 pounds. Basal length, 92.9–122.0; zygomatic breadth, 62.0–76.5. Females slightly smaller than males. Upper parts brown; underparts paler; lower jaw and throat whitish; whitish extending to level of ear in some individuals; underfur gray, tipped with brown. Skull flattened dorsally; mastoid processes weak; tympanic bullae flattened, not in contact with paroccipital processes; palate projecting not more than 15 mm. behind upper molars; postorbital constriction wider than in *L. longicaudis;* 14 pairs of ribs.

Some names listed below by van Zyll de Jong as synonyms of one or another of the seven subspecies of *Lutra canadensis* probably apply to valid subspecies unrecognized by him.

Lutra canadensis canadensis (Schreber)

1776. *Mustela lutra canadensis* Schreber, Die Säugthiere . . . , theil 3, heft 18, pl. 126b. Type locality, Eastern Canada [= Quebec].

1803. "*Mustela hudsonica* Lacép. [ède]," Desmarest, Nouv. Dict. Hist. Nat., XIII, p. 384. Preceding is from Rhoads, Trans. Amer. Philos. Soc., n.s., 19:424, 1898, who uses *hudsonica* as the specific name of the Canadian river otter, but see Allen, J. A., Bull. Amer. Mus. Nat. Hist., 18:459, 460, November 10, 1898, for use of *Lutra canadensis*.

1823. *Lutra canadensis*, Sabine, *in* Franklin, Narrative of a journey to the shores of the Polar Sea in . . . 1819–22, p. 653.

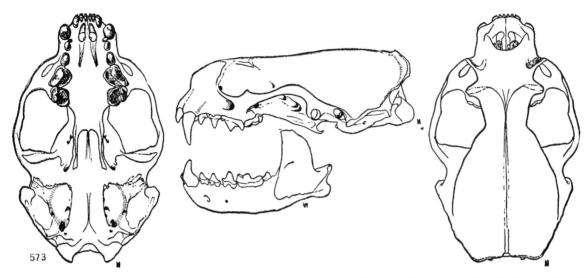

Fig. 573. *Lutra canadensis sonora*, 8 mi. up Colorado River from Needles, San Bernardino Co., California, No. 61451 M.V.Z., ♂, X ½.

1847. *Lutra americana* Wyman, Proc. Boston Soc., 2:249, a *nomen nudum*.
1863. *Lutra destructor* Barnston, Canadian Nat. Geol., 8:152, April, type from Michipicoten Island, Lake Superior, Ontario.
1898. *Lutra degener* Bangs, Proc. Biol. Soc. Washington, 12:35, March 24, type from Bay St. George, Newfoundland.
1945. *Lutra canadensis chimo* Anderson, Ann. Rept. Provancher Soc. Nat. Hist. Canada, Quebec, for 1944, p. 59, November 3, type from Chimo, Ungava Dist., Quebec, about 30 mi. S tip of Ungava Bay, Quebec.

MARGINAL RECORDS (van Zyll de Jong, 1972:97, unless otherwise noted).—Quebec: head Povungnituk River (Harper, 1961:124); Payne Bay (*ibid.*); Ungava Bay, Fort Chimo. Labrador: Okak (Harper, 1961:124); Hopedale; Black Bay (Harper, 1961:124). Newfoundland: Gander River; Bay St. George. Nova Scotia: Cape Breton Island (Hall and Kelson, 1959:945); 4 mi. S Halifax. Maine: Bucksport. Massachusetts: Kingston (Hall and Kelson, 1959:945). Connecticut: Liberty Hill; Portland. New York: Herkimer County. Ontario: Dalhousie Lake, Lanark Co. Michigan: Gladwin County; Missaukee County. Wisconsin: Waterford (Jackson, 1961:389); *Grant County* (*op. cit.*:384, Map 70); Pierce County (*ibid.*); Douglas County (*ibid.*). Ontario: Cedar Lake; Fort Severn. Quebec: Little Whale River (Hall and Kelson, 1959:945).

Lutra canadensis kodiacensis Goldman

1935. *Lutra canadensis kodiacensis* Goldman, Proc. Biol. Soc. Washington, 48:180, November 15, type from Uyak Bay, Kodiak Island, Alaska.

MARGINAL RECORDS (van Zyll de Jong, 1972:99).—Alaska: Shuyak Island; *Afognak Island*; type locality.

Lutra canadensis lataxina F. Cuvier

1823. *Lutra lataxina* F. Cuvier, *in* Dictionnaire des sciences naturelles . . . , 27:242. Type locality, South Carolina.
1898. *Lutra canadensis lataxina*, J. A. Allen, Bull. Amer. Mus. Nat. Hist., 10:460, November 10.
1843. *Lataxina mollis* Gray, List of the . . . Mammalia in the . . . British Museum, p. 70. Based on *Lutra lataxina* F. Cuvier.
1897. *Lutra rhoadsi* Cope, Proc. Acad. Nat. Sci. Philadelphia, for 1896, p. 385, 391, 392, type from Port Kennedy Bone Deposits, Pennsylvania.
1898. *Lutra hudisonica vaga* Bangs, Proc. Boston Soc. Nat. Hist., 28:224, March, type from Micco, Brevard Co., Florida.
1920. *Lutra canadensis interior* Swenk, Univ. Nebraska Stud., 18:2, May 15, type from Lincoln Creek, W of Seward, Seward Co., Nebraska.
1933. *Lutra parviceps* Gidley and Gazin, Jour. Mamm., 14:349, November 13, type from Pleistocene deposit, Cumberland Cave, Maryland. Regarded as indistinguishable from *L. c. lataxina* by Hall (Carnegie Inst. Washington Publ. 473:75–77, November 20, 1936).
1935. *Lutra canadensis texensis* Goldman, Proc. Biol. Soc. Washington, 48:184, November 15, type from 20 mi. W Angleton, Brazoria Co., Texas.

MARGINAL RECORDS (van Zyll de Jong, 1972:97, 98, unless otherwise noted).—Iowa: Mississippi River, 5 mi. N Lansing (Bowles, 1975:137). Indiana: Patoka Creek. Pennsylvania: Bushkill Creek, SW Pike Co. Connecticut: Killingsworth. New Jersey: Tuckerton. Virginia: Virginia Beach. North Carolina: Fort Caswell. South Carolina: Mount Pleasant. Georgia: Sapelo Island. Florida: Monroe County; Tarpon Springs. Alabama: Bayou La Batre (Hall and Kelson, 1959:945). Louisiana: 5 mi. W Holly Beach, 1 mi. N Hwy. 82 (Lowery, 1974:450). Texas: Victoria County

(Davis, 1966:92); Brownsville; region of Austin (Hall and Kelson, 1959:946); Mobeetie (*ibid.*). Oklahoma: Woodward County. Kansas: Trego County (Cockrum, 1952:261). Colorado: South Platte River, near Orchard (Armstrong, 1972:294). Nebraska (Jones, 1964c:298): Lodgepole Creek; Niobrara River, S of Cody.

Lutra canadensis mira Goldman

1935. *Lutra mira* Goldman, Proc. Biol. Soc. Washington, 48:185, November 15, type from Kasaan Bay, Prince of Wales Island, Alaska.
1968. *Lutra canadensis mira*, Harris, "Otters: A study of the Recent Lutrinae," Weidenfeld and Nicolson, London, p. 200.
1935. *Lutra vancouverensis* Goldman, Proc. Biol. Soc. Washington, 48:186, November 15, type from Quatsino, Vancouver Island, British Columbia. (Considered to be a synonym of *L. c. pacifica* by Cowan and Guiguet, Handbook No. 11, British Columbia Provincial Mus., p. 332, October 1965.)

MARGINAL RECORDS (van Zyll de Jong, 1972:99, unless otherwise noted).—Alaska: Anklin River, Yakutat; Taku. British Columbia: Stuie (Cowan and Guiguet, 1965:332); Sumas; Victoria (Cowan and Guiguet, 1965:332); Sooke (*ibid.*); Bamfield; Quatsino Sound. Alaska: Howkan, near Dall Island; Baronof Island (Hall and Kelson, 1959:946); Windfall Harbor.

Lutra canadensis pacifica Rhoads

1898. *Lutra hudsonica pacifica* Rhoads, Trans. Amer. Philos. Soc., n. s., 19:429, September, type from Lake Keechelus, 3000 ft., Kittitas Co., Washington.
1898. *Lutra canadensis pacifica*, J. A. Allen, Bull. Amer. Mus. Nat. Hist., 10:460, November 10.
1914. *Lutra canadensis brevipilosus* Grinnell, Univ. California Publ. Zool., 12:306, October 31, type from Grizzly Island, Suisun Bay, Solano Co., California.
1935. *Lutra canadensis preblei* Goldman, Proc. Biol. Soc. Washington, 48:178, November 15, type from near McTavish Bay, Great Bear Lake (on canoe route from Lake Hardisty), Mackenzie.
1935. *Lutra canadensis optiva* Goldman, Proc. Biol. Soc. Washington, 48:179, November 15, type from Zaikof Bay, Montague Island, Alaska.
1935. *Lutra canadensis yukonensis* Goldman, Proc. Biol. Soc. Washington, 48:180, November 15, type from Unalakleet, Norton Sound, Alaska.
1935. *Lutra canadensis extera* Goldman, Proc. Biol. Soc. Washington, 48:181, November 15, type from Nagai Island, Shumagin Islands, Alaska. Known only from Nagai Island, Alaska.
1935. *Lutra canadensis evexa* Goldman, Proc. Biol. Soc. Washington, 48:182, November 15, type from Stuart Lake, near headwaters Fraser River, British Columbia.
1935. *Lutra canadensis nexa* Goldman, Proc. Biol. Soc. Washington, 48:182, November 15, type from near Deeth, Humboldt River, Elko Co., Nevada.

MARGINAL RECORDS.—Alaska: Kanayut Lake, about 15 mi. NE Tolugak Lake. Mackenzie: Mackenzie

[River] Delta; Lower Anderson River; near McTavish Bay, Great Bear Lake; Lockhart River. Keewatin (Harper, 1956:72): Kazan River, outlet of Ennadai Lake; Windy River mouth. Manitoba: York Factory; Oxford House. Minnesota (van Zyll de Jong, 1972:99): Elk River; Wabasha County. Colorado: Greeley; Puebla (Armstrong, 1972:295); White River Valley. Utah: Provo River, near Heber; Raft River, 2 mi. S Utah–Idaho border. Nevada: Deeth, Humboldt River; 1 mi. E Golconda; S end Marlette Lake. California: Babcock Lake, 9000 ft., Yosemite National Park; Kaweah River, 4 mi. NE Lemon Cove; Tulare Lake; San Joaquin River near Mendota; 2½ mi. E Hollister (Morejohn, 1969:84, as sight record only); Price Creek, Humboldt Co. Oregon: Rogue River; Smith River near Gardiner (van Zyll de Jong, 1972:99). Washington (*ibid.*): Forks; Mt. Vernon. British Columbia: 40 mi. W Hope (*op. cit.*: 98); Anahim Lake (*ibid.*); Bear Lake; Teslin Lake, near Yukon. Alaska (van Zyll de Jong, 1972:98, unless otherwise noted): Hinchinbrook Island; Cape Elizabeth; Nagai Island; Frosty Peak, Alaska Peninsula (Hall and Kelson, 1959:946); Mission.

Lutra canadensis periclyzomae Elliot

1905. *Lutra periclyzomae* Elliot, Proc. Biol. Soc. Washington, 18:80, February 21, type from Gawi, W coast Moresby Island, Queen Charlotte Islands, British Columbia.
1909. *Lutra canadensis periclyzomae*, Heller, Univ. California Publ. Zool., 5:262, February 18.

MARGINAL RECORDS (van Zyll de Jong, 1972:99).—British Columbia: Henslung Bay, Langara Island; *Graham Island; type locality;* Skedans Bay, Louise Island.

Lutra canadensis sonora Rhoads

1898. *Lutra hudsonica sonora* Rhoads, Trans. Amer. Philos. Soc., n.s., 19:431, September, type from Montezuma Well, Beaver Creek, Yavapai Co., Arizona.
1898. *Lutra canadensis sonora*, J. A. Allen, Bull. Amer. Mus. Nat. Hist., 10:460, November 10.

MARGINAL RECORDS.—Colorado: Grand Junction; headwaters Arkansas River, Lake Co. (Armstrong, 1972:294). New Mexico: Rinconada; Canadian River near mouth Mora River; Gila River, near Arizona line. California: 8 mi. N Needles. Nevada: Colorado River at jct. Virgin River. Utah: Colorado River at Glen Canyon.

Lutra longicaudis
Southern River Otter

External measurements: 940–1240; 390–500. One specimen from Durango weighed 14.75 kg. (32½ pounds) (Pohle, 1920:95). Skull: basal length, 101.0–113.0; zygomatic breadth, 72.0–85.0. Upper parts sepia brown to pale reddish brown; underparts gray brown; throat, inguinal, and pectoral regions paler; lips and inner side of forelegs dirty-white. Skull slightly convex dor-

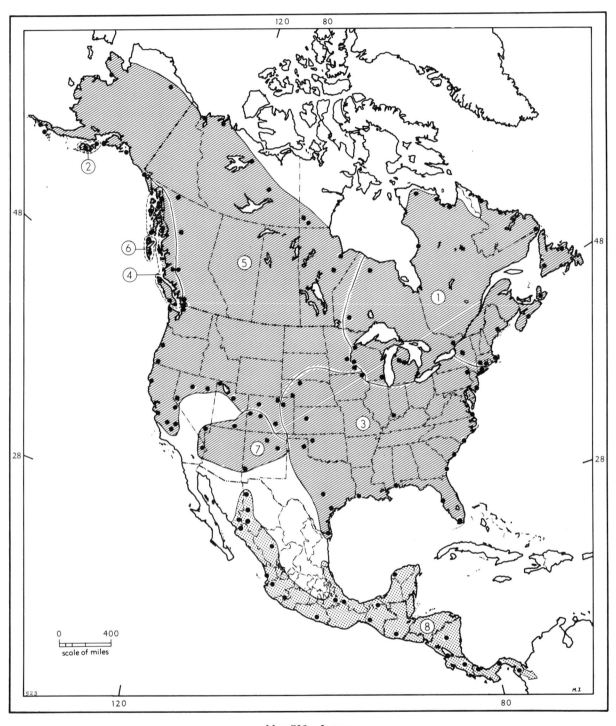

Map 523. *Lutra*.

Guide to kinds
1. *L. canadensis canadensis*
2. *L. canadensis kodiacensis*

3. *L. canadensis lataxina*
4. *L. canadensis mira*
5. *L. canadensis pacifica*

6. *L. canadensis periclyzomae*
7. *L. canadensis sonora*
8. *L. longicaudis annectens*

sally; mastoid processes strong; bullae long and highly inflated; postorbital constriction less than in *L. canadensis*; 15 or 16, rarely 14, pairs of ribs.

It is improbable that all the names listed below by van Zyll de Jong as synonyms of *Lutra longicaudis annectens* apply to a single subspecies.

Lutra longicaudis annectens Major

1897. *Lutra annectens* Major, Zoöl. Anzeiger, 20:142, April 26, type from Río de Tepic, Nayarit.
1904. *Lutra colombiana* J. A. Allen, Bull. Amer. Mus. Nat. Hist., 20:452, November 28, type from Bonda, Magdalena, Colombia.
1908. *Lutra latidens* J. A. Allen, Bull. Amer. Mus. Nat. Hist., 24:600, October 13, type from Lavala [= Savala], Matagalpa, Nicaragua.
1908. *Lutra emerita* Thomas, Ann. Mag. Nat. Hist., ser. 8, 1:390, May, type from Río Chama, 2000 m., Mérida, Venezuela.
1914. *Lutra parilina* Thomas, Ann. Mag. Nat. Hist., ser. 8, 14:59, July, type from San Juan, 870 ft., 15 mi. W Huigra, Ecuador.
1914. *Lutra repanda* Goldman, Smiths. Miscl. Coll., 63(5):3, March 14, type from Cana (Santa Cruz de Cana), 2000 ft., upper Río Tuyra, Darién, Panamá.
1924. *Lontra mesopetes* Cabrera, Bol. Real Soc. Española Hist. Nat., Madrid, 24:52, February, type from Costa Rica. Known only from type locality. Van Zyll de Jong, Royal Ontario Mus., Life Sci. Contrib., No. 80:87, June 15, 1972, stated that the position of *L. mesopetes* as a synonym of *L. l. annectens* was doubtful. Harris, "Otters: A study of the Recent Lutrinae," Weidenfeld and Nicolson, London, 1968, regards this taxon as a species.
1972. *Lontra longicaudis annectens*, van Zyll de Jong, Royal Ontario Mus., Life Sci. Contrib., No. 80:87, June 15.

MARGINAL RECORDS.—Chihuahua (Anderson, 1972:386): Río Gavilán, *ca.* 7 mi. W Pacheco (sight record only); Río Tutuaca, 20 mi. S Yaguarachic; Urique. Durango: Melchor, Río Nazas. Veracruz: Río Jamapa, near Huatusco (Hall and Dalquest, 1963:343); Río Blanco, 15 km. W Piedras Negras, 500 ft. Tabasco: Macuspana region (66221 KU). Yucatán: a small river, 40 mi. W Mérida. Nicaragua: Río Tuma. Costa Rica: Villa Colon. Panamá: Río Changuinola (Handley, 1966: 790); Gatún (*ibid.*), thence southeastward into South America and northwestward to Panamá: Cana (Santa Cruz de Cana), 2000 ft., upper Río Tuyra. Nicaragua: Peña Blanca. Guatemala: Santa Mixtan. Chiapas: Jiquipilas (66220 KU). Guerrero: Río Chapolapa, near Chapolapa [Chapalapa] (Davis and Lukens, 1958:359). Jalisco: 20 mi. S, 5 mi. W Pihuamo, 1100 ft. (111978 KU); 5 mi. SSE Mascota, 5400 ft. (100657 KU). Nayarit: type locality. Sinaloa: 8 km. N El Fuerte, 200 ft. (Armstrong, 1972:55). Sonora: Río Mayo, near San Bernardo (Cockrum, 1964:634).

Genus Enhydra Fleming—Sea Otter

1822. *Enhydra* Fleming, The philosophy of zoology . . . , 2:187. Type, *Mustela lutris* Linnaeus.
1816. *Pusa* Oken, Lehrbuch der Naturgeschichte, 3, 2:985. Type, *Pusa orientalis* Oken. Not of Scopoli, 1771. Oken's names not available.
1827. *Latax* Gloger, Nova Acta Phys.-Med. Acad. Caesar. Leop.-Carol., 13(2):511. Type, *Mustela lutris* Linnaeus. Under the mistaken impression that *Enhydra* Fleming was preoccupied by *Enhydris* Merrem, 1820, a reptile, *Latax*

has been used for the sea otters. *Enhydra* is not a homonym of *Enhydris* (see Article 36 of the International Code of Zoological Nomenclature, and Opinion 25 of the International Commission of Zoological Nomenclature).
1829. *Enydris* Fischer, Synopsis mammalium, p. 228, an emendation of *Enhydra* Fleming.

Head large, blunt; neck short, thick; ears short, pointed; legs and tail short; hind feet webbed, flattened into broad flippers, furred dorsally and ventrally, 5th digit longest; forefeet small, palms naked. Color varying from black to almost red, usually uniform dark brown, paler on throat and chest. Upper parts, underparts and head grizzled in some. Skull: high, inflated, flattened dorsally; facial region broad and flat; orbits relatively large; postorbital processes small; auditory bullae small; palate extending posterior to plane of M1 only a short distance. P4 and M1 wider than long; M1 and p4 oval; cusps broad, rounded. Dentition, i. $\frac{3}{2}$, c. $\frac{1}{1}$, p. $\frac{3}{3}$, m. $\frac{1}{2}$.

Enhydra lutris
Sea Otter

External measurements of an adult male and female, respectively, are: 1478, 360, 220; 1289, 273, 222. Weight of a male and a female, respectively, 34.4, 19.7 kg. Condylobasal length, 141–146, 131–133; zygomatic breadth, 103–106, 97–103. For additional characters see description of genus.

Roest (1973:1–17) has provided the best data available to date on geographic variation. From these I conclude that three subspecies are recognizable. Of these, *Enhydra lutris gracilis* (Beckstein) is extralimital (southern tip of Kamchatka Peninsula south originally to northern shores of Honshu). Not mapped.

Enhydra lutris lutris (Linnaeus)

1758. [*Mustela*] *lutris* Linnaeus, Syst. nat., ed. 10, 1:45. Type locality, Kamchatka, U.S.S.R.
1843. *Enhydra lutris*, Gray, List of the . . . Mammalia in the . . . British Museum, p. 72.
1777. [*Lutra*] *Marina* Erxleben, Systema regni animalis . . . , p. 445. Type locality, unknown.

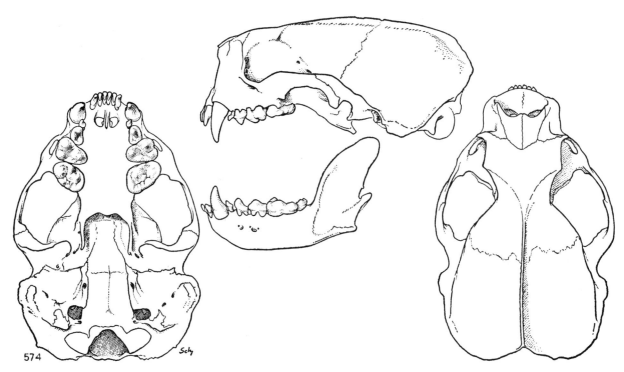

Fig. 574. *Enhydra lutris nereis*, shore ½ mi. N Bixby Creek, Monterey Co., California, No. 84812 M.V.Z., ♂, X ½.

1816. *Pusa orientalis* Oken, Lehrbuch der Naturgeschichte, 3(2):986. Oken's names are not available.
1827. *Lutra stelleri* Lesson, Manuel de mammalogie . . . , p. 156. Type locality, Kamchatka.
1922. *Enhydra lutris kamtschatica* Dybowski, Arch. Tow. Nauk. Lwow, 1:350, a *nomen nudum*.

RANGE.—Alaska: end of Aleutian Islands chain, along islands and coast to British Columbia: Vancouver.

Enhydra lutris nereis (Merriam)

1904. *Latax lutris nereis* Merriam, Proc. Biol. Soc. Washington, 17:159, October 6, type from San Miguel Island, Santa Barbara Islands, California.
1923. *Enhydra lutris nereis*, Grinnell, Univ. California Publ. Zool., 21:316, January 27.

RANGE.—Washington: Straits of Juan de Fuca, southward along coast to Baja California: Sebastián Vizcaíno Bay—[In 1970 and 1971 a number of sea otters, *E. l. lutris*, from Amchitka Island, Alaska, were released near Port Orford, Oregon. About 23 persisted there as of 1975 in what almost certainly was the natural range of *E. l. nereis* (Davis and Lidicker, 1975:434).]

FAMILY **VIVERRIDAE**—Viverrids

Dentition essentially as in the Mustelidae, but upper molars usually 2–2, and upper carnassial usually having 3 outer cusps; auditory bullae, when not rudimentary (1 African genus), inflated, thin-walled, divided by a distinct septum into 2 chambers; form usually slender, legs short; size moderate; feet digitigrade (after Miller, 1912: 440).

SUBFAMILY **HERPESTINAE**—Mongooses

Naked glandular pouch around anus and capable of being folded over anus; anal glands opening into mentioned pouch, outside anus proper; perineal perfume glands absent; claws long and nonretractile; ear lacking marginal bursa; penis short; external auditory meatus at end of a bony tube; orbit completely encircled with bone in most members of the subfamily (after Pocock, 1941:2).

Genus **Herpestes** Illiger
Mongooses

1811. *Herpertes* [sic] Illiger, Prodromus systematis mammalium et avium . . . , p. 135, a misprint, corrected to *Herpestes* on p. 302. Type, *Viverra ichneumon* Linnaeus.
1799. *Ichneumon* Lacépède, Tableau des divisions . . . des mammifères, des cétacés et des oiseaux, p. 7. Not *Ichneumon* Linnaeus, 1758, a genus of Hymenoptera.
1822. *Mangusta* Horsfield, Zoological researches in Java . . . , pt. 5, unpaged. Type, *Ichneumon javanicus* E. Geoffroy Saint-Hilaire.

1837. *Urva* Hodgson, Jour. Asiatic Soc. Bengal, 6:56. Type, *Gulo urva* Hodgson.

1841. *Mesobema* Hodgson, Jour. Asiatic Soc. Bengal, 10:910. Type, *Gulo urva* Hodgson.

1865. *Calogale* Gray, Proc. Zool. Soc. London, p. 560, October. Type, *Herpestes nepalensis* Gray [= *Mangusta auropunctata* Hodgson].

1865. *Calictis* Gray, Proc. Zool. Soc. London, p. 564, October. Type, *Herpestes smithii* Gray.

1865. *Taeniogale* Gray, Proc. Zool. Soc. London, p. 569, October. Type, *Herpestes smithii* Gray.

1865. *Onychogale* Gray, Proc. Zool. Soc. London, p. 570, October. Type, *Cynictis maccarthiae* Gray.

Total length, 585–865; tail, 230–440; guardhairs banded, imparting characteristic speckled appearance to coat; premolars $\frac{4}{4}$.

Herpestes auropunctatus
Indian Mongoose

External measurements: in males, 335; 270; 62; in females, 5–10 per cent less; condylobasal length of skull, 57 (females) to 66 (males); zygomatic breadth, 30–34; legs short. Reddish gray or yellowish gray, vermiculated or speckled. Not mapped.

Herpestes auropunctatus auropunctatus (Hodgson)

1836. *Mangusta auropunctata* Hodgson, Jour. Asiatic Soc. Bengal, 5:235. Type locality, Nepal.

1841. *H[erpestes]. auropunctatus*, Wagner, *in* Schreber, Die Säugthiere . . . , supplementband 2:310. See also Blanford, Mammals of British India, p. 121, 1888 *fide* Pocock, Jour. Bombay Nat. Hist. Soc., 39:241, April 15, 1937, and Ellerman and Morrison-Scott, Checklist of Palaearctic and Indian mammals 1758–1946, Publ. British Mus. (Nat. Hist.), p. 295, November 19, 1946.

RANGE.—Natural range of the subspecies *H. a. auropunctatus* is northern India. Introduced (Espeut, 1882:713) and widely established in West Indies. According to Varona (1974:53), occurring in Cuba, Hispaniola, Jamaica, Puerto Rico, Vieques, St. John, St. Thomas, St. Croix, Tortola, St. Martin, St. Kitts, Nevis, Antigua, Guadeloupe, Marie Galante, "Deseada?" [= Désirade ?], Martinique, Santa Lucia, St. Vincent, Grenada, and Barbados. Also introduced in Dominica but no longer living there.

FAMILY FELIDAE—Cats and Allies

Medium to large digitigrade carnivores; feet haired below, except for naked pads; plantar pad wide, indistinctly trilobate; claws short, curved, and retractile (except in *Acinonyx*); forefeet 5-toed; pollex small and set high above other toes; carpal pad single, small, and subconical; hind feet with 4 digits; metatarsal pads absent; baculum rudimentary.

Muzzle short, broad, not appreciably overlapping chin; ears with well-developed bursa, anterior flap of which is notched mesially; supratragus simple, with a clavate thickening; supercilliary and mystacial vibrissae well developed; interramal tuft of vibrissae absent.

Skull comparatively short and broad; muzzle wide; ethmoturbinals comparatively long, extending to anterior part of chamber and overlying reduced maxilloturbinals; auditory bulla high, rounded, and composed of ecto- and endo-tympanics divided by a partition; alisphenoid canal absent.

Upper and lower incisors forming straight, transverse line; canines pointed, grooved, somewhat compressed posteriorly; P2, when present, small, conical, single-rooted; P3 moderate-sized, crown triangular, compressed with 3 cusps set in a line; anterior cusp and posterior cusp small and set at base of large, pointed middle cusp; P4 large, crown compressed, with cusps set in a line, the 1st small and conical, the 2nd high and pointed, the 3rd low and bladelike; protocone well developed; M1 small, transversely set; p3 and p4 resembling P3; m1 with well-developed paraconid and protoconid; metaconid absent; talonid small. Dentition, i. $\frac{3}{3}$, c. $\frac{1}{1}$, p. $\frac{2}{2}$ or $\frac{3}{2}$, m. $\frac{1}{1}$.

KEY TO NORTH AMERICAN GENERA OF FELIDAE

1. Tail more than $\frac{1}{2}$ length of body; premolars $\frac{3}{2}$. . . . . *Felis*, p. 1035

1'. Tail less than $\frac{1}{2}$ length of body; premolars $\frac{2}{2}$. . . . . *Lynx*, p. 1049

Genus Felis Linnaeus—Cats

1758. *Felis* Linnaeus, Syst. nat., ed. 10, 1:41. Type, *Felis catus* Linnaeus.

1775. *Catus* Frisch, Das Natur-System der vierfüssigen Thiere . . . , Tab. 12. Type, *Felis catus* Linnaeus. Frisch's names, as well as Oken's, are not available.

1775. *Panthera* Frisch, Das Natur-System der vierfüssigen Thiere . . . , Tab. 13. Type, "Das Pantherthier." Frisch's names, as well as Oken's, are not available.

1775. *Tigris* Frisch, Das Natur-System der vierfüssigen Thiere . . . , Tab. 13. Type, *Felis tigris* Linnaeus. Frisch's names, as well as Oken's, are not available.

1829. *Leo* Brehm, Oken's Isis, p. 637.

1834. *Puma* Jardine, The naturalist's library, 2:266. Type, *Felis concolor* Linnaeus.

1841. *Otocolobus* Brandt, Bull. Acad. Sci. St. Pétersbourg, 9:38. Type, *Felis maul* Pallas.

1842. *Leopardus* Gray, Ann. Mag. Nat. Hist., ser. 1, 10:260, December. Type, *Leopardus griseus* Gray [= *Felis pardalis* Linnaeus].

1843. *Chaus* Gray, List of the . . . Mammalia in the . . . British Museum, p. 44. Type, *Felis chaus* Güldenstaedt.

1854. *Uncia* Gray, Ann. Mag. Nat. Hist., ser. 2, 14:394, November. Type, *Felis uncia* Schreber.

1858. *Catolynx* Severtzow, Revue et Mag. Zool., Paris, ser. 2, 10:385, September. Type, *Felis catus* Linnaeus.

1858. *Herpailurus* Severtzow, Revue et Mag. Zool., Paris, ser. 2, 10:385, September. Type, *Felis yaguarundi* Azara.

1858. *Jaguarius* Severtzow, Revue et Mag. Zool., Paris, ser. 2, 10:386, September. Type, *Felis onca* Linnaeus.

1858. *Oncoides* Severtzow, Revue et Mag. Zool., Paris, ser. 2, 10:386, September. Type, *Felis pardalis* Linnaeus.

1858. *Dendrailurus* Severtzow, Revue et Mag. Zool., Paris, ser. 2, 10:386, September. Type, *Felis strigilata* Wagner.

1858. *Oncifelis* Severtzow, Revue et Mag. Zool., Paris, ser. 2, 10:386, September. Type, *Felis geoffroyi* Gervais.

1858. *Profelis* Severtzow, Revue et Mag. Zool., Paris, ser. 2, 10:386, September. Type, *Felis celidogaster* Temminck [= *Felis aurata* Temminck].

1858. *Noctifelis* Severtzow, Revue et Mag. Zool., Paris, ser. 2, 10:386, September. Type, *Felis guigna* Molina.

1858. *Lynchailurus* Severtzow, Revue et Mag. Zool., Paris, ser. 2, 10:386, September. Type, *Felis pajeros* Desmarest.

1858. *Prionailurus* Severtzow, Revue et Mag. Zool., Paris, ser. 2, 10:387, September. Type, *Felis pardachrous* [*sic*] Hodgson [= *Felis bengalensis* Kerr].

1858. *Zibethailurus* Severtzow, Revue et Mag. Zool., Paris, ser. 2, 10:387, September. Type, *Felis viverrina* Bennett.

1858. *Catopuma* Severtzow, Revue et Mag. Zool., Paris, ser. 2, 10:387, September. Type, *Felis moormensis* Hodgson [= *Felis temminckii* Vigors and Horsfield].

1858. *Pardofelis* Severtzow, Revue et Mag. Zool., Paris, ser. 2, 10:387, September. Type, *Felis marmorata* Marin.

1858. *Ictailurus* Severtzow, Revue et Mag. Zool., Paris, ser. 2, 10:387, 388, September. Type, *Felis planiceps* Vigors and Horsfield.

1858. *Leptailurus* Severtzow, Revue et Mag. Zool., Paris, ser. 2, 10:389, September. Type, *Felis serval* Schreber.

1858. *Otailurus* Severtzow, Revue et Mag. Zool., Paris, ser. 2, 10:388, September. Type, *Felis megalotis* Müller.

1858. *Chrysailurus* Severtzow, Revue et Mag. Zool., Paris, ser. 2, 10:389, September. Type, *Felis neglecta* Gray [= *Felis celidogaster* Temminck].

1864. *Serval* Brehm, Fuhrer Z. Garten Hamburg, ed. 6, p. 52. Type, *Serval maculatus* Brehm.

1866. *Galeopardus* Heuglin and Fitzinger, Sitzungsb. k. Akad. Wiss., Wien, 54:557. Type, *Felis serval* Schreber.

1867. *Neofelis* Gray, Proc. Zool. Soc. London, p. 265, October. Type, *Felis macrocelis* Temminck.

1867. *Pardalina* Gray, Proc. Zool. Soc. London, p. 266. October. Type, *Felis himalayanus* Warwick [= *Felis geoffroyi* Gervais?].

1867. *Catolynx* Gray, Proc. Zool. Soc. London, p. 267, October. Type, *Felis marmorata* Martin. Not *Catolynx* Severtzow, 1858.

1867. *Viverriceps* Gray, Proc. Zool. Soc. London, p. 268, October. Type, *Felis viverrina* Bennett.

1867. *Pajeros* Gray, Proc. Zool. Soc. London, p. 269, October. Type, *Felis pajeros* Desmarest.

1868. *Pardus* Fitzinger, Sitzungsb. k. Akad. Wiss., Wien, 58:459. Type, *Felis pardus* Linnaeus.

1869. *Margay* Gray, Catalogue of the carnivorous, pachydermatous and edentate Mammalia in the British Museum, p. 21. Type, *Felis wiedii* Schinz.

1869. *Ailurogale* Fitzinger, Sitzungsb. k. Akad. Wiss., Wien, 60:249. Type, *Felis planiceps* Vigors and Horsfield.

1870. *Ailurinus* Gervais, Nouv. Arch. Mus. Paris, 6:159. Type, "l'Ailurin" Gervais, 1855, Histoire naturelle des mammifères . . . , 2:87 [= *Felis planiceps* Vigors and Horsfield].

1874. *Pyrofelis* Gray, Ann. Mag. Nat. Hist., ser. 4, 14:354, November. Type, *Felis temminckii* Vigors and Horsfield.

1882. *Plethaelurus* Cope, Proc. Amer. Philos. Soc., 20:475, November 20. Type, *Felis planiceps* Vigors and Horsfield.

1894. *Servalina* Grevé, Nova Acta Acad. Caesar. Leop.-Carol., 63:76. Type, *Felis serval* Schreber.

1905. *Trichaelurus* Satunin, Ann. Mus. Zool. St. Pétersbourg, 9:495, a renaming of *Otocolobus* Brandt, which was thought to be preoccupied.

1919. *Oncilla* J. A. Allen, Bull. Amer. Mus. Nat. Hist., 41:355. Type, *Felis pardinoides oncilla* Thomas.

1925. *Poliailurus* Lönnberg, Arkiv. Zool. Stockholm, 2(18A):2. Type, *Felis pallida* Buechner [= *Felis bieti* Milne-Edwards].

1926. *Microfelis* Roberts, Ann. Transvaal Mus., 11:250. Type, *Felis nigripes* Burchell.

1926. *Eremaelurus* Ognev, Ann. Mus. Zool. Leningrad, 27:356. Type, *Eremaelurus thinobius* Ognev [= a subspecies of *Felis margarita* Loche].

1932. *Badiofelis* Pocock, Proc. Zool. Soc. London, p. 749. Type, *Felis badia* Gray.

1933. *Mungofelis* Antonius, Zool. Gart., Leipzig, 6:13. Type, *Felis braccata* Cope.

Medium to large digitigrade carnivores; tail usually long; feet large; forefeet with 5 toes, hind feet with 4 toes; claws sharp, recurved, and retractile; eyes large; ears well developed, medium in size; coat tawny, spotted, or striped. Skull highly arched in frontal region; face flattened; rostrum short; carotid canal short or lacking; alisphenoid canal absent; tympanic bullae large, divided internally into 2 chambers; palate ending posterior to molars; clavicles small, floating; os penis rudimentary. M1 rudimentary; ml lacking talonid and metaconid. Dentition, i. $\frac{3}{3}$, c. $\frac{1}{1}$, p. $\frac{3}{2}$, m. $\frac{1}{1}$.

KEY TO NORTH AMERICAN SPECIES OF FELIS

1. Total length more than 1400; greatest length of skull more than 158.
    2. Upper parts not spotted; bregmatic processes of parietals extending anteromedially over frontals, approaching or reaching temporal crest and fusing with frontals in old animals.
        F. concolor, p. 1039
    2'. Upper parts spotted; bregmatic processes of parietals not extended as described in 2 above . . . . . *F. onca*, p. 1037
1'. Total length less than 1400; greatest length of skull less than 158.
    3. Upper parts not spotted; skull only slightly arched; nasals extending anteriorly beyond plane of anterior edge of palatine foramina; palatine foramina not visible in dorsal aspect.
        F. yagouaroundi, p. 1048

3'. Upper parts spotted; skull highly arched; nasals not extending anteriorly to plane of anterior edge of palatine foramina; palatine foramina visible in dorsal aspect.

    4. Hind foot more than 145; greatest length of skull more than 120; basilar length more than 90; length of P4 more than 12.7.   *F. pardalis*, p. 1043

    4'. Hind foot less than 145; greatest length of skull less than 120; basilar length less than 90; length of P4 less than 12.7.

        5. Hair on nape directed forward; mastoid breadth half or more (50.7–53.0%) of basilar length. . . . . . .*F. wiedii*, p. 1045

        5'. Hair on nape directed backward; mastoid breadth less (44.2–48.7%) than half of basilar length. . . . .*F. tigrina*, p. 1047

Felis onca
Jaguar

x 1/30

Revised by Nelson and Goldman, Jour. Mamm., 14:221, 240, August 17, 1933.

External measurements: (six males) 1727–2419; 523–665; (three males) 229–302; (five females) 1574–2190; 432–604; (one female) 225. Weight of one male from Brazil: 131.8 kg. Greatest length of skull, (males) 237.5–302.0, (females) 204.0–267.8; zygomatic breadth, (males) 166.5–207.0, (females) 143.8–175.5; crown length of upper carnassial, (males) 23.5–29.2, (females) 24.1–30.8. Size large; form robust; tail short and tapering; ears small, short, and rounded, without elongated terminal tufts; pelage short and rather bristly; upper parts profusely spotted at all ages.

Ground color of upper parts varying from light ochraceous buff or pale straw color to light golden tawny or tan, nearly uniform over median dorsal area from top of head to base of tail, becoming gradually paler to near cinnamon buff or light buff on cheeks, side of neck, lower parts of flanks and outer surfaces of legs; body heavily spotted with black; dorsum, except posterior median line and sides, marked with irregularly circular or crescentic or broken spots that tend to form rosettes or to enclose smaller spots in a field darker than the general ground color; posterior part of median line bearing elongated solid black spots that tend to present an irregular laterally paired arrangement or to become confluent; top and sides of head, neck, and feet marked by smaller black spots; underside of neck and inner sides of forelegs marked with more or less confluent black spots that tend to form transverse bars; upper surface of muzzle varying from pinkish buff to clay color, unspotted; cheeks, forehead, and feet with small, rounded black spots; upper and lower lips white near middle, becoming black toward angles of mouth; outer sides of ears deep black, with small buffy median spots; inner sides of ears thinly clothed with whitish hairs; tail with crowded, irregular black markings, separated on basal half above by narrow pinkish buffy or cinnamon buffy interspaces, becoming white below and toward tip, which is usually black (after Nelson and Goldman, 1933:223, 224).

Felis onca arizonensis Goldman

1932. *Felis onca arizonensis* Goldman, Proc. Biol. Soc. Washington, 45:144, September 9, type from near Cibecue, Navajo Co., Arizona.

MARGINAL RECORDS.—New Mexico: near Springer; Goldberg Ranch, Jornada Experiment Range, Caballo Mts. Sonora: western foothills of Sierra Madre, due W Casas Grandes, Chihuahua; 15 mi. W Alamo Ranch. Arizona: 20 mi. W Nogales. Baja California Norte: near S tip San Pedro Mártir Range (Leopold, 1959:466). California: *Tehachapi Mts.* Arizona: 4 mi. S Grand Canyon rim; type locality.

Felis onca centralis Mearns

1901. *Felis centralis* Mearns, Proc. Biol. Soc. Washington, 14:139, August 9, type from Talamanca, Limón, Costa Rica (probably near Sipurio, in valley of Río Sicsola).
1904. [*Felis*] *onca centralis*, Elliot, Field Columb. Mus., Publ. 95, Zool. ser., 4:446.

MARGINAL RECORDS.—Chiapas: near Comitán. Panamá (Handley, 1966:790, 791): Río Changuinola; Boca de Cupe, thence into South America.

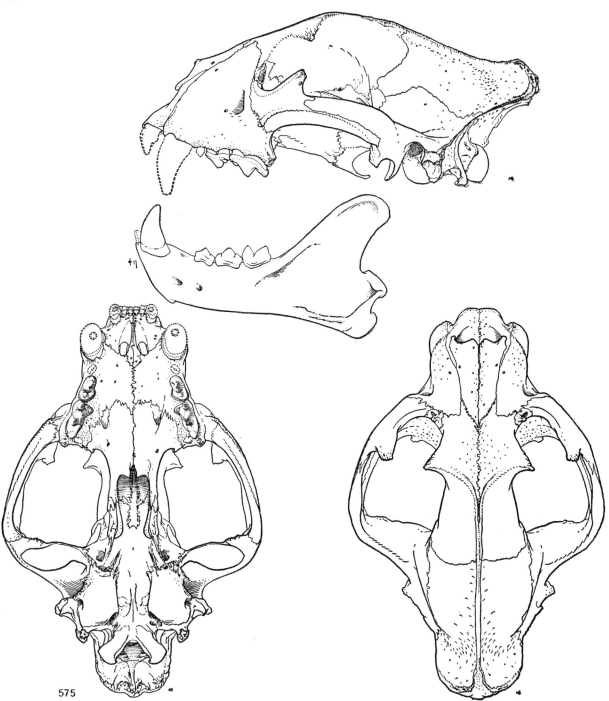

Fig. 575. *Felis onca hernandesii*, 10 mi. N Colotopec, Oaxaca, No. 35317 M.V.Z., sex?, X ⅖.

Felis onca goldmani Mearns

1901. *Felis hernandesii goldmani* Mearns, Proc. Biol. Soc. Washington, 14:142, August 9, type from Yohaltun, Campeche.

1932. *Felis onca goldmani*, Goldman, Proc. Biol. Soc. Washington, 45:144, September 9.

MARGINAL RECORDS.—Campeche: La Tuxpeña, thence around coast to *Belize*. Guatemala: La Libertad.

Felis onca hernandesii (Gray)

1858. *Leopardus hernandesii* Gray, Proc. Zool. Soc. London, for 1857, p. 278, January 28, type from Mazatlán, Sinaloa.

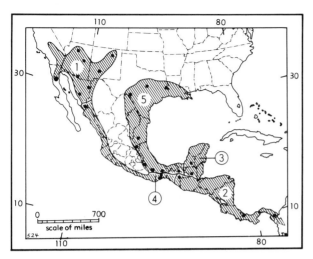

Map 524. *Felis onca.*

Guide to subspecies
1. *F. o. arizonensis*
2. *F. o. centralis*
3. *F. o. goldmani*
4. *F. o. hernandesii*
5. *F. o. veraecrucis*

1932. *Felis onca hernandesii*, Goldman, Proc. Biol. Soc. Washington, 45:144, September 9.

MARGINAL RECORDS.—Sonora: mts. near Alamos, thence southward along W slope Sierra Madre Occidental to Oaxaca: Chivela; Tehuantepec; thence northward along coast to point of beginning.

Felis onca veraecrucis Nelson and Goldman

1933. *Felis onca veraecrucis* Nelson and Goldman, Jour. Mamm., 14:236, August 17, type from San Andrés Tuxtla, Veracruz.

MARGINAL RECORDS.—Texas: Goldthwaite; S Jasper, thence southward along Gulf Coast to Chiapas: Palenque. Veracruz: 20 km. E Jesús Carranza, 300 ft. (Hall and Dalquest, 1963:344); Pérez; Orizaba. San Luis Potosí: Cueva de las Sabinas, 12 km. NE Ciudad Valles. Tamaulipas: 5 km. N Gómez Farías. Texas: near mouth Pecos River, Val Verde Co.

Felis concolor
Mountain Lion

X ¹/₂₀

Revised by Goldman, Classification of the races of the puma, pt. 2, pp. 177–302, Pls. 43–93, 1 fig. in text, *in* Young

and Goldman, The puma, mysterious American cat, pp. xiv + 358, 93 pls., 6 figs. in text, Amer. Wildlife Manag. Inst., Washington, D.C., November 16, 1946.

External measurements: (males) 1710–2743; 660–784; 240–292; (females) 1500–2332; 534–814; 220–267. Weight: (males) 67–103 kg.; (females) 36–60 kg. Greatest length of skull, (males) 172.0–237.0, (females) 158.3–203.0; zygomatic breadth, (males) 126.4–164.3, (females) 107.2–140.7; alveolar length of maxillary tooth-row, (males) 52.5–72.0, (females) 48.8–62.5. Form slender; ears small, rounded. Upper parts grizzled gray or dark brown to shades of buff, cinnamon, tawny, cinnamon-rufous, or ferruginous; color usually most intense along middorsal line from top of head to base of tail; shoulders and flanks lighter; underparts dull whitish, overlaid with buff across abdomen; sides of muzzle black; chin and throat white; ears black externally with grayish median patches on some; tail above like back, lighter below, distal 2–3 in. black. Young are spotted with black on a buffy ground color.

Skull short, broad; sagittal crest convex in outline; lambdoidal crest having a deep lateral concavity; spiculelike bregmatic processes of parietal extending anteromedially over frontals, approaching or reaching temporal ridge and fusing with frontals in old specimens; nasals expanded distally; palate wide; grooves on canines obsolescent or absent.

Felis concolor azteca Merriam

1901. *Felis hippolestes aztecus* Merriam, Proc. Washington Acad. Sci., 3:592, December 11, type from Colonia García, 6700 ft., about 60 mi. SW Casas Grandes, Chihuahua.
1929. *Felis concolor azteca*, Nelson and Goldman, Jour. Mamm., 10:347, November 11.

MARGINAL RECORDS.—New Mexico: 10 mi. W Tusas; Queen. Chihuahua: Gallego. Zacatecas: Chichimaquillas, 20 mi. E Fresnillo. San Luis Potosí: Platanito; Xilitla. Morelos: N of Yautepec. Jalisco: Los Masos, thence northward along coast to Sonora and inland to Arizona: Baboquivari Mts.; Wagoner; Burro Creek, Mohave Co.; Carro Canyon, near Kingman (Cockrum, 1961:245); Hilltop (Hoffmeister, 1971:84, as *F. concolor* only); *Indian Garden* (*ibid.*); Mt. Elden (Hoffmeister and Carothers, 1969: 188, as *F. concolor* only). New Mexico: Bluewater; head Santa Clara Creek, Jemez Mts.

Felis concolor browni Merriam

1903. *Felis aztecus browni* Merriam, Proc. Biol. Soc. Washington, 16:73, May 29, type from Lower Colorado River, 12 mi. S Yuma, Yuma Co., Arizona.
1929. *Felis concolor browni*, Nelson and Goldman, Jour. Mamm., 10:347, November 11.

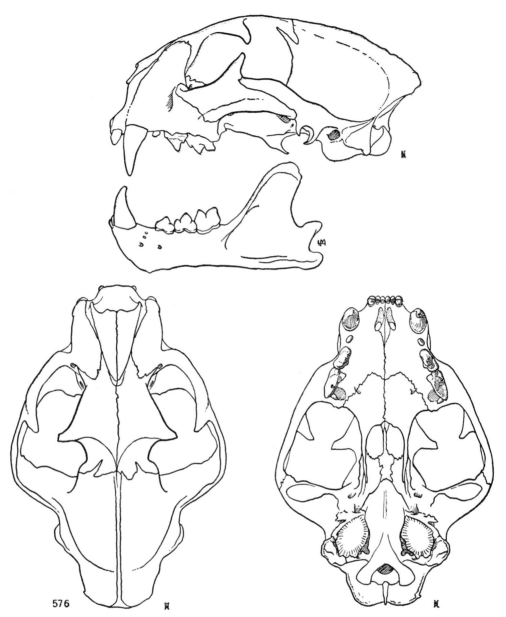

Fig. 576. *Felis concolor kaibabensis*, Potts, Nevada, No. 37295 M.V.Z., ♂, X ⅖.

MARGINAL RECORDS.—Arizona: Hualpai Mts.; vic. Squaw Creek, Kofa Mts., 50 mi. NE Yuma; type locality. Baja California: 6 mi. NW Cataviña; 18 mi. S Tres Pozos. California: Colorado River, 20 mi. N Picacho.

Felis concolor californica May

1896. *Felis californica* May, California game "marked down," p. 22. Type locality, Kern County, California.
1929. *Felis concolor californicus*, Nelson and Goldman, Jour. Mamm., 10:347, November 11.

MARGINAL RECORDS.—Oregon: 63 mi. NE Grants Pass; Rock Creek, Jackson Co. California: Genesee. Nevada: Spooner; vic. Wichman; Silver Peak Range. California: Marble Fork Bridge, Sequoia National Park; head Brush Creek, E side Kern River, 25 mi. above Kernville; San Bernardino Mts.; Strawberry Creek, San Jacinto Mts.; Buckman Springs. Baja California: Rancho San Antonio, San Pedro Mártir Mts.; vic. San Quintin, thence northward along coast to Oregon: 15 mi. W Waldo; 35 mi. W Selma.

Felis concolor coryi Bangs

1896. *Felis concolor floridana* Cory, Hunting and fishing in Florida. . . , p. 109, type from wilderness back of Sebastian, Brevard Co., Florida. Not *F. floridana* Desmarest, 1820.
1899. *Felis coryi* Bangs, Proc. Biol. Soc. Washington, 13:15, January 31, a renaming of *F. c. floridana* Cory.
1929. *Felis concolor coryi*, Nelson and Goldman, Jour. Mamm., 10:347, November 11.
1911. *Felis arundivaga* Hollister, Proc. Biol. Soc. Washington, 24:176, June 16, type from 12 mi. SW Vidalia, Concordia Parish, Louisiana.

MARGINAL RECORDS.—Arkansas: Greene County. Mississippi: Tunica County. Alabama: DeKalb County. Georgia: Bartow County; NW of Dahlonega; Glynn County, thence along Atlantic and Gulf coasts to Louisiana: Red Dirt Game Management Area, Nachitoches Parish (Goertz and Abegg, 1966:727); *highway between Robeline and Spanish Lake (ibid.)*; old Mansfield Road, *ca.* 2 mi. N De Soto Parish line, near Keithville (*ibid.*). Arkansas: *near Shady Lake Recreation Area* (Lewis, 1969:371, as *F. concolor* only, sight record only); near Mena (Lewis, 1969: 371, as *F. concolor* only); vic. Radding; Ozark National Forest, 25 mi. N Russellville (Lewis, 1969:371, as *F. concolor*, sight record).

Felis concolor costaricensis Merriam

1901. *Felis bangsi costaricensis* Merriam, Proc. Washington Acad. Sci., 3:596, December 11, type from Boquete, Chiriquí, Panamá.
1929. *Felis concolor costaricensis*, Nelson and Goldman, Jour. Mamm., 10:347, November 11.

MARGINAL RECORDS.—Costa Rica: Guapiles. Panamá (Handley, 1966:790): Cerro Punta; Río Bayano, 10 mi. above mouth Río Mamoní; Tacarcuna Village. Almost certainly occurs in adjacent Colombia.

Felis concolor cougar Kerr

1792. *Felis couguar* Kerr, The animal kingdom . . . , p. 151. Type locality, Pennsylvania.
1929. *Felis concolor couguar*, Nelson and Goldman, Jour. Mamm., 10:347, November 11.

MARGINAL RECORDS.—New Brunswick: headwaters Sevogle River. Nova Scotia: vic. Blue Mtn., Pictou Co., thence southward along coast to South Carolina: Santee River. Tennessee: Tellico River drainage, Monroe Co.; near Nashville. Indiana: Knox County; Porter County. Michigan: Mason County. Quebec: Lac des Quinze; near Sorel.

Felis concolor hippolestes Merriam

1897. *Felis hippolestes* Merriam, Proc. Biol. Soc. Washington, 11:219, July 15, type from Wind River Mts., near head Big Wind River, Fremont Co., Wyoming.
1929. *Felis concolor hippolestes*, Nelson and Goldman, Jour. Mamm., 10:347, November 11.

MARGINAL RECORDS.—South Dakota: jct. Grand and Missouri rivers. Nebraska (Jones, 1964c:301, 302): *Engineer Cantonment;* Cass County; Republican River, S of Franklin. Kansas: Trego County. Colorado: La Veta. New Mexico: *northern New Mexico.* Colorado: 18 mi. N Durango; *Mesa Verde National Park* (Anderson, 1961:61); Paradox; Gateway. Utah: White River, Uintah Co.; Spanish Fork Canyon, 4 mi. from mouth Lake Fork, 6 mi. SE Thistle; Provo; Park Valley. Idaho: 9 mi. NW Pocatello. Wyoming: type locality; Wolf; head Bear Creek, Bear Lodge Mts.

Felis concolor improcera Phillips

1912. *Felis improcera* Phillips, Proc. Biol. Soc. Washington, 25:85, May 4, type from Calmallí, Baja California.
1929. *Felis concolor improcera,* Nelson and Goldman, Jour. Mamm., 10:347, November 11.

MARGINAL RECORDS.—Baja California: type locality, thence southward to 10 mi. W Miraflores.

Felis concolor kaibabensis Nelson and Goldman

1931. *Felis concolor kaibabensis* Nelson and Goldman, Jour. Washington Acad. Sci., 21:209, May 19, type from Powell Plateau, 8700 ft., Grand Canyon National Park, Coconino Co., Arizona.

MARGINAL RECORDS.—Nevada: Cold Creek, Ruby Mts.; Spruce Mtn. Utah: Deep Creek Mts.; Antelope, near Manti; Thousand Lake Mtn. Arizona: Bright Angel, Grand Canyon National Park; *type locality.* Utah: Pine Valley Mts.; Modena. Nevada: 15 mi. W Sunnyside; Hot Creek, near Keystone; head left fork Shady Run, Stillwater Range.

Felis concolor mayensis Nelson and Goldman

1929. *Felis concolor mayensis* Nelson and Goldman, Jour. Mamm., 10:350, November 11, type from La Libertad, Petén, Guatemala.

MARGINAL RECORDS.—Quintana Roo: Pueblo Nuevo X-Can (92571 KU). Campeche: 103 km. SE Escárcega (93832 KU). Guatemala: *lowland forests of both slopes; mts. of Alta Verapaz and Quiché.* El Salvador: Lake Olomega (Burt and Stirton, 1961:49). Oaxaca: Limón (Goodwin, 1969:245). Guerrero (Davis and Lukens, 1958:360): Acahuizotla, 2800 ft.; Xaltianguis, 1600 ft. Veracruz: Catemaco.

Felis concolor missoulensis Goldman

1943. *Felis concolor missoulensis* Goldman, Jour. Mamm., 24:229, June 7, type from Sleeman Creek, about 10 mi. SW Missoula, Montana Co., Montana.

MARGINAL RECORDS.—Alberta: 59° 50' N, 112° 25' W (Kuyt, 1971:142). Saskatchewan: 3 mi. NW Squaw Creek Power Dam (Riome, 1973:100). Mani-

Map 525. *Felis concolor.*

Guide to subspecies	4. *F. c. coryi*	8. *F. c. improcera*	12. *F. c. oregonensis*
1. *F. c. azteca*	5. *F. c. costaricensis*	9. *F. c. kaibabensis*	13. *F. c. schorgeri*
2. *F. c. browni*	6. *F. c. couguar*	10. *F. c. mayensis*	14. *F. c. stanleyana*
3. *F. c. californica*	7. *F. c. hippolestes*	11. *F. c. missoulensis*	15. *F. c. vancouverensis*

toba (Soper, 1961:209): Makinak; Marquette; 35 mi. NE Winnipeg (Nero, 1974:56). North Dakota: Fort Union. Montana: Fort Keogh [= Keough]. Wyoming: Wapiti River; Buffalo River, *ca.* 20 mi. SE Jackson Lake. Idaho: Sheep Creek, Boise National Forest. Oregon: Upper Imnaha River. Washington: Blue Mts.; Republic. British Columbia: Nelson; near Horsefly (Lesowski, 1963:586, as *F. concolor* only); Big Muddy River, on Alaska Hwy. Several of the marginal occurrences are sight records only. At least some of half a dozen sight records in southern Yukon (Youngman, 1975:153) in the late 1960's and early 1970's indicate a more northern range than is shown on Map 525.

Felis concolor oregonensis (Rafinesque)

1832. *Felix* [*sic*] *oregonensis* Rafinesque, Atlantic Jour., 1:62, June 20. Type locality, Ohanapecosh River, Mount Rainier National Park, Pierce Co., Washington (by restriction, Nelson and Goldman, Proc. Biol. Soc. Washington, 45:105, July 15, 1932).

1904. [*Felis*] *concolor oregonensis*, Elliot, Field Columb. Mus., Publ. 95, Zool. Ser., 4:454, August 2.

1897. *Felis hippolestes olympus* Merriam, Proc. Biol. Soc. Washington, 11:220, July 15, type from Lake Cushman, Olympic Mts., Mason Co., Washington. (Regarded as inseparable from *oregonensis* by Dalquest, Univ. Kansas Publ., Mus. Nat. Hist., 2:237, April 9, 1948.)

MARGINAL RECORDS.—British Columbia: Bella Coola Inlet; Vernon. Washington: Loomis; Dollar Mtn., near Republic; Leavenworth; Goldendale. Oregon: Murderer Creek, Dayville; Rock Creek, 14 mi. NW Trail; 12 mi. E Agness; Pistol River, thence northward along coast, including Nelson Island, British Columbia (Cowan and Guiguet, 1965:339), to point of beginning.

Felis concolor schorgeri Jackson

1955. *Felis concolor schorgeri* Jackson, Proc. Biol. Soc. Washington, 68:149, October 31, type from near Appleton, Outagamie Co., Wisconsin.

MARGINAL RECORDS.—Ontario: Beardmore. Michigan: Marquette County. Wisconsin: type locality. Illinois: Cook County. Missouri: vic. St. Louis; N of Shirley (Lewis, 1969:372, as *F. concolor*, sight record only); near Willow Springs (*ibid.*, as *F. concolor*, sight record only). Oklahoma: Craig County (Lewis, 1969:371, as *F. concolor*, sight record only). Kansas: Harper County; Barber County; Comanche County; Catherine. Minnesota: 8½ mi. S Williams; Duluth. Ontario: *ca.* 43 mi. W Port Arthur.

Felis concolor stanleyana Goldman

1936. *Felis concolor youngi* Goldman, Proc. Biol. Soc. Washington, 49:137, August 22, type from Bruni Ranch, near Bruni, Webb Co., Texas. Not *Felis youngi* Pei, Palaeont. Sinica, ser. c, 8(1):133, May, 1934.

1938. *Felis concolor stanleyana* Goldman, Proc. Biol. Soc. Washington, 51:63, March 18, a renaming of *F.c. youngi* Goldman.

MARGINAL RECORDS.—Oklahoma: Mayes County (Lewis, 1969:371, as *F. concolor*, sight record only); 10 mi. W Checotah (*ibid.*, as *F. concolor* only). Texas: Brazos County (Davis, 1966:118, as *F. concolor* only); Brazoria County, thence along coast to Tamaulipas: Zamorina (Alvarez, 1963:463); Miquihuana, 6400 ft. Chihuahua: Sierra Almagre, 5400 ft., 12 mi. S Jaco. Texas: Presidio County; Hudspeth County (Davis, 1966:118, as *F. concolor* only); 25 mi. N

Van Horn; Burned Cabin, near head McKittrick Canyon, Culberson County; Memphis. Oklahoma: Tulsa County (Lewis, 1969:371, as *F. concolor*, sight record only).

Felis concolor vancouverensis Nelson and Goldman

1932. *Felis concolor vancouverensis* Nelson and Goldman, Proc. Biol. Soc. Washington, 45:105, July 15, type from Campbell Lake, Vancouver Island, British Columbia.

MARGINAL RECORDS.—British Columbia: *Vancouver Island; Quadra Island* (Cowan and Guiguet, 1965:339); Saltspring Island (*ibid.*).

Felis pardalis
Ocelot

North American subspecies revised by Goldman, Jour. Mamm., 24:375–380, August 17, 1943.

External measurements: (males) 950–1367; 280–400; 132–180; (females) 920–1209; 270–371; 130–173. Greatest length of skull, (males) 121.0–158.0, (females) 120.0–140.7; zygomatic breadth, (males) 81.0–108.0, (females) 80.0–96.4; crown length of maxillary tooth-row, (males) 35.5–47.1, (females) 35.1–39.6. Ground color of upper parts grayish to cinnamon; dark markings forming chainlike streaks, generally forming black-bordered elongated spots, which run obliquely down sides; area enclosed by black that usually is more intense than ground color; head with small black spots and two black stripes on each cheek; 4 or 5 parallel black stripes on neck; ground color of sides paler than dorsum; underparts and inner surfaces of limbs whitish; tail above marked with dark bars or blotches.

Felis pardalis albescens Pucheran

1855. *Felis albescens* Pucheran, *in* I. Geoffroy Saint-Hilaire, Mammifères, *in* Petit-Thoaurs, Voyage autour du monde sur la frégate la Venus . . . , Zoologie, p. 149. Type locality, Arkansas.

1906. *Felis pardalis albescens*, J. A. Allen, Bull. Amer. Mus. Nat. Hist., 22:219, July 25.

1901. *Felis limitis* Mearns, Proc. Biol. Soc. Washington, 14:146, August 9, types from Brownsville, Cameron Co., Texas.

1911. *Felis ludoviciana* Brass, Aus dem Reiche der Pelze, p. 411, April, a renaming of *F. limitis*.

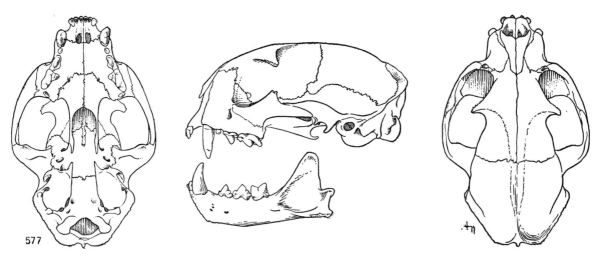

Fig. 577. *Felis pardalis albescens*, Brownsville, Texas, No. 2209 K.U., ♂, X ½.

MARGINAL RECORDS.—Texas: 2½ mi. NE Hedley. Arkansas: *type locality*. Texas: Jefferson County (Davis, 1966:116, as *F. pardalis* only); 18 mi. E Bay City, thence southward along Gulf Coast to Tamaulipas (Alvarez, 1963:464): 10 mi. N Altamira; Sierra San Carlos, San José. Texas: Brewster County (Davis, 1966:116, as *F. pardalis* only); Alamo de Cesarea Ranch, between Marfa and Terlingua.

Felis pardalis mearnsi J. A. Allen

1902. *Felis costaricensis*, Mearns, Proc. U.S. Nat. Mus., 25:245, September 17, type from Talamanca, Limón, Costa Rica. Not *Felis bangsi costaricensis* Merriam, 1901.
1904. *Felis mearnsi* J. A. Allen, Bull. Amer. Mus. Nat. Hist., 20:71, February 29, a renaming of *F. costaricensis* Mearns.
1909. *Felis pardalis mearnsi*, Lyon and Osgood, Bull. U.S. Nat. Mus., 62:208, January 28.
1860. *Felis pardalis minimus* Wilson, Proc. Acad. Nat. Sci. Philadelphia, p. 82. Type locality, Realjo, Nicaragua. Based on a kitten that was lost; if based on an ocelot, this name may apply to the subspecies for which we now use the name *F. p. mearnsi*.

MARGINAL RECORDS.—Nicaragua: Río Choco. Panamá (Handley, 1966:791): Sibube; Gatún; Cerro Pirre. Almost certainly occurs in adjacent Colombia.

Felis pardalis nelsoni Goldman

1925. *Felis pardalis nelsoni* Goldman, Jour. Mamm., 6:122, May 12, type from Manzanillo, 50 ft., Colima.

MARGINAL RECORDS.—Sinaloa: Mazatlán (Armstrong, *et al.*, 1972:56); *Los Limones*. Guerrero: Omilteme. Oaxaca: Puerto Angel, thence northward along Pacific Coast to Sinaloa: Escuinapa. Not found: Sinaloa: Los Pieles (J. A. Allen, 1906:219).

Felis pardalis pardalis Linnaeus

1758. [*Felis*] *pardalis* Linnaeus, Syst. nat., ed. 10, 1:42. Type locality, State of Veracruz (by restriction, J. A. Allen, Bull. Amer. Mus. Nat. Hist., 41:345, October 3, 1919).
1792. *Felis mexicana* Kerr, The animal kingdom . . . , p. 151, may refer to this subspecies.
1795. *Felis ocelot* Link, Beyträge zur Naturgeschichte, 2:90, based on *Felis pardalis* Linnaeus.
1830. *Felis griffithii* Fischer, Synopsis mammalium, p. 369 (= 569). Type locality, probably México.
1838. *Felis canescens* Swainson, Animals in menageries . . . , p. 118, a renaming of *F. griffithii* Fischer.
1842. *Leopardus griseus* Gray, Ann. Mag. Nat. Hist., ser. 1, 10:260, December, type from Guatemala.
1842. *Felis pseudopardalis* Boitard, Jardin des Plantes . . . , p. 187, may refer to this subspecies.
1911. ?*Felis buffoni* Brass, Aus dem Reiche der Pelze, p. 412, April. Type locality, México.
1911. ?*Felis mexicana* Brass, Aus dem Reiche der Pelze, p. 412, April. Type locality, México. Not *F. mexicana* Desmarest, 1816, nor *F. mexicana* Saussure, 1860.

MARGINAL RECORDS.—San Luis Potosí: 17 km. W Ebano. Campeche: 65 km. S, 128 km. E Escárcega (93835 KU). Honduras: Balfate; Archaga, 20 km. N Tegucigalpa. El Salvador: San Antonio (Burt and Stirton, 1961:49), thence northwestward along Pacific Coast to Oaxaca: Salina Garrapatera (Goodwin, 1969:245). Veracruz: Tierra Blanca (Hall and Dalquest, 1963:346); Mirador.

Felis pardalis sonoriensis Goldman

1925. *Felis pardalis sonoriensis* Goldman, Jour. Mamm., 24:378, August 17, type from Camoa, Río Mayo, Sonora.

MARGINAL RECORDS.—Arizona: Fort Verde. Sonora: Guirocoba; type locality.

Map 526. *Felis pardalis.*

1. *F. p. albescens*   2. *F. p. mearnsi*   3. *F. p. nelsoni*   4. *F. p. pardalis*   5. *F. p. sonoriensis*

Felis wiedii
Margay

North American subspecies revised by Goldman, Jour. Mamm., 24:380–385, August 17, 1943.

External measurements: (males) 862–1300; 331–510; 102–130; (females) 805–1029; 342–440; 89–132. Greatest length of skull, (males) 89.0–107.0, (females) 86.1–100.6; zygomatic breadth, (males) 61.3–72.2, (females) 59.6–65.8; crown length of maxillary tooth-row, (males) 27.2–29.4, (females) 24.7–29.4. "Ground color of upper parts Clay Color, slightly darker along middorsal line and slightly grayer on head; eye ring black; cheeks Cinnamon with narrow area of white between cheek and eye; top of muzzle dull Cinnamon; top of head near Tawny-Olive with two narrow lines of black extending posteriorly from above eyes to between ears and enclosing

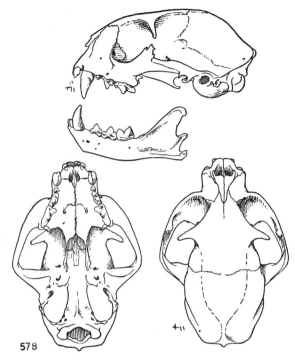

Fig. 578. *Felis wiedii nicaraguae*, Canas Gordas, Costa Rica, No. 141989 A.M.N.H., ♂, X ½.

numerous flecks of black; backs of ears black with round white spot, 7 mm. in diameter; back of neck with five parallel lines; median line narrow and not distinct; second lines heavy and distinct, 10 mm. wide; outer lines narrow, 6 mm. wide, but distinct; indistinct, broken area of black, enclosing narrow line of Clay Color or Tawny-Olive, extending along median line from shoulders to rump; rest of upper parts Clay Color with numerous irregular black spots, averaging 20 mm. in diameter, and enclosing small areas of Tawny-Olive; lower on sides, spots tending to break up and become elongated; pair of broad, black transverse bars on each shoulder surrounding areas of Tawny-Olive; front and back legs with numerous black spots, often oriented transversely; under parts white with numerous solid black spots; tail Tawny-Olive with numerous irregular black spots, forming complete rings near tip." (Dalquest, 1953:201, 202, description of specimen from San Luis Potosí.)

F. wiedii resembles a small, long-tailed *F. pardalis*. Differences from *pardalis* are: skull smaller; postorbital constriction shallower; orbit relatively larger; postorbital processes relatively longer and slenderer (fusing with jugals in some skulls); dorsal contour of skull evenly convex from nasals to occiput; sagittal crest absent; temporal crests lyriform.

Felis wiedii cooperi Goldman

1943. *Felis wiedii cooperi* Goldman, Jour. Mamm., 24:384, August 17, type from Eagle Pass, Maverick Co., Texas. Known only from type locality.

Felis wiedii glaucula Thomas

1903. *Felis glaucula* Thomas, Ann. Mag. Nat. Hist., ser. 7, 12:235, August, type from Beltrán, Jalisco.
1943. *Felis wiedii glaucula*, Goldman, Jour. Mamm., 24:384, August 17.

MARGINAL RECORDS.—Chihuahua: 40 km. N, 6 km. W Choix, Sinaloa (Anderson, 1972:388). Sinaloa: 44 km. ENE Sinaloa, 600 ft. (Armstrong, et al., 1972:57); Tatemales. Jalisco: Los Masos. Guerrero (Davis and Lukens, 1958:360): Acahuizotla, 2800–4000 ft.; *2 mi. W Omiltemi, 7900 ft.* Not found: Jalisco: type locality.

Felis wiedii nicaraguae (J. A. Allen)

1919. *Margay glaucula nicaraguae* J. A. Allen, Bull. Amer. Mus. Nat. Hist., 41:357, October 3, type from Volcán de Chinandega, Chinandega, Nicaragua.
1943. *Felis wiedii nicaraguae*, Goldman, Jour. Mamm., 24:382, August 17.

MARGINAL RECORDS.—Honduras: Santa Bárbara; Guaymaca. Nicaragua: Matagalpa. Costa Rica: Santa Clara; Cañas Gordas. Nicaragua: type locality.

Felis wiedii oaxacensis Nelson and Goldman

1860. *Felis mexicana* Saussure, Revue et Mag. Zool., Paris, ser. 2, 1:12, January. Type locality, Alvarado, Veracruz. Not *F. mexicana* Desmarest, 1816.
1931. *Felis glaucula oaxacensis* Nelson and Goldman, Jour. Mamm., 12:303, August 24, type from Cerro San Felipe, 10,000 ft., near Oaxaca, Oaxaca.
1943. *Felis wiedii oaxacensis*, Goldman, Jour. Mamm., 24:383, August 17.

MARGINAL RECORDS.—Tamaulipas: Rancho del Cielo. Veracruz: Alvarado. Oaxaca (Goodwin, 1969:245): La Gloria; Santa Lucía; type locality. Veracruz: Córdoba. San Luis Potosí: Apetsco.

Felis wiedii pirrensis Goldman

1914. *Felis pirrensis* Goldman, Smiths. Miscl. Coll., 63(5):4, March 14, type from Cana, 2000 ft., Darién, Panamá.
1943. *Felis wiedii pirrensis*, Goldman, Jour. Mamm., 24:380, August 17.

MARGINAL RECORDS.—Panamá: Sibube (Handley, 1966:791); Calcobevora; type locality, thence into South America.

Felis wiedii salvinia (Pocock)

1941. *Leopardus wiedii salvinia* Pocock, Field Mus. Nat. Hist., Publ. 511, Zool. Ser., 27:366, December 8, type from Vera Paz, Guatemala.

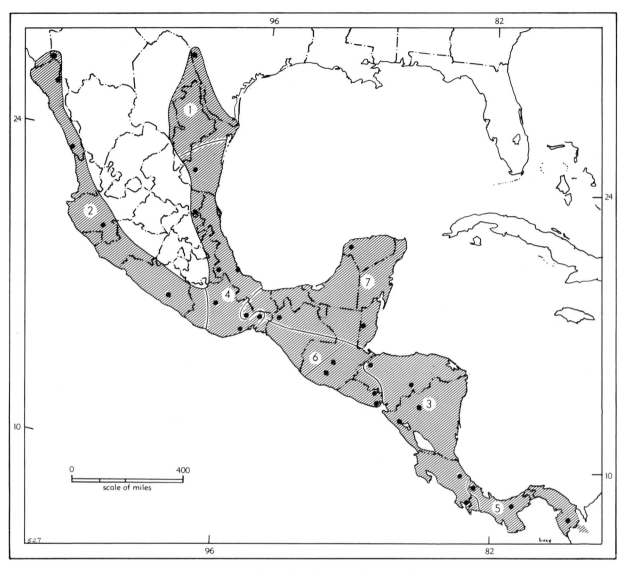

Map 527. *Felis wiedii.*

Guide to subspecies	2. *F. w. glaucula*	4. *F. w. oaxacensis*	6. *F. w. salvinia*
1. *F. w. cooperi*	3. *F. w. nicaraguae*	5. *F. w. pirrensis*	7. *F. w. yucatanica*

1943. *Felis wiedii salvinia*, Goldman, Jour. Mamm., 24:382, August 17.

MARGINAL RECORDS.—Guatemala: type locality. El Salvador (Burt and Stirton, 1961:49): Mt. Cacaguatique; Colinas de Jucuarán. Guatemala: Dueñas.

Felis wiedii yucatanica Nelson and Goldman

1931. *Felis glaucula yucatanica* Nelson and Goldman, Jour. Mamm., 12:304, August 24, type from Mérida, Yucatán.
1943. *Felis wiedii yucatanica*, Goldman, Jour. Mamm., 24:382, August 17.

MARGINAL RECORDS.—Yucatán: type locality, thence around peninsula to Belize: near Uaxactún.

Chiapas: Tecpactán. Oaxaca: Ocotal (Goodwin, 1969:245).

Felis tigrina
Little Spotted Cat

External measurements: 755–970; 255–420; 98–110; 38–41. Greatest length of skull, 83–88; zygomatic breadth, 49.5–59.3; palatal length, 30–31; length of P4, 9.6–10.7. Ground color of upper parts light fulvous, becoming lighter on sides and belly; face with white supraorbital line; black frontal lines interrupted; back of ears black, whitish patch at base small, but well defined; black median nuchal line present in some; 2 pairs

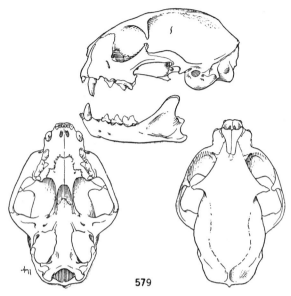

579

Fig. 579. *Felis tigrina pardinoides* [this subspecies does not occur in North America], Andalucia, Huila, Colombia, No. 33896 A.M.N.H., ♂, X ½.

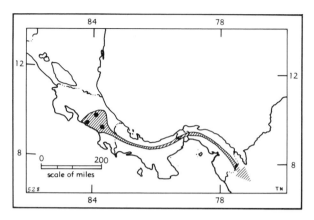

Map 528. *Felis tigrina oncilla.*

(*ibid.*); Pozo Azul. Also occurs in South America, but presence in Panamá is unverified.

Felis yagouaroundi
Jaguarundi

strong black lines and 3rd indistinct one on sides of neck; mid-dorsal line interrupted; laterally lines change to rosettes about 1 in. in diameter, with dark fulvous centers; belly and limbs spotted; chin, chest, and groins whitish, with little spotting; belly heavily spotted; limbs dull yellowish fulvous, spots becoming smaller distally, disappearing on digits; tail with 11–13 black rings, broader than light intervening spaces.

Skull small and delicate, smooth, and with only a poorly developed lambdoidal crest; face short; nasals not markedly convex above; postorbital processes small and weak; braincase only slightly constricted; bullae well inflated (after Thomas, 1903:236, 237). Mastoid breadth amounts to less (44.2–48.7%) than half of basilar length.

Many authors have used the name *F. pardinoides* Gray, 1867, for this species that seems to have been named *F. tigrina* by Schreber in 1775.

Felis tigrina oncilla Thomas

1902. *Felis pardinoides oncilla* Thomas, Ann. Mag. Nat. Hist., ser. 7, 12:237, August, type from Volcán de Irazú, Cartago, Costa Rica.
1959. *Felis tigrina oncilla*, Hall and Kelson, The mammals of North America, Ronald Press, p. 964, March 31.
1904. *Felis carrikeri* J. A. Allen, Bull. Amer. Mus. Nat. Hist., 20:47, February 29, type from Pozo Azul, Río Pirrís, San José Prov., Costa Rica.

MARGINAL RECORDS.—Costa Rica: type locality; La Piedra, 4 km. SW Cerro Chirripó, 10,500 ft. (Gardner, 1971:464); *Fila la Maquina, 7½ km. E Canaan*

External measurements: 888–1372; 330–609; 120–152; 25–40. Greatest length of skull, 89.3–115.5; zygomatic breadth, 60.0–74.5; crown length of maxillary tooth-row, 27.5–32.5. Body elongate, head long and low, ears small. Upper parts unspotted, dichromatic: blackish to brownish-gray or reddish-yellow to chestnut. Cranium long, never strongly convex longitudinally; facial area short; temporal crests almost always separated, forming a lyriform area; rarely complete sagittal crest; muzzle compressed above; narrowest part of interorbital region situated forward, just behind nasomaxillary line; maxilla broad above, with nearly horizontal upper edge, rounded anterosuperior angle, and nearly vertical anterior border. P1 minute or absent; mesopterygoid fossa narrowed anteriorly, with rounded angles, anterior border inflated, notched; inner chamber of auditory bulla never much inflated anteriorly, not projecting far beyond the low partition; occipital crest not appreciably concave above condyles.

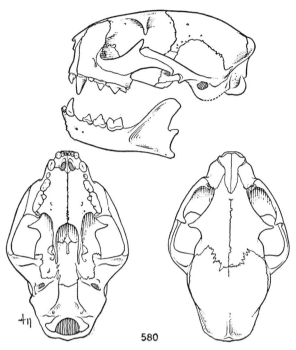

Fig. 580. *Felis yagouaroundi cacomitli*, Brownsville, Texas, No. 2219 K.U., ♂, X ½.

Felis yagouaroundi cacomitli Berlandier

1859. *Felis cacomitli* Berlandier, *in* Baird, Mammals of the boundary, *in* Emory, Rept. . . . U.S. and Mexican boundary survey, 2(2):12, January, type from Matamoros, Tamaulipas.
1905. *Felis jaguarondi cacomitli*, Elliot, Field Columb. Mus., Publ. 105, Zool. Ser., 6:370, December 6.
1901. *Felis apache* Mearns, Proc. Biol. Soc. Washington, 14:150, August 9, type from Matamoros, Tamaulipas.

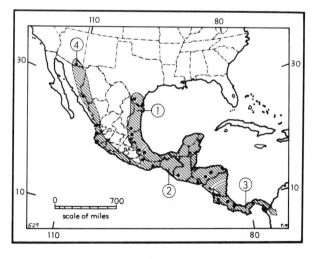

Map 529. *Felis yagouaroundi.*

1. *F. y. cacomitli* 3. *F. y. panamensis*
2. *F. y. fossata* 4. *F. y. tolteca*

MARGINAL RECORDS.—Texas: Starr County; Willacy County, thence southward along coast to Veracruz: Mirador. San Luis Potosí: near Ebano; vic. Xilitla. Tamaulipas: near Gómez Farías.

Felis yagouaroundi fossata Mearns

1901. *Felis fossata* Mearns, Proc. Biol. Soc. Washington, 14:150, August 9, type from Mérida, Yucatán.
1942. *Felis (yagouaroundi) fossata*, Goodwin, Bull. Amer. Mus. Nat. Hist., 79:186, May 29.

MARGINAL RECORDS.—Quintana Roo: 6½ km. NE Playa del Carmen (Birney, *et al.*, 1974:17); 81 km. W Chetumal (93838 KU). Honduras: La Flor Archaga; Muin. El Salvador (Burt and Stirton, 1961:50): Lake Olomega; *Puerto del Triunfo.* Guatemala: Finca Ciprés. Oaxaca (Goodwin, 1969:248): Salina Garrapatera; Chiltepec. Veracruz: Río Blanco, 20 km. WNW Piedras Negras.

Felis yagouaroundi panamensis J. A. Allen

1904. *Felis panamensis* J. A. Allen, Bull. Amer. Mus. Nat. Hist., 20:71, February 29, type from Boquerón, Chiriquí, Panamá.
1946. *Felis yaguarondi* [sic] *panamensis*, Goodwin, Bull. Amer. Mus. Nat. Hist., 87:443, December 31.

MARGINAL RECORDS.—Costa Rica: Catalina; San José. Panamá (Handley, 1966:791): Sibube; Empire; Cana, thence into South America.

Felis yagouaroundi tolteca Thomas

1898. *Felis yaguaroundi* [sic] *tolteca* Thomas, Ann. Mag. Nat. Hist., ser. 7, 1:41, January, type from Tatemales, Sinaloa.

MARGINAL RECORDS.—Arizona: *ca.* 3 mi. E Canelo Ranger Station, Coronado National Forest, Santa Cruz Co. Jalisco: vic. Chamela Bay (López-F., *et al.*, 1973:104). Nayarit: 1 mi. N San Blas, 10 ft. Sinaloa (Armstrong, *et al.*, 1972:58): 2 mi. E Palmito, Isla Palmito del Verde, 10 ft.; 15 mi. NE Choix.

Genus Lynx Kerr—Lynx and Bobcat

1792. *Lynx* Kerr, The animal kingdom . . . , 1:155. Type, *Lynx vulgaris* Kerr [= *Felis lynx* Linnaeus].
1821. *Lynceus* Gray, London Med. Repos., 15:302, April 1. Type, *Felis lynx* Linnaeus. Not *Lynceus* Müller, 1785, a crustacean.
1825. *Lyncus* Gray, Thompson's Ann. Philos., 26:339, November, a renaming of *Lynceus* Gray, 1821.
1829. *Pardina* Kaup, Skizzirte Entwickelungs–Geschichte und natürliches System der europäischen Thierwelt, 1:53. Type, *Felis pardina* Temminck.
1834. *Lynchus* Jardine, The naturalist's library, 2:274, an emendation of *Lynceus* Gray.
1843. *Caracal* Gray, List of the . . . Mammalia in the . . . British Museum, p. 46. Type, *Felis caracal* Schreber.
1858. *Urolynchus* Severtzow, Revue et Mag. Zool., Paris, ser. 2, 10:389, September. Type, *Felis caracal* Schreber.

1867. *Cervaria* Gray, Proc. Zool. Soc. London, p. 276, October. Type, *Lyncus pardinus* [= *Felis pardina* Temminck]. Not *Cervaria* Walker, 1866, a lepidopteran.

1903. *Eucervaria* Palmer, Science, n.s., 17:873, May 29, a renaming of *Cervaria* Gray, 1867.

Medium-sized cats with short tails, large, triangular, tufted ears, large paws, broad, short heads. Cranial characters essentially as in *Felis*, but with nasal branch of premaxillae slender and gradually attenuated; postorbital processes thinner, less depressed and sharper; notching of suborbital edge of palate shallower; and premolars $\frac{2}{2}$.

KEY TO NORTH AMERICAN SPECIES OF LYNX

1. Tail tipped with black above and below; tail less than ½ length of hind foot; anterior condyloid foramen separate from foramen lacerum posterius. . . . *L. canadensis*, p. 1050
1′. Tail tipped with black only above; tail more than ½ length of hind foot; anterior condyloid foramen confluent with foramen lacerum posterius. . . . . . . *L. rufus*, p. 1051

Lynx canadensis
Lynx

External measurements: 825–954; 95–125; 203–250. Weight, 6.8–18.1 kg. Greatest length of skull, 120–136; zygomatic breadth, 82–93; alveolar length of maxillary tooth-row, 38.1–41.6. Females are smaller than males. "Legs long; feet large; pads well furred. Color: Upper parts a mixture of Ochraceous-Tawny, dark brown and black sprinkled with white; all guard hairs white subterminally and black terminally and darkest middorsally; underfur Cinnamon Brown; top of head like back, but heavily suffused with white-tipped hairs; eyelids white; ears buffy brown at base, with central white spot; dorsal margin of ears brown, terminating in a long, slender black tuft; ears white inside; ruff well developed, consisting of white, black, and brown and white hairs; sides and flanks Ochraceous-Buff, sprinkled with white, guard hairs white subterminally, black terminally; tail Cinnamon above, white beneath, black-tipped; feet and legs buffy white; chin and throat white; underparts buffy white sparsely mottled with light brown. Skull: Large; postorbital processes of frontals small; presphenoid wide transversely, but constricted anteroposteriorly; posterior palatine foramina situated near orbital rim of palate; anterior margin of interpterygoid fossa semicircular; anterior condyloid foramen separate from foramen lacerum posterius." (Durrant, 1952:438, 439.)

Kurtén and Rausch (1959:41) regarded *Lynx canadensis* of North America as specifically different from *Lynx lynx* of Eurasia on the basis of carefully prepared data obtained from a comparison of skulls from Fennoscandia and Alaska. Comparison of Alaskan skulls with skulls from northeastern Asia would yield information more useful in deciding on specific *versus* subspecific relationship of Recent *Lynx* of northern Eurasia and northern North America. Until that information becomes available the conservative arrangement of recognizing two closely related species (*Lynx canadensis* in North America and *Lynx lynx* in the Old World) is followed here.

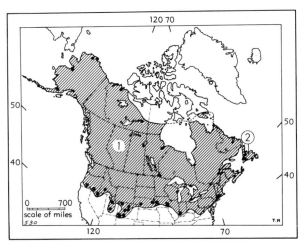

Map 530. *Lynx canadensis.*

1. *L. c. canadensis*   2. *L. c. subsolanus*

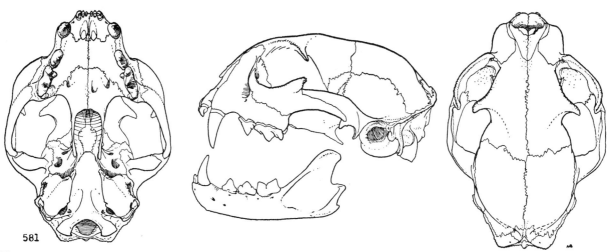

Fig. 581. *Lynx canadensis canadensis*, Kusilif, Alaska, No. 4214 M.V.Z., ♂. Lower jaw is of ♀ 48081 K.U., Wonder Lake, Mt. McKinley Dist., Alaska. X ½.

Lynx canadensis canadensis Kerr

1792. *Lynx canadensis* Kerr, The animal kingdom . . . , 1:157. Type locality, Eastern Canada [= Quebec].
1900. *Lynx canadensis mollipilosus* Stone, Proc. Acad. Nat. Sci. Philadelphia, 52:48, March 24, type from Wainwright Inlet [*sic*], Point Barrow, Alaska.

MARGINAL RECORDS.—Alaska: Point Barrow. Mackenzie: Franklin Bay; E shore Artillery Lake. Manitoba: York Factory. District of Franklin: *Baffin Island*, thence southward along Atlantic Coast of continental North America to Connecticut: Southington. Pennsylvania: "*southern borders.*" Indiana: Knox County. Iowa: island S of Muscatine; sec. 4, T. 80 N, R. 37 W, Shelby Co. (Rasmussen, 1969:371). Nebraska (Jones, 1964c:304): 3 mi. W Norfolk; Cedar Point, near Keystone. Colorado: Grand Lake; vic. Mt. Lincoln; mts. N Pagosa Springs; La Plata; region about Snow Mass Peak; Elkhead Mts. Utah: Bald Ridge, *ca.* 25 mi. W and a little N of Orangeville; N. Fork Provo River. Idaho: 22 mi. N Hill City. Nevada: Little Owyhee River. Oregon: Kiger Creek, Steens [Mts.]; Fort Klamath, thence northward along Pacific Coast to Alaska: western shore Brooks Lake (Cahalane, 1959:196); mts. near Unalakleet; barren country back of Wainwright.

Lynx canadensis subsolanus Bangs

1897. *Lynx subsolanus* Bangs, Proc. Biol. Soc. Washington, 11:49, March 16, type from Codroy, Newfoundland.
1901. [*Lynx canadensis*] *subsolanus*, Elliot, Field Columb. Mus., Publ. 45, Zool. Ser., 2:296, March 6.

MARGINAL RECORDS.—Newfoundland: Cormack (Cameron, 1959:102); type locality.

Lynx rufus
Bobcat

Eastern subspecies revised by Peterson and Downing, Contrib. Royal Ontario Mus. Zool. and Palaeont., 33:1–23, April 8, 1952; Californian subspecies revised by Grinnell and Dixon, Univ. California Publ. Zool., 21:339–354, January 24, 1924.

External measurements: (males) 787–1252; 130–195; 158–223; 61–77 (crown); (females) 710–1219; 95–170; 143–180; 63–76 (crown). Weight: (males) 7.2–31.0 kg., (females) 5.7–24.0 kg. Greatest length of skull, (males) 114.5–149.5, (females) 105.8–137.4; zygomatic breadth, (males) 82.2–107.1, (females) 75.2–90.5; alveolar length of maxillary tooth-row, (males) 35.9–43.1, (females) 33.7–35.6. Upper parts grayish, buffy, or reddish, usually with black spots; color most intense mid-dorsally, becoming lighter laterally; rump and hind legs buffy; ears blackish with white spot near tip, tufted; eyelids white; under-

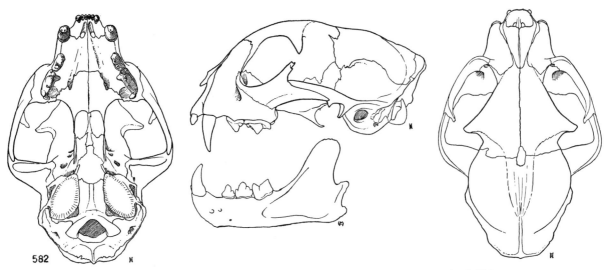

Fig. 582. *Lynx rufus pallescens*, Marble, Nevada, No. 24584 M.V.Z., ♂, X ⅓.

parts and inner side of legs whitish, with black spots; tail with indistinct black rings, tipped with black above, whitish below. Skull robust; sagittal crests lyrate; lambdoidal crest and inion well developed; tympanic bullae large; postorbital processes prominent; presphenoid elongate; anterior condyloid foramen confluent with foramen lacerum posterius.

Lynx rufus baileyi Merriam

1890. *Lynx baileyi* Merriam, N. Amer. Fauna, 3:79, September 11, type from Moccasin Spring, N of Colorado River, Coconino Co., Arizona.
1901. [*Lynx rufa*] *baileyi*, Elliot, Field Columb. Mus., Publ. 45, Zool. Ser., 2:297, March 6.
1897. *Lynx rufus eremicus* Mearns, Preliminary diagnoses of new mammals . . . from the Mexican boundary line, p. 1, January 12 (preprint of Proc. U.S. Nat. Mus., 20:458, December 24, 1897), type from New River, near Laguna Station, Colorado Desert, Imperial Co., California.

MARGINAL RECORDS.—Nevada: Carson River, 8 mi. W Fallon; Pahranagat Valley; Rainbow Canyon, Lincoln Co. Utah: Beaverdam Wash, 8 mi. N Utah-Arizona border, 2800 ft.; W Kolob Peak, Washington Co.; 8½ mi. SW Orderville, 5000 ft.; Sanford Canyon, Garfield Co.; Big Hollow, Wayne Co.; Thousand Lake Mtn., Wayne Co.; Harley Dome, Grand Co. Colorado (Armstrong, 1972:300): West Paradox Valley; Dolores, 7000 ft.; White Canyon, Mesa Verde; Durango; "Piedra River, Durango"; Chromo; La Jara; Monte Vista; Russell Springs; 4 mi. S, 1 mi. E Great Sand Dunes National Monument; 10 mi. E Gardner; southern part of Bent County. Kansas: Barber County. Oklahoma: 12 mi. NE Seiling; *Red River Valley*. Texas: 2 mi. S, 10 mi. E Pringle (Cutter, 1959:32). New Mexico: 12 mi. S, 2 mi. W Conchas Dam; 7 mi. NE Corona; Pecos River, 35

mi. N Roswell. Texas: Dawson County (Davis, 1966:122, as *L. rufus* only); Stanton; Sterling County (Davis, 1966:122, as *L. rufus* only); Crockett County (*ibid.*). Coahuila: 2 mi. SW Carricitos, 3500 ft. Durango: La Pila (Baker and Greer, 1962:142); Coyotes. Chihuahua: Urique (Anderson, 1972:388). Sonora: Costa Rica Ranch. Arizona: near Rancho Bonita Well, Organ Pipe Cactus National Monument. California: near Imperial; Vallecito; Victorville; E slope Mt. Baxter, 5500 ft.; Onion Valley; Cottonwood Lakes, 11,000 ft. Nevada: Desert Creek Ranch, S end [Lyon] Co.

Lynx rufus californicus Mearns

1897. *Lynx rufus californicus* Mearns, Preliminary diagnoses of new mammals . . . from the Mexican boundary line, p. 2, January 12 (preprint of Proc. U.S. Nat. Mus., 20:458, December 24, 1897), type from San Diego, San Diego Co., California.
1899. *Lynx (Cervaria) fasciatus oculeus* Bangs, Proc. New England Zool. Club, 1:23, March 31, type from Nicasio, Marin Co., California.

MARGINAL RECORDS.—California: Mt. Shasta, 6500 ft.; McCloud River, near Baird. Nevada: 2½ mi. NW Lakeview; Glenbrook. California: foot Leevining Creek grade; Piute Mts.; Riverside; San Jacinto Mts.; Santa Rosa Mts.; Campo. Baja California: Hanson Lagoon, [near] Hanson Laguna Mts. California: type locality, thence along Pacific Coast to Gualala; South Yolla Bolly Mtn.; Weaverville.

Lynx rufus escuinapae J. A. Allen

1903. *Lynx ruffus* [*sic*] *escuinapae* J. A. Allen, Bull. Amer. Mus. Nat. Hist., 19:614, November 14, type from Escuinapa, Sinaloa.

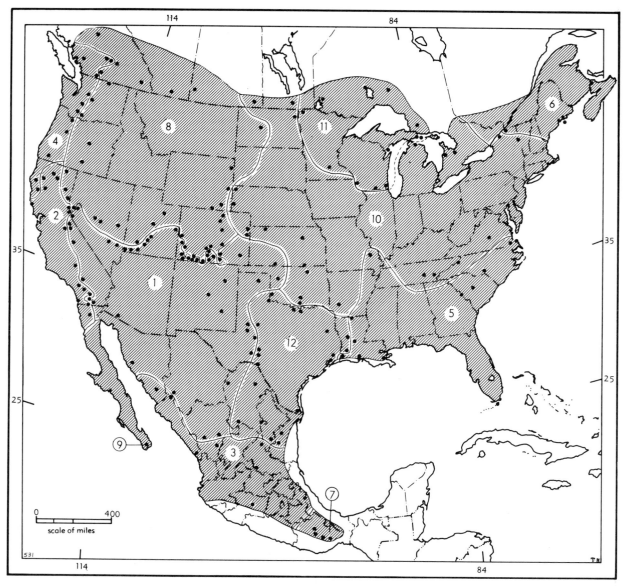

Map 531. *Lynx rufus.*

1. *L. r. baileyi*	4. *L. r. fasciatus*	7. *L. r. oaxacensis*	10. *L. r. rufus*
2. *L. r. californicus*	5. *L. r. floridanus*	8. *L. r. pallescens*	11. *L. r. superiorensis*
3. *L. r. escuinapae*	6. *L. r. gigas*	9. *L. r. peninsularis*	12. *L. r. texensis*

MARGINAL RECORDS.—Sonora: Camoa. Sinaloa: *1 mi. S El Cajón, 1800 ft.* (Armstrong, *et al.,* 1972:58); 18 km. NNE Choix (*ibid.*); type locality. Durango: 28 mi. S, 17 mi. W Vicente Guerrero, 8350 ft. (Baker and Greer, 1962:142). San Luis Potosí: Bledos; 6 km. S Matehuala. Veracruz: 3 km. W Limón, 7500 ft. Michoacán: Pátzcuaro, thence westward to Pacific Coast and northward to point of beginning.

Lynx rufus fasciatus Rafinesque

1817. *Lynx fasciatus* Rafinesque, Amer. Month. Mag., 2(1):46, November. Type locality, "Northwest Coast"; based on specimens from near mouth Columbia River, on Netul River [= Lewis and Clark River], near Astoria, Oregon.

1901. [*Lynx rufa*] *fasciata*, Elliot, Field Columb. Mus., Publ. 45, Zool. Ser., 2:297, March 6.

MARGINAL RECORDS.—British Columbia: Brigade Lake (Cowan and Guiguet, 1965:342); Westwold; Manning Park (Cowan and Guiguet, 1965:342). Washington: Skykomish; Mt. Rainier; headwaters Lewis River. Oregon: Multnomah County; Linn County; Jackson County. California: Carlotta; Laytonville, thence northward along Pacific Coast to British

Columbia: Halfmoon Bay (Cowan and Guiguet, 1965:342); Alta Lake.

Lynx rufus floridanus Rafinesque

1817. *Lynx floridanus* Rafinesque, Amer. Month. Mag., 2(1):46, November. Type locality, Florida.
1858. *Lynx rufus* var. *floridanus*, Baird, Mammals, *in* Repts. Expl. Surv. . . . , 8(1):91, July 14.

MARGINAL RECORDS.—Missouri: Banner. Tennessee: *western Tennessee.* Alabama: *no specific locality.* South Carolina: Aiken County; Richard County [= Richland County?]. North Carolina: Montgomery County. Virginia: Nansemond County; Dismal Swamp, thence southward along Atlantic Coast to Florida: south on keys to Lower Matecumbe (Layne, 1974:400). Thence along Gulf Coast to Louisiana (Lowery, 1974:470): Burbridge; Abbeville; 25 mi. SW Lake Charles; near Lecompte; *near Georgetown;* near Spearsville.

Lynx rufus gigas Bangs

1897. *Lynx gigas* Bangs, Proc. Biol. Soc. Washington, 11:50, March 16, type from 15 mi. back of Bear River, Nova Scotia.
1952. *Lynx rufus gigas*, Peterson and Downing, Contrib. Royal Ontario Mus. Zool. and Paleont., 33:11, April 8.

MARGINAL RECORDS.—Nova Scotia: Colchester County, thence along coast to Maine: *Hancock County;* Mount Desert Island (Manville, 1960:416, as *L. rufus* only). New York: Franklin County.

Lynx rufus oaxacensis Goodwin

1963. *Lynx rufus oaxacensis* Goodwin, Amer. Mus. Novit., 2139:1, May 7, type from Los Nanches, San Pedro Jilotepec, District of Tehuantepec, Oaxaca, Mexico, *ca.* 5000 ft.

MARGINAL RECORDS (Goodwin, 1969:249).— Oaxaca: Macuiltianguis; type locality; *Tequisistlán;* San Bartolo Yautepec; Mitla.

Lynx rufus pallescens Merriam

1899. *Lynx fasciatus pallescens* Merriam, N. Amer. Fauna, 16:104, October 28, type from S side Mt. Adams, near Trout Lake, Skamania Co., Washington.
1901. [*Lynx rufa*] *pallescens*, Elliot, Field Columb. Mus., Publ. 45, Zool. Ser., 2:297, March 6.
1902. *Lynx uinta* Merriam, Proc. Biol. Soc. Washington, 15:71, March 22, type from Bridger Pass, Carbon Co., Wyoming. (Regarded by Grinnell and Dixon, Univ. California Publ. Zool., 21:350, January 24, 1924, as inseparable from *pallescens.*)

MARGINAL RECORDS.—British Columbia: Williams Lake; Fort Steele. Alberta: Milk River between Writingonstone and Police Coulee. Saskatchewan: Maple Creek (Beck, 1958:22). North Dakota: near Garrison. Wyoming (Long, 1965a:709): Black Hills, Sand Creek; *Newcastle;* Manville; *5 mi. E Horse Creek;* Federal; Jelm. Colorado (Armstrong, 1972:300, 301):

20 mi. N Fort Collins; SW of Berthoud; Denver; 20 mi. S Colorado Springs; 2 mi. W La Veta; *ca.* 3 mi. E Villa Grove; 10 mi. NW Saguache; Conejos River, 22 mi. W Antonito; Gordon Creek, near jct. Piedra River; Coventry; Unaweep Canyon. Utah: S Myton; Joes Valley; W reservoir, Minersville. Nevada: Shingle Pass, Lincoln Co.; Monitor Valley, 8 mi. E Toquima Peak; Millett P.O.; *Marble;* 6 mi. NE Virginia City, 6000 ft. California: Plumas Junction; E side Eagle Lake; Straw. Oregon: Lake County; Crook County. Washington: type locality; Lake Keechelus; Oroville. British Columbia: Princeton (Cowan and Guiguet, 1965:342); Trinity Valley.

Lynx rufus peninsularis Thomas

1898. *Lynx rufus peninsularis* Thomas, Ann. Mag. Nat. Hist., ser. 7, 1:42, January, type from Santa Anita, Baja California. Known only from Baja California.

Lynx rufus rufus (Schreber)

1777. *Felis rufa* Schreber, Die Säugthiere . . . , thiel 3, heft 95, Pl. 109b. Type locality, New York. *Felis ruffus* Güldenstaedt, 1776, is not a scientific name according to J. A. Allen, Jour.. Mamm., 1:91, March 2, 1920.
1817. *Lynx rufus*, Rafinesque, Amer. Month. Mag., 2:46, November.
1817. *Lynx montanus* Rafinesque, Amer. Month. Mag., 2:46, November. Type locality, Catskill Mts., New York.

MARGINAL RECORDS.—Saskatchewan: Whitewood. Manitoba: La Rivière, Pembina Hills. North Dakota: Minto. Michigan: Presque Isle County. Ontario: Bruce County; Simcoe County. New York: Adirondack Mts. Vermont: Addison County, thence eastward to Atlantic Coast and then southward to Virginia: Buckingham County. North Carolina: near Mt. Mitchell. Tennessee: Cherokee National Forest; Walden Ridge, near Soddy. Arkansas: state game refuge near Graysonia Park, Clark Co. Oklahoma: Carter County; Murray County; Osage County. Kansas: 2½ mi. SW Silverdale; Junction City; Oberlin (4762 KU). Colorado (Armstrong, 1972:301): S. Fork Republican River, near Tuttle; along Chief Creek, near Wray. Nebraska (Jones, 1964c:306, 307): 10 mi. W Harrisburg; *SW corner Sioux Co., N of Mitchell; Harrison;* Squaw Canyon, N of Harrison; *NW corner Dawes Co. (Jim Moody's Ranch).*

Lynx rufus superiorensis Peterson and Downing

1952. *Lynx rufus superiorensis* Peterson and Downing, Contrib. Royal Ontario Mus. Zool. and Paleont., 33:1, April 8, type from McIntyre Township, near Port Arthur, Ontario.

MARGINAL RECORDS.—Ontario: Onaping; 20 mi. E English River Post; T. 22, R. 14, Algoma Dist.; Cockburn Island. Michigan: Chippewa County; Menominee County. Wisconsin (Jackson, 1961:403, Map 74): *Door County;* Racine County (pre-1900); Rock County (pre-1900); Crawford County (pre-1900). Minnesota: Wright County; Roseau County. Ontario: Sioux Narrows, Kenora Dist.

Lynx rufus texensis J. A. Allen

1829. *Felis maculata* Horsfield and Vigors, Zool. Jour., 4:381, pl. 13, type from México. Not *Felis (Lynx) vulgaris maculatus* Kerr, 1792.
1895. *Lynx texensis* J. A. Allen, Bull. Amer. Mus. Nat. Hist., 7:188, June 20, based on the description of a bobcat by Audubon and Bachman, The viviparous quadrupeds of North America, 2:293, 1851, from "the vicinity of Castroville, on the headwaters of the Medina [River]," Medina Co., Texas (see Audubon and Bachman, *op. cit.*:295).
1897. *Lynx rufus texensis,* Mearns, Preliminary diagnoses of new mammals . . . from the Mexican boundary line, p. 2, January 12 (preprint of Proc. U.S. Nat. Mus., 20:458, December 24).

MARGINAL RECORDS.—Oklahoma: 19 mi. N, 1 mi. E Hollis (Martin and Preston, 1970:56, as *L. rufus* only, sight record). Texas: Montague County; Cooke County; Rusk County (Davis, 1966:122, as *L. rufus* only). Louisiana (Lowery, 1974:470): *1 mi. W Flora;* Gorum; below Ragley. Texas: Sour Lake; Chambers County (Davis, 1966:122, as *L. rufus* only), thence along coast to Brownsville. Tamaulipas: [region of] San Carlos Mts.; Rancho Santa Rosa, 360 m. Nuevo León: 7 mi. S Aramberri. Coahuila: 3 mi. S, 13 mi. E San Antonio de las Alazanas, 8900 ft.; 4 mi. E Las Margaritas, 3100 ft.; La Gacha, 1600 ft. Texas: Devils River; Sutton County (Davis, 1966:122, as *L. rufus* only); Schleicher County (*ibid.*); Scurry County (*ibid.*); Hardeman County (Dalquest, 1968:20).

Order PINNIPEDIA

Reviewed by Scheffer, Seals, sea lions, and walruses . . . , Stanford Univ. Press, pp. x–179, illustrated, April 24, 1958.

North American pinnipeds monographed by J. A. Allen, U.S. Geol. and Geog. Surv. Territories, Miscl. Publ., 12:xvi+785, 60 figs., 1880.

Limbs modified for aquatic locomotion, enclosed to elbow or knee in integument; 1st digit of manus usually longest, 1st and 5th digits of pes usually longer than 3 middle digits; iliac part of pelvis short, anterior border everted; ischia barely meeting, symphysis short and ischia never ankylosed in males, ischia separated in females; skull narrow interorbitally; facial region short, broad; braincase expanded, lachrymal bones imperforate; palatine bones usually separated from frontal bones by a vacuity; tympanic bones separated from exoccipital bones; teeth simple and unspecialized except that upper canines are tremendously enlarged in *Rosmarus;* all cheekteeth essentially same shape in any one species; incisors reduced in number, never more than $\frac{3}{2}$; deciduous dentition lost early in life; Cowper's glands absent; mammae, 2 or 4, abdominal; tail vestigial.

Key to Families of Pinnipedia

1. Hind limbs capable of rotating forward; alisphenoid canals present.   Otariidae, p. 1056
1'. Hind limbs not capable of rotating forward; alisphenoid canals absent. Phocidae, p. 1063

Family OTARIIDAE
Eared Seals and Walrus

Front and hind limbs approx. same size; hind limbs capable of rotating forward; ventral surface of manus and pes and most of dorsal surface thereof naked; alisphenoid canals present.

Key to Recent Subfamilies of Otariidae

1. Pinnae present; postorbital processes present; upper canine shorter than length of upper tooth-row. . . . . . . .Otariinae, p. 1056
1'. Pinnae absent; postorbital processes absent; upper canine longer than length of upper tooth-row. . . . . . .Rosmarinae, p. 1061

Subfamily OTARIINAE—Eared Seals

Reviewed by Sivertsen, Sci. Results Norwegian Antarctic Exped. 1927–1928 . . . , 36:1–76, 46 figs., 10 pls., December 21, 1954. American species revised by King, Bull. British Mus., Zoology, 2(10):309–337, 3 figs., 2 pls., November, 1954.

Form slender, elongate; pinnae present, small external conch subcylindrical; nails frequently rudimentary; skull having well-developed postorbital processes; scalpula broad, large; mandibular symphysis not fused in adult; upper canine caniform; crista tympani having overhang of free ventral edge; pars petrosa, dorsal to meatus acousticus internus, and (according to Mitchell, 1968:1885) rounded and smooth; testes scrotal.

Key to North American Genera of Otariinae

1. Postorbital processes quadrate; diastema between P4 and M1 large, 5% or more of total length of skull. . . . .*Eumetopias*, p. 1058
1'. Postorbital processes triangular; diastema between P4 and M1 small, never more than 4% of total length of skull.
  2. Nasal bones short, broad, combined widths at anterior ends 80–90% of length; rostral length less than 25% of total length of skull.   *Callorhinus*, p. 1056
  2'. Nasal bones long, slender, combined widths at anterior ends 40–50% of length; rostral length more than 25% of total length of skull.
    3. Palate narrow, sides parallel between P1 and P3; tympanic bullae smooth, rounded; sagittal crest not greatly elevated in old males.
              Arctocephalus, p. 1058
    3'. Palate broad, sides increasing in width between P1 and P3; tympanic bullae irregular; sagittal crest greatly elevated in old males.
              Zalophus, p. 1059

Genus Callorhinus Gray—Northern Fur Seal

1859. *Callorhinus* Gray, Proc. Zool. Soc. London, p. 359, October, 1859–February, 1860. Type, *Phoca ursina* Linnaeus.

1817. *Otoes* G. Fischer, Mém. Soc. Imp. Nat. Moscou, 5:373, 445. Based on a composite species, *Phoca jubata* Gmelin, and therefore not available (see Palmer, N. Amer. Fauna, 23:487, 488, January 25, 1904).

1866. *Arctocephalus* Gill, Comm. Essex Inst., 5:7, 11, April 7. Type, *Phoca ursina* Linnaeus. Not *Arctocephalus* F. Cuvier, 1826.

1892. *Callotaria* Palmer, Proc. Biol. Soc. Washington, 7:156, July 27, a renaming of *Callorhinus* Gray.

Total length of males, 7–8 ft., weight to 700 pounds; total length of females, 4–5 ft., weight to 130 pounds. Males blackish to dark brownish dorsally, shoulders and neck gray, underparts and limbs reddish-brown; females gray to grayish-brown dorsally, reddish-brown or gray ventrally. Condylobasal length, (males) 221–275, (females) 168–200; zygomatic breadth, (males) 115–137, (females) 89–93; length of series of upper cheek-teeth, (males) 49–55, (females) 40–46. Facial part of skull short, broad, convex, only slightly depressed; palate short, narrow, deeply notched on posterior margin; interorbital region long, approx. 20 per cent of condylobasal length; nasals short and broad, combined widths at anterior ends 80–90 per cent of length; postorbital processes triangular. Dentition, i. $\frac{3}{2}$, c. $\frac{1}{1}$, p. $\frac{4}{4}$, m. $\frac{2}{1}$.

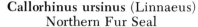

Callorhinus ursinus (Linnaeus)
Northern Fur Seal

X 1/38

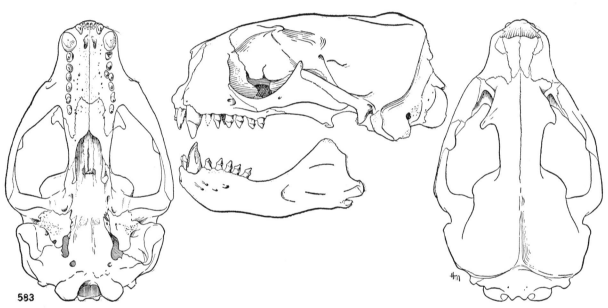

583

Fig. 583. *Callorhinus ursinus*, Pacific Grove, California, No. 2873 K.U., sex ?, X $\frac{1}{3}$.

1758. *Phoca ursina* Linnaeus, Syst. nat., ed. 10, 1:37, type locality, Bering Island, Commander Islands.

1859. *Callorhinus ursinus*, Gray, Proc. Zool. Soc. London, 1859:359.

1792. *Siren cynocephala* Walbaum, *in* Artedi, Petr. Artedi . . . ichthyologica, pt. 3, Genera piscium, p. 560. Type locality, North Pacific Ocean south of Alaska Peninsula at approx. lat. 53° N, long. 155° W (see Stejneger, Georg Wilhelm Steller . . . , Pl. 12 facing p. 278, August, 1936). (Based on the sea ape of Steller; see Stejneger, *op. cit.*:285.)

1835. *Phoca minima* Tilesius, Oken's Isis, 1835:715, type locality, Bay of Patience, Sakhalin Island.

1866. *Arctocephalus californianus* Gray, Catalogue of seals and whales in the British Museum, p. 51, type locality, Monterey, California.

1898. *Callorhinus alascanus* Jordan and Clark, *in* Jordan, The fur seals and fur-seal islands of the North Pacific Ocean, Publ. U.S. Treas. Dept., Spec. Agents Div., pt. 1, p. 45, type from Pribilof Islands, Alaska.

898. *Callorhinus curilensis* Jordan and Clark, *in* Jordan, The fur seals and fur-seal islands of the North Pacific Ocean, Publ. U.S. Treas. Dept., Spec. Agents Div., pt. 1, p. 45, type from Robben Island [= Seal Island, W of Sakhalin Island].

See characters of the genus. Scheffer (1958:82) regards the northern fur seal as a monotypic species.

RANGE.—Northwest Territories: Letty Harbour (Radvanyi, 1960:277). Alaska: E of Point Barrow, thence along coast southward to California: United States–Mexican border.

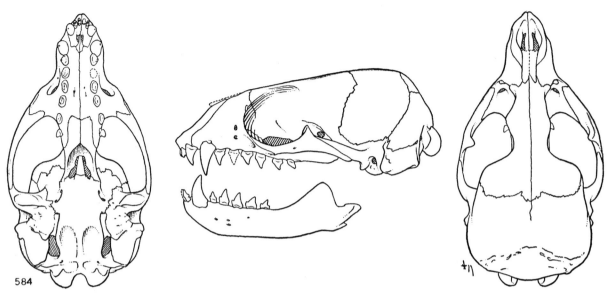

Fig. 584. *Arctocephalus philippii townsendi*, Guadalupe Island, Baja California, No. 83618 U.S.N.M., ♀, X ⅓.

Genus **Arctocephalus** É. Geoffroy St.-Hilaire and F. Cuvier
Southern Fur Seals

1826. *Arctocephalus* É. Geoffroy St.-Hilaire and F. Cuvier, Dictionaire des sciences naturelles, Phoques, vol. 39:553. Type, *Phoca pusilla* Schreber.
1866. *Arctophoca* Peters, Monatsb. preuss Akad. Wiss., Berlin, p. 276, pl. II, May. Type, *Otaria philippii* Peters.

Total length of males, 5–6 ft., females slightly smaller. Condylobasal length, (males) 235–256, (females) 197–217; zygomatic breadth, (males) 130–140, (females) 107–118; interorbital constriction, (males) 26.0–30.7, (females) 21.5–22.0. Skull narrow, interorbital breadth 10–12 per cent length of skull; postorbital processes triangular; nasal bones long, slender; rostrum long, rostral length averaging approx. 30 per cent of total length of skull. Skull resembles that of young *Zalophus* but differs in smaller size, more rounded tympanic bullae, narrower interorbital constriction, narrower palate, larger postorbital processes, longer maxillae, broader zygomatic roots of maxillae, and less elevated sagittal crest in males. Dentition, i. $\frac{3}{2}$, c. $\frac{1}{1}$, p. $\frac{4}{4}$, m. $\frac{2}{1}$.

This description may not apply equally well to all species in the genus, since it is based on the nominal genus *Arctophoca*, which includes only *A. philippii* in the present classification.

Arctocephalus philippii
Guadalupe Fur Seal

Blackish-gray dorsally, becoming yellowish-gray on head and neck; brownish-black ventrally;

lips, chin, and proximal parts of limbs rusty brown. Skull narrow; prolongations of palatine bones not thickened; zygomatic root of maxillae broad; zygomatic arch narrow; rostrum narrow; cheek-teeth lacking accessory cusps and widely spaced.

This species was nearly exterminated in the nineteenth century. Peterson, *et al.* (1968:665–675) show that the species has increased in numbers in the twentieth century but is restricted to Guadalupe Island, Baja California. The Chilean population may be extinct.

Arctocephalus philippii townsendi Merriam

1897. *Arctocephalus townsendi* Merriam, Proc. Biol. Soc. Washington, 11:178, July 1, type locality, Isla de Guadalupe, Baja California.
1958. *Arctocephalus philippii townsendi*, Scheffer, Seals, sea lions, and walruses . . . , Stanford Univ. Press, p. 80, April 24.

RANGE.—Formerly from California: Farallone Islands, thence southward to Baja California: Guadalupe Island (Rice, *et al.*, 1965:85); San Benito Islands (King, 1954:327, 328); Cedros Island (Rice, *et al.*, 1965:83).

Genus **Eumetopias** Gill
Northern Sea Lion or Steller's Sea Lion

1866. *Eumetopias* Gill, Comm. Essex Inst., 5:7, April 7. Type, *Arctocephalus monteriensis* Gray [= *Phoca jubata* Schreber].

Total length of males, 10–13 ft., weight to 2000 pounds; total length of females, 7–9 ft., weight to 600 pounds. Pelage harsh, no underfur; color buff or yellowish-tan, lighter immediately after molt; naked parts of skin black; males with mane on neck. Condylobasal length, (males) 374–400, (females) 294–298; zygomatic breadth, (males) 220–260, (females) 145–173; interorbital constriction, (males) 76–101.6 (21–27% of length of skull); diastema between P4 and M1, (males) 21–28 (5–8% of total length of skull). Skull large; palate long, contracted and truncate posteriorly; postorbital processes quadrate, placed anteriorly. Dentition, i. $\frac{3}{2}$, c. $\frac{1}{1}$, p. $\frac{4}{4}$, m. $\frac{1}{1}$.

This genus is monotypic.

Eumetopias jubatus (Schreber)
Northern Sea Lion or Steller's Sea Lion

1776. *Phoca jubata* Schreber, Die Säugthiere . . . , theil 3, heft 17, Pl. 83b and p. 300 (vernacular name and description). Type locality, North Pacific Ocean.
1902. *Eumetopias jubata*, J. A. Allen, Bull. Amer. Mus. Nat. Hist., 16:113, March 15.
1828. *Otaria stellerii* Lesson, Dict. Class. Hist. Nat., 13:420, a renaming of *Phoca jubata* Schreber.
1859. *Arctocephalus monteriensis* Gray, Proc. Zool. Soc. London, p. 358, 360, pl. 72. Type locality, Monterey, California. (Based on a skull of *Eumetopias* and a skin of *Callorhinus*.)

See characters of the genus.

RANGE.—Alaska: Bering Strait, thence southward to California: San Nicolas Island.

Genus Zalophus Gill—California Sea Lion

1866. *Zalophus* Gill, Comm. Essex Inst., 5:7, April 7. Type, *Otaria gillespii* MacBain [= *Otaria californiana* Lesson].

Total length of males to 8 ft., weight to 600 pounds; total length of females to 6 ft., weight to 200 pounds. Color blackish when wet, varying from light buff to deep sepia when dry. Muzzle broader and heavier than in *Callorhinus*; males with conspicuous crest on top of head; skull more slender and elongate than in other American otariids (except *Arctocephalus*). Condylobasal length, (males) 248–330, (females) 195–251; zygomatic breadth, (males) 145–180, (females) 78–128; length of series of upper cheek-teeth, (males) 53–58, (females) 46–52. Sagittal crest of skull high, rising steeply from frontals in supraorbital region, reaching greatest height at level of postorbital processes of jugal bone; palate relatively short (37–47% of condylobasal length); anterior nares narrow; tympanic bullae irregular in shape; zygomatic roots of maxillae narrow (5–7% of condylobasal length). Dentition, i. $\frac{3}{2}$, c. $\frac{1}{1}$, p. $\frac{4}{4}$, m. $\frac{2-1}{1}$.

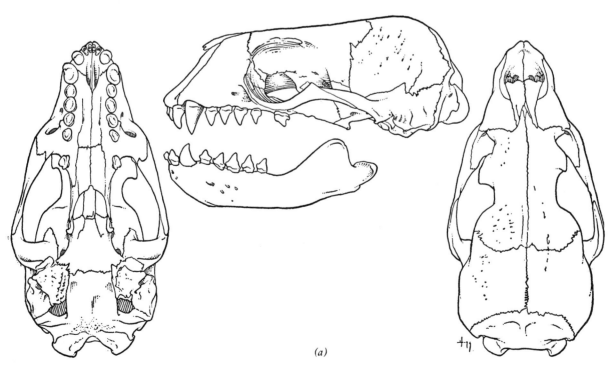

(a)

Fig. 585a. *Eumetopias jubatus*, Bolinas Bay, Marin Co., California, No. 48100 K.U., immature ♀, X $\frac{1}{4}$.

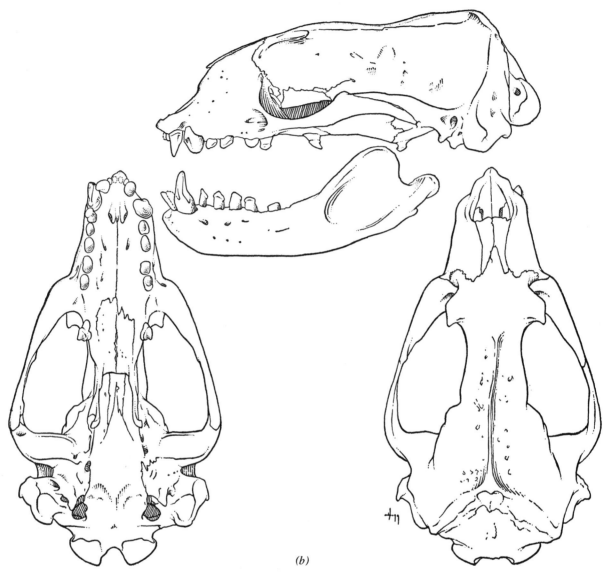

Fig. 585b. *Eumetopias jubatus*, California, No. 2762 K.U., adult ♀, X ⅓ [incorrectly identified as *Zalophus californianus* in Hall and Kelson, 1959: Fig. 515 on p. 977].

Zalophus californianus
California Sea Lion

See characters of genus. Average number of cheek-teeth varies geographically: 5 or 6. Ratio of breadth of braincase (above jugals) to breadth of postorbital processes varies from 73–88 per cent in the Galapagos to 93–123 per cent in North America. Considered polytypic by Scheffer (1958:60).

Zalophus californianus californianus (Lesson)

1828. *Otaria californiana* Lesson, Dict. Class. Hist. Nat.,

13:420. Type locality, vic. San Francisco Bay, California.
1880. *Zalophus californianus*, J. A. Allen, U.S. Geol. and Geog. Surv. Territories, Miscl. Publ., 12:276.
1858. *Otaria gillespii* MacBain, Proc. Royal Soc. Edinburgh, 1:422. Type locality, California.

RANGE.—British Columbia: Effingham Inlet (Cowan and Guiguet, 1965:351), thence southward along Pacific Coast and on adjacent islands, including Gulf of California (Turner Island, No. 95607 KU) possibly to Nayarit: Tres Marías Islands (probably according to Nelson, 1899:17, but improbably according to Van Gelder, 1960:23).

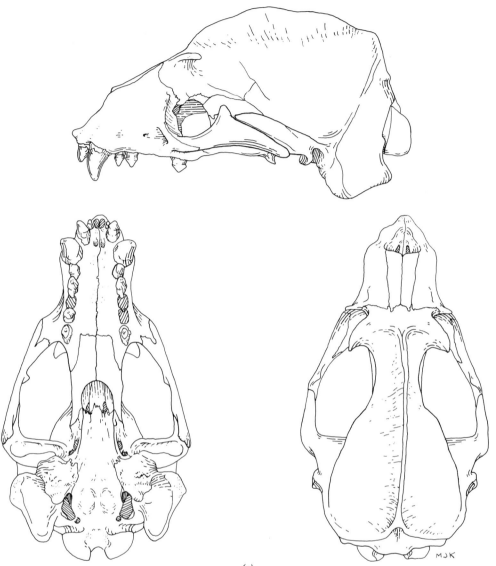

Fig. 585c. *Zalophus californianus californianus,* Turner Island, Gulf of California, Sonora, No. 95607 K.U., ♂, X ⅓.

SUBFAMILY **ROSMARINAE**—Walrus

The review of fossils by Mitchell, Jour. Fisheries Research Board of Canada, 25(9):1843–1900, 16 figs., 5 tables, September 1968, is basis for according subfamily instead of family rank to walruses.

Body thick and heavy; head relatively short; muzzle blunt, broad, with coarse bristles; pinnae absent; anterior portion of skull swollen; canines forming long tusks; postorbital processes lacking; surface of scapula long, narrow; mandibular symphysis fused in adults; crista tympani lacking overhang of free frontal edge; pars petrosa, dorsal to meatus acousticus internus, and (according to Mitchell, 1968:1885) flattened and slightly striated; testes abdominal.

Genus **Rosmarus** Brunnich—Walrus

1762. *Odobenus* Brisson, Regnum animale . . . , ed. 2, p. 30. Type, *Odobenus* Brisson [= *Phoca rosmarus* Linnaeus].

Brisson's names are non-Linnaean, not consistently binomial (see: Hopwood, Proc. Zool. Soc. London, 117:534, 535, October 30, 1947; Hershkovitz, Proc. Biol. Soc. Washington, 62:12, March 17, 1949; and International Comm. on Zool. Nomenclature Opinion 467-16 (pt. 16):73–88, May 31, 1957), and therefore are not to be used.

1766. *Trichechus* Linnaeus, Syst. nat., ed. 12, 1:49. Not *Trichechus* Linnaeus, 1758, ed. 10, p 34.

1772. *Rosmarus* Brunnich, Zoologiae fundamenta . . . , p. 38. Type, "Zavhaften" [= *Phoca rosmarus* Linnaeus].

Total length of males to 12 ft., weight to 3000 pounds; females approx. one-third smaller than males. Hair short, becoming sparse with age; skin thick, wrinkled, yellowish-brown, becoming reddish-brown or chestnut on belly and at base of appendages. Condylobasal length, (males) —387, (females) 322–373; zygomatic breadth, (males) 220–250, (females) 184–204; length of series of upper cheek-teeth, (males) 87–98, (females) 63–65. See characters of the family. Functional dentition, i. $\frac{1}{0}$, c. $\frac{1}{1}$, p. $\frac{3}{3}$, m. $\frac{0}{0}$.

Rosmarus rosmarus
Walrus

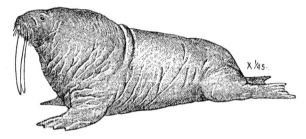

See characters of the genus and family.

Rosmarus rosmarus divergens (Illiger)

1815. *Trichechus obesus* Illiger, Abh. preuss. Akad. Wiss., Berlin, 1804–1811, p. 64, a *nomen nudun*. Cape, Alaska.
1932. *Odobenus r[osmarus]. divergens*, Pohle, Fauna arctica, 6(lief.2):78, March 15.
1815. *Trichechus obesus* Illiger, Abh. preuss. Akad. Wiss., Berlin, 1804–1811, p. 64, a *nomen nudum*.
1831. *Trichechus cookii* Fremery, Bijdrag. Nat. Vetensk., 6:385. Type locality, off Icy Cape, Alaska (lat. 70° 06′ N, long. 113° 18′ W).
1922. *Trichechus orientalis* Dybowski, Arch. Tow. Nauk. Lwow, 1:351, a *nomen nudum*.

RANGE.—Bering Sea, S along Alaska: Pribilof Islands and Bristol Bay. East in Canada (Harrington, 1966:508–512): to SW coasts of Banks and Victoria islands and mouth Bathurst Inlet.

Rosmarus rosmarus rosmarus (Linnaeus)

1758. *[Phoca] rosmarus* Linnaeus, Syst. nat., ed. 10, 1:38. Type locality, Arctic regions.

RANGE.—Atlantic and Arctic coasts south to Massachusetts: Plymouth (Manville and Favour, 1960:501); west (Harrington, 1966:508–512) in Canada: to Mould Bay of Prince Patrick Island, "just east of" Cape George Richards of Melville Island, and Spence Bay of mainland; east (G. M. Allen, 1930:139) to Greenland and Eurasia.

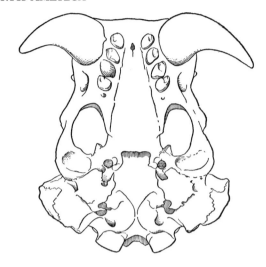

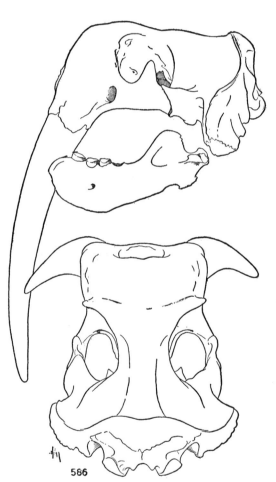

Fig. 586. *Rosmarus rosmarus rosmarus*, Ingelfield Gulf, North Greenland, No. 2654 K.U. ♂, X $\frac{1}{8}$.

top—ventral view
center—lateral view
bottom—dorsal view

FAMILY PHOCIDAE—Earless Seals

Hind limbs not capable of rotating forward; manus and pes hairy; nails frequently well developed; front limbs noticeably smaller than hind limbs; neck short; pinnae absent; postorbital processes rudimentary or absent; mastoid processes swollen, continuous with auditory bullae; alisphenoid canal absent; scapula small, trochanter minor of femur rudimentary; ilium short, broad, abruptly turned outward and recurved anteriorly; acetabulum opposite first sacral vertebra; testes abdominal.

KEY TO NORTH AMERICAN GENERA OF PHOCIDAE

1. Single incisor in each lower jaw; premaxillae not meeting nasals; males having inflatable pouch of skin on nose or having elongate, inflatable tubular proboscis.
    2. Auditory bullae concave on anterior border; males having elongate inflatable tubular proboscis; nails small, rudimentary on pes. . . . .*Mirounga,* p. 1069
    2'. Auditory bullae straight on anterior border; males having inflatable pouch of skin on nose; nails well developed on all digits. . . . .*Cystophora,* p. 1068
1'. Two incisors in each lower jaw; premaxillae meeting nasals; males lacking inflatable pouch of skin on nose or elongate, inflatable tubular proboscis.
    3. Incisors ⅖; nails rudimentary, frequently absent on pes. . . . .*Monachus,* p. 1071
    3'. Incisors ⅗; all digits having well-developed nails.
        4. Snout long (distance between tip of nose and eye nearly twice distance between eye and auditory meatus); nasals, frontals and parietals forming straight line in lateral view.
        Halichoerus, p. 1067
        4'. Snout short (distance between tip of nose and eye much less than twice distance between eye and auditory meatus); nasals, frontals and parietals forming convex outline in lateral view.
            5. Third digit of manus longer than 1st two; jugal bone short, deep (depth of jugal not less than ½ its length); mammae, 4. . . . .*Erignathus,* p. 1066
            5'. First and 2nd digits of manus longer than 3rd; jugal bone long, narrow (depth of jugal less than ½ its length); mammae, 2.
                6. Posterior margin of palate broadly U-shaped; bony nasal septum reaching, or nearly reaching, posterior margin of palate.
                    7. Cheek-teeth usually lacking accessory cusps; yellowish bands around neck, forelimbs and rump. . . . .*Histriophoca,* p. 1064
                    7'. Cheek-teeth having accessory cusps, yellowish bands absent. . . . *Pagophilus,* p. 1065
                6'. Posterior margin of palate V-shaped; bony nasal septum not reaching posterior margin of palate.
                    8. Cheek-teeth large, length of P2 6.8 or more; pelage usually having dark spots on pale background. . . . .*Phoca,* p. 1063
                    8'. Cheek-teeth small, length of P2 less than 6.8; pelage usually having whitish spots with dark centers. . . . .*Pusa,* p. 1065

SUBFAMILY PHOCINAE

Incisors ⅗; posterior 4 cheek-teeth double-rooted; digits with 5 well-developed nails; toes of hind feet subequal, 1st and 5th not greatly exceeding others in length; premaxillae prolonged posterodorsally, meeting nasals.

Genus Phoca Linnaeus—Harbor Seals

1758. *Phoca* Linnaeus, Syst. nat., ed. 10, 1:37. Type, *Phoca vitulina* Linnaeus.

Males 5–6 ft. long; P4 6.8 long or longer; for other details see key to genera and account of *P. vitulina.*

Phoca vitulina
Harbor Seal

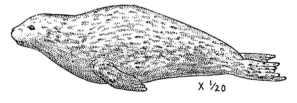

X ½₀

Total length of males, 5–6 ft., females smaller; weight to approx. 300 pounds. Color highly variable, usually yellowish-gray dorsally varied with irregular spots of dark brown or black, yellowish-white ventrally, usually with small dark brown spots; sometimes uniform brownish-yellow or grayish dorsally, paler ventrally, lacking spots (after J. A. Allen, 1880:562–565). Total length of skull, 157.3–221.5; mastoid breadth, 100.0–128.0; least interorbital breadth, 9.5–16.0; nasals broad, tapering gradually to level of maxillofrontal suture and then tapering abruptly; posterior margin of palate V-shaped; bony nasal septum reaching posterior margin of palate; cheek-teeth large, frequently set obliquely in jaw, 3rd upper cheek-tooth usually the largest; posterior palatine foramina situated anterior to maxillopalatine suture; mandible heavy and deep; manidbular symphysis short. Dentition, i. $\frac{3}{2}$, c. $\frac{1}{1}$, p. $\frac{4}{4}$, m. $\frac{1}{1}$.

See addenda.

Phoca vitulina concolor De Day

1842. *Phoca concolor* De Kay, Zoology of New-York . . . , 1(pt. 1, Mammalia):53, type from Long Island Sound, near Sands Point, Nassau Co., New York.
1904. [*Phoca (Phoca) vitulina*] *concolor*, Trouessart, Catalogus mammalium . . . , Suppl., p. 286.

RANGE.—Atlantic Coast from Franklin: Ellesmere Island, thence southward to South Carolina (Caldwell, 1961:425) and Florida: Ponce de Leon Inlet (Caldwell and Caldwell, 1969:379); rarely westward to Lake Ontario. Also Hudson Bay (Edwards, 1963:6, recorded only as *Phoca vitulina*).

Phoca vitulina mellonae Doutt

1942. *Phoca vitulina mellonae* Doutt, Ann. Carnegie Mus., 29:111, May 12, type from Lower Seal Lake, Quebec, approx. 90 mi. E Richmond Gulf, Hudson Bay, lat. 56° 30′ N, long. 74° 30′ W.

RANGE.—Quebec: type locality. See Harper (1961:127, Map 42) for several other places in the interior of the Ungava Peninsula at which *P. v. mellonae* and other land-locked populations of *P. vitulina* occur.

Phoca vitulina richardii (Gray)

1864. *Halicyon richardii* Gray, Proc. Zool. Soc. London, p. 28, July, type from Vancouver Island, British Columbia. (Ognev, The mammals of U.S.S.R. . . . , 3:524, 1935, regarded *richardii* as subspecifically inseparable from the earlier named *Phoca largha* of Pallas, Zoographia Rosso-Asiatica . . . , 1:113, 1811. Scheffer, Seals, sea lions and walruses . . . , 1958:93, Stanford Univ. Press, regarded *P. v. largha* as a subspecies occurring in the western part of the North Pacific. McLaren, Jour. Mamm., 47:466–471, August 29, 1966, regarded *P. largha* as specifically different from *P. vitulina richardii*.)
1942. *Phoca vitulina richardii*, Doutt, Ann. Carnegie Mus., 29:112, May 12.
1902. *Phoca richardii pribilofensis* J. A. Allen, Bull. Amer. Mus. Nat. Hist., 16:495, December 12, type from St. Paul Island, Pribilof Islands, Alaska. (Regarded by Schwarz,

Jour. Mamm., 23:222, May 14, 1942, as identical with the *Phoca largha* of Pallas, Zoographia Rosso-Asiatica . . . , 1:113, 1811.)
1902. *Phoca richardii geronimensis* J. A. Allen, Bull. Amer. Mus. Nat. Hist., 16:495, December 12, type from San Gerónimo Island, Baja California.

RANGE.—Arctic and Pacific coasts from Yukon: Herschel Island, thence westward and southward to Pribilof Islands (Kenyon, 1960:520), and along coast to northern Baja California, 28° N (Bartholomew and Boolootian, 1960:370).

Genus **Histriophoca** Gill—Ribbon Seal

1873. *Histriophoca* Gill, Amer. Nat., 7:179, March. Type, *Phoca fasciata* Zimmermann.

Males up to 6 ft.; cheek-teeth usually lacking accessory cusps; for other details see account of *H. fasciata*.

Histriophoca fasciata (Zimmermann)
Ribbon Seal

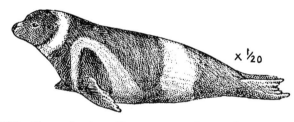

1783. *Phoca fasciata* Zimmermann, Geographische Geschichte . . . , 3:277. Type locality, Kurile Islands, Japan.
1932. *Histriophoca fasciata*, Pohle, Die Säugtiere. . . . *In* Fauna arctica . . . , Jena, G. Fischer, vol. 6:78.
1811. *Phoca equestris* Pallas, Zoographia Rosso-Asiatica . . . , 1:111, a renaming of *P. fasciata*.

Total length of males to 6 ft., females slightly smaller; weight of an adult male recorded as 209 pounds. Males dark brown with bands of yellowish encircling neck, rump and forelimbs; females grayish-yellow or grayish-brown with faint indication of yellowish bands, the one encircling rump most distinct. Total length of skull, 193.0–201.3; mastoid breadth, 122.0–133.4; least interorbital breadth, 9.8–14.6. Nasals narrow, tapering gradually from anterior to posterior; posterior margin of palate broadly U-shaped; auditory bullae large; cheek-teeth small, usually lacking accessory cusps; posterior palatine foramina situated on, or posterior to, maxillopalatine suture; mandible slender; mandibular symphysis long; masseteric fossa broad and deep; coronoid process large. Dentition as in *Phoca*.

RANGE.—Alaska: Point Barrow; edge of drift ice in Bering Sea (Roest, 1964:416). One extralimital occurrence in 1962 in California: Spooner's Cove, near mouth Islay Creek, 6½ mi. S Morrow Bay (*ibid.*).

Genus **Pusa** Scopoli—Ringed Seals

1777. *Pusa* Scopoli, Introductio ad historiam naturalem . . . , p. 490. Type, *Phoca foetida* Fabricius [= *Phoca hispida* Schreber].

1864. *Pagomys* Gray, Proc. Zool. Soc. London, p. 31, July. Type, *Phoca foetida* Fabricius [= *Phoca hispida* Schreber].

Males up to 4½ ft.; cheek-teeth small (P2 less than 6.8); for other details see key to genera and account of *P. hispida*.

Pusa hispida
Ringed Seal

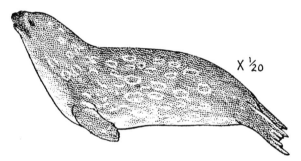

X ¹⁄₂₀

Total length to approx. 4½ ft.; weight to approx. 200 pounds. Color highly variable, resembling *P. vitulina;* generally white, silvery-white or yellowish-white ventrally; spotted, streaked or marbled with black dorsally, often having white spots with dark centers; hairs around lips, and often on head, axillae, ventral surface of tail and hind limbs, tawny or russet; pelage coarse and stiff, usually directed backward. Total length of skull, 155.5–177.0; mastoid breadth, 90.4–105.8; least interorbital breadth, 4.0–6.2. Nasals tapering as in *P. vitulina;* posterior margin of palate V-shaped; bony nasal septum not reaching posterior margin of palate; cheek-teeth small, 3rd the largest both above and below; posterior palatine foramina situated on, or posterior to, maxillopalatine suture; mandible slender; mandibular symphysis long; masseteric fossa narrow; coronoid process small.

Dentition as in *Phoca*.

Pusa hispida hispida (Schreber)

1775. *Phoca hispida* Schreber, Die Säugthiere . . . , theil 3, heft 13, Pl. 86. Type locality, coasts of Greenland and Labrador.

1958. *Pusa hispida hispida,* Scheffer, Seals, sea lions and walruses . . . , Stanford Univ. Press, p. 97, April 24.

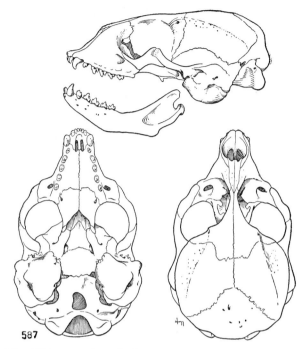

587

Fig. 587. *Pusa hispida hispida,* Holstenberg, Greenland, No. 2827 K.U., sex ?, X ⅓.

1776. *Phoca foetida* Fabricius, *in* Müller, Zoologiae Danicae prodromus . . . , p. viii. Type locality, Greenland (see Fabricius, Fauna Groenlandica, p. 13, 1780).

1820. *Phoca annellata* Nilsson, Skand. fauna . . . , 1:365, a renaming of *Phoca foetida* Fabricius.

1921. *Pusa hispida pygmaea* Zukowsky, Wiegmann's Arch. für Naturgesch., 10(87A):183. Type locality, Greenland or Novaya Zemlya, Russia.

1943. *Phoca hispida beaufortiana* Anderson, Ann. Rept. Provancher Soc. Nat. Hist. Canada, Quebec, p. 25 (for 1942), September 7, type from Cockburn Point, Dolphin and Union Strait, Mackenzie; lat. 68° 55′ 29″ N, long. *ca.* 115° 10′ W.

1943. *Phoca hispida soperi* Anderson, Ann. Rept. Provancher Soc. Nat. Hist. Canada, Quebec, p. 27 (for 1942), September 7, type from near mouth Takuirbing River, eastern end Nettilling Lake, *ca.* 85 ft., Baffin Island, Franklin; lat. 66° 16′ N, long. 74° 33′ 36″ W.

RANGE (Scheffer, 1958:98).—"Arctic Ocean, occasionally to the Pole in open leads, and on sea ice along northern Eurasia, Greenland, and North America; southward to . . . Labrador and Hudson Bay or perhaps James Bay."

Genus **Pagophilus** Gray—Harp Seal

1844. *Pagophilus* Gray, *in* The zoology of the voyage of H. M. S. Erebus and Terror . . . , 1(Mamm.):3. Type, *Phoca groenlandica* Erxleben.

1904. *Pagophoca* Trouessart, Catalogus mammalium . . . , Suppl., p. 287, a renaming of *Pagophilus* Gray.

Males up to 6 ft.; cheek-teeth small, having accessory cusps; for other details see key and account of *P. groenlandicus*.

Pagophilus groenlandicus groenlandicus (Erxleben)—Harp Seal

1777. [*Phoca*] *groenlandica* Erxleben, Systema regni animalis . . . , 1:588. Type locality, Greenland or Newfoundland.
1850. *Pagophilus Groenlandicus*, Gray, Catalogue of the . . . Mammalia in the . . . British Museum, part 2 seals:25.
1785. *Phoca semilunaris* Boddaert, Elenchus animalium, p. 170. Type locality, Greenland or Iceland.
1822. *Phoca albicauda* Desmarest, Mammalogie . . . , *in* Encyclopédie méthodique . . . , p. 541, a *nomen dubium*. Type locality, unknown.
1824. *Phoca leucopla* Thienemann, Nat. Bemerk. Reise Europe, 1:102, Pl. 13. Type locality, "A few miles north of Grimsey Island, north of Iceland." (Ellerman and Morrison-Scott, Checklist of Palaearctic and Indian mammals, 1758 to 1946. British Mus., London, p. 331, November 19, 1951.)
1851. *Phoca albini* Alessandrini, Mem. R. R. Accad. Bologna, 2:158. Type locality, unknown.

Total length of males, 5–6 ft., females usually smaller; weight to approx. 400 pounds, although some old males have been recorded as reaching 800 pounds. Males whitish to yellowish-white dorsally, face and head blackish; black elliptical band on side from shoulder nearly to tail, bands from each side meeting broadly at shoulder, barely meeting near tail; hind limbs generally irregularly spotted with black. Females similar in color to males but black markings absent, or if present irregular and indistinct. Total length of skull, 190.9–221.0; mastoid breadth, 114.6–123.6; least interorbital breadth, 8.3–20.0. Nasals long, narrow, tapering gradually from anterior to posterior; posterior margin of palate broadly U-shaped; cheek-teeth small, having accessory cusps; 3rd cheek-tooth the largest both above and below; posterior palatine foramina on, or posterior to, maxillopalatine suture; mandible as in *H. fasciata*. Dentition as in *Phoca*.

RANGE.—North Atlantic Ocean and adjoining waters of Arctic Ocean, west to Mackenzie: Aklavik, Mackenzie River Delta, and south occasionally as far as Virginia: near Cape Henry at Little Creek (Goodwin, 1954:258).

Genus Erignathus Gill—Bearded Seal

1866. *Erignathus* Gill, Comm. Essex Inst., 5:5, 9, April 7. Type, *Phoca barbata* Erxleben.

Total length of males, 8–12 ft., weight to 1000 pounds; total length of females, 7½–8 ft., weight to approx. 500 pounds. Body grayish to yellowish, darker grayish on back; muzzle broad, having vibrissae dense and straight; forehead high, convex; 3rd digit of manus longest. Condylobasal length, (males) 227–257, (females) 208–238; zygomatic breadth, (males) 134–138, (females) 128–137; length of upper series of cheek-teeth, (males) 83–85, (females) 72–81. Skull having small postorbital processes; palatal area broad, oval; posterior border of palate emarginate, not triangular; jugal short, deep (depth of jugal not less than half its length); incomplete narial septum; posterior margin of mandibular ramus forming a straight line between angular process and articular process; scapula lacking acromion process; upper border of ilium lacking abrupt eversion; mammae, 4. Dentition, i. $\frac{3}{2}$, c. $\frac{1}{1}$, p. $\frac{4}{4}$, m. $\frac{1}{1}$.

The genus is monotypic.

Erignathus barbatus
Bearded Seal

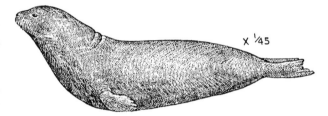

See characters of the genus.

Erignathus barbatus barbatus (Erxleben)

1777. [*Phoca*] *barbata* Erxleben, Systema regni animalis . . . , 1:590. Type locality, coasts of Scotland, or southern Greenland or Iceland.
1866. *Erignathus barbatus*, Gill, Comm. Essex Inst., 5(1):12, April 7.
1776. *Phoco barbata* Fabricius, *in* Müller, Zoologiae Danicae prodromus . . . , p. viii, a *nomen nudum*.
1778. *Phoca leporina* Lepechin, Acta. Ac. Petrop., 1:264, Pl. 8. Type locality, White Sea.
1828. *Phoca parsonii* Lesson, Dict. Class. Hist. Nat., 13:414. Type locality, Northern seas.
1828. *Phoca lepechenii* Lesson, Dict. Class. Hist. Nat., 13:415, a renaming of *Phoca leporina* Lepechin.

RANGE.—Canada: Hudson Bay, east to Greenland: Davis Strait, thence southward to Newfoundland.

Erignathus barbatus nauticus (Pallas)

1811. *Phoca nautica* Pallas, Zoographia Rosso-Asiatica . . . , 1:108. Type locality, Okhotsk Sea. (North American specimens regarded by Anderson, Canadian Field-Nat., 44:99, April 1, 1930, as identical with *E. b. barbatus*.)

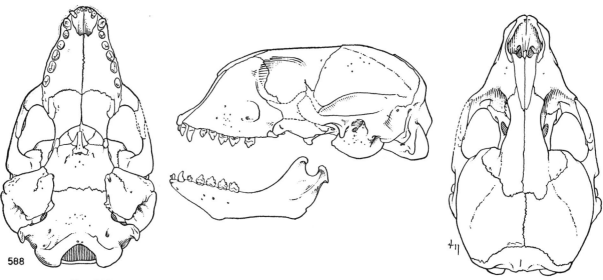

Fig. 588. *Erignathus barbatus barbatus*, Holstenberg, Greenland, No. 2881 K.U., sex ?, X ⅓.

1904. *Erignathus barbatus nauticus*, Osgood, N. Amer. Fauna, 24:47, November 23.
1811. *Phoca albigena* Pallas, Zoographia Rosso-Asiatica . . . , 1:109. Type locality, Kamchatka.

RANGE.—Franklin: "All coasts of Banks Island" (Manning and Macpherson, 1958:64). Yukon: Pauline Cove, Herschel Island (Youngman, 1975:158, as *E. barbatus* only because he regards *E. barbatus* as a monotypic species). Alaska: Point Barrow, thence southward along coast of Alaska to Bristol Bay.

Genus **Halichoerus** Nilsson—Gray Seal

1820. *Halichoerus* Nilsson, Skand. fauna . . . , 1:376. Type, *Halichoerus griseus* Nilsson [= *Phoca grypus* Fabricius].

Total length of males to 12 ft., weight to 800 pounds; total length of females, 5½–6½ ft., weight to 700 pounds. Color of body varying from silverish to dusky-gray, obscure black spots dorsally and laterally. Snout long (distance between tip of nose to eyes nearly twice that between eye and auditory meatus); muzzle broad, vibrissae crenulated. Total length of skull, 208–320; zygomatic breadth, 120–198; length of upper series of cheek-teeth, 73. Skull arched dorsally, profile of parietals, frontals, and nasals form straight line; cheek-teeth single-rooted, excepting 2 posteriormost teeth in upper jaw and single posteriormost tooth in lower jaw. Dentition, i. ³⁄₂, c. ¹⁄₁, p. ⁴⁄₄, m. ¹⁄₁.

The genus is monotypic.

Halichoerus grypus (Fabricius)
Gray Seal

1791. *Phoca grypus* Fabricius, Skrivter af Naturh.-Selskabet, Kjøbenhavn, 1(2):167, Pl. 13, Fig. 4. Type locality, Greenland.
1841. *Halichoerus grypus*, Nilsson, Wiegmann's Arch. für Naturgesch., Jahrg. 7, 1:318.
1820. *Halichoerus griseus* Nilsson, Skand. fauna . . . , 1:377. Type locality, Greenland.
1824. *Phoca halichoerus* Thienemann, Nat. Bermerk. Reise Europe, 1:142. Type locality, Norway.
1851. *Halichoerus macrorhynchus* Hornschuch and Schilling, Wiegmann's Arch. für Naturgesch., 17, 2:28. Type locality, Baltic Sea.
1851. *Halichoerus pachyrhynchus* Hornschuch and Schilling, Wiegmann's Arch. für Naturgesch., 17, 2:28. Type locality, Baltic Sea.
1886. *Halichoerus grypus* var. *atlantica* Nehring, Sitzungsb. Gesell. Naturforsch. Freunde, Berlin, p. 122. Type locality, west coast of Norway.
1886. *Halichoerus grypus* var. *baltica* Nehring, Sitzungsb. Gesell. Naturforsch. Freunde, Berlin, p. 122. Type locality, Baltic Sea.

See characters of the genus.

RANGE.—Greenland to Labrador, thence southward, including breeding in 1963 on Muskeget Island, Massachusetts (Standard Times, New Bedford, Massachusetts, September 1, 1963—p. 13 of 2nd sec.), to New Jersey.

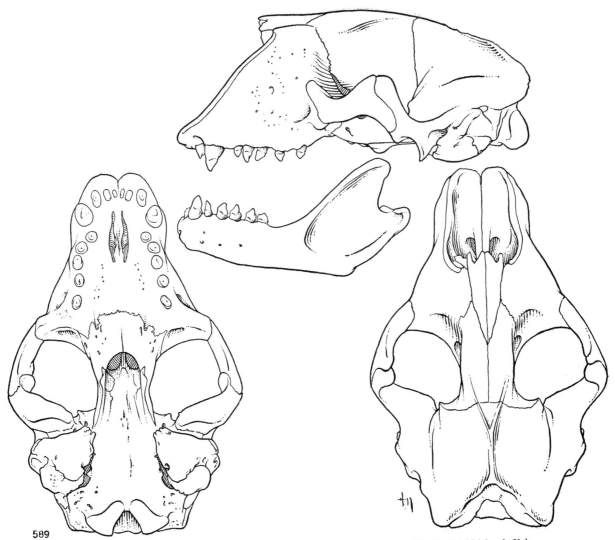

589

Fig. 589. *Halichoerus grypus*, SW Point Anticosti, Quebec, No. 35291 U.S.N.M., ♂, X ⅓.

SUBFAMILY CYSTOPHORINAE

Incisors ²⁄₁; cheek-teeth single-rooted, except 5th upper and occasionally 4th upper, which are double-rooted; males having an inflatable pouch of skin on nose or having an elongate, inflatable tubular proboscis; toes of hind feet subequal or having 1st and 5th toes much longer than others; nails well developed (*Cystophora*), or rudimentary (*Mirounga*); premaxillae not meeting nasals.

Genus **Cystophora** Nilsson—Hooded Seal

1820. *Cystophora* Nilsson, Skand. fauna . . . , 1:382. Type, *Cystophora borealis* Nilsson [= *Phoca cristata* Erxleben].
1826. *Stemmatopus* F. Cuvier, Dict. Sci. Nat., 39:550. Type, *Stemmatopus cristatus* Cuvier [= *Phoca cristata* Erxleben].

1911. *Cystophoca* Brass, Aus dem Reiche der Pelze, p. 668, April, a renaming, perhaps accidental, of *Cystophora* Nilsson, 1820.

Total length of males to 10½ ft., weight to 850 pounds; total length of females to 8 ft., weight to 400 pounds. Color slate grayish to blackish above, sides paler spotted with white. Males have bladderlike protuberance on nose, inflatable in times of danger, or anger; hind feet slightly emarginate on distal borders having distalmost digit of pes slightly longer than medial digits; digits of forefeet and hind feet with large claws. Condylobasal length, (males) 264–275, (female) 220; zygomatic breadth, (males) 210–220, (female) 165; length of series of upper cheek-teeth, (males) 90–91, (female) 74. Skull having pala-

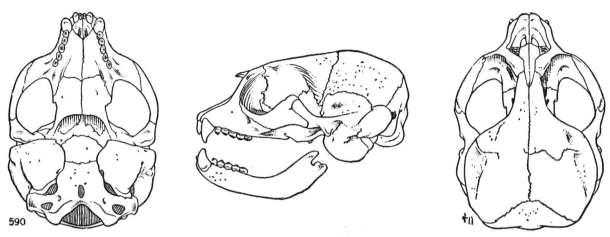

Fig. 590. *Cystophora cristata*, origin unknown [obtained from Swope Park Zoo, Kansas City, Missouri], No. 11094 K.U., sex ?, X ⅓.

tines slightly emarginate on posterior border; auditory bullae square in front; basisphenoid and basioccipital bones form broad, flat, not concave, interbullar space; superior notch of mandible deeply U-shaped. Dentition, i. $\frac{2}{1}$, c. $\frac{1}{1}$, p. $\frac{4}{4}$, m. $\frac{1}{1}$.

This genus is monotypic.

Cystophora cristata (Erxleben)
Hooded Seal

1777. [*Phoca*] *cristata* Erxleben, Systema regni animalis . . . , 1:590. Type locality, Southern Greenland or Newfoundland.
1841. *Cystophora cristata*, Nilsson, Wiegmann's Arch. für Naturgesch., Jahrg. 7, 1:326.
1820. *Cystophora borealis* Nilsson, Skand. fauna . . . , 1:383. Type locality, Southern Greenland or Newfoundland.

See characters of the genus.

RANGE.—Franklin: Cape Sabine, Ellesmere Island, thence southward to Florida: Cape Canaveral, Brevard Co.

Genus **Mirounga** Gray—Elephant Seals

1827. *Mirounga* Gray, *in* Griffith, The animal kingdom . . . by the Baron Cuvier . . . , 5:179. Type, *Phoca proboscidea* Péron [= *Phoca leonina* Linnaeus].
1826. *Macrorhinus* F. Cuvier, Dict. Sci. Nat., 39:552. Type, *Phoca proboscidea* Péron [= *Phoca leonina* Linnaeus]. Not *Macrorhinus* Latreille, 1825, a coleopteran.
1830. *Rhinophoca* Wagler, Natürliches System der Amphibien . . . , p. 27. Type, *Phoca proboscidea* Péron [= *Phoca leonina* Linnaeus].
1841. *Physorhinus* Gloger, Gemeinnütziges Hand- und Hilfs-buch Naturgeschichte, 1:xxxiv, 163. Type, the Mirounga [= *Phoca leonina* Linnaeus] of the southern seas. Not *Physorhinus* Eschscholtz, 1836, a genus of Coleoptera.
1843. *Morunga* Gray, List of the . . . Mammalia in the . . .

British Museum, p. xxiii, 103, May 13, an accidental renaming of *Mirounga* Gray, 1827.

Total length of males, 15–16 ft., weight to 5000 pounds; total length of females to 11 ft., weight to approx. 1700 pounds. Color of body dull yellowish-brown mixed with gray, darker on back, yellowish on venter; immature individuals grayish; males have a proboscislike inflatable snout; hind feet bilobed; claws, if present, rudimentary. Condylobasal length, (males) 480–510, (female) 320; zygomatic breadth, (males) 360–365, (female) 200; length of series of upper cheek-teeth, (males) 130–140, (female) 98. Palatine bones deeply emarginate on posterior border; auditory bullae concave anteriorly; interbullar space narrow and deeply hollowed; superior notch of mandible not deeply emarginate. Dentition, i. $\frac{2}{1}$, c. $\frac{1}{1}$, p. $\frac{4}{4}$, m. $\frac{1}{1}$.

Mirounga angustirostris (Gill)
Northern Elephant Seal

1070   MAMMALS OF NORTH AMERICA

1866. *Macrorhinus angustirostris* Gill, Comm. Essex Inst., 5(1):13, April 7, type from San Bartolomé Bay, Baja California.
1904. [*Mirounga*] *angustirostris*, Elliot, Field Columb. Mus., Publ. 95, Zool. Ser., 4:545, August 2.

See characters of the genus. The only other species, *M. leonina*, occurs in the Southern Hemisphere.

RANGE.—Alaska: Kasaan, Prince of Wales Island, thence southward to Baja California (Rice, *et al.*, 1965:33): Guadalupe Island; San Benito Islands; Cedros Island.

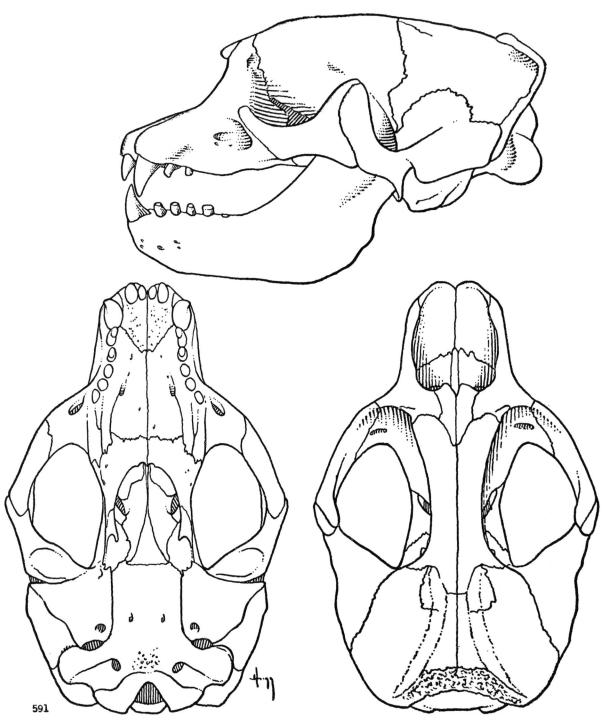

591

Fig. 591. *Mirounga angustirostris*, St. Bartholomews Bay, Baja California, No. 4704 U.S.N.M., ♀, (after J. A. Allen, Miscl Publ. No. 12, U.S. Geol. and Geog. Surv. Territories, Figs., 57–59, pp. 744–746, 1880), X ⅓.

SUBFAMILY MONACHINAE

Revised by King, Bull. British Mus. (Nat. Hist.), Zool., 3:201–256, 6 pls., January, 1956.

Incisors ²⁄₂; posterior 4 cheek-teeth double-rooted; 1st and 5th toes of hind feet greatly exceeding others in length; nails rudimentary or absent; premaxillae prolonged posterodorsally, meeting nasals.

Genus **Monachus** Fleming—Tropical Seals

1822. *Monachus* Fleming, The philosophy of zoology . . . , 2:187. Type, *Phoca monachus* Hermann.
1824. *Pelagios* F. Cuvier, Mém. Mus. Hist. Nat., 11:196, Pl. 2. Type, *Phoca monachus* Hermann.
1826. *Pelagius* F. Cuvier, Dict. Sci. Nat., 39:550, an emendation of *Pelagios* F. Cuvier, 1824.
1841. *Pelagocyon* Gloger, Gemeinnütziges Hand- und Hilfs-buch der Naturgeschichte, 1:xxxiv, 163. Type, *Phoca monachus* Hermann.
1848. *Rigoon* Gistel, Naturgeschichte Thierreichs für höhere Schulen, p. x, a renaming of *Pelagios* F. Cuvier, 1824.
1854. *Heliophoca* Gray, Ann. Mag. Nat. Hist., ser. 2, 13:201. Type, *Heliophoca atlantica* Gray [= *Phoca monachus* Hermann].

Total length of males, to 8½ ft. Dorsal surface of body brownish-gray without spots, white or yellowish-gray ventrally; muzzle broad, and slightly grooved between nostrils; vibrissae short, not dense; nails well developed on forefeet and rudimentary on hind feet; digits of hind feet becoming shorter proximally; underfur sparse or absent. Condylobasal length, (male subadult, 100369 USNM, from Campeche), 275; (female adult, 102524 USNM, from Campeche), 282; zygomatic breadth, (male) 183, (female) 172;

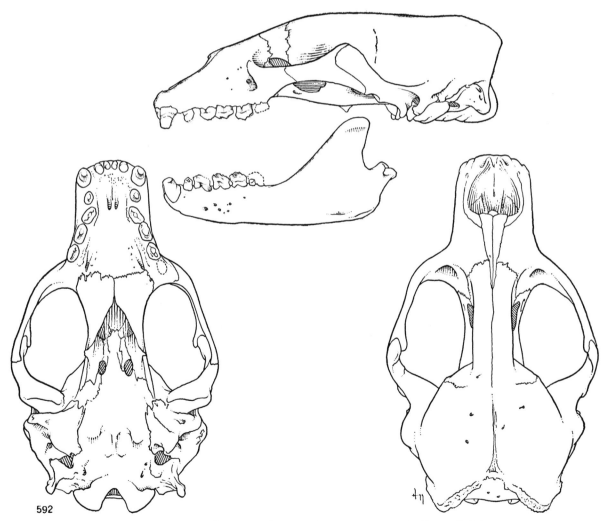

Fig. 592. *Monachus tropicalis*, Yucatán, No. 10421 A.M.N.H., sex ?, X ⅓.

1072 MAMMALS OF NORTH AMERICA

interorbital constriction, (male) 33.4, (female) 28.3. Rostrum elongate and depressed; palate notched posteriorly; canines large, conical; cheek-teeth conical, hollow, having strongly lobed cingulum on proximal side; cheek-teeth obliquely placed in respect to median palatine line, all double-rooted except anteriormost 2, in both jaws, which are single rooted; molars having slightly developed accessory cusps anteriorly and posteriorly. Dentition, i. $\frac{2}{2}$, c. $\frac{1}{1}$, p. $\frac{4}{4}$, m. $\frac{1}{1}$.

Three species are recognized: *M. monachus* of the Black, Adriatic, and part of the Mediterranean seas and northwestern Africa; *M. schauinslandi* of the Central Pacific (Sandwich [Hawaiian] Islands); and *M. tropicalis* of America (Caribbean region).

Monachus tropicalis (Gray)
West Indian Seal

1850. *Phoca tropicalis* Gray, Catalogue of the . . . Mammalia in the . . . British Museum, 2:28. Type locality, Jamaica.

M. tropicalis differs from *M. monachus* in slightly lesser size and cranial features (see Scheffer, 1958:113), and from *M. schauinslandi* in cranial features, for example posteriorly pointed (instead of rounded) nasals and V-shaped instead of U-shaped posterior margin of palate.

Lack of recent sightings suggests that this species is extinct.

RANGE.—Florida, and Bahama Islands, thence southwestward in Gulf of Mexico to Honduras (Moore, 1953:118–120; G. M. Allen, 1942:453).

ORDER SIRENIA
Manatees, Dugongs, and Sea Cow

Fusiform; entirely aquatic; no visible separation of head from body; eyes small, eyelids imperfectly developed; external auditory meatus small, pinnae absent; pectoral appendages paddle-shaped, digits not externally visible; carpal elements vary from 3 (Dugongidae) to 5 or 6 (Trichechidae); pelvic appendages absent; tail flattened and laterally expanded, rounded or forked posteriorly; skin tough and rugose, naked or sparsely haired; bones of skull and postcranial skeleton massive and dense; nasal bones absent (Dugongidae) or vestigial (Trichechidae); 6–8 cervical vertebrae; clavicles absent; sacrum absent; anterior caudal vertebrae having chevron bones; toothless or having incisors (I2 a tusk in Dugongidae) and molariform teeth that are separated by wide diastema; Recent genera having anterior portion of palate and corresponding portion of mandible covered with rough, horny plate; tongue small, fixed, rugose; single pair of pectoral mammae; stomach of several chambers.

FAMILY TRICHECHIDAE—Manatees

Vestigial nails present on pectoral appendages; skin nearly hairless excepting bristles on snout; upper lip cleft, each half capable of moving independently of other; nostrils valvelike; rostrum short, formed dorsally by union of premaxillae anterior to external nares; rostrum only slightly convex ventrally; nasal bones present; incisors $\frac{2}{2}$, rudimentary, concealed beneath horny plate and lost before maturity; molariform teeth $\frac{11}{11}$, selenodont, rarely more than $\frac{6}{6}$ at any one time; anterior molariform teeth wear down and fall out, replaced at posterior end of tooth-row; 2 tuberculate, transverse ridges on molariform teeth; 6 cervical vertebrae; 5 or 6 carpal elements.

The family Trichechidae contains a single genus.

Genus Trichechus Linnaeus—Manatees

Revised by Hatt, Bull. Amer. Mus. Nat. Hist., 66:533–566, 1 pl., 2 figs., September 10, 1934; see also Simpson, Bull. Amer. Mus. Nat. Hist., 59:419–503, 23 figs., September 6, 1932.

1758. *Trichechus* Linnaeus, Syst. nat., ed. 10, 1:34. Type, *Trichechus manatus* Linnaeus.

1772. *Manatus* Brünnich, Zoologiae fundamenta . . . , pp. 34, 38. Type, *Trichechus manatus* Linnaeus.
1803. *Oxystomus* G. Fischer, Das Nat. Mus. Naturgesch. zu Paris, 2:353. Type, *Trichechus manatus* Linnaeus.
1848. *Halipaedisca* Gistel, Naturgeschichte Thierreichs für höhere Schulen, p. 83. Type, *Manatus americanus* Desmarest [= *Trichechus manatus* Linnaeus].

See characters of the family. There are two species in the genus other than the North American *T. manatus*. These are: *T. inunguis* of the rivers of northeastern South America and *T. senegalensis* of West Africa.

Trichechus manatus
Caribbean Manatee

Total length to approx. 15 ft., usually 8–13. Weight to approx. 1500 pounds. Color dull grayish to blackish throughout. See characters of the family.

Trichechus manatus latirostris (Harlan)

1823. *Manatus latirostris* Harlan, Jour. Acad. Nat. Sci. Philadelphia, 3(1):394. Type locality, near the capes of East Florida.
1934. *Trichechus manatus latirostris*, Hatt, Bull. Amer. Mus. Nat. Hist., 66:538, September 10.

RANGE.—Coast and coastal rivers of southeastern United States from North Carolina: Beaufort, thence southward and westward to Texas: mouth Rio Grande, thence along coast to Veracruz (Hall and Dalquest, 1963:348): Alvarado; Coatzacoalcos.

Trichechus manatus manatus Linnaeus

1758. [*Trichechus*] *manatus* Linnaeus, Syst. nat., ed. 10, 1:34. Type locality, restricted to West Indies by Thomas (Proc. Zool. Soc. London, p. 132, March 22, 1911).
1784. *Manati Trichechus* Boddaert, Elenchus animalium, 1:173. Type locality, restricted to West Indies by Hatt (Bull. Amer. Mus. Nat. Hist., 66:535, September 10, 1934).
1795. *Trichechus antillarum* Link, Beyträge zur Naturgeschichte, 1(2):109. Type locality, restricted to West Indies by Hatt (Bull. Amer. Mus. Nat. Hist., 66:535, September 10, 1934).
1795. *Trichechus americanus* Link, Beyträge zur Naturgeschichte, 1(2):109. Type locality, restricted to West Indies by Hatt (Bull. Amer. Mus. Nat. Hist., 66:535, September 10, 1934).
1800. *Manatus Guyannensis* Bechstein, *in* Thomas Pennant's allgemeine Uebersicht der vierfüssigen Thiere, 2:732. Type locality, unknown (name applied to Pennant's "Guiana manatee").

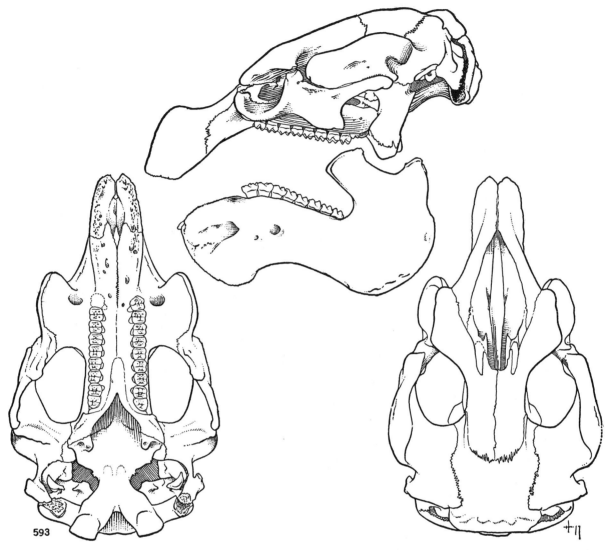

Fig. 593. *Trichechus manatus manatus*, Alvarado, Veracruz, No. 32229 K.U., sex ?, X ¼.

1800. *Manatus Oronocensis* Bechstein, *in* Thomas Pennant's allgemeine Uebersicht der vierfüssigen Thiere, 2:732. Type locality, restricted to West Indies by Hatt (Bull. Amer. Mus. Nat. Hist., 66:535, September 10, 1934).

1800. *Trichechus Clusii* Shaw, General zoology . . . , 1:246. Type locality, restricted to West Indies by Hatt (Bull. Amer. Mus. Nat. Hist., 66:536, September 10, 1934).

1800. *Trichechus Amazonicus* Shaw, General zoology . . . , 1:246. Type locality, restricted to West Indies by Hatt (Bull. Amer. Mus. Nat. Hist., 66:536, September 10, 1934).

1802. *Manatus minor* Daudin, Buffon's Histoire Naturelle, Didot ed., Quadrupeds, 14:134. Type locality, restricted to West Indies by Hatt (Bull. Amer. Mus. Nat. Hist., 66:536, September 10, 1934).

1838. *Manatus atlanticus* Oken, Allegemeine Naturgeschichte . . . , 2(7):1098. Type locality, restricted to West Indies by Hatt (Bull. Amer. Mus. Nat. Hist., 66:536, September 10, 1934).

1897. *Manatus Koellikeri* Kükenthal, Zoöl. Anzeiger, 20:40. Type locality, Surinam, northern South America.

RANGE.—Coasts and coastal rivers of the West Indies, for example, Puerto Rico (Erdman, 1970:638): 1 km. W Corcega Beach; Caracol Cay, thence southeastward to northern South America.

ORDER PERISSODACTYLA
Perissodactyls

Main axis of foot lies directly over 3rd digit (mesaxonic); 3rd digit on both forefeet and hind feet larger than any of the other digits; digit symmetrical with free border of ungual phalanx evenly rounded; never plantigrade; femur having 3rd trochanter; astragalus having pulleylike articulation with tibia but distal surface flattened; fibula not articulating with calcaneum. Alisphenoid canal present; nasals expanded posteriorly; thoracic vertebrae plus lumbar vertebrae never fewer than 22, usually 23; os penis absent; clavicle absent; horns as true bony outgrowths absent in Recent families. Premolars and molars in continuous series; crowns massive, quadrate, transversely ridged or complex; posterior premolars often resembling true molars in size and structure; crown of last molar commonly bilobed; teeth lophodont, brachydont to hypsodont. Stomach simple; caecum large; mammae inguinal; gall bladder absent; testes scrotal.

SUBORDER CERATOMORPHA

Cement absent on crowns of teeth; lower molars with two transverse ridges (Tapiroidea) or partially bicrescentic (Rhinoceratoidea); hypoconulid of m3 usually absent; lower molars asymmetrical with anterior crest of hypoconid, when present, attached to, or pointing toward, protoconid; manus tridactyl or tetradactyl, pes tridactyl; ulna and fibula not reduced to vestiges, not ankylosed with radius and tibia, respectively; body heavy; neck short; limbs short and stocky.

SUPERFAMILY TAPIROIDEA

Incisors chisel-shaped, reduced to no fewer than $\frac{3}{3}$; canines conical, never lost; cheek-teeth brachydont; protoloph and metaloph at right angles to ectoloph; parastyle and metacone well developed; never horned; postglenoid- and paroccipital-processes well developed.

FAMILY TAPIRIDAE—Tapirs

Legs and feet short; manus tetradactyl, pes tridactyl; canines well developed; canine-premolar diastema of moderate length; posterior 3 premolars molariform; cheek-teeth with simple pattern, uppers with lophs, lowers with 2 transverse crests; short proboscis; nasal bones shortened.

Genus Tapirus Brünnich—Tapirs

American species revised by Hershkovitz, Proc. U.S. Nat. Mus., 103:465–496, May 18, 1954.

1772. *Tapirus* Brünnich, Zoologiae fundamenta . . . , pp. 44, 45. Type, *Hippopotamus terrestris* Linnaeus. For use of Brünnich instead of Brisson as author of *Tapirus* see Hopwood, Proc. Zool. Soc. London, 117:534–536, October 30, 1947, and Hershkovitz, Proc. U.S. Nat. Mus., 103:466, May 18, 1954.

1779. *Tapir* Blumenbach, Handbuch der Natürgeschichte, 1:129. Type, *Tapir suillus* Blumenbach [= *Hippopotamus terrestris* Linnaeus].

1827. *Syspotamus* Billberg, Synop. faunae Scandinaviae, 1(1): Mammalia, conspectus A (before p. 1), a renaming of *Tapir* Gmelin, 1778.

1830. *Rhinochoerus* Wagler, Natürliches System der Amphibien . . . , p. 17, a renaming of *Tapirus* Brisson.

1865. *Elasmognathus* Gill, Proc. Acad. Nat. Sci. Philadelphia, 17:183. Type, *Elasmognathus bairdii* Gill. Not Fieber, 1844, an hemipteran.

1872. *Tapyra* Liais, Climats, géol., faune et géog. botanique, Brésil, p. 397, an emendation of *Tapirus* Brünnich.

1873. *Cinchacus* Gray, Hand-list of the edentate, thick-skinned, and ruminant mammals in the British Museum, p. 34. Type, *Tapirus leucogenys* Gray [= *Tapirus pinchaque* Roulin]. *Cinchacus* is regarded as a typographical error for *Pinchacus* by Hershkovitz (Proc. U.S. Nat. Mus., 103:469, May 18, 1954) and is so emended. (Valid as a subgenus according to Hershkovitz, *loc. cit.*)

1903. *Tapirella* Palmer, Science, n.s., 17:873, May 29, a renaming of *Elasmognathus* Gill.

1913. *Acrocodia* Goldman, Proc. Biol. Soc. Washington, 26:65. Type, *Tapirus indicus* Desmarest. (Valid as a subgenus according to Hershkovitz, Proc. U.S. Nat. Mus., 103:467, May 18, 1954.)

Body stocky, neck long; upper lip long, prehensile; ears short, tail short; legs short, stout; skin thick. Forefeet with 4 toes, hind feet with 3 toes; 3rd digit longest. Skull with no trace of postorbital bar; anterior nasal openings enlarged, recessed; nasal bones small, arched. Teeth brachydont, without cement; molars bilophodont, without extra cusps or crests; deciduous dentition well developed and long persisting. Dentition, i. $\frac{3}{3}$, c. $\frac{1}{1}$, p. $\frac{4}{4}$, m. $\frac{3}{3}$.

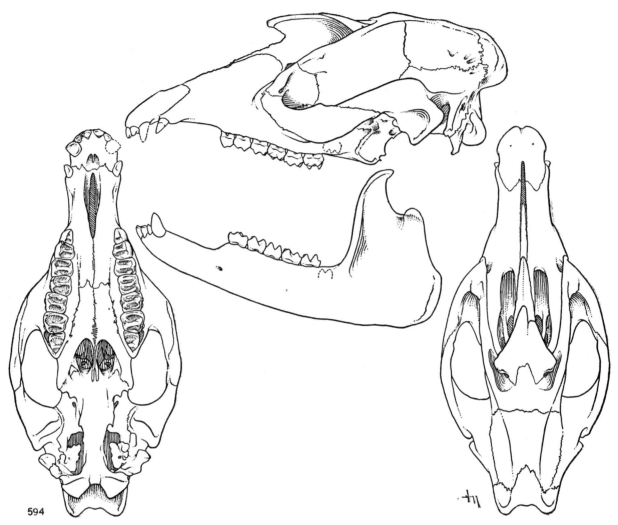

Fig. 594. *Tapirus bairdii*, 60 km. SE Jesús Carranza, 450 ft., Veracruz, No. 24558 K.U., ♀, X ¼.

Subgenus **Tapirella** Palmer

1865. *Elasmognathus* Gill, Proc. Acad. Nat. Sci. Philadelphia, 17:183. Type, *Elasmognathus bairdii* Gill. Not Fieber, 1844, an hemipteran.

1903. *Tapirella* Palmer, Science, n.s., 17:873, May 29, a renaming of *Elasmognathus* Gill.

"Head flattened dorsally, a low mane, not always well defined, extending from front of ears to withers; proboscis longer and bulkier than in other American species; pelage of lowland populations thin and not completely hiding skin, in highland populations longer, thicker, and completely hiding skin; hoofs broad, larger than in subgenus *Tapirus* with middle hoof always wider than long. Dorsal contour of skull . . . flattened or slightly rounded; median frontal line usually placed abruptly above level of nasals, roughly parallel to horizontal plane of nasals but strongly divergent from horizontal plane of maxillary tooth row; superior longitudinal parietal ridges separated by a broad flat table, not uniting to form a sagittal crest . . . ossification of vertical mesethmoid plate extending beyond tips of nasals to angle between premaxillae in old adults; outer anterodorsal surface of maxillae produced upward to form thin, parallel-sided plates embracing mesethmoid; posterolateral maxillary process projecting back to form inner lateral wall of narial meatus but not contacting nasal bone; posterolateral border of premaxilla rounded or angular and embraced by maxilla; nasal without descending process overlapping maxilla; two ossification centers of nasal sometimes persistent in fully ossified bone of adult. First upper premolar as in subgenus *Tapirus;* upper incisors orthodont." (Hershkovitz, 1954:488, 489.)

Tapirus bairdii (Gill)
Baird's Tapir

1865. *Elasmognathus bairdii* Gill, Proc. Acad. Nat. Sci. Philadelphia, 17:183, type from Isthmus of Panamá, Panamá.
1882. *Tapirus (Elasmognathus) bairdi* [*sic*], Sumichrast, La Naturaleza, 5:332.
1870. *Elasmognathus dowii* Gill, Amer. Jour. Sci., 50:142, type from Guatemala.

External measurements of an adult from Veracruz: 2020; 70; 375; 140. Skull of an adult female from Nicaragua: Greatest length, 420; condylobasal length, 395; zygomatic width, 195; palatal length, 201; ". . . entire margin or only upper borders and lower edges of ears either conspicuously trimmed or spotted with white or buff, or uniformly brown or black; lips edged white, gray or buff; cheeks paler than crown, the contrastingly paler color usually continuing onto throat, chest, and, to a varying degree, on belly; chin darker than cheeks and lips; upper parts of head and body, sides, and limbs light drab to brown in palest individuals, dark brown to black in darkest individuals. Juvenal pelage marked by a variable pattern of yellow and white spots and stripes covering entire body." (Hershkovitz, 1954:468.)

MARGINAL RECORDS.—Veracruz: Buena Vista. Oaxaca: Zanatepec (Goodwin, 1969:249), thence throughout Middle America to Panamá (Handley, 1966:792): upper Río Changena, 2400 ft.; Gatún; Cana, thence into South America.

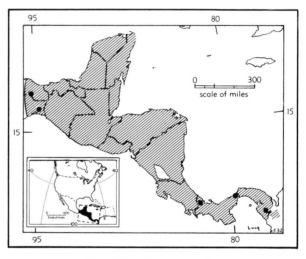

Map 532. *Tapirus bairdii.*

ORDER **ARTIODACTYLA**—Artiodactyls

Main axis of foot lies between 3rd and 4th digits (paraxonic); toes 2 or 4 (except Tayassuidae), 1st (inner) digit suppressed; 3rd and 4th digits subequal and larger than lateral pair (when present); ungual phalanges of 3rd and 4th digits flattened on inner surface and more or less symmetrical to vertical axis of foot; never plantigrade. Femur lacking 3rd trochanter; distal articular surface of astragalus divided into 2 nearly equal facets that rest on navicular and cuboid; magnum articulates freely with scaphoid; fibula articulating with calcaneum. Alisphenoid canal absent; nasals not markedly expanded posteriorly; clavicle generally absent; thoracic plus lumbar vertebrae total 19; os penis absent. Premolars generally simpler than molars; upper premolars usually with a single lobe, upper molars bilobed; last lower molar usually with 3 lobes; teeth bunodont to selenodont, brachydont to hypsodont. Stomach more or less complex; caecum small; cerebral hemispheres well convoluted; uterus bicornuate; mammae inguinal or abdominal; indeciduate placenta, diffuse or cotyledonary; testes scrotal.

The peccaries (family Tayassuidae) have fewer than four chambers in the stomach and therein differ from the other three North American artiodactyls (deer, pronghorn, and bovids), which have 4-chambered stomachs. Species of these three groups are ruminants. All ruminants, or "cud-chewers," graze grasses and woody material in which there is a relatively low amount of nutriments and swallow it rapidly with little chewing. When first swallowed, the food enters the rumen or paunch and, after undergoing a softening process, it is regurgitated into the mouth, where it is chewed again and further mixed with the salivary juices. The food is then swallowed a second time, entering the second compartment of the stomach. It then progresses to the third stomach, and then to the fourth stomach or digesting chamber (the reed or abomasum), where the greatest digestive activity takes place. In this manner, cud-chewing animals can quickly consume a large quantity of low-grade food and, impart to it the necessary grinding and chemical treatment. (After Walker, "Mammals of the World," 1964, Johns Hopkins Press, Baltimore.)

KEY TO NORTH AMERICAN FAMILIES OF ARTIODACTYLA

1. Frontal appendages absent; 3 hoofs (1 vestigial) present on hind foot; upper incisors present; canines well developed; lower canines not incisiform. . . . Tayassuidae, p. 1078
1'. Frontal appendages present (at least in males) 4 or 2 hoofs (2 or none vestigial) present on hind foot; upper incisors absent; upper canines absent or poorly developed; lower canines incisiform.
  2. Frontal appendages not deciduous, usually present in both sexes; lachrymal articulating with nasal.
                      Bovidae, p. 1108
  2'. Frontal appendages deciduous, absent in females of some species; lachrymal not articulating with nasal.
    3. Frontal appendages composed of bone. . . . . Cervidae, p. 1083
    3'. Frontal appendages composed of fused hair surrounding bony core. . . . . . . Antilocapridae, p. 1106

SUBORDER **SUIFORMES**

Upper incisors present; horns or antlers absent; metacarpal and metatarsal bones of main pair of toes not fused; feet short, usually tetradactyl. Dentition little reduced (44 teeth in one living family, Suidae); cheek-teeth bunodont or bunoselenodont (Hippopotamidae) in Recent genera; canines large, triangular in cross section; stomach simple or complex.

SUPERFAMILY **SUOIDEA**

Snout elongate, mobile, with expanded, truncate, nearly naked, flat, oval, terminal surface in which nostrils are located; manus tetradactyl, pes tetradactyl or tridactyl; incisors rooted; cheek-teeth bunodont; caecum present.

FAMILY **TAYASSUIDAE**—Peccaries

Genera revised by Woodburne, Mem. Southern California Acad. Sci., 7:1–48, July 5, 1968.

I3 absent; P $\frac{1}{1}$ absent; upper canines directed downward, not outward or upward; stomach complex; pes tridactyl; premolars and molars forming a continuous series, gradually increasing in size from first to last; P $\frac{4}{4}$ nearly as complex as molars; molars with square, quadricuspidate crowns. Dentition, i. $\frac{2}{3}$, c. $\frac{1}{1}$, p. $\frac{3}{3}$, m. $\frac{3}{3}$.

KEY TO GENERA OF TAYASSUIDAE

1. Whitish or yellowish collar extending from mane over shoulders to throat; diastema between p1 and c1 less than total length of premolars in lower tooth-row. *Dicotyles,* p. 1079
1'. Collar absent; diastema between p1 and c1 equal to or more than total length of premolars in lower tooth-row. . . . *Tayassu,* p. 1081

Genus **Dicotyles** G. Cuvier
Collared Peccaries or Javelina

1817. *Dicotyles* G. Cuvier, Le règne animal . . . , 1:237. Type, *Dicotyles torquatus* G. Cuvier, 1817 [= *Sus tajacu* Linnaeus, 1758].
1775. *Tagassu* Frisch, Das Natur-System der vierfüssigen Thiere . . . , p. 3. Type, "Das amerikanische einzige Schwein-Geschlechte," or *Sus tajacu* Linnaeus (see Palmer, N. Amer. Fauna, 23:955, January 23, 1904). Frisch's names are unavailable in zoology because the system he employed was non-Linnaean (see Sherborn, Index animalium . . . , sec. 1, p. xxv, 1902; Thomas and Miller, Ann. Mag. Nat. Hist., ser. 7, 16:461–464, October 1905; and Hershkovitz, Jour. Mamm., 29:272–275, August 31, 1948).
1828. *Adenonotus* Brookes, Prodromus synopsis animalium . . . collection of Joshua Brookes, p. 11, previous to May. A renaming of *Dicotyles* G. Cuvier. Type, *Sus tajassu* [= *tajacu*] Linnaeus, 1758.
1835. *Pecari* Reichenbach, Bildergalerie der Thierwelt . . . , heft 6, p. 1. Type, *Sus torquatus* [= *Dicotyles torquatus* G. Cuvier].

External measurements: 870–940; 19–55; 180–200; 84–100; sexes similar in size. Weight 40–65 pounds. Grizzled grayish to blackish throughout, excepting yellowish tinge on cheeks and whitish to yellowish collar extending from mane over shoulders to throat; feet blackish. Young yellowish-brown, grizzled with blackish; black stripe down back; grayish-yellow ventrally. Large, well-developed musk gland on mid-dorsal line of rump; mane of long, stiff hairs along mid-dorsal line from crown to rump. Greatest length of skull, 234–248; condylobasal length, 199–217; zygomatic breadth, 97.0–114.7. Rostrum narrow, slightly convex dorsally; sides of rostrum not flattened, divided by zygomatic ridge; maxillae not laterally expanded above maxillary tooth-row; distinct ridge on palate from C to P2; infraorbital canals rounded; molars having well-developed cingula, and cusps not closely connected by intermediate cusplets; wide diastema between c1 and p1, but less than total length of lower premolars.

Dicotyles tajacu
Collared Peccary or Javelina

x ¹/₂₂ C G Pritchard

See characters for the genus.

Dicotyles tajacu angulatus Cope

1889. *Dicotyles angulatus* Cope, Amer. Nat., 23:147, February, type from Guadalupe River, Texas.
1968. *Dicotyles tajacu angulatus,* Woodburne, Mem. Southern California Acad. Sci., 7:1–48, July 5.

MARGINAL RECORDS.—Texas: Montague County; McLennan County; Brazos County; Brazoria County; Aransas National Wildlife Refuge (Halloran, 1961:25, as *Tayassu tajacu angulatus*); mouth Rio Grande. Tamaulipas: approx. 10 mi. N Cues; *Rancho del Cielo; Alta Cima.* San Luis Potosí: Presa de Guadalupe. Guanajuato: approx. 15 mi. S San Bartolo de Berrio. San Luis Potosí: Bledos. Nuevo León: approx. 10 mi. E San Rafael. Coahuila: Jaral; 6 mi. NE Hda. La Mariposa, 1700 ft.; approx. 15 mi. W Los Alamos. Texas: vic. Mule Ear Peaks, Brewster Co.; Christmas Mts. New Mexico: eastern edge Pecos Valley. Texas: 10–15 mi. N Monahans.

Dicotyles tajacu bangsi (Goldman)

1917. *Pecari angulatus bangsi* Goldman, Proc. Biol. Soc. Washington, 30:109, May 23, type from Boca de Cupe, 250 ft., Darién, Panamá.

MARGINAL RECORDS.—Panamá: Gatún, Canal Zone; Tacarcuna Village (Handley, 1966:792), thence into South America.

Dicotyles tajacu crassus (Merriam)

1901. *Tayassu angulatus crassus* Merriam, Proc. Biol. Soc. Washington, 14:124, July 19, type from Metlatoyuca, Puebla.

MARGINAL RECORDS.—San Luis Potosí: 6½ mi. E Sabanito; Ebano, thence eastward to Caribbean Coast and southward along coast to Veracruz: *Arroyo Saoso, 7 km. S, 37 km. E Jesús Carranza; 32 km. ESE Jesús Carranza, 350 ft.* (Hall and Dalquest, 1963:350); *25 km. SE Jesús Carranza, 250 ft.* (op. cit.:351); 30 km.

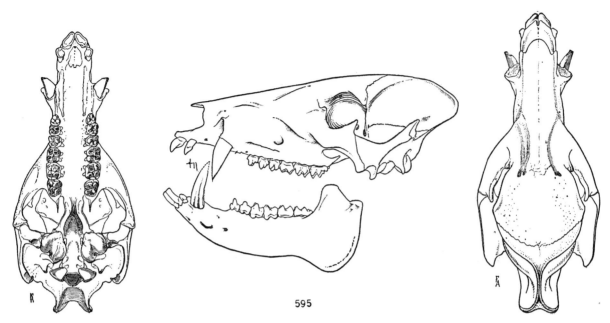

595

Fig. 595. *Dicotyles tajacu sonoriensis*, Ferguson Ranch, 35 mi. from Florence, Pinal Co., Arizona, No. 34177 M.V.Z., ♂, X ⅓.

SSE Jesús Carranza, 300 ft. Oaxaca: Tollosa. Puebla: type locality. *E Hidalgo. NE Querétaro.*

Dicotyles tajacu crusnigrum (Bangs)

1902. *Tayassu crusnigrum* Bangs, Bull. Mus. Comp. Zool., 39:20, April, type from Boquete, Chiriquí, Panamá.

MARGINAL RECORDS.—Nicaragua: Río Coco. Panamá: Sibube (Handley, 1966:792); Paracoté. Nicaragua: San Rafael del Norte.

Dicotyles tajacu humeralis (Merriam)

1901. *Tayassu angulatus humeralis* Merriam, Proc. Biol. Soc. Washington, 14:122, July 19, type from Armería, Colima.

MARGINAL RECORDS.—Michoacán: near Penjamillo de Degollado; approx. 5 mi. W Panindicuaro. Guerrero: near Chilpancingo. Oaxaca (Goodwin, 1969:251): Jayacatlán; Tolosa; Nizanda; mts. N of Zanatepec; San Antonio, thence northward along coast to Colima: type locality. Michoacán: near Los Reyes.

Dicotyles tajacu nanus (Merriam)

1901. *Tayassu nanus* Merriam, Proc. Biol. Soc. Washington, 14:102, July 19, type from Cozumel Island, Quintana Roo. Known only from Cozumel Island. Regarded by Hershkovitz (Fieldiana-Zool., 31:566, 567, July 10, 1951) as doubtfully distinct from *yucatanensis* and possibly based on an introduced population.

Dicotyles tajacu nelsoni (Goldman)

1926. *Pecari angulatus nelsoni* Goldman, Proc. Biol. Soc. Washington, 39:48, July 30, type from Huehuetán, 500 ft., Chiapas.

MARGINAL RECORDS.—Quintana Roo: La Vega, thence southward along coast to Belize: Stann Creek Valley. Guatemala: *Pacific slope.* Chiapas: type locality; 3 km. SW Mapastepec. Belize: Kate's Lagoon. Quintana Roo: Esmeralda. Yucatán: Tekom.

Dicotyles tajacu nigrescens (Goldman)

1926. *Pecari angulatus nigrescens* Goldman, Proc. Biol. Soc. Washington, 39:49, July 30, type from Chamelecón, Cortés, Honduras. Regarded as identical with *nelsoni* by Hershkovitz, Fieldiana-Zool., 31:455, July 10.

MARGINAL RECORDS.—Guatemala: *Caribbean slope.* Honduras: type locality; Cantoral. El Salvador (Burt and Stirton, 1961:66): *Lake Olomega;* Hda. San Pedro. Honduras: Las Flores.

Dicotyles tajacu sonoriensis Mearns

1897. *Dicotyles angulatus sonoriensis* Mearns, Preliminary diagnoses of new mammals . . . *Mephitis, Dorcelaphus,* and *Dicotyles,* from the Mexican border of the United States, p. 3, February 11 (preprint of Proc. U.S. Nat. Mus., 20:469, December 24), type from San Bernardino River, Sonora, near Monument 77, Mexican boundary.

MARGINAL RECORDS.—Arizona: Fossil Creek; Parker Creek, Sierra Ancha Experimental Forest (Johnson and Johnson, 1964:323). New Mexico: SW

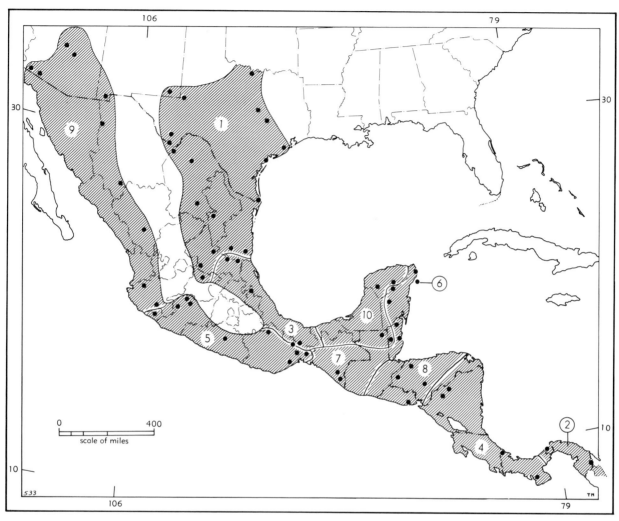

Map 533. *Dicotyles tajacu.*

1. *D. t. angulatus*	4. *D. t. crusnigrum*	8. *D. t. nigrescens*
2. *D. t. bangsi*	5. *D. t. humeralis*	9. *D. t. sonoriensis*
3. *D. t. crassus*	6. *D. t. nanus*	10. *D. t. yucatanensis*
	7. *D. t. nelsoni*	

Big Hatchet Peak. Chihuahua: 5 mi. SW Pacheco (Anderson, 1972:389); approx. 15 mi. S Tecorichic. Durango: approx. 5 mi. S Durango. Jalisco: Los Masos; *Volcán de Fuego;* La Ciénega, thence northward along coast. Arizona: W of Growler Mine, Organ Pipe Cactus National Monument; Tule Tank, 1250 ft. (Cockrum, 1961:250).

Dicotyles tajacu yucatanensis (Merriam)

1901. *Tayassu angulatus yucatanensis* Merriam, Proc. Biol. Soc. Washington, 14:123, July 19, type from Tunkás, Yucatán.

MARGINAL RECORDS.—Yucatán: type locality. Belize: Mountain Pine Ridge. Guatemala: Uaxacutún.

Yucatán: Ticul.—Introduced in 1930 in western and eastern Cuba (Varona, 1974:57).

Genus **Tayassu** G. Fischer
White-lipped Peccaries

1814. *Tayassu* G. Fischer, Zoognosia tabulis synopticis illustrata . . . , 3:284. Type by tautonomy, *Tayassu pecari* G. Fischer.

1817. *Notophorus* Fischer, Adversaria Zool., Mém. Soc. Imp. Nat. Moscou, V:373. A renaming of *Dicotyles* G. Cuvier; no generic or specific diagnoses given. Type, *Notophorous* [= *Tayassu*] *pecari* Fischer, 1814.

1901. *Olidosus* Merriam, proposed as a subgenus, Proc. Biol. Soc. Washington, 14:120, July 19. Type, *Sus albirostris* Illiger 1811 [= *Tayassu* (*Olidosus*) *albirostris* (Illiger)].

External measurements of a female of *T. p. spiradens* from Carillo, Costa Rica: 1040; ——; 190; 80. Blackish dorsally, frequently tinged with tawny; bases of hairs buffy; blackish ventrally, becoming grizzled in pectoral and inguinal regions; feet whitish; muzzle, chin, and cheeks white or yellowish-white. Young russet-colored, blackish mid-dorsally from crown to rump. Cranial measurements of the type specimen of *T. p. spiradens:* Greatest length of skull, 280.5; condylobasal length, 245.0; zygomatic breadth, 120.0. Skull large and heavy; rostrum broad, flattened or slightly concave dorsally; sides of rostrum flattened, not divided by zygomatic ridge; maxillae laterally expanded above maxillary tooth-row from level of canine posterior to level of M1; distinct ridge on palate from C to P2 lacking; infraorbital canals elliptical; molars lacking well-developed cingula, but having cusps closely connected by intermediate cusplets; wide diastema between c1 and p1, greater than total length of lower premolars.

Tayassu pecari
White-lipped Peccary

See characters for the genus.

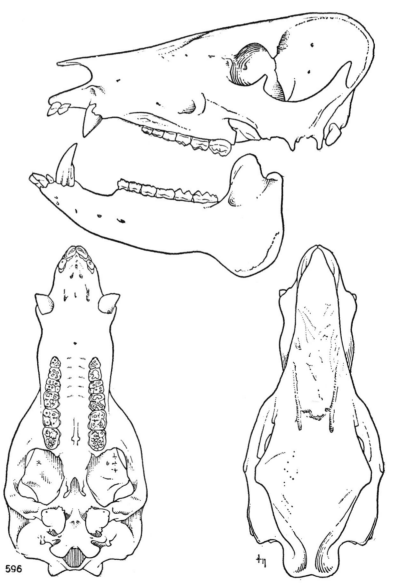

Fig. 596. *Tayassu pecari ringens*, 20 km. E Jesús Carranza, Veracruz, No. 24562 K.U., ♂, X ⅓.

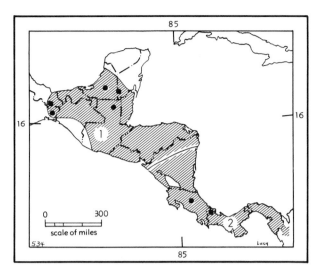

Map 534. *Tayassu pecari.*

1. *T. p. ringens* 2. *T. p. spiradens*

Tayassu pecari ringens Merriam

1901. *Tayassu albirostris ringens* Merriam, Proc. Biol. Soc. Washington, 14:121, July 19, type from Apazote, Campeche.
1902. *Tayassu pecari ringens*, J. A. Allen, Bull. Amer. Mus. Nat. Hist., 16:166, July 1.

MARGINAL RECORDS.—Campeche: type locality. Quintana Roo: 86 km. W Chetumal (93839 KU). Guatemala: Uaxactún; *tropical lowlands of both coasts.* N Chiapas. Oaxaca: Sierra Madre, N of Zanatepec (Goodwin, 1969:251). Veracruz: *20 km. E Jesús Carranza;* 20 km. ENE Jesús Carranza (Hall and Dalquest, 1963:352), thence along coast to point of beginning.—Introduced in 1930 in western and eastern Cuba (Varona, 1974:56).

Tayassu pecari spiradens Goldman

1912. *Tayassu albirostris spiradens* Goldman, Proc. Biol. Soc. Washington, 25:189, December 24, type from Talamanca [= Sipurio, Río Sixaola, Limón, near Caribbean Coast], Costa Rica.
1920. *Tayassu pecari spiradens* Goldman, Smiths. Miscl. Coll., 69(5):74, April 24.

MARGINAL RECORDS.—Costa Rica: Carillo, San José. Panamá: Río Changuinola (Handley, 1966:792), thence eastward into South America.

SUBORDER RUMINANTIA

Upper incisors absent; horns or antlers usually present, at least in males; metatarsal bones of main pair of toes fused into a single cannon-bone (same is true of metacarpal bones), terminating inferiorly in a pair of pulleylike condyles; metacarpals and metatarsals of lateral toes incomplete or absent; navicular and cuboid of tarsus united. Lower canines incisiform, forming, with incisors, a semicircle of spatulate pairs which bite against a callous pad in upper jaw; cheek-teeth selenodont. Stomach 4-chambered; placenta cotyledonous.

SUPERFAMILY CERVOIDEA

Chiefly browsers; antlers usually present in males only, shed and replaced annually, covered with "velvet" that atrophies and is rubbed off when antlers reach full growth.

FAMILY CERVIDAE—Cervids

Frontal appendages (when present) of (usually) branched, bony, solid, deciduous antlertissue, supported on permanent skin-covered pedicels. Upper canines long in antlerless species, but absent or poorly developed in North America cervids, all of which have antlers in at least males; lower canines incisiform; cheekteeth selenodont, generally brachydont. Lateral toes generally present, with constituent bones fully developed; lateral metacarpals lacking either upper (telemetacarpalian) or lower (plesiometacarpalian) ends. Gall bladder usually absent; 2 pair mammae present (except in *Moschus* ?); facial glands present (except in *Capreolus*); tarsal, metatarsal, and interdigital glands may be present; placenta with few cotyledons.

KEY TO NORTH AMERICAN GENERA OF CERVIDAE

1. Antlers more or less palmate.
    2. Antlers strongly palmate; length of skull more than 500. . . . . .*Alces*, p. 1099
    2'. Antlers slightly palmate; length of skull less than 500. . . . .*Rangifer*, p. 1101
1'. Antlers not palmate.
    3. Posterior narial cavity divided by vomer.
        4. Antlers unbranched, less than ½ length of head; auditory bullae slightly inflated. . . . .*Mazama*, p. 1097
        4'. Antlers generally branched, more than ½ length of head; auditory bullae moderately inflated. *Dama*, p. 1087
    3'. Posterior narial cavity not completely divided by vomer. . . . . .*Cervus*, p. 1083

Genus Cervus Linnaeus—Wapiti

1758. *Cervus* Linnaeus, Syst. nat., ed. 10, 1:66. Type, *Cervus elaphus* Linnaeus.

1799. *Alce* Blumenbach, Handbuch der Naturgeschichte, p. 697. Type, *Alce gigantea* Blumenbach [a Pleistocene fossil].

1827. *Elaphus* Hamilton-Smith, *in* Griffith, The animal kingdom . . . by the Baron Cuvier . . . , 5:307. Type, *Cervus elaphus* Linnaeus.

1827. *Rusa* Hamilton-Smith (*op. cit.*:309). Type, *Cervus unicolor* Kerr.

1836. *Hippelaphus* Bonaparte, Iconografia della fauna Italica . . . , 1 (fasc. xv, xvi):4, a renaming of *Rusa* Hamilton-Smith. Not Reichenbach, 1835.

1838. *Harana* Hodgson, Ann. Nat. Hist., 1:154. Type, *Cervus wallichii* Cuvier.

1838. *Rucervus* Hodgson, Ann. Nat. Hist., 1:154. Type, *Cervus elaphoides* Hodgson [= *Cervus duvaucelii* Cuvier].

1841. *Pseudocervus* Hodgson, Jour. Asiatic Soc. Bengal, 10:914, July–December. Type, *Cervus wallichii* Cuvier.

1843. *Panolia* Gray, List of the . . . Mammalia in the . . . British Museum, p. 180. Type, *Panolia acuticornis* Gray [= *Cervus eldii* M'Clelland].

1844. *Megaceros* Owen, Rept. British Assoc. Adv. Sci., for 1843, pp. 237–239. Type, *Megaceros hibernicus* Owen [a Pleistocene fossil].

1846. *Hippelaphus* Sundevall, Kongl. Svenska Vetensk.-Akad. Handl., for 1844, p. 177. Type, *Cervus hippelaphus* Cuvier. Not Reichenbach, 1835.

1846. *Strongyloceros* Owen, A history of British fossil mammals and birds, p. 470. Type, *Cervus elaphus* Linnaeus.

1846. *Recervus* Gray, Catalogue of the Mammalia . . . Nepal and Tibet in the British Museum, p. 33, an emendation of *Rucervus* Hodgson.

1870. *Sika* Sclater, Proc. Zool. Soc. London, p. 115, June. Type, *Cervus sika* Temminck [= *Cervus nippon* Temminck].

1870. *Recurvus* Jäger and Bessels, Petermann's Geog. Mitth., 16:87, a misprint.

1872. *Pseudaxis* Gray, Catalogue of the ruminant Mammalia British museum, p. 70. Type, *Cervus taiouanus* Blyth.

1874. *Elaphoceros* Fitzinger, Sitzungsb. k. Akad. Wiss., Wien, 68:347, 352, for 1873. Type, *Cervus sika* Temminck [= *Cervus nippon* Temminck].

1876. *Strangalioceros* Alston, Zool. Record, for 1874, 9:556, a *lapsus*?

1876. *Strangyloceros* Alston (*ibid.*), a *lapsus*?

1888. *Sambur* Heude, Mém. Hist. Nat. Empire Chinois, 2:8. Type, *Cervus aristotelis* Cuvier.

1888. *Roussa* Heude, Mém. Hist. Nat. Empire Chinois, 2:8, an emendation?

1897. *Russa* Jentink, Notes Leyden Mus., 19:63, an emendation?

1898. *Sikaillus* Heude, Mém. Hist. Nat. Empire Chinois, 4:98. Type, *Cervus sika* Temminck [= *Cervus nippon* Temminck].

1898. *Sica* Trouessart, Catalogus mammalium . . . , fasc. 4:878, a renaming of *Sika*.

1899. *Eucervus* Acloque, Faune de France, Mammals, p. 71. Type, *Cervus elaphus* Linnaeus. Not Gray, 1866.

1930. *Przewalskium* Flerov, Comptes Rendus Acad. Sci. U.R.S.S., p. 115. Type, *Cervus albirostris* Przewalski.

1943. *Thaocervus* Pocock, Jour. Bombay Nat. Hist. Soc., 43:554, 559. Type, *Rucervus schomburgki* Blyth.

Size small to large; antlers subcylindrical or somewhat flattened; brow tine and at least 2 other tines present; muzzle bare; tail medium to short; coat uniformly colored or spotted; main hoofs long, pointed, hind pair united through most of basal length by deep interungual membrane; no specialized gland or deep interdigital cleft on front of fore-pasterns or hind pasterns; skull long, narrow; cheek-teeth moderately short-crowned, wide; crowns of lower incisors varying in size; young generally spotted. Dentition, i. $\frac{0}{3}$, c. $\frac{1}{1}$, p. $\frac{3}{3}$, m. $\frac{3}{3}$.

Cervus elaphus
Wapiti

× 1/30

The three westernmost North American subspecies "revised" by McCullough, Univ. California Publ. Zool., 88:vii + 209, 8 pls., 41 figs., 30 tables, June 5, 1969. See remarks below.

External measurements: (males) 2030–2972; 80–213; 464–660. Weight to 1100 pounds, average about 650 pounds. Females average smaller and lighter. Other dimensions among males: basilar length of skull, 358–455; zygomatic breadth, 155–203; length of upper molar series, 128–143. Upper parts light grayish brown to dark brown; head, mane, neck, and legs dark brown, almost black; underparts darker than back; rump patch whitish to rich tawny; summer pelage more tawny. Lachrymal vacuity widely open, pit well developed; vomer low posteriorly, not dividing posterior nares into 2 chambers; base of pedicel extending conspicuously over posterior part of orbital cavity. Maxillary canine present in both sexes; lower incisors distinctly, but not excessively, differentiated in size and form. Antlers generally with at least 5 tines, including bez; brow tine forms obtuse angle with beam.

Wapiti formerly occurred in most of central North America, but have been extirpated in many parts of that region, although they have been reintroduced in some parts of it. Also they have been introduced in some places outside of their range as known to explorers from Europe and eastern North America. One such place is the southwestern part of the Yukon Territory, where wapiti were introduced in the early 1950s (see Youngman, 1975:159).

Barclay (1935:798) regarded all named kinds of the wapiti of North America as belonging to the same species as the named kinds of red deer in Eurasia. The earliest available specific name is [*Cervus*] *elaphus* Linnaeus. Ellerman and Morrison-Scott (1951:367) concurred with Barclay, as did Flerov (1952:175 [158 English ed.]–178 [161]). McCullough (1969:2, 3) states that red deer and wapiti introduced into separate areas of New Zealand extended their geographic ranges there, and that hybridization occurred where they met with the result that much of the population in the area of the "overlap" shows intermediate characteristics. If this crossbreeding in nature is interpreted as integradation, McCullough (1969:3) is perhaps correct in his feeling that red deer and wapiti are genetically a single species.

Cervus elaphus canadensis Erxleben

1777. [*Cervus elaphus*] *canadensis* Erxleben, Systema regni animalis . . . , p. 305. Type locality, eastern Canada [= Quebec].
1808. *Cervus wapiti* Barton, Jour. Med. Phys. Philadelphia, 3:36. Type locality, unknown.
1815. *Cervus major* Ord, *in* Guthrie, A new geog., hist. and coml. grammar . . . , Philadelphia, ed. 2, 2:292, description on p. 306. Type locality, unknown.

MARGINAL RECORDS.—Ontario: 200 mi. NE Lake Nipigon; Ottawa. New York: N branch Saranac. Pennsylvania: Pike County; Monroe County. Virginia: Clarke County; Bedford County. South Carolina: Fairfield County. Georgia: *Appalachian Mts.* Tennessee: near Nashville; Reelfoot Lake. Louisiana: near Mound, Madison Parish; [vic.] Colfax. Texas: *Panhandle*, thence northward in the plains to Nebraska: within 1½ mi. Sidney (Jones, 1964c:310), thence to Alberta, and eastward to North Dakota: "*all . . . North Dakota*." Minnesota: "throughout the State."

Cervus elaphus manitobensis Millais

1915. *Ć*[*ervus*]. *c*[*anadensis*]. *manitobensis* Millais, *in* The gun at home and abroad . . . , p. 281, type from Manitoba and eastern Saskatchewan.

MARGINAL RECORDS.—Saskatchewan: Prince Albert Park (Soper, 1961:39). Manitoba: Duck Mtn.; Red River; Pembina Hills; Turtle Mtn. Alberta: Cypress Hills.

Cervus elaphus merriami Nelson

1902. *Cervus merriami* Nelson, Bull. Amer. Mus. Nat. Hist., 16:7, January 16, type from head Black River, *ca.* 9000 ft., White Mts., Apache Co., Arizona.
1935. *Cervus elaphus merriami*, Barclay, Proc. Zool. Soc. London, p. 798, January.

MARGINAL RECORDS.—Arizona: near mouth Little Colorado River; vic. Camp Apache; type locality. New Mexico: Datil Mts.; Manzano Mts. Oklahoma: Wichita Mts.; Rainy Mtn., *ca.* 40 mi. W Lawton. Texas: Guadalupe Mts. New Mexico: Guadalupe Mts.; Sacramento Mts.; near Ruidoso; Gallina Mts. Chihuahua: *northwestern Chihuahua.* Arizona: head Oak Creek, San Francisco Mts.

Cervus elaphus nannodes Merriam

1905. *Cervus nannodes* Merriam, Proc. Biol. Soc. Washington, 18:24, February 2, type from Buttonwillow, Kern Co., California.
1935. *Cervus elaphus nannodes*, Barclay, Proc. Zool. Soc. London, p. 798, January.

MARGINAL RECORDS (McCullough, 1969:179–181).—California: Cow Creek, 10 mi. N confluence with Sacramento River; *Sacramento River, 6 mi. N Deer Creek;* near Oroville; *Bear River, 10 mi. above confluence with Feather River; Yuba River, 10 mi. above Feather River;* Grand Island; *Stanislaus River;* Merced River at foothills; *opposite side Merced River at foothills; left bank San Joaquin River, 37° N, 119°40′ W;* Tule Swamp S of Kaweah River; China Grade Bluffs, 4 mi. NE Bakersfield; Rincon Point, thence N along coast to Green Valley; *between Healdsburg and Geyserville;* S side Clear Lake; 2 mi. SE South Yolla Bolly Mtn.; *Cow Creek, 8 mi. N confluence with Sacramento River.*

Cervus elaphus nelsoni V. Bailey

1935. *Cervus canadensis nelsoni* V. Bailey, Proc. Biol. Soc. Washington, 38:188, November 15, type from Yellowstone National Park, Wyoming.
1969. *Cervus elaphus nelsoni*, McCullough, Univ. California Publ. Zool., 88:3, June 5.

MARGINAL RECORDS.—Alberta: Fort Smith. Manitoba: N end Lake Winnipeg. Saskatchewan: Cumberland House. Alberta: Fort Saskatchewan, near Edmonton; near head Pembina River; Waterton Lakes Park. Wyoming (Long, 1965a:711, 712): Slough Creek; Sheridan; *Bear Lodge Mts.;* Black Hills; Fort Laramie; Fort D. A. Russell [= Warren AF Base]; *Pole Mtn.; Medicine Bow Mts.* New Mexico: Pecos River Mts.; Jemez Mts. Colorado: Long Canyon (Anderson, 1961:63). Utah: Willis Creek, Bryce. Nevada: Stevenson's Canyon, Snake Range; Wild Bruneau Mts., vic.

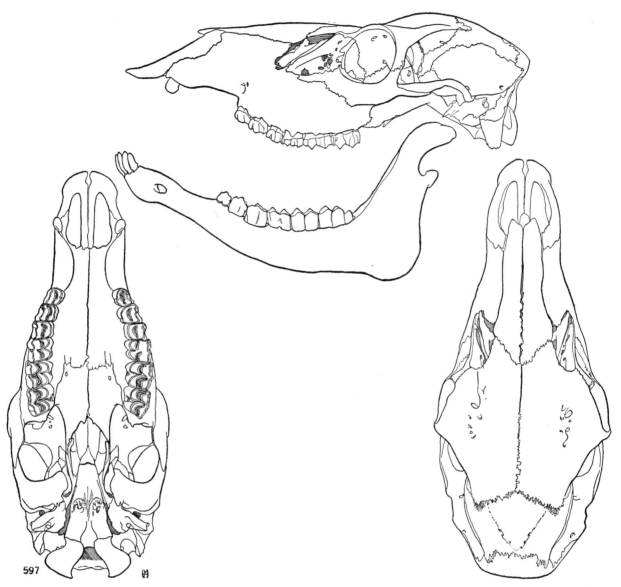

Fig. 597. *Cervus elaphus nelsoni*, 1 mi. N Green River Lakes, Sublette Co., Wyoming, No. 89267 M.V.Z., ♀, X ¼.

Mountain City. Oregon: Steens Mts. California (McCullough, 1969:181): on Pit River, 7 mi. N, 3 mi. W Fall River Mills; N bank Pit River S of Brock Mtn.; Scott Mts.; *Sisson; Mt. Shasta.* Oregon: headwaters Deschutes River; Burns; Blue Mts. Washington: Blue Mts.; *NE Washington.* British Columbia: Chilco Creek, 8 mi. E Vanderhoof; Lower Post (Cowan and Guiguet, 1965:359). In British Columbia introduced on *Graham Island (Queen Charlotte Islands),* and on mainland at *McNab Creek* (Cowan and Guiguet, 1965:361).

Cervus elaphus roosevelti Merriam

1897. *Cervus roosevelti* Merriam, Proc. Biol. Soc. Washington, 11:272, December 17, type from Mt. Elaine, on ridge between heads of Hoh, Elwha, and Soleduc rivers, near Mt. Olympus, Mason Co., Washington.

1969. *Cervus elaphus roosevelti,* McCullough, Univ. California Publ. Zool., 88:3, June 5.
1827. *C[ervus]. occidentalis* Hamilton-Smith, *in* Griffith, The animal kingdom . . . by the Baron Cuvier . . . , 4:101. Type locality, western North America (regarded by V. Bailey, N. Amer. Fauna, 55:81, August 29, 1936, as probably not identifiable as to genus or locality).

MARGINAL RECORDS (Cowan and Guiguet, 1965:362, unless otherwise noted).—British Columbia: Cape Scott, Vancouver Island; *Saltspring Island; Saturna Island; Saanich Peninsula, Vancouver Island.* Washington: headwaters Cispus River, S of Mt. Rainier. Oregon: near The Dalles. California (McCullough, 1969:181): Elk Lake, Marble Mts.; Mt. Sanhedrin; near Plantation, thence northward to point of beginning.

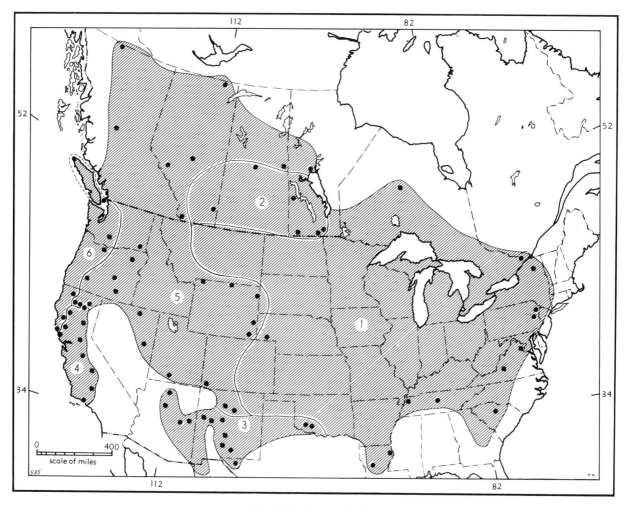

Map 535. *Cervus elaphus.*

| 1. *C. e. canadensis* | 3. *C. e. merriami* | 5. *C. e. nelsoni* |
| 2. *C. e. manitobensis* | 4. *C. e. nannodes* | 6. *C. e. roosevelti* |

Genus **Dama** Zimmermann—New World Deer

Reviewed by Kellogg (*D. virginiana*, pp. 31–55), and Cowan (*D. hemionus*, pp. 334–359), *in* The deer of North America (edited by W. P. Taylor), Amer. Wildlife Manag. Inst., Washington, D.C., xvi + 4 + 668 pp., illustrated, 1956.

1780. *Dama* Zimmermann, Geographische Geschichte . . . , 2:129. Type, *Dama virginiana* Zimmermann. Not *Dama* Frisch, 1775, nor *Cervus dama* Linnaeus, 1758.

1827. *Mazama* Hamilton-Smith, *in* Griffith, The animal kingdom . . . by the Baron Cuvier . . . , 5:314. No type designated. Not *Mazama* Rafinesque, 1817.

1832. *Odocoileus* Rafinesque, Atlantic Jour., 1:109. Type, *Odocoileus speleus* Rafinesque [= *Dama virginiana* Zimmermann, or a closely related subfossil form].

1841. *Dorcelaphus* Gloger, Gemeinnütziges Hand- und Hilfs-buch der Naturgeschichte, p. 140.

1842. *Cariacus* Lesson, Nouveau tableau du règne animal . . . , mammifères, p. 173. No type designated; based in part on *Mazama* Hamilton-Smith, 1827.

1842. *Aplacerus* Haldeman, Proc. Acad. Nat. Sci. Philadelphia, 1:188, a renaming of *Mazama* Hamilton-Smith.

1844. *Reduncina* Wagner, *in* Schreber, Die Säugthiere . . . , Suppl., 4:373. Based on *Mazama* Hamilton-Smith.

1855. *Macrotis* Wagner, *in* Schreber, Die Säugthiere . . . , Suppl., 5:368. Type, *Cervus macrotis* Say [= *Cervus hemionus* Rafinesque]. Not Dejean, 1833, a coleopteran, nor Reid, 1836.

1866. *Eucervus* Gray, Ann. Mag. Nat. Hist., ser. 3, 18:338, October. Type, *Cervus hemionus* Rafinesque.

1874. *Otelaphus* Fitzinger, Sitzungsb. k. Akad. Wiss., Wien, 68:356, for 1873, a renaming of *Macrotis* Wagner.

1879. *Gymnotis* Fitzinger, Sitzungsb. k. Akad. Wiss., Wien, 78:343. Type, *Gymnotis wiegmanni* Fitzinger [= *Cervus gymnotis* Wiegmann].

1901. *Odocoelus* J. A. Allen, Amer. Nat., 35:449, June, an emendation.

1902. *Odontocoelus* Sclater, Ann. Mag. Nat. Hist., ser. 7, 9:290, April 1, an emendation.

1902. *Dama* J. A. Allen, Bull. Amer. Mus. Nat. Hist., 16:19. Type, *Cervus virginianus* Boddaert. Not *Dama* Zimmermann, 1780.

Medium-sized deer; lateral metacarpals with only distal ends persisting; vomer high, dividing nostrils into 2 chambers posteriorly; antlers present only in males, large, with beam rising at marked angle to plane of face, generally dichotomously forked, with subbasal snag; anterior prong of main fork more or less developed at expense of hind prong; main prongs with secondary forking; face long and narrow; rhinarium well developed; ears variable in size and hairiness; tail long to moderate, with or without hair below; coat uniformly colored; tarsal and usually metatarsal glands present; interdigital glands present on both forefeet and hind feet; lachrymal pits and preorbital glands small; lachrymal fossa shallow; lachrymal ducts on rim of orbit; upper canines usually absent; naviculocuboid of tarsus free from cuneiform; young spotted; premolars moderately proportioned; p3 open; tendency toward closed anterior fossette in P4; diastema moderately long. Dentition, i. $\frac{0}{3}$, c. $\frac{0}{1}$, p. $\frac{3}{3}$, m. $\frac{3}{3}$.

KEY TO NORTH AMERICAN SPECIES OF DAMA

1. Tail brown above, fringed with white laterally, white below; ears approx. $\frac{1}{2}$ length of head; metatarsal gland less than 25 mm. long; antlers with 1 main beam from which tines rise vertically; lachrymal fossa shallow. . . . .*D. virginiana*, p. 1091
1'. Tail tipped with black, white or black above; ears approx. $\frac{3}{4}$ length of head; metatarsal gland more than 25 mm. long; antlers branching dichotomously, tines about equal in size; lachrymal fossa deep.
D. hemionus, p. 1088

Dama hemionus
Black-tailed Deer or Mule Deer

External measurements: (males) 1370–1800, 106–230, 330–585, 118–250 (crown); (females) 1160–1800, 115–200, 325–475, 118–243 (crown);

length of metatarsal gland (males) 25–150, (females) 35–116. Weight of males up to 456 pounds. Basilar length, (males) 207–288, (females) 202–266; zygomatic breadth, (males) 94–133, (females) 92–113; length of maxillary tooth-row, (males) 58–85, (females) 62–81. Upper parts (summer) reddish or yellowish tawny, (winter) dark brownish or rufous gray, speckled with whitish; dark brown patch extending nearly to eyes on forehead; brown patch on each side of nose; remainder of face white or gray; ears black on front border, whitish on inner surface; inner sides of buttocks and legs, abdomen, and throat white; remainder of underparts blackish brown; tail black-tipped, white elsewhere above and below, or black above and white below. Antlers with short subbasal snag, beam curving outward and upward, forking dichotomously, the prongs about equal in size.

Dama hemionus californica (Caton)

1876. *Cervus macrotis* var. *Californicus* Caton, Amer. Nat., 10:464, August. Type locality, near Gaviota Pass, 1050 ft., 40 mi. up coast (W) from Santa Barbara, Santa Barbara Co., California.
1902. *Dama h[emionus]. californica*, J. A. Allen, Bull. Amer. Mus. Nat. Hist., 16:20, February 1.

MARGINAL RECORDS.—California: Strawberry Valley, Yuba Co.; near N. Fork Silver Creek; Dardanelles; near Mirror Lake, Yosemite Valley; 6 mi. N Cedar Grove; Kennedy Meadows in Tulare Co.; Tehachapi; W end Big Bear Lake; Orange County; Casitas Pass; Salmon Creek, Monterey Co.; Alviso Canyon; Tejon Ranch; Greenhorn Mtn.; Dunlap; S. Fork, Merced River; near Bloods, Calaveras Co.

Dama hemionus cerrosensis (Merriam)

1898. *Odocoileus cerrosensis* Merriam, Proc. Biol. Soc. Washington, 12:101, April 30, type from Cerros [= Cedros] Island, off W coast Baja California. Known only from Cedros Island, Baja California.
1959. *Dama hemionus cerrosensis*, Hall and Kelson, Mammals of North America, Ronald Press, p. 1004, March 31.

Dama hemionus columbiana (Richardson)

1829. *Cervus macrotis* var. *Columbiana* Richardson, Fauna Boreali-Americana, 1:257. Type locality, mouth Columbia River, Oregon (= Cape Disappointment, Pacific Co., Washington; see Cowan, California Fish and Game, 22:218, 219, July, 1936).
1959. *Dama hemionus columbiana*, Hall and Kelson, Mammals of North America, Ronald Press, p. 1004, March 31.
1936. *Odocoileus hemionus columbianus*, Cowan, California Fish and Game, 22:215, July.
1848. *Cervus Lewisii* Peale, Mammalia and ornithology, *in* U.S. Expl. Exped. . . . , p. 39, applied to specimens from Feather River and San Francisco Bay, California.

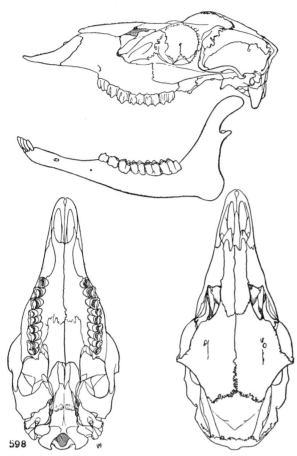

Fig. 598. *Dama hemionus hemionus*, 12-Mile Creek, Nevada, ½ mi. E California boundary, No. 74295 M.V.Z., ♀, X ¼.

1852. *Cariacus punctulatus* Gray, Proc. Zool. Soc. London, p. 239, for 1850, January 24. Type locality, California.
1853. *Cervus Richardsoni* Audubon and Bachman, The viviparous quadrupeds of North America, 3:27. Type locality, California.
1873. *Eucervus pusilla* Gray, Hand-list of the edentate, thick-skinned, and ruminant mammals in the British Museum, p. 157. Type locality, North America [= British Columbia?].
1898. *Odocoileus columbianus scaphiotus* Merriam, Proc. Biol. Soc. Washington, 12:101, April 30, type from Laguna Ranch, Gabilan Range, San Benito Co., California. (Regarded as indistinguishable from *columbianus* by Cowan, California Fish and Game, 22:215, July, 1936.)

MARGINAL RECORDS.—British Columbia: Rivers Inlet; Pemberton Meadows; Boston Bar; upper Skagit Valley. Washington: *summit Cascades westward*. California: E side Mt. Shasta; Eagle Lake; Lake Tahoe; Pigeon Roost Camp; Chinquapin; Elkhorn Station, Yolo Co.; Fourteen Mile House; Laguna Ranch, Gabilan Range; near Coalinga, thence northward along coast, including Vancouver Island, to point of beginning.—**See** addenda.

Dama hemionus crooki (Mearns)

1897. *Dorcelaphus crooki* Mearns, Preliminary diagnoses of new mammals of the genera *Mephitis*, *Dorcelaphus*, and *Dicotyles*, from the Mexican border . . ., p. 2, February 11 (preprint of Proc. U.S. Nat. Mus., 20:468, December 24, 1897), type from summit Dog Mts., 6129 ft., Hidalgo Co., New Mexico. Regarded as a hybrid by Seton, Lives of game animals . . ., vol. 3:327, 328, 1929, and Goldman and Kellogg, Jour. Mamm., 20:507, November 14, 1939.
1959. *Dama hemionus crooki*, Hall and Kelson, Mammals of North America, Ronald Press, p. 1004, March 31.
1901. *Odocoileus hemionus canus* Merriam, Proc. Washington Acad. Sci., 3:560, November 29, type from Sierra en Media, Chihuahua.

MARGINAL RECORDS.—Arizona: Colorado Canyon between Cataract Creek and Pine Springs; San Francisco Mts.; Bakers Butte, Mogollon Mts.; White Mts. (Hoffmeister, 1962:56). New Mexico: ½ mi. S Madre Mtn.; 15 mi. W Magdalena; Hell Canyon, Manzano Mts.; San Andres Mts.; Dark Canyon, Guadalupe Mts. Texas: Washburn; Mobeetie; Castle Mts.; Fort Lancaster; mouth Pecos River. Coahuila: Sierra de la Paila (Baker, 1956:320); *southwestern Coahuila*. Tamaulipas: Cerro del Tigre (Alvarez, 1963:465). Zacatecas: *northern Zacatecas*. Durango: 7 mi. NNW La Zarca (Baker and Greer, 1962:143). Chihuahua: Casas Grandes (Anderson, 1972:391); Sierra en Media. Sonora: San Bernardino Valley. Arizona: Woods Ranch (20 mi. SW Tucson); Castle Dome Mts. (Hoffmeister, 1962:56); 8 mi. E Ehrenberg (*ibid.*); Hualapai Mts. (*ibid.*).

Dama hemionus eremica (Mearns)

1897. *Dorcelaphus hemionus eremicus* Mearns, Preliminary diagnoses of new mammals of the genera *Mephitis*, *Dorcelaphus*, and *Dicotyles* from the Mexican border . . ., p. 4, February 11 (preprint of Proc. U.S. Nat. Mus., 20:470, December 24, 1897), type from Sierra Seri, near Gulf of California, Sonora.
1902. *Dama h[emionus]. eremica*, J. A. Allen, Bull. Amer. Mus. Nat. Hist., 16:20, February 1.

MARGINAL RECORDS.—California: vic. Coxcomb Mtn. Arizona: Colorado River opposite Blythe; valley of Ajo. Sonora: Pozo de Luis; La Libertad Ranch. Baja California: delta Colorado River. California: Indio. [Hoffmeister (1962:53) arranged *D. h. eremica* as a synonym of *D. h. crooki*, but, because in arriving at his conclusion, he failed to examine specimens from California, Sonora, and certain localities in southwestern Arizona, *D. h. eremica* is here listed as a subspecies.]

Dama hemionus fuliginata (Cowan)

1933. *Odocoileus hemionus fuliginatus* Cowan, Jour. Mamm., 14:326, November 13, type from Barona Ranch, 30 mi. E San Diego, San Diego Co., California.
1959. *Dama hemionus fuliginata*, Hall and Kelson, Mammals of North America, Ronald Press, p. 1006, March 31.

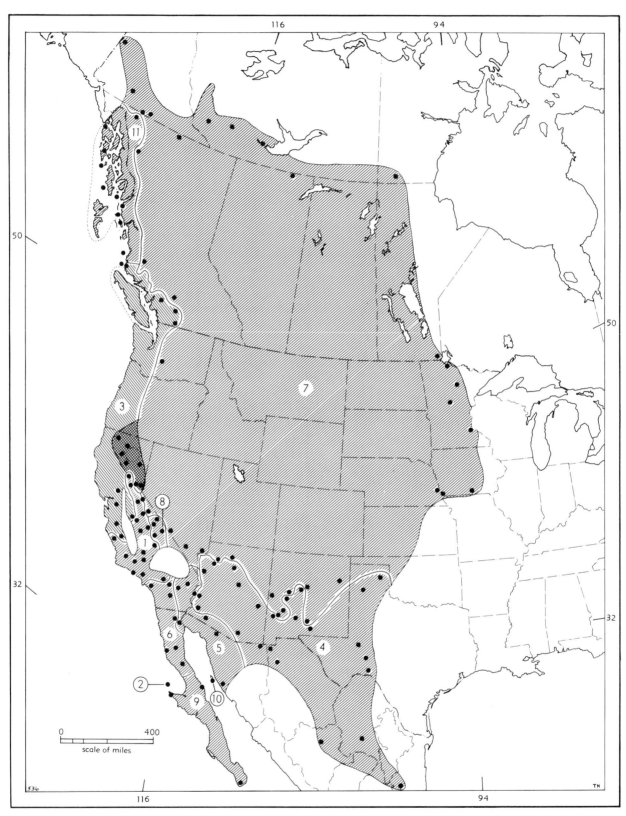

Map 536. *Dama hemionus.*

Guide to subspecies	3. *D. h. columbiana*	6. *D. h. fuliginata*	9. *D. h. peninsulae*
1. *D. h. californica*	4. *D. h. crooki*	7. *D. h. hemionus*	10. *D. h. sheldoni*
2. *D. h. cerrosensis*	5. *D. h. eremica*	8. *D. h. inyoensis*	11. *D. h. sitkensis*

1090

MARGINAL RECORDS.—California: Tahquitz Valley; Toro Peak, Santa Rosa Mts. Baja California: Hanson Lagoon, Laguna Hanson Mts.; Parral; Mission Santa María; San Quentín. California: San Mateo Valley, Orange Co.

Dama hemionus hemionus (Rafinesque)

1817. *Cervus hemionus* Rafinesque, Amer. Month. Mag., 1:436, October. Type locality, mouth Big Sioux River, South Dakota (see V. Bailey, N. Amer. Fauna, 49:41, January 8, 1927).

1902. *Dama hemionus,* J. A. Allen, Bull. Amer. Mus. Nat. Hist., 16:20, February 1.

1820. *Cervus auritus* Warden, Description . . . des États-Unis . . . , 5:640. Type locality, ". . . la contrée située à l'ouest des montagnes Rocky, près des bords de la rivière de Kooskooskée."

1823. *Cervus macrotis* Say, *in* Long, Account of an exped. . . . to the Rocky Mountains . . . , 2:88. Type locality, Mora River, near present town of Mora, New Mexico.

1881. *Cervus macrotis* var. *Montanus* Canton, The antelope and deer of America, ed. 2, p. 94. Type locality, lat. 42° on Missouri River.

1899. *Cariacus virgultus* Hallock, Forest and Stream, 52(21):404, May 27. Type locality, near Hallock, Kittson Co., Minnesota.

MARGINAL RECORDS.—Yukon: Dawson (Youngman, 1975:160). Mackenzie District: near confluence Flat and South Nahanni rivers (Scotter, 1975:489); Simpson; W end Great Slave Lake. Alberta: Wood Buffalo Park. Keewatin: edge of timber, Nueltin Lake (Harper, 1956:75). Manitoba: near town of Sprague. Minnesota: Zippel Twp., Lake of the Woods Co.; 25 mi. NE Grand Rapids; near Pillager; on U.S. Hwy. 10, near Valentine Lake. Iowa (Kline, 1959:148): near Leone; Missouri River, 1 mi. N, 4½ mi. W Hamburg. Nebraska: *ca.* 3 mi. S Bellevue (Jones, 1964c:314). Kansas: *western ⅔ of state.* New Mexico: northern escarpment Staked Plains; Sacramento Mts.; Manzano Mts.; Gallinas Mtn.; W of Chloride; along Gila River; Mogollon Mts. Arizona: Point Imperial Road (Hoffmeister, 1962:58); 5 mi. N Pine Springs, Hualapai Indian Reservation, 7000 ft. (*ibid.*). Nevada: Bonelli Peak; Charleston Peak; Grapevine Peak. California: Tuolumne Meadows; Dardanelles; Pigeon Roost Camp; Charles Meadows; Buck Butte; Red Rock Valley. Washington: Yakima Park. British Columbia: near Lillooet; Atnarko Valley (Cowan and Guiguet, 1965:372); Telegraph Creek area (*ibid.*). Yukon: Tarfu Lake area (Youngman, 1964:5); Richthofen Island, Lake Laberge (*ibid.*); Nisutlin River, near Teslin. British Columbia: Lower Liard Crossing.—**See** addenda.

Dama hemionus inyoensis (Cowan)

1933. *Odocoileus hemionus inyoensis* Cowan, Proc. Biol. Soc. Washington, 46:69, April 27, type from Kid Mtn., 11,000 ft., 10 mi. W Big Pine, Inyo Co., California.

1959. *Dama hemionus inyoensis,* Hall and Kelson, Mammals of North America, Ronald Press, p. 1007, March 31.

MARGINAL RECORDS.—California: Piute Monument, White Mts.; *Poison Creek, White Mts.; Black Mtn., Inyo Mts.;* Badger Flat, Inyo Mts.; Sawtooth Peak; *George Creek, NE Mt. Bernard [= Barnard]; S Kearsarge Peak; 7 mi. W Independence; type locality;* Bishop Creek Canyon.

Dama hemionus peninsulae (Lydekker)

1898. *Mazama (Dorcelaphus) hemionus peninsulae* Lydekker, Proc. Zool. Soc. London, for 1897, pp. 899, 900, April, type from between La Laguna and Victoria Mtn., about 6000 ft., Sierra Laguna, Baja California (see Cowan, California Fish and Game, 22:232, July, 1936).

1902. *Dama h[emionus]. peninsulae,* J. A. Allen, Bull. Amer. Mus. Nat. Hist., 16:20, February 1.

MARGINAL RECORDS.—Baja California: Santa Teresa Bay; Cape San Lucas; San Bartolomé Bay.

Dama hemionus sheldoni (Goldman)

1939. *Odocoileus hemionus sheldoni* Goldman, Jour. Mamm., 20:497, November 15, type from Tiburón Island, Sonora. Known only from Tiburón Island, Sonora.

1959. *Dama hemionus sheldoni,* Hall and Kelson, Mammals of North America, Ronald Press, p. 1007, March 31.

Dama hemionus sitkensis (Merriam)

1898. *Odocoileus columbianus sitkensis* Merriam, Proc. Biol. Soc. Washington, 12:100, April 30, type from Sitka, Alaska.

1959. *Dama hemionus sitkensis,* Hall and Kelson, Mammals of North America, Ronald Press, p. 1007, March 31.

MARGINAL RECORDS.—British Columbia: Atlin; Port Simpson; Porcher Island; Pitt Island (Cowan and Guiguet, 1965:373); Bella Bella; Calvert Island. Alaska: Duke Island; Dall Island; Egg Harbour, Coronation Island; type locality; Indian [= Inian] Island, Icy Straits (Anderson, 1947:176). Introduced on *Graham Island* and *Moresby Island,* British Columbia.

Dama virginiana
White-tailed Deer

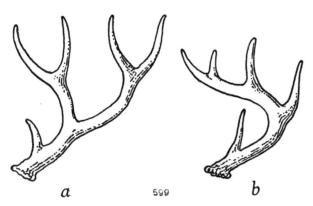

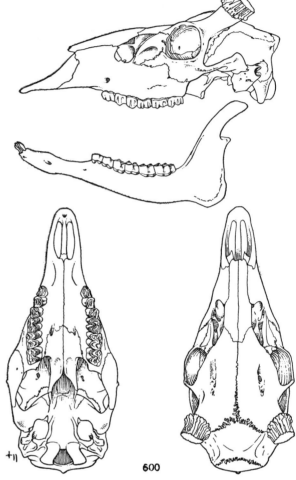

Fig. 599. Right antlers of deer, viewed from inside, to show difference between antler of Black-tailed Deer (*a*) and White-tailed Deer (*b*). Note that the antler of the White-tailed Deer (*b*) has one main beam with minor prongs branching off same, whereas the antler of the Black-tailed Deer (*a*) branches into two nearly equal parts. X approx. ⅛.

External measurements: 1340–2062; 152–330; 362–521; 140–229; height at shoulder, 660–1143. Weight, 50–350 pounds. Condylobasal length, 242.3–290; length of maxillary tooth-row, 68.0–88.5; width of orbit at frontojugal suture, 102.8–123.5. Females average smaller than males. Upper parts (summer) bright fulvous chestnut to gray, grayer in winter; underparts white, lower surface of tail, chin, throat, band around muzzle, and ring around eye, white; young spotted. Antlers with long subbasal snag, beam curving forward, with posterior prong upright, unbranched; prongs all appearing to arise vertically from main beam and generally are unbranched.

In the period 1790 to date *D. virginiana* from several mainland localities in the Americas has been introduced widely in the West Indies (Varona, 1974:57).

Fig. 600. *Dama virginiana texana*, Encinal, La Salle Co., Texas, No. 16538 K.U., ♂, X ¼.

Dama virginiana acapulcensis (Caton)

1877. *Cervus acapulcensis* Caton, The antelope and deer of America, p. 113. Type locality, Acapulco, Guerrero.
1959. *Dama virginiana acapulcensis*, Hall and Kelson, Mammals of North America, Ronald Press, p. 1008, March 31.

MARGINAL RECORDS.—Colima: Armería. Oaxaca: Huilotepec (Goodwin, 1969:253).

Dama virginiana borealis (Miller)

1900. *Odocoileus americanus borealis* Miller, Bull. New York State Mus. Nat. Hist., 8:83, November 21, type from Bucksport, Hancock Co., Maine.
1902. *Dama v[irginiana]. borealis*, J. A. Allen, Bull. Amer. Mus. Nat. Hist., 16:20, February 1.

MARGINAL RECORDS.—Ontario: Newpost, Abitibi River. Quebec: Anticosti Island; *Gaspé Penin-*

sula. New Brunswick. Cape Breton Island. Nova Scotia. Maine: Mount Desert Island (Manville, 1960:416, as *Odocoileus virginianus* only), thence southward along coast to Pennsylvania: Adams County. Maryland: Cumberland, thence along N shore Ohio River and then northward along E shore Mississippi River to Wisconsin: Nelson (Jackson, 1961:420). Minnesota: near Becker; Schroeder (Timm, 1975:36).

Dama virginiana carminis (Goldman and Kellogg)

1940. *Odocoileus virginianus carminis* Goldman and Kellogg, Proc. Biol. Soc. Washington, 53:81, June 28, type from Botellas Cañon, 6500 ft., Sierra del Carmen, Coahuila.
1959. *Dama virginiana carminis*, Hall and Kelson, Mammals of North America, Ronald Press, p. 1008, March 31.

MARGINAL RECORDS.—Texas: near Moss Well, 5400 ft. Coahuila: type locality; *10 mi. NW Las Margaritas, 3800 ft.*; 4 mi. E Las Margaritas, 4000 ft.; 4 mi. N, 21 mi. W Cuatro Ciénegas, 6500 ft.; Jaral; 26 mi. W

Santa Teresa, 7050 ft.; 3 mi. S, 8 mi. E Hechicero, Chihuahua (Baker, 1956:323). Texas: *Burro Mesa*.

Dama virginiana chiriquensis (J. A. Allen)

1910. *Odocoileus rothschildi chiriquensis* J. A. Allen, Bull. Amer. Mus. Nat. Hist., 28:95, April 30, type from Boquerón, Chiriquí, Panamá.
1959. *Dama virginiana chiriquensis*, Hall and Kelson, Mammals of North America, Ronald Press, p. 1008, March 31.

MARGINAL RECORDS.—Panamá: Bayano River; type locality.

Dama virginiana clavia (Barbour and G. M. Allen)

1922. *Odocoileus virginianus clavium* Barbour and G. M. Allen, Jour. Mamm., 3:73, May 9, type from Big Pine Key, Monroe Co., Florida.
1959. *Dama virginia clavia*, Hall and Kelson, Mammals of North America, p. 1008, Ronald Press, March 31.

MARGINAL RECORDS (Layne, 1974:400).—Formerly Matecumbe Keys to Key West; now limited to the Big Pine Group of Keys.

Dama virginiana couesi (Coues and Yarrow)

1875. *Cariacus virginianus* var. *couesi* Coues and Yarrow, Report upon the collections of mammals, p. 111, *in* Rept. Geog. Geol. Expl. and Surv., West of One Hundredth Merid. (Wheeler), 5(Zool.):72, type from Camp Crittenden [= Crittenden], on Sonoita Creek, between Santa Rita and Patagonia Mts., Santa Cruz Co., Arizona.
1959. *Dama virginiana couesi*, Hall and Kelson, Mammals of North America, Ronald Press, p. 1008, March 31.
1903. *Odocoileus battyi* J. A. Allen, Bull. Amer. Mus. Nat. Hist., 19:591, November 12, type from Rancho Santuario, an old "Spanish Grant" ranch on the plains, 7000 ft., Durango.
1915. *O[docoileus]. v[irginianus]. baileyi* Lydekker, Catalogue of the ungulate mammals in the . . . British Museum, 4:158, an accidental renaming of *O. battyi*.

MARGINAL RECORDS.—Arizona (Hoffmeister, 1962:61, 64, unless otherwise noted): near San Francisco Peaks (McCulloch, 1967:484, sight record only); *within 15 mi. of Flagstaff* (Hoffmeister and Carothers, 1969:188, as *Odocoileus virginianus* only, sight record only); 6 mi. NE Pine; *10 mi. SE Long Valley, on Blue Ridge;* W of Nutrioso. New Mexico: San Francisco Mts.; Datil Mts.; Gallinas Mts.; Animas Peak (Findley, *et al.*, 1975:332). Chihuahua: 20 mi. SW Gallego (Anderson, 1972:391). Durango (Baker and Greer, 1962:144): Rosario; Mt. San Gabriel. Zacatecas: Plateado. Jalisco: near Bolaños. Nayarit: Santa Teresa. Sinaloa (Armstrong, *et al.*, 1972:59): 20 km. E Badiraguato, 1800 ft.; *17 mi. NE San Benito, 6500 ft.;* 6 km. NE El Fuerte. Sonora: El Cobre; Sierra Seri; Sierra Cubabi. Arizona: Ehrenberg (Miller and Kellogg, 1955:803, but see Hoffmeister, 1962:61, who notes that *D. hemionus crooki* occurs 8 mi. E Ehrenberg, and that the westernmost occurrence of *D. virginiana* in Arizona known

to him is in the Ajo Mts. approx. 145 mi. to the SE); Bill Williams Mtn. (Hoffmeister, 1962:61).

Dama virginiana dacotensis (Goldman and Kellogg)

1940. *Odocoileus virginianus dacotensis* Goldman and Kellogg, Proc. Biol. Soc. Washington, 52:82, June 28, type from White Earth River, Mountrail Co., North Dakota.
1959. *Dama virginiana dacotensis*, Hall and Kelson, Mammals of North America, Ronald Press, p. 1010, March 31.

MARGINAL RECORDS.—Mackenzie: downstream from mouth Wrigley Creek, between Hells Gate on South Nahanni River and its confluence with Flat River, 61° 34′ N, 125° 28′ W (Scotter, 1975:488, as species only); a ridge below Twisted Mtn., N of South Nahanni River (*ibid.*); 8 mi. W Bell Rock (Kuyt, 1966:194, sight record only). Saskatchewan: Prince Albert Park (Soper, 1961:39, sight record only). Manitoba: Riding Mts.; *beach ridges of Lake Manitoba adjacent to Delta Marsh* (Tamsitt, 1962:77, as *Odocoileus virginiana* only); Whiteshell Forest Reserve (Soper, 1946:151, as *O. v. borealis*). Minnesota: Marshall County. Wyoming: Bear Lodge Mts. (Long, 1965a:714). Colorado: Julesberg; Denver; near summit of Floyd Hill (Armstrong, 1972:306); Larimer County; *Marvine Lodge* (Armstrong, 1972:306); Josephine Basin, *ca.* 4 mi. W Meeker (*ibid.*). Wyoming: Medicine Bow Creek (Long, 1965a:714). Montana: *east of mts.* Alberta: Edmonton.

Dama virginiana hiltonensis (Goldman and Kellogg)

1940. *Odocoileus virginianus hiltonensis* Goldman and Kellogg, Proc. Biol. Soc. Washington, 53:83, June 28, type from Hilton Head Island, Beaufort Co., South Carolina. Known only from Hilton Head Island.
1959. *Dama virginiana hiltonensis*, Hall and Kelson, Mammals of North America, Ronald Press, p. 1010, March 31.

Dama virginiana leucura (Douglas)

1829. *Cervus leucurus* Douglas, Zool. Jour., 4:330, October–January, 1829. Type locality, Lower Columbia River [= Willamette River Valley]; see V. Bailey, Proc. Biol. Soc. Washington, 45:44, April 2, 1932.
1959. *Dama virginia leucura*, Hall and Kelson, Mammals of North America, Ronald Press, p. 1010, March 31.

MARGINAL RECORDS.—Washington: Steilacoom; Cowlitz River. Oregon: foothills about Beaverton; near Sweet Home; near headwaters Umpqua River.

Dama virginiana macroura (Rafinesque)

1817. *Corvus* [sic] *macrourus* Rafinesque, Amer. Month. Mag., 1:436, October. Type locality, "Plains of the Kangar [= Kansas] River" [= plains near Wakarusa Creek, Douglas Co., Kansas].
1902. *Dama v[irginiana]. macroura*, J. A. Allen, Bull. Amer. Mus. Nat. Hist., 16:20, February 1.
1901. *Odocoelus virginianus louisianae* G. M. Allen, Amer. Nat., 35(414):449, June, type from Mer Rouge, Morehouse Parish, Louisiana.

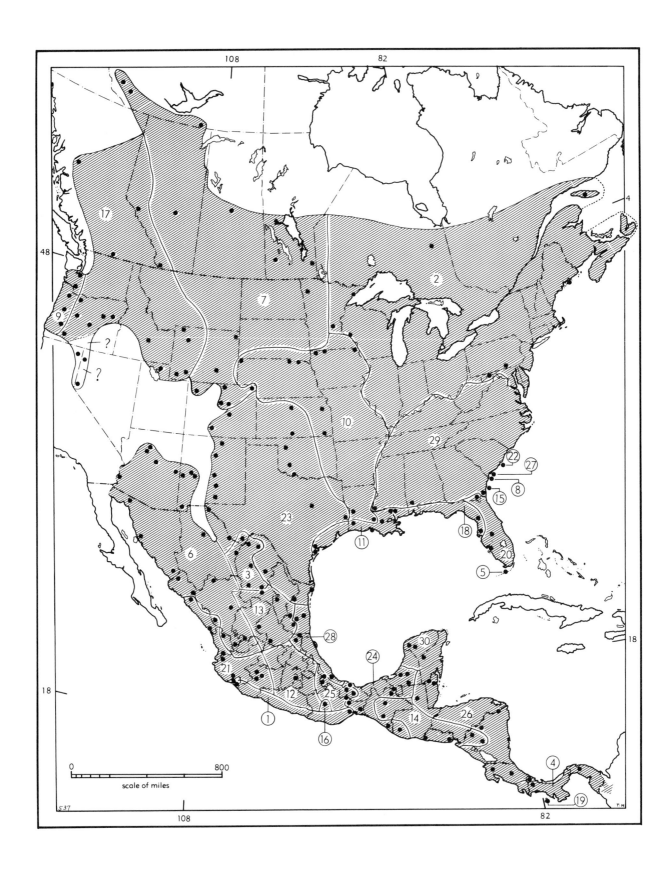

MARGINAL RECORDS.—Iowa (Bowles, 1975:146): Spirit Lake; upper Iowa River Valley, thence southward along W bank Mississippi River to Louisiana (Lowery, 1974:499, as *Odocoileus virginianus* only): 2 mi. W Rosedale; *6 mi. E Cecelia; 5 mi. S Elizabeth;* Fort Polk Game Management Area. Texas: *northeastern Texas.* Oklahoma: *eastern Oklahoma.* Kansas: Chetopa; type locality. Nebraska (Jones, 1964c:319, 320): Warbonnet Canyon; 3 mi. E Bristow; 3½ mi. W St. Helena; *9 mi. NW Ponca.* Iowa: T. 98 N, R. 44 W, Lyon Co. (Bowles, 1975:146).

Dama virginiana mcilhennyi (F. W. Miller)

1928. *Odocoileus virginianus mcilhennyi* F. W. Miller, Jour. Mamm., 9:57, February 9, type from near Avery Island, Iberia Parish, Louisiana.
1959. *Dama virginiana mcilhennyi,* Hall and Kelson, Mammals of North America, Ronald Press, p. 1010, March 31.

MARGINAL RECORDS.—Louisiana: 1 mi. S Angie (Lowery, 1974:499, as *Odocoileus virginianus* only); *Pearl River,* thence westward along Gulf Coast, including Marsh Island (*loc. cit.,* as *Odocoileus virginianus* only), to Texas: near Matagorda Bay; Aransas National Wildlife Refuge (Halloran, 1961:25). Louisiana (Lowery, 1974:499, as *Odocoileus virginianus* only): Calcasieu Parish; Sorrento.

Dama virginiana mexicana (Gmelin)

1788. [*Cervus*] *mexicanus* Gmelin, Syst. nat., ed. 13, 1:179. Type locality, Valley of Mexico, México; see Osgood, Proc. Biol. Soc. Washington, 15:88, April 25, 1902, and Jour. Mamm., 1:77, March 2, 1920.
1959. *Dama virginiana mexicana,* Hall and Kelson, Mammals of North America, Ronald Press, p. 1010, March 31.
1902. *Dama lichtensteini* J. A. Allen, Bull. Amer. Mus. Nat. Hist., 16:20, February 1, a renaming of *Cervus mexicanus* Lichtenstein.

MARGINAL RECORDS.—*Guanajuato. Puebla.* Morelos: *Barranca Tepeite* (Ramírez-P., 1971:283). Distrito Federal: Tlalpan.

Dama virginiana miquihuanensis (Goldman and Kellogg)

1940. *Odocoileus virginianus miquihuanensis* Goldman and Kellogg, Proc. Biol. Soc. Washington, 53:84, June 28, type from Sierra Madre Oriental, 6000 ft., near Miquihuana, Tamaulipas.

1959. *Dama virginiana miquihuanensis,* Hall and Kelson, Mammals of North America, Ronald Press p. 1010, March 31.

MARGINAL RECORDS.—Coahuila: Sierra Guadalupe. Nuevo León: Cerro Potosi. Tamaulipas (Alvarez, 1963:466): 15 km. W Rancho Santa Rosa, 4500 ft.; Ejido Santa Isabel, 2000 ft.; *Rancho Pano Ayuctle.* San Luis Potosi: mts. near Jesús María; Bledos.

Dama virginiana nelsoni (Merriam)

1898. *Odocoileus nelsoni* Merriam, Proc. Biol. Soc. Washington, 12:103, April 30, type from San Cristóbal, 8500 ft., Chiapas.
1959. *Dama virginiana nelsoni,* Hall and Kelson, Mammals of North America, Ronald Press, p. 1010, March 31.

MARGINAL RECORDS.—Chiapas: type locality. *Guatemala.* Nicaragua: [San] Rafael del Norte; Muy Muy. El Salvador (Burt and Stirton, 1961:66): Colinas de Jacuarán; Barra de Santiago. Chiapas: Nueva Palestina, 846 m., 6 km. S Jaltenango.

Dama virginiana nigribarbis (Goldman and Kellogg)

1940. *Odocoileus virginianus nigribarbis* Goldman and Kellogg, Proc. Biol. Soc. Washington, 53:85, June 28, type from Blackbeard Island, McIntosh Co., Goergia.
1959. *Dama virginiana nigribarbis,* Hall and Kelson, Mammals of North America, Ronald Press, p. 1010, March 31.

MARGINAL RECORDS.—Georgia: *Sapelo Island;* type locality.

Dama virginiana oaxacensis (Goldman and Kellogg)

1940. *Odocoileus virginianus oaxacensis* Goldman and Kellogg, Proc. Biol. Soc. Washington, 53:85, June·28, type from mts. 15 mi. W Oaxaca, Oaxaca.
1959. *Dama virginiana oaxacensis,* Hall and Kelson, Mammals of North America, Ronald Press, p. 1010, March 31.

MARGINAL RECORDS.—Oaxaca: Cerro San Felipe; *type locality.*

Dama virginiana ochroura (V. Bailey)

1932. *Odocoileus virginianus ochrourus* V. Bailey, Proc. Biol. Soc. Washington, 45:43, April 2, type from Coolin, S end Priest Lake, Bonner Co., Idaho.
1959. *Dama virginiana ochroura,* Hall and Kelson, Mammals of North America, Ronald Press, p. 1010, March 31.

Map 537. *Dama virginiana.*

1. *D. v. acapulcensis*
2. *D. v. borealis*
3. *D. v. carminis*
4. *D. v. chiriquensis*
5. *D. v. clavia*
6. *D. v. couesi*
7. *D. v. dacotensis*
8. *D. v. hiltonensis*
9. *D. v. leucura*
10. *D. v. macroura*
11. *D. v. mcilhennyi*
12. *D. v. mexicana*
13. *D. v. miquihuanensis*
14. *D. v. nelsoni*
15. *D. v. nigribarbis*
16. *D. v. oaxacensis*
17. *D. v. ochroura*
18. *D. v. osceola*
19. *D. v. rothschildi*
20. *D. v. seminola*
21. *D. v. sinaloae*
22. *D. v. taurinsulae*
23. *D. v. texana*
24. *D. v. thomasi*
25. *D. v. tolteca*
26. *D. v. truei*
27. *D. v. venatoria*
28. *D. v. veraecrucis*
29. *D. v. virginiana*
30. *D. v. yucatanensis*

1933. *Odocoileus virginianus ochrurus*, V. Bailey, Nature Mag., 21(3):126, a *lapsus*.

MARGINAL RECORDS.—British Columbia: Babine Mts. Alberta: Jasper; Waterton Lakes National Park. Montana: *western Montana*. Wyoming: Tower Falls (Long, 1965a:714); Valley (*ibid.*); Green River (Miller and Kellogg, 1955:802); Fort Bridger (*ibid.*). Utah: N of Ogden. Idaho: Mackay. Oregon: Powder River Valley; Buck Creek, Grant Co.; Davis Creek, Crook Co. Nevada (remarks of Adams, 1963:518–522, cast doubt on the validity of alleged occurrences in this state): Granite Peak; Verdi, Truckee River. California: head S fork Pit River (Adams, 1963:519, record in question). Oregon: Fort Klamath; Metolius River, W of Deschutes River; Mill Creek, NE base Mt. Hood. British Columbia: Oliver (Cowan and Guiguet, 1965:376).

Dama virginiana osceola (Bangs)

1896. *Cariacus osceola* Bangs, Proc. Biol. Soc. Washington, 10:26, February 25, type from Citronelle, Citrus Co., Florida.
1902. *Dama v[irginiana]. osceola*, J. A. Allen, Bull. Amer. Mus. Nat. Hist., 16:20, February 1.
1896. *Cariacus* (or *Damelaphus*) *fraterculus* [Coues], The Nation, 62:404, May 21. Type locality, Florida. (Described by Cory, Hunting and fishing in Florida . . . , p. 113.)

MARGINAL RECORDS.—Mississippi: Pearl River County (Kennedy, *et al.*, 1974:29, as species only). Alabama: *southern Alabama*. Florida: Blitches Ferry; N of Tampa Bay.

Dama virginiana rothschildi Thomas

1902. *Dama rothschildi* Thomas, Nov. Zool., 9:136, April 10, type from Coiba Island, Panamá. Known only from Coiba Island.
1959. *Dama virginiana rothschildi*, Hall and Kelson, Mammals of North America, Ronald Press, p. 1011, March 31.

Dama virginiana seminola (Goldman and Kellogg)

1940. *Odocoileus virginianus seminolus* Goldman and Kellogg, Proc. Biol. Soc. Washington, 53:86, June 28, type from 10 mi. NE Everglades, Collier Co., Florida.
1959. *Dama virginiana seminola*, Hall and Kelson, Mammals of North America, Ronald Press, p. 1011, March 31.

MARGINAL RECORDS.—Georgia: Okefinokee Swamp, thence along coast to Florida: Lee County; Lake Arbuckle.

Dama virginiana sinaloae (J. A. Allen)

1903. *Odocoileus sinaloae* J. A. Allen, Bull. Amer. Mus. Nat. Hist., 19:613, November 14, type from Escuinapa, Sinaloa.
1959. *Dama virginiana sinaloae*, Hall and Kelson, Mammals of North America, Ronald Press, p. 1011, March 31.

MARGINAL RECORDS.—Sinaloa: Culiacán. Durango: 6 mi. S Pueblo Nuevo, 3000 ft. (Baker and Greer,

1962:144). Michoacán: Nahuatzin, 8500 ft.; Pátzcuaro, 7000 ft.; Uruapan, 4500 ft. Colima: Paso del Río. Jalisco: Volcán de Fuego; Wakenakili Mts.; Estancia. Sinaloa: type locality.

Dama virginiana taurinsulae (Goldman and Kellogg)

1940. *Odocoileus virginianus taurinsulae* Goldman and Kellogg, Proc. Biol. Soc. Washington, 53:87, June 28, type from Bulls Island, Charleston Co., South Carolina. Known only from Bulls Island.
1959. *Dama virginiana taurinsulae*, Hall and Kelson, Mammals of North America, Ronald Press, p. 1011, March 31.

Dama virginiana texana (Mearns)

1898. *Dorcelaphus texanus* Mearns, Proc. Biol. Soc. Washington, 12:23, January 27, type from Fort Clark [north of Eagle Pass on Big Bend of Rio Grande], Kinney Co., Texas.
1902. *Dama v[irginiana]. texensis* [sic], J. A. Allen, Bull. Amer. Mus. Nat. Hist., 16:20, February 1.
1901. *Odocoileus texensis* Miller and Rehn, Proc. Boston Soc. Nat. Hist., 30:17, December 27, an accidental renaming of *texanus*.

MARGINAL RECORDS.—Kansas: forks of Solomon River; Barber County. Oklahoma: Woodward County; Cache; Beaver Creek, Texas–Oklahoma line. Texas: McLennan County; Jasper County. Tamaulipas: 33 mi. S Washington Beach (Alvarez, 1963:466); El Mulato. Nuevo León: *northern Nuevo León*. Coahuila: Monclova. Chihuahua: Ojinaga. Texas: San Elizario. New Mexico: Sacramento Mts.; *near Highrolls*; Ruidoso; Gallo Canyon, N of Jicarilla Mts.; E side Pecos Baldy; E slopes Taos Mts. Colorado (Armstrong, 1972:306): San Luis Valley; foothills W of Monument.

Dama virginiana thomasi (Merriam)

1898. *Odocoileus thomasi* Merriam, Proc. Biol. Soc. Washington, 12:102, April 30, type from Huehuetán, Chiapas.
1959. *Dama virginiana thomasi*, Hall and Kelson, Mammals of North America, Ronald Press, p. 1011, March 31.

MARGINAL RECORDS.—Campeche (as *D. virginiana* only): 7 km. N, 51 km. E Escárcega (93853 KU); 103 km. SE Escárcega (93855 KU). Tabasco: Montecristo. Chiapas: Palenque. Guatemala: Finca Ciprés. Chiapas: type locality. Oaxaca (Goodwin, 1969:255): Tapanatepec; Sarabia. Veracruz: Catemaco. Campeche: 7½ km. W Escárcega (92574 KU, as *D. virginiana* only).

Dama virginiana tolteca (Saussure)

1860. *Cervus toltecus* Saussure, Revue et Mag. Zool., Paris, ser. 2, 12:247. Type locality, near Orizaba, Veracruz.
1959. *Dama virginia tolteca*, Hall and Kelson, Mammals of North America, Ronald Press, p. 1011, March 31.

MARGINAL RECORDS.—Veracruz: Mirador; Pasa Nueva; 35 km. ENE Jesús Carranza (Hall and Dalquest, 1963:354); *20 km. E Jesús Carranza, 200 ft.* Oaxaca: Achotal. Veracruz: type locality.

Dama virginiana truei (Merriam)

1889. *Cariacus clavatus* True, Proc. U.S. Nat. Mus., 11:417, July 5. Type from Segovia River, about 50 mi. from sea, Comarca de El Cabo, Nicaragua. Not Hamilton-Smith, 1827, unidentifiable.

1898. *Odocoileus truei* Merriam, Proc. Biol. Soc. Washington, 12:103, April 30, a renaming of *Cariacus clavatus* True.

1959. *Dama virginiana truei*, Hall and Kelson, Mammals of North America, Ronald Press, p. 1011, March 31.

1901. *Odocoileus costaricensis* Miller, Proc. Biol. Soc. Washington, 14:36, April 25, type from Talamanca region, E side Costa Rica, between coast and foot of Cordilleras, Costa Rica.

MARGINAL RECORDS.—Belize: Kate's Lagoon; Belize. Nicaragua: type locality. Panamá: Sibube (Handley, 1966:792). Costa Rica: *Talamanca;* Cartago; *Carillo;* Bebedero. Nicaragua: Río Coco. Guatemala: 7 mi. S La Libertad.

Dama virginiana venatoria (Goldman and Kellogg)

1940. *Odocoileus virginianus venatorius* Goldman and Kellogg, Proc. Biol. Soc. Washington, 53:88, June 28, type from Hunting Island, Beaufort Co., South Carolina. Known only from Hunting Island.

1959. *Dama virginiana venatoria*, Hall and Kelson, Mammals of North America, Ronald Press, p. 1012, March 31.

Dama virginiana veraecrucis (Goldman and Kellogg)

1940. *Odocoileus virginianus veraecrucis* Goldman and Kellogg, Proc. Biol. Soc. Washington, 53:89, June 28, type from Chijol, 200 ft., Veracruz.

1959. *Dama virginianus veraecrucis*, Hall and Kelson, Mammals North America, Ronald Press, p. 1012, March 31.

MARGINAL RECORDS.—Tamaulipas: Soto la Marina. Veracruz: near port of Veracruz. San Luis Potosí: 5 km. NE Pujal; Ebano. Tamaulipas: Savinito Tierra Caliente (Alvarez, 1963:466).

Dama virginiana virginiana Zimmermann

1780. *Dama virginiana* Zimmermann, Geographische Geschichte . . . , 2:24, 129. Type from Virginia.

1919. *Cariacus wisconsinensis* Belitz, Wisconsin Conserv., 1:1, a renaming of *D. virginiana* Zimmermann.

MARGINAL RECORDS.—West Virginia: *northern boundary.* Virginia: *northern boundary.* Georgia: Cumberland Island. Alabama: Mount Vernon Barracks. Mississippi: Wilkinson County (Kennedy, *et al.*, 1974:29, as *O. virginiana* only), thence northward on E bank Mississippi River and eastward on S bank Ohio River.

Dama virginiana yucatanensis (Hays)

1872. *Cervus yucatanensis* Hays, Ann. Lyc. Nat. Hist., New York, 10(8):218, October–December. Type locality, Yucatán and southern part México.

1959. *Dama virginiana yucatanensis*, Hall and Kelson, Mammals of North America, Ronald Press, p. 1012, March 31.

MARGINAL RECORDS.—Yucatán: Calcehtok; Yokat. Quintana Roo: Esmeralda.

Genus Mazama Rafinesque—Brockets

1817. *Mazama* Rafinesque, Amer. Month. Mag., 1(5):363, September. Type, by subsequent designation (Merriam, Science, n.s., 1:208, February 22, 1895), *Mazama pita* Rafinesque [= *Moschus americanus* Erxleben].

1827. *Subulo* Hamilton-Smith, *in* Griffith, The animal kingdom . . . by the Baron Cuvier . . . , 5:318. Type not designated; included species were *Cervus rufus* Illiger, *C. simplicicornis* Illiger, and *C. nemorivagus* F. Cuvier.

1841. *Passalites* Gloger, Gemeinnütziges Hand- und Hilfs-buch der Naturgeschichte, p. xxxiii, 140. Type, *Cervus nemorigavus* F. Cuvier [= *Cervus gouazoubira* Fischer].

1842. *Subula* Lesson, Nouveau tableau du règne animal . . . mammifères, p. 174, a renaming of *Subulo* Hamilton-Smith. Not *Subula* Schumacher, 1817, a mollusk.

1843. *Coassus* Gray, List of the . . . Mammalia in the . . . British Museum, p. 174. Type not designated; included species were *Cervus rufus* Illiger, and *C. nemorivagus* F. Cuvier.

1872. *Homelaphus* Gray, Catalogue of the ruminant Mammalia . . . in the British Museum, p. 90. Type, *Homelaphus inornatus* Gray [= *Moschus americanus* Erxleben].

1874. *Nanelaphus* Fitzinger, Sitzungsb. k. Akad. Wiss., Wien (for 1873), 68(1):360. Type, *Nanelaphus namby* Fitzinger [= *Cervus gouazoubira* Fischer].

1874. *Doryceros* Fitzinger, Sitzungsb. k. Akad. Wiss., Wien (for 1873), 68(1):360. Type not designated; included species were *Cervus tschudii* Wagner, and *C. nemorivagus* F. Cuvier.

Size small; back arched; tail short; ears moderate; face elongated; facial glands small and exposed; toe-gland pits deep and triangular. Antlers simple spikes, less than half length of head. Skull resembles that of *Dama* but auditory bullae less inflated, facial profile more arched; upper border of orbits flat, nearly straight. Canines (deciduous ?) often present; molars sometimes with supplementary columns. Dentition, i. $\frac{0}{3}$, c. $\frac{0}{1}$ or $\frac{1}{1}$, p. $\frac{3}{3}$, m. $\frac{3}{3}$.

KEY TO NORTH AMERICAN SPECIES OF MAZAMA

1. Upper parts some shade of rufous; underparts same color as flanks. *M. americana*, p. 1098
1'. Upper parts some shade of brown; underparts white. . . . . . .*M. gouazoubira*, p. 1099

Mazama americana
Red Brocket

x 1/24

External measurements: (males) 1050–1350, 120; (females) 1050–1420, 50–180. Condylobasal length, (males) 177–210, (females) 148–222; zygomatic breadth, (males) 82–99, (females) 70–100; length of maxillary tooth-row, (males) 55–68; (females) 48–73.5. Upper parts varying from yellowish rufous to dark chestnut rufous; midline and back of neck varying from dull brownish to black; legs pale rufous to black; underparts same color as flanks; hair of nape reversed in some; hair coarser than in the brown brocket.

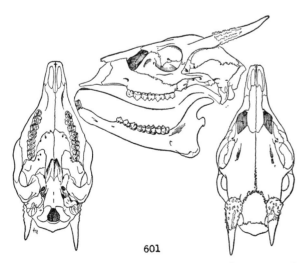

601

Fig. 601. *Mazama americana temama*, 35 km. SE Jesús Carranza, 400 ft., Veracruz, No. 24565 K.U., ♂, X ¼.

Mazama americana cerasina Hollister

1914. *Mazama tema cerasina* Hollister, Proc. Biol. Soc. Washington, 27:209, October 31, type from Talamanca [= Sipurio, Río Sixaola, Limón, near Caribbean Coast], Costa Rica.
1951. *Mazama americana cerasina,* Hershkovitz, Fieldiana-Zool., Field Mus. Nat. Hist., 31:567, July 10.

MARGINAL RECORDS.—Guatemala: Uaxactún. Belize: Mountain Pine Ridge, thence southward along coast to Costa Rica: type locality. Panamá (Handley, 1966:793): Sibube; *Almirante; 14 km. SSW Changuinola; Río Teribe.* Costa Rica: Pozo Azul, thence northwest along Pacific Coast to Guatemala: *Pacific Coast region.*

Mazama americana pandora Merriam

1901. *Mazama pandora* Merriam, Proc. Biol. Soc. Washington, 14:105, July 19, type from Tunkás, Yucatán.
1966. *Mazama americana pandora,* Hershkovitz, Ectoparasites of Panama, Field Mus. Nat. Hist., p. 743, November 22.

MARGINAL RECORDS.—Yucatán: type locality; Chichén-Itzá. Campeche: Dzibalchén (95115 KU).

Mazama americana reperticia Goldman

1913. *Mazama tema reperticia* Goldman, Smiths. Miscl. Coll., 60(22):2, February 28, type from Gatún, Canal Zone, Panamá.
1951. *Mazama americana reperticia,* Hershkovitz, Fieldiana-Zool., Chicago Nat. Hist. Mus., 31:567, July 10.

MARGINAL RECORDS (Handley, 1966:793).—Panamá: type locality; *Rio Chucunaque; El Real;* Tacarcuna; *Cana,* into South America, thence west (in Panamá) along Pacific Coast to Chepigana; *Río Trinidad;* Boquete.

Mazama americana temama (Kerr)

1792. *Cervus temama* Kerr, The animal kingdom . . . , p. 303. Type locality, restricted to Mirador, Veracruz, by Hershkovitz (Fieldiana-Zool., Chicago Nat. Hist. Mus., 31:567, July 10, 1951).
1951. *Mazama americana temama,* Hershkovitz, Fieldiana-Zool., Chicago Nat. Hist. Mus., 31:567, July 10.
1817. *Mazama tema* Rafinesque, Amer. Month. Mag., 2(1):44, November, a renaming of *Cervus temama* Kerr.
1860. *C[ervus]. sartorii* Saussure, Revue et Mag. Zool., Paris, ser. 2, 12:252, June. Type locality, Mirador, Veracruz.

MARGINAL RECORDS.—Tamaulipas: Rancho Pano Ayuctle (Alvarez, 1963:467); *Alta Cima,* thence southward along coast to Campeche: 42 km. E Escárcega (93857 KU). Quintana Roo: 86 km. W Chetumal (93856 KU). Belize: Bokowina. Campeche: 103 km. SE Escárcega (93859 KU). Chiapas: 1 km. S Finca Prusia, 1160 m. Oaxaca: San Antonio (Goodwin, 1969:255); Playa de Carrizo, 30 km. (by road) W Tuxtepec. Veracruz: type locality. San Luis Potosí: Xilitla; Platanito.

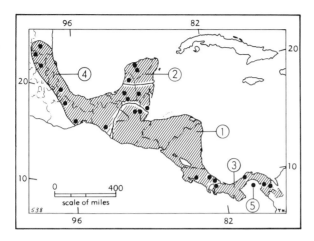

Map 538. *Mazama.*

1. *M. americana cerasina* 3. *M. americana reperticia*
2. *M. americana pandora* 4. *M. americana temama*
5. *M. gouazoubira permira*

Mazama gouazoubira
Brown Brocket

External measurements: (male) 1060, 85; (female) 1000, 80. Condylobasal length, (male) 171, (female) 157; zygomatic breadth, (males) 68.5–74.7, (female) 68.5; length of maxillary tooth-row, (males) 51.0–52.0, (female) 54.0 (after Kellogg, 1946:58). Upper parts varying from yellowish-gray brown to drab brown; underparts white; hair of nape never reversed; hair less coarse than in the red brocket.

Mazama gouazoubira permira Kellogg

1946. *Mazama permira* Kellogg, Proc. Biol. Soc. Washington, 59:57, March 11, type from Isla San José, Archipelago de las Perlas, Golfo de Panamá, Panamá. Known only from Isla San José.
1951. *Mazama gouazoubira permira*, Hershkovitz, Fieldiana-Zool., Chicago Nat. Hist. Mus., 31:567, July 10.

Genus Alces Gray—Moose

Revised by Peterson, Contrib. Royal Ontario Mus. Zool. and Palaeont., 34:1–30, 8 figs., 1 table, October 15, 1952.

1821. *Alces* Gray, London Med. Repos., 15:307. Type, *Cervus alces* Linnaeus.
1841. *Alcelaphus* Gloger, Gemeinnütziges Hand- und Hilfs-buch der Naturgeschichte, p. 143, a renaming of *Alces* Gray. Not Blainville, 1816.
1902. *Paralces* J. A. Allen, Bull. Amer. Mus. Nat. Hist., 16:160, a renaming of *Alces* Gray on the erroneous assumption that *Alces* Gray was a homonym of *Alce* Blumenbach, 1799 [= *Megaceros* Owen, 1844, a subgenus of *Cervus*].

Size large; body heavy; legs long; shoulder region high; tail short; muzzle broad and overhanging; ears large; neck short; dewlap present; nose with triangular bare spot between nostrils; antlers, present only in males, large, with palmate blades and projecting tines. Upper parts (new spring pelage) dark, varying from black to dark brown, reddish brown, or grayish brown; underparts and lower legs lighter, grayish or brownish, sometimes almost whitish; winter pelage grayer. Skull having premaxillary region greatly lengthened; nasals short, nasal aperture large; distance from front of nasal to front of premaxilla about equal to that from back of nasal to back of occiput; vomer low posteriorly, not dividing posterior nares into 2 chambers; lachrymal vacuity widely open, lachrymal pit well developed; maxillary canines usually absent in both sexes, lower canines incisiform; incisors little differentiated; molars and premolars broad and low crowned. Lateral metacarpals strongly attenuated, rudimentary, only distal ends developed as styloid vestiges. Metatarsal gland absent; hoofs long, narrow, pointed, forehoofs larger than hind hoofs. (After Peterson, 1952:13, 14.) Dentition, i. $\frac{0}{3}$, c. $\frac{0}{1}$, p. $\frac{3}{3}$, m. $\frac{3}{3}$.

Alces alces
Moose

External measurements of a male from Quebec: 2903, 63, 792, 254; height at shoulder, 1826. Greatest length of skull, (males) 521–686; mastoid breadth, (males) 132–183; crown length of maxillary tooth-row, (males) 136–162. Females average smaller than males. See characters of the genus.

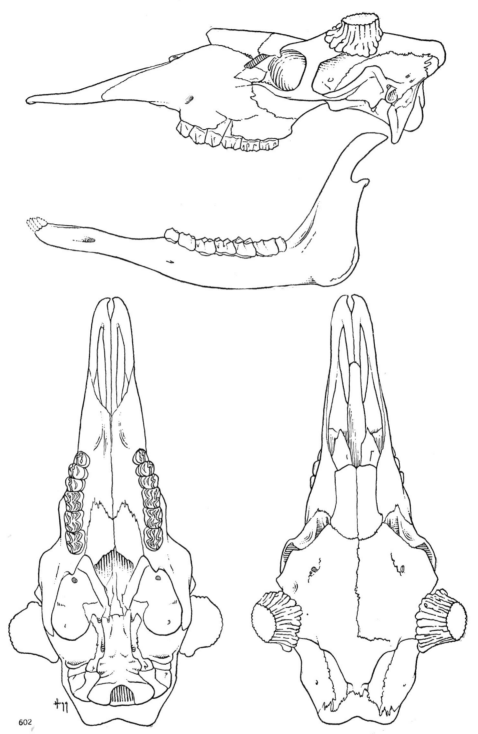

Fig. 602. *Alces alces andersoni*, Lake of the Woods, Marshall Co., Minnesota, No. 2477 K.U., ♂, X ⅓.

Alces alces americana (Clinton)

1822. *Cervus americanus* [Clinton], Letters on the natural history and internal resources . . . of New York, p. 193. Type locality, "Country north of Whitestown" [probably in the western Adirondack region], New York.

1915. *Alces alces americana*, Lydekker, Catalogue of the ungulate mammals in the . . . British Museum, 4:234.

1907. *Alces columbae* Lydekker, The Field, London, 109:182, February 2, type from Ontario (originally given as British Columbia, see Miller and Kellogg, Bull. U.S. Nat. Mus., 205:810, March 3, 1955). Regarded as indeterminable by

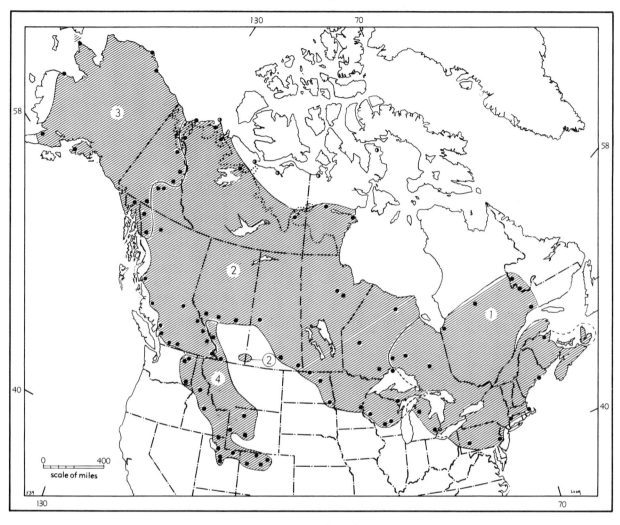

Map 539. *Alces alces.*

1. *A. a. americana* 2. *A. a. andersoni* 3. *A. a. gigas* 4. *A. a. shirasi*

Peterson, Royal Ontario Mus. Zool., Life Sci. Occas. Pap., 9:6, May 25, 1950.

MARGINAL RECORDS.—Labrador (Harper, 1961:128): Molson Lake; Ashuanipi Lake. Quebec: Moisie River (Harper, 1961:128). New Brunswick: Nepisiquit. Maine: *Mount Desert Island* (Manville, 1960:416); Little Cranberry Island (*ibid.*), thence along coast to Massachusetts: vic. Plymouth. Connecticut: Litchfield. Pennsylvania: Northampton County; between Altoona and Pittsburgh. Michigan: Oakland County; Missaukee County. Ontario: 250 mi. NW Sault Ste. Marie; Flint Lake, 35 mi. E Longlac; 15 mi. S Kapuskasing. Quebec (Harper, 1961:128): Nemiscau River; Lac Ochiltrie. Introduced (Cameron, 1959:66, 102) on Newfoundland (from New Brunswick) and *ca.* 1900 on Anticosti Island.

Alces alces andersoni Peterson

1950. *Alces americana andersoni* Peterson, Royal Ontario Mus. Zool., Life Sci. Occas. Pap., 9:1, May 25, type from sec. 27, T. 10, R. 16, Sprucewood Forest Reserve, 15 mi. E Brandon, Manitoba.
1952. *Alces alces andersoni* Peterson, Contrib. Royal Ontario Mus. Zool. and Palaeont., 34:24, October 15.

MARGINAL RECORDS.—Mackenzie: Ellice Island, mouth Mackenzie River; between head Eskimo Lakes and Liverpool Bay; Wilmot Horton River, near 69° N; Dease River, NE Great Bear Lake; Thelon River below jct. with Hanbury. Keewatin (Kelsall, 1972:133): near Schultz Lake; S of Chesterfield Inlet. Manitoba: Split Lake; Steel River, a few miles below Fox River. Ontario: Shamattawa River; Nungessor Lake; 40 mi. NW Port Arthur; Simpson Island. Michigan: Dickinson

County. Wisconsin (Schorger, 1956:7–9): Green Lake; Dellona; Lake Sapin. Minnesota: southern Pine County. North Dakota: Sargent County; around Devils Lake; Turtle Mts. Saskatchewan: Moose Mtn.; 1½ mi. N Estlin (Lahrman, 1971:141); Birch Lake. Alberta: Elk Island Park; Brazeau River, 175 mi. NW Calgary; between Smoky River and Jasper House. British Columbia: Indianpoint Lake; Wood River (Cowan and Guiguet, 1965:380); Golden; Invermere; Cummings Creek (Cowan and Guiguet, 1965:380); Similkameen Valley; Flood (Cowan and Guiguet, 1965:380); Squamish; Pemberton Meadows; Stuie; Kitimat; McDame Post, Dease River. Yukon: Teslin River District; Lapie River (Canol Road) mi. 120. Mackenzie: vic. Fort McPherson. Also Alberta: Cypress Hills (Soper, 1965:355).

Some far northern extralimital occurrences in the Northwest Territories, from west to east (Kelsall, 1972:129–138), include the following: Baillie Island; Cape Kendall; near Arctic Sound; half-day walk inland from Stewart Point at W base Adelaide Peninsula; near Igloolik.

[Youngman (1975:161) referred specimens from the southeastern part of the Yukon Territory to *A. a. gigas* instead of to *A. a. andersoni* as Peterson (1952) did. Peterson's arrangement is retained here.]

Alces alces gigas Miller

1899. *Alces gigas* Miller, Proc. Biol. Soc. Washington, 13:57, May 29, type from N side Tustumena Lake, Kenai Peninsula, Alaska.
1915. *Alces alces gigas*, Lydekker, Catalogue of the ungulate mammals in the . . . British Museum, 4:237.

MARGINAL RECORDS.—Alaska: 3 km. S Barrow Dewline Site (71° 17′ N, 156° 38′ W) (Chesemore, 1968:528, as *A. alces* only); mouth Colville River. Yukon: Peel River; upper waters Stewart River; N fork Macmillan River; Teslin River district. British Columbia: Atlin; Telegraph Creek; 15 mi. NW Kelsall Lake. Alaska: Chugachick Bay, Kenai Peninsula; Katmai, Alaska Peninsula; Yukon Delta; Cape Prince of Wales, Seward Peninsula.

Alces alces shirasi Nelson

1914. *Alces americanus shirasi* Nelson, Proc. Biol. Soc. Washington, 27:72, April 25, type from Snake River, 4 mi. S Yellowstone National Park, Teton Co., Wyoming.
1952. *Alces alces shirasi*, Peterson, Contrib. Royal Ontario Mus. Zool. and Palaeont., 34:23, October 15.

MARGINAL RECORDS.—British Columbia: *Fraser River near Alberta line*. Alberta: Banff; Waterton. Montana: *upper Missouri River*. Wyoming (Long, 1965a:715, unless otherwise noted): S Fort Shoshone River, near Cody (Peterson, 1952:24); South Pass; Moose Flat, Greys River. Colorado: Mica Basin, E of Hahn's Peak P.O. (Armstrong, 1972:307); Estes Park region; Grand County; Middle Fork Stuart Creek, 40 mi. SW Meeker (Armstrong, 1972:307, and Fig. 114, on p.

306). Utah: Uinta Mts.; Spanish Fork Canyon; American Fork Canyon; near Kaysville; Lewiston. Idaho: Salmon River Mts.; Locksa River, Idaho Co. Washington: Blue Mts.; Twin Lakes, Ferry Co.; 3 mi. N Boyd's. British Columbia: *Flathead Valley*.

Genus **Rangifer** Hamilton-Smith
Reindeer and Caribou

Revised by Banfield, Nat. Mus. Canada, Bull. 177 (Biol. Ser. 66): vi + 137, 12 pls., 15 figs., 23 tables, February 7, 1962.

1827. *Rangifer* Hamilton-Smith, *in* Griffith, The animal kingdom . . . by the Baron Cuvier . . . , 5:304. Type, *Cervus tarandus* Linnaeus.
1827. *Tarandus* Billberg, Synops. faunae Scandinaviae, p. 22. Type, *Cervus tarandus* Linnaeus.
1838. *Procerus* Serres, Essai sur les Cavernes à Ossements, ed. 3, p. 143. Type not designated; included species were *Cervus tarandus* Linnaeus and *Procerus caribaeus* Serres.
1840. *Procervus* Blainville, Comptes Rendus Acad. Sci. Paris, 11:392, a renaming of *Procerus* Serres.
1845. *Achlis* Reichenbach, Die vollständigste Naturgeschichte . . . , Säugethiere, 3:12. Type, *Cervus tarandus* Linnaeus.

Size moderate; ears and tail short; muzzle wide, hairy; mane on neck; lateral hoofs functional; main hoofs broad, deeply cleft, almost circular when held together; tarsal glands present, metatarsal glands absent; antlers large, variable, with brow and bez tines, semipalmate distally, usually present in both sexes. Lachrymal vacuity large, pit small; vomer high posteriorly, completely dividing posterior nares; orbital cavities not extended backward under bases of horn pedicles. Maxillary canine usually present in both sexes; lower incisors relatively small; posterior lobe of last molar small. Dentition, i. $\frac{0}{3}$, c. $\frac{1}{1}$, p. $\frac{3}{3}$, m. $\frac{3}{3}$.

Some mammalogists having knowledge of geographic variation in *Rangifer* have felt that Banfield (1962a:vi + 137) recognized too few subspecies or at least wished they knew his philosophy and the conceptual basis on which his classification was made. Under the title of "The post glacial dispersal of American caribou" he explained his basis more concisely in 1963 than in the monograph of the preceding year. Banfield's later (1963:206) explanation follows.

"A quantitative study of 704 recent and 57 sub-fossil caribou specimens led to the recognition of six Western Hemisphere subspecies of the Holarctic reindeer, *Rangifer tarandus* (Linné). These allop[a]tric subspecies displayed a broad pattern of concordant characters and . . . intergradation [in the marginal parts of their geographic ranges]. The whole population also exhibited a mosaic of individual characters which were considered indicative of local demes below the subspecific level.

"The subspecies can be arranged in three supra-subspecies groups of approximate equal taxonomic rank: mainland tundra taxa (*R. t. granti* and *R. t. groenlandicus*); Arctic insular taxa (*R. t. pearyi* and *R. t. eogroenlandicus*); and forest taxa: (*R. t. caribou* and *R. t. dawsoni*). When the distributions of the subspecies were plotted on a map of the Last (Wisconsin) Glaciation, it was noticed that an unglaciated refuge was available in the heart of the range of each group: Beringia, the continental tundra refuge; Pearyland for the Arctic insular group; and the periglacial Continental forested refuge for the woodland group. The finding of the late-Pleistocene caribou remains in each of these refugia confirms their occupation during the Wisconsin stage. It is also noteworthy that present areas of intergradation centre on the boundaries of these refugia.

"Although the evidence points to the evolution of most of these subspecies *in situ*, there are certain anomalies in the distribution pattern that can best be explained by emigration into corridors which are known to have become ice-free early in the deglaciation process.

"Such reasoning explains the northern extension of woodland caribou in the Mackenzie Valley, along the Cordilleran–Laurentide Ice Sheet border; the evolution of *dawsoni* in relative isolation on the Queen Charlotte Islands after reaching the islands soon after their exposure; and the eastward expansion of *groenlandicus* to the Keewatin tundra, Baffin Island and southwestern Greenland, along the northern boundary of the shrinking Laurentide Ice Sheet while much of northeastern Canada was unavailable to southern populations of caribou."

Understandably some taxonomists may choose to apply subspecific names to certain of the geographic variants that Banfield has relegated to the lower category of demes.

Domesticated strains of the reindeer from Eurasia more than once were introduced into North America (Alaska, Canada, and Greenland) after 1890, but by 1975 most or all of them had died out in America. According to Banfield (1962a:43, 44) all the introduced animals were subspecies *Rangifer tarandus tarandus* (Linnaeus 1758). For leads to the voluminous literature on these introductions, see Banfield (1962a:102).

As of 1977 the fate of native herds is uncertain. Early decision on state or federal authority to manage the natural rangelands is desirable.

Rangifer tarandus
Caribou

External measurements: 1369–2536, 102–218, 381–700. Height at shoulder, 680–1397. Weight, to 600 pounds. Females average approx. 10 per cent smaller than males. Length of skull, 295–435; occipitonasal length, 221–302; zygomatic breadth, 110–149; length of series of upper cheek-teeth, 76.9–110. Color varying from almost pure white to blackish brown; neck usually with white throat-fringe; rump patch white; white band above hoofs.

Banfield (Preprint, Nat. Mus. Canada Bull. 185:60–71, 6 plates, May, 1963) designated neotypes for six subspecies of *Rangifer tarandus: arcticus, caribou, fennicus, groenlandicus, platyrhynchus,* and *tarandus.*

Rangifer tarandus caribou (Gmelin)

1788. [*Cervus tarandus*] *caribou* Gmelin, Syst. nat., ed. 13, 1:177. Type locality, eastern Canada [= restricted to Quebec City by Banfield, Nat. Mus. Canada Bull. 177 (Biol. Ser. 66):78, February 7, 1962].

1884. *Rangifer tarandus caribou,* True, Proc. U.S. Nat. Mus., 7:592, November 29.

1829. *Cervus tarandus* var. *sylvestris* Richardson, Fauna Boreali-Americana, 1:250. Type locality, southwestern shores of Hudson Bay.

1847. *Cervus hastalis* Agassiz, Proc. Boston Soc. Nat. Hist., 2:188, for North American counterpart of Old World *Cervus tarandus, auct.,* but no type locality, type specimen, or description. Here considered to be a *nomen nudum.* If not a *nomen nudum,* probably is a junior synonym of [*Cervus tarandus*] *caribou* Gmelin 1788.

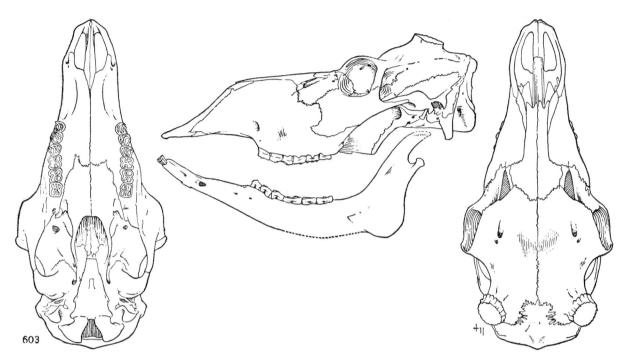

Fig. 603. *Rangifer tarandus groenlandicus*, Holstenberg, Greenland, No. 2266 K.U., sex ?, X ¼.

1896. *Rangifer terraenovae* Bangs, Preliminary description of the Newfoundland caribou, Boston, p. 1, November 11, type from Codroy, Newfoundland.

1899. *Rangifer montanus* Thompson-Seton, Ottawa Nat., 13(5):129, August, type from Illecillewaet watershed, near Revelstoke, Selkirk Range, British Columbia.

1901. *Rangifer stonei* J. A. Allen, Bull. Amer. Mus. Nat. Hist., 14:143, May 28, type from Kenai Peninsula, Alaska.

1902. *Rangifer osborni* J. A. Allen, Bull. Amer. Mus. Nat. Hist., 16:149, April 16, type from Cassiar Mts. [60 mi. SE Dease Lake], British Columbia.

1912. *Rangifer fortidens* Hollister, Smiths. Miscl. Coll., 56(35):3, February 7, type from head Moose Pass branch Smoky River, Alberta.

1914. *Rangifer arcticus caboti*, G. M. Allen, Proc. New England Zool. Club, 4:104, March 24, type from 30 mi. N Nachvak, Labrador.

1915. *Tarandus rangifer keewatinensis* Millais, *in* The gun at home and abroad . . . , p. 257. Type locality, central or northern Manitoba, or Keewatin, or north or south Saskatchewan or as far north as Peace River or Lake Athabasca (see Miller and Kellogg, Bull. U.S. Nat. Mus., 205:815, March 3, 1955).

1915. *Tarandus rangifer labradorensis* Millais, *in* The gun at home and abroad . . . , p. 259. Type locality, ". . . horns brought into Nain, Davis Inlet, and Fort Chimo."

1915. *Tarandus rangifer ogilvyensis* Millais, *in* The gun at home and abroad . . . , p. 263, type from Ogilvie Mts., N of Dawson, Yukon.

1919. *Rangifer mcguirei* Figgins, Proc. Colorado Mus. Nat. Hist., 3(1):1, December 28, type from Kletson Creek, a tributary of the White River, 4 mi. E Alaska–Yukon boundary, Yukon.

1935. *Rangifer montanus selousi* Barclay, Proc. Zool. Soc. London, p. 306, July 4, type from mountains S of South Fork Macmillan River, 5000 ft., Yukon.

MARGINAL RECORDS (Banfield, 1962a:86, 88, unless otherwise noted).—Mackenzie: Keith Bay, Great Bear Lake; Fort Rae; Fort Reliance. Manitoba: Cranberry Portage; Wabowden; Nelson River. Ontario: 65 mi. E Winisk (Hall and Kelson, 1959:1018, as *R. t. arcticus*). Quebec: Rupert River (Harper, 1961:144); Cape Jones (*op. cit.*:135); Cape Wolstenholme (*op. cit.*: 134); Belle Isle Strait (Jacobi, 1931:110). Newfoundland: Gander Lake (Cameron, 1959:106); Codroy. Quebec: Gaspé, thence along Atlantic Coast, including Prince Edward Island, to Nova Scotia: Victoria Co., Cape Breton Island, and remainder of *Nova Scotia* (Banfield, 1962a:73, 86). New Brunswick: Woodstock (J. A. Allen, 1894:100). Maine: Calais; Upton. New Hampshire: *Connecticut Lakes* (Goodwin, 1935:178). Vermont: *northern part* (G. M. Allen, 1904:8). Michigan (Burt, 1942:214): Spitza Farm, near Minden City; Beaver Islands. Wisconsin (Cory, 1912:81): near White River, Price Co.; near La Pointe. Minnesota: Cook County (Swanson, 1945:100); Red Lake. North Dakota: Turtle Mts. (V. Bailey, 1927:33). Saskatchewan: Meadow Lake. Alberta: Fort McMurray; Lesser Slave Lake; Thoral Creek (Carter *in* Crowe, 1943:409); near Banff (Rand, 1948:214). British Columbia: Moyie Range, E of Kootenay Lake (Anderson, 1947:179). Idaho: neighborhood of Elk City (Davis, 1939:371). Washington (Anderson, 1947:179): Usk; Okanogan County. British Columbia: Okanagan Lake; Wells Gray Park; Jarvis Pass; *Ithca Mtn. in Coast Range* (Banfield, 1962a:75);

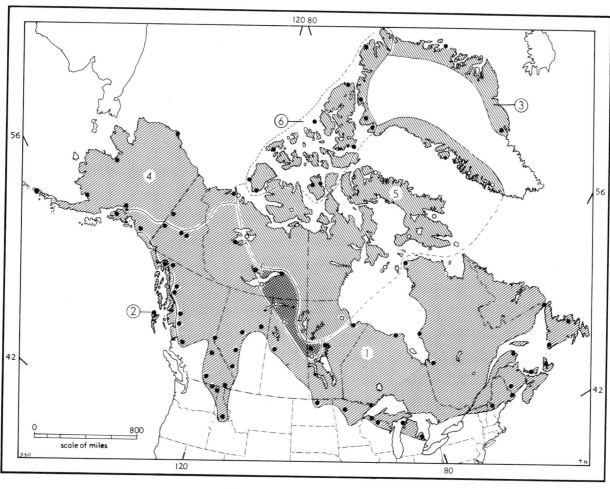

Map 540. *Rangifer tarandus*.

1. *R. t. caribou*   3. *R. t. eogroenlandicus*   5. *R. t. groenlandicus*
2. *R. t. dawsoni*   4. *R. t. granti*   6. *R. t. pearyi*

Cariboo [Caribou] Mtn., in Coast Range (*ibid.*); Morice Lake in Coast Range (*ibid.*); Suskwa River (Cowan and Guiguet, 1965:384); *Skeena River*; Glenora (Cowan and Guiguet, 1965:384); *Glenora Mtn.* (Jacobi, 1931:92); Level Mtn. (J. A. Allen, 1903:526); Atlin (Cowan and Guiguet, 1965:384). Alaska: Coast Range above Skagway (O. J. Murie, 1935:77); Copper River (Banfield, 1962a:76); Kenai Peninsula. Yukon: Kletson Creek, a tributary of White River, 4 mi. E Alaska–Yukon boundary (this is the type locality of *R. mcguirei* placed as a synonym of *R. t. caribou* by Banfield, 1962a:72); Pelly River; Macmillan River.

Rangifer tarandus dawsoni Thompson-Seton

1900. *Rangifer dawsoni* Thompson-Seton, Ottawa Nat., 13:260, February, type from foothills W of Virago Sound, N end Graham Island, Queen Charlotte Islands, British Columbia.

1915. *Rangifer tarandus dawsoni*, Lydekker, Catalogue of the ungulate mammals in the . . . British Museum, 4:251.

MARGINAL RECORDS.—British Columbia, Graham Island: type locality (close to Naden Harbor); *valley of river running into Virago Sound; near Frederick Island*. According to Cowan and Guiguet (1965:386), *R. dawsoni* occurred only on the part of Graham Island W of Naden Harbor and may be extinct as no definite evidence of its presence has been recorded since 1920. Banfield (1963:48) thought the time of extinction was "shortly after 1910."·

Rangifer tarandus eogroenlandicus Degerbøl

1957. *Rangifer tarandus eogroenlandicus* Degerbøl, Acta Arctica, Fasc. 10:45, prior to August 16, type from head Scoresby Sund, Greenland.

MARGINAL RECORDS.—Greenland: Germanialand; Angmagssalik.

Rangifer tarandus granti J. A. Allen

1902. *Rangifer granti* J. A. Allen, Bull. Amer. Mus. Nat. Hist.,
16:122, March 31, type from W end Alaska Peninsula, opposite Popof Island, Alaska.
1910. *Rangifer tarandus granti*, Ward, Records of big game,
ed. 6, p. 85.
1912. *Rangifer excelsifrons* Hollister, Smiths. Miscl. Coll.,
56(35):5, February 7, type from Meade River, near Point
Barrow, Alaska.

MARGINAL RECORDS (Banfield, 1962a:59, unless
otherwise noted).—Alaska: Meade River, thence east-
ward along coast to Northwest Territories, including
Richards Island (Porsild, 1945:20, under *R. arcticus*) to
Anderson River. Yukon: Clinton Road. Alaska: Susitna
River, thence southwestward on Alaska Peninsula to
Unimak Island. Thence northeastward to Nushagak
(Osgood, 1900b:23), up coast to Unalakleet (*ibid.*), and
along coast to point of beginning.

Rangifer tarandus groenlandicus (Linnaeus)

1767. *Capra groenlandica* Linnaeus, Syst. nat., 12th ed., 1:98,
from Alpine America. In synonymy of *Cervus tarandus*.
1898. *Rangifer tarandus groenlandicus*, Lydekker, Deer of
all lands, p. 47.
1780. *Cervus grönlandicus* Borowski, Gemeinnüzzige
Naturgeschichte des Thierreichs . . . , 1(3):72. Type locality, Greenland.
1829. *Cervus tarandus* var. *arctica* Richardson, Fauna
Boreali-Americana, 1:241. Type locality, Fort Enterprise,
about 64° 30′ N, 113° W, Mackenzie, Northwest Territories,
Canada (see J. A. Allen, Bull. Amer. Mus. Nat. Hist., 24:584,
September 11, 1908).

MARGINAL RECORDS (Banfield, 1962a:55, 56, un-
less otherwise noted).—Franklin: Pond Inlet, Baffin Is-
land. Greenland: Holsteinborg; Godthaab. Franklin
(Baffin Island): Cumberland Sound; NE of Chorkbak
Inlet (Manning, 1943:50). Keewatin: Mansel Island;
Coats Island. Manitoba: McClintock; The Pas. Sas-
katchewan: Cree Lake; Lake Athabasca. Mackenzie:
Fort Resolution; Great Bear Lake; Horton River.
Franklin: Southern Victoria Island; Milne Inlet, Baffin
Island.

Rangifer tarandus pearyi J. A. Allen

1902. *Rangifer pearyi* J. A. Allen, Bull. Amer. Mus. Nat. Hist.,
16:409, October 31, type from Ellesmere Land, 79° N,
Franklin.
1915. *Rangifer tarandus pearyi*, Lydekker, Catalogue of the
ungulate mammals in the . . . British Museum, 4:256.

MARGINAL RECORDS (Banfield, 1962a:64, unless
otherwise noted).—Greenland: North Greenland; N of
Kane Basin (Anderson, 1947:180); Marshall Bay; *In-
glefield Bay*; Olriks Fiord. Franklin: Craig Harbour;
Browne Bay, Prince of Wales Island; Ommanney Bay,
Prince of Wales Island; De Salis Bay; Cape Kellett;
Mould Bay, Prince Patrick Island; Meighen Island
(Jacobi, 1931:84); Cape Hecla, Ellesmere Island (*op.
cit.*:112).

SUPERFAMILY **BOVOIDEA**

Chiefly grazers; horns, usually present on both
sexes, permanent bony outgrowths from skull,
covered with highly keratinized horny sheath
that is permanent (Bovidea) or shed and replaced
annually (Antilocapridae).

FAMILY **ANTILOCAPRIDAE**—Pronghorn

Horn-cores simple or forked; sheath sometimes
forked when core is simple; sheath shed annually.
Of the several genera, only one persists into Recent time.

Genus **Antilocapra** Ord—Pronghorn

1818. *Antilocapra* Ord, Jour. Phys. Chim. Hist. Nat. et Arts,
87:149, August. Type, *Antilope Americanus* Ord.
1827. *Dicranoceros* Hamilton-Smith, *in* Griffith, The animal
kingdom . . . by the Baron Cuvier . . . , 4:169, Type, *Antilope Americanus* Ord.

For characters see account of *Antilocapra
americana*.

Antilocapra americana
Pronghorn

External measurements of males: 1245–1472;
89–178; 400–432. Weight to approx. 138 pounds.
Females average approx. 10 per cent smaller than
males. Basilar length, 240–256; occipitonasal
length, 216–240; alveolar length of maxillary
tooth-row, 67–72. Upper parts warm tan, neck
with black mane; underparts, rump, and 2 bands
across neck white. Horns in both sexes, deciduous, composed of fused hairs, branched about
two-thirds distance between base and tip; bony

horn-core unbranched, flattened laterally, sharp anteriorly, rounded posteriorly. Dew-hoofs absent. Frontal sinuses open to outside by two large, longitudinal fossae in dorsal surface of frontal bones; orbits large, situated ventral to horn-cores; lachrymal bone not articulating with nasal bone; cheek-teeth hypsodont, selenodont, and rootless; p4 with closed anterior fossette. Dentition, i. $\frac{0}{3}$, c. $\frac{0}{1}$, p. $\frac{3}{3}$, m. $\frac{3}{3}$.

Antilocapra americana americana (Ord)

1815. *Antilope Americanus* Ord, *in* Guthrie, A new geog., hist. and coml. grammar . . . , Philadelphia, ed. 2, 2:292 (described on p. 308). Type locality, plains and highlands of the Missouri River.
1818. *Antilocapra americana* Ord, Jour. Phys. Chim. Hist. Nat. et Arts, 87:149.
1855. *Antilocapra anteflexa* Gray, Proc. Zool. Soc. London, p. 10, March 27, type "most probably from America." (Based on an anomolous pair of horns.)

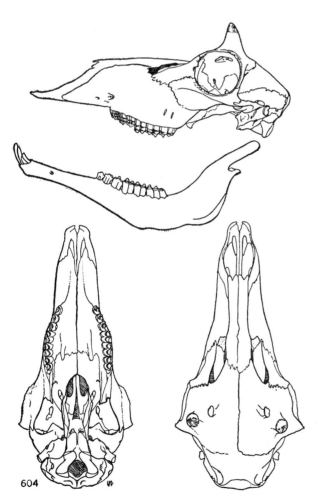

Fig. 604. *Antilocapra americana americana,* 13 mi. W, 6 mi. N Deep Hole, Washoe Co., Nevada, No. 93881 M.V.Z., ♀, X ¼.

MARGINAL RECORDS.—Saskatchewan: Carlton House. Manitoba: *southwestern Manitoba.* Minnesota: Kittson County; Lyon County. Iowa: northwestern Harrison County (Jones, 1960:249, sight record only). Nebraska: vic. Elkhorn (Jones, 1964c:324). Oklahoma: Kay County. Texas: Hutchinson County (Cutter, 1959:33). New Mexico: Portales; Eddy; *ca.* Jarilla; Sandia Mts. (Findley, *et al.,* 1975:334, as *A. americana* only); Beaver Dam Creek, E base Mogollon Mts. Arizona: SE of Springerville; Anderson Mesa, 7000 ft., 30 mi. SE Flagstaff (Cockrum, 1961:256); Walnut [Walnut Grove, SW of Prescott] (*ibid.*); Antelope Valley. California: W side Colorado Desert, near Mexican line; Campo; San Jacinto Valley; Antelope Valley; Salinas Valley; Contra Costa County; Shasta Valley; vic. Mt. Dome. Nevada: 12-Mile Creek, Washoe Co.; High Rock Canyon, 5000 ft. Idaho: forks of Owyhee River, Owyhee Co.; Antelope Valley; Lemhi Valley. Montana: Red Rock [Deer] Creek; Nine-mile Prairie (Hoffmann, *et al.,* 1969:599, as *A. americana* only); *Ovando* (*ibid.*). Alberta: near Rocky Mountain House.—Bowles, 1971:426, mentions the possibility that *A. americana* occurred on the prairies "about ten years before the first sttlers [*sic*]" as far eastward as *Jefferson County,* which is in southeastern Iowa.

Antilocapra americana mexicana Merriam

1901. *Antilocapra americana mexicana* Merriam, Proc. Biol. Soc. Washington, 14:31, April 5, type from Sierra en Media [about 10 mi. S New Mexico border], Chihuahua.

MARGINAL RECORDS.—Arizona: near Peach Springs; Big Chino Valley; Mogollon Mtn., 25 mi. NE Fort Verde. New Mexico: 15 mi. W Las Palomas; valley E of Socorro; Tularosa Valley; W of Organ Mts. Texas: 5 mi. N, 5 mi. E Draw (Garner, 1967:286, as *A. americana* only); western Fannin County; central Kaufman County; western Robertson County; western Bastrop County; northeastern Kennedy County. Hidalgo: Cazadero. San Luis Potosí: Tanque de Dolores. Zacatecas: Llanos de Palula, between boundary of San Luis Potosi and 60 km. NNW Catorce. Chihuahua: 60 km. W Ciudad Chihuahua (Anderson, 1972:393). Sonora: Llanos de Guaje, 108 km. NW Ocampo. Arizona: base Huachuca Mts.; Lechuguilla Desert, thence northward, W of Colorado River, to point of beginning.

Antilocapra americana oregona V. Bailey

1932. *Antilocapra americana oregona* V. Bailey, Proc. Biol. Soc. Washington, 45:45, April 2, type from Hart Mtn. (Warner Mts.), Lake Co., Oregon.

MARGINAL RECORDS.—Washington: *much of plains country of eastern Washington.* Oregon: warm springs near sources of John Dea's [Day] River, about 100 mi. from Fort Dalles; near Antone; southern Malheur County; Warner Lake Valley; Rogue River Valley; Pumice Desert between Crater Lake and Diamond Lake; head N fork Umpqua River; S of The Dalles.

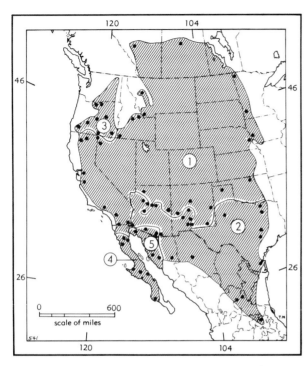

Map 541. *Antilocapra americana.*

Guide to subspecies	3. *A. a. oregona*
1. *A. a. americana*	4. *A. a. peninsularis*
2. *A. a. mexicana*	5. *A. a. sonoriensis*

Antilocapra americana peninsularis Nelson

1912. *Antilocapra americana peninsularis* Nelson, Proc. Biol. Soc. Washington, 25:107, June 29, type from 45 mi. S Calmallí, Baja California.

MARGINAL RECORDS.—Baja California: Cañón Esperanza; San Felipe Bay; inland from Santa Rosalía Bay; S beyond Magdalena Bay; type locality; Vizcaíno Bay; San Quentín.

Antilocapra americana sonoriensis Goldman

1945. *Antilocapra americana sonoriensis* Goldman, Proc. Biol. Soc. Washington, 58:3, March 21, type from 40 mi. N Costa Rica, a ranch on N side Río de Sonora, SW of Hermosillo, Sonora.

MARGINAL RECORDS.—Arizona: near Cipriano Well, Organ Pipe Cactus National Monument; Crittenden. Sonora: Hermosillo; type locality; Santa Rosa Valley, near Monument 183. The well-marked, diagnostic, cranial characters for both sexes of this subspecies are described by Paradiso and Nowak (1971:855–858).

FAMILY BOVIDAE—Bovids

Frontal appendages hollow, nondeciduous, horny sheaths supported by bony cores, frequently present in both sexes. Upper canines absent; lower canines with simple crowns; cheek-teeth frequently hypsodont. Lateral toes always incomplete, sometimes absent; lower ends of lateral metacarpals absent. Gall bladder usually present; placenta with numerous cotyledons.

KEY TO NORTH AMERICAN GENERA OF BOVIDAE

1. Total length more than 2000; length of skull more than 350; length of maxillary tooth-row more than 120.
    2. Tail longer than 300; horns smooth, conical; accessory column on inner side of molars not reduced; paroccipital processes widely separated from condyles. . . . .*Bison,* p. 1108
    2'. Tail shorter than 150; horns rugose, flattened at base; accessory column on inner side of molars reduced; paroccipital processes not widely separated from condyles. . . . .*Ovibos,* p. 1112
1'. Total length less than 2000; length of skull less than 350; length of maxillary tooth-row less than 120.
    3. Tail longer than 150; lachrymal pits absent; horns less than 6 in. in circumference at base. . . . . .*Oreamnos,* p. 1110
    3'. Tail shorter than 150; lachrymal pits present; horns more than 6 in. in circumference at base. . . . . . .*Ovis,* p. 1113

Genus **Bison** Hamilton-Smith—Bison

1827. *Bison* Hamilton-Smith, *in* Griffith, The animal kingdom . . . by the Baron Cuvier . . . , 5:373. Type, *Bos bison* Linnaeus.

Forehead short, broad; muzzle narrowed; nasals pointed, not reaching premaxillae; nasal opening composed of premaxillae, maxillae, and nasals; orbits tubular, composed of frontals, lachrymals, and jugals; molars with style between anterior and posterior lobes; both sexes having permanent horns. Dentition, i. $\frac{0}{3}$, c. $\frac{0}{1}$, p. $\frac{3}{3}$, m. $\frac{3}{3}$.

Bison bison
Bison

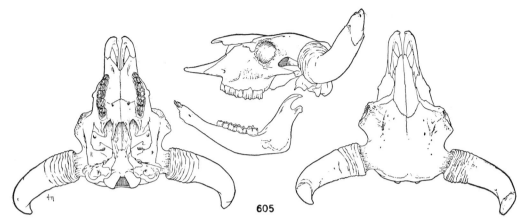

Fig. 605. *Bison bison bison*, Buffalo National Park, Wainwright, Alberta, No. 48115 K.U., ♂, X 1/12.

External measurements: (males) 3042–3803; 330–913; 584–661; (females) 2132–2896; 304–508; 508. Weight of large male approx. 818 kg.; large female, 545 kg. Head heavy; neck short; shoulders high, humped; tail short, tufted at tip; fur on head, neck, and shoulders brownish black, long, wooly, forming beard on chin; fur of body short, brown to light brown. Greatest length of skull, (males) 491–595; basilar length, (males) 454–555; alveolar length of molars, (males) 82–98; length of maxillary tooth-row, (males) 142–150 (135–147.5 in females).

As of 1974 a pure blooded remnant of *Bison bison athabascae* persists in the District of Mackenzie in the Nyarling River–Big Buffalo Lake area of Wood Buffalo Park. After 1957 stock from there was transported more than 100 mi. WNW and established in an area near Fort Providence (Banfield, 1974:407). Herds of *B. b. bison* have been established in several wildlife preserves in the United States and Canada.

Bison bison athabascae Rhoads

1898. *Bison bison athabascae* Rhoads, Proc. Acad. Nat. Sci. Philadelphia, 49:498, January 18, type from within 50 mi. SW Fort Resolution, Mackenzie.
1899. *Bison occidentalis* Lucas, Proc. U.S. Nat. Mus., 21:758, type from St. Michael, Alaska.
1932. *Bison bison oregonus* V. Bailey, Proc. Biol. Soc. Washington, 45:48, April 2, type from Malheur Lake, Harney Co., Oregon.
1933. *Bison bison haningtoni* Figgins, Proc. Colorado Mus. Nat. Hist., 12(4):30, December 5, type from head Rock Creek, NE of South Park, Park Co., Colorado.

MARGINAL RECORDS.—Alaska?: 30 mi. above mouth Tanana River. Yukon?: Peel River. Mackenzie: vic. Great Marten Lake; type locality; 40 mi. NE Fort Smith. Alberta: E side Athabaska above and below Fort McMurray; between Methye Portage and Lake La Biche. Montana: Glacier National Park. Idaho: Agency Creek, Lemhi Co. Wyoming: Gardiner River, Yel-

lowstone National Park. Montana: Pryor Mts. Wyoming (Long, 1965a:719, unless otherwise noted): W slope Bighorn Mts. (Fryxfell, 1928:138); 22 mi. W Cody; *Hoback River*; Gross Ventre Range, E Jackson. Colorado: Lost Park, near Marvin Lake; Alma, 10,232 ft. Wyoming: Snowy Range (Long, 1965a:719). Colorado: Pikes Peak region; Elk Mts., Gunnison Co.; top of Book Cliffs. Utah: Twelve Mile Creek Canyon, near Gunnison. Idaho: 20 mi. W Raft River, Cassia Co. Oregon: Barren Valley, E of Steens Mts. California: near Honey Lake; near Eagle Lake; Alturas Valley. Oregon: Izee. Washington: Grand Coulee. Oregon: near Joseph (Olterman and Verts, 1972:10, as *B. bison* only). Alberta: 15 mi. S Henry House, Buffalo Prairie. British Columbia: Fort St. John region; Cypress Creek; Fort Nelson. Yukon: near Fort Liard. British Columbia: Atlin

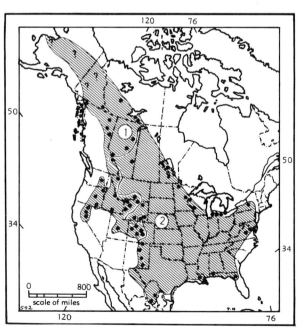

Map 542. *Bison bison.*

1. *B. b. athabascae*   2. *B. b. bison*

(Cowan and Guiguet, 1965:388). Yukon: 12 mi. above mouth Pelly River, near Selkirk?. Alaska?: near St. Michael.

Bison bison bison (Linnaeus)

1758. [*Bos*] *bison* Linnaeus, Syst. nat., ed. 10, 1:72. Type locality, ancient "Quivira," central Kansas (Hershkovitz, Proc. Biol. Soc. Washington, 70:32, June 28, 1957).

1888. *B*[*ison*]. *bison*, Jordan, Manual of the vertebrate animals . . . , ed. 5, p. 337.

1762. *Bison americanus* Brisson, Regnum animale . . . , p. 56. Type locality, America.

1915. *Bison sylvestris* Hay, Proc. U.S. Nat. Mus., 48:515, April 8, type from Huron County, Ohio (Pleistocene).

1915. *Bison americanus pennsylvanicus* Shoemaker, A Pennsylvania bison hunt, p. 9. Type locality, Pennsylvania (based on hearsay and regarded as invalid by Skinner and Kaisen, Bull. Amer. Mus. Nat. Hist., 89:163, October 31, 1947; see Opinion 2, International Commission on Zoological Nomenclature).

1933. *Bison bison septemtrionalis* [*sic*] Figgins, Proc. Colorado Mus. Nat. Hist., 12:28, December 5, type from 6 mi. NE Palmer, Merrick Co., Nebraska.

MARGINAL RECORDS.—Saskatchewan: *northern edge Great Plains region.* Manitoba: Lake Winnipeg. Minnesota: Lake of the Woods; SE Pine County. Wisconsin: St. Croix River; Tremepleau [= Trempealeau] River. Michigan: Kalamazoo County; Jackson County; Wayne County. New York: salt lakes, vic. Syracuse. District of Columbia: [Wahington]. Virginia: Goochland County. Florida: Newman's Lake; vic. Pensacola. Louisiana (Lowery, 1974:502, 504): Bayou Terre aux Boeufs; Manchacq [Bayou Manchac], thence westward to Coahuila: Monclova; Parras. Chihuahua: *NW Chihuahua.* New Mexico: Carlsbad (Findley, *et al.*, 1975:335); 15 mi. S Fort Sumner (*ibid.*); near Ribera; upper valley of the Vermejo. Colorado: *plains east of mts.* Wyoming (Long, 1965a:719, 720): Laramie Plains; Snake River, mouth Muddy Creek (Baggs); Powder River, Sheridan Co. Montana: *east of continental divide.* British Columbia: Howse Pass (Cowan and Guiguet, 1965:388). Alberta: Wainwright; *Hand Hills, 500 mi. W Manitoba.*—See Reed (1955:130) on early occurrence in *Arizona* and New Mexico. See also Dalquest (1961:73), who records the remains of the extinct *Bison antiquus* along with remains of the modern bison at Wichita Falls, Texas, in a deposit "considered to be of early Recent age." Some persons think *B. b. bison* occurred in Tamaulipas of northeasternmost México in the sixteenth century (see Coopwood, 1900), but I am not convinced that the animal did occur there.

Genus Oreamnos Rafinesque—Mountain Goat

1817. *Oreamnos* Rafinesque, Amer. Month. Mag., 2:44, November. Type, *Mazama dorsata* Rafinesque [= *Rupicapra americana* Blainville].

1827. *Aplocerus* Hamilton-Smith, *in* Griffith, The animal kingdom . . . by the Baron Cuvier . . . , 5:354. Type, *Antilope lanigera* Hamilton-Smith [= *Rupicapra americana* Blainville].

1844. *Haplocerus* Wagner, *in* Schreber, Die Säugthiere . . . , Suppl., 4:462, a renaming of *Aplocerus* Hamilton-Smith.

1884. *Haploceros* Flower and Garson, Cat. Osteol. Mus. Roy. Coll. Surg., pt. 2, p. 257, an accidental renaming of *Haplocerus* Wagner.

1901. *Oreamnus* Elliot, A synopsis of the mammals of North America . . . (Zool. Publ., Field Mus.), 2:44, an accidental renaming of *Oreamnos* Rafinesque.

Body stocky, hump on withers, muzzle small, "beard" present, tail short; interdigital cleft deep, foot-glands rudimentary or absent, cannon bones broad; facial glands and lachrymal depression absent; horns present in both sexes, moderate in size, unbranched, conical, ridged at base, diverging laterally, inclining posteriorly; a pair of bare black glands present behind horns; hair white or yellowish white, long, with thick, wooly underfur, mammae, 4. Dentition, i. $\frac{0}{3}$, c. $\frac{0}{1}$, p. $\frac{3}{3}$, m. $\frac{3}{3}$.

Oreamnos americanus
Mountain Goat

External measurements: (males) 1521–1787; 152–203; 330–368. Weight, 150–300 pounds. Females are 10–20 per cent smaller than males in some dimensions. Basal length of skull, 270–293; occipitonasal length, 299–336; interorbital breadth, 95–98; palatilar length, 176–184; length of maxillary tooth-row, 69–77. Other characters as for the genus.

Cowan and McCrory (1970:60–73) studied skulls and concluded "that there are no valid reasons for recognizing subspecies within *Oreamnos americanus* (Blainville)." Some other investigators have thought differences in external measurements (*e.g.*, length of tail vertebrae) provide basis for recognizing subspecies. If valid subspecies exist, their geographic ranges may be approx. as shown on Map 543.

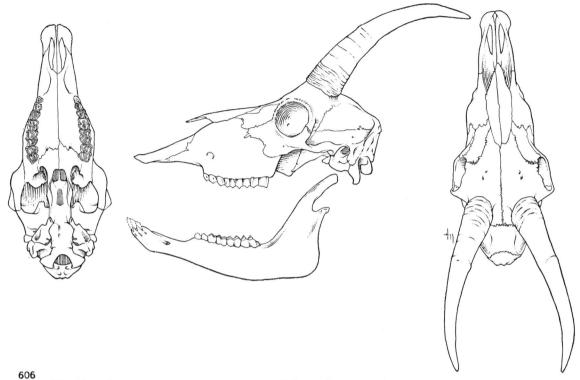

606

Fig. 606. *Oreamnos americanus americanus*, Tcho Park Mts., Washington, No. 1885 K.U., ♀, X ¼.

Oreamnos americanus americanus (Blainville)

1816. *R[upicapra]. americana* Blainville, Bull. Soc. Philom., Paris, p. 80. Type locality, restricted to Cascade Range, near Columbia River, in Oregon or Washington by Hollister, Proc. Biol. Soc. Washington, 25:185, December 24, 1912; regarded as probably Mt. Adams, Washington, by Dalquest, Univ. Kansas Publ., Mus. Nat. Hist., 2:409, April 9, 1948.

1912. *Oreamnos americanus americanus*, Hollister, Proc. Biol. Soc. Washington, 25:186, December 24.

1815. *Ovis montanus* Ord, *in* Guthrie, A new geog., hist. and coml. grammar . . . , Philadelphia, ed. 2, 2:292. Not Schreber, 1804, or Tiedemann, 1808.

1817. *Mazama dorsata* Rafinesque, Amer. Month. Mag., 2:44, November, a renaming of *Ovis montanus* Ord.

1817. *Mazama sericea* Rafinesque, Amer. Month. Mag., 2:44, November, a renaming of *Rupicapra americana* Blainville.

1822. *Antilope lanigera* Hamilton-Smith, Trans. Linnaean Soc., 13:38, a renaming of *Rupicapra americana* Blainville.

1823. *Capra columbiana* Desmoulins, Dict. Class Hist. Nat., 3:580, a renaming of *Rupicapra americana* Blainville.

MARGINAL RECORDS.—British Columbia: Tesla Lake; Texas Creek; Manning Park. Washington: Lake Chelan; E of Mt. Stuart, Wenatchee Range; Mt. Adams. Oregon: Mt. Jefferson. Washington: Mt. St. Helens; Mt. Index; Mt. Higgins; Mt. Baker, thence northward to British Columbia: Kynoch Inlet.

Oreamnos americanus columbiae Hollister

1904. *Oreamnos montanus columbianus* J. A. Allen, Bull. Amer. Mus. Nat. Hist., 20:20, February 10, type from Shesley Mts., British Columbia. Not *Capra columbiana* Desmoulins, 1823.

1912. *Oreamnos americanus columbiae* Hollister, Proc. Biol. Soc. Washington, 25:186, December 24, a renaming of *Oreamnos montanus columbianus* J. A. Allen.

MARGINAL RECORDS.—Mackenzie: Glacier Lake, Iron Mts., 4500 ft.; Nahanni Mts. British Columbia: N of Keily Creek; banks Peace River; Omineca Mts. Alaska: Boca de Quadra; Taku River. Yukon: St. Elias Range; head Pelly River.—The often cited occurrence of *Oreamnos* at *Lapierre House* in northern Yukon at ca. 67° 23', 137° 00', seems to have been in error as explained by Youngman (1975:168).

Oreamnos americanus kennedyi Elliot

1900. *Oreamnus [sic] kennedyi* Elliot, Field Columb. Mus., Publ. 46, Zool. Ser., 3:3, June 20, type from mts. at mouth Copper River, opposite Kayak Island, Alaska.

1913. *Oreamnos americanus kennedyi*, Lydekker, Catalogue of the ungulate mammals in the . . . British Museum, 1:221.

MARGINAL RECORDS.—Alaska: Cook Inlet; Chugach Mts., near Thompson Pass; type locality.

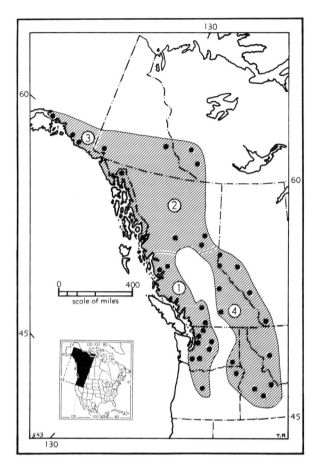

Map 543. *Oreamnos americanus.*

1. *O. a. americanus* 3. *O. a. kennedyi*
2. *O. a. columbiae* 4. *O. a. missoulae*

Oreamnos americanus missoulae J. A. Allen

1904. *Oreamnos montanus missoulae* J. A. Allen, Bull. Amer. Mus. Nat. Hist., 20:20, February 10, type from Missoula, Missoula Co., Montana.
1912. *Oreamnos americanus missoulae*, Hollister, Proc. Biol. Soc. Washington, 25:186, December 24.

MARGINAL RECORDS—British Columbia: *nearly to Peace River.* Alberta: mts. bordering Torrens River; Thoral Creek; Banff, Spray Creek; Waterton Park. Montana: type locality. Idaho: Salmon River Mts., Lemhi Co.; Sawtooth Mts., near Alturus Lake; near Stanley Lake. Oregon: Wallowa County. Washington: Blue Mts.; *northeastern Washington.* British Columbia: Shorts Creek; Wells Gray Park. Alberta: Mt. Robson.

Introduced. *Montana: Beartooth Plateau* (Long, 1965a:720). *Wyoming: 30 mi. E Yellowstone National Park, on Montana-Wyoming state line* (ibid.). *Colorado: Collegiate Range, ca. 20 mi. NW Salida* (Hibbs, 1967:242). *South Dakota: Mt. Rushmore–Needles–Harney Peak area of Black Hills* (Turner, 1974:145).

Genus **Ovibos** Blainville—Muskox

Revised by J. A. Allen, Mem. Amer. Mus. Nat. Hist., n.s., 1:101–226, 8 pls., 46 figs., March, 1913.

1816. *Ovibos* Blainville, Bull. Soc. Philom., Paris, p. 76, May. Type, *Bos moschatus* Zimmermann.
1911. *Bosovis* Kowarzik, Zoöl. Anzeiger, 37:107, February 14. Type, *Bos moschatus* Zimmermann.

For characters see account of *Ovibos moschatus.* This genus is monotypic.

Ovibos moschatus
Muskox

X¹/30   C.G. Pritchard

External measurements: (males) 2028–2434; 89–102; 432–508. Weight to 900 pounds. Females are smaller. Greatest length of skull, (males) 422–516, (females) 385–430; mastoid breadth, (males) 149–198, (females) 144–165; length of maxillary tooth-row, (males) 130–150, (females) 122–146. Horns of old males having broad base covering postorbital region of skull, decurved abruptly, recurved apically; roughened basally, smooth proximally; horns of females shorter, more slender, not so broad at base as in males. Legs and tail short; ears pointed, nearly concealed in hair. Upper parts dark brown, light patch in mid-dorsal region; chin, breast, belly black; forelegs and hind legs dirty white; upper side of tail black, underside white. Underfur thick, wooly, overhair coarse, long, nearly reaching the ground; space between nostrils and upper lip covered with short, close hair; mammae, 4. Orbits tubular; nasals short, not reaching premaxillae; dentition resembling that of *Bison* but less hypsodont and accessory column on inner side of molars reduced; paroccipital processes less widely separated from condyles. Dentition, i. $\frac{0}{3}$, c. $\frac{0}{1}$, p. $\frac{3}{3}$, m. $\frac{3}{3}$.

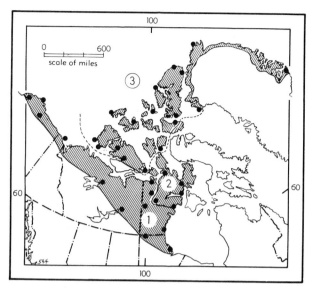

Map 544. *Ovibos moschatus.*

1. *O. m. moschatus*   2. *O. m. niphoecus*
3. *O. m. wardi*

Ovibos moschatus moschatus (Zimmermann)

1780. *Bos moschatus* Zimmermann, Geographische Geschichte . . . , 2:86. Type locality, between Seal and Churchill rivers, Manitoba.
1822. *Ovibos moschatus*, Desmarest, *in* Encyclopédie méthodique . . . , Mammalogie . . . , pt. 2, p. 492.
1908. *O*[*vibos*]. *moschatus mackenzianus* Kowarzik, Zoöl. Anzeiger, 33:617, November 10, type from Great Slave Lake, Mackenzie.

MARGINAL RECORDS.—Alaska: Point Barrow. Yukon: Herschel Island (Youngman, 1975:169), thence eastward along Arctic Coast to Keewatin: near Sherman Inlet; Back River; vic. Baker Lake; NW of Cape Eskimo. Manitoba: halfway between York Factory and Fort Churchill; between Seal and Kazan rivers. Mackenzie: Artillery Lake; 40–50 mi. E Dease Bay, Great Bear Lake. Alaska: near Tulugak Creek; Kunk [Kuk] River.

Ovibos moschatus niphoecus Elliot

1905. *Ovibos moschatus niphoecus* Elliot, Proc. Biol. Soc. Washington, 18:135, April 18, type from head Wager River, Keewatin.

MARGINAL RECORDS.—Keewatin: Murchison River region; Repulse Bay; Wager Inlet; Chesterfield Inlet; N side Baker Lake.

Ovibos moschatus wardi Lydekker

1900. *Ovibos moschatus wardi* Lydekker, Nature, 63:157, December 13, type from Clavering Island, Greenland.
1909. *O*[*vibos*]. *moschatus melvillensis* Kowarzik, Zoöl. Anzeiger, 33:617, November 10, type from Melville Island, Franklin.

1780. *Bos grunniens* Fabricius, Fauna Groenlandica, p. 28. Type locality, ice floe off W coast Greenland. Not Linnaeus, 1766.

MARGINAL RECORDS.—Greenland: Cape Morris Jesup, southward along E coast to Cape Dalton; southward along W coast to Thule region. Ellesmere Island: Cape Columbia; Fort Conger; Feilden Peninsula. Devon Island: Cape Sparbo. Somerset Island: Cape Garry. Taylor Island. Victoria Island: N end Prince Albert Sound; NE Minto Inlet. Banks Island: Nelson Head; near Cape Kellett. Prince Patrick Island. Melville Island: Cape Mudge. Cornwall Island. Axel Heiberg Island: within 50 mi. N end. Introduced on *Nunivak Island, Alaska.*

Genus **Ovis** Linnaeus—Sheep

American species revised by Cowan, Amer. Midland Nat., 24:505–580, November, 1940.

1758. *Ovis* Linnaeus, Syst. nat., ed. 10, 1:70. Type, *Ovis aries* Linnaeus.
1762. *Aries* Brisson, Regnum animale . . . , p. 12. Type, *Aries* Brisson [= *Ovis aries* Linnaeus].
1766. *Musimon* Pallas, Spicilegia zoologica, 11:8. Type, *Musimon asiaticus* Pallas [= *Capra ammon* Linnaeus].
1798. *Musmon* Schrank, Fauna Boica, 1:78, a renaming of *Ovis* Linnaeus.
1816. *Ammon* Blainville, Bull. Soc. Philom., Paris, p. 76, a renaming of *Ovis* Linnaeus.
1847. *Caprovis* Hodgson, Jour. Asiatic Soc. Bengal, 16:702. Type, *Ovis musimon* Pallas.
1852. *Argali* Gray, Catalogue of the . . . Mammalia in the . . . British Museum, 3:174. Type, *Aegoceros argali* Pallas [= *Ovis ammon* Linnaeus].
1936. *Pachyceros* Gromova, Neue Forsch. in Tierz. u. Abstamm. (Festschr. z. 60 Geburstag Dr. Duerst, Bern), p. 84. Type, *Ovis nivicola* Eschscholtz.

Body stout; nose narrow, pointed; ears small, pointed; tail, including hair, shorter than ear; lateral hoofs present; no beard on chin, horns of males either spiraled with tips directed outward or bent in an arc with tips pointing either forward or toward each other behind the head; pedal glands and preorbital glands present, subcaudal glands absent. Coronal suture projecting forward in an angle; infraorbital foramen small, with well-defined rim, about equal to length of P4; upper ends of premaxillae not usually wedged between nasals and maxillae; lambdoidal suture forming more or less straight line. Dentition, i. $\frac{0}{3}$, c. $\frac{0}{1}$, p. $\frac{3}{3}$, m. $\frac{3}{3}$.

Much of the literature on the systematic position of the three living species of *Ovis* and their immediate ancestors for which the subgeneric name *Pachyceros* Gromova has been proposed is listed by Korobitsyna, *et al.* (1974:244, 245). The three are the two North American species and *O. nivicola* of eastern Asia.

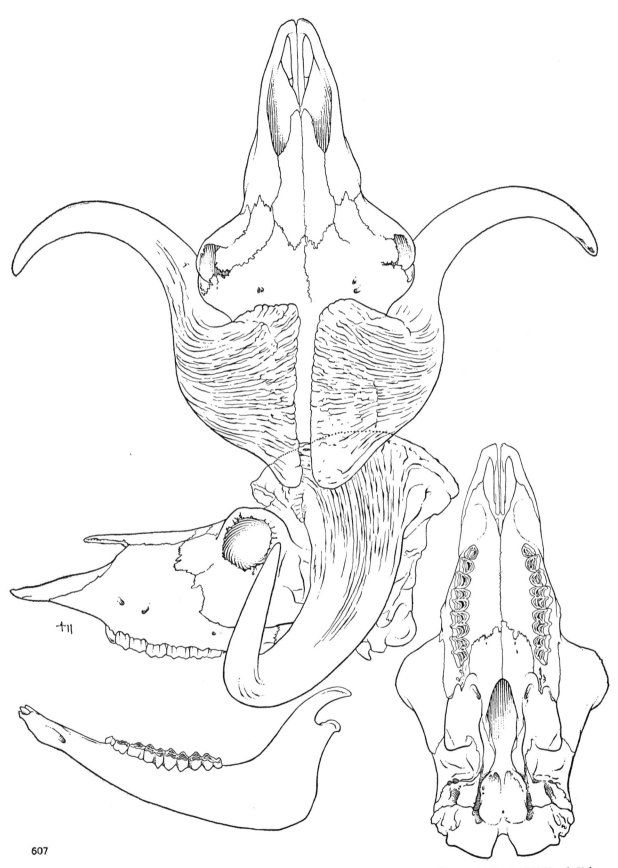

607

Fig. 607. *Ovibos moschatus* subsp. ?, near head Wagner Inlet, Hudson Bay, Canada, No. 19490 A.M.N.H., ♂, X ¼.

KEY TO NORTH AMERICAN SPECIES OF OVIS

1. Upper parts some shade of brown; nasals in males usually longer than 105, in females usually longer than 85.
O. canadensis, p. 1115
1'. Upper parts whitish (except in O. d. stonei, which is brownish-black); nasals in males usually shorter than 105, in females usually shorter than 85. . . . . O. dalli, p. 1117

Ovis canadensis
Mountain Sheep

X ¹/₂₈

External measurements: (males) 1326–1953, 70–150, 357–482; (females) 1166–1887, 77–130, 276–420. Weights to 156 kg. Basilar length of skull, (males) 247–285, (females) 226–261; zygomatic breadth, (males) 117–135, (females) 107–123; mastoid breadth, (males) 84–96, (females) 71–87; length of upper molar series, (males) 80–95, (females) 77–96. Upper parts Natal Brown or Mars Brown to Vinaceous Buff or Pale Vinaceous Fawn, underparts light brown to white, rump yellowish white to white. Horns of adult males massive, broad at base (to 18½ in. in circumference), not widespread (to 26 in. from tip to tip). From O. dalli, O. canandensis differs in average larger external and cranial measurements, more massive, less rugose, and less widely expanded horns, and in having some shade of brown, rather than white or blackish.

Ovis canadensis auduboni Merriam

1901. *Ovis canadensis auduboni* Merriam, Proc. Biol. Soc. Washington, 14:31, April 5, type from Upper Missouri, probably Badlands between Cheyenne and White rivers, South Dakota.

MARGINAL RECORDS.—North Dakota: jct. Yellowstone and Missouri rivers; Missouri River between Little Knife and White Earth rivers; Mandan. South Dakota: Cave Hills, Harding Co.; Imlay; between Yellow Medicine and Porcupine creeks. Nebraska (Jones, 1964c:330): near Long Pine; near Hay Springs; 23 mi. S Agate; Birdwood Creek; *5 mi. S Chimney Rock;* Courthouse Rock. Wyoming (Long, 1965a:722): Spoon Butte; Black Hills. Montana: *eastern Montana.*

Ovis canadensis californiana Douglas

1829. *Ovis californianus* Douglas, Zool. Jour., 4:332, January. Type locality, near Mt. Adams, Yakima Co., Washington (see J. A. Allen, Bull. Amer. Mus. Nat. Hist., 31:25, March 4, 1912) or Falls of the Columbia, near mouth Deschutes River (see V. Bailey, N. Amer. Fauna, 55:64, 65, August 29, 1936).
1912. *Ovis canadensis californiana,* Miller, Bull. U.S. Nat. Mus., 79:396, December 31.
1912. *Ovis cervina sierrae* Grinnell, Univ. California Publ. Zool., 10:144, May 9, type from E slope Mt. Baxter, 11,000 ft., Sierra Nevada, Inyo Co., California.
1915. *Ovis canadensis samilkameenensis* Millais, *in* The gun at home and abroad . . . , p. 324, type from Similkameen Mts., British Columbia.
1913. *Ovis dalli ellioti* Kowarzik, Zoöl. Anzeiger, 41:444, a *nomen nudum.*

MARGINAL RECORDS.—British Columbia: S of head Riske Creek (Cowan and Guiguet, 1965:397); Tranquille Creek; N end Okanagan Lake; E of Vaseux Lake (Cowan and Guiguet, 1965:397). Washington: Hellgate dist., SE Ferry Co.; Grant County. Oregon: Blue Mts.; Mahogany Mts. Nevada: Division Peak; E side Pyramid Lake; along Truckee River. California: Sonora Peak; Tin Mtn.; vic. Olancha Peak; Glenville; Kaweah Peaks; Big Pine; Scott Mtn.; Siskiyou Mts. Oregon: vic. Bend; Maupin. Washington: Klickitat Valley; type locality; Stehekin; Bauerman Ridge. British Columbia: Chilko Lake; 120 mi. NW Lillooet.

Ovis canadensis canadensis Shaw

1804. *Ovis canadensis* Shaw, Naturalists' Miscl., 51: text to Pl. 610 (December, 1803?). Type locality, mts. on Bow River, near Exshaw, Alberta.
1900. *Nemorhoedus palmeri* Cragin, Bull. Geol. Soc. America, 11:611, August 10, type from Cave on Glen Eyrie Estate, about 5 mi. NW Colorado Springs, El Paso Co., Colorado.

MARGINAL RECORDS.—British Columbia: Mt. Quintette, Kiskatenaw River. Alberta: Smoky River; Thoral Creek; Red Deer River; Waterton Lakes Park. Montana: *Sun River country* (Forrester and Senger, 1965:524); Scattering Springs area, Sun River (*ibid.*);

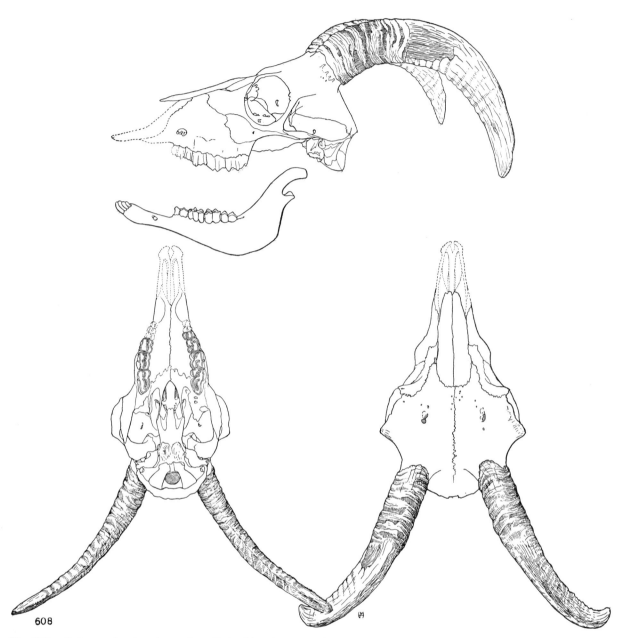

608

Fig. 608. *Ovis canadensis nelsoni*, 4 mi. E Boulder City, Nevada, No. 71803 M.V.Z., ♀. (Lower jaw is No. 2320 M.V.Z., ♀, from Deep Canyon, Santa Rosa Mts., Riverside Co., California. Its teeth are: permanent 1st incisor, deciduous 2nd and 3rd incisors, deciduous canine, 3 deciduous premolars, and 2 permanent molars. The 3rd permanent molar has not erupted.) X ¼.

Three Buttes. Wyoming (Long, 1965a:722): Crystal Creek; Fort Laramie. Colorado: Kenosha Range; Denver Mountain Park; Pikes Peak. New Mexico: Truchas Peaks; Santa Fe Baldy; Frijoles Canyon, Jemez Mts. Arizona: Bright Angel Canyon (Hoffmeister, 1971:155, as *O. canadensis* only). Utah: *Virgin River Canyon, Utah–Arizona boundary.* Nevada: Granite Mtn., 11,200 ft.; near head Willow Creek Canyon; King Lea Peak, Jackson Mts. Idaho: near Silver City. Oregon: upper Sheep Creek, Wallowa Mts., SE Joseph. Washington: upper Tucannon River; Pend Oreille Mts. British Columbia: head Kootenay River, 35 mi. SW Banff; vic. Wapiti Pass.

Ovis canadensis cremnobates Elliot

1904. *Ovis cervina cremnobates* Elliot, Field Columb. Mus., Publ. 87, Zool. Ser., 3(14):239, January 7, type from Mattomi, Sierra San Pedro Mártir, Baja California.

1912. *Ovis canadensis cremnobates*, Miller, Bull. U.S. Nat. Mus., 79:396, December 31.

MARGINAL RECORDS.—California: Chuckwalla Mts. Baja California: San Felipe Bay; type locality; Laguna Hanson. California: Julian. Not found: Baja California: 20 mi. W Fort St. Louis; 40 mi. NW Fort San Luis; 8 mi. NW Fort San Luis; Aroyo [*sic*] Grande; Saltillo del Rey.

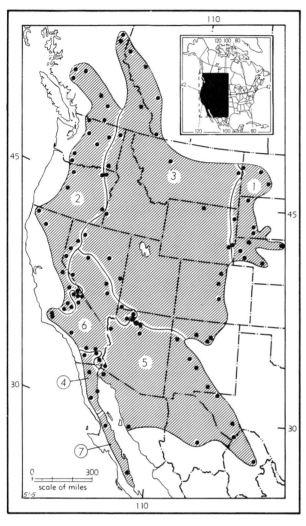

Map 545. *Ovis canadensis.*

Guide to subspecies
1. *O. c. auduboni*
2. *O. c. californiana*
3. *O. c. canadensis*
4. *O. c. cremnobates*
5. *O. c. mexicana*
6. *O. c. nelsoni*
7. *O. c. weemsi*

Ovis canadensis mexicana Merriam

1901. *Ovis mexicanus* Merriam, Proc. Biol. Soc. Washington, 14:30, April 5, type from Lago de Santa María, Chihuahua.

1901. *Ovis canadensis mexicanus*, Lydekker, The great and small game of Europe, western and northern Asia and America, p. 11.

1907. *Ovis canadensis gaillardi* Mearns, Bull. U.S. Nat. Mus., 56:240, April 13, type from Gila Mts., between Tinajas Altas and Mexican boundary, Yuma Co., Arizona.

1912. *Ovis canadensis texianus* V. Bailey, Proc. Biol. Soc. Washington, 25:109, June 29, type from Guadalupe Mts., El Paso Co., Texas.

1916. *Ovis sheldoni* Merriam, Proc. Biol. Soc. Washington, 29:130, September 6, type from El Rosario, northern Sonora.

MARGINAL RECORDS.—Arizona: mouth Cataract (= Havasu) Canyon (Hoffmeister, 1971:177); *Hava-supai Point* (*op. cit.*: 155, as *O. canadensis* only); Desert View (*op. cit.*: 177); San Francisco Mtn. New Mexico: Zuñi Mts.; Magdalena Mts.; Sheep Mtn.; Guadalupe Mts. Texas: E rim above Frijole; Glass Mts.; Boquillas Canyon. Coahuila: approx. 5 mi. E San Lazaro. Chihuahua: Chihuahua. Sonora: Costa Rica Ranch; Rosario. Arizona: Gila Mts.; Black Mts.; *Colorado Canyon, Vitz's Crossing, 17 mi. N Pine Springs.* Not found: Arizona: Black Mesa of The Alamo.

Ovis canadensis nelsoni Merriam

1897. *Ovis nelsoni* Merriam, Proc. Biol. Soc. Washington, 11:218, July 15, type from Grapevine Mts. [high limestone ridge forming middle of range, about 5 mi. southerly from Grapevine Peak], on boundary between Inyo Co., California, and Esmeralda Co., Nevada, just S of lat. 37° N.

1898. *Ovis canadensis nelsoni*, Lydekker, Wild oxen, sheep, and goats of all lands, living and extinct, p. 208.

MARGINAL RECORDS.—Nevada: 28 mi. N Austin; 6 mi. NE Italian Spring, Quinn Canyon Mts., 7000 ft.; Irish Mtn. Utah: Beaverdam Mts. (Stock, 1970:433). Arizona: Kanab Canyon; Toroweap, thence along W side Colorado River to California: 20 mi. W Picacho; Canyon Springs, Chuckwalla Mts.; Santa Rosa Mts.; head Rock Creek, San Gabriel Mts.; Cuyama Plain, base White Mtn.; Caliente Peak, Cuyama Valley; W slope Inyo Range; W slope Sheep Mtn., White Mts.; Tin Mts.; Grapevine Mts. Nevada: Middle Fork Cat Creek, 8900 ft.

Ovis canadensis weemsi Goldman

1937. *Ovis canadensis weemsi* Goldman, Proc. Biol. Soc. Washington, 50:30, April 2, type from Cajon de Tecómaja, 2000 ft., Sierra de la Giganta, about 30 mi. S Cerro de la Giganta, Baja California.

MARGINAL RECORDS.—Baja California: Sierra San Borjas, 20 mi. N Calmallí; type locality.

Dall's Sheep

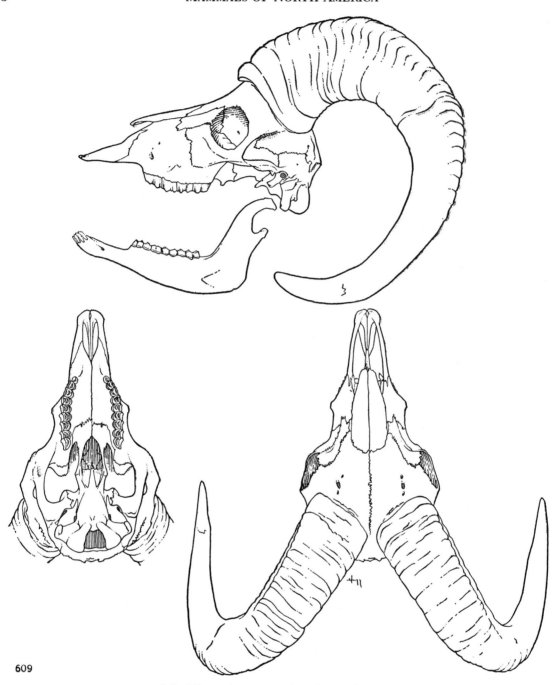

609

Fig. 609. *Ovis dalli dalli*, Knik River, Cooks Inlet, Alaska, No. 1845 K.U., ♂, X ¼.

External measurements: (males) 1300–1780, 75–115, 395–450; (females) 1324–1377, 77–99. Weights to 91 kg. Basilar length of skull, (males) 235–274, (females) 215–238; zygomatic breadth, (males) 114–134, (females) 107–117; mastoid breadth, (males) 82–97, (females) 70–79; length of upper molar series, (males) 68–88, (females) 68–74. Upper parts pure white or Seal Brown, or Bone Brown, or Blackish Brown; underparts white; neck white or grizzled; face white to dirty white. Horns of adult males narrow at base to 15½ in. in circumference, widespread to 34½ in. from tip to tip. For comparison see *O. canadensis*.

Ovis dalli dalli Nelson

1884. *Ovis montana dalli* Nelson, Proc. U.S. Nat. Mus., 7:13, June 3, type from mts. S of Fort Yukon on W bank Yukon River, Alaska; probably Tanana Hills.

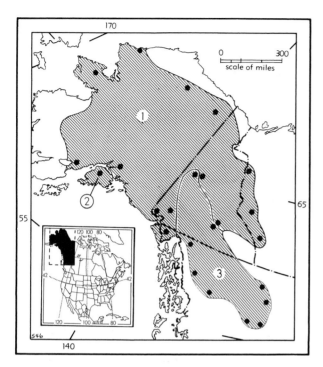

Map 546. *Ovis dalli.*

1. *O. d. dalli*   2. *O. d. kenaiensis*
3. *O. d. stonei*

1897. *Ovis dalli*, J. A. Allen, Bull. Amer. Mus. Nat. Hist., 9:112, April 8.

MARGINAL RECORDS.—Alaska: Cape Beaufort; Anaktuvuk Pass; Romanzoff Mts. Mackenzie: Rockies, 66° N lat.; mts. above Sekwi River and Dodo Creek; South Nahanni River. Yukon: Rose–Lapie Pass area; Klondike River; Kluane Lake. British Columbia: St.

Elias Range. Yukon: St. Elias Alps, head White River, Mt. Logan. Alaska: Knik River; mts. near Iliamna Bay; *Kuskokwim River;* Berry Creek.

Ovis dalli kenaiensis J. A. Allen

1902. *Ovis dalli kenaiensis* J. A. Allen, Bull. Amer. Mus. Nat. Hist., 16:145, April 23, type from head Sheep Creek, Kenai Peninsula, Alaska. Regarded by Osgood, N. Amer. Fauna, 30:51, October 7, 1909, as identical with *O. d. dalli;* regarded as a valid subspecies by Cowan, Amer. Midland Nat., 24:528, November, 1940.

MARGINAL RECORDS.—Alaska: Skilak Lake, Kenai Peninsula, thence throughout Kenai Peninsula.

Ovis dalli stonei J. A. Allen

1897. *Ovis stonei* J. A. Allen, Bull. Amer. Mus. Nat. Hist., 9:111, April 8, type from Che-on-nee Mts., headwaters Stikine River, British Columbia.
1912. *Ovis dalli stonei*, J. A. Allen, Bull. Amer. Mus. Nat. Hist., 31:28, March 4.
1898. *Ovis canadensis liardensis* Lydekker, Wild oxen, sheep, and goats of all lands, living and extinct, p. 215, type from Liard River, British Columbia.
1901. *Ovis fannini* Hornaday, Ann. Rept. New York Zool. Soc., 1900, App. 1:2, January 8, type from Dawson City, Yukon.
1907. *Ovis cowani* Rothschild, Proc. Zool. Soc. London, p. 238, August 1, type from near Mt. Logan, British Columbia.
1915. *Ovis canadensis niger* Millais, *in* The gun at home and abroad . . . , p. 324, type from mts. at head Skeena River, British Columbia.

MARGINAL RECORDS.—Yukon: Dawson; Pelly River. British Columbia: Prophet River (Cowan and Guiguet, 1965:400); Sikanni Chief River; Halfway River (Cowan and Guiguet, 1965:400); Peace River on slopes bordering Nabesche River; mts. at head Skeena River; Telegraph Creek; Lake Atlin.

SUGGESTIONS FOR COLLECTING AND PREPARING STUDY SPECIMENS OF MAMMALS

In the first place the value of specimens is greatly enhanced by having them in excellent condition, and, of course, they must be accompanied by adequate data. Often a specimen may be well prepared and adequate data for it recorded, but subsequently, through faulty packing, incorrect method of labeling, or improper care and protection against pests and climatic conditions, it may be rendered worthless. Proper care at all stages of field work, therefore, is required to insure good quality in specimens. More details than are provided here on the techniques of preparing mammalian specimens are given in other authors' publications that can profitably be consulted. One of the best publications in this field is R. M. Anderson's "Methods of Collecting and Preserving Vertebrate Animals" (Bulletin No. 69, pages 1–162, illustrated, second edition, National Museum of Canada, March 22, 1944).

What to Collect

In preparing mammals caught, divide your efforts between common and rare species. As a rule, even with the more common species, save a pair from each locality. Series of up to 30 from one locality should be saved in each center of suspected differentiation—these to include young of different ages as well as adults.

Weathered skeletons or skulls of animals found may be desirable as specimens. For these, care must be taken to obtain all bones, and also teeth, as these frequently will have fallen out.

Contents of cheek pouches and samples of feces may be saved dry.

If lice, earwigs, flies, fleas, and other ectoparasites are to be preserved, it is desirable to examine not only the mammals but also their nests.

Plants to be saved should be placed in presses according to usual botanical practice. Seeds may be collected dry in any adequate containers.

Equipment and Properties

Mouse traps of the variety that snap shut on the animal can be purchased in almost every hardware store. These traps are satisfactory for shrews. For mice, however, the larger "Museum Special," manufactured by the Animal Trap Company of Lititz, Pennsylvania, is best because the wire that strikes and kills the mouse is far enough from the treadle to keep the head of the mouse from being struck and crushed. Broken skulls are less desirable than unbroken skulls. The still larger rat trap is stocked in most hardware stores and is suitable for taking animals the size of wood rats and small ground squirrels. Steel traps in sizes 0–4 are used in many areas to secure other animals. McAbee gopher traps are the best yet devised for taking pocket gophers. A variety of mole traps are on the market; the stabbing variety is preferred by most of the collectors that we know.

Many specimens are most effectively taken by shooting. For smaller and medium-sized kinds a shotgun is recommended, and, furthermore, shot of small size should be used to avoid unnecessary mutilation of the animal. For many years the double-barreled 16-gauge shotgun has been a favorite of collectors; it is customary to carry a standard 16-gauge shell in the left barrel and a metal bushing in the right barrel. The bushing is drilled to chamber a 32-caliber metal shell loaded with No. 12 shot. The collector is ready to shoot animals of fox- or eagle-size with the charge from the left barrel, or mice, chipmunks, or small birds with the charge from the right barrel.

Nets of silk, such as are manufactured in Japan, are useful to the mammal collector, especially in capturing several kinds of bats.

Metal forceps, 10–14 in. long, frequently come in handy when bats or other small mammals have to be removed from crevices.

For safely storing prepared specimens in accessible fashion, a museum cabinet that excludes insects, dust, and light is essential. A visit to the nearest museum maintaining a collection of study specimens of mammals, or a letter of inquiry addressed there, will yield all needed information on the type of container best suited to the needs of the collector. Advice concerning the cleaning of skulls can be obtained from the same source.

Every state has its own laws relating to hunting, and the collector should obtain and read the laws so as to carry on his collecting in conformance with the law. The State Fish and Game office can provide a copy of the laws, and that office, in most states, is the place to apply for a "scientific collecting permit." The collector should also find and inform the local game protector of proposed collecting, because this can be of mutual benefit; the collector often receives valuable advice as to where certain species occur locally, and the Game Protector needs to know who is afield in his area and oftentimes obtains information valuable to him from the catches of the collector.

Wherever animals are collected on privately owned lands the permission of the landowner or his tenant must be obtained in advance. Application of the Golden Rule with landowners and the Game Protector will avoid trouble for the collector and bring him welcome assistance.

NOTE-TAKING

Field notes can usefully be divided into: (1) catalogue of specimens, (2) itinerary or journal, (3) accounts of species. Enter the name of the collector and the year in the upper left-hand corner of every page, but far enough from the margin to permit binding of the pages. Each page should be filled before another page is started. Use only Higgins Eternal Black Ink. For convenience all three sections of the notes ordinarily are kept in a single binder, but separate binders may be used.

CATALOGUE

In the catalogue, all specimens of vertebrate animals should be given consecutive numbers. Never repeat a number; for instance, do not begin a new series each year. One line of the notebook page should be devoted to the precise locality. Include distance in airline miles from some well-established landmark. Include also elevation, county, and state. Devote one line to each specimen. If a specimen is not a conventional one, indicate the nature by entry directly above the field number, whether (if) skeleton, skull-only, skin-only, or alcoholic. Use the vernacular name of the species if you are not *sure* of the scientific name.

On the first line of the itinerary enter date and locality. Follow with a concise account of route and travel area and habitats studied, and record number and kinds of traps set, distance between traps, number of vertebrates collected, as well as other pertinent information—for example, record number of traps set in each type of vegetation and numbers and kinds of animals caught therein. Section, township, and range comprise useful information.

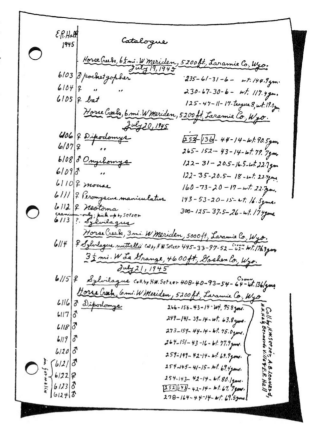

Fig. 610. A page from a field catalogue, slightly less than one-half natural size. Blue horizontal lines fail to show in this illustration.

ACCOUNTS OF SPECIES

Accounts of species should be headed with either the scientific or common name, as preferred. The date and locality for the account should be given on the first line. Only one species should be written about on a single page. Information in the account should not be a repetition of material given in the itinerary or journal. Do not only include facts, but make interpretations and generalizations. The accounts should be written in a style suitable for quoting in any publication. Accounts of species need not be restricted to kinds collected. If the account is about animals collected, it is wise to refer to the animals by your field numbers.

Head each notebook page with collector's name and year, page number (if number system is used), locality (in *detail* the first time used), and date.

Write *full* notes, even at risk of entering much information of seemingly little value. One cannot anticipate the needs of the future, when notes and collections are worked up. The following are

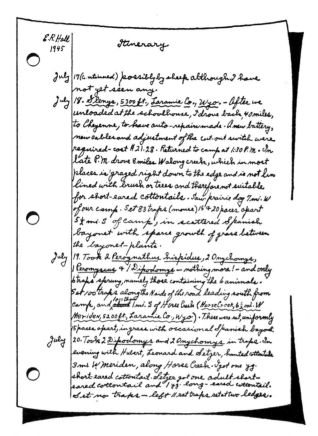

Fig. 611. A page from a field notebook, slightly less than one-half actual size. Blue horizontal lines fail to show in this illustration.

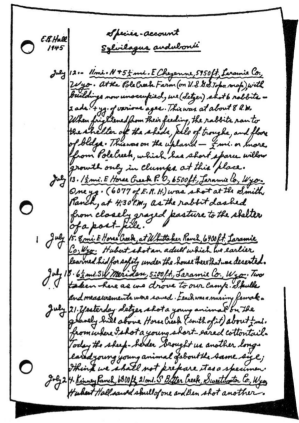

Fig. 612. A page from a field notebook, slightly less than one-half natural size. Blue horizontal lines fail to show in this illustration.

suggested topics, but do not restrict yourself to these alone. Be alert for new ideas and new facts.

Describe vegetation (saving plant-press samples of species not positively known), nature of ground, slope exposure, and drainage in each belt of animal life sampled. Describe exact location of trap lines, referring to your topographic maps, and also enter a sketch, in profile or surface view or both, to illustrate the location and relations of the different habitats crossed. Properly marked maps for each region worked should ultimately be bound in with the field notes of at least one member of your field party.

Keep record of closeness of settings of traps, distance covered, and results of each night's trapping; give number and type of traps put out in each habitat and number of animals of each species captured in each habitat (whether or not preserved). In some instances it is advisable to record the sex, age, and breeding condition of each animal.

Keep full record of breeding data: number and approximate size (length) of embryos or of young

found in nests. Dig out burrows if practicable; make drawings to scale, showing plan, and elevation; describe fully.

Record food plants; keep specimens for identification where not known by a definite name; preserve contents of cheek pouches and stomachs. If these are not saved, identify and record contents.

Note regularly in notebook all "pick-ups," that is, odd skulls or fragments of animals of whatever sort or source, serially numbered along with specimens of the more usual sort. Give full information, as with odd skulls secured from trappers. Label all such specimens adequately, as elsewhere described.

When leaving a well-worked locality, enter a summary of species observed, with remarks of a general nature, relating to local conditions of terrain, human activities, and other pertinent conditions.

Where feasible, interview old residents, trappers, National Forest and National Park rangers in each locality visited. Always record accurately the name, official position or occupation, and address of each person giving information; also give your opinion as to his reliability. Note general attitude of persons interviewed as to game laws, conservation, and effects of settlement by Man, and record specific comments, complaints, or criticisms.

Ascertain present numbers and distribution of large mammals as compared with their former status. So far as is possible get definite statements expressing ratio of present abundance compared with a definite number of years back. Seek such information, where feasible, by indirect query. Do not risk influencing your informant's statements by leading questions. Record fully all evidence as to human influence upon original or "natural" balance. Record present economic relations of vertebrate animal life, that is, effect on agriculture and stockraising, with full details. Note opinions of persons interviewed as to whether species should be protected or destroyed. Describe local methods of capture or destruction; give your opinion as to their effectiveness and justification.

Opportunity offering, record detailed observations on effects upon vertebrate animals of: severe storms; floods; forest, brush, or prairie fires; overgrazing; tree-cutting; road-building; or tree-planting.

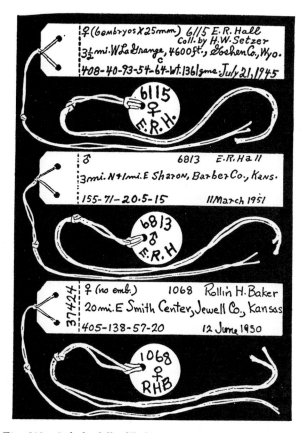

Fig. 613. Labels, fully filled out. The larger labels are for stuffed skins; the smaller labels are for uncleaned skulls. X ¾.

LABELING OF SPECIMENS

Use one serial set of field numbers for *all* specimens (including "pick-ups," wet preservations, and ectoparasites).

Of course every specimen is to be fully labeled at the time it is prepared—and before the preparation of another is begun. A complete, authentic label for a mammal in most instances is scientifically more valuable than the skin to which it is attached. Beginners may not realize the prejudice engendered in careful zoologists by sight of a specimen incompletely labeled or of a specimen with a complete label unattached or insecurely attached to the specimen.

For each specimen always give altitude and county as well as exact place; for example: "3 mi. NE Lone Star, 850 ft., Douglas Co., Kansas." Attend minutely to proper punctuation. If not true NE, give miles north and miles east. Distances always to be airline. Locality data to be given in notebook precisely as on specimen label.

Record data on labels for skins on one side only (the opposite side is left blank for entering the name in eternal ink after laboratory studies have been completed). Enter data on three or four

lines as may be necessary, and in the order indicated below.

On the first (top) line record sex (if female, record number and size of embryos or absence of embryos), collector's field number, and collector's name. Immediately following the field number record the name of the person to whom the number pertains, and preface the name of the second person with "Coll. by" or with "Prep. by" for the purpose of insuring that the field number will be associated only with the name to which it pertains and also to show which person was the collector and which one was the preparator. This information is important and especially so when necessity arises for tracing back through the field notes (catalogue, species account, or itinerary) to obtain supplementary information.

On the second line, or on the second and third lines, record the locality exactly as in the catalogue and other parts of the field notes—even to punctuation and abbreviation.

On the last line, record total length, length of tail, length of hind foot, height of ear from notch, weight in grams, and date. Use dashes (not commas, periods, colons, or semicolons) to separate the measurements. Use the abbreviation "gms." after the weight to show that the weight is recorded in grams instead of in grains or in ounces. Write out the names of months that are no longer than five letters; abbreviate the names of the others. Never use a numeral to represent the month, because doing so increases the chances of error in later transcriptions of the month and day. For example, depending on the country from which a person comes, the date of June 9, 1953, is written in at least the following forms: VI-9-1953, 6-IX-1953, 6-9-1953, VI-IX-1953, 9-6-1953; still other combinations are possible and may be used; even within different parts of one country (the United States) the form varies. Any one of the above combinations of numerals means Sept. 6, 1953, to some persons and June 9, 1953, to other persons; consequently it should be written "June 9, 1953" or "9 June 1953" to avoid error. Also use four numerals for the year (1953) and not an apostrophe and two numerals ('53) because the apostrophe can waste valuable time of the zoologist in ascertaining whether 1753, 1853, or 1953 is correct.

On a skull-tag enter field number, initials, and sex symbol with Higgins Eternal Ink; write 'large and heavy' to prevent fading of the number and consequent loss of a specimen.

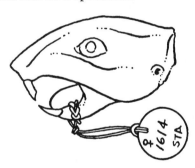

Fig. 614.   Skull-label correctly tied to one ramus of the lower jaw of a skull freshly skinned out, and immediately before immersion in water. X ¾.

Attach tag to skull by running the string between the *lower jaws*. Tie with about ¼-in. slack in the string. Do not tie tightly around the lower jaw; there is danger of breaking the jaws of bats or shrews. Also, a loose string allows the larval dermestids to eat the meat directly between the string and jaw bone. Cut off the loose ends of the string.

All miscellaneous material should bear labels with notebook references by specimen numbers, and initials of collector. Nests should be "threaded" and have labels attached to them (not only to container). Folders for plants should each be inscribed with full data, according to usual botanical practice.

LABELS

Labels for skins. Paper used as labels should last for centuries, take ink well, hold ink well even when labels are immersed in liquids, and retain its strength in water or formalin or alcohol. At the Museum of Natural History of the University of Kansas we use, as best meeting the requirements mentioned above, "Resistall Index, Bristol 100 per cent rag." This paper has a weight of 110 lb. From sheets 25½ × 30½ in. we cut out the labels, after printing, that are to be attached to skins.

Stacks of 50 labels are firmly held in a metal frame of appropriate size, and two holes are drilled by means of a bit in a drill press near one end of the stack of labels. See Fig. 613 for exact position of the two holes.

For threading the labels we use Star Brand, white, 6-strand, mercerized, size-10 cotton thread manufactured by the American Thread Company. This thread comes in spools of 150 yd. to the spool. For stringing one label a piece of thread 330 mm. long is used. After the thread is strung through the two holes in the label, the loose ends of the thread are strung through a half-turn loop taken in the thread of the "backside" of the label. The threads are pulled taut, and a knot is tied in the two threads exactly one inch from the end of the label. This label and thread are recommended as satisfactory in all respects save one. The size of our label, in our opinion, is too large (see Fig. 613). A width of 15 mm. and a length of 70 mm. would suffice. Predecessors established the size now used at the University of Kansas, and we continue to use it because the advantages of uniformity within a single collection outweigh the advantages that would be gained by changing to another size.

Threading of the labels is done by hand. A smooth flat board, say, 4 × 8 in., with the point of a finishing nail showing on one side at a distance of 1 in. from an ink mark facilitates tying the knot at the correct distance from the edge of the label. A skillful person can thread and tie 200 to 300 labels an hour.

Labels for skulls and skeletons are of a diameter (19 mm.) that fits loosely in the bottom of the smaller of the two sizes of glass vials used for housing skulls. These round labels are "Fiber Water Proof Stock, 10 pt." purchased from the Dennison Manufacturing Company of St. Louis, Missouri. The labels are perforated at one side and are strung with No. 5 linen twine, in which a knot is tied exactly 1 in. from edge of label (see Figs. 613 and 614). If pressure is exerted when writing on these labels, the writing will be retained even though the labels are immersed in water, ammonia, or alcohol. This fiber stock yields labels that are resistant to dermestid beetles and their larvae—insects used to remove flesh from the osteological specimens to which these round labels are attached. Some other kinds of paper are eaten by dermestids if the paper becomes stained with blood.

Notebook paper that is high in rag content is used because its lasting qualities are thought to be superior to most other kinds of paper. Field notes containing observations of behavior of animals written "on the spot," and census counts of animals under natural conditions in areas that later are greatly modified by Man, have increasing value with the passage of time. Such records, therefore, should be on paper that will not deteriorate with age.

SKINNING AND STUFFING SMALL MAMMALS

Tools and supplies normally used are as follows:

Labels for skins
Labels for skulls
Higgins Eternal Ink (not Higgins Drawing Ink or Higgins India Ink)
Dip pen or fountain pen in which Higgins Eternal Ink can be used
30-cm. rule graduated in millimeters
Sewing needles
White (not colored) cotton thread, two sizes
Pliers with clean-cutting jaws and wire-gripping ends on jaws
Forceps with fine sharp points but having arms strong enough to permit gripping solidly
Scalpel or razor-sharp knife blade
Carborundum sharpening stone
Scissors, surgical-type, with two sharp points
Cotton, long-fibered, resilient, in smooth bats
Arsenical powder or soap in wide-mouthed container
One shallow pie tin
White corn meal or fine hardwood sawdust

Other supplies and equipment are optional and may include scraper for removing fat and a three-cornered file for working metal.

Measure any small mammal in millimeters and weigh it in grams. Experienced preparators take two measurements, write them on the label, take two more measurements, write them on the label,

weigh the animal, and write the weight on the label.

The standard measurements for a study specimen are taken as follows:

Total length. Manipulate mammal so that it lies out straight (do not stretch it; guard against error that can result from a broken vertebral column), and measure distance from tip of nose-pad to tip of fleshy part of tail, excluding hairs that project beyond tip.

Fig. 615. Measuring total length of a small mammal. X ½.

Length of tail. Bend tail up at right-angle with body and measure from bend on back to tip of fleshy part of tail, excluding hairs that project beyond tip.

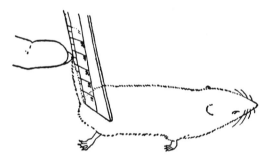

Fig. 616. Measuring length of tail of a small mammal. X ½.

Length of hind foot. With its toes out straight measure the distance from tip of longest claw to heel—in the same way that the overall length of a person's foot would be measured. Outside North America the claw is excluded from the measurement of the length of the hind foot, and only the fleshy part of the foot is measured in mammals from continents other than North America (Greenland, Panamá, and Central America are parts of North America).

Height of ear from notch. Insert end of rule in notch at bottom of ear and measure to distalmost border of fleshy part of ear.

Next make out the labels (for skin and skull), and then make the entry in the field catalogue. It may be necessary to enter the sex (and certainly number and size of embryos if any) after the animal is skinned.

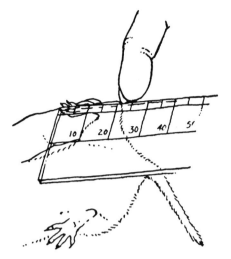

Fig. 617. Measuring length of hind foot of a small mammal
by means of a transparent (plastic) rule. X ¾.

Fold a square piece of cotton or other stuffing to desired size and shape, and lay it to one side. A body too large or too small will cause trouble. A body wider than high and wider at the middle than at either end, to begin with, gives best results. Make the body one-fourth longer than the body of the animal (the extra fourth will be cut off later). The back and sides of the body should be smooth; small irregularities in the body make larger-appearing irregularities in the dried specimen. The outside edge of the cotton should be feathered; that is to say, enough of the fiber should be removed to make the cotton taper to a thin edge. When the last fold is made, drag the body across the skinning board or table top so as to cause the thin edge of the cotton to adhere to the previous layer. So doing will prevent the body from unrolling while the skin is being turned over it; partial unrolling twists the skin, making it unsightly. The aim, whatever method is adopted for forming the body, is to obtain a symmetrical, firm, yet resilient body that will retain its shape while the skin is drying.

After the body is made, select Monel metal wire of appropriate gauge; with pliers holding one end of the wire stretch a piece until it is perfectly straight. Cut and lay to one side five pieces, one piece for the tail and four pieces for the four legs. A leg-wire should be as long as the bony structure of the limb; wires for the two hind legs should be longer than those for the front legs. The tail-wire should be a half longer than the tail. For pocket gophers and wood rats use No. 20 wire; for a larger *Peromyscus* use No. 22; for small pocket mice use No. 24; for tails of the smallest bats use No. 24 or even No. 26 wire. Monel metal wire, or annealed tin wire, is to be used. Do not substitute other kinds of wire (not even copper, brass, or galvanized wire), because those eventually corrode or rust and destroy the specimen. If the correct kind of wire is not available, use split bamboo instead, and in any event use bamboo or other straight-grained, tough wood in place of wire in the legs of mammals larger than wood rats. Striped skunks and other mammals of equal and smaller size are to be stuffed; badgers, foxes, and larger animals are skinned so that the hides can be properly tanned and preserved unstuffed.

Have ruler, threaded needle, scalpel, forceps, scissors, arsenic container, and pie tin containing a double handful of sawdust or corn meal before you on a table or on a flat skinning board supported on the arms of a camp chair.

With the mammal held, back down, on the sawdust in the pie tin, use the scissors to cut the skin, and if possible not the body wall, on the midline of the belly from immediately in front of the anal opening foward one-fourth the distance (no more) toward the chest. Experience will enable you to make this cut by means of no more than two snips of the scissors, and to cut only the skin and not the body wall. Retain the scissors on the

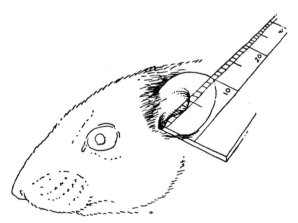

Fig. 618. Measuring height of ear from notch in a small
mammal. X 1½.

Fig. 619. Making initial incision in skinning a small
mammal. X ⅔.

thumb and finger of one hand while proceeding so as not to waste time in laying down and picking up the instrument each time that it is needed.

With the free hand sweep sawdust onto the cut area. With the third and fourth fingers of the hand holding the mammal, thrust the knee toward the midline of the body, meantime pushing the skin of the belly and flank away from the midline by means of the thumb of the same hand. Grasp the exposed knee with the scissors hand, and with the forefinger and thumb of the other hand separate the leg from the shin and push the skin all the way down to the ankle. Insert a blade of the scissors at the ankle in such a way that the bones of the lower leg are between the blades of the scissors, and in one motion by means of the blades of the scissors strip the flesh from the lower leg and free the lower leg by cutting it immediately below the knee joint. Repeat the process on the other hind leg.

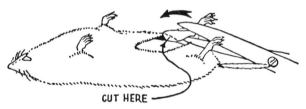

Fig. 620. Skinning out hind leg. X ⅔.

One reason for leaving the bones of the lower leg attached to the skin, instead of severing the hind leg by cutting through the ankle, is to meet the need that sometimes arises to measure the length of the dry hind foot. If the bones of the lower leg are present, the heel can quickly and certainly be located. If the bones of the lower leg are not present, there will be uncertainty concerning the heel: Was it cut away with the bones of the lower leg? If the heel is located, is all of it present? To answer such a question, it is necessary to relax the skin of the hind foot, cut the skin, and visually inspect the bones. These time-consuming operations can be avoided by leaving the bones of the leg in the skin.

Meanwhile, and throughout the whole of the skinning, keep the specimen buried as much as possible in sawdust, because the sawdust absorbs fat and any other body fluids, thus preventing them from getting on the fur or on the hands of the preparator. If the fat gets on the hands of the preparator, it will be transferred to the fur. Fat on the fur makes the pelage unsightly and spreads eventually to the label and other specimens; fat anywhere on the preparation oxidizes and in

time destroys the skin. Keeping the specimen immersed in sawdust (or white corn meal) requires that the preparator learn to work as much as possible by touch instead of by sight. Beginners seldom use enough sawdust. The thought that "if some is good more is better" really applies to the use of sawdust in skinning mammals. After eight to ten small mammals that are exceptionally fat are skinned in a small quantity of sawdust, it becomes saturated with grease. Discard it and use fresh sawdust.

When the two hind legs have been freed and skinned out in the manner described above, separate the skin from the body wall by working a finger, or the handle of a scalpel, between the skin and the body. With the fingernails sever the gut and associated connections of the skin to the body. Scissors or scalpel can be used, but time is saved by using the fingernails. When the skin is free all around the base of the tail, place three fingernails behind the skin and push it off the tail. One motion will do the job. Take care to keep the nails against the tail vertebrae and behind the skin so as to avoid the misfortune of having the skin of the tail turn inside out; this misfortune slows down even the experienced preparator and is one that may be insurmountable for the beginner.

When the skin is free from the tail and hind legs, separate the skin from the body wall at the anterior end of the initial incision that was made on the belly. Then turn the skin down all around the body and push (don't pull) it off the body. Be sure that the skin is reflected (turned down) on the belly as well as elsewhere; otherwise the skin will gradually tear forward from the front end of the initial incision on the belly as you proceed with the skinning, and that misfortune causes the skin to fit improperly on the cotton body, with the result that the dried skin of the specimen will be misshapen. Push the skin down until the mammal's elbows are in sight. Tear or cut the thin skin-muscle that shows up. Push the skin down to the wrist. Straddle the forearm near the wrist with the blades of the scissors; in bringing the blades of the scissors up from the wrist almost to the elbow, strip the flesh from the lower leg and cut the bones immediately below the elbow joint. Do the same with the other foreleg. Drop the scissors, because you will not need them again in skinning this mammal. Push the skin farther until the bases of the ears come into view. By means of the fingernails securely grasp the ear tube on two sides where it emerges from the bone and pull the tube free. When the two ears are free, push the skin forward until the eyes are reached. Pick up the scalpel or knife and make two cuts on each eye:

one cut straight down across the eye to sever most of the attachments of the skin to the skull, and a second cut on a transverse plane, with the blade of the scalpel against the bone so as to sever the lower (front) angle of the eye from the skull. Care has to be used in making the last cut, or the lids at the angle of the eye will be cut, with the result that the eye opening will be unnaturally large in the stuffed skin.

Push the skin to the tip of the lower jaws, sever the skin of each cheek, and then separate the skin from the front of the jaw by means of three slices of the scalpel or by means of three tears made with the fingernails. By means of the thumbnail push the skin on each side of the rostrum (forepart of the skull) to the tip of the nose. Cut the skin free by a stroke of the scalpel, taking care to cut well ahead of the tips of the nasal bones. Beginners often cut off the ends of the nasal bones and are especially apt to do this if scissors are used instead of a scalpel.

With experience, less than a minute should be required to skin a mouse; that is to say, no more than a minute need elapse from the instant when the first incision (cut) is made on the venter until the skin is freed from the tip of the nose.

Note the sex. If the specimen is female, note the absence or presence of embryos. If embryos are present, note the number and length from crown to rump. Complete entries on labels and in catalogue. Attach the skull-label to the skull and drop it in water.

With scalpel in hand go over the skin, laid flat on the sawdust, to remove any fat. Remove it all. Fascia or small bits of red muscle are not worth removing. Grasp threaded needle; sew up mouth with a triangular stitch, and secure with a knot to prevent the stitch from coming loose. Cut the thread above the knot. Tie a knot with one hand at the end of the thread, and place threaded needle where you will reach for it the next time it is

needed. To all parts of the flesh-side of the skin now apply arsenical soap by means of a brush, or apply powdered arsenic by means of a wad of cotton held in forceps.

In each foreleg thrust a wire along the bone of the foreleg into the palm and to the base of the nail of the middle finger without perforating the skin. Begin at the wrist with a thin wisp of cotton, and wrap the bone of the lower leg and wire firmly together. Moistened fingers will secure the cotton at the upper end of the wrapping. The foreleg shaped from cotton should be smaller than the original foreleg.

With the forceps pick up a cotton body. Place the outer seam down. By means of the closed forceps press down on the center of one end of the roll of cotton; with a finger and the thumb of the free hand press inward each side of the cotton at the same time; maintain pressure of the thumb and finger on the two sides of the end of the body;

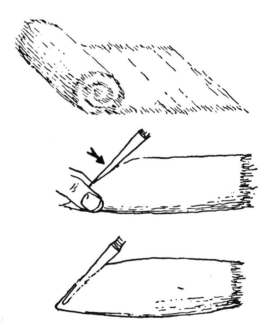

Fig. 622. Folding cotton to make a body, and shaping the end of the body by means of forceps and thumb and forefinger so that the body will fit properly in the nose of the skin. X ⅓.

transfer the forceps, now opened, to the outside of the finger and thumb, and by means of the forceps constrict the cotton. Repetition of the maneuver forms a firm, sharp-pointed, symmetrical end, pointed downward. Holding this pointed end by means of the forceps, place the nose-pad of the skin (still fleshside out) against the pointed end of the cotton body; turn the skin over the cotton body as far as the forelegs, meanwhile maintain-

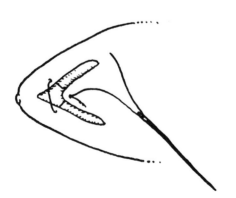

Fig. 621. Sewing of lips. X 1.

ing the grip with the forceps on the pointed end of the cotton; set in place the skin of the head and neck by tugging (not strongly enough to tear it) at the skin in the area of each eye, the skin of the throat, the skin around each ear, and the skin of the chest, making certain that the hard pointed nose of cotton is all the way to the tip of the nose, that the two eye openings are symmetrically set, that the ears are exactly opposite each other, and that the skin of the head is fully stuffed. Then release the grip of the forceps. If the head end of the cotton body was correctly formed the cotton will expand and thus fill out the skin of the head and cause it to be bilaterally symmetrical.

Handling the body and skin as little as possible, turn the remaining part of the skin over the body, which should be slightly longer than the natural body. With scissors cut off the surplus end of the body but leave a thin extension on the back (top) side. See that the cut is exactly vertical and exactly transverse. Let the thin extension of cotton from the back cover the cut end.

Rotate each hind foot one-half turn outward. Then wire each hind leg in the fashion described above for the forelegs. Pull the hind feet out behind. The soles should be down. The skin of the hind leg should be stuffed so that the circumference of the lower leg and thigh is the same as it was in life. The wrapping that binds the wire and bone of the lower leg together should be long enough and frayed enough at the upper (proximal) end to make a gradual blend with the cotton body.

Pick up the tail-wire, moisten one end, twirl (by rotating the wire) on a thin strand of cotton, continue twirling so as to wrap a strand of cotton of gradually increasing diameter on the wire. The tail on the skinned body is a pattern for size. Remember that projecting ends of fibers cause the artificial tail to be functionally larger than it appears to be. Consequently, the wrapped tail-wire should seem to be of a slightly lesser diameter than the actual tail that has been slipped out of its skin. If the cotton is firmly anchored at each end of the tail–wire and if the diameter is exactly correct, all will go nicely. Hang the unstuffed tail down over the near edge of the skinning board (or table), dust some arsenic on the artificial tail, place the tip of the tail-wire into the open base of the tail, and then in one continuous movement thrust the tail-wire all the way to the tip of the skin of the tail—almost to the tip is not sufficient, because any unstuffed part will wither and be broken off.

With wire cutters snip off only as much of the free end of the wire as can not be got through the slit in the skin of the belly. The wire should lie on

the midline of the underside of the cotton body between it and the skin.

Adjust the four legs so that the pairs are symmetrically placed and so that the leg-wires lie parallel to the midline and as close to it as tension on the skin will permit. See that the tail-wire is lined up properly—parallel with the leg-wires and on the midline of the belly midway between the

Fig. 623.   Sewing a stuffed skin. X ⅔.

wires of the two hind legs. All this should be done without picking up the stuffed skin; unnecessary handling at this stage makes the body misshapen. Next sew up the slit in the belly. This is to be done by first catching the very edge of the cut edge of the skin with the needle; three or four diagonal cross-stitches should suffice. After the last one is taken, pull the whole lot tight; throw a loop in the thread, and by means of the points (closed) of the scissors run the loop down to the fur side of the skin to form a knot that prevents the stitches from pulling out as the skin dries. Cut the thread close, say, 2 mm. above the knot. Before returning the needle to the place selected for it, tie a knot in the thread so that it will be ready for instant use to sew together the lips of the next specimen that will be stuffed.

Lay the stuffed skin belly-down with hind feet projecting over the near edge of the skinning board. Tie on the label above right heel; take double turn (on the thread, not the leg), pull the thread almost as tight as possible without breaking it, complete as a square knot or tie four alternate granny knots. Cut off the loose ends of the string. The aims are first to tie the knot so snug that it will not slip off after the diameter of the heel and foot has decreased owing to drying out, and, second, to tie the knot so secure that it will have to be cut to remove the label.

Use a toothbrush having soft bristles to comb the pelage.

Select 8 glass-headed pins no longer than the depth of the drying tray. Pin down the skin in the form shown in Fig. 624. Pin down the forefeet first, seeing that each foot has the palm flat down on the board. The forefeet should be underneath

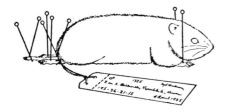

Fig. 624. Completed skin pinned down to dry. X ½.

the head well toward the midline of the body. Slant the pins outward so that they will not crease the skin or fur of the sides of the head. Next pin down the base of the tail by crossing two pins over it, in the fur at its base instead of over the bare part of the tail. Line up the tip of the nose, base of the tail, and tip of the tail; then pin down the tip by means of two crossed pins. Finally pin down the two hind feet. Inspect the stuffed animal from the rear to see that the two hind feet are equidistant from the tail; inspect from the front to see that the two forefeet are equidistant from the nose, and inspect from each side to see that the tips of the toes of each pair of feet are correctly arranged in anterior to posterior position.

By means of a pair of forceps, or by means of a thumb and forefinger, compress each external ear so that the two ears will be creased exactly alike. This helps in causing them to lie down flat as they dry, and if the treatment is repeated on the following day, the ears are more likely to be symmetrical when thoroughly dry.

Use a metal pin, the back of a scalpel, or the sharp edge of forceps to smooth the pelage. Set the tray where the skin can dry in a shady, airy place safe from animals and storms. Two days to two weeks later, whenever the skin is thoroughly dry, it can be unpinned. The amount of humidity has most to do with the time required for drying; in a hot, dry desert, rodent skins may dry in 24 hours, but in a humid, hot, tropical setting, the skins may never dry unless artificial heat is used to drive out the moisture.

With practice the average preparator can complete a specimen in 12–15 minutes. The most rapid preparator I know did the job in slightly less than 5 minutes. He, however, could not maintain that pace and actually turned out approximately eight specimens per hour. Self-analysis reveals to the beginner that he is making unnecessary movements. Attention to the elimination of superfluous movements probably is the best method of increasing the speed of preparation.

Many persons have suggested means of saving the skins of small mammals without stuffing them, or means of partly stuffing them, as for ex-

ample with only a piece of flat cardboard. Such skins are less useful for studying geographic variation and speciation than those prepared by the method described above, because such studies require comparisons of the new specimens with old specimens—specimens prepared in earlier years. Satisfactory comparisons are made only with specimens prepared in the same form. Since nearly all the specimens from the past were prepared according to the method described above, specimens to be saved now are most useful if they also are prepared as described here.

An exception of a sort is made for rabbits in that the artificial bodies are prepared in a fashion different from that described above. The approved method is the following: cut a piece of corrugated cardboard for the center of the body; attach a straight stick to the cardboard at two places by means of wire or twine so that the stick strengthens the cardboard and projects far behind the cardboard to form a paddle; cover the cardboard with a thin (¼-in.) layer of smooth cotton bringing the two edges of the cotton together on the midline of the belly-side of the body along the stick; turn the skin, prepared in the usual way even to the legs, over the flattened body; with heavy shears cut off the posterior end of the cardboard to an appropriate shape; by means of Monel metal wire or a large sacking needle and twine sew through each hind foot around the stick at two places so that the stick will support the two hind legs; with a single stitch secure each forefoot to the skin of the throat; sew up the slit in the belly; thrust a single pin (to be removed when the skin is dry) through the tip of each ear to secure the ears in the desired position; tie the label on the right hind leg; lay the skin away to dry. The skin need not be pinned down. Some advantages of this method are that it provides maximum support for the long hind legs which are wobbly and subject to breakage if otherwise prepared; a flattened body that is strong allows the specimen to be stored in less space than would be required if the body were deeper; and the method requires a minimum of time for stuffing.

Flattening the bodies of all specimens is desirable, and the larger the specimen the thinner the body should be in relation to its width. Convenience in storing is one reason for flattening the bodies. In many collections of study specimens of mammals—for example, those in the University of Kansas Museum of Natural History—the storage cases have the runners so spaced that the distance between the topside of one tray bottom and the underside of the tray bottom next above is $1\frac{13}{16}$ in. Sectional cardboard trays are used for specimens of shrew size and mouse size and reduce the available vertical space to 1⅞ in. or slightly less. (The overall depth of boxes to house skulls was the factor governing the vertical space between trays.) Because there is only 1⅞ in. of vertical space between trays, the artificial bodies of the smaller mammals, say, chipmunks, are made in such a size that the overall height of a specimen is less than 1⅞ in. The pinning trays in the collecting-chests (part of the field equipment) have end pieces 1½ in. high. When the trays of freshly stuffed specimens are stacked in the chest, any specimen that is slightly more than the specified height is at once compressed to 1½ in. Once the skin has dried in that space the height will not increase much if any.

Specimens as large as jack rabbits and opossums may need to be slightly deeper than 1½ in. even if the artificial body is much flattened. The height of such a specimen should not exceed 2¾ in., because that height of body is almost the maximum that can be accommodated in a storage case by separating succeeding trays by one additional case-runner. The means of caring for these specimens in the trays of the collecting chest is to stack one empty pinning tray upside down on the pinning tray that contains the jack rabbits or opossums. In actual practice, however, some collectors keep the heights of even opossums and jack rabbits in 1½ in.

In preparing study specimens of mammals, just as in doing other work, different methods will be found for gaining the same result. By slightly altering their methods from year to year, two preparators who used the same methods to begin with will employ appreciably different methods after the lapse of several years. It is understandable therefore that no two preparators of experience use exactly the same methods—and this, I think, is as it should be, for each preparator should constantly strive to improve the quality of his product. The aim should be firm, symmetrical skins, free of all fat.

It seems to me that every preparator takes pride in his specimens and derives genuine personal gratification from the contemplation of one of his mammal specimens if well prepared, firmly made, and indelibly and accurately labeled. At such moments he may even consider the possibility that one or another of several particularly well-stuffed mice a century hence will be as deservingly spoken of and achieve more of favorable remembrance for him than some other accomplishment at the moment better appreciated by his associates.

A few collectors have immersed skins in a salt-alum bath or treated them only with salt to preserve the skins until it was convenient to stuff them. Salt, or salt-alum solution, does alter the color of the pelage. I recommend against using these preserving agents even at a sacrifice in quantity of specimens. Also, it seems to me that the least that should be done for study skins subjected to the salt-alum treatment or to *any* treatment offering fair likelihood of resultant color change is to label the skins at the time of preparation with clear indication of departure from the conventional technique.

A departure that I regard as the least of several evils is the soaking of autumn-taken skins of ground squirrels in white gasoline to remove surplus fat that may have been released by scraping the flesh side of the skin: 12–18 hours in gasoline removes the fat. Longer immersion tends to dehydrate the skin, making it difficult to shape properly and making it so brittle after it is dried that it may break. For the skins of some ground squirrels almost ready for hibernation, we know of no substitute for gasoline, carbon tetrachloride, or other compound that removes fat. Such skins that we degreased only by scraping and applying corn meal or fine sawdust to absorb the fat, although seemingly free of fat at the time of stuffing, later showed some greasiness that spread to several parts of the skin and down the string or thread of the label onto the label. Gross inspection of skins of mammals almost ready to hibernate suggests that there are fat cells *in* the skin of some as well as on the flesh side of the skin. Scraping and using an absorbent such as corn meal removes most of the fat but not all of it. In spring, when the same kinds of mammals are lean, the soaking of the skin in gasoline or in some other liquid compound is unnecessary—undesirable, in fact.

When a skin has to be washed in soap and water for the purpose of removing dirt, bloodstains, and the like, it can be dried more quickly if its final bath is in white gasoline instead of in water. This is because the gasoline displaces the water and the more volatile gasoline is quickly removed by corn meal or sawdust in which the skin is buried or with which the skin is repeatedly dusted. Such a skin must be thoroughly dried until the fur is everywhere fluffy before being stuffed; otherwise the fur will always cling together in patches and be unsightly.

Mammal skins to be tanned. Skins that are to be tanned or stuffed at a later date should be cased. Slit the skin from hind feet, down inside of hind legs to base of tail, and split tail full length. Do not open pads of feet on carnivores. In deer, open front legs from "elbow" to hoof.

Fat on skins should be scraped off before skins

are stretched for drying. It is not necessary to soak such skins in gasoline.

Do not apply salt, alum, or formalin to skins that are to be relaxed later. Stretch skins to dry, flesh-side out, over a board, cardboard, frame of wire, or two poles. When nearly dry, skins as large as those of deer may be rolled up for packing; skins of carnivores should be packed flat, with the tail folded on the body if desired, with a wisp of excelsior between the skins. Be sure that skins dry and that no fly eggs are on the skins when they are packed. Label dried skins with skull tags only.

Preparation of skulls. Skulls should be severed from the vertebral column, using extreme care not to injure the skull. Skulls the size of those of *Spermophilus franklinii* or larger should have the major part of the masseter muscle snipped off to allow the skull to dry quickly.

As soon as possible, skulls should be put in a glass container of cold water to soak for 12 hours, to remove the blood and loosen the brain. In very hot weather it may be necessary to change the water to prevent fermentation.

After removing the skulls from the water, blow out the brains with the aid of a hypodermic fitted with a blunt needle, or with an atomizer bulb fitted with a short rubber tube and blunt hypodermic needle.

Large and small skulls should not be strung on the same wire to dry. If, for example, squirrel and mouse skulls are strung together, some of the smaller skulls certainly will be more or less broken.

Above all, do not allow skulls to become fly blown. This is apt to occur when they are hung up to dry and while soaking, as some will float and thus be exposed to flies. Maggots do much damage by discoloring the bone, loosening the sutures, and obliterating data on tags. Never hang skulls in the sun—always in the shade and, if possible, where there is a breeze. When skulls are quickly dried, any fly eggs deposited will not hatch. If, due to damp weather, the skulls are apt to remain soggy, protect them by cheesecloth (when hung up) to exclude flies. When packing skulls for shipment, or when moving camp, use a container with plenty of air holes. *Never* put damp or even dry skulls in airtight containers; this causes sweating and maceration. For directions on cleaning mammal skulls, see Jour. Mamm., 14(4):372–374, 1933.

Preparation of skeletons. When preparing skeletons, skin the body completely, which means to the tip of the tail and to the claws of the feet. The pads of the feet of mammals and the skin on the tarsi of birds are nearly impervious to the dermestid beetles. Always "draw" the animal and cut off all large muscles. Tag skeletons (each separate piece) in the same way as skulls.

Take care to save baculum, hyoid bones, marsupial bones, patellas, tip of tail, pygostyle, and alula, as these are easily lost. Do not remove tongue or eyes; they contain important bones.

When a skeleton has been roughed out, wrap it with thread or string so that the head and extremities will not be broken off when they are dry and brittle. The legs are pulled up along the body and the head brought back. A few wrappings of thread will suffice for small skeletons. Do not use so much thread that the beetles have difficulty in getting to the meat in the cleaning process. Do not wrap too tightly, as fresh bones are easily bent.

The higher the humidity, the more thoroughly skeletons should be roughed out. Be sure to remove heart and lungs. A mammal skeleton the size of a squirrel's or larger should have its skull detached and its brains removed. Be sure to tie the skull to the body. A skeleton with its skull and leg bones packed in rib-basket is slow in drying. Keep sawdust off skeletons. Tie an additional label on skeletons of extremely young animals calling attention to them, so that they will receive extra care in cleaning. See that each separate part has a tag.

Packing and shipping. Keep skins of mammals with you until thoroughly dry. In preparing for shipment, take special pains to pack specimens tight so that they will not move. Much of the smoothness and symmetry may be lost through loose packing.

Final Suggestions

Read suggestions above every few days when on a field trip, devoting half an hour or so to thoughtful consideration of the objects of your field work, which probably are: to ascertain everything possible in regard to the natural history of the mammals of the regions traversed, and to make careful records of the facts gathered in the form of specimens and notes to be preserved for all time. All this is for the *information* of others; strive to make your records in all respects clearly intelligible. Remember that the value of your notes increases as the years go by and faunal changes take place. Some earlier notebooks describe conditions now vanished in the localities with which they dealt.

GLOSSARY

Some technical and several semitechnical terms used in the preceding text.

abdomen, *n.*—That part of the body (excepting the back) between the thorax (rib-basket) and the pelvis.

alveolar, *adj.*—Of or pertaining to an alveolus (plural, alveoli), a small cavity or pit, as a socket for a tooth. Alveolar length of a tooth-row therefore denotes the length of the row of the teeth, taken from the posteriormost place where the back tooth emerges from the bone to the anteriormost point where the front tooth in the row emerges from the bone—the overall length of the bony sockets for the row of teeth.

annulation, *n.*—A circular or ringlike formation, as of the dermal scales on the tail of a mammal where one ring of scales that extends entirely around the tail is succeeded, posteriorly, by other rings.

arboreal, *adj.*—Inhabiting or frequenting trees—contrasted with fossorial, aquatic, and cursorial.

auditory bulla (plural, auditory bullae).—A hollow, bony prominence of rounded form (in most mammals formed by the tympanic bone) partly enclosing structures of the middle and inner ear. See Fig. 626.

basal length.—Distance on skull from the anteriormost inferior border of the foramen magnum to a line connecting the anteriormost parts of the premaxillary bones. See Fig. 632.

basilar length.—Distance on skull from the anteriormost inferior border of the foramen magnum to a line connecting the posteriormost margins of the alveoli of the first upper incisors. See Fig. 632.

bead, *n.*—A salient, rounded cordlike projecting ridge of bone, as in certain rodents where the superior border of the orbit is beaded.

braincase, *n.*—The part of the skull enclosing the brain.

calcar, *n.*—In bats a process connected with the calcaneum (heel bone), helping to support the edge of the fold of skin that extends between the leg and tail.

cancellous, *adj.*—Having a spongy or porous structure.

canine, *adj.* & *n.*—Of, pertaining to, or designating the tooth next to the incisors in mammals. See Fig. 632. Of or pertaining to dogs or to the family Canidae.

carnivore, *n.*—An animal that preys on other animals; an animal that eats the flesh of other animals; especially any mammal of the Order Carnivora.

cheek-teeth, *n.*—Teeth behind the canines.

conch (plural, conchs), *n.*—The external ear of a mammal; sometimes the spelling is concha (plural, conchae); the origin of both spellings is conch or konch, originally a bivalve shell of a marine mollusk.

condylar (articular) process.—On a mandible, the process ending in the articular condyle.

condylobasal length.—Least distance on skull from a line connecting the posteriormost projections of the exoccipital condyles to a line connecting the anteriormost projections of the premaxillary bones. See Fig. 632.

coronoid process.—The upward projecting process of the posterior part of the mandible, giving attachment on its outward side to the masseter muscle and on its inner side to the temporal muscle. See Fig. 630.

dental formula (plural, formulae).—A brief method for expressing the number and kind of teeth of mammals. The abbreviations i. (incisor), c. (canine), p. or pm. (premolar), and m. (molar) indicate the kinds in the permanent dentition, and the number in each jaw is written like a fraction, the figures above the horizontal line showing the number in the upper jaw, and those below, the number in the lower jaw. The dental formula of an adult coyote is i. $\frac{3}{3}$; c. $\frac{1}{1}$; p. $\frac{4}{4}$; m. $\frac{2}{3}$. [d.i., d.c. and d.p. (in place of i., c. and p.) designate deciduous ("milk") teeth.]

dentine, *n.*—A calcareous material, harder and denser than bone, which composes the principal mass of a tooth.

dentition, *n.*—The teeth, considered collectively, of an animal.

deuterocone, *n.*—One of the cusps of a premolar tooth of a mammal corresponding in position (anteromedial) to the protocone of a true molar. See Fig. 633.

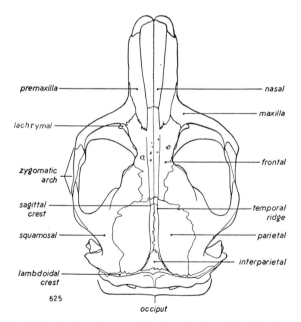

Fig. 625. Parts of the skull in dorsal view of the Southern Pocket Gopher, *Thomomys umbrinus nevadensis*, from Malloy Ranch, 5 mi. W Austin, Nye County, Nevada, No. 37073 M.V.Z., ♂, X 1½. Figures 625–634 are, with slight modifications, from "Mammals of Nevada" by E. Raymond Hall, Univ. California Press, Berkeley, 1946.

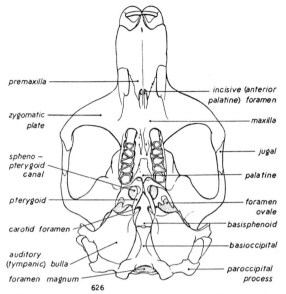

Fig. 626. Parts of the skull in ventral view of the Southern Pocket Gopher, X 1½ (same specimen shown in Fig. 625).

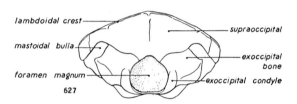

Fig. 627. Posterior view of cranium of Southern Pocket Gopher, X 1½ (same specimen shown in Fig. 625).

diastema, *n.*—A vacant place or gap between teeth in a jaw. See Fig. 628.

diurnal, *adj.*—Active by day—opposed to nocturnal.

enamel, *n.*—Of teeth, the hardest substance of the mammalian body and forming a thin layer that caps or partly covers a tooth.

faeces (singular and plural), *n.*—Intestinal excrement.

feces (see faeces).

femur (plural, femora), *n.*—The proximal bone of the hind limb.

foramen magnum.—The large opening in the back of a skull through which the spinal cord passes to become the medulla oblongata of the brain. See Fig. 627.

forearm, *n.*—The part of the forelimb between the elbow and wrist.

fossorial, *adj.*—Fitted for digging.

frontal, *adj.* & *n.*—Pertaining to or designating the bone (paired) immediately in front of the parietal bone and behind the nasal. See Fig. 625.

gestation period.—The period of carrying young in the uterus, as applied to placental mammals; the period of pregnancy.

guard-hairs.—The stiffer, longer hairs that grow up through the limber, shorter hairs (fur) of a mammal's pelage.

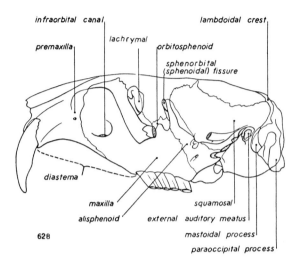

Fig. 628. Lateral view of left side of skull of Southern Pocket Gopher, X 1½ (same specimen shown in Fig. 625).

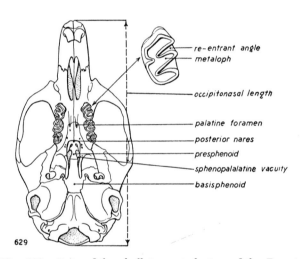

Fig. 629. Parts of the skull in ventral view of the Desert Wood Rat, *Neotoma lepida lepida,* Baker Creek, 7300 ft., White Pine Co., Nevada, No. 42031 M.V.Z., ♀, X 1½.

habitat, *n.*—The kind of environment in which a species of organism is normally found.

hamular process of pterygoid.—A hooklike process on the pterygoid bone. See Fig. 626 for pterygoid bone.

hibernation, *n.*—Of an animal, torpidity especially in winter; the bodily temperature approximates that of the surroundings; the rate of respiration and the heart beat ordinarily are much slower than in an active mammal.

horizontal ramus.—In a lower jaw, the ramus bearing the teeth, and anterior to the vertical ramus.

incisive foramina.—The anterior palatine foramina (singular, foramen), of which there are two, in the bony roof of the anterior part of the cavity of the mouth at the juncture of the premaxillary bones and maxillary bones; transmit nasal branches of palatine arteries and nasopalatine ducts of Jacobson. See Fig. 626.

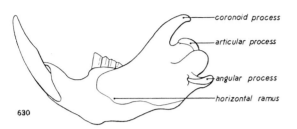

Fig. 630. Parts of the left lower mandible in lateral view of the Southern Pocket Gopher, X 1½ (same specimen shown in Fig. 625).

incisor, *adj.* & *n.*—Pertaining to or designating one of the teeth in front of the canine tooth; those in the upper jaw invariably are in the premaxillary bone. See Fig. 632.

infraorbital canal.—A canal through the maxillary bone from the orbit to the face. See Fig. 628.

inguinal, *adj.*—Pertaining to or in the region of the groin.

insectivorous, *adj.*—Eating insects; preying or feeding on insects.

interfemoral membrane.—In a bat the fold of skin stretching from hind legs to tail. The uropatagium.

interorbital constriction.—The least distance across the top of the skull between the orbits (eye sockets). See Fig. 631.

interorbital region.—The region between the eye sockets; the region of the skull between the rostrum and the braincase.

interparietal, *adj.* & *n.*—Pertaining to or designating the bone (rarely paired) immediately in front of the supraoccipital bone and between the two parietal bones. See Fig. 625.

litter, *n.*—The two or more young brought forth at one birth by a female mammal.

loph, *n.*—A combining form used as the terminal part of certain words and denoting the ridges (or areas) composed of several cusps and styles on the occlusal face of a tooth, as protoloph. See metaloph in Fig. 629.

M2.—Designation of the second true molar in the upper jaw of a mammal.

mammae (singular, mamma), *n.*—The glandular organs for secreting milk.

mastoid, *adj.* & *n.*—Designating or pertaining to the mastoid bone (paired) or its process. This bone is bounded by the squamosal bone, the exoccipital bone, and the tympanic bone. See Figs. 628 and 631.

maxillary breadth.—Width of skull from some designated place on the lateral face of the right maxillary bone (maxilla) to the corresponding place on the left maxillary bone; in shrews, across the ends of the zygomatic processes of the two maxillary bones.

maxillary tooth-row.—The row of teeth in one maxillary bone; in most mammals all the premolars and molars on one side of the upper jaw.

metabolic water.—Water formed as an end product of combustion of foodstuffs in an animal's body.

metacarpal, *adj.* & *n.*—Of or pertaining to a metacarpal bone. A bone of the hand or forefoot between the wrist and fingers; when all the digits are present there are five more or less elongated metacarpal bones, one at the base of each digit.

molar, *adj.* & *n.*—Of or pertaining to a molar tooth. One of the teeth behind the premolar teeth; for example, in the opossum three on each side in upper jaw and in lower jaw, making 12 in all; a molar tooth is not preceded in embryological development by a deciduous (milk) tooth.

molt (moult), *n.* & *v.*—In a mammal, the act or process of shedding or casting off the hair, or outer layer of skin or horns; most mammals shed the hair once, twice, or three times annually. The castoff covering (*obsolete*). As a verb: to be shed (*intransitive*) or to shed (*transitive*).

nasal, *adj.* & *n.*—Of or pertaining to the nose, as a nasal bone (paired) on the dorsal surface of the skull at its anterior end. A nasal bone. See Fig. 625.

nocturnal, *adj.*—Active by night—opposed to diurnal.

occipitonasal length.—Least distance between two vertical lines, one touching the posteriormost part of the skull above the foramen magnum (opening for the spinal cord) and the other touching the anteriormost part of the nasal bones or a nasal bone.

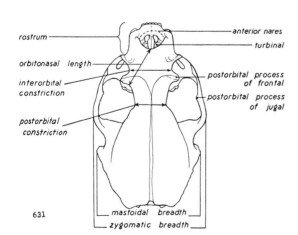

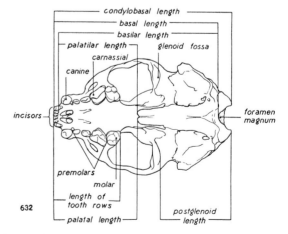

Figs. 631 and 632. Parts of the skull with points and parallels between which several "standard" measurements are taken, of the River Otter, *Lutra canadensis sonora*, Colorado River, 8 mi. upriver from Needles, San Bernardino Co., California, ♂ No. 61451 M.V.Z., X ⅘. Fig. 631 is dorsal view and Fig. 632 is ventral view.

occlusal, *adj.*—Of or pertaining to the grinding or biting (occluding) surface of a tooth.

opposable, *adj.*—Capable of being placed opposite something else; said of the first toe of an opossum in the sense that it can be placed opposite each of the other toes on that same foot.

orbit, *n.*—The cavity in the skull in which the eye and its appendages are situated; the eye socket.

osseous, *adj.*—Composed of, or resembling, bone; bony. Osseous tissue is bony tissue.

overhairs, *n.*—The longer hairs of the pelage of a mammal that project above the fur (shorter hairs).

P3.—Designation of the third (next to last) premolar in the upper jaw of a mammal. Capital letters designate teeth in the upper jaw and lowercase letters designate teeth in the lower jaw. See Fig. 633.

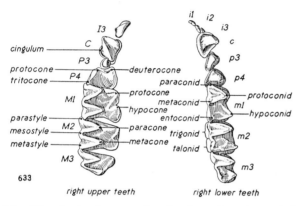

Fig. 633. Occlusal views of teeth of Brazilian Free-tailed Bat, *Tadarida brasiliensis mexicana*, X 6 (same specimen shown in Fig. 195).

palatal, *adj.*—Of or pertaining to the palate (as used in the foregoing account, the bony roof of the mouth made up of two palatine bones, two maxillary bones, and two premaxillary bones).

palate, *n.*—The roof of the mouth, consisting of the structures that separate the mouth from the nasal cavity. The bony palate is composed of the following bones: premaxillae, maxillae, and palatines.

parietal, *adj. & n.*—Pertaining to or designating the parietal bone (paired) roofing the braincase. This bone is behind the frontal bone and in front of the occipital bones. See Fig. 625.

pectoral, *adj.*—Of, pertaining to, or situated or occurring in or on, the chest.

pencil, *n.*—Tuft of fur or hair, as a black pencil on the end of the tail of a mammal.

phalanx (plural, phalanges), *n.*—A bone, in a finger, distal to the metacarpus or a bone, in a toe, distal to the metatarsus.

pinna (plural, pinnae), *n.*—The projecting part of an ear.

postauricular, *adj.*—Situated behind the auricle (pinna) of the ear, as a postauricular patch (ordinarily referring to a patch of fur differing in color from surrounding fur).

postorbital, *adj.*—Situated behind the eye, as postorbital process of the frontal bone or postorbital process of the jugal bone. See Fig. 631.

premaxillary, *adj.*—Of or referring to the premaxilla, a bone (paired), in the mammalian skull bearing the incisor teeth of

the upper jaw; the premaxilla is situated in front of the maxilla. See Fig. 625.

premolar, *adj. & n.*—Designating or pertaining to one of the teeth (a maximum of 4 on each side of upper jaw and lower jaw of placental mammals, or 16 in all) in front of the true molars. When canine teeth are present, premolars are behind these teeth; premolars are preceded by decidious teeth, and in the upper jaw are confined to the maxillary bone. See Fig. 632.

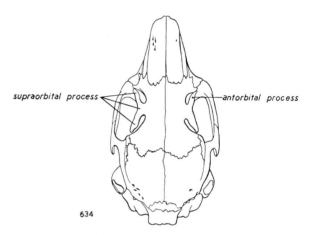

Fig. 634. Parts of the skull of the Pygmy Rabbit, *Sylvilagus idahoensis*, Millett P.O., Nye Co., Nevada, X 1¼.

re-entrant angle.—An infold of the enamel layer on the side, front, or back of a cheek-tooth, as in a molar of a muskrat or wood rat. See Fig. 629.

rostrum, *n.*—Of a mammalian skull, the part projecting in front of the orbits.

rut, *n.*—The breeding period, as in deer.

sagittal crest.—The ridge of bone at the juncture of the two parietal bones resulting from the coalescence of the temporal ridges; in old individuals of many species of mammals the crest extends from the middle of the lambdoidal crest anteriorly onto the frontal bones and divides there into two temporal ridges, each of which extends anterolaterally on the posterior edge of the postorbital process of the frontal bone. See Fig. 625.

saliva, *n.*—The fluid secreted by the glands discharging into the mouth.

subauricular spot.—A spot, patch of hair, distinctively colored immediately below the ear.

supraorbital process of frontal.—The process of the frontal bone on the top rim of the orbit, as in a rabbit. See Fig. 634.

tarsus, *n.*—The ankle.

temporal ridge (paired).—A curved, raised line on the side of the braincase marking the upper limit of attachment of the fascia of the temporal muscle. The temporal ridge is prominent on the parietal bone, frequently extends forward onto the frontal bone, and in some kinds of mammals extends backward onto the interparietal bone. When present, the sagittal crest is formed by the coalescence of the two temporal ridges. See Fig. 625.

terrestrial, *adj.*—Inhabiting the land, rather than the water, trees, or air.

tibia (plural, tibiae), *n.*—The inner and usually the larger of the two bones of the hind limb (leg) between the knee and the ankle.

torpid, *adj.*—Having lost most of the power of exertion; dormant. A ground squirrel is torpid when it is hibernating.

tricolor, *adj. & n.*—Having three colors. Said of hair on back of mammal when hair has three bands each of a different color.

type locality.—The place where a type specimen was obtained.

underfur, *n.*—The short hair of a mammal; in temperate and boreal climates the underfur ordinarily is denser, made up of more hairs, than the longer and coarser overhair.

underparts, *n.*—The underneath (ventral) side of a mammal (not the back or sides), as of a woods mouse with white *underparts*.

upper parts.—The top (dorsal) surface and all of the sides (not the belly, chest, or throat), as of a woods mouse with reddish-brown *upper parts*.

uropatagium, *n.*—The interfemoral membrane of a bat; that is to say, the fold of skin that stretches from the hind legs to the tail.

zygomatic breadth.—Greatest distance across zygomatic arches of cranium at right angles to long axis of skull. See Fig. 631.

LITERATURE CITED

A letter of the Roman alphabet is added to the year of publication if the page numbers overlap in two publications of the same author(s) in the same year.

ADAMS, L. 1963. Do white-tailed deer occur in northeastern California? Jour. Mamm., 44:518–522, 1 fig., December 13.

ADAMS, W. H. 1965. New locality records of two North Carolina mammals. Jour. Mamm., 46:499, August 26.

ALLEN, G. M. 1904. Fauna of New England. 3. List of the Mammalia. Occas. Pap. Boston Soc. Nat. Hist., 7:1–35, June.

———. 1911. Mammals of the West Indies. Bull. Mus. Comp. Zool., 54:173–263, July.

———. 1917. Two undescribed West Indian bats. Proc. Biol. Soc. Washington, 30:165–170, October 23.

———. 1919. Bats from Mount Whitney, California. Jour. Mamm., 1:1–5, November 28.

———. 1930. The walrus in New England. Jour. Mamm., 11:139–145, May 9.

———. 1942. Extinct and vanishing mammals *of the* Western Hemisphere *with the* marine species of all the oceans. Amer. Committee Internat. Wild Life Protection, spec. publ. 11:xvi + 620, frontispiece, 24 figs.

ALLEN, G. M., and SANBORN, C. C. 1937. Notes on bats from the Bahamas. Jour. Mamm., 18:226–228, 1 fig., May 12.

ALLEN, H. 1891. Description of a new species of Vampyrops. Proc. Acad. Nat. Sci. Philadelphia, 43:400–405, September 22.

———. 1892. Description of a new genus of phyllostome bats. Proc. U.S. Nat. Mus., 15:441–442, 2 figs.

ALLEN, J. A. 1869. Catalogue of the mammals of Massachusetts: with a critical revision of the species. Bull. Mus. Comp. Zool., 1:143–252, 10 tables.

———. 1874. Notes on the mammals of portions of Kansas, Colorado, Wyoming and Utah. Bull. Essex Inst., 6:43–66, May.

———. 1879. On the species of the genus Bassaris. Bull. U.S. Geol. and Geog. Surv. Territories, 5:331–340, November 30.

———. 1880. History of North American pinnipeds, a monograph of the walruses, sea-lions, sea-bears and seals of North America. U.S. Geol. and Geog. Surv. Territories, Miscl. Publ., 12:xvi + 785, 60 figs.

———. 1894. Notes on mammals from New Brunswick, with description of a new species of Evotomys. Bull. Amer. Mus. Nat. Hist., 6:99–106, April 14.

———. 1897. Additional notes on Costa Rican mammals, with descriptions of new species. Bull. Amer. Mus. Nat. Hist., 9:31–44, 1 pl., March 11.

———. 1898. Revision of the chickarees, or North American red squirrels (subgenus Tamiasciurus). Bull. Amer. Mus. Nat. Hist., 10:249–298, July 22.

———. 1900. Note on the generic names Didelphis and Philander. Bull. Amer. Mus. Nat. Hist., 13:185–190, October 12.

———. 1903. Mammals collected in Alaska and northern British Columbia by the Andrew J. Stone Expedition of 1902. Bull. Amer. Mus. Nat. Hist., 19:521–567, October 10.

———. 1906. Mammals from the states of Sinaloa and Jalisco, Mexico, collected by J. H. Batty during 1904 and 1905. Bull. Amer. Mus. Nat. Hist., 22:191–262, pls. 20–33, July 25.

———. 1910. Additional mammals from Nicaragua. Bull. Amer. Mus. Nat. Hist., 28:87–115, April 30.

———. 1915. Review of the South American Sciuridae. Bull. Amer. Mus. Nat. Hist., 34:147–309, 14 pls., 25 figs., 8 tables, May 17.

ALLRED, D. M., and BECK, D. E. 1963. Range of movement and dispersal of some rodents at the Nevada Atomic Test Site. Jour. Mamm., 44:190–200, 10 figs., 1 table, May 20.

ALSTON, E. R. 1879–1882. Biologia Centrali-Americana. Mammalia, xx + 220 pp., 22 pls.

ALVAREZ, T. 1958. Roedores colectados en el territorio de la Baja California. Acta Zool. Mexicana, 2:1–7, November 20.

———. 1961. Sinopsis de las especies Mexicanas del genero Dipodomys. Rev. Soc. Mexicana Hist. Nat., 21:391–424, 8 maps, 1 fig. (unnumbered), May 19.

———. 1961. Taxonomic status of some mice of the Peromyscus boylii group in eastern Mexico, with description of a new subspecies. Univ. Kansas Publ., Mus. Nat. Hist., 14:111–120, 1 fig., December 29.

———. 1962. A new subspecies of ground squirrel (Spermophilus spilosoma) from Tamaulipas, México. Univ. Kansas Publ., Mus. Nat. Hist., 14:121–124, March 7.

———. 1962. A new subspecies of wood rat (Neotoma) from northeastern Mexico. Univ. Kansas Publ., Mus. Nat. Hist., 14:139–143, April 30.

———. 1963. The Recent mammals of Tamaulipas, México. Univ. Kansas Publ., Mus. Nat. Hist., 14:363–473, 5 figs., May 20.

———. 1968. Notas sobre una colección de mamíferos de la región . . . costera del Río Balsas entre Michoacán y Guerrero. Rev. Soc. Mexicana Hist. Nat., 29:21–35, December.

ALVAREZ, T., and AVIÑA, C. E. 1963. Notas acerca de algunas especies Mexicanas de ardillas del genero *Sciurus* (Rodentia: Sciuridae). Rev. Soc. Mexicana Hist. Nat., 24:33–39, December.

———. 1964. Nuevos registros en México de la familia Molossidae. Rev. Soc. Mexicana Hist. Nat., 25:243–254, 2 maps, December.

ALVAREZ, T., and RAMÍREZ-P., J. 1968. Descripción de una nueva subespecie de *Spermophilus adocetus* (Rodentia, Sciuridae) de Michoacán, México y estado taxonómico de S. *a. arceliae* (Villa-R., 1942). Rev. Soc. Mexicana Hist. Nat., 29:181–189, 2 figs., 1 table, December.

———. 1972. Notas acerca de murciélagos Mexicanos. Anal.

Escuela Nac. Cienc. Biol., México, 19:167–178, 1 fig., February.

ALVAREZ DEL TORO, M. 1952. Los animales silvestres de Chiapas. Ed. Gob. Estado, Tuxtla Gutiérrez. Pp. 247—2, 119 figs., May 31.

AMEGHINO, F. 1910. Montaneia anthropomorpha. Un género de monos hoy extinguido de la isla de Cuba.—nota preliminar. Anal. Mus. Nac. Buenos Aires, 20 (Ser. 3ª, t. 13):317–318, September 16.

ANDERSEN, K. 1906. *Brief diagnoses of a new genus and ten new forms of stenodermatous bats.* Ann. Mag. Nat. Hist., ser. 7, 18:419–423, December.

———. 1908. A monograph of the chiropteran genera *Uroderma, Enchisthenes,* and *Artibeus.* Proc. Zool. Soc. London, pp. 204–319, 59 figs., September 7.

ANDERSEN, K. W., and FLEHARTY, E. D. 1964. Additional fox records for Kansas. Trans. Kansas Acad. Sci., 67:193–194, June 12.

ANDERSEN, K. W., and JONES, J. K., JR. 1971. Mammals of northwestern South Dakota. Univ. Kansas Publ., Mus. Nat. Hist., 19:361–393, 8 figs., January 18.

ANDERSON, R. M. 1942. Six additions to the list of Quebec mammals with descriptions of four new forms. Ann. Report for 1941, Provancher Soc. Nat. Hist. Canada, Quebec, pp. 31–43 (English), 45–57 (French), July 14.

———. 1945. Three mammals of the weasel family (Mustelidae) added to the Quebec list with descriptions of two new forms. Ann. Report for 1944, Provancher Soc. Nat. Hist. Canada, Quebec, pp. 56–61 (English), 62–68 (French), November 2.

———. 1947. Catalogue of Canadian Recent mammals. Bull. Nat. Mus. Canada, 102:v + 238, 1 fig., January 24.

ANDERSON, R. M., and RAND, A. L. 1945. The varying lemming (genus Dicrostonyx) in Canada. Jour. Mamm., 26:301–306, November 14.

ANDERSON, S. 1956. Extensions of known ranges of Mexican bats. Univ. Kansas Publ., Mus. Nat. Hist., 9:347–351, August 15.

———. 1959. Mammals of the Grand Mesa, Colorado. Univ. Kansas Publ., Mus. Nat. Hist., 9:405–414, 1 fig., May 20.

———. 1959. Distribution, variation, and relationships of the montane vole, Microtus montanus. Univ. Kansas Publ., Mus. Nat. Hist., 9:415–511, 12 figs., August 1.

———. 1960. The baculum in microtine rodents. Univ. Kansas Publ., Mus. Nat. Hist., 12:181–216, 49 figs., February 19.

———. 1960. Neotropical bats from western México. Univ. Kansas Publ., Mus. Nat. Hist., 14:1–8, October 24.

———. 1961. A relict population of *Microtus pennsylvanicus* in southwestern New Mexico. Amer. Mus. Novit., 2034:1–3, 1 fig., May 26.

———. 1961. Mammals of Mesa Verde National Park, Colorado. Univ. Kansas Publ., Mus. Nat. Hist., 14(3):29–67, 2 pls., 3 figs., July 24.

———. 1962. Tree squirrels (*Sciurus colliaei* group) of western Mexico. Amer. Mus. Novit., 2093:1–13, 5 figs., June 13.

———. 1962. A new northern record of *Synaptomys borealis* in Ungava. Jour. Mamm., 43:421–422, August 20.

———. 1964. The systematic status of *Perognathus artus* and *Perognathus goldmani* (Rodentia). Amer. Mus. Novit., 2184:1–27, 60 figs., July 9.

———. 1965. Conspecificity of *Plagiodontia aedium* and P.

hylaeum (Rodentia). Proc. Biol. Soc. Washington, 78:95–97, October 8.

———. 1966. Taxonomy of gophers, especially *Thomomys* in Chihuahua, Mexico. Systematic Zool., 15:189–198, 6 figs., September 26.

———. 1969. Taxonomic status of the woodrat, Neotoma albigula, in southern Chihuahua, México. Univ. Kansas Mus. Nat. Hist., Miscl. Publ., 51:25–50, 7 figs., 2 tables, July 11.

———. 1972. Mammals of Chihuahua taxonomy and distribution. Bull. Amer. Mus. Nat. Hist., 148:149–410, 366 figs., 15 tables, September 8.

ANDERSON, S., and BERG, W. N. 1959. Extension of the known range of the cotton rat, Sigmodon hispidus, in New Mexico. Southwestern Nat., 4:40–42, July 22.

ANDERSON, S., and GAUNT, A. S. 1962. A classification of the white-sided jack rabbits of Mexico. Amer. Mus. Novit., 2088:1–16, 3 figs., May 24.

ANDERSON, S., and HADARY, S. 1965. A kit fox from southern Zacatecas. Jour. Mamm., 46:343, May 20.

ANDERSON, S., and HUBBARD, J. P. 1971. Notes on geographic variation of *Microtus pennsylvanicus* (Mammalia, Rodentia) in New Mexico and Chihuahua. Amer. Mus. Novit., 2460:1–8, 2 figs., April 21.

ANDERSON, S., and JONES, J. K., JR. 1960. Records of harvest mice, Reithrodontomys, from Central America, with description of a new subspecies from Nicaragua. Univ. Kansas Publ., Mus. Nat. Hist., 9:519–529, January 14.

ANDERSON, S., and LONG, C. A. 1961. Small mammals in pellets of barn owls from Miñaca, Chihuahua. Amer. Mus. Novit., 2052:1–3, 1 table, October 12.

ANDERSON, S., and NELSON, B. C. 1958. Additional records of mammals of Kansas. Trans. Kansas Acad. Sci., 61:302–312, November 21.

ANDERSON, S., and NELSON, C. E. 1960. Birds and mammals in barn owl pellets from near Laguna, Chihuahua, Mexico. Southwestern Nat., 5:99–101, 1 table, August 10.

———. 1965. A systematic revision of *Macrotus* (Chiroptera). Amer. Mus. Novit., 2212:1–39, 19 figs., 2 tables, March 17.

ANDERSON, S., and OGILVIE, P. W. 1957. Vertebrates found in owl-pellets from northeastern Chihuahua. Southwestern Nat., 2:33–37, October 15.

ANDREWS, R. D. 1963. The golden mouse in southern Illinois. Nat. Hist. Misc., Chicago Acad. Sci., 179:1–3, December 29.

ANTHONY, H. E. 1917. A new rabbit and a new bat from Neotropical regions. Bull. Amer. Mus. Nat. Hist., 37:335–337, Pl. 34, May 28.

———. 1918. The indigenous land mammals of Porto Rico, living and extinct. Mem. Amer. Mus. Nat. Hist., n.s., 2:331–435, Pls. 55–74, 55 figs., numerous unnumbered tables, June.

———. 1919. Mammals collected in eastern Cuba in 1917. With descriptions of two new species. Bull. Amer. Mus. Nat. Hist., 41:625–643, Pls. 35–37, December 30.

———. 1923. Preliminary report on Ecuadorean mammals. Amer. Mus. Novit., 55:1–14, 4 figs., January 31.

———. 1932. A new genus of rodents from Yucatán. Amer. Mus. Novit., 586:1–3, 1 fig., November 16.

ARMSTRONG, D. M. 1969. Noteworthy records of bats from Costa Rica. Jour. Mamm., 50:808–810, November 28.

———. 1971. Notes on variation in *Spermophilus tridecemlineatus* (Rodentia: Sciuridae) in Colorado and adjacent states, and description of a new subspecies. Jour. Mamm., 52:528–536, 1 table, August 26.

———. 1972. Distribution of mammals in Colorado. Univ. Kansas Mus. Nat. Hist., Monog. no. 3, x + 415 pp., 8 pls., 33 figs., 36 tables, July 20.

———. 1974. Second record of the Mexican big-eared bat in Utah. Southwestern Nat., 19:114–115, March 25.

ARMSTRONG, D. M., and JONES, J. K., JR. 1971. Mammals from the Mexican state of Sinaloa. I. Marsupialia, Insectivora, Edentata, Lagomorpha. Jour. Mamm., 52:747–757, 6 figs., December 16.

ARMSTRONG, D. M., JONES, J. K., JR., and BIRNEY, E. C. 1972. Mammals from the Mexican state of Sinaloa. III. Carnivora and Artiodactyla. Jour. Mamm., 53:48–61, 1 fig., March 24.

ARMSTRONG, F. H. 1957. Notes on Sorex preblei in Washington State. Murrelet, 38:6, August.

ARMSTRONG, F. H., and JOHNSON, M. L. 1969. Noctilio leporinus in Hispaniola. Jour. Mamm., 50:133, February 26.

ARNOLD, J. R., and SCHONEWALD, J. 1972. Notes on the distribution of some bats in southern México. Wasmann Jour. Biol., 30:171–174.

ARREDONDO, O. 1958. Los roedores Cubanos extinguidos. El Cartero Cubano, ano 17 (no. 12):8–11, 48, 7 figs., December.

———. 1970. Dos nuevas especies subfosiles de mamíferos (Insectivora: Nesophontidae) del Holoceno Precolombino de Cuba. Mem. Soc. Cienc. Nat. La Salle, 30:122–152, 11 figs., 5 tables [Rec. Univ. Kansas Library Sept. 13, 1971].

ASDELL, S. A. 1946. Patterns of mammalian reproduction. Comstock Publishing Co., Inc., Ithaca, New York, x + (4) + 437 pp., 12 pls., numerous figs. and tables.

AXTELL, R. W. 1961. An additional record for the bat Tadarida molossa from Trans-Pecos Texas. Southwestern Nat., 6:52–53, June 10.

———. 1962. An easternmost record for the bat Choeronycteris mexicana from Coahuila, Mexico. Southwestern Nat., 7:76, June 1.

BACCUS, J. T. 1968. Two noteworthy records of rodents from Baylor County. Southwestern Nat., 13:362, December 10.

———. 1971. The mammals of Baylor County, Texas. Texas Jour. Sci., 22:177–185, April 30.

BACCUS, J. T., GREER, R. E., and RAUN, G. G. 1971. Additional records of Baiomys taylori (Rodentia: Cricetidae) for northern Texas. Texas Jour. Sci., 23:148–149, October.

BACKUS, R. H. 1961. Stranded killer whale in the Bahamas. Jour. Mamm., 42:418–419, August 21.

BADER, R. S., and HALL, J. S. 1960. Mammalian remains from an Indiana cave. Jour. Mamm., 41:111–112, February 20.

BAILEY, V. 1900. Revision of American voles of the genus Microtus. N. Amer. Fauna, 17:1–88, 5 pls., 17 figs., June 6.

———. 1905. Biological survey of Texas. N. Amer. Fauna, 25:1–222, 16 pls., 24 figs., October 24.

———. 1915. Revision of the pocket gophers of the genus Thomomys. N. Amer. Fauna, 39:1–136, 8 pls., 10 figs., November 15.

———. 1927. A biological survey of North Dakota. N. Amer. Fauna, 49:vi + 226, 21 pls., 8 figs., 1 map (frontispiece), January 8.

———. 1932. Mammals of New Mexico. N. Amer. Fauna, 53:1–412, 22 pls., 58 figs., March 1.

———. 1936. The mammals and life zones of Oregon. N. Amer. Fauna, 55:1–416, 51 pls., 102 figs., 1 map (frontispiece), August 29.

BAIRD, S. F. 1858. Mammals: Explorations and surveys for a railroad route . . . , pp. xlviii + 757, Pls. 17–60, Figs. 1–35, July 14.

BAKER, R. H. 1956. Mammals of Coahuila, México. Univ. Kansas Publ., Mus. Nat. Hist., 9:125–335, 75 figs., June 15.

———. 1959. Substitute name for Lepus americanus seclusus Baker and Hankins. Jour. Mamm., 40:145, February 20.

———. 1962. Additional records of Notiosorex crawfordi from Mexico. Jour. Mamm., 43:283, May 29.

———. 1964. Myotis lucifugus lucifugus (Leconte) and Pipistrellus hesperus maximus Hatfield in Knox County, new to north-central Texas. Southwestern Nat., 9:205, August 20.

———. 1966. Further notes on the mammals of Durango, México. Jour. Mamm., 47:344–345, May 26.

———. 1967. A new subspecies of pallid bat (Chiroptera: Vespertilionidae) from northeastern Mexico. Southwestern Nat., 12:329–330, October 31.

———. 1968. Records of opossum and kit fox from Zacatecas. Jour. Mamm., 49:318, May 20.

———. 1969. Cotton rats of the Sigmodon fulviventer group. Univ. Kansas Mus. Nat. Hist., Miscl. Publ., 51:177–232, 21 figs., 6 tables, July 11.

BAKER, R. H., and GREER, J. K. 1960. Notes on Oaxacan mammals. Jour. Mamm., 41:413–415, August 15.

———. 1962. Mammals of the Mexican state of Durango. Publ. Mus. Michigan State Univ., Biol. Ser., 2:25–154, 4 pls., 6 figs., August 27.

BAKER, R. H., and PETERSEN, M. K. 1965. Notes on a climbing rat, Tylomys, from Oaxaca, México. Jour. Mamm., 46:694–695, 1 fig., November 30.

———. 1969. Records of the badger from Mexico. Southwestern Nat., 14:251–252, October 10.

BAKER, R. H., and PHILLIPS, C. J. 1965. Peromyscus ochraventer in San Luis Potosí. Jour. Mamm., 46:337–338, May 20.

———. 1965. Mammals from El Nevado de Colima, México. Jour. Mamm., 46:691–693, November 30.

BAKER, R. H., and WEBB, R. G. 1967. Notas acerca de los anfibios, reptiles y mamíferos de La Pesca, Tamaulipas. Rev. Soc. Mexicana Hist. Nat., 27:179–190, 1 table, June 29.

BAKER, R. H., WEBB, R. G., and DALBY, P. 1967. Notes on reptiles and mammals from southern Zacatecas. Amer. Midland Nat., 77(1):223–226, January.

BAKER, R. H., WEBB, R. G., and STERN, E. 1973. Amphibians, reptiles and mammals from north-central Chiapas. Anal. Inst. Biol., Univ. Nac. Autó. México, 42 (for 1971, Ser. Zool. 1):77–85, 1 fig., 1 table, January 31.

BAKER, R. H., and WOMOCHEL, D. 1966. Mammals from southern Oaxaca. Southwestern Nat., 11:306, June 30.

BAKER, R. J. 1967. Karyotypes of bats of the family Phyllostomidae and their taxonomic implications. Southwestern Nat., 12:407–428, 18 figs., 1 table, December 31.

BAKER, R. J., and CHRISTIANSON, L. 1966. Notes on bats from Sonora, Mexico. Southwestern Nat., 11:310–311, June 30.

BAKER, R. J., and COCKRUM, E. L. 1966. Geographic and ecological range of the long-nosed bat, *Leptonycteris*. Jour. Mamm., 47:329–331, 1 fig., May 26.

BAKER, R. J., and GENOWAYS, H. H. 1975. A new subspecies of Geomys bursarius (Mammalia: Geomyidae) from Texas and New Mexico. Occas. Pap. Mus. Texas Tech Univ., 29:1–18, 5 figs., 2 tables, April 25.

BAKER, R. J., and GOMEZ, G. 1968. Notes on some bats of Tamaulipas. Southwestern Nat., 13:361–362, December 10.

BAKER, R. J., and JONES, J. K., JR. 1972. *Tadarida aurispinosa* in Sonora, Mexico. Southwestern Nat., 17:308–309, October 6.

———. 1975. Additional records of bats from Nicaragua, with a revised checklist of Chiroptera. Occas. Pap. Mus. Texas Tech Univ., 32:1–13, 1 fig., July 18.

BAKER, R. J., JONES, J. K., JR., and CARTER, D. C. 1976. Biology of bats of the New World family Phyllostomatidae. Part I. Special Publ. Mus. Texas Tech Univ., 10:1–218, 94 figs., 12 tables, June 25.

BAKER, R. J., and LOPEZ, G. 1968. Notes on some bats of Tamaulipas. Southwestern Nat., 13:361–362, December 10.

BAKER, R. J., and MASCARELLO, J. T. 1969. Chromosomes of some vespertilionid bats of the genera *Lasiurus* and *Plecotus*. Southwestern Nat., 14:249–251, 3 figs., October 10.

BAKER, R. J., and MCDANIEL, V. R. 1972. A new subspecies of Uroderma bilobatum (Chiroptera: Phyllostomatidae) from Middle America. Occas. Pap. Mus. Texas Tech Univ., 7:1–4, 1 fig., 1 table, November 3.

BAKER, R. J., MOLLHAGEN, T., and LOPEZ, G. 1971. Notes on *Lasiurus ega*. Jour. Mamm., 52:849–852, 1 fig., 2 tables, December 16.

BAKER, R. J., and PATTON, J. L. 1967. Karyotypes and karyotypic variation of North American vespertilionid bats. Jour. Mamm., 48:270–286, 15 figs., 1 table, May 20.

BAKER, R. J., and WARD, C. M. 1967. Distribution of bats in southeastern Arkansas. Jour. Mamm., 48:130–132, February 20.

BAKER, W. W. 1967. *Myotis leibii leibii* in Georgia. Jour. Mamm., 48:142, February 20.

BALSER, D. S., and LONGLEY, W. H. 1966. Increase of the fisher in Minnesota. Jour. Mamm., 47:547–550, 1 fig., 1 table, August 29.

BANFIELD, A. W. F. 1958. The mammals of Banff National Park, Alberta. Nat. Mus. Canada, Bull. 159:1–53, 5 figs.

———. 1960. Some noteworthy accessions to the National Museum mammal collection. Nat. Hist. Pap., Nat. Mus. Canada, 6:1–2, June 2.

———. 1961. A red bat on Southampton Island, Northwest Territories. Canadian Field-Nat., 75:264, December 29.

———. 1962a. A revision of the reindeer and caribou, genus *Rangifer*. Nat. Mus. Canada, Bull. 177 (Biol. Ser. 66):vi + 137, 12 pls., 15 figs., 23 tables, February 7.

———. 1962b. Notes on the mammals of Pelee Island, Ontario. Bull. Nat. Mus. Canada, 183:118–122, December 3.

———. 1963. The disappearance of the Queen Charlotte Islands' caribou. Preprint Nat. Mus. Canada, Bull. 185:40–49, 1 pl., 1 fig., 2 tables, March.

———. 1963. The post glacial dispersal of American caribou. Proc. XVI International Congress of Zool., 1:206, August 20–27.

———. 1974. The mammals of Canada. National Museums of Canada, Univ. Toronto Press, xxv + 438 pp., 46 colored pls., 113 figs., 176 maps, November 6.

BANGS, O. 1902. Chiriqui mammalia. Bull. Mus. Comp. Zool., 39:15–51, 27 figs., April.

BANKS, R. C. 1964. Range extensions for three bats in Baja California, Mexico. Jour. Mamm., 45:489, September 15.

———. 1964. The mammals of Cerralvo Island, Baja California. Trans. San Diego Soc. Nat. Hist., 13:397–404, 1 fig., 2 tables, December 30.

———. 1967a. The *Peromyscus guardia–interparietalis* complex. Jour. Mamm., 48:210–218, 3 figs., 1 table, May 20.

———. 1967b. A new insular subspecies of spiny pocket mouse (Mammalia, Rodentia). Proc. Biol. Soc. Washington, 80:101–103, 1 table, July 28.

———. 1967c. Birds and mammals of La Laguna, Baja California. Trans. San Diego Soc. Nat. Hist., 14:205–232, 4 figs., August 18.

BANKS, R. C., and BROWNELL, R. L. 1969. Taxonomy of the common dolphins of the eastern Pacific Ocean. Jour. Mamm., 50:262–271, 3 figs., 1 table, June 12.

BARBOUR, R. W., and DAVIS, W. H. 1969. Bats of America. Univ. Press of Kentucky, Lexington, 286 pp., 24 pls., 131 figs. (numbered), 73 figs. (unnumbered), December 20.

BARCLAY, E. N. 1935 (for 1934). The red deer of the Caucasus. Proc. Zool. Soc. London, pp. 789–798, 3 pls., 5 figs., January.

BARKALOW, F. S. 1967. Range extension and notes on the least weasel in North Carolina. Jour. Mamm., 48:488, August 21.

BARKALOW, F. S., JR., and FUNDERBURG, J. B., JR. 1960. Probable breeding and additional records of the seminole bat in North Carolina. Jour. Mamm., 41:394–395, August 15.

BARKER, M. S., JR., and BEST, T. L. 1976. The wolverine (*Gulo luscus*) in Nevada. Southwestern Nat., 21:133, March 20.

BARTHOLOMEW, G. A., and BOOLOOTIAN, R. A. 1960. Numbers and population structure of the pinnipeds on the California Channel Islands. Jour. Mamm., 41:366–375, 1 pl., 1 fig., 3 tables, August 15.

BATEMAN, G. C. 1967. Abnormal coloration in the pocket mouse, *Perognathus baileyi*. Jour. Mamm., 48:493, August 21.

BEASLEY, L. E., and SEVERINGHAUS, W. D. 1973. A survey of the cricetine rodents of west Tennessee. Jour. Tennessee Acad. Sci., 48:106–111, 8 figs., July.

BECK, W. H. 1958. A guide to Saskatchewan mammals. Saskatchewan Nat. Hist. Soc., Special Publ., 1:1–52, 20 figs., several maps, October 17.

———. 1961. Range extension of the wandering shrew. Blue Jay, 19:185, December.

———. 1964. Records of mammals in the Lake Athabasca area, Saskatchewan. Blue Jay, 22:165–172, 3 figs., 2 tables, December.

BECK, W. H., and WILSON, A. E. 1969. The star-nosed mole in Manitoba. Blue Jay, 27:93–95, 2 figs., 1 table, June.

BEE, J. W., and HALL, E. R. 1956. Mammals of northern Alaska on the Arctic Slope. Univ. Kansas Mus. Nat. Hist., Miscl. Publ., 8:1–309, colored frontispiece, 4 pls., 127 figs., 5 tables, March 10.

BEER, J. R. 1961. Hibernation in *Perognathus flavescens*. Jour. Mamm., 42:103, February 20.

BEST, T. L. 1971. Notes on the distribution and ecology of five eastern New Mexico mammals. Southwestern Nat., 16:210–211, November 15.

———. 1972. Mound development by a pioneer population of the banner-tailed kangaroo rat, Dipodomys spectabilis baileyi Goldman, in eastern New Mexico. Amer. Midland Nat., 87:201–206, 4 figs., February 23.

———. 1973. Ecological separation of three genera of pocket gophers (Geomyidae). Ecology, 54:1311–1319, 1 fig., 2 tables, autumn.

BEST, T. L., GREER, J. K., and ELDER, F. F. B. 1972. Two bat records from Zacatecas, Mexico. Southwestern Nat., 17:97–98, May 1.

BEST, T. L., and KENNEDY, M. L. 1972. The porcupine (Erethizon dorsatum Linnaeus) in the Texas Panhandle and adjacent New Mexico and Oklahoma. Texas Jour. Sci., 24:351, December.

BIGGERS, C. J., and DAWSON, W. D. 1971. Serum protein polymorphisms in Peromyscus polionotus of South Carolina. Jour. Mamm., 52:376–385, 2 figs., 4 tables, May 28.

BIGGERS, J. D. 1967. Notes on reproduction of the woolly opossum (Caluromys derbianus) in Nicaragua. Jour. Mamm., 48:678–680, 4 tables, November 20.

BIRKENHOLZ, D. E. 1967. The harvest mouse (Reithrodontomys megalotis) in central Illinois. Trans. Illinois State Acad. Sci., 60:49–53, 1 table, March 1.

BIRNEY, E. C. 1974. Twentieth century records of wolverine in Minnesota. Loon, 46:78–81, summer.

BIRNEY, E. C., BOWLES, J. B., TIMM, R. M., and WILLIAMS, S. L. 1974. Mammalian distributional records in Yucatán and Quintana Roo, with comments on reproduction, structure, and status of peninsular populations. Bell Mus. Nat. Hist., Univ. Minnesota, Occas. Pap., 13:ii + 25, 2 figs., 3 tables, August 27.

BIRNEY, E. C., and JONES, J. K., JR. 1972. Woodrats (genus Neotoma) of Sinaloa, México. Trans. Kansas Acad. Sci., 74:197–211, 3 figs., 2 tables, April 7.

BIRNEY, E. C., JONES, J. K., JR., and MORTIMER, D. M. 1971. The yellow-faced pocket gopher, Pappogeomys castanops, in Kansas. Trans. Kansas Acad. Sci., 73:368–375, 1 fig., 2 tables, April 30.

BIRNEY, E. C., and LAMPE, R. P. 1972. Sagebrush vole (Lagurus curtatus) in South Dakota. Amer. Midland Nat., 88:466, October 18.

BIRNEY, E. C., and RISING, J. D. 1968. Notes on distribution and reproduction of some bats from Kansas, with remarks on incidence of rabies. Trans. Kansas Acad. Sci., 70:519–524, 1 fig., March 12.

BLACK, H. L. 1970. Occurrence of the Mexican big-eared bat in Utah. Jour. Mamm., 51:190, May 9.

BLACK, J. D. 1936. Mammals of northwestern Arkansas. Jour. Mamm., 17:29–35, February 17.

———. 1937. Mammals of Kansas. Kansas State Board Agric., 30th Biennial Report, 35:116–217, Figs. 17–36.

BLAIR, W. F. 1939. Faunal relationships and geographic distribution of mammals in Oklahoma. Amer. Midland Nat., 22:85–133, 1 fig., 2 tables, July.

———. 1942. Systematic relationships of Peromyscus and several related genera as shown by the baculum. Jour. Mamm., 23:196–204, 2 figs., 1 table, June 3.

———. 1952. Bats of the Edwards Plateau in central Texas. Texas Jour. Sci., 4:95–98, March 30.

———. 1954. Mammals of the Mesquite Plains Biotic District in Texas and Oklahoma, and speciation in the central grasslands. Texas Jour. Sci., 6:235–264, 1 fig., September.

BLANKS, G. A., and SHELLHAMMER, H. S. 1968. Chromosome polymorphism in California populations of harvest mice. Jour. Mamm., 49:726–731, 2 figs., November 26.

BLEAKNEY, J. S. 1965. First specimens of eastern pipistrelle from Nova Scotia. Jour. Mamm., 46:528–529, August 26.

———. 1965. Notes on the migratory tree bats of Nova Scotia. Canadian Field-Nat., 79:154–155, October 20.

BLEICH, V. C. 1974. Muskrats (Ondatra zibethicus) in Amargosa Canyon, Inyo and San Bernardino counties, California. Murrelet, 55:7–8, July.

BLEICH, V. C., and SCHWARTZ, O. A. 1974. Western range extension of Stephens' kangaroo rat (Dipodomys stephensi), a threatened species. California Fish and Game, 60:208–210, 1 fig., October.

BOGAN, M. A. 1974. Identification of Myotis californicus and M. leibii in southwestern North America. Proc. Biol. Soc. Washington, 87:49–56, 3 figs., 4 tables, April 25.

———. 1975. Geographic variation in Myotis californicus in the southwestern United States and Mexico. U.S. Fish Wildlife Serv., Wildlife Research Report, 3:iv + 31, 10 figs., prior to May 8.

BOGAN, M. A., and JONES, C. 1975. Observations on Lepus callotis in New Mexico. Proc. Biol. Soc. Washington, 88:45–49, April 23.

BOGAN, M. A., and WILLIAMS, D. F. 1970. Additional records of some Chihuahuan bats. Southwestern Nat., 15:131–134, June 1.

———. 1975. Neotoma micropus in Chihuahua. Southwestern Nat., 20:278–279, June 27.

BOLE, B. P., JR., and MOULTHROP, P. N. 1942. The Ohio Recent mammal collection in the Cleveland Museum of Natural History. Sci. Publs., Cleveland Mus. Nat. Hist., 5:83–181, September 11.

BOLEY, R. B., and KENNERLY, T. E., JR. 1969. Cellulolytic bacteria and reingestion in the plains pocket gopher, Geomys bursarius. Jour. Mamm., 50:348–349, June 12.

BONGARDT, H., RICHENS, V. B., and HOWARD, W. E. 1968. Serum protein patterns in pocket gophers. Jour. Mamm., 49:544–547, 3 figs., August 20.

BOOTH, E. S. 1968. Mammals of Southern California. Univ. California Press, 99 pp., illustrated.

BORELL, A. E. 1961. Fox squirrel attacks mourning dove. Jour. Mamm., 42:101, February 20.

BORRECCO, J. E., and HOOVEN, E. F. 1972. Northern distribution record of the California meadow mouse (Microtus californicus) in Oregon. Murrelet, 53:32–33, August.

BOWEN, W. W. 1968. Variation and evolution of Gulf Coast populations of beach mice, Peromyscus polionotus. Bull. Florida State Mus. (Biol. Sci.), 12(1):1–91, 34 figs., April 19.

BOWERS, J. H. 1974. Genetic compatibility of Peromyscus maniculatus and Peromyscus melanotis, as indicated by breeding studies and morphometrics. Jour. Mamm., 55:720–737, 2 figs., 6 tables, November 20.

BOWERS, J. H., and JUDD, F. W. 1969. Notes on the distribution of Didelphis marsupialis and Citellus spilosoma in western Texas. Texas Jour. Sci., 20(3):277, February 10.

BOWLES, J. B. 1971. Historical record of some Iowan mammals. Trans. Kansas Acad. Sci., 73:419–430, July 28.
———. 1975. Distribution and biogeography of mammals of Iowa. Special Publ. Mus. Texas Tech Univ., 9:1–184, 62 figs., 10 tables, July 18.
BRADLEY, W. G. 1966. Status of the cotton rat in Nevada. Jour. Mamm., 47:349–350, May 26.
BRADLEY, W. G., AUSTIN, G. T., and O'FARRELL, M. 1965. Lasionycteris noctivagans, Lasiurus cinereus and Tadarida molossa in Clark County, Nevada. Southwestern Nat., 10:220, July 1.
BRADLEY, W. G., and COCKRUM, E. L. 1968. A new subspecies of the meadow vole (Microtus pennsylvanicus) from northwestern Chihuahua, México. Amer. Mus. Novit., 2325:1–7, 1 table, June 19.
BRADLEY, W. G., and HANSEN, C. G. 1965. Observations on the distribution of the ring-tailed cat in southern Nevada. Southwestern Nat., 10:310–311, November 15.
BRADLEY, W. G., and MAUER, R. A. 1965. A collection of bats from Chihuahua, Mexico. Southwestern Nat., 10:74–75, March 15.
BRADLEY, W. G., and O'FARRELL, M. J. 1967. The mastiff bat, Eumops perotis, in southern Nevada. Jour. Mamm., 48:672, November 20.
BRADSHAW, G. VR. 1961. New Arizona locality for the dwarf shrew. Jour. Mamm., 42:96, February 20.
BRADSHAW, G. VR., and HAYWARD, B. 1960. Mammal skulls recovered from owl pellets in Sonora, Mexico. Jour. Mamm., 41:282–283, May 20.
BRADSHAW, W. N., and GEORGE, W. A. 1969. The karyotype in Peromyscus maniculatus nubiterrae. Jour. Mamm., 50:822–824, 1 fig., November 28.
BRAND, D. D. 1937. The natural landscape of northwestern Chihuahua. Univ. New Mexico Bull., Geol. Ser., 5(2):1–74, 10 pls., 5 maps, 3 charts, November 1.
BRAND, L. R., and RYCKMAN, R. E. 1968. Laboratory life histories of Peromyscus eremicus and Peromyscus interparietalis. Jour. Mamm., 49:495–501, 3 figs., August 20.
BROADBOOKS, H. E. 1958. Northward extension of the kangaroo rat in Washington. Jour. Mamm., 39:299–301, May 20.
———. 1961. California ground squirrel invades Washington. Jour. Mamm., 42:257–258, May 20.
———. 1961. The funnel-eared bat in Sonora. Jour. Mamm., 42:403, August 21.
———. 1965. Ecology and distribution of the pikas of Washington and Alaska. Amer. Midland Nat., 73:299–335, 15 figs., 9 tables, April.
BROWN, J. H. 1968. Activity patterns of some neotropical bats. Jour. Mamm., 49:754–757, November 26.
BROWN, J. H., and WELSER, C. F. 1968. Serum albumin polymorphisms in natural and laboratory populations of Peromyscus. Jour. Mamm., 49:420–426, 2 figs., 1 table, August 20.
BROWN, L. N. 1961. Sorex longirostris in southwestern Missouri. Jour. Mamm., 42:527, November 20.
———. 1963. Notes on the distribution of Peromyscus nuttalli flammeus and Peromyscus maniculatus ozarkium. Jour. Mamm., 44:424–425, August 22.
———. 1964. Breeding records and notes on Phenacomys silvicola in Oregon. Jour. Mamm., 45:647–648, January 25.
———. 1964. Ecology of three species of Peromyscus from southern Missouri. Jour. Mamm., 45:189–202, 3 figs., 4 tables, May 20.
———. 1964. First record of Microtus ochrogaster from Arkansas. Jour. Mamm., 45:471–472, September 15.
———. 1965a. Status of opossum, Didelphis marsupialis, in Wyoming. Southwestern Nat., 10:142–143, April 30.
———. 1965b. A fisher, Martes pennanti, in Sheridan County, Wyoming. Southwestern Nat., 10:143, April 30.
———. 1965c. Abert's squirrel in southern Wyoming. Jour. Mamm., 46:516, August 26.
———. 1966. First record of the pygmy shrew in Wyoming and description of a new subspecies (Mammalia: Insectivora). Proc. Biol. Soc. Washington, 79:49–51, May 23.
———. 1967. Ecological distribution of six species of shrews and comparison of sampling methods in the central Rocky Mountains. Jour. Mamm., 48:617–623, 1 fig., 2 tables, November 20.
———. 1968. Smallness of mean litter size in the Mexican vole. Jour. Mamm., 49:159, February 20.
BROWN, L. N., and METZ, D. 1966. First record of Perognathus flavescens in Wyoming. Jour. Mamm., 47:118, March 12.
BRYANT, M. D. 1945. Phylogeny of Nearctic Sciuridae. Amer. Midland Nat., 33:257–390, 8 pls., 48 figs., 2 tables, March.
BUCHANAN, O. M., and HOWELL, T. R. 1965. Observations on the natural history of the thick-spined rat, Hoplomys gymnurus, in Nicaragua. Ann. Mag. Nat. Hist., ser. 13, 8:549–559, 2 figs., 1 pl., September.
———. 1967. Zoogeography of Scotinomys in Middle America, with the description of a new subspecies from Nicaragua. Jour. Mamm., 48:414–419, August 21.
BUCKNER, C. H. 1961. The pale big brown bat in Manitoba. Jour. Mamm., 42:529, November 20.
BUDEN, D. W. 1975. Monophyllus redmani Leach (Chiroptera) from the Bahamas, with notes on variation in the species. Jour. Mamm., 56:369–377, 1 fig., 2 tables, May 30.
———. 1975. A taxonomic and zoogeographic appraisal of the big-eared bat (Macrotus waterhousii Gray) in the West Indies. Jour. Mamm., 56:758–769, 2 figs., 3 tables, November 18.
———. 1976. A review of the bats of the endemic West Indian genus Erophylla. Proc. Biol. Soc. Washington, 89:1–15, 3 figs., 3 tables, June 25.
———. 1977. First records of bats of the genus Brachyphylla from the Caicos Islands, with notes on geographic variation. Jour. Mamm., 58:221–225, 2 figs., 1 table, May 31.
BUECHNER, H. K. 1944. Helminth parasites of the gray fox. Jour. Mamm., 25:185–188, 1 table, May 26.
BURNS, J. J. 1964. Comparison of two populations of mink from Alaska. Canadian Jour. Zool., 42:1071–1079, 4 figs., 2 tables, November.
BURT, W. H. 1932. Descriptions of heretofore unknown mammals from islands in the Gulf of California, Mexico. Trans. San Diego Soc. Nat. Hist., 7:161–182, 1 table, October 31.
———. 1933. Additional notes on the mammals of southern Arizona. Jour. Mamm., 14:114–122, May 15.
———. 1938. Faunal relationships and geographic distribution of mammals in Sonora, Mexico. Miscl. Publ. Mus.

Zool., Univ. Michigan, 39:1–77, 26 maps, 2 tables, February 14.

———. 1942. A caribou antler from the lower peninsula of Michigan. Jour. Mamm., 23:214, June 3.

———. 1946. The mammals of Michigan. Univ. Michigan Press, Ann Arbor, xv + 288 pp., 13 pls., 107 figs., 67 maps.

BURT, W. H., and BARKALOW, F. S., JR. 1942. A comparative study of the bacula of wood rats (subfamily Neotominae). Jour. Mamm., 23:287–297, 3 figs., 1 table, August 13.

BURT, W. H., and STIRTON, R. A. 1961. The mammals of El Salvador. Miscl. Publ. Mus. Zool., Univ. Michigan, 117:1–69, 2 figs., September 22.

BUTTERWORTH, B. B. 1964. Parental relations and the behavior of juvenile kangaroo rats (Dipodomys) in captivity. Southwestern Nat., 8:213–220, 1 table, February 5.

CABRERA, A. 1925. Genera mammalium: Insectivora, Galeopithecia. Mus. Nac. de Cienc. Nat., Madrid, 1–232, 18 pls., November 29.

———. 1958. Catálogo de los mamíferos de America del Sur. Rev. Mus. Argentino de Cienc. Nat., 4:iv + 1–308, March 27. Also, on August 25, 1961, as part 2 of vol. 4: nota, title page, advertencia, photo of deceased author, biographical account (xix–xxii), catalogue (309–625), and index (627–732).

CAHALANE, V. H. 1959. A biological survey of Katmai National Monument. Smiths. Miscl. Coll., 138(5), iii + 246 pp., 17 pls., 3 figs., 1 map, August 20.

CALDWELL, D. K. 1961. The harbor seal in South Carolina. Jour. Mamm., 42:425, August 21.

CALDWELL, D. K., and CALDWELL, M. C. 1969. The harbor seal, Phoca vitulina concolor, in Florida. Jour. Mamm., 50:379–380, June 12.

———. 1969. Gray's dolphin, Stenella styx, in the Gulf of Mexico. Jour. Mamm., 50:612–614, 1 fig., August 22.

———. 1969. Addition of the leatherback sea turtle to the known prey of the killer whale, Orcinus orca. Jour. Mamm., 50:636, August 22.

———. 1971. Sounds produced by two rare cetaceans stranded in Florida. Cetology, No. 4:1–6, 10 figs., May 14.

———. 1971. Beaked whales, Ziphius cavirostris, in the Bahamas. Quart. Jour. Florida Acad. Sci., 34:157–160.

———. 1975. Pygmy killer whales and short-snouted spinner dolphins in Florida. Cetology, No. 18:1–5, 3 figs., September 26.

CALDWELL, D. K., CALDWELL, M. C., and ARRINDELL, G. 1973. Dwarf sperm whales, Kogia simus, from the Lesser Antillean island of St. Vincent. Jour. Mamm., 54:515–517, May 31.

CALDWELL, D. K., CALDWELL, M. C., RATHJEN, W. F., and SULLIVAN, J. R. 1971. Cetaceans from the Lesser Antillean island of St. Vincent. Fishery Bull., 69:303–312, 3 figs., 1 table.

CALDWELL, D. K., RATHJEN, W. F., and CALDWELL, M. C. 1971. Cuvier's beaked whale, Ziphius cavirostris, from Barbados. Bull. Southern California Acad. Sci., 70:52–53.

CALDWELL, L. D. 1966. Marsh rabbit development and ectoparasites. Jour. Mamm., 47:527–528, 1 fig., August 29.

CALLAHAN, J. R. 1975. Status of the peninsula chipmunk. Jour. Mamm., 56:266–269, 2 figs., February 20.

———. 1977. Diagnosis of Eutamias obscurus (Rodentia: Sciuridae). Jour. Mamm., 58:188–201, 8 figs., 2 tables, May 31.

CAMERON, A. W. 1953. The mammals of southeastern Quebec. Ann. Report Provancher Soc. Nat. Hist. Canada, Quebec, for 1950–1951, pp. 20–86, illustrated.

———. 1958. New mammal records for Prince Edward Island. Jour. Mamm., 39:291–292, May 20.

———. 1959. Mammals of the islands in the Gulf of St. Lawrence. Nat. Mus. Canada, Bull. 154:iii + 165, 8 figs., 28 maps, January 12.

———. 1962. Mammalian zoogeography of the Magdalen Islands Archipelago, Quebec. Jour. Mamm., 43:505–514, December 7.

CAMERON, A. W., and MORRIS, W. A. 1951. The mammals of the Lake Mistassini and Lake Albanel regions, Quebec. Bull. Nat. Mus. Canada, 123:120–130.

CAMP, C. L., ALLISON, H. J., and NICHOLS, R. H. 1964. Bibliography of fossil vertebrates 1954–1958. Geol. Soc. Amer., Mem. 92:xxv + 647, prior to October 19.

CAMPBELL, R. W. 1972. Range extension of the big brown bat on Vancouver Island, British Columbia. Murrelet, 53:12, May.

CARLETON, M. D. 1977. Interrelationships of populations of the Peromyscus boylii species group (Rodentia, Muridae) in western Mexico. Occas. Pap. Mus. Zool., Univ. Michigan, 675:1–47, 16 figs., 4 tables, 3 appendices, March 1.

CARPENTER, C. R. 1935. Behavior of red spider monkeys in Panama. Jour. Mamm., 16:171–180, 1 table, August 12.

CARTER, D. C. 1968. A new species of Anoura (Mammalia: Chiroptera: Phyllostomidae) from South America. Proc. Biol. Soc. Washington, 81:427–430, 1 table, December 30.

CARTER, D. C., and DAVIS, W. B. 1961. Tadarida aurispinosa (Peale) (Chiroptera: Molossidae) in North America. Proc. Biol. Soc. Washington, 74:161–165, August 11.

CARTER, D. C., PINE, R. H., and DAVIS, W. B. 1966. Notes on Middle American bats. Southwestern Nat., 11:488–499, December 31.

CASEBEER, R. S., LINSKY, R. B., and NELSON, C. E. 1963. The phyllostomid bats, Ectophylla alba and Vampyrum spectrum, in Costa Rica. Jour. Mamm., 44:186–189, 1 pl., May 20.

CHAPMAN, F. M. 1901. A revision of the genus Capromys. Bull. Amer. Mus. Nat. Hist., 14:313–323, Pls. 39–40, 3 figs., November 12.

CHAPMAN, J. A., and MORGAN, R. P. II. 1973. Systematic status of the cottontail complex in western Maryland and nearby West Virginia. Wildlife Monog., Wildlife Soc., 36:1–54, frontispiece, 36 figs., 22 tables, December.

CHAPMAN, J. A., and PARADISO, J. L. 1972. First records of the New England cottontail (Sylvilagus transitionalis) from Maryland. Chesapeake Sci., 13:149, June.

CHERMOCK, R. L., and WHITE, J. S. 1953. Myotis keenii septentrionalis, a new bat record for Alabama. Jour. Alabama Acad. Sci., 25:24, December.

CHERNYAVSKY, F. B. 1972. [The distribution and geographical variability of the American long-tailed ground squirrel (Citellus parryi Rich., 1827) in northeastern Siberia.] Trans. Moscow Soc. Naturalists, 48:199–214, 2 figs.

CHESEMORE, D. L. 1968. Occurrence of moose near Barrow, Alaska. Jour. Mamm., 49:528–529, August 20.

CHEW, R. M., and BUTTERWORTH, B. B. 1964. Ecology of rodents in Indian Cove (Mojave Desert). Joshua Tree National Monument, California. Jour. Mamm., 45:203–225, 7 figs., 11 tables, May 20.

CHEW, R. M., LINDBERG, R. G., and HAYDEN, P. 1965. Circadian rhythm of metabolic rate in pocket mice. Jour. Mamm., 46:477–494, 7 figs., 2 tables, August 26.

CHILDS, H. E. 1969. Birds and mammals of the Pitmegea River region, Cape Sabine, northwestern Alaska. Biol. Pap., Univ. Alaska, 10:1–76, 2 figs., 4 tables, May.

CHOATE, J. R. 1970. Systematics and zoogeography of Middle American shrews of the genus Cryptotis. Univ. Kansas Publ., Mus. Nat. Hist., 19:195–317, 20 figs., 1 table, December 30.

———. 1972. Notes on geographic distribution and habitats of mammals eaten by owls in southern New England. Trans. Kansas Acad. Sci., 74:212–216, April 7.

CHOATE, J. R., and BIRNEY, E. C. 1968. Sub-Recent Insectivora and Chiroptera from Puerto Rico, with the description of a new bat of the genus Stenoderma. Jour. Mamm., 49:400–412, 5 figs., 3 tables, August 20.

CHOATE, J. R., and CLIFTON, P. L. 1970. Noteworthy records of bats from Tamaulipas, Mexico. Southwestern Nat., 14:358–360, January 15.

CHOATE, J. R., and DUBOS, R. E. 1971. Distributional status of four species of Connecticut mammals. Univ. Connecticut Occas. Pap., 2:17–20, December.

CHOATE, J. R., and FLEHARTY, E. D. 1975. Synopsis of native, Recent mammals of Ellis County, Kansas. Occas. Pap. Mus. Texas Tech Univ., 37:1–80, 5 figs., December 19.

CHOATE, J. R., and GENOWAYS, H. H. 1967. Notes on some mammals from Nebraska. Trans. Kansas Acad. Sci., 69:238–241, April 25.

CHOATE, J. R., PHILLIPS, C. J., and GENOWAYS, H. H. 1967. Taxonomic status of the brush mouse, Peromyscus boylii cansensis Long, 1961. Trans. Kansas Acad. Sci., 69:306–313, 2 figs., 1 table, April 25.

CLARK, W. K. 1953. The baculum in the taxonomy of Peromyscus boylei and P. pectoralis. Jour. Mamm., 34:189–192, 1 fig., 1 table, May 14.

CLEVELAND, A. G. 1970. The current geographic distribution of the armadillo in the United States. Texas Jour. Sci., 22:90–92, 1 fig., September 15.

CLOTHIER, R. R. 1960. Banner-tailed kangaroo rat in central Arizona. Jour. Mamm., 41:517, November 11.

———. 1969. Reproduction in the gray-necked chipmunk. Jour. Mamm., 50:642, August 22.

CLOUGH, G. C. 1959. Extension of range of the woodland jumping mouse. Jour. Mamm., 40:449, August 20.

———. 1974. Additional notes on the biology of the Bahamian hutia, Geocapromys ingrahami. Jour. Mamm., 55:670–672, 1 fig., August 20.

COCKRUM, E. L. 1952. Mammals of Kansas. Univ. Kansas Publ., Mus. Nat. Hist., 7:1–303, 73 figs., 37 tables, August 25.

———. 1961. The Recent mammals of Arizona: their taxonomy and distribution. Univ. Arizona Press, pp. viii + 276, 112 figs., February 11.

———. 1964. Southern river otter, Lutra annectens, from Sonora, Mexico. Jour. Mamm., 45:634–635, January 25.

COCKRUM, E. L., and BRADSHAW, G. VAN R. 1963. Notes on mammals from Sonora, México. Amer. Mus. Novit., 2138:1–9, May 7.

COCKRUM, E. L., and GARDNER, A. L. 1960. Underwood's mastiff bat in Arizona. Jour. Mamm., 41:510–511, November 11.

COCKRUM, E. L., and MUSGROVE, B. 1964. Cave myotis, Myotis velifer, from southern Nevada. Jour. Mamm., 45:636–637, January 25.

———. 1965. Extension of known range of the pocketed free-tailed bat. Jour. Mamm., 46:509, August 26.

COCKRUM, E. L., and ORDWAY, E. 1959. Bats of the Chiricahua Mountains, Cochise County, Arizona. Amer. Mus. Novit., 1938:1–35, 10 figs., June 5.

COLLINS, L. R. 1973. Monotremes and marsupials [.] A reference for zoological institutions. Smiths. Inst. Publ. 4888:v + 323, frontispiece, 56 figs., 37 numbered tables, 2 charts, August 10.

CONLEY, W. H. 1970. Geographic variation in the least chipmunk, Eutamias minimus, in New Mexico and eastern Arizona. Jour. Mamm., 51:695–702, 1 fig., 1 table, November 30.

CONNOR, P. F. 1960. The small mammals of Otsego and Schoharie counties, New York. New York State Mus. and Sci. Serv., Bull. 382:1–84, 15 figs., 10 tables, July.

———. 1966. The mammals of the Tug Hill Plateau, New York. New York State Mus. and Sci. Serv., Bull. 406:5 unnumbered + 1–82, 14 figs., 3 maps, 1 table, May.

———. 1971. The mammals of Long Island, New York. New York State Mus. and Sci. Serv., Bull. 416:i–v + 1–78, 26 figs., 3 maps, July.

CONSTANTINE, D. G. 1959. Pteronotus davyi in northwestern Mexico. Jour. Mamm., 40:442, August 20.

———. 1961. Locality records and notes on western bats. Jour. Mamm., 42:404–405, August 21.

———. 1966. New bat locality records from Oaxaca, Arizona and Colorado. Jour. Mamm., 47:125–126, March 12.

COOPWOOD, B. 1900. Route of Cabeza de Vaca. Quart. Texas State Hist. Assoc., 3:229–264, April.

CORTHUM, K. W., JR. 1967. Reproduction and duration of placental scars in the prairie vole and the eastern vole. Jour. Mamm., 48:287–292, May 20.

CORY, C. B. 1912. The mammals of Illinois and Wisconsin. Field Mus. Nat. Hist., Publ. 153, Zool. Ser., 11:1–505, numerous pls. and figs.

COSGROVE, G. E., and O'FARRELL, T. P. 1965. Papillomas and other lesions in the stomachs of pine mice. Jour. Mamm., 46:510–513, 2 pls., 1 table, August 26.

COULOMBE, H. N., and BANTA, B. H. 1964. The distribution and ecology of the Crawford desert shrew, Notiosorex crawfordi, in Saline Valley, Inyo County, California. Wasmann Jour. Biol., 22:277–297, 5 figs., 3 tables, fall.

COWAN, I. McT., and GUIGUET, C. J. 1956. The mammals of British Columbia. British Columbia Provincial Mus., Handbook No. 11: 1–413, illustrated, July 15.

———. 1960. The mammals of British Columbia. British Columbia Provincial Mus., Handbook No. 11, 2nd ed.: 1–413, 103 pls., 29 figs., 53 maps, June 8.

———. 1965. The mammals of British Columbia. British Columbia Provincial Mus., Handbook No. 11, 3rd ed.: 1–414, 103 pls., 29 figs., 53 maps, October.

COWAN, I. McT., and MCCRORY, W. 1970. Variation in the mountain goat, Oreamnos americanus (Blainville). Jour. Mamm., 51:60–73, 2 figs., 7 tables, February 20.

COX, T. J. 1965. Behavior of the mastiff bat. Jour. Mamm., 46:687–688, November 30.

CRAIN, J. L., and CLIBURN, J. W. 1965. A preliminary study

of the mammals of southeastern Mississippi. Jour. Mississippi Acad. Sci., 11:271–280.

CRAIN, J. L., and PACKARD, R. L. 1966. Notes on mammals from Washington Parish, Louisiana. Jour. Mamm., 47:323–325, May 26.

CREEL, G. C., and THORNTON, W. A. 1970. Extensions of the known ranges of two species of Texas mammals. Texas Jour. Sci., 21:481, June 12.

CRONAN, J. M., and BROOKS, A. 1962. The mammals of Rhode Island. Rhode Island Dept. Agric. and Conservation, Div. Fish and Game, Wildlife Pamphlet No. 6:ix + 133, 39 figs., May.

CROSSIN, R. S., SOULE, O. H., WEBB, R. G., and BAKER, R. H. 1973. Biotic relationships in the canon del Rio Mezquital, Durango, Mexico. Southwestern Nat., 18:187–200, 5 figs., 2 tables, June 29.

CROWE, P. E. 1943. Notes on some mammals of the southern Canadian Rocky Mountains. Bull. Amer. Mus. Nat. Hist., 80:391–410, Pls. 32–34, 1 map, February 4.

CSUTI, B. A. 1971. Distribution of some Southern California kangaroo rats. Bull. Southern California Acad. Sci., 70:50–51, August 27.

CUTTER, W. L. 1959. Notes on some mammals from northern Texas. Southwestern Nat., 4:30–34, June 4.

———. 1959. The Warren wood rat in Texas. Jour. Mamm., 40:448–449, August 20.

DALBY, P. L., and BAKER, R. H. 1967. Crawford's desert shrew first reported from Nuevo Leon. Southwestern Nat., 12:195–196, August 4.

DALE, F. H. 1940. Geographic variation in the meadow mouse in British Columbia and southeastern Alaska. Jour. Mamm., 21:332–340, 1 table, August 13.

DALQUEST, W. W. 1948. Mammals of Washington. Univ. Kansas Publ., Mus. Nat. Hist., 2:1–444, 140 figs., frontispiece, April 9.

———. 1953. Mammals of the Mexican state of San Luis Potosí. Louisiana State Univ. Studies, Biol. Ser., 1:1–229, 1 fig., December 28.

———. 1961. Two species of bison contemporaneous in early Recent deposits in Texas. Southwestern Nat., 6:73–78, September 1.

———. 1968. Mammals of north-central Texas. Southwestern Nat., 13:13–21, May 28.

———. 1975. The montane vole in northeastern New Mexico and adjacent Colorado. Southwestern Nat., 20:138–139, May 15.

DALQUEST, W. W., and COLLIER, G. 1964. Notes on Dipodomys elator, a rare kangaroo rat. Southwestern Nat., 9:146–150, August 20.

DALQUEST, W. W., and KILPATRICK, W. 1973. Dynamics of pocket gopher distribution on the Edwards Plateau of Texas. Southwestern Nat., 18:1–9, 1 fig., March 30.

DALQUEST, W. W., and WERNER, H. J. 1954. Histological aspects of the faces of North American bats. Jour. Mamm., 35:147–160, 1 pl., 1 fig., 2 tables, May 26.

DAVIS, B. L., and BAKER, R. J. 1974. Morphometrics, evolution, and cytotaxonomy of mainland bats of the genus Macrotus (Chiroptera: Phyllostomatidae). Systematic Zool., 23:26–39, 5 figs., 4 tables, May 21.

DAVIS, D. D. 1964. The giant panda: a morphological study of evolutionary mechanisms. Fieldiana: Zool. Mem. (Chicago Nat. Hist. Mus.), 3:1–339, frontispiece, 159 figs., 25 tables, December 7.

DAVIS, J., and LIDICKER, W. Z., JR. 1975. The taxonomic status of the southern sea otter. Proc. California Acad. Sci., 4th ser., 40:429–437, October 3.

DAVIS, J. A. 1965. The mammals of northwestern New Jersey. New Jersey Nature News, 20:150–155.

DAVIS, J. A., JR. 1957. A new shrew (Sorex) from Florida. Amer. Mus. Novit., 1844:1–9, 1 fig., 3 tables, October 10.

DAVIS, W. B. 1939. The Recent mammals of Idaho. Caxton Printers, Ltd., Caldwell, Idaho, 400 pp., 2 pls., 33 figs., April 5.

———. 1940. Distribution and variation of pocket gophers (genus Geomys) in the southwestern United States. Bull. Texas Agric. Exper. Sta., 590:1–38, 6 figs., 4 tables, October 23.

———. 1941. Additional records of Notiosorex crawfordi from Texas. Jour. Mamm., 22:319–320, 1 fig., August 14.

———. 1958. Review of Mexican bats of the Artibeus cinereus complex. Proc. Biol. Soc. Washington, 71:163–166, 1 fig., December 31.

———. 1961. The female of Mustela frenata texensis. Jour. Mamm., 42:273, May 20.

———. 1965. Review of the Eptesicus brasiliensis complex in Middle America with the description of a new subspecies from Costa Rica. Jour. Mamm., 46:229–240, 5 figs., 2 tables, May 20.

———. 1966. Review of South American bats of the genus Eptesicus. Southwestern Nat., 11:245–274, 5 figs., 4 tables, July 20.

———. 1966. The mammals of Texas. Bull. Texas Parks and Wildlife Dept., 41:1–267, 77 figs., 69 maps.

———. 1968. Review of the genus Uroderma (Chiroptera). Jour. Mamm., 49:676–698, 9 figs., November 26.

———. 1969. A review of the small fruit bats (genus Artibeus) of Middle America. Southwestern Nat., 14:15–29, 4 figs., 2 tables, May 16.

———. 1970. A review of the small fruit bats (genus Artibeus) of Middle America. Southwestern Nat., 14:389–402, 2 figs., 3 tables, February 16.

———. 1970. The large fruit bats (genus Artibeus) of Middle America, with a review of the Artibeus jamaicensis complex. Jour. Mamm., 51:105–122, 5 figs., 3 tables, February 20.

———. 1973. Geographic variation in the fishing bat, Noctilio leporinus. Jour. Mamm., 54:862–874, 4 figs., 2 tables, December 14.

DAVIS, W. B., and CARTER, D. C. 1962. Notes on Central American bats with description of a new subspecies of Mormoops. Southwestern Nat., 7:64–74, 1 fig., June 1.

———. 1962. Review of the genus Leptonycteris (Mammalia: Chiroptera). Proc. Biol. Soc. Washington, 75:193–198, August 28.

———. 1964. A new species of fruit-eating bat (genus Artibeus) from Central America. Proc. Biol. Soc. Washington, 77:119–122, 1 fig., June 26.

DAVIS, W. B., CARTER, D. C., and PINE, R. H. 1964. Noteworthy records of Mexican and Central American bats. Jour. Mamm., 45:375–387, September 15.

DAVIS, W. B., and LUKENS, P. W., JR. 1958. Mammals of the Mexican sate of Guerrero, exclusive of Chiroptera and Rodentia. Jour. Mamm., 39:347–367, 10 tables, August 20.

DAVIS, W. B., and ROBERTSON, J. L., JR. 1944. The mammals of Culberson County, Texas. Jour. Mamm., 25:254–273, 1 pl., 2 figs., September 8.

DAVIS, W. H. 1959. Taxonomy of the eastern pipistrel. Jour. Mamm., 40:521–531, 2 figs., November 20.

———. 1966. Arizona expedition. Bat Research News, 7:28, July. [Processed Literature.]

DAVIS, W. H., and RIPPY, C. L. 1968. Distribution of *Myotis lucifugus* and *Myotis austroriparius* in the southeastern United States. Jour. Mamm., 49:113–117, 2 figs., February 20.

DAWBIN, W. H., NOBLE, B. A., and FRASER, F. C. 1970. Observations on the electra dolphin, *Peponocephala electra*. Bull. British Mus. Nat. Hist. Zool., 20:173–201, 13 figs., 3 tables, December 21.

DEACON, J. E., BRADLEY, W. G., and LARSEN, K. M. 1964. Ecological distribution of the mammals of Clark Canyon, Charleston Mountains, Nevada. Jour. Mamm., 45:397–409, 2 tables, September 15.

DEGERBØL, M. 1935. Report of the mammals collected by the Fifth Thule Expedition [1921–24] to Arctic North America: Zoology I. Mammals. Vol. II, no. 4:1–67, 12 figs., 9 tables.

———. 1957. The extinct reindeer of East-Greenland. . . . Acta Arctica, Fasc. 10:1–66, 11 pls., 12 figs., 1 map, 12 tables, prior to August 16.

DE LA TORRE, L. 1952. An additional record of the bat, *Sturnira ludovici*, in Mexico. Chicago Acad. Sci., Nat. Hist. Miscl., 105:1–2, May 2.

———. 1958. The status of the bat *Myotis velifer cobanensis* Goodwin. Proc. Biol. Soc. Washington, 71:167–170, December 31.

DELLINGER, S. C., and BLACK, J. D. 1940. Notes on Arkansas mammals. Jour. Mamm., 21:187–191. May 16.

DEMOTT, S. L. 1975. Pygmy shrew, *Microsorex hoyi*, in Gunnison County, Colorado. Southwestern Nat., 20:417–418, October 10.

DICKERMAN, R. W. 1960. "Davian behavior complex" in ground squirrels. Jour. Mamm., 41:403, August 15.

DIERSING, V. E. 1976. An analysis of *Peromyscus difficilis* from the Mexican–United States boundary area. Proc. Biol. Soc. Washington, 89:451–466, 5 figs., 2 tables, October 12.

DIERSING, V. E., and HOFFMEISTER, D. F. 1974. The rock mouse, *Peromyscus difficilis*, in western Texas. Southwestern Nat., 19:213, July 26.

DIMMICK, R. W. 1969. The distribution of *Microtus ochrogaster* in Tennessee. Jour. Mamm., 50:126, February 26.

DOBSON, G. E. 1878. Catalogue of the Chiroptera in the collection of the British Museum. British Museum, London, xlii + 567 pp., 30 pls., numerous unnumbered tables, after May 20.

DOUGLAS, C. L. 1967. New records of mammals from Mesa Verde National Park, Colorado. Jour. Mamm., 48:322–323, May 20.

———. 1969. Comparative ecology of pinyon mice and deer mice in Mesa Verde National Park, Colorado. Univ. Kansas Publ., Mus. Nat. Hist., 18:421–504, 20 figs., 13 tables, August 20.

DOUTT, J. K., HEPPENSTALL, C. A., and GUILDAY, J. E. 1973. Mammals of Pennsylvania, 3rd ed., Pennsylvania Game Commission, 288 pp., illustrated.

DUNMIRE, W. W. 1961. Breeding season of three rodents on White Mountain, California. Jour. Mamm., 42:489–493, 3 tables, November 20.

DUNNAWAY, P. B., and KAYE, S. V. 1961. Cotton rat mortality during severe winter. Jour. Mamm., 42:265–268, 1 pl., May 20.

DUNNIGAN, P. B. 1967. Pocket gophers of the genus Thomomys of the Mexican state of Sinaloa. Radford Review, 21(3):139–168, 4 figs., 5 tables.

DURRANT, S. D. 1952. Mammals of Utah. Univ. Kansas Publ., Mus. Nat. Hist., 6:1–549, 91 figs., 30 tables, August 10.

DURRANT, S. D., and DEAN, N. K. 1959. Mammals of Glen Canyon. Univ. Utah Anthro. Pap., 40:73–103, 1 fig., 2 tables, June.

DURRANT, S. D., and HANSEN, R. M. 1954. A new rock squirrel (*Citellus variegatus*) from the Great Basin with critical comments on related subspecies. Proc. Biol. Soc. Washington, 67:263–271, 2 tables, November 15.

DUSBÁBEK, F. 1969. Macronyssidae (Acarina: Mesostigmata) of Cuban bats. Folia Parasitologica (Prague), 16:321–328, 2 figs.

EASTERLA, D. A. 1967. First specimens of plains pocket mouse from Missouri. Jour. Mamm., 48:479–480, August 21.

———. 1968a. First records of the pocketed free-tailed bat for Texas. Jour. Mamm., 49:515–516, August 20.

———. 1968b. Range extension of the southern bog lemming in Arkansas. Southwestern Nat., 13:364, December 10.

———. 1968c. Hispid cotton rat north of the Missouri River. Southwestern Nat., 13:364–365, December 10.

———. 1968d. First records of *Blarina brevicauda minima* in Missouri and Arkansas. Southwestern Nat., 13:448–449, December 31.

———. 1970. First records of the spotted bat in Texas and notes on its natural history. Amer. Midland Nat., 83:306–308, February 24.

———. 1970. First records of the least weasel, *Mustela nivalis*, from Missouri and southwestern Iowa. Jour. Mamm., 50:333–340, 3 figs., 2 tables, May 20.

EASTERLA, D. A., and BACCUS, J. 1973. A collection of bats from the Fronteriza Mountains, Coahuila, Mexico. Southwestern Nat., 17:424–427, January 10.

EASTERLA, D. A., and DAMMAN, D. L. 1977. The masked shrew and meadow vole in Missouri. Northwestern Missouri State Univ. Studies, 37(2):1–25, 2 figs., 2 tables, May.

EASTERLA, D. A., and WATKINS, L. C. 1969. Pregnant *Myotis sodalis* in northwestern Missouri. Jour. Mamm., 50:372–373, June 12.

EDWARDS, R. L. 1963. Observations on the small mammals of the southeastern shore of Hudson Bay. Canadian Field-Nat., 77:1–12, 3 figs., April 26.

EGER, J. L. 1974. A new subspecies of the bat *Eumops auripendulus* (Chiroptera: Molossidae), from Argentina and eastern Brazil. Royal Ontario Mus., Life Sci. Occas. Pap., 25:1–8, 2 figs., 2 tables, June 19.

———. 1977. Systematics of the genus *Eumops* (Chiroptera: Molossidae). Royal Ontario Mus., Life Sci. Contrib., 110:1–69, 28 figs., 22 tables, June 13.

EGOSCUE, H. J. 1961. Small mammal records from western Utah. Jour. Mamm., 42:122–124, February 20.

———. 1964. The kit fox in southwestern Colorado. Southwestern Nat., 9:40, April 10.

———. 1964. Ecological notes and laboratory life history of

the canyon mouse. Jour. Mamm., 45:387–396, 3 figs., 2 tables, September 15.

———. 1965. Records of shrews, voles, chipmunks, cottontails and mountain sheep from Utah. Jour. Mamm., 46:685–687, November 30.

———. 1966. Gray fox records from western Utah. Southwestern Nat., 11:143–144, March 30.

———. 1968. *Eutamias dorsalis grinnelli* Burt in extreme western Utah. Jour. Mamm., 49:145–146, February 20.

EGOSCUE, H. J., and LEWIS, T. J. 1968. An albino long-tailed pocket mouse from Utah. Jour. Mamm., 49:319, May 20.

ELDER, W. H., and SHANKS, C. E. 1962. Age changes in tooth wear and morphology of the baculum in muskrats. Jour. Mamm., 43:144–150, 1 fig., 2 tables, May 29.

ELLERMAN, J. R., HAYMAN, R. W., and HOLT, G. W. C. 1941. The families and genera of living rodents, with a list of named forms (1758–1936). British Museum (Nat. Hist.), London, 2:xii + 690, 50 figs., March 21.

ELLERMAN, J. R., and MORRISON-SCOTT, T. C. S. 1951. Checklist of Palaearctic and Indian mammals . . ., pp. 1–810, November 19.

ELLIOT, D. G. 1907. A catalogue of the collection of mammals in the Field Columbian Museum. Field Columb. Mus. Publ. 115, Zool. Ser., 8:viii + 694, 92 figs., February 9.

ELLIS, D. V. 1957. Some observations on mammals in the area between Coppermine and Pond Inlet, N.W.T., during 1954 and 1955. Canadian Field-Nat., 71:1–6, 1 map, July 17.

ENDERS, R. K. 1932. Mammal distribution in Saline and Camden counties, Missouri. Amer. Midland Nat., 13:114–123, May.

ERDBRINK, D. P. 1953. A review of fossil and Recent bears of the Old World. . . . Deventer–Drukkerij Jan De Lange. xii + 597 + 12 pp., 22 pls., 61 figs., 7 tables, several maps.

ERDMAN, D. S. 1962. Stranding of a beaked whale, *Ziphius cavirostris* Cuvier, on the south coast of Puerto Rico. Jour. Mamm., 43:276–277, 2 tables, May 29.

———. 1970. Marine mammals from Puerto Rico to Antigua. Jour. Mamm., 51:636–639, August 28.

ERNST, C. H., and ERNST, E. M. 1972. The eastern chipmunk, *Tamias striatus*, in southwestern Minnesota, U.S.A. Canadian Field-Nat., 86:377, December 16.

ESPEUT, W. B. 1882. On the acclimatization of the Indian mungoos in Jamaica. Proc. Zool. Soc. London, pp. 712–714.

FARNEY, J. P. 1975. Natural history and northward dispersal of the hispid cotton rat in Nebraska. Platte Valley Review, No. 3:11–16, 2 figs. + 1 unnumbered, 2 tables.

FARNEY, J. P., and JONES, J. K., JR. 1975. Noteworthy records of bats from Nebraska. Mammalia, 39:327–330, December 20.

FASSLER, D. J. 1974. Mammals of Pulaski County, Kentucky. Trans. Kentucky Acad. Sci., 35:37–43, 1 fig., June.

FEDYK, S. 1970. Chromosomes of *Microtus* (*Stenocranius*) *gregalis major* (Ognev, 1923) and phylogenetic connections between Sub-arctic representatives of the genus *Microtus* Schrank, 1798. Acta Theriologica, 15:143–152, 2 pls., 1 fig., April 30.

FEIST, J. D., and MCCOURT, K. H. 1973. A northern range extension for the pika in the Northwest Territories. Canadian Field-Nat., 87:317–318, 1 fig., October 17.

FELTEN, H. 1956. Fledermäuse (Mammalia, Chiroptera) aus El Salvador. Teil 3. Senckenbergiana Biol., 37:179–212, Pls. 24–27, 7 figs., April 15.

———. 1956. Fledermäuse (Mammalia, Chiroptera) aus El Salvador. Teil 4. Senckenbergiana Biol., 37:341–367, Pls. 45–48, 4 figs., September 15.

———. 1957. Fledermäuse (Mammalia, Chiroptera) aus El Salvador. Teil 5. Senckenbergiana Biol., 38:1–22, Pls. 1–2, 3 figs., January 15.

———. 1958. Nagetiere (Mammalia, Rodentia) aus El Salvador. Senckenbergiana Biol., 39:133–144, August 30.

———. 1958. Weitere Säugetiere aus El Salvador (Mammalia: Marsupialia, Insectivora, Primates, Edentata, Lagomorpha, Carnivora und Artiodactyla). Senckenbergiana Biol., 39:213–228, 1 fig., December 15.

FENTON, M. B. 1966. *Myotis sodalis* in caves near Watertown, New York. Jour. Mamm., 47:526, August 29.

———. 1972. Distribution and overwintering of *Myotis leibii* and *Eptesicus fuscus* (Chiroptera: Vespertilionidae) in Ontario. Royal Ontario Mus., Life Sci. Occas. Pap., 21:1–8, 3 figs., 1 table, July 28.

FERGUSON, D. E. 1962. The woodchuck in Mississippi. Jour. Mamm., 43:107, February 28.

FERTIG, D. S., and LAYNE, J. M. 1963. Water relationships in the Florida mouse. Jour. Mamm., 44:322–334, 3 figs., August 22.

FINDLEY, J. S. 1955. Speciation of the wandering shrew. Univ. Kansas Publ., Mus. Nat. Hist., 9:1–68, 18 figs., 1 table, December 10.

———. 1957. The hog-nosed bat in New Mexico. Jour. Mamm., 38:513–514, November 20.

———. 1959. The Bailey pocket mouse in New Mexico. Jour. Mamm., 40:145, February 20.

———. 1960. Identity of the long-eared Myotis of the southwest and Mexico. Jour. Mamm., 41:16–20, 1 fig., 1 pl., February 20.

———. 1961. Geographic variation in New Mexican chickarees. Jour. Mamm., 42:313–322, 3 figs., August 21.

FINDLEY, J. S., Harris, A. H., Wilson, D. E., and JONES, C. 1975. Mammals of New Mexico. Univ. New Mexico Press, Albuquerque, xxii + 360, 131 figs., 3 tables, April 1.

FINDLEY, J. S., and JONES, C. J. 1960. Geographic variation in the yellow-nosed cotton rat. Jour. Mamm., 41:462–469, 1 fig., November 11.

———. 1962. Distribution and variation of voles of the genus *Microtus* in New Mexico and adjacent areas. Jour. Mamm., 43:154–166, 5 figs., May 29.

———. 1965. Northernmost records of some neotropical bat genera. Jour. Mamm., 46:330–331, May 20.

———. 1965. Comments on spotted bats. Jour. Mamm., 46:679–680, 1 table, November 30.

———. 1967. Taxonomic relationships of bats of the species *Myotis fortidens*, *M. lucifugus*, and *M. occultus*. Jour. Mamm., 48:429–444, 8 figs., August 21.

FINDLEY, J. S., and POORBAUGH, J. 1957. Another dwarf shrew from New Mexico. Jour. Mamm., 38:513, November 20.

FINDLEY, J. S., and PULLEN, B. E. 1958. The fulvous harvest mouse in New Mexico. Jour. Mamm., 39:306–307, May 20.

FINDLEY, J. S., and TRAUT, G. L. 1970. Geographic variation in *Pipistrellus hesperus*. Jour. Mamm., 51:741–765, 10 figs., 5 tables, November 30.

FINLEY, R. B., JR. 1953. A new subspecies of wood rat (*Neotoma mexicana*) from Colorado. Univ. Kansas Publ., Mus. Nat. Hist., 5:527–534, 2 figs., August 15.

——. 1958. The wood rats of Colorado: distribution and ecology. Univ. Kansas Publ., Mus. Nat. Hist., 10:213–552, 34 pls., 8 figs., November 7.

FISCHER, J. B. 1829. Synopsis mammalium. Stuttgardt, xlii + 725 pp.

FISCUS, C. H., RICE, D. W., and JOHNSON, A. M. 1969. New records of *Mesoplodon stejnegeri* and *Ziphius cavirostris* from Alaska. Jour. Mamm., 50:127, February 26.

FISLER, G. F. 1965. Adaptation and speciation in harvest mice of the marshes of San Francisco Bay. Univ. California Publ. Zool., 77:1–108, 23 figs., numerous tables, November 3.

FITCH, H. S., GOODRUM, P., and NEWMAN, C. 1952. The armadillo in the southeastern United States. Jour. Mamm., 33:21–37, 2 figs., 1 table, February 18.

FIVUSH, B., PARKER, R., and TAMARIN, R. H. 1975. Karyotype of the beach vole, *Microtus breweri*, an endemic island species. Jour. Mamm., 56:272–273, February 20.

FLEHARTY, E. D. 1960. The status of the gray-necked chipmunk in New Mexico. Jour. Mamm., 41:235–242, 6 figs., May 20.

FLEHARTY, E. D., and ANDERSON, K. W. 1964. The meadow vole, Microtus pennsylvanicus (Ord) in Kansas. Trans. Kansas Acad. Sci., 67:129–130, June 12.

FLEHARTY, E. D., and STADEL, D. L. 1968. Distribution of Peromyscus leucopus (woods mouse) in western Kansas. Trans. Kansas Acad. Sci., 71:231–233, 1 fig., August 19.

FLEMING, T. H., HOOPER, E. T., and WILSON, D. E. 1972. Three Central American bat communities: structure, reproductive cycles, and movement patterns. Ecology, 53:555–569, 9 figs., 9 tables.

FLEROV, K. K. 1952. Fauna of USSR: mammals, musk deer and deer, vol. I, no. 2. English translation: Israel Program for Scientific Translations, pp. 1–257, 79 figs.

FORBES, R. B. 1962. Notes on food of silky pocket mice. Jour. Mamm., 43:278–279, May 29.

——. 1966. Notes on a litter of least chipmunks. Jour. Mamm., 47:159–161, 1 table, March 12.

——. 1967. Some aspects of the water economics of two species of chipmunks. Jour. Mamm., 48:466–468, 2 tables, August 21.

FORMAN, G. L., BAKER, R. J., and GERBER, J. D. 1968. Comments on the systematic status of vampire bats (family Desmodontidae). Systematic Zool., 17:417–425, 8 figs., December.

FORRESTER, D. J., and SENGER, C. M. 1965. Fetal measurements and milk characteristics of bighorn sheep. Jour. Mamm., 46:524–525, 2 tables, August 26.

FORSMAN, E., and MASER, C. 1970. Saw-whet owl preys on red tree mice. Murrelet, 51:10, June.

FORSYTH MAJOR, C. I. 1901. *The musk-rat of Santa Lucia (Antilles)*. Ann. Mag. Nat. Hist., ser. 7, 7:204–206, February.

FOSTER, J. B. 1961. Life history of the phenacomys vole. Jour. Mamm., 42:181–198, 1 pl., 3 figs., 1 table, May 20.

FRANCIS, C. S. 1959. Raccoons in Saskatchewan. Blue Jay, 17:30, March.

FRYXELL, F. M. 1928. The former range of the bison in the Rocky Mountains. Jour. Mamm., 9:129–139, May 9.

FUNDERBURG, J. B. 1961. Virginia muskrat in southeastern North Carolina. Jour. Mamm., 42:268, May 20.

——. 1961. Erythristic raccoons from North Carolina. Jour. Mamm., 42:270–271, May 20.

GALYON, A. 1968. Desert denizen now calls Memphis home. Tennessee Conservationist, 34:14–15, 3 figs.

GARDNER, A. L. 1962. Bat records from the Mexican states of Colima and Nayarit. Jour. Mamm., 43:102–103, February 28.

——. 1962. A new bat of the genus *Glossophaga* from Mexico. Los Angeles Co. Mus. Contrib. Sci., 54:1–7, 4 figs., May 11.

——. 1963. Nota acerca de la distribucion de dos murcielagos en Mexico. Rev. Soc. Mexicana Hist. Nat., 24:41–44, 1 table, December.

——. 1965. New bat records from the Mexican state of Durango. Proc. Western Foundation Vert. Zool., 1:99–106, 1 fig., June.

——. 1966. A new subspecies of the Aztec mastiff bat, *Molossus aztecus* Saussure, from southern Mexico. Los Angeles Co. Mus. Contrib. Sci., 111:1–5, November 9.

——. 1971. Notes on the little spotted cat, *Felis tigrina oncilla* Thomas, in Costa Rica. Jour. Mamm., 52:464–465, 1 table, May 27.

——. 1973. The systematics of the genus Didelphis (Marsupialia: Didelphidae) in North and Middle America. Special Publ. Mus. Texas Tech Univ., 4:1–81, 14 figs., 7 tables, July 3.

GARDNER, A. L., LaVal, R. K., and WILSON, D. E. 1970. The distributional status of some Costa Rican bats. Jour. Mamm., 51:712–729, 1 fig., November 30.

GARDNER, A. L., and PATTON, J. L. 1972. New species of *Philander* (Marsupialia: Didelphidae) and *Mimon* (Chiroptera: Phyllostomidae) from Perú. Occas. Pap. Mus. Zool., Louisiana State Univ., 43:1–12, 3 figs., 3 tables, September 13.

——. 1976. Karyotypic variation in Oryzomyine rodents (Cricetinae) with comments on chromosomal evolution in the neotropical Cricetine complex. Occas. Pap. Mus. Zool., Louisiana State Univ., 49:1–48, 10 figs., 2 tables, April 15.

GARNER, H. W. 1967. An ecological study of the brush mouse, *Peromyscus boylii*, in western Texas. Texas Jour. Sci., 19:285–291, 3 figs., 2 tables, November.

GASHWILER, J. S. 1976. A new distribution record of Merriam's shrew in Oregon. Murrelet, 57:13–14, summer.

GAUMER, G. F. 1917. Monographia de los mamiferos de Yucatán. Dept. de Talleres Gráficos de la Secretaría de Fomento, México, xli + 332 pp., 57 pls., 2 photographs, 1 map.

GEIST, O. W., BUCKLEY, J. L., and MANVILLE, R. H. 1960. Alaskan records of the narwhal. Jour. Mamm., 41:250–253, 1 pl., May 20.

GELUSO, K. N. 1972. Western spotted skunk in Oklahoma. Southwestern Nat., 16:457–458, February 18.

GENNARO, A. L. 1968. Northern geographic limits of four desert rodents of the genera Peromyscus, Perognathus, Dipodomys, and Onychomys in the Rio Grande Valley. Amer. Midland Nat., 80:477–493, 8 figs., October 29.

GENNARO, A. L., and Salb, T. J. 1969. The hispid pocket mouse in San Luis Potosi. Southwestern Nat., 14:251, October 10.

GENOWAYS, H. H. 1967. Second record of Myotis volans from North Dakota. Trans. Kansas Acad. Sci., 69:355, April 25.

——. 1971. A new species of spiny pocket mouse (genus

Liomys) from Jalisco, Mexico. Univ. Kansas Mus. Nat. Hist., Occas. Pap., 5:1–7, 2 figs., 1 table, June 18.

———. 1973. Systematics and evolutionary relationships of spiny pocket mice, genus Liomys. Special Publ. Mus. Texas Tech Univ., 5:1–368, 66 figs., 49 tables, December 7.

GENOWAYS, H. H., and BAKER, R. J. 1975. A new species of Eptesicus from Guadeloupe, Lesser Antilles (Chiroptera: Vespertilionidae). Occas. Pap. Mus. Texas Tech Univ., 34:1–7, 3 figs., 1 table, July 18.

GENOWAYS, H. H., and BIRNEY, E. C. 1974. Neotoma alleni. Mammalian Species, 41:1–4, 4 figs., June 28.

GENOWAYS, H. H., and CHOATE, J. R. 1970. Additional notes on some mammals from eastern Nebraska. Trans. Kansas Acad. Sci., 73:120–122, November 30.

———. 1972. A multivariate analysis of systematic relationships among populations of the short-tailed shrew (genus *Blarina*) in Nebraska. Systematic Zool., 21:106–116, 5 figs., 1 table, March 31.

GENOWAYS, H. H., and JONES, J. K., JR. 1967. Notes on distribution and variation in the Mexican big-eared bat, *Plecotus phyllotis*. Southwestern Nat., 12:477–480, December 31.

———. 1968. A new mouse of the genus *Nelsonia* from southern Jalisco, Mexico. Proc. Biol. Soc. Washington, 81:97–100, April 30.

———. 1968. Notes on bats from the Mexican state of Zacatecas. Jour. Mamm., 49:743–745, November 26.

———. 1969. Taxonomic status of certain long-eared bats (genus *Myotis*) from the southwestern United States and Mexico. Southwestern Nat., 14:1–13, 5 figs., 1 table, May 16.

———. 1969. Notes on pocket gophers from Jalisco, México, with descriptions of two new subspecies. Jour. Mamm., 50:748–755, 3 figs., 1 table, November 28.

———. 1971. Systematics of southern banner-tailed kangaroo rats of the *Dipodomys phillipsii* group. Jour. Mamm., 52:265–287, 5 figs., 6 tables, May 28.

———. 1971. Notes on spotted skunks (genus *Spilogale*) from western México. Anal. Inst. Biol., Univ. Nac. Autó. México, 39:123–131, 2 figs., 2 tables, September 11.

———. 1971. Second specimen of *Oryzomys dimidiatus*. Jour. Mamm., 52:833–834, December 16.

———. 1972. Mammals from southwestern North Dakota. Occas. Pap. Mus. Texas Tech Univ., 6:1–36, 1 fig., 5 tables, November 3.

———. 1973. Notes on some mammals from Jalisco, México. Occas. Pap. Mus. Texas Tech. Univ., 9:1–22, 1 fig., 3 tables, January 26.

———. 1975. Additional records of the stenodermine bat, *Sturnira thomasi*, from the Lesser Antillean island of Guadeloupe. Jour. Mamm., 56:924–925, 1 table, November 18.

GENOWAYS, H. H., and SCHLITTER, D. A. 1967. Northward dispersal of the hispid cotton rat in Nebraska and Missouri. Trans. Kansas Acad. Sci., 69:356–357, April 25.

GENTRY, J. B. 1966. Invasion of a one-year abandoned field by *Peromyscus polionotus* and *Mus musculus*. Jour. Mamm., 47:431–439, 2 figs., 1 table, August 29.

GENTRY, J. B., ODUM, E. P., MASON, M., NABHOLZ, V., MARSHALL, S., and MCGINNIS, J. T. 1968. Effect of altitude and forest manipulation on relative abundance of small mammals. Jour. Mamm., 49:539–541, 1 fig., 1 table, August 20.

GEORGES, S. 1976. A range extension of the coyote in Quebec. Canadian Field-Nat., 90:78–79, 1 fig., March 19.

GETZ, L. L. 1961. New locality records of some Kansas mammals. Jour. Mamm., 42:282–283, May 20.

GHISELIN, J. 1965. *Thomomys bottae* in Granite Springs Valley, Pershing County, Nevada. Jour. Mamm., 46:525, August 26.

GILEVA, E. A., and LOBANOVA, N. A. 1974. Complex chromosome variability in Arctic lemming and taxonomy of the genus Dichrostonyx. First International Theriological Cong., 1:191–192, June.

GLASS, B. P. 1958. The seminole bat in Oklahoma. Jour. Mamm., 39:587, November 20.

———. 1960. The taxonomic status of Oklahoma beavers, *Castor canadensis*. Southwestern Nat., 5:21–24, 1 fig., April 15.

———. 1961. Two noteworthy records of bats for Oklahoma. Southwestern Nat., 6:200–201, December 29.

GLASS, B. P., and BAKER, R. J. 1965. *Vespertilio subulatus* Say, 1823: proposed supression under the plenary powers (Mammalia, Chiroptera). Bull. Zool. Nomenclature, 22:204–205, August.

———. 1968. The status of the name *Myotis subulatus* Say. Proc. Biol. Soc. Washington, 81:257–260, August 30.

GLASS, B. P., and HALLORAN, A. F. 1961. The small mammals of the Wichita Mountains Wildlife Refuge, Oklahoma. Jour. Mamm., 42:234–239, May 20.

GLASS, B. P., and HALLORAN, A. H. 1960. Status and distribution of the red fox, (*Vulpes vulpes*) in Oklahoma. Southwestern Nat., 5:71–74, 1 fig., August 10.

GLASS, B. P., and MORSE, R. C. 1959. A new pipistrel from Oklahoma and Texas. Jour. Mamm., 40:531–534, November 20.

GLASS, B. P., and WARD, C. M. 1959. Bats of the genus Myotis from Oklahoma. Jour. Mamm., 40:194–201, May 20.

GOERTZ, J. W. 1964. Range extension of the muskrat in California. Jour. Mamm., 45:480, September 15.

GOERTZ, J. W., and ABEGG, R. 1966. Pumas in Louisiana. Jour. Mamm., 47:727, December 2.

GOLDMAN, E. A. 1910. Revision of the wood rats of the genus Neotoma. N. Amer. Fauna, 31:1–124, 8 pls., 14 figs., October 19.

———. 1911. Revision of the spiny pocket mice (genera Heteromys and Liomys). N. Amer. Fauna, 34:1–70, 3 pls., 6 figs., September 7.

———. 1912a. Descriptions of twelve new species and subspecies of mammals from Panama. Smiths. Miscl. Coll., 56(36):1–11, February 19.

———. 1912b. New mammals from eastern Panama. Smiths. Miscl. Coll., 60(2):1–18, September 20.

———. 1917. New mammals from North and Middle America. Proc. Biol. Soc. Washington, 30:107–116, May 23.

———. 1918. The rice rats of North America (genus Oryzomys). N. Amer. Fauna, 43:1–100, 6 pls., 11 figs., September 23.

———. 1920. Mammals of Panama. Smiths. Miscl. Coll., 69:1–309, 39 pls., 24 figs., April 24.

———. 1937. *New rodents from Middle America*. Jour. Washington Acad. Sci., 27:418–423, October 15.

———. 1938. List of the gray foxes of Mexico. Jour. Washington Acad. Sci., 28:494–498, November 15.

————. 1941. A new western subspecies of golden mouse. Proc. Biol. Soc. Washington, 54:189–191, December 8.

————. 1944. Classification of wolves, pt. 2, pp. 387–507, Pls. 88–131, 1 fig., in Young, S. P., and Goldman, E. A., The wolves of North America. Amer. Wildlife Inst., Washington, D.C., xx + 636 pp., 131 pls., 15 figs., 7 tables.

————. 1950. Raccoons of North and Middle America. N. Amer. Fauna, 60:vi + 153, 22 pls., 2 figs., November 7.

GOLLEY, F. B. 1962. Mammals of Georgia. Univ. Georgia Press, Athens, xii + 218 pp., 118 figs., 5 tables.

————. 1966. South Carolina mammals. Contrib. Charleston Mus., 15:xiv + 181 pp., 90 figs., 4 tables.

GOLLEY, F. B., GENTRY, J. B., CALDWELL, L. D., and DAVENPORT, L. B., JR. 1965. Number and variety of small mammals on the AEC Savannah River Plant. Jour. Mamm., 46:1–18, 5 figs., 7 tables, February 20.

GOODPASTER, W. W., and HOFFMEISTER, D. F. 1952. Notes on the mammals of western Tennessee. Jour. Mamm., 33:362–371, 1 table, August 19.

————. 1954. Life history of the golden mouse, Peromyscus nuttalli, in Kentucky. Jour. Mamm., 35:16–27, 1 pl., 2 figs., 3 tables., February 10.

————. 1968. Notes on Ohioan mammals. Ohio Jour. Sci., 68:116–117, March.

GOODWIN, G. G. 1929. Mammals of the Cascapedia Valley, Quebec. Jour. Mamm., 10:239–246, August 10.

————. 1932. New records and some observations on Connecticut mammals. Jour. Mamm., 13:36–40, February 9.

————. 1932. Three new Reithrodontomys and two new Peromyscus from Guatemala. Amer. Mus. Novit., 560:1–5, September 16.

————. 1935. The mammals of Connecticut. Bull. Connecticut State Geol. and Nat. Hist. Surv., 53:1–221, 33 pls., 19 figs.

————. 1942. Mammals of Honduras. Bull. Amer. Mus. Nat. Hist., 79:107–195, May 29.

————. 1946. Mammals of Costa Rica. Bull. Amer. Mus. Nat. Hist., 87:271–473, Pl. 17, 50 figs., 1 map, December 31.

————. 1954. Southern records for Arctic mammals and a northern record for Alfaro's rice rat. Jour. Mamm., 35:258, May 26.

————. 1955a. Two new white-footed mice from Oaxaca, México. Amer. Mus. Novit., 1732:1–5, June 10.

————. 1955b. Mammals of Guatemala, with the description of a new little brown bat. Amer. Mus. Novit., 1744:1–5, August 12.

————. 1958. Bats of the genus Rhogeëssa. Amer. Mus. Novit., 1923:1–17, 1 table, December 31.

————. 1959a. Renaming of Sigmodon planifrons minor Goodwin. Jour. Mamm., 40:447, August 20.

————. 1959b. Descriptions of some new mammals. Amer. Mus. Novit., 1967:1–8, 2 figs., October 29.

————. 1959c. Bats of the subgenus Natalus. Amer. Mus. Novit., 1977:1–22, December 22.

————. 1960. The status of Vespertilio auripendulus Shaw, 1800, and Molossus ater Geoffroy, 1805. Amer. Mus. Novit., 1994:1–6, 1 fig., March 8.

————. 1961a. Flying squirrels (Glaucomys volans) of Middle America. Amer. Mus. Novit., 2059:1–22, 7 figs., November 29.

————. 1961b. The murine opossums (Genus Marmosa) of the West Indies, and the description of a new subspecies of Rhipidomys from Little Tobago. Amer. Mus. Novit., 2070:1–20, 10 figs., December 29.

————. 1963. A new subspecies of bobcat (Lynx rufus) from Oaxaca, México. Amer. Mus. Novit., 2139:1–7, 3 figs., May 7.

————. 1969. Mammals from the state of Oaxaca, México, in the American Museum of Natural History. Bull. Amer. Mus. Nat. Hist., 141:1–270, 40 pls., 97 figs., 43 tables, April 30.

GOODWIN, G. G., and GREENHALL, A. M. 1961. A review of the bats of Trinidad and Tobago. Bull. Amer. Mus. Nat. Hist., 122:187–301, Pls. 7–46, 113 figs., 2 tables, 2 maps, June 26.

————. 1962. Two new bats from Trinidad, with comments on the status of the genus Mysophylla. Amer. Mus. Novit., 2080:1–18, April 24.

GOODWIN, R. E. 1970. The ecology of Jamaican bats. Jour. Mamm., 51:571–579, August 28.

GORDON, R. E., and BAILEY, J. R. 1963. The occurrence of Parascalops breweri on the Highlands (North Carolina) Plateau. Jour. Mamm., 44:580–581, 1 table, December 13.

GORHAM, S. W., and JOHNSTON, D. H. 1963. Notes on New Brunswick bats. Canadian Field-Nat., 76:228, January 7.

GOTTSCHANG, J. L. 1965. Winter populations of small mammals in old fields of southwestern Ohio. Jour. Mamm., 46:44–52, 6 tables, February 20.

GRAY, J. E. 1873. Notes on the rats; with the description of some new species from Panama and the Aru Islands. Ann. Mag. Nat. Hist., ser. 4, 12:416–419, 2 figs., November.

GRAY, R. L. 1977. Extension of red fox distribution in California. California Fish and Game, 63:58, January.

GREER, J. K. 1960. Southern yellow bat from Durango, México. Jour. Mamm., 41:511, November 11.

GREER, J. K., and GREER, M. 1970. Record of the pygmy spotted skunk (Spilogale pygmaea) from Colima, México. Jour. Mamm., 51:629–630, August 28.

GRIMM, W. C., and WHITEBREAD, R. 1952. Mammal survey of northeastern Pennsylvania. Pennsylvania Game Comm., 82 pp., 5 maps, 2 numbered tables.

GRINNELL, J. 1912. A new member of the Perognathus parvus group of pocket mice. Proc. Biol. Soc. Washington, 25:127–128, July 31.

————. 1922. A geographical study of the kangaroo rats of California. Univ. California Publ. Zool., 24:1–124, 7 pls., 24 figs., June 17.

————. 1933. Review of the Recent mammal fauna of California. Univ. California Publ. Zool., 40:71–234, September 26.

GRINNELL, J., DIXON, J. S., and LINSDALE, J. M. 1930. Vertebrate natural history of a section of northern California through the Lassen Peak region. Univ. California Publ. Zool., 35:v + 594, 181 figs., October 10.

GROMOV, I. M., BIBIKOV, D. I., KALABUKHOV, N. I., and MEYER, M. N. 1965. Fauna SSSR Mlekopitayushchie [Fauna of the U.S.S.R., mammals]. Nauka, Acad. Sci., Moscow and Leningrad, vol. 3, no. 2, pp. 1–466 + 2, 102 figs., 18 tables, November 30.

GROMOV, I. M., GUREEV, A. A., NOVIKOV, G. A., SOKOLOV, I. I., STRELKOV, P. P., and CHAPSKII, K. K. 1963. Mlekopitayushchie fauni SSSR [Mammals of the fauna of the U.S.S.R.]. Acad. Sci., Moscow and Leningrad, 2 vols., 1–638, 639–1100, 540 figs., June 27 (vol. 1), July 26 (vol. 2).

GUIGUET, C. J. 1955. A record of Baird's dolphin (*Delphinus bairdii Dall*) in British Columbia. Canadian Field-Nat., 68:136, 1 fig., May 4.

GUILDAY, J. E., and MAYER-OAKES, W. J. 1952. An occurrence of the rice rat (*Oryzomys*) in West Virginia. Jour. Mamm., 33:253–255, 1 fig., May 14.

GUNDERSON, H. L., and BEER, J. R. 1953. The mammals of Minnesota. Occas. Pap. Univ. Minnesota, Mus. Nat. Hist., 6:xii+190, illustrated.

GUNIER, W. J., and ELDER, W. H. 1973. New records of Myotis leibii from Missouri. Amer. Midland Nat., 89:489–490, May 11.

GUNTER, G. 1955. Blainville's beaked whale, *Mesoplodon densirostris*, on the Texas coast. Jour. Mamm., 36:573–574, December 14.

GUNTER, G., and OVERSTREET, R. 1974. Cetacean Notes. I. Sei and rorqual whales on the Mississippi coast, a correction. II. A dwarf sperm whale in Mississippi Sound and its helminth parasites. Gulf Research Reports, 4:479–481, December.

HAHN, D. E. 1966. The nine-banded armadillo, *Dasypus novemcinctus*, in Colorado. Southwestern Nat., 11:303, June 30.

HAHN, W. L. 1907. A review of the bats of the genus Hemiderma. Proc. U.S. Nat. Mus., 32:103–118, February 8.

HALL, E. R. 1943. Criteria for vertebrate subspecies, species and genera: the mammals. Ann. New York Acad. Sci., 44:141–144, June 8.

———. 1946. Mammals of Nevada. Univ. California Press, Berkeley, xi + 710 pp., frontispiece, 11 pls., 485 figs., 27 tables, July 1.

———. 1951. American weasels. Univ. Kansas Publ., Mus. Nat. Hist., 4:1–446, 41 pls., 31 figs., December 27.

———. 1955. Handbook of mammals of Kansas. Univ. Kansas Mus. Nat. Hist., Miscl. Publ., 7:1–303, illustrated, December 13.

———. 1960. Oryzomys couesi only subspecifically different from the marsh rice rat, Oryzomys palustris. Southwestern Nat., 5:171–173, November 1.

———. 1971. Variation in the blackish deer mouse, Peromyscus furvus. Anal. Inst. Biol., Univ. Nac. Autó. México, 39 (for 1968, Ser. Zool. 1):149–154, 1 fig., 1 table, September 11.

HALL, E. R., and ALVAREZ, T. 1961. A new species of mouse (*Peromyscus*) from northwestern Veracruz, México. Proc. Biol. Soc. Washington, 74:203–205, August 11.

———. 1961. A new subspecies of the black myotis (bat) from eastern Mexico. Univ. Kansas Publ., Mus. Nat. Hist., 14:69–72, 1 fig., December 29.

HALL, E. R., and COCKRUM, E. L. 1952. Comments on the taxonomy and geographic distribution of North American microtines. Univ. Kansas Publ., Mus. Nat. Hist., 5:293–312, November 17.

———. 1953. A synopsis of the North American microtine rodents. Univ. Kansas Publ., Mus. Nat. Hist., 5:373–498, 149 figs., January 15.

HALL, E. R., and DALQUEST, W. W. 1950. A synopsis of the American bats of the genus Pipistrellus. Univ. Kansas Publ., Mus. Nat. Hist., 1:591–602, 1 fig., 1 table, January 20.

———. 1963. The mammals of Veracruz. Univ. Kansas Publ., Mus. Nat. Hist., 14:165–362, 2 figs., May 20.

HALL, E. R., and GENOWAYS, H. H. 1970. Taxonomy of the Neotoma albigula-group of woodrats in central Mexico. Jour. Mamm., 51:504–516, 4 figs., 2 tables, August 28.

HALL, E. R., and JONES, J. K., JR. 1961. North American yellow bats, "Dasypterus," and a list of the named kinds of the genus Lasiurus Gray. Univ. Kansas Publ., Mus. Nat. Hist., 14:73–98, 4 figs., December 29.

HALL, E. R., and KELSON, K. R. 1952. Comments on the taxonomy and geographic distribution of some North American marsupials, insectivores and carnivores. Univ. Kansas Publ., Mus. Nat. Hist., 5:319–341, 1 table, December 5.

———. 1959. The mammals of North America. Ronald Press Co., New York, 1:xxx + 546 + 79, and 2:ix + 547–1083 + 79; 1231 illustrations, March 31.

HALL, E. R., and TAMSITT, J. R. 1968. A new subspecies of the red fig-eating bat from Puerto Rico. Royal Ontario Mus., Life Sci. Occas. Pap., 11:1–5, 2 figs., August 9.

HALL, J. G. 1968. Spotted skunk on Shiva Temple in Grand Canyon. Plateau, 40:98–100, 1 fig., winter.

HALL, J. S. 1961. *Myotis austroriparius* in central Kentucky. Jour. Mamm., 42:399–400, August 21.

HALL, J. S., and WILSON, N. 1966. Seasonal populations and movements of the gray bat in the Kentucky area. Amer. Midland Nat., 75:317–324, 2 figs., 3 tables, April 20.

HALLORAN, A. F. 1961. The carnivores and ungulates of the Aransas National Wildlife Refuge, Texas. Southwestern Nat., 6:21–26, June 10.

———. 1962. An Arizona specimen of the red fox. Jour. Mamm., 43:432, August 20.

HALLORAN, A. F., and TABER, F. E. 1965. Carnivore notes from the Navajo Indian Reservation. Southwestern Nat., 10:139–140, April 30.

HAMILTON, W. J., JR. 1950. The prairie deer mouse in New York and Pennsylvania. Jour. Mamm., 31:100, February 21.

———. 1963. Distribution of the fox squirrel in New York. Jour. Mamm., 44:124–125, February 20.

HAMMER, E. W. 1971. A southwestern range extension of the sagebrush vole in Oregon. Murrelet, 52:26, November.

HANDLEY, C. O., JR. 1959. A review of the genus *Hoplomys* (thick-spined rats), with description of a new form from Isla Escudo de Veraguas, Panamá. Smiths. Miscl. Coll., 139(4):1–10, 1 fig., July 3.

———. 1959. A revision of American bats of the genera Euderma and Plecotus. Proc. U.S. Nat. Mus., 110:95–246, 27 figs., September 3.

———. 1960. Descriptions of new bats from Panama. Proc. U.S. Nat. Mus., 112:459–479, October 6.

———. 1966. Descriptions of new bats (*Chiroderma* and *Artibeus*) from Mexico. Anal. Inst. Biol., Univ. Nac. Autó. México, 36:297–301, June 20.

———. 1966. Checklist of mammals of Panama. Pp. 753–795 of "Ectoparasites of Panama," Field Mus. Nat. Hist., November 22.

———. 1966. A synopsis of the genus *Kogia* (pygmy sperm whales). Pp. 62–69 of "Whales, dolphins, and porpoises," Univ. California Press.

———. 1976. Mammals of the Smithsonian Venezuelan project. Brigham Young Univ., Sci. Bull., Biol. Series, 20(5):1–89 + 1 fig. + facing list of localities, 1 table, June 30.

HANDLEY, C. O., JR., and FERRIS, K. C. 1972. Descriptions of new bats of the genus *Vampyrops*. Proc. Biol. Soc. Washington, 84:519–523, February 29.

HANDLEY, C. O., JR., and PATTON, C. P. 1947. Wild mammals of Virginia. Commonwealth of Virginia Comm. Game and Inland Fisheries, Richmond, vi+220 pp., frontispiece, 103 figs.

HANSEN, E. L. 1965. Muskrat distribution in south-central Oregon. Jour. Mamm., 46:669–671, November 30.

HANSEN, R. M. 1960. Age and reproductive characteristics of mountain pocket gophers in Colorado. Jour. Mamm., 41:323–335, 8 tables, August 15.

———. 1962. Dispersal of Richardson ground squirrel in Colorado. Amer. Midland Nat., 68:58–66, 1 fig., August 1.

HARPER, F. 1956. The mammals of Keewatin. Univ. Kansas Mus. Nat. Hist., Miscl. Publ., 12:1–94, 6 pls., 8 figs., 1 map, October 26.

———. 1961. Land and fresh-water mammals of the Ungava Peninsula. Univ. Kansas Mus. Nat. Hist., Miscl. Publ., 27:1–178, 8 pls., 3 figs., 45 maps, August 11.

HARRINGTON, C. R. 1966. Extralimital occurrences of walruses in the Canadian Arctic. Jour. Mamm., 47:506–513, 1 fig., August 29.

HARRIS, A. H. 1974. *Myotis yumanensis* in interior southwestern North America, with comments on *Myotis lucifugus.* Jour. Mamm., 55:589–607, 7 figs., 6 tables, August 20.

HARRIS, A. H., and FINDLEY, J. S. 1962. Status of *Myotis lucifugus phasma* and comments on variation in *Myotis yumanensis.* Jour. Mamm., 43:192–199, 4 figs., 1 table, May 29.

HARRIS, W. P., JR. 1932. Four new mammals from Costa Rica. Occas. Pap. Mus. Zool., Univ. Michigan, 248:1–6, August 4.

———. 1937. Revision of *Sciurus variegatoides,* a species of Central American squirrel. Miscl. Publ. Mus. Zool., Univ. Michigan, 38:1–39, 3 pls., 3 figs., 1 map, September 7.

———. 1943. A list of mammals from Costa Rica. Occas. Pap. Mus. Zool., Univ. Michigan, 476:1–15, 1 map, October 8.

HART, B. J. 1972. Distribution of the pigmy mouse, *Baiomys taylori,* in north-central Texas. Southwestern Nat., 17:213–214, 1 fig., September 15.

HATFIELD, D. M. 1935. A natural history study of Microtus californicus. Jour. Mamm., 16:261–271, 1 fig., 1 table, November 15.

HATT, R. T. 1929. The red squirrel: its life history and habits, with special reference to the Adirondacks of New York and the Harvard Forest. Bull. New York State Coll. Forestry, Syracuse Univ., Roosevelt Wildlife Annals, 2:1–146, 52 figs., 1 map, March.

HAUGEN, A. O. 1961. Wolverine in Iowa. Jour. Mamm., 42:546–547, November 20.

HAVEMAN, J. R., and ROBINSON, W. L. 1976. Northward range extension of the southern flying squirrel in Michigan. Jack-Pine Warbler, 54:40–41, March.

HAYS, H. A., and IRELAND, P. H. 1967. A big free-tailed bat (*Tadarida macrotis*) taken in southeastern Kansas. Southwestern Nat., 12:196, August 4.

HAYWARD, B. J. 1970. The natural history of the cave bat *Myotis velifer.* WRI-SCI, Western New Mexico Univ., 1:1–74, 21 figs., 11 tables, February 15.

HAYWARD, B. J., and JOHNSON, R. R. 1961. Notes on *Plecotus phyllotis* from Arizona. Jour. Mamm., 42:402, August 21.

HAYWARD, C. L., and KILLPACK, M. L. 1958. Distribution

and variation of the Utah population of the Great Basin pocket mouse. Great Basin Nat., 18:26–30, May 31.

HAZARD, E. B. 1963. Records of the opossum in northern Minnesota. Jour. Mamm., 44:118, February 20.

HEANEY, L. R., and BIRNEY, E. C. 1975. Comments on the distribution and natural history of some mammals in Minnesota. Canadian Field-Nat., 89:29–34, 1 table, April 18.

———. 1977. Distribution and natural history notes on some mammals from Puebla, Mexico. Southwestern Nat., 21:543–545, February 10.

HENDRICKS, L. J. 1963. Observation of armadillo in east-central New Mexico. Jour. Mamm., 44:581, December 13.

HENSON, O. W., JR., and NOVICK, A. 1966. An additional record of the bat, *Phyllonycteris aphylla.* Jour. Mamm., 47:351–352, May 26.

HEPTNER, V. G. 1952. The systematic position of the water rat group (*Arvicola* Lac., Mammalia, Muridae) and the superspecific grouping of voles. Bull. Moscow Soc. Naturalists, Ser. Biol., 57(2):58–61.

HERSHKOVITZ, P. 1944. A systematic review of the neotropical water rats of the genus *Nectomys* (Cricetinae). Miscl. Publ. Mus. Zool., Univ. Michigan, 58:1–101, 4 pls., 5 figs., 2 maps, 19 tables, January 4.

———. 1948. Mammals of northern Colombia, preliminary report no. 3: water rats (genus Nectomys), with supplemental notes on related forms. Proc. U.S. Nat. Mus., 98:49–56, June 30.

———. 1949. Generic names of the four-eyed pouch opossum and the woolly opossum (*Didelphidae*). Proc. Biol. Soc. Washington, 62:11–12, March 17.

———. 1949. Mammals of northern Colombia, preliminary report no. 4: monkeys (Primates), with taxonomic revisions of some forms. Proc. U.S. Nat. Mus., 98:323–427, Pls. 15–17, Figs. 52–59, May 10.

———. 1949. Mammals of northern Colombia, preliminary report no. 5: bats (Chiroptera). Proc. U.S. Nat. Mus., 99:429–454, Fig. 38, May 10.

———. 1954. Mammals of northern Colombia, preliminary report no. 7: tapirs (genus *Tapirus*), with a systematic review of American species. Proc. U.S. Nat. Mus., 103:465–496, Figs. 58–62.

———. 1960. Mammals of northern Colombia, preliminary report no. 8: arboreal rice rats, a systematic revision of the subgenus Oecomys, genus Oryzomys. Proc. U.S. Nat. Mus., 110 (3420):513–568, 12 pls., 6 figs., 10 tables, February 4.

———. 1961. On the nomenclature of certain whales. Fieldiana, Zool., 39:547–565, July 25.

———. 1966. Catalog of living whales. Bull. U.S. Nat. Mus., 246:viii + 259, February 28.

———. 1966. Mice, land bridges and Latin American faunal interchange. Pp. 725–751 of "Ectoparasites of Panama," Field Mus. Nat. Hist., November 22.

———. 1966. Taxonomic notes on tamarins, genus *Saguinus* (Callithricidae, Primates), with descriptions of four new forms. Folia Primat., 4:381–395, 4 figs.

———. 1970. Supplementary notes on Neotropical *Oryzomys dimidiatus* and *Oryzomys hammondi* (Cricetinae). Jour. Mamm., 51:789–794, 2 figs., 2 tables, November 30.

———. 1976. Comments on generic names of four-eyed opossums (Family Didelphidae). Proc. Biol. Soc. Washington, 89:295–303, 1 fig., October 12.

HEWSTON, J. 1962. Albino mammals in North Dakota. Jour. Mamm., 43:269–270, May 29.

HIBBS, L. D. 1967. Food habits of the mountain goat in Colorado. Jour. Mamm., 48:242–248, 3 tables, May 20.

HILL, J. E. 1937. Morphology of the pocket gopher mammalian genus Thomomys. Univ. California Publ. Zool., 42:81–171, 26 figs., 1 table, August 25.

HINTON, M. A. C. 1926. Monograph of the voles & Lemmings (Microtinae) living and extinct. British Museum (Nat. Hist.). Vol. 1. pp. xvi + 488, 15 pls., 110 figs., June 26.

HOFFMANN, R. S. 1971. Relationships of certain Holarctic shrews, genus Sorex. Zeitschr. f. Säugetierkunde, 36:193–200, 4 figs., 2 tables, August.

HOFFMANN, R. S., and JONES, J. K., JR. 1970. Influence of late-glacial and post-glacial events on the distribution of Recent mammals on the northern Great Plains. Pleistocene and Recent Environments . . . Dept. Geol. Univ. Kansas Special Publ., 3:355–394, 17 figs., 2 tables.

HOFFMANN, R. S., and NADLER, C. F. 1968. Chromosomes and systematics of some North American species of the genus Marmota (Rodentia: Sciuridae). Experientia, 24:740–742, July 15.

HOFFMANN, R. S., PATTIE, D. L., and BELL, J. F. 1969. The distribution of some mammals in Montana. II. Bats. Jour. Mamm., 50:737–741, 5 figs., November 28.

HOFFMANN, R. S., and TABER, R. D. 1960. Notes on Sorex in the Rocky Mountain Alpine Zone. Jour. Mamm., 41:230–234, May 20.

HOFFMANN, R. S., WRIGHT, P. L., and NEWBY, F. E. 1969. The distribution of some mammals in Montana I. Mammals other than bats. Jour. Mamm., 50:579–604, 16 figs., 2 tables, August 22.

HOFFMEISTER, D. F. 1951. A taxonomic and evolutionary study of the piñon mouse, Peromyscus truei. Illinois Biol. Monog., 21:x+104, 5 pls., 24 figs., 7 tables, November 12.

———. 1954. Distribution of some Illinois mammals. Chicago Acad. Sci., Nat. Hist. Miscl., 128:1–4, February 10.

———. 1956. Mammals of the Graham (Pinaleno) Mountains, Arizona. Amer. Midland Nat., 55:257–288, 7 figs., 1 table, April.

———. 1957. Review of the long-nosed bats of the genus Leptonycteris. Jour. Mamm., 38:454–461, 1 fig., November 20.

———. 1959. A new spotted ground squirrel from Mexico. Proc. Biol. Soc. Washington, 72:37–38, May 1.

———. 1959. Distributional records of certain mammals from southern Arizona. Southwestern Nat., 4:14–19, 1 fig., July 22.

———. 1962. The kinds of deer, Odocoileus, in Arizona. Amer. Midland Nat., 67:45–64, 2 tables, 6 figs., January.

———. 1968. Pygmy shrew, Microsorex hoyi winnemana, in Great Smoky Mountains National Park. Jour. Mamm., 49:331, May 20.

———. 1969. The species problem in the Thomomys bottae–Thomomys umbrinus complex of pocket gophers in Arizona. Univ. Kansas Mus. Nat. Hist., Miscl. Publ., 51:75–91, 10 figs., 2 tables, July 11.

———. 1971. Mammals of Grand Canyon. Univ. Illinois Press, pp. 1–183, illustrated, May.

———. 1974. The taxonomic status of Perognathus penicillatus minimus Burt. Southwestern Nat., 19:213–214, July 26.

———. 1977. Status of the cotton mouse, Peromyscus gossypinus, in southern Illinois. Amer. Midland Nat., 97:222–224, 2 figs., 1 table, January.

———. 1977. Noteworthy range extensions of mammals in northern Mexico and Arizona. Southwestern Nat., 22:150–151, March 1.

HOFFMEISTER, D. F., and CAROTHERS, S. W. 1969. Mammals of Flagstaff, Arizona. Plateau, 41:184–188, 1 fig., 1 table.

HOFFMEISTER, D. F., and DE LA TORRE, L. 1960. A revision of the wood rat, Neotoma stephensi. Jour. Mamm., 41:476–491, 1 pl., 4 figs., 4 tables, November 11.

———. 1961. Geographic variation in the mouse, Peromyscus difficilis. Jour. Mamm., 42:1–13, 2 figs., February 20.

HOFFMEISTER, D. F., and DIERSING, V. E. 1973. The taxonomic status of Peromyscus merriami goldmani Osgood, 1904. Southwestern Nat., 18:354–357, 1 fig., October 5.

HOFFMEISTER, D. F., and DURHAM, F. E. 1971. Mammals of the Arizona Strip including Grand Canyon National Monument. Mus. Northern Arizona, Tech. Ser., 11:1–44, 15 figs.

HOFFMEISTER, D. F., and GOODPASTER, W. W. 1954. The mammals of the Huachuca Mountains, southeastern Arizona. Illinois Biol. Monog., 24:v+152, 27 figs., 4 tables, December 31.

———. 1962. Observations on a colony of big-eared bats, Plecotus rafinesquii. Trans. Illinois State Acad. Sci., 55:87–89, 1 fig., no date (pers. comm.).

HOFFMEISTER, D. F., and KRUTZSCH, P. H. 1955. A new subspecies of Myotis evotis (H. Allen) from southeastern Arizona and Mexico. Chicago Acad. Sci., Nat. Hist. Miscl., 151:1–4, December 28.

HOFFMEISTER, D. F., and LEE, M. R. 1963. Eastern cottontail in Sonora, Mexico. Jour. Mamm., 44:120–121, February 20.

———. 1963. The status of the sibling species Peromyscus merriami and Peromyscus eremicus. Jour. Mamm., 44:201–213, 4 figs., 3 tables, May 20.

———. 1963. Taxonomic review of cottontails, Sylvilagus floridanus and Sylvilagus nuttallii, in Arizona. Amer. Midland Nat., 70:138–148, 2 figs., 1 table, July 10.

———. 1963. Revision of the desert cottontail, Sylvilagus audubonii, in the southwest. Jour. Mamm., 44:501–518, 2 figs., 3 tables, December 13.

———. 1967. Revision of the pocket mice, Perognathus penicillatus. Jour. Mamm., 48:361–380, 5 figs., 4 tables, August 21.

HOFFMEISTER, D. F., and NADER, I. A. 1963. Distributional notes on Arizona mammals. Illinois State Acad. Sci., 56:92–93, October 7.

HOLLISTER, N. 1915. The genera and subgenera of raccoons and their allies. Proc. U.S. Nat. Mus., 49:143–150, 2 pls., August 13.

HOLSINGER, J. R. 1964. The gray myotis in Virginia. Jour. Mamm., 45:151–152, March 12.

HOOPER, D. 1973. Gray squirrels at Weekes, Saskatchewan. Blue Jay, 31:238, December.

HOOPER, E. T. 1938. Geographical variation in wood rats of the species Neotoma fuscipes. Univ. California Publ. Zool., 42:213–246, Pls. 7–8, 2 figs., March 1.

———. 1943. Geographic variation in harvest mice of the

species *Reithrodontomys humulis*. Occas. Pap. Mus. Zool., Univ. Michigan, 477:1–19, 1 map, November 5.

———. 1947. Notes on Mexican mammals. Jour. Mamm., 28:40–57, February 17.

———. 1952. A systematic review of the harvest mice (genus *Reithrodontomys*) of Latin America. Miscl. Publ. Mus. Zool., Univ. Michigan, 77:1–255, 9 pls., 24 figs., 12 maps, 7 tables, January 16.

———. 1955. Notes on mammals of western Mexico. Occas. Pap. Mus. Zool., Univ. Michigan, 565:1–26, March 31.

———. 1957. Records of Mexican mammals. Occas. Pap. Mus. Zool., Univ. Michigan, 586:1–9, April 30.

———. 1957. Record of the Mexican harvest mouse (*Reithrodontomys mexicanus*) from Michoacán, México. Jour. Mamm., 38:521–522, November 20.

———. 1958. The male phallus in mice of the genus *Peromyscus*. Miscl. Publ. Mus. Zool., Univ. Michigan, 105:1–24, 14 pls., 1 fig., 1 table, December 29.

———. 1960. The glans penis in *Neotoma* (Rodentia) and allied genera. Occas. Pap. Mus. Zool., Univ. Michigan, 618:1–21, 11 pls., 1 table, October 17.

———. 1961. Notes on mammals from western and southern Mexico. Jour. Mamm., 42:120–122, February 20.

———. 1968. Habitats and food of amphibious mice of the genus *Rheomys*. Jour. Mamm., 49:550–553, 2 tables, August 20.

———. 1972. A synopsis of the rodent genus *Scotinomys*. Occas. Pap. Mus. Zool., Univ. Michigan, 665:1–32, 12 figs., 1 table, April 11.

HOOPER, E. T., and HART, B. S. 1962. A synopsis of Recent North American microtine rodents. Miscl. Publ. Mus. Zool., Univ. Michigan, 120:1–68, 11 figs., 2 tables, June 26.

HOOPER, E. T., and MUSSER, G. G. 1964. Notes on classification of the rodent genus *Peromyscus*. Occas. Pap. Mus. Zool., Univ. Michigan, 635:1–13, 2 figs., 1 table. April 22.

HOUSTON, C. S., and HOUSTON, M. I. 1973. A history of raccoons in Saskatchewan. Blue Jay, 31:103–104, 1 fig., June.

HOWELL, A. B. 1926. Voles of the genus Phenacomys. N. Amer. Fauna, 48:iv + 66, 7 pls., 11 figs., 2 tables, October 16.

HOWELL, A. H. 1929. Revision of the American chipmunks (genera Tamias and Eutamias). N. Amer. Fauna, 52:1–157, 10 pls., 9 figs., 2 tables, November 30.

———. 1938. Revision of the North American ground squirrels, with a classification of the North American Sciuridae. N. Amer. Fauna, 56:1–256, 32 pls., 20 figs., May 18.

———. 1943. Two new cotton rats from Florida. Proc. Biol. Soc. Washington, 56:73–75, June 16.

HSU, T. C., and ARRIGHI, F. E. 1968. Chromosomes of *Peromyscus* (Rodentia, Cricetidae) I. Evolutionary trends in 20 species. Cytogenetics, 7:417–446, 17 figs.

HSU, T. C., and JOHNSON, M. L. 1970. Cytological distinction between *Microtus montanus* and *Microtus canicaudus*. Jour. Mamm., 51:824–826, 2 figs., November 30.

HUBBARD, J. P. 1972. Hooded skunk on the Mogollon Plateau, New Mexico. Southwestern Nat., 16:458, February 18.

HUBBS, C. L. 1960. The marine vertebrates of the outer coast [of Baja California]. Systematic Zool., 9:134–147, September–December.

HUBBS, C. L., PERRIN, W. F., and BALCOMB, K. C.

1973. *Stenella coeruleoalba* in the eastern and central Tropical Pacific. Jour. Mamm., 54:549–552, 4 figs., May 31.

HUEY, L. M. 1932. A new species of pocket gopher from southern Arizona. Trans. San Diego Soc. Nat. Hist., 7:157–160, July 28.

———. 1951. The kangaroo rats (Dipodomys) of Baja California, Mexico. Trans. San Diego Soc. Nat. Hist., 11:205–255, 2 figs., 3 tables, Maps 3–6, April 30.

———. 1955. A new race of *Dipodomys* and a new race of *Thomomys* from Arizona. Trans. San Diego Soc. Nat. Hist., 12:99–101, February 10.

———. 1957. A new race of wood rat (Neotoma) from the gulf side of central Baja California, Mexico. Trans. San Diego Soc. Nat. Hist., 12:287–288, September 25.

———. 1960. Two new races of Perognathus spinatus from Baja California, Mexico. Trans. San Diego Soc. Nat. Hist., 12:409–412, February 1.

———. 1960. Comments on the pocket mouse, Perognathus fallax, with descriptions of two new races from Baja California, Mexico. Trans. San Diego Soc. Nat. Hist., 12:415–419, February 1.

———. 1961. Further northern extension of the hog-nosed skunk. Jour. Mamm., 42:421, August 21.

———. 1964. The mammals of Baja California, Mexico. Trans. San Diego Soc. Nat. Hist., 13:85–168, January 15.

HUSSON, A. M. 1962. The bats of Suriname. Zool. Verhand. Rijksmuseum Nat. Hist. Leiden, 58:1–282, 30 pls., 39 figs., 31 tables, November 13.

HUSSON, A. M., and HOLTHUIS, L. B. 1974. *Physeter macrocephalus* Linnaeus, 1758, the valid name for the sperm whale. Zool. Med. Rijksmuseum Nat. Hist. Leiden, 48:205–217, 3 pls., October 24.

INGLES, L. G. 1959. Notas acerca de los mamíferos Mexicanos. Anal. Inst. Biol., Univ. Nac. Autó. México, 19:319–408, March 31.

IRWIN, D. W., and BAKER, R. J. 1967. Additional records of bats from Arizona and Sinaloa. Southwestern Nat., 12:195, August 4.

IVERSON, S. L., and TURNER, B. N. 1973. Ecological notes on Manitoba *Napaeozapus insignis*. Canadian Field-Nat., 87:15–19, 1 fig., 2 tables, March 30.

IVEY, R. D. 1957. Ecological notes on the mammals of Bernalillo County, New Mexico. Jour. Mamm., 38:490–502, 1 fig., November 20.

JACKSON, C. F. 1938. Notes on the mammals of southern Labrador. Jour. Mamm., 19:429–434, November 14.

JACKSON, H. H. T. 1951. Classification of the races of the coyote, pt. 2, pp. 227–341, Pls. 58–81, Figs. 20–28, *in* Young, S. P., and Jackson, H. H. T., The clever coyote. Stackpole Co., Harrisburg, Pa., and Wildlife Manag. Inst., Washington, D.C., xv + 411 pp., 81 pls., 28 figs., 11 tables, November 29.

———. 1961. Mammals of Wisconsin. Univ. Wisconsin Press, xiii + 504 pp., illustrations (approx. 459 including 81 maps).

JACOBI, A. 1931. Das Rentier, eine zoologische Monographie der Gattung Rangifer. Zool. Anzeiger, 96 (suppl.), vii + 264 pp., 6 pls., 32 figs.

JAMES, P., and HAYSE, A. 1963. Albinism in the spotted ground squirrel. Jour. Mamm., 44:574–575, December 13.

JAMESON, D. K. 1959. A survey of the parasites of five species of bats. Southwestern Nat., 4:61–65, September 15.

JAMESON, E. W., JR., and MEAD, R. A. 1964. Seasonal

changes in body fat, water and basic weight in *Citellus lateralis, Eutamias speciosus* and *E. amoenus*. Jour. Mamm., 45:359–365, 3 figs., 1 table, September 15.

JANES, D. W., and GIER, H. T. 1966. Distribution, numbers, and hunting of foxes in Kansas. Trans. Kansas Acad. Sci., 69:23–31, 1 fig., 2 tables, June 15.

JANNETT, F. J., JR., and JANNETT, J. Z. 1974. Drum-marking by Arvicola richardsoni and its taxonomic significance. Amer. Midland Nat., 92:230–234, July.

JARRELL, G. H. 1965. A correction on the range of *Cryptotis parva* in New England. Jour. Mamm., 46:671, November 30.

JIMÉNEZ-G., A. 1971. Nuevos registros de murciélagos para Nuevo León, México. Anal. Inst. Biol., Univ. Nac. Autó. México, 39:133–144, 3 figs., 1 table, 3 maps, September 11.

JOHANSEN, K. 1962. Buoyancy and insulation in the muskrat. Jour. Mamm., 43:64–68, 2 figs., 1 table, February 28.

JOHNSON, D. H. 1943. Systematic review of the chipmunks (genus Eutamias) of California. Univ. California Publ. Zool., 48:63–148, Pl. 6, 12 figs., 6 tables, December 24.

———. 1948. A rediscovered Haitian rodent, Plagiodontia aedium, with a synopsis of related species. Proc. Biol. Soc. Washington, 61:69–74, June 16.

JOHNSON, D. R. 1967. Diet and reproduction of Colorado pikas. Jour. Mamm., 48:311–315, May 20.

JOHNSON, G. L., and PACKARD, R. L. 1974. Electrophoretic analysis of Peromyscus comanche Blair, with comments on its systematic status. Occas. Pap. Mus. Texas Tech Univ., 24:1–16, 5 figs., 8 tables, May 24.

JOHNSON, M. L. 1962. *Another Washington record of the fringed myotis*. Murrelet, 42:44, May.

———. 1968. Application of blood protein electrophoretic studies to problems in mammalian taxonomy. Systematic Zool., 17:23–30, March 27.

———. 1973. Characters of the heather vole, *Phenacomys*, and the red tree vole, *Arborimus*. Jour. Mamm., 54:239–244, 2 tables, April 26.

JOHNSON, M. L., and BENSON, S. B. 1960. Relationship of the pocket gophers of the Thomomys mazama–talpoides complex in the Pacific Northwest. Murrelet, 41:16–22, 1 fig., December.

JOHNSON, M. L., and OSTENSON, B. T. 1959. Comments on the nomenclature of some mammals of the Pacific Northwest. Jour. Mamm., 40:571–577, November 20.

JOHNSON, R. R., and JOHNSON, J. E. 1964. Notes on distribution of bats and other mammals in Arizona. Jour. Mamm., 45:322–324, May 20.

JOHNSON, W. E., and SELANDER, R. K. 1971. Protein variation and systematics in kangaroo rats (genus *Dipodomys*). Systematic Zool., 20:377–405, 7 figs., 11 tables, December 30.

JOHNSTON, D. W. 1974. Food of the barn owl on Grand Cayman, B.W. I. Quart. Jour. Florida Acad. Sci., 35:171–172, March 15.

JONES, C. J. 1961. Additional records of shrews in New Mexico. Jour. Mamm., 42:399, August 21.

———. 1961. Additional records of bats in New Mexico. Jour. Mamm., 42:538–539, November 20.

JONES, C. [J.], and FINDLEY, J. S. 1963. The long-nosed bat in New Mexico. Southwestern Nat., 8:174–175, November 14.

JONES, C. [J.], and FISHER, R. D. 1973. Comments on the type-specimen of *Neotoma desertorum sola* Merriam 1894 (Mammalia: Rodentia). Proc. Biol. Soc. Washington, 86:435–437, December 14.

JONES, C. J., FLEHARTY, E. D., and HARRIS, A. H. 1960. Unusual habits of grasshopper mice in New Mexico. Jour. Mamm., 41:275–276, May 20.

JONES, C. [J.], and SUTTKUS, R. D. 1972. Notes on netting bats for eleven years in western New Mexico. Southwestern Nat., 16:261–266, 3 tables, February 18.

JONES, E. T. 1974. Sixteen Alberta bat records, 1971–1974. Blue Jay, 32:244–245, 2 figs., December.

JONES, F. L. 1955. Records of southern wolverine, *Gulo luscus luteus*, in California. Jour. Mamm., 36:569, December 14.

JONES, J. K., JR. 1960. The pronghorn, Antilocapra americana, in western Iowa. Amer. Midland Nat., 63:249, January.

———. 1960. The hispid cotton rat in Nebraska. Jour. Mamm., 41:132, February 20.

———. 1964a. Additional records of mammals from Durango, México. Trans. Kansas Acad. Sci., 66:750–753, February 25.

———. 1964b. A new subspecies of harvest mouse, *Reithrodontomys gracilis*, from Isla del Carmen, Campeche. Proc. Biol. Soc. Washington, 77:123–124, June 26.

———. 1964c. Distribution and taxonomy of mammals of Nebraska. Univ. Kansas Publ., Mus. Nat. Hist., 16:1–356, 4 pls., 82 figs., October 1.

———. 1964d. Bats new to the fauna of Nicaragua. Trans. Kansas Acad. Sci., 67:506–508, December 11.

———. 1964e. Bats from western and southern Mexico. Trans. Kansas Acad. Sci., 67:509–516, 2 tables, December 11.

———. 1966. Recent records of the shrew, *Megasorex gigas* (Merriam), from western México. Amer. Midland Nat., 75:249–250, February 8.

———. 1966. Bats from Guatemala. Univ. Kansas Publ., Mus. Nat. Hist., 16:439–472, 3 tables, April 18.

———. 1970. The red bat in Baja California, Mexico. Southwestern Nat., 14:361, January 15.

———. 1976. Note on the distribution of *Myotis leibii* (Mammalia: Chiroptera) in Nebraska. Trans. Kansas Acad. Sci., 78:89, September 30.

JONES, J. K., JR., and ALVAREZ, T. 1962. Taxonomic status of the free-tailed bat, Tadarida yucatanica Miller. Univ. Kansas Publ., Mus. Nat. Hist., 14:125–133, 1 fig., March 7.

———. 1964. Additional records of mammals from the Mexican state of San Luis Potosí. Jour. Mamm., 45:302–303, May 20.

JONES, J. K., JR., ALVAREZ, T., and LEE, M. R. 1962. Noteworthy mammals from Sinaloa, México. Univ. Kansas Publ., Mus. Nat. Hist., 14:145–159, May 18.

JONES, J. K., JR., and ANDERSON, S. 1958. Noteworthy records of harvest mice in México. Jour. Mamm., 39:446–447, August 20.

———. 1959. The eastern harvest mouse, Reithrodontomys humulis, in Oklahoma. Southwestern Nat., 4:153–154, October 24.

JONES, J. K., JR., and BLEIER, W. J. 1974. Sanborn's long-tongued bat, *Leptonycteris sanborni*, in El Salvador. Mammalia, 38:144–145, June 15.

JONES, J. K., JR., and CARTER, D. C. 1972. *Cyclopes didac-*

tylus in British Honduras. Mammalia, 36:535–536, September.

JONES, J. K., JR., CHOATE, J. R., and CADENA, A. 1972. Mammals from the Mexican state of Sinaloa. II. Chiroptera. Univ. Kansas Mus. Nat. Hist., Occas. Pap., 6:1–29, 1 fig., 4 tables, January 25.

JONES, J. K., JR., and CORTNER, G. L. 1961. The subspecific identity of the gray squirrel (Sciurus carolinensis) in Kansas and Nebraska. Trans. Kansas Acad. Sci., 63:285–288, 1 table, January 11.

———. 1965. The least weasel, Mustela nivalis, in Kansas. Amer. Midland Nat., 73:247, January.

JONES, J. K., JR., FLEHARTY, E. D., and DUNNIGAN, P. B. 1967. The distributional status of bats in Kansas. Univ. Kansas Mus. Nat. Hist., Miscl. Publ., 46:1–33, 11 figs., May 1.

JONES, J. K., JR., and GENOWAYS, H. H. 1966. Records of bats from western North Dakota. Trans. Kansas Acad. Sci., 69:88–90, June 15.

———. 1967. A new subspecies of the fringe-tailed bat, *Myotis thysanodes* from the Black Hills of South Dakota and Wyoming. Jour. Mamm., 48:231–235, 2 figs., May 20.

———. 1967. Notes on the Oaxacan vole, *Microtus oaxacensis* Goodwin, 1966. Jour. Mamm., 48:320–321, May 20.

———. 1967. Annotated checklist of bats from South Dakota. Trans. Kansas Acad. Sci., 70:184–196, September 22.

———. 1967. A new subspecies of the free-tailed bat, *Molossops greenhalli*, from western Mexico (Mammalia; Chiroptera). Proc. Biol. Soc. Washington, 80:207–210, 1 table, December 1.

———. 1968. Distribution of the porcupine, *Erethizon dorsatum*, in Mexico. Mammalia, 32:709–711, 1 fig., December.

———. 1970. Harvest mice (genus *Reithrodontomys*) of Nicaragua. Occas. Pap. Western Found. of Vert. Zool., 2:1–16, 4 figs., 2 tables, July 20.

JONES, J. K., JR., GENOWAYS, H. H., and WATKINS, L. C. 1971. Bats of the genus Myotis from western Mexico, with a key to the species. Trans. Kansas Acad. Sci., 73:409–418, 1 table, July 28.

JONES, J. K., JR., and GLASS, B. P. 1960. The short-tailed shrew, *Blarina brevicauda*, in Oklahoma. Southwestern Nat., 5:136–142, 2 figs., 1 table, November 1.

JONES, J. K., JR., and HENDERSON, F. R. 1963. Noteworthy records of foxes from South Dakota. Jour. Mamm., 44:283, May 20.

JONES, J. K., JR., and LAWLOR, T. E. 1965. Mammals from Isla Cozumel, Mexico, with description of a new species of harvest mouse. Univ. Kansas Publ., Mus. Nat. Hist., 16:409–419, 1 fig., April 13.

JONES, J. K., JR., and LEE, M. R. 1962. Three species of mammals from western Texas. Southwestern Nat., 7:77–78, June 1.

JONES, J. K., JR., and LONG, C. A. 1961. *Sorex longirostris* in Mississippi. Jour. Mamm., 42:252, May 20.

JONES, J. K., JR., and MURSALOǦLU, B. 1961. Geographic variation in the harvest mouse, *Reithrodontomys megalotis*, on the central Great Plains and in adjacent regions. Univ. Kansas Publ., Mus. Nat. Hist., 14:9–27, 1 fig., July 24.

JONES, J. K., JR., and PHILLIPS, C. J. 1970. Comments on systematics and zoogeography of bats in the Lesser Antilles.

Studies on the fauna of Curaçao and other Caribbean islands, 32:131–145, 1 fig., 2 tables, March.

JONES, J. K., JR., and PHILLIPS, G. L. 1964. A new subspecies of the fruit-eating bat, Sturnira ludovici, from western Mexico. Univ. Kansas Publ., Mus. Nat. Hist., 14:475–481, 1 fig., March 2.

JONES, J. K., JR., and SCHWARTZ, A. 1967. Bredin-Archbold-Smithsonian biological survey of Dominica. 6. Synopsis of bats of the Antillean genus *Ardops*. Proc. U.S. Nat. Mus., 124(3634):1–13, 3 figs., 1 table, December.

JONES, J. K., JR., SMITH, J. D., and ALVAREZ, T. 1965. Notes on bats from the cape region of Baja California. Trans. San Diego Soc. Nat. Hist., 14:53–56, 1 table, August 4.

JONES, J. K., JR., SMITH, J. D., and GENOWAYS, H. H. 1973. Annotated checklist of mammals of the Yucatán Peninsula, México. I. Chiroptera. Occas. Pap. Mus. Texas Tech Univ., 13:1–31, 2 figs., 1 table, May 16.

JONES, J. K., JR., SMITH, J. D., and TURNER, R. W. 1971. Noteworthy records of bats from Nicaragua, with a checklist of the chiropteran fauna of the country. Univ. Kansas Mus. Nat. Hist., Occas. Pap., 2:1–35, 1 fig., 4 tables, April 29.

JONES, J. K., JR., and STANLEY, W. C. 1962. *Myotis subulatus* in North Dakota. Jour. Mamm., 43:263, May 29.

JONES, J. K., JR., and VAUGHAN, T. A. 1959. The evening bat in Nebraska. Jour. Mamm., 40:246–247, May 20.

JORGENSEN, C. D., and HAYWARD, C. L. 1963. Notes on shrews from southern Nevada. Jour. Mamm., 44:582, December 13.

———. 1965. Mammals of the Nevada Test Site. Brigham Young Univ. Sci. Bull., Biol. Ser., 6:1–81, 48 figs., 20 tables, March.

JUDD, F. W. 1970. Geographic variation in the deer mouse, *Peromyscus maniculatus*, on the Llano Estacado. Southwestern Nat., 14:261–282, January 15.

JUDD, F. W., and SCHMIDLY, D. J. 1969. Distributional notes for some mammals from western Texas and eastern New Mexico. Texas Jour. Sci., 20:381–383, April 25.

KASPAR, T. C., and PARRISH, J. F. III. 1974. The porcupine, *Erethizon dorsatum* Linnaeus, (Rodentia), in Texas. Southwestern Nat., 19:214–215, July 26.

KAUFMANN, J. H., LANNING, D. V., and POOLE, S. E. 1976. Current status and distribution of the coati in the United States. Jour. Mamm., 57:621–637, 4 figs., 1 table, December 10.

KEARNY, S. W. 1912. An expedition across Iowa in 1820. A journal by S. W. Kearny. Annals of Iowa, 10(3rd ser.):343–371, January–April.

KEE, D. T., and ENRIGHT, M. J. 1970. Southern records of *Microtus ochrogaster* in Arkansas. Southwestern Nat., 14:358, January 15.

KEITH, A. R. 1969. The mammals of Martha's Vineyard. Dukes County Intelligencer, 11:47–98, 15 figs., November.

KEITH, L. B., and MESLOW, E. C. 1966. Animals using runways in common with snowshoe hares. Jour. Mamm., 47:541, 1 table, August 29.

KELLOGG, R. 1939. Annotated list of Tennessee mammals. Proc. U.S. Nat. Mus., 86:245–303, February 14.

———. 1946. Three new mammals from the Pearl Islands, Panama. Proc. Biol. Soc. Washington, 59:57–62, March 11.

———. 1946. Mammals of San José Island, Bay of Panamá. Smiths. Miscl. Coll., 106(7):1–4, July 18.

KELSALL, J. P. 1971. A range-extension for the bushy-tailed wood rat. Canadian Field-Nat., 85:326, 1 fig., December.

———. 1972. The northern limits of moose (*Alces alces*) in western Canada. Jour. Mamm., 53:129–138, 1 fig., March 24.

KENNEDY, M. L., RANDOLPH, K. N., and BEST, T. L. 1974. A review of Mississippi mammals. Nat. Sci. Research Inst., Eastern New Mexico Univ., 2:ii + 36, December.

KENNERLY, T. E., JR. 1958. Comparisons of morphology and life history of two species of pocket gophers. Texas Jour. Sci., 10:133–146, 3 figs., 1 table, June.

———. 1959. Contact between the ranges of two allopatric species of pocket gophers. Evolution, 13:247–263, June 5.

KENYON, K. W. 1960. A ringed seal from the Pribilof Islands, Alaska. Jour. Mamm., 41:520–521, November 11.

KILPATRICK, C. W. 1971. Distribution of the brush mouse, *Peromyscus boylii*, and the encinal mouse, *Peromyscus pectoralis*, in north-central Texas. Southwestern Nat., 16:211–213, 1 fig., November 15.

KILPATRICK, C. W., and CAIRE, W. 1973. First record of the encinal mouse, *Peromyscus pectoralis*, for Oklahoma, and additional records for north-central Texas. Southwestern Nat., 18:351, October 5.

KILPATRICK, C. W., and ZIMMERMAN, E. G. 1975. Genetic variation and systematics of four species of mice of the *Peromyscus boylii* species group. Systematic Zool., 24:143–162, 7 figs., 6 tables, September 15.

KING, J. E. 1954. The otariid seals of the Pacific Coast of America. Bull. British Mus. (Nat. Hist.), Zool., 2:309–337, 2 pls., 3 figs., November.

KINSELLA, J. M. 1967. Unusual habitat of the water shrew in western Montana. Jour. Mamm., 48:475–477, 1 fig., August 21.

KIRKLAND, G. L., JR. 1975. Taxonomy and geographic distribution of *Peromyscus maniculatus nubiterrae* Rhoads (Mammalia: Rodentia). Ann. Carnegie Mus., 45:213–229, 1 fig., 8 tables, July 11.

KIRKPATRICK, R. D. 1961. New observations of Indiana swamp rabbits. Jour. Mamm., 42:99–100, February 20.

KIRKPATRICK, R. D., CARTWRIGHT, A. M., BRIER, J. C., and SPICKA, E. J. 1975. Additional mammal records for Belize. Mammalia, 39:330–331, December 20.

KLIMSTRA, W. D., and SCOTT, T. G. 1956. Distribution of the rice rat in southern Illinois. Chicago Acad. Sci. Nat. Hist. Miscl., 154:1–3, 1 fig., April 30.

KLINE, P. D. 1959. Additional mule deer records for Iowa. Jour. Mamm., 40:148–149, February 20.

KLINGENER, D. 1964. Notes on the range of *Napaeozapus* in Michigan and Indiana. Jour. Mamm., 45:644–645, January 25.

KOMAREK, E. V., and KOMAREK, R. 1938. Mammals of the Great Smoky Mountains. Bull. Chicago Acad. Sci., 5:137–162, 6 figs., 1 map, August 15.

KOOPMAN, K. F. 1955. A new subspecies of *Chilonycteris* from the West Indies and a discussion of the mammals of La Gonave. Jour. Mamm., 36:109–113, 1 pl., 1 table, February 28.

———. 1958. Does *Pygoderma* occur in North America? Jour. Mamm., 39:584–585, November 20.

———. 1961. A collection of bats from Sinaloa, with remarks on the limits of the neotropical region in northwestern Mexico. Jour. Mamm., 42:536–537, November 20.

———. 1965. A northern record of the yellow bat. Jour. Mamm., 46:695, November 30.

———. 1968. Taxonomic and distributional notes on Lesser Antillean bats. Amer. Mus. Novit., 2333:1–13, 1 table, July 19.

———. 1974. Eastern limits of *Plecotus* in Mexico. Jour. Mamm., 55:872–873, November 20.

———. 1975. Bats of the Virgin Islands in relation to those of the Greater and Lesser Antilles. Amer. Mus. Novit., 2581:1–7, June 20.

KOOPMAN, K. F., and JONES, J. K., JR. 1970. Classification of bats. Pp. 22–28, *in* About bats: a chiropteran symposium (B. H. Slaughter and D. W. Walton, eds.), Southern Methodist Univ. Press, Dallas, Texas, vii + 339 pp., December.

KOOPMAN, K. F., and RUIBAL, R. 1955. Cave-fossil vertebrates from Camaguey, Cuba. Breviora, Mus. Comp. Zool., 46:1–8, June 24.

KOOPMAN, K. F., and WILLIAMS, E. E. 1951. Fossil Chiroptera collected by H. E. Anthony in Jamaica, 1919–1920. Amer. Mus. Novit., 1519:1–29, 6 figs., 4 tables, June 6.

KOROBITSYNA, K. V., NADLER, C. F., VORONTSOV, N. N., and HOFFMANN, R. S. 1974. Chromosomes of the Siberian snow sheep, *Ovis nivicola*, and implications concerning the origin of Amphiberingian wild sheep (subgenus *Pachyceros*). Quaternary Research, 4:235–245, September.

KORTLUCKE, S. M. 1973. Morphological variation in the kinkajou, *Potos flavus* (Mammalia: Procyonidae), in Middle America. Univ. Kansas Mus. Nat. Hist., Occas. Pap., 17:1–36, 10 figs., 5 tables, October 26.

KRUTZSCH, P. H. 1954. North American jumping mice (genus Zapus). Univ. Kansas Publ., Mus. Nat. Hist., 7:349–472, 47 figs., 4 tables, April 21.

KRUTZSCH, P. H., and HEPPENSTALL, C. A. 1955. Additional distributional records of bats in Utah. Jour. Mamm., 36:126–127, February 28.

KUKALOVÁ-PECK, J. 1973. A phylogenetic tree of the animal kingdom. National Mus. Natural Sciences, Ottawa, Publ. Zool., 8:1–79, 30 charts with facing linear classifications of higher systematic categories.

KUNZ, T. H. 1965. Notes on some Nebraskan bats. Trans. Kansas Acad. Sci., 68:201–203, May 14.

KUNZ, T. H., and SCHLITTER, D. A. 1968. An annotated checklist of bats from Iowa. Trans. Kansas Acad. Sci., 71:166–175, August 19.

KURTÉN, B. 1964. The evolution of the polar bear, Ursus maritimus Phipps. Acta Zool. Fennica, 108:1–30, 4 pls., 4 figs., 9 tables, December.

———. 1966. Pleistocene bears of North America: 1. Genus Tremarctos, spectacled bears. Acta Zool. Fennica, 115:1–120, 24 pls., 64 figs., 37 tables, March.

———. 1967. Pleistocene bears of North America: 2. Genus Arctodus, short-faced bears. Acta. Zool. Fennica, 117:1–60, 30 figs., 30 tables, March.

KURTÉN, B., and RAUSCH, R. 1959. A comparison between Alaskan and Fennoscandian wolverine (Gulo gulo Linnaeus). Acta Arctica, Fasc. 11:5–20, 1 pl., 7 figs., 5 tables.

———. 1959. A comparison between the northern lynxes of Fennoscandia and Alaska. Acta Arctica, Fasc. 11:21–44, 10 figs., 5 tables.

KUYT, E. 1966. White-tailed deer near Fort Smith, N.W.T. Blue Jay, 24:194, December.

———. 1971. Possible occurrence of cougar near Fort Smith, N.W.T. Blue Jay, 29:142–143, September.

LACKEY, J. A. 1967. Biosystematics of *heermanni* group kangaroo rats in southern California. Trans. San Diego Soc. Nat. Hist., 14:313–343, 8 figs., 11 tables, December 13.

———. 1970. Distributional records of bats from Veracruz. Jour. Mamm., 51:384–385, 1 table, May 20.

———. 1977. Small mammals of the St. Lawrence River System. Pp. 189–194, *in* Preliminary Report Biol. Characteristics of the St. Lawrence River System, State Univ. College of Environmental Sci. and Forestry . . . Syracuse, New York 13210, February.

LAHRMAN, F. W. 1971. Moose at Estlin. Blue Jay, 29:141, 1 fig., September.

LAMPE, R. P., JONES, J. K., JR., HOFFMANN, R. S., and BIRNEY, E. C. 1974. The mammals of Carter County, southeastern Montana. Univ. Kansas Mus. Nat. Hist., Occas. Pap., 25:1–39, 1 fig., 2 tables, June 17.

LANGE, A. L. 1956. Woodchuck remains in northern Arizona caves. Jour. Mamm., 37:289–291, June 9.

LANGE, K. I. 1959. Soricidae of Arizona. Amer. Midland Nat., 61:96–108, 8 figs., 1 table, January.

———. 1959. Taxonomy and nomenclature of some pocket gophers from southeastern Arizona. Proc. Biol. Soc. Washington, 72:127–132, November 4.

———. 1960. Mammals of the Santa Catalina Mountains, Arizona. Amer. Midland Nat., 64:436–458, 1 table, October.

LARSON, E. A. 1964. The peninsula chipmunk in Baja California. Jour. Mamm., 45:634, January 25.

LAUGHLIN, J. M., and COOPER, A. L. 1973. A range extension of the kit fox in Oregon. Murrelet, 54:23, October.

LAURIE, E. M. O. 1953. Rodents from British Honduras, Mexico, Trinidad, Haiti and Jamaica collected by Mr. I. T. Sanderson. Ann. Mag. Nat. Hist., ser. 12, 6:382–394, May.

LAVAL, R. K. 1967. Records of bats from the southeastern United States. Jour. Mamm., 48:645–648, November 20.

———. 1969. Records of bats from Honduras and El Salvador. Jour. Mamm., 50:819–822, November 28.

———. 1970. Banding returns and activity periods of some Costa Rican bats. Southwestern Nat., 15:1–10, 1 table, 6 figs., June 1.

———. 1970. Infraspecific relationships of bats of the species *Myotis austroriparius*. Jour. Mamm., 51:542–552, 3 figs., 5 tables, August 28.

———. 1972. Distributional records and band recoveries of bats from Puebla, Mexico. Southwestern Nat., 16:449–451, February 18.

———. 1973a. A revision of the neotropical bats of the genus *Myotis*. Nat. Hist. Mus. Los Angeles Co. Sci. Bull., 15:1–54, 23 figs., 14 tables, February 14.

———. 1973b. Systematics of the genus *Rhogeessa* (Chiroptera: Vespertilionidae). Univ. Kansas Mus. Nat. Hist., Occas. Pap., 19:1–47, 14 figs., 5 tables, November 9.

———. 1977. Notes on some Costa Rican bats. Brenesia, Dept. Hist. Nat. Mus. Nac. Costa Rica, Nos. 10/11:77–83, Marzo.

LAVAL, R. K., and SHIFFLETT, W. A. 1972. *Choeronycteris mexicana* from Texas. Bat Research News, 12:40, February. [Processed Literature.]

LAVROV, L. S., and ORLOV, V. N. 1973. Karyotypes and taxonomy of modern beavers (Castor, Castoridae, Mammalia). Zool. Jour., 52(5):734–742, 6 figs., 2 tables.

LAWLOR, T. E. 1965. The Yucatan deer mouse, Peromyscus yucatanicus. Univ. Kansas Publ., Mus. Nat. Hist., 16:421–438, 2 figs., 2 tables, July 20.

———. 1969. A systematic study of the rodent genus *Ototylomys*. Jour. Mamm., 50:28–42, 5 figs., 3 tables, February 26.

———. 1971. Evolution of *Peromyscus* on northern islands in the Gulf of California, Mexico. Trans. San Diego Soc. Nat. Hist., 16:91–124, 19 figs., 4 tables, February 24.

———. 1971. Distribution and relationships of six species of *Peromyscus* in Baja California and Sonora, Mexico. Occas. Pap. Mus. Zool., Univ. Michigan, 661:1–22, 2 tables, 8 figs., March 1.

LAWRENCE, B. 1934. New *Geocapromys* from the Bahamas. Occas. Pap. Boston Soc. Nat. Hist., 8:189–196, 3 figs., 1 table, November 7.

LAWRENCE, B., and BOSSERT, W. H. 1969. The cranial evidence for hybridization in New England Canis. Breviora, Mus. Comp. Zool., 330:1–13, 5 figs., 2 tables, September 15.

LAY, D. M. 1963. Seis mamíferos nuevos para la fauna de México. Anal. Inst. Biol., Univ. Nac. Autó. México, 33:373–377, February 6.

LAYNE, J. N. 1958. Notes on mammals of southern Illinois. Amer. Midland Nat., 60:219–254, 1 fig., July.

———. 1960. The growth and development of young golden mice, Ochrotomys nuttalli. Quart. Jour., Florida Acad. Sci., 23:36–58, 4 figs., 1 table, March.

———. 1965. Observations on marine mammals in Florida waters. Bull. Florida State Mus., 9:131–181, 4 figs., 10 tables, May 19.

———. 1974. The land mammals of South Florida. Miami Geol. Soc. Mem., 2:386–413, 2 figs., November.

LAYNE, J. N., and SHOOP, C. R. 1971. Records of the water shrew (*Sorex palustris*) and smoky shrew (*Sorex fumeus*) from Rhode Island. Jour. Mamm., 52:215, February 26.

LEATHERWOOD, S., CALDWELL, D. K., and WINN, H. E. 1976. Whales, dolphins, and porpoises of the western North Atlantic a guide to their identification. National Oceanic and Atmospheric Administration, National Marine Fisheries Service, iv + 176 pp., 182 figs., August.

LEE, D. S. 1974. A second pigmy shrew from Maryland. Chesapeake Sci., 15:60, March.

LEE, M. R., and DURRANT, S. D. 1960. A new long-tailed vole (*Microtus longicaudus* (Merriam)) from Utah. Proc. Biol. Soc. Washington, 73:167–170, 1 table, December 30.

———. 1960. A new jumping mouse (*Zapus princeps* Allen) from Utah. Proc. Biol. Soc. Washington, 73:171–174, December 30.

LEE, M. R., and HOFFMEISTER, D. F. 1963. Status of certain fox squirrels in Mexico and Arizona. Proc. Biol. Soc. Washington, 76:181–190, 2 figs., 1 table, August 2.

LEE, M. R., and ZIMMERMAN, E. G. 1969. Robertsonian polymorphism in the cotton rat, *Sigmodon fulviventer*. Jour. Mamm., 50:333–339, 3 figs., 1 table, June 12.

LEOPOLD, A. S. 1959. Wildlife of Mexico—The game birds and mammals. Univ. California Press, xvi+568 pp., 194 figs., 18 tables.

LEOPOLD, A. S., and HALL, E. R. 1945. Some mammals of Ozark County, Missouri. Jour. Mamm., 26:142–145, July 19.

LESOWSKI, J. 1963. Two observations of cougar cannibalism. Jour. Mamm., 44:586, December 13.

LEWIS, H. F., and DOUTT, J. K. 1942. Records of the Atlantic walrus and the polar bear in or near the northern part of the Gulf of St. Lawrence. Jour. Mamm., 23:365–375, 1 fig., December 30.

LEWIS, J. C. 1969. Evidence of mountain lions in the Ozarks and adjacent areas, 1948–1968. Jour. Mamm., 50:371–372, June 12.

LIBBY, W. L. 1959. The distribution of some small mammals in eastern interior Alaska. Jour. Mamm., 40:607–609, November 20.

LIDICKER, W. Z., JR. 1960. An analysis of intraspecific variation in the kangaroo rat Dipodomys merriami. Univ. California Publ. Zool., 67:125–218, Pls. 9–12, 20 figs., August 4.

———. 1960. A new subspecies of the cliff chipmunk from central Chihuahua. Proc. Biol. Soc. Washington, 73:267–273, 3 tables, December 30.

———. 1961. Two southern range extensions in Chihuahua. Jour. Mamm., 42:281, May 20.

———. 1971. Corrections and additions to our knowledge of the pocket gopher Thomomys mazama pugetensis. Murrelet, 52:12–13, April.

LINDSAY, D. M. 1960. Mammals of Ripley and Jefferson counties, Indiana. Jour. Mamm., 41:253–262, May 20.

LINSLEY, J. H. 1842. A catalogue of the mammalia of Connecticut, arranged according to their natural families. Amer. Jour. Sci., 43:345–354, October.

LINZEY, A. V., and LAYNE, J. N. 1969. Comparative morphology of the male reproductive tract in the rodent genus Peromyscus (Muridae). Amer. Mus. Novit., 2355:1–47, 11 figs., 3 tables, January 27.

LINZEY, A. V., and LINZEY, D. W. 1967. Microtus pennsylvanicus in North Carolina and Tennessee. Jour. Mamm., 48:310, May 20.

LINZEY, A. V., LINZEY, D. W., and PERKINS, S. E., JR. 1976. The Peromyscus leucopus species group in Alabama. Jour. Alabama Acad. Sci., 47:109–113, 1 fig., 1 table, April.

LINZEY, D. W., and LINZEY, A. V. 1967. Maturational and seasonal molts in the golden mouse, Ochrotomys nuttalli. Jour. Mamm., 48:236–241, 1 fig., 1 table, May 20.

———. 1967. Growth and development of the golden mouse, Ochrotomys nuttalli nuttalli. Jour. Mamm., 48:445–458, 8 figs., 3 tables, August 21.

———. 1969. First record of the yellow bat in Alabama. Jour. Mamm., 50:845, November 28.

LONG, C. A. 1961. Woodchuck in Oklahoma and southeastern Kansas. Jour. Mamm., 42:255–256, May 20.

———. 1961. Reithrodontomys montanus griseus in Missouri. Jour. Mamm., 42:417–418, August 21.

———. 1963. Taxonomic comments on some aquatic mammals of San Juan Island, Washington. Murrelet, 44:25–26, December.

———. 1964. The baculum in pocket gophers of southwestern Wyoming. Trans. Kansas Acad. Sci., 66:754–756, 1 fig., 1 table, February 25.

———. 1965a. The mammals of Wyoming. Univ. Kansas Publ., Mus. Nat. Hist., 14:493–758, 82 figs., July 6.

———. 1965b. Fulvous harvest mouse in Missouri. Jour. Mamm., 46:506, August 26.

———. 1971. A new subspecies of chipmunk from the Door Peninsula, Wisconsin (Mammalia: Rodentia). Proc. Biol. Soc. Washington, 84:201–202, November 4.

———. 1972. Taxonomic revision of the mammalian genus Microsorex Coues. Trans. Kansas Acad. Sci., 74:181–196, 1 fig., 1 table, April 7.

———. 1972. Taxonomic revision of the North American badger, Taxidea taxus. Jour. Mamm., 53:725–759, 11 figs., 4 tables, November 30.

———. 1976. Microtus ochrogaster minor in Wisconsin. Univ. Wisconsin Mus. Nat. Hist., Reports on the Fauna and Flora of Wisconsin, 11:1, processed literature, May 20.

———. 1976. The occurrence, status and importance of bats in Wisconsin with a key to the species. Trans. Wisconsin Acad. Sci., Arts and Letters, 64:62–82, 9 figs., 1 table.

LONG, C. A., and COPES, F. A. 1968. Note on the rate of dispersion of the opossum in Wisconsin. Amer. Midland Nat., 80:283–284, July 25.

LONG, C. A., and HAYS, H. A. 1962. Yellow mutant in the muskrat, Ondatra zibethicus. Jour. Mamm., 43:104, February 28.

LONG, C. A., and LONG, C. F. 1964a. Geographic records of the swift fox, Vulpes velox. Southwestern Nat., 9:108, May 15.

———. 1964b. Range extension of the gray fox, Urocyon cinereoargenteus. Southwestern Nat., 9:108–109, May 15.

LONG, C. A., and SEVERSON, R. G. 1969. Geographical variation in the big brown bat in the north-central United States. Jour. Mamm., 50:621–624, 1 fig., August 22.

LOOMIS, R. B., and DAVIS, R. M. 1965. The vampire bat in Sonora, with notes on other bats from southern Sonora. Jour. Mamm., 46:497, August 26.

LOOMIS, R. B., and JONES, J. K., JR. 1964. The northern yellow bat in Sinaloa, Mexico. Bull. Southern California Acad. Sci., 63:32, April 21.

LOOMIS, R. B., and STEPHENS, R. C. 1962. Marmosa canescens in Sonora, Mexico. Jour. Mamm., 43:111, February 28.

———. 1962. Neotoma phenax in Sinaloa, Mexico. Jour. Mamm., 43:112, February 28.

———. 1965. Cliff chipmunk from the vicinity of Guaymas, Sonora. Jour. Mamm., 46:501, August 26.

LÓPEZ-F., W., SÁNCHEZ-H., C., and VILLA-R., B. 1973. Algunos mamíferos de la región de Chamela, Jalisco, México. Anal. Inst. Biol., Univ. Nac. Autó. México, Ser. Zool., 42:99–106, January 31.

LOWE, C. E. 1958. Ecology of the swamp rabbit in Georgia. Jour. Mamm., 39:116–127, 2 figs., 6 tables, February 20.

LOWERY, G. H., JR. 1936. A preliminary report on the distribution of the mammals of Louisiana. Proc. Louisiana Acad. Sci., 3:11–39, 4 pls., 2 figs., March.

———. 1943. Check-list of the mammals of Louisiana and adjacent waters. Occas. Pap. Mus. Zool., Louisiana State Univ., 13:213–257, 5 figs., 4 tables, November 22.

———. 1974. The mammals of Louisiana and its adjacent waters. Illustrated by H. D. Pratt. Louisiana State Univ. Press, published for the Louisiana Wild Life and Fisheries Commission, xxiii + 565 pp., 14 colored pls., 247 numbered figs., 70 maps, 86 unnumbered illustrations, 10 tables, June 7.

LYNCH, G. M. 1967. Long-range movement of a raccoon in Manitoba. Jour. Mamm., 48:659–660, 1 fig., November 20.

———. 1971. Raccoons increasing in Manitoba. Jour. Mamm., 52:621–622, 1 table, August 26.

MACHADO-ALLISON, C. E. 1967. The systematic position of the bats Desmodus and Chilonycteris, based on host-parasite relationships (Mammalia; Chiroptera). Proc. Biol. Soc. Washington, 80:223–226, December 1.

MacLeod, C. F., and Cameron, A. W. 1961. Distribution records for mammals in west-central Quebec. Jour. Mamm., 42:281–282, May 20.

Macpherson, A. H. 1964. A northward range extension of the red fox in the eastern Canadian Arctic. Jour. Mamm., 45:138–140, 1 fig., March 12.

Maher, W. J. 1960. Recent records of the California grey whale (Eschrichtius glaucus) along the north coast of Alaska. Arctic, 13:257–265, 2 figs., 1 table.

———. 1972. Two new records of long-eared myotis in Saskatchewan. Blue Jay, 30:236, December.

Maher, W. J., and Wilimovsky, N. J. 1963. Annual catch of bowhead whales by Eskimos at Point Barrow, Alaska, 1928–1960. Jour. Mamm., 44:16–20, 2 tables, February 20.

Manning, T. H. 1943. Notes on the mammals of south and central West Baffin Island. Jour. Mamm., 24:47–59, 1 fig., February 20.

———. 1971. Geographical variation in the polar bear Ursus maritimus Phipps. Canadian Wildlife Serv. Report Ser., 13:1–27, frontispiece, 2 figs., 10 tables, received at Univ. Kansas May 24.

Manning, T. H., and Macpherson, A. H. 1958. The mammals of Banks Island. Arctic Inst. of North America, Tech. Pap. 2, Pp. 1–74, 15 figs., 1 map, September 18.

———. 1961. A biological investigation of Prince of Wales Island, N.W.T. Trans. Royal Canadian Inst., 33:116–239, 13 figs., 20 tables, October.

Manville, R. H. 1950. The mammals of Drummond Island, Michigan. Jour. Mamm., 31:358–359, August 21.

———. 1960. Recent changes in the mammal fauna of Mount Desert Island, Maine. Jour. Mamm., 41:415–416, August 15.

———. 1961. The entepicondylar foramen and Ochrotomys. Jour. Mamm., 42:103–104, February 20.

———. 1961. Notes on behavior of marten. Jour. Mamm., 42:112, February 20.

———. 1961. Notes on some mammals of the Gaspé Peninsula, Quebec. Canadian Field-Nat., 75:108–109, June 9.

Manville, R. H., and Favour, P. G., Jr. 1960. Southern distribution of the Atlantic walrus. Jour. Mamm., 41:499–503, 1 fig., November 11.

Manville, R. H., and Young, S. P. 1965. Distribution of Alaska mammals. Bur. Sport Fisheries and Wildlife, circular 211, iv + 74 pp., 62 pls., 63 figs.

Marshall, L. G., and Weisenberger, G. J. 1971. A new dwarf shrew locality for Arizona. Plateau, 43:132–137, 2 figs., winter.

Martell, A. M. 1974a. A northern range extension for the bushy-tailed wood rat, Neotoma cinerea (Ord). Canadian Field-Nat., 88:348, 1 fig., October 31.

———. 1974b. A northern range extension for the northern bog lemming, Synaptomys borealis borealis (Richardson). Canadian Field-Nat., 88:348, October 31.

———. 1975. Taxonomic status of Microtus pennsylvanicus arcticus Cowan. Jour. Mamm., 56:255–257, February 20.

Marti, C. D. 1972. Notes on the least shrew in Colorado. Southwestern Nat., 16:447–448, February 18.

Martin, R. A. 1971. New records of the dwarf shrew from South Dakota. Jour. Mamm., 52:835–836, December 16.

Martin, R. E., and Matocha, K. G. 1972. Distributional status of the kangaroo rat, Dipodomys elator. Jour. Mamm., 53:873–877, 2 figs., November 30.

Martin, R. E., and Preston, J. R. 1970. The mammals of Harmon County, Oklahoma. Proc. Oklahoma Acad. Sci., 49:42–60, 2 figs.

Martin, R. L. 1966. Redescription of the type locality of Sorex dispar. Jour. Mamm., 47:130–131, March 12.

Maser, C. 1974. The sage vole, Lagurus curtatus (Cope, 1868), in the Crooked River National Grassland, Jefferson County, Oregon. A contribution to its life history and ecology. Säugetierkund. Mitteil., 22:193–222, 17 figs., 2 tables, September.

Maser, C., and Johnson, M. L. 1968. Notes on the white-footed vole (Phenacomys albipes). Murrelet, 48:23–27, 1 table, 1 map, March.

Matson, J. O., and Patten, D. R. 1975. Notes on some bats from the state of Zacatecas, Mexico. Los Angeles Co. Mus. Contrib. Sci., 263:1–12, 2 tables, May 16.

Maxwell, M. H., and Brown, L. N. 1968. Ecological distribution of rodents on the high plains of eastern Wyoming. Southwestern Nat., 13:143–158, 5 tables, September 10.

McCarley, [W.] H. 1959. The mammals of eastern Texas. Texas Jour. Sci., 11:385–426, 2 figs., 2 tables, December.

———. 1960. The rice rat in southern Oklahoma. Jour. Mamm., 41:130–131, February 20.

———. 1961. New locality records for some Oklahoma mammals. Southwestern Nat., 6:108–109, September 1.

———. 1966. Annual cycle, population dynamics and adaptive behavior of Citellus tridecemlineatus. Jour. Mamm., 47:294–316, 3 figs., 8 tables, May 26.

McCarley, [W.] H., and Free, P. 1962. A new record of the woodchuck from Oklahoma. Jour. Mamm., 43:271–272, May 29.

McCoy, C. J., Jr., and Miller, P. H. 1964. Ecological distribution of the subspecies of Ammospermophilus leucurus in Colorado. Southwestern Nat., 9:89–93, 1 table, May 15.

McCulloch, C. Y. 1967. Recent records of white-tailed deer in northern Arizona. Southwestern Nat., 12:482–484, 1 table, December 31.

McCullough, D. R. 1969. The tule elk: its history, behavior, and ecology. Univ. California Publ. Zool., 88:vii + 209 pp., frontispiece, 8 pls., 41 figs., 30 tables, June 5.

McCully, H. 1967. The broad-handed mole, Scapanus latimanus, in a marine littoral environment. Jour. Mamm., 48:480–482, 1 fig., August 21.

McDaniel, L. L. 1967. Merriam's shrew in Nebraska. Jour. Mamm., 48:493, August 21.

McGee, L. E. 1965. Extension of the range eastward for Taxidea taxus (Carnivora: Mustelidae). Southwestern Nat., 10:78, March 15.

McKeever, S. 1960. Food of the northern flying squirrel in northeastern California. Jour. Mamm., 41:270–271, May 20.

McKenna, M. C. 1975. Toward a phylogenetic classification of the Mammalia, in Phylogeny of the Primates, Plenum Publ. Corp., New York, 2:21–46, 3 figs., December.

McLaren, I. A. 1966. Taxonomy of harbor seals of the western North Pacific and evolution of certain other hair seals. Jour. Mamm., 47:466–473, August 29.

McLaughlin, C. A. 1958. A new race of pocket gopher Geomys bursarius from Missouri. Los Angeles Co. Mus. Contrib. Sci., 19:1–4, January 29.

McPherson, A. B. 1972. A new subspecies of the squirrel

Sciurus variegatoides Ogilby from Costa Rica. Rev. Biol. Trop., 19:191–194, February 22.

MEDEIROS, J. L., and HECKMANN, R. A. 1971. *Euderma maculatum* from California infected with rabies virus. Jour. Mamm., 52:858, December 16.

MERRIAM, C. H. 1892. Description of a new genus and species of murine rodent (*Xenomys nelsoni*) from the state of Colima, western Mexico. Proc. Biol. Soc. Washington, 7:159–163, September 29.

———. 1897. Notes on the chipmunks of the genus *Eutamias* occurring west of the east base of the Cascade–Sierra System, with descriptions of new forms. Proc. Biol. Soc. Washington, 11:189–212, July 1.

———. 1897. *Nelsonia neotomodon*, a new genus and species of murine rodent from Mexico. Proc. Biol. Soc. Washington, 11:277–279, Figs. 14–15, December 17.

———. 1901. Seven new mammals from Mexico, including a new genus of rodents. Proc. Washington Acad. Sci., 3:559–563, November 29.

———. 1903. Four new mammals, including a new genus (*Teanopus*), from Mexico. Proc. Biol. Soc. Washington, 16:79–82, May 29.

MEYERRIECKS, A. J. 1963. Florida otter preys on common gallinule. Jour. Mamm., 44:425–426, August 22.

MEYLAN, A., and HAUSSER, J. 1973. Les chromosomes des *Sorex* du groupe *araneus-arcticus* (Mammalia, Insectivora). Zeitschr. f. Säugetierkunde, 38:143–158, 7 figs., 1 table, June.

MICHAEL, E. D., and BIRCH, J. B. 1967. First Texas record of *Plecotus rafinesquii*. Jour. Mamm., 48:672, November 20.

MICHAEL, E. D., WHISENNAND, R. L., and ANDERSON, G. 1970. A recent record of *Myotis austroriparius* from Texas. Jour. Mamm., 51:620, August 28.

MICKEY, A. B. 1961. Record of the spotted bat from Wyoming. Jour. Mamm., 42:401–402, August 21.

MILLER, A. H., and STEBBINS, R. C. 1964. The lives of desert animals in Joshua Tree National Monument. Univ. California Press, vi + 452, frontispiece, 7 pls., 141 figs.

MILLER, D. H. 1964. Northern records of the pine mouse in Vermont. Jour. Mamm., 45:627–628, January 25.

MILLER, G. S., JR. 1896. The genera and subgenera of voles and lemmings. N. Amer. Fauna, 12:1–84, 3 pls., 40 figs., July 23.

———. 1899. Preliminary list of the mammals of New York. Bull. New York State Mus., 6:271–390, several tables, November 18.

———. 1900. Three new bats from the Island of Curaçao. Proc. Biol. Soc. Washington, 13:123–127, April 6.

———. 1902. Twenty new American bats. Proc. Acad. Nat. Sci. Philadelphia, pp. 389–412, September 12.

———. 1904. Notes on the bats collected by William Palmer in Cuba. Proc. U.S. Nat. Mus., 27:337–348, Pl. 9, 5 tables, January 23.

———. 1907. The families and genera of bats. Bull. U.S. Nat. Mus., 57:xvii + 282, 14 pls., 49 figs., June 29.

———. 1912. Catalogue of the mammals of western Europe . . . . British Mus. (Nat. Hist.). xv + 1019 pp., 213 figs., November 23.

———. 1913. Notes on the bats of the genus Molossus. Proc. U.S. Nat. Mus., 46:85–92, August 23.

———. 1916. Bones of mammals from Indian sites in Cuba and Santo Domingo. Smiths. Miscl. Coll., 66(12):1–10. 1 pl., December 7.

———. 1917. A hooded seal in Florida. Proc. Biol. Soc. Washington, 30:121, July 27.

———. 1918. Three new bats from Haiti and Santo Domingo. Proc. Biol. Soc. Washington, 31:39–40, May 16.

———. 1918. Mammals and reptiles collected by Theodoor de Booy in the Virgin Islands. Proc. U.S. Nat. Mus., 54:507–511, Pl. 81, October 15.

———. 1922. Remains of mammals from caves in the Republic of Haiti. Smiths. Miscl. Coll., 74(3):1–8, October 16.

———. 1924a. List of North American Recent mammals, 1923. Bull. U.S. Nat. Mus., 128:xvi + 673, April 29.

———. 1924b. A Pollack whale from Florida presented to the National Museum by the Miami Aquarium Association. Proc. U.S. Nat. Mus., 66(art. 9):1–15, 22 pls.

———. 1927. A Pollack whale on the coast of Virginia. Proc. Biol. Soc. Washington, 40:111–112, September 26.

———. 1928. The Pollack whale in the Gulf of Campeche. Proc. Biol. Soc. Washington, 41:171, October 15.

———. 1929a. A second collection of mammals from caves near St. Michel, Haiti. Smiths. Miscl. Coll., 81(9):1–30, 10 pls., March 30.

———. 1929b. The characters of the genus Geocapromys Chapman. Smiths. Miscl. Coll., 82(4):1–3, 1 pl., December 9.

———. 1929c. Mammals eaten by Indians, owls, and Spaniards in the coast region of the Dominican Republic. Smiths. Miscl. Coll., 82(5):1–16, 2 pls., December 11.

———. 1931. The red bats of the Greater Antilles. Jour. Mamm., 12:409–410, November 11.

MILLER, G. S., JR., and ALLEN, G. M. 1928. The American bats of the genera Myotis and Pizonyx. Bull. U.S. Nat. Mus., 144:viii + 218, 1 pl., 1 fig., 13 maps, numerous tables, May 25.

MILLER, G. S., JR., and GIDLEY, J. W. 1918. *Synopsis of the supergeneric groups of rodents.* Jour. Washington Acad. Sci., 8:431–448, July 19.

MILLER, G. S., JR., and KELLOGG, R. 1955. List of North American Recent mammals. Bull. U.S. Nat. Mus., 205:xii + 954, March 3.

MILLER, P. H., and McCoy, C. J., JR. 1965. Kit fox in Colorado. Jour. Mamm., 46:342–343, May 20.

MILLS, R. S. 1971. Second record of the Mexican free-tailed bat in Ohio. Jour. Mamm., 52:479, May 28.

MILSTEAD, W. W., and TINKLE, D. W. 1959. Seasonal occurrence and abundance of bats (Chiroptera) in northwestern Texas. Southwestern Nat., 4:134–142, October 24.

MITCHELL, E. 1968. The Mio-Pliocene pinniped *Imagotaria*. Jour. Fisheries Research Board Canada, 25:1843–1900, 16 figs., 5 tables, September.

———. [Special Editor.] 1975. Report of the meeting on smaller cetaceans Montreal, April 1–11, 1974. Jour. Fisheries Research Board Canada, 32 (No. 7, special issue):887–983, 43 figs., July.

MITCHELL, E., and KOZICKI, V. M. 1975. Autumn stranding of a northern bottlenose whale (*Hyperoodon ampullatus*) in the Bay of Fundy, Nova Scotia. Jour. Fisheries Research Board Canada, 32(No. 7, special issue):1019–1040, 7 figs., 5 tables, July.

MITCHELL, G. C. 1965. First record of *Eumops underwoodi*

from the state of Jalisco, Mexico. Jour. Mamm., 46:100, February 20.

MOHLHENRICH, J. S. 1961. Distribution and ecology of the hispid and least cotton rats in New Mexico. Jour. Mamm., 42:13–24, 3 figs., February 20.

MOHR, E. 1939. Die Baum- und Ferkelratten-Gattungen Capromys Desmarest (sens. ampl.) und Plagiodontia Cuvier. Mitteil. Hamburgischen Zool. Mus. und Inst. in Hamburg, 48:48–118, 53 figs., several tables, May.

MOLLHAGEN, T. 1971. Checklist of bats in caves in the regions of the Sierra de Guatemala and Sierra de El Abra, northeastern Mexico. Assoc. Mexican Cave Studies, Bull., 4:19–22.

———. 1973. Distributional and taxonomic notes on some west Texas bats. Southwestern Nat., 17:427–430, January 10.

MOLLHAGEN, T., and BAKER, R. H. 1972. Myotis volans interior in Knox County, Texas. Southwestern Nat., 17:97, May 1.

MOOJEN, J. 1948. Speciation in the Brazilian spiny rats (genus Proechimys, family Echimyidae). Univ. Kansas Publ., Mus. Nat. Hist., 1:301–406, 140 figs., 1 table, December 10.

MOORE, J. C. 1945. Life history notes on the Florida weasel. Proc. Florida Acad. Sci. (for 1944), 7:247–263.

———. 1953. Distribution of marine mammals to Florida waters. Amer. Midland Nat., 49:117–158, 19 figs., January.

———. 1956. Variation in the fox squirrel in Florida. Amer. Midland Nat., 55:41–65, 7 figs., 7 tables, January.

———. 1958. A beaked whale from the Bahama Islands and comments on the distribution of Mesoplodon densirostris. Amer. Mus. Novit., 1897:1–12, 6 figs., 2 tables, July 22.

———. 1959. Relationships among the living squirrels of the Sciurinae. Bull. Amer. Mus. Nat. Hist., 118:153–206, 7 figs., 3 tables, July 27.

———. 1960. The relationships of the gray squirrel, Sciurus carolinensis, to its nearest relatives. Southeastern Assoc. Game and Fish Commissioners, 13th Ann. Conference, pp. 356–363.

———. 1966. Diagnoses and distributions of beaked whales of the genus Mesoplodon known from North American waters. Pp. 32–61, 12 figs., in Whales, dolphins, and porpoises, Univ. California Press, pp. xv + 789, illustrated. Edited by Kenneth S. Norris.

———. 1968. Relationships among the living genera of beaked whales with classifications, diagnoses and keys. Fieldiana, Zool., 53:iv + 209–298 pp., 24 figs., 1 table, December 13.

MOORE, J. C., and CLARK, E. 1963. Discovery of right whales in the Gulf of Mexico. Science, 141:269, July 19.

MOORE, J. C., and GILMORE, R. M. 1965. A beaked whale new to the Western Hemisphere. Nature, 205(4977):1239–1240, March 20.

MORAN, R. J., and OZOGA, J. J. 1965. Elk calf pursued by coyotes in Michigan. Jour. Mamm., 46:498, August 26.

MOREJOHN, G. V. 1969. Evidence of river otter feeding on freshwater mussels and range extension. California Fish and Game, 55:83–84, 1 fig., January.

MOREJOHN, G. V., LOEB, V., and BALTZ, D. M. 1973. Coloration and sexual dimorphism in the Dall porpoise. Jour. Mamm., 54:977–982, 3 figs., 1 table, December 14.

MOREJOHN, G. V., and RICE, D. W. 1973. First record of Bryde's whale (Balaenoptera edeni) off California. California Fish and Game, 59:313–315, October [received December 12].

MORSE, R. C., and GLASS, B. P. 1960. The taxonomic status of Antrozous bunkeri. Jour. Mamm., 41:10–15, 2 figs., February 20.

MÖRZER BRUYNS, W. F. J. 1971. Field guide of whales and dolphins. uitgeverij tor/n. v. uitgeverij v.h.c.a. mees zieseniskade 14^II Amsterdam, Netherlands (ISBN 90 70055 090), 258 pp., 96 figs. (colored), 19 charts, 3 graphs.

MUIR, T. J., and POLDER, E. 1960. Notes on hibernating bats in Dubuque County caves. Proc. Iowa Acad. Sci., 67:602–606, December 27.

MUMFORD, R. E. 1969. Distribution of the mammals of Indiana. Indiana Acad. Sci., Monog., 1:vii + 114, 17 figs., 3 tables, October 1.

MUMFORD, R. E., OAKLEY, L. L., and ZIMMERMAN, D. A. 1965. June bat records from Guadalupe Canyon, New Mexico. Southwestern Nat., 9:43–45, April 10.

MUMFORD, R. E., and ZIMMERMAN, D. A. 1962. Notes on Choeronycteris mexicana. Jour. Mamm., 43:101–102, 1 pl., February 28.

MUNRO, J. A. 1958. Distribution of the harvest mouse in British Columbia. Canadian Field-Nat., 72:146, August 1.

MURIE, J. O., and DICKINSON, D. 1973. Behavioral interactions between two species of red-backed vole (Clethrionomys) in captivity. Canadian Field-Nat., 87:123–129, 1 fig., 2 tables, June 4.

MURIE, O. J. 1935. Alaska-Yukon caribou. N. Amer. Fauna, 54:1–93, 10 pls., 16 figs., 4 tables, August 8.

———. 1959. Fauna of the Aleutian Islands and Alaska Peninsula. N. Amer. Fauna, 61:xiv + 364 pp., 43 figs., 2 tables, November 4.

MURRAY, K. F., and BARNES, A. M. 1969. Distribution and habitat of the woodrat, Neotoma fuscipes, in northeastern California. Jour. Mamm., 50:43–48, 1 fig., February 26.

MUSSER, G. G. 1961. A new subspecies of flying squirrel (Glaucomys sabrinus) from southwestern Utah. Proc. Biol. Soc. Washington, 74:119–125, 1 table, August 11.

———. 1964. Notes on geographic distribution, habitat, and taxonomy of some Mexican mammals. Occas. Pap. Mus. Zool., Univ. Michigan, 636:1–22, 1 fig., June 17.

———. 1968. A systematic study of the Mexican and Guatemalan gray squirrel, Sciurus aureogaster F. Cuvier (Rodentia: Sciuridae). Misc. Publ. Mus. Zool., Univ. Michigan, 137:1–112, 34 figs., 6 tables, November 1.

———. 1969. Notes on Peromyscus (Muridae) of Mexico and Central America. Amer. Mus. Novit., 2357:1–23, 3 figs., 2 tables, January 27.

MUUL, I. 1969. Photoperiod and reproduction in flying squirrels, Glaucomys volans. Jour. Mamm., 50:542–549, 1 fig., 3 tables, August 22.

MUUL, I., and CARLSON, F. W. 1963. Red-back vole in trees. Jour. Mamm., 44:415–416, August 22.

NADER, I. A. 1965. Two new subspecies of kangaroo rats, genus Dipodomys. Proc. Biol. Soc. Washington, 78:49–54, 1 fig., July 21.

NADLER, C. F. 1966. Chromosomes and systematics of the American ground squirrels of the subgenus Spermophilus. Jour. Mamm., 47:579–596, 9 figs., 4 tables, December 2.

————. 1968. The chromosomes of *Spermophilus townsendi* (Rodentia: Sciuridae) and report of a new subspecies. Cytogenetics, 7:144–157, 5 figs., 2 tables, April.

NADLER, C. F., and HOFFMANN, R. S. 1977. Patterns of evolution and migration in the arctic ground squirrel, *Spermophilus parryii* (Richardson). Canadian Jour. Zool., 55:748–758, 2 figs., 2 tables, April.

NADLER, C. F., HOFFMANN, R. S., and GREER, K. R. 1971. Chromosomal divergence during evolution of ground squirrel populations (Rodentia: *Spermophilus*). Systematic Zool., 20:298–305, 3 figs., September.

NADLER, C. F., and HUGHES, C. E. 1966. Chromosomes and taxonomy of the ground squirrel subgenus *Ictidomys*. Jour. Mamm., 47:46–53, 5 figs., 1 table, March 12.

NADLER, C. F., VORONTSOV, N. N., HOFFMANN, R. S., FORMICHOVA, I. I., and NADLER, C. F., JR. 1973. Zoogeography of transferrins in arctic and long-tailed ground squirrel populations. Comp. Biochem. Physiol., 44B:33–40, 2 figs., 1 table, prior to March 21.

NADLER, C. F., and YOUNGMAN, P. M. 1969. Transferrin polymorphism among populations of the arctic ground squirrel, *Spermophilus undulatus* (Pallas). Canadian Jour. Zool., 47:1051–1057, 5 figs., prior to October 28.

NAGORSEN, D. W., and PETERSON, R. L. 1975. Karyotypes of six species of bats (Chiroptera) from the Dominican Republic. Royal Ontario Mus., Life Sci. Occas. Pap., 28:1–8, 2 figs., August 25.

NECKER, W. L., and HATFIELD, D. M. 1941. Mammals of Illinois. Bull. Chicago Acad. Sci., 6:17–60, 15 figs., May 15.

NEGUS, N. C., and DUNDEE, H. A. 1965. The nest of *Sorex longirostris*. Jour. Mamm., 46:495, August 26.

NELSON, A. L. 1934. Notes on Wisconsin mammals. Jour. Mamm., 15:252–253, August 10.

NELSON, E. W. 1899. Revision of the squirrels of Mexico and Central America. Proc. Washington Acad. Sci., 1:15–110, 2 pls., May 9.

————. 1909. The rabbits of North America. N. Amer. Fauna, 29:1–314, 13 pls., 19 figs., numerous tables, August 31.

NELSON, E. W., and GOLDMAN, E. A. 1929. Six new pocket mice from Lower California and notes on the status of several described species. Proc. Biol. Soc. Washington, 42:103–111, March 25.

————. 1930. *A new pocket mouse from southern Lower California*. Jour. Washington Acad. Sci., 20:223–224, June 19.

————. 1933. Revision of the jaguars. Jour. Mamm., 14:221–240, 1 table, August 17.

NERO, R. W. 1958. Additional pocket mouse records. Blue Jay, 16:176–179, December.

————. 1960. Short-tailed shrew north of the North Saskatchewan River. Blue Jay, 18:41–42, 2 figs., March.

————. 1964. Sagebrush vole range extension and other records. Blue Jay, 22:119–120, September.

————. 1965. Recent pocket mouse records for Saskatchewan. Blue Jay, 23:36–38, 1 table, March.

————. 1974. Cougars in Manitoba. Blue Jay, 32:55–56, March.

NEWBY, F. E., and McDOUGAL, J. J. 1964. Range extension of the wolverine in Montana. Jour. Mamm., 45:485–487, 1 fig., 1 table, September 15.

NEWBY, F. E., and WRIGHT, P. L. 1955. Distribution and status of the wolverine in Montana. Jour. Mamm., 36:248–253, 1 fig., 3 tables, May 26.

NEWSOM, W. M. 1937. Mammals on Anticosti Island. Jour. Mamm., 18:435–442, November 22.

NORRIS, K. S., and McFARLAND, W. N. 1958. A new harbor porpoise of the genus *Phocoena* from the Gulf of California. Jour. Mamm., 39:22–39, 4 pls., 2 tables, February 20.

NOVICK, A. 1960. Successful breeding in captive *Artibeus*. Jour. Mamm., 41:508–509, November 11.

————. 1963. Orientation in neotropical bats. II. Phyllostomatidae and Desmodontidae. Jour. Mamm., 44:44–56, 2 pls., February 20.

NUGENT, R. F., and CHOATE, J. R. 1970. Eastward dispersal of the badger, *Taxidea taxus*, into the northeastern United States. Jour. Mamm., 51:626–627, August 28.

O'FARRELL, M. J., and BRADLEY, W. G. 1969. A new bat record, *Plecotus phyllotis*, from Nevada. Jour. Mamm., 50:128, February 26.

O'FARRELL, M. J., BRADLEY, W. G., and JONES, G. W. 1967. Fall and winter bat activity at a desert spring in southern Nevada. Southwestern Nat., 12:163–171, 2 figs., 4 tables, August 4.

OGNEV, S. I. 1944. Contribution to the taxonomy of *Microtus ratticeps* Keyserl. et Blas. Comptes Rendus (Doklady) Acad. Sci. U.R.S.S, Nouv. Sér., 44:165–168, August 10.

————. 1948. The mammals of the U.S.S.R. and adjacent countries. . . . Publ. Acad. Sci. U.S.S.R., 6:1–559, illustrated. [In Russian.]

————. 1950. The mammals of the U.S.S.R. and adjacent countries. . . . Publ. Acad. Sci. U.S.S.R., 7:1–706, illustrated. [In Russian.]

————. 1964. Mammals of the U.S.S.R. and adjacent countries. U.S. Dept. Commerce, Washington, D.C., 7:ix + 1–626, 10 pls., 347 figs., 15 maps. [English translation of Ognev, 1950.]

OJASTI, J., and LINARES, O. J. 1972. Adiciónes a la fauna de murciélagos de Venezuela con notas sobre las especies del genero *Diclidurus* (Chiroptera). Acta Biol. Venezuelica, 7:421–441, 1 fig., 2 tables, May.

OLSEN, R. W. 1969. Agonistic behavior of the short-tailed shrew (*Blarina brevicauda*). Jour. Mamm., 50:494–500, 2 figs., August 22.

OLTERMAN, J. H., and VERTS, B. J. 1972. Endangered plants and animals of Oregon. IV. *Mammals*. Agricultural Experiment Station, Oregon State Univ., Special Report, 364:1–47, 38 figs., 2 tables, December.

OMURA, H., OHSUMI, S., NEMOTO, T., NASU, K., and KASUYA, T. 1969. Black right whales in the North Pacific. Sci. Reports Whales Research Inst., No. 21:1–78, 18 pls., 27 figs., 42 tables, June.

ORR, R. T. 1940. The rabbits of California. Occas. Pap. California Acad. Sci., 19:1–227, 10 pls., 30 figs., May 25.

————. 1963. A northern record for the Pacific bottlenose dolphin. Jour. Mamm., 44:424, August 22.

————. 1965. Longevity in *Pizonyx vivesi*. Jour. Mamm., 46:497, August 26.

————. 1966. Cuvier's beaked whale in the Gulf of California. Jour. Mamm., 47:339, May 26.

————. 1966. Risso's dolphin on the Pacific Coast of North America. Jour. Mamm., 47:341–343, 1 fig., May 26.

————. 1967. A second specimen of *Ziphius cavirostris* from Mexico. Jour. Mamm., 48:328, May 20.

ORR, R. T., and BANKS, R. C. 1964. Bats from islands in the

Gulf of California. Proc. California Acad. Sci., 30:207–210, December 31.

ORR, R. T., and SILVA-T., G. 1960. A new species of bat of the genus *Antrozous* from Cuba. Proc. Biol. Soc. Washington, 73:83–86, 1 table, August 10.

OSGOOD, W. H. 1900a. Revision of the pocket mice of the genus Perognathus. N. Amer. Fauna, 18:1–72, 4 pls., 15 figs., 1 table, September 20.

———. 1900b. Mammals of the Yukon region, pp. 21–46, Pls. 4–7, *in* Osgood, W. H., and Bishop, L. B., Results of a biological reconnaissance of the Yukon River region. N. Amer. Fauna, 19:1–100, 7 pls., October 6.

———. 1904. A biological reconnaissance of the base of the Alaska Peninsula. N. Amer. Fauna, 24:1–86, 7 pls., November 23.

———. 1909. Revision of the mice of the American genus Peromyscus. N. Amer. Fauna, 28:1–285, 8 pls., 12 figs., several tables, April 17.

PACKARD, R. L. 1960. Speciation and evolution of the pygmy mice, genus Baiomys. Univ. Kansas Publ., Mus. Nat. Hist., 9:579–670, 4 pls., 12 figs., 4 tables, June 16.

———. 1961. Additional records of mammals from eastern Texas. Southwestern Nat., 6:193–195, December 29.

———. 1963. Distribution of the black-tailed jackrabbit in eastern Texas. Texas Jour. Sci., 15:107–110, 1 fig., March.

———. 1965. Range extension of the hooded skunk in Texas and Mexico. Jour. Mamm., 46:102, February 20.

———. 1966. *Myotis austroriparius* in Texas. Jour. Mamm., 47:128, March 12.

———. 1969. Taxonomic review of the golden mouse, *Ochrotomys nuttalli*. Univ. Kansas Mus. Nat. Hist., Miscl. Publ., 51:373–406, 3 figs., 2 tables, July 11.

PACKARD, R. L., and BOWERS, J. H. 1970. Distributional notes on some foxes from western Texas and eastern New Mexico. Southwestern Nat., 14:450–451, February 16.

PACKARD, R. L., and GARNER, H. W. 1964. Records of some mammals from the Texas high plains. Texas Jour. Sci., 16:387–390, September.

PACKARD, R. L., and JUDD, F. W. 1968. Comments on some mammals from western Texas. Jour. Mamm., 49:535–538, August 20.

PAGELS, J. F., and ADLEMAN, R. G. 1971. A note on the cotton rat in central Virginia. Virginia Jour. Sci., 22:195.

PARADISO, J. L. 1958. The common blackfish in Virginia coastal waters. Jour. Mamm., 39:440–441, August 20.

———. 1959. A new star-nosed mole (*Condylura*) from the southeastern United States. Proc. Biol. Soc. Washington, 72:103–107, 1 fig., 1 table, July 24.

———. 1960. A new white-footed mouse (*Peromyscus leucopus*) from southeastern Virginia. Proc. Biol. Soc. Washington, 73:21–23, August 10.

———. 1963. The type locality of *Tamias striatus fisheri*. Jour. Mamm., 44:579–580, December 13.

———. 1965. Recent records of red wolves from the Gulf Coast of Texas. Southwestern Nat., 10:318–319, November 15.

———. 1966. Recent records of coyotes, *Canis latrans*, from the southeastern United States. Southwestern Nat., 11:500–501, December 21.

———. 1967. A review of the wrinkle-faced bats (*Centurio senex* Gray) with description of a new subspecies. Mammalia, 31:595–604, 1 pl., 1 fig., December.

———. 1969. Mammals of Maryland. N. Amer. Fauna, 66:iv + 193, 54 figs., August 21.

PARADISO, J. L., and MANVILLE, R. H. 1961. Taxonomic notes on the tundra vole (*Microtus oeconomus*) in Alaska. Proc. Biol. Soc. Washington, 74:77–92, 1 map, May 19.

PARADISO, J. L., and NOWAK, R. M. 1971. Taxonomic status of the Sonoran pronghorn. Jour. Mamm., 52:855–858, 1 table, December 16.

———. 1972. A report on the taxonomic status and distribution of the red wolf. U.S. Bur. Sport Fisheries and Wildlife, Special Sci. Report—Wildlife No. 145:1–36, 15 figs., 1 table, January 18.

PARK, H., and HALL, E. R. 1951. The gross anatomy of the tongues and stomachs of eight New World bats. Trans. Kansas Acad. Sci., 54:64–72, 34 figs., March 17.

PARMALEE, P. W. 1962. A second porcupine record for Illinois. Trans. Illinois State Acad. Sci., 55:90–91.

PARMALEE, P. W., BIERI, R. A., and MOHRMAN, R. K. 1961. Mammal remains from an Illinois cave. Jour. Mamm., 42:119, February 20.

PARMALEE, P. W., and GUILDAY, J. E. 1966. A recent record of porcupine from Tennessee. Jour. Tennessee Acad. Sci., 41:81–82.

PATRIC, E. F. 1962. Reproductive characteristics of red-backed mouse during years of differing population densities. Jour. Mamm., 43:200–205, 3 figs., 2 tables, May 29.

PATTEN, D. R., and FINDLEY, L. T. 1970. Observations and records of *Myotis* (*Pizonyx*) *vivesi* Menegaux. Los Angeles Co. Mus. Contrib. Sci., 183:1–9, April 17.

PATTERSON, B. 1962. An extinct solenodontid insectivore from Hispaniola. Breviora, 165:1–11, 4 figs., August 22.

PATTON, J. L. 1967. Chromosome studies of certain pocket mice, genus *Perognathus* (Rodentia: Heteromyidae). Jour. Mamm., 48:27–37, February 20.

———. 1967. Chromosomes and evolutionary trends in the pocket mouse subgenus *Perognathus* (Rodentia: Heteromyidae). Southwestern Nat., 12:429–438, 6 figs., 1 table, December 31.

PATTON, J. L., and DINGMAN, R. E. 1968. Chromosome studies of pocket gophers, genus *Thomomys*. I. The specific status of *Thomomys umbrinus* (Richardson) in Arizona. Jour. Mamm., 49:1–13, 8 figs., 4 tables, February 20.

PATTON, J. L., and HSU, T. C. 1967. Chromosomes of the golden mouse, *Peromyscus* (*Ochrotomys*) *nuttalli* (Harlan). Jour. Mamm., 48:637–639, 1 fig., November 20.

PATTON, J. L., and JONES, J. K., JR. 1972. First records of *Perognathus baileyi* from Sinaloa, México. Jour. Mamm., 53:371–372, June 23.

PATTON, J. L., and SOULE, O. H. 1967. *Natural hybridization in pocket mice, genus Perognathus.* Mammalian Chromosomes Newsletter, 8:263–265, 1 fig., October, processed literature, + 1 p. MS and 1 fig.

PATTON, T. H. 1966. Occurrence of fossil vertebrates on Cayman Brac, B.W.I. Caribbean Jour. Sci., 6:181, December.

PAUL, J. R. 1965. *Blarina telmalestes* in North Carolina. Jour. Mamm., 46:496, August 26.

———. 1968. Risso's dolphin, *Grampus griseus*, in the Gulf of Mexico. Jour. Mam., 49:746–748, 2 figs., November 26.

PAUL, J. R., and CORDES, C. L. 1969. New distributional records for five mammals of the coastal plain of North

Carolina. Trans. Illinois State Acad. Sci., 62:372–374, December 1.

PAUL, J. R., and QUAY, T. L. 1963. Notes on the mammalian fauna of the Toxaway River Gorge, North Carolina. Jour. Elisha Mitchell Sci. Soc., 79(2):124–126, 2 tables, November.

PAYNE, N. F. 1975. Range extension of the marsh rabbit in Virginia. Chesapeake Sci., 16:77–78, 1 table, March.

———. 1976. Red squirrel introduction to Newfoundland. Canadian Field-Nat., 90:60–64, 2 figs., 1 table, March 19.

PEACOCK, D. B. 1967. Some notes on Virginia and North Carolina mammals. Jour. Washington Acad. Sci., 57:242–244, December.

PEACOCK, R., and PEACOCK, D. 1961. The harvest mouse in northern Virginia. Jour. Mamm., 42:543–544, November 20.

PEFAUR, J. E., and HOFFMANN, R. S. 1971. Merriam's shrew and hispid pocket mouse in Montana. Amer. Midland Nat., 86:247–248, July 21.

PERRIN, W. F. 1975. Distribution and differentiation of populations of dolphins of the genus Stenella in the eastern Tropical Pacific. Jour. Fish Research Board Canada, 32:1059–1067, 9 figs., 3 tables.

PERRIN, W. F., BEST, P. B., DAWBIN, W. H., BALCOMB, K. C., GAMBELL, R., and ROSS, G. J. B. 1973. Rediscovery of Fraser's Dolphin Lagenodelphis hosei. Nature, 241:345–350, 4 figs., 2 tables, February 2.

PERRIN, W. F., and HUBBS, C. L. 1969. Observations on a young pygmy killer whale (Feresa attenuata Gray) from the eastern tropical Pacific Ocean. Trans. San Diego Soc. Nat. Hist., 15:297–308, 8 figs., November 24.

PERRIN, W. F., and WALKER, W. A. 1975. The rough-toothed porpoise, Steno bredanensis, in the eastern Tropical Pacific. Jour. Mamm., 56:905–907, November 18.

PETERSON, R. L. 1952. A review of the living representatives of the genus Alces. Contrib. Royal Ontario Mus. Zool. and Palaeont., 34:1–30, 8 figs., 1 table, October 15.

———. 1962. Notes on the distribution of Microtus chrotorrhinus. Jour. Mamm., 43:420, August 20.

———. 1965. The genus Vampyressa recorded from British Honduras. Jour. Mamm., 46:676, November 30.

———. 1966. Notes on the Yucatan vesper rat, Otonyctomys hatti, with a new record, the first from British Honduras. Canadian Jour. Zool., 44:281–284, 3 figs., 1 table, March 21.

———. 1968. A new bat of the genus Vampyressa from Guyana, South America, with a brief systematic review of the genus. Royal Ontario Mus., Life Sci. Contrib., 73:1–17, 6 figs., September 20.

———. 1968. Cryptotis mayensis recorded from British Honduras. Jour. Mamm., 49:796, November 26.

PETERSON, R. L., and KIRMSE, P. 1969. Notes on Vampyrum spectrum, the false vampire bat, in Panama. Canadian Jour. Zool., 47:140–142. [Rec. February 1969.]

PETERSON, R. S. 1967. The land mammals of Unalaska Island: present status and zoogeography. Jour. Mamm., 48:119–129, 1 fig., 1 table, February 20.

PETERSON, R. S., HUBBS, C. L., GENTRY, R. L., and DeLONG, R. L. 1968. The Guadalupe fur seal: habitat, behavior, population size, and field identification. Jour. Mamm., 49:665–675, 8 figs., November 26.

PETERSON, R. S., and SYMANSKY, A. 1963. First record of the Gaspé shrew from New Brunswick. Jour. Mamm., 44:278–279, May 20.

PETTER, F. 1961. Eléments d'une révision des lièvres européens et asiatiques du sous-genre Lepus. Zeitschr. f. Säugetierkunde, 26:30–40, 3 figs., March.

PFEIFER, W. K., and HIBBARD, E. A. 1970. A recent record of the swift fox (Vulpes velox) in North Dakota. Jour. Mamm., 51:835, November 30.

PFEIFFER, C. J., and GASS, G. H. 1963. Note on the longevity and habits of captive Cryptotis parva. Jour. Mamm., 44:427–428, August 22.

PHILLIPS, C. J., and JONES, J. K., JR. 1968. Additional comments on reproduction in the woolly opossum (Caluromys derbianus) in Nicaragua. Jour. Mamm., 49:320–321, 1 table, May 20.

———. 1969. Notes on reproduction and development in the four-eyed opossum, Philander opossum, in Nicaragua. Jour. Mamm., 50:345–348, 1 fig., 1 table, June 12.

———. 1971. A new subspecies of the long-nosed bat, Hylonycteris underwoodi, from México. Jour. Mamm., 52:77–80, 2 figs., February 26.

PINE, R. H. 1969. Three rodents new to the fauna of Honduras. Jour. Mamm., 50:643, August 22.

———. 1971. A review of the long-whiskered rice rat, Oryzomys bombycinus Goldman. Jour. Mamm., 52:590–596, 1 fig., 1 table, August 26.

———. 1972. The bats of the genus Carollia. Tech. Monog., Texas A&M Univ., Texas Agric. Exp. Station, 8:1–125, 3 figs., 7 maps, 8 tables, September 26.

———. 1973. Anatomical and nomenclatural notes on opossums. Proc. Biol. Soc. Washington, 86:391–402, December 14.

PINE, R. H., and CARTER, D. C. 1970. Distributional notes on the thick-spined rat (Hoplomys gymnurus) with the first records from Honduras. Jour. Mamm., 51:804, November 30.

PINE, R. H., CARTER, D. C., and LaVAL, R. K. 1971. Status of Bauerus Van Gelder and its relationships to other nyctophiline bats. Jour. Mamm., 52:663–669, 2 figs., December 16.

PIRLOT, P. 1962. Mammifères de la Province de Québec. Le Naturaliste Canadien, 89:129–147, April.

PIZZIMENTI, J. J., and NADLER, C. F. 1972. Chromosomes and serum proteins of the Utah prairie dog, Cynomys parvidens (Sciuridae). Southwestern Nat., 17:279–286, 3 figs., October 6.

PLATT, A. P. 1968. Differential trap mortality as a measure of stress during times of population increase and decrease. Jour. Mamm., 49:331–335, 1 fig., 2 tables, May 20.

POCOCK, R. I. 1932. The black and brown bears of Europe and Asia. Jour. Bombay Nat. Hist. Soc. Part 1, 35:771–823, 11 figs., July 15. Part 2, 36:101–138, 2 pls., 12 figs., November 15.

———. 1941. The fauna of British India, including Ceylon and Burma. Mammalia.—Vol. II. Carnivora. Taylor and Francis Ltd., London, xii + 503, 12 pls., 115 figs., 1 map, numerous tables, September 15.

PORSILD, A. E. 1945. Mammals of the Mackenzie Delta. Canadian Field-Nat., 59:4–22, August 16.

POURNELLE, G. H., and BARRINGTON, B. A. 1953. Notes on mammals of Anastasia Island, St. Johns County, Florida. Jour. Mamm., 34:133–135, February 9.

PRESTON, J. R., and SEALANDER, J. A. 1969. Unusual second

record of *Notiosorex* from Arkansas. Jour. Mamm., 50:641–642, August 22.

PRINGLE, L. P. 1960. Notes on coyotes in southern New England. Jour. Mamm., 41:278, May 20.

QUAY, W. B. 1962. Apocrine sweat glands in the angulus oris of microtine rodents. Jour. Mamm., 43:303–310, 3 pls., August 20.

RACEY, K., and COWAN, I. McT. 1935. Mammals of the Alta Lake region of south-western British Columbia. Report of Provincial [British Columbia] Museum, 1935: pp. H15–H29, 5 pls., 1 fig., 2 tables.

RADVANYI, A. 1960. Off-course migration of a northern fur seal. Jour. Mamm., 41:277, May 20.

RALSTON, G. L., and CLARK, W. H. 1971. Occurrence of *Mustela frenata* in northern Baja California, Mexico. Southwestern Nat., 16:209, November 15.

RAMÍREZ-P., J. 1971. Nuevos registros de murciélagos para el estado de Morelos, México. Anal. Inst. Biol., Univ. Nac. Autó. México, 40 (for 1969, Ser. Zool. 1):123–127, December 15.

——. 1971. Contribución al estudio de los mamíferos del Parque Nacional "Lagunas de Zempoala," Morelos, México. Anal. Inst. Biol., Univ. Nac. Autó. México, 40 (for 1969, Ser. Zool. 2):253–290, 6 figs., 3 cuadros, 2 tables, December 30.

RAMÍREZ-P., J., and ALVAREZ, T. 1972. Notas sobre los murciélagos del genero *Leptonycteris* en México, con la designación del lectotipo de *L. yerbabuenae* Martínez y Villa, 1940. Southwestern Nat., 16:249–259, 5 figs., 1 table, February 18.

RAMÍREZ-P., J., and PHILLIPS, A. R. 1971. Primer registro de comadreja (*Mustela*) en Quintana Roo, México. Anal. Inst. Biol., Univ. Nac. Autó. México, 39 (for 1968, Ser. Zool. 1):145–148, September 11.

RAMÍREZ-P., J., and SÁNCHEZ-H., C. 1971. *Tylomys nudicaudus* from the Mexican states of Puebla and Guerrero. Jour. Mamm., 52:481, May 28.

——. 1974. Regurgitaciones de lechuza, procedentes de la Cueva del Cañón del Zopilote, Guerrero, México. Rev. Soc. Mexicana Hist. Nat., 33 (for December 1972):107–112, 1 table, July 18.

RAMSEY, P. R., and CARLEY, C. J. 1970. Additions to the known range and ecology of three species of *Dipodomys*. Southwestern Nat., 14:351–353, January 15.

RAND, A. L. 1945. Mammal investigations on the Canol Road, Yukon and Northwest Territories, 1944. Bull. Nat. Mus. Canada, 99:1–52, 20 pls., 1 fig., several tables, June.

——. 1945. Mammals of the Ottawa District. Canadian Field-Nat., 59:111–132, December 21.

——. 1948. Mammals of the eastern Rockies and western plains of Canada. Bull. Nat. Mus. Canada, 108:vii + 237, 89 figs.

RASMUSSEN, D. L., and CHAMBERLAIN, N. B. 1959. A new Richardson's meadow mouse from Utah. Jour. Mamm., 40:53–56, 1 fig., February 20.

RASMUSSEN, J. L. 1969. A recent record of the lynx in Iowa. Jour. Mamm., 50:370–371, June 12.

RAUN, G. G. 1959. A new Texas locality for *Dipodomys ordi*. Jour. Mamm., 40:146–147, February 20.

——. 1960. Barn owl pellets and small mammal populations near Mathis, Texas, in 1956 and 1959. Southwestern Nat., 5:194–200, 2 tables, December 20.

——. 1965. The subspecific status of the cottontail, *Sylvilagus floridanus*, in northern Coahuila, Mexico. Jour. Mamm., 46:519–521, 2 tables, August 26.

RAUN, G. G., and WILKS, B. J. 1961. Noteworthy records of the hog-nosed skunk (*Conepatus*) from Texas. Texas Jour. Sci., 13:204–205, July.

RAUSCH, R. 1953. On the status of some Arctic mammals. Arctic, 6:91–148, 17 figs., 5 tables, July.

——. 1962. Notes on the collared pika, *Ochotona collaris* (Nelson), in Alaska. Murrelet, 42:22–24, 1 fig., May.

——. 1964. The specific status of the narrow-skulled vole (subgenus *Stenocranius* Kashchenko) in North America. Zeitschr. f. Säugetierkunde, 29:343–358, 7 figs., 1 table, November.

——. 1967. New records of the pigmy shrew *Microsorex hoyi* (Baird) in Alaska. Murrelet, 48:9–10, September.

RAUSCH, R. L., and RAUSCH, V. R. 1968. On the biology and systematic position of *Microtus abbreviatus* Miller, a vole endemic to the St. Matthew Islands, Bering Sea. Zeitschr. f. Säugetierkunde, 33:65–99, 20 figs., 4 tables, April.

——. 1971. The somatic chromosomes of some North American marmots (Sciuridae), with remarks on the relationships of *Marmota broweri* Hall and Gilmore. Mammalia, 35:85–101, 6 figs., March.

——. 1975. Taxonomy and zoogeography of *Lemmus* spp. (Rodentia: Arvicolinae), with notes on laboratory-reared lemmings. Zeitschr. f. Säugetierkunde, 40:8–34, 11 figs., February.

RAVEN, H. C. 1937. Notes on the taxonomy and osteology of two species of *Mesoplodon* (*M. europaeus* Gervais, *M. mirus* True). Amer. Mus. Novit., 905:1–30, 15 figs., 6 tables, January 14.

——. 1942. On the structure of *Mesoplodon densirostris*, a rare beaked whale. Bull. Amer. Mus. Nat. Hist., 80:23–50, 26 figs., 6 tables, September 1.

RAY, C. E. 1964a. A new capromyid rodent from the Quaternary of Hispaniola. Breviora, 203:1–4, 1 fig., April 10.

——. 1964b. The taxonomic status of *Heptaxodon* and dental ontogeny in *Elasmodontomys* and *Amblyrhiza* (Rodentia: Caviomorpha). Bull. Mus. Comp. Zool., 131:107–127, 2 figs., May 25.

——. 1965. The relationships of Quemisia gravis (Rodentia: ?Heptaxodontidae). Smiths. Miscl. Coll., 149(3):1–12, 1 pl., 2 figs., 1 table, April 28.

REDDEL, J. R. 1968. The hairy-legged vampire, *Diphylla ecaudata*, in Texas. Jour. Mamm., 49:769, November 26.

REED, E. K. 1955. Bison beyond the Pecos. Texas Jour. Sci., 7:130–135, June.

REEDER, W. G. 1965. Occurrence of the big brown bat in southwestern Alaska. Jour. Mamm., 46:332–333, May 20.

REHN, J. A. G. 1904. A revision of the mammalian genus Macrotus. Proc. Acad. Nat. Sci. Philadelphia, 56:427–446, June 27.

REICHSTEIN, H. 1958. Schädelvariabilität europäischer Mauswiesel (*Mustela nivalis* L.) und Hermeline (*Mustela erminea* L.) in Beziehung zu Verbreitung und Geschlecht. Zeitschr. f. Säugetierkunde, 22:151–182, 7 figs., 16 tables.

REIG, O. A. 1955. Noticia preliminar sobre la presencia de microbiotherinos vivientes en la fauna Sudamericana. Investigaciones Zool. Chilenas, 2:121–130, 5 figs., April.

REPENNING, C. A. 1967. Subfamilies and genera of the

Soricidae. Geol. Surv. Professional Pap., 565:iv + 74 pp., 42 figs.

REYNOLDS, H. G. 1966. Abert's squirrels feeding on pinyon pine. Jour. Mamm., 47:550–551, August 29.

REYNOLDS, T. E., KOOPMAN, K. F., and WILLIAMS, E. E. 1953. A cave faunule from western Puerto Rico with a discussion of the genus Isolobodon. Breviora, 12:1–8, 1 pl., April 23.

RICE, D. W. 1957. Life history and ecology of Myotis austroriparius in Florida. Jour. Mamm., 38:15–32, 3 figs., 5 tables, February 25.

———. 1965. Bryde's whale in the Gulf of Mexico. Norsk Hvalfangst-Tidende, 54:114–115, 1 fig., May.

———. 1977. A list of the marine mammals of the world. U.S. Dept. Commerce, National Oceanographic and Atmospheric Administration Tech. Report. NMFS SSRF-711, pp. iii + 15, April.

RICE, D. W., KENYON, K. W., and LLUCH, B. D. 1965. Pinniped populations at Islas Guadalupe, San Benito, and Cedros, Baja California, in 1965. Trans. San Diego Soc. Nat. Hist., 14:73–84, 5 figs., 3 tables, December 28.

RICE, D. W., and SCHEFFER, V. B. 1968. A list of the marine mammals of the World. U.S. Fish and Wildlife Serv. Special Sci. Report, Fisheries, 579:iii + 16, December.

RICHMOND, E. A. 1962. The fauna and flora of Horn Island, Mississippi. Gulf Research Reports, 1:57–106, 27 figs., April.

RICHMOND, N. D., and MCDOWELL, R. D. 1952. The least weasel (Mustela rixosa) in Pennsylvania. Jour. Mamm., 33:251–253, 1 fig., 1 table, May 14.

RICHMOND, N. D., and ROSLUND, H. R. 1949. Mammal survey of northwestern Pennsylvania. Pennsylvania Game Comm., and U.S. Fish and Wildlife Serv., 67 pp., 5 maps.

RICK, A. M. 1965. Otonyctomys hatti in Guatemala. Jour. Mamm., 46:335–336, May 20.

———. 1968. Notes on bats from Tikal, Guatemala. Jour. Mamm., 49:516–520, August 20.

RICKART, E. A. 1972. An analysis of barn owl and great horned owl pellets from western Nebraska. Prairie Nat., 4:35–38, 2 tables, June.

RÍMOLI, R. O. 1977. Roedores fósiles de la Hispaniola. Univ. Central del Este, República Dominicana. Ser. Científica III. 96 pp., 19 pls., 16 tables, January 10.

———. 1977. Una nueva especie de monos (Cebidae: Saimirinae: Saimiri) de la Hispaniola. Univ. Autó. Santo Domingo, "No. 1 Vol. CCXLII," pp. 1–14, 2 pls., 2 tables, prior to December 29.

RINKER, G. C. 1960. The entepicondylar foramen in Peromyscus. Jour. Mamm., 41:276, May 20.

RIOME, L. S. 1968. Short-tailed shrew at Nipawin. Blue Jay, 26:201–203, 3 figs., December.

RIOME, S. D. 1973. Evidence of cougars near Nipawin, Saskatchewan. Blue Jay, 31:100–102, 2 figs., June.

RIPPY, C. L., and HARVEY, M. J. 1965. Notes on Plecotus townsendii virginianus in Kentucky. Jour. Mamm., 46:499, August 26.

ROBERTSON, P. B., and MUSSER, G. G. 1976. A new species of Peromyscus (Rodentia: Cricetidae), and a new specimen of P. simulatus from southern Mexico, with comments on their ecology. Univ. Kansas Mus. Nat. Hist., Occas. Pap., 47:1–8, 2 figs., 1 table, March 1.

ROBINS, J. D. 1971. Movement of Franklin's ground squirrel into northeastern Minnesota. Jour. Minnesota Acad. Sci., 37:30–31.

ROBINSON, J. W., and HOFFMANN, R. S. 1975. Geographical and interspecific cranial variation in big-eared ground squirrels (Spermophilus): a multivariate study. Systematic Zool., 24:79–88, 5 figs., 4 tables, March.

ROBINSON, T. S., and QUICK, F. W. 1965. The cotton rat in Kentucky. Jour. Mamm., 46:100, February 20.

ROEST, A. I. 1964. A ribbon seal from California. Jour. Mamm., 45:416–420, 1 fig., 2 tables, September 15.

———. 1970. Kogia simus and other cetaceans from San Luis Obispo County, California. Jour. Mamm., 51:410–417, 1 fig., 2 tables, May 20.

———. 1973. Subspecies of the sea otter, Enhydra lutris. Los Angeles Co. Mus. Contrib. Sci., 252:1–17, 5 figs., 9 tables, December 21.

ROHWER, S. A., and KILGORE, D. L., JR. 1973. Interbreeding in the arid-land foxes, Vulpes velox and V. macrotis. Systematic Zool., 22:157–165, 4 figs., 3 tables, July 30.

ROUK, C. S., and GLASS, B. P. 1970. Comparative gastric histology of five North and Central American bats. Jour. Mamm., 51:455–472, 19 figs., August 28.

ROWLAND, R. H., and TURNER, F. B. 1964. Correlation of the local distributions of Dipodomys microps and D. merriami and of the annual grass Bromus rubens. Southwestern Nat., 9:56–61, 2 tables, May 15.

ROWLETT, R. A. 1972. First records of Eumops perotis and Microtus ochrogaster in New Mexico. Jour. Mamm., 53:640, September 27.

RUDD, R. L. 1955. Population variation and hybridization in some California shrews. Systematic Zool., 4:21–34, 7 figs., 2 tables, March.

RUFFER, D. G. 1965. Burrows and burrowing behavior of Onychomys leucogaster. Jour. Mamm., 46:241–247, 2 figs., May 20.

RUSSELL, R. J. 1968. Evolution and classification of the pocket gophers of the subfamily Geomyinae. Univ. Kansas Publ., Mus., Nat. Hist., 16:473–579, 9 figs., August 5.

———. 1968. Revision of pocket gophers of the genus Pappogeomys. Univ. Kansas Publ., Mus. Nat. Hist., 16:581–776, 10 figs., August 5.

———. 1969. Intraspecific population structure of the species Pappogeomys castanops. Univ. Kansas Mus. Nat. Hist., Miscl. Publ., 51:337–371, 17 figs., July 11.

RUST, C. C. 1966. Notes on the star-nosed mole (Condylura cristata). Jour. Mamm., 47:538, August 29.

RYAN, R. M. 1960. Mamiferos colectados en Guatemala en 1954. Acta Zool. Mexicana, 4:1–19, June 30.

RYSER, F. A. 1964. Spiny pocket mouse in Nevada. Jour. Mamm., 45:301–302, May 20.

ST. ROMAIN, P. A. 1976. Variation in the cotton mouse (Peromyscus gossypinus) in Louisiana. Southwestern Nat., 21:79–88, 2 figs., 2 tables, March 20.

SALT, J. R., and WERSHLER, C. R. 1975. A range and elevation extension for the sagebrush vole in Alberta. Canadian Field-Nat., 89:184, May 28.

SANBORN, C. C. 1932. The bats of the genus Eumops. Jour. Mamm., 13:347–357, November 2.

———. 1933. Bats of the genera Anoura and Lonchoglossa. Field Mus. Nat. Hist., Zool. Ser., 20:23–27, December 11.

———. 1937. American bats of the subfamily Emballonurinae. Field Mus. Nat. Hist., Zool. Ser., 20:321–354, Figs. 37–48, 1 table, December 28.

———. 1941. Descriptions and records of Neotropical bats. Field Mus. Nat. Hist., Zool. Ser., 27:371–387, December 8.

SANDS, J. L. 1960. The opossum in New Mexico. Jour. Mamm., 41:393, August 15.

SCHALDACH, W. J., JR. 1965. Notas breves sobre algunos mamíferos del sur de México. Anal. Inst. Biol., Univ. Nac. Autó. México, 35:129–137, August 25.

———. 1966. New forms of mammals from southern Oaxaca, México, with notes on some mammals of the coastal range. Säugetierkund. Mitteil., 14:286–297, October.

SCHALDACH, W. J., [JR.], and McLAUGHLIN, C. A. 1960. A new genus and species of glossophagine bat from Colima, Mexico. Los Angeles Co. Mus. Contrib. Sci., 37:1–8, 3 figs., May 19.

SCHEFFER, V. B. 1958. Seals, sea lions, and walruses, a review of the Pinnipedia. Stanford Univ. Press, Stanford, California, x + 179 pp., 32 pls., 15 figs., April 24.

SCHEFFER, V. B., and RICE, D. W. 1963. A list of the marine mammals of the world. U.S. Fish and Wildlife Serv. Special Sci. Report, Fisheries, 431:1–12.

SCHEFFER, V. B., and SLIPP, J. W. 1948. The whales and dolphins of Washington state with a key to the cetaceans of the west coast of North America. Amer. Midland Nat., 39:257–337, 50 figs., 6 tables, April 22.

SCHILLER, E. L., and RAUSCH, R. 1956. Mammals of the Katmai National Monument. Arctic, 9:191–201.

SCHINZ, H. R. 1821 and 1825. Das Thierreich eingetheilt nach dem Bau der Thiere als Grundlage ihrer Naturgeschichte und der vergleichenden Anatomie von dem Herrn Ritter von Cuvier. Vols. 1 and 4. Stuttgart and Tübingen.

SCHLITTER, D. A. 1973. *Notiosorex crawfordi evotis* from Nayarit. Southwestern Nat., 17:423, January 10.

SCHLITTER, D. A., and BOWLES, J. B. 1968. Noteworthy distributional records of some mammals in Iowa. Trans. Kansas Acad. Sci., 70:525–529, March 12.

SCHMIDLY, D. J. 1972. Geographic variation in the white-ankled mouse, *Peromyscus pectoralis*. Southwestern Nat., 17:113–138, 7 figs., 2 tables, September 15.

———. 1973. Geographic variation and taxonomy of *Peromyscus boylii* from Mexico and the southern United States. Jour. Mamm., 54:111–130, 6 figs., 4 tables, April 26.

———. 1973. The systematic status of *Peromyscus comanche*. Southwestern Nat., 18:269–278, 3 figs., 2 tables, October 5.

SCHMIDLY, D. J., BELEAU, M. H., and HILDEBRAN, H. 1972. First record of Cuvier's dolphin from the Gulf of Mexico with comments on the taxonomic status of *Stenella frontalis*. Jour. Mamm., 53:625–628, 1 table, September 27.

SCHMIDLY, D. J., and HENDRICKS, F. S. 1976. Systematics of the southern races of Ord's kangaroo rat, *Dipodomys ordii*. Bull. Southern California Acad. Sci., 75:225–237, 6 figs., 5 tables, November.

SCHMIDLY, D. J., HENDRICKS, F. S., and LIEB, C. S. 1974. Noteworthy additions to the bat fauna of the San Carlos Mountains, Tamaulipas, México. Texas Jour. Sci., 25:87–88, March.

SCHMIDLY, D. J., and MARTIN, C. O. 1973. Notes on bats from the Mexican state of Querétaro. Bull. Southern California Acad. Sci., 72:90–92, August.

SCHMIDLY, D. J., MARTIN, C. O., and COLLINS, G. F.

1972. First occurrence of a black right whale (*Balaena glacialis*) along the Texas coast. Southwestern Nat., 17:214–215, 1 fig., September 15.

SCHMIDLY, D. J., and MELCHER, B. A. 1974. Annotated checklist and key to the cetaceans of Texas waters. Southwestern Nat., 18:453–464, 2 figs., January 31.

SCHORGER, A. W. 1956. The moose in early Wisconsin. Wisconsin Acad. Sci., Arts and Letters, 45:1–10, 1 fig.

———. 1973. The mammals of Dane County. Trans. Wisconsin Acad. Sci., Arts and Letters, 61:75–85.

SCHULTZ, V. 1955. Status of the coyote and related forms in Tennessee. Jour. Tennessee Acad. Sci., 30:44–46, January.

SCHULZ, T. A., RADOVSKY, F. J., and BUDWISER, P. D. 1970. First insular record of *Notiosorex crawfordi*, with notes on other mammals of San Martin Island, Baja California, Mexico. Jour. Mamm., 51:148–150, February 20.

SCHWARTZ, A. 1956. The cottontail rabbits (*Sylvilagus floridanus*) of peninsular Florida. Proc. Biol. Soc. Washington, 69:145–151, 1 table, September 12.

SCHWARTZ, A., and JONES, J. K., JR. 1967. Review of bats of the endemic Antillean genus *Monophyllus*. Proc. U.S. Nat. Mus., 124(3635):1–20, 1 fig., 2 tables, December.

SCHWARZ, E., and SCHWARZ, H. K. 1943. The wild and commensal stocks of the house mouse, *Mus musculus* Linnaeus. Jour. Mamm., 24:59–72, February 20.

SCOTTER, G. W. 1975. White-tailed deer and mule deer observations in southwestern district of Mackenzie, Northwest Territories. Canadian Field-Nat., 88:487–489, 1 fig., February 17.

SCOTTER, G. W., and SIMMONS, N. M. 1975. Range extensions for the bushy-tailed wood rat in the Northwest Territories. Canadian Field-Nat., 88:489–490, February 17.

SEALANDER, J. A., JR. 1956. A provisional check-list and key to the mammals of Arkansas (with annotations). Amer. Midland Nat., 56:257–296, 1 fig., October.

———. 1960. Some noteworthy records of Arkansas mammals. Jour. Mamm., 41:525–526, November 11.

———. 1961. Hematological values in deer mice in relation to botfly infection. Jour. Mamm., 42:57–60, 1 table, February 20.

———. 1967. First record of small-footed Myotis in Arkansas. Jour. Mamm., 48:666, November 20.

———. 1977. New marginal records for the eastern harvest mouse and southeastern shrew in Arkansas. Southwestern Nat., 22:148–149, March 1.

SEALANDER, J. A., JR., and FORSYTH, B. J. 1966. Occurrence of the badger in Arkansas. Southwestern Nat., 11:134, March 30.

SEALANDER, J. A., [JR.], and GIPSON, P. S. 1972. Range extension of ringtail cat into Arkansas. Southwestern Nat., 16:458–459, February 18.

SEALANDER, J. A., [JR.], GIPSON, P. S., and MANLEY, J. M. H. 1975. The distribution of the prairie vole (*Microtus ochrogaster*) and the southern bog lemming (*Synaptomys cooperi*) in Arkansas. Texas Jour. Sci., 26:421–430, 2 figs., 2 tables, August.

SEALANDER, J. A., [JR.], and PRICE, J. F. 1964. Free-tailed bat in Arkansas. Jour. Mamm., 45:152, March 12.

SEALEY, S. 1961. Northern flying squirrel reported at Battleford. Blue Jay, 19:184, December.

SERGEANT, D. E., and BRODIE, P. F. 1969. Body size in white whales, *Delphinapterus leucas*. Jour. Fisheries Research Board Canada, 26:2561–2580, 15 figs., 3 tables, October 31.

Seton, E. T. 1929. Lives of game animals. . . . 2:xvii + 746, illustrated. Doubleday, Doran & Co., Inc., Garden City, New York.

Setzer, H. W. 1949. Subspeciation in the kangaroo rat, Dipodomys ordii. Univ. Kans. Publ., Mus. Nat. Hist., 1:473–573, 27 figs., 7 tables, December 27.

Severaid, J. H. 1950. The gestation period of the pika (Ochotona princeps). Jour. Mamm., 31:356–357, August 21.

Severinghaus, W. D. 1977. Description of a new subspecies of prairie vole, Microtus ochrogaster. Proc. Biol. Soc. Washington, 90:49–54, 1 fig., 2 tables, June 16.

Severinghaus, W. D., and Beasley, L. E. 1973. A survey of the microtine and zapodid rodents of west Tennessee. Jour. Tennessee Acad. Sci., 48:129–133, 2 figs., 2 tables, October.

Shamel, H. H. 1931. Notes on the American bats of the genus Tadarida. Proc. U.S. Nat. Mus., 78:1–27, numerous tables, May 6.

Shaw, J. H. 1962. The bushy-tailed wood rat in southeastern Alaska. Jour. Mamm., 43:431–432, August 20.

Sheppe, W., [Jr.]. 1960. Systematic relations of Clethrionomys in the Pacific Northwest. Canadian Field-Nat., 74:171–173, 1 table, November 28.

———. 1961. Systematic and ecological relations of Peromyscus oreas and P. maniculatus. Proc. Amer. Philos. Soc., 105:421–446, 12 figs., 3 tables, 1 map, August.

Sherman, H. B. 1937. A list of the Recent wild land mammals of Florida. Proc. Florida Acad. Sci., for 1936, 1:102–128.

Shippee, E. A., and Egoscue, H. J. 1958. Additional mammal records from the Bonneville Basin, Utah. Jour. Mamm., 39:275–277, May 20.

Silva-T., G. 1976. Historia y Actualización Taxonómica de Algunas Especies Antillanas de Murciélagos de los Géneros Pteronotus, Brachyphylla, Lasiurus, y Antrozous (Mammalia: Chiroptera). Poeyana, No. 153:1–24, 1 fig., 5 tables, November 30.

Silva-T., G., and Koopman, K. F. 1964. Notes on the occurrence and ecology of Tadarida laticaudata yucatanica in eastern Cuba. Amer. Mus. Novit., 2174:1–6, 3 figs., April 14.

Silva-T., G., and Pine, R. H. 1969. Morphological and behavioral evidence for the relationships between the bat genus Brachyphylla and the Phyllonycterinae. Biotropica, 1:10–19, 4 figs., June.

Silver, J. 1924. The European hare (Lepus europaeus Pallas) in North America. Jour. Agric. Research, Washington, 28:1133–1137, 1 fig., June 14.

Simkin, D. W. 1966. Extralimital occurrences of raccoons in Ontario. Canadian Field-Nat., 80:144–146, 1 fig., 1 table, December 8.

Simmons, N. M. 1966. Observations of mammals in the Cabeza Prieta Game Range area, Arizona. Jour. Mamm., 47:122, March 12.

Singh, R. P., and McMillan, D. B. 1966. Karyotypes of three subspecies of Peromyscus. Jour. Mamm., 47:261–266, 3 figs., May 26.

Smith, E., and Goodpaster, W. 1960. A free-tailed bat found in Ohio. Jour. Mamm., 41:117, February 20.

Smith, H. C. 1969. Recent Alberta pocket mouse records. Blue Jay, 27:227, December.

———. 1972. Some recent records of Alberta mammals. Blue Jay, 30:53–54, March.

Smith, J. D. 1964. Second record of the eastern harvest mouse from Oklahoma. Trans. Kansas Acad. Sci., 67:204–205, June 12.

———. 1970. The systematic status of the black howler monkey, Alouatta pigra Lawrence. Jour. Mamm., 51:358–369, 3 figs., 1 table, May 20.

———. 1972. Systematics of the chiropteran family Mormoopidae. Univ. Kansas Mus. Nat. Hist., Miscl. Publ., 56:1–132, 40 figs., 5 tables, March 10.

Smith, J. D., and Jones, J. K., Jr. 1967. Additional records of the Guatemalan vole, Microtus guatemalensis Merriam. Southwestern Nat., 12:189–191, 1 fig., August 4.

Smith, J. D., and Lawlor, T. E. 1964. Additional records of the armadillo in Kansas. Southwestern Nat., 9:48–49, April 10.

Smith, M. H. 1968. A comparison of different methods of capturing and estimating numbers of mice. Jour. Mamm., 49:455–462, 1 fig., 1 table, August 20.

Smith, P. B. 1971. Occurrence of Sylvilagus floridanus in Vermont. Jour. Mamm., 52:624–625, August 26.

Smith, R. W. 1940. The land mammals of Nova Scotia. Amer. Midland Nat., 24:213–241, 1 fig., July 31.

Soper, J. D. 1961. Field data on the mammals of southern Saskatchewan. Canadian Field-Nat., 75:23–41, 5 figs., March 9.

———. 1961. The mammals of Manitoba. Canadian Field-Nat., 75:171–219, 30 figs., December 29.

———. 1965. The mammals of Alberta. Hamly Press, Edmonton, 402 pp. + index, 67 pls., 41 figs., March 15.

Sparks, D. R. 1968. Occurrence of milk in stomachs of young jackrabbits. Jour. Mamm., 1968:324–325, May 20.

Spencer, A. W., and Pettus, D. 1966. Habitat preferences of five sympatric species of long-tailed shrews. Ecology, 47:677–683, 2 figs., 4 tables, summer.

Spenrath, C. A., and LaVal, R. K. 1970. Records of bats from Querétaro and San Luis Potosí, México. Jour. Mamm., 51:395–396, May 20.

Stains, H. J., and Turner, R. W. 1963. Harvest mice south of the Illinois River in Illinois. Jour. Mamm., 44:274–275, May 20.

Starrett, A. 1960. The penis of Scotinomys longipilosus and a second specimen from Costa Rica. Jour. Mamm., 41:492–495, 1 fig., November 11.

Starrett, A., and Casebeer, R. S. 1968. Records of bats from Costa Rica. Los Angeles Co. Mus. Contrib. Sci., 148:1–21, June 28.

Starrett, A., and de la Torre, L. 1964. Notes on a collection of bats from Central America, with the third record for Cyttarops alecto Thomas. Zoologica, New York Zool. Soc., 49:53–63.

Starrett, A., and Rolle, F. J. 1963. A record of the genus Lasiurus from Puerto Rico. Jour. Mamm., 44:264, May 20.

Stephens, F. 1906. California mammals. West Coast Publishing Co., San Diego, 351 pp., illustrated.

Stevenson, H. M. 1962. Occurrence and habits of the eastern chipmunk in Florida. Jour. Mamm., 43:110–111, February 28.

Stewart, G. R. 1969. A western yellow bat in Los Angeles County, California. Bull. Southern California Acad. Sci., 68:194–195.

STEWART, R. E. A., STEPHEN, J. R., and BROOKS, R. J. 1975. Occurrence of muskrat, *Ondatra zibethicus albus*, in the District of Keewatin, Northwest Territories. Jour. Mamm., 56:507, May 30.

STOCK, A. D. 1970. Notes on mammals of southwestern Utah. Jour. Mamm., 51:429–433, May 20.

———. 1974. Chromosome evolution in the genus *Dipodomys* and its taxonomic and phylogenetic implications. Jour. Mamm., 55:505–526, 7 figs., 2 tables, August 20.

STONES, C. S., and HABER, G. C. 1965. Eastern pipistrelle in Michigan. Jour. Mamm., 46:688, November 30.

STORER, T. I. 1937. The muskrat as native and alien. Jour. Mamm., 18:443–460, 6 figs., November 22.

STREETER, R. G., and BRAUN, C. E. 1968. Occurrence of pine marten, *Martes americana*, (Carnivora: Mustelidae) in Colorado alpine areas. Southwestern Nat., 13:449–451, 1 table, December 31.

STROUD, R. K. 1968. Risso dolphin in Washington State. Jour. Mamm., 49:347–348, 2 tables, May 20.

STRUHSAKER, P. 1967. An occurrence of the minke whale, *Balaenoptera acutorostrata*, near the northern Bahama Islands. Jour. Mamm., 48:483, August 21.

STUBBE, M. 1970. Fledermausnachweise aus Gewöllen europäischer und kubanischer Eulen. Beiträge Vogelkunde, 16:393–398, 2 tables, June.

STUDIER, E. H. 1968. Fringe-tailed bat in northeast New Mexico. Southwestern Nat., 13:362, December 10.

STUDIER, E. H., BECK, L. R., and LINDEBORG, R. G. 1967. Tolerance and initial metabolic response to ammonia intoxication in selected bats and rodents. Jour. Mamm., 48:564–572, 4 figs., 2 tables, November 20.

STUDIER, E. H., and BACA, T. P. 1968. Atmospheric conditions in artificial rodent burrows. Southwestern Nat., 13:401–410, 3 figs., December 31.

STUPKA, A. 1960. Second specimen of least weasel from North Carolina. Jour. Mamm., 41:519–520, November 11.

STUPKA, R. C., BROWER, J. E., and HENRIKSEN, J. 1973. New northeastern Illinois locality records for the western harvest mouse (*Reithrodontomys megalotis dychei*). Trans. Illinois Acad. Sci., 65(3–4):112–114, 1 fig., August.

SUTTON, D. A., and NADLER, C. F. 1969. Chromosomes of the North American chipmunk genus *Eutamias*. Jour. Mamm., 50:524–535, 2 figs., 2 tables, August 22.

———. 1974. Systematic revision of three Townsend chipmunks (*Eutamias townsendii*). Southwestern Nat., 19:199–211, 4 figs., 3 tables, July 26.

SUTTON, R. W. 1958. A Manitoba record of the gray fox. Jour. Mamm., 39:439–440, August 20.

———. 1964. Range extension of raccoon in Manitoba. Jour. Mamm., 45:311–312, 1 table, May 20.

SWANSON, G. 1945. A systematic catalog of the mammals of Minnesota, pp. 52–102, *in* Swanson, G., Surber, T., and Roberts, T. S., The mammals of Minnesota. Minnesota Dept. Conservation, Tech. Bull., 2:1–108, illustrated.

TAMSITT, J. R. 1959. *Peromyscus nasutus* in northeastern New Mexico. Jour. Mamm., 40:611–613, November 20.

———. 1960. Some mammals of Riding Mountain National Park, Manitoba. Canadian Field-Nat., 74:147–150, August 30.

———. 1962. Mammals of the Delta Marsh Region of Lake Manitoba, Canada. Canadian Field-Nat., 76:71–78, August 29.

TAMSITT, J. R., and VALDIVIESO, D. 1966. Taxonomic comments on *Anoura caudifer, Artibeus literatus* and *Molossus molossus*. Jour. Mamm., 47:230–238, 1 fig., 1 table, May 26.

TATE, G. H. H. 1933. A systematic revision of the Marsupial genus *Marmosa*. . . . Bull. Amer. Mus. Nat. Hist., 66:1–250, 26 pls., 29 figs., 9 tables, August 10.

———. 1935. The taxonomy of the genera of neotropical hystricoid rodents. Bull. Amer. Mus. Nat. Hist., 68:295–447, June 12.

TAYLOR, J. 1964. Noteworthy predation on the guano bat. Jour. Mamm., 45:300–301, May 20.

———. 1965. Additional mammals for Bryan County, Oklahoma. Jour. Mamm., 45:640–642, January 25.

TAYLOR, W. P. 1915. Description of a new subgenus (*Arborimus*) of *Phenacomys*, with a contribution to knowledge of the habits and distribution of *Phenacomys longicaudus* True. Proc. California Acad. Sci. 4th ser., 5(5):111–161, 1 pl., 4 figs., several tables, December 30.

TENAZA, R. R. 1966. Migration of hoary bats on South Farallon Island. Jour. Mamm., 47:533–535, August 29.

TERREL, T. L. 1972. The swamp rabbit (Sylvilagus aquaticus) in Indiana. Amer. Midland Nat., 87:283–295, 1 fig., 5 tables, May 8.

THAELER, C. S., JR. 1961. Variation in some salt-marsh populations of Microtus californicus. Univ. California Publ. Zool., 60:67–94, 21 figs., 3 tables, November 28.

———. 1968. Karyotypes of sixteen populations of the *Thomomys talpoides* complex of pocket gophers (Rodentia-Geomyidae). Chromosoma, 25:172–183, 8 figs., 1 table, July 1.

———. 1968. An analysis of the distribution of pocket gopher species in northeastern California (genus *Thomomys*). Univ. California Publ. Zool., 86:1–46, 13 figs., 1 table, September 19.

———. 1968. An analysis of three hybrid populations of pocket gophers (genus *Thomomys*). Evolution, 22:543–555, 3 figs., 5 tables, September 20.

———. 1972. Taxonomic status of the pocket gophers, *Thomomys idahoensis* and *Thomomys pygmaeus* (Rodentia, Geomyidae). Jour. Mamm., 53:417–428, 3 figs., 1 table, September 27.

THOMA, B., and GUNDERSON, H. L. 1963. A new county record for the grasshopper mouse. Flicker, 35:27–28, March.

THOMAS, O. 1901. *On a collection of mammals from the Kanuku Mountains, British Guiana*. Ann. Mag. Nat. Hist., ser. 7, 8:139–154, August.

———. 1901. *New insular forms of* Nasua *and* Dasyprocta. Ann. Mag. Nat. Hist., ser. 7, 8:271–273, October.

———. 1902. On some mammals from Coiba Island, off the west coast of Panama. Novitates Zool., 9:135–137, April.

———. 1902. *On the geographical races of the kinkajou*. Ann. Mag. Nat. Hist., ser. 7, 9:266–269. April.

———. 1902. Diagnosis of a new Central-American porcupine. Ann. Mag. Nat. Hist., ser. 7, 10:169, August.

———. 1903. *Notes on neotropical mammals of the genera* Felis, Hapale, Oryzomys, Akodon, *and* Ctenomys, *with descriptions of new species*. Ann. Mag. Nat. Hist., ser. 7, 12:234–243, August.

———. 1913. *The geographical races of the woolly opossum* (Philander laniger). Ann. Mag. Nat. Hist., ser. 8, 12:358–361, October.

THOMAS, R., and SCHEFFER, V. B. 1962. Records of ringed seals from the Pribilof Islands. Jour. Mamm., 43:428, August 20.

THOMPSON, J. N., and GREER, J. K. 1969. Oklahoma range of Dipodomys ordii (Rodentia: Heteromyidae). Proc. Oklahoma Acad. Sci., 48:108–109.

THOMPSON, L. S. 1977. Dwarf shrew (Sorex nanus) in north-central Montana. Jour. Mamm., 58:248–250, May 31.

THRELFALL, W. 1969. Further records of helminths from Newfoundland mammals. Canadian Jour. Zool., 47:197–201, 2 tables, March.

TIMM, R. M. 1975. Distribution, natural history, and parasites of mammals of Cook County, Minnesota. Bell Mus. Nat. Hist., Univ. Minnesota, Occas. Pap., 14:ii + 56, 7 figs., 15 tables, December 30.

TOLL, J. E., BASKETT, T. S., and CONAWAY, C. H. 1960. Home range, reproduction, and foods of the swamp rabbit in Missouri. Amer. Midland Nat., 63:398–412, 6 figs., 3 tables, April.

TOWNSEND, C. H. 1912. Mammals collected by the "Albatross" Expedition in Lower California in 1911, with descriptions of new species. Bull. Amer. Mus. Nat. Hist., 31:117–130, Pls. 8–9, June 14.

TRETHEWEY, D. E. C., and VERTS, B. J. 1971. Reproduction in eastern cottontail rabbits in western Oregon. Amer. Midland Nat., 86:463–476, 6 figs., 3 tables, October.

TRUE, F. W. 1885. On the occurrence of Loncheres armatus, (Geoff.) Wagner, in the island of Martinique, West Indies. Proc. U.S. Nat. Mus., 7:550–551.

———. 1889. Contributions to the natural history of the cetaceans. A review of the family Delphinidae. Bull. U.S. Nat. Mus., 36:1–191, 47 pls., numerous tables.

———. 1913. Description of Mesoplodon mirum, a beaked whale recently discovered on the coast of South Carolina. Proc. U.S. Nat. Mus., 45:651–657, Pls. 52–57, 1 fig., November 29.

TURNER, R. W. 1968. The olive-backed pocket mouse in southern Colorado. Jour. Mamm., 49:524, August 20.

———. 1974. Mammals of the Black Hills of South Dakota and Wyoming. Univ. Kansas Mus. Nat. Hist., Miscl. Publ., 60:1–178, 14 figs., 29 tables, April 3.

TURNER, R. W., and DAVIS, W. H. 1970. Bats from the Black Hills of South Dakota. Trans. Kansas Acad. Sci., 72:360–364, March 3.

TURNER, R. W., and JONES, J. K., JR. 1968. Additional notes on bats from western South Dakota. Southwestern Nat., 13:444–447, December 31.

TUTTLE, M. D. 1964. Additional record of Sorex longirostris in Tennessee. Jour. Mamm., 45:146–147, March 12.

———. 1964. Myotis subulatus in Tennessee. Jour. Mamm., 45:148–149, March 12.

———. 1968. First Tennessee record of Mustela nivalis. Jour. Mamm., 49:133, February 20.

TUTTLE, M. D., and ROBERTSON, P. B. 1969. The gray bat, Myotis grisescens, east of the Appalachians. Jour. Mamm., 50:370, June 12.

TYSON, E. L. 1964. Two new records of bats (Molossidae) from Panama. Jour. Mamm., 45:495–496, September 15.

ULMER, F. A. 1963. Northward extension of the range of the cotton mouse in Virginia. Jour. Mamm., 44:273–274, May 20.

URBAN, E. K., and WIMMER, R. B. 1959. Reithrodontomys megalotis and Urocyon cinereoargenteus from western Oklahoma. Jour. Mamm., 40:450, August 20.

VALDEZ, R., and LAVAL, R. K. 1971. Records of bats from Honduras and Nicaragua. Jour. Mamm., 52:247–250, February 26.

VAN BREE, P. J. H. 1971. On Globicephala sieboldii Gray, 1846, and other species of pilot whales (Notes on Cetacea, Delphinoidea III). Beaufortia, Inst. Taxonomic Zool. (Zool. Mus.), Univ. Amsterdam, Ser. Miscl. Publs., No. 249, Vol. 19:79–87, 3 figs., 1 table, September 17.

VAN BREE, P. J. H., and CADENAT, J. 1968. On a skull of Peponocephala electra (Gray, 1846) (Cetacea, Globicephalinae) from Sénégal. Beaufortia, 14:193–202, 3 figs., 3 tables, July 10.

VAN BREE, P. J. H., and PURVES, P. E. 1972. Remarks on the validity of Delphinus bairdii (Cetacea, Delphinidae). Jour. Mamm., 53:372–374, 1 table, June 23.

VAN CURA, N. J., and HOFFMEISTER, D. F. 1966. A taxonomic review of the grasshopper mice, Onychomys, in Arizona. Jour. Mamm., 47:613–630, 6 figs., 3 tables, December 2.

VAN GELDER, R. G. 1959. A taxonomic revision of the spotted skunks (genus Spilogale). Bull. Amer. Mus. Nat. Hist., 117:229–392, 47 figs., 32 tables, June 15.

———. 1960. Results of the Puritan-American Museum of Natural History Expedition to western Mexico. 10. Marine mammals from the coasts of Baja California and the Tres Marías Islands, México. Amer. Mus. Novit., 1992:1–27, 10 figs., 3 tables, March 8.

———. 1960. Carnivores in Puebla, Mexico. Jour. Mamm., 41:519, November 11.

VAN GELDER, R. G., and WINGATE, D. B. 1961. The taxonomy and status of bats in Bermuda. Amer. Mus. Novit., 2029:1–9, 4 figs., 1 table, May 8.

VAN ZYLL DE JONG, C. G. 1972. A systematic review of the Nearctic and Neotropical river otters (genus Lutra, Mustelidae, Carnivora). Royal Ontario Mus., Life Sci. Contrib., No. 80:1–104, 3 appendices, 38 figs., 3 tables, June 15.

VAN ZYLL DE JONG, C. G., CAMPBELL, D. C., and MERILEES, W. J. 1975. A new distribution record for the California Myotis (Myotis californicus) in British Columbia. Canadian Field-Nat., 89:322–323, 1 fig., September 5.

VARONA, L. S. 1965. Balaenoptera borealis Lesson (Mammalia: Cetacea) capturada en Cuba. Poeyana, Inst. Biol., ser. A, 7:1–4, 1 table, June 22.

———. 1970a. Nueva especie y nuevo subgénero de Capromys (Rodentia: Caviomorpha) de Cuba. Poeyana, ser. A, 73:1–18, 4 figs., 4 tables, July 30.

———. 1970b. Descripción de una nueva especie de Capromys del sur de Cuba (Rodentia: Caviomorpha). Poeyana, ser. A, 74:1–16, 4 figs., 4 tables, July 30.

———. 1974. Catálogo de los mamíferos vivientes y extinguidos de las Antillas. Acad. Cienc. Cuba, viii + 139, 1 map.

VARONA, L. S., and GARRIDO, O. H. 1970. Vertebrados de los cayos de San Felipe, Cuba, incluyendo una nueva especie de jutía. Poeyana, ser. A, 75:1–26, 3 figs., 3 tables, August 12.

VAUGHAN, T. A. 1959. Functional morphology of three bats: Eumops, Myotis, Macrotus. Univ. Kansas Publ., Mus. Nat. Hist., 12:1–153, 4 pls., 17 figs., July 8.

———. 1961. Vertebrates inhabiting pocket gopher burrows in Colorado. Jour. Mamm., 42:171–174, May 20.

———. 1969. Reproduction and population densities in a montane small mammal fauna. Univ. Kansas Mus. Nat. Hist., Miscl. Publ., 51:51–74, 8 figs., 4 tables, July 11.

VAUGHAN, T. A., and BATEMAN, G. C. 1970. Functional morphology of the forelimb of mormoopid bats. Jour. Mamm., 51:217–235, 7 figs., May 20.

VERTS, B. J. 1965. Extension of the range of the cotton mouse in Arkansas. Jour. Mamm., 46:100, February 20.

———. 1975. New records for three uncommon mammals in Oregon. Murrelet, 56(3):22–23, December.

VIBE, C. 1967. Arctic animals in relation to climatic fluctuations. Meddel. om Gronlend, 170(5):1–227, 101 figs., 3 tables, September 6.

VILLA-R., B. 1958. El mono araña (Ateles geoffroyi) encontrado en la costa de Jalisco y en la región central de Tamaulipas. Anal. Inst. Biol., Univ. Nac. Autó. México, 28:345–347, June 14.

———. 1967. Los murciélagos de México. Anal. Inst. Biol., Univ. Nac. Autó. México, xvi + 491, 171 figs., 38 tables, 98 maps, February 6.

VILLA-R., B., and COCKRUM, E. L. 1962. Migration in the guano bat Tadarida brasiliensis mexicana (Saussure). Jour. Mamm., 43:43–64, 2 figs., February 28.

VILLA-R., B., and RAMÍREZ-P., J. 1971. Diclidurus virgo Thomas, el murciélago blanco, en la costa de Nayarit, México. Anal. Inst. Biol., Univ. Nac. Autó. México, 39:155–158, 1 fig., July 14.

VOGT, D. W., and ARAKAKI, D. T. 1971. Karyotype of the American red fox (Vulpes fulva). Jour. Hered., 62:318–319, 1 fig., September.

VON BLOEKER, J. C., JR. 1967. The land mammals of the southern California islands. Proc. Symp. Biol. California islands, Santa Barbara Bot. Garden, California, pp. 245–263, August 18.

VORHIES, C. T., and TAYLOR, W. P. 1922. Life history of the kangaroo rat, Dipodomys spectabilis spectabilis Merriam. Bull. U.S. Dept. Agric., 1091:1–40, 9 pls., 3 figs., 2 tables, September 13.

VORONTSOV, N. N., and LIAPUNOVA, E. A. 1973. Genetics and the problems of transberingian connections of Holarctic mammals, pp. 146–149, in "The Bering land bridge and its role for the history of Holarctic floras and faunas in the late Cenozoic"—theses of the reports of all-union symposium. Acad. Sci. U.S.S.R., far-eastern scientific centre, Khabarovsk, pp. 1–222.

WALLACE, J. T., and HOUP, R. 1969. Marginal record of Parascalops breweri (Bachman) from Kentucky. Trans. Kentucky Acad. Sci., 29:9.

WALLEY, H. D. 1970. A Brazilian free-tailed bat (Tadarida brasiliensis) taken in north-central Illinois. Trans. Illinois Acad. Sci., 63:113.

WALTERS, R. D., and ROTH, V. D. 1950. Faunal nest study of the woodrat, Neotoma fuscipes monochroura Rhoads. Jour. Mamm., 31:290–292, August 21.

WALTON, D. W. 1963. A collection of the bat Lonchophylla robusta Miller from Costa Rica. Tulane Studies in Zool., 10:87–90, May 3.

WARD, F. D., and LEONARD, R. G. 1968. Rodents drown in improvised mast traps. Jour. Mamm., 49:530, August 20.

WARNER, D. W., and BEER, J. R. 1957. Birds and Mammals of the Mesa de San Diego, Puebla, Mexico. Acta Zool. Mexicana, 2(4–5):1–21, 1 table, December 15.

WATERHOUSE, G. R. 1848. A natural history of the Mammalia. Hippolyte Baillière, London, 2 (Rodentia):1–500, illustrated.

WATERS, J. H. 1962. Range of Peromyscus maniculatus in southern New England. Jour. Mamm., 43:102, February 28.

———. 1969. The systematic position of white-footed mice, genus Peromyscus, of Nantucket, Massachusetts. Jour. Mamm., 50:129–132, 2 tables, February 26.

WATERS, J. H., and RAY, C. E. 1961. Former range of the sea mink. Jour. Mamm., 42:380–383, 1 fig., August 21.

WATERS, J. H., and RIVARD, C. J. 1962. Terrestrial and marine mammals of Massachusetts and other New England states. Privately published, pp. vi + 151, 33 figs., 1 table, 47 maps.

WATERS, J. H., and STOCKLEY, B. H. 1965. Hibernating meadow jumping mouse on Nantucket Island, Massachusetts. Jour. Mamm., 46:67–76, 3 figs., 1 table, February 20.

WATKINS, L. C., JONES, J. K., JR., and GENOWAYS, H. H. 1972. Bats of Jalisco, México, Special Publ. Mus. Texas Tech Univ., 1:1–44, 3 figs., 3 tables, December 8.

WAUER, R. H. 1961. Peculiar actions of coyote and kit fox. Jour. Mamm., 42:109, February 20.

———. 1965. Genus and species of shrew new for Utah. Jour. Mamm., 46:496, August 26.

WEAVER, R. L. 1940. Notes on a collection of mammals from the southern coast of the Labrador Peninsula. Jour. Mamm., 21:417–422, November 14.

WEBB, R. G., and BAKER, R. H. 1971. Vertebrados terrestres del suroeste de Oaxaca. Anal. Inst. Biol., Univ. Nac. Autó. México, 40:139–151, 1 fig., 1 table, December 15.

WENTWORTH, F. A., and SUTTON, D. A. 1969. Chromosomes of the Townsend pocket gopher, Thomomys townsendii. Southwestern Nat., 14:157–161, 3 figs., October 10.

WEST, G. C., and HUTCHISON, V. H. 1964. The southern bog lemming in Rhode Island. Jour. Mamm., 45:479–480, September 15.

WETZEL, R. M., and KOCK, D. 1973. The identity of Bradypus variegatus Schinz (Mammalia, Edentata). Proc. Biol. Soc. Washington, 86:25–34, May 31.

WETZEL, R. M., and SHELAR, E. 1964. The water shrew in southern Connecticut. Jour. Mamm., 45:311, May 20.

WHARTON, C. H. 1968. First records of Microsorex hoyi and Sorex cinereus from Georgia. Jour. Mamm., 49:158, February 20.

WHARTON, C. H., and WHITE, J. J. 1967. The red-backed vole, Clethrionomys gapperi, in north Georgia. Jour. Mamm., 48:670–672, 1 table, November 20.

WHITAKER, J. O., JR., and GOODWIN, R. E. 1960. Additional record of Peromyscus maniculatus bairdi in New York. Jour. Mamm., 41:518, November 11.

WHITAKER, J. O., JR., and SLY, G. R. 1970. First record of Reithrodontomys megalotis in Indiana. Jour. Mamm., 51:381, May 20.

WHITAKER, J. O., JR., and ZIMMERMAN, E. G. 1965. Additional Mustela nivalis records for Indiana. Jour. Mamm., 46:516, August 26.

———. 1968. Microtus ochrogaster in Alabama and Tennessee. Jour. Mamm., 49:328, May 20.

WHITE, J. A. 1953. Genera and subgenera of chipmunks.

Univ. Kansas Publ., Mus. Nat. Hist., 5:543–561, 12 figs., December 1.

———. 1953. Taxonomy of the chipmunks, Eutamias quadrivittatus and Eutamias umbrinus. Univ. Kansas Publ., Mus. Nat. Hist., 5:563–582, 6 figs., December 1.

———. 1953. Geographic distribution and taxonomy of the chipmunks of Wyoming. Univ. Kansas Publ., Mus. Nat. Hist., 5:583–610, 3 figs., December 1.

———. 1953. The baculum in the chipmunks of western North America. Univ. Kansas Publ., Mus. Nat. Hist., 5:611–631, 19 figs., December 1.

WIEGERT, R. G., and MAYENSCHEIN, J. C. 1966. Distribution and trap response of a small wild population of cotton rats (*Sigmodon h. hispidus*). Jour. Mamm., 47:118–120, 1 table, March 12.

WILLIAMS, D. F. 1968. *A new record of* Myotis thysanodes *from Washington.* Murrelet, 49:26–27, December.

WILLIAMS, D. F., and FINDLEY, J. S. 1968. The plains pocket mouse (*Perognathus flavescens*) in New Mexico. Jour. Mamm., 49:771, November 26.

WILLIAMS, E. E., and KOOPMAN, K. F. 1951. A new fossil rodent from Puerto Rico. Amer. Mus. Novit., 1515:1–9, 2 figs., June 4.

———. 1952. West Indian fossil monkeys. Amer. Mus. Novit., 1546:1–16, 4 figs., March 3.

WILLIAMS, S. L., and BAKER, R. J. 1976. Vagility and local movements of pocket gophers (Geomyidae: Rodentia). Amer. Midland Nat., 96:303–316, 3 figs., 2 tables, November 23.

WILSON, A. D. 1908. Hunting bears on horseback. National Geographic, 19:350–356, 4 photos, May.

WILSON, D. E. 1968. Ecological distribution of the genus *Peromyscus.* Southwestern Nat., 13:267–273, 4 tables, December 10.

———. 1973. The systematic status of *Perognathus merriami* Allen. Proc. Biol. Soc. Washington, 86:175–191, May 31.

———. 1976. The subspecies of *Thyroptera discifera* (Lichtenstein and Peters). Proc. Biol. Soc. Washington, 89:305–311, 1 fig., 1 table, October 12.

WILSON, J. W. III. 1974. Analytical zoogeography of North American mammals. Evolution, 28:124–140, 16 figs., 4 tables, April 4.

WILSON, N. 1960. A northernmost record of Plecotus rafinesquii Lesson (Mammalia, Chiroptera). Amer. Midland Nat., 64:500, October.

WING, E. S., HOFFMAN, C. A., JR., and RAY, C. E. 1968. Vertebrate remains from Indian sites on Antigua, West Indies. Caribbean Jour. Sci., 8:123–139, 4 figs., 4 tables, September–December.

WINKELMANN, J. R. 1962. Mammal records from Guerrero and Michoacán, México. Jour. Mamm., 43:108–109, February 28.

———. 1962. Additional records of *Mimon cozumelae.* Jour. Mamm., 43:112, February 28.

WOLFE, J. L. 1968. Armadillo distribution in Alabama and northwest Florida. Quart. Jour. Florida Acad. Sci., 31:209–212, 1 fig., September.

WOLFE, J. L., and ROGERS, D. T. 1969. Old field mammals in western Alabama. Jour. Mamm., 50:609–612, 1 fig., 1 table, August 22.

WOLFRAM, G. 1964. Coyotes: the silent invaders. Canadian Audubon, 26:112–115, illustrated, September–October.

WOOD, A. E. 1935. Evolution and relationships of the heteromyid rodents with new forms from the Tertiary of western North America. Ann. Carnegie Mus., 24:73–262, 157 figs., 5 tables, May 13.

WOOD, J. E., and ODUM, E. P. 1964. A nine-year history of furbearer populations on the AEC Savannah River Plant area. Jour. Mamm., 45:540–551, 3 figs., 4 tables, January 25.

WOOD, T. J., and TESSIER, G. D. 1974. First records of eastern flying squirrel (*Glaucomys volans*) from Nova Scotia. Canadian Field-Nat., 88:83–84, June.

WOOD, W. 1974. Muskrat origin, distribution, and range extension through the coastal areas of Del Norte Co. Calif. & Curry Co., Oregon. Murrelet, 55:1–4, July.

WOOLSEY, N. G. 1959. New northern record for *Lepus alleni* in Arizona. Jour. Mamm., 40:250, May 20.

WRIGLEY, R. E. 1969. Ecological notes on the mammals of southern Quebec. Canadian Field-Nat., 83:201–211, November 13.

———. 1972. Systematics and biology of the woodland jumping mouse, *Napaeozapus insignis.* Illinois Biol. Monog., 47:8 + 118, 22 figs., 8 tables, December.

———. 1974. Ecological notes on animals of the Churchill Region of Hudson Bay. Arctic, 27:201–214, 6 figs., 2 tables, September.

WRIGLEY, R. E., DRESCHER, H.-E., and DRESCHER, S. 1973. First record of the fox squirrel in Canada. Jour. Mamm., 54:782–783, August 31.

WRIGLEY, R. E., and DUBOIS, J. E. 1973. Distribution of the pocket gophers *Geomys bursarius* and *Thomomys talpoides* in Manitoba. Canadian Field-Nat., 87:167–169, 2 figs., June 4.

YARBROUGH, J. W., and STUDIER, E. H. 1968. Mink in northeast New Mexico. Southwestern Nat., 13:105, May 28.

YATES, T. L., and SCHMIDLY, D. J. 1977. Systematics of Scalopus aquaticus (Linnaeus) in Texas and adjacent states. Occas. Pap. Mus. Texas Tech Univ., 45:1–36, 8 figs., 5 tables, June 3.

YOAKUM, J. 1966. Ringtail in northern Nevada. Jour. Mamm., 47:351, May 26.

YOCOM, C. F., and ELEY, T. 1972. Extension of golden-mantled ground squirrel range west to the Pacific coastal area of Humboldt and Del Norte counties, northwestern California. Murrelet, 53:29–30, August.

YOUNG, W. A. 1962. Range extension of dusky-footed wood rat. Murrelet, 42:6, February.

YOUNGMAN, P. M. 1958. Geographic variation in the pocket gopher, Thomomys bottae, in Colorado. Univ. Kansas Publ., Mus. Nat. Hist., 9:363–384, 7 figs., February 21.

———. 1962. The spread of the European hare to the Ottawa region of Ontario. Canadian Field-Nat., 76:223, January 7.

———. 1964. Range extensions of some mammals from northwestern Canada. Nat. Mus. Canada Nat. Hist. Pap., 23:1–6, April 28.

———. 1967. A new subspecies of varying lemming, Dicrostonyx torquatus (Pallas), from Yukon Territory (Mammalia, Rodentia). Proc. Biol. Soc. Washington, 80:31–33, March 24.

———. 1967. Insular populations of the meadow vole, Microtus pennsylvanicus, from northeastern North America, with descriptions of two new subspecies. Jour. Mamm., 48:579–588, 4 figs., 1 table, November 20.

——. 1968. Notes on mammals of southeastern Yukon Territory and adjacent Mackenzie District. Bull. Nat. Mus. Canada, 223:70–86, 3 pls., 1 fig., April.

——. 1975. Mammals of the Yukon Territory. Nat. Mus. Nat. Sci. Canada, Publ. Zool., No. 10:1–192, colored frontispiece, 8 figs., 39 tables, 60 maps, September 4.

YOUNGMAN, P. M., and GILL, D. A. 1968. First record of the southern flying squirrel, *Glaucomys volans volans,* from Quebec. Canadian Field-Nat., 82:227–228, 1 fig., October 15.

ZIMMERMAN, E. G. 1970. Karyology, systematics and chromosomal evolution in the rodent genus, *Sigmodon.* Publ. Michigan State Univ., Biol. Ser., 4:385–454, 16 pls., 9 figs., 4 tables, December 11.

ZIMMERMAN, E. G., and COTHRAN, E. G. 1976. Hybridization in the Mexican and 13-lined ground squirrels, *Spermophilus mexicanus* and *Spermophilus tridecemlineatus.* Experientia, 32:704–706, 6 figs., June 15.

ZIMMERMANN, K. 1955. Die Gattung *Arvicola* Lac. im System der Microtinae. Säugetierkund. Mitteil., 3:110–112, 2 abb., July 1.

ADDENDA

The following information was published or came to the author's attention after June 30, 1977, when the manuscript was completed.

P. xv. "Calibration of the great American interchange . . . of terrestrial [mammalian] faunas between the Americas" by L. G. Marshall, *et al.* (Science, 204:272–279, April 20, 1979) provides pertinent new information on minimum time (before the present) when several genera of American mammals were in South America.

P. 27. *Sorex cinereus.* van Zyll de Jong (Canadian Jour. Zool., 54:963–973, 1976), without deciding on subspecific versus specific status of taxa of the *S. cinereus* group, found *S. c. ugyunak* along with the insular *S. hydrodromus* (probable synonym, *S. pribilofensis*) and *S. jacksoni* in the Bering Sea, to be closely related as "tundra forms," but distinct from the "woodland forms." *S. c. haydeni*, the "prairie form," resembled the tundra forms. *S. c. hollisteri* resembled the woodland forms.

P. 29. *Sorex cinereus fontinalis.* Kirkland (Proc. Pennsylvania Acad. Sci., 51(1):43–46, December 1977), on basis of differences in size of specimens in southeastern Pennsylvania and southward slightly alters the geographic ranges there of *S. c. fontinalis* and *S. c. cinereus.*

P. 29. *Sorex cinereus haydeni* in Canada does not intergrade with *Sorex cinereus cinereus* according to van Zyll de Jong (Jour. Mamm., 61:66, February 20, 1980). If the two taxa do not intergrade south of Canada *Sorex haydeni* Baird 1858 will stand as a species.

P. 29. *Sorex cinereus lesueurii.* The occurrence just west of U.S. 41, 7/10 mi. S of Ohio River in Henderson Co., Kentucky (see French, Trans. Kentucky Acad. Sci., 39:78, March 1978) provides a marginal record within the shaded area on Map 14.

P. 31. *Sorex preblei.* Specimen No. 186678 USNM from Fort Custer, Montana, and No. 133547 KU from sec. 4, T. 46 N, R. 21 E, in Nevada (manuscript of R. S. Hoffmann and Robert D. Fisher, 1978) provide marginal records within the shaded area on Map 15.

P. 32. Hennings and Hoffmann (Occas. Pap. Mus. Nat. Hist. Univ. Kansas, 68:1–35, July 15, 1977) list several unsolved problems for *Sorex vagrans,* and propose new name combinations most of which do not meet the conditions imposed by the International Code of Zoological Nomenclature.

P. 46. *Sorex gaspensis.* The occurrence on South Mountain, in the Grande Anse Valley, and in the Cheticamp River Valley (Roscoe and Majka, Canadian Field-Nat., 90(4):497, 1976) of Cape Breton Island, Nova Scotia, are precise localities not plotted on Map 26.

P. 47. *Sorex dispar dispar.* The occurrence at Beaver Brook, Aroostook Co., Maine (A. J. Godin, Wild mammals of New England, The Johns Hopkins Univ. Press, p. 28, 1977) pro-vides the northernmost record of this taxon. The specimen from $5\frac{3}{10}$ km. N, $3\frac{1}{2}$ km. W of Riverside-Albert, Albert Co., New Brunswick (Kirkland, *et al.,* Canadian Field-Nat., 93(2):196, 1979) provides the easternmost record. See also Kirkland and Van Deusen (The shrews of the *Sorex dispar* Group . . . , Amer. Mus. Novit., 2675:1–21, 5 figs., 6 tables, May 7, 1979).

P. 48. *Sorex merriami* (likely *leucogenys*). An easternmost occurrence for Colorado is 1 mi. S, 2 mi. W Ramah (Spencer, Jour. Colorado-Wyoming Acad. Sci., 7(6):48, 1975, who listed two specimens not identified to subspecies). Ramah is *ca.* 42 mi. NE of the "Black Forest" locality on Map 28 and outside the shaded area.

P. 50. **Sorex arizonae** Diersing and Hoffmeister (Jour. Mamm., 58:330, August 20, 1977) is named as a new species from SE Arizona, possibly related to *Sorex oreopolus emarginatus,* or to *Sorex merriami* that they regard (p. 328) as monotypic species. Conway and Schmitt (Jour. Mamm., 59:631, August 21, 1978) record *S. arizonae* from New Mexico: Turkey Spring, Animas Mts., Hidalgo Co. Caire, Vaughan, and Diersing (Southwestern Nat., 22(2):532, August 10, 1978) record *S. arizonae* from *ca.* 60 km. by road W Tomochic, Chihuahua, Mexico.

P. 51. *Microsorex* is reduced from generic to subgeneric rank by Diersing (Jour. Mamm., 61:80, February 20, 1980). He places *Sorex alnorum* Preble 1902, *Microsorex hoyi intervectus* Jackson 1925, and *Microsorex hoyi washingtoni* Jackson 1925 as synonyms of *Sorex hoyi hoyi* Baird 1858.

P. 65. *Notiosorex crawfordi evotis.* The occurrence "near Plateado, Zacatecas" recorded by Fisher and Bogan (Proc. Biol. Soc. Washington, 90:826, December 6, 1977) is a marginal record within the shaded area on Map 38.

P. 66. *Notiosorex gigas.* A specimen listed from 3 mi. N Coapan, Nayarit, by Fisher and Bogan (Proc. Biol. Soc. Washington, 90:827, December 6, 1977) is approximately 220 km. northward from the shaded area on Map 38.

P. 104. *Macrotus waterhousii californicus.* A pre-Columbian occurrence, recorded as *M. californicus,* by Ray and Wilson (Occas. Pap. Mus. Texas Tech Univ., 57:2, 3, 7, March 30, 1979) from $2\frac{1}{2}$ mi. W of Terlingua, Texas, suggests a former distribution "across the continent in the region of the present international boundary."

P. 110. Davis and Carter (Occas. Pap. Mus. Texas Tech Univ., 53:7, 8, September 29, 1978) name as new **Tonatia silvicola centralis** with type locality at El Castillo, 40 m., department of Río San Juan, Nicaragua, and **Tonatia evotis** with type locality at 25 km. SSW Puerto Barrios, 75 m., department of Izabel, Guatemala.

P. 126. *Monophyllus plethodon luciae.* Baker, *et al.* (Occas. Pap. Mus. Texas Tech Univ., 50:4, May 26, 1978) report specimens from Guadeloupe, Lesser Antilles.

P. 151. *Chiroderma villosum jesupi.* The occurrence at 4 km. E San Felipe Orizatlán, *ca.* 500 m., Hidalgo, reported by Carter and Jones (Occas. Pap. Mus. Texas Tech Univ., 54:7, September 29, 1978) provides the northernmost record for this species in the lowlands of eastern México.

P. 151. *Chiroderma improvisum.* Jones and Baker (Occas. Pap. Mus. Texas Tech Univ., 60:2, September 21, 1979) record this species from Montserrat, 55 km. NW of the type locality.

P. 155. **Artibeus jamaicensis schwartzi** Jones, J. K., Jr. (Occas. Pap. Mus. Texas Tech Univ., 51:2, August 11, 1978) is known only from St. Vincent, Lesser Antilles.

Pp. 164, 165. *Ardops nichollsi annectens* and *Ardops nichollsi montserratensis.* Jones and Baker (Occas. Pap. Mus. Texas Tech Univ., 60:2–5, August 11, 1978) provide new information on the range in size of these two subspecies.

P. 169. **Brachyphylla cavernarum intermedia** Swanepoel and Genoways (Bull. Carnegie Mus. Nat. Hist., No. 12:38, December 27, 1978), type from 1 mi. W Corozal, Puerto Rico, is named as a new subspecies occurring in Puerto Rico and the Virgin Islands excluding St. Croix.

P. 170. *Brachyphylla nana* Miller is raised from subspecific rank to specific rank by Swanepoel and Genoways (Bull. Carnegie Mus. Nat. Hist., No. 12:49, December 27, 1978).

P. 170. *Brachyphylla pumila* Miller is arranged as a synonym of *Brachyphylla nana* by Swanepoel and Genoways (Bull. Carnegie Mus. Nat. Hist., No. 12:49, December 27, 1978).

P. 171. *Erophylla bombifrons santacristobalensis* (Elliot) is arranged as a synonym of *Erophylla sezekorni bombifrons* (Miller) by Buden (Proc. Biol. Soc. Washington, 89:14, June 25, 1976).

P. 174. *Desmodus rotundus* possibly *murinus.* Osteological remains of pre-Columbian age of this species from a fissure deposit 2½ mi. W of Terlingula, Texas (Ray and Wilson, Occas. Pap. Mus. Texas Tech Univ., 57:1,2, March 30, 1979) are from a place not shown on Map 138.

P. 178. *Natalus stramineus stramineus.* Baker, *et al.* (Occas. Pap. Mus. Texas Tech Univ., 50:11, May 26, 1978) report a specimen from Guadeloupe, Lesser Antilles.

P. 187. **Myotis findleyi** Bogan, Jour. Mamm., 59:524, August 21, 1978, type from Isla María Magdalena, Islas Tres Marías, Nayarit, is named as a new species.

P. 189. *Myotis yumanensis lutosus.* The occurrence at 5 km. E Zacualtipán, 2100 m., Hidalgo, reported by Carter and Jones (Occas. Pap. Mus. Texas Tech Univ., 54:10, September 29, 1978) provides the easternmost record for this species in México.

P. 199. *Myotis carteri* is accorded specific, not subspecific, rank by Bogan (Jour. Mamm., 59:519, August 21, 1978).

P. 199. *Vespertilio exiguus* H. Allen, Proc. Acad. Nat. Sci. Philadelphia, 18:281, 1866, with type from Aspinwall [= Colón, Panamá] may be a synonym of *Myotis nigricans nigricans.*

P. 199. *Myotis punensis* J. A. Allen, Bull. Amer. Mus. Nat. Hist., 33:383, July 9, 1914, type from Puna Island, Ecuador, is arranged by Bogan (Jour. Mamm., 59:524, August 21, 1978) as a synonym of *M. nigricans nigricans.*

P. 199. *Myotis nigricans extremus* is recognized as a valid

subspecies by Bogan (Jour. Mamm., 59:524, August 21, 1978), who records specimens from only the type locality.

P. 200. *Myotis dominicensis.* Carter and Dolan (Special Publ. Mus. Texas Tech Univ., 15:75, July 28, 1978) call attention to the name *Myotis splendidus* (Wagner) proposed as *Vespertilio splendidus* (Arch. Naturgesch., 11(1):148, 1845) for the *Myotis* of St. Thomas Island, Virgin Islands.

P. 200. *Vespertilio leucogaster.* Carter and Dolan (Special Publ. Mus. Texas Tech Univ., 15:73, July 28, 1978) convincingly maintain that *Vesp[ertilio]. leucogaster* is one of the names of which Schinz, not Wied-Neuwied, is the author.

P. 205. *Myotis keenii septentrionalis* (Trouessart). van Zyll de Jong (Canadian Jour. Zool., 57:993, May 1979) "suggests specific [instead of subspecific] status for" this taxon, and records it from as far west and north as: Cadomin in Alberta, Hudson Hope in British Columbia, and Nahanni National Park in the Northwest Territories.

P. 208. *Myotis evotis evotis.* The specimen from Corral Draw, South Dakota, and the one from Warbonnet Canyon, Nebraska, are identified (Jones and Choate, The Prairie Naturalist, 10(2):50, 51, June 1978) as *Myotis thysanodes pahasapensis.* This does not alter Map 161 of *M. thysanodes* but does alter Map 164 of *M. evotis* because the latter species now is not known to occur in Nebraska or southwestern South Dakota.

P. 221. *Nycteris intermedia intermedia.* The occurrence at 4 km. E San Felipe Orizatlán, *ca.* 500 m., Hidalgo, reported by Carter and Jones (Occas. Pap. Mus. Texas Tech Univ., 54:10, September 29, 1978) provides a marginal record for this subspecies.

P. 227. *Nycticeius humeralis humeralis.* A specimen from 3 mi. S Red Cloud, Nebraska (3 mi. N Kansas boundary), is barely within the shaded area on Map 177 and the westernmost record in Nebraska (Farney and Jones, Mammalia, 39:329, December 20, 1975).

P. 232. *Idionycteris* is accorded generic instead of subgeneric rank by Williams, Druecker, and Black (Jour. Mamm., 51:605, August 28, 1970).

P. 239. *Molossops planirostris planirostris.* Carter and Dolan (Special Publ. Mus. Texas Tech Univ., 15:85, 86, July 28, 1978) designate a lectotype from Cayenne, French Guiana.

P. 241. *Tadarida brasiliensis cynocephala.* The occurrence at four places in North Carolina as far north as Campbell College, Buies Creek, Harnett Co. extends the known range 258 km. northeastward from Columbia, Sumter Co., South Carolina according to Lee and Marsh (Amer. Midland Nat., 100:240, July 1978).

P. 242. *Tadarida brasiliensis mexicana.* A specimen from 1 mi. S and 18 mi. E Valentine, in Keya Paha County, of north-central Nebraska, extends the known geographic range of this subspecies to within 10 miles of the South Dakota boundary (Farney and Jones, Mammalia, 39:329, December 20, 1975).

P. 248. *Eumops perotis gigas.* Carter and Dolan (Special Publ. Mus. Texas Tech Univ., 15:91, July 28, 1978) list the holotype or syntype as from Rio Negro [in state of Amazonas], Brazil.

P. 301. *Sylvilagus floridanus chapmani.* Schmidly (The Mammals of Trans-Pecos Texas, 1977:180, November 14) reports specimens from 24 mi. SW Ft. Stockton and 23 mi. SW McCamey. Both are marginal records to the west.

P. 301. **Sylvilagus floridanus macrocarpus** Diersing and Wilson (Smithsonian Contrib. Zool. No. 297:8, January 8, 1980), type from Estanzuela, 1372 m., Nayarit.

P. 304. S[ylvilagus]. f[loridanus]. restrictus is a synonym of S[ylvilagus]. f[loridanus]. orizabae according to Diersing and Wilson, Smithsonian Contrib. Zool. No. 297:9, January 8, 1980.

P. 305. S[ylvilagus]. f[loridanus]. subcinctus is a synonym of S[ylvilagus]. f[loridanus]. orizabae according to Diersing and Wilson, Smithsonian Contrib. Zool. No. 297:14, January 8, 1980.

P. 313. According to Diersing and Wilson, Smithsonian Contrib. Zool. No. 297:5, January 8, 1980, materials (2 specimens) here recorded as Sylvilagus cunicularius insolitus (on authority of Armstrong and Jones, Jour. Mammalogy, 52:755, December 16, 1971) from San Ignacio, actually are Sylvilagus floridanus holzneri as is a third specimen recorded (op. cit.: 754) from San Ignacio.

P. 314. **Sylvilagus graysoni badistes** Diersing and Wilson, Smithsonian Contrib. Zool. No. 297:15, January 8, 1980), type from San Juanito Island of the Tres Marias Islands, Nayarit.

P. 318. Lepus othus othus. Anderson, H. L. (The Murrelet, 59:73, 1978) believes that L. othus no longer occurs north of the "Selawik-Kotzebue area" and therefore is not now to be found on any part of the Arctic Slope of Alaska.

P. 338. Tamias striatus pipilans Lowery is regarded as a synonym of Tamias striatus striatus (Linnaeus) by C. Jones and R. D. Suttkus (Proc. Biol. Soc. Washington, 91:837, February 23, 1979).

P. 340. Tamias striatus venustus. The occurrence 9 mi. SE Shawnee, Pottawatomie Co., Oklahoma, reported by Black, J. H., et al. (Proc. Oklahoma Acad. Sci., 57:163, September 1, 1977) extends the known geographic range approximately 45 miles westward from the longitude of occurrences previously reported in Oklahoma.

Pp. 411, 415. Pizzimenti (Occas. Pap. Mus. Nat. Hist. Univ. Kansas, 39:1–73, September 10, 1975) supports Hollister's classification of species but questions the validity of two subspecies: C. l. arizonensis and C. g. zuniensis.

P. 418. Sciurus carolinensis pennsylvanicus. The occurrence in Rock County, Minnesota, reported by Ernst and French (Trans. Kansas Acad. Sci., 79:50, 1977) provides a marginal record within the shaded area on Map 279.

P. 436. According to Hoffmeister and Diersing (Jour. Mamm., 59:402–413, 5 figs., May 30, 1978) Sciurus kaibabensis Merriam 1904 is a subspecies of S. aberti Woodhouse 1853; S. a. mimus Merriam 1904 is a synonym of S. a. ferreus True 1900; S. a. navajo Durrant and Kelson 1947 is a synonym of S. a. aberti Woodhouse 1853; and S. a. phaeurus J. A. Allen 1904 is a synonym of S. a. durangi Thomas 1893.

P. 458. Thomomys talpoides confinis becomes Thomomys idahoensis confinis and occurs in the Bitterroot Valley, Montana, from Corvallis southward probably for no more than 40 km. according to Thaeler (Murrelet, 58:49–51, summer [= August 10], 1977).

P. 459. Thomomys talpoides idahoensis Merriam is arranged as a species (Thomomys idahoensis, consisting of T. i. confinis and T. i. idahoensis) distinct from Thomomys talpoides (Richardson) by Thaeler (Murrelet, 58:49–51, summer [= August 10], 1977).

Pp. 499, 500, 501. Geomys bursarius brazensis, Geomys bursarius dutcheri, Geomys bursarius ludemani, Geomys bursarius pratincola, and Geomys bursarius terricolus are regarded by Honeycutt and Schmidly (Occas. Pap. Mus. Texas Tech Univ., 58:43, June 1, 1979) as synonyms of Geomys bursarius sagittalis Merriam.

P. 529. Perognathus infraluteus Thomas 1893, Perognathus fasciatus litus Cary 1911, and Perognathus flavescens olivaceogriseus Swenk 1940 are arranged as synonyms of Perognathus fasciatus fasciatus Wied 1839 by Williams and Genoways (Ann. Carnegie Mus. Nat. Hist., 48:95, March 6, 1979).

P. 530. Perognathus flavescens cleomophila Goldman is arranged as a synonym of Perognathus flavescens apache by Williams (Bull. Carnegie Mus. Nat. Hist., No. 10:45, December 18, 1978).

P. 531. Perognathus flavescens gypsi Dice is arranged as a synonym of Perognathus flavescens melanotis by Williams (Bull. Carnegie Mus. Nat. Hist., No. 10:54, December 18, 1978).

P. 531. Perognathus flavescens melanotis. The occurrence at 3 mi. SE Willcox, Arizona (reported by Williams, Bull. Carnegie Mus. Nat. Hist., No. 10:55, December 18, 1978) provides a marginal record for this subspecies.

P. 570. Dipodomys ordii richardsoni. B. P. Glass (Southwestern Nat., 15(4):497–499, May 15, 1971) shows the type locality to be in Cimarron (not Harper) County, Oklahoma and localizes it to the confluence of Cienquilla and Currumpaw creeks, in sec. 32, T. 2 N, R. 2 E.

Pp. 576, 577. Best (Jour. Mamm., 59:174, February 17, 1978) arranged Dipodomys peninsularis peninsularis, D. p. australis, D. p. eremoecus, and D. p. pedinomus as valid subspecies of the species Dipodomys agilis. He arranged Dipodomys paralius as a synonym of Dipodomys agilis plectilis, and Dipodomys antiquarius as a synonym of Dipodomys agilis pedionomus.

P. 609. Oryzomys palustris texensis J. A. Allen and Oryzomys aquaticus J. A. Allen. On the basis of less than ideal cranial material Hall (Southwestern Nat., 5: 171–173, November 1, 1960) reported intergradation in southern Texas between O. palustris texensis and the more southern O. couesi aquaticus (of authors) and so arranged the two taxa as subspecies of one species.

On the basis of careful habitat study and more abundant material from other sites in southern Texas Benson and Gehlback (Jour. Mamm., 60:225–228. February 20, 1979) found no intergradation in size and coloration and so arranged the two taxa as separate species.

Additional study in southern Texas where the two taxa occur to learn if they intergrade at some place(s) or at no place(s) would be worthwhile.

P. 609. **Oryzomys argentatus** Spitzer and Lazell (Jour. Mamm., 59:787, November 24, 1978) is a species known only from the type locality, Cudjoe Key, Monroe Co., Florida.

P. 695. Peromyscus boylii cordillerae becomes Peromyscus aztecus cordillerae according to Carleton (Jour. Mamm., 60:293, May 29, 1979).

P. 697. Peromyscus boylii spicilegus becomes Peromyscus spicilegus according to Carleton (Jour. Mamm., 60:289, May 29, 1979).

P. 699. *Peromyscus evides* becomes *Peromyscus aztecus evides* according to Carleton (Jour. Mamm., 60:294, May 29, 1979).

P. 699. *Peromyscus aztecus* becomes *Peromyscus aztecus aztecus* according to Carleton (Jour. Mamm., 60:293, May 29, 1979).

P. 700. *Peromyscus oaxacensis* becomes *Peromyscus aztecus oaxacensis* according to Carleton (Jour. Mamm., 60:295, May 29, 1979).

P. 701. *Peromyscus hylocetes* becomes *Peromyscus aztecus hylocetes* according to Carleton (Jour. Mamm., 60:294, May 29, 1979).

P. 727. *Onychomys leucogaster albescens*. Schmidly (The Mammals of Trans-Pecos Texas, 1977:197, November 14) records two specimens from 4 mi. W Sierra Blanca which provide a marginal record outside the shaded area on Map 421. Also he refers specimens from Hudspeth County and 7½ mi. E El Paso City Hall to *O. l. albescens* instead of to *O. l. ruidosae.*

P. 730. *Onychomys leucogaster ruidosae.* See addenda under *O. l. albescens.*

P. 731. *Onychomys torridus longicaudus.* The occurrence 22 mi. W Winnemucca, Nevada (Engstrom and Fleharty, Great Basin Nat., 38:180, 1978), extends the known range *ca.* 65 mi. NNE.

P. 732. *Onychomys torridus arenicola* Mearns 1896 is recognized by Hinesley (Jour. Mamm., 60:117–128, February 20, 1979) as a monotypic species, *Onychomys arenicola.* Its geographic range so far as known is in New Mexico and extreme W Texas, and therefore wholly within that shown on Map 422 for *Onychomys torridus torridus.*

P. 736. *Sigmodon hispidus.* Figure 1 of Severinghaus and Hoffmeister (Jour. Mamm., 59:868, November 24, 1978) shows cranial differences between *S. hispidus* and *S. arizonae.*

P. 736. *Sigmodon hispidus*, probably the subspecies *alfredi*, has been reported by Carey, A. B. (Jour. Mamm., 59:624, August 21, 1978) from the Pueblo Army Depot, Colorado— *ca.* 38° 18' N, 104° 21' W.

P. 740. *Sigmodon mascotensis.* Severinghaus and Hoffmeister (Jour. Mamm., 59:869, 870, November 24, 1978) mention specimens from southern Nayarit south of the Río Grande de Santiago as "probably referable to" *S. mascotensis.* These are north of the geographic range shown on Map 427.

P. 741. *Sigmodon arizonae.* Figure 1 of Severinghaus and Hoffmeister (Jour. Mamm., 59:868, November 24, 1978) shows cranial differences between *S. arizonae* and *S. hispidus.*

P. 749. *Neotoma floridana illinoensis.* Details of contraction of geographic range in the present century in southern Illinois are documented by Nawrot and Klimstra (Nat. Hist. Miscellanea, Chicago Acad. Sci., No. 196:1–12, January 15, 1976).

P. 811. *Microtus chrotorrhinus* (unidentified to subspecies). The occurrence on a north-facing slope of Grande Anse Valley (Roscoe and Majka, Canadian Field-Nat., 90(4):498, 1976) of Cape Breton Island, Nova Scotia, is a precise locality not plotted on Map 459.

P. 842. *Zapus hudsonius* (unidentified to subspecies). The capture of a specimen about 25 mi. NW of Yellowknife, Mackenzie, by Krapu and Traugher (Amer. Midland Nat., 88:467, October 1972) extends the known range approximately 150 miles (along longitude 115°) north of the shaded area on Map 474.

P. 863. Varona (Poeyana, No. 195:1–51, December 12, 1979) named 12 additional new species of *Capromys* from Cuba. They are: **C. latus** (p. 5), **C. antiquus** (p. 8), **C. robustus** (p. 10), **C. megas** (p. 12), **C. jaumei** (p. 15), **C. kraglievichi** (p. 19), **C. barbouri** (p. 21), **C. delicatus** (p. 26), **C. beatrizae** (p. 28), **C. silvai** (p. 31), **C. gracilis** (p. 17), and **C. minimus** (p. 24). All are of Pleistocene age and the four last named occur also in deposits of Holocene age. **Brachycapromys** (p. 15), **Palaeocapromys** (p. 5) and **Stenocapromys** (p. 17) are names for three subgenera that Varona recognizes as heretofore unnamed.

Pp. 863, 864. Varona (Poeyana, No. 194:1–33, November 28, 1979), of the genus *Capromys*, named (page 5) **Pygmaeocapromys** as a new subgenus from the Holocene of Cuba with **Capromys (Pygmaeocapromys) angelcabrerai** new species (described on page 6) as the type species of the new subgenus. *Capromys nana* G. M. Allen 1917 also from the Holocene of Cuba is referred by Varona (*op. cit.:* 7) to the subgenus *Pygmaeocapromys.*

P. 871. *Echimys armatus armatus* (I. Geoffroy St.-Hilaire). 1830. *Mus hispidus* Lichtenstein, Darstellung neue . . . Säugethiere, pl. 35, fig. 1 and pl. 36, with text, type from Cayenne, French Guiana. Not *Echimys hispidus* Desmarest, 1817 [= *Mesomys hispidus*], type from South America.—Several lists of North American mammals (for example, Miller and Kellogg, 1955:648, Hall and Kelson, 1959:802, and Varona, 1974:73) have recorded *Echimys armatus* or *E. a. armatus* from the island of Martinique. Presumably the basis is True's (1885:550) account. He listed, described, and identified as *Loncheres armatus* a skin (13039 USNM) from the island of Martinique, received in 1878 from F. A. Ober, among specimens of West Indian animals. True speculated that the individual had been brought from South America on a sailing vessel, and had perished on the island without opportunity to reproduce. If such was the case, the species is not properly to be included in the North American list.

Pp. 880, 882. *Stenella longirostris* and *Stenella clymene.* According to Perrin, W. F., Mitchell, E. D., van Bree, P. J. H., and Caldwell, D. K. (in a typed, possibly published, paragraph), "It has been generally thought that spinner dolphins worldwide belong to a single species, *Stenella longirostris* (Gray, 1828). New data show that two closely related species exist in the tropical Atlantic, *S. longirostris* & *S. clymene* (Gray, 1850). The two forms differ in color pattern, external proportions & skull dimensions & proportions. *S. clymene* closely resembles *S. longirstris* in color pattern and tooth shape and size and is obviously closely related to it, but has a short, broad rostrum, is more robust in build, and has fewer teeth. Its skull is very much like that of *S. coeruleoalba* in proportions, but smaller, and some specimens have been erroneously referred to that species, as well as to *S. frontalis. S. longirostris* and *S. clymene* are sympatric. Both have been recorded from the Southeast, the Gulf of Mexico, the Caribbean, and the coast of Africa. The two species may represent successive waves of invasion from the Indopacific during interglacial periods" (Abstract, Sec. 1, Proc. Second Conference on the biology of marine mammals, San Diego, California, 12–15 December 1977).

See also van Bree and Perrin ("On the diagnosis of the spinner dolphin, *Stenella longirostris* (Gray, 1828) and its holotype," Zool. Mededelingen . . . Rijksmuseum . . . Leiden, 52(21): 255–259, 15 December 1977).

P. 958. *Ursus maritimus.* Don E. Wilson (paper 44:447–453, 8 figs., 3rd International Conference on bears, International Union for Conservation of Nature and Natural Resources, Morges, Switzerland—date of publication prior to April 19, 1978) analyzed data on significant geographic variation in North American Holocene polar bears but chose not to make taxonomic use of his data "until the Old World populations have been similarly analyzed."

P. 1064. *Phoca vitulina largha* Pallas 1811, is recorded by Shaughnessy and Fay (Jour. Zool., London, 182:397, July 1977) in the eastern Pacific from North America as follows: Herschel Island, Yukon Territory and Point Barrow, Alaska, south to the Pribilof Islands, Bristol Bay, and the eastern Aleutian Islands. *P. v. largha* inhabits principally the southern edge of the Arctic ice pack. Most of its geographic range is in the western Pacific. Differences between the three Pacific taxa of *Phoca* in color, size, and cranial shape, are well illustrated by photographs of skins and line drawings of the three subspecies. [Although listed as a subspecies on p. 395 (*op. cit.*), *largha* is identified as a species on p. 416 (*op. cit.*).]

Pp. 1089, 1091. In the dark area of Map 536, in northern California *Dama hemionus columbiana* and *Dama hemionus hemionus* occur together in spring, summer and early autumn, but in late autumn *columbiana* moves to lower elevations and westward, and *hemionus* moves to lower elevations eastward and northeastward. Because the two kinds are not together in the breeding season they do not interbreed in the darkened area although they do so farther north and south and therefore are arranged as subspecies of a single species. For details see Cowan, California Fish and Game, 22(3): 155–246, July 1936).

INDEX TO VERNACULAR NAMES

INDEX TO TECHNICAL NAMES

azarae *(Continued)*
Didelphis, 3
azteca,
Carollia, 137
Felis, 1039
aztecus,
Artibeus, 159
Arvicola, 795
Caluromys, 12
Felis, 1039
Hesperomys, 699
Lepus, 301
Molossus, 255
Myotis, 203
Oryzomys, 609
Peromyscus, 699, 1179
Philander, 12
Potos, 978
Reithrodontomys, 638
Sylvilagus, 301
azuerensis,
Ateles, 268
Oryzomys, 609
azulensis,
Peromyscus, 715

B

bachmani,
Lepus, 296
Musaraneus, 32
Sciurus, 428
Sylvilagus, 296
Thomomys, 477
badia,
Felis, 1036
Badiofelis, 1036
badistes,
Sylvilagus, 1179
badius,
Peromyscus, 710
Spermophilus, 394
Thomomys, 465
Baeodon, 228
alleni, 230
meyeri, 238
bahamensis,
Eptesicus, 214
Nyctinomus, 241
Tadarida, 241
Vespertilio, 214
baileyi,
Canis, 930
Castor, 601
Dipodomys, 579
Lepus, 308
Lynx, 1052
Microtus, 807
Myotis, 193
Neotoma, 748
Odocoileus, 1093
Perognathus, 544

baileyi *(Continued)*
Sciurus, 441
Sigmodon, 744
Sylvilagus, 308
Tamiasciurus, 441
Thomomys, 477
Baiomys, 723
allex, 725
analogus, 725
ater, 725
brunneus, 723
canutus, 726
fuliginatus, 726
grisescens, 723
handleyi, 723
infernatis, 724
musculus, 724
nebulosus, 724
nigrescens, 724
pallidus, 724
paulus, 726
pullus, 725
subater, 726
taylori, 725
Baiosciurus, 416
bairdi,
Elasmognathus, 1077
Gulo, 1008
Microtus, 810
Peromyscus, 671
Sorex, 34
Tapirus, 1077
Ursus, 954
bairdii,
Berardius, 904
Delphinus, 884
Elasmognathus, 1075
Hesperomys, 679
Lepus, 315
Mus, 671
Peromyscus, 671
Tapirus, 1077
balaclavae,
Peromyscus, 672
Balaena, 919
albicans, 899
aleoutiensis, 921
ampullata, 910, 911
angulata, 920
antarctica, 919
antiquorum, 916
arctica, 916, 921
atlanticus, 919
australis, 919, 920
biscayensis, 920
boops, 913, 916, 918
borealis, 914, 921
britannica, 920
cisarctica, 920
cullamach, 921
eubalaena, 920
euskariensis, 920

Balaena *(Continued)*
gibbosa, 912, 914
glacialis, 919, 920
groenlandica, 921
islandica, 920
japonica, 921
kuliomoch, 921
lalandii, 918
longimana, 918
lunulata, 921
major, 916
mediterranea, 920
minimus, 914
musculus, 914, 917
mysticetus, 919, 921
nodosa, 918
nordcaper, 920
novae angliae, 918
physalus, 914, 916
pitlekajensis, 921
robusta, 916
rostrata, 911, 914, 916
roysii, 921
sieboldi, 921
sulcata, 916
tarentina, 920
van benedeniana, 920
van benediana, 920
Balaenidae, 919
Balaenoptera, 913
acutorostrata, 914
acuto-rostrata, 914
alba, 916
antarctica, 919
aragous, 916
arctica, 916
australis, 918
blythii, 917
bonaerensis, 915
boops, 913
borealis, 915
brevicauda, 917
brydei, 915
capensis, 918
carolinae, 918
copei, 917
davidsoni, 915
edeni, 915
eschrichtii, 914
gibbar, 913, 916
gigas, 918
huttoni, 914
indica, 918
intermedia, 917
iwasi, 916
jubartes, 917
laticeps, 914, 916
leucopteron, 918
mediterraneensis, 916
microcephala, 915
miramaris, 917
musculus, 917

Balaenoptera *(Continued)*
patachonicus, 916
physalus, 916
quoyi, 916
robusta, 912
rorqual, 916
rostrata, 913
schlegelli, 915
swinhoei, 914
swinhoii, 916
syncondylus, 919
tenuirostris, 916
thalmaha, 915
velifera, 917
Balaenopteridae, 913
Balantiopteryx, 83
io, 84
ochoterenai, 84
pallida, 83
plicata, 83
Balena, 919
baliolus,
Peromyscus, 669
Sciurus, 422
Baloena, 919
balteatus,
Delphinus, 885
baltica,
Halichoerus, 1067
banderanus,
Peromyscus, 711, 721
bangsi,
Dicotyles, 1079
Felis, 1041, 1044
Glaucomys, 450
Mustela, 989
Pecari, 1079
Perognathus, 537
Sciuropterus, 450
Sciurus, 423
Vulpes, 939
bangsii,
Lepus, 320
banksianus,
Canis, 930
banksicola,
Lepus, 320
barabensis,
Glis, 367
barang,
Leptonyx, 1029
Barangia, 1029
barbara,
Eira, 1004
Galictis, 1005
Mustela, 1004
Tayra, 1005
barbata,
Phoca, 1066
barbatus,
Erignathus, 1066
Promops, 248

griseimembria *(Continued)*
  Aotus, 260
griseoalbus,
  Canis, 930
griseo-albus,
  Canis, 930
griseobracatus,
  Peromyscus, 668
griseocaudatus,
  Sciurus, 425
griseoflavus,
  Macroxus, 420
  Reithrodontomys, 644
  Sciurus, 420
griseoventer,
  Neotoma, 762
griseoventris,
  Cryptotis, 60
grisescens,
  Ateles, 269
  Baiomys, 723
  Eutamias, 344
  Myotis, 197
griseus,
  Arctopithecus, 280
  Bradypus, 280
  Canis, 930
  Delphinus, 891, 892
  Grampus, 892
  Halichoerus, 1067
  Heteromys, 597
  Leopardus, 1044
  Lepus, 328
  Peromyscus, 706
  Pteronotus, 92
  Reithrodontomys, 636
  Sciurus, 433
  Sigmodon, 738
  Tamias, 337
  Ursus, 954
Grison, 1006
  allamandia, 1006
Grisonella, 1006
Grisonia, 1006
groenlandica,
  Balaena, 921
  Capra, 1106
  Phoca, 1066
  Pterobalaena, 914
groenlandicus,
  Alopex, 935
  Anarak, 910, 911
  Canis, 935
  Dicrostonyx, 835
  Lepus, 320
  Mus, 836
  Pagophilus, 1066
  Rangifer, 1106
  Thalassarctos, 959
grönlandicus,
  Cervus, 1106
growlerensis,

growlerensis *(Continued)*
  Thomomys, 483
grunniens,
  Bos, 1113
Grymaeomys, 12
grypus,
  Halichoerus, 1067
  Phoca, 1067
gryphus,
  Pterobalaena, 918
  Vespertilio, 192
guadaloupensis,
  Globiocephalus, 894
guadalupensis,
  Microtus, 814
  Thomomys, 483
guadeloupensis,
  Eptesicus, 217
Guandira, 114
  cayanensis, 114
guardia,
  Peromyscus, 661
guardiae,
  Perognathus, 558
guatemalae,
  Ototylomys, 629
  Urocyon, 943
guatemalensis,
  Microtus, 720, 803
  Peromyscus, 714
gubernator,
  Lagenorhynchus,
    888
Guerlinguetus, 436
guerlinguetus,
  Sciurus, 436
guerrerensis,
  Cryptotis, 60
  Hodomys, 772
  Liomys, 591
  Neotoma, 772
  Orthogeomys, 508
  Oryzomys, 616
  Potos, 978
  Sigmodon, 744
guianensis,
  Armadillo, 282
  Vespertilio, 118
guigna,
  Felis, 1036
gularis,
  Bradypus, 279
  Procyon, 970
Gulo, 1006
  auduboni, 1008
  bairdi, 1008
  gulo, 1008
  hylaeus, 1008
  katschemakensis, 1008
  luscus, 1007
  luteus, 1009
  niediecki, 1008

Gulo *(Continued)*
  sibiricus, 1006
  urva, 1035
  vancouverensis, 1009
gulo,
  Gulo, 1008
  Mustela, 1006
gulosa,
  Mustela, 986
gulosus,
  Perognathus, 537
gundlachi,
  Capromys, 864
gunnisoni,
  Cynomys, 414
  Peromyscus, 683
  Spermophilus, 414
gunnisonii,
  Cynomys, 414
güntheri,
  Mesoplodon, 905
guyannensis,
  Manatus, 1073
gyas,
  Ursus, 954
gymnicus,
  Sciurus, 442
  Tamiasciurus, 442
gymnonotus,
  Chilonycteris, 90
  Pteronotus, 95
Gymnotis, 1087
  wiegmanni, 1087
gymnotis,
  Cervus, 1087
  Peromyscus, 712
gymnurus,
  Cratogeomys, 525
  Dasypus, 282
  Geomys, 525
  Hoplomys, 871
  Pappogeomys, 524
  Platygeomys, 525
  Tylomis, 627
  Tylomys, 627
gypsi,
  Perognathus, 531, 1179

H

Habromys, 718
Haematonycteris, 175
haematoreia,
  Neotoma, 749
haidarum,
  Mustela, 989
  Putorius, 989
haitiensis,
  Ardops, 165
  Phyllops, 165
Halibalaena, 920
Halichoerus, 1067

Halichoerus *(Continued)*
  atlantica, 1067
  baltica, 1067
  griseus, 1067
  grypus, 1067
  macrorhynchus, 1067
  pachyrhynchus, 1067
halichoerus,
  Phoca, 1067
halicoetes,
  Reithrodontomys, 642
  Sorex, 34
Halicyon, 1064
  richardii, 1064
Halipaedisca, 1073
hallensis,
  Alopex, 935
  Vulpes, 935
halli,
  Microtus, 807
  Taxidea, 1010
hallorani,
  Taxidea, 1010
halophilus,
  Microtus, 802
hamatus,
  Delphinus, 886
hamiltoni,
  Ursus, 950
hammondi,
  Nectomys, 608
handleyi,
  Baiomys, 723
haningtoni,
  Bison, 1109
hansae,
  Eumops, 250
Hapale,
  geoffroyi, 271
Haploceros, 1110
Haplocerus, 1110
Haplodon,
  californicus, 335
  leporinus, 335
Haplomylomys, 656
Harana, 1084
harbisoni,
  Peromyscus, 662
hardyi,
  Zapus, 841
harlani,
  Brachysorex, 61
  Cryptotis, 61
  Delphinus, 893
harquahalae,
  Thomomys, 484
harrimani,
  Vulpes, 938
harrisi,
  Reithrodontomys, 649
  Scotinomys, 734
harrisii,

Nycteris *(Continued)*
cinerea, 225
degelida, 221
ega, 221
egregia, 225
floridana, 220
frantzii, 223
insularis, 220
intermedia, 220, 1178
mexicana, 224
minor, 222
panamensis, 221
pfeifferi, 221
seminola, 224
teliotis, 224
xanthina, 221
Nycticea, 226
crepuscularis, 227
cynocephala, 241
Nycticeius, 226
cubanus, 227
humeralis, 227, 1178
mexicanus, 227
subtropicalis, 227
Nycticejus, 226
noveboracensis, 222
ruepellii, 226
Nycticeus, 226
Nycticeyx, 226
Nyctiellus, 179
Nyctinomops, 240
yucatanicus, 245
Nyctinomus,
aequatorialis, 245
antillularum, 241
bahamensis, 241
brasiliensis, 241
californicus, 242
depressus, 245
femorosaccus, 240, 243
jugularis, 246
macrotis, 245
megalotis, 245
minutus, 246
mohavensis, 242
murinus, 242
musculus, 243
nevadensis, 245
orthotis, 250
Nyctipithecus, 259
felinus, 259
rufipes, 260
vociferans, 259
Nyctiplanus, 138
rotundatus, 138
Nyctiptennis, 213
Nyctomys, 630
colimensis, 630
costaricensis, 630
decolorus, 630
florencei, 630
nitellinus, 630

Nyctomys *(Continued)*
pallidulus, 631
salvini, 631
sumichrasti, 630
venustulus, 631
Nyctophilinae, 236
nyctor,
Myotis, 202
nymphaea,
Vampyressa, 148
Nystactes, 183

O

oaxacae,
Dipodomys, 582
Marmosa, 16
Sorex, 50
oaxacensis,
Bassariscus, 965
Dama, 1095
Eumops, 248
Felis, 1046
Glaucomys, 448
Jentinkia, 965
Lynx, 1054
Microtus, 800
Odocoileus, 1095
Peromyscus, 1179
obesus,
Trichechus, 1062
obliquidens,
Lagenorhynchus, 889
obliquus,
Elasmodontomys, 855
obscura,
Blarina, 59
Cryptotis, 59
Lichonycteris, 134
Marmota, 373
Ochotona, 290
obscuroides,
Sorex, 35
obscurus,
Antrozous, 237
Clethrionomys, 785
Dipodomys, 569
Eutamias, 355
Evotomys, 785
Fiber, 827
Liomys, 593
Molossus, 256
Ondatra, 827
Perodipus, 569
Perognathus, 552
Pipistrellus, 213
Procyon, 970
Reithrodontomys, 641
Sorex, 35
Tamias, 355
Vespertilio, 190
obsidianus,

obsidianus *(Continued)*
Spermophilus, 397
obsoletus,
Spermophilus, 396
obtusa,
Electra, 895
Phyllonycteris, 172
obtusus,
Delphinus, 887
obvelatus,
Sigmodon, 739
occidentalis,
Arvicola, 805
Bison, 1109
Canis, 933
Cervus, 1086
Chincha, 1022
Clethrionomys, 782
Dipodomys, 572
Evotomys, 782
Lonchorhina, 107
Mephitis, 1020, 1022
Neotoma, 769
Sturnira, 140
occipitalis,
Fiber, 827
Ondatra, 827
Thomomys, 477
occisor,
Mustela, 997
Putorius, 997
occulta,
Feresa, 896
occultus,
Myotis, 193
Perognathus, 559
Scapanus, 70
oceanicus,
Peromyscus, 684
ocelot,
Felis, 1044
Ochetodon, 632
ochoterenai,
Balantiopteryx, 84
Ochotona, 286
albata, 287
albatus, 287
barnesi, 287
brooksi, 287
brunnescens, 287
cinnamomea, 287
clamosa, 287
collaris, 286
cuppes, 289
fenisex, 289
figginsi, 289
fumosa, 289
fuscipes, 289
goldmani, 289
howelli, 289
incana, 289
jewetti, 289

Ochotona *(Continued)*
lasalensis, 290
lemhi, 290
levis, 290
littoralis, 290
lutescens, 290
moorei, 290
muiri, 290
nevadensis, 290
nigrescens, 290
obscura, 290
princeps, 290
saturata, 291
saturatus, 291
saxatilis, 291
schisticeps, 291
septentrionalis, 291
sheltoni, 291
taylori, 291
tutelata, 291
uinta, 291
utahensis, 291
ventorum, 291
wasatchensis, 292
Ochotonidae, 286
ochracea,
Marmota, 370
Neotoma, 762
ochraceus,
Clethrionomys, 783
Eutamias, 349
Evotomys, 783
Oryzomys, 623
Peromyscus, 688
Procyon, 970
ochraventer,
Peromyscus, 709
ochrogaster,
Hypudaeus, 789, 813
Microtus, 812
ochrogenys,
Eutamias, 351
ochrognathus,
Sigmodon, 744
ochropus,
Canis, 928
Ochrotomys, 721
aureolus, 722
flammeus, 722
floridanus, 722
lisae, 722
nuttalli, 722
ochroura,
Dama, 1095
ochrourus,
Odocoileus, 1095
ochrurus,
Odocoileus, 1096
ochrus,
Perognathus, 557
ocius,
Thomomys, 462

ordi,
  Perodipus, 566
ordii,
  Dipodomys, 565
  Perodipus, 563
Oreamnos, 1110
  americanus, 1110
  columbiae, 1111
  columbianus, 1111
  kennedyi, 1111
  missoulae, 1112
  montanus, 1111
Oreamnus, 1110
  kennedyi, 1111
oreas,
  Microtus, 819
  Peromyscus, 681
oregona,
  Antilocapra, 1107
oregonensis,
  Felis, 1043
  Felix, 1043
  Glaucomys, 453
  Mustela, 997
  Pteromys, 453
  Putorius, 997
  Vespertilio, 186
oregoni,
  Arvicola, 789, 810
  Microtus, 809
oregonus,
  Bassariscus, 964
  Bison, 1109
  Citellus, 388
  Lepus, 316
  Microdipodops, 561
  Spermophilus, 388
  Thomomys, 466
  Zapus, 845
oreinus,
  Sorex, 38
oreocetes,
  Cratogeomys, 522
  Eutamias, 346
oreoecus,
  Thomomys, 489
Oreolagus, 286
oreopolus,
  Sorex, 50
Oreosciurus, 416
oresterus,
  Heteromys, 600
  Peromyscus, 683
orgiloides,
  Ursus, 956
orgilos,
  Ursus, 956
oribasa,
  Mustela, 998
oribasus,
  Mustela, 998
  Putorius, 998

oribasus (Continued)
  Ursus, 956
oribates,
  Perognathus, 559
oricolus,
  Spermophilus, 396
orientalis,
  Armadillo, 282
  Peromyscus, 713
  Pusa, 1034
  Trichechus, 1062
origenes,
  Martes, 984
  Mustela, 984
orinomus,
  Bassariscyon, 978
  Myotis, 188
  Urocyon, 943
orinus,
  Oryzomys, 617
  Reithrodontomys, 651
orion,
  Canis, 933
  Scoteinus, 226
orizabae,
  Lepus, 304
  Neotoma, 763
  Neotomodon, 746
  Peromyscus, 711
  Reithrodontomys, 642
  Sorex, 36
  Sylvilagus, 304
  Thomomys, 489
ornatus,
  Ateles, 269
  Dipodomys, 582
  Lasiurus, 224
  Sorex, 37
  Vampyrodes, 146
orolestes,
  Neotoma, 771
oronocensis,
  Manatus, 1074
orophila,
  Blarina, 61
  Cryptotis, 61
orophilus,
  Phenacomys, 786
örstedii,
  Chrysothrix, 266
orthodon,
  Physeter, 903
Orthogeomys, 506
  alleni, 507
  alvarezi, 508
  annexus, 508
  carbo, 508
  carlosensis, 513
  cartagoensis, 512
  cavator, 512
  cayoensis, 509
  cherriei, 513

Orthogeomys (Continued)
  chiapensis, 509
  concavus, 510
  costaricensis, 514
  cuniculus, 507
  dariensis, 513
  dolichocephalus, 512
  engelhardi, 508
  felipensis, 508
  grandis, 507
  guerrerensis, 508
  heterodus, 512
  hispidus, 509
  hondurensis, 511
  huixtlae, 508
  isthmicus, 511
  lanius, 512
  latifrons, 509
  latirostris, 511
  matagalpae, 514
  negatus, 511
  nelsoni, 509
  nigrescens, 513
  pansa, 513
  pluto, 509
  pygacanthus, 509
  scalops, 509
  soconuscensis, 509
  teapensis, 511
  tehuantepecus, 511
  torridus, 511
  underwoodi, 513
  vulcani, 509
  yucatanensis, 511
orthotis,
  Nyctinomus, 250
Orthriomys, 790
Ortotheriinae, 274
Oryctogale, 1024
Oryctolagus, 292
Oryctomys, 455
  bottae, 455
oryzivora,
  Arvicola, 610
Oryzomys, 607
  agrestis, 615
  albigularis, 617
  albiventer, 608
  alfari, 623
  alfaroi, 615
  alleni, 617
  angusticeps, 615
  antillarum, 609
  apatelius, 609
  aphrastus, 618
  argentatus, 1179
  aquaticus, 609, 1179
  aztecus, 609
  azuerensis, 609
  bicolor, 619
  bombycinus, 616
  bulleri, 610

Oryzomys (Continued)
  caliginosus, 622
  capito, 618
  carrikeri, 618
  carrorum, 614
  caudatus, 615
  chapmani, 616
  cherriei, 733
  chrysomelas, 622
  colimensis, 614
  coloratus, 609
  concolor, 619
  costaricensis, 620
  couesi, 609, 1179
  cozumelae, 610
  creper, 621
  crinitus, 610
  dariensis, 616
  devius, 617
  dilutior, 616
  dimidiatus, 612
  efficax, 624
  engraciae, 621
  floridanus, 609
  frontalis, 619
  fulgens, 612
  fulvescens, 620
  gatunensis, 610
  gloriaensis, 616
  goldmani, 609
  guerrerensis, 616
  huastecae, 616
  hylocetes, 616
  idoneus, 622
  incertus, 615
  jalapae, 609
  lambi, 610
  lenis, 622
  longicaudatus, 620
  luciae, 625
  mayensis, 622
  megadon, 614
  melanotis, 613
  mexicanus, 610
  molestus, 609
  natator, 610
  navus, 620
  nelsoni, 612
  nicaraguae, 622
  ochraceus, 623
  orinus, 617
  pacificus, 622
  palatinus, 616
  palustris, 608
  panamensis, 618
  peninsulae, 610
  peragrus, 611
  phaeopus, 622
  phaeotis, 618
  pinicola, 612
  pirrensis, 617
  planirostris, 612

planifrons,
  Cratogeomys, 520
  Erophylla, 171
  Heteromys, 597
  Hyperoodon, 911
  Pappogeomys, 520
  Phyllonycteris, 171
  Sigmodon, 743
planirostris,
  Molossops, 239, 1178
  Molossus, 239
  Oryzomys, 612
  Thomomys, 491
planorum,
  Thomomys, 491
plantinarensis,
  Liomys, 593
platycephalus,
  Clethrionomys, 779
  Dipodomys, 586
platyceps,
  Micronycteris, 102
Platygeomys, 515
  angustirostris, 524
  gymnurus, 525
  imparilis, 525
  inclarus, 525
  neglectus, 523
  planiceps, 524
  tylorhinus, 524
  varius, 524
  zinseri, 524
platyrhina,
  Chincha, 1022
platyrhinus,
  Otisorex, 25, 27
Platyrhinus, 144
  dorsalis, 145
  helleri, 144
  vittatus, 145
Plecotus, 232
  australis, 234
  ingens, 235
  lecontii, 234
  macrotis, 234
  mexicanus, 233
  pallescens, 235
  phyllotis, 232
  rafinesquii, 234
  townsendii, 234
  virginianus, 236
plectilis,
  Dipodomys, 576
Pleisothomomys,
  potomacensis, 455
pleistocenica,
  Geocapromys, 866
pleistocenicus,
  Capromys, 866
plenus,
  Sigmodon, 742
plerus,

plerus (Continued)
  Perognathus, 532
plesius,
  Citellus, 389
  Spermophilus, 391
Plethaelurus, 1036
plethodon,
  Monophyllus, 126
plicata,
  Balantiopteryx, 83
plumbea,
  Blarina, 56
plumbeus,
  Peromyscus, 681
pluto,
  Orthogeomys, 509
pluvialis,
  Peromyscus, 681
poadromus,
  Lepus, 318
poasensis,
  Sciurus, 439
  Syntheosciurus, 439
Podomys, 720
Poecilolagus, 314
poeppingii,
  Mysateles, 863
Poescopia, 918
poeskop,
  Megaptera, 918
poeyanus,
  Solenodon, 22
poeyi,
  Capromys, 865
  Phyllonycteris, 172
polaris,
  Mustela, 991
  Putorius, 991
  Thalarctos, 947
  Ursus, 959
Poliailurus, 1036
Poliocitellus, 397
polionotus,
  Microdipodops,
    561
  Mus, 669
  Peromyscus, 667
poliopus,
  Sciurus, 420
polius,
  Peromyscus, 698
polycephus,
  Physeter, 904
polypolius,
  Peromyscus, 660
Polyprotodontia, 2
pomeegra,
  Delphinus, 884
ponticus,
  Tursiops, 887
poolei,
  Hexolobodon, 867

popei,
  Perognathus, 553
popofensis,
  Microtus, 806
porsildi,
  Lepus, 322
porteri,
  Scalopus, 75
portoricensis,
  Chilonycteris, 91
  Isolobodon, 869
  Monophyllus, 125
  Pteronotus, 91
potomacensis,
  Pleisothomomys, 455
Potomogale, 20
Potomogalidae, 20
Potos, 977
  arborensis, 977
  aztecus, 978
  boothi, 977
  campechensis, 977
  caucensis, 978
  chiriquensis, 977
  dugesii, 977
  flavus, 977
  guerrerensis, 978
  isthmicus, 978
  mansuetus, 978
  megalotis, 978
  prehensilis, 978
  tolimensis, 978
potosinus,
  Pipistrellus, 212
  Thomomys, 491
potrerograndei,
  Reithrodontomys, 651
powelli,
  Thomomys, 491
praeceps,
  Artibeus, 155
Praopus, 283
pratensis,
  Mynomes, 789, 795
  Pappogeomys, 520
  Spermophilus, 397
pratincola,
  Geomys, 501, 1179
  Microtus, 799
pratincolus,
  Geomys, 501
  Microtus, 799
preblei,
  Dipodomys, 572
  Lutra, 1031
  Perodipus, 572
  Peromyscus, 704
  Phenacomys, 786
  Sorex, 30
  Tamiasciurus, 445
  Zapus, 844
preblorum,

preblorum (Continued)
  Marmota, 370
prehensilis,
  Capromys, 864
  Coendou, 853
  Hystrix, 853
  Potos, 978
  Viverra, 978
premaxillaris,
  Thomomys, 466
pretiosa,
  Neotoma, 758
pretiosus,
  Liomys, 591
  Molossus, 253
prevostensis,
  Peromyscus, 681
  Sorex, 36
pribilofensis,
  Alopex, 936
  Phoca, 1064
  Sorex, 30
  Vulpes, 936
pricei,
  Eutamias, 355
  Perognathus, 549
  Tamias, 355
prietae,
  Perognathus, 559
Primates, 258
primulina,
  Mustela, 998
primus,
  Natalus, 178
princeps,
  Lepus, 290
  Ochotona, 290
  Zapus, 844
principalis,
  Microtus, 820
Priodontini, 281
Prionailurus, 1036
priscus,
  Dipodomys, 569
  Rhynchiscus, 78
problematicus,
  Arctopithecus, 279
Proboscidea, 78
  rivalis, 78
  saxatilis, 78
  villosa, 78
proboscidea,
  Phoca, 1069
Procerus, 1102
  caribaeus, 1102
Procervus, 1102
Prochilus, 947
Procyon, 966
  annulatus, 970
  auspicatus, 968
  brachyurus, 970, 973
  californicus, 972